ENVIRONM

ENCYCLC

FOURTH ED

ENVIRONMENTAL ENCYCLOPEDIA

FOURTH EDITION

VOLUME

1

A–K

GALE
CENGAGE Learning

Detroit • New York • San Francisco • New Haven, Conn • Waterville, Maine • London

Environmental Encyclopedia Fourth Edition

Project Editor: Deirdre S. Blanchfield

Rights Acquisition and Management:
 Robyn Young

 Composition: Evi Abou-El-Seoud

Manufacturing: Wendy Blurton

Imaging: John Watkins

Product Design: Kristine Julien and Jennifer Wahi

© 2011 Gale, Cengage Learning

ALL RIGHTS RESERVED. No part of this work covered by the copyright herein may be reproduced, transmitted, stored, or used in any form or by any means graphic, electronic, or mechanical, including but not limited to photocopying, recording, scanning, digitizing, taping, Web distribution, information networks, or information storage and retrieval systems, except as permitted under Section 107 or 108 of the 1976 United States Copyright Act, without the prior written permission of the publisher.

For product information and technology assistance, contact us at
Gale Customer Support, 1-800-877-4253.
For permission to use material from this text or product,
submit all requests online at **www.cengage.com/permissions.**
Further permissions questions can be emailed to
permissionrequest@cengage.com

Cover photograph reproduced by permission of Spanishsnapper/Dreamstime.com (picture of Pelican); Raphoto/Dreamstime.com (picture of desert); Aero 17/ Dreamstime.com (picture of glacier); Fotoplanner/Dreamstime.com (picture of acorn); Smithore/Dreamstime.com (picture of plastic on beach); Dehooks/Dreamstime.com (picture of oil spill); and WDG Photo, 2010/Shutterstock.com (picture of a butterfly).

While every effort has been made to ensure the reliability of the information presented in this publication, Gale, a part of Cengage Learning, does not guarantee the accuracy of the data contained herein. Gale accepts no payment for listing; and inclusion in the publication of any organization, agency, institution, publication, service, or individual does not imply endorsement of the editors or publisher. Errors brought to the attention of the publisher and verified to the satisfaction of the publisher will be corrected in future editions.

Library of Congress Cataloging-in-Publication Data

Environmental encyclopedia / project editors, Deirdre S. Blanchfield. —4th ed.
 p. cm.
 Includes bibliographical references and index.
 ISBN 978-1-4144-8737-3 (volume 1) — ISBN 978-1-4144-8738-0 (volume 2) —
ISBN 978-1-4144-8736-6 (set)
 1. Environmental sciences—Encyclopedias. 2. Ecology—Encyclopedias. 3. Earth sciences—Encyclopedias. I. Blanchfield, Deirdre S. II. Title.
 GE10.E38 2010
 363.7003—dc22

Gale
27500 Drake Rd.
Farmington Hills, MI, 48331-3535

ISBN-13: 978-1-4144-8736-6 (set)   ISBN-10: 1-4144-8736-3 (set)
ISBN-13: 978-1-4144-8737-3 (vol. 1)   ISBN-10: 1-4144-8737-1 (vol. 1)
ISBN-13: 978-1-4144-8738-0 (vol. 2)   ISBN-10: 1-4144-8738-X (vol. 2)

This title is also available as an e-book.
ISBN-13: 978-1-4144-8739-7 ISBN-10: 1-4144-8739-8
Contact your Gale, a part of Cengage Learning sales representative for ordering information.

Printed in China
1 2 3 4 5 6 7 15 14 13 12 11

CONTENTS

LIST OF ENTRIES

A

Abbey, Edward
Absorption
Acclimation
Accounting for nature
Accuracy
Acetone
Acid and base
Acid deposition
Acid mine drainage
Acid rain
Acidification
Activated sludge
Acute effects
Adams, Ansel
Adaptation
Adaptive management
Adirondack Mountains
Adsorption
Aeration
Aerobic
Aerobic sludge digestion
Aerobic/anaerobic systems
Aerosol
Aflatoxin
African Wildlife Foundation
Africanized bees
Agency for Toxic Substances and
    Disease Registry
Agent Orange
Agglomeration
Agricultural chemicals

Agricultural environmental
    management
Agricultural pollution
Agricultural Research Service
Agricultural revolution
Agricultural Stabilization and
    Conservation Service
Agroecology
Agroforestry
AIDS
Air and Waste Management
    Association
Air pollution
Air pollution control
Air pollution index
Air quality
Air quality control region
Air quality criteria
Air-pollutant transport
Airshed
Alar
Alaska Highway
Alaska National Interest Lands
    Conservation Act (1980)
Albedo
Algal bloom
Algicide
Allelopathy
Allergen
Alligator, American
Alpha ptopic
Alternative energy sources
Aluminum
Amazon basin
Ambient air

Amenity value
American box turtle
American Cetacean Society
American Committee for
    International Conservation
American Farmland Trust
American Forests
American Indian Environmental
    Office
American Oceans Campaign
American Wildlands
Ames test
Amoco Cadiz
Amory, Cleveland
Anaerobic
Anaerobic digestion
Anemia
Animal cancer tests
Animal Legal Defense Fund
Animal rights
Animal waste
Animal Welfare Institute
Antarctic Treaty (1961)
Antarctica
Anthrax
Anthropogenic
Antibiotic resistance
Aquaculture
Aquarium trade
Aquatic chemistry
Aquatic microbiology
Aquatic toxicology
Aquatic weed control
Aquifer
Aquifer depletion

Effluent tax
E_h
Ehrlich, Paul R.
El Niño
Electric utilities
Electromagnetic field
Electron acceptor and donor
Electrostatic precipitation
Elemental analysis
Elephants
Elton, Charles S.
Emergency Planning and
  Community Right-to-Know Act
  (1986)
Emergent diseases (human)
Emergent ecological diseases
Emission
Emission standards
Emphysema
Endangered species
Endangered Species Act (1973)
Endemic species
Endocrine disruptors
Energy and the environment
Energy conservation
Energy efficiency
Energy flow
Energy path, hard vs. soft
Energy policy
Energy recovery
Energy Reorganization Act
  (1973)
Energy taxes
Enteric bacteria
Environment
Environment Canada
Environmental accounting
Environmental aesthetics
Environmental auditing
Environmental chemistry
Environmental Defense
Environmental degradation
Environmental design
Environmental dispute resolution
Environmental economics
Environmental education
Environmental enforcement

Environmental engineering
Environmental estrogens
Environmental ethics
Environmental health
Environmental history
Environmental impact assessment
Environmental Impact Statement
Environmental law
Environmental Law Institute
Environmental liability
Environmental literacy and
  ecocriticism
Environmental monitoring
Environmental Monitoring and
  Assessment Program
Environmental policy
Environmental Protection
  Agency (EPA)
Environmental racism
Environmental refugees
Environmental resources
Environmental science
Environmental stress
Environmental Working Group
Environmentalism
Environmentally preferable
  purchasing
Environmentally responsible
  investing
Enzyme
Ephemeral species
Epidemiology
Erodible
Erosion
Escherichia coli
Essential fish habitat
Estuary
Ethanol
Ethnobotany
Eurasian milfoil
European Union
Eutectic
Evapotranspiration
Everglades
Evolution
Exclusive economic zone
Exotic species

Experimental Lakes Area
Exponential growth
Externality
Extinction
Exxon Valdez

F

Family planning
Famine
Fauna
Fecundity
Federal Energy Regulatory
  Commission
Federal Insecticide, Fungicide
  and Rodenticide Act (1972)
Federal Land Policy and
  Management Act (1976)
Federal Power Commission
Feedlot runoff
Feedlots
Fertilizer
Fibrosis
Field capacity
Filters
Filtration
Fire ants
First world
Fish and Wildlife Service
Fish kills
Fisheries and Oceans Canada
Floatable debris
Flooding
Floodplain
Flora
Florida panther
Flotation
Flu pandemic
Flue gas
Flue-gas scrubbing
Fluidized bed combustion
Fluoridation
Fly ash
Flyway
Food additives
Food and Drug Administration

Multiple chemical sensitivity
Multiple Use-Sustained Yield Act (1960)
Municipal solid waste
Municipal solid waste composting
Mutagen
Mutation
Mutualism
Mycorrhiza
Mycotoxin

N

Nader, Ralph
Naess, Arne
Nagasaki, Japan
National Academy of Sciences
National Ambient Air Quality Standard
National Audubon Society
National Emission Standards for Hazardous Air Pollutants
National Environmental Policy Act (1969)
National Estuary Program
National forest
National Forest Management Act (1976)
National Institute for the Environment
National Institute for Occupational Safety and Health
National Institute for Urban Wildlife
National Institute of Environmental Health Sciences (Research Triangle Park, North Carolina)
National lakeshore
National Mining and Minerals Act (1970)
National Oceanic and Atmospheric Administration (NOAA)
National park
National Park Service
National Parks and Conservation Association

National pollutant discharge elimination system
National Priorities List
National Recycling Coalition
National Research Council
National seashore
National Wildlife Federation
National wildlife refuge
Native landscaping
Natural gas
Natural resources
Natural Resources Defense Council
Nature
The Nature Conservancy
Nearing, Scott
Nekton
Neoplasm
Neotropical migrants
Neritic zone
Neurotoxin
Neutron
Nevada Test Site
New Madrid, Missouri
New Source Performance Standard
New York Bight
Niche
Nickel
Nitrates and nitrites
Nitrification
Nitrogen
Nitrogen cycle
Nitrogen fixation
Nitrogen oxides
Nitrogen waste
Nitrous oxide
Noise pollution
Nonattainment area
Noncriteria pollutant
No degradable pollutant
Nongame wildlife
Nongovernmental organization
Nonpoint source
Nonrenewable resources
Non-timber forest products

No-observable-adverse-effect-level
North American Association for Environmental Education
North American Free Trade Agreement
Northern spotted owl
Not In My Backyard
Nuclear fission
Nuclear fusion
Nuclear power
Nuclear Regulatory Commission
Nuclear test ban
Nuclear weapons
Nuclear winter
Nucleic acid
Nutrient

O

Oak Ridge, Tennessee
Occupational Safety and Health Act (1970)
Occupational Safety and Health Administration
Ocean Conservatory, The
Ocean dumping
Ocean Dumping Ban Act (1988)
Ocean farming
Ocean outfalls
Ocean thermal energy conversion
Octane rating
Odén, Svante
Odor control
Odum, Dr. Eugene P.
Office of Civilian Radioactive Waste Management
Office of Management and Budget
Office of Surface Mining
Off-road vehicles
Ogallala Aquifer
Oil drilling
Oil embargo
Oil shale
Oil spills

Public Health Service
Public interest group
Public land
Public Lands Council
Public trust
Puget Sound/Georgia Basin
  International
  Task Force
Pulp and paper mills
Purple loosestrife

Q

Quaamen, David

R

Rabbits in Australia
Rachel Carson Council
Radiation exposure
Radiation sickness
Radioactive decay
Radioactive fallout
Radioactive pollution
Radioactive waste
Radioactive waste management
Radioactivity
Radiocarbon dating
Radioisotope
Radiological Emergency
  Response Team
Radionuclides
Radiotracer
Radon
Rails-to-Trails Conservancy
Rain forest
Rain shadow
Rainforest Action Network
Rangelands
Raprenox (nitrogen scrubbing)
Rare species
Recharge zone
Reclamation
Record of Decision
Recreation
Recyclables

Recycling
Red tide
Redwoods
Refuse-derived fuels
Regan, Tom
  [Thomas Howard]
Regulatory review
Rehabilitation
Reilly, William K.
Relict species
Religion and the environment
Remediation
Renewable energy
Renewable Natural Resources
  Foundation
Reserve Mining Corporation
Reservoir
Residence time
Resilience
Resistance (inertia)
Resource Conservation and
  Recovery Act
Resource recovery
Resources for the Future
Respiration
Respiratory diseases
Restoration ecology
Retention time
Reuse
Rhinoceroses
Ribonucleic acid
Richards, Ellen H.
Right-to-know
Riparian Land
Riparian rights
Risk analysis
Risk assessment (public health)
River basins
River blindness
River dolphins
Rocky Flats nuclear plant
Rocky Mountain Arsenal
Rocky Mountain Institute
Rodale Institute
Rolston, Holmes
Ronsard, Pierre
Roosevelt, Theodore

Roszak, Theodore
Rowland, Frank Sherwood
Ruckleshaus, William Doyle
Runoff

S

Safe Drinking Water
  Act (1974)
Sagebrush Rebellion
Sahel
St. Lawrence Seaway
Sale, Kirkpatrick
Saline soil
Salinity
Salinization
Salinization of soils
Salmon
Salt, Henry S.
Salt (road)
Salt water intrusion
Sand dune ecology
Sanitary sewer overflows
Sanitation
Santa Barbara oil spill
Saprophyte
Savanna
Savannah River site
Save the Whales
Save-the-Redwoods League
Scarcity
Scavenger
Schistosomiasis
Schumacher, Ernst E.
Schweitzer, Albert
Scientific Committee on
  Problems of the Environment
Scotch broom
Scrubbers
Sea level change
Sea otter
Sea Shepherd Conservation
  Society
Sea turtles
Seabed disposal
Seabrook Nuclear Reactor

W

War, environmental effects of
Waste exchange
Waste Isolation Pilot Plan
Waste management
Waste reduction
Waste stream
Wastewater
Water allocation
Water conservation
Water diversion projects
Water Environment Federation
Water hyacinth
Water pollution
Water quality
Water quality standards
Water reclamation
Water resources
Water rights
Water table
Water table draw-down
Water treatment
Waterkeeper Alliance
Waterlogging
Watershed
Watershed management
Watt, James Gaius
Wave power
Weather modification
Weathering
Wells

Werbach, Adam
Wet scrubber
Wetlands
Whale strandings
Whales
Whaling
White, Gilbert
White Jr., Lynn Townsend
Whooping crane
Wild and Scenic Rivers Act (1968)
Wild river
Wilderness
Wilderness Act (1964)
Wilderness Society
Wilderness Study Area
Wildfire
Wildlife
Wildlife management
Wildlife refuge
Wildlife rehabilitation
Wilson, Edward O.
Wind energy
Windscale (Sellafield) plutonium
   reactor
Winter range
Wise use movement
Wolman, Abel
Wolves
Woodwell, George M.
World Bank
World Conservation Strategy
World Resources Institute

World Trade Organization
   (WTO)
World Wildlife Fund
Wurster, Charles F.

X

X-ray
Xenobiotic
Xylene

Y

Yard waste
Yellowstone National Park
Yokkaichi asthma
Yosemite National Park
Yucca Mountain

Z

Zebra mussel
Zebras
Zero discharge
Zero population growth
Zero risk
Zone of saturation
Zoo
Zooplankton

ADVISORY BOARD

Over several editions of the *Environmental Encyclopedia*, a number of experts in journalism, library science, law, environmental policy, and environmental science communities have provided invaluable assistance in the development of topics and content areas related to their expertise. We would like to express our sincere appreciation to:

Dean Abrahamson: Hubert H. Humphrey Institute of Public Affairs, University of Minnesota, Minneapolis, Minnesota

Joseph Patterson Hyder, J.D.: Independent scholar. Managing partner for the Hyder Law Group in Jacksonville, Florida

Alexander I. Ioffe, Ph.D.: Senior Scientist, Russian Academy of Sciences. Moscow, Russia

Maria Jankowska: Library, University of Idaho, Moscow, Idaho

Adrienne Wilmoth Lerner, J.D.: Independent scholar. Partner, Hyder Law Group in Jacksonville, Florida

Terry Link: Library, Michigan State University, East Lansing, Michigan

Holmes Rolston: Department of Philosophy, Colorado State University, Fort Collins, Colorado

Frederick W. Stoss: Science and Engineering Library, State University of New York—Buffalo, Buffalo, New York

Hubert J. Thompson: Conrad Sulzer Regional Library, Chicago, Illinois

CONTRIBUTORS

Susan Aldridge: Independent scholar and science writer, London, United Kingdom

Margaret Alic, Ph.D.: Freelance writer, Eastsound, Washington

William G. Ambrose Jr., Ph.D.: Department of Biology, East Carolina University, Greenville, North Carolina

James L. Anderson, Ph.D.: Soil Science Department, University of Minnesota, St. Paul, Minnesota

Monica Anderson: Freelance Writer, Hoffman Estates, Illinois

Bill Asenjo M.S., CRC: Science writer, Iowa City, Iowa

William Arthur Atkins, M.S.: Independent scholar and science writer, Normal, Illinois

Terence Ball, Ph.D.: Department of Political Science, University of Minnesota, Minneapolis, Minnesota

Brian R. Barthel, Ph.D.: Department of Health, Leisure and Sports, The University of West Florida, Pensacola, Florida

Stuart Batterman, Ph.D.: School of Public Health, University of Michigan, Ann Arbor, Michigan

Eugene C. Beckham, Ph.D.: Department of Mathematics and Science, Northwood Institute, Midland, Michigan

Milovan S. Beljin, Ph.D.: Department of Civil Engineering, University of Cincinnati, Cincinnati, Ohio

Julie Berwald, Ph.D.: Geologist and writer, Austin, Texas

Heather Bienvenue: Freelance writer, Fremont, California

Lawrence Biskowski, Ph.D.: Department of Political Science, University of Georgia, Athens, Georgia

E. K. Black: University of Alberta, Edmonton, Alberta, Canada

Paul R. Bloom, Ph.D.: Soil Science Department, University of Minnesota, St. Paul, Minnesota

Gregory D. Boardman, Ph.D.: Department of Civil Engineering, Virginia Polytechnic Institute and State University, Blacksburg, Virginia

Marci L. Bortman, Ph.D.: The Nature Conservancy, Huntington, New York

Pat Bounds: Freelance writer

Peter Brimblecombe, Ph.D.: School of Environmental Sciences, University of East Anglia, Norwich, United Kingdom

Kenneth N. Brooks, Ph.D.: College of Natural Resources, University of Minnesota, St. Paul, Minnesota

Peggy Browning: Freelance writer

Marie Bundy: Freelance Writer, Port Republic, Maryland

Ted T. Cable, Ph.D.: Department of Horticulture, Forestry and Recreation Resources, Kansas State University, Manhattan, Kansas

John Cairns Jr., Ph.D.: University Center for Environmental and Hazardous Materials Studies, Virginia Polytechnic Institute and State University, Blacksburg, Virginia

Liane Clorfene Casten: Freelance journalist, Evanston, Illinois

Ann S. Causey: Prescott College, Prescott, Arizona

Ann N. Clarke: Eckenfelder Inc., Nashville, Tennessee

David Clarke: Freelance journalist, Bethesda, Maryland

Sally Cole-Misch: Freelance writer, Bloornfleld Hills, Michigan

Edward Cooney: Patterson Associates, Inc., Chicago, Illinois

Terence H. Cooper, Ph.D.: Soil Science Department, University of Minnesota, St. Paul, Minnesota

Contributors

Deborah L. Swackhammet, Ph.D.: School of Public Health, University of Minnesota, Minneapolis, Minnesota

Liz Swain: Freelance writer, San Diego, California

Ronald D. Taskey, Ph.D.: Soil Science Department, California Polytechnic State University, San Luis Obispo, California

Mary Jane Tenerelli, M.S.: Freelance writer, East Northport, New York

Usha Vedagiri: IT Corporation, Edison, New Jersey

Donald A. Villeneuve,, Ph.D.: Ventura College, Ventura, California

Nikola Vrtis: Freelance writer, Kentwood, Michigan

Eugene R. Wahl: Freelance writer, Coon Rapids, Minnesota

Terry Watkins: Indianapolis, Indiana

Ken R. Wells: Freelance writer, Laguna Hills, California

Roderick T. White Jr.: Freelance writer, Atlanta, Georgia

T. Anderson White, Ph.D.: University of Minnesota, St. Paul, Minnesota

Kevin Wolf: Freelance writer, Minneapolis, Minnesota

Angela Woodward: Freelance writer, Madison, Wisconsin

Gerald L. Young, Ph.D.: Program in Environmental Science and Regional Planning, Washington State University, Pullman, Washington

Melanie Barton Zoltán, M.S.: Independent scholar and science writer, Amherst, Massachusetts

INTRODUCTION

The third edition of the *Environmental Encyclopedia*, edited by William P. Cunningham carried important thoughts still applicable to this, the fourth edition.

"As you might imagine, choosing what to include and what to exclude from this collection has been challenging. Almost everything has some environmental significance, so our task has been to select a limited number of topics we think are of greatest importance in understanding our environment and our relation to it. Undoubtedly, we have neglected some topics that interest you and included some you may consider irrelevant, but we hope that overall you will find this new edition helpful and worthwhile."

"The word environment is derived from the French environ, which means to "encircle" or "surround." Thus, our environment can be defined as the physical, chemical, and biological world that envelops us, as well as the complex of social and cultural conditions affecting an individual or community. This broad definition includes both the natural world and the "built" or technological environment, as well as the cultural and social contexts that shape human lives. You will see that we have used this comprehensive meaning in choosing the articles and definitions contained in this volume."

"Among some central concerns of environmental science are:

• How did the natural world on which we depend come to be as it is, and how does it work?

• What have we done and what are we now doing to our environment—both for good and ill?

• What can we do to ensure a sustainable future for ourselves, future generations, and the other species of organisms on which—although we may not be aware of it—our lives depend?"

"The articles in this volume attempt to answer those questions from a variety of different perspectives. Historically, environmentalism is rooted in natural history, a search for beauty and meaning in nature. Modern environmental science expands this concern, drawing on almost every area of human knowledge including social sciences, humanities, and the physical sciences. Its strongest roots, however, are in ecology, the study of interrelationships among and between organisms and their physical or nonliving environment. A particular strength of the ecological approach is that it studies systems holistically; that is, it looks at interconnections that make the whole greater than the mere sum of its parts. You will find many of those interconnections reflected in this book. Although the entries are presented individually so that you can find topics easily, you will notice that many refer to other topics that, in turn, can lead you on through the book if you have time to follow their trail. This series of linkages reflects the multilevel associations in environmental issues."

The fourth edition of the *Environmental Encyclopedia*, updated by a team of scientists and scholars, attempts to preserve the structure and relationships established in the third edition with updated information and resources that articulate concerns about global warming, climate change and other environmental perils that have become overwhelmingly clearer since the publication of the third edition in 2002.

Accordingly, to this solid foundation and structure, the fourth edition of the *Environmental Encyclopedia* updates, reinforces, and clarifies essential environmental science concepts, with an emphasis on topics increasingly the subject of economic and geopolitical news. For example, this revision incorporates, highlights, and further updates some of the most fundamental environmental data and analysis contained in the most recent Intergovernmental Panel on Climate Change (IPCC) report.

In addition to clarification of existing material, the fourth edition contains information and data distilled from more than 800 peer reviewed journal reports published since the last edition. Along with hundreds of new photos, approximately 5,000 new or updated references are included.

Intended for a wide and diverse audience, every effort has been made to update the *Environmental Encyclopedia* entries in everyday language and to provide accurate and generous explanations of the most important scientific terms. Entries are designed to instruct, challenge, and excite less experienced students, while providing a solid foundation and reference for more advanced students. Although certainly not a substitute for in-depth study of important topics, the fourth edition of the *Environmental Encyclopedia* is designed to better provide students and readers with the basic information, resources, and insights that will enable a greater understanding of the news and stimulate critical thinking regarding current events.

Appropriate to the diversity of environmental sciences, the fourth edition gives special attention to the contributions by women and scientists of diverse ethnic and cultural backgrounds. In addition, the editors have included special contributions written by respected writers and experts. New entries include in-depth information and analysis related to climate change, climate change controversies (including the 2009 and 2010 "Climategate" investigations), and the 2010 Gulf Oil Spill along with recent revisions to environmental law and policy. The fourth edition of *The Gale Environmental Encyclopedia* also expands upon the content provided in earlier editions to include a broader range and treatment of topics of international and global concern.

Environmental science is, of course, not static. In some cases, including the *Deepwater Horizon* oil spill, the full environmental impact will take years and decades to fully manifest and investigate. At the time this book went to press, debates about the size of the *Deepwater Horizon* spill still raged, new regulations regarding offshore drilling were pending, and the formal investigation regarding the cause of the spill remained open. Regardless, the *Environmental Encyclopedia* serves a solid base from which to begin a journey toward an in-depth understanding of topics critical to understanding the complexities of environmental issues. With ongoing issues we have made a special effort to set the context and provide a base of understanding that will allow students and readers to more critically understand the updated information contained in the reliable resources provided.

Thus far Earth is the only known planet with blue skies, warm seas, and life. It is our most tangible and insightful laboratory. Because Earth is our only home, environmental studies also offer a profound insight into delicate balance and the tenuousness of life. As Carl Sagan wrote in *Pale Blue Dot: A Vision of the Human Future in Space*: "The Earth is a very small stage in a vast cosmic arena." For humans to play wisely upon that stage, to secure a future for the children who shall inherit the Earth, we owe it to ourselves to become players of many parts, so that our repertoire of scientific knowledge enables us to use reason and intellect in our civic debates, and to understand the complex harmonies of Earth.

K. Lee Lerner & Brenda Wilmoth Lerner, Editors
Paris, France
November, 2010

A

Abbey, Edward

1927–1989
American environmentalist and writer

Novelist, essayist, white-water rafter, and self-described "desert rat," Abbey wrote of the wonders and beauty of the American West that was fast disappearing in the name of development and progress. Often angry, frequently funny, and sometimes lyrical, Abbey recreated for his readers a region that was unique in the world. The American West was perhaps the last place where solitary selves could discover and reflect on their connections with wild things and with their fellow human beings.

Abbey was born in Indiana, Pennsylvania, in 1927. He received his BA from the University of New Mexico in 1951. After earning his master's degree in 1956, he joined the National Park Service, where he served as park ranger and as a firefighter. He later taught writing at the University of Arizona.

Abbey's books and essays, such as *Desert Solitaire* (1968) and *Down the River* (1982), had their angrier fictional counterparts—most notably, *The Monkey Wrench Gang* (1975) and *Hayduke Lives!* (1990)—in which he gave voice to his outrage over the destruction of deserts and rivers by dam-builders and developers of all sorts. In *The Monkey Wrench Gang* Abbey weaves a tale of three "ecoteurs" who defend the Wild West by destroying the means and machines of development—dams, bulldozers, logging trucks—which would otherwise reduce forests to lumber and raging rivers to irrigation channels.

This aspect of Abbey's work inspired some radical environmentalists, including Dave Foreman and other members of Earth First!, to practice "monkey-wrenching" or "ecotage" to slow or stop such environmentally destructive practices as strip mining, the clear-cutting of old-growth forests on public land, and the damming of wild rivers for flood control, hydroelectric power, and

what Abbey termed "industrial tourism." Although Abbey's description and defense of such tactics have been widely condemned by many environmental groups, he remains a revered figure among many who believe that gradualist tactics have not succeeded in slowing, much less stopping, the destruction of North American wilderness. Abbey also had an oceangoing ship named after him. The activist Sea Shepherd Conservation Society purchased the former U.S. Coast Guard patrol vessel *Cape Knox*, renamed it the *Edward Abbey*, and used it for a number of expeditions.

Abbey died on March 14, 1989. He is buried in a desert in the southwestern United States.

Resources

BOOKS

Abbey, Edward. *Desert Solitaire*. New York: McGraw-Hill, 1968.
Abbey, Edward. *Down the River*. Boston: Little, Brown, 1982.
Abbey, Edward. *Hayduke Lives!* Boston: Little, Brown, 1990.
Abbey, Edward. *The Monkey Wrench Gang*. Philadelphia: Lippincott, 1975.
Berry, W. "A Few Words in Favor of Edward Abbey." In *What Are People For?* San Francisco: North Point Press, 1991.
Bowden, C. "Goodbye, Old Desert Rat." In *The Sonoran Desert*. New York: Abrams, 1992.
Manes, C. *Green Rage: Radical Environmentalism and the Unmaking of Civilization*. Boston: Little, Brown, 1990.

Terence Ball

Absorption

Absorption, or more generally "sorption," is the process by which one material (the sorbent) takes up and retains another (the sorbate) to form a homogenous concentration at equilibrium.

Sorption is defined as the adhesion of gas molecules, dissolved substances, or liquids to the surface of solids with which they are in contact. In soils, three types of mechanisms, often working together, constitute sorption. They can be grouped into physical sorption, chemiosorption, and penetration into the solid mineral phase. Physical sorption (also known as adsorption) involves the attachment of the sorbent and sorbate through weak atomic and molecular forces. Chemiosorption involves chemical bonds similar to holding atoms in a molecule. Electrostatic forces operate to bond minerals via ion exchange, such as the replacement of sodium, magnesium, potassium, and aluminum cations ($+$) as exchangeable bases with acid ($-$) soils. While cation (positive ion) exchange is the dominant exchange process occurring in soils, some soils have the ability to retain anions (negative ions) such as nitrates, chlorine, and to a larger extent, oxides of sulfur.

Absorption and wastewater treatment

In on-site wastewater treatment, the soil absorption field is the land area where the wastewater from the septic tank is spread into the soil. One of the most common types of soil absorption field has porous plastic pipes extending away from the distribution box in a series of two or more parallel trenches, usually 1.5 to 2 feet (45.7–61 cm) wide. In conventional, below-ground systems, the trenches are 1.5 to 2 feet deep. Some absorption fields must be placed at a shallower depth than this to compensate for some limiting soil condition, such as a hardpan or high water table. In some cases they may even be placed partially or entirely in fill material that has been brought to the lot from elsewhere.

The porous pipe that carries wastewater from the distribution box into the absorption field is surrounded by gravel that fills the trench to within a foot or so of the ground surface. The gravel is covered by fabric material or building paper to prevent plugging. Another type of drainfield consists of pipes that extend away from the distribution box, not in trenches but in a single, gravel-filled bed that has several such porous pipes in it. As with trenches, the gravel in a bed is covered by fabric or other porous material.

Usually the wastewater flows gradually downward into the gravel-filled trenches or bed. In some instances, such as when the septic tank is lower than the drainfield, the wastewater must be pumped into the drainfield. Whether gravity flow or pumping is used, wastewater must be evenly distributed throughout the drainfield. It is important to ensure that the drainfield is installed with care to keep the porous pipe level, or at a very gradual downward slope away from the distribution box or pump chamber, according to specifications stipulated by public health officials. Soil beneath the gravel-filled trenches or bed must be permeable so that wastewater and air can move through it and come in contact with each other. Good aeration is necessary to ensure that the proper chemical and microbiological processes will be occurring in the soil to cleanse the percolating wastewater of contaminants. A well-aerated soil also ensures slow travel and good contact between wastewater and soil.

How common are septic systems with soil absorption systems?

According to the data published by the Environmental Protection Agency (EPA) in 2008, usage of septic systems ranges from 7 percent of households in urban areas to 61 percent use in small communities and rural areas. Approximately 19.8 million homes in the United States use septic tanks or cesspools.

According to a study conducted by the EPA's Office of Technology Assessment, virtually all septic tank waste is discharged to subsurface soils, which can impact groundwater quality. The EPA recommends that to decrease environmental risks and save money on septic system repairs, homeowners using septic systems should inspect and pump systems on a regular basis, use water efficiently, and avoid disposing household hazardous wastes into sinks and toilets. Landscape planning near the septic system and feeder lines is also important. Septic tank users are cautioned not to plant tress or shrubs that may have root systems that interfere either with tank integrity or that can hinder absorption from subsurface drainage lines.

See also Pollution control.

Resources

BOOKS
Bitton, Gabriel. *Wastewater Microbiology*. Hoboken, NJ: Wiley-Liss and John Wiley & Sons, 2005.
Eaton, Andrew D., and M. A. H. Franson. *Standard Methods for the Examination of Water & Wastewater*. Washington, DC: American Public Health Association, 2005.
Russell, David L. *Practical Wastewater Treatment*. New York: Wiley-Interscience, 2006.

OTHER
United States Environmental Protection Agency (EPA). "Water: Wastewater: Municipal Wastewater Treatment." http://www.epa.gov/ebtpages/watewastewater municipalwastewatertreatment.html (accessed November 9, 2010).
United States Environmental Protection Agency (EPA). "Water: Wastewater: Wastewater Systems." http://www. epa.gov/ebtpages/watewastewaterwastewatersystems. html (accessed November 9, 2010).

United States Environmental Protection Agency (EPA). "Water: Water Pollution Control: Wastewater Treatment." http://www.epa.gov/ebtpages/watewaterpollution wastewatertreatment.html (accessed November 9, 2010).

Carol Steinfeld

Acclimation

Acclimation is the process by which an organism adjusts to a change in its environment. It generally refers to the ability of living things to adjust to changes in climate, and may occur rapidly as in the case of color adaptation, or slowly as with physiological acclimatization to cold and altitude.

Some scientists draw a sharper distinction between acclimation and acclimatization, defining the latter adjustment as made under natural conditions when the organism is subject to the full range of changing environmental factors. Acclimation, however, refers to a change in only one environmental factor under laboratory conditions.

In an acclimation experiment, adult frogs (*Rana temporaria*) maintained in the laboratory at a temperature of either 50°F (10°C) or 86°F (30°C) were tested in an environment of 32°F (0°C). It was found that the group maintained at the higher temperature was inactive at freezing temperatures. The group maintained at the lower temperature of 50°F (10°C), however, was active at freezing temperatures and thus showed acclimation to lower temperatures.

Acclimation and acclimatization can have profound effects upon behavior, inducing shifts in preferences and in mode of life. The golden hamster (*Mesocricetus auratus*) prepares for hibernation when the environmental temperature drops below 59°F (15°C). Temperature preference tests in the laboratory show that the hamsters develop a marked preference for cold environmental temperatures during the pre-hibernation period. Following arousal from a simulated period of hibernation, the situation is reversed, and the hamsters actively prefer the warmer environments.

An acclimated microorganism is any microorganism that is able to adapt to environmental changes such as a change in temperature or a change in the quantity of oxygen or other gases. Many organisms that live in environments with seasonal changes in temperature make physiological adjustments that permit them to continue to function properly, even though their environmental temperature goes through a definite annual temperature cycle.

Acclimatization usually involves a number of interacting physiological processes. For example, in acclimatizing to high altitudes, the first response of human beings is to increase their breathing rate. After about forty hours, changes have occurred in the oxygen-carrying capacity of the blood, which makes it more efficient in extracting oxygen at high altitudes. Full acclimatization, as measured by blood gases and breathing rate, may take weeks. Such acclimatization is usually accompanied by increased red blood cell counts as the body attempts to increase the oxygen carrying capacity of the body.

Athletes often train at altitude to acclimatize to high-altitude competition, or to increase performance at lower altitude. The acclimatization processes are measurable physiologically and some sports regulatory agencies limit the periods athletes may stay at high altitude prior to competition at lower altitudes. At higher altitudes, the body compensates by increasing its production of red blood cells (erythrocytes). In the body this increase is mediated by the hormone erythropoietin (EPO), and so some athletes may attempt to artificially boost red blood cell numbers (thereby increasing the oxygen-carrying capacity of the blood) by taking EPO supplements. For this reason EPO, a drug that then mimics acclimatization processes, is generally banned in athletic competition.

Resources

BOOKS

Gerday, Charles, and Nicolas Glansdorff. *Physiology and Biochemistry of Extremophiles*. Washington, DC: ASM Press, 2007.
Hill, Richard W. *Animal Physiology*. Sunderland, MA: Sinauer Associates, 2004.
Hillman, Stanley S. *Ecological and Environmental Physiology of Amphibians*. Oxford, UK: Oxford University Press, 2009.

Linda Rehkopf

Accounting for nature

Accounting for nature is an approach to national income accounting in which the degradation and depletion of natural resource stocks and environmental amenities are explicitly included in the calculation of net national product (NNP). NNP is equal to gross national product (GNP) minus capital depreciation, and GNP is equal to the value of all final goods and

Accuracy

services produced in a nation in a particular year. It is recognized that natural resources are economic assets that generate income, and that just as the depreciation of buildings and capital equipment are treated as economic costs and subtracted from GNP to get NNP, depreciation of *natural capital* should also be subtracted when calculating NNP. In addition, expenditures on environmental protection, which at present are included in GNP and NNP, are considered defensive expenditures in accounting for nature that should not be included in either GNP or NNP.

Resources

OTHER

United States Environmental Protection Agency (EPA). "Economics: Environmental Accounting." http://www. epa.gov/ebtpages/econenvironmentalaccounting.html (accessed September 3, 2010).

United States Environmental Protection Agency (EPA). "Economics: Environmental Accounting: Full Cost Accounting." http://www.epa.gov/ebtpages/econenvironmentala fullcostaccounting.html (accessed September 3, 2010).

Accuracy

Accuracy is the closeness of an experimental measurement to the true value (i.e., actual or specified) of a measured quantity. A true value (within measurable limits) can be determined by an experienced analytical scientist who performs repeated analyses of a sample of known purity or concentration using reliable, well-tested methods.

Measurement is inexact, and the magnitude of that exactness is referred to as the error. Error is inherent in measurement and is a result of such factors as the precision of the measuring tools, their proper adjustment, the method, and competency of the analytical scientist.

Statistical methods are used to evaluate accuracy by predicting the likelihood that a result varies from the true value. The analysis of probable error is also used to examine the suitability of methods or equipment used to obtain, portray, and utilize an acceptable result. Highly accurate data can be difficult to obtain and costly to produce. However, different applications can require lower levels of accuracy that are adequate for a particular study.

Resources

BOOKS

Freedman, David; Robert Pisani; and Roger Purves. *Statistics*. 4th ed. New York: W. W. Norton, 2007.

Manly, Bryan F. J. *Statistics for Environmental Science and Management*. London: Chapman & Hall, 2008.

McCleery, Robin H.; Trudy A. Watt; and Tom Hart. *Introduction to Statistics for Biology*. 3rd ed. London: Chapman and Hall, 2007.

Judith L. Sims

Acetone

Acetone (C_3H_6O) is a colorless liquid that is used as a solvent in products, such as in nail polish and paint, and in the manufacture of other chemicals such as plastics and fibers. It is a naturally occurring compound that is found in plants and is released during the metabolism of fat in the body. It is also found in volcanic gases, and is manufactured by the chemical industry (sometimes under the label "2-propanone," a chemical synonym). Acetone is also found in the atmosphere as an oxidation product of both natural and anthropogenic volatile organic compounds (VOCs). It has a strong smell and taste, and is soluble in water. The evaporation point of acetone is quite low compared to water, and the chemical is highly flammable. Because it is so volatile, the acetone manufacturing process results in a large percentage of the compound entering the atmosphere. Ingesting acetone can cause damage to the tissues in the mouth and can lead to unconsciousness. Breathing acetone can cause irritation of the eyes, nose, and throat; headaches; dizziness; nausea; unconsciousness; and possible coma and death. Women may experience menstrual irregularity. However, despite concern about the carcinogenic potential of acetone, laboratory studies and studies of workers routinely exposed to acetone show no evidence that acetone causes cancer.

As of 2010 the National Institute for Occupational Safety and Health's Registry of Toxic Effects of Chemical Substances continues monitoring for suspected acetone exposure-related contributions to respiratory, gastrointestinal, kidney, and liver diseases.

Resources

OTHER

United States Environmental Protection Agency (EPA). "Pollutants/Toxics: Soil Contaminants: Acetone." http://www.epa.gov/ebtpages/pollsoilcacetone.html (accessed September 3, 2010).

Marie H. Bundy

Acid and base

In chemistry, an acid is a substance that increases the hydrogen ion (H^+) concentration in a solution and a base is a substance that removes hydrogen ions from a solution. In water, removal of hydrogen ions results in an increase in the hydroxide ion (OH^-) concentration. Water with a pH of 7 is neutral, while lower pH values are acidic and higher pH values are basic.

The acidity of a liquid (aqueous solution) is measured as its concentration of hydrogen ions. The pH scale expresses this concentration in logarithmic units, ranging from very acidic solutions of pH 0, through the neutral value of pH 7, to very alkaline (or basic) solutions of pH 14. A one-unit difference in pH (for example, from pH 3 to pH 4) represents a ten fold difference in the concentration of hydrogen ions.

Changes in the pH of soil and water (e.g., soil and water becoming more acidic or basic) can have devastating impacts on habitat. For example, marine biologists contend that climate change and pollution are driving changes in pH in ocean waters that threaten both algae and corals.

Resources

BOOKS

Lew, Kristi. *Acids and Bases*. New York: Chelsea House Publications, 2008.

Petheram, Louise. *Acid Rain (Our Planet in Peril)*. Mankato, MI: Capstone Press, 2006.

Acid deposition

Acid precipitation from the atmosphere, whether in the form of dryfall (finely divided acidic salts), rain, or snow results in acid deposition.

Naturally occurring carbonic acid normally makes rain and snow mildly acidic (approximately 5.6 pH). Human activities often introduce much stronger and more damaging acids. Sulfuric acids formed from sulfur oxides (SO_2) released in coal or oil combustion or smelting of sulfide ores predominate as the major atmospheric acid in industrialized areas. Nitric acid created from oxides of nitrogen (NO and NO_2), formed by oxidizing atmospheric nitrogen when any fuel is burned in an oxygen-rich environment, constitutes the major source of acid precipitation in cities such as Los Angeles, California, with little industry but large numbers of trucks and automobiles. The damage caused to building materials, human health, crops, and natural ecosystems by atmospheric acids amounts to billions of dollars per year in the United States. Particulates of ammonium sulfate ($NH_{42}SO_4$) and ammonium nitrate (NH_4NO_3) also impact acid base balances.

Dry deposition results from atmospheric particulate matter, as well as the uptake of gaseous sulfur dioxide and nitric oxides by plants, soil, and water. Once they are dry deposited, certain chemicals can generate important quantities of acidity as plants decompose in the ecosystem. In relatively polluted environments close to emissions sources, dry depositons account for a greater percentage of acidifying pollution than wet depostions (e.g., acid rain).

For example, within a 25-mile (40-km) radius of a smelter, even though only 1 percent of the total sulfur dioxide is deposited within that area, about 50 percent of the total input of acidifying sulfur dioxide from the atmosphere found on the ground is due to dry deposition.

Resources

BOOKS

Petheram, Louise. *Acid Rain (Our Planet in Peril)*. Mankato, MI: Capstone Press, 2006.

OTHER

United States Environmental Protection Agency (EPA). "Air: Air Pollutants: Sulfur Oxides (SO2)." http://www.epa.gov/ebtpages/airairpollutantssulfuroxidesso2.html (accessed September 4, 2010).

Acid mine drainage

The process of mining the earth for coal and metal ores has a long history of rich economic rewards—and a high level of environmental impact to the surrounding aquatic and terrestrial ecosystems. Acid mine drainage is the highly acidic, sediment-laden discharge from exposed mines that is released into the ambient aquatic environment. The bright orange seeps of acid mine drainage threaten aquatic life in streams and ponds that receive mine discharge. In the Appalachian coal mining region, almost 7,500 miles (12,000 km) of streams and almost 30,000 acres (12,000 ha) of land are estimated to be seriously affected by the discharge of uncontrolled acid mine drainage.

Acid mine drainage in Spain. *(Ashiga/Shutterstock.com)*

In the United States, coal-bearing geological strata occur near the surface in large portions of the Appalachian mountain region. The relative ease with which coal could be extracted from these strata led to a type of mining known as strip mining that was practiced heavily in the nineteenth and early twentieth centuries. In this process, large amounts of earth, called the overburden, were physically removed from the surface to expose the coal-bearing layer beneath. The coal was then extracted from the rock as quickly and cheaply as possible. Once the bulk of the coal had been mined, and no more could be extracted without a huge additional cost, the sites were usually abandoned. The remnants of the exhausted coal-bearing rock and soil are called the mine spoil waste.

Acid mine drainage is not generated by strip mining itself but by the nature of the rock where it takes place. Three conditions are necessary to form acid mine drainage: pyrite-bearing rock, oxygen, and iron-oxidizing bacteria. In the Appalachians, the coal-bearing rocks usually contain significant quantities of pyrite (iron). This compound is normally not exposed to the atmosphere because it is buried underground within the rock; it is also insoluble in water. The iron and the sulfide are said to be in a reduced state, that is, the iron atom has not released all the electrons that it is capable of releasing. When the rock is mined, the pyrite is exposed to air. It then reacts with oxygen to form ferrous iron and sulfate ions, both of which are highly soluble in water. This leads to the formation of sulfuric acid and is responsible for the acidic nature of the drainage. But the oxidation can only occur if the bacteria *Thiobacillus ferrooxidans* are present. These activate the iron-and-sulfur oxidizing reactions and use the energy released during the reactions for their own growth. They must have oxygen to carry these reactions through. Once the maximum oxidation is reached, these bacteria can derive no more energy from the compounds and all reactions stop.

The acidified water may be formed in several ways. It may be generated by rain falling on exposed mine spoil wastes or when rain and surface water (carrying dissolved oxygen) flow down and seep into rock fractures and mine shafts, coming into contact with pyrite-bearing rock. Once the acidified water has been formed, it leaves the mine area as seeps or small streams.

Characteristically bright orange to rusty red in color due to the iron, the liquid may be at a pH of between 2.0 and 4.0. These are extremely low pH values and signify a very high degree of acidity. Vinegar, for example, has a pH of about 4.7 and the pH associated with acid rain is in the range of between 4.0 and 6.0. Thus, acid mine drainage with a pH of 2 is more acidic than almost any other naturally occurring liquid released in the environment (with the exception of some volcanic lakes that are pure acid). Usually, the drainage is also very high in dissolved iron, manganese, aluminum, and suspended solids.

The acidic drainage released from the mine spoil wastes usually follows the natural topography of its area and flows into the nearest streams or wetlands where its effect on the water quality and biotic community is unmistakable. The iron coats the stream bed and its vegetation as a thick orange coating that prevents sunlight from penetrating leaves and plant surfaces. Photosynthesis stops and the vegetation (both vascular plants and algae) dies. The acid drainage eventually also makes the receiving water acid. As the pH drops, the fish, the invertebrates, and algae die when their metabolism can no longer adapt. Eventually, there is no life left in the stream with the possible exception of some bacteria that may be able to tolerate these conditions. Depending on the number and volume of seeps entering a stream and the volume of the stream itself, the area of impact may be limited and improved conditions may exist downstream, as the acid drainage is diluted. Abandoned mine spoil areas also tend to remain barren, even after decades. The colonization of the acidic mineral soil by plant species is a slow and difficult process, with a few lichens and aspens being the most hardy species to establish.

While many methods have been tried to control or mitigate the effects of acid mine drainage, very few have been successful. Federal mining regulations (Surface Mining Control and Reclamation Act of 1978) now require that when mining activity ceases, the mine spoil wastes should be buried and covered with the overburden and vegetated topsoil. The intent is to restore the area to premining condition and to prevent the generation of acid mine drainage by limiting the exposure of pyrite to oxygen and water. Although some minor seeps may still occur, this is the single most effective way to minimize the potential scale of the problem. Mining companies are also required to monitor the effectiveness of their restoration programs and must post bonds to guarantee the execution of abatement efforts, should any become necessary in the future.

There are, however, numerous abandoned sites exposing pyrite-bearing spoils. Cleanup efforts for these sites have focused on controlling one or more of the three conditions necessary for the creation of the acidity: pyrite, bacteria, and oxygen. Attempts to remove bulk quantities of the pyrite-bearing mineral and store it somewhere else are extremely expensive and difficult to execute. Inhibiting the bacteria by using detergents, solvents, and other bactericidal agents are temporarily effective, but usually require repeated application. Attempts to seal out air or water are difficult to implement on a large scale or in a comprehensive manner.

Since it is difficult to reduce the formation of acid mine drainage at abandoned sites, one of the most promising new methods of mitigation treats the acid mine drainage after it exits the mine spoil wastes. The technique channels the acid seeps through artificially created wetlands, planted with cattails or other wetland plants in a bed of gravel, limestone, or compost. The limestone neutralizes the acid and raises the pH of the drainage, while the mixture of oxygen-rich and oxygen-poor areas within the wetland promote the removal of iron and other metals from the drainage. As of 2010, many agencies, universities, and private firms are working to improve the design and performance of these artificial wetlands. A number of additional treatment techniques may be strung together in an interconnected system of anoxic limestone trenches, settling ponds, and planted wetlands. This provides a variety of physical and chemical microenvironments so that each undesirable characteristic of the acid drainage can be individually addressed and treated; for example, acidity is neutralized in the trenches, suspended solids are settled in the ponds, and metals are precipitated in the wetlands. In the United States, the research and treatment of acid mine drainage continues to be an active field of study in the Appalachians and in the metal-mining areas of the Rocky Mountains.

Mine drainage and discharge, even if unintentional and at seemingly low levels, can have devastating environmental impacts. For example, cyanide leakage from gold mines in northern Idaho caused gradual acidification in areas of the South Fork Salmon River for years. One area so contaminated was where chinook salmon breed, so the change in pH threatened the entire chinook salmon population.

Resources

BOOKS

National Research Council. *Superfund and Mining Megasites: Lessons from the Coeur D'alene River Basin.* Washington, DC: National Academies Press, 2006.

OTHER

United States Environmental Protection Agency (EPA). "Industry: Industrial Processes: Mining." http://www.epa.gov/ebtpages/induindustmining.html (accessed August 27, 2010).

Usha Vedagiri

Acid rain

Acid rain is the term generally used in the popular press that is equivalent to wet acidic deposition as used in the scientific literature. Acid deposition results from the deposition of airborne acidic pollutants on land and in bodies of water. These pollutants can cause damage to forests as well as to lakes and streams.

The major pollutants that cause acidic deposition are sulfur dioxide (SO_2) and nitrogen oxides (NO_x) produced during the combustion of fossil fuels. In the atmosphere these gases oxidize to sulfuric acid (H_2SO_4) and nitric acid (HNO_3) that can be transported long distances before being returned to the earth dissolved in rain drops (wet deposition), deposited on the surfaces of plants as cloud droplets, or directly on plant surfaces (dry deposition).

Electrical utilities contribute the greatest percentage of SO_2 added to the atmosphere. Most of this is from the combustion of coal. Electric utilities also contribute approximately one-third of the NO_x added to the atmosphere (internal combustion engines used in automobiles, trucks, and buses contribute about half). Natural sources such as forest fires, swamp gases, volcanoes, lightning, and microbial processes in soils contribute only 5 percent and 15 percent of the atmospheric SO_2 and NO_x.

In response to air quality regulations, electrical utilities have switched to coal with lower sulfur content and installed scrubbing systems to remove SO_2. This has resulted in a steady decrease in SO_2 emissions in the United States since 1970, with an 18–20 percent decrease between 1975 and 1988. Emissions of NO_x

Acid rain damage on forest, Mount Mitchell, North Carolina.
(Will & Deni McIntyre/Getty Images)

have also decreased from the peak in 1975, with a 9–15 percent decrease from 1975 to 1988. With still tougher (Phase II) standards implemented in 2000, by 2009 the U.S. Environmental Protection Agency (EPA) characterized its acid-rain reduction program as a success.

A commonly used indicator of the intensity of acid rain is the pH of this rainfall. The pH of nonpolluted rainfall in forested regions is in the range between 5.0 and 5.6. The upper limit is 5.6, not neutral (7.0), because of carbonic acid that results from the dissolution of atmospheric carbon dioxide. The contribution of naturally occurring nitric and sulfuric acid, as well as organic acids, reduces the pH somewhat to less than 5.6. In arid and semiarid regions, rainfall pH values can be greater than 5.6 because of the effect of alkaline soil dust in the air. Nitric and sulfuric acids in acidic rainfall (wet deposition) can result in pH values for individual rainfall events of less than 4.0.

In North America, the lowest acid rainfall is in the northeastern United States and southeastern Canada. The lowest mean pH in this region is 4.15. Even lower pH values are observed in central and northern Europe. Generally, the greater the population density and density of industrialization, the lower the rainfall pH. Long distance transport, however, can result in low pH rainfall even in areas with low population and low density of industries, as in parts of New England, eastern Canada, and in Scandinavia.

A very significant portion of acid deposition occurs in the dry form. In the United States, it is estimated that 30–60 percent of acidic deposition occurs as dry fall. This material is deposited as sulfur dioxide gas and very finely divided particles (aerosols) directly on the surfaces of plants (needles and leaves). The rate of deposition depends not only on the concentration of acid materials suspended in the air, but also on the nature and density of plant surfaces exposed to the atmosphere and the atmospheric conditions (e.g., wind speed and humidity).

Direct deposition of acid cloud droplets can be very important especially in some high-altitude forests. Acid cloud droplets can have acid concentrations of five to twenty times that in wet deposition. In some high elevation sites that are frequently shrouded in clouds, direct droplet deposition is three times that of wet deposition from rainfall.

Acid deposition has the potential to adversely affect sensitive forests as well as lakes and streams. Agriculture is generally not included in the assessment of the effects of acidic deposition because experimental evidence indicates that even the most severe episodes of acid deposition do not adversely affect the growth of agricultural crops, and that any long-term soil acidification can readily be managed by addition of agricultural lime. In fact, the acidifying potential of the fertilizers normally added to cropland is much greater than that of acidic deposition. In forests, however, long-term acidic deposition on sensitive soils can result in the depletion of important nutrient elements (e.g., calcium, magnesium, and potassium) and in soil acidification. Also, acidic pollutants can interact with other pollutants (e.g., ozone) to cause more immediate problems for tree growth. Acid deposition can also result in the acidification of sensitive lakes and the loss of biological productivity.

Long-term exposure of acid-sensitive materials used in building construction and in monuments (e.g., zinc, marble, limestone, and some sandstone) can result in surface corrosion and deterioration. Monuments tend to be the most vulnerable because they are usually not as protected from rainfall as most building materials. In particular, buildings made of limestone and marble contain vulnerable calcium carbonate. Dry deposition reacts with the calcium carbonate to damage surface features. A number of famous buildings and sculptures, especially in Europe, have been damaged by acid deposition and remain the focus of restoration efforts.

Nutrient depletion due to acid deposition on sensitive soils is a long-term (decades to centuries) consequence of acidic deposition. Acidic deposition greatly accelerates the very slow depletion of soil nutrients because of natural weathering processes. Soils that contain less plant-available calcium, magnesium, and potassium are less buffered with respect to degradation due to acidic deposition. The most sensitive soils are shallow sandy soils over hard bedrock. The least vulnerable soils are the deep clay soils that are highly buffered against changes because of acidic deposition.

The more immediate possible threat to forests is the forest decline phenomenon that has been observed in forests in northern Europe and North America. Acidic deposition in combination with other stress factors such as ozone, disease, and adverse weather conditions can lead to decline in forest productivity and, in certain cases, to dieback. Acid deposition alone cannot account for the observed forest decline, and acid deposition probably plays a minor role in the areas where forest decline has occurred. Ozone is a much more serious threat to forests, and it is a key factor in the decline of forests in the Sierra Nevada and San Bernardino mountains in California.

The greatest concern for adverse effects of acidic deposition is the decline in biological productivity in lakes. When a lake has a pH less than 6.0, several species of minnows, as well as other species that are part of the food chain for many fish, cannot survive. At pH values less than about 5.3, lake trout, walleye, and smallmouth bass cannot survive. At pH less than about 4.5, most fish cannot survive (largemouth bass are an exception).

Many small lakes are naturally acidic due to organic acids produced in acid soils and acid bogs. These lakes have chemistries dominated by organic acids, and many have brown-colored waters due to the organic acid content. These lakes can be distinguished from lakes acidified by acidic deposition, because lakes strongly affected by acidic deposition are dominated by sulfate.

Lakes that are adversely affected by acidic deposition tend to be in steep terrain with thin soils. In these settings the path of rainwater movement into a lake is not influenced greatly by soil materials. This contrasts to most lakes where much of the water that collects in a lake flows first into the groundwater before entering the lake via subsurface flow. Due to the contact with soil materials, acidity is neutralized and the capacity to neutralize acidity is added to the water in the form of bicarbonate ions (bicarbonate alkalinity). If more than 5 percent of the water that reaches a lake is in the form of groundwater, a lake is not sensitive to acid deposition.

Resources

BOOKS

Brimblecombe, Peter. *Acid Rain: Deposition to Recovery*. Dordrecht, Netherlands: Springer, 2007.

Morgan, Sally. *Acid Rain*. London: Watts Publishing Group, 2005.

Petheram, Louise. *Acid Rain (Our Planet in Peril)*. Mankato, MI: Capstone Press, 2006.

Visgilio, Gerald R. *Acid in the Environment: Lessons Learned and Future Prospects*. New York: Springer Science+ Business Media, 2007.

OTHER

National Geographic Society. "Acid Rain." http://environment.nationalgeographic.com/environment/global-warming/acid-rain-overview.html (accessed August 31, 2010).

United States Environmental Protection Agency (EPA). "Air: Air Pollution Effects: Acid Rain." http://www.epa.gov/ebtpages/airairpollutionefacidrain.html (accessed August 31, 2010).

Paul R. Bloom

Acoustics *see* **Noise pollution.**

Acquired immune deficiency *see* **AIDS.**

Acidification

The process of becoming more acidic. The common measure of acidification is a decrease in pH, reflecting an increase in hydrogen ion (proton) concentration. Acidification of soils and natural waters by acid rain or acidic wastes can result in reduced biological productivity. Normal rainfall is slightly acidic, with a pH of about 5.6. Rain with a pH below 5.6 is considered to be acid rain.

Acidity *see* **pH.**

Activated sludge

The activated sludge process is an aerobic (oxygen-rich), continuous-flow biological method for the treatment of domestic and biodegradable industrial wastewater, in which organic matter is utilized by microorganisms for life-sustaining processes, that is, for energy for reproduction, digestion, movement, and so forth, and as a food source to produce cell growth and more microorganisms. During these activities of utilization and degradation of organic materials, degradation products of carbon dioxide and water are also formed. The activated sludge process is characterized by the suspension of microorganisms in the wastewater, a mixture referred to as the mixed liquor. Activated sludge is used as part of an overall treatment system, which includes primary treatment of the wastewater for the removal of particulate solids before the use of activated sludge as a secondary treatment process to remove suspended and dissolved organic solids.

The conventional activated sludge process consists of an aeration basin, with air as the oxygen source, where treatment is accomplished. Soluble (dissolved) organic materials are absorbed through the cell walls of the microorganisms and into the cells, where they are broken down and converted to carbon dioxide, water, energy, and the production of more microorganisms. Insoluble (solid) particles are adsorbed on the cell walls, transformed to a soluble form by enzymes (biological catalysts) secreted by the microorganisms, and absorbed through the cell wall, where they are also digested and used by the microorganisms in their life-sustaining processes.

The microorganisms that are responsible for the degradation of the organic materials are maintained in suspension by mixing induced by the aeration system. As the microorganisms are mixed, they collide with other microorganisms and stick together to form larger particles called *floc*. The large flocs that are formed settle more readily than individual cells. These flocs also collide with suspended and colloidal materials (insoluble organic materials), which stick to the flocs and cause the flocs to grow even larger. The microorganisms digest these adsorbed materials, thereby reopening sites for more materials to stick.

The aeration basin is followed by a secondary clarifier (settling tank), where the flocs of microorganisms with their adsorbed organic materials settle out. A portion of the settled microorganisms, referred to as sludge, are recycled to the aeration basin to maintain an active population of microorganisms and an adequate supply of biological solids for the adsorption of organic materials. Excess sludge is wasted by being piped to separate sludge-handling processes. The liquids from the clarifier are transported to facilities for disinfection and final discharge to receiving waters, or to tertiary treatment units for further treatment.

Activated sludge processes are designed based on the mixed liquor suspended solids (MLSS) and the organic loading of the wastewater, as represented by the biochemical oxygen demand (BOD) or chemical oxygen demand (COD). The MLSS represents the quantity of microorganisms involved in the treatment of the organic materials in the aeration basin, while the organic loading determines the requirements for the design of the aeration system.

Modifications to the conventional activated sludge process include:

- *Extended aeration.* The mixed liquor is retained in the aeration basin until the production rate of new cells is the same as the decay rate of existing cells, with no excess sludge production. In practice, excess sludge is produced, but the quantity is less than that of other activated sludge processes. This process is often used for the treatment of industrial wastewater that contains complex organic materials requiring long detention times for degradation.

- *Contact stabilization.* This process is based on the premise that as wastewater enters the aeration basin (referred to as the contact basin), colloidal and insoluble organic biodegradable materials are removed rapidly by biological sorption, synthesis, and flocculation during a relatively short contact time. This method uses a reaeration (stabilization) basin before the settled sludge from the clarifier is returned to the contact basin. The concentrated flocculated and adsorbed organic materials are oxidized in the reaeration basin, which does not receive any addition of raw wastewater.

- *Plug flow.* Wastewater is routed through a series of channels constructed in the aeration basin; wastewater flows through and is treated as a plug as it winds its way through the basin. As the "plug" passes through the tank, the concentrations of organic materials are gradually reduced, with a corresponding decrease in oxygen requirements and microorganism numbers.

- *Step aeration.* Influent wastewater enters the aeration basin along the length of the basin, while the return sludge enters at the head of the basin. This process results in a more uniform oxygen demand in the basin and a more stable environment for the microorganisms; it also results in a lower solids loading on the clarifier for a given mass of microorganisms.

- *Oxidation ditch.* A circular aeration basin (racetrack-shaped) is used, with rotary brush aerators that extend across the width of the ditch. Brush aerators aerate the wastewater, keep the microorganisms in suspension, and drive the wastewater around the circular channel.

Resources

OTHER

United States Environmental Protection Agency (EPA). "Wastes: Solid Waste - Nonhazardous: Sewage Sludge." http://www.epa.gov/ebtpages/wastsolidwas tesewagesludge.html (accessed November 7, 2010).

United States Environmental Protection Agency (EPA). "Water: Wastewater." http://www.epa.gov/ebtpages/ watewastewater.html (accessed November 7, 2010).

Judith Sims

Acute effects

Acute effects are effects that manifest quickly, often dramatically. For example, an acute infection is one of rapid onset and of short duration, which either resolves or becomes chronic (long-term).

Environmental stresses and changes may be characterized as acute (short-term) or chronic (long-term). For example, global climate change imposes both acute and chronic stress on ecosystems.

With regard to toxicity, acute toxicity usually refers to the impact of a short-term exposure to a toxic chemical.

Adams, Ansel

1902–1984
American photographer and conservationist

Ansel Adams is best known for his stark black-and-white photographs of nature and the American landscape. He was born and raised in San Francisco, California. Schooled at home by his parents, he received little formal training except as a pianist. A trip to Yosemite Valley as a teenager had a profound influence on him, and Yosemite National Park and the Sierra "range of light" attracted him back many times and inspired two great careers: photographer and conservationist. As he observed, "Everybody needs something to believe in [and] my point of focus is conservation." He used his photographs to make that point more vivid and turned it into an enduring legacy.

Adams was a painstaking artist, and some critics have chided him for an overemphasis on technique and for creating in his work "a mood that is relentlessly

optimistic." Adams *was* a careful technician, making all of his own prints (reportedly hand-producing over 13,000 in his lifetime), sometimes spending a whole day on one print. He explained, "I have made thousands of photographs of the natural scene, but only those images that were most intensely felt at the moment of exposure have survived the inevitable winnowing of time."

He did winnow, ruthlessly, and the result was a collection of work that introduced millions of people to the majesty and diversity of the American landscape. Not all of Adams's pictures were uplifting or optimistic images of scenic wonders; he also documented scenes of overgrazing in the arid Southwest and of incarcerated Japanese Americans in the Manzanar internment camp.

From the beginning, Adams used his photographs in the cause of conservation. His pictures played a major role in the late 1930s in establishing Kings Canyon National Park. Throughout his life, he remained an active, involved conservationist; for many years he was on the board of the Sierra Club and strongly influenced the club's activities and philosophy.

Ansel Adams's greatest bequest to the world will remain his photographs and advocacy of wilderness and the national park ideals. Through his work he not only generated interest in environmental conservation, he also captured the beauty and majesty of nature for all generations to enjoy.

On August 20, 2007, Adams was posthumously named for induction to the California Hall of Fame.

Resources

BOOKS

Adams, Ansel, and Andrea Gray Stillman. *Ansel Adams in the National Parks: Photographs from America's Wild Places.* New York: Little, Brown, 2010.

Nash, Eric Peter. *Ansel Adams: The Spirit of Wild Places.* New York: New Line Books, 2006.

Gerald L. Young

Adaptation

From the Latin *ad* ("toward") plus *aptus* ("fit for some role"), adaptation refers to any structural, physiological, or behavioral trait that aids an organism's survival and ability to reproduce in its existing environment.

For example, all members of a population share many characteristics in common, such as, all finches in a particular forest being alike in many ways. But if many hard-to-shell seeds are found in the forest, those finches with stronger, more conical bills will have better rates of reproduction and survival than finches with thin bills. Therefore, a conical, stout bill can be considered an adaptation to that forest environment.

Successful genetically-based adaptations are more likely to be passed from generation to generation through the survival of better-adapted organisms.

Adaptive management

Adaptive management is taking an idea, implementing it, and then documenting and learning from any mistakes or benefits of the experiment. The concept can apply in business, but is also relevant and important in environmental science.

The basic idea behind adaptive management is that, for several reasons, natural systems are not predictable. Management policies and procedures must therefore become more adaptive and capable of change to cope with unpredictable systems. Put another way, adaptive management allows decisions to be made when confronted with uncertainty.

Adaptive management was developed in the late 1970s and mid-1980s. Advocates suggest treating management policies as experiments, which are then designed to maximize learning, rather than focusing on immediate resource yields. If the environmental and resource systems on which human beings depend are constantly changing, then societies who utilize that learning cannot rely on those systems to sustain continued use. Adaptive management mandates a continual experimental process, an ongoing process of reevaluation and reassessment of planning methods and human actions, and a constant long-term monitoring of environmental impacts and change. This would keep up with the constant change in the environmental systems to which the policies or ideas are to be applied.

The Grand Canyon Protection Act of 1992 is one example of adaptive management at work. It entailed the study and monitoring of the Glen Canyon Dam and the operational effects on the surrounding environment, both ecological and biological.

A more recent example occurred in 2007. Then, a nonprofit organization called Foundations of Success in concert with the Ocean Conservancy applied adaptive management principles to encourage the end of overfishing. Another example is the 2009 publication of the Louisiana Coastal Protection and Restoration Technical Report by the United States Army Corps of Engineers. The report outlines a strategy to deal with the declining wetlands, marshes, and beaches along Louisiana's coast.

Resources

BOOKS

Aguado, Edward, and James E. Burt. *Understanding Weather and Climate*. Upper Saddle River, NJ: Pearson/Prentice Hall, 2009.

Allan, Catherine, and George Henry Stankey. *Adaptive Environmental Management: A Practitioner's Guide*. New York: Springer, 2009.

Mann, Michael E., and Lee R. Kump. *Dire Predictions: Understanding Global Warming*. Boston: Beacon Press, 2007.

Gerald L. Young

Adirondack Mountains

The Adirondacks are a range of mountains in northeastern New York, containing Mt. Marcy (5,344 ft; 1,644 m), the state's highest point. Bounded by the Mohawk Valley on the south, the St. Lawrence Valley on the northeast, and by the Hudson River and Lake Champlain on the east, the Adirondack Mountains form the core of Adirondack Park. This park is one of the earliest and most comprehensive examples of regional planning in the United States. The regional plan attempts to balance conflicting interests of many users at the same time as it controls environmentally destructive development. Although the plan remains controversial, it has succeeded in largely preserving one of the last and greatest wilderness areas in the East.

The Adirondacks serve a number of important purposes for surrounding populations. Vacationers, hikers, canoeists, and anglers use the area's 2,300 wilderness lakes and extensive river systems. The state's greatest remaining forests stand in the Adirondacks, providing animal habitat and serving recreational visitors. Timber and mining companies, employing much of the area's resident population, also rely on the forests, some of which contain the East's most ancient old-growth groves. Containing the headwaters of numerous rivers, including the Hudson, Adirondack Park is an essential source of clean water for farms and cities at lower elevations.

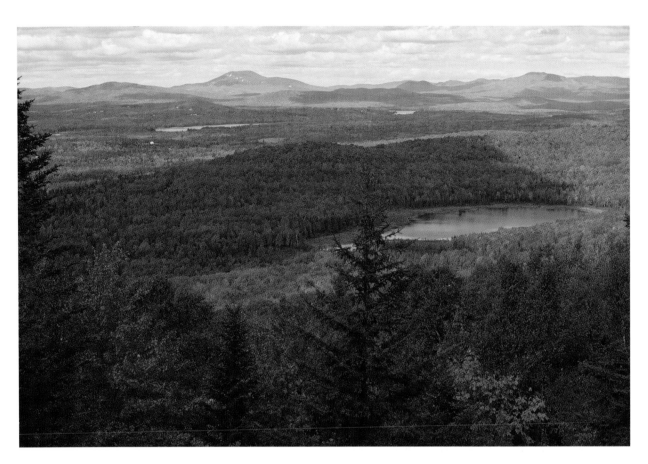

View of the Adirondack Mountains in upstate New York during autumn. *(Roy Whitehead / Photo Researchers, Inc.)*

Adirondack Park was established by the New York State Constitution of 1892, which mandates that the region shall remain "forever wild." Encompassing six million acres (2.4 million ha), this park is the largest wilderness area in the eastern United States—nearly three times the size of Yellowstone National Park. Only one-third of the land within park boundaries, however, is owned by the state of New York. Private mining and timber concerns, public agencies, several towns, thousands of private cabins, and 107 units of local government occupy the remaining property.

Because the development interests of various user groups and visitors conflict with the state constitution, a comprehensive regional land use plan was developed in 1972 and 1973. The novelty of the plan lay in the large area it covered and in its jurisdiction over land uses on private land as well as public land. According to the regional plan, all major development within park boundaries must meet an extensive set of environmental safeguards drawn up by the state's Adirondack Park Agency. Stringent rules and extensive regulations frustrate local residents and commercial interests, who complain about the plan's complexity and resent

"outsiders" ruling on what Adirondackers are allowed to do. Nevertheless, this plan has been a milestone for other regions trying to balance the interests of multiple users. By controlling extensive development, the park agency has preserved a wilderness resource that has become extremely rare in the eastern United States. The survival of this century-old park, surrounded by extensive development, demonstrates the value of preserving wilderness in spite of ongoing controversy.

In recent decades forestry and recreation interests in the Adirondacks have encountered a new environmental problem in acid precipitation. Evidence of deleterious effects of acid rain and snow on aquatic and terrestrial vegetation began to accumulate in the early 1970s. Studies revealed that about one-half of the Adirondack lakes situated above 3,300 feet (1,000 m) have pH levels so low that all fish have disappeared. Prevailing winds put these mountains directly downstream of urban and industrial regions of western New York and southern Ontario. Because they form an elevated obstacle to weather patterns, these mountains capture a great deal of precipitation carrying acidic sulfur and nitrogen oxides from upwind industrial cities. Clean air legislation passed in 1990 has had only a limited impact on the Adirondack lakes. According to a 2009 survey carried out by the Adirondack Lakes Survey Corporation some local pollutant loads have improved but others are getting worse.

Resources

OTHER

National Geographic Society. "Mountains." http://science. nationalgeographic.com/science/earth/surface-of-the-earth/mountains-article.html (accessed October 2, 2010).

United Nations System-Wide EarthWatch. "Mountains." http://earthwatch.unep.net/mountains/index.php (accessed October 2, 2010).

Mary Ann Cunningham

Adsorption

The process where ions or molecules from solutions become bound to solid surfaces. Adsorption is commonly confused with absorption, which occurs when molecules diffuse into a liquid or solid. A widely-used adsorbent used for removal of undesirable particulates or chemicals is activated carbon or charcoal. Activated carbon, either granulated activated carbon (GAC) or powdered activated carbon (PAC), is used in water purification systems to extract chemicals or organic contaminants from the water through adsorption. Phosphorus (P) is removed from water flowing through soils by adsorption on soil particles. Some pesticides adsorb strongly on soil particles. Adsorption by suspended solids is also an important process in natural waters. Adsorption can also be used for removal of pollutants from air. Pollution-control systems use components referred to as scrubbers equipped with adsorbents to remove pollutants such as sulfur dioxide (SO_2) in order to reduce harmful air-pollutant emissions.

Resources

BOOKS

Inglezakis, Vassilis J., and Stavros G. Poulopoulos. *Adsorption, Ion Exchange and Catalysis: Design of Operations and Environmental Applications.* Amsterdam: Elsevier, 2006.

Yaws, Carl L. *Yaws Handbook of Properties for Environmental and Green Engineering: Adsorption Capacity, Water Solubility, Henry's Law Constant.* Houston, TX: Gulf Publishing, 2008.

AEC *see* **Atomic Energy Commission.**

AEM *see* **Agricultural environmental management.**

Aeration

With regard to plant growth, aeration refers to an exchange that takes place in soil or another medium allowing oxygen to enter and carbon dioxide to escape into the atmosphere. Crop growth is often reduced when aeration is poor. In geology, particularly with reference to groundwater, aeration is the portion of Earth's crust where the pores are only partially filled with water. In relation to water treatment, aeration is the process of exposing water to air in order to remove such undesirable substances in drinking water as iron and manganese.

Aerobic

Aerobic refers to either an environment that contains molecular oxygen gas (O_2); an organism or tissue that requires oxygen for its metabolism; or a chemical or biological process that requires oxygen. Aerobic organisms use molecular oxygen in respiration, releasing carbon dioxide (CO_2) in return. These organisms include mammals, fish, birds, and green plants, as well as many of the lower life forms such as fungi, algae, and sundry bacteria and actinomycetes. Many, but not all, organic decomposition processes are aerobic; a lack of oxygen halts or greatly slows these processes.

Aerobic sludge digestion

Wastewater treatment plants produce organic sludge as wastewater is treated; this sludge must be further treated before ultimate disposal. Sludges are generated from primary settling tanks, which are used to remove settable, particulate solids, and from secondary clarifiers (settling basins), which are used to remove excess biomass production generated in secondary biological treatment units.

Disposal of sludges from wastewater treatment processes is a costly and difficult problem. The processes used in sludge disposal include: (1) reduction in sludge volume, primarily by removal of water, which constitutes 97–98 percent of the sludge; (2) reduction of the volatile (organic) content of the sludge, which eliminates nuisance conditions by reducing putrescibility and reduces threats to human health by reducing levels of microorganisms; and (3) ultimate disposal of the residues.

Aerobic sludge digestion is one process that may be used to reduce both the organic content and the volume of the sludge. Under aerobic conditions, a large portion of the organic matter in sludge may be oxidized biologically by microorganisms to carbon dioxide and water. The process results in approximately 50 percent reduction in solids content. Aerobic sludge digestion facilities may be designed for batch or continuous flow operations. In batch operations, sludge is added to a reaction tank, while the contents are continuously aerated. Once the tank is filled, the sludges are aerated for two to three weeks, depending on the types of sludge. After aeration is discontinued, the solids and liquids are separated. Solids at concentrations of 2–45 percent are removed, and the clarified liquid supernatant is decanted and recycled to the wastewater treatment plant. In a continuous flow system, an aeration tank is utilized, followed by a settling tank.

Aerobic sludge digestion is usually used only for biological sludges from secondary treatment units, in the absence of sludges from primary treatment units. The most commonly used application is for the treatment of sludges wasted from extended aeration systems (which is a modification of the activated sludge system). Since there is no addition of an external food source, the microorganisms must utilize their own cell contents for metabolic purposes in a process called endogenous respiration. The remaining sludge is a mineralized sludge, with remaining organic materials comprised of cell walls and other cell fragments that are not readily biodegradable.

The advantages of using aerobic digestion, as compared to the use of anaerobic digestion, include (1) simplicity of operation and maintenance; (2) lower capital costs; (3) lower levels of biochemical oxygen demand (BOD) and phosphorus in the supernatant; (4) fewer effects from upsets such as the presence of toxic interferences or changes in loading and pH; (5) less odor; (6) nonexplosive; (7) greater reduction in grease and hexane solubles; (8) greater sludge fertilizer value; (9) shorter retention periods; and (10) an effective alternative for small wastewater treatment plants.

Disadvantages include: (1) higher operating costs, especially energy costs; (2) highly sensitive to ambient temperature (operation at temperatures below 59°F [15°C] may require excessive retention times to achieve stabilization; if heating is required, aerobic digestion may not be cost-effective); (3) no useful by-product such as methane gas that is produced in anaerobic digestion; (4) variability in the ability to dewater to reduce sludge volume; (5) less reduction in volatile solids; and (6) unfavorable economics for larger wastewater treatment plants.

Resources

OTHER

United States Environmental Protection Agency (EPA). "Wastes: Solid Waste - Nonhazardous: Sewage Sludge." http://www.epa.gov/ebtpages/wastsolidwastesewagesludge.html (accessed November 7, 2010).
United States Environmental Protection Agency (EPA). "Water: Wastewater." http://www.epa.gov/ebtpages/watewastewater.html (accessed November 7, 2010).

Judith Sims

Aerobic/anaerobic systems

Most living organisms require oxygen to function normally, but a few forms of life exist exclusively in the absence of oxygen and some can function both in the presence of oxygen (aerobically) and in its absence (anaerobically). Examples of anaerobic organisms are found in bacteria of the genus *Clostridium*, found in parasitic protozoans from the gastrointestinal tract of humans and other vertebrates, and in ciliates associated with sulfide-containing sediments. Organisms capable of switching between aerobic and anaerobic existence are found in forms of fungi known as yeasts. The ability of an organism to function both aerobically and anaerobically increases the variety of sites in which it is able to exist and conveys some advantages over organisms with less adaptive potential.

Microbial decay activity in nature can occur either aerobically or anaerobically. Aerobic decomposers of compost and other organic substrates are generally preferable because they act more quickly and release fewer noxious odors. Large sewage treatment plants use a two-stage digestion system in which the first stage is anaerobic digestion of sludge that produces flammable methane gas that may be used as fuel to help operate the plant. Sludge digestion continues in the aerobic second stage, a process which is easier to control but more costly because of the power needed to provide aeration. Although most fungi are generally aerobic organisms, yeasts used in bread-making and in the production of fermented beverages such as wine and beer can metabolize anaerobically. In the process, they release ethyl alcohol and the carbon dioxide that causes bread to rise.

Tissues of higher organisms may have limited capability for anaerobic metabolism, but they need elaborate compensating mechanisms to survive even brief periods without oxygen. For example, human muscle tissue is able to metabolize anaerobically when blood cannot supply the large amounts of oxygen needed for vigorous activity. Muscle contraction requires an energy-rich compound called adenosine triphosphate (ATP). Muscle tissue normally contains enough ATP for twenty to thirty seconds of intense activity. ATP must then be metabolically regenerated from glycogen, the muscle's primary energy source. Muscle tissue has both aerobic and anaerobic metabolic systems for regenerating ATP from glycogen. Although the aerobic system is much more efficient, the anaerobic system is the major energy source for the first minute or two of exercise. The carbon dioxide released in this process causes the heart rate to increase. As the heart beats faster and more oxygen is delivered to the muscle tissue, the more efficient aerobic system for generating ATP takes over. A person's physical condition is important in determining how well the aerobic system is able to meet the needs of continued activity. In fit individuals who exercise regularly, heart function is optimized, and the heart is able to pump blood rapidly enough to maintain aerobic metabolism. If the oxygen level in muscle tissue drops, anaerobic metabolism will resume. Toxic products of anaerobic metabolism, including lactic acid, accumulate in the tissue, and muscle fatigue results.

Other interesting examples of limited anaerobic capability are found in the animal kingdom. Some diving ducks have an adaptation that allows them to draw oxygen from stored oxyhemoglobin and oxymyoglobin in blood and muscles. This adaptation permits them to remain submerged in water for extended periods. To prevent desiccation, mussels and clams close their shells when out of the water at low tide, and their metabolism shifts from aerobic to anaerobic. When once again in the water, the animals rapidly return to aerobic metabolism and purge themselves of the acid products of anaerobiosis accumulated while they were dry.

Resources

BOOKS

Betsy, Tom, and James Keogh. *Microbiology Demystified.* New York: McGraw-Hill Professional, 2005.

Leadbetter, Jared R., ed. *Environmental Microbiology.* Amsterdam; Boston: Elsevier Academic Press, 2005.

Pasteur, Louis. *Louis Pasteur's Studies on Fermentation: The Diseases of Beer, Their Causes, and the Means of Preventing Them.* New York: MacMillan, 2005 [1879].

Douglas C. Pratt

Aerosol

An aerosol is a suspension of particles, liquid or solid, in a gas. The term implies a degree of permanence in the suspension, with an average particle size of about 0.00004 in (1 micrometer). Thus, in proper use the term connotes the ensemble of the particles and the suspending gas.

The atmospheric aerosol has two major components, generally referred to as coarse and fine particles, with different sources and different composition. Coarse particles result from mechanical processes, such as grinding. The smaller the particles are ground, the more surface they have per unit of mass. Creating new surface requires energy, so the smallest average size that can be created by such processes is limited by the available energy. It is rare for such mechanically-generated particles to be less than 0.00004 inches in diameter. Fine particles, by contrast, are formed by condensation from the vapor phase. For most substances, condensation is difficult from a uniform gaseous state; it requires the presence of pre-existing particles on which the vapors can deposit. Alternatively, very high concentrations of the vapor are required, compared with the concentration in equilibrium with the condensed material. These very small particles penetrate deep into the lung and can have serious health effects.

Hence, fine particles form readily in combustion processes when substances are vaporized. The gas is then quickly cooled. These can then serve as nuclei for the formation of larger particles, still in the fine particle size range, in the presence of condensable vapors. However, in the atmosphere such particles become rapidly more scarce with increasing size, and are relatively rare in sizes much larger than a few micrometers. At about 0.00008 inches (2 micrometers), coarse and fine particles are about equally abundant. In recent years there has been an interest in ultrafines, particles about 10–100 nanometers in diameter. These are produced by diesel engines, for example, but their implications for health are not well-understood.

Using the term strictly, one rarely samples the atmospheric aerosol, but rather the particles out of the aerosol. The presence of aerosols is generally detected by their effect on light. Aerosols of a uniform particle size in the vicinity of the wavelengths of visible light can produce rather spectacular optical effects. In the laboratory, such aerosols can be produced by condensation of the heated vapors of certain oils on nuclei made by evaporating salts from heated filaments. If the suspending gas is cooled quickly, particle size is governed by the supply of vapor compared with the supply of nuclei, and the time available for condensation to occur. Since these can all be made nearly constant throughout the gas, the resulting particles are quite uniform. It is also possible to produce uniform particles by spraying a dilute solution of a soluble material, then evaporating the solvent. If the spray head is vibrated in an appropriate frequency range, the drops will be uniform in size, with the size controlled by the frequency of vibration and the rate of flow of the spray. Obviously, the final particle size is also a function of the concentration of the sprayed solution.

One form of aerosol pollution is soot: tiny, dark particles consisting mostly of carbon that are released by burning many fuels. Some aerosols are light-colored and so tend to cool the earth by reflecting sunlight, thus reducing radiative forcing (energy imbalance between the earth and sunlight); but soot absorbs sunlight and so tends to warm the earth. In 2008, researchers studying soot released by cooking fires in South Asia and coal-burning in East Asia concluded that soot is a more important contributor to climate change than had been hitherto known, edging out methane as the second-largest greenhouse pollutant after carbon dioxide. The melting of glaciers in the Himalayan Mountains may have more to do with soot than carbon dioxide. Dimming of the skies by soot reduces evaporation and regional temperature differences over the Indian Ocean, causing less rain to fall during the monsoon season and so contributing to drought in Southeast Asia. The new estimate of the radiative forcing (energy imbalance between the earth and sunlight) caused by soot was about double that cited by the Intergovernmental Panel on Climate Change (IPCC) in its influential 2007 summary of the state of climate science. Replacing dung and wood as cooking fuels in South Asia with biogas, natural gas, and solar cookers could significantly reduce the problem. The aerosol haze produced by human activities in Asia may be affecting the climate in other regions. Research has shown that the presence of Asian aerosols alters temperature and pressure gradients over the Indian Ocean, which results in the direction of monsoon weather to Australia, thus increasing rainfall and cloud cover in that region. Studies in recent decades show that these climate changes caused by Asian aerosols may be shifting climate patterns southward. More research is being carried out to uncover the affects of human-generated atmospheric aerosols on climate change.

Resources

BOOKS

Harley, Naomi H., and Lev S. Ruzer. *Aerosols Handbook: Measurement, Dosimetry, and Health Effects.* Boca Raton, FL: CRC Press, 2005.
Integrated Land Ecosystem-Atmosphere Processes Study. *Special Issue on Aerosols, Clouds, Precipitation, Climate.* Helsinki, Finland: iLEAPS International Project Office, 2008.
Obernberger, Ingwald. *Aerosols in Biomass Combustion: Formation, Characterisation, Behavior, Analysis, Emissions, Health Effects.* Graz, Austria: BIOS Bioenergiesysteme, 2005.

OTHER

Centers for Disease Control and Prevention (CDC). "Aerosols." http://www.cdc.gov/niosh/topics/aerosols/ (accessed September 17, 2010).
Centers for Disease Control and Prevention (CDC). "Infectious Aerosols." http://www.cdc.gov/niosh/topics/infectaero/ (accessed September 17, 2010).
NASA Earth Observatory. http://earthobservatory.nasa.gov/Features/Aerosols/ (accessed September 17, 2010).
NASA Facts Online. "Atmospheric Aerosols: What Are They, and Why Are They So Important?" http://oea.larc.nasa.gov/PAIS/Aerosols.html (accessed September 17, 2010).
United States Environmental Protection Agency (EPA). "Air: Air Pollutants: Aerosols." http://www.epa.gov/ebtpages/airairpollutantsaerosols.html (accessed September 17, 2010).

James P. Lodge Jr.

Aflatoxin

Aflatoxin is a toxic compound produced by some fungi and among the most potent naturally occurring carcinogens for humans and animals. Aflatoxin intake is positively related to high incidence of liver cancer in humans in many developing countries. In many farm animals aflatoxin can cause acute or chronic diseases. Aflatoxin is a metabolic by-product produced by the fungi *Aspergillus flavus* and the closely-related species *Aspergillus parasiticus* growing on grains and decaying organic compounds. There are four naturally occurring aflatoxins: B_1, B_2, G_1, and G_2. All of these compounds will fluoresce under a UV (black) light around 425 to 450 nanometers providing a qualitative test for the presence of aflatoxins. In general, starch grains, such as corn, are infected in storage when the moisture content of the grain reaches 17 to 18 percent and the temperature is 79° to 99°F (26° to 37°C). However, the fungus may also infect grain in the field under hot, dry conditions.

African Wildlife Foundation

The African Wildlife Foundation (AWF), headquartered in Washington, DC, was established in 1961 to promote the protection of animals native to Africa. In addition to AWF's Washington, DC, headquarters, the organization maintains an office and conservation center in Nairobi, Kenya. AWF also has field offices in the following cities: Johannesburg, South Africa; Arusha, Tanzania; Kinshasa, Democratic Republic of Congo; and Livingstone, Zambia. The African offices promote the idea that Africans themselves are best able to protect the wildlife of their continent. AWF also established the Mweka College of Wildlife Management in Tanzania to provide professional training to rangers and park and reserve wardens. Conservation education, especially as it relates to African wildlife, has always been a major AWF goal—in fact, it has been the association's primary focus since its inception.

AWF carries out its mandate to protect Africa's wildlife through a wide range of projects and activities. Since 1961, AWF has provided a radio communication network in Africa, as well as several airplanes and jeeps for anti poaching patrols. These were instrumental in facilitating the work of Richard Leakey in the Tsavo National Park, Kenya. In 1999, the African Heartlands project was set up to try to connect large areas of wildland, which is home to wild animals. The project also attempts to involve people who live adjacent to protected wildlife areas by asking them to take joint-responsibility for natural resources. The program demonstrates that land conservation and the needs of neighboring people and their livestock can be balanced, and the benefits shared.

Another highly successful AWF program is the Elephant Awareness Campaign. Its slogan, "only elephants should wear ivory," has become extremely popular, both in Africa and the United States, and is largely responsible for bringing the plight of the African elephant (*Loxodonta africana*) to public awareness.

Although AWF is concerned with all the wildlife of Africa, in recent years the group has focused on saving African elephants, black rhinoceroses (*Diceros bicornis*), and mountain gorillas (*Gorilla gorilla berengei*). These species are seriously endangered and are benefiting from AWF's work to aid these and other animals in critical danger.

From its inception, AWF has supported education centers, wildlife clubs, national parks, and reserves. AWF also involves teachers and community leaders in its endeavors with a series of publications and classes. AWF publications, written in Swahili, have been used in both elementary schools and adult literacy classes in African villages.

Resources

BOOKS

Bolen, Eric, and William Robinson. *Wildlife Ecology and Management*. New York: Benjamin Cummings, 2008.

Fulbright, Timothy E., and David G. Hewitt. *Wildlife Science: Linking Ecological Theory and Management Applications*. Boca Raton, FL: CRC Press, 2008.

Skinner, J. D., and Christian T. Chimimba. *The Mammals of the Southern African Subregion*. Cambridge, UK: Cambridge University Press, 2005.

ORGANIZATIONS

African Wildlife Foundation., 1400 16th Street, NW, Suite 120, Washington, DC, USA, 20036, (202) 939-3333, (202) 939-3332, africanwildlife@awf.org, http://www. awf.org

Cathy M. Falk

Africanized bees

The Africanized bee (*Apis mellifera scutellata*), or killer bee, is an extremely aggressive honeybee. This bee evolved when African honeybees were brought to Brazil to mate with other bees to increase honey production. The imported bees were accidentally released and they have since spread northward, traveling at a rate of 300 miles (483 km) per year. The bees first appeared in the United States at the Texas-Mexico border in late 1990, but have now spread across the southern United States.

The bees get their killer title because of their vigorous defense of colonies or hives when disturbed. Aside from temperament, they are much like their counterparts now in the United States, which are European in lineage. Africanized bees are slightly smaller than their more passive cousins. Although Africanized honeybees

An Africanized bee collecting grass pollen in Brazil.
(Photograph by Scott Camazine. Photo Researchers Inc.)

are more adapted to warmer climates and are intolerant of harsh winters, global warming and climate change may contribute to their invasive tendencies and promote their colonization of the northern United States, thus displacing the dwindling number of domestic honeybee colonies.

Honeybees are social insects and live and work together in colonies. When bees fly from plant to plant, they help pollinate flowers and crops. Africanized bees, however, seem to be more interested in reproducing than in honey production or pollination. For this reason they are constantly swarming and moving around, while domestic bees tend to stay in local, managed colonies. Swarming is defined as a group of bees that separate from their present colony and move to a new location or hive. European or domestic honeybees swarm about once a year, while Africanized bees may swarm every six weeks. Africanized bees are also not as particular as domestic honeybees about choosing a new location. Thus, finding a new home is easier for Africanized bees. Because Africanized bees are also much more aggressive than domestic honeybees when their colonies are disturbed, they can be harmful to people who are allergic to bee stings.

More problematic than the threat to humans, however, is the impact the bees will have on fruit and vegetable industries in the southern parts of the United States. Many fruit and vegetable growers depend on honeybees for pollination, and in places where the Africanized bees have appeared, honey production has fallen by as much as 80 percent. Beekeepers in this country are experimenting with re-queening

their colonies (removing and replacing Africanized queens with European queens) regularly to ensure that the colonies reproduce gentle offspring.

Another danger is the propensity of the Africanized bee to mate with honeybees of European lineage, a kind of infiltration of the gene pool of more domestic bees. Researchers from the U.S. Department of Agriculture (USDA) are watching for the results of this interbreeding, particularly for those bees that display European-style physiques and African behaviors, or vice versa.

When Africanized bees first appeared in southern Texas, researchers from the USDA's Honeybee Research Laboratory in Weslaco, Texas, destroyed the colony, estimated at 5,000 bees. Some of the members of the 3-pound (1.4-kg) colony were preserved in alcohol and others in freezers for future analysis. Researchers are also developing management techniques, including the annual introduction of young mated European queens into domestic hives, in an attempt to maintain gentle production stock and ensure honey production and pollination. Other methods of generating more gentle colonies include injecting Africanized queens with European honeybee sperm and re-queening highly aggressive colonies.

Africanized bees are now found in much of the South according to the USDA's Agricultural Research Service. The latest state to be affected is Georgia, where a swarm of the bees killed a man in October 2010. There have also been reported colonies in Puerto Rico and the Virgin Islands. Southern Nevada bees were almost 90 percent Africanized in June of 2001. Most of Texas has been labeled as a quarantine zone, and beekeepers are not able to move hives out of these boundaries. The largest colony found to date was in southern Phoenix, Arizona. The hive was almost 6 feet (1.8 m) long and held about 50,000 Africanized bees.

Resources

BOOKS

Green, Jen. *Ants, Bees, Wasps, and Termites.* London: Southwater, 2004.

Horn, Tammy. *Bees in America: How the Honey Bee Shaped a Nation.* Lexington: University of Kentucky, 2006.

Tautz, Jürgen; David C. Sandeman; and Helga R. Heilmann. *The Buzz about Bees: Biology of a Superorganism.* Berlin: Springer, 2008.

Linda Rehkopf

Agency for Toxic Substances and Disease Registry

The Agency for Toxic Substances and Disease Registry (ATSDR) studies the health effects of hazardous substances in general and at specific locations. As indicated by its title, the agency maintains a registry of people exposed to toxic chemicals. Along with the Environmental Protection Agency (EPA), ATSDR prepares and updates profiles of toxic substances. In addition, ATSDR assesses the potential dangers posed to human health by exposure to hazardous substances at Superfund sites. The agency will also perform health assessments when petitioned by a community. Though ATSDR's early health assessments have been criticized, the agency's later assessments and other products are considered more useful.

ATSDR was created in 1980 by the Comprehensive Environmental Response, Compensation, and Liability Act (CERCLA), also known as the Superfund, as part of the U.S. Department of Health and Human Services. As originally conceived, ATSDR's role was limited to performing health studies and examining the relationship between toxic substances and disease. The Superfund Amendments and Reauthorization Act (SARA) of 1986 codified ATSDR's responsibility for assessing health threats at Superfund sites. ATSDR, along with the national Centers for Disease Control and state health departments, conducts health surveys in communities near locations that have been placed on the Superfund's National Priorities List for cleanup.

ATSDR's first assessments were harshly criticized. The General Accounting Office (GAO), a congressional agency that reviews the actions of the federal administration, charged that most of these assessments were inadequate. Some argued that the agency was underfunded and poorly organized. ATSDR only receives about 5 percent of the money appropriated for the Superfund project.

Subsequent health assessments have generally been more complete, but they still may not be adequate in informing the community and the EPA of the dangers at specific sites. In general, ATSDR identifies a local agency to help prepare the health surveys. Unlike many of the first assessments, more recent surveys now include site visits and face-to-face interviews. However, other data on environmental effects are limited. ATSDR only considers environmental information provided by the companies that created the hazard or data collected by the EPA. In addition, ATSDR only assesses health risks from illegal emissions, not from "permitted" emissions. Some scientists contend that not enough is known about the health effects of exposure to hazardous substances to make conclusive health assessments.

Reaction to the performance of ATSDR's other functions has been generally more positive. As mandated by SARA, ATSDR, and the EPA have prepared hundreds of toxicological profiles of hazardous substances. These profiles have been judged generally helpful, and the GAO praised ATSDR's registry of people who have been exposed to toxic substances.

ATSDR still receives criticism from politicians and environmental groups, however, for underreporting the impact of toxic substances on human health. In 2008 and 2009, the U.S. House of Representatives Science Committee held hearings on ATSDR's issuance of a flawed health consultation on the impact of formaldehyde in trailers used by the Federal Emergency Management Agency (FEMA) to house evacuees of hurricanes Katrina and Rita in 2005.

Resources

BOOKS

Kumar, C.S.S.R. *Nanomaterials: Toxicity, Health and Environmental Issues.* Weinheim, Germany: Wiley-VCH, 2006.

Lippmann, Morton, ed. *Environmental Toxicants: Human Exposures and Their Health Effects.* Hoboken, NJ: Wiley-Interscience, 2006.

Rapp, Doris. *Our Toxic World: A Wake-up Call.* Buffalo, NY: Environmental Research Foundation, 2004.

ORGANIZATIONS

Agency for Toxic Substances and Disease Registry, 4770 Buford Highway, NE, Atlanta, GA, USA, 30341, (800) 232-4636, ATSDRIC@cdc.gov, http://www.atsdr.cdc.gov

Alair MacLean

Agent Orange

Agent Orange is a herbicide recognized for its use during the Vietnam War. It is composed of equal parts of two chemicals: 2,4-D and 2,4,5-T. A less potent form of the herbicide has also been used for clearing heavy growth on a commercial basis for a number of years. However, it does not contain 2,4-D. On a commercial level, the herbicide was used in forestry control as early as the 1930s. In the 1950s through the 1960s, Agent Orange was also exported. For example, New Brunswick, Canada, was the scene of major Agent Orange spraying to control forests for industrial development.

Air force aircraft spray Agent Orange in Vietnam. *(© Archive Image / Alamy)*

In Malaysia in the 1950s, the British used compounds with the chemical mixture 2,4,5-T to clear communication routes.

In the United States, herbicides were considered for military use toward the end of World War II, during the action in the Pacific. However, the first American military field tests were actually conducted in Puerto Rico, Texas, and Fort Drum, New York, in 1959.

That same year—1959—the Crops Division at Fort Detrick, Maryland, initiated the first large-scale military defoliation effort. The project involved the aerial application of Agent Orange to about 4 square miles (10.4 km^2) of vegetation. The experiment proved highly successful; the military had found an effective tool. By 1960, the South Vietnamese government, aware of these early experiments, had requested that the United States conduct trials of these herbicides for use against guerrilla forces. Spraying of Agent Orange in Southeast Asia began in 1961. South Vietnam

President Ngo Dinh Diem stated that he wanted this powder in order to destroy the rice and the food crops that would be used by the Vietcong. Thus began the use of herbicides as a weapon of war.

The United States military became involved, recognizing the limitations of fighting in foreign territory with troops who were not accustomed to jungle conditions. The military wanted to clear communication lines and open up areas of visibility in order to enhance their opportunities for success. Eventually, the United States military took complete control of the spray missions. Initially, there were to be restrictions: the spraying was to be limited to clearing power lines and roadsides, railroads and other lines of communications, and areas adjacent to depots. Eventually, the spraying was used to defoliate the thick jungle brush, thereby obliterating enemy hiding places.

Once under the authority of the military, and with no checks or restraints, the spraying continued to

OTHER

United States Congress. *Our Forgotten Responsibility: What Can We Do to Help Victims of Agent Orange?: Hearing before the Subcommittee on Asia, the Pacific, and the Global Environment of the Committee on Foreign Affairs, House of Representatives, One-hundred Tenth Congress, Second Session, May 15, 2008*. Washington, DC: U.S. Government Printing Office.

Martin, M. F. *Vietnamese Victims of Agent Orange and U.S.-Vietnam Relations*. Ft. Belvoir: Defense Technical Information Center, 2008. Available at http://handle.dtic.mil/100.2/ADA490443

Liane Clorfene Casten
Paula Anne Ford-Martin

Agglomeration

Agglomeration refers to a process by which a group of individual particles is clumped together into a single mass. The term has a number of specialized uses. Some types of rocks are formed by the agglomeration of particles of sand, clay, or some other material. In geology, an agglomerate is a rock composed of volcanic fragments. One technique for dealing with air pollution is ultrasonic agglomeration (U.S. agglomeration), also referred to as ultrasound-assisted agglomeration. A source of very high frequency sound is attached to a smokestack, and the ultrasound produced by this source causes tiny particulate matter in waste gases to agglomerate into particles large enough to be collected. U.S. agglomeration has been employed for environmental remediation to collect by-products of coal combustion, such as fly ash, and to remove particles from diesel exhaust. Agglomeration with petroleum coke, a solid formed during oil refinement, as an adsorbent has also been used for soil remediation for cleanup of soil contaminated with oil. This is also a technical term in the European Union to define an air quality management area.

Resources

BOOKS

Kaufmann R., and C. Cleveland. *Environmental Science*. New York: McGraw-Hill International Edition, 2008.

Luque de Castro, M. D., and F. Priego Capote. *Analytical Applications of Ultrasound, Vol. 26, Techniques and Instrumentation in Analytical Chemistry*. Amsterdam, Netherlands: Elsevier, 2007.

Pfafflin, J. R. *The Dictionary of Environmental Science and Engineering*. New York: Routledge, 2008.

Agricultural chemicals

The term agricultural chemical refers to any substance involved in the growth or utilization of any plant or animal of economic importance to humans. An agricultural chemical may be a natural product, such as urea, or a synthetic chemical, such as dichlorodiphenyltrichloroethane (DDT). The agricultural chemicals used in more recent decades include fertilizers, pesticides, growth regulators, animal feed supplements, and raw materials for use in chemical processes.

In the broadest sense, agricultural chemicals can be divided into two large categories: those that promote the growth of a plant or animal and those that protect plants or animals. Plant fertilizers and animal food supplements constitute the first group, while pesticides, herbicides, animal vaccines, and antibiotics belong to the second group.

In order to remain healthy and grow normally, crops require a number of nutrients, some in relatively large quantities termed "macronutrients," and others in relatively small quantities termed "micronutrients." Nitrogen (N), phosphorus (P), and potassium (K) are considered macronutrients, and boron (B), calcium (Ca), chlorine (Cl), copper (Cu), iron (Fe), magnesium (Mg), and manganese (Mn), among others, are micronutrients.

Farmers have long-understood the importance of replenishing the soil, and they have traditionally done so by natural means, using such materials as manure, dead fish, or compost. Synthetic fertilizers were first available in the early twentieth century, but they became widely used only after World War II (i.e., 1945). By 1990, farmers in the United States were using about 20 million tons (18.1 million metric tons)

Agricultural chemical treatments in spring vineyard, Oltrepo Pavese, Italy. (*scattoselvaggio/Shutterstock.com.*)

of these fertilizers a year. The average fertilizer use from 2005 to 2006 for all crops was 21.7 million tons (19.7 million metric tons) per year. Total use of the three major commercial fertilizers has increased 194 percent from 1960 to 2006 based on fertilizer sales.

Synthetic fertilizers are designed to provide either a single nutrient or some combination of nutrients. Examples of single-component or straight fertilizers are urea (NH_2CONH_2), which supplies nitrogen, or potassium chloride (KCl), which supplies potassium. The composition of mixed fertilizers, those containing more than one nutrient, is indicated by the analysis printed on their container. An 8–10–12 fertilizer, for example, contains 8 percent nitrogen by weight, 10 percent phosphorus, and 12 percent potassium.

Synthetic fertilizers can be designed to release nutrients almost immediately (quick-acting) or over longer periods of time (time-released). They may also contain specific amounts of one or more trace nutrients needed for particular types of crops or soil. Controlling micronutrients is one of the most important problems in fertilizer compounding and use; the presence of low concentrations of some elements can be critical to a plant's health, whereas higher levels can be toxic to the same plants or to animals that might ingest the micronutrient.

Plant growth patterns can also be influenced by direct application of certain chemicals. For example, the gibberellins are a class of compounds that can dramatically affect the rate at which plants grow and fruits and vegetables ripen. They have been used for a variety of purposes ranging from the hastening of root development to the delay of fruit ripening. Delaying ripening is most important for marketing agricultural products because it extends the time a crop can be transported and stored on grocery shelves. Other kinds of chemicals used in the processing, transporting, and storage of fruits and vegetables include those that slow down or speed up ripening (e.g., maleic hydrazide, $C_4H_4N_2O_2$; ethylene oxide, C_2H_4O; potassium permanganate, $KMnO_4$; ethylene, C_2H_4; and acetylene, HC_2H), that reduce weight loss (e.g., chlorophenoxyacetic acid, $C_9H_9ClO_3$), retain green color (e.g., cycloheximide, $C_{15}H_{23}NO_4$), and control firmness (e.g., ethylene oxide).

The term "agricultural chemical" is most likely to bring to mind the range of chemicals used to protect plants against competing organisms: pesticides and herbicides. These chemicals disable or kill bacteria, fungi, rodents, worms, snails and slugs, insects, mites, algae, termites, or any other species of plant or animal that feeds upon, competes with, or otherwise interferes with the growth of crops. Such chemicals are named according to the organism against which they are designed to act. Some examples are fungicides (designed to kill fungi), insecticides (used against insects), nematicides (to kill round worms), avicides (to control birds), and herbicides (to combat plants).

The introduction of synthetic pesticides in the years following World War II, which ended in 1945, produced spectacular benefits for farmers. More than fifty major new products appeared between 1947 and 1967, resulting in yield increases in the United States ranging from 400 percent for corn to 150 percent for sorghum and 100 percent for wheat and soybeans. Similar increases in less developed countries, resulting from the use of both synthetic fertilizers and pesticides, eventually became known as the Green Revolution.

By the 1970s, however, the environmental consequences of using synthetic pesticides became apparent. Chemicals were becoming less effective as pests developed resistances to them, and their toxic effects on other organisms had grown more apparent. Farmers were also discovering drawbacks to chemical fertilizers as they found that they had to use larger and larger quantities each year in order to maintain crop yields, resulting in increased material and labor costs. One solution to the environmental hazards posed by synthetic pesticides is the use of natural chemicals such as juvenile hormones, sex attractants, and anti-feedant compounds. The development of such natural pest-control materials has, however, been relatively modest; the vast majority of agricultural companies and individual farmers continue to use synthetic chemicals that have served them so well for over a half century.

Chemicals are also used to maintain and protect livestock. At one time, farm animals were fed almost exclusively on readily available natural foods. They grazed on rangelands or were fed hay or other grasses. In recent decades, carefully blended chemical supplements are commonly added to the diet of most farm animals. These supplements have been determined on the basis of extensive studies of the nutrients that contribute to the growth or milk production of cows, sheep, goats, and other types of livestock. A typical animal supplement diet consists of various vitamins, minerals, amino acids, and nonprotein (simple) nitrogen compounds. The precise formulation depends primarily on the species; a vitamin supplement for cattle, for example, tends to include A, D, and E, while swine and poultry diets would also contain Vitamin K, riboflavin (B_2), niacin (B_3), pantothenic acid (B_5), and choline ($C_5H_{14}NO^+$).

A number of chemicals added to animal feed serve no nutritional purpose but provide other benefits. For example, the addition of certain hormones to the feed of dairy cows can significantly increase their output of milk. Genetic engineering is also becoming increasingly important in the modification of crops and livestock. Cows injected with a genetically modified chemical, bovine somatotropin, produce a significantly larger quantity of milk.

It is estimated that infectious diseases cause the death of 15-20 percent of all farm animals each year. Just as plants are protected from pests by pesticides, livestock are protected from disease-causing organisms by immunization, antibiotics, and other techniques. Animals are vaccinated against species-specific diseases, and farmers administer antibiotics, sulfonamides, nitrofurans, arsenicals, and other chemicals that protect against disease-causing organisms. Some additives are simply adulterants. In 2008, the surplus of melamine ($C_3H_6N_6$) produced in China led it to be added to cattle feed, to boost the apparent protein content. As a result of melamine-contaminated milk ingestion, children died and hundreds of thousands of people became sick. The ensuing scandal led to several death sentences from the Chinese courts.

The use of chemicals with livestock can have deleterious effects, just as crop chemicals have. In the 1960s, for example, the hormone diethylstilbestrol (DES) was widely used to stimulate the growth of cattle, but scientists found that detectable residues of the hormone remained in meat sold from the slaughtered animals. DES is now considered a carcinogen, and the U.S. Food and Drug Administration banned its use in cattle feed as of 1979.

Because of increased health and environmental risks presented by agricultural chemicals, crop production utilizing either integrated pest management (IPM) or organic methods is being increasingly implemented. IPM practices focus on using biological controls to minimize the use of chemicals. Biological controls consist of crop rotation, use of cover crops, pest predator introduction, use of pest-resistant varieties, and tillage or cultivation to prevent weed growth. Timing of planting and chemical application, if any, is based on monitoring pest numbers and pest mating cycles. No synthetic pesticides are used for organic methods of crop production. Biological controls and natural pesticides are utilized for pest control in organic agriculture.

Resources

BOOKS

Allen, Will. *The War on Bugs*. White River Junction, VT: Chelsea Green Publishing, 2008.

Clark, John Marshall, and Hideo Ohkawa. *Environmental Fate and Safety Management of Agrochemicals*. Washington, DC: American Chemical Society, 2005.

Hardy, Sandra, and Mark Scott. *Spray Sense: Safe and Effective Use of Farm Chemicals*. Orange, NSW, Australia: NSW Department of Primary Industries, 2006.

Kennedy, I. R. *Rational Environmental Management of Agrochemicals: Risk Assessment, Monitoring, and Remedial Action*. Washington, DC: American Chemical Society, 2007.

Milne, George W. A. *Pesticides: An International Guide to 1800 Pest Control Chemicals*. Aldershot, Hampshire, England: Ashgate, 2004.

Tressaud, Alain. *Fluorine and the Environment: Agrochemicals, Archaeology, Green Chemistry & Water*. Vol. 2. Amsterdam, Netherlands: Elsevier, 2006.

Uri, Noel D. *Agriculture and the Environment*. New York: Novinka, 2006.

Washington State Library. *Best Management Practices for Agricultural Chemicals: A Guide for Pesticide and Fertilizer Storage and Operation Area Facilities*. Olympia: Washington State Department of Ecology, 2005.

OTHER

European Commission. "From Mad Cows to Melamine: Reliable Food Safety Testing." http://ec.europa.eu/dgs/jrc/downloads/jrc_leaflet_food_safety_en.pdf (accessed August 2, 2010).

United States Environmnetal Protection Agency (EPA). "Fertilizer Applied for Agricultural Purposes." http://cfpub.epa.gov/eroe/index.cfm?fuseaction=detail.viewInd&ch=48 &subtop=228&&lv=list.listByChapter &r=188231 (accessed August 2, 2010).

United States Environmental Protection Agency (EPA). "Pollutants/Toxics: Agricultural Chemicals." http://www.epa.gov/ebtpages/pollagriculturalchemicals.html (accessed August 2, 2010).

United States Environmental Protection Agency (EPA) and States Pesticide Data Management and Subscription Service. "2005 Pesticide Industry Report." http://www.know tify.net/2005USPestIndReptExecSum.pdf (accessed August 2, 2010).

David E. Newton

Agricultural environmental management

The complex interaction of agriculture and environment has been an issue since the origins of humans. Humans grow food to eat and also hunt animals that depend on natural resources for healthy ongoing habitats. Therefore, the world's human population must balance farming activities with maintaining natural

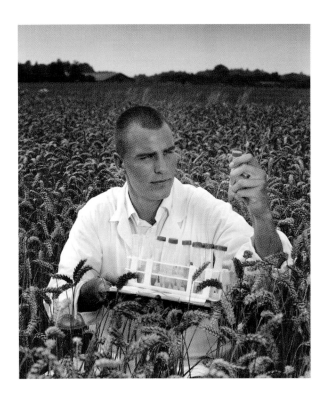
Agriculture scientist examining samples in rye field. *(Noam Armonn/Shutterstock.com)*

farming methods that use fewer synthetic (man-made) pesticides and fertilizers and encourage organic farming. They also work to conserve energy and water.

Soil erosion, converting virgin land to agricultural use, introduction of fertilizer and pesticides, animal wastes, and irrigation are parts of farming that can lead to changes in quality or availability of water. An expanding human population has lead to increased farming and accelerated soil erosion. When soil has a low capacity to retain water, farmers must pump groundwater up and spray it over crops. After years of doing so, the local water table will eventually fall. This water depletion can impact native vegetation in the area.

The industry calls the balance of environment and lessening of agricultural effects sustainability or sustainable development. In some parts of the world, such as the High Plains of the United States or parts of Saudi Arabia, populations and agriculture are depleting water aquifers faster than the natural environment can replenish them. Sustainable development involves dedicated, scientifically-based plans to ensure that agricultural activity is managed in such a way that aquifers are not prematurely depleted.

Agroforestry is a method of cultivating both crops and trees on the same land. Between rows of trees, farmers plant agricultural crops that generate income during the time it takes the trees to grow mature enough to produce earnings from nuts or lumber.

Increased modernization of agriculture also impacts the environment. Traditional farming practice, which continues in underdeveloped countries today, consists of subsistence agriculture. In subsistence farming, just enough crops and livestock are raised to meet the needs of a particular family. However, today large farms produce food for huge populations. More than one-half of the world's working population is employed by some agricultural or agriculturally-associated industry. Almost 40 percent of the world's land area is devoted to agriculture (including permanent pasture). The growing use of machines, pesticides, and man-made fertilizers have all seriously impacted the environment.

For example, the use of pesticides such as dichloro-diphenyltrichloroethane (DDT) in the 1960s was identified as leading to the deaths of certain species of birds. Most Western countries banned use of such harmful pesticides, and some bird populations began to recover. As of 2010, use of pesticides is strictly regulated in the United States.

Many more subtle effects of farming occur on the environment. When grasslands and wetlands or forests are converted to crops, and when crops are not

resources. The term agriculture originally meant the act of cultivating fields or growing crops. However, its meaning has expanded to include raising livestock as well.

When early settlers began farming and ranching in the United States, they faced pristine wilderness and open prairies. There was little cause for concern about protecting the environment or wildlife populations, and for two centuries, the country's land and water were aggressively used to create a healthy supply of ample food for Americans. In fact, many American families settled in rural areas and made a living as farmers and ranchers, passing family businesses down through generations. By the 1930s, the federal government began requiring farmers to idle certain acres of land to prevent oversupply of food and to protect exhausted soil.

Since that time, agriculture has become a complex science, as farmers must carefully manage soil and water to lessen the risk of degrading the soil and its surrounding environment or depleting water tables beneath the land's surface. In fact, farming and ranching present several environmental challenges that require careful management by farmers and local and federal regulatory agencies that guide their activities. The science of applying principles of ecology to agriculture is called agroecology. Those involved in agroecology develop rotated, eventually the land changes to the point that entire species of plants and animals can become threatened. Urbanization also encroaches onto farmland and cuts the amount of land available for farming.

Throughout the world, countries and organizations develop strategies to protect the environment, natural habitats, and resources while still supplying the food human populations require. In 1992, the United Nations Conference on Environment and Development (also known as the Earth Summit) in Rio de Janeiro focused on how to sustain the world's natural resources, with the goal of balancing good policies on environment and community vitality. In the United States, the Department of Agriculture has published its own policy on sustainable development, which works toward balancing economics, environment, and social needs concerning agriculture. In 1993, an Executive Order formed the President's Council on Sustainable Development (PCSD), which developed new approaches for achieving economic and environmental goals for public policy in agriculture. Guiding principles included sections on agriculture, forestry, and rural community development.

According to the United States Environmental Protection Agency (EPA), Agricultural Environmental Management (AEM) is one of the most innovative programs in New York state. Though the state of New York is perceived as mostly urban, in reality most of the state is composed of farmland, forests, mountainous areas, and waterways, with agriculture forming a large part of the state's economy. The AEM program serves important environmental functions and helps keep New York's farms economically viable. As of 2007, more than 10,000 farms were participating in the program.

The program began in June 2000 when Governor George Pataki introduced legislation to the state's Senate and Assembly proposing a partnership to promote farming's good stewardship of land and to provide the funding and support of farmers' efforts. The bill was passed and signed into law by the governor that same year. The purpose of the law is to help farmers develop agricultural environmental management plans that control agricultural pollution and comply with federal, state, and local regulations on use of land, water quality, and other environmental concerns. New York's AEM program brings together agencies from state, local, and federal governments, conservation representatives, businesses from the private sector, and farmers. The program is voluntary and offers education, technical assistance, and financial incentives to farmers to participate.

An example of a successful AEM project occurred at a dairy farm in central New York. The farm composted animals' solid wastes, which reduced the amount of waste spread on the fields. This in turn reduced pollution in the local watershed. The New York State Department of Agriculture and Markets oversees the program. It begins when a farmer expresses interest in AEM. Next, the farmer completes a series of five tiers of the program.

In Tier I, the farmer completes a short questionnaire that surveys current farming activities and future plans to identify potential environmental concerns. Tier II involves worksheets that document current activities that promote stewardship of the environment and help prioritize any environmental concerns. In Tier III, a conservation plan is developed that is tailored specifically for the individual farm. The farmer works together with an AEM coordinator and several members of the cooperating agency staff.

Under Tier IV of the AEM program, agricultural agencies and consultants provide the farmer with educational, technical, and financial assistance to implement the best management practices for preventing pollution to water bodies in the farm's area. The plans use Natural Resources Conservation Service standards and guidance from cooperating professional engineers. Finally, farmers in the AEM program receive ongoing evaluations to ensure that the plan they have devised helps protect the environment and also ensures viability of the farm business.

Funding for the AEM program comes from a variety of sources, including New York's Clean Water/Clean Air Bond Act and the State Environmental Protection Fund. Local soil and water conservation districts (SWCDs) also partner in the effort, and farmers can access funds through these districts. The EPA says involvement of the SWCDs has likely been a positive factor in farmers' acceptance of the program.

Agricultural organizations and businesses in other states and countries are following programs for environmental management and protection. The EPA encourages organizations to implement Environmental Management Systems (EMSs) for which the International Organization of Standardization (ISO) set guidelines in 1996, and updated in 2004. The ISO defines an EMS as "that part of the overall management system which includes organizational structure, planning activities, responsibilities, practices, procedures, processes, and resources for developing, implementing, achieving, reviewing, and maintaining the environmental policy." The EMS standard set forth by the ISO is referred to as ISO 14001 and has seventeen key elements including objectives and targets; environmental policy; training, awareness, and competence; and monitoring and measurement. Development and adherence to EMS procedures plays a significant role in maintaining environmental quality through sustainable agriculture.

Resources

BOOKS

Holthaus, Gary H. *From the Farm to the Table: What All Americans Need to Know About Agriculture.* Lexington: University Press of Kentucky, 2006.

Molden, D. *Water for Food, Water for Life: A Comprehensive Assessment of Water Management in Agriculture.* London: Earthscan, 2007.

OTHER

United States Environmental Protection Agency (EPA). "Agriculture." http://www.epa.gov/agriculture (accessed November 10, 2010).

United States Environmental Protection Agency (EPA). "Ecosystems: Agroecosystems: Agriculture." http://www.epa.gov/ebtpages/ecosagroecosystemsagriculture.html (accessed November 10, 2010).

Teresa G. Norris

Agricultural pollution

The development of modern agricultural practices is one of the great success stories of applied sciences. Improved plowing techniques, new pesticides and fertilizers, and better strains of crops are among the factors that have resulted in significant increases in agricultural productivity.

Yet these improvements have not come without cost to the environment and sometimes to human health. Modern agricultural practices have contributed to the pollution of air, water, and land. Air pollution may be the most memorable, if not the most significant, of these consequences. During the 1920s and 1930s, huge amounts of fertile topsoil were blown away across vast stretches of the Great Plains, an area that eventually became known as the "dust bowl." The problem occurred because of soil erosion; farmers either did not know about or chose not to use techniques for protecting and conserving their soil. The soil then blew away during droughts, resulting not only in the loss of valuable farmland, but also in the pollution of the surrounding atmosphere.

Summertime satellite observations of ocean color in the Gulf of Mexico show highly turbid waters, which may include large blooms of phytoplankton because of agricultural runoff carried by rivers to the sea. *(Courtesy of NASA)*

Soil conservation techniques developed rapidly in the 1930s, including contour plowing, strip cropping, crop rotation, windbreaks, and minimum- or no-tillage farming, and thereby greatly reduced the possibility of erosion on such a scale. However, such events, though less dramatic, have continued to occur, and in recent decades they have presented new problems. When topsoil is blown away by winds, they can carry with them the pesticides, herbicides, and other crop chemicals now so widely used. In the worst cases, these chemicals have contributed to the collection of air pollutants that endanger the health of plants and animals, including humans. Ammonia (NH_3), released from the decay of fertilizers, is one example of a compound that may cause minor irritation to the human respiratory system and more serious damage to the health of other animals and plants.

A more serious type of agricultural pollution are the solid waste problems resulting from farming and livestock practices. Authorities estimate that slightly over one-half of all the solid wastes produced in the United States each year—a total of about 2 billion tons (1.8 billion metric tons)—come from a variety of agricultural activities. Some of these wastes pose little or no threat to the environment. Crop residue left on cultivated fields and animal manure produced on rangelands, for example, eventually decay, returning valuable nutrients to the soil.

Some modern methods of livestock management, however, tend to increase the risks posed by animal wastes. Farmers are raising a larger variety of animals, as well as larger numbers of them, in smaller and smaller areas such as feedlots or huge barns. In such cases, large volumes of wastes are generated in these areas. Many livestock managers attempt to sell these waste products or dispose of them in a way that poses no threat to the environment. Yet in many cases the wastes are allowed to accumulate in massive dumps where soluble materials are leached out by rain. Some of these materials then find their way into groundwater or surface water, such as lakes and rivers. Some are harmless to the health of animals, though they may contribute to the eutrophication (excess primary productivity as a result of excess nutrient concentrations) of lakes and ponds. Other materials, however, may have toxic, carcinogenic, or genetic effects on humans and other animals.

The leaching of hazardous materials from animal-waste dumps contributes to perhaps the most serious form of agricultural pollution: the contamination of water supplies. Many of the chemicals used in agriculture today can be harmful to plants and animals. Pesticides and herbicides are the most obvious of these; used by farmers to disable or kill plant and animal pests, they may also cause problems for beneficial plants and animals as well as humans.

Runoff from agricultural land is another serious environmental problem posed by modern agricultural practices. Runoff constitutes a nonpoint source of pollution. Rainfall leaches out and washes away pesticides, fertilizers, and other agricultural chemicals from a widespread area, not a single source such as a sewer pipe. Maintaining control over nonpoint sources of pollution is an especially difficult challenge. In addition, agricultural land is more easily leached out than is nonagricultural land. When lands are plowed, the earth is broken up into smaller pieces, and the finer the soil particles, the more easily they are carried away by rain. Studies have shown that the nitrogen (N) and phosphorus (P) in chemical fertilizers are leached out of croplands at a rate about five times higher than from forest woodlands or idle lands.

The accumulation of nitrogen and phosphorus in waterways from chemical fertilizers has contributed to the acceleration of eutrophication of lakes and ponds. Scientists believe that the addition of human-made chemicals such as those in chemical fertilizers can increase the rate of eutrophication by a factor of at least ten. A more deadly effect is the poisoning of plants and animals by toxic chemicals leached off of farmlands. The biological effects of such chemicals are commonly magnified many times as they move up a food chain. The best-known example of this phenomenon involved a host of biological problems—from reduced rates of reproduction to malformed animals to increased rates of death—

attributed to the use of dichlorodiphenyltrichloro-ethane (DDT) in the 1950s and 1960s.

Sedimentation also results from the high rate of erosion on cultivated land, and increased sedimentation of waterways poses its own set of environmental problems. Some of these are little more than cosmetic annoyances. For example, lakes and rivers may become murky and less attractive, losing potential as recreation sites. However, sedimentation can block navigation channels, and other problems may have fatal results for organisms. Aquatic plants may become covered with sediments and die; marine animals may take in sediments and be killed; and cloudiness from sediments may reduce the amount of sunlight received by aquatic plants so extensively that they can no longer survive.

Environmental scientists are especially concerned about the effects of agricultural pollution on groundwater. Groundwater is polluted by much the same mechanisms as is surface water, and evidence for that pollution has accumulated rapidly in the past decade. Groundwater pollution tends to persist for long periods of time. Water flows through an aquifer much more slowly than it does through a river, and agricultural chemicals are not flushed out quickly.

Many solutions are available for the problems posed by agricultural pollution, but many of them are not easily implemented. Chemicals that are found to have serious toxic effects on plants and animals can be banned from use, such as DDT in the 1970s, but this kind of decision is seldom easy. Regulators must always assess the relative benefit of using a chemical, such as increased crop yields, against its environmental risks. Such a risk-benefit analysis means that some chemicals known to have certain deleterious environmental effects remain in use because of the harm that would be done to the agricultural economy if they were banned.

Another way of reducing agricultural pollution is to implement better farming techniques. In the practices of minimum-farming or no-tillage farming, for example, plowing is reduced or eliminated entirely. Ground is left essentially intact, reducing the rate at which soil and the chemicals it contains are eroded away. This method also prevents the breaking up of soil into increasingly smaller particles that would be easily eroded. Reducing the amount of chemicals necessary for efficient crop production can be achieved through integrated pest management (IPM) processes and crop rotation, key methods in the implementation of sustainable agriculture. Sustainable agriculture focuses on the profitable and efficient production of crops while minimally affecting the ecosystem. Through IPM, increased crop monitoring to obtain information about pest numbers and status can help determine the optimal time for pesticide application, reducing the need for future applications. Both sustainable agriculture and organic farming have become more popular methods for crop production due to their minimal environmental impact.

Agriculture has a significant influence on the environment and climate change. Agricultural emissions can have a global impact through the emission of greenhouse gases, such as methane (CH_4) from cattle and paddy fields or ozone depleting nitrous oxide (N_2O) from fertile farmland. The growing global population has pushed humans, in terms of food production, into previously uninhabited areas, resulting in a large amount of deforestation. In developing countries, slash-and-burn agriculture is employed to quickly gain new area for crop production. This process releases pent-up carbon from the plants into the atmosphere as well as removes plants that were previously able to process carbon in the air, resulting in an increase in greenhouse gas emissions.

Resources

BOOKS

Brubaker, Elizabeth. *Greener Pastures: Decentralizing the Regulation of Agricultural Pollution.* Toronto: University of Toronto Press, 2007.

Cook, Christopher D. *Diet for a Dead Planet: How the Food Industry Is Killing Us.* New York: New Press, 2004.

McDowell, Richard W. *Grazed Pastures and Surface Water Quality.* New York: Nova Science Publishers, 2008.

Reddy, Rajendra, and J. P. Abhay Shankar. *Agricultural Pollution.* New Delhi: Commonwealth Publishers, 2008.

PERIODICALS

Workshop on Agricultural Air Quality: State of the Science, and Viney P. Aneja. "Agricultural Air Quality: State of the Science." *Journal of the Air and Waste Management Association,* 58 (9) (2008).

David E. Newton

Agricultural Research Service

A branch of the U.S. Department of Agriculture charged with the responsibility of agricultural research on a regional or national basis, the Agricultural Research Service (ARS) has a mission to develop new knowledge and technology needed to solve agricultural problems of broad scope and high national priority in order to ensure adequate production of high-quality food and agricultural products for the United States. The national research center of the ARS is located at Beltsville, Maryland, consisting of laboratories, land, and other facilities. In addition, there are many other research centers located throughout the United States. Scientists of the ARS are also located at land grant universities throughout the country where they conduct cooperative research with state scientists.

Agricultural revolution

The development of agriculture has been a fundamental part of the march of civilization. It is an ongoing challenge, for as long as population growth continues, mankind will need to improve agricultural production.

The agricultural revolution is actually a series of four major advances, closely linked with other key historical periods. The first, the *Neolithic* or New Stone Age, marks the beginning of sedentary (settled) farming. Much of this history is lost in antiquity, dating back perhaps 10,000 years or more. Still, humans owe an enormous debt to those early pioneers who so painstakingly nourished the best of each year's crop. Archaeologists have found corn cobs a mere two inches (5.1 cm) long, so different from today's much larger ears.

The second major advance came as a result of Genoese explorer Christopher Columbus's (c. 1451–1506) voyages to the New World. Isolation had fostered the development of two completely independent agricultural systems in the New and Old Worlds. A short list of interchanged crops and animals clearly illustrates the global magnitude of this event; furthermore, the current population explosion began its upswing during this period. From the New World came maize, beans, the Irish potato, squash, peanuts, tomatoes, and tobacco. From the Old World came wheat, rice, coffee, cattle, horses, sheep, and goats. Maize is now a staple food in Africa. Several Indian tribes in America adopted new lifestyles, notably the Navajo as sheepherders and the Cheyenne as nomads using the horse to hunt buffalo.

The Industrial Revolution both contributed to and was nourished by agriculture. The greatest agricultural advances came in transportation, where first canals, then railroads and steamships made the shipment of food from areas of surplus possible. This in turn allowed more specialization and productivity, but most importantly, it reduced the threat of starvation. The steamship ultimately brought refrigerated meat to

Europe from distant Argentina and Australia. Without these massive increases in food shipments, the exploding populations and greatly increased demand for labor by newly emerging industries could not have been sustained.

In turn the Industrial Revolution introduced major advances in farm technology, such as the cotton gin, the mechanical reaper, improved plows, the combine harvester, and, in the twentieth century, tractors and trucks. These advances enabled fewer and fewer farmers to feed larger and larger populations, freeing workers to fill demands for factory labor and the growing service industries. The fixation of atmospheric nitrogen via the Haber process enhanced the potential crop yield from farmland by enabling the production of nitrogen fertilizers. The development of pesticides, which include insecticides, fungicides, and herbicides, has made it possible to increase crop yields with less damage by insects and fungal pests as well as reduced competition from weeds.

Finally, agriculture has fully participated in the scientific advances of the twentieth and twenty-first centuries. Key developments include hybrid corn, the high responders in tropical lands, described as the Green Revolution, and modern genetic research. Agriculture has benefited enormously from scientific advances in biology, and the future here is bright for applied research, especially involving genetics. Great potential exists for the development of crop strains with greatly improved dietary characteristics, such as higher protein or reduced fat. Crops have been genetically modified to be resistant to herbicide so that the production fields can be sprayed to control competing weeds without damaging crops. The development of irrigation systems that pump water with diesel and electric motors also made it possible to water crops without depending on rain to supply enough water to sustain crop production.

Growing populations, made possible by these food surpluses, have forced agricultural expansion onto less and less desirable lands. Because agriculture radically simplifies ecosystems and greatly amplifies soil erosion, many areas such as the Mediterranean Basin and tropical forest lands have suffered severe degradation.

Major developments in civilization are directly linked to the agricultural revolution. A sedentary lifestyle, essential to technological development, was both mandated and made possible by farming. Urbanization flourished, which encouraged specialization and division of labor. Large populations provided the energy for massive projects, such as the Egyptian pyramids and the colossal engineering efforts of the Romans.

The plow represented the first lever, both lifting and overturning the soil. The draft animal provided the first in a long line of nonhuman energy sources. Plant and animal selection for desired characteristics and breeding are likely the first application of science and technology toward specific goals. A number of important crops bear little resemblance to the ancestors from which they were derived. Animals such as the fat-tailed sheep represent thoughtful cultural control of their lineage.

Climate dominates agriculture, second only to irrigation. Farmers are especially vulnerable to variations, such as late or early frosts, heavy rains, or drought. Rice, wheat, and maize have become the dominant crops globally because of their high caloric yield, versatility within their climate range, and their cultural status as the staff of life. Many would not consider a meal complete without rice, bread, or tortillas. This cultural influence is so strong that even starving peoples have rejected unfamiliar food. China provides a good example of such cultural differences, with a rice culture in the south and a wheat culture (noodles) in the north.

These crops all need a wet season for germination and growth, followed by a dry season to allow spoilage-free storage. Rice was domesticated in the monsoonal lands of Southeast Asia, whereas wheat originated in the Fertile Crescent of the Middle East. Historically, wheat was planted in the fall, and harvested in late spring, coinciding with the cycle of wet and dry seasons in the Mediterranean region. Maize needs the heavy summer rains provided by the Mexican highland climate.

Other crops predominate in areas with less suitable climates. These include barley in semiarid lands; oats and potatoes in cool, moist lands; rye in colder climates with short growing seasons; and dry rice on hillsides and drier lands where paddy rice is impractical.

Although food production is the main emphasis in agriculture, more and more industrial applications have evolved. Cloth fibers have been a mainstay, but paper products and many chemicals now come from cultivated plants. With shortages in fossil fuels and the threat of climate change amplified by the use of fossil fuels, agricultural research has focused on using plant biomass as a renewable source of fuel.

The agricultural revolution is also associated with some of mankind's darker moments. In the tropical and subtropical climates of the New World, slave labor was extensively exploited. Close, unsanitary living conditions among workers and in cities, made possible by the agricultural revolution, fostered plagues. Overdependence on limited crop sources periodically led to famines.

Resources

BOOKS

Conkin, Paul Keith. *A Revolution Down on the Farm: The Transformation of American Agriculture since 1929.* Lexington: University Press of Kentucky, 2008.

Dewar, James A. *Perennial Cornucopia: Planning the Next Agricultural Revolution.* Santa Monica, CA: RAND, 2007.

Francis, Charles A.; Raymond P. Poincelot; and George W. Bird. *Developing and Extending Sustainable Agriculture: A New Social Contract.* New York: Haworth Food & Agricultural Products Press, 2006.

Herren, Ray V. *Introduction to Biotechnology: An Agricultural Revolution.* Clifton Park, NY: Delmar Learning, 2005.

Kerridge, Eric. *The Agricultural Revolution.* London: Routledge, 2006.

Paarlberg, Don, and Philip Paarlberg. *The Agricultural Revolution of the 20th Century.* New York: Blackwell Publishing, 2007.

Nathan H. Meleen

Agricultural Stabilization and Conservation Service

For the past half century, agriculture in the United States has faced the somewhat unusual and enviable problem of overproduction. Farmers have produced more food than United States citizens can consume, and, as a result, per capita farm income has decreased as the volume of crops has increased. To help solve this problem, the Secretary of Agriculture established the Agricultural Stabilization and Conservation Service on June 5, 1961. The purpose of the service was to administer commodity and land-use programs designed to control production and to stabilize market prices and farm income. The service operated through state committees of three to five members each and committees consisting of three farmers in approximately 3,080 agricultural counties in the nation. In 1994, the service became part of the Consolidated Farm Service Agency, following a reorganization of the U.S. Department of Agriculture.

Agriculture and energy conservation *see* **Environmental engineering.**

Agriculture, drainage *see* **Runoff.**

Agriculture, sustainable *see* **Sustainable agriculture.**

Agroecology

Agroecology is an interdisciplinary field of study that applies ecological principles to the design and management of agricultural systems. Agroecology concentrates on the relationship of agriculture to the biological, economic, political, and social systems of the world.

The combination of agriculture with ecological principles such as biogeochemical cycles, energy conservation, and biodiversity has led to practical applications that benefit the whole ecosystem rather than just an individual crop. For instance, research into integrated pest management has developed ways to reduce reliance on pesticides. Such methods include biological or biotechnological controls such as genetic engineering, cultural controls such as changes in planting patterns, physical controls such as quarantines to prevent entry of new pests, and mechanical controls such as physically removing weeds or pests.

Sustainable agriculture is another goal of agroecological research. Sustainable agriculture views farming as a total system and stresses the long-term conservation of resources. It balances the human need for food with concerns for the environment and maintains that agriculture can be carried on without reliance on pesticides and fertilizers. Sustainable agriculture integrates methods to protect the environment, to be profitable for the farmers, and to promote successful farming communities. Implementing crop rotation to inhibit pest numbers and using livestock manure as fertilizer are two practices employed to reduce input of insecticides and synthetic fertilizers into the environment, while also effectively reducing the crop maintenance costs for the farmer.

Agroecology advocates the use of biological controls rather than pesticides to minimize agricultural damage from insects and weeds. Biological controls use natural enemies to control weeds and pests, such as ladybugs that kill aphids. Biological controls include the disruption of the reproductive cycles of pests and the introduction of more biologically diverse organisms to inhibit overpopulation of different agricultural pests.

Agroecological principles shift the focus of agriculture from food production alone to wider concerns, such as environmental quality, food safety, the quality of rural life, humane treatment of livestock, and conservation of air, soil, and water. Agroecology also studies how agricultural processes and technologies will be impacted by wider environmental problems such as global warming, desertification, or salinization. Desertification of land occurs in arid, semiarid, dry subhumid areas and is caused by deforestation, overgrazing, and water depletion resulting from improper irrigation. Salinization is the accumulation of salt in soils affected by excess irrigation, deforestation, and climate change. These processes can degrade the quality and fertility of the soil, making it difficult for successful plant growth and crop cultivation and harvest.

The entire world population depends on agriculture, and, as the number of people continues to grow, agroecology is becoming more important, particularly in developing countries. Agriculture is the largest economic activity in the world, and in areas such as sub-Saharan Africa about 75 percent of the population is involved in some form of it. As population pressures on the world food supply increase, the application of agroecological principles is expected to stem the ecological consequences of traditional agricultural practices such as pesticide poisoning and erosion.

The World Agroforestry Centre works to improve the livelihoods of poor farm shareholders and improve the sustainability and productivity of agriculture. *(Wendy Stone / Corbis)*

Resources

BOOKS

Altieri, Miguel A. *Agroecology for Food Sovereignty.* Oakland, CA: Food First Books, 2009.

Clements, David, and Anil Shrestha. *New Dimensions in Agroecology.* Binghamton, NY: Food Products Press, 2004.

Gliessman, Stephen R. *Agroecology: The Ecology of Sustainable Food Systems.* Boca Raton, FL: CRC Press, 2007.

Warner, Keith. *Agroecology in Action: Extending Alternative Agriculture through Social Networks.* Cambridge, MA: MIT, 2007.

Wojtkowski, Paul A. *Introduction to Agroecology: Principles and Practices.* New York: Food Products Press, 2006.

Linda Rehkopf

Agroforestry

Agroforestry is a land use system in which woody perennials (trees, shrubs, vines, palms, bamboo, etc.) are intentionally combined on the same land management unit with crops and sometimes animals, either in a spatial arrangement or a temporal sequence. It is based on the premise that woody perennials in the landscape can enhance the productivity and sustainability of agricultural practice. The approach is especially pertinent in tropical and subtropical areas where improper land management and intensive, continuous cropping of land have led to widespread devastation. Agroforestry recognizes the need for an alternative

agricultural system that will preserve and sustain productivity. The need for both food and forest products has led to an interest in techniques that combine production of both in a manner that can halt and may even reverse the ruin caused by existing practices.

Although the term agroforestry has come into widespread use only in the last twenty to twenty-five years, environmentally sound farming methods similar to those now proposed have been known and practiced in some tropical and subtropical areas for many years. As an example, one type of intercropping found on small rubber plantations (less than 25 acres [10 ha]), in Malaysia, Thailand, Nigeria, India, and Sri Lanka involves rubber plants intermixed with fruit trees, pepper, coconuts, and arable crops such as soybeans, corn, banana, and groundnut. Poultry may also be included. Unfortunately, in other areas the pressures caused by expanding human and animal populations have led to increased use of destructive farming practices. In the process, inhabitants have further reduced their ability to provide basic food, fiber, fuel, and timber needs and contributed to even more environmental degradation and loss of soil fertility.

The successful introduction of agroforestry practices in problem areas requires the cooperative efforts of experts from a variety of disciplines. Along with specialists in forestry, agriculture, meteorology, ecology, and related fields, it is often necessary to enlist the help of those familiar with local culture and heritage to explain new methods and their advantages. Usually, techniques must be adapted to local circumstances, and research and testing are required to develop viable systems for a particular setting. Intercropping combinations that work well in one location may not be appropriate for sites only a short distance away because of important meteorological or ecological differences. Despite apparent difficulties, agroforestry has great appeal as a means of arresting problems with deforestation and declining agricultural yields in warmer climates. The practice is expected to grow significantly in the next several decades. Some areas of special interest include intercropping with coconuts as the woody component, and mixing tree legumes with annual crops.

This intercropping allows for a wider diversity of inhabitants compared to an area that focuses on the production of one type of crop (termed "monoculture"). With a great diversity of plants, new niches or habitats are created for more wildlife species including insects and birds. With ever-increasing amounts of deforestation, it is important to develop new habitats for biodiversity to flourish. Additionally, adding trees to the landscape in intercropping may lessen the effects of global warming because more carbon can be taken up by these trees, thus reducing the accumulation of carbon in the atmosphere.

Agroforestry does not seem to lend itself to mechanization as easily as the large-scale grain, soybean, and vegetable cropping systems used in industrialized nations because practices for each site are individualized and usually labor-intensive. For these reasons they have had less appeal in areas such as the United States and Europe. Nevertheless, temperate zone applications have been developed or are under development. Examples include small-scale organic gardening and farming, mining wasteland reclamation, and biomass energy crop production on marginal land.

Resources

BOOKS

Alavalapati, Janaki R. R., and D. Evan Mercer. *Valuing Agroforestry Systems Methods and Applications*. Vol. 2 of *Advances in Agroforestry*. Dordrecht, Netherlands: Kluwer Academic Publishers, 2004.

Batish, D. *Ecological Basis of Agroforestry*. Boca Raton, FL: CRC Press, 2008.

Schroth, G. *Agroforestry and Biodiversity Conservation in Tropical Landscapes*. Washington, DC: Island Press, 2004.

United States National Agroforestry Center. *Agroforestry: Working Trees for Agriculture*. Lincoln, NE: Author, 2008.

World Agroforestry Centre. *Defying the Odds: Agroforestry Helps African Farmers Meet Food Security Goals*. Nairobi, Kenya: World Agroforestry Centre, 2006.

Douglas C. Pratt

AIDS

AIDS (acquired immune deficiency syndrome) is an infectious and fatal viral disease. AIDS is *pandemic*, which means that it is worldwide in distribution. A sufficient understanding of AIDS can be gained only by examining its causation (etiology), symptoms, treatments, and the risk factors for transmitting and contracting the disease.

AIDS occurs as a result of infection with the HIV (human immunodeficiency virus). HIV is a ribonucleic acid (RNA) virus that targets and kills T-lymphocytes, specialized white blood cells that are important in immune protection. Depletion of helper T-lymphocytes leaves the AIDS-infected person with a disabled immune system and at risk for infection by organisms that ordinarily pose no special hazard to the individual. Infection by these organisms is thus opportunistic and is frequently fatal.

The initial infection with HIV may entail no symptoms or relatively benign symptoms of short duration that may mimic infectious mononucleosis. This initial period is followed by a longer period (usually from one to as many as ten years) when the infected person is in apparent good health. The HIV-infected person, despite the outward image of good health, is in fact capable of spreading the virus to others, and appropriate care must be exercised to prevent spread of the virus at this time. During the course of the infection, the effects of the depletion of helper T cells eventually become evident. Symptoms include weight loss, persistent cough, persistent colds, diarrhea, periodic fever, weakness, fatigue, enlarged lymph nodes, and malaise. Following this, the

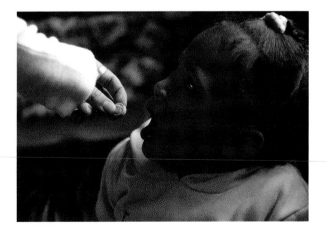

Natasha is given her antiretroviral medication at Bowy House, a home for orphans and sick children, in Paarl, near Cape Town, South Africa, June 14, 2007. *(AP Photo/Obed Zilwa)*

person with AIDS becomes vulnerable to chronic infections by opportunistic pathogens. These include, but are not limited to, oral yeast infections (thrush), pneumonia caused by the fungus *Pneumocystis carinii*, and infection by several kinds of herpes viruses. The AIDS patient is vulnerable to Kaposi's sarcoma, which is a cancer seldom seen except in those individuals with depressed immune systems. Death of the AIDS patient may be accompanied by confusion, dementia, and coma.

There is no cure for AIDS, but with modern combination therapy, the disease often remains suppressed for years. Opportunistic infections are treated with antibiotics, and combinations of antiretroviral drugs, which slow the progress of the HIV infection, are available. Research to find a vaccine for AIDS has not yet yielded satisfactory results, but scientists have been encouraged by the development of a vaccine for feline leukemia—a viral disease in cats that has similarities to AIDS. Unfortunately, this does not provide hope of a cure for those already infected with the HIV virus.

Prevention is crucial for a lethal disease with no cure. Thus, modes of transmission must be identified and avoided. Everyone is at risk: Globally, men and women are equally infected with the disease. According to the Joint United Nations Program on HIV and AIDS (UNAIDS), in 2008 there were approximately 34.4 million people infected worldwide with HIV or AIDS. Some 2.7 million new HIV infections occurred in 2008, with sub-Saharan Africa accounting for over two-thirds (1.9 million) of all new infections worldwide. Overall, according to the UNAIDS, the HIV virus took the lives of two million people in 2008 alone; between 1981 and 2008, a total of 25 million lives were lost to the disease.

According to the U.S. Centers for Disease Control and Prevention (CDC), about 55,000 new cases are diagnosed each year in the United States. Worldwide, AIDS cases in heterosexual males and women are on the increase, and no sexually active person can be considered safe from AIDS any longer. Therefore, everyone who is sexually active should be aware of the principal modes of transmission of the HIV virus—infected blood, semen from the male, and genital tract secretions of the female—and use appropriate means to prevent exposure. While the virus has been identified in tears, saliva, and breast milk, infection by exposure to those substances seems to be significantly less.

Worldwide increase in infectious diseases

In August 2007, the World Health Organization's (WHO) 2007 annual report, *A Safer Future*, reported an overall increase in risk from infectious disease epidemics. Report findings indicated that major infectious diseases such as AIDS, avian flu, cholera, Ebola and other hemorrhagic fevers, polio, and severe acute respiratory syndrome (SARS) were spreading faster than at any time in modern history and that the rate of emerging infectious diseases discoveries is "historically unprecedented." The report also noted alarming increases in diseases such as drug-resistant tuberculosis, which is often associated with AIDS, related to improper use of antibiotics that can lead to antimicrobial resistance. Other factors cited included increased global airline travel and lack of full international cooperation in sharing of data, technology, and other efforts to develop effective vaccines and other countermeasures.

Research continues to unravel the history of HIV, with possible benefits for clinical medical practice. In October 2007, a study published by the U.S. National Academy of Sciences reported that, based on the genetic analysis of blood samples, the strain of the HIV virus (the specific subtype of the virus known as HIV-1 group M subtype B) that is now most common in the United States, Europe, and parts of Australia, South America, and Japan, first traveled to the United States from Haiti around 1969. The report stated that the first transmission was via a single carrier. The study also stated that HIV traveled from Africa to Haiti in the mid-1960s. Scientists have predicted that studies of the history of HIV mutation would greatly increase the chances of predicting future mutation and facilitate vaccine development.

Resources

BOOKS

World Health Organization. *2008 Report on the Global AIDS Epidemic*. Geneva, Switzerland: World Health Organization, 2008.

Barnett, Tony, and Alan Whiteside. *AIDS in the Twenty-first Century, Fully Revised and Updated Edition: Disease and Globalization*. 2nd ed. New York: MacMillan, 2006.

Itano, Nicole. *No Place Left to Bury the Dead: Denial, Despair, and Hope in the African AIDS Pandemic*. New York: Atria Book, 2007.

Whiteside, Alan. *HIV/AIDS: A Very Short Introduction*. Oxford, UK: Oxford University Press, 2007.

World Health Organization. *Preventing HIV/AIDS in Young People*. Geneva, Switzerland: Author, 2006.

PERIODICALS

Hecht, R., et al. "Putting It Together: AIDS and the Millennium Development Goals." *PLoS Medicine*. 2006, 3(11).

Sempere, J. M.; V. Soriano; and J. M. Benito. "T Regulatory Cells and HIV Infection." *AIDS Rev*. January-March 2007, 9 (1):54–60.

OTHER

Centers for Disease Control and Prevention (CDC). "Acquired Immune Deficiency Syndrome (AIDS)." http://www.cdc.gov/hiv/ (accessed November 8, 2010).

Centers for Disease Control and Prevention (CDC). "Global
    HIV/AIDS." http://www.cdc.gov/globalaids (accessed
    November 8, 2010).
National Institutes of Health (NIH). "AIDS." http://health.
    nih.gov/topic/AIDS (accessed November 8, 2010).
World Health Organization (WHO). "HIV/AIDS." http://
    www.who.int/topics/hiv_aids/en (accessed November 8,
    2010).

Robert G. McKinnell

Ailuropoda melanoleuca *see* **Giant panda.**

Air and Waste Management Association

Founded in 1907 as the International Association for the Prevention of Smoke, this group changed its name several times as the interests of its members changed, becoming the Air and Waste Management Association (A&WMA) in the late 1980s. Although an international organization for environment professionals in more than sixty-five countries, the association is most active in North America and most concerned with North American environmental issues. Among its main concerns are air pollution control, environmental management, and waste processing and control.

A nonprofit organization that promotes the basic need for a clean environment, the A&WMA seeks to educate the public and private sectors of the world by conducting seminars, holding workshops and conferences, and offering continuing education programs for environmental professionals in the areas of pollution control and waste management. One of its main goals is to provide "a neutral forum where all viewpoints of an environmental management issue (technical, scientific, economic, social, political and public health) receive equal consideration." Approximately ten to twelve specialty conferences are held annually, as well as five or six workshops. The topics continuously revolve and change as new issues arise.

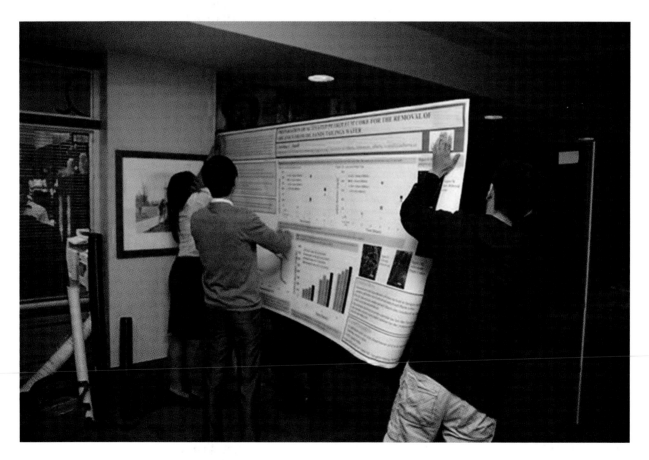

Air and Waste Management Association office workers prepare a presentation. *(Courtesy of Air and Waste Management Association)*

Education is so important to the A&WMA that it funds projects and activities at all levels, from the professional to the general public realm, including scholarships for graduate students pursuing careers in fields related to waste management and pollution control. Although A&WMA members are all professionals, they seek to educate even the very young by sponsoring essay contests, science fairs, and community activities, and by volunteering to speak to elementary, middle school, and high school audiences on environmental management topics.

The association's 12,000 members, all of whom are volunteers, are involved in virtually every aspect of every A&WMA project. There are thirty-two association sections across the world, facilitating meetings at regional and even local levels to discuss important issues. Training seminars are an important part of A&WMA membership, and members are taught the skills necessary to run public outreach programs designed for students of all ages and the general public.

The A&WMA's publications deal primarily with air pollution and waste management, and include the *Journal of the Air & Waste Management Association*, a scientific monthly; a bimonthly newsletter; a wide variety of technical books; and numerous training manuals and educational videotapes.

ORGANIZATIONS

Air and Waste Management Association, 420 Fort Duquesne Blvd, One Gateway Center, Pittsburgh, PA, USA, 15222, (412) 232-3444, (412) 232-3450, info@awma.org, http://www.awma.org

Cathy M. Falk

Air pollution

Air pollution is a general term that covers a broad range of contaminants in the atmosphere. Pollution can occur from natural causes or from human activities. Discussions about the effects of air pollution have focused mainly on human health but attention

Cyclists ride past a traditional Chinese gateway during a day murky from fog and pollution in Beijing, China, with less than ten months to go before the 2008 Olympics. Air pollution emerged as one of Beijing's biggest headaches. *(AP Photo/Ng Han Guan)*

is being directed to environmental quality and amenity as well. Air pollutants are found as gases or particles, and on a restricted scale they can be trapped inside buildings as indoor air pollutants. Urban air pollution has long been an important concern for civic administrators, but increasingly, air pollution has become an international problem.

The most characteristic sources of air pollution have always been the combustion processes. Here the most obvious pollutant is smoke. However, the widespread use of fossil fuels have made sulfur (S) and nitrogen oxides (NO_x) pollutants of great concern. With increasing use of petroleum-based fuels, a range of organic compounds have become widespread in the atmosphere.

In urban areas, air pollution has been a matter of concern since historical times. Indeed, there were complaints about smoke in ancient Rome. The use of coal throughout the centuries has caused cities to have dense smoke. Along with smoke, large concentrations of sulfur dioxide (SO_2) were produced. It was this mixture of smoke and sulfur dioxide that typified the foggy streets of Victorian London. Such situations are far less common in the cities of North America and Europe today. However, until recently, they have been evident in other cities, such as Ankara, Turkey, and Shanghai, China, that rely heavily on coal.

Coal is still burnt in large quantities to produce electricity or to refine metals, but these processes are frequently undertaken outside cities. Within urban areas, fuel use has shifted toward liquid and gaseous hydrocarbons (petrol and natural gas). These fuels typically have a lower concentration of sulfur, so the presence of sulfur dioxide has declined in many urban areas. However, the widespread use of liquid fuels in automobiles has meant increased production of carbon monoxide (CO), nitrogen oxides, and volatile organic compounds (VOCs).

Primary pollutants such as sulfur dioxide or smoke are the direct emission products of the combustion process. Many of the key pollutants in the urban atmospheres are secondary pollutants, produced by processes initiated through photochemical reactions. The photochemical smog in Los Angeles, California, is now characteristic of urban atmospheres dominated by secondary pollutants.

Although the automobile is the main source of air pollution in contemporary cities, there are other equally significant sources. Stationary sources are still important and the oil-burning furnaces that have replaced the older coal-burning ones are still responsible for a range of gaseous emissions and fly ash. Incineration is also an important source of complex combustion products, especially where this incineration burns a wide range of refuse. These emissions can include chlorinated hydrocarbons such as dioxin. When plastics, which often contain chlorine (Cl), are incinerated, hydrochloric acid (HCl) results in the waste gas stream. Metals, especially where they are volatile at high temperatures, can migrate to smaller, respirable particles. The accumulation of toxic metals, such as cadmium (Cd), on fly ash gives rise to concern over harmful effects from incinerator emissions. In specialized incinerators designed to destroy toxic compounds such as polychlorinated biphenyls (PCBs), many questions have been raised about the completeness of this destruction process. Even under optimum conditions where the furnace operation has been properly maintained, great care needs to be taken to control leaks and losses during transfer operations (fugitive emissions).

The enormous range of compounds used in modern manufacturing processes have also meant that there has been an ever-widening range of emissions from both the industrial processes and the combustion of their wastes. Although the amounts of these exotic compounds are often rather small, they add to the complex range of compounds found in the urban atmosphere. Again, it is not only the deliberate loss of effluents through discharge from pipes and chimneys that needs attention. Fugitive emissions of volatile substances that leak from valves and seals often warrant careful control.

Air pollution control procedures are increasingly an important part of civic administration, although their goals are far from easy to achieve. It is also noticeable that although many urban concentrations of primary pollutants (e.g., smoke and sulfur dioxide) are on the decline in developed countries, this is not always true in the developing countries. Here the desire for rapid industrial growth has often lowered urban air quality. Secondary air pollutants are generally proving a more difficult problem to eliminate than primary pollutants such as smoke.

Urban air pollutants have a wide range of effects, with health problems being the most enduring concern. In the classical polluted atmospheres filled with smoke and sulfur dioxide, a range of bronchial diseases was enhanced. While respiratory diseases are still the principal problem, the issues are somewhat subtler in atmospheres where the air pollutants are not so obvious. In photochemical smog, eye irritation from the secondary pollutant peroxyacetyl nitrate (PAN) is one on the most characteristic direct effects of the smog. High concentrations of carbon monoxide in cities where automobiles operate at high density means that the human heart has to work harder to

raised concerns that athletes would be at risk. In the month preceding the competition, traffic was reduced and air polluting industries were shuttered. As a result, the air quality was acceptable during the Olympics.

Resources

BOOKS

Allan, Catherine, and George Henry Stankey. *Adaptive Environmental Management: A Practitioner's Guide.* New York: Springer, 2009.

Aguado, Edward, and James E. Burt. *Understanding Weather and Climate.* Upper Saddle River, NJ: Pearson/Prentice Hall, 2009.

Mann, Michael E., and Lee R. Kump. *Dire Predictions: Understanding Global Warming.* Boston: Beacon Press, 2007.

David Clarke
Jeffrey Muhr

Air quality control region

The Clean Air Act defines an intrastate or interstate air quality control region (AQCR) as a contiguous area where air quality, and thus air pollution, is relatively uniform. In those cases where topography is a factor in air movement, AQCRs often correspond with airsheds. AQCRs may consist of two or more cities, counties, or other governmental entities, and each region is required to adopt consistent pollution control measures across the political jurisdictions involved. AQCRs may even cross state lines and, in these instances, the states must cooperate in developing pollution control strategies. Each AQCR is treated as a unit for the purposes of pollution reduction and achieving National Ambient Air Quality Standards. As of 1993, most AQCRs had achieved national air quality standards; however, the remaining AQCRs, where standards had not been achieved, were a significant group, where a large percentage of the United States population dwelled. AQCRs involving major metro areas such as Los Angeles, California, New York City, New York, Houston, Texas, Denver, Colorado, and Philadelphia, Pennsylvania, were not achieving air quality standards because of smog, motor vehicle emissions, and other pollutants.

Resources

BOOKS

Schwartz, Joel. *Air Quality in America: A Dose of Reality on Air Pollution Levels, Trends, and Health Risks.* Washington, DC: AEI Press, 2008.

OTHER

United States Environmental Protection Agency (EPA). "Air: Air Pollution Control." http://www.epa.gov/ebt pages/airairpollutioncontrol.html (accessed September 3, 2010).

Air quality criteria

The relationship between the level of exposure to air pollutant concentrations and the adverse effects on health or public welfare associated with such exposure, air quality criteria are critical in the development of ambient air quality standards, which define levels of acceptably safe exposure to an air pollutant.

Air-pollutant transport

Air-pollutant transport is the advection or horizontal convection of air pollutants from an area where emission occurs to a downwind receptor area by local or regional winds. It is sometimes referred to as atmospheric transport of air pollutants. This movement of air pollution is often simulated with computer models for point sources as well as for large diffuse sources, such as urban regions.

In some cases, strong regional winds or low-level nocturnal jets can carry pollutants hundreds of miles from source areas of high emissions. The possibility of transport over such distances can be increased through topographic channeling of winds through valleys. Air-pollutant transport over such distances is often referred to as long-range transport.

Air-pollutant transport is an important consideration in air quality planning. Where such impact occurs, the success of an air quality program may depend on the ability of air pollution control agencies to control upwind sources.

Airshed

An airshed is a geographical region, usually a topographical basin, that tends to have uniform air quality. The air quality within an airshed is influenced predominantly by emission activities native to that airshed because the elevated topography around the

Smog on the outskirts of Moscow. *(iStockphoto.com/Mordolff)*

basin constrains horizontal air movement. Pollutants move from one part of an airshed to other parts fairly quickly, but are not readily transferred to adjacent airsheds. An airshed tends to have a relatively uniform climate and relatively uniform meteorological features at any given point in time.

Alar

Alar is the trade name for the chemical compound daminozide, manufactured by the Uniroyal Chemical Company. The compound has been used since 1968 to keep apples from falling off trees before they are ripe and to keep them red and firm during storage. As late as the early 1980s, up to 40 percent of all red apples produced in the United States were treated with Alar.

In 1985, the Environmental Protection Agency (EPA) found that N,N-dimethylhydrazine (UDMH), a compound produced during the breakdown of daminozide, was a carcinogen. UDMH was routinely produced during the processing of apples, as in the production of apple juice and apple sauce, and the EPA suggested a ban on the use of Alar by apple growers. An outside review of the EPA studies, however, suggested that they were flawed, and the ban was not instituted. Instead, the agency recommended that Uniroyal conduct further studies on possible health risks from daminozide and UDMH.

Even without a ban, Uniroyal felt the impact of the EPA's research well before its own studies were concluded. Apple growers, fruit processors, legislators, and the general public were all frightened by the possibility that such a widely used chemical might be carcinogenic. Many growers, processors, and store owners pledged not to use the compound nor to buy or sell apples on which it had been used. By 1987, sales of Alar had dropped by 75 percent.

In 1989, two new studies again brought the subject of Alar to the public's attention. The consumer research organization Consumers' Union found that, using a very sensitive test for the chemical, eleven of twenty red apples they tested contained Alar. In addition, twenty-three of forty-four samples of apple juice tested

Apple harvest crop, Okanagan Valley, British Columbia.
(© iStockphoto.com/laughingmango)

caused blood-vessel tumors in mice. The agency once more declared its intention to ban Alar, and within a month, Uniroyal announced it would end sales of the compound in the United States.

Resources

Levine, Marvin J. *Pesticides: A Toxic Time Bomb in Our Midst*. New York: Praeger, 2007.
Matthews, Graham. *Pesticides: Health, Safety, and the Environment*. New York: Wiley-Blackwell, 2006.

David E. Newton

contained detectable amounts of the compound. The Natural Resources Defense Council (NRDC) announced their findings on the compound at about the same time. The NRDC concluded that Alar and certain other agricultural chemicals pose a threat to children about 240 times higher than the one-in-a-million risk traditionally used by the EPA to determine the acceptability of a product used in human foods.

The studies by the NRDC and the Consumers' Union created a panic among consumers, apple growers, and apple processors. Many stores removed all apple products from their shelves, and some growers destroyed their whole crop of apples. The industry suffered millions of dollars in damage. Representatives of the apple industry continued to question how much of a threat Alar truly posed to consumers, claiming that the carcinogenic risks identified by the EPA, the NRDC, and Consumers' Union were greatly exaggerated. But in May of that same year, the EPA announced interim data from its most recent study, which showed that UDMH

Alaska Highway

The Alaska Highway, sometimes referred to as the Alcan (*Al*aska-*Can*ada) Highway, is the final link of a binational transportation corridor that provides an overland route between the lower United States and Alaska. The first, all-weather, 1,680-miles (2,700 km) Alcan Military Highway was hurriedly constructed during 1942 and 1943 to provide land access between Dawson Creek, a Canadian village in northeastern British Columbia, and Delta Junction, Alaska, which connected to Fairbanks, a town on the Yukon River in central Alaska. Construction of the road was motivated by perception of a strategic, but ultimately unrealized, Japanese threat to maritime supply routes to Alaska during World War II. Today, the Alaska

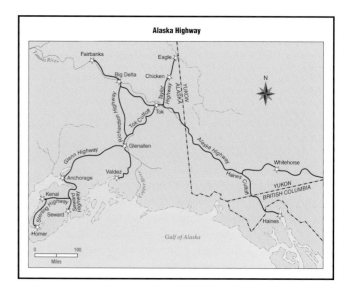

Map of the Alaska (Alcan) Highway. *(Reproduced by permission of Gale, a part of Cengage Learning)*

Highway runs approximately 1,390 miles (2,237 km) following the rerouting of some sections.

The route of the Alaska Highway extended through what was then a wilderness. An aggressive technical vision was supplied by the United States Army Corps of Engineers and the civilian U.S. Public Roads Administration using the labor of approximately 11,000 American soldiers and 16,000 American and Canadian civilians. In spite of the extraordinary difficulties of working in unfamiliar and inhospitable terrain, the route was opened for military passage in less than two years. Among the formidable challenges faced by the workers was a need to construct 133 bridges and thousands of smaller culverts across energetic watercourses, the infilling of alignments through a boggy muskeg capable of literally swallowing bulldozers, and working in winter temperatures that were so cold that vehicles were not turned off for fear they would not restart (steel dozer-blades became so brittle that they cracked upon impact with rock or frozen ground).

In hindsight, the planning and construction of the Alaska Highway could be considered an unmitigated environmental debacle. The enthusiastic engineers were almost totally inexperienced in the specialized techniques of Arctic construction, especially about methods dealing with permafrost, or permanently frozen ground. If the integrity of permafrost is not maintained during construction, then this underground, ice-rich matrix will thaw and become unstable, and its water content will run off. An unstable morass could be produced by the resulting erosion, mudflow, slumping, and thermokarst-collapse of the land into subsurface voids left by the loss of water. Repairs were very difficult, and reconstruction was often unsuccessful, requiring abandonment of some original alignments. Physical and biological disturbances caused terrestrial landscape scars that persist to this day and will continue to be visible (especially from the air) for centuries. Extensive reaches of aquatic habitat were secondarily degraded by erosion or sedimentation. In the decades following World War II, large construction projects in the Arctic, such as the Trans-Alaska Pipeline System (TAPS), are much more intensively scrutinized and planned regarding their ecological impacts. Though the TAPS has experienced its share of environmental problems, such as occasional oil spills and seepages on land, it was planned and built with the special conditions of the Arctic in mind—by the end of 2010, the TAPS had delivered approximately 16 billion barrels of oil to the United States since beginning operations in 1977. Such careful planning and construction is in marked contrast with the unfettered and freewheeling engineering associated with the initial construction of the Alaska Highway.

The Alaska Highway has been more or less continuously upgraded since its initial completion and was opened to unrestricted traffic in 1947. Nonmilitary benefits of the Alaska Highway include provision of access to a great region of the interior of northwestern North America. This access fostered economic development through mining, forestry, trucking, and tourism, as well as helping to diminish the perception of isolation felt by many northern residents living along the route.

Compared with the real dangers of vehicular passage along the Alaska Highway during its earlier years, the route safely provides one of North America's most spectacular ecotourism opportunities. Landscapes range from alpine tundra to expansive boreal forest, replete with abundantly cold and vigorous streams and rivers. There are abundant opportunities to view large mammals such as moose (*Alces alces*), caribou (*Rangifer tarandus*), and bighorn sheep (*Ovis canadensis*), as well as charismatic smaller mammals and birds and a wealth of interesting Arctic, boreal, and alpine species of plants.

Resources

BOOKS

Dalby, Ron. *Guide to the Alaskan Highway*. Birmingham, AL: Menasha Ridge Press, 2008.

Bill Freedman

Alaska National Interest Lands Conservation Act (1980)

The Alaska National Interest Lands Conservation Act (ANILCA), commonly referred to as the Alaska Lands Act, protected 104 million acres (42 million ha), or 28 percent, of the state's 375 million acres (152 million ha) of land. The law added 44 million acres (18 million ha) to the national park system, 55 million acres (22.3 million ha) to the fish and wildlife refuge system, 3 million acres (1.2 million ha) to the national forest system, and made twenty-six additions to the national wild and scenic rivers system. The law also designated 56.7 million acres (23 million ha) of land as wilderness, with the stipulation that 70 million acres (28.4 million ha) of additional land be reviewed for possible wilderness designation.

The genesis of this Act can be traced to 1959, when Alaska became the forty-ninth state. As part of the statehood act, Alaska could choose 104 million acres (42.1 million ha) of federal land to be transferred to the state. This selection process was halted in 1966 to clarify land claims made by Alaskan indigenous peoples. In 1971, the Alaska Native Claims Settlement Act (ANSCA) was passed to satisfy the native land claims and allow the state-selection process to continue. This Act stipulated that the secretary of the interior could withdraw 80 million acres (32.4 million ha) of land for protection as national parks and monuments, fish and wildlife refuges, and national forests, and that these lands would not be available for state or native selection. Congress would have to approve these designations by 1978. If Congress failed to act, the state and the natives could select any lands not already protected. These lands were referred to as national interest, or d-2 lands, after Section 17(d)(2) of the ANCSA.

Secretary of the Interior Rogers Morton recommended 83 million acres (33.6 million ha) for protection in 1973, but this did not satisfy environmentalists. The ensuing conflict over how much and which lands should be protected, and how these lands should be protected, was intense. The environmental community formed the Alaska Coalition, which by 1980 included over 1,500 national, regional, and local organizations with a total membership of 10 million people. Meanwhile, the state of Alaska and development-oriented interests launched a fierce and well-financed campaign to reduce the area of protected land.

In 1978, the House of Representatives passed a bill protecting 124 million acres (50.2 million ha). The Senate passed a bill protecting far less land, and House-Senate negotiations over a compromise broke down in October. Thus, Congress would not act before the December 1978 deadline. In response, the executive branch acted. Department of the Interior Secretary Cecil Andrus withdrew 110 million acres (44.6 million ha) from state selection and mineral entry. President Jimmy Carter then designated 56 million acres (22.7 million ha) of these lands as national monuments under the authority of the Antiquities Act. Forty million additional acres (16.2 million ha) were withdrawn as fish and wildlife refuges, and 11 million acres (4.5 million ha) of existing national forests were withdrawn from state selection and mineral entry. Carter indicated that he would rescind these actions once Congress had acted.

In 1979, the House passed a bill protecting 127 million acres (51.4 million ha). The Senate passed a bill designating 104 million acres (42.1 million ha) as national interest lands in 1980. Environmentalists and the House were unwilling to reduce the amount of land to be protected. In November of 1980, however, Ronald Reagan was elected president, and the environmentalists and the House decided to accept the Senate bill rather than face the potential for much less land under a president who would side with development interests. President Carter signed the ANILCA into law on December 2, 1980.

The ANILCA also mandated that the U.S. Geological Service (USGS) conduct biological and petroleum assessments of the coastal plain section of the Arctic National Wildlife Refuge (ANWR), 19.8 million acres (8 million ha) known as area 1002. While the USGS did determine a significant quantity of oil reserves in the area, they also reported that petroleum development would adversely impact many native species, including caribou (*Rangifer tarandus*), snow geese (*Chen caerulescens*), and muskoxen (*Ovibos moschatus*).

In 2001, the George W. Bush's administration unveiled a new energy policy that would open up this area to oil and natural gas exploration. The House passed bills authorizing drilling in the ANWR in 2000 and 2005. In 2002, the Senate rejected the 2000 bill from the House, and the ANWR drilling provisions of the 2005 bill were removed in the House and Senate conference committee. In December 2005, Alaska senator. Ted Stevens, a strong supporter of drilling in the ANWR, attached an amendment that would allow drilling in the ANWR to a defense appropriations bill. Senate Democrats filibustered the appropriations bill until the ANWR amendment was withdrawn. U.S. President Barack Obama opposes drilling in the ANWR, citing the damage that it would cause to wildlife in the area and the negligible impact of the ANWR drilling on global oil supplies and prices.

Resources

BOOKS

Waterman, Jonathan. *Where Mountains Are Nameless: Passion and Politics in the Arctic Wildlife Refuge*. New York: W.W. Norton, 2007.

PERIODICALS

Biello, David. "No Arctic Oil Drilling? How About Selling Parks?" *San Francisco Chronicle* (September 24, 2005).

Revkin, Andrew C. "Scientists Report Severe Retreat of Arctic Ice." *New York Times* (September 21, 2007).

OTHER

"Alaska National Interest Lands Conservation Act." 16 USC 3101–3223; Public Law 96–487. [June 2002].http://www.access.gpo.gov/uscode/title16/ chapter51;us.html (accessed July 30, 2009).

ORGANIZATIONS

The Alaska Coalition, 419 6th St, #328, Juneau, AK, USA, 99801, (907) 586-6667, (907) 463-3312, info@alaska coalition.org, http://www.alaskacoalition.org

Christopher McGrory Klyza
Paula Anne Ford-Martin

Alaska National Wildlife Refuge *see* **Arctic National Wildlife Refuge.**

Alaska pipeline *see* **Trans-Alaska pipeline.**

Albedo

The reflecting power of a surface, expressed as a ratio of reflected radiation to incident or incoming radiation; it is sometimes expressed as a percentage. Albedo is also called the "reflection coefficient" and derives from the Latin root word *albus*, which means whiteness. Sometimes expressed as a percentage, albedo is more commonly measured as a fraction on a scale from zero to one, with a value of one denoting a completely reflective, white surface, whereas a value of zero would describe an absolutely black surface that reflects no light rays.

Albedo varies with surface characteristics such as color and composition, as well as with the angle of the sun. The albedo of natural earth surface features such as oceans, forests, deserts, and crop canopies varies widely. Some measured values of albedo for various surfaces are shown below.

The albedo of clouds in the atmosphere is important to life on earth because extreme levels of radiation would make the planet uninhabitable; at any moment

Types of Surface	Albedo
Fresh, dry snow cover	0.80–0.95
Aged or decaying snow cover	0.40–0.70
Oceans	0.07–0.23
Dense clouds	0.70–0.80
Thin clouds	0.25–0.50
Tundra	0.15–0.20
Desert	0.25–0.29
Coniferous forest	0.10–0.15
Deciduous forest	0.15–0.20
Field crops	0.20–0.30
Bare dark soils	0.05–0.15

Albedo varies with surface characteristics. *(Reproduced by permission of Gale, a part of Cengage Learning)*

in time about 50 percent of the planet's surface is covered by clouds. The mean albedo for Earth, called the planetary albedo, is about 30–35 percent.

Mark W. Seeley

Algal bloom

Algae are simple, filamentous aquatic organisms that can be unicellular (single-celled) or multicellular (having more than one cell); they grow in colonies and are commonly found floating in ponds, lakes, and oceans. Populations of algae fluctuate with the availability of nutrients; a sudden increase in nutrients often results in a profusion of algae known as algal bloom.

The growth of a particular algal species can be both sudden and massive. Algal cells can increase to very high densities in the water, often thousands of cells per milliliter, and the water itself can be colored brown, red, or green. Algal blooms occur in freshwater systems and in marine environments, and they usually disappear in a few days to a few weeks. These blooms consume oxygen, increase turbidity, and clog lakes and streams. Some algal species release water-soluble compounds that may be toxic to fish and shellfish, resulting in fish kills and poisoning episodes.

Algal groups are generally classified on the basis of the pigments that color their cells. The most common algal groups are cyanobacteria (commonly referred to as blue-green algae), green algae, red algae, and brown algae. Algal blooms in freshwater lakes and ponds tend to be caused by cyanobacteria and green algae. The excessive amounts of nutrients that cause these blooms are often the result of human activities. For example, nitrates and phosphates introduced into a lake from fertilizer runoff during a storm can cause rapid algal growth. Accumulation of nitrates and phosphates in an ecosystem is termed "eutrophication." Some common cyanobacteria known to cause blooms as well as release nerve toxins (neurotoxins) are *Microcystis*, *Nostoc*, and *Anabaena*. These types of blooms are referred to as Harmful Algal Bloom (HABs). Algal blooms can occur in drinking water and are able to survive typical water purification procedures.

Red tides in coastal areas are a type of algal bloom. They are common in many parts of the world, including the New York Bight, the Gulf of California, and the Red Sea. The causes of algal blooms are not as well-understood in marine environments as they are in

Algal bloom in a lake caused by the bluegreen alga *Microcystis*. *(© iStockphoto.com/Heike Kampe)*

freshwater systems. Although human activities may well have an effect on these events, weather conditions probably play a more important role: turbulent storms that follow long, hot, dry spells have often been associated with algal blooms at sea. Toxic red tides most often consist of genera from the dinoflagellate algal group such as *Gonyaulax* and *Gymnodinium*. The potency of the toxins has been estimated to be ten to fifty times higher than cyanide or curare, and people who eat exposed shellfish may suffer from paralytic shellfish poisoning within thirty minutes of consumption. Some species such as *Karenia brevis* (commonly referred to as Florida red tide) have induced respiratory effects or symptoms caused by their airborne (aerosol form) toxins. A number of cyanobacteria genera such as *Oscillatoria* and *Trichodesmium* have also been associated with red blooms, but they are not necessarily toxic in their effects. Some believe that the blooms caused by these genera gave the Red Sea its name.

The economic and health consequences of algal blooms can be sudden and severe, but the effects are generally not long-lasting. In the past, there has been little evidence that algal blooms have long-term effects on water quality or ecosystem structure. However, some scientists argue that a factor in the development of algal blooms is climate change and that global warming may result in an increasing occurrence of algal blooms. An increase in the number of HABs may have a dramatic impact on the seafood industry as well as the entire ecosystem.

Resources

BOOKS

Evangelista, Valtere. *Algal Toxins: Nature, Occurrence, Effect, and Detection*. Dordrecht, Netherlands: Springer, 2008.

Graham, Linda E.; James M. Graham; and Lee Warren Wilcox. *Algae*. San Francisco: Benjamin Cummings, 2009.

Gualtieri, Paolo, and L. Barsanti. *Algae: Anatomy, Biochemistry, and Biotechnology*. Boca Raton, FL: Taylor & Francis, 2006.

Hagen, Kristian N. *Algae: Nutrition, Pollution, Control, and Energy Sources*. New York: Nova Science, 2008.

Waaland, J. Robert, and Carole A.. *Algae and Human Affairs*. Cambridge, UK: Cambridge University Press, 2007.

Washington State Department of Health, Office of Environmental Health Assessments. *Toxic Blue-Green Algae Blooms*. Olympia, WA: Washington State Dept. of

Health, Division of Environmental Health, Office of Environmental Health Assessments, 2007.

OTHER

Centers for Disease Control and Prevention (CDC). "Harmful Algal Blooms (HABs)." http://www.cdc.gov/hab/ (accessed August 29, 2010).

University of California Museum of Paleontology. "Algae and Seaweeds: Cyanobacteria." http://www.ucmp.berkeley.edu/bacteria/cyanointro.html (accessed August 29, 2010).

Usha Vedagiri
Douglas Smith

Algicide

The presence of nuisance algae can cause unsightly appearance, odors, slime, and coating problems in aquatic media. Algicides are chemical agents used to control or eradicate the growth of algae in aquatic media such as industrial tanks, swimming pools, and lakes. These agents used may vary from simple inorganic compounds, such as copper sulphate, which are broad-spectrum in effect and control a variety of algal groups, to complex organic compounds that are targeted to be species-specific in their effects. Algicides usually require repeated application or continuous application at low doses in order to maintain effective control.

Aline, Tundra *see* **Tundra.**

Allelopathy

Derived from the Greek words *allelo* (other) and *pathy* (causing injury to), allelopathy is a form of competition among plants. One plant produces and releases a chemical into the surrounding soil that inhibits the germination or growth of other species in the immediate area. These chemical substances are both acids and bases and are called secondary compounds. For example, black walnut (*Jugans nigra*) trees release a chemical called juglone that prevents other plants, such as tomatoes, from growing in the immediate area around each tree. In this way, plants such as black walnut reduce competition for space, nutrients, water, and sunlight.

Allergen

An allergen is any substance that can bring about an allergic response in an organism. Hay fever and asthma are two common allergic responses. The allergens that evoke these responses include pollen, fungi, and dust. Allergens can be described as host-specific agents in that a particular allergen may affect some individuals, but not others. A number of air pollutants are known to be allergens. Formaldehyde, thiocyanates, and epoxy resins are examples. People who are allergic to natural allergens, such as pollen, are more inclined to be sensitive also to synthetic allergens, such as formaldehyde.

Alligator, American

The American alligator (*Alligator mississippiensis*) is a member of the reptilian family Crocodylidae, which consists of twenty-one species found in tropical and subtropical regions throughout the world. It is a species that has been reclaimed from the brink of extinction.

Historically, the American alligator ranged in the Gulf and Atlantic coast states from Texas to the Carolinas, with rather large populations concentrated in the swamps and river bottomlands of Florida and Louisiana. From the late nineteenth century into the middle of the twentieth century, the population of this species decreased dramatically. With no restrictions on their activities, hunters killed alligators as pests or to harvest their skin, which was highly valued in the leather trade. The American alligator was killed in such great numbers that biologists predicted its probable extinction. It has

An American alligator (*Alligator mississippiensis*).
(FlordiaStock/Shutterstock.com)

been estimated that about 3.5 million of these reptiles were slaughtered in Louisiana between 1880 and 1930. The population was also impacted by the fad of selling young alligators as pets, principally in the 1950s.

States began to take action in the early 1960s to save the alligator from extinction. In 1963 Louisiana banned all legalized trapping, closed the alligator hunting season, and stepped up enforcement of game laws against poachers. By the time the Endangered Species Act was passed in 1973, the species was already experiencing a rapid recovery. Because of the successful re-establishment of alligator populations, its endangered classification was downgraded in several southeastern states, and there are now strictly regulated seasons that allow alligator trapping. Because of the persistent demand for its hide for leather goods and an increasing market for the reptile's meat, alligator farms are now both legal and profitable. According to the nonprofit National Parks Conservation Association (NPCA), by 2010 there were over a million alligators in the United States, most of which could be found in the southern coastal states of Texas, Louisiana, Georgia, and Florida.

Human fascination with large, dangerous animals, along with the American alligator's near extinction, have made it one of North America's best-studied reptile species. Population pressures, primarily resulting from being hunted so ruthlessly for decades, have resulted in a decrease in the maximum size attained by this species. The growth of a reptile is indeterminate, and they continue to grow as long as they are alive, but old adults used to attain larger sizes than their modern counterparts. The largest recorded American alligator was an old male killed in January 1890, in Vermilion Parish, Louisiana, which measured 19.2 feet (6 m) long. The largest female ever taken was only about half that size.

Alligators do not reach sexual maturity until they are about 6 feet (1.3 m) long and nearly ten years old. Females construct a nest mound in which they lay about thirty-five to fifty eggs. The nest is usually 5 to 7 feet (1.5–2.1 m) in diameter and 2 to 3 feet (0.6–0.9 m) high, and decaying vegetation produces heat, which keeps the eggs at a fairly constant temperature during incubation. The young stay with their mother through their first winter, striking out on their own when they are about 1.5 feet (0.5 m) in length.

Resources

OTHER

U.S. Government; science.gov. "Reptiles and Amphibians." http://www.science.gov/browse/w_115A10.htm (accessed November 7, 2010).

Eugene C. Beckham

Alligator mississippiensis *see* **Alligator, American.**

All-terrain vehicle *see* **Off-road vehicles.**

Alpha particle

Emitted by radioactive materials, an alpha particle is identical to the nucleus of a helium atom, consisting of two protons and two neutrons. Some common alpha-particle emitters are uranium-235, uranium-238, radium-226, and radon-222. Alpha particles have relatively low penetrating power. They can be stopped by a thin sheet of paper or by human skin. The inhalation of alpha-emitting radon gas escaping from bedrock into houses in some areas is a health hazard. The U.S. Centers for Disease Control and Prevention (CDC) and the Environmental Protection Agency (EPA) estimate that exposure to radon in the home is responsible for 20,000 lung cancer deaths per year.

Alternative energy sources

Energy consumption is most clearly understood if final energy consumption is distinguished from primary energy consumption. Final energy consumption is energy consumed, not wasted at the point of generation. For example, of all the energy released by burning coal at a coal-fired power plant, only about 40 percent is turned into electricity; the rest is wasted up the chimney as heat. A further 5 percent or so is lost in power lines; so, of the total energy released by fuel-burning at such a plant (usually termed "primary energy"), only about 35 percent reaches end uses and can be counted as final energy consumption. In a heating system for a building, only the energy that heats the building is calculated in final energy consumption; energy that goes up the vent-stack is primary energy consumption but does not heat the house.

Fossil fuels, nuclear power, and large-scale hydroelectric power from dams are all considered conventional energy sources and are contributors to climate change because of carbon dioxide (CO_2) emissions in the atmosphere. Other sources are termed "alternative." The main fossil fuels—coal, oil, and natural gas (methane)—provided the vast majority (over 75 percent) of final energy consumption worldwide in the

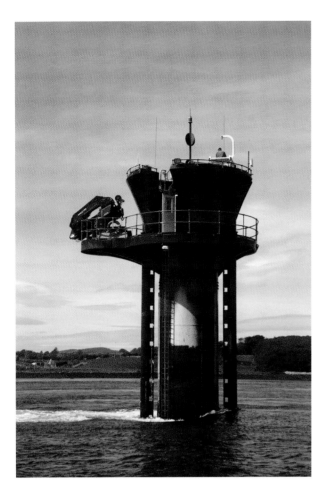

Experimental tidal generator, Strangford Lough, County Down, Ulster, Northern Ireland. (© *Robert Harding Picture Library / SuperStock*)

power, and around 16 percent was from hydropower and alternative (renewable) sources such as wind turbines and solar cells.

As for primary energy (as opposed to final energy), the exact contribution alternative energy sources make to the total primary energy used around the world is not known. Conservative estimates place their share at 3 to 4 percent, but some energy experts dispute these figures. U.S. energy expert Amory Lovins (1947–), a prominent figure in expert debates around nuclear power, alternative energy, and energy efficiency starting in the 1970s, has argued that the statistics collected are based primarily on large electric utilities and the regions they serve. They fail to account for areas remote from major power grids, which are more likely to use solar energy, wind energy, or other sources. When these areas are taken into consideration, Lovins, who is also the chief scientist at the Rocky Mountain Institute, has stated that alternative energy sources contribute as much as 11 percent to the total primary energy used in the United States. Animal manure, furthermore, is widely used as an energy source in India, parts of China, and much of Africa, and when this is taken into account the percentage of the worldwide contribution that alternative sources make to primary energy production could rise as high as 10 to 15 percent.

Now considered an alternative energy source, wind power is one of the earliest forms of energy harvested mechanically by humankind. Wind is caused by the uneven heating of the earth's surface by the sun, and so (like wave and tidal power) is considered an indirect form of solar power. The kinetic energy of wind—the energy that a moving air mass has by virtue of its motion—is proportional to the square of wind velocity, and the energy that a wind turbine (windmill generator) can *extract* from wind is proportional to the cube of velocity, so the ideal location for a windmill generator is an area with constant and relatively fast winds. In most locations, wind speed rises (up to a point) with height, and turbulence from ground roughness decreases—also a crucial consideration. Modern turbines are therefore placed on high towers, typically about 200 feet (60 m) above the ground.

An efficient wind turbine can produce 175 watts per square meter of propeller blade area at a height of 75 feet (25 m); in fact, turbines being installed today are much higher than this, and so intercept more wind and produce more power per unit of blade area. In 2008, the cost of generating one kilowatt hour of wind power was about eight cents, as compared to five cents for hydropower and fifteen cents for nuclear power. (A kilowatt-hour is the amount of energy consumed by a 1,000-watt load during one hour; a 1,000-watt load can be imagined

late 2000s—and it is expected to continue to do so into the future. Nuclear reactors provided around 3 percent of final energy consumption, and hydroelectric power from dams also provided about 3 percent. Alternative energy sources, including wind power, active and passive solar systems, biofuel, and geothermal energy, supplied approximately 5 percent of final energy consumption. Due to the depletion of fossil fuel sources as well as the pollution caused by all conventional energy sources, the share of worldwide energy consumption originating from alternative energy sources is growing.

Power generated with electricity is only one energy source, although it is a particularly versatile one. Because much fossil fuel is consumed for nonelectric forms of energy (heat and transport), the breakdown of the world electricity supply is differently stated than that for the world energy supply. As of the end of the 2000s, about 69 percent of world electricity came from fossil fuels, approximately 15 percent was from nuclear as ten 100-watt light bulbs.) Because of low capital costs, low fuel costs, rapid construction, and zero pollution, global wind-generation capacity was growing exponentially from the late 1990s to the end of the 2000s, and continues a rapid increase in the early 2010s. From 2000 to 2007, global wind capacity increased by a factor of about five. For instance, about 20 percent of Denmark's electricity was generated from wind, about 10 percent of Germany's, and about 1 percent of the United States'. However, because the United States is so large, even at this low percentage it was the world's third-largest producer of wind power, with capacity about equal to that of Spain and about one-half that of Germany.

Solar energy can be utilized either directly as heat or indirectly by converting it to electrical power using photovoltaic cells or steam-turbine plants. Windows and collectors alone are considered passive systems; an active solar system uses a fan, pump, or other machinery to transport the heat generated from the sun. Greenhouses and solariums are the most common examples of the passive use of solar energy, with glass windows allowing entrance to sunlight but restricting the heat from escaping. In most climates, buildings can be made entirely self-sufficient in heating and cooling through use of superinsulation and passive solar heat collection. Flat-plate collectors that use fluid circulating over a dark surface to collect solar energy are another direct solar-thermal method, often mounted on rooftops. These collectors can provide from one-third to all of the energy required for space heating, depending on climate and overall building efficiency. In 2007, about 50 million homes worldwide, mostly in China, were using flat-plate collectors to supply most or all of their domestic hot water.

Photovoltaic cells are made of semiconductor materials such as silicon. These cells are capable of absorbing part of sunlight to produce a direct electric current with about 14 percent efficiency, which is to say about 14 percent of the energy in the sunlight is converted to electricity. In the laboratory, efficiencies of over 40 percent have been achieved, and such work will eventually raise the efficiency of commercially available cells. The cost of producing photovoltaic power is about four dollars a watt: that is, a system capable of producing 1,000 watts in bright sunshine costs about $4,000 to install. However, thin-film cells and other technologies are being perfected that are steadily reducing photovoltaic costs. Photovoltaics are now being used economically in lighthouses, boats, cars, and in rural villages and other remote areas. Large solar systems (producing approximately 20 million watts) are being installed in Europe, the United States, and elsewhere as quickly as the solar industry can meet the demand for photovoltaic cells. Rooftop photovoltaics are also being installed rapidly. These have the advantage that they produce power where it is to be used and use land that has already been developed for buildings, rather than covering wild land with solar cell arrays. As of 2007, Germany was consuming the largest amount of photovoltaic electricity.

Geothermal energy is the heat generated in the interior of Earth by radioactive elements. Earth's interior radioactive material is so dilute that it does not pose a radiation threat. Like solar energy, it can be used directly as heat or indirectly to generate electricity. Geothermal steam is classified as either dry (without water droplets), or wet (mixed with water). When it is generated in certain areas containing corrosive sulfur (S) compounds, it is known as sour steam; and when generated in areas that are free of sulfur, it is known as sweet steam. Geothermal energy can be used to generate electricity by the flashed steam method, in which high temperature geothermal brine is used as a heat exchanger to convert injected water into steam. The produced steam is used to turn a turbine. When geothermal wells are not hot enough to create steam, a fluid that evaporates at a much lower temperature than water, such as isobutane (C_4H_{10}) or ammonia (NH_{13}), can be placed in a closed system in which the geothermal heat provides the energy to evaporate the fluid and run the turbine.

There are at least seventy countries worldwide that utilize geothermal energy, including the United States, Mexico, Italy, Iceland, Japan, and Russia. Geothermal energy can, in some cases, have more environmental impact than solar or wind energy. Depending on the type of geothermal resource being exploited, it can contribute to air pollution; it can also emit dissolved salts and, in some cases, toxic heavy metals such as mercury (Hg) and arsenic (As).

Though there are several ways of extracting energy from the ocean, the most promising are the harnessing of tidal power, wave power, and ocean thermal energy conversion. The power of ocean tides is based on the difference between high and low water. Traditionally, it was thought that for tidal power to be effective the differences in height need to be very great, more than 15 feet (4.6 m), and there are only a few places in the world where such differences exist. These include the Bay of Fundy (on Canada's east coast) and a few sites in China. However, most tide-power projects as of 2010 do not rely on such extreme geography. More modest tides can be harnessed by impounding high tides behind dams and then letting the water run out through turbines at low tide, or by installing underwater turbines resembling windmills in places where coastal and seafloor shapes cause strong tidal currents to run. For instance, since early 2007, Verdant Power, headquartered in Arlington, Virginia, has operated a pilot project in the East River of New York City. It is considered the first major tidal-power project in the United States. Technologies are also being developed for harvesting energy from ocean waves—not the crashing waves seen at the shore, but swells out to sea. The rise-and-fall motion of a float, for example, can be used to compress air that runs a generator, whose power output can be conveyed to shore through cables.

Ocean thermal energy conversion utilizes temperature differences rather than tides and waves. Ocean temperature is stratified, especially near the tropics, and the process takes advantage of this fact by using a fluid with a low boiling point, such as ammonia. The vapor from the fluid drives a turbine, and cold water from lower depths is pumped up to condense the vapor back into liquid. The electrical power generated by this method can be shipped to shore or used to operate a floating plant such as a cannery.

Other sources of alternative energy are being exploited at various scales as of 2010. These include harnessing the energy in biomass through the production of wood from trees or the production of ethanol (C_2H_6O) from crops such as sugarcane, corn, or switchgrass (*Panicum virgatium*). Scientific research studies are being employed to determine efficient and cost-effective ways of deriving energy from biomass. Methane (CH_4) gas can be generated from the anaerobic breakdown of organic waste in sanitary landfills and from wastewater treatment plants. With the cost of garbage disposal rapidly increasing, the burning of garbage is becoming a viable option as an energy source. For instance, in Denmark, garbage is being treated as a clean and viable alternative energy source as the country incinerates much of its trash in waste-to-energy plants. The *New York Times* reports that Denmark, as of April 2010, has twenty-nine such plants that burn garbage in ninety-eight communities totaling 5.5 million citizens. The trend continues in the future with ten more plants planned or under construction. In addition, about 400 additional waste-to-energy plants are scattered throughout Europe. Whether trash incineration can be rendered safe by air pollution controls is, however, disputed.

Ethanol and methanol (CH_3OH) are produced from biomass and used in transportation; in Brazil, all gasoline is sold with at least 20 percent ethanol from sugar cane (in the United States, a 10 percent corn-ethanol blend is common). However, production of fuel ethanol has been criticized both as a net-energy consumer (or poor net-energy producer) and also as injurious to the world's poor population. In early 2008, for example, food shortages gripped much of the world, with riots against high prices becoming more common in some poor countries. Food prices worldwide had risen 83 percent since 2005, according to the World Bank. Although there were a number of causes for rising prices, including a drought in Australia that lowered production of rice, the main food of over one-half the world's population, food experts agreed that the push to raise crops for ethanol biofuel for vehicles was part of the problem. Biofuel manufacturers compete directly with food buyers in the market for corn, and high demand for biofuel crops causes growers to switch acreage away from food production. The result is higher food prices, which some of the world's poorest simply cannot pay. In Haiti, for example, by late 2007 many of the rural poor were routinely eating mud, which has no food value but eases hunger pangs. Expert estimates of how much of the rise in global food prices was due to biofuel programs ranged from 10 to 75 percent.

Hydrogen (H) could be a valuable aid in utilizing energy from renewable sources if problems of supply and storage can be solved. Its only combustion by-products are water and heat, and it can be combined with oxygen in battery-like devices termed fuel cells to generate electricity directly with high efficiency. Also, despite a bad reputation, it is not nearly as explosive as gasoline vapor; hydrogen disperses rapidly when released, rather than pooling near the ground. Some alternative-energy advocates visualize a hydrogen economy in which windmills, solar cells, and other renewable energy sources produce hydrogen by cracking water molecules (hydrogen and oxygen), a process known as electrolysis. The hydrogen would then provide clean fuel for vehicles as well as electricity from fuel cells to smooth the output of variable sources such as wind and sun. To date, the difficulty of producing affordable fuel cells and hydrogen slows the deployment of hydrogen technologies.

Of all alternative energy sources, energy conservation or efficiency is perhaps the most important. In practice, improving efficiency is in most settings a far cheaper way to provide an additional unit of energy service than is buying the output of new generating capacity, whether wind, nuclear, coal, or other. Efficiency improvements such as thicker insulation and redesign of industrial machines also entails relatively small levels of pollution or material use compared to generating the energy that the efficiency displaces. Therefore, energy efficiency is the best way to meet energy demands without adding to air and water pollution and the other harms entailed by generating various kinds of energy.

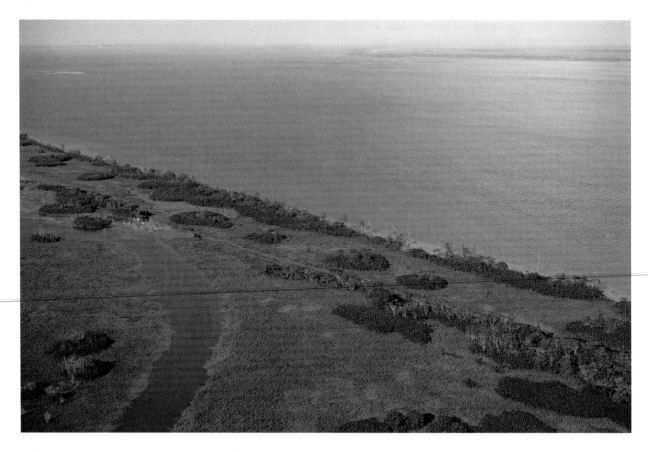

Aerial view of the lower Amazon River with floodplain grasslands (Campos de varzea) in Para, Brazil. *(Jacques Jangoux/Photo Researchers, Inc.)*

increased by 600 percent over the past forty years, and this has led to the clearing of over 65 percent of the region's forests for agriculture. In 1970, the percentage of the Amazon rainforest that was deforested or degraded was about 1 percent compared with the 2005 estimate of 16–20 percent. The mean annual deforestation rate in the Amazon rainforest from 2000 to 2005 was substantially higher than in the previous five years. Such rates would mean that in two decades the entire forest will be reduced by 40 percent, which is of significant concern to environmentalists.

In Brazil and other parts of the Amazon basin, cattle ranchers, plantation owners, and small landowners clear the forest by setting it on fire (termed "slash-and-burn" agriculture), which causes an estimated 23–43 percent increase in carbon dioxide (CO_2) levels worldwide and spreads smoke over millions of square miles. In Brazil, up to 70 percent of the deforestation is tied to cattle ranching. In the past, large governmental subsidies and tax incentives have encouraged this practice, which had little or no financial success and caused widespread environmental damage.

Tropical soils rapidly lose their fertility, and this land degradation allows only limited annual meat production. It is often only 300 pounds (136 kg) per acre, compared to over 3,000 pounds (1,360 kg) per acre in North America.

Further damage to the tropical forests of the Amazon basin is linked to commercial logging. Although only five of the approximately 1,500 tree species of the region are extensively logged, tremendous damage is done to the surrounding forest as these are selectively removed. When loggers build roads to move in heavy equipment, they may damage or destroy half of the trees in a given area.

The deforestation taking place in the Amazon basin has a wide range of environmental effects. The clearing and burning of vegetation produces smoke or air pollution, which at times has been so abundant that it is clearly visible from space. Removal of trees, which are significant carbon sinks, decreases the amount of carbon dioxide that can be extracted from the air, and in the case of slash-and-burn methods, the carbon dioxide that was stored in the tree biomass is released to the atmosphere during combustion. The result of deforestation is an accumulation of carbon dioxide in the air that can lead to climate change. Global rising surface and ocean temperatures and deforestation are contributing to climate change and inducing extreme changes in weather patterns. Flooding in the Amazon basin in 2009 left over 400,000 people homeless. Clearing also leads to increased soil erosion after heavy rains, and can result in water pollution through siltation as well as increased water temperatures from increased exposure. Yet the most alarming, and definitely the most irreversible, environmental problem facing the Amazon basin is the loss of biodiversity. Through the irrevocable process of extinction, this may cost humanity more than the loss of species. It may cost them the loss of potential discoveries of medicines and other beneficial products derived from these species. As of early 2009, the Brazilian government had added 19.3 million acres (7.8 million ha) of the Amazon rainforest as protected areas to ensure conservation of biodiversity and prevention of climate change.

Resources

BOOKS

Barthem, R.B. *Global International Waters Assessment: Amazon Basin.* Nairobi, Kenya: UNEP, 2004.

Bush, Mark B., and John Flenley. *Tropical Rainforest Responses to Climatic Change.* Berlin: Springer, 2007.

Fraser, Lauchlan H., and Paul A. Keddy. *The World's Largest Wetlands: Ecology and Conservation.* Cambridge, UK: Cambridge University Press, 2005.

London, Mark, and Brian Kelly. *The Last Forest: The Amazon in the Age of Globalization.* New York: Random House, 2007.

Rudel, Thomas K. *Tropical Forests: Regional Paths of Destruction and Regeneration in the Late Twentieth Century.* New York: Columbia University Press, 2005.

Souter, David. *Amazon Basin.* Kalmar, Sweden: University of Kalmar, 2004.

PERIODICALS

Asner, Gregory P., et al. "Condition and Fate of Logged Forests in the Brazilian Amazon." *Proceedings of the National Academy of Sciences* 103 (2006): 12947–12950.

OTHER

Posey, Darrell Addison, and Michael J. Balick. *Human Impacts on Amazonia: The Role of Traditional Ecological Knowledge in Conservation and Development.* New York: Columbia University Press, 2006.

Eugene C. Beckham

Ambient air

Ambient air is the air, external to buildings and other enclosures, found in the lower atmosphere over a given area, usually near the surface. Air pollution standards normally refer to ambient air.

Amenity value

Amenity value is the value something has because of the pleasant feelings it generates to those who use or view it. This value is often used in cost-benefit analysis, particularly in shadow pricing, to determine the worth of natural resources that will not be harvested for economic gain. A virgin forest will have amenity value, but its value will decrease if the forest is harvested.

American alligator *see* **Alligator, American.**

American box turtle

Box turtles are in the Order Chelonia, Family Emydidae, and genus *Terrapene*. There are two major species in the United States: *carolina* (Eastern box turtle) and *ornata* (Western or ornate box turtle).

Box turtles are easily recognized by their dome-shaped upper shell (carapace) and by their lower shell (plastron), which is hinged near the front. This hinging allows them to close up tightly into the "box" when in danger (hence their name).

Box turtles are fairly small, having an adult maximum length of 4 to 7 inches (10–18 cm). Their range is restricted to North America, with the Eastern species located over most of the eastern United States and the Western species located in the central and southwestern United States and into Mexico, but not as far west as California. Both species are highly variable in coloration and pattern, ranging from a uniform tan to dark brown or black, with yellow spots or streaks. They prefer a dry habitat such as woodlands, open brushlands or prairie. They typically inhabit sandy soil, but are sometimes found in springs or ponds during hot weather. During the winter, they hibernate in the soil below the frost line, often as deep as 2 feet (60 cm). Their home range is usually fairly small, and they often live within areas less than 300 square yards (240 m^2).

Eastern box turtle. *(Photograph by Robert Huffman. Fieldmark Publications)*

Indians used to eat box turtles and incorporated their shells into their ceremonies as rattles.

Resources

BOOKS

Bonin, Franck; Bernard Devaux; and Alain Dupre. *Turtles of the World.* Baltimore: Johns Hopkins University Press, 2006.

OTHER

University of California Museum of Paleontology. "Marine Vertebrates: Turtles." http://www.ucmp.berkeley.edu/anapsids/testudines/testudines.html (accessed November 5, 2010).

John Korstad

Box turtles are omnivorous, feeding on living and dead insects, earthworms, slugs, fruits, berries (particularly blackberries and strawberries), leaves, and mushrooms. They have been known to ingest some mushrooms which are poisonous to humans, and there have been reports of people eating box turtles and getting sick. Other than this, box turtles are harmless to humans and are commonly collected and sold as pets (although this should be discouraged because they are now a threatened species). They can be fed raw hamburger, canned pet food, or leafy vegetables.

Box turtles normally live as long as thirty to forty years. Some have been reported with a longevity of more than 100 years, and this makes them the longest-lived land turtle. They are active from March until November and are diurnal, usually being more active in the early morning. During the afternoons they typically seek shaded areas. They breed during the spring and autumn, and the females build nests from May until July, typically in sandy soil where they dig a hole with their hind feet. The females can store sperm for several years. They typically hatch three to eight eggs that are elliptically-shaped and about 1.5 inches (4 cm) in diameter. Male box turtles have a slight concavity in their plastron that aids in mounting females during copulation. All four toes on the male's hind feet are curved, which aids in holding down the posterior portion of the female's plastron during copulation. Females have flat plastrons, shorter tails, and yellow or brown eyes. Most males have bright red or pink eyes. The upper jaw of both sexes ends in a down-turned beak.

Predators of box turtles include skunks, raccoons, foxes, snakes, and other animals. Native American

American Cetacean Society

The American Cetacean Society (ACS), located in San Pedro, California, is dedicated to the protection of whales and other cetaceans, including dolphins and porpoises. Principally an organization of scientists and teachers (though its membership does include students and laypeople) the ACS was founded in 1967 and claims to be the oldest whale conservation group in the world.

The ACS argues that the best protection for whales, dolphins, and porpoises is better public awareness about "these remarkable animals and the problems they face in their increasingly threatened habitat." The organization is committed to political action through education, and much of its work has been in improving communication between marine scientists and the general public.

The ACS has developed several educational resource materials on cetaceans, which are widely available for use in classrooms. There is a cetacean research library at the national headquarters in San Pedro, California, and the organization responds to thousands of inquiries every year. The ACS supports marine mammal research and sponsors a biennial conference on whales. It also assists in conducting whale-watching tours.

The organization also engages in more traditional and direct forms of political action. A representative in Washington, DC, monitors legislation that might affect cetaceans, attends hearings at government agencies, and participates as a member of the International Whaling Commission. The ACS also networks with other conservation groups. In addition, the ACS directs letter-writing campaigns, sending out "Action Alerts" to citizens and politicians. The organization is currently emphasizing the threats to marine life posed by oil spills, toxic wastes from industry and agriculture, the use of sound and sonar equipment, and particular fishing practices (including commercial whaling).

The ACS publishes a national newsletter on whale research, conservation, and education, called *Spyhopper*, and a quarterly journal of scientific articles on the same subjects, called *Whalewatcher*.

Resources

ORGANIZATIONS

American Cetacean Society, PO Box 1391, San Pedro, CA, USA, 90733, (310) 548-6279, (310) 548-6950, acsoffice @acsonline.org, http://www.acsonline.org

Douglas Smith

American Committee for International Conservation

The American Committee for International Conservation (ACIC), located in Washington, DC, is an association of nongovernmental organizations (NGOs) that is concerned about international conservation issues. The ACIC, founded in 1930, includes twenty-one member organizations. It represents conservation groups and individuals in forty countries. While ACIC does not fund conservation research, it does promote national and international conservation research activities. Specifically, ACIC promotes conservation and preservation of wildlife and other natural resources, and encourages international research on the ecology of endangered species.

Formerly called the American Committee for International Wildlife Protection, ACIC assists the International Union for Conservation of Nature (IUCN), an independent organization of nations, states, and NGOs, in promoting natural resource conservation. ACIC also coordinates its members' overseas research activities.

Member organizations of the ACIC include the African Wildlife Leadership Foundation, National Wildlife Federation, World Wildlife Fund, Caribbean Conservation Corporation, National Audubon Society, Natural Resources Defense Council, Nature Conservancy, International Association of Fish and Wildlife Agencies, and National Parks and Conservation Association. Members also include The Conservation Foundation, International

Institute for Environment and Development; Massachusetts Audubon Society; Chicago Zoological Society; Wildlife Preservation Trust; Wildfowl Trust; School of Natural Resources, University of Michigan; World Resources Institute; Global Tomorrow Coalition; and The Wildlife Society.

ACIC holds no formal meetings or conventions, nor does it publish magazines, books, or newsletters. Contact: American Committee for International Conservation, c/o Center for Marine Conservation, 1725 DeSales Street, NW, Suite 500, Washington, DC 20036.

Resources

BOOKS

Akcakaya, H. Resit. *Species Conservation and Management: Case Studies.* New York: Oxford University Press, 2004.
Chiras, Daniel D.; John P. Reganold; and Oliver S. Owen. *Natural Resource Conservation: Management for a Sustainable Future.* New York: Prentice-Hall, 2004.
Frankham, Richard; Jonathan D. Ballou; David A. Briscoe; and Karina H. McInnes. *A Primer of Conservation Genetics.* Cambridge, UK: Cambridge University Press, 2004.
Ladle, Richard J. *Biodiversity and Conservation: Critical Concepts in the Environment.* London: Routledge, 2009.

PERIODICALS

Willis, K. J., and H. J. B. Birks. "What Is Natural? The Need for a Long-Term Perspective in Biodiversity Conservation." *Science* 314 (2006): 1261–1265.

OTHER

Conservation International. "Conservation International." http://www.conservation.org/ (accessed October 12, 2010).
International Union for Conservation of Nature (IUCN). "The IUCN Red List of Threatened Species." http://www.iucnredlist.org/ (accessed October 12, 2010).

Linda Rehkopf

American Farmland Trust

Headquartered in Washington, DC, the American Farmland Trust (AFT) is an advocacy group for farmers and farmland. It was founded in 1980 to help reverse or at least slow the rapid decline in the number of productive acres nationwide, and it is particularly concerned with protecting land held by private farmers. The principles that motivate the AFT are perhaps best summarized in a line from William Jennings Bryan of the late nineteenth-early twentieth-century period that the organization has often quoted: "Destroy our farms, and the grass will grow in the streets of every city in the country."

Over 1 million acres (404,700 ha) of farmland in the United States is lost each year to development, according to the AFT, and in Illinois one and a half bushels of topsoil are lost for every bushel of corn produced. The AFT argues that such a decline poses a serious threat to the future of the American economy. As farmers are forced to cultivate increasingly marginal land, food will become more expensive, and the United States could become a net importer of agricultural products, damaging its international economic position. The organization believes that a declining farm industry would also affect American culture, depriving the country of traditional products such as cherries, cranberries, and oranges and imperiling a sense of national identity that is still in many ways agricultural.

The AFT works closely with farmers, business people, legislators, and environmentalists "to encourage sound farming practices and wise use of land." The group directs lobbying efforts in Washington, DC, working with legislators and policymakers and frequently testifying at congressional and public hearings on issues related to farming. In addition to mediating between farmers and state and federal government, the trust is also involved in political organizing at the grassroots level, conducting public opinion polls, contesting proposals for incinerators and toxic waste sites, and drafting model conservation easements. They conduct workshops and seminars across the country to discuss farming methods and soil conservation programs, and they worked with the state of Illinois to establish the Illinois Sustainable Agriculture Society. The group produces kits for distribution to schoolchildren in both rural and urban areas called "Seed for the Future," which teach the benefits of agriculture and help each child grow a plant.

The AFT has a reputation for innovative and determined efforts to realize its goals, and former Secretary of Agriculture John R. Block has said that "this organization has probably done more than any other to preserve the American farm." The foundation claims that between 1980 (the year it was founded) and 2010, it has "helped to save more than three million acres of farmland and led the way for the adoption of conservation practices on millions more." The AFT continues to battle urban sprawl in areas such as California's Central Valley and Berks County, Pennsylvania, as well as working to support farms in states such as Vermont, which are threatened not so much by development but by a poor agricultural economy. The AFT promotes a wetland policy that is fair to farmers, while meeting environmental standards, and won a national award from the Soil and Water Conservation Society for its publication *Does Farmland Protection Pay?*

The AFT publishes a quarterly magazine called *American Farmland*, a newsletter called *Farmland Update*, and a variety of brochures and pamphlets that offer practical information on soil erosion, the cost of community services, and estate planning. They also distribute videos, including *The Future of America's Farmland*, which explains the sale and purchase of development rights. In March 2010 the AFT sponsored its first-ever *No Farms No Food* rally in Albany, New York, the state's capital. A variety of groups were represented at the rally, including farmers, conservationists, and food advocates. Topics addressed included the loss of farmland to urban sprawl, the need to help farmers keep the water supply clean, and the economic benefits of locally grown food. Some 100 New York state legislators met with various advocacy leaders as part of the *No Farms No Food* rally.

Resources

ORGANIZATIONS

The American Farmland Trust (AFT), 1200 18th Street, NW, Suite 800, Washington, DC, USA, 20036, (202) 331-7300, (202) 659-8339, info@farmland.org, http://www.farmland.org

Douglas Smith

American Forests

Located in Washington, DC, American Forests was founded in 1875, during the early days of the American conservation movement, to encourage forest management. Originally called the American Forestry Association, the organization was renamed in the latter part of the twentieth century. The group is dedicated to promoting the wise and careful use of all natural resources, including soil, water, and wildlife, and it emphasizes the social and cultural importance of these resources as well as their economic value.

Although benefiting from increasing national and international concern about the environment, American Forests takes a balanced view on preservation, and it has worked to set a standard for the responsible harvesting and marketing of forest products. American Forests sponsors the Trees for People program, which is designed to help meet the national demand for wood and paper products by increasing the productivity of private woodlands. It provides educational and technical information to individual forest owners and makes recommendations to legislators and policymakers in Washington, DC.

To draw attention to the greenhouse effect, American Forests inaugurated their Global ReLeaf program in October 1988. Global ReLeaf is what American Forests calls "a tree-planting crusade." The message is, "Plant a tree, cool the globe," and Global ReLeaf has organized a national campaign challenging Americans to plant millions of trees. American Forests has gained the support of government agencies and local conservation groups for this program, as well as many businesses, including such Fortune 500 companies as Texaco, McDonald's, and Ralston-Purina. The goal of the project is to plant 100 million trees by the year 2020. By 2010, there had been 33 million trees planted, distributed throughout all fifty U.S. states, as well as in twenty other nations. Global ReLeaf also launched a cooperative effort with the American Farmland Trust called Farm ReLeaf, and it has also participated in the campaign to preserve Walden Woods in Massachusetts. American Forests has brought Global ReLeaf to eastern Europe, running a workshop in Budapest, Hungary, for environmental activists from many former communist countries.

American Forests has been extensively involved in the controversy over the preservation of old-growth forests in the American Northwest. They have been working with environmentalists and representatives of the timber industry, and consistent with the history of the organization, American Forests is committed to a compromise that both sides can accept: "If we have to choose between preservation and destruction of old-growth forests as our only options, neither choice will work." American Forests supports an approach to forestry known as *New Forestry*, where the priority is no longer the quantity of wood or the number of board feet that can be removed from a site, but the vitality of the ecosystem the timber industry leaves behind. The organization advocates the establishment of an Old Growth Reserve in the Pacific Northwest, which would be managed by the principles of New Forestry under the supervision of a Scientific Advisory Committee.

American Forests publishes the *National Registry of Big Trees,* which celebrated its seventieth anniversary in 2010. The registry is designed to encourage the appreciation of some of the largest and oldest (and oftentimes, most unusual) trees in the United States. The group also publishes *American Forests,* a

bimonthly magazine, and *Resource Hotline*, a biweekly newsletter, as well as *Urban Forests: The Magazine of Community Trees*. It presents the Annual Distinguished Service Award, the John Aston Warder Medal, and the William B. Greeley Award, among others. According to its 2008 annual report, American Forests had revenue in 2008 of $5,186,515 (including gifts, donations, and membership fees).

Resources

ORGANIZATIONS

American Forests, PO Box 2000, Washington, DC, USA, 20013, (202) 955-4500, (202) 955-4588, info@amfor. org, http://www.americanforests.org

Douglas Smith

American Indian Environmental Office

The American Indian Environmental Office (AIEO) was created to increase the quality of public health and environmental protection on Native American land and to expand tribal involvement in running environmental programs.

Native Americans are the second-largest landholders in the United States besides the government. Their land is often threatened by environmental degradation such as strip mining, clear-cutting, and toxic storage. The AIEO, with the help of former President Ronald Reagan's Federal Indian Policy (January 24, 1983), works closely with the U.S. Environmental Protection Agency (EPA) to prevent further degradation of the land. The AIEO has received grants from the EPA for environmental cleanup and obtained a written policy that requires the EPA to continue with the trust responsibility, a clause expressed in certain treaties that requires the EPA to notify the tribe when performing any activities that may affect reservation lands or resources. This involves consulting with tribal governments, providing technical support, and negotiating EPA regulations to ensure that tribal facilities eventually comply.

The pollution of Dine Reservation land is an example of an environmental injustice that the AIEO wants to prevent in the future. The reservation has over 1,000 abandoned uranium mines that leak radioactive contaminants and is also home to the largest coal strip mine in the world. The cancer rate for the Dine people is seventeen times the national average. To help tribes with pollution problems similar to the Dine, several offices now exist that handle specific environmental projects. They include the Office of Water, Air, Environmental Justice, Pesticides and Toxic Substances; Performance Partnership Grants; Solid Waste and Emergency Response; and the Tribal Watershed Project. Each of these offices reports to the National Indian Headquarters in Washington, DC.

At the Rio Earth Summit in 1992, the Biodiversity Convention was drawn up to protect the diversity of life on the planet. Many Native American groups believe that the convention also covered the protection of indigenous communities, including Native American land. In addition, the groups demand that prospecting by large companies for rare forms of life and materials on their land must stop.

Tribal environmental concerns

Tribal governments face both economic and social problems dealing with the demand for jobs, education, health care, and housing for tribal members. Often the reservations' largest employer is the government, which owns the stores, gaming operations, timber mills, and manufacturing facilities. Therefore, the government must deal with the conflicting interests of protecting both economic and environmental concerns. Many tribes are becoming self-governing and manage their own natural resources along with claiming the reserved right to use natural resources on portions of public land that border their reservation. As a product of the reserved treaty rights, Native Americans can use water, fish, and hunt anytime on nearby federal land.

Robert Belcourt, Chippewa-Cree tribal member and director of the Natural Resources Department in Montana stated:

"We have to protect nature for our future generations. More of our Indian people need to get involved in natural resource management on each of our reservations. In the long run, natural resources will be our bread and butter by our developing them through tourism and recreation and just by the opportunity they provide for us to enjoy the outdoor world."

Belcourt has fought to destroy the negative stereotypes of conservation organizations that exist among Native Americans who believe, for example, that conservationists are extreme tree huggers and insensitive to Native American culture. These stereotypes are a result of cultural differences in philosophy, perspective, and communication. To work together effectively, tribes and conservation groups need to learn about one another's cultures, and this means they must listen both at meetings and in one-on-one exchanges.

The AIEO also addresses the organizational differences that exist between tribal governments and conservation organizations, such as differences in style, motivation, and the pressures each group faces. Pressures on the Wilderness Society, for example, include fending off attempts in Washington, DC, to weaken key environmental laws or securing members and raising funds. Pressures on tribal governments more often are economic and social in nature and have to do with the need to provide jobs, health care, education, and housing for tribal members. Because tribal governments are often the reservations' largest employers and may own businesses such as gaming operations, timber mills, manufacturing facilities, and stores, they function as both governors and leaders in economic development.

Native Americans currently occupy and control over 52 million acres (21.3 million ha) in the continental United States and 45 million more acres (18.5 million ha) in Alaska, yet this is only a small fraction of their original territories.

In the nineteenth century, many tribes were confined to reservations that were perceived to have little economic value, although valuable natural resources have subsequently been found on some of these lands. Pointing to their treaties and other agreements with the federal government, many tribes assert that they have reserved rights to use natural resources on portions of public land.

In previous decades these natural resources on tribal lands were managed by the Bureau of Indian Affairs (BIA). Now many tribes are becoming self-governing and are taking control of management responsibilities within their own reservation boundaries. In addition, some tribes are pushing to take back management over some federally-managed lands that were part of their original territories. For example, the Confederated Salish and Kootenai tribes of the Flathead Reservation are taking steps to assume management of the National Bison Range, which lies within the reservation's boundaries and is, as of 2010, managed by the U.S. Fish and Wildlife Service.

Another issue concerns Native American rights to water. There are legal precedents that support the practice of reserved rights to water that are within or bordering a reservation. In areas where tribes fish for food, mining pollution has been a continous threat to maintaining clean water. Mining pollution is monitored, but the amount of fish that Native Americans consume is higher than the government acknowledges when setting health guidelines for their consumption. This is why the AIEO is asking that stricter regulations be imposed on mining companies. As tribes increasingly exercise their rights to use and consume water and fish, their roles in natural resource debates will increase.

Many tribes are establishing their own natural resource management and environmental quality protection programs with the help of the AIEO. Tribes have established fisheries, wildlife, forestry, water quality, waste management, and planning departments. Some tribes have prepared comprehensive resource management plans for their reservations while others have become active in the protection of particular species. The AIEO is uniting tribes in their strategy and involvement level with improving environmental protection on Native American land.

Resources

ORGANIZATIONS

American Indian Environmental Office, 1200 Pennsylvania Avenue, NW, Washington, DC, USA, 20460, (202) 564-0303, (202) 564-0298, http://www.epa.gov/aieo/index.htm

Nicole Beatty

American Oceans Campaign

Located in Los Angeles, California, the American Oceans Campaign (AOC) was founded in 1987 as a political interest group dedicated primarily to the restoration, protection, and preservation of the health and vitality of coastal waters, estuaries, bays, wetlands, and oceans. In 2002, the group joined other groups to form Oceana, a nonprofit environmental advocacy group. The focus of this partnership is the Oceans at Risk program that concentrates on the impact that wasteful fisheries have on the marine environment.

More national and conservationist (rather than international and preservationist) in its focus than other groups with similar concerns, the AOC tended to view the oceans as a valuable resource whose use should be managed carefully. As cofounder Ted Danson asserted, the oceans must be regarded as far more than a natural preserve by environmentalists; rather, healthy oceans "sustain biological diversity, provide us with leisure and recreation, and contribute significantly to our nation's GNP."

The AOC's main political efforts reflected this focus. Central to the AOC's lobbying strategy was a desire to build cooperative relations and consensus among the general public, public interest groups, private sector corporations and trade groups, and governmental authorities around responsible management of ocean resources. As Oceana, the group continues AOC's interests in grassroots public awareness campaigns through mass media and community outreach programs.

As a lobbying organization, the AOC developed contacts with government leaders at all levels from local to national, attempting to shape and promote a variety of legislation related to clean water and oceans. It was active in lobbying for strengthening various aspects of the Clean Water Act, the Safe Drinking Water Act, the Oil Pollution Act, and the Ocean Dumping Ban Act. The AOC regularly provided consultation services, assistance in drafting legislation, and occasional expert testimony on matters concerning ocean ecology.

Also very active at the grassroots level, AOC organized numerous cleanup operations, which drew attention to the problems caused by ocean dumping and made a practical contribution to reversing the situation. Concentrating its efforts in California and the Pacific Northwest, the AOC launched a "Dive for Trash" program in 1991.

Lawrence J. Biskowski

American Wildlands

American Wildlands (AWL) is a nonprofit wildland resource conservation and education organization founded in 1977. AWL is dedicated to protecting and promoting proper management of America's publicly-owned wild areas and to securing wilderness designation for public land areas. The organization has played a key role in gaining legal protection for many wilderness and river areas in the U.S. interior West and in Alaska.

Founded as the American Wilderness Alliance, AWL is involved in a wide range of wilderness resource issues and programs including timber management policy reform, habitat corridors, rangeland management policy reform, riparian and wetlands restoration, and public land management policy reform. AWL promotes ecologically sustainable uses of public wildlands resources including forests, wilderness, wildlife, fisheries, and rivers. It pursues this mission through grassroots activism, technical support, public education, litigation, and political advocacy.

AWL maintains three offices: the central Rockies office in Lakewood, Colorado; the northern Rockies office in Bozeman, Montana; and the Sierra-Nevada office in Reno, Nevada. The organization's 2009 budget from income and other revenue (such as donations) was about $580,000.

The central Rockies office in Bozeman considers its main concern timber management reform. It has launched the Timber Management Reform Policy Program, which monitors the U.S. Forest Service and works toward a better management of public forests. Since initiation of the program in 1986, the program includes resource specialists, a wildlife biologist, forester, water specialist, and an aquatic biologist who all report to an advisory council. A major victory of this program was stopping the sale of 4.2 million board feet (1.3 million m) of timber near the Electric Peak Wilderness Area.

Other programs coordinated by the central Rockies office include: (1) *Corridors of Life Program*, which identifies and maps wildlife corridors, land areas essential to the genetic interchange of wildlife that connect roadless lands or other wildlife habitat areas. Areas targeted are in the interior West, such as Montana, North and South Dakota, Wyoming, and Idaho. The AWL contributed to the language embodied in a 2007 resolution—the *Wildlife Corridors Policy Initiative*—approved by the Western Governors' Association, which is composed of the governors of nineteen western U.S. states. The resolution aims to conserve various wildlife corridors in the western United States; (2) The *Rangeland Management Policy Reform Program* monitors grazing allotments and files appeals as warranted. An education component teaches citizens to monitor grazing allotments and to use the appeals process within the U.S. Forest Service and Bureau of Land Management; (3) The *Recreation-Conservation Connection*, through newsletters and travel-adventure programs, teaches the public how to enjoy the outdoors without destroying nature. Many hundreds of travelers have participated in ecotourism trips through AWL.

AWL is also active internationally. The AWL/Leakey Fund has aided Richard Leakey's wildlife habitat conservation and elephant poaching elimination efforts in Kenya. A partnership with the Island Foundation has helped fund wildlands and river protection efforts in Patagonia, Argentina. AWL also is an active member of Canada's Tatshenshini International Coalition to protect that river and its 2.3 million acres (930,780 ha) of wilderness.

Resources

ORGANIZATIONS

American Wildlands, 40 East Main #2, Bozeman, MT, USA, 59715, (406) 586-8175, (406) 586-8242, info@wildlands.org, http://www.wildlands.org

Linda Rehkopf

Ames test

A laboratory test developed by biochemist Bruce N. Ames to determine the possible carcinogenic nature of a substance, the Ames test involves using a particular strain of the bacteria *Salmonella typhimurium* that lacks the ability to synthesize histidine and is therefore very sensitive to mutation. The bacteria are inoculated into a medium deficient in histidine but containing the test compound. If the compound results in deoxyribonucleic acid (DNA) damage with subsequent mutations, some of the bacteria will regain the ability to synthesize histidine and will proliferate to form colonies. The culture is evaluated on the basis of the number of mutated bacterial colonies it produced. The ability to replicate mutated colonies leads to the classification of a substance as probably carcinogenic.

The Ames test is a test for mutagenicity, not carcinogenicity. However, approximately nine out of ten mutagens are indeed carcinogenic. Therefore, a substance that can be shown to be mutagenic by being subjected to the Ames test can be reliably classified as a suspected carcinogen and thus recommended for further study.

Resources

BOOKS

Bignold, Leon P. *Cancer: Cell Structures, Carcinogens, and Genomic Instability*. New York: Birkhauser, 2005.

Brian R. Barthel

Amoco Cadiz

This shipwreck in March 1978 off the Brittany coast was the first major supertanker accident since the *Torrey Canyon* eleven years earlier. Ironically, this spill, more than twice the size of the *Torrey Canyon*,

Off the coast of Portsall, the prow of the *Amoco Cadiz* sticks out of the water like a giant shark. *(© Vauthey Pierre/Corbis Sygma)*

blackened some of the same shores and was one of four substantial oil spills occurring there since 1967. It received great scientific attention because it occurred near several renowned marine laboratories.

The cause of the wreck was a steering failure as the ship entered the English Channel off the northwest Brittany coast, and failure to act swiftly enough to correct it. During the next twelve hours, the *Amoco Cadiz* could not be extricated from the site. In fact, three separate lines from a powerful tug broke trying to remove the tanker before it drifted onto rocky shoals. Eight days later the *Amoco Cadiz* split in two.

Seabirds seemed to suffer the most from the spill, although the oil devastated invertebrates within the extensive, 20 to 30 feet (6-9 m) high intertidal zone. Thousands of birds died in a bird hospital described by one oil spill expert as a bird morgue. Thirty percent of France's seafood production was threatened, as well as an extensive kelp crop, harvested for fertilizer, mulch, and livestock feed. However, except on oyster farms located in inlets, most of the impact was restricted to the few months following the spill.

In an extensive journal article, Erich Grundlach and others reported studies on where the oil went and summarized the findings of biologists. Of the 245,300 tons (223,000 metric tons) released, 13.5 percent was incorporated within the water column, 8 percent went into subtidal sediments, 28 percent washed into the intertidal zone, 20–40 percent evaporated, and 4 percent was altered while at sea. Much research was done on chemical changes in the hydrocarbon fractions over time, including that taken up within organisms. Researchers found that during early phases, biodegradation was occurring as rapidly as evaporation.

The cleanup efforts of thousands of workers were helped by storm and wave action that removed much of the stranded oil. High-energy waves maintained an adequate supply of nutrients and oxygenated water, which provided optimal conditions for biodegradation. This is important because most of the biodegradation was done by aerobic organisms. Except for protected inlets, much of the impact was gone three years later, but some effects were expected to last a decade.

Resources

PERIODICALS

Grove, N. "Black Day for Brittany: *Amoco Cadiz* Wreck." *National Geographic* 154 (July 1978): 124–135.

Grundlach, Erich R., et al. "The Fate of *Amoco Cadiz* Oil." *Science* 221 (July 8, 1983): 122–129.

Spooner, M. F., ed. *Amoco Cadiz Oil Spill.* New York: Pergamon, 1979.

Wang, Zhendi, and Scott Stout. *Oil Spill Environmental Forensics: Fingerprinting and Source Identification.* New York: Academic, 2006.

Nathan H. Meleen

Amory, Cleveland

1917–1998
American Activist and writer

Amory was known both for his series of classic social history books and his work with the Fund for Animals. Born in Nahant, Massachusetts, to an old Boston family, Amory attended Harvard University, where he became editor of *The Harvard Crimson.*

Amory was hired by *The Saturday Evening Post* after graduation, becoming the youngest editor ever to join that publication. He worked as an intelligence officer in the United States Army during World War II, and in the years after the war, wrote a trilogy of social commentary books, now considered to be classics. *The Proper Bostonians* was published to critical acclaim in 1947, followed by *The Last Resorts* (1948), and *Who Killed Society?* (1960), all of which became best sellers.

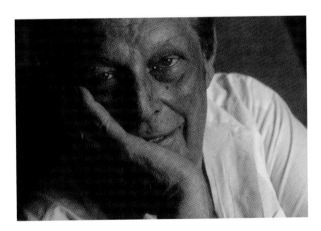

Author Cleveland Amory poses at Black Beauty Ranch July 24, 1991, in New York City. Amory was a social historian, television critic, and animal rights advocate. *(Photo by Paul S. Howell/ Liaison)*

Beginning in 1952, Amory served for eleven years as social commentator on NBC's "The Today Show." The network fired him after he spoke out against cruelty to animals used in biomedical research. From 1963 to 1976, Amory served as a senior editor and columnist for *Saturday Review* magazine, while doing a daily radio commentary titled "Curmudgeon-at-Large." He was also chief television critic for *TV Guide,* where his biting attacks on sport hunting angered hunters and generated bitter but unsuccessful campaigns to have him fired.

In 1967, Amory founded The Fund for Animals in his words, "to speak for those who can't," and served as its unpaid president. Animal protection became his passion and his life's work, and he was considered one of the most outspoken and provocative advocates of animal welfare. Under his leadership, the fund became a highly activist and controversial group, engaging in such activities as confronting hunters of whales and seals, and rescuing wild horses, burros, and goats. The fund, and Amory in particular, are well-known for their campaigns against sport hunting and trapping, the fur industry, abusive research on animals, and other activities and industries that engage in or encourage what they consider cruel treatment of animals.

In 1975, Amory published *ManKind? Our Incredible War on Wildlife,* using humor, sarcasm, and graphic rhetoric to attack hunters, trappers, and other exploiters of wild animals. The book was praised by *The New York Times* in a rare editorial. His next book, *AniMail* (1976), discussed animal issues in a question-and-answer format. In 1987, he wrote *The Cat Who Came for Christmas,* a book about a stray cat he rescued from the streets of New York, which became a national best seller. This was followed in 1990 by its sequel, also a best seller, *The Cat and the Curmudgeon.* Amory had been a senior contributing editor of *Parade* magazine since 1980, where he often profiled famous personalities.

Amory died of an aneurysm at the age of 81 on October 14, 1998. He remained active until his death, spending the day in his office at the Fund for Animals and then passing away in his sleep later that evening.

See also Fund for Animals.

Resources

BOOKS

Amory, Cleveland. *The Cat and the Curmudgeon.* New York: G. K. Hall, 1991.

Amory, Cleveland. *The Cat Who Came for Christmas.* New York: Little, Brown, 1987.

Lewis G. Regenstein

Anaerobic

Anaerobic refers to an environment lacking in molecular oxygen (O_2), or to an organism, tissue, chemical reaction, or biological process that does not require oxygen. Anaerobic organisms can use a molecule other than O_2 as the terminal electron acceptor in respiration. These organisms can be either obligate, meaning that they cannot use O_2, or facultative, meaning that they do not require oxygen but can use it if it is available.

Organic matter decomposition in poorly aerated environments, including water-logged soils, septic tanks, and anaerobically-operated waste treatment facilities, produces large amounts of methane gas. The methane can become an atmospheric pollutant, or it may be captured and used for fuel, as in biogas-powered electrical generators.

Anaerobic digestion

Anaerobic digestion refers to the biological degradation of sludge or solid waste under anaerobic conditions (i.e., an environment lacking oxygen). In the digestive process, solids are converted to noncellular end products.

In the anaerobic digestion of sludges, the goals are to reduce sludge volume, ensure the remaining solids are chemically stable, reduce disease-causing pathogens, and enhance the effectiveness of subsequent dewatering methods, sometimes recovering methane as a source of energy. Anaerobic digestion is commonly used to treat sludges that contain primary sludges, such as that from the first settling basins in a wastewater treatment plant, because the process is capable of stabilizing the sludge with little biomass production, a significant benefit over aerobic sludge digestion, which would yield more biomass in digesting the relatively large amount of biodegradable matter in primary sludge.

The microorganisms responsible for digesting the sludges anaerobically are often classified in two groups: the acid formers and the methane formers. The acid formers are microbes that create, among others, acetic and propionic acids from the sludge. These chemicals generally make up about one-third of the by-products initially formed based on a chemical oxygen demand (COD) mass balance; some of the propionic and other acids are converted to acetic acid.

The methane formers convert the acids and by-products resulting from prior metabolic steps (e.g., alcohols, hydrogen, carbon dioxide) to methane. Often, approximately 70 percent of the methane formed is derived from acetic acid, about 10-15 percent from propionic acid.

Anaerobic digesters are designed as either standard- or high-rate units. The standard-rate digester has a solids retention time of thirty to ninety days, as opposed to ten to twenty days for the high-rate systems. The volatile solids loadings of the standard- and high-rate systems are in the area of 0.5 to 1.6 and 1.6 to 6.4 $Kg/m^3/d$, respectively. The amount of sludge introduced into the standard-rate is therefore generally much less than the high-rate system. Standard-rate digestion is accomplished in single-stage units, meaning that sludge is fed into a single tank and allowed to digest and settle. High-rate units are often designed as two-stage systems in which sludge enters into a completely mixed first stage that is mixed and heated to approximately 98°F (35°C) to speed digestion. The second-stage digester, which separates digested sludge from the overlying liquid and scum, is not heated or mixed.

With the anaerobic digestion of solid waste, the primary goal is generally to produce methane, a valuable source of fuel that can be burned to provide heat or used to power motors. There are basically three steps in the process. The first involves preparing the waste for digestion by sorting the waste and reducing its size. The second consists of constantly mixing the sludge, adding moisture, nutrients, and pH neutralizers while heating it to about 143°F (60°C) and digesting the waste for a week or longer. In the third step, the generated gas is collected and sometimes purified, and digested solids are disposed of. For each pound of undigested solid, about 8 to 12 cubed feet (0.24-0.36 cubic m) of gas is formed, of which about 60 percent is methane.

Resources

BOOKS

Bitton, Gabriel. *Wastewater Microbiology*. Hoboken, NJ: Wiley-Liss and John Wiley & Sons, 2005.

Leadbetter, Jared R., ed. *Environmental Microbiology*. Amsterdam, Netherlands; Boston: Elsevier Academic Press, 2005.

PERIODICALS

Kniemeyer, O., et al. "Anaerobic Oxidation of Short-chain Hydrocarbons by Marine Sulphate-reducing Bacteria." *Nature* 449 (2007): 898–901.

Gregory D. Boardman

Anemia

Anemia is a medical condition in which the red cells of the blood are reduced in number or volume or are deficient in hemoglobin, their oxygen-carrying pigment. Almost 100 different varieties of anemia are known. Iron deficiency is the most common cause of anemia worldwide. Other causes of anemia include ionizing radiation, lead poisoning, vitamin B_{12} deficiency, folic acid deficiency, certain infections, and pesticide exposure. Some 350 million people worldwide—mostly women of childbearing age—suffer from anemia.

The most noticeable symptom is pallor of the skin, mucous membranes, and nail beds. Symptoms of tissue oxygen deficiency include pulsating noises in the ear, dizziness, fainting, and shortness of breath. The treatment varies greatly depending on the cause and diagnosis, but may include supplying missing nutrients, removing toxic factors from the environment, improving the underlying disorder, or restoring blood volume with transfusion.

Aplastic anemia is a disease in which the bone marrow fails to produce an adequate number of blood cells. It is usually acquired by exposure to certain drugs, to toxins such as benzene, or to ionizing radiation. Aplastic anemia from radiation exposure is well-documented from the Chernobyl experience. Bone marrow changes typical of aplastic anemia can occur several years after the exposure to the offending agent has ceased.

Aplastic anemia can manifest itself abruptly and progress rapidly; more commonly it is insidious and chronic for several years. Symptoms include weakness and fatigue in the early stages, followed by headaches, shortness of breath, fever, and a pounding heart. Usually a waxy pallor and hemorrhages occur in the mucous membranes and skin. Resistance to infection is lowered and becomes the major cause of death. While spontaneous recovery occurs occasionally, the treatment of choice for severe cases is bone marrow transplantation.

Marie Curie, who discovered the element radium and did early research into radioactivity, died in 1934 of aplastic anemia, most likely caused by her exposure to ionizing radiation.

While lead poisoning, which leads to anemia, is usually associated with occupational exposure, toxic amounts of lead can leach from imported ceramic dishes. Other consumer products produced in the developing world also contain unacceptably high concentrations of lead. For instance, in a report issued on October 11, 2010, the International Tobacco Control (ITC) Project found that Chinese cigarettes, when compared to Canadian cigarettes, contained three times the concentration of lead and other heavy metals. Additional environmental sources of lead exposure include old paint or paint dust, and drinking water pumped through lead pipes or lead-soldered pipes.

Cigarette smoke is known to cause an increase in the level of hemoglobin in smokers, which leads to an underestimation of anemia in smokers. Studies suggest that carbon monoxide (a by-product of smoking) chemically binds to hemoglobin, causing a significant elevation of hemoglobin values. Compensation values developed for smokers can now detect possible anemia.

Resources

OTHER

Centers for Disease Control and Prevention (CDC). "Anemia." http://www.cdc.gov/nccdphp/dnpa/nutrition/nutrition_for_everyone/iron_deficiency/index.htm (accessed November 8, 2010).
Mayo Clinic. "Anemia." http://www.mayoclinic.com/health/anemia/DS00321 (accessed November 8, 2010).
National Institutes of Health (NIH). "Anemia." http://health.nih.gov/topic/Anemia (accessed November 8, 2010).

Linda Rehkopf

Animal cancer tests

Cancer causes more loss of life-years than any other disease in the United States. According to a Centers for Disease Control and Prevention (CDC) report issued in January 2010, cancer and heart disease still accounted for nearly one-half (48.5 percent) of all U.S. deaths in 2007. However, many deaths from heart attack and stroke occur in the elderly. The loss of life-years of an eighty-five-year-old person (whose life expectancy at the time of his or her birth was between fifty-five and sixty) is, of course, zero. However, the loss of life-years of a child of ten who dies of a pediatric leukemia is between sixty-five to seventy years. This comparison of youth with the elderly is not meant in any way to demean the *value* that reasonable people place on the lives of the elderly. Rather, the comparison is made to emphasize the great decrease in life span due to malignancies, especially in the young.

The chemical causation of cancer is not a simple process. Many, perhaps most, chemical carcinogens do not in their usual condition have the potency to cause cancer. The noncancer causing form of the chemical is called a "procarcinogen." Procarcinogens are frequently complex organic compounds that the human body attempts to dispose of when ingested. Hepatic enzymes chemically change the procarcinogen in several steps to yield a chemical that is more easily excreted. The chemical changes result in modification of the procarcinogen (with no cancer forming ability) to the ultimate carcinogen (with cancer causing competence). Ultimate carcinogens have been shown to have a great affinity for deoxyribonucleic acid (DNA), ribonucleic acid (RNA), and cellular proteins, and it is the interaction of the ultimate carcinogen with the cell macromolecules that causes cancer. It is unfortunate indeed that one cannot look at the chemical structure of a potential carcinogen and predict whether or not it will cause cancer. There is no computer program that will predict what hepatic enzymes will do to procarcinogens and how the metabolized end product(s) will interact with cells.

Great strides have been made in the development of chemotherapeutic agents designed to cure cancer. The drugs have significant efficacy with certain cancers (these include but are not limited to pediatric acute lymphocytic leukemia, choriocarcinoma, Hodgkin's disease, and testicular cancer), and some treated patients attain a normal life span. While this development is heartening, the cancers listed are, for the most part, relatively infrequent. More common cancers such as colorectal carcinoma, lung cancer, breast cancer, and ovarian cancer remain intractable with regard to treatment.

These several reasons are why animal testing is used in cancer research. The majority of Americans support the effort of the biomedical community to use animals to identify potential carcinogens with the hope that such knowledge will lead to a reduction of cancer prevalence. Similarly, they support efforts to develop more effective chemotherapy. Animals are used under terms of the Animal Welfare Act of 1966 and its several amendments. The Act designates that the U.S. Department of Agriculture is responsible for the humane care and handling of warm-blooded creatures and other animals used for biomedical research. The Act also calls for inspection of research facilities to ensure that adequate food, housing, and care are provided. It is the belief of many that the constraints of the law have enhanced the quality of biomedical research. Poorly maintained animals do not provide quality research. The law also has enhanced the care of animals used in cancer research.

Resources

BOOKS

Bunz, Fred. *Principles of Cancer Genetics*. Baltimore: Springer, 2008.

Colditz, Graham A. *Encyclopedia of Cancer and Society*. Los Angeles: Sage Publications, 2007.
King, Roger, J. B., and Mike W. Robins. *Cancer Biology*. Harlow, UK; New York: Pearson/Prentice Hall, 2006.
Weinberg, Robert A. *The Biology of Cancer*. New York: Garland Science, 2007.

OTHER

Centers for Disease Control and Prevention (CDC). "Cancer." http://www.cdc.gov/cancer/ (accessed November 7, 2010).
World Health Organization (WHO). "Cancer." http://www.who.int/entity/mediacentre/factsheets/fs297/en/index.html (accessed November 7, 2010).

Robert G. McKinnell

Animal Legal Defense Fund

Originally established in 1979 as Attorneys for Animal Rights, this organization changed its name to Animal Legal Defense Fund (ALDF) in 1984, and is known as the law firm of the animal rights movement. Their motto is "we may be the only lawyers on earth whose clients are all innocent." ALDF contends that animals have a fundamental right to legal protection against abuse and exploitation. Hundreds of attorneys work for ALDF, and the organization has more than 100,000 supporting members who help the cause of animal rights by writing letters and signing petitions for legislative action. The members are also strongly encouraged to work for animal rights at the local level.

ALDF's work is carried out in many places including research laboratories, large cities, small towns, and the wild. ALDF attorneys try to stop the use of animals in research experiments, and continue to fight for expanded enforcement of the Animal Welfare Act. ALDF also offers legal assistance to humane societies and city prosecutors to help in the enforcement of anti-cruelty laws and the exposure of veterinary malpractice. The organization attempts to protect wild animals from exploitation by working to place controls on trappers and sport hunters. In California, ALDF successfully stopped the hunting of mountain lions and black bears. ALDF is also active internationally bringing legal action against elephant poachers as well as against animal dealers who traffic in endangered species.

ALDF's clear goals and swift action have resulted in many court victories. In 1992 alone, the organization won cases involving cruelty to dolphins, dogs, horses, birds, and cats. It has also blocked the importation of over 70,000 monkeys from Bangladesh for research purposes, and has filed suit against the National Marine Fisheries Services to stop the illegal gray market in dolphins and other marine mammals. In 2009, the ALDF filed an amicus curiae brief in the case *U.S. v. Stevens*, the first animal cruelty case to come before the Supreme Court of the United States in fifteen years. In *Stevens*, the Court ruled that a federal law that prohibited the production, sale, and distribution of videos depicting animal cruelty violated the First Amendment's right to freedom of speech.

Resources

ORGANIZATIONS

Animal Legal Defense Fund, 170 East Cotati Ave., Cotati, CA, USA, 94931, (707) 795-2533, (707) 795-7280, info@aldf.org, http://www.aldf.org

Cathy M. Falk

Animal rights

Concerns about the way humans treat non-human animals has spawned a powerful social and political movement driven by the conviction that humans and certain animals are similar in morally significant ways, and that these similarities oblige humans to extend to those animals serious moral consideration, including rights. Though animal welfare movements, concerned primarily with humane treatment of pets, date back to the 1800s, modern animal rights activism, also called animal liberationism, has developed primarily out of concern about the use and treatment of domesticated animals in agriculture and in medical, scientific, and industrial research. The rapid growth in membership of animal rights organizations testifies to the increasing momentum of this movement. The leading animal rights group today, People for the Ethical Treatment of Animals (PETA), was founded in 1980 with 100 individuals; in 2010, it has over two million members. The nonprofit PETA, with about 300 employees, is headquartered in Norfolk, Virginia. According to its Web site: "Animals are not ours to eat. Animals are not ours to wear. Animals are not ours to experiment on. Animals are not ours to use for entertainment. Animals are not ours to abuse in any way." The animal rights activist movement has closely followed and used the work of modern philosophers who seek to establish a firm logical foundation for the extension of moral considerability beyond the human community into the animal community.

The nature of animals and appropriate relations between humans and animals have occupied Western thinkers for millennia. Traditional Western views, both religious and philosophical, have tended to deny that humans have any moral obligations to nonhumans. The rise of Christianity and its doctrine of personal immortality, which implies a qualitative gulf between humans and animals, contributed significantly to the dominant Western paradigm. When seventeenth-century philosopher René Descartes (1596–1650) declared animals mere biological machines, the perceived gap between humans and nonhuman animals reached its widest point. English philosopher Jeremy Bentham (1748–1832), the father of ethical utilitarianism, challenged this view and fostered a widespread anticruelty movement and exerted powerful force in shaping legal and moral codes. Its modern legacy, the animal welfare movement, is reformist in that it continues to accept the legitimacy of sacrificing animal interests for human benefit, provided animals are spared any suffering that can conveniently and economically be avoided.

In contrast to the conservatively reformist platform of animal welfare crusaders, a new radical movement began in the late 1970s. This movement, variously referred to as animal liberation or animal rights, seeks to put an end to the routine sacrifice of animal interests for human benefit. In seeking to redefine the issue as one of rights, some animal protectionists organized around the well-articulated and widely disseminated utilitarian perspective of Australian philosopher Peter Singer (1946–). In his 1975 classic *Animal Liberation*, Singer argued that because some animals can experience pleasure and pain, they deserve moral consideration by humans. While not actually a rights position, Singer's work nevertheless uses the language of rights and was among the first to abandon welfarism and to propose a new ethic of moral considerability for all sentient (conscious) creatures.

To assume that humans are inevitably superior to other species simply by virtue of their species membership is an injustice, which Singer terms "speciesism," an injustice parallel to racism and sexism.

Singer does not claim all animal lives to be of equal worth, nor that all sentient beings should be treated identically. In some cases, human interests may outweigh those of nonhumans, and Singer's utilitarian calculus would allow people to engage in practices that require the use of animals in spite of their pain, where those practices can be shown to produce an overall balance of pleasure over suffering.

Some animal advocates thus reject utilitarianism on the grounds that it allows the continuation of morally

People for Ethical Treatment of Animals (PETA) demonstrate for their cause nearly nude in winter during rush hour on a main commuter route into Washington, DC, 1993. *(Wally McNamee)*

abhorrent practices. Lawyer Christopher Stone and American philosophers Joel Feinberg (1926–2004) and Tom Regan (1938–) have focused on developing cogent arguments in support of rights for certain animals. Regan's 1983 book *The Case For Animal Rights* developed an absolutist position that criticized and broke from utilitarianism. It is Regan's arguments, not reformism or the pragmatic principle of utility, which have come to dominate the rhetoric of the animal rights crusade. In 2005, Regan wrote the book *Empty Cages*, where he continues his views on animal rights.

The question of which animals possess rights then arises. Regan asserts it is those who, like humans, are subjects experiencing their own lives. By "experiencing" Regan means conscious creatures aware of their environment and with goals, desires, emotions, and a sense of their own identity. These characteristics give an individual inherent value, and this value entitles the bearer to certain inalienable rights, especially the right to be treated as an end in itself, and never merely as a means to human ends.

The environmental community has not embraced animal rights; in fact, the two groups have often been at odds. A rights approach focused exclusively on animals does not cover all the entities such as ecosystems that many environmentalists feel ought to be considered morally. Yet a rights approach that would satisfy environmentalists by encompassing both living and nonliving entities may render the concept of rights philosophically and practically meaningless. Regan accuses environmentalists of environmental fascism, insofar as they advocate the protection of species and ecosystems at the expense of individual animals. Most animal rightists advocate the protection of ecosystems only as necessary to protect individual animals, and do not assign any more value to the individual members of a highly-endangered species than to those of a common or domesticated species. Thus, because of its focus on the individual, animal rights cannot offer a realistic plan for managing natural systems or for protecting ecosystem health, and may at times hinder the efforts of resource managers to effectively address these issues.

For most animal activists, the practical implications of the rights view are clear and uncompromising. The rights view holds that all animal research, factory farming, and commercial or sport hunting and trapping should be abolished. This change of moral status necessitates a fundamental change in contemporary Western moral attitudes toward animals, for it requires humans to treat animals as inherently valuable beings with lives and interests independent of human needs and wants. While this change is not likely to occur in the near future, the efforts of animal rights advocates may ensure that wholesale slaughter of these creatures for unnecessary reasons is no longer routinely the case, and that when such sacrifice is found to be necessary, it is accompanied by moral deliberation.

Resources

BOOKS

Bekoff, Marc, ed. *Encyclopedia of Animal Rigths and Animal Welfare*. Santa Barbara, CA: Greenwood Press, 2010.

Evans, Kim Masters. *Animal Rights*. Detroit, MI: Gale Cengage Learning, 2010.

Regan, Tom. *Emtpy Cages: Facing the Challenge of Animal Rights*. Lanham, MD: Rowman & Littlefield, 2004.

Regan, Tom, and Peter Singer. *Animal Rights and Human Obligations*. 2nd ed. Englewood Cliffs, NJ: Prentice-Hall, 1989.

Singer, Peter. *Animal Liberation: The Definitive Classic of the Animal Movement*. New York: Harper Perennial, 2009.

Zimmerman, M. E., et al., eds. *Environmental Philosophy: From Animal Rights to Radical Ecology*. Upper Saddle Road, NJ: Pearson/Prentice-Hall, 2005.

OTHER

AnimalsVoice.com (The Culture and Animals Foundation and Tom Regan). "Empty Cages." http://www.animals voice.com/TomRegan/home.html (accessed October 25, 2010).

PETA. "Home Web Page." http://www.peta.org/ (accessed October 25, 2010).

Ann S. Causey

Animal waste

Animal wastes most commonly refer to the excreted materials from live animals. However, under certain production conditions, the waste may also include straw, hay, wood shavings, or other sources of organic debris. It has been estimated that there may be as much as 2 billion tons (1.8 billion metric tons) of animal wastes produced in the United States annually. Application of excreta to soil brings benefits such as improved soil tilth, increased water-holding capacity, and some plant nutrients. Concentrated forms of excreta or high application rates to soils without proper management may lead to high salt concentrations in the soil and cause serious on-site or off-site pollution.

Animal Welfare Institute

Founded in 1951, the Animal Welfare Institute (AWI) is a nonprofit organization that works to educate the public and to secure needed action to protect animals. AWI is a highly respected, influential, and effective group that works with Congress, the public, the news media, government officials, and the conservation community on animal protection programs and projects. Its major goals include improving the treatment of laboratory animals and a reduction in their use; eliminating cruel methods of trapping wildlife; saving species from extinction; preventing painful experiments on animals in schools and encouraging humane science teaching; improving shipping conditions for animals in transit; banning the importation of parrots and other exotic wild birds for the pet industry; and improving the conditions under which farm animals are kept, confined, transported, and slaughtered.

In 1971, AWI launched the Save the Whales Campaign to help protect whales. The organization provides speakers and experts for conferences and meetings around the world, including Congressional hearings and international treaty and commission meetings. Each year, the institute awards its prestigious Albert Schweitzer medal to an individual for outstanding achievement in the advancement of animal welfare. Its publications include *The AWI Quarterly*; books such as *Animals and Their Legal Rights*; *Facts about Furs*; and *The Endangered Species Handbook*. Booklets, brochures, and other educational materials are distributed to schools, teachers, scientists, government officials, humane societies, libraries, and veterinarians. The AWI also releases an annual report describing its ongoing goals and efforts toward animal welfare. The fifty-eighth annual report describes the institute's efforts from July 1, 2008, through June 30, 2009. The annual reports can be downloaded for free from the AWI official Web site.

The AWI works closely with its associate organization, The Society for Animal Protective Legislation (SAPL), a lobbying group based in Washington, Founded in 1955, SAPL devotes its efforts to supporting legislation to protect animals, often mobilizing its 14,000 correspondents in letter-writing campaigns to members of Congress. SAPL has been responsible for the passage of more animal protection laws than any other organization in the country, and perhaps the world, and it has been instrumental in securing the enactment of fourteen federal laws. Principal goals of SAPL include enacting legislation to end the use of cruel steel-jaw, leg-hold traps and to secure proper enforcement, funding, administration, and reauthorization of existing animal protection laws.

Major federal legislation that the SAPL has promoted includes the first federal Humane Slaughter Act in 1958 and its strengthening in 1978; the 1959 Wild Horse Act; the 1966 Laboratory Animal Welfare Act and its strengthening in 1970, 1976, 1985, and 1990; the 1969 Endangered Species Act and its strengthening in 1973; a 1970 measure banning the crippling or "soring" of Tennessee walking horses; measures passed in 1971 prohibiting hunting from aircraft, protecting wild horses, and resolutions calling for a moratorium on commercial whaling; the 1972 Marine Mammal Protection Act; negotiation of the 1973 Convention on International Trade in Endangered Species of Fauna and Flora (CITES); the 1979 Packwood-Magnuson Amendment protecting whales and other ocean creatures; the 1981 strengthening of the Lacey Act to restrict the importation of illegal wildlife; the 1990 Pet Theft Act; the 1992 Wild Bird Conservation Act, protecting parrots and other exotic wild birds; the International Dolphin Conservation Act, restricting the killing of dolphins by tuna fishermen; and the Driftnet Fishery Conservation Act, protecting whales, seabirds, and other ocean life from being caught and killed in huge, 30-mile-long (48-km-long) nets.

Both AWI and SAPL were long headed by their chief volunteer and founder, Christine Stevens (1918–2002), a prominent Washington, DC, humanitarian and community leader. Stevens founded the AWI in 1951 and led the organization for over fifty years. Both nationally and internationally, Stevens used her access to business and government leaders to bring about positive change toward the humane treatment of animals. Her contributions to animal welfare have been ably described by Jane Goodall, the famed British primatologist, who said of Stevens after her passing, "Christine Stevens was a giant voice for animal welfare. Passionate, yet always reasoned, she took up one cause after another and she never gave up. Millions of animals are better off because of Christine's quiet and very effective advocacy. She will sorely be missed by all of us."

Resources

ORGANIZATIONS

Animal Welfare Institute, 900 Pennsylvania Ave., Washington, DC, USA, 20003, (202) 337-2332, awi@awionline.org, http://www.awionline.org

Society for Animal Protective Legislation, PO Box 3719, Washington, DC, USA, 20007, (202) 337-2334, (202) 338-9478, sapl@saplonline.org, http://www.saplonline.org

Lewis G. Regenstein

Anion *see* **Ion.**

Antarctic Treaty (1961)

The Antarctic Treaty, which entered into force in 1961, established an international administrative system for the continent. The impetus for the treaty was the international geophysical year, during 1957 and 1958, which had brought scientists from many nations together to study Antarctica. The political situation in Antarctica was complex at the time, with the following seven nations having made sometimes overlapping territorial claims to the continent: Argentina, Australia, Chile, France, New Zealand, Norway, and the United Kingdom. Several other nations, most notably the former Soviet Union and the United States, had been active in Antarctic exploration and research and were concerned with how the continent would be administered.

Negotiations on the treaty began in June 1958 with Belgium, Japan, and South Africa joining the original nine countries. The treaty was opened for signature in December 1959 and took effect in June 1961. It begins by "recognizing that it is in the interest of all mankind that Antarctica shall continue forever to be used exclusively for peaceful purposes." The key to the treaty was the nations' agreement to disagree on territorial claims. Signatories of the treaty are not required to renounce existing claims; nations without claims have an equal voice as those with claims, and no new claims or claim enlargements can take place while the treaty is in force. This agreement defused the most controversial and complex issue regarding Antarctica in an unorthodox way. Among the other major provisions of the treaty are that the continent will be demilitarized; nuclear explosions and the storage of nuclear wastes are prohibited; the right of unilateral inspection of all facilities on the continent to ensure that the provisions of the treaty are being honored is guaranteed; and scientific research can continue throughout the continent.

The treaty runs indefinitely and can be amended only by the unanimous consent of the signatory nations. Provisions were also included for other nations to become parties to the treaty. These additional nations can either be acceding parties, which do not conduct significant research activities but agree to abide by the terms of the treaty, or consultative parties, which have acceded to the treaty and undertake substantial scientific research on the continent. Sixteen nations have joined the original twelve in becoming consultative parties, including Brazil, China, Finland, Germany, India, Italy, Peru, Poland, South Korea, Spain, Sweden, and Uruguay.

Under the auspices of the treaty, the Convention on the Conservation of Antarctic Marine Living Resources was adopted in 1982. This regulatory regime is an effort to protect the Antarctic marine ecosystem from severe damage caused by overfishing. Following this convention, negotiations began on an agreement for the management of Antarctic mineral resources. The Convention on the Regulation of Antarctic Mineral Resource Activities was concluded in June 1988, but in 1989 Australia and France rejected the convention, urging that Antarctica be declared an international wilderness closed to mineral development. The Protocol on Environmental Protection to the Antarctic Treaty was drafted 1991 and also in 1998. This agreement prevents development and provides for the protection of the Antarctic environment through five specific annexes on marine pollution, fauna and flora, environmental impact assessments, waste management, and protected areas. It prohibits all activities relating to mineral resources except scientific. A sixth annex addressing liability arising from environmental emergencies was adopted in 2005. However, as of 2010, the sixth annex had yet to take effect because all consultative parties to the Antarctic Treaty (meaning all twenty-eight nations with consultative status) must first implement its terms under their respective national laws before it can go into effect.

Resources

BOOKS

Cioc, Mark. *The Game of Conservation: International Treaties to Protect the World's Migratory Animals.* Athens: Ohio University Press, 2009.

Fox, William. *Terra Antarctica: Looking into the Emptiest Continent.* San Antonio, TX: Trinity University Press, 2005.

French, Duncan; Matthew Saul; and Nigel D. White, eds. *International Law and Dispute Settlement: New Problems and Techniques.* Oxford, UK: Hart, 2010.

Riffenburgh, Beau, ed. *Encyclopedia of the Antarctic.* New York: Routledge, 2007.

Triggs, Gillian, and Anna Riddell. *Antarctica: Legal and Environmental Challenges for the Future.* London: British Institute of International and Comparative Law, 2007.

OTHER

"Secretariat of the Antarctic Treaty." http://www.ats.aq/index_e.htm (accessed November 8, 2010).

Christopher McGrory Klyza

Antarctica

Antarctica is Earth's fifth largest continent, centered asymmetrically around the South Pole. Ninety-eight percent of this land mass, which covers approximately 5.4 million square miles (13.8 million km^2), is covered by ice sheets to an average depth of 1.25 miles (2 km). This continent receives very little precipitation, less than 5 inches (12 cm) annually, and the world's coldest temperature was recorded here, at $-128°F$ ($-89°C$). Exposed shorelines and inland mountaintops support life only in the form of lichens, two species of flowering plants, and several insect species. In sharp contrast, the ocean surrounding the Antarctic continent is one of the world's richest marine habitats. Cold water rich in oxygen and nutrients supports teeming populations of phytoplankton and shrimp-like Antarctica krill, the food source for the region's numerous whales, seals, penguins, and fish. During the nineteenth and early twentieth centuries, whalers and sealers severely depleted Antarctica's marine mammal populations. In recent decades the whale and seal populations have begun to recover, but interest has grown in new resources, especially oil, minerals, fish, and tourism.

The Antarctic region's outer bound is a band of turbulent ocean currents and high winds that circle the continent at about sixty degrees south latitude. This ring is known as the Antarctic convergence zone. Ocean turbulence in this zone creates a barrier marked by sharp differences in salinity and water temperature. Antarctic marine habitats, including the limit of krill populations, are bounded by the convergence.

Since 1961 the Antarctic Treaty has formed a framework for international cooperation and compromise in the use of Antarctica and its resources. The treaty reserves the Antarctic continent for peaceful scientific research and bans all military activities. Nuclear explosions and radioactive waste are also banned, and the treaty neither recognizes nor establishes territorial claims in Antarctica. However, neither does the treaty deny pre–1961 claims, of which seven exist. Furthermore, some signatories to the treaty, including the United States, reserve the right to make claims at a

Melting icebergs are seen in Antarctica during a 2007 visit of U.N. Secretary General Ban Ki-moon as part of several environment-related visits to see the effects of global warming and the impact of melting glaciers. *(AP Photo/Roberto Candia)*

later date. At present the United States has no territorial claims, but it does have several permanent stations, including one at the South Pole. Questions of territorial control could become significant if oil and mineral resources were to become economically recoverable. The primary resources exploited are fin fish and krill fisheries. Interest in oil and mineral resources has risen in recent decades, most notably during the 1973 "oil crisis." The expense and difficulty of extraction and transportation have so far made exploitation uneconomical, however.

Human activity has brought an array of environmental dangers to Antarctica. A growing and largely uncontrolled fishing industry may be depleting both fish and krill populations in Antarctic waters. The parable of the "tragedy of the commons" seems ominously appropriate to Antarctica fisheries, which have already nearly eliminated many whale, seal, and penguin species. Solid waste and oil spills associated with research stations and with tourism pose an additional threat. Although Antarctica remains free of permanent settlement, year-round scientific research stations are maintained on the

continent. Oil and mineral extraction could seriously threaten marine habitat and onshore penguin and seal breeding grounds. In 1989, the Antarctic had its first oil spill when an Argentine supply ship, carrying tourists and 170,000 gallons (643,500 l) of diesel fuel, ran aground. Spilled fuel destroyed a nearby breeding colony of Adele penguins (*Pygoscelis adeliae*). Tourists themselves present a further threat to penguins and seals. Visitors have been accused of disturbing breeding colonies, thus endangering the survival of young penguins and seals. In 2008 more than 45,000 tourists visited Antarctica during its summer season. In 2009, the twenty-eight nations considered as consultative parties to the Antarctic Treaty announced an intention to restrict tourism in order to limit environmental damage. Consultative parties are nations that have signed and ascribe to the terms of the treaty, as well as being obliged to perform substantial research in the Antarctic on an ongoing basis.

The West Antarctic Peninsula, where melting has accelerated so notably, is one of the fastest-warming regions in the world. Jutting out into the Antarctic Ocean, the peninsula is fringed by ice shelves, which are vast, floating masses of ice anchored to the coast and slowing the emptying of glaciers into the sea. As climate warms in the region, some of these shelves are weakening and breaking up, allowing glaciers to flow faster. Such an event occurred in March 2008, when the Wilkins ice shelf began to rapidly disintegrate. A 160-square mile (414-square km) area of ice broke away, leaving the rest of the shelf held in place only by a narrow strip of stable ice anchored to an island at each end. The Wilkins shelf breakup was the latest sign of dramatic climate change in that part of Antarctica. In the last thirty years, six West Antarctic ice shelves have collapsed, and the shedding of over 300 West Antarctic glaciers has been confirmed. Collapsing ice shelves do not raise sea level significantly, since floating ice has about the same effect on sea level whether it melts or not, but inland glaciers add new water to the ocean, raising its level.

In late November 2009, a study released by the Scientific Committee on Antarctic Research (SCAR) warned that elevated sea temperatures are accelerating ice shelf melting along the western coast of Antarctica. At these rates, the melting will significantly increase prior estimates of global sea level rise by 2100. Although temperatures across the Antarctic as a whole have remained stable over recent decades, the coastal areas along the western Antarctic peninsula have warmed by approximately 5.4° (3° Centigrade) since 1960. In contrast to the warming of the western Antarctic peninsula, other regions of Antarctica have cooled, and have experienced increases in snowpack and ice thickness. Climatologists attribute the temporary cooling of Antarctic regions to temporary decreases in ozone layer thickness caused by the use of chlorofluorocarbons (CFC) prior to their ban under the Montreal Protocol. Other climate models predict increased snowfall because of higher levels of atmospheric humidity as global temperatures rise.

Resources

BOOKS

Fox, William. *Terra Antarctica: Looking into the Emptiest Continent*. San Antonio, TX: Trinity University Press, 2005.
Triggs, Gillian, and Anna Riddell. *Antarctica: Legal and Environmental Challenges for the Future*. London: British Institute of International and Comparative Law, 2007.

PERIODICALS

Eilperin, Juliet. "Antarctic Ice Sheet Is Melting Rapidly." *Washington Post* (March 3, 2006).
Shepherd, Andrew, and Duncan Wingham. "Recent Sea-Level Contributions of the Antarctic and Greenland Ice Sheets." *Science* 315 (2007): 1529–1532.

OTHER

United States Central Intelligence Agency (CIA). "Antarctica." *CIA World Factbook*. https://www.cia.gov/library/publications/the-world-factbook/geos/ay.html (accessed November 10, 2010).

Mary Ann Cunningham

Anthracite coal *see* **Coal.**

Anthrax

Anthrax is a bacterial infection caused by *Bacillus anthracis*. It usually affects cloven-hoofed animals, such as cattle, sheep, and goats, but it can occasionally spread to humans. Anthrax is almost always fatal in animals, but it can be successfully treated in humans if antibiotics are given soon after exposure. In humans, anthrax is usually contracted when spores are inhaled or come in contact with the skin. It is also possible for people to become infected by eating the meat of contaminated animals. Anthrax, a deadly disease in nature, gained worldwide attention in 2001 after it was used as a bioterrorism agent against the United States. Until the 2001

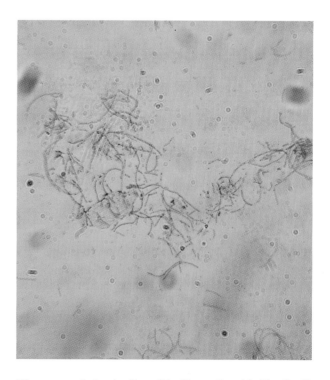

Microscope photo of anthrax (*Bacillus anthracis*). *(Carolina K. Smith, M.D./Shutterstock.com.)*

attack, only eighteen cases of anthrax had been reported in the United States in the previous 100 years.

Anthrax occurs naturally. The first reports of the disease date from around 1500 BCE, when it was thought to be the cause of an Egyptian plague. Robert Koch first identified the anthrax bacterium in 1876 and Louis Pasteur developed an anthrax vaccine for sheep and cattle in 1881. Anthrax bacteria are found in nature in South and Central America, southern and eastern Europe, Asia, Africa, the Caribbean, and the Middle East. Anthrax cases in the United States are rare, probably due to widespread vaccination of animals and the standard procedure of disinfecting animal products such as cowhide and wool. Reported cases occur most often in Texas, Louisiana, Mississippi, Oklahoma, and South Dakota.

Anthrax spores can remain dormant (inactive) for years in soil and on animal hides, wool, hair, and bones. There are three forms of the disease, each named for its means of transmission: cutaneous (through the skin), inhalation (through the lungs), and intestinal (caused by eating anthrax-contaminated meat). Symptoms appear within several weeks of exposure and vary depending on how the disease was contracted.

Cutaneous anthrax is the mildest form of the disease. Initial symptoms include itchy bumps, similar to insect bites. Within two days, the bumps become inflamed and a blister forms. The centers of the blisters are black because of dying tissue. Other symptoms include shaking, fever, and chills. In most cases, cutaneous anthrax can be treated with antibiotics such as penicillin. Intestinal anthrax symptoms include stomach and intestinal inflammation and pain, nausea, vomiting, loss of appetite, and fever, all becoming progressively more severe. Once the symptoms worsen, antibiotics are less effective, and the disease is usually fatal.

Inhalation anthrax is the deadliest form of the disease. The initial symptoms of inhalation anthrax are flu-like, but breathing becomes progressively more difficult. Inhalation anthrax can be treated successfully if antibiotics are given before symptoms develop. Once symptoms develop, the disease is usually fatal.

The inhalation form of the disease was involved in the bioterrorism attacks of October and November 2001 in the eastern United States. Five people died after being exposed to anthrax through contaminated mail. At least seventeen other people contracted the disease but survived. Media representatives in Florida and New York received envelopes containing anthrax, including Tom Brokaw. Anthrax-contaminated letters also were sent to the Washington, DC, offices of two senators.

In the aftermath, an intensive investigation by U.S. federal agents had failed to identify the person or group. However, with time, evidence accumulated and, according to the Federal Bureau of Investigation (FBI), pointed to Bruce Ivins, a microbiologist and anthrax researcher at the U.S. Army biodefense lab at Fort Detrick, Maryland. Ivins committed suicide as he was about to be charged. In February 2010, the FBI issued its final report on the 2001 anthrax attacks, in which Ivins was identified as the sole person who sent the anthrax-tainted letters.

While the case has been officially closed, skeptics still contend that Ivins did not act alone because his laboratory did not have the capacity to weaponize the anthrax traced to his lab.

The only natural outbreak of anthrax among people in the United States occurred in Manchester, New Hampshire, in 1957. Nine workers in a textile mill that processed wool and goat hair contracted the disease, five with inhalation anthrax and four with cutaneous anthrax. Four of the five people with inhalation anthrax died. By coincidence, workers at the mill were participating in a study of an experimental anthrax vaccine. No workers who had been vaccinated contracted the disease.

Following this outbreak, the study was stopped, all workers at the mill were vaccinated, and vaccination became a condition of employment. After that, no mill workers contracted anthrax. The mill closed in 1968. However, in 1966 a man who worked across the street from the mill died from inhalation anthrax. He is thought to have contracted it from anthrax spores carried from the mill by the wind. The United States Food and Drug Administration approved the anthrax vaccine in 1970. It is used primarily for military personnel and some health care workers. During the 2001 outbreak, thousands of postal workers were offered the vaccine after anthrax spores from contaminated letters were found at several post office buildings.

The largest outbreak worldwide of anthrax in humans occurred in the former Soviet Union in 1979, when anthrax spores released from a military laboratory infected seventy-seven people, sixty-nine of whom died. Anthrax is an attractive weapon to bioterrorists. It is easy to transport and is highly lethal. The World Health Organization (WHO) estimates that 110 pounds (50 kg) of anthrax spores released upwind of a large city would kill tens of thousands of people, with thousands of others ill and requiring medical treatment.

The Geneva Convention, which established a code of conduct for war, outlawed the use of anthrax as a weapon in 1925. However, Japan developed anthrax weapons in the 1930s and used them against civilian populations during World War II. During the 1980s, Iraq mass-produced anthrax as a weapon.

Resources

BOOKS

Coen, Bob, and Eric Nadler. *Dead Silence: Fear and Terror on the Anthrax Trail.* Berkeley, CA: Counterpoint, 2009.

Cole, Leonard A. *The Anthax Letters: A Bioterrorism Expert Investigates the Attack that Shocked America.* New York: Skyhorse Publishing, 2009.

Johnstone, William. *Bioterror: Anthrax, Influenza, and the Future of Public Health Security.* New York: Praeger, 2008.

Jones, Susan D. *Death in a Small Package: A Short History of Anthrax.* Baltimore: The Johns Hopkins University Press, 2010.

ORGANIZATIONS

Centers for Disease Control and Prevention, 1600 Clifton Road, Atlanta, GA, USA, 30333, (404) 639-3534, (888) 246-2675, cdcresponse@ashastd.org, http://www.cdc.gov

Ken R. Wells

Anthropocentrism *see* **Environmental ethics.**

Anthropogenic

Anthropogenic refers to changes in the natural world due to the activities of people. Such changes may be positive or negative. For example, anthropogenic changes in soils can occur because of plowing, fertilizing, using the soil for construction, or long-continued manure additions. When manure is added to soils, the change is considered beneficial, but when soils are compacted for use as parking lots, the change is considered negative. Other examples of anthropogenic effects on the environment include oil spills, acid rain, logging of old-growth forests, creation of wetlands, and preservation of endangered species.

Antibiotic resistance

Antibiotics are drugs principally derived from naturally occurring fungi and microorganisms that kill bacteria and can cure patients with bacterial diseases. Before the advent of antibiotics in the 1940s, many common diseases were lethal or incurable. Tuberculosis, pneumonia, scarlet fever, staph and

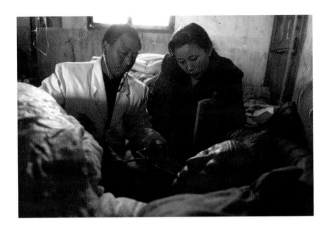

A doctor examines an ill woman at her home in Anjui Province, China. While access to antibiotics such as penicillin have done wonders to improve health care in China, experts say these powerful drugs are being overused, and with dangerous consequences. *(AP Photo/Eugene Hoshiko)*

strep infections, typhoid fever, gonorrhea, and syphilis were all dreaded diseases until the development of penicillin and other antibiotics in the middle of the twentieth century. Yet almost as soon as antibiotics came into common use, scientists noticed that some strains of disease-causing bacteria developed resistance to the antibiotic used most often against it.

People infected with an antibiotic-resistant bacteria must be treated with different antibiotics, often more potent and toxic than the commonly-used drug. In some cases, bacteria may be resistant to several antibiotics. Tuberculosis, once the leading killer in the United States at the beginning of the nineteenth century, seemed defeated with the introduction of streptomycin (derived from the actinobacterium *Streptomyces griseus*) and PAS (scientifically known as 4-aminosalicylic acid) in the 1940s and early 1950s. However, tuberculosis resurged in the United States and worldwide in the 1990s as people came down with antibiotic-resistant strains of the disease. Bacteria that cause salmonella, a foodborne illness, had become increasingly resistant to antibiotics by the early twenty-first century—as had the bacteria that commonly cause early childhood ear infections. Misuse and overuse of antibiotics contribute to the rise of resistant strains.

Bacteria can become resistant to antibiotics relatively quickly. Bacteria multiply rapidly, producing a new generation in as little as one-half hour. Hence, evolutionary pressures can produce bacteria with new characteristics in very little time. When a person takes an antibiotic, the drug will typically kill almost all the bacteria it is designed to destroy, plus other beneficial bacteria. Some small percentage of the disease bacteria, possibly as little as 1 percent, may have a natural ability to resist the antibiotic. Therefore, a small number of resistant bacteria may survive drug treatment. When these resistant bacteria are all that are left, they are free to multiply, passing the resistance to their offspring. Physicians warn people to take the fully prescribed course of antibiotics even if symptoms of the disease disappear in a day or two. This is to limit the danger of resistant bacteria flourishing. Bacteria can also develop resistance by contact with other species of bacteria that are resistant. Neighboring bacteria can pass genetic material back and forth by swapping bits of deoxyribonucleic acid (DNA) called plasmids. If bacteria that normally live on the skin and bacteria that live in the intestine should come into contact with each other, they may make a plasmid exchange, and spread antibiotic resistant qualities. Antibiotic-resistant bacteria are often resistant to a whole class of antibiotics, that is, a group of antibiotics that function in a similar way. People afflicted with a resistant strain of bacteria must be treated with a different class of antibiotics.

Antibiotic resistance was evident in the 1940s, though penicillin had only become available in 1941. By 1946, one London hospital reported that 14 percent of patients with staph infections had penicillin-resistant strains, and that number rose precipitously over the next decade. In 1943, scientists brought out streptomycin, a new antibiotic that fought tuberculosis (penicillin was found not to work against that disease). However, streptomycin-resistant strains of tuberculosis developed rapidly, and other drugs had to be found. In 1959, physicians in Japan found a virulent strain of dysentery that was resistant to four different classes of antibiotic. Some troubling cases of antibiotic resistance have been isolated incidents. Nevertheless, by the 1990s it was clear that antibiotic resistance was a widespread and growing problem. A few cases around the world in 1999 found deadly bacteria resistant to vancomycin, a powerful antibiotic described as a drug of last resort because it is only used when all other antibiotics fail. By this time, scientists in many countries were deeply alarmed about the growing public health threat of antibiotic resistance. In the mid-2000s, the Global Health Council estimated that 9.5 million people die each year from infectious diseases, most that are preventable with the use of antibiotics and other such medicines. In the United States, the Institute of Medicine stated that antimicrobial (antibiotic) resistance was likely to add $50 billion annually to the health care system in the late 2000s in the form of longer hospital stays and more expensive medicines to treat such problems. The U.S. Centers for Disease Control and Prevention (CDC) claimed in 2001 that antibiotic resistance had spread to "virtually all important human pathogens treatable with antibiotics." In 2010, the CDC further stated that these antibiotic resistance organisms make it more likely that humans will die as a result of an infection.

Antibiotic resistance makes treatment of infected patients difficult. The sexually transmitted disease gonorrhea was easily cured with a single dose of penicillin in the middle of the twentieth century. By the 1970s, penicillin-resistant strains of the disease had become prevalent in Asia, and migrated from there to the rest of the world. Penicillin was no longer used to treat gonorrhea in the United States after 1987. Standard treatment was then a dose of either of two classes of antibiotics, fluoroquinolones or cephalosporins. By the late 1990s, strains of gonorrhea resistant to fluoroquinolones had been detected in Asia. The resistant strains showed up in California in 2001. The California CDC soon recommended not using fluoroquinolones to treat gonorrhea, fearing that use of these drugs would actually strengthen the antibiotic resistance. If patients were only partially cured by fluoroquinolones, yet some infection lingered, they could pass the resistant strain to others. In addition, the resistance could become stronger as only the most resistant bacteria survived exposure to the drug. So public health officials and doctors were left with cephalosporins to treat gonorrhea, more costly drugs with more risk of side effects.

Overuse of antibiotics contributes to antibiotic resistance. The number of antibiotic prescriptions for children rose almost 50 percent in the United States between 1980 and 1992. Two decades later, in the early 2010s, children, the elderly, and those with compromised immune systems are the most likely to receive antibiotic prescriptions. That same decade, the medical community estimates that about 150 million prescriptions are issued each year in the United States. Of those, about 90 million are for antibiotics; and well over one-half of them are considered "absolutely unnecessary or inappropriate." Antibiotics work only against bacterial diseases, and are useless against viral infections. Yet physicians frequently prescribe antibiotics for coughs and colds.

An article on the problem in *American Family Physician* found that most doctors understood the inappropriateness of their prescriptions, yet feared that patients were unsatisfied with their care unless they received a drug. The CDC launched various state and national initiatives to educate both doctors and their patients about overuse of antibiotics. Other groups took on specific diseases, such as the overprescribing of antibiotics for childhood ear infections. The common ailment was known to be treatable without antibiotics, but many doctors continue to give antibiotics anyway. In fact, many children in day care centers who had ear infections had an antibiotic-resistant form of the condition. Consequently, pediatricians and parents are advised to use antibiotics only when necessary. Most bacteria live in the body without causing harm, but can make people ill if they build up to certain levels, or if a person's immune system is weakened. People carrying Cipro-resistant bacteria could potentially come down with a resistant form of pneumonia or some other bacterial illness later in life.

People are also exposed to antibiotics through meat and other food. About one-half the antibiotics used in the United States go to farm animals, and some are also sprayed on fruits and vegetables. Some farm animals are given antibiotics to cure a specific disease. However, other antibiotics are given as preventives, and to promote growth. Animals living in crowded and dirty conditions are more susceptible to disease, and the preventive use of antibiotics keeps such animals healthier than they would otherwise be. The antibiotics prescribed by veterinarians are similar or the same as drugs used in humans. Farm animals in the United States are routinely given penicillin, amoxicillin, tetracycline, ampicillin, erythromycin, and neomycin, among others, and studies have shown that antibiotic resistance is common in contaminated meat and eggs. Salmonella bacteria are killed when meat is cooked properly, and most cases of salmonella disease get better without treatment. However, for the small percent of cases of more serious infection, multiple antibiotic resistance could make treatment very difficult. Twenty percent of the urinary tract infections studied were resistant to Bactrim (a branded name for trimethoprim-sulfamethoxazole), meaning that in most cases physicians would be advised to treat with a stronger antibiotic with more side effects.

Antibiotics in preventive doses or for growth promotion of farm animals were banned by the European Union in 1998. Many groups in the United States concerned with antibiotic resistance recommend the United States follow suit. However, as of 2010, the United States has not done so. The plan released by the CDC, the Food and Drug Administration (FDA), and the National Institutes of Health (NIH) in 2001 to combat antibiotic resistance called for increased monitoring of antibiotic use in agriculture and in human health. The plan also called for public education on the risks of overuse and improper use of antibiotics, and for more research in combating drug-resistant diseases. As of 2010, the FDA, the NIH, and the CDC continue to investigate links between agricultural use of antibiotics and human health.

Resources

BOOKS

Germ Wars: Battling Killer Bacteria and Microbes. New York: Rosen, 2008.
Sachs, Jessica Snyder. *Good Germs, Bad Germs: Health and Survival in a Bacterial World.* New York: Hill and Wang, 2007.
Spellberg, Brad. *Rising Plague: The Global Threat from Deadly Bacteria and Our Dwindling Arsenal to Fight Them.* Amherst, NY: Prometheus Books, 2009.
Zimmerman, Barry E., and David J. Zimmerman. *Killer Germs: Microbes and Diseases that Threaten Humanity.* Chicago: Contemporary Book, 2003.

OTHER

Centers for Disease Control and Prevention. "Antibiotic/Antimicrobial Resistance." http://www.cdc.gov/drugresistance/index.html (accessed November 8, 2010).
Global Health Council. "The Impact of Infectious Diseases." http://www.globalhealth.org/infectious_diseases/ (accessed November 8, 2010).
Infectious Diseases Society of America. "Facts about Antibiotic Resistance." http://www.idsociety.org/Content.aspx?id = 5650 (accessed November 8, 2010).

ORGANIZATIONS

Alliance for the Prudent Use of Antibiotics, 75 Kneeland Street, Boston, MA, USA, 02111, (617) 636-0966, (617) 636-3999, http://www.tufts.edu/med/apua/

Angela Woodward

Ants *see* **Fire ants.**

ANWR *see* **Arctic National Wildlife Refuge.**

Apis mellifera scutellata *see* **Africanized bees.**

AQCR *see* **Air Quality Control Region.**

Aquaculture

Aquaculture is the husbandry or rearing of aquatic organisms under controlled or semicontrolled conditions. Stated another way, it is the art of cultivating natural plants and animals in water for human consumption or use. It can be considered aquatic agriculture or, as some people wish to call it, underwater agriculture. It is sometimes incorrectly termed "aquiculture." Aquaculture involves production in both freshwater and saltwater. Mariculture is aquaculture in saline (brackish and marine) water and is usually accomplished through nearshore or offshore cages. Hydroponics is the process of growing plants in water. Organisms that are grown in aquaculture include fish, shellfish (crustaceans such as crawfish and shrimp, and mollusks such as oysters and clams), algae, and aquatic plants. More people in the United States are eating seafood for the added health benefits. Not only can these organisms be raised for human consumption, but they can also be reared for the lucrative baitfish, health food, aquarium, and home garden-pond industries.

Aquaculture dates back more than 3,500 years, when carp were spawned and reared in China. There have also been records of aquaculture practices being performed in Egypt and Japan nearly that long ago. Mariculture is thought to have been brought to Hawaii about 1,500 years ago. Modern aquaculture finds its roots in the 1960s in the culturing of catfish in the United States and salmon in Europe.

There are two general methods used in aquaculture, extensive and intensive. Extensive aquaculture involves the production of low densities of organisms. For example, fingerling fish can be raised in ponds where they feed on the natural foods such as phytoplankton and zooplankton, and the matured fish are harvested at the end of the growing season. These ponds can also be fertilized through the addition of nutrients to enhance the food chain, thus increasing fish or shellfish production. Typically, several hundred to several thousand pounds of fish are raised per acre annually. Intensive practices utilize much higher densities of organisms. This method of aquaculture requires better water quality, which necessitates circulation, oxygenation, added commercial foods, and biological filters (with bacteria) to remove toxic wastes. These systems can produce more than 1 million pounds of fish per acre (45,000) ha. Intensive aquaculture often utilizes tanks to grow the fish or shellfish, either indoor or outdoor. The water can be recirculated as long as toxic wastes such as ammonia (NH_3) are removed. Some processors run the effluent from the fish tanks through greenhouses with plants raised hydroponically to remove these chemicals. In this way, added revenue is gained. Some aquaculture operations raise only one type of organism (termed monoculture), whereas others grow several species together (termed "polyculture"). An example of polyculture would be growing tilapia and catfish together. Another method of aquaculture consists of raising crops in the open ocean. The National Offshore Aquaculture Act, submitted in 2007 by the National Oceanic and Atmospheric Association (NOAA) and the Department of Commerce, would

regulate commercial offshore aquaculture operations to ensure the maintenance of the natural ecosystem.

In 2004, 32 percent of the world production of fish was attributed to aquaculture. About 86 percent of the world's aquaculture production comes from Asia. China is the leading producer nation, accounting for about 70 percent (as of 2005), with most of their production coming from carp, and secondarily shrimp. India is the second-largest producer nation, followed by Japan, Taiwan, and the Philippines. Salmon is another major fish produced by aquaculture, with most coming from Norway. Great Britain, Canada, Chile, and Iceland are also major producers of farm-raised salmon. Other important aquaculture species and some of the countries where they are raised include tilapia (Caribbean countries, Egypt, India, Israel, and the Philippines); milkfish (the Philippines); Nile perch (Egypt); sea bass and sea bream (Egypt and Israel); mullet (Egypt, India, and Israel); dolphin fish, also known as mahi-mahi (Egypt); grass and silver carp (Israel); halibut and other flatfishes (France and Norway); prawns (Caribbean countries, Israel, and the Philippines); mussels and oysters (the Philippines); crabs (India); seaweeds (India and Japan); *Spirulina*, a type of cyanobacteria (India); and many other examples.

Although people in the United States consume less fish and shellfish than people in other parts of the world, they spend billions of dollars importing edible fish each year. Total production from aquaculture in the United States in 2006 was about 397 thousand tons (360 thousand metric tons), which earned about $1.2 billion. This comprises only about 1 percent of the world's total aquaculture production. Domestically, catfish is the major aquaculture species of the United States. Trout (primarily raised in Idaho) is the second major species raised in aquaculture and accounts for a $55 million per year industry. Shrimp, hybrid-striped bass, and tilapia are other species commonly raised in the United States.

In 1986 the U.S. Congress placed aquaculture under the jurisdiction of the U.S. Department of Agriculture (USDA) and appropriated $3 million to establish four regional aquaculture centers. These centers are located at the University of Washington, Southeastern Massachusetts University, Mississippi State University, and a Center for Tropical and Subtropical Aquaculture jointly administered by the University of Hawaii and the Oceanic Institute. The following year Congress created a fifth center located at Michigan State University. The mandate for these centers was, and still is, to promote aquaculture in each region and to solve some of the problems facing this industry. More recent legislation has created mariculture

Starting in 1984 international conservation groups have worked to introduce net fishing as a less harmful fishing alternative that still allows divers to retain their income from live fish. To catch fish with nets, divers may use a stick to drive fish from their hiding places in the reef and then trap the fish with a fine mesh net. Reports indicate that long-term survival rates of net-caught fish may be as high as 90 percent, compared to as little as 10 percent among cyanide-caught fish. In Hawaii and Australia, where legal controls are more effective than in Southeast Asia, nets are used routinely. Although the aquarium industry has provided little aid in the effort to increase net fishing, local communities in the Philippines and Indonesia, with the aid of international conservationist organizations, have worked to encourage net fishing. Increased local control of reef fisheries is also important in helping communities to control cyanide fishing in their nearby reefs. Local communities that have depended on reefs and their fish for generations understand that sustainable use is both possible and essential for their own survival, so they often have a greater incentive to prevent reef damage than either state governments or transient fishing enterprises. In an effort to help coastal villages help themselves, the government of the Philippines has recently granted villagers greater rights to patrol and control nearby fishing areas.

Another step toward the control of cyanide fishing is the development of cyanide-detection techniques that allow wholesalers to determine whether fish are tainted with residual cyanide. Simply by sampling the water a fish is carried in, the test can detect if the fish is releasing cyanide from its tissues or metabolizing cyanide. Ideally this test could help control illegal and harmful fish trade, but thus far it is not widely used. Attempts have also begun to establish cyanide-free certification, but this has been slow to take effect because the market is dispersed and there are many different exporters.

Coastal villages also suffer from reef damage. Most coastal communities in coral reef regions have traditionally relied on the rich fishery supported by the reef as a principal protein and food source. As reefs suffer, fisheries deteriorate. Some researchers estimate that just 1,300 feet (400 m) of healthy reef can support 800 people, while the same amount of damaged reef can support only 200 people. Other ecological benefits also disappear as reefs deteriorate, since healthy coral reefs perform critical water clarification functions. Corals, along with sea anemones and other lifeforms they shelter, filter floating organic matter from the water column, cycling it into the food chain that supports fish, crustaceans, birds, and humans. By

breaking ocean waves offshore, healthy and intact corals also control beach erosion and reduce storm surges, the unusually large waves and tides associated with storms and high winds.

Divers collecting the fish suffer as well. Exposure to cyanide and inadequate or unsafe breathing hoses are common problems. In addition divers must search deeper waters as more accessible fish are depleted. In 1993, forty divers in a single small village were reported to have been injured and ten killed by the bends, which results from rapid changes of pressure as divers rise from deep water.

Unfortunately, there has been relatively little breeding of tropical fish in aquaria, at least in part because these fish often have specialized habitat needs and lifestyle requirements that are difficult to produce under controlled or domestic conditions. In addition the aquarium trade is specialized and limited in volume, and gearing up to produce fish can be an expensive undertaking for which a reliable market must be assured. Equally unfortunate is the fact that American and European dealers in aquarium products tend to be elusive about the details of where their fish came from and how they were caught. If pet shops do not mention the source, many aquarium owners are able to ignore the implications of the fish trade they are participating in.

In addition to aquarium fishing, cyanide has now been introduced to the live food-fish market, especially in Asia, where live fish are an expensive delicacy. The live food/fish trade, centered in China and Hong Kong, is estimated to exceed $1 billion per year. The Hong Kong cyanide fleet alone has hundreds of vessels, each employing up to twenty-five divers to catch fish with cyanide for the Chinese and Hong Kong restaurant market.

Another harmful fishing technique is blast fishing—releasing a small bomb that breaks up coral masses in which fish hide. A single explosive might destroy corals in a circle from 10 to 33 feet (3–10 m) wide. The fish, briefly stunned by the blast, are easily retrieved from the rubble. Half the countries in the South Pacific have seen coral damage from blasting, including Guam, Indonesia, Malaysia, and Thailand. The technique has also spread to Africa, where it is used in Tanzania and other countries bordering the Indian Ocean.

Some environmentalists stress that simply eliminating the live fish trade is not an adequate solution. Because live fish are so lucrative, much more valuable than the same weight in dead fish, fishermen may be able to produce a better income with fewer fish when they catch live fish. Steering the fish capture methods to a safer alternative, especially the use of nets, could do more to save reef communities than eliminating the trade altogether.

Resources

BOOKS

Cote, I. M., and J. D. Reynold, eds. *Coral Reef Conservation.* Cambridge, UK: Cambridge University Press, 2006.

PERIODICALS

Dean, Cornelia. "Coral Reefs and What Ruins Them." *New York Times* (February 26, 2008).

Mary Ann Cunningham

Aquatic chemistry

Water can exist in various forms within the environment, including (1) liquid water of oceans, lakes and ponds, rivers and streams, soil interstices, and underground aquifers; (2) solid water of glacial ice and more-ephemeral snow, rime, and frost; and (3) vapor water of cloud, fog, and the general atmosphere. More than 97 percent of the total quantity of water in the hydrosphere occurs in the oceans, whereas about 2 percent is glacial ice, and less than 1 percent is groundwater. Only about 0.01 percent occurs in freshwater lakes, and the quantities in other compartments are even smaller.

Each compartment of water in the hydrosphere has its own characteristic chemistry. Seawater has a relatively large concentration of inorganic solutes (about 3.5 percent), dominated by the ions chloride (Cl^-) (1.94 percent), sodium (Na^+) (1.08 percent), sulfate (SO_4^{2-}) (0.27 percent), magnesium (Mg^{2+}) (0.13 percent), calcium (Ca^{2+}) (0.041 percent), potassium (K^+) (0.049 percent), and bicarbonate (HCO_3^-) (0.014 percent).

Surface waters such as lakes, ponds, rivers, and streams are highly variable in their chemical composition. Saline and soda lakes of arid regions have total salt concentrations that can substantially exceed that of seawater. Lakes such as the Great Salt Lake in Utah and the Dead Sea in Israel can have salt concentrations that exceed 25 percent. The shores of such lakes are caked with a crystalline rime of evaporate minerals, which are sometimes mined for industrial use.

The most chemically dilute surface waters are lakes in watersheds with hard, slowly weathering bedrock and soils. Such lakes can have total salt concentrations of less than 0.001 percent. For example, Beaverskin Lake in Nova Scotia has very clear, dilute water that is chemically dominated by chloride, sodium, and sulfate, in concentrations of two-thirds of the norm for surface water or less, with only traces of calcium, usually most abundant, and no silica (SiO_2). A nearby body of water, Big Red Lake, has a similarly dilute concentration of inorganic ions but, because it receives drainage from a bog, its chemistry also includes a large concentration of dissolved organic carbon, mainly comprised of humic or fulvic acids that stain the water a dark brown and greatly inhibit the penetration of sunlight.

Water in the form of precipitation is considerably more dilute than that of surface waters, with concentrations of sulfate, calcium, and magnesium of ranging from one-fortieth to one-hundredth of surface water levels, but adding small amounts of nitrate and ammonium (NH_4^+). Chloride and sodium concentrations depend on proximity to saltwater. For example, precipitation at a remote site in Nova Scotia, only 31 miles (50 km) from the Atlantic Ocean, will have six to ten times as much sodium and chloride as a similarly remote location in northern Ontario, Canada.

Acid rain is associated with the presence of relatively large concentrations of sulfate and nitrate (NO_3^-) in precipitation water. If the negative electrical charges of the sulfate and nitrate anions cannot be counterbalanced by positive charges of the cations sodium, calcium, magnesium, and ammonium, then hydrogen (H) ions go into solution, making the water acidic. Hubbard Brook Experimental Forest, New Hampshire, within an airshed of industrial, automobile, and residential emissions from the northeastern United States and eastern Canada, receives a substantially acidic precipitation, with an average pH of about 4.1. At Hubbard Brook, sulfate and nitrate together contribute 87 percent of the anion-equivalents in precipitation. Because cations other than the hydrogen ion can only neutralize about 29 percent of those anion charges, hydrogen ions must go into solution, making the precipitation acidic.

Fogwaters can have much larger chemical concentrations, mostly because the inorganic chemicals in fogwater droplets are less diluted by water than in rain and snow. For example, fogwater on Mount Moosilauke, New Hampshire, has average sulfate and nitrate concentrations about nine times more than in rainfall there, with ammonium eight times more, sodium seven times more, and potassium and the hydrogen ion three times more.

The above descriptions deal with chemicals present in relatively large concentrations in water. Often, however, chemicals that are present in much smaller concentrations can be of great environmental importance.

For example, in freshwaters phosphate (PO_4^{3-}) is the nutrient that most frequently limits the productivity of plants, and therefore, of the aquatic ecosystem. If the average concentration of phosphate in lake water is less than about 10μ/l, then the algae productivity will be very small, and the lake is classified as oligotrophic. Lakes with phosphate concentrations ranging from about 10 to 35μ/l are mesotrophic, those with 35 to 100μ/l are eutrophic, and those with more than 100μ/l are very productive, and very green, hypertrophic bodies of water. In a few exceptional cases, the productivity of freshwater may be limited by nitrogen, silica, or carbon, and sometimes by unusual micronutrients. For example, the productivity of phytoplankton in Castle Lake, California, has been shown to be limited by the availability of the trace metal, molybdenum (Mo).

Sometimes, chemicals present in trace concentrations in water can be toxic to plants and animals, causing substantial ecological changes. An important characteristic of acidic waters is their ability to solubilize aluminum (Al) from minerals, producing ionic aluminum (Al^{3+}). In nonacidic waters, ionic aluminum is generally present in minute quantities, but in very acidic waters when pH is less than 2.0, attainable by acid mine drainage, soluble-aluminum concentrations can rise drastically. Although some aquatic biota are physiologically tolerant of these aluminum ions, other species, such as fish, suffer toxicity and may disappear from acidified bodies of water. Many aquatic species cannot tolerate even small quantities of ionic aluminum. Many ecologists believe that aluminum ions are responsible for most of the toxicity of acidic waters and also of acidic soils.

Some chemicals can be toxic to aquatic biota even when present in ultra trace concentrations. Many species within the class of chemicals known as chlorinated hydrocarbons are insoluble in water, but are soluble in biological lipids such as animal fats. These chemicals often remain in the environment because they are not easily metabolized by microorganisms or degraded by ultraviolet radiation or other inorganic processes. Examples of chlorinated hydrocarbons are the insecticides dichlorodiphenyltrichloroethane (DDT), dichlorodiphenyldichloroethane (DDD), dieldrin, and methoxychlor, the class of dielectric fluids known as polychlorinated biphenyls (PCBs), and the chlorinated dioxin, tetrachlorodibenzo-p-dioxin (TCDD).

These chemicals are so dangerous because they collect in biological tissues, and accumulate progressively as organisms age through a process termed "bioaccumulation." They also accumulate into especially large concentrations in organisms at the top of the ecosystem's food chain in a process termed "biomagnification." In some cases, older individuals of top predator species have been found to have very large concentrations of chlorinated hydrocarbons in their fatty tissues. The toxicity caused to raptorial birds and other predators as a result of their accumulated doses of DDT, PCBs, and other chlorinated hydrocarbons is a well-recognized environmental problem. If humans eat aquatic organisms with high concentrations of bioaccumulated chemicals, then the chemicals can accumulate in their bodies if they are not able to metabolize them, ultimately posing a health risk.

Water pollution can also be caused by the presence of hydrocarbons. Accidental spills of petroleum from disabled tankers are the highest profiled causes of oil pollution, but smaller spills from tankers disposing of oily bilge waters, chronic discharges from refineries, and urban runoff are also significant sources of oil pollution. Hydrocarbons can also be present naturally, as a result of the release of chemicals synthesized by algae or during decomposition processes in anaerobic sediment. In a few places, there are natural seepages from near-surface petroleum reservoirs, as occurs in the vicinity of Santa Barbara, California. In general, the typical, naturally occurring concentration of hydrocarbons in seawater is quite small. Beneath a surface slick of spilled petroleum, however, the concentration of soluble hydrocarbons can be multiplied several times, sufficient to cause toxicity to some biota. This dissolved fraction does not include the concentration of finely suspended droplets of petroleum, which can become incorporated into an oil-in-water emulsion toxic to organisms that become coated with it. In general, within the very complex mix of hydrocarbons found in petroleum, the smallest molecules are the most soluble in water.

Agricultural practices result in a large amount of water pollution. Chemicals used as pesticides and fertilizers collect in nearby bodies of water and groundwater due to runoff from rainwater or irrigation water. Additionally, continued soil tillage and cultivation breaks soil particles down and increases soil erosion. These particles, typically accompanied by agricultural chemicals, collect in adjacent bodies of water and affect water quality. Regulations such as the Clean Water Act and the implementation of organic or sustainable methods of agriculture such as integrated pest management (IPM) focus on reducing the chemical input into the environment.

Resources

BOOKS

Crompton, T. R. *Toxicants in Aqueous Ecosystems: A Guide for the Analytical and Environmental Chemist.* Berlin: Springer, 2007.

Howard, Alan G. *Aquatic Environmental Chemistry.* Oxford, UK: Oxford University Press, 2004.

Manahan, Stanley E. *Environmental Chemistry*. Boca Raton, FL: CRC Press, 2005.

Weiner, Eugene R. *Applications of Environmental Aquatic Chemistry: A Practical Guide*. 2nd ed. Boca Raton, FL: CRC Press, 2008.

Bill Freedman

Aquatic microbiology

Aquatic microbiology is the science that deals with microscopic living organisms in fresh- or saltwater systems. While aquatic microbiology can encompass all microorganisms, including microscopic plants and animals, it more commonly refers to the study of bacteria, viruses, and fungi and their relation to other organisms in the aquatic environment.

Bacteria are quite diverse in nature. The scientific classification of bacteria divides them into nineteen major groups based on their shape, cell structure, staining properties (used in the laboratory for identification), and metabolic functions. Bacteria occur in many sizes as well, ranging from 0.1 micrometers to greater than 500 micrometers. Some are motile and have flagella, which are tail-like structures used for movement.

Although soil is the most common habitat of fungi, they are also found in aquatic environments. Aquatic fungi are collectively called *water molds* or *aquatic Phycomycetes*. They are found on the surface of decaying plant and animal matter in ponds and streams. Some fungi are parasitic and prey on algae and protozoa.

Viruses are the smallest group of microorganisms and usually are viewed only with the aid of an electron microscope. They are disease-causing organisms that are very different than bacteria, fungi, and other cellular lifeforms. Viruses are infectious nucleic acid enclosed within a coat of protein. They penetrate host cells and use the nucleic acid of other cells to replicate.

Bacteria, viruses, and fungi are widely distributed throughout aquatic environments. They can be found in freshwater rivers, lakes, and streams, in the surface waters and sediments of the world's oceans, and even in hot springs. They have even been found supporting diverse communities at hydrothermal vents in the depths of the oceans.

Microorganisms living in these diverse environments must deal with a wide range of physical conditions, and each has specific adaptations to live in the particular place it calls home. For example, some have adapted to live in freshwater with very low salinity, whereas others live in the saltiest parts of the ocean. Some must deal with the harsh cold of Arctic waters, whereas those in hot springs are subjected to intense heat. In addition, aquatic microorganisms can be found living in environments where there are extremes in other physical parameters such as pressure, sunlight, organic substances, dissolved gases, and water clarity.

Aquatic microorganisms obtain nutrition in a variety of ways. For example, some bacteria living near the surface of either fresh or marine waters, where there is often abundant sunlight, are able to produce their own food through the process of photosynthesis. Bacteria living at hydrothermal vents on the ocean floor where there is no sunlight can produce their own food through a process known as chemosynthesis, which depends on preformed organic carbon as an energy source. Many other microorganisms are not able to produce their own food. Rather, they obtain necessary nutrition from the breakdown of organic matter, such as dead organisms.

Aquatic microorganisms play a vital role in the cycling of nutrients within their environment, and thus are a crucial part of the food chain. Many microorganisms obtain their nutrition by breaking down organic matter in dead plants and animals. As a result of this process of decay, nutrients are released in a form usable by plants. These aquatic microorganisms are especially important in the cycling of the nutrients nitrogen, phosphorus, and carbon. Without this recycling, plants would have few, if any, organic nutrients to use for growth.

In addition to breaking down organic matter and recycling it into a form of nutrients that plants can use, many of the microorganisms become food themselves. There are many types of animals that graze on bacteria and fungi. For example, some deposit-feeding marine worms ingest sediments and digest numerous bacteria and fungi found there, later expelling the indigestible sediments. Therefore, these microorganisms are intimate members of the food web in at least two ways.

Humans have taken advantage of the role these microorganisms play in nutrient cycles. At sewage treatment plants, microscopic bacteria are cultured and then used to break down human wastes. However, in addition to the beneficial uses of some aquatic microorganisms, others may cause problems for people because they are pathogens, which can cause serious diseases. For example, viruses such as *Salmonella typhi*, *S. paratyphi*, and the Norwalk virus are found in water contaminated by sewage and can cause illness. Fecal coliform (*E. coli*) bacteria and Enterococcus bacteria are two types of microorganisms that are used to indicate the presence of disease-causing microorganisms in aquatic environments.

Marci L. Bortman

Aquatic toxicology

Aquatic toxicology is the study of the adverse effects of toxins and their activities on aquatic ecosystems. Aquatic toxicologists assess the condition of aquatic systems, monitor trends in conditions over time, diagnose the cause of damaged systems, guide efforts to correct damage, and predict the consequences of proposed human actions so the ecological consequences of those actions can be considered before damage occurs. Aquatic toxicologists study adverse effects at different spatial, temporal, and organizational scales. Because aquatic systems contain thousands of species, each of these species can respond to toxicants in many ways, and interactions between these species can be affected.

Consequently, a virtually unlimited number of responses could be produced by chemicals. Scientists study effects as specific as a physiological response of a local fish species or as inclusive as the biological diversity of a large river basin. Generally, attention is first focused on responses that are considered important either socially or biologically. For example, if a societal goal is to have fishable waters, responses to toxicants of game fishes and the organisms they depend on would be of interest.

The two most common tools used in aquatic toxicology are the field survey and the toxicity test. A field survey characterizes the indigenous biological community of an aquatic system or watershed. Chemistry, geology, and land use are also essential components of field surveys. Often, the characteristics of the community are compared to other similar systems that are in good condition. Field surveys provide the best evidence of the existing condition of natural systems. However, when damaged communities and chemical contamination co-occur, it is difficult to establish a cause-and-effect relationship using the field survey alone. Toxicity tests can help to make this connection. The toxicity test excises some replicable piece of the system of interest, perhaps fish or microbial communities

of lakes, and exposes it to a chemical in a controlled, randomized, and replicable manner. The most common aquatic toxicity test data provide information about the short-term survival of fish exposed to one chemical. Such tests demonstrate whether a set of chemical conditions can cause a specified response. The tests also provide information about what concentrations of the chemical are of concern. Toxicity tests can suggest a threshold chemical concentration below which the adverse effect is not expected to occur; they can also present an index of relative toxicity used to rank the toxicity of two chemicals. By using information from the field survey and toxicity tests, along with other tools such as tests on the fate of chemicals released into aquatic systems, aquatic toxicologists can develop models predicting the effects of proposed actions.

Chemical and thermal pollution caused by human activities can have a dramatic impact on aquatic ecosystems. Chemicals such as pesticides, herbicides, and detergents can drain into rivers and ponds, affecting water quality and thus the inhabiting organisms. These chemicals can negatively affect aquatic organisms, compromising their physiological systems. Some chemicals, such as the banned pesticide dichlorodiphenyltrichloroethane (DDT), act as endocrine disruptors (exogenous chemicals that can mimic the function of endogenous hormones) leading to altered reproductive development. Fish living in a polluted aquatic ecosystem can accumulate these chemicals in their bodies. If humans or other animals consume these fish, the consumers are then exposed to the chemicals. This exposure may induce harmful effects in the consumers, especially in cases involving a developing fetus. Thermal pollution results from either unnaturally warm or cold water being introduced into an aquatic ecosystem. This water can be from industrial sources, reservoirs, or urban runoff (storm water heated from passing over pavement, sidewalks, or rooftops). Water that is unnaturally warmer than the normal temperatures of a body of water can deplete the amount of available oxygen, thus killing fish and other aquatic organisms. Thermal pollution can disrupt the natural biodiversity and chemical composition of the water. Testing bodies of water to obtain information about water quality helps assess the state of an aquatic ecosystem to identify possible sources of contamination and plan preventive measures against further contamination.

The most important challenge to aquatic toxicology will be to develop methods that support sustainable use of aquatic and other ecosystems. Sustainable use is intended to ensure that those now living do not deprive future generations of the essential natural resources necessary to maintain a quality lifestyle.

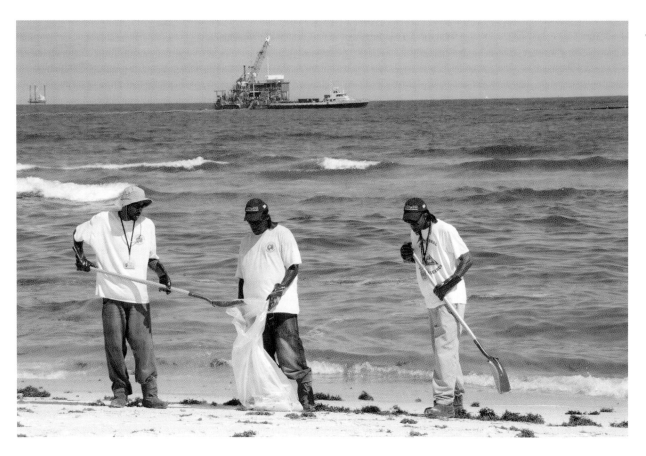

Oil spill workers collect tainted debris and dark oil patches along the beach as oil washes ashore at Perdido Pass, Alabama, June 2010. *(Cheryl Casey/Shutterstock.com)*

Achieving the goal of sustainable use will require aquatic toxicologists to have a much longer temporal perspective than is now had. Additionally, in a more crowded and increasingly affluent world, cumulative effects will become extremely important.

Resources

BOOKS

Carson, Rachel. *Silent Spring*. Boston: Houghton Mifflin, 1962.

Dietrich, Daniel R.; Simon F. Webb; and Thomas Petry. *Hot Spot Pollutants: Pharmaceuticals in the Environment*. Amsterdam: Elsevier, 2005.

Giesy, John P. *Aquatic Toxicology Lab Exercises*. Albany, GA: Lewis Publishing, 2008.

Kumar, Arvind. *Ecobiology of Polluted Waters*. Delhi: Daya Publishing House, 2006.

Ostrander, Gary Kent. *Techniques in Aquatic Toxicology*. Boca Raton, FL: Taylor & Francis, 2005.

Paquin, Paul R. *Bioavailability and Effects of Ingested Metals on Aquatic Organisms*. Alexandria, VA: Water Environment Research Foundation, 2006.

Rand, Gary M. *Fundamentals of Aquatic Toxicology: Effects, Environmental Fate, and Risk Assessment*. Boca Raton, FL: CRC, 2008.

Svensson, Elias P. *Aquatic Toxicology Research Focus*. New York: Nova Science Publishers, 2008.

John Cairns Jr.

Aquatic weed control

A simple definition of an aquatic weed is a plant that grows (usually too densely) in an area such that it hinders the usefulness or enjoyment of that area. Some common examples of aquatic plants that can become weeds are the water milfoils, ribbon weeds, and pondweeds. They may grow in ponds, lakes, streams, rivers, navigation channels, and seashores, and the growth may be because of a variety of factors such as excess nutrients in the water or the introduction of rapidly growing exotic species. The problems caused by aquatic weeds are many, ranging from

Aquifer

unsightly growth and nuisance odors to clogging of water-ways, damage to shipping and underwater equipment, and impairment of water quality.

It is difficult and usually unnecessary to eliminate weeds completely from a lake or stream. Therefore, aquatic weed control programs usually focus on controlling and maintaining the prevalence of the weeds at an acceptable level. The methods used in weed control may include one or a combination of the following: physical removal, mechanical removal, habitat manipulation, biological controls, or chemical controls.

Physical removal of weeds involves cutting, pulling, or raking weeds by hand. It is time-consuming and labor-intensive, and is most suitable for small areas or for locations that cannot be reached by machinery. Mechanical removal is accomplished by specialized harvesting machinery equipped with toothed blades and cutting bars to cut the vegetation, collect it, and haul it away. It is suitable for offshore weed removal or to supplement chemical control. Repeated harvesting is usually necessary and often the harvesting blades may be limited in the depth or distance that they can reach. Inadvertent dispersal of plant fragments may also occur and lead to weed establishment in new areas. Operation of the harvesters may disturb fish habitat.

Habitat manipulation involves a variety of innovative techniques to discourage the establishment and growth of aquatic weeds. Bottom liners of plastic sheeting placed on lake bottoms can prevent the establishment of rooted plants. Artificial shading can discourage the growth of shade-intolerant species. Draw down of the water level can be used to eliminate some species by desiccation. Dredging to remove accumulated sediments and organic matter can also delay colonization by new plants.

Biological control methods generally involve the introduction of weed-eating fish, insects, competing plant species, or weed pathogens into an area of high weed growth. While there are individual success stories (for example, stocking lakes with grass carp), it is difficult to predict the long-term effects of the introduced species on the native species and ecology and, therefore, biological controls should be used with caution.

Chemical control methods consist of the application of herbicides that may be either systemic or contact in nature. Systemic herbicides are taken up into the plant and cause plant death by disrupting its metabolism in various ways. Contact herbicides only kill the directly exposed portions of the plant, such as the leaves. While herbicides are convenient and easy to use, they must be selected and used with care at the appropriate times and in the correct quantities. Sometimes, they may also kill nontarget plant species and, in some cases, toxic residues from the degrading herbicide may be ingested and transferred up the food chain.

In regions of the globe that experience change in aquatic conditions because of climate change in the coming century, the invasion by nonindigenous weed species may be more problematic, necessitating more dedicated weed-control efforts.

Resources

BOOKS

Contreras, Sofia A. *Effects of Climate Change on Aquatic Invasive Species.* Boston: Nova Science, 2010.
Rozema, Jelte; Rien Aerts; and Hans Cornelissen. *Plants and Climate Change.* New York: Springer, 2010.

Usha Vedagiri

Aquifer

Natural zones below the surface that yield water in sufficient quantities to be economically important for industrial, agricultural, or domestic purposes, aquifers can occur in a variety of geologic materials, ranging from glacial-deposited outwash to sedimentary beds of limestone and sandstone, and fractured zones in dense igneous rocks. Composition and characteristics are almost infinite in their variety.

Aquifers can be confined or unconfined. Unconfined aquifers are those where direct contact can be made with the atmosphere, whereas confined aquifers are separated from the atmosphere by impermeable materials. Confined aquifers are also artesian aquifers. Though "artesian" was originally a term applied to water in an aquifer under sufficient pressure to produce flowing wells, the term is now generally applied to all confined situations.

Aquifer depletion

An aquifer is a water-saturated geological layer that easily releases water to wells or springs for use as a water supply. Also called groundwater reservoirs or water-bearing formations, aquifers are created and replenished when excess precipitation (rain and snowfall)

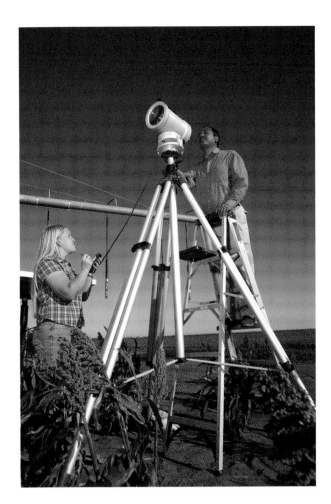

Agricultural engineer (on ladder) and hydrologic technician determining the water loss from a field, by measuring the sensible heat flux with a large-aperture scintillometer. *(Science Source/Photo Researchers, Inc.)*

is held in the soil. This water is not released through runoff, nor is it removed by the surface flows of rivers or streams. Plants have used what they need (transpiration) and little is evaporated from nonliving surfaces, such as soil. The remaining excess water slowly percolates downward through the soil and through the air spaces and cracks of the surface overburden of rocks into the bedrock. As water collects in this saturated area or recharge zone, it becomes groundwater. The uppermost level of the saturated area is called the water table.

Groundwater is especially abundant in humid areas where the overburden is relatively thick and the bedrock is porous or fractured, particularly in areas of sedimentary rocks such as sandstone or limestone. Aquifers are extremely valuable natural resources in regions where lakes and rivers are not abundant. Groundwater is usually accessed by drilling a well and then pumping the water to the surface.

In less-moist environments, however, the quantity of precipitation available to recharge groundwater is much smaller. Slowly recharging aquifers in arid environments are easily depleted if their groundwater is used rapidly by humans. In some cases, groundwater sources may exist in water-bearing geologic areas that make pumping nearly impossible. Moreover, increased irrigation use has led to heavy pumping that is draining aquifers and lowering water tables around the world. Aquifer depletion is a growing problem as world populations increase and with them the need for increased food supplies.

Large, rapidly recharging aquifers underlying humid landscapes can sustain a high rate of pumping of their groundwater. As such, they can be sustainably managed as a renewable resource. Aquifers that recharge very slowly, however, are essentially filled with old, so-called "fossil" water that has accumulated over thousands of years or longer. This kind of aquifer has little capability of recharging as the groundwater is used because the groundwater is depleted so rapidly for human use. Therefore, slowly recharging aquifers are essentially nonrenewable resources whose reserves are mined by excessive use.

In 1999 and again in 2006, the Worldwatch Institute reported that water tables were falling on every continent in the world, mainly because of excessive human consumption. Groundwater in India, in particular, is being pumped at double the rate of the aquifer's ability to recharge from rainfall. The aquifer under the North China Plain is seeing its water table fall at 5 feet (1.5 m) a year.

In the United States, the situation is similar. The largest aquifer in the world, known as the Ogallala Aquifer, is located beneath the arid lands of the western United States. The Ogallala aquifer is very slowly recharged by underground seepage that mostly originates with precipitation falling on a distant recharge zone in mountains located in its extreme western range. Much of the groundwater presently in the Ogallala is fossil water that has accumulated during tens of thousands of years of extremely slow infiltration. Although the Ogallala aquifer is an enormous resource, it is being depleted alarmingly by pumping at more than 150,000 wells. Most of the groundwater being withdrawn by the wells is used in irrigated agriculture, and some for drinking and other household purposes. In recent years, the level of the Ogallala aquifer has been decreasing by as much as 3.2 feet (1 m) per year in intensively utilized zones, while the recharge rate is only of the order of one-thirty-second of an inch (1 mm) per year. Obviously, the Ogallala aquifer is being mined on a large scale.

Aquifer depletion brings with it more than the threat of water scarcity for human use. Serious environmental consequences can occur when large amounts of water are pumped rapidly from groundwater reservoirs. Commonly, the land above an aquifer will subside or sink as the water is drained from the geologic formation and the earth compacts. In 1999, researchers noted that portions of Bangkok, Thailand, and Mexico City, Mexico, were sinking as a result of overexploitation of their aquifers. This can cause foundations of buildings to shift and may even contribute to earthquake incidence. Large cities in the United States such as Albuquerque, New Mexico, Phoenix, Arizona, and Tucson, Arizona, lie over aquifers that are being rapidly depleted.

Unfortunately, no solution to aquifer depletion has went beyond drilling deeper wells or abandoning agriculture and import foods. Both are costly choices for any country, both in dollars and in economic independence.

Resources

BOOKS

Bear, Jacob, and A. H.D. Cheng. *Modeling Groundwater Flow and Contaminant Transport.* Dordrecht, Netherlands: Springer, 2010.

Glennon, Robert J. *Water Follies: Groundwater Pumping and the Fate of America's Fresh Waters.* Washington, DC: Island Press, 2004.

Taniguchi, Makoto, and I. P. Holman. *Groundwater System Response to a Changing Climate.* Boca Raton, FL: CRC Press, 2010.

Tóth, József. *Gravitational Systems of Groundwater Flow: Theory, Evaluation, Utilization.* Cambridge, UK: Cambridge University Press, 2009.

PERIODICALS

Gardner, Gary. "From Oasis to Mirage: The Aquifers that Won't Replenish." *World Watch* 8 (2005): 30–37.

OTHER

United States Department of the Interior, United States Geological Survey (USGS). "Aquifers." http://www.usgs.gov/science/science.php?term=48&type=feature (accessed October 26, 2010).

ORGANIZATIONS

World Resources Institute, 10 G Street, NE (Suite 800), Washington, DC, USA, 20002, (202) 729-7600, (202) 729-7610, front@wri.org, http://www.wri.org/

Worldwatch Institute, 1776 Massachusetts Ave., NW, Washington, DC, USA, 20036–1904, (202) 452–1999, (202) 296-7365, worldwatch@worldwatch.org, http://www.worldwatch.org/

Bill Freedman

Aquifer restoration

Once an aquifer is contaminated, the process of restoring the quality of water is generally time-consuming and expensive, and it is often more cost-effective to locate a new source of water. For these reasons, the restoration of an aquifer is usually evaluated on the basis of these criteria: (1) the potential for additional contamination; (2) the time period over which the contamination has occurred; (3) the type of contaminant; and (4) the hydrogeology of the site. Restoration techniques fall into two major categories, in situ methods and conventional methods of withdrawal, treatment, and disposal.

Remedies undertaken within the aquifer involve the use of chemical or biological agents that either reduce the toxicity of the contaminants or prevent them from moving any further into the aquifer, or both. One such method requires the introduction of biological cultures or chemical reactants and sealants through a series of injection wells. This action will reduce the toxicity, form an impervious layer to prevent the spread of the contaminant, and clean the aquifer by rinsing. However, a major drawback to this approach is the difficulty and expense of installing enough injection wells to assure a uniform distribution throughout the aquifer.

In situ degradation, another restoration method, can theoretically be accomplished through either biological or chemical methods. Biological methods involve placing microorganisms in the aquifer that are capable of utilizing and degrading the hazardous contaminant. A great deal of progress has been made in the development of microorganisms that will degrade both simple and complex organic compounds. It may be necessary to supplement the organisms introduced with additional nutrients and substances to help them degrade certain insoluble organic compounds. Before introduction of these organisms, it is also important to evaluate the intermediate products of the degradation to carbon dioxide and water for toxicity. Chemical methods of in situ degradation fall into three general categories: (1) injection of neutralizing agents for acid or caustic compounds; (2) addition of oxidizing agents such as chlorine or ozone to destroy organic compounds; and (3) introduction of amino acids to reduce polychlorinated biphenyls (PCBs).

There are also methods for stabilizing an aquifer and preventing a contaminant plume from extending. One stabilizing alternative is the conversion of a contaminant to an insoluble form. This method is limited to use on inorganic salts, and even those compounds

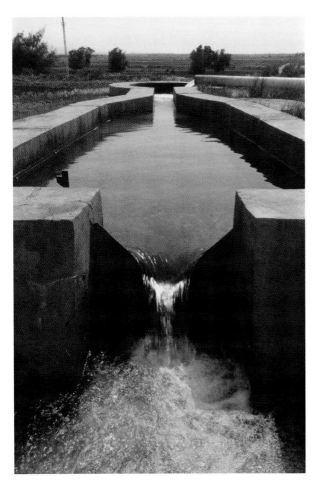

An aquifer drainage channel at Bahariya Oasis, Egypt, 2001.
(© Ron Watts/Corbis)

usually made of concrete and are put in place by digging a trench, which is then filled with the slurry. The depth at which these walls can be used is limited to 80 to 90 feet (24–27 m). Grout curtains are formed by injecting a cement grout under pressure into the aquifer through a series of injection wells, but it is difficult to know how effective a barrier this is and whether it has uniformly penetrated the aquifer. Depth is again a limiting factor, and this method has a range of 50 to 60 feet (15–18 m). Steel sheets can be driven to a depth of 100 feet (30 m) but no satisfactory technique for forming impermeable joints between the individual sheets has been developed. Floor seals are grouts installed horizontally, and they are used where the contaminant plume has not penetrated the entire depth of the aquifer. None of these methods offers a permanent solution to the potential problems of contamination; in addition, some type of continued monitoring and maintenance is necessary.

Conventional methods of restoration involve removal of the contaminant followed by withdrawal of the water, and final treatment and disposal. Site characteristics that are important include the topography of the land surface, characteristics of the soil, depth to the water table and how this depth varies across the site, and depth to impermeable layers. The options for collection and withdrawal can be divided into five groups: (1) collection wells; (2) subsurface gravity collection drains; (3) impervious grout curtains (as described above); (4) cutoff trenches; or (5) a combination of these options. Collection wells are usually installed in a line, and are designed to withdraw the contaminant plume and to keep movement of other clean water into the wells at a minimum. Various sorts of drains can be effective in intercepting the plume, but they do not work in deep aquifers or hard rock. Cutoff trenches can be dug if the contamination is not too deep, and water can then be drained to a place where it can be treated before final discharge into a lake or stream.

Water taken from a contaminated aquifer requires treatment before final discharge and disposal. To what degree it can be treated depends on the type of contaminant and the effectiveness of the available options. Reverse osmosis uses pressure at a high temperature to force water through a membrane that allows water molecules, but not contaminants, to pass. This process removes most soluble organic chemicals, heavy metals, and inorganic salts. Ultrafiltration also uses a pressure-driven method with a membrane. It operates at lower temperatures and is not as effective for contaminants with smaller molecules. An ion exchange uses a bed or series of tubes filled with a resin to remove selected

can dissolve if the physical or chemical properties of the aquifer change. A change in pH, for instance, might allow the contaminant to return to solution. Like other methods of in situ restoration, conversion requires the contaminant to be contained within a workable area.

The other important stabilizing alternatives are containment methods, which enclose the contaminant in an insoluble material and prevent it from spreading through the rest of the aquifer. There are partial and total containment methods. For partial containment, a clay cap can be applied to keep rainfall from moving additional contaminants into the aquifer. This method has been used quite often where there are sanitary landfills or other types of hazardous waste sites. Total containment methods are designed to isolate the area of contamination through the construction of some kind of physical barrier, but these have limited usefulness. These barriers include slurry walls, grout curtains, sheet steel, and floor seals. Slurry walls are compounds from the water and replace them with harmless ions. The system has the advantage of being transportable; and depending on the type of resin, it can be used more than once by flushing the resin with either an acid or salt solution. Both organic and inorganic substances can be removed using this method.

Wet-air oxidation is another important treatment method. It introduces oxygen into the liquid at a high temperature, which effectively treats inorganic and organic contaminants. Combined ozonation or ultraviolet radiation is a chemical process in which the water containing toxic chemicals is brought into contact with ozone and ultraviolet radiation to break down the organic contaminants into harmless parts. Chemical treatment is a general name for a variety of processes that can be used to treat water. They often result in the precipitation of the contaminant, and will not remove soluble organic and inorganic substances. Aerobic biological treatments are processes that employ microorganisms in the presence of dissolved oxygen to convert organic matter to harmless products. Another approach uses activated carbon in columns—the water is run over the carbon and the contaminants become attached to it. Contaminants begin to cover the surface area of the carbon over time, and these filters must be periodically replaced.

These treatment processes are often used in combination. The methods used depend on the ultimate disposal plan for the end products. The three primary disposal options are (1) discharge to a sewage treatment plant; (2) discharge to a surface water body; and (3) land application. Each option has positive and negative aspects. Any discharge to a municipal treatment plant requires pretreatment to standards that will allow the plant to accept the waste. Land application requires an evaluation of plant nutrient supplying capability and any potentially harmful side effects on the crops grown. Discharge to surface water requires that the waste be treated to the standard allowable for that water. For many organics the final disposal method is burning.

Resources

BOOKS

Bear, Jacob, and A. H.D. Cheng. *Modeling Groundwater Flow and Contaminant Transport.* Dordrecht, Netherlands: Springer, 2010.

Glennon, Robert J. *Water Follies: Groundwater Pumping and The Fate of America's Fresh Waters.* Washington, DC: Island Press, 2004.

Taniguchi, Makoto, and I. P. Holman. *Groundwater System Response to a Changing Climate.* Boca Raton, FL: CRC Press, 2010.

Tóth, József. *Gravitational Systems of Groundwater Flow: Theory, Evaluation, Utilization.* Cambridge, UK: Cambridge University Press, 2009.

PERIODICALS

Gardner, Gary. "From Oasis to Mirage: The Aquifers that Won't Replenish." *World Watch* 8 (2005): 30–37.

OTHER

United States Department of the Interior, United States Geological Survey (USGS). "Aquifers." http://www.usgs.gov/science/science.php?term=48&type=feature (accessed October 26, 2010).

James L. Anderson

Arable land

Arable land has soil and topography suitable for the economical and practical cultivation of crops. These range from grains, grasses, and legumes to fruits and vegetables. Permanent pastures and rangelands are not considered arable. Forested land is also nonarable, although it can be farmed if the area is cleared and the soil and topography are suitable for cultivation.

Aral Sea

The Aral Sea is a large, shallow saline lake hidden in the remote deserts of the republics of Uzbekistan and Kazakhstan in the south-central region of the former Soviet Union. Once the world's fourth largest lake in area (smaller only than North America's Lake Superior, Siberia's Lake Baikal, and East Africa's Lake Victoria), in 1960 the Aral Sea had a surface area of 26,250 square miles (68,000 km^2), and a volume of 260 cubic miles (1,090 cubic km). Its only water sources are two large rivers, the Amu Darya and the Syr Darya. Flowing northward from the Pamir Mountains on the Afghan border, these rivers pick up salts as they cross the Kyzyl Kum and Kara Kum deserts. Evaporation from the landlocked sea's surface (it has no outlet) makes the water even saltier.

The Aral Sea's destruction began in 1918 when plans were made to draw off water to grow cotton, a badly needed cash crop for the newly formed Soviet

Aral Sea disaster, Kazakhstan. *(gopixgo/Shutterstock.com.)*

Union. The amount of irrigated cropland in the region was expanded greatly (from 7.2–18.8 million acres [2.9–7.6 million ha]) in the 1950s and 1960s with the completion of the Kara Kum canal. Annual water flows in the Amu Darya and Syr Darya dropped from about 13 cubic miles (55 cubic km) to less than 1 cubic mile (5 cubic km). In some years the rivers were completely dry when they reached the lake. Water extraction has similarly much reduced the volume of the Dead Sea, which is situated between the nations of Israel and Jordan.

Soviet authorities were warned that the sea would die without replenishment, but sacrificing a remote desert lake for the sake of economic development seemed an acceptable tradeoff. Inefficient irrigation practices drained away the lifeblood of the lake. Dry years in the early 1970s and mid-1980s accelerated water shortages in the region. More recently, in a disaster of unprecedented magnitude and rapidity, the Aral Sea seems to be quickly disappearing.

Until 1960 the Aral Sea was fairly stable, but by 1990, it had lost 40 percent of its surface area and two-thirds of its volume. Surface levels dropped 42 feet (914 m), turning 11,580 square miles (30,000 km^2, about the size of the state of Maryland) of former seabed into a salty, dusty desert. Fishing villages that were once at the sea's edge are now 25 miles (40 km) from water. Boats trapped by falling water levels lie abandoned in the sand. Salinity of the remaining water has tripled with devastating effects to aquatic life. Commercial fishing that brought in 52,800 tons (48,000 metric tons) in 1957 was completely gone in 1990.

Winds whipping across the dried-up seabed pick up salty dust, poisoning crops and causing innumerable health problems for residents. An estimated 47.3 million tons (43 million metric tons) of salt are blown onto nearby fields and cities each year. Eye irritations, intestinal diseases, skin infections, asthma, bronchitis, and a variety of other health problems have risen sharply in the past thirty years, especially among children. Infant mortality in the Karakalpak Autonomous Republic, adjacent to the Aral Sea, is seventy-five per 1,000, twice as high as in other former Soviet republics.

Among adults, throat cancers have increased fivefold in thirty years. Many physicians believe that heavy doses of pesticides used on the cotton fields and transported by runoff water to the lake sediments

ORGANIZATIONS

Arctic Council Secretariat, Polarmiljøsenteret NO-9296Tromsø, Norway, +47 77 75 01 40, +47 77 75 05 01, ac-chair@arctic-council.org, http://www.arctic-council.org

Nicole Beatty

Arctic haze

Arctic haze refers to the dry aerosol present in arctic regions during much of the year and responsible for substantial loss of visibility through the atmosphere. The arctic regions are, for the most part, very low in precipitation, qualifying on that basis as deserts. Ice accumulates because even less water evaporates than is deposited. Hence, particles that enter the arctic atmosphere are only very slowly removed by precipitation, a process that removes a significant fraction of particles from the tropical and temperate atmospheres. Thus, relatively small sources can lead to appreciable final atmospheric concentrations.

Studies have been conducted on the chemistry of the particles in the haze, and those trapped in the snow and ice. Much of the time the mix of trace elements in the particles is very close to that found in the industrial emissions from northern Europe and Siberia and quite different from that in such emissions from northern North America. Concentrations decrease rapidly with depth in the ice layers, indicating that these trace elements began to enter the atmosphere within the past few centuries. It is now generally conceded that most of the haze particles are derived from human activities, primarily—though not exclusively—in northern Eurasia.

Since the haze scatters light, including sunlight, it decreases the solar energy received at the ground level in polar regions and may, therefore, have the potential to decrease arctic temperatures. Arctic haze also constitutes a nuisance because it decreases visibility. The trace elements found in arctic haze apparently are not yet sufficiently concentrated in either atmosphere or precipitation to constitute a significant toxic hazard.

One form of aerosol pollution that contributes to Arctic haze is soot, tiny, dark particles consisting mostly of carbon that are released by burning many fuels. Soot, whether in the air or lying on icy surfaces, absorbs sunlight and so tends to warm the earth and accelerate melting of snow and ice. Not only the Arctic but other ice-covered regions are strongly affected by soot; in 2008, researchers studying soot released by cooking fires in South Asia and coal-burning in East Asia concluded that soot is a more important contributor to global warming than had been hitherto known, edging out methane as the second-largest greenhouse pollutant after carbon dioxide. Melting of glaciers in the Himalaya Mountains may have as much to do with soot as with carbon dioxide. Dimming of the skies by soot reduces evaporation and regional temperature differences over the Indian Ocean, causing less rain to fall during the monsoon season and so contributing to drought in Southeast Asia. The new estimate of the radiative forcing (energy imbalance between the earth and sunlight) caused by soot was about double that cited by the Intergovernmental Panel on Climate Change (IPCC) in its influential 2007 summary of the state of climate science. Replacing dung and wood as cooking fuels in South Asia with biogas, natural gas, and solar cookers could significantly reduce the problem.

As the Arctic continues to warm in this century, predictions are that the increasing amounts of sea ice will melt. If this occurs, the pollution in the region could increase because of the increased shipping and mining. This, combined with pollution drifting northward, could increase Arctic haze.

Resources

BOOKS

Baker, Stuart. *Climate Change in the Arctic*. New York: Benchmark Books, 2009.

Kolbert, Elizabeth. *Field Notes from a Catastrophe: Man, Nature, and Climate Change*. New York: Bloomsbury, 2006.

Martin, James. *Planet Ice: A Climate for Change*. Seattle, WA: Mountaineers Books, 2009.

Zellen, Barry Scott. *Arctic Doom, Arctic Boom: The Geopolitics of Climate Change in the Arctic*. New York: Praeger, 2009.

James P. Lodge Jr.

Arctic National Wildlife Refuge

Beyond the jagged Brooks Range in Alaska's far northeastern corner lies one of the world's largest nature preserves, the 19.8 million-acre (8 million-ha) Arctic National Wildlife Refuge (ANWR). It is the largest wildlife refuge in the United States. A narrow strip of treeless coastal plain in the heart of the refuge presents one of nature's grandest spectacles, as well as one of the longest-running environmental battles of the past century. For a few months during the brief

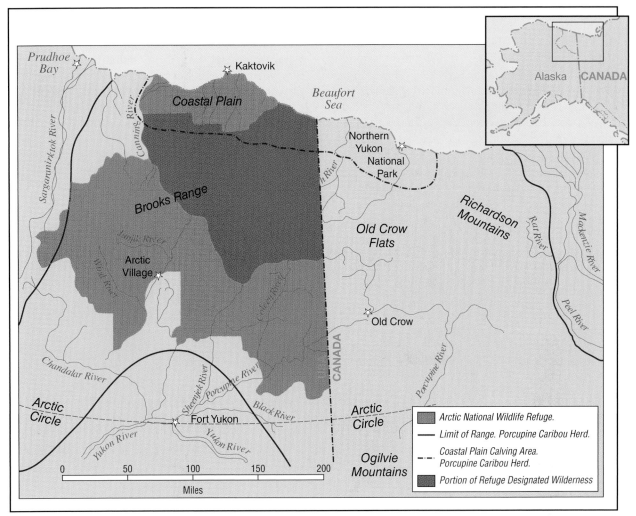

A map of the Arctic National Wildlife Refuge (ANWR) in Alaska. *(Reproduced by permission of Gale, a part of Cengage Learning)*

arctic summer, the tundra teems with wildlife. This is the calving ground of the 130,000 caribou of the Porcupine herd (*Rangifer tarandus granti*), which travels up to 400 miles (640 km) each summer to graze and give birth along the Arctic Ocean shore. It also is an important habitat for tens of thousands of snow geese (*Chen caerulescens*), tundra swans (*Cygnus columbianus*), shorebirds, and other migratory waterfowl; a denning area for polar bears (*Ursus maritimus*), arctic foxes (*Alopex lagopus*), and arctic wolves (*Canis lupus arctos*); and a year-round home to about 350 shaggy musk ox (*Ovibos moschatus*). In wildlife density and diversity, it rivals Africa's Serengeti.

When Congress established the Wildlife Refuge in 1980, a special exemption was made for about 1.5 million acres (600,000 ha) of coastline between the mountains and the Beaufort Sea, where geologists think sedimentary strata may contain billions of

barrels of oil and trillions of cubic feet of natural gas. Termed the 1002 area for the legislative provision that put it inside the wildlife refuge but reserved the right to drill for fossil fuels, this narrow strip of tundra may be the last big, onshore, liquid petroleum field in North America. It also is a critical habitat for one of the richest biological communities in the world. The possibility of extracting the fossil fuels without driving away the wildlife and polluting the pristine landscape is supported by oil industry experts and disputed by biologists and environmentalists.

The amount of oil and gas beneath the tundra is uncertain. Only one seismic survey has been done and a few test wells drilled. Industry geologists claim that there may be sixteen billion barrels of oil under the ANWR, but guessing the size and content of the formation from this limited evidence is as much an art as a science. Furthermore, the amount of oil it is economical to recover depends on market prices and shipping costs. When wholesale prices were $25 per barrel, the U.S. Geological Survey estimated that seven billion barrels might be pumped profitably. If prices drop below $10 per barrel, as they did in the early 1990s, the economic resource might be only a few hundred million barrels; but if they soar to over $100 per barrel, as they did in 2008, it could be quite lucrative to think in terms of billions of barrels again.

Energy companies are extremely interested in ANWR because any oil found there could be pumped out through the existing Trans-Alaska pipeline, thus extending the life of their multibillion-dollar investment. The state of Alaska hopes that revenues from ANWR will replenish dwindling coffers as the oil supply from nearby Prudhoe Bay wells dry up. Automobile companies, tire manufacturers, filling stations, and others who depend on the continued use of petroleum argue that domestic oil supplies are rapidly being depleted, leaving the United States dependent on foreign countries for more than one-half of its energy consumption.

Oil drilling proponents point out that prospecting will occur only in winter when the ground is covered with snow and most wildlife is absent or hibernating. Once oil is located, they claim, it will take only four or five small drilling areas, each occupying no more than a few hundred hectares, to extract it. Heavy equipment would be hauled to the sites during the winter on ice roads built with water pumped from nearby lakes and rivers. Each 2.2-yards-thick (2-m), gravel drilling pad would hold up to fifty closely spaced wells, which would penetrate the permafrost (soil or rock at temperatures of 32° F [0° C] or below) and then spread out horizontally to reach pockets of oil up to 6 miles (10 km) away from the wellhead. A central processing facility would strip water and gas from the oil, which would then be pumped through elevated, insulated pipelines to join oil flowing from Prudhoe Bay.

Opponents of this project argue that the noise, pollution, and construction activity accompanying this massive operation will drive away wildlife and leave scars on the landscape that could last for centuries. Pumping the millions of gallons of water needed to build ice roads could dry up local ponds on which wildlife depends for summer habitat. Every day six to eight aircraft—some as big as the C–130 Hercules—would fly into ANWR. The smell of up to 700 workers and the noise of numerous trucks and enormous power-generating turbines, each as large and loud as a jumbo aircraft engine, would waft out over the tundra. Pointing to the problems of other Arctic oil drilling operations where drilling crews dumped garbage, sewage, and toxic drilling waste into surface pits,

environmentalists predict disaster if drilling is allowed in the refuge. Pipeline and drilling spills at Prudhoe Bay have contaminated the tundra and seeped into waterways. And scars from bulldozer tracks made fifty years ago can still be seen clearly today. While serving as president of the Natural Resources Defense Council, John Adams claimed that the once pristine wildlife habitat of Prudhoe Bay has become a toxic, industrial wasteland.

Oil companies planning to drill in ANWR, by contrast, claim that old, careless ways are no longer permitted in their operations. Wastes are collected and either burned or injected into deep wells. Although some animals do seem to have been displaced at Prudhoe Bay, the Central Arctic herd with 27,000 caribou is five times larger now than it was in 1978 when drilling operations began there. But in ANWR the Porcupine herd has five times as many animals crowded into one-fifth the area and may be much more sensitive to disturbance than their cousins to the west.

Native people are divided on this topic. The coastal Inupiat people, many of whom work in the oil fields, support opening the refuge to oil exploration. They hope to use their share of the oil revenues to build new schools and better housing. The Gwich'in people, who live south of the refuge, would gain nothing from oil exploitation. They worry that the migrating caribou populations on which they depend might decline as a result of drilling on critical calving grounds.

Even if ANWR contains seven billion barrels of oil, it will take at least a decade to begin to get it to market and the peak production rate will probably be about one million barrels of oil per day. Flow from the 1002 area would then meet less than 4 percent of the U.S. daily oil consumption. Improving the average fuel efficiency of all cars and light trucks in America by just 1 mile (1.6 km) per gallon would save more oil than is ever likely to be recovered from ANWR, and it would do so far faster and cheaper than extracting and transporting crude oil from the arctic. Cutting fossil fuel consumption also is vital if catastrophic global climate change is to be avoided.

In 1995, Congress passed a budget that included a provision to allow drilling in ANWR, but President Bill Clinton vetoed the bill. In 2002, the Republican-controlled House of Representatives once again passed an energy bill that authorized oil and gas exploration in ANWR. President George W. Bush strongly supported this bill and promised to sign it if given an opportunity. The Senate energy bill, however, rejected ANWR drilling by a vote of fifty-four to forty-six, and the measure was abandoned by its sponsors. Drilling in Alaska was a controversial topic of debate in the 2008 United States presidential election. As a candidate for president, Barack Obama stated that he opposed drilling for oil in ANWR, a position he still held when he entered office in 2009.

Undoubtedly, this debate will not disappear anytime soon. With a potential of billions of dollars to be made from oil and gas formations, the industry will not give up easily. However, with the uncertainties about the amount of obtainable oil and oil prices voiced by the U.S. Department of Energy in 2008, drilling may not be as profitable as many hoped. The impact of drilling on the well-being and survival of the inhabitants of the wildlife refuge, as well as the potential effects on climate change resulting from the use of fossil fuels from the ANWR combined, may be a strong enough argument to prevent Alaskan drilling. For many political conservatives, this issue has become a matter of principle. They believe that the United States has both a right and a duty to exploit the resources available to them. Environmental groups feel equally strongly about the value of wildlife, wilderness, and one of the last remaining undisturbed places in the world. Protecting this harsh but beautiful land garnered more donations and public passions than any other environmental issue in the past decade. Clearly, however ANWR ends up being managed will be a landmark in environmental history.

Complying with a federal judge's order to decide whether polar bears are a threatened species, in May 2008 the U.S. Interior Department designated the polar bear a threatened species. The bear is therefore entitled to Endangered Species Act protections. Federal scientists had previously stated that a two-thirds reduction of the current population level (about 25,000) was likely by 2050. Among the reasons cited for the bears' declining security was loss of the Arctic sea ice that is an essential part of its habitat. Global warming has been found to be the cause of the diminishing ice, but the government's May 2008 statement emphasized that U.S. climate policy would not change to protect the bears. The announcement also claimed that oil and gas exploration in the Arctic (including the ANWR, which supports an unusually high density of polar-bear land dens) does not threaten the bears and stated that the bears' new status would have no affect on such activities. In fact, no changes in policy or regulation were to be promulgated, reducing the announcement, environmental organizations and many scientists said, to a mere formality. As of 2010, the International Union for Conservation of Nature (IUCN) lists polar bears as "vulnerable."

Resources

BOOKS

Arctic National Wildlife Refuge: Issues and Legislation. New York: Nova Science, 2009.

Brown, S. C. *Arctic Wings: Birds of the Arctic National Wildlife Refuge.* Seattle, WA: Mountaineers Books, 2006.

Corn, M. L., and B. A. Roberts. *Arctic National Wildlife Refuge (ANWR) Votes and Legislative Actions, Ninety-fifth Congress through 110th Congress.* Washington, DC: Congressional Research Service, Library of Congress, 2008.

Haugen, D. M. *Should Drilling be Permitted in the Arctic National Wildlife Refuge?* Detroit, MI: Greenhaven Press, 2008.

Kaye, R. *Last Great Wilderness: The Campaign to Establish the Arctic National Wildlife Refuge.* Fairbanks: University of Alaska Press, 2006.

Kotchen, M. J., and N. E. Burger. *Should We Drill in the Arctic National Wildlife Refuge?: An Economic Perspective.* Cambridge, MA: National Bureau of Economic Research, 2007.

Lieland, B. T. *Arctic National Wildlife Refuge: A Review.* New York: Nova Science, 2009.

Standlea, D. M. *Oil, Globalization, and the War for the Arctic Refuge.* Albany: State University of New York Press, 2006.

Waterman, Jonathan. *Where Mountains Are Nameless: Passion and Politics in the Arctic Wildlife Refuge.* New York: W.W. Norton, 2007.

OTHER

International Union for Conservation of Nature and Natural Resources (IUCN). "IUCN Red List of Threatened Species, 2009." http://www.iucnredlist.org/details/22823/0 (accessed September 17, 2010).

William P. Cunningham

Arid

Arid lands are dry areas or deserts where a shortage of rainfall prevents permanent rain-fed agriculture. They are bordered by, and interact with, marginal or semiarid lands where the annual rainfall (still only 10–15 inches [25–38 cm]) allows limited agriculture and light grazing. However, in many parts of the world human mismanagement, driven by increasing populations, has degraded these areas into unusable arid lands. For example, clearing natural vegetation and overgrazing have led to soil erosion, reduced land productivity, and ultimately reduced water availability in the region. This degradation of semiarid lands to deserts, which can also occur naturally, is known as desertification.

when he entered the fifth-grade class at the Cathedral School in Uppsala. After graduating in 1876, Arrhenius enrolled at the University of Uppsala.

At Uppsala, Arrhenius concentrated on mathematics, chemistry, and physics and passed the candidate's examination for the bachelor's degree in 1878. He then began a graduate program in physics at Uppsala, but left after three years of study. He was said to be dissatisfied with his physics adviser and felt no more enthusiasm for the only adviser available in chemistry, Per Theodor Cleve. As a result, he obtained permission to do his doctoral research in absentia with the physicist Eric Edlund at the Physical Institute of the Swedish Academy of Sciences in Stockholm.

The topic Arrhenius selected for his dissertation was the electrical conductivity of solutions. In 1884 Arrhenius submitted his thesis on this topic. In the thesis he hypothesized that when salts are added to water, they break apart into charged particles now known as ions. What was then thought of as a molecule of sodium chloride, for example, would dissociate into a charged sodium atom (a sodium ion) and a charged chlorine atom (a chloride ion). The doctoral committee that heard Arrhenius's presentation in Uppsala was totally unimpressed by his ideas. Among the objections raised was the question of how electrically charged particles could exist in water. In the end, the committee granted Arrhenius his PhD, but with a score so low that he did not qualify for a university teaching position.

Convinced that he was correct, Arrhenius had his thesis printed and sent it to a number of physical chemists on the continent, including Rudolf Clausius, Jacobus van't Hoff, and Wilhelm Ostwald. These men formed the nucleus of a group of researchers working on problems that overlapped chemistry and physics, developing a new discipline that would ultimately be known as physical chemistry. From this group, Arrhenius received a much more encouraging response than he had received from his doctoral committee. In fact, Ostwald came to Uppsala in August 1884 to meet Arrhenius and to offer him a job at Ostwald's Polytechnikum in Riga, Latvia. Arrhenius was flattered by the offer and made plans to leave for Riga, Latvia, but eventually declined for two reasons. First, his father was gravely ill (he died in 1885), and second, the University of Uppsala decided at the last moment to offer him a lectureship in physical chemistry.

Arrhenius remained at Uppsala only briefly, however, as he was offered a travel grant from the Swedish Academy of Sciences in 1886. The grant allowed him to spend the next two years visiting major scientific laboratories in Europe, working with Ostwald in Riga, Friedrich Kohlrausch in Wurzburg, Ludwig Boltzmann in Graz, and van't Hoff in Amsterdam. After his return to Sweden, Arrhenius rejected an offer from the University of Giessen, Germany, in 1891, in order to take a teaching job at the Technical University in Stockholm. Four years later he was promoted to professor of physics there. In 1903, during his tenure at the Technical University, Arrhenius was awarded the Nobel Prize in chemistry for his work on the dissociation of electrolytes.

Arrhenius remained at the Technical University until 1905 when, declining an offer from the University of Berlin, he became director of the physical chemistry division of the Nobel Institute of the Swedish Academy of Sciences in Stockholm. He continued his association with the Nobel Institute until his death in Stockholm on October 2, 1927.

Although he will always be remembered best for his work on dissociation, Arrhenius was a man of diverse interests. In the first decade of the twentieth century, for example, he became especially interested in the application of physical and chemical laws to biological phenomena. In 1908, Arrhenius published a book titled *Worlds in the Making* in which he theorized about the transmission of life-forms from planet to planet in the universe by means of spores.

Arrhenius's name has also surfaced in recent years because of the work he did in the late 1890s on the greenhouse effect. He theorized that carbon dioxide in the atmosphere has the ability to trap heat radiated from the earth's surface, causing a warming of the atmosphere. Changes over time in the concentration of carbon dioxide in the atmosphere would then, he suggested, explain major climatic variations such as the glacial periods. In its broadest outlines, the Arrhenius theory sounds similar to speculations about climate changes resulting from global warming.

Among the honors accorded Arrhenius in addition to the Nobel Prize were the Davy Medal of the Royal Society (1902), the first Willard Gibbs Medal of the Chicago section of the American Chemical Society (1911), and the Faraday Medal of the British Chemical Society (1914).

A chemical equation and a crater on the moon have been named after him.

Resources

BOOKS

Arrhenius, Svante. *Worlds in the Making: The Evolution of the Universe*. New York: Harper, 1908.

Arrhenius, Svante. *Theories of Solutions*. New Haven, CT: Yale University Press, 1912.

Arrhenius, Svante. *Chemistry in Modern Life*. New York: Van Nostrand, 1925.

Arsenic

Arsenic (As) is an element having an atomic number of thirty-three and an atomic weight of 74.9216 that is listed by the U.S. Environmental Protection Agency (EPA) as a hazardous substance and as a carcinogen. *The Merck Index* states that the symptoms of acute poisoning following arsenic ingestion are irritation of the gastrointestinal tract, nausea, vomiting, and diarrhea that can progress to shock and death. According to the 2001 Update Board on Environmental Studies and Toxicology, such toxic presentations generally require weeks to months of exposure to arsenic at high doses, as much as 0.02 mg/lb/day (0.04 mg/ kg/day). Furthermore, long-term, or chronic poisoning can result in skin thickening, exfoliation, and hyperpigmentation. Continuing exposure also has been associated with development of herpes, peripheral neurological manifestations, and degeneration of the liver and kidneys. Of primary importance, however, is the association of chronic arsenic exposure with increased risk of developing high blood pressure, cardiovascular disease, and skin cancer, diseases that are increasing public health concerns.

Safe drinking water standards are regulated by the EPA in the United States. Arsenic enters drinking water supplies naturally from deposits and from agricultural and industrial processes. In February 2002, the EPA revised the former standard for acceptable levels of arsenic in drinking water. The revision of the Safe Drinking Water Act reduced the standard maximum contaminant level (MCL) from fifty parts per billion (ppb) to ten ppb of arsenic as the acceptable level for drinking water. The regulation required that all states comply with the new standard by January 2006. Sources of arsenic contamination of drinking water include erosion of natural deposits and run off of waste from glassmaking and electronics industries. Also, arsenic is used in insecticides and rodenticides, although much less widely than it once was due to new information regarding its toxicity.

Perhaps one of the greatest environmental problems from arsenic has occurred in Bangladesh, where the provision of clean drinking water led to the construction of numerous wells in the 1970s. Prior to this time, drinking water was obtained from surface waters, which were contaminated with bacteria and other disease-causing agents. The geology of the area is such that arsenic concentration in much of the ground water is at toxic levels, resulting in high concentrations in well water. It is now estimated that between 35 and 77 million Bangladeshis may be drinking water containing more arsenic than is permissible. The high variability of arsenic in water from different wells means there is no simple overall solution to what is a very important problem. Until the contaminated wells are tested and identified, the only solution is to collect rainwater for drinking or to use treated surface waters. As of late 2006, there were 40,000 diagnosed cases of arsenic poisoning in Bangladesh. Results of a ten-year study published in 2010 has indicated that upward of 77 million people have well water with at least periodic toxic levels of arsenic, and that tens of millions of people have been put at risk of early death because of their exposure. Estimates are that 90 percent of the population of Bangladesh uses potentially contaminated groundwater as their source of drinking water.

Resources

BOOKS

Henke, Kevin R. *Arsenic: Environmental Chemistry, Health Threats, and Waste Treatment*. Chichester, UK: Wiley, 2009.

Ravenscroft, Peter; H. Brammer; and K. S. Richards. *Arsenic Pollution: A Global Synthesis*. Chichester, UK: Wiley-Blackwell, 2009.

United States, and Syracuse Research Corporation. *Toxicological Profile for Arsenic*. Atlanta, GA.: U.S. Department of Health and Human Services, Public Health Service, Agency for Toxic Substances and Disease Registry, 2007.

OTHER

Centers for Disease Control and Prevention (CDC). "Arsenic." http://emergency.cdc.gov/agent/arsenic/index.asp (accessed October 26, 2010).

Russell, Louise. "Reducing Arsenic Poisoning in Bangladesh." United Nations International Children's Emergency Fund (UNICEF). September 11, 2006. http://www.unicef.org/infobycountry/bangladesh_35701.html (accessed October 26, 2010).

World Health Organization (WHO). "Arsenic in drinking water." http://www.who.int/entity/mediacentre/factsheets/fs210/en/index.html (accessed October 26, 2010).

Arsenic-treated lumber

Arsenic-treated wood is wood that has been pressure-treated with a pesticide containing inorganic arsenic (i.e., the arsenic compound does not contain carbon) to protect it from dry rot, fungi, molds, termites, and other pests. The arsenic can be a part of a chromated copper arsenate (CCA) chemical mixture consisting of three pesticidal compounds, consisting of copper, chromate, and arsenic; the most commonly used type of CCA contains 34 percent arsenic as arsenic pentoxide. Less commonly used wood preservatives containing arsenic include the pesticide ammoniacal copper arsenate (ACA), which contains ammonium, copper, and arsenic, and the pesticide ammoniacal copper zinc arsenate (ACZA), which contains ammonia, copper, zinc, and arsenic.

Inorganic arsenic in CCA has been used since the 1940s. CAA is injected into wood through a process that uses high pressure to saturate wood products with the chemicals. Preserved wood products, such as utility poles, highway noise barriers, signposts, retaining walls, boat bulkheads, dock pilings, and wood decking, are used in the construction, railroad, and utilities industries. Historically CCA has been the principal chemical used to treat wood for outdoor uses around a home. Residential uses of arsenic-treated woods include play structures, decks, picnic tables, gazebos, landscaping timbers, residential fencing, patios, and walkways or boardwalks. After wood is pressure-treated with arsenic compounds, residues of the preservatives can remain on the surface. The initial residues wash off, but as the wood weathers, new layers of treated wood and pesticides are exposed. Arsenic is also present in paints that are used to cover the cut ends of treated wood. Freshly arsenic-treated wood, if not coated, has a greenish tint, which fades over time.

Arsenic is acutely toxic. Contact with arsenic may cause irritation of the stomach, intestines, eyes, nose, and skin; blood vessel damage; and reduced nerve function. In addition, according to the National Academy of Sciences and the National Research Council, exposure to arsenic increases the risk of human lung, bladder, and skin cancer over a lifetime and is suspected as a cause of kidney, prostate, and nasal passage cancer. The National Academy of Sciences and the Science Advisory Board of the United States Environmental Protection Agency has also reported that arsenic may cause high blood pressure, cardiovascular disease, and diabetes.

Arsenic may enter the body through the skin, by ingestion, or by inhalation. Ingestion occurs most frequently when contaminated hands are put in the mouth, or when contaminated hands are used for eating food. Repeated exposure will increase risks of adverse health effects. Splinters of wood piercing the skin may also be a means of entry of arsenic into the body, but the importance of this route has not been well-studied.

The use of most pesticides containing arsenic had been banned by the United States Environmental Protection Agency, but in 1985 CCA was designated as a restricted-use pesticide. However, CCA-treated wood products were not regulated like the pesticides the wood products contained because it was assumed that the pesticides would stay in the wood. Unfortunately adequate information was not available on whether arsenic is fixed in the wood permanently and whether the wood product is safe. It is known that fixation of the chemicals in the wood matrix is enhanced if the treated wood is wrapped in tarps and stored for a sufficient length of time, which varies with the temperature. For example, to achieve fixation, the wood must be stored for thirty-six days in 50°F (21°C) weather and for twelve days in 70°F (10°C) weather. At the freezing temperature point, no fixation occurs.

However, research by the Florida Center for Solid and Hazardous Waste Management and the Connecticut Agricultural Experiment Station suggests that leaching of arsenic into the soils or into surface water under CCA-treated structures occurs at greater-than-safe levels. Florida studies showed that arsenic was found in soils underneath the eight pressure-treated decks that were investigated. Of seventy-three samples taken, sixty-one samples had levels of arsenic higher than the Florida cleanup levels for industrial sites. One sample had arsenic levels 300 times higher than the state's mandated cleanup level.

Based on the potential for adverse health effects, the United States Protection Agency announced in February 2002 that the use of consumer products made with CCA, including play structures, decks, picnic tables, landscaping timbers, patios, walkways, and boardwalks, will be phased out voluntarily on the wood treating industry on December 31, 2003. Wood treated prior to December 31, 2003, can still be used, and already-built structures using CCA-treated wood and the soil surrounding the structures will not have to be replaced or removed. As of August 2001, Switzerland, Vietnam, and Indonesia had already banned the use of CCA- treated wood, while Germany, Sweden, Denmark, Japan, Australia, and New Zealand restricted its use.

The United States Environmental Protection Agency has issued cautionary recommendations to reduce consumer exposure to arsenic-contaminated wood. What follows are some of those recommendations. Treated wood should not be used where the preservative can become a component of water, food, or animal feed. Treated wood should not be used where it may come into direct or indirect contact with drinking water, except for uses where there is incidental contact, such as docks and bridges. Treated wood should not be used for cutting boards or countertops, and food should not be placed directly on treated wood. Children and others should wash their hands after playing or working outdoors, as arsenic may be swallowed from hand-to-mouth activity. Children and pets should be kept out from under-deck areas. Edible plants should not be grown near treated decks or other structures constructed of treated wood. A plastic liner should be placed on the inside of arsenic-treated boards used to frame garden beds. Sawdust from treated wood should not be used as animal bedding, and treated boards should not be used to construct structures for storage of animal feed or human food. Treated wood should not be used in the construction of the parts of beehives that may come into contact with honey.

Only wood that is visibly clean and free of surface residues (e.g., wood that does not show signs of crystallization or resin on its surface) should be used for patios, decks, and walkways. Consumers working with arsenic-treated wood should reduce their exposure by only sawing, sanding, and machining treated wood outdoors and by wearing a dust mask, goggles, and gloves when performing these types of activities. They should also wash all exposed areas of their bodies thoroughly with soap and water before eating, drinking, or using tobacco products. Work clothes should be washed separately from other household clothing before being worn again. Sawdust, scraps, and other construction debris from arsenic-treated wood should not be composted or used as mulch, but caught on tarps for disposal off-site. Pressure-treated wood has an exemption from hazardous waste regulations and can be disposed of in municipal landfills without a permit. Pressure-treated wood should not be burned in open fires or in stoves, fireplaces, or residential boilers, as both the smoke and the ash may contain toxic chemicals. Treated wood from commercial or industrial uses, such as from construction sites, may only be burned in commercial or industrial incinerators or boilers in accordance with state and federal regulations.

The United States Consumer Product Safety Commission has recommended that playground equipment be painted or sealed with a double coat of a penetrating, nontoxic and non slippery sealant (e.g., oil-based or semitransparent stains) every one or two years, depending upon wear and weathering. However, available data are limited on whether sealants will reduce the migration of wood preservative chemicals from CCA-treated wood. Other potential treatments include the use of polyurethane or other hard lacquer, spar varnish, or paint. However, the use of film-forming or nonpenetrating stains are not recommended, as subsequent flaking and peeling may later increase exposure to preservatives in the wood. Structures constructed of arsenic-treated wood should be inspected regularly for wood decay or structural weakness. If the treated wood cracks expose the interior of the wood to be still structurally sound, the affected area should be covered with a double coat of a sealant.

Alternatives to the use of arsenic-treated lumber include the use of painted metal; stones or brick; recycled plastic lumber, which may be all plastic or a composite of plastic and wood fiber; lumber that has been treated with the alternative pesticide ACQ (alkaline copper quaternary), which does not contain arsenic; or untreated rot-resistant wood such as cedar and redwood.

Arsenic compounds are referred to as arsenicals.

Resources

BOOKS

Cooper, Chris. *Arsenic*. New York: Benchmark Books, 2006.

OTHER

Centers for Disease Control and Prevention (CDC). "Arsenic." http://emergency.cdc.gov/agent/arsenic/index.asp (accessed November 11, 2010).

United States Environmental Protection Agency (EPA). "Pollutants/Toxics: Multimedia Pollutants: Arsenic." http://www.epa.gov/ebtpages/pollmultimediapollarsenic.html (accessed November 11, 2010).

United States Environmental Protection Agency (EPA). "Pollutants/Toxics: Soil Contaminants: Arsenic." http://www.epa.gov/ebtpages/pollsoilcontaminanarsenic.html (accessed November 11, 2010).

United States Environmental Protection Agency (EPA). "Water: Water Pollutants: Arsenic." http://www.epa.gov/ebtpages/watewaterpollutantarsenic.html (accessed November 11, 2010).

World Health Organization (WHO). "Arsenic in drinking water." http://www.who.int/entity/mediacentre/factsheets/fs210/en/index.html (accessed November 11, 2010).

World Health Organization (WHO). "Arsenic." http://www.who.int/topics/arsenic/en (accessed November 11, 2010).

Judith L. Sims

Artesian well

A well that discharges water held in a confined aquifer, artesian wells are usually thought of as wells whose water is free-flowing at the land surface.

However, there are many other natural systems that can result in such wells. The classic concept of artesian flow involves a basin with a water-intake area above the level of groundwater discharge. These systems can include stabilized sand dunes; fractured zones along bedrock faults; horizontally-layered rock formations; and the intermixing of permeable and impermeable materials along glacial margins.

Asbestos

Asbestos is a fibrous mineral silicate (a compound containing a negatively-charged form of silicon) that occurs in numerous forms.

A form called amosite [$Fe_5Mg_2(Si_8O_{22})(OH)_2$] has been shown to cause mesothelioma, squamous cell carcinoma, and adenocarcinoma of the lung after inhalation over a long period of time. This substance has been listed by the Environmental Protection Agency (EPA). Lung cancer is most likely to occur in those individuals who are exposed to high airborne doses of asbestos and who also smoke. The potential of asbestos to cause disease appears to be related to the length-to-diameter ratio of particles. Particles less than 2 micrometers in length are the most hazardous.

Asbestos exposure causes thickening and calcified hardened areas (plaques) on the lining of the chest cavity. When inhaled, it forms "asbestos bodies" in the lungs, yellowish-brown particles created by reactions between the fibers and lung tissue. This disease was first described by W. E. Cooke in 1921 and given the name asbestosis. Asbestosis does not develop immediately after inhalation of asbestos. Rather, there is usually a time lag. This is referred to as latency. The latency period is generally longer than twenty years—the heavier the exposure, the more likely and the earlier is the onset of the disease.

In 1935 an association between asbestos and cancer was noted by Kenneth M. Lynch. However, it was not until 1960 that Christopher Wagner demonstrated a particularly lethal association between cancer of the lining of the lungs and asbestos. By 1973 the National Institute of Occupational Safety and Health recommended adoption of an occupational standard of two asbestos fibers per cubic centimeter of air.

During this time, many cases of lung cancer began to surface, especially among asbestos workers who had been employed in shipbuilding during World War II. The company most impacted by lawsuits was the Manville Corporation, which had been the supplier to the United States government. Manville Corporation eventually sought Title 11 Federal Bankruptcy protection as a result of these lawsuits.

The Reserve Mining Company, a taconite (iron ore) mining operation in Silver Bay, Minnesota, was involved in litigation over the dumping of tailings (wastes) from their operations into Lake Superior. These tailings contained amosite asbestos particles that appeared to migrate into the Duluth, Minnesota, water supply. In an extended lawsuit, the Reserve Mining Company was ordered to shut down their operations. One controversial question raised during the legal action was whether cancer could be caused by drinking water containing asbestos fibers. In other cancer cases related to asbestos, the asbestos was inhaled rather than ingested. Federal courts held that there is reasonable cause to believe that asbestos in food and drink is dangerous—even in small quantities—and ordered Reserve Mining to stop dumping tailings in the lake.

Other examples of communities affected by asbestos include Baryulgil, New South Wales, and Libby, Montana. The latter was the site of a $60 million settlement in 2008 that involved homeowners and business people who had used products containing an ore called vermiculite that was mined at Libby. The vermiculite was contaminated with asbestos.

A significant industry has developed for removing asbestos materials from private and public buildings as a result of the tight standards placed on asbestos concentrations by the Occupational Safety and Health Administration. At one time, many steel construction materials, especially horizontal beams, were sprayed with asbestos to enhance their resistance to fires. Wherever these materials are now exposed to ambient air in buildings, they have the potential to create a hazardous condition. The asbestos must either be covered or removed, and another insulating material substituted. Removal, however, causes its own problems, releasing high concentrations of fibers into the air. Many experts regard covering asbestos in place with a plastic covering to be the best option in most cases. What was once considered a lifesaving material for its flame retardancy, now has become a hazardous substance that must be removed and sequestered in sites specially certified for holding asbestos building materials.

In May 2008, researchers reported that nanotubes—microscopic, artificial, and needle-like tubes, usually composed of carbon and used in many new materials because of their strength and conductivity—can have affects on lung tissue similar to those of asbestos. The research indicated that the nanotubes are not dangerous if sealed inside plastics or other materials; like asbestos fibers, they must be inhaled before they can do harm. Nanotubes are therefore mostly a threat to those handling them in manufacturing, not to the general public. Being forewarned of likely health hazards from nanotubes, manufacturers can take steps to protect workers. Nanoparticles (particles smaller than 100 nanometers across) have also been found to be a cancer hazard, causing deoxyribonucleic acid (DNA) damage similar to that caused by ionizing radiation.

Resources

BOOKS

Craighead, John E. *Asbestos and its Diseases.* New York: Oxford University Press 2008.

Newman, Michael C. *Fundamentals of Ecotoxicology, Third Edition.* Boca Raton, FL: CRC Press, 2009.

Webster, Peter. *White Dust Black Death: The Tragedy of Asbestos Mining at Baryulgil.* New York: Trafford Publishing, 2006.

Malcolm T. Hepworth

Asbestos removal

Asbestos is a naturally occurring mineral, used by humans since ancient times but not extensively until the 1940s. After World War II and for the next thirty years, it was widely used as a construction material in schools and other public buildings. The United States Environmental Protection Agency (EPA) estimates that asbestos-containing materials were installed in most of the primary and secondary schools, as well as in most public and commercial buildings, in the nation. It is estimated that 27 million Americans had significant occupational exposure to asbestos between 1940 and 1980. Asbestos has been popular because it is readily available, low in cost, and has very useful properties. It does not burn, conducts heat and electricity poorly, strengthens concrete products into which it is incorporated, and is resistant to chemical corrosion. It has been used in building materials as a thermal and electrical insulator and has been sprayed on steel beams in buildings for protection from heat in fires. Asbestos has also been used as an acoustical plaster. In 1984, an EPA survey found that approximately

Asbestos removal at the Jussieu campus of the Univeritsy of Paris. *(Thomas Jouanneau/Sygma/Corbis)*

for treatment. That is why asbestos could be referred to as a silent killer.

Exposure to asbestos not only affects factory personnel working with asbestos, but individuals who live in areas surrounding asbestos emissions. In addition to asbestosis, exposure may result in a rare form of cancer called mesothelioma, which affects the lining of the lungs or stomach. Approximately 5-10 percent of all workers employed long-term in asbestos manufacturing or mining operations die of mesothelioma.

Asbestosis is characterized by dyspnea (labored breathing) on exertion, a nonproductive cough, hypoxemia (insufficient oxygenation of the blood), and decreased lung volume. Progression of the disease may lead to respiratory failure and cardiac complications. Asbestos workers who smoke have a marked increase in the risk for developing bronchogenic cancer.

Increased risk is not confined to the individual alone, but there is an extended risk to workers' families, as asbestos dust is carried on clothes and in hair. Consequently, in the fall of 1986, President Ronald Reagan signed into law the Asbestos Hazard Emergency Response Act, requiring that all primary and secondary schools be inspected for the presence of asbestos; if such materials are found, the school district must file and carry out an asbestos abatement plan. The Environmental Protection Agency was charged with the oversight of the project.

In May 2008, researchers reported that nanotubes—microscopic, artificial, and needle-like tubes, usually composed of carbon and used in many new materials because of their strength and conductivity—can have affects on lung tissue similar to those of asbestos. The researchers, whose results were published in the journal *Nature Nanotechnology*, emphasized that nanotubes are not dangerous if sealed inside plastics or other materials; like asbestos fibers, they must be inhaled before they can do harm. Nanotubes are therefore mostly a threat to those handling them in manufacturing, not to the general public.

Resources

BOOKS

Asbestos: Selected Cancers. Washington, DC: National Academies Press, 2006.

Bartrip, P. W. J. *Beyond the Factory Gates: Asbestos and Health in Twentieth-Century America.* London: Continuum, 2006.

Craighead, John E., and A. R. Gibbs. *Asbestos and Its Diseases.* New York: Oxford University Press, 2008.

Schneider, Andrew, and David McCumber. *An Air That Kills: How the Asbestos Poisoning of Libby, Montana, Uncovered a National Scandal.* New York: Putnam, 2004.

PERIODICALS

"Asbestos in Canada–Hazardous Hypocrisy." *The Economist* 389, *8603* (2008): 65.

LaDou, J., et al. "The Case for a Global Ban on Asbestos." *Environmental Health Perspectives* 118, 7 (2010): 897–901.

R, J. "Yes, It's Asbestos." *Science News.* 171, 2 (2007): 29.

Wagner, G.R. "The Fallout from Asbestos." *Lancet* 369, *9566* (2007): 973–974.

OTHER

U.S. Centers for Disease Control and Prevention: Agency for Toxic Substances and Disease Registry. "Asbestos." http://www.atsdr.cdc.gov/asbestos/asbestos/health_effects/ (accessed November 8, 2010).

Brian R. Barthel

Ash, fly *see* **Fly ash.**

Asian longhorn beetle

The Asian longhorn beetle (*Anoplophora glabripennis*) is classified as a pest in the United States and its homeland of China. According to the United States Department of Agriculture (USDA), the beetles have the potential to destroy millions of hardwood trees.

Longhorn beetles live for one year. They are 1 to 1.5 inches (2.5–3.8 cm) long, and their backs are black with white spots. The beetles' long antennae are black and white and extend up to 1 inch (2.5 cm) beyond the length of their bodies.

Female beetles chew into tree bark and lay from thirty-five to ninety eggs. After hatching, larvae tunnel into the tree, staying close to the sapwood, and eat tree tissue throughout the fall and winter. After pupating, adult beetles leave the tree through drill-like holes. Beetles feed on tree leaves and young bark for two to three days and then mate. The cycle of infestation continues as more females lay eggs.

Beetle activity can kill trees such as maples, birch, horse chestnut, poplar, willow, elm, ash, and black locust. According to the USDA, beetles came to the United States in wooden packaging material and pallets from China. The first beetles were discovered in 1996 in Brooklyn, New York. Other infestations were found in other areas of the state. Subsequent infestations of beetles, differing in origin from that of the

Stag beetle mating. *(iStockphoto.com / Lingbeek)*

Brooklyn beetles, were discovered in Chicago and New Jersey.

Liz Swain

Asian (Pacific) shore crab

The Asian (Pacific) shore crab (*Hemigrapsus sanguineus*) is a crustacean also known as the Japanese shore crab or Pacific shore crab. It was probably brought from Asia to the United States in ballast water. When a ship's hold is empty, it is filled with ballast water to stabilize the vessel.

The first Asian shore crab was seen in 1988 in Cape May, New Jersey. By 2001, Pacific shore crabs colonized the East Coast, with populations located from New Hampshire to North Carolina. Crabs live in the sub-tidal zone where low-tide water is several feet deep.

The Asian crab is 2 to 3 inches (5–7.7 cm) wide. Their shell color is pink, green, brown, or purple. There are three spines on each side of the shell. The crab has two claws and bands of light and dark color on its six legs.

A female produces 56,000 eggs per clutch. Asian Pacific crabs have three or four clutches per year. Other crabs produce one or two clutches annually.

At the start of the twenty-first century, there was concern about the possible relationship between the rapidly growing Asian crab population and the decline in native marine populations, such as the lobster population in Long Island Sound.

Liz Swain

Asiatic black bear

The Asiatic black bear or moon bear (*Ursus thibetanus*) ranges through southern and eastern Asia, from Afghanistan and Pakistan through the Himalayas to Indochina, including most of China, Manchuria, Siberia, Korea, Japan, and Taiwan. The usual habitat of this bear is angiosperm forests, mixed hardwood-conifer forests, and brushy areas, occurring in mountainous areas up to the tree line, which can be as high as 13,000 feet (4,000 m) in parts of the Himalayan range.

The Asiatic black bear has an adult body length of 4.3 to 6.5 feet (1.3–2.0 m), a tail of 2.5 to 3.5 feet (75–105 cm), a height at the shoulder of 2.6 to 3.3 feet (80–100 cm), and a body weight of 110 to 440 pounds (50–200 kg). Male animals are considerably larger than females. Their weight is greatest in late summer and autumn, when the animals are fat in preparation for winter. Their fur is most commonly black, with white patches on the chin and a crescent- or Y-shaped patch on their chest. The base color of some individuals is brownish rather than black.

Female Asiatic black bears, or sows, usually give birth to two small cubs that are 12 to 14 ounces (350–400 g) in a winter den, although the litter can range from one to three. The gestation period is six to eight months. After leaving their birth den, the cubs follow their mother closely for about two and one-half years, after which they are able to live independently. Asiatic black bears become sexually mature at an age of three to four years, and they can live for as long as thirty-three years.

stigmata—wounds resembling those Jesus suffered on the cross. He died in 1226 and was canonized in 1228.

Francis's unconventional life has made him attractive to many who have questioned the direction of their own societies. Francis loved nature. In his book "Canticle of the Creatures," he praises God for the gifts, among others, of "Brother Sun," "Sister Moon," and "Mother Earth, Who nourishes and watches us." But this is not to say that Francis was a pantheist or nature worshipper. He loved nature not as a whole, but as the assembly of God's creations. As G.K. Chesterton remarked, Francis "did not want to see the wood for the trees. He wanted to see each tree as a separate and almost a sacred thing, being a child of God and therefore a brother or sister of man."

For White, Francis was "the greatest radical in Christian history since Christ" because he departed from the traditional Christian view in which humanity stands over and against the rest of nature—a view, White charges, that is largely responsible for modern ecological crises. Against this view, Francis "tried to substitute the idea of the equality of all creatures, including man, for the idea of man's limitless rule of creation."

Resources

BOOKS

Sorrell, Roger D. *St. Francis of Assisi and Nature: Tradition and Innovation in Western Christian Attitudes Toward the Environment.* New York: Oxford University Press, 2009.

Hart, John. *Sacramental Commons: Christian Ecological Ethics.* Lanham, MD: Rowman & Littlefield, 2006.

Richard K. Dagger

▌Asthma

Asthma is a condition characterized by unpredictable and disabling shortness of breath, wheezing, coughing, and a feeling of tightness in the chest. It features episodic attacks of bronchospasm (prolonged contractions of the bronchial smooth muscle), and is a complex disorder involving biochemical, autonomic, immunologic, infectious, endocrine, and psychological factors to varying degrees in different individuals. Asthma affects about 300 million people worldwide.

Asthma occurs in families, suggesting that there is a genetic predisposition for the disorder, although the exact mode of genetic transmission remains

unclear. The environment appears to play an important role in the expression of the disorder. For example, asthma can develop when *predisposed* individuals become infected with viruses or are exposed to allergens or pollutants. On occasion, foods or drugs may precipitate an attack. Psychological factors have been investigated but have yet to be identified with any specificity.

The severity of asthma attacks varies among individuals, over time, and with the degree of exposure to the triggering factors. Approximately one-half of all cases of asthma develop during childhood. Another one-third of cases develop before the age of forty. There are two basic types of asthma—allergic or intrinsic asthma, and non allergic or extrinsic asthma. Allergic asthma is triggered by allergens, while intrinsic asthma is not. Allergic asthma is classified as Type I or Type II, depending on the type of allergic response involved. Type I allergic asthma is the classic allergic asthma, which is common in children and young adults who are highly sensitive to dust and pollen, and is often seasonal in nature. It is characterized by sudden, brief, intermittent attacks of bronchospasms that readily respond to bronchodilators (inhaled medication that opens swollen airways). Type II allergic asthma, or allergic alviolitis, develops in adults under age thirty-five after long exposure to irritants. Attacks are more prolonged than Type I and are more inflammatory. Fever and infiltrates, which are visible on chest X-rays, often accompany bronchospasm.

Non allergic asthma has no known immunologic cause and no known seasonal variation. It usually occurs in adults over the age of thirty-five, many of whom are sensitive to aspirin and have nasal polyps. Attacks are often severe and do not respond well to bronchodilators.

A third type of asthma that occurs in otherwise normal individuals is called exercise-induced asthma. Individuals with exercise-induced asthma experience mild to severe bronchospasms during or after moderate to severe exertion. They have no other occurrences of bronchospasms when not involved in physical exertion. Although the cause of this type of asthma has not been established, it is readily controlled by using a bronchodilator prior to beginning exercise.

Resources

BOOKS

Bellenir, Karen. *Asthma Sourcebook: Basic Consumer Health Information about the Causes, Symptoms, Diagnosis, and Treatment of Asthma in Infants, Children, Teenagers, and Adults.* Holmes, PA: Omnigraphics, 2006.

Berger, William, and Jackie Joyner-Kersee. *Asthma for Dummies.* New York: For Dummies, 2004.

Wolf, Raoul L. *Essential Pediatric Allergy, Asthma, and Immunology*. New York: McGraw-Hill Medical Publishing Division, 2004.

PERIODICALS

Silverstein, Alvin; Virginia Silverstein; and Laura Silverstein Nunn. *The Asthma Update*. Berkeley Heights, NJ: Enslow Publishers, 2006.

OTHER

Centers for Disease Control and Prevention (CDC). "Asthma and Allergies." http://www.cdc.gov/health/asthma.htm (accessed November 6, 2010).
Mayo Clinic. "Asthma." http://www.mayoclinic.com/health/asthma/DS00021 (accessed November 6, 2010).
National Institutes of Health (NIH). "Asthma." http://health.nih.gov/topic/Asthma (accessed November 6, 2010).
United States Environmental Protection Agency (EPA). "Human Health: Health Effects: Asthma." http://www.epa.gov/ebtpages/humahealtheffectsasthma.html (accessed November 6, 2010).
World Health Organization (WHO). "Asthma." http://www.who.int/entity/mediacentre/factsheets/fs307/en/index.html (accessed November 6, 2010).

Brian R. Barthel

Aswan High Dam

A heroic symbol and an environmental liability, this dam on the Nile River was built as a central part of modern Egypt's nationalist efforts toward modernization and industrial growth. Begun in 1960 and completed by 1970, the High Dam lies near the town of Aswan, which sits at the Nile's first cataract, or waterfall, 200 river miles (322 km) from Egypt's southern border. The dam generates urban and industrial power, controls the Nile's annual flooding, ensures year-round, reliable irrigation, and has boosted the country's economic development as its population climbed from 20 million in 1947 to 58 million in 1990. The Aswan High Dam is one of a generation of huge dams built on the world's major rivers between 1930 and 1970 as both functional and symbolic monuments to progress and development. It also represents the hazards of large-scale efforts to control nature. Altered flooding, irrigation, and sediment deposition patterns have led to the displacement of villagers and farmers, a costly dependence on imported fertilizer, water quality problems and health hazards, and erosion of the Nile Delta.

Aswan attracted international attention in 1956 when planners pointed out that flooding behind the new dam would drown a number of ancient Egyptian tombs and monuments. A worldwide plea went out for assistance in saving the 4,000-year-old monuments, including the tombs and colossi of Abu Simbel and the temple at Philae. The United Nations Educational, Scientific, and Cultural Organization (UNESCO) headed the epic project, and over the next several years the monuments were cut into pieces, moved to higher ground, and reassembled above the waterline.

The High Dam, built with international technical assistance and substantial funding from the former Soviet Union, was the second dam to be built near Aswan. English and Egyptian engineers built the first Aswan dam between 1898 and 1902. Justification for the first dam was much the same as that for the second, larger dam, namely flood control and irrigation. Under natural conditions the Nile experienced annual floods of tremendous volume. Fed by summer rains on the Ethiopian Plateau, the Nile's floods could reach sixteen times normal low season flow. These floods carried terrific silt loads, which became a rich fertilizer when flood waters overtopped the river's natural banks and sediments settled in the lower surrounding fields. This annual soaking and fertilizing kept Egypt's agriculture prosperous for thousands of years. But annual floods could be wildly inconsistent. Unusually high peaks could drown villages. Lower than usual floods might not provide enough water for crops. The dams at Aswan were designed to eliminate the threat of high water and ensure a gradual release of irrigation water through the year.

Flood control and regulation of irrigation water supplies became especially important with the introduction of commercial cotton production. Cotton was introduced to Egypt by 1835, and within fifty years it became one of the country's primary economic assets. Cotton required dependable water supplies; but with reliable irrigation, up to three crops could be raised in a year. Full-year commercial cropping was an important economic innovation, vastly different from traditional seasonal agriculture. By holding back most of the Nile's annual flood, the first dam at Aswan captured 65.4 billion cubic yards (50 billion cubic m) of water each year. Irrigation canals distributed this water gradually, supplying a much greater acreage for a much longer period than did natural flood irrigation and small, village-built waterworks. But the original Aswan dam allowed 39.2 billion cubic yards (30 billion cubic m) of annual flood waters to escape into the Mediterranean. As Egypt's population, agribusiness, and development needs grew, planners decided this was a loss that the country could not afford.

Aswan High Dam, Aswan, Egypt. *(Jon Arnold Images Ltd./Alamy)*

The High Dam at Aswan was proposed in 1954 to capture escaping floods and to store enough water for long-term drought, something Egypt had seen repeatedly in history. Three times as high and nearly twice as long as the original dam, the High Dam increased the reservoir's storage capacity from an original 6.5 billion cubic yards (5 billion cubic m) to 205 billion cubic yards (157 billion cubic m). The new dam lies 4.3 miles (7 km) upstream of the previous dam, stretches 2.2 miles (3.6 km) across the Nile, and is nearly 0.6 miles (1 km) wide at the base. Because the dam sits on sandstone, gravel, and comparatively soft sediments, an impermeable screen of concrete was injected 590 feet (180 m) into the rock, down to a buried layer of granite. In addition to increased storage and flood control, the new project incorporates a hydropower generator. The dam's turbines, with a capacity of 8 billion kilowatt hours per year, doubled Egypt's electricity supply when they began operation in 1970.

Lake Nasser, the reservoir behind the High Dam, now stretches 311 miles (500 km) south to the Dal cataract in Sudan. Averaging 6.2 miles (10 km) wide, this reservoir holds the Nile's water at 558 feet (170 m) above sea level. Because this reservoir lies in one of the world's hottest and driest regions, planners anticipated evaporation at the rate of 13 billion cubic yards (10 billion cubic m) per year. Dam engineers also planned for siltation, since the dam would trap nearly all the sediments previously deposited on downstream flood plains. Expecting that Lake Nasser would lose about 5 percent of its volume to siltation in 100 years, designers anticipated a volume loss of 39.2 billion cubic yards (30 billion cubic m) over the course of five centuries.

An ambitious project, the Aswan High Dam has not turned out exactly according to sanguine projections. Actual evaporation rates today stand at approximately 19 billion cubic yards (15 billion cubic m) per year, or one-half of the water gained by constructing the new dam. Another 1.3 to 2.6 cubic yards (1–2 billion cubic m) are lost each year through seepage from unlined irrigation canals. Siltation is also more severe than expected. With 60 to 180 million tons (54-162 metric tons) of silt deposited in the lake each year, more recent projections suggest that the reservoir will be completely filled in 300 years. The dam's effectiveness in flood control, water storage, and power generation will decrease much sooner. With the river's silt load trapped behind the dam, Egyptian farmers have had to turn to chemical fertilizer, much of it imported at substantial cost. While this strains commercial cash crop producers, a need for fertilizer application seriously troubles local food growers who have less financial backing than agribusiness ventures.

A further unplanned consequence of silt storage is the gradual disappearance of the Nile Delta. The Delta has been a site of urban and agricultural settlement for millennia, and a strong local fishing industry exploited the large schools of sardines that gathered near the river's outlets to feed. Longshore currents sweep across the Delta, but annual sediment deposits counteracted the erosive effect of these currents and gradually extended the delta's area. Now that the Nile's sediment load is negligible, coastal erosion is causing the delta to shrink. The sardine fishery has collapsed because river discharge and nutrient loads have been so severely depleted. Decreased freshwater flow has also cut off water supply to a string of freshwater lakes and underground aquifers near the coast. Saltwater infiltration and soil salinization have become serious threats.

Water quality in the river and in Lake Nasser have suffered as well. The warm, still waters of the reservoir support increasing concentrations of phytoplankton, or floating water plants. These plants, most of them microscopic, clog water intakes in the dam and decrease water quality downstream. Salt concentrations in the river are also increasing as a higher percentage of the river's water evaporates from the reservoir.

While the High Dam has improved the quality of life for many urban Egyptians, it has brought hardship to much of Egypt's rural population. Most notably, severe health risks have developed in and around irrigation canal networks. These canals used to flow only during and after flood season; once the floods dissipated the canals would again become dry. Now that they are year-round, irrigation canals have become home to a common tropical snail that carries schistosomiasis, a debilitating disease that severely weakens its victims. Malaria may also be spreading, as moist mosquito breeding spots have multiplied. Farm fields, no longer washed clean each year, are showing high salt concentrations in the soil. Perhaps most tragic is the displacement of villagers, especially Nubians, who are ethnically distinct from their northern Egyptian neighbors and who lost most of their villages to Lake Nasser. Resettled in apartment blocks and forced to find work in the cities, Nubians are losing their traditional culture.

The Aswan High Dam was built as a symbol of national strength and modernity. By increasing industrial and agricultural output the Aswan High Dam helps bolster the nation's economic growth and raise the standard of living in the Nile River Valley. Lake Nasser now supports a fishing industry that partially replaces jobs lost in the delta fishery, and tourists contribute to the national income when they hire cruise boats on the lake. The country's expanded population needs a great deal of water. The Egyptian population is at 81 million, as of 2010, and projected to rise to 90 million by the year 2035.

Resources

BOOKS

Hile, Kevin. *Dams and Levees. Our environment.* Detroit, MI: KidHaven Press, 2007.
International Commission on Large Dams. *Dams and the World's Water: An Educational Book that Explains How Dams Help to Manage the World's Water.* Paris: CIGB ICOLD, 2007.

OTHER

National Geographic Society. "Egypt." http://travel.nationalgeographic.com/places/countries/country_egypt.html (accessed October 2, 2010).
United States Central Intelligence Agency (CIA). "Egypt." *CIA World Factbook.* https://www.cia.gov/library/publications/the-world-factbook/geos/eg.html (accessed October 2, 2010).

Mary Ann Cunningham

Atmosphere

The atmosphere is the envelope of gas surrounding Earth, which is for the most part permanently bound to Earth by its gravitational field. It is composed primarily of nitrogen (N) (78% by volume) and oxygen (O) (21% by volume). There are also small amounts of argon (Ar), carbon dioxide (CO_2), and water vapor, as well as trace amounts of other gases and particulate matter.

Trace components of the atmosphere can be very important in atmospheric functions. Ozone (O_3) accounts on average for two parts per million (ppm) of the atmosphere but is more concentrated in the stratosphere. This stratospheric ozone is critical to the existence of terrestrial life on the planet as it serves to block the sun's harmful ultraviolet radiation from reaching Earth's surface. Particulate matter is another important trace component. Aerosol loading of the atmosphere, as well as changes in the carbon dioxide composition of the atmosphere, can be responsible for significant changes in climate. Greenhouse gases such as carbon dioxide can absorb and trap radiation in the lower atmosphere, thus causing a heating of Earth.

The composition of the atmosphere changes over time and space. Outside of water vapor, which can vary in the lower atmosphere from trace amounts in desert regions to 4 percent over large bodies of water, the concentrations of the major components vary little in time. Approximately 50 miles (80 km) above sea level, however, the relative proportions of component gases change significantly. As a result, the atmosphere is divided into two compositional layers: the homosphere and the heterosphere. Below 50 miles altitude is the homosphere, and above 50 miles altitude is the heterosphere. In the homosphere, the ratio of the gaseous components remains fairly constant with no gross change in the proportions of components.

The atmosphere is also divided according to its thermal behavior. Based on this criteria, the atmosphere can be divided into several layers. The lowest layer is the *troposphere*, which extends about 6 to 12 miles (10-20 km). The temperature decreases with height in the troposphere, which is the domain of weather. The troposphere contains large amounts of greenhouse gases such as carbon dioxide and ozone, which contribute to climate change. The tropopause is the boundary, or transition zone, between the troposphere and the stratosphere. The stratosphere is the stable layer above the troposphere and extends from about 7 to 31 miles (11-31 km) altitude. The major components of the stratosphere are nitrogen, oxygen, and ozone. This layer is important because it contains much of the ozone, which filters ultraviolet light out of the incident solar radiation. The next layer is the *mesosphere*, which is much less stable, exhibits decreasing temperature with increasing height, and extends to about 53 to 59 miles (85-95 km) altitude. Finally there is the *thermosphere*, which is another very stable zone, but its contents are barely dense enough to cause a visible degree of solar radiation scattering. The thermosphere begins at about 62 miles (100 km) altitude and is characterized by increasing temperature with increasing height.

Resources

BOOKS

Kaufmann, Robert, and Cutler Cleveland. *Environmental Science*. New York: McGraw-Hill, 2007.

Keeling, R. F., ed. *The Atmosphere*. Boston: Elsevier, 2006.

OTHER

United States Environmental Protection Agency (EPA). "Air: Atmosphere." http://www.epa.gov/ebtpages/air atmosphere.html (accessed November 9, 2010).

Robert B. Giorgis Jr.

Atmospheric (air) pollutants

Atmospheric pollutants are substances that are present in the air to a degree that is harmful to living organisms or to materials exposed to the air. Common air pollutants include smoke, smog, and gases such as carbon monoxide (CO), nitrogen and sulfur oxides (NO_x, SO_x), and hydrocarbon fumes. While gaseous pollutants are generally invisible, solid or liquid pollutants in smoke and smog are easily seen. One particularly noxious form of air pollution occurs when oxides of sulfur and nitrogen react with water and oxygen in the atmosphere to produce sulfuric and nitric acid. When the acids are brought to Earth in the form of acid rain, damage is inflicted on lakes, rivers, vegetation, buildings, and other objects. Because sulfur and nitrogen oxides can be carried for long distances in the atmosphere before they are deposited with precipitation, damage may occur far from pollution sources. Large emissions from fossil fuel combustion of air pollutants such as carbon dioxide (CO_2) in the lower atmosphere can absorb and trap heat, thus resulting in increased temperatures on Earth. These air pollutants are referred to as greenhouse gases because they contribute to the greenhouse effect, which is causing global warming of the earth commonly referred to as climate change.

Smoke is an ancient environmental pollutant, but increased use of fossil fuels in recent centuries has resulted in increased accumulation of air pollutants. Smoke can induce respiratory illness, aggravate symptoms of asthma, bronchitis, and emphysema; and long-term exposure can lead to lung cancer. Smoke toxicity increases when fumes of sulfur dioxide (SO_2), commonly released by coal combustion, are inhaled with it. One particularly bad air pollution incident occurred in London in 1952 when more than 4,000 deaths resulted from high smoke and sulfur dioxide levels that accumulated in the metropolitan area during an atmospheric inversion. For four days, a dense fog settled in London with temperatures near freezing and extremely low visibility. The concentration of smoke in the air ranged from 0.1 parts per million (ppm) to over 1.0 ppm (0.3 to more than 4 mg per m^3). The United

States' primary standard for sulfur dioxide, set as part of the Environmental Protection Agency (EPA) National Ambient Air Quality Standards (NAAQS), is 0.14 ppm (365 mg per m^3), which is a maximum for a twenty-four-hour period and is not to be exceeded more than one time in a year. The fog did not disperse until mild weather and wind advanced into the area.

Sources of air pollutants are particularly abundant in industrial population centers. Major sources include power and heating plants, industrial manufacturing plants, and transportation vehicles. Mexico City is a particularly bad example of a very large metropolitan area that has not adequately controlled harmful emissions into the atmosphere; and, as a result, many area residents suffer from respiratory ailments. The Mexican government has been forced to shut down a large oil refinery and ban new polluting industries from locating in the city. More difficult for the Mexican government to control are the emissions from over 3.5 million automobiles, trucks, and buses.

Smog (a combination of smoke and fog) is formed from the condensation of moisture (fog) on particulate matter (smoke) in the atmosphere. Although smog has been present in urban areas for a long time, photochemical smog is an exceptionally harmful and annoying form of air pollution found in large urban areas. It was first recognized as a serious problem in Los Angeles, California, in the late 1940s. Photochemical smog forms by the catalytic action of sunlight on some atmospheric pollutants, including unburned hydrocarbons evaporated from automobile and other fuel tanks. Products of the photochemical reactions include ozone (O_3), aldehydes, ketones, peroxyacetyl nitrate (PAN), and organic acids. Photochemical smog causes serious eye and lung irritation and other health problems.

Because of the serious damage caused by atmospheric pollution, control has become a high priority in many countries. Several approaches have proven beneficial. An immediate control measure involves a system of air alerts announced when air pollution monitors disclose dangerously high levels of contamination. Industrial sources of pollution are forced to curtail activities until conditions return to normal. These emergency situations are often caused by atmospheric inversions that limit the upward mixing of pollutants. Wider dispersion of pollutants by taller smokestacks can alleviate local pollution. However, dispersion does not lessen overall amounts of pollution; instead, it can create problems for communities downwind from the source. Greater energy efficiency in vehicles, homes, electrical power plants, and industrial plants lessens pollution from these sources. This reduction in fuel use and the use of alternative sources of energy such as solar power can reduce pollution created as a result of energy use; in addition, more energy-efficient operation reduces costs. Efficient and convenient mass transit also helps to mitigate urban air pollution by cutting the need for personal vehicles.

Soot is also a contributor to air pollution. Tiny, dark particles consisting mostly of carbon, which are released by burning many fuels, soot absorbs sunlight and so tends to warm Earth. In 2008, researchers studying soot released by cooking fires in South Asia and coal-burning in East Asia concluded that soot is a more important contributor to climate change than had been suspected, edging out methane (CH_4) as the second-largest greenhouse pollutant after carbon dioxide. Dimming of the skies by soot reduces evaporation and regional temperature differences over the Indian Ocean, causing less rain to fall during the monsoon season and so contributing to drought in Southeast Asia. The new estimate of the radiative forcing (energy imbalance between Earth and sunlight) caused by soot was about double that cited by the Intergovernmental Panel on Climate Change (IPCC) in its influential 2007 summary of the state of climate science. Replacing dung and wood as cooking fuels in South Asia with alternative fuel sources such as biogas, natural gas, and solar cookers could significantly reduce the problem.

Resources

BOOKS

Desonie, Dana. *Atmosphere: Air Pollution and Its Effects.* New York: Chelsea House Publishers, 2007.

Manahan, Stanley E. *Environmental Chemistry.* Boca Raton, FL: CRC Press, 2005.

Prinn, Ronald G. *Effects of Air Pollution Control on Climate.* Cambridge, MA: MIT Joint Program on the Science and Policy of Global Change, 2005.

Vallero, Daniel A. *Fundamentals of Air Pollution.* Amsterdam: Elsevier, 2008.

PERIODICALS

Law, Kathy S. "Arctic Air Pollution: Origins and Impacts." *Science* 315 (2007): 1537–1540.

OTHER

Centers for Disease Control and Prevention (CDC). "Air Pollution and Respiratory Health." http://www.cdc.gov/nceh/airpollution/default.htm (accessed October 27, 2010).

National Geographic Society. "Air Pollution." http://environment.nationalgeographic.com/environment/global-warming/pollution-overview.html (accessed October 27, 2010).

National Institutes of Health (NIH). "Air Pollution." http://health.nih.gov/topic/AirPollution (accessed October 27, 2010).

United States Environmental Protection Agency (EPA). "Air: Air Pollution Control." http://www.epa.gov/ebt pages/airairpollutioncontrol.html (accessed October 27, 2010).

Douglas C. Pratt

Atmospheric deposition

Many kinds of suspended particles and gases are deposited from the atmosphere to the surfaces of terrestrial and aquatic ecosystems. Wet deposition refers to deposition occurring while it is raining or snowing, whereas dry deposition occurs in the time intervals between precipitation events.

Relatively large particles suspended in the atmosphere, such as dust entrained by strong winds blowing over fields or emitted from industrial smokestacks, may settle gravitationally to nearby surfaces at ground level. Particulates smaller than about 0.5 microns in diameter, however, do not settle in this manner because they behave aerodynamically like gases. Nevertheless, they may be impaction-filtered from the atmosphere when an air mass passes through a physically complex structure. For example, the large mass of foliage of a mature conifer forest provides an extremely dense and complex surface. As such, a conifer canopy is relatively effective at removing particulates of all sizes from the atmosphere, including those smaller than 0.5 microns. Forest canopies dominated by hardwood (or angiosperm) trees are also effective at doing this, but to a lesser degree compared with conifers.

Dry deposition also includes the removal of certain gases from the atmosphere. For example, carbon dioxide (CO_2) gas is absorbed by plants and fixed by photosynthesis into simple sugars. Plants are also rather effective at absorbing certain gaseous pollutants such as sulfur dioxide (SO_2), nitric oxide (NO), nitrogen dioxide (NO_2), and ozone (O_3). Most of this gaseous uptake occurs by absorption through the numerous tiny pores in leaves known as stomata. These same gases are also dry-deposited by absorption by moist soil, rocks, and water surfaces.

Wet deposition involves substances that are dissolved in rainwater and snow. In general, the most abundant substances dissolved in precipitation water are sulfate (SO_4^{2-}), nitrate (NO_3^-), calcium (Ca^{2+}), magnesium (Mg^{2+}), and ammonium (NH_4^+). At places close to the ocean, sodium (Na^+) and chloride (Cl^-) derived from sea-salt aerosols are also abundant in precipitation water. Acidic precipitation occurs whenever the concentrations of the anions (negatively charged ions) sulfate, and nitrate occur in much larger concentrations than the cations (positively charged ions) calcium, magnesium, and ammonium. In such cases, the cation deficit is made up by hydrogen ions going into solution, creating solutions that may have an acidic pH of less than 4.

The deposition of acidified snow or rain can result in the acidification of vulnerable surface waters. This is particularly the case of streams, rivers, ponds, and lakes that have low concentrations of alkalinity and consequently little capacity for neutralizing inputs of acidity. Acid rain has devastated many populations of fish and other aquatic organisms that are unable to tolerate the increased acidity. The low pH also causes the release of toxic metals such as aluminum (Al) that can adversely affect aquatic organisms. The biodiversity in and around some lakes has changed from native fish and plants to more acid-tolerant species. The National Atmospheric Deposition Program (NADP) was established by the United States to sample and monitor atmospheric deposition. The environmental effects of acid deposition caused the U.S. Environmental Protection Agency (EPA) to propose a limit of 22 to 44 pounds (10-20 kg) of sulfate emissions per approximately 2.5 acres (1 ha) each year.

Acidification may also be caused by the dry deposition of certain gases, especially sulfur dioxide and oxides of nitrogen. After they are dry-deposited, these gases become oxidized in soil or water into sulfate and nitrate ions, respectively, a process accompanied by the production of equivalent amounts of hydrogen ions. In some environments the gaseous concentrations of sulfur dioxide and oxides of nitrogen are relatively high, particularly in areas polluted by gaseous emissions from large numbers of automobiles, power plants, or other industries. In those circumstances the dry deposition of acidifying substances will be a much more important cause of environmental acidification than the wet deposition of acidic precipitation.

By implementing alternative sources of energy and electricity such as natural gas, wind energy, solar energy, or hydropower, reductions in sulfur and nitrogen oxide emissions can be accomplished, thus decreasing the amounts of atmospheric deposition of these pollutants. Congress established the Acid Rain Program as part of the Clean Air Act (CAA) to reduce these emissions, initiating a cap on sulfur oxide emissions from power plants to 8.95 million tons (8.12 million metric tons) annually in 2010.

Resources

BOOKS

Brimblecombe, Peter. *Acid Rain: Deposition to Recovery.* Dordrecht, Netherlands: Springer, 2007.

National Atmospheric Deposition Program (U.S.). *National Atmospheric Deposition Program: 2004 Annual Summary.* Champaign, IL: 2005.

Vallero, Daniel A. *Fundamentals of Air Pollution.* Amsterdam: Elsevier, 2008.

Yadav, P. R., and Shubhrata R. Mishra. *Environmental Ecology.* New Delhi: Discovery Publishing House, 2004.

OTHER

National Geographic Society. "Acid Rain." http://environment.nationalgeographic.com/environment/global-warming/acid-rain-overview.html (accessed August 17, 2010).

United States Environmental Protection Agency (EPA). "Reducing Acid Rain." http://www.epa.gov/acidrain/reducing/index.html (accessed August 17, 2010).

ORGANIZATIONS

United States Environmental Protection Agency, Clean Air Markets Division, 1200 Pennsylvania Avenue, NW, Washington, DC, USA, 20460, (202) 564-9150, http://www.epa.gov/airmarkets/acidrain/

Worldwatch Institute, 1776 Massachusetts Ave., NW, Washington, DC, USA, 20036–1904, (202) 452–1999, (202) 296-7365, worldwatch@worldwatch.org, http://www.worldwatch.org/

Bill Freedman

Atmospheric inversion

Atmospheric inversions are horizontal layers of air that increase in temperature with height. Such warm, light air often lies over air that is cooler and heavier. As a result, the air has a strong vertical stability, especially in the absence of strong winds.

Atmospheric inversions play an important role in transient air quality. They can trap air pollutants below or within them, causing high concentrations in a volume of air that would otherwise be able to dilute air pollutants throughout a large portion of the troposphere.

Atmospheric inversions are quite common, and there are several ways in which they are formed. Surface inversions can form during the evening when the radiatively cooling ground becomes a heat sink at the bottom of an air mass immediately above it. As a result, heat flows down through the air, which sets up a temperature gradient that coincides with an inversion. Alternatively, relatively warm air may flow over a cold surface with the same results.

Elevated atmospheric inversions can occur when vertical differences in wind direction allow warm air to set up over cold air. However, it is more common in near-subtropical latitudes, especially on the western sides of continents, to get subtropical subsidence inversions. Subtropical subsidence inversions often team with mountainous topography to trap air both horizontally and vertically.

Atmospheric pollutants *see* **Carbon monoxide; Nitrogen oxides; Particulate; Smog; Sulfur dioxide; Volatile organic compound.**

Atomic bomb *see* **Nuclear weapons.**

Atomic bomb testing *see* **Bikini atoll; Nevada Test Site.**

Atomic energy *see* **Nuclear power.**

Atomic Energy Commission

The Atomic Energy Commission (AEC) was established by the United States Congress in 1946. It originally had a dual function: to promote the development of nuclear power and to regulate the nuclear power industry. These two functions were occasionally in conflict, however. Critics of the AEC claimed that the commission sometimes failed to apply appropriate safety regulations because doing so would have proved burdensome to the industry and hindered its growth. In 1975, the AEC was dissolved. Its regulatory role was assigned to the Nuclear Regulatory Commission and its research and development role to the Energy Research and Development Administration.

Atomic fission *see* **Nuclear fission.**

Atomic fusion *see* **Nuclear fusion.**

Atrazine

Atrazine is used as a selective preemergent herbicide on crops, including corn, sorghum, and sugarcane, and as a nonselective herbicide along fence lines, right-of-ways, and roadsides. It is used in a variety of formulations, both alone and in combination with other herbicides. In the United States it is the single most widely-used herbicide, accounting for about 15 percent of total herbicide application by weight. Its popularity stems from its effectiveness, its selectivity (it inhibits photosynthesis in plants lacking a detoxification mechanism), and its low mammalian toxicity (similar to that of table salt). Concern over its use stems from the fact that a small percentage of the amount applied can be carried by rainfall into surface waters (streams and lakes), where it may inhibit the growth of plants and algae or contaminate drinking-water supply. Conflicting independent studies have shown adverse impacts on amphibian populations, but other studies (especially those relied upon by the Environmental Protection Agency [EPA]) show no definitive adverse impacts. Although setting limits on allowable concentrations, the EPA has essentially ruled the use of atrazine safe. As of 2010, Atrazine still faces broader restrictions and opposition in the European Union.

Attainment area

An attainment area is a politically- or geographically-defined region that is in compliance with the National Ambient Air Quality Standards (NAAQS) established by the Clean Air Act (CAA), as set by the Environmental Protection Agency (EPA). A region is considered to be an attainment area if it does not exceed the specified thresholds for air pollutants that may be a threat to public health or the environment. An area that has concentrations of a pollutant above the set standards is referred to as a nonattainment area. The standards are set for a variety of ambient air criteria pollutants, including particulates, ozone (O_3), carbon monoxide (CO), sulfur oxides (SO_x), lead (Pb), and nitrogen oxides (NO_x). There are two sets of standards for these criteria pollutants: Primary standards are limits set to protect public health, whereas secondary standards are in place to prevent damage to the environment and property. A region can be an attainment area for one pollutant and a nonattainment area for another. If an area is designated as nonattainment for a given pollutant, the local and state governments must take action to reduce these concentrations to meet the NAAQS and be considered in attainment. Under the CAA, the federal government has the power to withhold funds from areas that do not meet the NAAQS for air pollutants. As of 2008, many metropolitan areas were designated as nonattainment for one or more of the criteria pollutants, with approximately 90 million Americans inhabiting these areas.

Resources

BOOKS

Vallero, Daniel A. *Fundamentals of Air Pollution.* Amsterdam: Elsevier, 2008.

Audubon, John James

1785–1851
American naturalist, artist, and ornithologist

John James Audubon, the most renowned artist and naturalist in nineteenth-century America, left a legacy of keenly observant writings as well as a portfolio of exquisitely rendered paintings of the birds of North America.

Born April 26, 1785, Audubon was the illegitimate son of a French naval captain and a domestic servant girl from Santo Domingo (now Haiti). Audubon spent his childhood on his father's plantation in Santo Domingo and most of his late teens on the family estate in Mill Grove, Pennsylvania—a move intended to prevent him from being conscripted into the Napoleonic army.

Audubon's early pursuits centered around natural history, and he was continuously collecting and drawing plants, insects, and birds. At a young age he developed a habit of keeping meticulous field notes of his observations. At Mill Grove, Audubon, in order to learn more of the movements and habits of birds, tied bits of colored string onto the legs of several Eastern Phoebes and proved that these birds returned to the same nesting sites the following year. Audubon was the first to use banding to study the movement of birds.

While at Mill Grove, Audubon began courting their neighbor's eldest daughter, Lucy Bakewell, and they were married in 1808. They made their first home in Louisville, Kentucky, where Audubon tried being a storekeeper. He could not stand staying inside, and so

John James Audubon. *(Corbis-Bettmann)*

he spent most of his time afield to "supply fowl for the table," thus dooming the store to failure.

In 1810, Audubon met, by chance, Alexander Wilson, who is considered the father of American ornithology. Wilson had finished much of his nine-volume *American Ornithology* at this time, and it is believed that his work inspired Audubon to embark on his monumental task of painting the birds of North America.

The task that Audubon undertook was to become *The Birds of America*. Because Audubon decided to depict each species of bird life-size, thus rendering each on a 36.5 by 26.5 inch (93 cm x 67 cm) page, this was the largest book ever published up until that time. He was able to draw even larger birds such as the whooping crane life-size by depicting them with their heads bent to the ground. Audubon pioneered the use of fresh models instead of stuffed museum skins for his paintings. He would shoot birds and wire them into lifelike poses to obtain the most accurate drawings possible. Even though his name is affixed to a modern conservation organization, the National Audubon Society, it must be

remembered that little thought was given to the conservation of birds in the early nineteenth century. It was not uncommon for Audubon to shoot a dozen or more individuals of a species to get what he considered the perfect one for painting.

Audubon solicited subscribers for his *Birds of America* to finance the printing and hand-coloring of the plates. The project took nearly twenty years to complete, but the resulting double elephant folio, as it is known, was truly a work of art, as well as the ornithological triumph of the time. Later in his life, Audubon worked on a book of the mammals of North America with his two sons, but failing health forced him to let them complete the work. He died in 1851, leaving behind a remarkable collection of artwork that depicted the natural world he loved so much.

Resources

BOOKS

Audubon, John James. *Birds of America*. Philadelphia: J. B. Chevalier, 1842.

Eugene C. Beckham

Audubon Society *see* **National Audubon Society.**

Australia

The Commonwealth of Australia, a country in the South Pacific Ocean, occupies the smallest continent and covers an area of 2,966,200 square miles (7,682,000 km²). Along the northeast coast of Australia, for a distance of 1,200 miles (2,000 km) in length and offshore as much as 100 miles (160 km) into the Pacific Ocean, lies the Great Barrier Reef, the greatest assemblage of coral reefs in the world. The island of Tasmania, which is a part of Australia, lies off the southeast coast of the mainland. Australia consists mainly of plains and plateaus, most of which average 2,000 feet (600 m) above sea level. There are several low mountain ranges in the country.

A national census is conducted every five years. As of 2010, the latest census figures available were gathered in 2006. According to the Australian Bureau of Statistics, in 2006 there were nearly 20 million people living in Australia (not counting visitors). Around 450,000 people described their ancestry as Aboriginal,

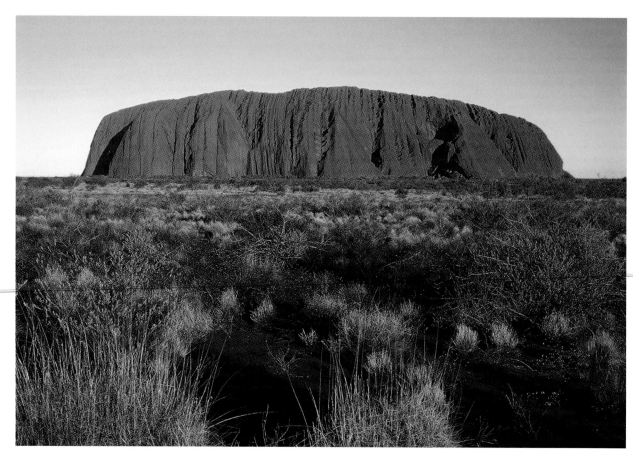

Uluru (Ayers Rock), Northern Territory, Australia. *(Steve Lovegrove/Shutterstock.com)*

which refers to the various peoples native to Australia before the arrival of Europeans.

More than one-third of Australia receives under 10 inches (25 cm) of annual rainfall, while less than one-third receives over 20 inches (51 cm). Many areas experience prolonged drought and frequent heat waves. The Outback is one such area. It is comprised of dry, barren land and vast interior deserts. Soils generally are poor, with fertile land found only in the lowlands and valleys near the east and southeast coast. The largest river is the Murray River in southeastern Australia. In the inland area, few rivers have much water due to the lack of rainfall. In areas with coastal highlands, rivers flow only a short distance to the sea. Lakes frequently dry up and become beds of salt. In contrast, there are lush rain forests in the north and snow-covered mountains in the south.

The most common tree is the eucalyptus, of which there are at least 600 different kinds. Since Australia is the driest of the inhabited continents, there are few forests. The acacia, or wattle, is Australia's national flower. Many of the mammals of Australia (kangaroos, wallabies, and koala bears) are marsupials, and are unlike animals found anywhere else in the world (but marsupials are not confined to Australia; there are some sixty different kinds of marsupial opossums native to South America). The young of marsupials are born premature and require continued development outside the womb in pouches on the parent's body. The egg-laying mammal, or monotreme, is also only found in Australia. There are two species of monotreme: the duck-billed platypus and the echidna, or spiny anteater. The most famous Australian bird is the kookaburra. There are no hoofed animals in Australia. Two flight-less birds, the emu and the cassowary, also are found only in Australia. Snakes are common and include many poisonous species. Two kinds of crocodiles are found in the northern part of Australia. All of these unique fauna and flora species have evolved due to the long separation of the Australian continent from Southeast Asia and a general lack of predators.

The goal of environmental protection and management in Australia is integration of the principles of ecologically-sustainable development into policies and programs encompassing all aspects of environmental issues.

Australia's land resources form the basis of its unique natural heritage and its agricultural industry. However, problems of land degradation are widespread. Land degradation due to dryland salinity is a national problem that threatens both biodiversity and agricultural productivity. Dryland salinity affects about 5,434,000 acres (2.2 million ha) of once-productive agricultural land and costs an estimated $243 million per year in lost agricultural production. Dryland salinity originates from salt deposited in the landscape over millions of years due to marine sedimentation. Land clearing over the last 200 years and the replacement of deep-rooted perennial native vegetation with shallow-rooted annuals has caused water tables to rise, bringing saline water close to the surface into the root zones of plants. Excessive irrigation has also led to increases in levels of water tables. Strategies to address this problem include research for the prediction, prevention, and reversal of dryland salinity. An example of a research effort to ameliorate the effects of salinity is a project that addresses the identification of salt-tolerant plant species and development of appropriate establishment and growth management techniques to ensure their survival. Australia's naturally fire-adapted ecosystem has frequent brushfires that now threaten many new developments and even major cities. Other land degradation problems being addressed include the rehabilitation of mining sites and remediation of sites contaminated by railway operations.

Australia contains about 10 percent of the world's biological diversity. The Commonwealth is addressing the issues of loss of biodiversity through a program referred to as the National Vegetation Initiative (NVI). About 988 million acres (400 million ha) of land is protected within a terrestrial reserves system, but 1,235 million acres (500 million ha), or more than two-thirds of Australia's land area, are managed by private landholders. Biodiversity outside the reserves has been affected by vegetation clearance and modification. The NVI will attempt to reverse the long-term decline in the quality and extent of Australia's native vegetation by developing programs and incentives to conserve, enhance, and manage remnant native vegetation; to increase revegetation activities; and to encourage the integration of native vegetation into conventional farming systems.

The state of Australia's rivers and other aquatic systems has been declining due to a range of factors. Unsustainable water extractions for agricultural production have resulted in reduced flows in rivers that result in blue-green algae outbreaks, declines in native fish populations, increases in salinity, loss of wetlands, and loss of the beneficial aspects of small floods. Research is being conducted to define environmental flow requirements so that water allocations or entitlements will include environmental needs as a legitimate use. Other factors affecting the decline in aquatic systems include poor land and vegetation management, agricultural and urban pollution, salinity, destruction of native habitat, and spread of exotic pests such as the European carp and the water weed *Salvinia*.

The Australian government's primary air quality objective is to provide all its citizens with equivalent protection from air pollution, which will be achieved by the development of national air quality strategies and standards that minimize the adverse environmental impacts of air pollution. Since over 60 percent of the Australian population lives in cities, the government's first priority is to develop measures to reduce the impact of air pollution on urban areas.

The government of Australia considers climate changes as one of its most important environmental issues. The Department of Climate Change was established by the Australian government in December 2007; its name was changed to the Department of Climate Change and Energy Efficiency in March 2008. The main goals of the department are to reduce the nation's greenhouse gas emissions, to formulate plans for adapting to climate change, and to work with other nations to find global solutions to the climate change problem.

Regarding greenhouse gas emissions, the Australian government has developed cooperative agreements with industry to abate and reduce such emissions. Large-scale vegetation plantings for improvement of land resources will reduce carbon dioxide releases, a major greenhouse gas, into the atmosphere. Methane emission from intensive agriculture is being reduced through livestock waste treatment. The use of alternative transport fuels such as liquefied petroleum gas, compressed natural gas, ethanol, and other alcohol blends is encouraged through the use of a fuel excise exemption.

Although some ozone layer depletion in the atmosphere occurs over most of the Earth, the most dramatic changes are seen when an ozone "hole" (more precisely, a thinning of the ozone layer) forms over Antarctica each spring. The ozone layer protects life on Earth from harmful ultraviolet radiation from the sun, but it can be depleted by some widely-used, ozone-depleting chemicals. The ozone hole sometimes drifts over Australia and exposes residents to dangerous levels of ultraviolet radiation. Australia, because of its proximity to

Antarctica, has taken a leading role in ozone protection and has been influential in the development and implementation of cost-effective mechanisms to ensure the phase out of use of ozone-depleting substances by all countries. The government also monitors solar ultraviolet radiation levels and assesses the consequences for public health.

Australia's marine and coastal environments are rich in natural and cultural resources, are adjacent to most of the nation's population (about 86 percent), and are a focus of much of Australia's economic, social, tourism, and recreational activity. Australia's Exclusive Economic Zone (EEZ) is one of the largest in the world, comprising 4.25 million square miles (11 million km^2) of marine waters. Uncoordinated and ad hoc development has been identified as a major contributing factor to the decline of coastal water quality and marine and estuarine habitats. Programs that include cooperation of the Commonwealth government with state and local governments, industry groups, nongovernmental organizations, and the community have been developed to protect and rehabilitate these environments. These programs include addressing threats to coastal water quality and marine biodiversity from land-based and marine pollution; developing integrated management plans for the conservation and sustainable use of coastal resources, including fisheries resources; supporting capital works and improvement of technologies to reduce the impacts of sewage and stormwater; controlling the introduction and spread of exotic marine pests in Australian waters; protecting and restoring fish habitats; and supporting comprehensive and consistent coastal monitoring.

The Great Barrier Reef is facing growing pressures from the reef-based tourism industry, commercial and recreational fishing (which is worth about $1.3 billion per year), expanding coastal urban areas, and downstream effects of land use from agricultural activities. Several programs are being developed and implemented to ensure that activities can continue on an ecologically-sustainable basis. For example, tourist and recreational impacts are being reduced, while providing diverse tourist activities is increasing. Fishing catches and efforts are being monitored, major or critical habitats are being identified, and fishing bycatch is being reduced through the development of new methods. Sediment, nutrient, and other land-based runoff that impacts the health of adjacent marine areas are being controlled in coastal developments. Also, spill contingency planning and other responses to prevent pollution from ships, navigational aids, and ship reporting systems are all being improved. In addition, the state of water quality is being monitored in long-term programs throughout the reef and threats from pollution are being assessed. Finally, all planning exercises include consultation with Aboriginal and Torres Strait Islander (ATSI) communities to ensure that their interests are considered.

The Commonwealth, through its National and World Heritage programs, is protecting elements of Australia's natural and cultural heritage that are of value for this and future generations. The National Estate comprises natural, historic, and indigenous places that have aesthetic, historic, scientific or social significance, or other special value. The Register of the National Estate had, by 2007 (when the Registry list was frozen, that is no additional entries allowed) some 13,000 places listed. The Register educates and alerts Australians to places of heritage significance. World Heritage properties are areas of outstanding universal cultural or national significance that are included in the United Nations Educational, Scientific, and Cultural Organization (UNESCO) World Heritage List. In 2010, there were 18 Australian properties on the World Heritage List, including the Great Barrier Reef.

See also Exclusive economic zone; Great Barrier Reef.

Resources

BOOKS

Australian National Audit Office; Peter McVay; Cameron Mathie; and David Crossley. *The Conservation and Protection of National Threatened Species and Ecological Communities: Department of the Environment and Water Resources.* Canberra, ACT: 2007.

Chapman, Arthur D. *Numbers of Living Species in Australia and the World.* Canberra, ACT: Australian Department of the Environment and Heritage, 2006.

Clarke, Philip. *Where the Ancestors Walked: Australia as an Aboriginal Landscape.* San Diego, CA: Allen & Unwin Academic, 2004.

Guile, Melanie. *World Issues Come to Australia: Drought and El Niño.* Port Melbourne, Victoria, Australia: Heinemann Library, 2008.

Lindenmayer, David. *On Borrowed Time: Australia's Environmental Crisis and What We Must Do About It.* Camberwell, Victoria, Australia: Penguin, 2007.

Markus, Nicola. *On Our Watch: The Race to Save Australia's Environment.* Carlton, Victoria, Australia: Melbourne University Press, 2009.

Natural Heritage Trust. *Threatened Australian Plants.* Canberra, Australian Capital Territory: Department of the Environment and Heritage, 2004.

New, T. R. *Conservation Biology in Australia: An Introduction.* South Melbourne, Victoria, Australia: Oxford University Press, 2006.

Thomas, Ian. *Environmental Management Processes and Practices in Australia.* Annandale, New South Wales, Australia: Federation Press, 2005.

OTHER

National Geographic Society. "Australia." http://travel.nation algeographic.com/places/countries/country_australia.html (accessed November 7, 2010).

United States Central Intelligence Agency (CIA). "Australia." *CIA World Factbook.* https://www.cia.gov/library/publications/the-world-factbook/geos/as.html (accessed November 7, 2010).

Judith L. Sims

Autecology

Autecology is a branch of ecology emphasizing the interrelationships among individual members of the same species and their environment. It includes the study of the life history or behavior of a particular species in relation to the environmental conditions that influence its activities and distribution. Autecology also includes studies on the tolerance of a species to critical physical factors (e.g., temperature, salinity, oxygen level, light) and biological factors (e.g., predation, symbiosis) thought to limit its distribution. Such data are gathered from field measurement or from controlled experiments in the laboratory. Autecology contrasts with *synecology*, the study of interacting groups (i.e., communities) of species.

Automobile

The development of the automobile at the end of the nineteenth century fundamentally changed the structure of society in the developed world and has had wide-ranging effects on the environment, the most notable being the increase of air pollution in cities. The piston-type internal combustion engine is responsible for the peculiar mix of pollutants that it generates. There are a range of other engines suitable for automobiles, but they have yet to displace engines using rather volatile petroleum derivatives.

The simplest and most successful way of improving gaseous emissions from automobiles is to find alternative fuels. Diesel fuels have always been popular for larger vehicles, although a large portion of private vehicles in Europe are also diesel-powered. Compressed natural gases have been widely-used as fuel in some countries (e.g., New Zealand), while ethanol has had success in places such as Brazil, where it can be produced relatively cheaply from sugarcane. There is limited, but growing, enthusiasm for the use of alternative fuels in the United States. A good example is the fuel blend E85, which, after a slow start, is becoming an increasingly popular fuel alternative. E85 is a mixture of up to 85 percent fuel ethanol (denatured grain alcohol) combined with some other liquid hydrocarbon, usually gasoline. According to the U.S. Department of Energy (DOE), by mid–2010 there were nearly 2,500 stations selling E85, mostly in the Midwestern United States where ethanol is widely derived from corn.

Others have suggested that fundamental changes to the engine itself can lower the impact of automobiles on air quality. The Wankel rotary engine is a promising power source that offers both low vibration and pollutant emissions from a relatively lightweight engine. Although Wankel engines are found on a number of exotic cars, there are still doubts about long-term engine performance and durability in the urban setting. Steam and gas turbines have many of the advantages of the Wankel engine, but questions of their expense and suitability for automobiles have restricted their use.

Traditionally, electric vehicles have had some positive impact for special sectors of the market. They have proved ideal for small vehicles within cities where frequent stop-start operation is required (e.g., delivery vans). Over the years, a few small, one-seat vehicles have been available at times, but they have failed to achieve any enduring popularity. The electric vehicle has traditionally suffered from low range, low speed and acceleration, and heavy batteries. However, they produce none of the conventional combustion-derived pollutants during operation, although the electricity to recharge the batteries requires the use of an electricity supply, which may or may not come from a power plant burning fossil fuels. From 1993 through 1996, General Motors (GM) Corporation produced and leased the EV1 all-electric car. GM cancelled its electric vehicle program in 2002, taking back its leased EV1s and destroying most of them (a few ended up in museums). GM claimed to have spent around one billion dollars developing and producing the EV1. GM management claimed that the market for electric cars was too limited and that profit margins were either too small, or nonexistent. In contrast to the American-based and managed General Motors, Japan-based Nissan announced that it would begin selling its all-electric compact car in the United States in December 2010, with sales in Europe planned for 2011.

In addition to the combustion engine and battery-powered electric cars, there are other technologies that have been explored. Fuel cells are an alternate source of electricity for electric automobiles. Fuel cells produce electricity directly from a chemical reaction, by

automobile emissions from malfunctioning units and from leaks. These can be from the evaporation of fuel, especially leaded fuels where the volatile tetraethyl lead is present. Carbon monoxide, from the exhaust system, can cause drowsiness and impair judgment. However, in many cases, the interior of a properly functioning automobile, without additional sources such as smoking, can have somewhat better air quality than the air outside. In general, pedestrians, cyclists, and those who work at road sites are likely to experience the worst of automotive pollutants such as carbon monoxide and potentially carcinogenic hydrocarbons.

Although huge quantities of fossil fuels are burnt in power generation and a range of industrial processes, automobiles make a significant and growing contribution to carbon dioxide emissions, which enhances the greenhouse effect. Ethanol, made from sugarcane, is a renewable source of energy and has the advantage of not making as large a contribution to the greenhouse gases as gasoline. Automobiles are not large emitters of sulfur dioxide and thus do not contribute greatly to the regional sulfric acid rain problem. Nevertheless, the nitrogen oxides emitted by automobiles are ultimately converted to nitric acid, and these are making an increasing contribution to rainfall acidity. Diesel-powered vehicles use fuel of a higher sulfur content and can contribute to the sulfur compounds in urban air, although regulations often require the removal of sulfur from modern diesel fuel. Newer ultra-low sulfur diesel (ULSD) fuel has greatly reduced the emissions of sulfur compounds. ULSD contains 15 to 50 parts per million (ppm), which represents a dramatic decrease from low-sulfur diesel that contained up to 500 ppm.

Despite the enormous problems created by the automobile, few propose its abolition, although some regulators in Europe have attempted to reduce usage. The ownership of a car carries with it powerful statements about personal freedom and power. Beyond this, the structure of many modern cities requires the use of a car. Thus while air pollution problems might well be cured by a wide range of sociological changes, a technological fix has been favored, such as the use of catalytic converters. Despite this and other devices, cities still face daunting air quality problems. In some areas, most notably the Los Angeles Basin in California, it is clear that there will have to be a wide range of changes if air quality is to improve. Although much attention is being given to lowering emissions of volatile organic compounds, it is likely that non polluting vehicles will have to be manufactured and a better mass transit system created.

Automakers will have to employ fuel-efficiency technologies to meet new, more stringent fuel economy standards in the United States. In 2009, U.S. President Barack Obama (1961–) announced a policy to increase the average fuel economy standard of vehicles in the United States to 35.5 miles per gallon (56.8 km per gallon) by 2016 from approximately 25 miles per gallon (40 km per gallon) in 2010. The new fuel economy standard will save an estimated 1.8 billion barrels of oil over the life of the program and is equivalent in emissions reductions to removing 58 million cars from the road for a year.

In April 2010, the United States set new greenhouse gas emissions standards for automobiles that included the most substantial increase in fuel efficiency standards since the 1970s. Canada also set new and similar emissions standards. Starting with 2012 model automobiles, the U.S. Environmental Protection Agency (EPA) and U.S. Department of Transportation cooperated in finalizing standards that will require cars and trucks to average 35.5 miles per gallon (15 km per l) by 2016, an increase of 42 percent from 2010 standards. Average vehicle emissions also cannot exceed 250 grams (8.75 oz) of carbon dioxide per mile by 2016. New measures intended for industrial tractor-trailer trucks are anticipated by the end of 2010. The shared standards of the United States and Canada reflect more than a shared environmental concern. The automotive industry has component factories on both sides of the border, and so shared standards are vital to economic interests. Several industries, including several in the energy and refining sectors, have joined to take legal action to challenge the EPA's authority to impose such limits.

In Europe, a manufacturer's cars must meet Europe's 130 grams per kilometer carbon dioxide emissions target by 2015. Thus far, European consumers have favored transition to diesel cars over transition to hybrid gas or electric vehicles.

See also Automobile.

Resources

OTHER

United States Environmental Protection Agency (EPA). "Air: Mobile Sources: Vehicle Emissions." http://www.epa.gov/ebtpages/airmobilesourcesvehicleemissions.html (accessed October 15, 2010).

Peter Brimblecombe

Autotroph

An autotroph is an organism that derives its carbon for building body tissues from carbon dioxide (CO_2) or carbonates and obtains its energy for bodily functions from radiant sources, such as sunlight, or from the oxidation of certain inorganic substances. The leaves of green plants and the bacteria that oxidize sulfur, iron, ammonium, and nitrite are examples of autotrophs. The oxidation of ammonium to nitrite, and of nitrite to nitrate, a process called nitrification, is a critical part of the nitrogen cycle. Moreover, the creation of food by photosynthetic organisms is largely an autotrophic process.

Avalanche

An avalanche is a sudden slide of snow and ice, usually in mountainous areas where there is heavy snow accumulation on moderate to steep slopes. Snow avalanches flow at an average speed of 80 miles per hour (130 km/h), and their length can range from less than 300 feet (100 m) to 2 miles (3.2 km) or more. Generally the term "avalanche" refers to sudden slides of snow and ice, but it can also be used to describe catastrophic debris slides consisting of mud and loose rock. Debris avalanches are especially associated with volcanic activity in which melted snow, earthquakes, and clouds of flowing ash can trigger movement of rock and mud. Snow avalanches generally consist either of loose, fresh snow or of slabs of accumulated snow and ice that move in large blocks. Snow avalanches occur most often where the snow surface has melted under the sun and then refreezes, forming a smooth surface of snow. Later snow falling on this smooth surface tends to adhere poorly, and it may slide off the slick plane of recrystallized snow when it is shaken by any form of vibration—including sound waves, earthquakes, or the movement of skiers.

Several factors contribute to snow avalanches, including snow accumulation, hill slope angle, slope shape (profile), and weather. Avalanches are most common where there is heavy snow accumulation on slopes of twenty-five to sixty-five degrees, and they occur most often on slopes between thirty and forty-five degrees. On slopes steeper than sixty-five degrees snow tends to slough off rather than accumulate. On shallow slopes, avalanches are likely to occur only in wet (melting) conditions, when accumulated snow

may be heavy, and when snowmelt collecting along a hardened old snow surface within the snowpack can loosen upper layers, allowing them to release easily. Slab avalanches may be more likely to start on convex slopes, where snow masses can be fractured into loose blocks, but they rarely begin on tree-covered slopes. However, loose snow avalanches often start among trees, gathering speed and snow as they cross open slopes. Weather can influence avalanche probability by changing the stability and cohesiveness of the snowpack. Many avalanches occur during storms when snow accumulates rapidly, or during sustained periods of cold weather when new snow remains loose. Like snowmelt, rainfall can increase chances of avalanche by lubricating the surface of hardened layers within the snowpack. Sustained winds increase snow accumulation on the leeward side of slopes, producing snow masses susceptible to slab movement. When conditions are favorable, an avalanche can be triggered by the weight of a person or by loud noises, earthquakes, or other sources of vibration. Avalanches tend to be most common in mid-winter, when snow accumulation is high, and in late spring, when melting causes instability in the snowpack.

Avalanches play an ecological role by keeping slopes clear of trees, thus maintaining openings vegetated by grasses, forbs, and low brush. They are also a geomorphologic force because they maintain bare rock surfaces, which are susceptible to erosion.

Most research into the dynamics and causes of avalanches has occurred in populous mountain regions such as the Alps, the Cascades, and the Rocky Mountains, where avalanches cause damage and fatalities by crushing buildings and vehicles, and covering highways and railways. Avalanches are very powerful. They can destroy buildings, remove full-grown trees from hillsides, and even sweep railroad trains from their tracks. One of the greatest avalanche disasters on record occurred in 1910 in the Cascades near Seattle, Washington, when a passenger train, trapped in a narrow valley in a snowstorm for several days, was caught in an avalanche and swept to the bottom of the valley. Ninety-six passengers died as the cars were crushed with snow. Although avalanches are among the more dangerous natural hazards, they have caused just over 600 recorded mortalities in North America from 1985 to 2004, and most avalanche victims in North America are caught in slides they triggered themselves by walking or skiing across open slopes with accumulated snow.

Reducing the size and likelihood of avalanches can be accomplished using explosives and support structures. Areas where many people are at risk, such as resort areas, may employ explosives to initiate small avalanches as snow builds up. This process effectively reduces the magnitude of an avalanche that may occur by removing snow buildup gradually. Snow fences or nets are structures used to prevent avalanche initiation in the starting zone, where unstable snow gives way and begins to move. These supports give external support to the accumulating snow, limit avalanche size by introducing discontinuity in the snow surface, and slow momentum for small avalanches.

Resources

BOOKS

American Avalanche Association, and National Avalanche Center (U.S.). *Snow, Weather, and Avalanches: Observational Guidelines for Avalanche Programs in the United States.* Pagosa Springs, CO: American Avalanche Association, 2004.

Bolognesi, Robert. *Avalanche!: Understand and Reduce the Risks from Avalanches.* Milnthorpe, UK: Cicerone, 2007.

McClung, David, and P. A. Schaerer. *The Avalanche Handbook.* Seattle, WA: Mountaineers Books, 2006.

Moynier, John. *Avalanche Aware: The Essential Guide to Avalanche Safety.* Guilford, CT: Falcon, 2006.

Pudasaini, Shiva P., and Kolumban Hutter. *Avalanche Dynamics: Dynamics of Rapid Flows of Dense Granular Avalanches.* Berlin: Springer, 2007.

Mary Ann Cunningham

B

Bacillus thuringiensis

Bacillus thuringiensis, or *B.t.*, is a family of bacterial-based, biological insecticides. Specific strains of *B.t.* are used against a wide variety of leaf-eating lepidopteran pests such as European corn borer, tomato hornworms, and tobacco moths, and some other susceptible insects such as blackflies and mosquitoes. The active agent in *B.t.* is toxic organic crystals that bind to the gut of an insect and poke holes in cell membranes, literally draining the life from the insect. *B.t.* can be applied using technology similar to that used for chemical insecticides, such as high-potency, low-volume sprays of *B.t.* spores applied by aircraft. The efficacy of *B.t.* is usually more variable and less effective than that of chemical insecticides, but the environmental effects of *B.t.* are argued to be more acceptable because there is little nontarget toxicity.

Background radiation

Ionizing radiation has the potential to kill cells or cause somatic (affecting the body) or germinal (relating to reproductive cells) mutations. It has this ability by virtue of its power to penetrate living cells and produce highly reactive charged ions. The ions (electrically charged atoms), within the ionizing radiation, directly cause cell damage. Radiation accidents and the potential for radiation from nuclear bombs (nuclear fission weapons, sometimes also called atomic bombs) and hydrogen bombs (nuclear fusion weapons) create a fear of radiation release around human activity.

However, people are subjected to diagnostic and therapeutic radiation each day. In addition, many older Americans were exposed to radioactive fallout from atmospheric testing of nuclear weapons in the middle part of the twentieth century. In the twenty-first century, there is a small amount of environmental contamination from nuclear fuel used in power plants. Accordingly, there is still considerable interest in radiation effects on biological systems and the sources of radiation in the environment.

Concern for radiation safety is certainly justified and most individuals seek to minimize their exposure to human-generated radiation. However, for most people, exposure levels to radiation from natural sources far exceed exposure to radiation produced by humans. In the United States, current estimates of human exposure levels of ionizing radiation suggest that less than 20 percent is of human origin. The remaining radiation (80 percent) is from natural sources and is referred to as "background radiation." While radiation doses vary tremendously from person to person, the average human has an annual exposure to ionizing radiation of about 360 millirem. Millirem or mrem (one-thousandth of one rem, where rem stands for roentgen equivalent in man) is a measure of radiation absorbed by tissue multiplied by a factor that takes into account the biological effectiveness of a particular type of radiation and other factors such as the competence of radiation repair. One mrem is equal to 10 μSv; where μSv is an abbreviation for microsievert (or one millionth of one sievert), a unit that is used internationally.

Some radiation has little biological effect. Visible light and infrared radiation, two types of electromagnetic radiation, do not cause ionization, are not mutagenic (mutation causing) and are not carcinogenic (cancer causing). Consequently, background radiation refers to ionizing radiation that is derived from cosmic radiation, terrestrial radiation, and radiation from sources internal to the body. (Background radiation has the potential for producing inaccurate counts from devices such as Geiger counters. For example, cosmic rays will be recorded when measuring the radioactive decay of a sample. This background "noise" must be subtracted from the indicated count level to give a true indication of activity of the sample.)

Cosmic rays are of galactic origin, entering Earth's atmosphere from outer space. Solar activity in the form of solar flares (explosions on the Sun) and sunspots (dark spots on the Sun) affects the intensity of cosmic rays. The atmosphere of Earth serves as a protective layer for humans and anything that damages that protective layer will increase the radiation exposure of those who live under it. The dose of cosmic rays doubles at 4,920 feet (1,500 m) above sea level. Because of this, citizens of Denver, Colorado, near the Rocky Mountains, receive more than twice the dose of radiation from cosmic rays as do citizens of coastal cities such as New Orleans, Louisiana. The aluminum shell of a jet airplane provides little protection from cosmic rays, and for this reason passengers and crews of high-flying jet airplanes receive more radiation than their earth-traveling compatriots. Even greater is the cosmic radiation encountered at 60,000 feet (18,300 m) where supersonic jets fly. The level of cosmic radiation there is 1,000 times that at sea level. While the cosmic ray dose for occasional flyers is minimal, flight and cabin crews of ordinary jet airliners receive an additional exposure of 160 mrem per year, an added radiation burden to professional flyers of more than 40 percent. Cosmic sources for nonflying citizens at sea level are responsible for about 8 percent (29–30 mrem) of background radiation exposure per annum.

Another source of background radiation is terrestrial radioactivity from naturally occurring minerals, such as uranium, thorium, and cesium, in soil and rocks. The abundance of these minerals differs greatly from one geographic area to another. Residents of the Colorado plateau receive approximately double the dose of terrestrial radiation as those who live in Iowa or Minnesota. The geographic variations are attributed to the local composition of Earth's crust and the kinds of rock, soil, and minerals present. Houses made of stone are more radioactive than houses made of wood. Limestones and sandstones are low in radioactivity when compared with granites and some shales. Naturally occurring radionuclides in soil may become incorporated into grains and vegetables and thus gain access to the human body. Radon is a radioactive gas produced by the disintegration of radium, which is produced from uranium. Radon escapes from Earth's crust and becomes incorporated into all living matter including humans. It is the largest source of inhaled radioactivity and comprises about 55 percent of total human radiation exposure (both background and human generated). Energy efficient homes, which do not leak appreciable amounts of air, may have a higher concentration of radon inside than is found in outside air. This is especially true of basement air. The radon in the home decays into radioactive "daughters" that become attached to aerosol particles which, when inhaled, lodge on lung and tracheal surfaces. Obviously, the level of radon in household air varies with construction material and with geographic location. Is radon in household air a hazard? Many people believe it is, since radon exposure (at a much higher level than occurs breathing household air) is responsible for lung cancer in nonsmoking uranium miners.

Naturally occurring radioactive carbon (carbon–14) similarly becomes incorporated into all living material. Thus, external radiation from terrestrial sources often becomes internalized via food, water, and air. Radioactive atoms (radionuclides) of carbon, uranium, thorium, and actinium and radon gas provide much of the terrestrial background radiation. The combined annual exposure to terrestrial sources, including internal radiation and radon, is about 266 mrem and far exceeds other, more feared sources of radiation.

Life on Earth evolved in the presence of ionizing radiation. It seems reasonable to assume that mutations can be attributed to this chronic, low level of radiation. Mutations are usually considered to be detrimental, but over the long course of human and other organic evolution, many useful mutations occurred, and it is these mutations that have contributed to the evolution of higher forms.

Nevertheless, it is to an organism's advantage to resist the deleterious effects associated with most mutations. The forms of life that inhabit Earth today are descendants of organisms that existed for millions of years on Earth. Inasmuch as background ionizing radiation has been on Earth longer than life, humans and all other organisms obviously cope with chronic low levels of radiation. Survival of a particular species is not due to a lack of genetic damage by background radiation. Rather, organisms survive because of a high degree of redundancy of cells in the body, which enables organ function even after the death of many cells (e.g., kidney and liver function, essential for life, does not fail with the loss of many cells; this statement is true for essentially all organs of the human body). Further, stem cells in many organs replace dead and discarded cells. Naturally occurring antioxidants are thought to protect against free radicals produced by ionizing radiation. Finally, repair mechanisms exist that can, in some cases, identify damage to the double helix and effect DNA (deoxyribonucleic acid) repair. Hence, while organisms are vulnerable to background radiation, mechanisms are present which assure survival.

Resources

BOOKS

Benarde, M. A. *Our Precarious Habitat.* New York: Wiley-Interscience, 2007.

OTHER

Cornell Lab of Ornithology. "Citizen Science." http://www. birds.cornell.edu/NetCommunity/Page.aspx?pid = 708 (accessed October 12, 2010).

Jefferson Laboratory. "Radiation Sources." http://www. jlab.org/div_dept/train/rad_guide/sources.html (accessed October 12, 2010).

Scandia National Laboratories. "Radiation." http://www.san dia.gov/ciim/ISA/1rad.html (accessed October 12, 2010).

Robert G. McKinnell

Bacon, Sir Francis

1561–1626
English statesman, author, and philosopher

Sir Francis Bacon, philosopher and lord chancellor of England, was one of the key thinkers involved in the development of the procedures and epistemological standards of modern science. Bacon thus has also played a vital role in shaping modern attitudes towards nature, human progress, and the environment. He inspired many of the great thinkers of the Enlightenment, especially in England and France. Moreover, Bacon laid the intellectual groundwork for the mechanistic view of the universe characteristic of eighteenth and nineteenth century thought and for the explosion of technology in the same period.

In *The Advancement of Learning* (1605) and *Novum Organum* (1620), Bacon attacked all teleological ways of looking at nature and natural processes (i.e., the idea found in Aristotle and in medieval scholasticism that there is an end or purpose which somehow guides or shapes such processes). For Bacon, this way of looking at nature resulted from the tendency of human beings to make themselves the measure of the outer world, and thus to read purely human ends and purposes into physical and biological phenomena. Science, he insisted, must guard against such prejudices and preconceptions if it was to arrive at valid knowledge.

Instead of relying on or assuming imaginary causes, science should proceed empirically and inductively, continuously accumulating and analyzing data through observation and experiment. Empirical observation and the close scrutiny of natural phenomena allow the

Sir Francis Bacon. *(Painting by Paul Somer. Corbis-Bettmann)*

scientist to make inferences, which can be expressed in the form of hypotheses. Such hypotheses can then be tested through continued observation and experiment, the results of which can generate still more hypotheses. Advancing in this manner, Bacon proposed that science would come to more and more general statements about the laws which govern nature and, eventually, to the secret nature and inner essence of the phenomena it studied.

As Bacon rather famously argued, "Knowledge is power." By knowing the laws of nature and the inner essence of the phenomena studied, human beings can remake things as they desire. Bacon believed that science would ultimately progress to the point that the world itself would be, in effect, merely the raw material for whatever future ideal society human beings decided to create for themselves.

The possible features of this future world are sketched out in Bacon's unfinished utopia, *The New Atlantis* (1627). Here Bacon developed the view that the troubles of his time could be solved through the construction of a community governed by natural scientists and the notion that science and technology indeed could somehow redeem mankind. Empirical science would unlock the secrets of nature thus providing for technological advancement. With technological development would come material abundance and, implicitly, moral and political progress.

Bacon's utopia is ruled by a "Solomon's House"—an academy of scientists with virtually absolute power to decide which inventions, institutions, laws, practices, and so forth will be propitious for society. Society itself is dedicated to advancing the human mastery of nature: "The End of Our Foundation is the Knowledge of Causes and secret motions of things; and the enlarging of the bounds of the human empire, to the effecting of all things possible."

Resources

BOOKS

Bacon, Francis. *The Advancement of Learning.* 1605.
Bacon, Francis. *Novum Organum.* 1620.
Bacon, Francis. *The New Atlantis.* 1627.

Lawrence J. Biskowski

BACT *see* **Best Available Control Technology.**

Baghouse

A baghouse is an air pollution control device normally using a collection of long, cylindrical, fabric filters to remove particulate matter from an exhaust air stream. The filter arrangement is normally designed to overcome problems of cleaning and handling large exhaust volumes. In most cases, exhaust gas enters long (usually 33–50 ft [10–15 m]), vertical, cylindrical filters on the inside from the bottom. The bags are sealed at the top. As the exhaust air passes through the fabric filter, particles are separated from the air stream by sticking either to the filter fabric or to the cake of particles previously collected on the inside of the filter. The exhaust then passes to the atmosphere free of most of its original particulate-matter loading; collection efficiency usually increases with particle size.

The buildup of particles on the inside of the bags is removed periodically by various methods, such as rapping the bags, pulsing the air flow through the bags, or shaking. The particles fall down the long cylindrical bags and are normally caught in a collection bin, which is unloaded periodically. A baghouse system is usually much cheaper to install and operate than a system using electrostatic precipitation to remove particulates.

Balance of nature

The ideal of a balance of nature is based on a view of the natural world that is largely an artifact created by the temporal, spatial, and cultural filters through which humans respond to the natural world. For a variety of reasons, people tend to interpret the natural course of events in the world around them as maintaining equilibrium and seeking to return it to equilibrium when disturbed.

There are three components to nature's balance: ecological, evolutionary, and population. In an ecological sense, communities are thought to proceed through successional stages to a steady state climax. When disturbed, the community tends to return to that climax state. Stability is assumed to be an endpoint, and once reached the community becomes a partly closed homeostatic system. In an evolutionary sense, the current compliment of species is interpreted as the ultimate product of evolution rather than a temporary expression of a continually changing global taxa. In the population sense, concepts such as carrying capacity and the constant interplay between environmental resistance and biotic potential are interpreted as creating a balance of numbers in a population and between the population and its environment. Three ideas are fundamental to the above: that nature undisturbed is constant; that when nature is disturbed it returns to the constant condition; and that constancy in nature is the desired endpoint.

This interpretation of nature may be so strongly filtered by one's cultural interpretation and idealization of balance that one tends to produce conclusions not in keeping with one's observations of nature. Assumptions of human centrality may be sufficiently strong to bend the usually clear lens supplied by science in this case. Although the concept of balance in nature has been formally criticized in ecology for over sixty-five years (since Frederick Clements and Henry Gleason focused the argument in the 1920s), the core of the science did not change until about 1985. Since that time, a dynamic approach that pays no special attention to equilibrium processes has taken center stage in ecological theorizing.

The primary alternatives are part of the group of ideas termed intermediate disturbance hypotheses. These ideas offer a different view of how communities assemble, suggesting that disturbance is more frequent and/or more influential than performing a routine return to an equilibrium state. Furthermore, disturbance and nonequilibrium situations are responsible for the most diverse communities, for example, tropical rain forests and coral reefs, through the reduction in competition caused by disturbance factors.

Few theorists, however, suggest that nonequilibrium settings are the single most powerful explanation or are mutually exclusive with communities that do have an equilibrium. There are situations that seem to seek equilibrium and smaller subsystems that appear virtually closed. In local situations, certain levels of resources and disturbance may create long-term stability and niche differentiation or other mechanisms may be the principal cause of a species diverse situation.

Although some theorists have been working to verify, revise, and examine new developments in ecology, very little attention has been given to alternative, more complex theoretical interpretations of nature, in terms of time, space, and dynamism in resource and environmental management. The implications of accepting a non-equilibrium orientation for environmental management are significant. Most of the underpinnings of resource management include steady state carrying capacity, succession, predator-prey balance, and community equilibrium as foundations.

There are three major implications in the shift from equilibrium to nonequilibrium approaches to the environment. First, until a more realistic theory is used, the rate of resource extraction from nature will be subject to considerably higher uncertainty than many people suspect. Since communities are expected to seek equilibrium, populations and species numbers are likely to be predicted to increase more than may be warranted. Second, people perceive that areas of high biodiversity are due to long- and short-term stability, when in fact the forces may be just the opposite. Therefore, management that attempts to maintain stability is the reverse of what is actually needed. Third, the kinds of disturbance people create in these diverse communities (deforestation, introduction of nonnative species, or oil pollution) may not mimic anything natural and a species may have little defense against them. A characteristic of communities with high species diversity is small population size, thus human

disturbance may cause exceptionally high rates of extinction. A central facet of the burgeoning practice of ecological restoration should be an ability to accurately mimic disturbance regimes.

Balance in nature has a strong appeal and has anchored natural resource management practices. Now practice stands well behind theoretical developments, and increased integration of more modern ecological science is required to avoid costly resource management mistakes.

Resources

BOOKS

Des Jardins, Joseph R. *Environmental Ethics: An Introduction to Environmental Philosophy*. Belmont, CA: Wadsworth, 2005.

Falk, C. L. *New Mexico Organic Conference; Cultivating an Ecological Conscience; Essays from a Farmer Philosopher*. 2010.

Hull, David L., and Michael Ruse. *The Cambridge Companion to the Philosophy of Biology*. Cambridge, UK: Cambridge University Press, 2007.

Nelson, William M. *Oxford Handbook of Green Chemistry: From Philosophy to Industrial Applications*. New York: Oxford University Press, 2010.

Sideris, Lisa H., and Kathleen Dean Moore. *Rachel Carson: Legacy and Challenge*. Albany: State University of New York Press, 2008.

Dave Duffus

Bald eagle

The bald eagle (*Haliaeetus leucocephalus*), one of North America's largest birds of prey with a wingspan of up to 7.5 feet (2.3 m), is a member of the family Accipitridae. Adult bald eagles are dark brown to black with a white head and tail; immature birds are dark brown with mottled white wings and are often mistaken for golden eagles (*Aquila chrysaetos*). Bald eagles feed primarily on fish, but also eat rodents, other small mammals, and carrion. The bald eagle is the national emblem for the United States, adopted as such in 1782 because of its fierce, independent appearance. This characterization is unfounded, however, as this species is usually rather timid.

Formerly occurring over most of North America, the bald eagle's range—particularly in the lower forty-eight states—had been drastically reduced by a variety of reasons, one being its exposure to DDT and related

Bald eagle (*Haliaeetus leucocephalus*). *(FlordiaStock/ Shutterstock.com)*

pesticides, which are magnified in the food chain/web. This led to reproductive problems, in particular, thin-shelled eggs that were crushed during incubation. The banning of DDT use in the United States in 1972 may have been a turning point in the recovery of the bald eagle. Eagle populations also were depleted due to leadpoisoning. Estimates are that for every bird that hunters shot and carried out with them, they left behind about a half pound of lead shot, which affects the wildlife in that ecosystem long after the hunters are gone. Since 1980, more than sixty bald eagles have died from lead poisoning. Other threats facing their populations include habitat loss or destruction, human encroachment, collisions with high power lines, and shooting.

In 1982, the population in the lower forty-eight states had fallen to less than 1,500 pairs, but by 1988, their numbers had risen to about 2,400 pairs. Due to strict conservation laws, the numbers have continued to rise and there are now 6,000 pairs. On July 4, 2000, the bald eagle was removed from the Endangered listing and is now listed as Threatened in the lower forty-eight states. The bald eagle is not endangered in the state of Alaska, since a large, healthy population of about 35,000 birds exists there. During the annual salmon run, up to 4,000 bald eagles congregate along the Chil-kat River in Alaska to feed on dead and dying salmon.

Bald eagles, which typically mate for life and build huge platform nests in tall trees or cliff ledges, have been aided by several recovery programs, including the construction of artificial nesting platforms. They will reuse, add to, or repair the same nest annually, and some pairs have been known to use the same nest for over thirty-five years. Because the bald eagle has been listed as either endangered or threatened throughout most of the United States, the federal government

has provided some funding for its conservation and recovery projects. In 1989, the federal government spent $44 million on the conservation of threatened and endangered species. Of the 554 species listed, $22 million, half of the total allotment, was spent on the top twelve species on a prioritized list. The bald eagle was at the top of that list and received $3 million of those funds.

Resources

BOOKS

Bald Eagle (Haliaeetus Leucocephalus). Washington, DC: U.S. Fish & Wildlife Service, 2006.

Guidry, Jeff. *An Eagle Named Freedom: My True Story of a Remarkable Friendship.* New York: William Morrow, 2010.

Rogers, Denny, and Lori Corbett. *The Illustrated Bald Eagle.* East Petersburg, PA: Fox Chapel, 2006.

OTHER

International Union for Conservation of Nature and Natural Resources. "IUCN Red List of Threatened Species: *Haliaeetus leucocephalus.*"http://www.iucnredlist.org/ apps/redlist/details/144341/0 (accessed November 9, 2010).

ORGANIZATIONS

American Eagle Foundation, P.O. Box 333, Pigeon Forge, TN, USA, 37868, (865) 429-0157, (865) 429-4743, (800) 2EAGLES, EagleMail@Eagles.Org., http://www. eagles.org

Eugene C. Beckham

Barrier island

An elongated island that lies parallel to, but mostly separate from, a coastline. Barrier islands are composed of sediments, mainly sand, deposited by longshore currents, wind, and wave action. Both marine and terrestrial plants and animals find habitat on barrier islands or along their sandy beach shore-lines. These islands also protect coastal lagoons from ocean currents and waves, providing a warm, quiet environment for species that cannot tolerate more violent wind and wave conditions. In recent decades these linear, sandy islands, with easy access from the mainland, have proven a popular playground for vacationers. These visitors now pose a significant threat to breeding birds and other coastal species. Building houses, roads, and other disruptive human activities can destabilize dunes and expose barrier islands to disastrous storm damage. In some cases, whole islands are swept away, exposing protected lagoons and delicate wetlands to further damage. Major barrier island formations in North America include those along the eastern coasts of the mid-Atlantic states, Florida, and the Gulf of Mexico.

Basel Convention

The Basel Convention on the Control of Transboundary Movements of Hazardous Wastes and their Disposal is a global treaty that was adopted in 1989 and became effective on May 5, 1992. The Basel Convention represents a response by the international community to problems caused by the international shipment of wastes. Of the over 400 million tons of hazardous wastes generated globally every year, an unknown amount is subject to transboundary movement. In the 1980s, several highly publicized "toxic ships" were accused of trying to dump hazardous wastes illegally in developing countries. The uncontrolled movement and disposal of hazardous waste, especially in developing countries that often lacked the know-how and equipment to safely manage and treat hazardous wastes, became a significant problem due to high domestic costs of treating or disposing of wastes.

Through its Secretariat in the United Nations Environment Programme, the Convention aims to control the transboundary movement of wastes, monitor and prevent illegal traffic, provide technical assistance and guidelines, and promote cooperation. The convention imposes obligations on treaty signatories to ensure that wastes are managed and disposed of in an environmentally sound manner. The main principles of the Basel Convention are to (1) reduce transboundary movements of hazardous wastes to a minimum consistent with environmentally sound management; (2) treat and dispose of hazardous wastes as close as possible to their source of generation; and (3) reduce and minimize the generation of hazardous waste. The convention generally prohibits parties to the convention from importing or exporting hazardous wastes or other wastes from or to a noncontracting party. However, a party to the convention may allow such import or export if the party has a separate bilateral or multilateral agreement regarding the transboundary movement of hazardous wastes with a nonparty and that agreement provides for "environmentally sound management." To date, the convention has defined environmentally sound management practices for a number of wastes and technologies, including organic solvents, waste oils, pentachloraphenol (PCBs), household wastes, landfills, incinerators, and oil recycling.

The Conference of the Parties (COP) is the governing body of the Basel Convention and is composed of all governments that have ratified or acceded to it. The COP has met nine times since the convention entered into force in May 1992. At the Third Meeting of the COP (COP3), the parties adopted an amendment to the convention that will ban the export of hazardous wastes from developed countries to developing ones. The Fourth Meeting of COP (COP4), held in Kuala Lumpur in October 1997, incorporated work on lists of wastes into the system of the Basel Convention. These lists should help to mitigate practical difficulties in determining exactly what is a waste, a problem that has been encountered by a number of parties.

As of 2010, the European Union and 173 states were parties to the Basel Convention. Though the United States was among the original signatories to the Basel Convention in 1989, the United States was still not a party to the convention since Congress had not yet passed implementing legislation. The United States was participating in planning and technical aspects of the convention.

Stuart Batterman

Bass, Rick

1958–
American writer

Rick Bass is an environmental activist and writer.

Bass was born in south Texas and grew up there, absorbing stories and family lore from his grandfather during deer-hunting trips. These early forays into Texas hill country form the basis of the author's first book, *The Deer Pasture*, published when he was twenty-seven years old. Bass has lived in Texas, Mississippi, Vermont, Utah, Arkansas, and Montana.

Bass received a degree in geology from Utah State University in 1979 and went to work as a petroleum geologist in Mississippi, prospecting for new oil wells. This experience informed one of his better-known nonfiction books, *Oil Notes*. Written in journal form, *Oil Notes* offers meditations on the art and science of finding energy in the ground, as well as reflections on the author's personal life and his outdoor adventures.

Author Rick Bass writes about wilderness and woodsmen from his home in the small, isolated burg and has carved out a niche as one of Montana's most recognized writers. *(AP Photo/Independent Record, Jason Mohr)*

nature themes. *Publisher's Weekly* said *Colter* was as much "a book about appreciating nature and life" as it was the story of Bass and his dog.

Although most of Bass's publications are nonfiction, the author has written both short stories and novels. Perhaps not surprisingly, the masculine pursuits of hunting, fishing, and drinking are central to many of Bass's stories. In 1998, Bass published his first novel, *Where the Sea Used to Be*. This book was followed by a collection of short stories in 2002, called *The Hermit's Story*. In 2006, he published a short fiction collection titled *The Lives of Rocks*, which was nominated for a 2007 Story Prize. In 2007, he published *The New Wolves: The Return of the Mexican Wolf to the American Southwest*. His autobiography *Why I Came West* was one of the finalists for the 2008 National Book Critics Award.

Bass has continued to publish essays and stories in many popular magazines, including *Sports Afield*, *Audubon*, *National Geographic Traveler*, *Atlantic Monthly*, *Sierra*, and others.

Bass is a passionate environmentalist whose nonfiction in particular celebrates efforts to reclaim a wilder America. Books such as *The Lost Grizzlies: A Search for Survivors* demonstrate his conviction that America's larger predators should be allowed to survive and thrive. Bass highlights the plight of the wolf in 1992's *The Ninemile Wolves* and 1998's *The New Wolves*.

Bass features his adopted Montana homeland in some of his publications, including his nonfiction titles *Winter: Notes from Montana*, a 1991 release, and *The Book of Yaak*, which was published in 1996. Bass's essays are often linked by comments about his dog, a German shorthaired pointer named Colter. Bass gave readers a more exclusive look at Colter with his 2000 book *Colter: The True Story of the Best Dog I Ever Had*. The work, however, also stays true to Bass's characteristic

Bats

Bats, the only mammals that fly, are among nature's least understood and unfairly maligned creatures. Bats are extremely valuable animals, responsible for consuming huge numbers of insects and pollinating and dispersing the seeds of fruit-bearing plants and trees, especially in the tropics. Yet, superstitions about and fear of these nocturnal creatures have led to their persecution and elimination from many areas, and several species of bats are now threatened with extinction.

There are over 900 species of bats, representing almost a quarter of all mammal species, and they are found on every continent except Antarctica. Most types of bats live in the tropics, and some forty species are found in the United States and Canada. The largest bats, flying foxes, found on Pacific islands, have wingspreads of 5 feet (1.5 m). The smallest bats, bamboo bats, are the size of the end of a person's thumb.

Bats commonly feed on mosquitoes and other night-flying insects, especially over ponds and other bodies of water. Some bats consume half of their weight in insects a night, eating up to 5,000 gnat-sized mosquitoes one hour, thus helping to keep insect populations under control. Some bats hunt ground-dwelling species, such as spiders, scorpions, large insects, and beetles, and others prey on frogs, lizards, small birds, rodents,

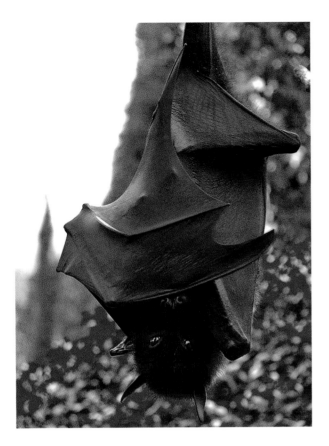

Flying fox bat, Bali, Indonesia. (Dominque Capelle/ Shutterstock.com)

fish, and even other bats. The infamous vampire bat of Central and South America does actually feed on blood, daily consuming about a tablespoon from cattle and other animals, but it does not generally bother humans.

Bats that live in tropical areas, such as fruit bats (also called flying foxes), often feed on and pollinate plants. Bats are thus extremely important in helping flowers and fruit-bearing plants to reproduce. In tropical rain forests, for example, bats are responsible for pollinating most of the fruit trees and plants.

Bats are usually social animals. Some colonies consist of millions of bats and use the same roost for centuries. Bat manure (guano) is often collected from caves and used as fertilizer. Most bats come out only at night and spend their days in dark roosts, hanging upside down, sleeping, nursing, and tending their young, or grooming their wings and fur. Bats become active an hour or so before dark, and at dusk they leave their roosting areas and fly out to feed, returning home before dawn. Many bats flying at night navigate and locate food, such as flying insects, by echolocation, emitting continuous high frequency sounds that echo

or bounce off of nearby objects. Such sounds cannot be heard by humans. Most bats have just one or two young a year, though some have up to four offspring at a time. The newborn must hold onto its mother, sometimes for several weeks, and be nursed for six to eight weeks. Some species of bats live up to twenty-five years. Most bats in North America migrate or hibernate in caves during the winter, when food is scarce and temperatures reach freezing point. Superstitions about and prejudice against bats have existed for hundreds of years, but most such tales are untrue. Bats do not carry bedbugs or become entangled in women's hair; they are not blind and indeed do not even have poor vision. In fact, except for the occasional rabid bat, these creatures are not dangerous to humans and are quite timid and will try to escape if confronted. In recent years, public education programs and conservationists, such as American ecologist Merlin Tuttle (b. 1941), head of Bat Conservation International in Austin, Texas, have helped correct these misconceptions about bats and have increased appreciation for the valuable role these creatures play in destroying pests and pollinating crops. Bracken Cave, located between San Antonio and Austin, is owned by Bat Conservation International and with some twenty million Mexican freetailed bats residing there in the spring and summer, the cave is said to shelter the world's largest bat colony and the largest collection of mammals anywhere on the planet. The pregnant females migrate there in early March from central Mexico to nurse and raise their young, and the colony can consume 250 tons of insects a night.

According to Tuttle, a colony of just 150 big brown bats can eat almost 40,000 cucumber beetles in a summer, which "means that they've protected local farmers from 18 million root worms, which cost American farmers $1 billion a year," including crop damage and pesticide costs. Tuttle and his organization suggest that people attract the creatures and help provide habitat for them by constructing or buying bathouses. Nevertheless, bats continue to be feared and exterminated throughout the world. Major threats to the survival of bats include intentional killing, loss of habitat (such as old trees, caves, and mines), eviction from barns, attics, and house eaves, pesticide poisoning, and vandalism and disturbance of caves where they roost. According to Tuttle, "Bats are among the most endangered animals in America. Nearly 40 percent of America's 43 species are either endangered or candidates for the list."

Over a dozen species of bats worldwide are listed by the U.S. Department of the Interior as endangered species, including the gray bat (*Myotis grisescens*) of

Low tide at Bay of Fundy, Nova Scotia, Canada. *(© Susan E. Degginger / Alamy)*

and Nova Scotia and the U.S. state of Maine. It encompasses about 62.5-thousand mi^2 (180-thousand km^2) of marine coastal-shelf habitat, mostly less than about 660 ft (200 m) deep. The bay is renowned for its exceptionally high tides, which can exceed 53 ft (16 m) in its upper reaches in the Minas Basin. These are higher tides than occur anywhere else in the world. During the peak tidal flooding of the bay the flow of water is about 880-million ft^3/s (25-million m^3/s), equivalent to about 2000 times the average flow of the Saint Lawrence River.

The astonishing tides of the Bay of Fundy occur because its long shape, great size, and increasing up-bay shallowness result in its tidal waters "piling up" to great depths. This effect is amplified by the natural period of tidal oscillation of the bay of about 13 hours, which further pushes against the natural tidal cycle of 12.4 hours. This rare physical phenomenon is known as a "near-resonant response." The tidal heights are particularly extreme in the upper reaches of the bay, but even in its lower areas small boats are commonly left high and dry during the twice-daily low tides, and

rivers may have a reversing tidal bore (or advancing wave) moving upstream with each tide. The "Reversing Falls" near the mouth of the Saint John River is another natural phenomenon associated with the great tides of the Bay of Fundy.

The huge tidal flows of the Bay of Fundy result in great upwellings of nutrient-rich bottom waters at some places, allowing high rates of ecological productivity to occur. The high productivity of marine phytoplankton supports a dense biomass of small crustaceans known as zooplankton, which are fed upon by great schools of small fishes such as herring. The zooplankton and fishes attract large numbers of such seabirds as gulls, phalaropes, and shearwaters to the bay during the summer and autumn months, and also abundant fin whales, humpback whales, northern right whales (this is the most endangered species of large whale), harbor porpoises, and white-sided dolphins.

The high productivity of the bay once also supported large stocks of commercial marine species, such as cod, haddock, scallop, and others. While overfishing has been a problem, the bay still supports large commercial fisheries of lobster and herring. There has also been a huge development of aquaculture in the lower bay, especially in the Passamaquoddy Bay area of New Brunswick.

In shallow areas, the extreme tidal ranges of the bay expose extensive mudflats at low tide. In some parts of the upper bay these mudflats are utilized by immense numbers of shorebirds during their autumn migration. The most abundant of these is the semi-palmated sandpiper, one of the most abundant shorebirds in the world. During its autumn migration, hundreds of thousands of these birds feed on mud shrimp in exposed mudflats at low tide, and then aggregate in dense numbers on shingle beaches at high tide. The sandpipers greatly increase their body weight during the several weeks they spend in the upper Bay of Fundy, and then leave for a nonstop flight to South America, fuelled by the fat laid down in the bay.

During the early 1970s there was a proposal to develop a huge tidal-power facility at the upper Bay of Fundy, to harvest commercial energy from the immense, twice-daily flows of water. The tidal barrage would have extended across the mouth of Minas Basin, a relatively discrete embayment with a gigantic tidal flow. Partly because of controversy associated with the enormous environmental damages that likely would have been caused by this ambitious development, along with the extraordinary construction costs and untried technology, this tidal-power facility was never built. A much smaller, demonstration project of 20 MW was commissioned in 1984 at Annapolis Royal in the upper bay, and even this facility has caused significant local damages.

In 2007, the use of Bay of Fundy tidal power regained momentum, when Nova Scotia Power contracted a Scottish tidal power company, OpenHydro, to explore the feasibility of installing seabed turbines in the bay. The project was approved in 2009, and two other candidate companies were selected for the testing of power generation tidal-driven turbines.

Nova Scotia has set a target for tidal power as supplying 20 percent of the total energy generated in the province by 2013, with the bulk coming from the Bay of Fundy.

Resources

BOOKS

Davis, Scott. *Serious Microhydro: Water Power Solutions from the Experts*. Gabriola Island, BC: New Society Publishers, 2010.

Hardisty, Jack. *The Analysis of Tidal Stream Power*. New York: Wiley, 2009.

Thurston, Harry. *A Place Between the Tides: A Naturalist's Reflections on the Salt Marsh*. Vancouver, BC: Greystone Books, 2004.

Bill Freedman

Beach renourishment

Beach renourishment, also called beach recovery or replenishment, is the act of rebuilding eroded beaches with offshore sand and gravel that is dredged from the sea floor. Renourishment projects are sometimes implemented to widen a beach for more recreational capacity, or to save structures built on an eroding sandy shoreline. The process is one that requires ongoing maintenance; the shoreline created by a renourished beach will eventually erode again. According to the National Oceanic and Atmospheric Administration (NOAA), as of 2010, the estimated cost of long-term restoration of a beach is between $3.3 and $17.5 million per mile.

The process itself involves dredging sand from an offshore site and pumping it onto the beach. The sand dredging site (sometimes referred to as a sand borrow) must also be carefully selected to minimize any negative environmental impact. For instance, dredging near the shoreline can induce erosion. The dredging process can also be lethal to aquatic organisms that are in the vicinity of the borrow and may disturb or kill organisms that are buried or attached to the

Public beach restoration on the Atlantic shore of Miami Beach, Florida. *(Jeffrey Greenberg / Photo Researchers, Inc.)*

seafloor such as submerged aquatic vegetation (SAV). Removal of parts or layers of the seafloor alters the existing characteristics and habitats and disrupts the seafloor ecosystem. Dredging can stir up silt and bottom sediment, which increases water turbidity and cuts off oxygen and light to marine flora and fauna. The placement of the dredged sand into beach areas can cover nesting sites for turtles, plants, SAV, and other organisms burrowed in the sand, affecting their survival. However, placing sand in eroded areas can supply increased area for habitats.

Regulations that involve beach renourishment planning, permits, and implementation include the Clean Water Act, the National Environmental Policy Act (NEPA), and the Coastal Zone Management Act (CZMA). These laws ensure that the beach erosion amendment and prevention procedures are carried out in an environmentally responsible manner.

Under the Energy and Water Development Appropriations Act of 2002, 65 percent of beach renourishment costs are paid for with federal funds, and the remaining 35 percent with state and local monies. Beach renourishment programs are often part of a state's coastal zone management program. In some cases, state and local government work with the U.S. Army Corps of Engineers to evaluate and implement an erosion control program such as beach renourishment.

Environmentalists argue that renourishment is for the benefit of commercial and private development, not for the benefit of the beach. Erosion is a natural process governed by weather and sea changes, and critics charge that tampering with it can permanently alter the ecosystem of the area in addition to threatening endangered sea turtle and seabird nesting habitats. The renourishment practices may sometimes be temporary solutions, which can lead to continued intervention.

In some cases, it is the coastal development that brings about the need for costly renourishment projects. Coastal development can hasten the erosion of beaches, displacing dunes and disrupting beach grasses and other natural barriers. Other human constructions, such as sea walls and other armoring, can also alter the shoreline. All of these alterations can disrupt the natural ecosystem, thus displacing animals, degrading habitat, and reducing biodiversity.

Results of a U.S. Army Corps of Engineers Study completed in 2001—the Biological Monitoring Program for Beach Nourishment Operations in Northern New Jersey (Manasquan Inlet to Asbury Park Section)—found that although dredging in the borrow

area had a negative impact on local marine life, most species had fully recovered within 24–30 months. Further studies are needed to determine the full long--range impact of beach renourishment programs on biodiversity and local coastal habitats.

Resources

BOOKS

Bird, E. C. F. *Coastal Geomorphology: An Introduction.* Chichester, UK: Wiley, 2008.

OTHER

National Oceanic and Atmospheric Administration (NOAA). "Beach Nourishment: Law and Policy." http://www.csc.noaa.gov/beachnourishment/html/human/law/index.htm (accessed September 2, 2010).
National Oceanic and Atmospheric Administration (NOAA). "NOAA Office of Ocean and Coastal Resource Management." http://www.ocrm.nos.noaa.gov/ (accessed September 2, 2010).

Paula Anne Ford-Martin

Bear *see* **Grizzly bear.**

Beattie, Mollie

1947–1996
American forester and conservationist

Mollie Hanna Beattie was trained as a forester, worked as a land manager and administrator, and ended her brief career as the first woman to serve as director of the U.S. Fish and Wildlife Service. Beattie's bachelor's degree was in philosophy, followed by a master's degree in forestry from the University of Vermont in 1979. In 1991, Beattie used a Bullard Fellowship at Harvard University to add a master's degree in public administration.

Early in her career, Mollie Beattie served in several conservation administrative posts and land management positions at the state level. She was commissioner of the Vermont Department of Forests, Parks, and Recreation (1985–1989), and deputy secretary of the state's Agency of Natural Resources (1989–1990). She also worked for private foundations and institutes: as program director and lands manager (1983–1985) for the Windham Foundation, a private, nonprofit organization concerned with critical issues facing Vermont, and later (1990–1993) as executive director of the Richard A. Snelling Center for Government in Vermont, a public policy institute.

Before becoming director of the Fish and Wildlife Service, Beattie was perhaps most widely known (especially in New England) for the book she coauthored on managing woodlots in private ownership, an influential guide reissued in a second edition in 1993. As a reviewer noted about the first edition, for the decade it was in print "thousands of landowners and professional foresters [recommended the book] to others." Especially noteworthy is the book's emphasis on the responsibility of private landowners for effective stewardship of the land. This background was reflected in her thinking as Fish and Wildlife director by making the private lands program—conservation in partnership with private land-owners—central to the agency's conservation efforts.

As director of the Fish and Wildlife Service, Beattie quickly became nationally known as a strong voice for conservation and as an advocate for thinking about land and wildlife management in ecosystem terms. One of her first actions as director was to announce that "the Service [will] shift to an ecosystem approach to managing our fish and wildlife resources." She emphasized that people use natural ecosystems for many purposes, and "if we do not take an ecosystem approach to conserving biodiversity, none of [those uses] will be long lived."

Her philosophy as a forester, conservationist, administrator, and land manager was summarized in her repeated insistence that people must start making better connections—between wildlife and habitat health and human health; between their own actions and "the destiny of the ecosystems on which both humans and wildlife depend"; and between the well-being of the environment and the well-being of the economy. She stressed "even if not a single job were created, wildlife must be conserved" and that the diversity of natural systems must remain integral and whole because "we humans are linked to those systems and it is in our immediate self interest to care" about them. She reiterated that in any frame greater than the short-term, the economy and the environment are "identical considerations."

Though Beattie's life was relatively short, and her tenure at the Fish and Wildlife Service cut short by illness, her ideas were well received in conservation circles and her influence continues after her death, helping to create what she called "preemptive conservation," anticipating crises before they appear and using that foresight to minimize conflict, to maintain biodiversity and sustainable economies, and to prevent extinctions.

Resources

BOOKS

Beattie, M., C. Thompson, and L. Levine. *Working with Your Woodland: A Land-Owner's Guide*, rev. ed. Hanover, NH: University Press of New England, 1993.

Gerald L. Young

Bees *see* **Africanized bees.**

Bellwether species

Bellwether species are monitored as an indicator of larger and more complex changes in locations, systems, populations, or ecological strata. Bellwether species are also called indicator species or sentinel species and are used as early warning signs of environmental damage and ecosystem change.

Ecologists have identified many bellwether species in various ecosystems. Stresses observable in Arctic polar bear populations are, for example, considered as indicators of ecosystem stress and change in the Arctic.

Resources

OTHER

United States Environmental Protection Agency (EPA). "Ecosystems: Ecological Monitoring: Environmental Indicators." http://www.epa.gov/ebtpages/ecosecological monienvironmentalindicators.html (accessed October 24, 2010).

Douglas Dupler

Below Regulatory Concern

Large populations all over the globe continue to be exposed to low-level radiation. Sources include natural background radiation, widespread medical uses of ionizing radiation, and releases and leakages from nuclear power and weapons manufacturing plants and waste storage sites. In the late–1980s, the Nuclear Regulatory Commission (NRC) pushed a proposal that could have added potential hazard to public health in the United States. The NRC proposal would have allowed low-level radioactive waste generated in industry, research, and hospitals to be mixed with general household trash and industrial waste in unprotected dump sites.

Successful control of erosion and sedimentation from construction and mining activities involves a system of BMPs that targets each stage of the erosion process. The first stage involves minimizing the potential sources of sediment by limiting the extent and duration of land disturbance to the minimum needed, and protecting surfaces once they are exposed. The second stage of the BMP system involves controlling the amount of runoff and its ability to carry sediment by diverting incoming flows and impeding internally generated flows. The third stage involves retaining sediment that is picked up on the project site through the use of sediment-capturing devices. Acid drainage from mining activities requires even more complex BMPs to prevent acids and associated toxic pollutants from harming surface waters.

Other pollutant sources for which BMPs have been developed include atmospheric deposition, boats and marinas, habitat degradation, roads, septic systems, underground storage tanks, and wastewater treatment.

Resources

OTHER

United States Environmental Protection Agency (EPA). "Pollution Prevention: Best Management Practices." http://www.epa.gov/ebtpages/pollbestmanagement practices.html (accessed November 9, 2010).

United States Environmental Protection Agency (EPA). "Water: Water Pollution." http://www.epa.gov/ebtpages/watewaterpollution.html (accessed November 9, 2010).

United States Environmental Protection Agency (EPA). "Water: Water Pollution Control." http://www.epa.gov/ebtpages/watewaterpollutioncontrol.html (accessed November 9, 2010).

United States Environmental Protection Agency (EPA). "Water: Water Pollution: Nonpoint Sources." http://www.epa.gov/ebtpages/watewaterpollutionnonpointsources.html (accessed November 9, 2010).

Judith L. Sims

Best practical technology

Best practical technology (BPT) refers to any of the categories of technology-based effluent limitations pursuant to Sections 301(b) and 304(b) of the Clean Water Act as amended. These categories are the best practicable control technology currently available (BPT); the best available control technology (BAT) economically feasible (BAT); and the best conventional pollutant control technology (BCT).

Section 301(b) of the Clean Water Act specifies that "in order to carry out the objective of this Act there shall be achieved—(1)(A) not later than July 1, 1977, effluent limitations for point sources, other than publicly owned treatment works (i) which shall require the application of the best practicable control technology currently available as defined by the Administrator pursuant to Section 304(b) of this Act, or (ii) in the case of discharge into a publicly owned treatment works which meets the requirements of subparagraph (B) of this paragraph, which shall require compliance with any applicable pretreatment requirements and any requirements under Section 307 of this Act…"

The BPT identifies the current level of treatment and is the basis of the current level of control for direct discharges. BACT improves on the BPT, and it may include operations or processes not in common use in industry. BCT replaces BACT for the control of conventional pollutants, such as biochemical oxygen demand (BOD), total suspended solids (TSS), fecal coliform, and pH. Details such as the amount of constituents, and the chemical, physical, and biological characteristics of pollutants, as well as the degree of effluent reduction attainable through the application of the selected technology can be found in the development documents published by the Environmental Protection Agency (EPA). These development documents cover different industrial categories such as dairy products processing, soap and detergents manufacturing, meat products, grain mills, canned and preserved fruits and vegetables processing, and asbestos manufacturing.

In accordance with Section 304(b) of the Clean Water Act, the factors to be taken into account in assessing the BPT include the total cost of applying the technology in relation to the effluent reductions to the results achieved from such an application, the age of the equipment and facilities involved, the process employed, the engineering aspects of applying various types of control technologies and process changes, and calculations of environmental impacts other than water quality (including energy requirements). As far as evaluating the BCT is concerned, the factors are mostly the same. By they include consideration of the reasonableness of the relationship between the costs of attaining a reduction in effluents and the benefits derived from that reduction, and the comparison of the cost and level of reduction of such pollutants from the discharge from publicly owned treatment works to the cost and level of reduction of such pollutants from a class or category of industrial sources. Control technologies may include in-plant control and preliminary treatment, and end-of-pipe treatment, examples of which are water conservation and reuse, raw materials substitution, screening, multimedia filtration, and activated carbon absorption.

James W. Patterson

Beta particle

A beta particle is an electron emitted by the nucleus of a radioactive atom. The beta particle is produced when a neutron within the nucleus decays into a proton and an electron. Beta particles have greater penetrating power than alpha particles but less than x-ray or gamma rays. Although beta particles can penetrate skin, they travel only a short distance in tissue. Beta rays pose relatively little health hazard, therefore, unless they are ingested into the body. Naturally radioactive materials such as potassium-40, carbon-14, and strontium-90 emit beta particles, as do a number of synthetic radioactive materials.

See also Radioactivity.

Beyond Pesticides

Founded in 1981, Beyond Pesticides (originally called the National Coalition Against the Misuse of Pesticides) is a nonprofit, grassroots network of groups and individuals concerned with the dangers of pesticides. Members of Beyond Pesticides include individuals, such as "victims" of pesticides, physicians, attorneys, farmers and farmworkers, gardeners, and former chemical company scientists, as well as health, farm, consumer, and church groups. All want to limit pesticide use through Beyond Pesticides, which publishes information on pesticide hazards and alternatives, monitors and influences legislation on pesticide issues, and provides seed grants and encouragement to local groups and efforts.

Administered by a fifteen-member board of directors and a small full-time staff, including a toxicologist and an ecologist, Beyond Pesticides is now the most prominent organization dealing with the pesticide issue. It was established on the premise that much is unknown about the toxic effects of

pesticides and the extent of public exposure to them. Because such information is not immediately forthcoming, members of Beyond Pesticides assert that the only available way of reducing both known and unknown risks is by limiting or eliminating pesticides. The organization takes a dual-pronged approach to accomplish this. First, Beyond Pesticides draws public attention to the risks of conventional pest management; second, it promotes the least-toxic alternatives to current pesticide practices.

An important part of Beyond Pesticides's overall program is the Center for Community Pesticide and Alternatives Information. The center is a clearinghouse of information, providing a 2,000-volume library about pest control, chemicals, and pesticides. To concerned individuals it sells inexpensive brochures and booklets, which cover topics such as alternatives to controlling specific pests and chemicals; the risks of pesticides in schools, to food, and in reproduction; and developments in the Federal Insecticide, Fungicide and Rodenticide Act (FIFRA), the national law governing pesticide use and registration in the United States. Through the center Beyond Pesticides also publishes *Pesticides and You* (*PAY*) five times a year. It is a newsletter sent to approximately 4,500 people, including Beyond Pesticides members, subscribers, and members of Congress. The center also provides direct assistance to individuals through access to Beyond Pesticides's staff ecologist and toxicologist.

In 1991 Beyond Pesticides also established the Local Environmental Control Project after the Supreme Court decision affirming local communities' rights to regulate pesticide use. Although Beyond Pesticides supported bestowing local control over pesticide use, it needed a new program to counteract the subsequent mobilization of the pesticide industry to reverse the Supreme Court decision. The Local Environmental Control Project campaigns first to preserve the right accorded by the Supreme Court decision and second to encourage communities to take advantage of this right.

Beyond Pesticides marked its tenth anniversary in 1991 with a forum titled "A Decade of Determination: A Future of Change." It included workshops on wildlife and groundwater protection, cancer risk assessment, and the implications of GATT and free trade agreements. Beyond Pesticides has also established the annual National Pesticide Forum. Through such conferences, its aid to victims and groups, and its many publications, Beyond Pesticides above all encourages local action to limit pesticides and change the methods of controlling pests.

Resources

ORGANIZATIONS

Beyond Pesticides, 701 E Street, SE, Suite 200, Washington,
D.C., USA, 20003, (202) 543-5450, (202) 543-4791,
info@beyond pesticides.org, http://www.beyond
pesticides.org

Andrea Gacki

Bhopal, India

On December 3, 1984, one of the world's worst industrial accidents occurred in Bhopal, India. Along with Three Mile Island and Chernobyl, Bhopal stands as an example of the dangers of industrial development without proper attention to environmental health and safety.

A large industrial and urban center in the state of Madhya Pradesh, Bhopal was the location of a plant owned by the American chemical corporation Union Carbide, Inc. and its Indian subsidiary Union Carbide India, Ltd. The plant manufactured pesticides, primarily the pesticide carbaryl (marketed under the name Sevin), which is one of the most widely used carbamate-class pesticides in the United States and throughout the world. Among the intermediate chemical compounds used together to manufacture Sevin is methyl isocyanate (MIC)—a lethal substance that is reactive, toxic, volatile, and flammable. It was the uncontrolled release of MIC from a storage tank in the Bhopal facility that caused the deaths of many thousands of people. The actual number who died from the Bhopal disaster may never be known exactly, as various organizations and commissions have arrived at different figures. The best estimates (as of 2010) are that around 3,800 people died more-or-less immediately from exposure to the lethal gas cloud of MIC, with many more thousands dying in the coming days, weeks, and months. Moreover, well over 100,000 people were seriously injured. Many observers argue that the numbers cited above are substantial underestimates.

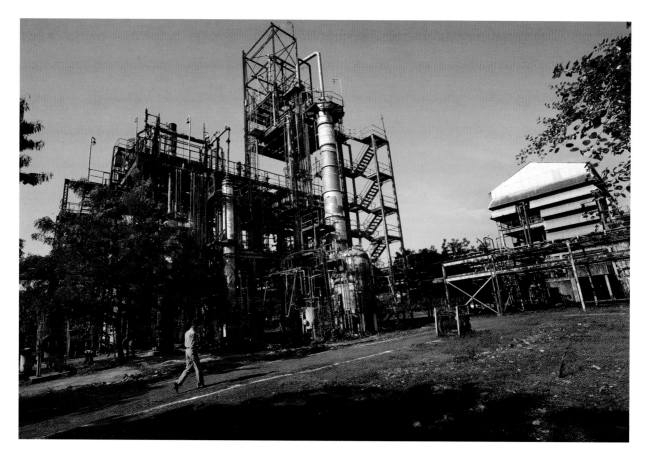

Indian policeman passing by the Union Carbide plant in the Indian city of Bhopal, 2009, over 25 years after a gas leak from the plant. *(Harish Tyagi)*

MIC (CH3-N = C = O) is highly volatile and has a boiling point of 89°F (39.1°C). In the presence of trace amounts of impurities such as water or metals, MIC reacts to generate heat, and if the heat is not removed, the chemical begins to boil violently. If relief valves, cooling systems, and other safety devices fail to operate in a closed storage tank, the pressure and heat generated may be sufficient to cause a release of MIC into the atmosphere. Because the vapor is twice as heavy as air, the vapors, if released, remain close to the ground where they can do the most damage, drifting along prevailing wind patterns. As set by the Occupational Health and Safety Administration (OSHA), the standards for exposure to MIC are set at 0.02 ppm over an eight-hour period. The immediate effects of exposure, inhalation, and ingestion of MIC at high concentrations (above 2 ppm) are burning and tearing of the eyes, coughing, vomiting, blindness, massive trauma of the gastrointestinal tract, clogging of the lungs, and suffocation of bronchial tubes. When not immediately fatal, the long-term health consequences include permanent blindness, permanently impaired lung functioning, corneal ulcers, skin damage, and potential birth defects.

Many explanations for the disaster have been advanced, but the most widely accepted theory is that trace amounts of water entered the MIC storage tank and initiated the hydrolysis reaction, which was followed by MIC's spontaneous reactions. The plant was not well-designed for safety, and maintenance was especially poor. Four key safety factors should have contained the reaction, but it was later discovered that they were all inoperative at the time of the accident. The refrigerator that should have slowed the reaction by cooling the chemical was shut off, and, as heat and pressure built up in the tank, the relief valve blew. A vent gas scrubber designed to neutralize escaping gas with caustic soda failed to work. Also, the flare tower that would have burned the gas to harmless by-products was under repair. Yet even if all these features had been operational, subsequent investigations found them to be poorly designed and insufficient for the capacity of the plant. Once the runaway reaction started, it was virtually impossible to contain.

The poisonous cloud of MIC released from the plant was carried by the prevailing winds to the south and east of the city—an area populated by highly congested communities of poorer people, many of whom worked as laborers at the Union Carbide plant and other nearby industrial facilities. Released at night, the silent cloud went undetected by residents who remained asleep in their homes, thus possibly ensuring a maximal degree of exposure. Many hundreds died in their sleep, others choked to death on the streets as they ran out in hopes of escaping the lethal cloud. Thousands more died in the following days and weeks. The Indian government and numerous volunteer agencies organized a massive relief effort in the immediate aftermath of the disaster consisting of emergency medical treatment, hospital facilities, and supplies of food and water. Medical treatment was often ineffective, for doctors had an incomplete knowledge of the toxicity of MIC and the appropriate course of action.

In the weeks following the accident, the financial, legal, and political consequences of the disaster unfolded. In the United States, Union Carbide's stock dipped 25 percent in the week immediately following the event. Union Carbide India Ltd. (UCIL) came forward and accepted moral responsibility for the accident, arranging some interim financial compensation for victims and their families. However its parent company, Union Carbide Inc., which owned 50.9 percent of UCIL, refused to accept any legal responsibility for their subsidiary. The Indian government and hundreds of lawyers on both sides pondered issues of liability and the question of a settlement. While Union Carbide hoped for out-of-court settlements or lawsuits in the Indian courts, the Indian government ultimately decided to pursue class action suits on behalf of the victims in the United States courts in the hope of larger settlements. The United States courts refused to hear the case, and it was transferred to the Indian court system.

In 1989, Union Carbide and the government of India entered into a settlement of damage claims on behalf of the victims and their families. Under the terms of the settlement, which was mediated by the Supreme Court of India, Union Carbide acknowledged its responsibility for the disaster and agreed to pay the Indian government the sum of $470 million in compensation to be dispersed to several hundred thousand claimants. The 1989 settlement was considered to be final, although many of the living victims, their families, and victims' advocacy groups contend that the compensation awarded was generally far too low. In June 2010, an Indian court found seven Indian nationals guilty of death due to negligence, which carries a maximum penalty of two years in prison. The seven men found guilty included the chairman of the now-defunct UCIL. American corporate officers, such as former Union Carbide chairman Warren Anderson, refused to answer to indictments from the Indian courts.

The disaster in Bhopal has had far-reaching political consequences in the United States. A number of congressional hearings were called and the Environmental Protection Agency (EPA) and OSHA initiated inspections and investigations. A Union Carbide plant in McLean, Virginia, that uses processes and products similar to those in Bhopal was repeatedly inspected by officials. While no glaring deficiencies in operation or maintenance were found, it was noted that several small leaks and spills had occurred at the plant in previous years that had gone unreported. These added weight to growing national concern about workers' right-to-know provisions and emergency response capabilities. In the years following the Bhopal accident, both state and federal environmental regulations were expanded to include mandatory preparedness to handle spills and releases on land, water, or air. These regulations include measures for emergency response such as communication and coordination with local health and law enforcement facilities, as well as community leaders and others. In addition, employers are now required to inform any workers in contact with hazardous materials of the nature and types of hazards to which they are exposed; they are also required to train workers in emergency health and safety measures.

The disaster at Bhopal raises a number of critical issues and highlights the wide gulf between developed and developing countries in regard to design and maintenance standards for health and safety. Management decisions allowed the Bhopal plant to operate in an unsafe manner and for a shantytown to develop around its perimeter without appropriate emergency planning. The Indian government, like many other developing nations in need of foreign investment, appeared to sacrifice worker safety in order to attract and keep Union Carbide and other industries within its borders. While a number of environmental and occupational health and safety standards existed in India before the accident, their inspection and enforcement was cursory or nonexistent. Often understaffed, the responsible Indian regulatory agencies were rife with corruption as well. The Bhopal disaster also raised questions concerning the moral and legal responsibilities of American companies abroad, and the willingness of those corporations to apply stringent United States safety and environmental standards to their operations in the Third World despite the relatively free hand given them by local governments.

Although worldwide shock at the Bhopal accident has largely faded, the suffering of many victims continues. While many national and international safeguards on the manufacture and handling of hazardous chemicals have been instituted, few expect that lasting improvements will occur in developing countries without a gradual recognition of the economic and political values of stringent health and safety standards.

Resources

BOOKS

Diamond, Arthur. *The Bhopal Chemical Leak*. San Diego, CA: Lucent, 1990.
D'Silva, Themistodles. *The Black Box of Bhopal: A Closer Look at the World's Deadliest Industrial Disaster*. Victoria, BC: Trafford, 2006.
Kurzman, Dan. *A Killing Wind: Inside the Bhopal Catastrophe*. New York: McGraw-Hill, 1987.

OTHER

PBS. "Seven Convicted in Deadly 1984 Bhopal Gas Leak in India." http://www.pbs.org/newshour/rundown/2010/06/seven-convicted-in-india-gas-leak.html (accessed November 8, 2010).
Union Carbide. "Bhopal Information Center." http://www.bhopal.com/ (accessed November 8, 2010).

Usha Vadagiri

Bikini atoll

Bikini Atoll was selected in late 1945 as the site for a number of tests of fission weapons, to experiment with different designs for the bomb and to test its effects on ships and the natural environment.

At that time, 161 people belonging to eleven families lived on Bikini. Since the Bikinians have no written history, little is known about their background. According to their oral tradition, the original home of their ancestors is nearby Wotje Atoll. Until the early 1900s, they had relatively little contact with strangers and were regarded with some disdain even by other Marshall Islanders. After the arrival of missionaries early in the twentieth century, the Bikinians became devout Christians. People lived on coconuts, breadfruits, arrowroot, fish, turtle eggs, and birds, all available in abundance on the atoll. The Bikinians were expert sailors and fishermen. Land ownership was important in the culture, and anyone who had no land was regarded as lacking in dignity.

President Truman had signed an order on January 10, 1946, authorizing the transfer of everyone living on Bikini Atoll to the nearly uninhabited Bongerik Atoll. The United States Government asked the Bikinians to give up their native land to allow experiments that would bring benefit to all humankind. Such an action, the Americans argued, would earn for the Bikinians special glory in heaven. The islanders agreed to the request and, along with their

A massive column of water rises from the sea as the second atom bomb test at Bikini Atoll explodes underwater July 25, 1946.
(AP Photo/Joint Task Force One)

homes, church, and community hall, were transported by the United States Navy to Rongerik.

In June and July of 1946, two tests of atomic bombs were conducted at Bikini as part of "Operation Crossroads." More than ninety vessels, including captured German and Japanese ships along with surplus cruisers, destroyers, submarines, and amphibious craft from the United States Navy, were assembled. Following these tests, however, the Navy concluded that Bikini was too small and moved future experiments to Eniwetok Atoll.

The first test of a fusion device occurred on October 31, 1952, at Eniwetok Atoll in the Marshall Islands in the Pacific Ocean. This was followed by a series of six more tests, code-named "Operation Castle," at Bikini Atoll in 1954. Two years later, on May 20, 1956, the first nuclear fusion bomb was dropped from an airplane over Bikini Atoll. The testing of a nuclear fusion device in Operation Castle thus marked the return of bomb testing to Bikini. The most memorable test of that series took place in 1954 and was code-named "Bravo." Experts expected a yield of 6 megatons from the hydrogen bomb used in the test, but measured instead a yield of 15 megatons, 250 percent greater. Bravo turned out to be the largest single explosion in all of human history, producing an explosive force greater than all of the bombs used in all the previous wars in history.

Fallout from Bravo was consequently much larger than had been anticipated. In addition, because of a shift in wind patterns, the fallout spread across an area of about 50,000 mi^2 (11.5 km^2), including three inhabited islands—Rongelap, Itrik, and Rongerik. A number of people living on these islands developed radiation burns and many were evacuated from their homes temporarily. Farther to the east, a Japanese fishing boat which had accidentally sailed into the restricted zone was showered with fallout. By the time the boat returned to Japan, twenty-three crew members had developed radiation sickness. One eventually died of infectious hepatitis, probably because of the numerous blood transfusions he received.

The value of Bikini as a test site ended in 1963 when the United States and the Soviet Union signed the Limited Test Ban Treaty which outlawed nuclear weapons testing in the atmosphere, the oceans, and outer space. Five years later, the United States government decided that it was safe for the Bikinians to return home. By 1971, some had once again taken up residence on their home island. Their return was short-lived. In 1978, tests showed that returnees had ingested quantities of radioactive materials much higher than the levels considered to be safe. The Bikinians were relocated once again, this time to the isolated and desolate island of Kili, 500 miles (804 km) from Bikini.

The primary culprit on Bikini was the radioactive isotope cesium–137. It had become so widely distributed in the soil, the water, and the crops on the island that no one living there could escape from it. With a half life of 30 years, the isotope is likely to make the island uninhabitable for another century.

Two solutions for this problem have been suggested. The brute-force approach is to scrape off the upper 12 inches (30 cm) of soil, transport it to some uninhabited island, and bury it under concrete. A similar burial site, the "Cactus Crater," already exists on Runit Island. It holds radioactive wastes removed from Eniwetok Atoll. The cost of clearing off Bikini's 560 acres (227 ha) and destroying all its vegetation (including 25,000 trees) has been estimated at more then $80 million. A second approach is more subtle and makes use of chemical principles. Since potassium replaces cesium in soil, scientists hope that adding potassium-rich fertilizers to Bikini's soil will leach out the dangerous cesium–137.

At the thirtieth anniversary of Bravo, the Bikinians had still not returned to their home island. Many of the original 116 evacuees had already died. A majority of the 1,300 Bikinians who then lived on Kili no longer wanted to return to their native land. If given the choice, most wanted to make Maui, Hawaii, their new home. But they did not have that choice. The United States continues to insist that they remain somewhere in the Marshall Islands. The only place there they cannot go, at least within most of their lifetimes, is Bikini Atoll.

See also Nuclear fission; Nuclear fusion; Radiation sickness.

Resources

BOOKS

Delgado, James P. *Nuclear Dawn from the Manhattan Project to Bikini Atoll.* Oxford, UK: Osprey, 2009.

Gunn, Angus M. *Encyclopedia of Disasters: Environmental Catastrophes and Human Tragedies.* Westport, CT: Greenwood Press, 2008.

David E. Newton

Bioaccumulation

Bioaccumulation is the general term for describing the accumulation of chemicals in the tissue of organisms. The chemicals that bioaccumulate are most often organic chemicals that are very soluble in fat and lipids and are slow to degrade, such as the pesticide dichlorodiphenyltrichloroethane (DDT). Usually used in reference to aquatic organisms, bioaccumulation occurs from exposure to contaminated water through respiration (e.g., gill uptake by fish), skin contact with a substance, or by consuming food that has accumulated the chemical (e.g., food chain/web transfer). Bioaccumulation has occurred when an organism has a higher concentration of the substance in its tissue compared with the concentration in the organism's environment. How much a certain chemical is bioaccumulated depends on several factors such as the uptake rate; the means of uptake (e.g., skin, ingestion); and environmental, physical, and biological factors. The concentration of the chemical in the organism is dependent on the metabolism of the organism. For instance, the speed with which the chemical can be excreted or eliminated from the organism, the chemical modification of the substance in the organism, and the concentration of fat or lipids in the organism are all factors that play a role in the extent of bioaccumulation of a substance.

DDT was a widely used synthetic pesticide to combat mosquitoes and in agriculture in the United States prior to banning use of the chemical in 1972. The chemical is still used in tropical regions to control malaria. It is an organochlorine chemical that readily bioaccumulates in organisms and is listed as a persistent organic pollutant (POP) in the "dirty dozen" chemicals identified at the Stockholm Convention on POPs. Many research studies have shown that DDT is toxic to organisms, especially affecting the reproductive systems of animals such as birds. DDT is thought to be an endocrine disruptor, meaning that it can mimic endogenous hormones and affect reproductive development.

Bioaccumulation of chemicals in fish has resulted in public health consumption advisories in some areas, and has affected the health of certain fish-eating wildlife including eagles, cormorants, terns, and mink.

Resources

BOOKS

Dunlap, Thomas R. *DDT, Silent Spring, and the Rise of Environmentalism: Classic Texts.* Weyerhaeuser Environmental Classics. Seattle: University of Washington Press, 2008.

Flinders, Camille. *The Potential for Bioaccumulation, Bioconcentration, and Biomagnification of Selected Metals in an Aquatic Food Chain: A Literature Review.* Research Triangle Park, NC: NCASI, 2006.

Great Britain Environment Agency. *Review of Bioaccumulation Models for Use in Environmental Standards.* Bristol, UK: Environment Agency, 2007.

OTHER

United States Geological Survey (USGS). "Bioaccumulation." http://toxics.usgs.gov/definitions/bioaccumulation.html (accessed July 14, 2009).

United States Geological Survey (USGS). "DDT." http://toxics.usgs.gov/definitions/ddt.html (accessed July 14, 2009).

Johanning, Eckardt. *Bioaerosols, Fungi, Mycotoxins and Human Health.* Albany, NY: Fungal Research Group Foundation, 2005.
National Research Council (U.S.). *A Framework for Assessing the Health Hazard Posed by Bioaerosols.* Washington, DC: National Academies Press, 2008.

Bioaerosols

Bioaerosols are airborne particles derived from plants and animals, or are themselves living organisms, including viruses, bacteria, fungi, and mammal and bird antigens. Bioaerosols can range in size from roughly 0.000004 inch (0.01 micrometer) (virus) to 0.004 inch (100 micrometer) (pollen).

These particles can be inhaled and can cause many types of health problems, including allergic reactions (specific activation of the immune system), infectious diseases (pathogens that invade human tissues), and toxic effects (due to biologically-produced chemical toxins). The most common outdoor bioaerosols are pollens from grasses, trees, weeds, and crops. The most common indoor biological pollutants are animal dander (minute scales from hair, feathers, or skin), dust mite and cockroach parts, fungi (molds), infectious agents (bacteria and viruses), and pollen.

The main concern regarding bioaerosols are their roles as allergens (e.g., mold in buildings) and the pathogenicity of certain types. It has been estimated that 40 percent of U.S. homes have a mold-related issue. These mold particles have the potential to induce allergic reactions and affect the respiratory system. Some molds produce toxins termed mycotoxins that can cause neurological effects and can be lethal. Bioaerosols pose a threat as potential agents for bioterrorism attacks, such as the anthrax spore distribution through the U.S. postal system in 2001 that killed five people. After years of investigation into the 2001 anthrax attacks, the U.S. Federal Bureau of Investigation (FBI) formally closed the case in February 2010, claiming that their prime suspect (since deceased) had acted alone. The FBI concluded that a microbiologist, who had worked in biodefense for the U.S. Army, had mailed envelopes containing anthrax spores. The attacks aroused widespread fears among the public at the time, and prompted the installation of costly equipment by the U.S. Postal Service and other government agencies in an attempt to foil any similar attacks using pathogens.

Resources

BOOKS

Bailey, Hollace S. *Fungal Contamination: A Manual for Investigation, Remediation and Control.* Jupiter, FL: Building Environment Consultants, Inc. (BECi), 2005.

Bioassay

A bioassay, or biological assay, refers to an evaluation of the effect of an effluent or other material on living organisms such as fish, rats, insects, bacteria, or other life forms. The bioassay may be used for many purposes, including the determination of: (1) permissible wastewater discharge rates; (2) the relative sensitivities of various animals; (3) the effects of physicochemical parameters on toxicity; (4) the compliance of discharges with effluent guidelines; (5) the suitability of a drug; (6) the safety of an environment; and (7) possible synergistic or antagonistic effects.

Bioassays can be qualitative or quantitative. A qualitative bioassay refers to testing the physical effect of a substance on an organism that cannot be measured, such as abnormal development. A quantitative bioassay refers to rating the toxicity or potency of a substance by measurement of the biological response it elicits within the organism.

There are those who wish to reserve the term simply for the evaluation of the potency of substances such as drugs and vitamins, but the term is commonly used as described above. Of course, there are times when it is inappropriate to use bioassay and evaluation of toxicity synonymously, as when the goal of the assay is not to evaluate toxicity.

Bioassays are conducted as static, renewal, or continuous-flow experiments. In static tests, the medium (air or water) about the test organisms is not changed; in renewal tests the medium is changed periodically; and in continuous-flow experiments the medium is renewed continuously. When testing chemicals or wastewaters that are unstable, continuous-flow testing is preferable. Examples of instability include the rapid degradation of a chemical, significant losses in dissolved oxygen, problems with volatility, and precipitation.

Bioassays are also classified on the basis of duration. The tests may be short-term (acute), intermediate-term, or long-term, also referred to as chronic. In addition, aquatic toxicologists speak of partial- or complete-life-cycle assessments. The experimental design of a bioassay is in part reflected in such labels as range-finding, which

African elephant (*Loxodonta africana*) and impala (*Aepyceros melampus*) drinking water in the Savuti Area of Chobe National Park, Botswana. *(Martin Harvey / Photo Researchers, Inc.)*

the fraction of global species that live in the tropics would increase to at least 86 percent.

Invertebrates comprise the largest number of described species, with insects making up the bulk of that total and beetles (*Coleoptera*) comprising most of the insects. Biologists believe that there still is a tremendous number of undescribed species of insects in the tropics, possibly as many as another 30 million species. This remarkable conclusion has emerged from experiments conducted by the American entomologist Terry Erwin (1940–), in which scientists fogged tropical forest canopies and then collected the rain of dead arthropods. This research suggests that: (1) a large fraction of the insect biodiversity of tropical forests is undescribed; (2) most insect species are confined to a single type of forest, or even to particular plant species, both of which are restricted in distribution; and (3) most tropical forest insects have a very limited dispersal ability.

The biodiversity and endemism (referring to a species being unique to a particular geographic area) of

other tropical forest biota are better known than that of arthropods. For example, a plot of only 0.2 acres (0.1 ha) in an Ecuadorian forest had 365 species of vascular plants. The richness of woody plants in tropical rain forest can approach 300 species per hectare, compared with fewer than twelve to fifteen tree species in a typical temperate forest, and thirty to thirty-five species in the Great Smokies of the United States, the richest temperate forest in the world.

There have been few systematic studies of all of the biota of particular tropical communities. In one case, American biologist D. H. Janzen (1939–) studied a savanna-like, 67 mi^2 (108 km^2) reserve of dry tropical forest in Costa Rica for several years. Janzen estimated that the site had at least 700 plant species, 400 vertebrate species, and a remarkable 13,000 species of insects, including 3,140 species of moths and butterflies.

Some might wonder why there is concern about the extinction of so many rare species of tropical insects, or of many other rare species of plants and animals. There are three classes of reasons why extinctions are regrettable:

1. There are important concerns in terms of the ethics of extinction. Central questions are whether humans have the right to act as the exterminator of unique and irrevocable species of wildlife and whether the human existence is somehow impoverished by the tragedy of extinction. These are philosophical issues that cannot be scientifically resolved, but it is certain that few people would applaud the extinction of unique species.

2. There are utilitarian reasons. Humans must take advantage of other organisms in myriad ways for sustenance, medicine, shelter, and other purposes. If species become extinct, their unique services, be they biological, ecological, or otherwise, are no longer available for exploitation.

3. The third class of reasons is in terms of ecology and involves the roles of species in maintaining the stability and integrity of ecosystems, that is, in terms of preventing erosion and controlling nutrient cycling, productivity, trophic dynamics, and other aspects of ecosystem structure and function. Because we rarely have sufficient knowledge to evaluate the ecological importance of particular species, it is likely that an extraordinary number of species will disappear before their ecological roles are understood.

There are many cases where research on previously unexploited species of plants and animals has revealed the existence of products of great utility to humans, such as food or medicinals. One example is the rosy periwinkle (*Catharantus roseus*), a plant native to Madagascar. During a screening of many plants for possible anticancer properties, an extract of rosy periwinkle was found to counteract the reproduction of cancer cells. Research identified the active ingredients as several alkaloids, which are now used to prepare the important anticancer drugs vincristine and vinblastine. This once obscure plant now allows treatment of several previously incurable cancers and is the basis of a multimillion-dollar economy.

Undoubtedly, there is a tremendous, undiscovered wealth of other biological products that are of potential use to humans. Many of these natural products are present in the biodiversity of tropical species that has not yet been discovered by taxonomists.

It is well known that extinction can be a natural process. In fact, most of the species that have ever lived on Earth are now extinct, having disappeared naturally for some reason or other. Perhaps they could not cope with changes in their inorganic or biotic environment, or they may have succumbed to some catastrophic event, such as a meteorite impact.

The rate of extinction has not been uniform over geological time. Long periods characterized by a slow and uniform rate of extinction have been punctuated by about nine catastrophic events of mass extinction. The most intense mass extinction occurred some 250 million years ago, when about 96 percent of marine species became extinct. Another example occurred 65 million years ago, when there were extinctions of many vertebrate species, including the reptilian orders Dinosauria and Pterosauria, but also of many plants and invertebrates, including about one half of the global fauna that existed at that time.

In modern times, however, humans are the dominant force causing extinction, mostly because of: (1) overharvesting; (2) effects of introduced predators, competitors, and diseases; and (3) habitat destruction. During the last 200 years, a global total of perhaps 100 species of mammals, 160 birds, and many other taxa are known to have become extinct through some human influence, in addition to untold numbers of undescribed, tropical species.

Even preindustrial human societies caused extinctions. Stone-age humans are believed to have caused the extinctions of large-animal fauna in various places, by the unsustainable and insatiable hunting of vulnerable species in newly discovered islands and continents. Such events of mass extinction of large animals, coincident with human colonization events, have occurred at various times during the last 10,000–50,000 years in Madagascar, New Zealand, Australia, Tasmania, Hawaii, North and South America, and elsewhere.

In more recent times, overhunting has caused the extinction of other large, vulnerable species, for example the flightless dodo of Mauritius. Some North American examples include Labrador duck (*Camptorhynchus labradorium*), passenger pigeon, Carolina parakeet (*Conuropsis carolinensis*), great auk, and Steller's sea cow (*Hydrodamalis stelleri*). Many other species have been brought to the brink of extinction by overhunting and loss of habitat. Some North American examples include eskimo curlew, plains bison, and a variety of marine mammals, including manatee (*Trichechus manatus*), right whales (*Eubalaena glacialis*), bowhead whales (*Balaena mysticetus*), and blue whales (*Balaenoptera musculus*).

Island biotas are especially prone to both natural and anthropogenic extinction. This syndrome can be illustrated by the case of the Hawaiian Islands, an ancient volcanic archipelago in the Pacific Ocean, about 994 mi (1,600 km) from the nearest island group and 2,484 mi

Biodiversity

(4,000 km) from the nearest continental landmass. At the time of colonization by Polynesians, there were at least sixty-eight endemic species of Hawaiian birds, out of a total richness of land birds of eighty-six species. Of the initial sixty-eight endemics, twenty-four are now extinct and twenty-nine are perilously endangered. Especially hard hit has been an endemic family, the Hawaiian honeycreepers (Drepanididae), of which thirteen species are believed extinct, and twelve endangered. More than fifty alien species of birds have been introduced to the Hawaiian Islands, but this gain hardly compensates for the loss and endangerment of specifically evolved endemics. Similarly, the native flora of the islands is estimated to have been comprised of 1,765–2,000 taxa of angiosperm plants, of which at least 94 percent were endemic. During the last two centuries, more than 100 native plants have become extinct, and the survival of at least an additional 500 taxa is threatened or endangered, some now being represented by only single individuals. The most important causes of extinction of Hawaiian biota have been the conversion of natural ecosystems to agricultural and urban landscapes, the introduction of alien predators, competitors, herbivores, and diseases, and to some extent, aboriginal overhunting of some species of bird.

Overhunting has been an important cause of extinction, but in modern times habitat destruction is the most important reason for the event of mass extinction that Earth's biodiversity is now experiencing. As was noted previously, most of the global biodiversity is comprised of millions of as yet undescribed taxa of tropical insects and other organisms. Because of the extreme endemism of most tropical biota, it is likely that many species will become extinct as a result of the clearing of natural tropical habitats, especially forest, and its conversion to other types of habitat.

The amount and rate of deforestation in the tropics are increasing rapidly, in contrast to the situation at higher latitudes where forest cover is relatively stable. Between the mid-1960s and the mid-1980s there was little change (less than 2 percent) in the forest area of North America, but in Central America forest cover decreased by 17 percent, and in South America by 7 percent (but by a larger percentage in equatorial countries of South America). The global rate of clearing of tropical rain forest in the mid-1980s was equivalent to 6–8 percent of that biome per year, a rate that if projected into the future would predict a biome half-life of only 9 to 12 years. Some of the cleared forest will regenerate through secondary succession, which would ultimately produce another mature forest. Little is known, however, about the rate and biological character of succession in tropical rainforests, or how long it would take to restore a fully biodiverse ecosystem after disturbance.

The present rate of disturbance and conversion of tropical forest predicts grave consequences for global biodiversity. Because of a widespread awareness and concern about this important problem, much research and other activity has recently been directed towards the conservation and protection of tropical forests. Many tropical forests are considered biodiversity hotspots, which are areas containing a high percentage of unique or endemic species under threat of extinction. Currently more than 15 percent of tropical forests have received some sort of protection. Of course the operational effectiveness of the protected status varies greatly, depending on the commitment of governments to these issues. Important factors include: (1) political stability; (2) political priorities; (3) finances available to mount effective programs to control poaching of animals and lumber and to prevent other disturbances; (4) the support of local peoples and communities for biodiversity programs; (5) the willingness of relatively wealthy nations to provide a measure of debt relief to impoverished tropical countries and thereby reduce their short term need to liquidate natural resources in order to raise capital and provide employment; and (6) local population growth, which also generates extreme pressures to overexploit natural resources.

The biodiversity crisis is a very real and very important aspect of the global environmental crisis. All nations have a responsibility to maintain biodiversity within their own jurisdictions and to aid nations with less economic and scientific capability to maintain their biodiversity on behalf of the entire planet. The modern biodiversity crisis focuses on species-rich tropical ecosystems, but the developed nations of temperate latitudes also have a large stake in the outcome and will have to substantially subsidize global conservation activities if these are to be successful. Much needs to be done, but an encouraging level of activity in the conservation and protection of biodiversity is beginning in many countries, including an emerging commitment by many nations to the conservation of threatened ecosystems in the tropics. The Tropical Forest Conservation Act was passed in the United States in 1998 in an effort to protect tropical forest areas worldwide. According to this act, the United States will provide funding for struggling developing countries (debt-for-nature) in exchange for protection of tropical forest territories within those cooperating countries. A new version of this act that includes conservation protection for coral reef habitats (renamed the Tropical Forest and

170

ENVIRONMENTAL ENCYCLOPEDIA 4

Coral Conservation Act of 2009) was being ushered through Congress in early 2009.

Although some species are lost to extinction, scientists are still discovering new species. A forest survey conducted from January to March 2007 in a remote corner of the Democratic Republic of Congo discovered six animal species previously unknown to scientists. A bat, a rodent, two shrew species, and two frog species were discovered by researchers from the Wildlife Conservation Society, the Field Museum of Chicago, the World Wildlife Fund, and the National Centre of Research and Science in Lwiro in a region that includes the Misotshi-Kabogo Forest and the Marunga Massif near Lake Tanganyika.

Determinations of species extinction and other changes in biodiversity often require years of observation. Sightings after species are thought to be extinct do occur. For example, after a fruitless search late in 2006, scientists failed to find a single Chinese river dolphin (*Lipotes vexillifer*) in its natural habitat in China. Following the search they proposed reclassifying the species as possibly extinct because none of these dolphins currently survives in captivity. In August 2007, however, Chinese media reported a confirmed sighting of the Chinese river dolphin (baiji) in the Yangtze River in east China. A big white animal filmed by a Chinese businessman in Tongling City, Anhui Province, with a digital camera, was later confirmed to be a Chinese river dolphin by Professor Wang Ding of the Chinese Academy of Sciences' Institute of Hydrobiology. This sighting provides a last hope that this species may be brought back from the brink of extinction by conservation efforts.

The plight of the baiji is, unfortunately, increasingly common. Declining populations of various wild species, a precursor to biodiversity loss, have been observed worldwide. In May 2008, the Zoological Society of London and World Wide Fund for Nature (WWF; formerly the World Wildlife Fund) released results of a study indicating a dramatic decline in global wildlife populations during the last four decades. The data showed that between 1970 and 2005, the populations of land-based species fell by 25 percent, of marine species by 28 percent, and of freshwater species by 29 percent. Biodiversity loss was also reported in the form of extinction of about 1 percent of the world's species each year. The two groups, in agreement with most biologists, stated that one of the great extinction episodes in Earth's history is now under way. Species whose populations shrink can be at greater risk of extinction than species with large populations.

The losses were attributed to pollution, urbanization, overfishing, and hunting, but both groups emphasized that climate change would play an increasing role in species decline over coming decades. Coral reefs, which contain about 25 percent of ocean biodiversity, are highly susceptible to climate change, and much of the existing coral reef habitat may be lost at the current rate of global warming. The Zoological Foundation and WWF based their conclusion upon population estimates and other data related to 1,400 species tracked as part of their joint Living Planet index. Data were obtained from publications in scientific journals.

Invasive plants, animals, and insects are second only to habitat loss as a cause of ongoing biodiversity loss around the world. In May 2008, biologists and botanists with the Global Invasive Species Group, a global organization dedicated to fighting the threat of invasive species, reported that many of the plants now being touted as ideal biofuel crops are invasive species when grown in many parts of the world. For example, the plant termed rapeseed (*Brassica napus*) in Europe and canola in North America is invasive in Australia, while a wetland plant called the giant reed (*Arundo donax*), scheduled for introduction as a biofuel crop in Florida, threatens to invade what is left of the Everglades swamp ecosystem. The overlap between second-generation biofuel crops (i.e., crops other than corn and sugar cane) and invasive plant species occurs because the same qualities that make a plant desirable for biofuel—fast growth, low maintenance, few enemies—also make it more likely to become invasive when introduced outside its native environmental niche.

In February 2009, ninety nations attended a United Nations conference on biodiversity in Norway. Evidence reviewed by the conference suggested that reducing biodiversity loss is also important to sustainability and the overall health of the world's economy. Diminishing biodiversity poses a growing threat to economic sectors such as agriculture, medical research, biofuel production, and tourism. As a consequence of failures to achieve a 2010 UN target of slowing extinctions, the report argues that the current extinction rate, the highest rate since the extinction of the dinosaurs some 65 million years, now poses both environmental and economic hazards. Developing nations without a diverse economic infrastructure, or those nations heavily dependant on nature-based tourism, are especially vulnerable.

At the UN Convention on Biological Diversity (CBD) meeting held in Nagoya, Japan, during October 2010, negotiators from about 190 countries agreed to new goals aimed at slowing and ultimately reversing the accelerating extinction rates for an array of plant and animal species. As part of an official accord, they agreed to profit-sharing agreements ensuring that both developed and developing nations would share in profits from pharmaceuticals and other products derived from unique genomes. A new biodiversity, the Nagoya Protocol, establishes a goal of reducing current extinction rates by 50 percent by 2020. New targets also call for increasing protection of land areas and financial agreements between rich and poorer countries to finance species protection programs.

Reports at the conference concluded that approximately half of Earth's readily available natural resources were already used, in use, or designated for human use. Human activities impacted nearly every ecosystem.

The UN's assessment report—the Global Biodiversity Outlook—concluded that despite local successes, almost all measurable trends indicated an accelerating biodiversity loss on a global scale. The report described the world as nearing a "tipping point" beyond which biodiversity loss might become irreversible or at a cost too great to make economically feasible. The Economics of Ecosystems and Biodiversity (TEEB) project already estimates the global economic costs of biodiversity loss at $2 to $5 trillion per year.

As of 2010, the United States has not ratified the underlying United Nations Convention on Biodiversity.

Resources

BOOKS

Bernstein, Aaron. *Sustaining Life: How Human Health Depends on Biodiversity*. New York: Oxford USA, 2008.
Hulot, Nicholas. *One Planet: A Celebration of Biodiversity*. New York: Abrams, 2006.
Ladle, Richard J. *Biodiversity and Conservation: Critical Concepts in the Environment*. London: Routledge, 2009.
Ninan, K. N. *Conserving and Valuing Ecosystem Services and Biodiversity: Economic, Institutional and Social Challenges*. London: Earthscan, 2009.

PERIODICALS

Araúo, Miguel B., and Carsten Rahbek. "How Does Climate Change Affect Biodiversity?" *Science* 313 (2006): 1,396–1,397.
Higgins, Paul A. T. "Biodiversity Loss Under Existing Land Use and Climate Change: An Illustration Using Northern South America." *Global Ecology and Biogeography* 16 (2007): 197–204.
Willis, K. J., and H. J. B. Birks. "What Is Natural? The Need for a Long-Term Perspective in Biodiversity Conservation." *Science* 314 (2006): 1,261–1,265.

OTHER

Nature Conservancy. "Nature Conservancy Biodiversity Page." http://www.nature.org/initiatives/
climatechange/strategies/art21202.html (accessed October 13, 2010).
United Nations System-Wide EarthWatch. "Biodiversity assessment." http://earthwatch.unep.net/emergingissues/biodiversity/assessment.php (October 13, 2010).

Bill Freedman

Biofilms

Biofilms refer to surface-adhering microorganisms that grow and divide, often within a protective layer of sugar (polysaccharide) produced by the microbes. Biofilms occur very commonly in nature and in infections in humans and other creatures. Within a biofilm, infectious bacteria are usually more resistant to antibiotics and other antibacterial agents than their bacterial counterparts that do not exist in a biofilm. Thus, biofilms are recognized as an important component of infections such as cystic fibrosis.

Marine microbiology began with the investigations of marine microfouling by L. E. Zobell and colleagues in the 1930s and 1940s. Their interests focused primarily on the early stages of settlement and growth of microorganisms, primarily bacteria on solid substrates immersed in the sea. The interest in the study of marine microfouling was sporadic from that time until the early 1960s when interest in marine bacteriology began to increase. Since 1970 the research on the broad problems of bioadhesion and specifically the early stages of microfouling has expanded tremendously.

The initial step in marine fouling is the establishment of a complex film. This film, which is composed mainly of bacteria and diatoms plus secreted extracellular materials and debris, is most commonly referred to as the primary film but may also be called the bacterial fouling layer or slime layer. The latter name is aptly descriptive since the film ultimately becomes thick enough to feel slippery or slimy to touch. In addition to the bacteria and diatoms that comprise most of the biota, the film may also include yeasts, fungi, and protozoans.

The settlement sequence in the formation of primary films is dependent upon a number of variables which may include the location of the surface, season of the year, depth, and proximity to previously fouled surfaces and other physiochemical factors.

Many studies have demonstrated the existence of some form of ecological succession in the formation of fouling communities, commencing with film forming microorganisms and reaching a climax community of macrofouling organisms such as barnacles, tunicates, mussels, and seaweeds.

Establishment of primary films in marine fouling has two functions: (1) to provide a surface favoring the settlement and adhesion of animal larvae and algal cells, and (2) to provide a nutrient source that could sustain or enhance the development of the fouling community.

Formation of a primary film is initiated by a phenomenon known as molecular fouling or surface conditioning. The formation of this molecular film was first demonstrated by Zobell in 1943 and since has been confirmed by many other investigators. The molecular film forms by the sorption to solid surfaces of organic matter dissolved or suspended in seawater. The sorption of this dissolved material creates surface changes in the surface of the substrate which are favorable for establishing biological settlement. These dissolved organic materials originate from a variety of sources such as end-products of bacterial decay, excretory products, dissolution from seaweeds, and so forth, and consist principally of sugars, amino acids, urea, and fatty acids.

This molecular film has been observed to form within minutes after any clean, solid surface is immersed in natural seawater. The role of this film in biofouling has been shown to modify the critical surface tension or wetability of the immersed surface, which then facilitates the strong bonding of the microorganisms through the mucopolysaccharides exuded by film-forming bacteria.

Bacteria have been found securely attached to substrates immersed in seawater after just a few hours. Initial colonization is by rod-shaped bacteria followed by stalked forms within 24–72 hours. As many as forty to fifty species have been isolated from the surface of glass slides immersed in seawater for a few days.

Following the establishment of the initial film of bacteria and their secreted extracellular polymer on a solid substrate, additional bacteria and other microorganisms may attach. Most significant in this population are benthic diatoms but there are also varieties of filamentous microorganisms and protozoans. These organisms, together with debris and other organic particular matter that adhere to the surface, create an intensely active biochemical environment and form the primary stage in the succession of a typical macrofouling community.

Considering the enormous economic consequences of marine fouling it is not at all surprising that

there continues to be intense interest in the results of recent research, particularly in the conditions and processes of molecular film formation.

Resources

BOOKS

Costerton, J. William. *The Biofilm Promer*. New York: Springer, 2010.
Krumbein, W. F., D. M. Paterson, and G. A. Zavarzin, eds. *Fossil and Recent Biofilms: A Natural History of Life on Earth*. New York: Springer, 2010.
Shirtliff, Mark, and Jeff G. Leid. *The Role of Biofilms in Device-Relayed Infections*. Berlin: Springer Berlin Heidelberg, 2009.
St. Clair, Larry, and Mark Seaward. *Biodeterioration of Stone Surfaces: Lichens and Biofilms as Weathering Agents of Rocks and Cultural Heritage*. New York: Springer, 2010.

Donald A. Villeneuve

Biofiltration

Biofiltration refers to the removal and oxidation of organic gases such as volatile organic compounds (abbreviated VOCs) from contaminated air in beds (biofilters) of compost, soil, or other materials such as municipal waste, sand, bark peat, volcanic ash, or diatomaceous earth. As contaminated air (such as air from a soil vapor extraction process) flows through the biofilter, the VOCs attach onto the surfaces of the pile and are degraded by microorganisms. Nutrient blends or exogenous microbial cultures can be added to a biofilter to enhance its performance. Moisture needs to be continually supplied to the biofilter to counteract the drying effects of the gas stream. The stationary support media that make up the biofiltration bed should be porous enough to allow gas to flow through the biofilter and should provide a large surface area with a high capacity to absorb fluid and absorb gas. This support media should also provide adequate buffering capacity and may also serve as a source of inorganic nutrients. Biofilters, also used to treat odors as well as organic contaminants, have been used in Europe for over twenty years. Compared to incineration and carbon adsorption, biofilters do not require landfilling of residuals or regeneration of spent materials.

Waste gases are moved through the units by induced or forced draft. Biofilters are capable of handling rapid air flow rates (e.g., up to 90,000 cubic ft (2,700 cubic m) per minute (cfm) in filters up to 20,000 sq ft (1,800 sq m) in wetted area) and VOC concentrations greater than 1,000 ppm. However, biofiltration removal of more highly halogenated compounds such as trichloroethylene (TCE) or carbon tetrachloride, which biodegrade very slowly under aerobic conditions, may require very long residence times (i.e., very large biofilters) or treatment of very low flow rates of air containing the contaminants.

The soil-type biofilter is similar in design to a soil compost pile. Fertilizers are preblended into the compost pile to provide nutrients for indigenous microorganisms, which accomplish the biodegradation of the VOCs. In the treatment bed type of biofilter, the waste air stream is humidified as it is passed through one or more beds of compost, municipal waste, sand, diatomaceous earth, or other materials. Another type of biofilter is the disk biofilter, which consists of a series of humidified, compressed disks placed inside a reactor shell. These layered disks contain activated charcoal, nutrients, microbial cultures, and compost material. The waste air stream is passed through the disk system. Collected water condensate from the process is returned to the humidification system for reuse.

Biofiltration can be an effective means of removing pollutants from the air. Even airborne pollutants can be degraded by microorganisms adhering to the biofilter surface as the air passes. A type of biofilter known as a trickling filter has been used as a system to remove some impurities from water for over 200 years.

Resources

BOOKS

Bellinger, Edward, and David D. Sigee. *Freshwater Algae: Identification and Use as Bioindicators.* New York: Wiley, 2010.

Meuser, Helmut. *Contaminated Urban Soils.* New York: Springer, 2010.

Pieribone, Vincent, David F. Gruber, and Sylvia Nasar. *Aglow in the Dark: The Revolutionary Science of Bioluminescence.* Cambridge, MA: Belknap Press, 2006.

Judith L. Sims

Biofouling

The term fouling, or more specifically biofouling, is used to describe the growth and accumulation of living organisms on the surfaces of submerged artificial structures as opposed to natural surfaces. Concern over and interest in fouling arises from practical considerations including the enormous costs resulting from fouling of ships, buoys, floats, pipes, cables, and other underwater man-made structures.

From its first immersion in the sea, an artificial structure or surface changes through time as a result of a variety of influences including location, season, and other physical and biological variables. Fouling communities growing on these structures are biological entities and must be understood developmentally. The development of a fouling community on a bare, artificial surface immersed in the sea displays a form of succession, similar to that seen in terrestrial ecosystems, which culminates in a community which may be considered a climax stage. Scientists have identified two distinct stages in fouling community development: (1) the primary or microfouling stage, and (2) the secondary or macrofouling stage.

Microfouling: When a structure is first submerged in seawater, microorganisms, primarily bacteria and diatoms, appear on the surface and multiply rapidly. Together with debris and other organic particulate matter, these microorganisms form a film on the surface. Although the evidence is not conclusive, it appears that the development of this film is a prerequisite to initiation of the fouling succession.

Macrofouling: The animals and plants that make up the next stages of succession in fouling communities are primarily the attached or sessile forms of animals and plants that occur naturally in shallow waters along the local coast. The development of fouling communities in the sea depends upon the ability of locally occurring organisms to live successfully in the new artificial habitat. The first organisms to attach to the microfouled surface are the swimming larvae of species present at the time of immersion. The kinds of larvae present vary with the season. Rapidly growing forms that become established first may ultimately be crowded out by others that grow more slowly. A comprehensive list of species making up fouling communities recorded from a wide variety of structures identified 2,000 species of animals and plants. Although the variety of organisms identified seems large, it actually represents a very small proportion of the known marine species. Further, only about 50 to 100 species are commonly encountered in fouling, including bivalve mollusks (primarily oysters and mussels), barnacles, aquatic invertebrates in the phylum Bryozoa, tubeworms and other organisms in the class Polychaeta, and green and brown algae.

Control of fouling organisms has long been a formidable challenge resulting in the development and application of a wide variety of toxic paints and greases, or the use of metals which give off toxic ions as they corrode. The most notable has probably been the use of tributyltin-based materials to coat the hulls of ships. The problem remains that none of the existing methods provide permanent control. Furthermore, the recognition of the potential environmental hazards attendant with the use of materials that leach toxins into the marine environment has led to the ban of some of the most widely used materials. This has stimulated efforts to develop alternative materials or methods of controlling biofouling that are environmentally safe.

Resources

BOOKS

Dürr, Simone, and Jeremy Thomason. *Biofouling*. Oxford, UK: Blackwell, 2008.

Flemming, Hans-Curt. *Marine and Industrial Biofouling*. Springer series on biofilms, Vol. 4. Berlin: Springer, 2009.

Railkin, Alexander I. *Marine Biofouling: Colonization Processes and Defenses*. Boca Raton, FL: CRC Press, 2004.

Donald A. Villeneuve

Biogeochemistry

Biogeochemistry refers to the quantity and cycling of chemicals in ecosystems. Biogeochemistry can be studied at various spatial scales, ranging from communities, landscapes (or seascapes), and over Earth as a whole. Biogeochemistry involves the study of chemicals in organisms, and also in nonliving components of the environment.

An important aspect of biogeochemistry is the fact that elements can occur in various molecular forms that can be transformed among each other, often as a result of biological reactions. Such transformations are an especially important consideration for nutrients, that is, those chemicals that are required for the healthy functioning of organisms. As a result of biogeochemical cycling, nutrients can be used repeatedly. Nutrients contained in dead biomass can be recycled through inorganic forms, back into living organisms, and so on. Biogeochemistry is also relevant to the movements and transformations of potentially toxic chemicals in ecosystems, such as metals, pesticides, and certain gases.

Nutrient cycles

Ecologists have a good understanding of the biogeochemical cycling of the most important nutrients. These include carbon, nitrogen, phosphorus, potassium, calcium, magnesium, and sulfur. Some of these can occur variously as gases in the atmosphere, as ions dissolved in water, in minerals in rocks and soil, and in a great variety of organic chemicals in the living or dead biomass of organisms. Ecologists study nutrient cycles by determining the quantities of the various chemical forms of nutrients in various compartments of ecosystems, and by determining the rates of transformation and cycling among the various compartments.

The nitrogen cycle is particularly well understood, and it can be used to illustrate the broader characteristics of nutrient cycling. Nitrogen is an important nutrient, being one of the most abundant elements in the tissues of organisms and a component of many kinds of biochemicals, including amino acids, proteins, and nucleic acids. Nitrogen is also one of the most common limiting factors to primary productivity, and the growth rates of plants in many ecosystems will increase markedly if they are fertilized with nitrogen. This is a fairly common characteristic of terrestrial and marine environments, and to a lesser degree of freshwater ones.

Plants assimilate most of their nitrogen from the soil environment, as nitrate (NO_3^-) or ammonium (NH_4^+) dissolved in the water that is taken up by roots. Some may also be taken up as gaseous nitrogen oxides (such as NO or NO_2) that are absorbed from the atmosphere. In addition, some plants live in a beneficial symbiosis with microorganisms that have the ability to fix atmospheric dinitrogen gas (N_2) into ammonia (NH_3), which can be used as a nutrient. In contrast, almost all animals satisfy their nutritional needs by eating plants or other animals and metabolically breaking down the organic forms of nitrogen, using the products (such as amino acids) to synthesize the necessary biochemicals of the animal. When plants and animals die, microorganisms active in the detrital cycle metabolize organic nitrogen in the dead biomass into simpler compounds, ultimately to ammonium.

The nitrogen cycle has always occurred naturally, but in modern times some of its aspects have been greatly modified by human influences. These include the fertilization of agricultural ecosystems, the dumping of nitrogen-containing sewage into lakes and other waterbodies, the emission of gaseous forms of nitrogen into the atmosphere, and the cultivation of nitrogen-fixing legumes. In some cases, human effects on nitrogen biogeochemistry result in increased productivity of crops, but in other cases serious ecological damages occur.

Toxic chemicals in ecosystems

Some human activities result in the release of toxic chemicals into the environment, which under certain conditions can pose risks to human health and cause serious damages to ecosystems. These damages are called pollution, whereas the mere presence of chemicals which cause no damage in the environment is referred to as contamination. Biogeochemistry is concerned with the emissions, transfers, and quantities of these potentially toxic chemicals in the environment and ecosystems.

Certain chemicals have a great ability to accumulate in organisms rather than in the nonliving (or inorganic) components of the environment. This tendency is referred to as bioconcentration. Chemicals that strongly bioconcentrate include methylmercury and all of the persistent organochlorine compounds, such as dichlorodiphenyl-trichloroethane (DDT), pentachlorophenol (PCBs), dioxins, and furans. Methylmercury bioconcentrates because it is rather tightly bound in certain body organs of animals. The organochlorines bioconcentrate because they are extremely insoluble in water but highly soluble in fats and lipids, which are abundant in the bodies of organisms but not in nonliving parts of the environment.

In addition, persistent organochlorines tend to occur in particularly large concentrations in the fat of top predators, that is, in animals high in the ecological food web, such as marine mammals, predatory birds, and humans. This happens because these chemicals are not easily metabolized into simpler compounds by these animals, so they accumulate in increasingly larger residues as the animals feed and age. This phenomenon is known as food-web magnification (or biomagnification). Food-web magnification causes chemicals such as DDT to achieve residues of tens or more parts per million (ppm) in the fatty tissues of top predators, even though they occur in the inorganic environment (such as water) in concentrations smaller than one part per billion (ppb). These high body residues can lead to ecotoxicological problems for top predators, some of which have declined in abundance because of their exposure to chlorinated hydrocarbons.

Because a few species of plants have an affinity for potentially toxic elements, they may bioaccumulate them to extremely high concentrations in their tissues. These plants are genetically adapted ecotypes which are themselves little affected by the residues, although they can cause toxicity to animals that might feed on their biomass. For example, some plants that live in environments in which the soil contains a mineral known as serpentine accumulate nickel to concentrations which may exceed thousands of ppm. Similarly, some plants (such as locoweed) growing in semi-arid environments can accumulate thousands of ppm of selenium in their tissues, which can poison animals that feed on the plants.

Resources

BOOKS

Reddy, K. R., and R. D. DeLaune. *Biogeochemistry of Wetlands Science and Applications*. Boca Raton, FL: CRC, 2008.

Walther, J. V. *Essentials of Geochemistry*. Sudbury, MA: Jones and Bartlett, 2005.

OTHER

United States Department of the Interior, United States Geological Survey (USGS). "Geochemistry." http://www.usgs.gov/science/science.php?term=437 (accessed September 4, 2010).

Bill Freedman

Biogeography

Biogeography is the study of the spatial distribution of plants and animals, both today and in the past. Developed during the course of nineteenth-century efforts to explore, map, and describe Earth, biogeography asks questions about regional variations in the numbers and kinds of species: Where do various species occur and why? What physical and biotic factors limit or extend the range of a species? In what ways do species disperse (expand their ranges), and what barriers block their dispersal? How has species distribution changed over centuries or millennia, as shown in the fossil record? What controls the makeup of a biotic community (the combination of species that occur together)? Biogeography is an interdisciplinary science: many other fields, including paleontology, geology, botany, oceanography, and climatology, both contribute to biogeography and make use of ideas developed by biogeographers.

Because physical and biotic environments strongly influence species distribution, the study of ecology is closely tied to biogeography. Precipitation, temperature ranges, soil types, soil or water salinity, and insolation (exposure to the sun) are some elements of the physical environment that control the distribution of plants and animals. Biotic limits to distribution, constraints imposed by other living things, are equally important. Species interact in three general ways: competition with other species (for space, sunlight, water, or food), predation (e.g., an owl species relying on rodents for food), and mutualism (e.g., an insect pollenizing a plant while the plant provides nourishment for the insect). The presence or absence of a key plant or animal may function as an important control on another species' spatial distribution. Community ecology, the ways in which an assemblage of species coexist, is also important. Biotic communities have a variety of niches, from low to high trophic levels, from generalist roles to specialized ones. The presence or absence of species filling one of these roles influences the presence or survival of a species filling another role.

Two other factors that influence a region's biotic composition or the range of a particular species are dispersal, or spreading, of a species from one place to another; and barriers, environmental factors that block dispersal. In some cases a species can extend its range by gradually colonizing adjacent, hospitable areas. In other cases a species may cross a barrier, such as a mountain range, an ocean, or a desert, and establish a colony beyond that barrier. The cattle egret (*Bubulcus ibis*) exemplifies both types of movement. Late in the nineteenth century these birds crossed the formidable barrier of the Atlantic Ocean, perhaps in a storm, and established a breeding colony in Brazil. During the past 100 years this small egret has found suitable habitat and gradually expanded its range around the coast of South America and into North America, so that by 1970 it had been seen from southern Chile to southern Ontario.

The study of dispersal has special significance in island biogeography. The central idea of island biogeography, proposed in 1967 by R. H. MacArthur and Edward O. Wilson, is that an island has an equilibrium number of species that increases with the size of the land mass and its proximity to other islands. Thus species diversity should be extensive on a large or nearshore island, with enough complexity to support large carnivores or species with very specific food or habitat requirements. Conversely, a small or distant island may support only small populations of a few species, with little complexity or niche specificity in the biotic community.

Principles of island biogeography have proven useful in the study of other "island" ecosystems, such as isolated lakes, small mountain ranges surrounded by deserts, and insular patches of forest left behind by clearcut logging. In such threatened areas as the Pacific Northwest and the Amazonian rain forests, foresters are being urged to leave larger stands of trees in closer proximity to each other so that species at high trophic levels and those with specialized food or habitat requirements (e.g., Northern spotted owls and Amazonian monkeys) might survive. In such areas as Yellowstone National Park, which national policy designates as an insular unit of habitat, the importance of adjacent habitat has received increased consideration. Recognition that clearcuts and farmland constitute barriers has led some planners to establish forest corridors to aid dispersal, enhance genetic diversity, and maintain biotic complexity in unsettled islands of natural habitat.

Resources

BOOKS

Sax, Dov F., John J. Stachowicz, Steven D. Gaines, eds. *Species Invasions: Insights into Ecology, Evolution, and Biogeography*. Sunderland, MA: Sinaur Associates, 2005.

Lomolino, Mark, and Lawrence Heaney. *Frontiers of Biogeography*. Sunderland, MA: Sinauer Associates, 2004.

Lomolino, Mark, Brett Riddle, and James Brown. *Biogeography*, 3rd ed. Sunderland, MA: Sinauer Associates, 2005.

Staller, John, et al. *Histories of Maize: Multidisciplinary Approaches to the Prehistory, Linguistics, Biogeography, Domestication, and Evolution of Maize*. San Diego, CA: Academic Press, 2006.

PERIODICALS

Higgins, Paul A. T. "Biodiversity Loss Under Existing Land Use and Climate Change: An Illustration Using Northern South America." *Global Ecology and Biogeography* 16 (2007): 197–204.

OTHER

United States Department of the Interior, United States Geological Survey (USGS). "Biogeography." http://www.usgs.gov/science/science.php?term=96 (accessed September 4, 2010).

Mary Ann Cunningham

Biohydrometallurgy

Biohydrometallurgy is a technique by which microorganisms are used to recover certain metals from ores. The technique was first used over 300 years ago to extract copper from low-grade ores. In recent years, its use has been extended to the recovery of uranium

In the environmental field, bioindicators are commonly used in field investigations of contaminated sites to document impacts on the biological community and ecosystem. These studies are then followed up with focused laboratory tests to pinpoint the source of toxicity or stress. After cleanup and remedial actions have been implemented at a site, bioindicators are also used to track the effectiveness of the remediation activity. In the future, bioindicators may be used more widely as investigative and decision-making tools from the initial pollution and impact assessment stage to the remediation and post-remediation monitoring stages.

Resources

BOOKS

Bellinger, Edward, and David D. Sigee. *Freshwater Algae: Identification and Use as Bioindicators.* New York: Wiley, 2010.
Meuser, Helmut. *Contaminated Urban Soils.* New York: Springer, 2010.
Pieribone, Vincent, David F. Gruber, and Sylvia Nasar. *Aglow in the Dark: The Revolutionary Science of Bioluminescence.* Cambridge, MA: Belknap Press, 2006.

Usha Vedagiri

In 2010, the globe is beginning to more evidently experience the effects of atmospheric warming. The changing climates in different regions of the Earth are already affecting biological communities. Even if the warming of the atmosphere could be stopped instantaneously, global changes will occur for at least another century, according to the predictive data from a number of different climate models.

While some changes may be relatively innocuous, changes that affect a species that are important to the food chain, for example to phytoplankton in the ocean, would be very influential and likely deleterious.

Resources

BOOKS

Ballesta, Laurent, Pierre Deschamp, and Jean-Michel Cousteau. *Planet Ocean: Voyage to the heart of the Marine Realm.* Washington, DC: National Geographic, 2007.
Lippson, Alice Jane, and Robert L. Lippson. *Life in the Chesapeake Bay.* Baltimore: Johns Hopkins University Press, 2006.
Molles, Manuel C. *Ecology: Concepts and Applications.* New York: McGraw Hill Science/Engineering/Math, 2009.
Schindler, David W., and John R. Vallentyne. *The Algal Bowl: Overfertilization of the World's Freshwaters and Estuaries.* Edmonton: University of Alberta Press, 2008.

Gerald L. Young

Biological community

A biological community is an association or group of populations of organisms living in a localized area or habitat. The community is a level of organization incorporating individual organisms, species, and populations. A population is a group of one species, and the community can consist of one or more populations. Communities may be large or small, ranging from the microscopic to the continent level of a biome and the global level of a biosphere.

In contrast to an ecosystem, a community does not necessarily include consideration of the physical environment or the habitat of a particular group of organisms. The term ecosystem was coined to incorporate study of a community together with its physical environment. Still, communities are adaptive systems, inseparable from and evolving in response to changing environmental conditions. So, the supply and availability of resources in the environment and also time are considerations in the dynamics of community structure and relationships. Individual communities may be relatively stable or constantly changing.

Biological fertility

Biological fertility is a term that refers to the number of offspring produced by a female organism. In a population, biological fertility is measured as the general fertility rate (the birth rate multiplied by the number of sexually productive females) or as the total fertility rate (the lifetime average number of offspring per female). Fertility and fecundity are synonyms. In population biology, biological fertility refers to the number of offspring actually produced, while fecundity is merely the biological ability to reproduce. Fecund individuals that fail to mate do not produce offspring (that is, are not biologically fertile), and do not contribute to population growth.

Biological fertility also applies to soil, and considers the chemical, physical, and biological components of soil. Knowledge of how these three parameters affect the capability of the soil to support crops is becoming a very important facet of agriculture, especially as the growing population of the globe is facing increasing pressure to produce enough food. Biological fertility of the soil can also be deleteriously affected by climate change, especially in regions where drought will be exacerbated.

Resources

BOOKS

Bruges, James. *The Biochar Debate: Charcoal's Potential to reverse Climate Change and Build Soil Fertility*. White River Junction, VT: Chelsea Green, 2010.

Chivian, Eric, and Aaron Burnstein. *Sustaining Life: How Human Health Depends on Biodiversity*. New York: Oxford University Press, 2008.

Cribb, Julian. *The Coming Famine: The Global Food Crisis and What We Can Do to Avoid It*. Berkeley: University of California Press, 2010.

Biological integrity *see* **Ecological integrity.**

Biological magnification *see* **Biomagnification.**

Biological methylation

Methylation is the process by which a methyl radical (-CH$_3$) is chemically combined with some other substance through the action of a living organism. One of the most environmentally important examples of this process is the methylation of mercury in the sediments of lakes, rivers, and other bodies of water. Elementary mercury and many of its inorganic compounds have relatively low toxicity because they are insoluble. However, in sediments, bacteria can convert mercury to an organic form, methylmercury, that is soluble in fat. When ingested by animals, methylmercury accumulates in body fat and exerts highly toxic, sometimes fatal, effects.

Biological methylation also refers to methylation that occurs in living organisms, and which is important in regulating activities of the organisms. The addition of a methyl group has been shown to be vital in the regulation of some genes, determining whether the transcription of the gene occurs or not. Methylation can also regulate the production of proteins in the process of translation.

Resources

BOOKS

Alberts, Bruce. *Molecular Biology of the Cell*. New York: Garland Science, 2008.

Carson, Susan, and Dominique Robertson. *Manipulation and Expression of Recombinant DNA: A Laboratory Manual*. Burlington, MA: Elsevier Academic Press, 2006.

Pisano, Gary P. *The Science Business: The Promise, the Reality, and the Future of Biotech*. Cambridge, MA: Harvard Business School Press, 2006.

Biological oxygen demand *see* **Biochemical oxygen demand.**

Biological Resources Discipline

Created to assess, monitor, and research biological resources in United States, the Biological Resources Discipline (BRD) of the United States Geological Survey (USGS) is the nonregulatory biological research component of the United States Department of the Interior. First created in 1994 as the National Biological Survey (NBS), an independent agency within the U.S. Department of the Interior, the NBS was merged with the USGS (also part of the Interior Department) in 1996.

The BRD is the principal biological research and monitoring agency of the federal government. It is responsible for gathering, analyzing, and disseminating biological information in order to support sound management and stewardship of the nation's biological and natural resources. It is also directed to foster understanding of biological systems and their benefits to society, and to make biological information available to the public. Although it was created mainly on the impetus of environmental and scientific organizations, the BRD also supports commercial and economic interests in that it seeks to identify opportunities for sustainable resource use. Agriculture and biotechnology are among the industries that stand to benefit from BRD research on new sources of food, fiber, and medicines.

Because it is independent of regulatory agencies—which are responsible for enforcing laws—the BRD has no formal regulatory, management, or enforcement roles. Therefore the BRD does not enforce laws such as the Endangered Species Act. Instead the BRD is responsible for gathering data that are scientifically sound and unbiased. The BRD also fosters public-private cooperation. For example, it worked with the International Paper Company in Alabama to develop a management plan for two species of pitcher plants found on company land that were candidates for the Endangered Species List. If it succeeds, this management plan will both preserve the pitcher plant populations—preventing the legal complications of having it listed as a federally endangered species—and allow continued judicious use of the land and resources.

The Biological Resources Discipline of the USGS was created in 1994 as an independent agency with the name National Biological Survey. The Survey was established on November 11, 1994, on the recommendations of President Bill Clinton and Interior Secretary Bruce Babbit. Renamed the National Biological Service (NBS) shortly after its creation, the agency was created by combining the biological research, inventory, and monitoring programs of seven agencies within the Department of the Interior. Built on the model of the Geological Survey, the NBS was established to provide accurate baseline data about ecosystems and species in United States territory. The mission of the NBS was to provide information to support sound stewardship of natural resources on public lands under the administration of the Department of the Interior. Part of its mission was also to foster cooperation among other entities involved in managing, monitoring, and researching natural resources. On October 1, 1996, the NBS was renamed the Biological Resources Discipline and merged with the United States Geological Survey. The USGS-BRD retains the research and information provision mandates of the NBS. Appointed as the first NBS/BRD director was H. Ronald Pulliam, a professor and research ecologist from of the Institute of Ecology and the University of Georgia in Athens.

The National Biological Service had a predecessor in the Bureau of Biological Survey, which operated as part of the Department of Agriculture from 1885 to 1939. The bureau's first director, C. Hart Merriam, revolutionized biological collection techniques and in 15 years nearly quadrupled the number of known American mammal species. In 1939, the bureau was transferred to the Department of the Interior, where it was the predecessor to the Fish and Wildlife Service. After its transfer to the Department of the Interior, the Bureau of Biological Survey's baseline data gathering and survey functions gradually diminished. The Fish and Wildlife Service now has regulatory and management responsibilities as well as research programs.

In recent years the need for a biological survey, and especially for the production of reliable baseline data, has resurfaced. The resurgence of interest in baseline data results in part from an interest in managing ecosystems, with a focus on stewardship and ecosystem restoration, in place of previous emphasis on individual species management. Managing and

restoring ecosystems requires basic data and information on biological indicators, which are often unavailable. Interest in threatened and endangered species, and public participation in environmental organizations probably also contributed to the widespread support for establishing a biological survey.

The BRD is important because it is the only federal agency whose principal mission is to perform basic scientific, biological, and ecological research. As an explicitly scientific research body, the BRD works to incorporate current ecological theory in its research and monitoring programs. Central to the BRD's mission are ideas such as long-term stewardship of resources, maintaining biodiversity, identifying ecological services of biological resources, anticipating climate change, and restoring populations and habitats. Because of its emphasis on basic research the BRD can address current scientific themes that pertain to public policy such as fire ecology, endocrine disruptors (chemicals such as pentachlorophenols [PCBs] that interfere with endocrine and reproductive functions in animals), ecological roles of wetlands, and habitat restoration. The primary purpose of this research activity is to provide land managers, especially within the Department of the Interior, with sound information to guide their management policies.

The BRD has two broad categories of research activities. One focus is on species for which the Department of the Interior has trust responsibilities, including endangered species, marine mammals, migratory birds, and anadromous fish such as salmon. Research on these organisms involves studies of physiology, behavior, population dynamics, and mortality. The second general focus is on ecosystems. This class of research is directed at using experimentation, modeling, and observation to produce practical information concerning the complex interactions and functions of ecosystems, as well as human-induced changes in ecosystems.

In addition to these ecological research questions, the BRD is mandated to perform basic classification, mapping, and description functions. The division is responsible for classifying and mapping species within the United States, as well as prospecting for new species. It is also directed to develop a standard classification system for ecological units, biological indicators for ecosystem health, protocols for managing pollution, and guidelines for ecosystem restoration.

Part of this basic assessment function is a series of National Status and Trends Reports, to be released by the BRD every two years. These reports are to assess the health of biological resources and to report trends in their decline or improvement.

The USGS BRD also provides a nationwide coordinated research agenda and a set of priorities for biological research. For example, it has undertaken a set of coordinated regional studies of the wide-ranging biological and ecological impacts of wetland distributions in Colorado, California, Texas, and other regions of the United States. Collectively these studies can provide a picture of nationwide trends and conditions, as well as producing coordinated information on restoration techniques, biodiversity issues, and biological indicators, at the same time as regional wetland problems are being addressed. Similar sets of coordinated studies have begun concerning the effects of endocrine-disrupting chemicals in the environment and on the use of prescribed fire as a management technique in a variety of ecosystems.

In addition to performing its own research the BRD provides coordination and a network of communication between federal agencies, state governments, universities, museums, and private conservation organizations. The National Partnership for Biological Survey, coordinated by the BRD, provides a network for sharing information and technology between federal, state, and other agencies. Participants in the Partnership include the United States Forest Service, the Natural Resources Conservation Service, the National Oceanic and Atmospheric Administration, the Environmental Protection Agency, the National Science Foundation, the Department of Defense, and the Army Corps of Engineers, as well as the Natural Heritage data centers and programs maintained by many states. The network of Natural Heritage programs is coordinated by The Nature Conservancy, a private organization dedicated to preserving habitat and species diversity. Also included in the Partnership are museums and universities nationwide that serve as repositories of information on biological resources and that conduct biological research, a range of non-governmental organizations, international conservation and research groups, Native American groups, private land holders, and resource user groups.

Resources

OTHER

United States Department of the Interior, United States Geological Survey (USGS). "Biological Resources Division Partnerships." http://biology.usgs.gov/partnership.html (accessed November 11, 2010).

ORGANIZATIONS

US Geological Survey (USGS), Biological Resources Division (BRD), Western Regional Office (WRO), 909 First Ave., Suite #800, Seattle, WA, USA, 98104, (206) 220-4600, (206) 220-4624, brd;uswro@usgs.gov, http://biology.usgs.gov

Mary Ann Cunningham

Biological treatment *see* **Bioremediation.**

Bioluminescence

Bioluminescence ("living light") is the production of light by living organisms through a biochemical reaction. The general reaction involves a substrate called luciferin and an enzyme called luciferase, and requires oxygen. Specifically, luciferin is oxidized by luciferase and the chemical energy produced is transformed into light energy. In nature, bioluminescence is fairly widespread among a diverse group of organisms such as bacteria, fungi, sponges, jellyfish, mollusks, crustaceans, some worms, fireflies, and fish. It is totally lacking in vertebrate animals. Fireflies are probably the most commonly recognized examples of bioluminescent organisms, using the emitted light for mate recognition.

Other probable functions of bioluminescence include communication between bacteria and a form of repulsion (similar to the release of inky compounds by squid).

Bioluminescence has been harnessed as an attribute of biotechnology. Fusion of the gene that encodes the production of bioluminescence with a target gene enables the transcription of the gene to be detected by the production of luminescence.

Resources

BOOKS

Pieribone, Vincent, David F. Gruber, and Sylvia Nasar. *Aglow in the Dark: The Revolutionary Science of Bioluminescence.* Cambridge, MA: Belknap Press, 2006.
Pisano, Gary P. *The Science Business: The Promise, the Reality, and the Future of Biotech.* Cambridge, MA: Harvard Business School Press, 2006.
Zimmer, Marc. *Glowing Genes: A Revolution in Biotechnology.* Amherst, NY: Prometheus Books, 2005.

Biomagnification

The bioaccumulation of chemicals in organisms beyond the concentration expected if the chemical was in equilibrium between the organism and its

DDT in fish-eating
birds 25 ppm

DDT in large
fish 2 ppm

DDT in small
fish (minnows)
0.5 ppm

DDT in zooplanton
0.04 ppm

DDT in water
0.000003 ppm
or 0.003 ppb

Biomagnification of the pesticide DDT in the food chain. DDT travels via runoff up through the food chain, accumulating in all exposed species. *(Reproduced by permission of Gale, a part of Cengage Learning)*

surroundings. Biomagnification can occur in both terrestrial and aquatic environments, but it is generally used in relation to aquatic situations. Most often, biomagnification occurs in the higher trophic levels of the food chain/web, where exposure to chemicals takes place mostly through food consumption rather than water uptake.

Biomagnification is a specific case of bioaccumulation and is different from bioconcentration. Bioaccumulation describes the accumulation of contaminants in the tissue of organisms. Typical examples of this include the elevated levels of many chlorinated pesticides and mercury in fish tissue. Bioconcentration is used to describe the concentration of a chemical in an organism from water uptake alone. This is quantitatively described by the bioconcentration factor, or BCF, which is the chemical concentration in tissue divided by the chemical concentration in water, expressed in equivalent units, at equilibrium. The vast majority of chemicals that bioaccumulate are aromatic organic compounds, particularly those with chlorine substituents. For organic compounds, the mechanism of bioaccumulation is thought to be the partitioning or solubilization of the chemical into the lipids of the organism. Thus the BCF should be proportional to the lipophilicity of the chemical, which is described by the octanol-water partition coefficient, Kow. The latter is a physical-chemical property of the compound describing its relative solubility in an organic phase and is the ratio of its solubility in octanol to its solubility in water at equilibrium. It is constant at a given temperature. If one assumes that a chemical's solubility in octanol is similar to its solubility in lipid, then we can approximate the lipid-normalized BCF as equal to the Kow. This assumption has been shown to be a reasonable first approximation for most chemicals accumulation in fish tissue.

However, animals are exposed to contaminants by other routes in addition to passive partitioning from water. For instance, fish can take up chemicals from the food they eat. It has been noted in field collections that for certain chemicals, the observed fish-water ratio (BCF) is significantly greater than the theoretical BCF, based on Kow. This indicates that the chemical has accumulated to a greater extent than its equilibrium concentration. This is defined as biomagnification. This condition has been documented in aquatic animals, including fish, shellfish, seals and sea lions, whales, and otters, and in birds, mink, rodents, and humans in both laboratory and field studies.

The biomagnification factor, BMF, is usually described as the ratio of the observed lipid-normalized BCF to Kow, which is the theoretical lipid-normalized BCF. This is equivalent to the multiplication factor above the equilibrium concentration. If this ratio is equal to or less than one, then the compound has not biomagnified. If the ratio is greater than one, then the chemicals are biomagnified by that factor. For instance, if a chemical's Kow were 100,000, then its lipid normalized BCF should be 100,000 if the chemical were in equilibrium in the organism's lipids. If the fish tissue concentration (normalized to lipids) were 500,000, then the chemical would be said to have biomagnified by a factor of five.

Biomagnification in the aquatic food chain often leads to biomagnification in terrestrial food chains,

particularly in the case of bird and wildlife populations that feed on fish. Consider the following example that demonstrates the results of biomagnification. The concentrations of the insecticide dieldrin in various trophic levels are determined to be the following: water, 0.1 ng/L; phytoplankton, 100 ng/g lipid; zooplankton, 200 ng/g lipid; fish, 600 ng/g lipid; terns, 800 ng/g lipid. If the Kow were equal to one million, then the phytoplankton would be in equilibrium with the water, but the zooplankton would have magnified the compound by a factor of 2, the fish by a factor of 6, and the terns by a factor of 8.

The mechanism of biomagnification is not completely understood. To achieve a concentration of a chemical greater than its equilibrium value indicates that the elimination rate is slower than for chemicals that reach equilibrium. Transfer efficiencies of the chemical would affect the relative ratio of uptake and elimination. There are many factors that control the uptake and elimination of a chemical from the consumption of contaminated food, and these include factors specific to the chemical as well as factors specific to the organism. The chemical properties include solubility, Kow, molecular weight and volume, and diffusion rates between organism gut, blood, and lipid pools. The organism properties include the feeding rate, diet preferences, assimilation rate into the gut, rate of chemical's metabolism, rate of egestion, and rate of organism growth. It is thought that the chemical's properties control whether biomagnification will occur, and that it is the transfer rate from lipid to blood that allows the chemical to attain a lipid concentration greater than its equilibrium value. Thus it follows that the chemicals that biomagnify have similar properties. They typically are organic; they have molecular weights between 200 and 600 daltons; they have Kows between 10,000 and 10 million; they are resistant to metabolism by the organism; they are non-ionic, neutral compounds; and they have molecular volumes between 260 and 760 cubic angstroms, a cross sectional width of less than 9.5 angstoms and a molecular surface area between 200 and 460 square angstroms. The latter dimensions allow them to more easily pass through lipid bilayers into cells but perhaps do not allow them to leave the cell easily due to their high lipophilicity. Since this disequilibrium would occur at each trophic level, it results in more and more biomagnification at each higher trophic level. Because humans occupy a very high trophic level, we are particularly vulnerable to adverse health effects as a result of exposure to chemicals that biomagnify.

Resources

BOOKS

de Ruiter, Peter C., Volkmar Wolters, and John C. Moore, eds. *Dynamic Food Webs: Multispecies Assemblages, Ecosystem Development and Environmental Change.* Burlington, MA: Academic Press, 2005.

Morgan, Kevin, Terry Marsden, and Jonathan Murdoch. *Worlds of Food: Place, Power, and Provenance in the Food Chain.* Oxford geographical and environmental studies. Oxford, UK: Oxford University Press, 2006.

PERIODICALS

Washam, Cynthia. "A Whiff of Danger: Synthetic Musks May Encourage Toxic Bioaccumulation." *Environmental Health Perspectives.* 113 (2005): A50–A51.

OTHER

United States Environmental Protection Agency (EPA). "Pollutants/Toxics: Toxic Substances: Persistent Bioaccumulative Toxic Pollutants (PBTs)." http://www.epa.gov/ebtpages/polltoxicpersistentbio accumulativetox.html (accessed September 19, 2010).

Deborah L. Swackhammer

Biomass

Biomass is a measure of the amount of biological substance minus its water content found at a given time and place on Earth's surface. Although sometimes defined strictly as living material, in actual practice the term often refers to living organisms, or parts of living organisms, as well as waste products or non-decomposed remains. It is a distinguishing feature of ecological systems and is usually presented as biomass density in units of dry weight per unit area. The term is somewhat imprecise in that it includes autotrophic plants, referred to as phytomass, heterotrophic microbes, and animal material, or zoomass. In most settings, phytomass is by far the most important component. A square meter of the planet's land area has, on average, about 22.05–26.46 lb (10–12 kg) of phytomass, although values may vary widely depending on the type of biome. Tropical rain forests average about 99.2 lb/3.28 ft^2 (45 kg/m^2) while a desert biome may have a value near zero. The global average for heterotrophic biomass is approximately 0.1 kg/m^2, and the average for human biomass has been estimated at 0.5 g/m^2 if permanently glaciated areas are excluded.

The nature of biomass varies widely. Density of fresh material ranges from a low of 0.14 g/cm^3 for floats of aquatic plants to values greater that 1 g/cm^3 for very dense hardwood. The water content of fresh material may be as low as 5 percent in mature seeds or as high as 95 percent in fruits and young shoots. Water levels for living plants and animals run from 50 to 80 percent, depending on the species, season, and growing conditions. To ensure a uniform basis for comparison, biomass samples are dried at 221°F (105°C) until they reach a constant weight.

Organic compounds typically constitute about 95 percent by weight of the total biomass, and nonvolatile residue, or ash, about 5 percent. Carbon is the principal element in biomass and usually represents about 45 percent of the total. An exception occurs in species that incorporate large amounts of inorganic elements such as silicon (Si) or calcium (Ca), in which case the carbon content may be much lower and nonvolatile residue several times higher. Another exception is found in tissues rich in lipids (oil or fat), where the carbon content may reach values as high as 70 percent.

Photosynthesis is the principal agent for biomass production. Light energy is used by chlorophyll-containing green plants to remove (or fix) carbon dioxide (CO$_2$) from the atmosphere and convert it to energy-rich organic compounds or biomass. It has been estimated that on the face of Earth approximately 200 billion tons of carbon dioxide are converted to biomass each year. Carbohydrates are usually the primary constituent of biomass, and cellulose is the single most important component. Starches are also important and predominate in storage organs such as tubers and rhizomes. Sugars reach high levels in fruits and in plants such as sugar cane and sugar beet. Lignin is a very significant noncarbohydrate constituent of woody plant biomass.

Recently, depletion of fossil fuel sources has led to the search for alternative sources of fuel. Plant biomass (a type of biofuel) is a potential source of fuel that is renewable (the source can be regenerated and replenished). Many researchers are conducting studies to determine the most suitable crops and the most efficient means of deriving energy from biomass. Switchgrass (*Panicum virgatum*), corn (*Zea mays*), and rapeseed (*Brassica napus*) plants are currently being grown and harvested for biofuel generation. However, in terms of the environmental impact of biofuels compared with fossil fuels, burning biomass does contribute to levels of carbon dioxide in the atmosphere. Two main issues are of concern with regard to biofuels. The amount of processing necessary to obtain energy from biomass is significant and relatively expensive. More dedicated research may play a part in resolving this issue. The other concern is that the amount of biofuels needed to replace fossil fuels is so substantial that there may not be enough area to dedicate to biomass production to meet worldwide energy requirements. Land for production of biofuel biomass may displace food crops, and thus lead to shortages of food supplies.

Biomass crops also have the potential to accumulate carbon in the root and shoot. This carbon sequestration may reduce the effects of global warming by removing amounts of carbon from the atmosphere with the crops acting as carbon sinks.

Resources

BOOKS

De la Garza, Amanda. *Biomass: Energy from Plants and Animals.* Detroit, MI: Greenhaven Press, 2007.
Goettemoeller, Jeffrey, and Adrian Goettemoeller. *Sustainable Ethanol: Biofuels, Biorefineries, Cellulosic Biomass, Flex-Fuel Vehicles, and Sustainable Farming for Energy Independence.* Maryville, MO: Prairie Oak Pub, 2007.

PERIODICALS

Dunn, David. "Utility Turns Biomass into Renewable Energy." *Biocycle* (September 2004).
Tilman, David, et al. "Carbon-Negative Biofuels from Low-Input High Diversity Grassland Biomass." *Science* 314 (2006): 1598–1600.

OTHER

World Energy Council. "Survey of Energy Resources 2007: BioEnergy: Biomass for Electricity Generation." http://www.worldenergy.org/publications/survey_of_energy_resources_2007/bioenergy/713.asp (accessed September 4, 2010).
World Energy Council. "Survey of Energy Resources 2007: BioEnergy: Biomass Resources. http://www.worldenergy.org/publications/survey_of_energy_resources_2007/bioenergy/717.asp (accessed September 4, 2010).

Douglas C. Pratt

Biomass fuel

A biomass fuel is an energy source derived from living organisms. Most commonly it is plant residue, harvested, dried and burned, or further processed into solid, liquid, or gaseous fuels. The most familiar and widely used biomass fuel is wood. Agricultural waste, including materials such as cereal straw, seed hulls, corn stalks and cobs, is also a significant source. Native shrubs and herbaceous plants are potential sources.

Virgin Atlantic's 747 is pushed out of the Virgin Hanger at Heathrow Airport February 24, 2008, to take off for Amsterdam in a flight by an airline using biofuels. *(Steve Parsons/PA Wire)*

Animal waste, although much less abundant overall, is a bountiful source in some areas.

Wood accounted for 25 percent of all energy used in the United States at the beginning of the twentieth century. With increased use of fossil fuels, its significance rapidly declined. By 1976, only 1 to 2 percent of United States energy was supplied by wood, and burning of tree wastes by the forest products industry accounted for most of it. Although the same trend has been evident in all industrialized countries, the decline has not been as dramatic everywhere. Sweden, Finland, and Austria meet up to 17 percent of their national energy needs with wood.

In industrialized countries, it is estimated that biomass supplies about 10 percent of total energy, and it continues to be a very important energy source for many developing countries, in which it amounts to 20 to 30 percent of total energy. Interest in biomass has greatly increased even in countries where its use has drastically declined. In the United States rising fuel prices led to a large increase in the use of wood-burning stoves and furnaces for space heating. Impending fossil fuel shortages have greatly increased research on its use in the United States and elsewhere. Because biomass is a potentially renewable resource, it is recognized as a possible replacement of petroleum and natural gas.

Historically, burning has been the primary mode for using biomass, but because of its large water content it must be dried to burn effectively. In the field, the energy of the sun may be all that is needed to sufficiently lower its water level. When this is not sufficient, another energy source may be needed.

Biomass is not as concentrated an energy source as most fossil fuels even when it is thoroughly dry. Its density may be increased by milling and compressing dried residues. The resulting briquettes or pellets are also easier to handle, store, and transport. Compression has been used with a variety of materials including crop residues, herbaceous native plant material, sawdust, and other forest wastes.

Solid fuels are not as convenient or versatile as liquids or gases, and this is a drawback to the direct use of biomass. Fortunately, a number of techniques are known for converting it to liquid or gaseous forms.

Partial combustion is one method. In this procedure, biomass is burned in an environment with restricted oxygen. Carbon monoxide (CO) and hydrogen (H) are formed instead of carbon dioxide (CO_2) and water. This mixture is called synthetic gas or syngas. It can serve as fuel although its energy content is lower than natural gas (methane). Syngas may also be converted to methanol (CH_3OH), a one carbon-alcohol that can be used as a transportation fuel. The conversion efficiency of syngas to methanol is over 99 percent. Because methanol is a liquid, it is easy to store and transport.

Anaerobic digestion is another method for forming gases from biomass. It uses microorganisms, in the absence of oxygen, to convert organic materials to methane. This method is particularly suitable for animal and human waste. Animal feedlots faced with disposal problems may install microbial gasifiers to convert waste to gaseous fuel used to heat farm buildings or generate electricity.

Biodiesel is diesel fuel that is formed from vegetable oil, animal fats, or waste from pulp and paper processing. Over 80 percent of commercial trucks in the United States are powered by biodiesel. Any diesel-powered vehicle can run with a mix of biodiesel (around 15 to 20 percent) and mineral diesel, but newer vehicles are able to run on pure biodiesel. The environmental impact of biodiesel on greenhouse gas emissions is a highly debated subject that depends on many factors such as the processing and production, deforestation for available space for crop production, and efficiency differences compared with mineral diesel.

For materials rich in starch and sugar, fermentation is an attractive alternative. Through acid hydrolysis or enzymatic digestion, starch can be extracted and converted to sugars. Sugars can be fermented to produce ethanol (C_2H_6O), a liquid biofuel with many potential uses.

Cellulose is the single most important component of plant biomass. Like starch, it is made of linked sugar components that may be easily fermented when separated from the cellulose polymer. The complex structure of cellulose makes separation difficult, but enzymatic means are being developed to do so. Perfection of this technology will create a large potential for ethanol production using plant materials that are not human foods.

The efficiency with which biomass may be converted to ethanol or other convenient liquid or gaseous fuels is a major concern. Conversion generally requires appreciable energy. If an excessive amount of expensive fuel is used in the process, costs may be prohibitive. Corn (*Zea mays*) has been a particular focus of efficiency studies. Inputs for the corn system include energy for production and application of fertilizer and pesticide, tractor fuel, on-farm electricity, and so forth, as well as those more directly related to fermentation. A recent estimate puts the industry average for energy output at 133 percent of that needed for production and processing. This net energy gain of 33 percent includes credit for co-products such as corn oil and protein feed as well as the energy value of ethanol. The most efficient production and conversion systems are estimated to have a net energy gain of 87 percent. Although it is too soon to make an accurate assessment of the net energy gain for cellulose-based ethanol production, it has been estimated that a net energy gain of 145 percent is possible.

Biomass-derived gaseous and liquid fuels share many of the same characteristics as their fossil fuel counterparts. Once formed, they can be substituted in whole or in part for petroleum-derived products. Gasohol, a mixture of 10 percent ethanol in gasoline, is an example. Ethanol contains about 35 percent oxygen, much more than gasoline, and a gallon contains only 68 percent of the energy found in a gallon of gasoline. For this reason, motorists may notice a slight reduction in gas mileage when burning gasohol. However, automobiles burning mixtures of ethanol and gasoline have a lower exhaust temperature. This results in reduced toxic emissions, one reason that clean air advocates often favor gasohol use in urban areas.

Biomass is called as a renewable resource since green plants are essentially solar collectors that capture and store sunlight in the form of chemical energy. Its renewability assumes that source plants are grown under conditions where yields are sustainable over long periods of time. Obviously, this is not always the case, and care must be taken to ensure that growing conditions are not degraded during biomass production.

A number of studies have attempted to estimate the global potential of biomass energy. Although the amount of sunlight reaching Earth's surface is substantial, less than a tenth of a percent of the total is actually captured and stored by plants. About half of it is reflected back to space. The rest serves to maintain global temperatures at life-sustaining levels. Other factors that contribute to the small fraction of the sun's energy that plants store include Antarctic and Arctic zones where little photosynthesis occurs, cold winters in temperate belts when plant growth is impossible, and lack of adequate water in arid regions. The global total net production of biomass energy has been estimated at 100 million megawatts per year per year. Forests and woodlands account for about 40 percent of the total, and oceans about 35 percent. Approximately 1 percent of all biomass is used as food by humans and other animals.

Soil requires some organic content to preserve structure and fertility. The amount required varies widely depending on climate and soil type. In tropical rain forests, for instance, most of the nutrients are found in living and decaying vegetation. In the interests of preserving photosynthetic potential, it is probably inadvisable to remove much if any organic matter from the soil. Likewise, in sandy soils, organic matter is needed to maintain fertility and increase water retention. Considering all the constraints on biomass harvesting, it has been estimated that about six million MWyr/yr of biomass are available for energy use. This represents about 60 percent of human society's total energy use and assumes that the planet is converted into a global garden with a carefully managed photosphere.

Although biomass fuel potential is limited, according to its advocates it provides a basis for significantly reducing society's dependence on nonrenewable reserves. Yet its potential is seriously diminished by factors that degrade growing conditions either globally or regionally. Thus, the impact of factors like climate change and acid rain must be taken into account to assess how well that potential might eventually be realized. It is in this context that one of the most important aspects of biomass fuel should be noted. Growing plants remove carbon dioxide from the atmosphere that is released back to the atmosphere when biomass fuels are used. Thus the overall concentration of atmospheric carbon dioxide should not change, and climate change should not result. Another environmental advantage arises from the fact that biomass contains much less sulfur (S) than most fossil fuels. As a consequence, biomass fuels should reduce the impact of acid rain.

The production of ethanol has been criticized both as a net energy consumer (or poor net-energy producer) and as injurious to the world's poor. In early 2008, food shortages gripped much of the world, with riots against high prices becoming more common in some poor countries. Food prices worldwide had risen 83 percent since 2005, according to the World Bank. Although there were a number of causes for rising prices, including a drought in Australia that lowered production of rice, the main food of over half the world's population, food experts agreed that the push to raise crops for ethanol biofuel for vehicles was part of the problem. Biofuel manufacturers compete directly with food buyers in the market for corn, and high demand for biofuel crops causes growers to switch acreage away from food production. The result is higher food prices, which some the world's poorest simply cannot pay. In Haiti, for example, by late 2007 many of the rural poor were routinely eating mud, which has no food value but eases hunger pangs. Expert estimates of how much of the rise in global food prices was due to biofuel programs ranged from 10 percent to over 30 percent.

See also Anaerobic digestion; Animal waste.

Resources

BOOKS

Goettemoeller, Jeffrey, and Adrian Goettemoeller. *Sustainable Ethanol: Biofuels, Biorefineries, Cellulosic Biomass, Flex-Fuel Vehicles, and Sustainable Farming for Energy Independence.* Maryville, MO: Prairie Oak, 2007.

Letcher, T. M. *Future Energy: Improved, Sustainable and Clean Options for Our Planet.* Amsterdam: Elsevier, 2008.

Mousdale, David M. *Biofuels: Biotechnology, Chemistry, and Sustainable Development.* Boca Raton, FL: CRC Press, 2008.

Welborne, Victor I. *Biofuels in the Energy Supply System.* New York: Novinka Books, 2006.

Worldwatch Institute. *Biofuels for Transport: Global Potential and Implications for Energy and Agriculture.* London: Earthscan Publications, Ltd., 2007.

PERIODICALS

Cressey, Daniel. "Advanced Biofuels Face an Uncertain Future." *Nature* 452 (2008): 670–71.

Cunningham, A. "Going Native: Diverse Grassland Plants Edge Out Crops as Biofuel." *Science News* 170, no. 24 (December 9, 2006): 372.

Fargione, Joseph. "Land Clearing and the Biofuel Carbon Debt." *Science* 319 (2008): 1235–37.

Righelato, Renton, and Dominick V. Spracklen. "Carbon Mitigation by Biofuels or by Saving and Restoring Forests?" *Science* 317 (2007): 902.

Rosner, Hillary. "Cooking Up More Uses for the Leftovers of Biofuel Production." *New York Times* (August 8, 2007).

Scharlemann, Jörn P. W., and William F. Laurance. "How Green Are Biofuels?" *Science* 319 (2008): 43–44.

Searchinger, Timothy. "Use of U.S. Croplands for Biofuels Increases Greenhouse Gases through Emissions from Land-Use Change." *Science* 319 (2008): 1598–1600.

OTHER

Food and Agriculture Organization of the United Nations (FAO). "Wood Energy." http://www.fao.org/forestry/14011/en/ (accessed September 5, 2010).

National Geographic Society. "Biofuels ." http://environment.nationalgeographic.com/environment/global-warming/biofuel-profile.html (accessed September 5, 2010).

World Energy Council. "Survey of Energy Resources 2007: BioEnergy: Biofuels. http://www.worldenergy.org/publications/survey_of_energy_resources_2007/bioenergy/714.asp (accessed September 5, 2010).

Douglas C. Pratt

Biome

A biome is a large terrestrial ecosystem characterized by distinctive kinds of plants and animals and maintained by a distinct climate and soil conditions. To illustrate, the desert biome is characterized by low annual rainfall and high rates of evaporation, resulting in dry environmental conditions. Plants and animals that thrive in such conditions include cacti, brush, lizards, insects, and small rodents. Special adaptations, such as waxy plant leaves, allow organisms to survive under low moisture conditions. Other examples of biomes include tropical rain forest, arctic tundra, grasslands, temperate deciduous forest, coniferous forest, tropical savanna, and Mediterranean chaparral.

Biophilia

The term biophilia was coined by the American biologist, Edward O. Wilson (1929–), to mean "the innate tendency [of human beings] to focus on life and lifelike processes" and "connections that human beings subconsciously seek with the rest of life." Wilson first used the word in 1979 in an article published in the *New York Times Book Review*, and then in 1984 he wrote a short book (157 pages) that explored the notion in more detail.

Clearly, humans have coexisted with other species throughout our evolutionary history. In Wilson's view, this relationship has resulted in humans developing an innate, genetically based, or "hard-wired" need to be close to other species and to be empathetic to their needs. The presumed relationship is reciprocal, meaning other animals are also to varying degrees empathetic with the needs of humans. The biophilia hypothesis seems intuitively reasonable, although it is likely impossible that it could ever be proven universally correct. For example, while many people may feel and express biophilia, some people may not.

There is a considerable body of psychological and medical evidence in support of the notion of biophilia, although most of it is anecdotal. There are, for example, many observations of sick or emotionally distressed people becoming well more quickly when given access to a calming natural environment, or when comforted by such pets as dogs and cats. There are also more general observations that many people are more comfortable in natural environments than in artificial ones—it can be much more pleasant to sit in a garden, for instance, than in a windowless room. The aesthetics and comfort of that room can, however, be improved by hanging pictures of animals or landscapes, by watching a nature show on television, or by providing a window that looks out onto a natural scene.

Wilson is a leading environmentalist and a compelling advocate of the need to conserve the threatened biodiversity of the world, such as imperiled natural ecosystems and endangered species. To a degree, his notion of biophilia provides a philosophical justification for conservation actions, by suggesting emotional

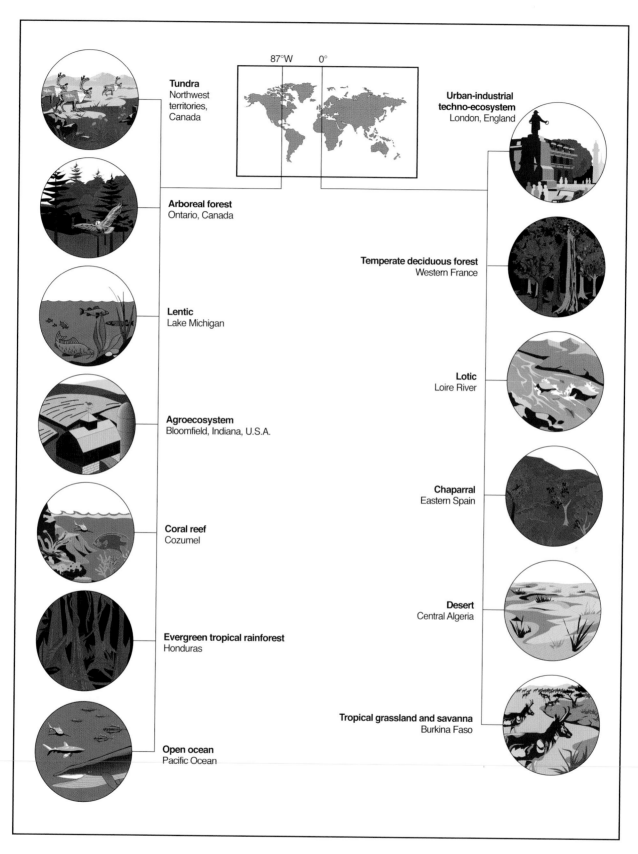

Tundra
Northwest territories, Canada

Urban-industrial techno-ecosystem
London, England

87°W   0°

Arboreal forest
Ontario, Canada

Temperate deciduous forest
Western France

Lentic
Lake Michigan

Lotic
Loire River

Agroecosystem
Bloomfield, Indiana, U.S.A.

Chaparral
Eastern Spain

Coral reef
Cozumel

Desert
Central Algeria

Evergreen tropical rainforest
Honduras

Tropical grassland and savanna
Burkina Faso

Open ocean
Pacific Ocean

Different biomes worldwide. *(Reproduced by permission of Gale, a part of Cengage Learning)* and even spiritual dimensions to the human relationship with other species and the natural world. If biophilia is a real phenomenon and an integral part of what it is to be human, then our affinity for other species means that willful actions causing endangerment and extinction can be judged to be wrong and even immoral. Our innate affinity for other species provides an intrinsic justification for doing them no grievous harm, and in particular, for avoiding their extinction. It provides a justification for enlightened stewardship of the natural world.

Humans also utilize animals in agriculture (we raise them as food), in research, and as companions. Ethical considerations derived from biophilia provide a rationale for treating animals as well as possible during such use, and also as a philosophical foundation for vegetarianism and anti-vivisectionism.

Resources

BOOKS

Darwin, Charles, and Edward O. Wilson. *From So Simple a Beginning: The Four Great Books of Charles Darwin.* New York: Norton, 2006.

Hölldobler, Bert, and Edward O. Wilson. *The Superorganism: The Beauty, Elegance, and Strangeness of Insect Societies.* New York: Norton, 2009.

Lumsden, Charles J., and Edward O. Wilson. *Genes, Mind, and Culture: The Coevolutionary Process.* Hackensack, NJ: World Scientific, 2005.

Wilson, Edward O. *Sociobiology: The New Synthesis.* Cambridge, MA: Belknap Press, 1975.

Wilson, Edward O. *The Diversity of Life.* Cambridge: Harvard University Press, 1992.

Wilson, Edward O. *Consilience: The Unity of Knowledge.* New York: Alfred A. Knopf, 1999.

Wilson, Edward O. *The Creation: An Appeal to Save Life on Earth.* New York: Norton, 2006.

Wilson, Edward O. *Nature Revealed: Selected Writings, 1949–2006.* Baltimore: Johns Hopkins University Press, 2006.

Bill Freedman

Bioregionalism

Drawing heavily upon the cultures of indigenous peoples, bioregionalism is a philosophy of living that stresses harmony with nature and the integration of humans as part of the natural ecosystem. The keys to bioregionalism involve learning to live off the land, without damaging the environment or relying on heavy industrial machines or products. Bioregionalists argue that if the relationship between nature and humans improves, the society as a whole will benefit.

Environmentalists who practice this philosophy "claim" a bioregion or area. For example, one's place might be a watershed, a small mountain range, a particular area of the coast, or a specific desert. To develop a connection to the land and a sense of place, bioregionalists try to understand the natural history of the area as well as how it supports human life. For example, they study the plants and animals that inhabit the region, the geological features of the land, as well as the cultures of the people who live or have lived in the area.

Bioregionalism also stresses community life where participation, self-determination, and local control play important roles in protecting the environment. Various bioregional groups exist throughout the United States, ranging from the Gulf of Maine to the Ozark Mountains to the San Francisco Bay area.

Resources

BOOKS

Sale, K. *Dwellers in the Land: The Bioregional Vision.* San Francisco: Sierra Club Books, 1985.

Snyder, G. *The Practice of the Wild.* San Francisco: North Point Press, 1990.

Christopher McGrory Klyza

Bioremediation

The increased use of chemicals in industrial processes and agriculture has resulted in a significant amount of these substances persisting and accumulating in our environment. Bioremediation is the process of removing contaminants or pollutants from the environment to return it to its original state through the use of biological mechanisms. These chemicals can be found in soil, sediments, or in bodies of water.

Plants can be utilized in bioremediation through root uptake and bioaccumulation of the chemicals from the soil and groundwater. Plants have been used for this process, termed phytoremediation, to amend problems in agricultural soil and for removal of environmental pollutants. This method of bioremediation can be accomplished through removal, sequestration, or degradation of contaminants by plants. Aquatic plants have been used to remove contaminants from bodies

Bioremediation pond for soil contaminated by crude oil beside an oil well in the Amazon. *(© Morley Read / Alamy)*

of water. These plants must be capable of growing in contaminated soil or water and must be able to survive with the chemicals accumulated in their living tissues. Some plants are able to process or degrade certain chemicals. Both vegetation and earthworms have been used for mitigation of pollutants, but another agent that has been utilized for this process is bacteria.

Bioremediation consists of a number of processes for remediating contaminated soils and groundwater based on the use of microorganisms to convert contaminants to less hazardous substances, termed biotransformation. Most commercial bioremediation processes are intended to convert organic substances to carbon dioxide (CO_2) and water, although processes for addressing metals are under development. Many bacteria ubiquitously found in soils and groundwater are able to biodegrade a range of organic compounds. Compounds found in nature, and ones similar to those, such as petroleum hydrocarbons, are most readily biodegraded by these bacteria. Bioremediation of chlorinated solvents, polychlorinated biphenyl (PCBs), pesticides, and many munitions compounds, while of great interest, are more difficult and have thus been much slower to reach commercialization.

In most bioremediation processes the bacteria use the contaminant as a food and energy source and thus survive and grow in numbers at the expense of the contaminant. In order to grow new cells, bacteria, like other biological species, require numerous minerals as well as carbon sources. These minerals are typically present in sufficient amounts except for phosphorus (P) and nitrogen (N), which are commonly added during bioremediation. If contaminant molecules are to be transformed, species called electron acceptors must also be present. By far the most commonly used electron acceptor is oxygen. Other electron acceptors include nitrate (NO_3^-), sulfate (SO_4^{2-}), carbon dioxide, and iron (Fe). Processes that use oxygen are called aerobic biodegradation. Processes that use other electron acceptors are commonly lumped together as anaerobic biodegradation.

The vast majority of commercial bioremediation processes use aerobic biodegradation and thus include some method for providing oxygen. The amount of oxygen that must be provided depends not only on the mass of contaminant present but also on the extent of conversion of the contaminants to carbon dioxide and water, other sources of oxygen, and the extent to which the contaminants are physically removed from the soils or groundwater. Typically, designs are based on adding two to three pounds of oxygen for each pound of biodegradable contaminant.

The processes generally include the addition of nutrient (nitrogen and phosphorus) sources. The amount of nitrogen and phosphorus that must be provided is quite variable and frequently debated. In general, this amount is less than the 100:10:1 ratio of carbon to nitrogen to phosphorus of average cell compositions. It is also important to maintain the soil or groundwater pH near neutral (pH 6–8.5), moisture levels at or above 50 percent of field capacity, and temperatures between 39°F (4°C) and 95°F (35°C), preferably between 68°F (20°C) and 86°F (30°C).

Bioremediation can be applied in situ and ex situ by several methods. Each of these processes is basically an engineering solution to providing oxygen (or alternate electron acceptors) and possibly, nutrients to the contaminated soils, which already contain the bacteria. The addition of other bacteria is not typically needed or beneficial.

In situ processes have the advantage of causing minimal disruption to the site and can be used to address contamination under existing structures. In situ bioremediation to remediate aquifers contaminated with petroleum hydrocarbons, such as gasoline, was pioneered in the 1970s and early 1980s by Richard L. Raymond and coworkers. These systems used groundwater recovery wells to capture contaminated water, which was treated at the surface and reinjected after amendment with nutrients and oxygen. The nutrients consisted of ammonium chloride (NH_4Cl) and phosphate salts and sometimes contained magnesium (Mg), manganese (Mn), and iron salts. Oxygen was introduced by sparging (bubbling) air into the reinjection water. As the injected water swept through the aquifer, oxygen and nutrients were carried to the contaminated soils and groundwater where the indigenous bacteria converted the hydrocarbons to new cell material, carbon dioxide, and water. Variations of this technology include the use of hydrogen peroxide (H_2O_2) as a source of oxygen and direct injection of air into the aquifer.

Bioremediation of soils located between the ground surface and the water table is most commonly practiced through bioventing. In this method, oxygen is introduced into the contaminated soils by either injecting air or extracting air from wells. The systems used are virtually the same as for vapor extraction. The major difference is in mode of operation and in the fact that nutrients are sometimes added by percolating nutrient-amended water through the soil from the ground surface or buried horizontal pipes. Systems designed for bioremediation operate at low air flow rates to replace oxygen consumed during biodegradation and to minimize physical removal of volatile contaminants.

Bioremediation can be applied to excavated soils by landfarming, soil cell techniques, or in soil slurries. The simplest method is landfarming. In this method soils are spread to a depth of 12–18 inches (30–46 cm). Nutrients, usually commercial fertilizers with high nitrogen and low phosphorous content, are added periodically to the soils, which are tilled or plowed frequently. In most instances, the treatment area is prepared by grading, laying down an impervious layer (clay or a synthetic liner), and adding a six-inch (15.2-cm) layer of clean soil or sand. Provisions for treating rainwater runoff are typically required. The frequent tilling and plowing breaks up soil clumps and exposes the soils and thus bacteria to air. This method is more suitable for treating high-silt and high-clay soils than are most of the other methods. It is not generally appropriate for soils contaminated with volatile contaminants such as gasoline, because vapors cannot be controlled unless the process is conducted within a closed structure.

Excavated soils can also be treated in cells or piles. A synthetic liner is placed on a graded area and covered with sand or gravel to permit collection of runoff water. The sands or gravel are covered with a permeable fabric and nutrient-amended soils are added. Slotted PVC pipe is added as the pile is built. The soils are covered with a synthetic liner and the PVC pipes are connected to a blower. Air is slowly extracted from the soils and, if necessary, treated before being discharged to the atmosphere. This method requires

less room than landfarming and less maintenance during operations, and it can be used to treat volatile contaminants because the vapors can be controlled.

Excavated soils can also be treated in soil/water slurries in either commercial reactors or in impoundments or lagoons. Soils are separated from oversize materials and mixed with water, nutrients are added, and the slurry is aerated to provide oxygen. In some cases additional sources of bacteria and/or surfactants are added. These systems are usually capable of attaining more rapid rates of biodegradation than other systems but have limited throughput.

Selection and design of a particular bioremediation method requires that the site be carefully investigated to define the lateral and horizontal extent of contamination including the total mass of biodegradable substances. Understanding the soil types and distribution and the site hydrogeology is as important as identifying the contaminants and their distribution in both soils and groundwater. Designing bioremediation systems requires the integration of microbiology, chemistry, hydrogeology, and engineering.

Bioremediation is generally viewed favorably by regulatory agencies and is actively supported by the U.S. Environmental Protection Agency (EPA). The mostly favorable publicity and the perception of bioremediation as a natural process has led to greater acceptance by the public compared to other technologies, such as incineration. It is expected that the use of bioremediation to treat soils and groundwater contaminated with petroleum hydrocarbons and other readily biodegradable compounds will continue to grow. Continued improvements in the design and engineering of bioremediation systems will result from the expanding use of bioremediation in a competitive market. It is anticipated that processes for treating the more recalcitrant organic compounds and metals will become commercial through greater understanding of microbiology and specific developments in the isolation of special bacteria and genetic engineering.

See also Bioventing; Groundwater; Microbes (microorganisms).

Resources

BOOKS

Atlas, Ronald M., and Jim Philp, eds. *Bioremediation: Applied Microbial Solutions for Real-world Environmental Cleanup.* Washington, DC: ASM Press, 2005.

Singh, Ajay, and Owen P. Ward. *Applied Bioremediation and Phytoremediation.* Soil Biology, 1. Berlin: Springer, 2004.

Singh, Ajay, and Owen P. Ward, eds. *Biodegradation and Bioremediation.* Berlin and New York: Springer, 2004.

Singh, Shree N., and R. D. Tripathi. *Environmental Bioremediation Technologies*. Berlin: Springer, 2007.

Trivedi, Pravin Chandra. *Pollution and Bioremediation*. Jaipur, India: Aavishkar, 2008.

Wani, Altaf H. *Bioremediation*. Practice Periodical of Hazardous, Toxic, and Radioactive Waste Management, v. 10, no. 2. Reston, VA: American Society of Civil Engineers, 2006.

OTHER

United States Environmental Protection Agency (EPA). "Treatment/Control: Treatment Technologies: Bioremediation." http://www.epa.gov/ebtpages/ treatreatmenttechnbioremediation.html (accessed October 13, 2010).

Robert D. Norris

Biosequence

Biosequence refers to the sequence of soils that contain distinctly different soil horizons (a horizon is a layer in the land that differs from the layers above and below it) because of the influence that vegetation had on the soils during their development. A typical biosequence would be the prairie soils in a dry environment, oak-savannah soils as a transition zone, and forested soils in a wetter environment. Prairie soils have dark, thick surface horizons while forested soils have a thin, dark surface with a light- colored zone below.

Resources

BOOKS

Brady, Nyle C., and Ray R. Weil. *Elements of the Nature and Properties of Soil*. New York: Prentice Hall, 2009.

Conkin, Keith. *A Revolution Down on the Farm: The Transformation of American Agriculture since 1929*. Lexington: University of Kentucky Press, 2009.

Logan, William Bryant. *Dirt: The Ecstatic Skin of the Earth*. New York: Norton, 2007.

Biosphere

The biosphere is the largest organic community on Earth. It is a terrestrial envelope of life, or the total global biomass of living matter. The biosphere incorporates every individual organism and species on the face of the earth.

Bios is the Greek word for life; "sphere" is from the Latin *sphaera*, which means essentially the "circuit or range of action, knowledge or influence," the "place or scene of action or existence," the "natural, normal or proper place."

Life began in a very different environment than found today: the atmosphere, for example, was mostly methane, ammonia, and carbon dioxide. As life evolved, it changed the atmosphere (and other abiotic components of Earth's surface), transforming it into the present oxygen-rich mixture of gases vital to life as it now exists. And those life-forms maintain that critical mixture in a complex, fluctuating system of global cycles.

The diversity and complexity of the biosphere is staggering. There are about 1.5 million known species. However, conservative estimates of the actual number of species range up to 5 million. Less conservative estimates that may more accurately reflect reality range up to a possible 100 million species.

Many species might be extinguished before they are even known. Human activities, especially destruction of habitat, are increasing the normal rate of species extinction. The diversity of the biosphere may be diminishing rapidly.

Taxonomically, the biosphere is organized into five kingdoms: monera, protista, fungi, animalia, and plantae, and a multitude of subsets of these, including the multiple millions of species mentioned above. Of the 1,200 to 1,800 billion tons dry weight of the biosphere, most of it—some 99 percent—is plant material. All the life-forms in the other four taxons, including animals and obviously the five billion-plus humans alive today, are part of that less than 1 percent.

The biosphere can also be subdivided into biomes: a biome incorporates a set of biotic communities within a particular region exposed to similar climatic conditions and which have dominant species with similar life cycles, adaptations, and structures. Deserts, grasslands, temperate deciduous forests, coniferous forests, tundra, tropical rain forests, tropical seasonal forests, freshwater biomes, estuaries, wetlands, and marine biomes are examples of specific terrestrial or aquatic biomes.

Another indication of the complexity of the biosphere is a measure of the processes that take place within it, especially the essential processes of photosynthesis and respiration. The sheer size of the biosphere is indicated by the amount of biomass present. The net primary production of Earth's biosphere has been estimated as 224.5×10^{15} grams.

The biosphere interacts in constant, intricate ways with other global systems: the atmosphere, lithosphere, hydrosphere, and pedosphere. Maintenance of life in the biosphere depends on this complex network of biological-biological, physical-physical, and biological-physical interactions. All the interactions are mediated by an equally complex system of positive and negative feedbacks—and the total makes up the dynamics of the whole system. Since each and all interpenetrate and react on each other constantly, outlining a global ecology is a major challenge.

Normally biospheric dynamics are in a rough balance. The carbon cycle, for example, is usually balanced between production and decomposition, the familiar equation of photosynthesis and respiration. As Piel notes: "The two planetary cycles of photosynthesis and aerobic metabolism in the biomass not only secure renewal of the biomass but also secure the steady-state mixture of gases in the atmosphere. Thereby, these life processes mediate the inflow and outflow of solar energy through the system; they screen out lethal radiation, and they keep the temperature of the planet in the narrow range compatible with life." But human activities, especially the combustion of fossil fuels, contribute to increases in carbon dioxide, distorting the balance and in the process changing other global relationships such as the nature of incoming and outgoing radiation and differentials in temperature between poles and tropics.

Humans are the dominant species in the biosphere. The transformation of radiant energy into useable biological energy is increasingly being diverted by humans to their own use. A common estimate is that humans are now diverting huge amounts of the net primary production of the globe to their own use: perhaps 40 percent of terrestrial production and close to 25 percent of all production is either utilized or wasted through human activity. Net primary production is defined as the amount of energy left after subtracting the respiration of primary producers, or plants, from the total amount of energy. It is the total amount of "food" available from the process of photosynthesis—the amount of biomass available to feed organisms, such as humans, that do not acquire food through photosynthesis.

Resources

BOOKS

Kaufman, Donald G. *The Biosphere: Protecting Our Global Environment*. Dubuque, IA: Kendall Hunt, 2008.

Poynter, Jane. *The Human Experiment: Two Years and Twenty Minutes Inside Biosphere 2*. New York: Basic Books, 2006.

Shugart, H. H. *Global Change and the Terrestrial Biosphere: Achievements and Challenges*. New York: Wiley, 2011.

Gerald L. Young

Biosphere reserve

A biosphere reserve is an area of land recognized and preserved for its ecological significance. Ideally biosphere reserves contain undisturbed, natural environments that represent some of the world's important ecological systems and communities. Biosphere reserves are established in the interest of preserving the genetic diversity of these ecological zones, supporting research and education, and aiding local, sustainable development. Official declaration and international recognition of biosphere reserve status is intended to protect ecologically significant areas from development and destruction. Since 1976 an international network of biosphere reserves has developed, with the sanction of the United Nations. Each biosphere reserve is proposed, reviewed, and established by a national biosphere reserve commission in the home country under United Nations guidelines. Communication among members of the international biosphere network helps reserve managers share data and compare management strategies and problems.

The idea of biosphere reserves first gained international recognition in 1973, when the United Nations Educational and Scientific Organization's (UNESCO) Man and the Biosphere Program (MAB) proposed

Placard in Manu National Park, Peru. (© *Aivar Mikko / Alamy*)

that a worldwide effort be made to preserve islands of the world's living resources from logging, mining, urbanization, and other environmentally destructive human activities. The term derives from the ecological word biosphere, which refers to the zone of air, land, and water at the surface of the earth that is occupied by living organisms. Growing concern over the survival of individual species in the 1970s and 1980s led increasingly to the recognition that endangered species could not be preserved in isolation. Rather, entire ecosystems, extensive communities of interdependent animals and plants, are needed for threatened species to survive. Another idea supporting the biosphere reserve concept was that of genetic diversity. Generally ecological systems and communities remain healthier and stronger if the diversity of resident species is high. An alarming rise in species extinctions in recent decades, closely linked to rapid natural resources consumption, led to an interest in genetic diversity for its own sake. Concern for such ecological principles as these led to UNESCO's proposal that international attention be given to preserving Earth's ecological systems, not just individual species.

The first biosphere reserves were established in 1976. In that year, eight countries designated a total of fifty-nine biosphere reserves representing ecosystems from tropical rain forest to temperate sea coast. The following year twenty-two more countries added another seventy-two reserves to the United Nations list, and by 2002 there was a network of 408 reserves established in ninety-four different countries. As of the most recent MAB meeting in May 2009, there are 553 reserves in 107 countries, collectively referred to as the World Network of Biosphere Reserves.

Like national parks, wildlife refuges, and other nature preserves, the first biosphere reserves aimed to protect the natural environment from surrounding populations, as well as from urban or international exploitation. To a great extent this idea followed the model of United States national parks, whose resident populations were removed so that parks could approximate pristine, undisturbed natural environments.

But in smaller, poorer, or more crowded countries than the United States, this model of the depopulated reserve made little sense. Around most of the world's nature preserves, well-established populations—often indigenous or tribal groups—have lived with and among the area's flora and fauna for generations or centuries. In many cases, these groups exploit local resources—gathering nuts, collecting firewood, growing food—without damaging their environment. Sometimes, contrary to initial expectations, the activity of indigenous peoples proves essential in maintaining

habitat and species diversity in preserves. Furthermore, local residents often possess an extensive and rare understanding of plant habitat and animal behavior, and their skills in using resources are both valuable and irreplaceable. At the very least, the cooperation and support of local populations is essential for the survival of parks in crowded or resource- poor countries. For these reasons, the additional objectives of local cooperation, education, and sustainable economic development were soon added to initial biosphere reserve goals of biological preservation and scientific research. Attention to humanitarian interests and economic development concerns today sets apart the biosphere reserve network from other types of nature preserves, which often garner resentment from local populations who feel excluded and abandoned when national parks are established. United Nations MAB guidelines encourage local participation in management and development of biosphere reserves, as well as in educational programs. Ideally, indigenous groups help administer reserve programs rather than being passive recipients of outside assistance or management.

In an attempt to mesh the diverse objectives of biosphere reserves, the MAB program has outlined a theoretical reserve model consisting of three zones, or concentric rings, with varying degrees of use. The innermost zone, the legally protected core, should be natural or minimally disturbed, essentially without human presence or activity. Ideally this is where the most diverse plant and animal communities live and where natural ecosystem functions persist without human intrusion. Surrounding the core is a buffer zone, mainly undisturbed but containing research sites, monitoring stations, and habitat rehabilitation experiments. The outermost ring of the biosphere reserve model is the transition zone. Here there may be sparse settlement, areas of traditional use activities, and tourist facilities. The development and maintenance of these areas involves programs for training and education as well as monitoring and ongoing research. In 1991, UNESCO launched a program termed Biosphere Reserve Integrated Monitoring (BRIM) to collect and incorporate data and results from "abiotic, biodiversity, socio-economic, and integrated monitoring" for the World Network of Biosphere Reserves.

Many biosphere reserves have been established in previously existing national parks or preserves. This is especially common in large or wealthy countries where well-established park systems existed before the biosphere reserve idea was conceived. In 1991 most of the United States' forty-seven biosphere reserves lay in national parks or wildlife sanctuaries. In countries with few such preserves, nomination for United Nations biosphere reserve status can sometimes attract international assistance and funding. In some instances debt for nature swaps have aided biosphere reserve establishment. In such an exchange, international conservation organizations purchase part of a country's national debt for a portion of its face value, and in exchange that country agrees to preserve an ecologically valuable region from destruction. Bolivia's Beni Biosphere Reserve came about this way in 1987 when Conservation International, a Washington-based organization, paid $100,000 to Citicorp, an international lending institution. In exchange, Citicorp forgave $650,000 in Bolivian debt, loans the bank seemed unlikely to ever recover, and Bolivia agreed to set aside a valuable tropical mahogany forest. This process has also produced other reserves, including Costa Rica's La Amistad, and Ecuador's Yasuni and Galapagos Biosphere Reserves.

In practice, biosphere reserves function well only if they have adequate funding and strong support from national leaders, legislatures, and institutions. Without legal protection and long-term support from the government and its institutions, reserves have no real defense against development interests.

National parks can provide a convenient institutional niche, defended by national laws and public policing agencies, for biosphere reserves. Pre-existing wildlife preserves and game sanctuaries likewise ensure legal and institutional support. Infrastructure—management facilities, access roads, research stations, and trained wardens—is usually already available when biosphere reserves are established in or adjacent to ready-made preserves.

Funding is also more readily available when an established national park or game preserve, with a pre-existing operating budget, provides space for a biosphere reserve. With intense competition from commercial loggers, miners, and developers, money is essential for reserve survival. Especially in poorer countries, international experience increasingly shows that unless there is a reliable budget for management and education, nearby residents do not learn cooperative reserve management, nor do they necessarily support the reserve's presence. Without funding for policing and legal defense, development pressures can easily continue to threaten biosphere reserves. Logging, clearing, and destruction often continue despite an international agreement on paper that resource extraction should cease. Turning parks into biosphere reserves may not always be a good idea. National park administrators in some less wealthy countries fear that biosphere reserve guidelines, with their compromising objectives and strong humanitarian interests, may weaken the mandate of national parks and wildlife sanctuaries set aside to protect endangered species from population pressures and development. In some cases, they argue, there exists a legitimate need to exclude people if rare species such as tigers or rhinoceroses are to survive.

Because of the expense and institutional difficulties of establishing and maintaining biosphere reserves, about two-thirds of the world's reserves exist in the wealthy and highly developed nations of North America and Europe. Poorer countries of Africa, Asia, and South America have some of the most important remaining intact ecosystems, but wealthy countries can more easily afford to allocate the necessary space and money. Developed countries also tend to have more established administrative and protective structures for biosphere reserves and other sanctuaries. An increasing number of developing countries are working to establish biosphere reserves, though. A significant incentive, aside from national pride in indigenous species, is the international recognition given to countries with biosphere reserves. Possession of these reserves grants smaller and less wealthy countries some of the same status as that of more powerful countries such as the United States, Germany, and Russia.

Some difficult issues surround the biosphere reserve movement. One question that arises is whether reserves are chosen for reasons of biological importance or for economic and political convenience. In many cases national biosphere reserve committees overlook critical forests or endangered habitats because logging and mining companies retain strong influence over national policy makers. Another problem is that in and around many reserves, residents are not yet entirely convinced, with some reason, that management goals mesh with local goals. In theory sustainable development methods and education will continue to encourage communication, but cooperation can take a long time to develop. Among reserve managers themselves, great debate continues over just how much human interference is appropriate, acceptable, or perhaps necessary in a place ideally free of human activity. Despite these logistical and theoretical problems, the idea behind biosphere reserves seems a valid one, and the inclusiveness of biosphere planning, both biological and social, is revolutionary.

Resources

BOOKS

Alfsen-Norodom, Christine, Benjamin D. Lane, and Melody Corry. *Urban Biosphere and Society: Partnership of Cities.* Annals of the New York Academy of Sciences, Vol. 1023. New York: New York Academy of Sciences, 2004.

Loneragan, Owen. *Biosphere.* Gardners Books, 2009.

Poynter, Jane. *The Human Experiment: Two Years and Twenty Minutes Inside Biosphere 2*. New York: Thunder's Mouth Press, 2006.

Sound, Christine. *Explaining Biosphere Reserves*. Paris, France: UNESCO, 2004.

Mary Ann Cunningham

Biota *see* **Biotic community.**

Biotechnology

Few developments in science have had the potential for such a profound impact on research, technology, and society in general as has biotechnology. Yet authorities do not agree on a single definition of this term. Sometimes writers have limited the term to techniques used to modify living organisms and, in some instances, the creation of entirely new kinds of organisms.

In most cases, however, a broader, more general definition is used. The Industrial Biotechnology Association, for example, uses the term to refer to any "development of products by a biological process." These products may indeed be organisms or they may be cells, components of cells, or individual and specific chemicals. A more detailed definition is that of the European Federation of Biotechnology, which defines biotechnology as the "integrated use of biochemistry, microbiology, and engineering sciences in order to achieve technological (industrial) application

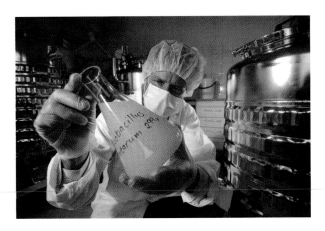

Researcher examining a flask containing a culture of the probiotic bacteria *Lactobacillus plantarum*. It is claimed that this bacterium has a beneficial effect on the stomach, and also reduces the risk of cardiovascular disease. *(Thierry Berrod, Mona Lisa Production / Photo Researchers, Inc.)*

of the capabilities of microorganisms, cultured tissue cells, and parts thereof." Biotechnology is an interdisciplinary field incorporating biochemistry, information technology, molecular biology, microbiology, pathology, embryology, and cell biology with other fields being incorporated over time as technology advances.

By almost any definition, biotechnology has been used by humans for thousands of years, long before modern science existed. Some of the oldest manufacturing processes known to humankind make use of biotechnology. Beer, wine, and breadmaking, for example, all occur because of the process of fermentation. During fermentation, microorganisms such as yeasts, molds, and bacteria are mixed with natural products, which they use as food. In the case of winemaking, for example, yeasts live on the sugars found in some type of fruit juice, most commonly, grape juice. They digest those sugars and produce two new products, alcohol and carbon dioxide (CO_2).

The alcoholic beverages produced by this process have been a mainstay of human civilization for untold centuries. In breadmaking, the products of fermentation are responsible for the odor (the alcohol) and texture (the carbon dioxide) of freshly-baked bread. Cheese and yogurt are two other products formed when microorganisms act on a natural product, in this case milk—changing its color, odor, texture, and taste.

Biotechnology has long been used in a variety of industrial processes. As early as the seventeenth century, bacteria were used to remove copper (Cu) from its ores. Around 1910, scientists found that bacteria could be used to decompose organic matter in sewage, thus providing a mechanism for dealing efficiently with such materials in solid waste. A few years later, a way was found to use microorganisms to produce glycerol ($C_3H_5[OH]_3$) synthetically. That technique soon became very important commercially, since glycerol is used in the manufacturing of explosives and World War I was in the beginning stages.

Not all forms of biotechnology depend on microorganisms. Hybridization is an example. Farmers long ago learned that they could control the types of animals bred by carefully selecting the parents. In some cases, they actually created entirely new animal forms that do not occur in nature. The mule, a hybrid of horse and donkey, is such an animal.

Hybridization has also been used in agriculture for centuries. Farmers found that they could produce food plants with any number of special qualities by carefully selecting the seeds they plant and by controlling growing conditions. As a result of this kind of process, the 2–3 inch (5.1–7.6 centimeter) vegetable known as maize has evolved over the years into the 12-inch (thirty centimeter), robust product termed corn. Indeed, there is hardly a fruit or vegetable in our diet today that has not been altered by long decades of hybridization.

Until the late nineteenth century, hybridization was largely a trial-and-error process. Then the work of Austrian botanist Gregor Mendel (1822–1884) started to become known. Mendel's research on the transmission of hereditary characteristics soon gave agriculturists a solid factual basis on which to conduct future experiments in cross-breeding.

Modern principles of hybridization have made possible a greatly expanded use of biotechnology in agriculture and many other areas. One of the greatest successes of the science has been in the development of new food crops that can be grown in a variety of less-than-optimal conditions. The dramatic increase in harvests made possible by these developments has become known as the agricultural revolution or green revolution.

Three decades after the green revolution first changed agriculture in many parts of the world, a number of problems with its techniques have become apparent. The agricultural revolution forced a worldwide shift from subsistence farming to cash farming, and many small farmers in developing countries lack the resources to negotiate this shift. A farmer must make significant financial investments in seed, agricultural chemicals such as fertilizers and pesticides, and machinery to make use of new farming techniques. In developing countries, peasants do not have and cannot borrow the necessary capital. The seed, chemicals, machinery, and oil to operate the equipment must commonly be imported, adding to already crippling foreign debts. In addition, the new techniques often have harmful effects on the environment. In spite of problems such as these, however, the green revolution has clearly made an important contribution to the lessening of world hunger.

Modern methods of hybridization have application in many fields besides agriculture. For example, scientists are now using controlled breeding techniques and other methods from biotechnology to ensure the survival of species that are threatened or endangered.

The nature of biotechnology has undergone a dramatic change in the last half century. That change has come about with the discovery of the role of deoxyribonucleic acid (DNA) in living organisms. DNA is a complex molecule consisting of strands of connected nucleotides (nitrogenous base, five-carbon

sugar, phosphate) arranged in a particular order. Two strands of DNA are connected by chemical bonds and run in antiparallel (opposite) directions. The many potential sequences of nucleotides allow DNA to store a large amount of genetic information. That information provides cells with the instructions they need to carry out all the functions they have to perform in a living organism. It also provides a mechanism by which that information is transmitted efficiently from one generation to the next.

As scientists learned more about the structure of the DNA molecule, they discovered precisely and in chemical terms how genetic information is stored and transmitted. With that knowledge, they have also developed the ability to modify DNA, creating new or modified instructions that direct cells to perform new and unusual functions. The process of DNA modification has come to be known as genetic engineering. Since genetic engineering normally involves combining two different DNA molecules, it is also referred to as recombinant DNA research. The sequencing of entire organism genomes (the full genetic content of an organism) has advanced the field of biotechnology, allowing scientists to uncover the functions of genes and their interactions as well as enabling researchers to manipulate single or multiple genes for a specific end purpose. The sequencing of the human genome was completed in 2003. The human genome sequence project and advancements in biotechnology has made gene therapy, which is the process of adding or replacing abnormally functioning genes with normally functioning genes to treat cancer and diseases such as AIDS.

There is little doubt that genetic engineering is the best known form of biotechnology today. Indeed, it is easy to confuse the two terms. However, the two terms are different in the respect that genetic engineering is only one type of biotechnology.

In theory, the steps involved in genetic engineering are relatively simple. First, scientists decide what kind of changes they want to make in a specific DNA molecule. They might, in some cases, want to alter a human DNA molecule to correct some error that results in a disease such as diabetes. In other cases, a researcher might want to add instructions to a DNA molecule that it does not normally carry. For example, a scientist might want to include instructions for the manufacture of a chemical such as insulin in the DNA of bacteria that normally lack the ability to make insulin.

Second, scientists find a way to modify existing DNA to correct errors or add new information. Such methods are now well developed. In one approach, enzymes (termed restriction enzymes) that recognize certain specific parts of a DNA molecule are used to cut open the molecule and then insert the new portion.

Third, scientists look for a way to insert the correct DNA molecule into the organisms in which it is to function. Once inside the organism, the new DNA molecule may provide corrected instructions to cells in humans (to avoid genetic disorders), in bacteria (resulting in the production of new chemicals), or in other types of cells for specific end purposes. *Agrobacterium* species are tumor-inducing bacteria that are used to transfer genetic information into plants. These bacteria are modified to remove the tumor-causing genetic component, which is replaced with the desired gene for transfer. Genetically modified organisms (GMOs) are produced through this process and may have improved growth, altered colors, or disease resistance.

Accomplishing these steps in practice is not always easy. One problem is ensuring the expression of an altered DNA molecule in the new host cells. Even if the molecule is able to enter a cell, it may not be able to operate and function (initiate expression) as planned. It is now possible for scientists to equip the altered DNA with the genetic information and mechanism necessary for expression. The expression of this DNA can be occurring all of the time in all parts of the organism or it can be initiated at certain times or under specific conditions.

In spite of problems, genetic engineering has already resulted in a number of impressive accomplishments. Dozens of products that were once available only from natural sources and in limited amounts are now manufactured in abundance by genetically engineered microorganisms at relatively low cost. Insulin, human growth hormone, tissue plasminogen activator, and alpha interferon are examples. In addition, the first trials with the alteration of human DNA to cure a genetic disorder were begun in 1991.

The prospects offered by genetic engineering have not been greeted with unanimous enthusiasm by everyone. Many people believe that the hope of curing or avoiding genetic disorders is a positive advance. But they question the wisdom of making genetic changes that are not related to life-threatening disorders. Some groups argue that these types of procedures should not be used for helping short children become taller or for making new kinds of tomatoes. Indeed, there are some critics who oppose *all* forms of genetic engineering, arguing that humans never have the moral right to modify an organism for any reason. As the technology available for genetic engineering continues to improve, debates over the use of these techniques in practical

settings are almost certainly going to continue—and to escalate—in the future.

As the field of genetic engineering progresses, so do other forms of biotechnology. The discovery of monoclonal antibodies is an example. Monoclonal antibodies are cells formed by the combination of tumor cells with animal cells that make one and only one kind of antibody. When these two kinds of cells are fused, they result in a cell that reproduces almost infinitely and that recognizes one specific kind of antigen. Such cells are extremely valuable in a vast array of medical, biological, and industrial applications, including the diagnosis and treatment of disease, the separation and purification of proteins, and the monitoring of pregnancy.

Biotechnology became a point of contention in 1992 during planning for the United Nations Earth Summit in Rio de Janeiro. In draft versions of the treaty on biodiversity, developing nations insisted on provisions that would force biotechnology companies in the developed world to pay fees to developing nations for the use of their genetic resources (the plants and animals growing within their boundaries). Companies had free access to most of these raw materials used in the manufacture of new drugs and crop varieties. U.S. President George Bush (1924–) argued that this provision would place an unfair burden on biotechnology companies in the United States, and refused to sign the biodiversity treaty that contained this clause. The Clinton administration subsequently endorsed the provisions of the biodiversity treaty that was signed by Madeleine Albright (1937–), U.S. ambassador to the United Nations, on June 4, 1993.

Another controversial topic regarding the use of biotechnology is stem cell research. Stem cells are undifferentiated cells and can be derived from adults and embryos. Researchers assert that embryonic stem cells have the potential to treat many human diseases, such as cancer. Opponents to stem cell research argue that embryos should not be sacrificed for this purpose. The Obama administration signed legislation in 2009 allowing broader use of federal funds to be used for embryonic stem cell research.

In what scientists broadly describe as a fundamental and far-reaching advance in synthetic biology, in May 2010, scientists announced the artificial creation of a genome capable of replacing the natural genome in a living cell.

Researchers from the J. Craig Venter Institute, located in Rockville, Maryland, recreated the genome of *Mycoplasma mycoides* (*M. mycoides*) bacteria and inserted the artificially created genome into a different species of bacteria, *Mycoplasma capricolum* (*M. capricolum cells*). Following the switch in genetic material, the transformed bacterial cell then began to function as though it were a native *Mycoplasma mycoides* bacterial cell.

Resources

BOOKS

Grace, Eric S. *Biotechnology Unzipped: Promises and Realities.* New York: Joseph Henry Press, 2006.

Scragg, Alan H. *Environmental Biotechnology.* Oxford, UK: Oxford University Press, 2005.

Slater, Adrian; Nigel W. Scott; and Mark R. Fowler. *Plant Biotechnology: The Genetic Manipulation of Plants,* 2nd ed. Oxford, UK: Oxford University Press, 2008.

Stewart, C. Neal. *Plant Biotechnology and Genetics: Principles, Techniques and Applications.* Hoboken, NJ: John Wiley and Sons, 2008.

Walker, Sharon. *Biotechnology Demystified.* New York: McGraw-Hill Professional, 2006.

Zimmer, Marc. *Glowing Genes: A Revolution In Biotechnology.* Loughton, UK: Prometheus Books, 2005.

OTHER

United Nations System-Wide EarthWatch. "Biotechnology." http://earthwatch.unep.net/biotechnology/index.php (accessed October 3, 2010).

United States Environmental Protection Agency (EPA). "Treatment/Control: Technology: Biotechnology." http://www.epa.gov/ebtpages/treatechnology biotechnology.html (accessed October 3, 2010).

David E. Newton

Bioterrorism

Bioterrorism refers to the use of lethal biological agents to wage terror against a civilian population. It differs from biological warfare in that it also thrives on public fear, which can demoralize a population. An example of bioterrorism is provided by the anthrax outbreak which occurred during September-November 2001 in the United States. Anthrax spores intentionally spread in the mail distribution system caused five deaths and a total of twenty-two infections. The Centers for Disease Control (CDC) classifies bioterror agents into three categories:

- Category A Diseases/Agents can be easily disseminated or transmitted from person to person and that can result in high mortality rates while causing public panic and social disruption. Anthrax, botulism,

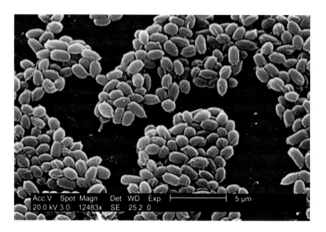

Electron micrograph of *bacillus anthracis*, the bacteria that causes anthrax. *(Centers for Disease Control)*

plague, smallpox, tularemis, and viral hemorrhagic fever viruses belong to this category.

- Category B Diseases/Agents are moderately easy to disseminate and that can result in low mortality rates. Brucellosis, food and water safety threats, melioidosis, psittacosis, staphylococcal enterotoxin B, and typhus belong to this category.

- Category C Diseases/Agents include emerging pathogens that could be engineered for mass dissemination in the future because of availability or ease of production and dissemination and that have potential for high mortality rates.

The anthrax attacks of 2001 were very limited in scope compared to the potential damage that could result from large-scale bioterrorism. A large-scale bioterrorism attack on the United States could threaten vital national security interests. Massive civilian casualties, a breakdown in essential services, violation of democratic processes, civil disorder, and a loss of confidence in government could compromise national security, according to a report prepared by four nonprofit analytical groups, including the Center for Strategic and International Studies and the John Hopkins Center for Civilian Biodefense Studies.

Probably the first sign of a bioterrorism attack is when people infected during the attack start developing symptoms and showing up in hospital emergency departments, urgent care centers, and doctors' offices. By this time, people infected in the initial attack will have begun spreading it to others.

An added concern is that most physicians have never treated a case of a bioterrorism agent such as smallpox or Ebola. This is likely to cause a delay in diagnosis, further promoting the spread of the contagious agent. For example, based on past smallpox history, it is estimated that each person infected during the initial attack will infect another ten to twelve persons. In the case of smallpox, only a few virus particles are needed to cause infection. One ounce of the smallpox virus could infect 3,000 persons if distributed through an aerosol attack, according to William Patrick (1927–), senior scientist in the United States biological weapons program before its official termination in 1969, in a 2001 *Washington Post Magazine* interview. Given these numbers, a terrorist with enough smallpox virus to fill a soda can could potentially infect 36,000 people in the initial attack who could then infect another 360,000–432,000. Of these, an estimated 30 percent or 118,800–140,400 would likely die.

Using disease as a weapon is not a new idea. It goes back at least hundreds of years and possibly much further. One account of the beginning of the great plague epidemic which occurred in Europe in the fourteenth century and killed a third of the population states that it started with an act of bioterrorism, as reported by A. Daniels in *National Review*. The Tartars were attacking a Genoan trading post on the Crimean coast in 1346 when the plague broke out among them. Turning the situation into a weapon, the Tartars catapulted the dead and diseased bodies over the trading post walls. The Genoans soon developed the deadly disease and took it back with them to Genoa, where it soon engulfed all of Europe. Another example from early North American history is provided by the British soldiers who deliberately gave smallpox-infected blankets to Native Americans in the 1700s.

The Hague Conventions of 1899 and 1907 included clauses outlawing the deliberate spread of a deadly disease. However, during World War I, German soldiers attempted to infect sheep destined for Russia with anthrax. After the war, forty members of the League of Nations, the precursor of the United Nations, outlawed biological weapons. But many countries continued biological warfare research. During World War II, the Japanese mass-produced a number of deadly biological agents, including anthrax, typhoid, and plague. They infected water supplies in China with typhoid, killing thousands, including 1,700 Japanese soldiers. Bioterrorism entered popular literature more than a century ago when British science fiction writer H. G. Wells (1866–1946) wrote "The Stolen Bacillus", a novel in which a terrorist tries to infect the London water supply with cholera, an acute and often deadly disease.

Throughout the Cold War era, several nations, including the United States and Soviet Union, developed sophisticated facilities to produce large amounts of biological agents to be used as weapons. Most nations have renounced the manufacture, possession, or use of biological weapons. However, a few rogue nations, including Iran, Iraq, and North Korea, still have active biological warfare programs according to the United States military. Many experts in the field believe that terrorists could obtain deadly biological agents from these rogue nations, or from other terrorist or criminal groups active in nations of the former Soviet Union.

Among the Category A Diseases/Agents, six highly lethal biological agents are most likely to be used by terrorists, according to the CDC. Depending on the biological agent, disease could be spread through the air, or by contaminating the food or water supply. Scientists are conducting research to develop methods of detection for bioterrorist attacks. Methods of real-time outbreak detection of diseases are referred to as biosurveillance. Researchers at the University of Pittsburgh Center for Biomedical Informatics developed an automated detection system in 1999. The system, Real-Time Outbreak Disease Surveillance (RODS), is able to access and collect data from hospitals, clinic, laboratories, and pharmacies for early detection of a potential bioterrorism attack. U.S. President George W. Bush issued a proposal to provide biosurveillance detection systems for all fifty states as part of the Public Health Security and Bioterrorism Preparedness and Response act of 2002. Other software programs have been developed for advanced detection such as Electronic Surveillance System Early Notification for Community-Based Epidemics (ESSENCE), Early Aberration Response System (EARS), Biowatch, and Bioshield. Biosense is a surveillance system used by the U.S. Center for Disease control. As of 2008, about more than one hundred sites in the U.S. were employing at least one method of biosurveillance.

- Anthrax, caused by *Bacillus anthracis*, is an acute infectious disease that most commonly occurs in hoofed animals but can also infect humans. Initial symptoms are flu-like and can occur up to several weeks after exposure. Treatment with antibiotics after exposure but before symptoms develop is usually successful in preventing infection. There is an anthrax vaccine used by the military but it is not available for civilian use. About 90 percent of people who are infected die.

- Botulism is a muscle-paralyzing disease caused by a toxin produced by a bacterium termed *Clostridium botulinum*. The botulinum toxin is the single most poisonous substance known, according to the Center for Civilian Biodefense Strategies. It is a major bioterrorism threat because of its extreme potency and

air from trenches or shallow wells that are screened within the unsaturated soils. The systems are similar to those used in vapor extraction. The main difference is that air extraction rates are low to minimize physical removal (stripping) of volatile organic compounds (VOCs), reducing the need for expensive treatment of the off-gases. The process may include the addition of nutrients such as common fertilizers, to provide nitrogen and phosphate for the bacteria. Bioventing is particularly attractive around buildings and actively used areas because it is relatively nonintrusive and results in minimal disturbance during installation and operation. The process is most suitable for petroleum hydrocarbon blends such as gasoline, jet fuel, and diesel oil, for petroleum distillates such as toluene, and for nonchlorinated solvents.

See also Biodegradable; Bioremediation; Vapor recovery system.

BirdLife International

"From research . . . to action. From birds . . . to people." So reads the cover of BirdLife International's annual report. This statement perfectly describes the beliefs of BirdLife International, a group founded under the original name International Council for Bird Preservation in 1922 by well-known American and European bird enthusiasts for the conservation of birds and their habitats.

Under the leadership of Director and Chief Executive Dr. Marco Lambertini, the group works to protect endangered birds worldwide and to promote public awareness of their ecological importance. BirdLife International has grown from humble beginnings in England to a federation of over 300 member organizations representing approximately 2.5 million members over 100 countries and territories. This includes developing tropical countries where few, if any, conservation movements existed prior to BirdLife International. There is also a worldwide network of enthusiastic volunteers.

BirdLife International is a key group in international efforts to protect bird migration routes, and also works to educate the public about endangered species and their ecological importance. The BirdLife International gathers and disseminates information about birds, maintaining a computerized data bank from which it generates reports. It conducts periodic symposiums on bird-related issues, runs the World Bird Club, maintains a Conservation Fund, runs special campaigns when opportunities such as the Migratory Bird Campaign present themselves, and develops and carries out priority projects in their Conservation Program.

The BirdLife International Conservation Program has undertaken many projects on behalf of endangered birds. BirdLife International began a captive breeding program for the pink pigeon (*Nesoenas mayeri*), a native of the island of Mauritius in the Indian Ocean, because the total population of this species had dwindled to just ten birds in the wild in 1990. As a result of these efforts, as of April 2007, 380 pink pigeons lived in the wild. Several pairs were released at Mauritius' Botanic Gardens of Pamplemousses. BirdLife International has focused on other seriously endangered birds as well, such as the imperial parrot (*Amazona imperialis*). In an attempt to protect its threatened habitat, BirdLife International has helped buy a forest reserve in Dominica. Conservation efforts have increased the imperial parrots numbers from sixty parrots to about 250 parrots. With the help of local citizens and educational facilities, BirdLife International hopes that their efforts to save the pink pigeon and imperial parrot will continue to be successful.

Another important BirdLife International project is the group's work to save the red-tailed parrot (*Amazona brasiliensis*) of southeastern Brazil. This project involves support of an extensive plan to convert an entire nearby island into a refuge for the parrots, which exist in only a very few isolated parts of Brazil. BirdLife International has also focused on islands in other conservation projects. The council purchased Cousin Island (famous for its numerous seabirds), in an effort to save the Seychelles brush warbler (*Acrocephalus sechellensis*). Native only to Cousin Island, this entire brush warbler species numbered only thirty individuals before BirdLife International bought their island. Today, there are more than 2,500 bush warblers, and BirdLife International continues to be actively involved in helping to breed more.

BirdLife International's publications are many and varied. Quarterly, it issues *Bird Conservation International*, a scholarly journal, and *World Birdwatch* magazine. It also publishes the well-respected series *Bird Red Data Books*, and such monographs as *Important Bird Areas in Europe* and *Key Forests for Threatened Birds in Africa*. BirdLife International produces numerous technical publications and study reports, and, occasionally, Conservation Red Alert pamphlets on severely threatened birds.

Resources

ORGANIZATIONS

BirdLife International, Wellbrook Court, Girton Road, Cambridge, United Kingdom, CB3 0NA, +44 (0) 1 223

277 318, +44 (0) 1 223 277 200, birdlife@birdlife.
org.uk, http://www.birdlife.org

Cathy M. Falk

Birth control *see* **Family planning; Male
contraceptives.**

Birth defects

Birth defects, also known as congenital malformations, are structural or metabolic abnormalities present at birth. While subtle variations normally occur in about half of all individuals in the United States, significant congenital defects are found in about 3 percent of live births. Fortunately, only about half of these require medical attention.

Birth defects may result from genetic causes or environmental insult. Defective genes are not easily repaired and thus are perhaps less interesting than teratogenic substances (substances able to cause defects in the developing fetus) to environmentalists. It is theoretically possible to limit exposure to teratogens by elimination of the agent in the environment or by modification of behavior to prevent contact. It should be noted, however, that the causes of more than half of congenital malformations remain unknown.

Birth defects of genetic origin may be due to aberrant chromosome number or structure, or to a single gene defect. Normal humans have 46 chromosomes, and variation from this number is referred to as aneuploidy. Down's syndrome, an example of aneuploidy, is usually characterized by an extra chromosome designated number 21. The individual with Down's thus has a total of 47 chromosomes, and the presence of the extra chromosome results in multiple defects. These often include developmental delays and physical characteristics comprising a small round head, eyes that slant slightly upward, a large and frequently protruding tongue, low set ears, broad hands with short fingers and short stature. People with Down's syndrome are particularly vulnerable to leukemia. Children with this condition are rarely born to mothers less than 25 years of age (less than one in 1,500), but the prevalence of Down's syndrome babies increases with mothers older than 45 (about one in twenty-five). Down's syndrome can be detected during pregnancy by chromosomal analysis of fetal cells. Fetal chromosomes may be studied by chorionic villus sampling or by amniocentesis.

Other congenital abnormalities with a chromosomal basis include Klinefelter's syndrome, a condition of male infertility associated with an extra X chromosome, and Turner's syndrome, a condition wherein females fail to mature sexually and are characterized by the aneuploid condition of a missing X chromosome.

Achondroplasia is a birth defect due to a dominant mutation of a Mendelian gene that results in dwarfism. Leg and arm bones are short but the trunk is normal and the head may be large. Spontaneous mutation accounts for most achondroplasia. The mortality rate for affected individuals is so high that the gene would be lost if it were not for mutation. Albinism, a lack of pigment in the skin, eyes, and hair, is another congenital defect caused by a single gene, which in this case is recessive.

The developing fetus is at risk for agents which can pass the placental barrier, such as infectious microbes, drugs and other chemicals, and ionizing radiation. Transplacental teratogens exert their effect on incompletely formed embryos or fetuses during the first three months of pregnancy. Organs and tissues in older and full term fetuses appear much as they will throughout life. It is not possible to alter the development of a fully formed structure. However, prior to the appearance of an organ or tissue, or during the development of that structure, teratogenic agents may have a profoundly deleterious effect.

Perhaps the best known teratogen is the sedative thalidomide which induces devastating anatomical abnormalities. The limb bones are either shortened or entirely lacking leading to a condition known as phocomelia. Intellectual development of thalidomide babies is unaffected. The experience with this drug, which started in 1959 and ended when it was withdrawn in 1961, emphasizes the fact that medications given to pregnant mothers generally cross the placenta and reach the developing embryo or fetus. Another drug that effects developmental abnormalities is warfarin, which is used in anticoagulant therapy. It can cause fetal hemorrhage, mental retardation, and a multiplicity of defects to the eyes and hands when given to pregnant women.

The teratogenic effects of alcohol, or the lifestyle that may accompany alcohol abuse, serves to illustrate that the term environment includes not only air and water, but the personal environment as well. Alcoholism during pregnancy can result in fetal alcohol syndrome with facial, limb, and heart defects accompanied by growth retardation and reduced intelligence. The effects of alcohol may be magnified by factors associated with alcoholism such as poor diet, altered metabolism and inadequate medical care. Because neither the time of vulnerability nor the toxic level of alcohol is known, the best advice is to eschew alcohol as a dietary constituent altogether during pregnancy.

Disease of the mother during pregnancy can present an environmental hazard to the developing fetus. An example of such a hazard is the viral disease German measles, also known as rubella. The disease is characterized by a slight increase in temperature, sore throat, lethargy and a rash of short duration. Greatest hazard to the fetus is during the second and third month. Children born of mothers who had rubella during this period may exhibit cataracts, heart defects, hearing loss and mental retardation. Obviously, the virus transverses the placenta to infect the embryo or fetus and that infection may persist in the newborn. Birth defects associated with rubella infection have decreased since the introduction of a rubella vaccine.

The most common viral infection that occurs in human fetuses is that of a herpes virus known as cytomegalovirus. The infection is detected in about 1–2 percent of all live births. Most newborns, fortunately, do not manifest symptoms of the infection. However, for a very small minority, the effects of congenital cytomegalovirus are cruel and implacable. They include premature birth or growth retardation prior to birth, frequently accompanied by hepatitis, enlarged spleen, and reduction in thrombocytes (blood cells important for clotting). Abnormally small heads, mental retardation, cerebral palsy, heart and cerebral infection, bleeding problems, hearing loss, and blindness can also occur. Exposure of the fetus to the virus occurs during infection of the pregnant woman or possibly from the father, as cytomegalovirus has been isolated from human semen.

Other infections known to provoke congenital defects include herpes simplex virus type II, toxoplasmosis, and syphilis.

Methylmercury is an effective fungicide for seed grain. Accidental human consumption of food made from treated seeds has occurred. Industrial pollution of sea water with organic mercury resulted in the contamination of fish, consumed by humans, from Minamata Bay in Japan. It has been established that organic mercury passes the placental barrier with effects that include mental retardation and a cerebral palsy-like condition due to brain damage. Anatomical birth defects, engendered by organic mercury, include abnormal palates, fingers, eyes, and hearts. The toxicity of methylmercury affects both early embryos and developing fetuses. Exclusion of mercury from human food can be effected by not using organic mercury as a

fungicide and by ending industrial discharge of mercury into the environment.

Of course other chemicals may be hazardous to the offspring of pregnant women. Polychlorinated biphenyl (PCB)s, relatively ubiquitous, but low level oily contaminants of the environment, cause peculiar skin pigmentation, low birth weights, abnormal skin and nails, and other defects in offspring when accidentally ingested by pregnant woman.

Uncharacterized mixtures of toxic chemicals that contaminate the environment, are thought to be potential teratogens. Cytogenetic (chromosomal) abnormalities and increased birth defects were detected among the residents of the contaminated Love Canal, New York. Cigarette smoke is the most common mixture of toxic substance to which fetuses are exposed. Tobacco smoke is associated with reduced birth weight but not specific birth anatomical abnormalities.

Much concern has arisen over the damaging effects of ionizing radiation, particularly regarding diagnostic x-rays and radiation exposure from nuclear accidents. The latter concern was given international attention following the explosion at Ukraine's Chernobyl Nuclear Power Station in 1986. Fear that birth defects would occur as a result was fueled by reports of defects in Japanese children whose mothers were exposed to radiation at Hiroshima. Radiation exposures to the fetus can result in many defects including various malformations, mental retardation, reduced growth rate, and increased risk for leukemia. Fortunately, however, the risk of these effects is exceptionally low. Fetal abnormalities caused by factors other than radiation are thought to be about ten times greater than those attributed to radiation during early pregnancy. However small the risk, most women choose to limit or avoid exposure to radiation during early pregnancy. This may be part of the reason for the increased popularity of diagnostic ultrasound as opposed to x-ray.

While concern is expressed for particular teratogenic agents or procedures, the etiology of most birth defects is unknown. Common defects, with unknown etiology, include cleft lip and cleft palate, extra fingers and toes, fused fingers, extra nipples, various defects in the heart and great vessels, narrowing of the entrance to the stomach, esophageal abnormalities, spina bifida, clubfoot, hip defects, and many, many others. Since the majority of birth defects are not caused by known effects of disease, drugs, chemicals or radiation, much remains to be learned.

A campaign launched in the 1990s to make women aware of the importance of folic acid in the diet before becoming pregnant is credited with reducing the prevalence of some neural tube birth defects including spina bifida by 23 percent by 2005. The National Institute of Health recommends that women of childbearing age receive 400 micrograms (mcg) of folic acid, a B vitamin, in their daily diet. According to the March of Dimes, 400 mcg of folic acid consumed daily at least one month before becoming pregnant reduces the risk of spina bifida and anencephaly in the fetus by 50-70 percent.

Resources

BOOKS

Ferretti, Patrizia. *Embryos, Genes, and Birth Defects.* Chichester, UK: Wiley, 2006.

March of Dimes Data Book for Policy Makers: Maternal, Infant, and Child Health in the United States, 2010. Washington, DC: March of Dimes, 2009.

Moore, Keith L., T. V. N. Persaud, and Mark G. Torchia. *Before We Are Born: Essentials of Embryology and Birth Defects.* Philadelphia: Saunders-Elsevier, 2008.

Wynbrandt, James, and Mark D. Lundman. *The Encyclopedia of Genetic Disorders and Birth Defects*, 3rd ed. New York: Facts On File, 2008.

PERIODICALS

Cleves, M. A., and C. A. Hobbs. "Collaborative Strategies for Unraveling the Complexity of Birth Defects." *Journal of Maternal-Fetal and Neonatal Medicine* 15, no. 1 (2004): 35-38.

Meyer, R. E., and A. B. Brown. "Folic Acid and Birth Defects Prevention: A Public Health Success Story." *North Carolina Medical Journal* 65, no. 3 (2004): 157.

Rasmussen, Sonja A., and Cynthia A. Moore. "Public Health Approach to Birth Defects, Developmental Disabilities, and Genetic Conditions." *American Journal of Medical Genetics* 125, no. 1 (2004): 1.

OTHER

Centers for Disease Control and Prevention (CDC). "Birth Defects." http://www.cdc.gov/ncbddd/bd/default.htm (accessed November 8, 2010).

March of Dimes.http://www.marchofdimes.com/ (accessed November 8, 2010).

National Institutes of Health (NIH). "Birth Defects." http://health.nih.gov/topic/BirthDefects (accessed November 8, 2010).

Robert G. McKinnell

Bison

The American bison (*Bison bison*) or buffalo is one of the most famous animals of the American West. Providing food and hides to the early Indians, it was

American bison (*Bison bison*). *(Oliver Le Qu/Shutterstock.com)*

almost completely eliminated by hunters, and now only remnant populations exist though its future survival seems assured.

Scientists do not consider the American bison a true buffalo (like the Asian water buffalo or the African buffalo), since it has a large head and neck, a hump at the shoulder, and fourteen pairs of ribs instead of thirteen. In United States, however, the names are used interchangeably. A full-grown American bison bull stands 5.5–6 feet (1.7–1.8 m) at the shoulder, extends 10–12.25 feet (3–3.8 m) in length from nose to tail, and weighs 1,600–3,000 pounds (726–1,400 kg). Cows usually weigh about 900 pounds (420 kg) or less. Bison are brown-black with long hair which covers their heads, necks, and humps, forming a "beard" at the chin and throat. Their horns can have a spread as large as 35 inches (89 cm). Bison can live for 30 or more years, and they are social creatures, living together in herds. Bison bulls are extremely powerful; a charging bull has been known to shatter wooden blanks 2 inches (5 cm) thick and 12 inches (30 cm) wide.

The American bison is one of the most abundant animals ever to have existed on the North American continent, roaming in huge herds between the Appalachians and the Rockies and as far south as

Florida. One herd seen in Arkansas in 1870 was described as stretching "from six to 10 miles (9.7 to 16.1 km) in almost every direction." In the far West, the herds were even larger, stretching as far as the eye could see, and in 1871 a cavalry troop rode for six days through a herd of bison.

The arrival of Europeans in America sealed the fate of the American bison. By the 1850s massive slaughters of these creatures had eliminated them from Illinois, Indiana, Kentucky, Ohio, New York, and Tennessee. After the end of the Civil War in 1865, railroads began to bring a massive influx of settlers to the West and bison were killed in enormous numbers. The famous hunter "Buffalo Bill" Cody was able to bag 4,280 bison in just eighteen months, and between 1854 and 1856, an Englishman named Sir George Gore killed about 6,000 bison along the lower Yellowstone River. Shooting bison from train windows became a popular recreation during the long trip west; there were contests to see who could kill the most animals on a single trip, and on one such excursion a group accompanying Grand Duke Alexis of Russia shot 1,500 bison in just two days. When buffalo tongue became a delicacy sought after by gourmets in the east, even more bison were killed for their tongues and their carcasses left to rot.

In the 1860s and 1870s extermination of the American bison became the official policy of the United States government in order to deprive the Plains Indians of their major source of food, clothing, and shelter. During the 1870s, two to four million bison were shot each year, and 200,000 hides were sold in St. Louis in a single day. Inevitably, the extermination of the bison helped to eliminate not only the Plains Indians, but also the predatory animals dependent on it for food, such as plains wolves. By 1883, according to some reports, only one wild herd of bison remained in the West, consisting of about 10,000 individuals confined to a small part of North Dakota. In September of that year, a group of hunters set off to kill the remaining animals and by November the job was done.

By 1889 or 1890 the entire North American bison population had plummeted to about 500 animals, most of which were in captivity. A group of about 20 wild bison remained in Yellowstone National Park, and about 300 wood bison (*Bison bison athabascae*) survived near Great Slave Lake in Canada's Northwest Territories. At that time, naturalist William Temple Hornaday led a campaign to save the species from complete extinction by the passage of laws and other protective measures. Canada enacted legislation to protect its remnant bison population in 1893 and the United States took similar action the following year.

In the early 2000s, thousands of bison are found in several national parks, private ranches, and game preserves in the United States. About 15,000 are estimated to inhabit Wood Bison National Park and other locations in Canada. The few hundred wood bison originally saved around Great Slave Lake also continued to increase in numbers until the population reached around 2,000 in 1922. But in the following years, the introduction of plains bison to the area caused hybridization, and pure specimens of wood bison probably disappeared around Great Slave Lake. Fortunately, a small, previously unknown herd of wood bison was discovered in 1957 on the upper North Yarling River, buffered from hybridization by 75 miles (121 km) of swampland. From this herd (estimated at about 100 animals in 1965) about twenty-four animals were successfully transplanted to an area near Fort Providence in the Northwest Territories and forty-five were relocated to Elk Island National Park in Alberta. Despite these rebuilding programs, the wood bison is still considered endangered and is listed as such by the U.S. Department of the Interior. It is also listed in Appendix I of the Convention on International Trade in Endangered Species of Fauna and Flora (CITES) treaty.

Controversy still surrounds the largest herd of American bison (5,000–6,000 animals in the early 1990s) in Yellowstone National Park. The free-roaming bison often leave the park in search of food in the winter, and Montana cattle ranchers along the park borders fear that the bison could infect their herds with brucellosis, a contagious disease that can cause miscarriages and infertility in cows. In an effort to prevent any chance of brucellosis transmission, the National Park Service (NPS) and the Montana Department of Fish, Wildlife and Parks, along with sport hunters acting in cooperation with these agencies, killed 1,044 bison between 1984 and 1992. Montana held a lottery-type hunt, and 569 bison were killed in the winter of 1988–1989, and 271 were killed in the winter of 1991–1992. The winter of 1996–1997 was exceptionally harsh, and some 850 buffalo of the park's remaining 3,500 starved or froze to death. In addition, the NPS, the U.S. Department of Agriculture (USDA), and the Montana Department of Livestock cooperated in a stepped-up buffalo killing program, in which some 1,080 were shot or shipped off to slaughterhouses. In all, more than half of Yellowstone's bison herd perished that winter.

Wildlife protection groups, such as the Humane Society of the United States and the Fund for

Animals, have protested the hunting of these bison—which usually consists of walking up to an animal and shooting it. Animal protection organizations have offered alternatives to the killing of the bison, including fencing certain areas to prevent them from coming into contact with cattle. Conversely, Montana state officials and ranchers, as well as the USDA, have long pressured the National Park Service to eradicate many or all of the Yellowstone bison herd or at least test the animals and eliminate those showing signs of brucellosis. Such an action, however, would mean the eradication of most of the Yellowstone herd, even though bison have not been known to infect cattle.

There is also a species of European bison called the wisent (*Bison bonasus*) which was once found throughout much of Europe. It was nearly exterminated in the early 1900s, but today a herd of about 1,600 animals can be found in a forest on the border between Poland and Russia. The European bison is considered vulnerable by IUCN—The World Conservation Union.

See also Endangered species; Endangered Species Act (1973); Overhunting; Rare species; Wildlife management.

Resources

BOOKS

Allen, J. A. *History of the American Bison: Bison Americanus*. Ann Arbor, MI: ProQuest, 2007.

Brodie, Jedediah F. *A Review of American Bison (Bos Bison): Demography and Population Dynamics*. Gardiner, MT: Wildlife Conservation Society & Pennsylvania State University, 2008.

Franke, Mary Ann. *To Save the Wild Bison: Life on the Edge in Yellowstone*. Norman: University of Oklahoma Press, 2005.

Harkin, Michael Eugene, and David Rich Lewis. *Native Americans and the Environment: Perspectives on the Ecological Indian*. Lincoln: University of Nebraska Press, 2007.

Isenberg, Andrew C. *The Destruction of the Bison: An Environmental History, 1750–1920*. Cambridge, UK: Cambridge University Press, 2008.

Winner, Cherie, and John F. McGee. *Bison*. Paw Prints, 2008.

Zontek, Ken. *Buffalo Nation: American Indian Efforts to Restore the Bison*. Lincoln: University of Nebraska Press, 2007.

Lewis G. Regenstein

Bituminous coal *see* **Coal.**

Black lung disease

Black lung disease, also known as anthracosis or coal workers' pneumoconiosis (CWP), is a chronic, fibrotic lung disease of coal miners. It is caused by inhaling coal dust, which accumulates in the lungs and forms black bumps or coal macules on the bronchioles. These black bumps in the lungs give the disease its common name. Lung disease among coal miners was first described by German mineralogist Georgius Agricola (1494–1555) in the sixteenth century, and it is now a widely recognized occupational illness.

Black lung disease occurs most often among miners of anthracite (hard) coal, but it is found among soft coal miners and graphite workers as well. The disease is characterized by gradual onset—the first symptoms usually appear only after ten to twenty years of exposure to coal dust. The extent and severity of the disease is clearly related to the length of this exposure. The disease also appears to be aggravated by cigarette smoking. The more advanced forms of black lung disease are frequently associated with emphysema or chronic bronchitis. There is no real treatment for this disease, but it may be controlled or its development arrested by avoiding exposure to coal dust.

Black lung disease is among the best-known occupational illnesses in the United States. Although there have been considerable reductions in the incidence of the disease since the mid-twentieth century, in some regions, more than 50 percent of coal miners develop the disease after thirty or more years on the job. The Federal Coal Mine Health and Safety Act (FCMHSA)

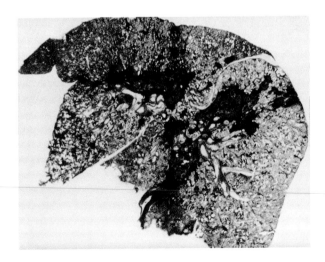

Thin section of whole lung showing the fibrosis characteristic of coalworker's pneumoconiosis, also known as black lung disease. *(Biophoto Associates / Photo Researchers, Inc.)*

of 1969 (amended in 1977) resulted in a reduction of CWP cases. The FCMHSA included provision for limits on coal mine dust and for a periodic chest radiograph program for mine workers. The Mine Safety and Health Administration (MSHA) coordinates with the National Institute for Occupational Safety and Health (NIOSH) to oversee the Coal Workers' Health Surveillance Program (CWHSP), which provides early detection and preventative measures against CWP. The rate of CWP in underground coal miners with twenty-five years or greater experience decreased from about 30 percent in the early 1970s to less than 5 percent in the late 1990s. Since 1995, however, the prevalence of black lung disease in CWHSP participants has doubled. Experts note the increase is especially associated with certain higher-risk or contract mining jobs, smaller mines, and mines in specific geographic areas.

Black-footed ferret (*Mustela nigripes*). (© All Canada Photos/ Alamy)

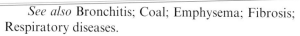

See also Bronchitis; Coal; Emphysema; Fibrosis; Respiratory diseases.

Resources

BOOKS

Levine, Linda. *Coal Mine Safety and Health*. CRS report for Congress, RL34429. Washington, DC: Congressional Research Service, Library of Congress, 2008.
McIvor, Arthur, and Ronald Johnston. *Miners' Lung: A History of Dust Disease in British Coal Mining*. Aldershot, UK: Ashgate, 2007.

OTHER

Medline Plus. "Coal Workers Pneumoconiosis." http:// www.nlm.nih.gov/medlineplus/ency/article/ 000130.htm (accessed November 7, 2010).
National Institute for Occupational Safety and Health (NIOSH). "Occupational Respiratory Disease Surveillance Program: Coal Workers' Health Surveillance Program." http://www.cdc.gov/niosh/ topics/surveillance/ORDS/CoalWorkersHealth SurvProgram.html (accessed November 7, 2010).

Linda Rehkopf

Black-footed ferret

A member of the Mustelidae (weasel) family, the black-footed ferret (*Mustela nigripes*) is the only ferret native to North America. It has pale yellow fur, an off-white throat and belly, a dark face, black feet, and a black tail. The black-footed ferret usually grows to a length of 18 inches (46 cm) and weighs 1.5–3 pounds (0.68–1.4 kg), though the males are larger than the females. These ferrets have short legs and slender bodies, and lope along by placing both front feet on the ground followed by both back feet.

Ferrets live in prairie dog burrows and feed primarily upon prairie dogs, mice, squirrels, and gophers, as well as small rabbits and carrion. Ferrets are nocturnal animals; activity outside the burrow occurs after sunset until about two hours before sunrise. They do not hibernate and remain active all year long.

Breeding takes place once a year, in March or early April, and during the mating season males and females share common burrows. The gestation period lasts approximately six weeks, and the female may have from one to five kits per litter. The adult male does not participate in raising the young. The kits remain in the burrow where they are protected and nursed by their mother until about four weeks of age, usually sometime in July, when she weans them and begins to take them above ground. She either kills a prairie dog and carries it to her kits or moves them into the burrow with the dead animal. During July and early August, she usually relocates her young to new burrows every three or four days, whimpering to encourage them to follow her or dragging them by the nape of their neck. At about eight weeks old the kits begin to play above ground. In late August and early September the mother positions her young in separate burrows, and by mid-September her offspring have left to establish their own territories.

Black-footed ferrets, like other members of the mustelid family, establish their territories by scent marking. They have well-developed lateral and anal scent glands. The ferrets mark their territory by either wiggling back and forth while pressing their pelvic scent glands against the ground, or by rubbing their lateral scent glands against shrubs and rocks. Urination is a third form of scent marking. Males establish large territories that may encompass one or more females of the species and exclude all other males. Females establish smaller territories.

Historically, the black-footed ferret was found from Alberta, Canada, southward throughout the Great Plains states. The decline of this species began in the 1800s with the settling of the west. Homesteaders moving into the Great Plains converted the prairie into agricultural lands, which led to a decline in the population of prairie dogs. Considering them a nuisance species, ranchers and farmers undertook a campaign to eradicate the prairie dog. The black-footed ferret is dependent upon the prairie dog: it takes 100–150 acres (40–61 ha) of prairie-dog colonies to sustain one adult. Because it takes such a large area to sustain a single adult, one small breeding group of ferrets requires at least 10 mi^2 (26 km^2) of habitat. As the prairie dog colonies became scattered, the groups were unable to sustain themselves.

In 1954, the National Park Service began capturing black-footed ferrets in an attempt to save them from their endangered status. These animals were released in wildlife sanctuaries that had large prairie dog populations. Black-footed ferrets, however, are highly susceptible to canine distemper, and this disease wiped out the animals the park service had relocated.

In September 1981, scientists located the only known wild population of black-footed ferrets near the town of Meeteetse in northwestern Wyoming. The colony lived in twenty-five prairie dog towns covering 53 mi^2 (137 km^2). But in 1985, canine distemper decimated the prairie dog towns around Meeteetse and spread among the ferret population, quickly reducing their numbers. Researchers feared that without immediate action the black-footed ferret would become extinct. The only course of action appeared to be removing them from the wild. If an animal had not been exposed to canine distemper, it could be vaccinated and saved. Some animals from the Meeteetse population did survive in captivity.

There is a breeding program and research facility called the National Black-footed Ferret Conservation Center in Wyoming, and in 1987, the Wyoming Fish and Game Department implemented a plan for preserving the black-footed ferret within the state. Researchers identified habitats where animals bred in captivity could be relocated. The program began with the eighteen animals from the wild population located at Meeteetse. In 1987, seven kits were born to this group. The following year thirteen female black-footed ferrets had

litters and thirty-four of the kits survived. In 1998, about 330 kits survived. Captive propagation efforts have improved the outlook for the black-footed ferret with over 7,000 kits born so far in the program. Captive populations will continue to be used to reestablish ferrets in the wild. About 2,300 black-footed ferrets that were bred and raised in captivity have been released into the wild.

Resources

BOOKS

Aronin, Miriam. *Black-Footed Ferrets: Back from the Brink*. New York: Bearport, 2008.

Debnam, Betty. *A "Tail" of Hope!: Black-Footed Ferret*. USA: Universal Press Syndicate, 2006.

Goodall, Jane, Thane Maynard, and Gail E. Hudson. *Hope for Animals and Their World: How Endangered Species Are Being Rescued from the Brink*. New York: Grand Central, 2009.

PERIODICALS

Cubie, D. "A Rare Species Gets a Second Chance Twenty-Five Years After the Black-Footed Ferret Was Rediscovered Surviving in the Wild, a Successful Captive-Breeding Program Is Giving the Endangered Animal a New Lease on Life." *National Wildlife* 45, no. 1 (2007): 12–13.

Dobson, A., and A. Lyles. "Ecology: Black-Footed Ferret Recovery." *Science* no. 5428 (2000): 985–87.

OTHER

Black-Footed Ferret Recovery Program. http://www.black footedferret.org/ (accessed November 9, 2010).

Debra Glidden

Blackout/brownout

A blackout is a total loss of electrical power. A blackout is usually defined as a drop in line voltage below 80 volts (V) (the normal voltage is 120V), since most electrical equipment will not operate below these levels. A blackout may be due to a planned interruption, such as limiting of loads during power shortages by rotating power shutoffs through different areas, or due to an accidental failure caused by human error, a failure of generating or transmission equipment, or a storm. Blackouts can cause losses of industrial production, disturbances to commercial activities, traffic and transportation difficulties, disruption of municipal services, and personal inconveniences. In the summer of 1977, a blackout caused by transmission line losses during a storm affected the New York City area. About nine million people were affected by the blackout, with some areas without power for more than 24 hours. The blackout was accompanied by looting and vandalism. A blackout in northeastern United States and eastern Canada due to a switching relay failure in November of 1965 affected 30 million people and resulted in improved electric utility power system design.

A brownout is a condition (usually temporary, but which may last longer, i.e., from periods ranging from fractions of a second to hours) when the alternating current (AC) electrical utility voltage is lower than normal. If the brownout lasts less than a second, it is called a *sag*. Brownouts may be caused by overloaded circuits, but are sometimes caused intentionally by a utility company in order to reduce the amount of power drawn by users during peak demand periods, or unintentionally when demand for electricity exceeds generating capacity. A sag can also occur when line switching is employed to access power from secondary utility sources. Equipment such as shop tools, compressors, and elevators starting up on a shared power line can cause a sag, which can adversely affect other sensitive electronic equipment such as computers. Generally, electrical utility customers do not notice a brownout except when it does affect sensitive electronic equipment.

Measures to protect against effects of blackouts and brownouts include efficient design of power networks, interconnection of power networks to improve stability, monitoring of generating reserve needs during periods of peak demand, and standby power for emergency needs. An individual piece of equipment can be protected from blackouts and brownouts by the use of an uninterruptible power source (UPS). A UPS is a device with internal batteries that is used to guarantee that continuous power is supplied to equipment even if the power supply stops providing power or during line sags. Commonly the UPS will boost voltage if the voltage drops to less than 103V and will switch to battery power at 90V and below. Some UPS devices are capable of shutting down the equipment during extended blackouts.

Resources

BOOKS

Grigsby, Leonard L. *Electric Power Generation, Transmission, and Distribution*. The electric power engineering series, 2. Boca Raton, FL: CRC Press, 2007.
Klein, Maury. *The Power Makers: Steam, Electricity, and the Men Who Invented Modern America*. New York: Bloomsbury Press, 2009.
Leckebusch, Gregor, and Tina Tin. *Stormy Europe: The Power Sector and Extreme Weather*. Gland, Switzerland: World Wildlife Fund for Nature, 2006.
Lenk, Ron. *Practical Design of Power Supplies*. New York: Wiley/IEEE, 2005.
Mazer, Arthur. *Electric Power Planning for Regulated and Deregulated Markets*. Hoboken, NJ: IEEE Press, 2007.

OTHER
National Geographic Society. "Future Power: Where Will the World Get Its Next Energy Fix?." http://environment.nationalgeographic.com/environment/global-warming/powering-the-future.html (accessed November 9, 2010).
U.S. Government; science.gov. "Electric Power Grid." http://www.science.gov/browse/w_121K.htm (accessed November 9, 2010).

Judith L. Sims

BLM *see* **Bureau of Land Management.**

Blow-out

A blow-out occurs where the soil is left unprotected to the erosive force of the wind. Blow-outs commonly occur as depressional areas, once enough soil has been removed. They most often occur in sandy soils, where vegetation is sparse.

Blue-baby syndrome

Blue-baby syndrome (or infant cyanosis) occurs in infants who drink water with a high concentration of nitrate or are fed formula prepared with water containing high nitrate levels. Excess nitrate can result in methemoglobinemia, a condition in which the oxygen-carrying capacity of the blood is impaired by an oxidizing agent such as nitrite, which can be reduced from nitrate by bacterial metabolism in the human mouth and stomach. Infants in the first three to six months of life, especially those with diarrhea, are particularly susceptible to nitrite-induced methemoglobinemia.

Adults convert about 10 percent of ingested nitrates into nitrites, and excess nitrate is excreted by the kidneys. In infants, however, nitrate is transformed to nitrite with almost 100 percent efficiency. The nitrite and remaining nitrate are absorbed into the body through the intestine. Nitrite in the blood reacts with hemoglobin to form methemoglobin, which does not transport oxygen to the tissues and body organs. The skin of the infant appears blue due to the lack of oxygen in the blood supply, which may lead to asphyxia, or suffocation.

Normal methemoglobin levels in humans range from 1 to 2 percent; levels greater than 3 percent are defined as methemoglobinemia. Methemoglobinemia is rarely fatal, readily diagnosed, and rapidly reversible with clinical treatment.

In adults, the major source of nitrate is dietary, with only about 13 percent of daily intake from drinking water. Nitrates occur naturally in many foods, especially vegetables, and are often added to meat products as preservatives. Only a few cases of methemoglobinemia have been associated with foods high in nitrate or nitrite. Nitrate is also found in air, but the daily respiratory intake of nitrate is small compared with other sources. Nearly all cases of the disease have resulted from ingestion by infants of nitrate in private well water that has been used to prepare infant formula. Levels of nitrate of three times the Maximum Contaminant Levels (MCLs) and above have been found in drinking water wells in agricultural areas. Federal MCL standards apply to all public water systems, though they are unenforceable recommendations. Insufficient data are available to determine whether subtle or chronic toxic effects may occur at levels of exposure below those that produce clinically obvious toxicity. If water has or is suspected to have high nitrate concentrations, it should not be used for infant feeding, nor should pregnant women or nursing mothers be allowed to drink it.

Domestic water supply wells may become contaminated with nitrate from mineralization of soil organic nitrogen, septic tank systems, and some agricultural practices, including the use of fertilizers and the disposal of animal wastes. Since there are many potential sources of nitrates in groundwater, the prevention of nitrate contamination is complex and often difficult.

Nitrates and nitrites can be removed from drinking water using several types of technologies. The Environmental Protection Agency (EPA) has designated reverse osmosis, anion exchange, and electrodialysis as the Best Available Control Technology (BAT) for the removal of nitrate, while recommending reverse osmosis and anion exchange as the BAT for nitrite. Other technologies can be used to meet MCLs for nitrate and nitrite if they receive approval from the appropriate state regulatory agency.

Resources

BOOKS

Addiscott, T. M. *Nitrate, Agriculture and the Environment.* Wallingford, UK: CABI, 2004.

Mozdzen, Miguel A. *Potential Bioremediation System for Nitrate Removal from Plant Nursery Runoff Water.* Gainesville, FL: University of Florida, 2007.

PERIODICALS

Faust, B. "Nitrate and the Blue Baby Syndrome." *Education in Chemistry* 41 (2004): 44–46.

Powlson, D. S., T. M. Addiscott, N. Benjamin, et al. "When Does Nitrate Become a Risk for Humans?" *Journal of Environmental Quality* 37, no. 2 (2008).

OTHER

MedLine Plus. "Methemoglobinemia." http://www.nlm.nih.gov/medlineplus/ency/article/000562.htm (accessed November 7, 2010).

Judith L. Sims

BMP *see* **Best management practices.**

BOD *see* **Biochemical oxygen demand.**

Bogs *see* **Wetlands.**

Bonn Convention *see* **Convention on the Conservation of Migratory Species of Wild Animals (1979).**

Bookchin, Murray

1921–2006
American social critic, environmentalist, and writer

Born in New York in 1921, Murray Bookchin was a writer, social critic, and founder of "social ecology." He had had a long and abiding interest in the environment, and as early as the 1950s he was concerned with the effects of human actions on the environment. In 1951 he published an article titled "The Problem of Chemicals," which exposed the detrimental effects of chemicals on nature and on human health. This work predates Rachel Carson's famous *Silent Spring* by over ten years.

In developing his theory of social ecology, Bookchin expounded the view that sound ecological practices in nature were not possible without having sound social practices in society. Put another way, harmony in society required harmony with nature. Bookchin described himself as an anarchist, contending that there was a natural relationship between natural ecology and anarchy.

Botanical garden

A botanical garden is a place where collections of plants are grown, managed, and maintained. Plants are normally labeled and available for scientific study by students and observation by the public. An arboretum is a garden composed primarily of trees, vines, and shrubs. Gardens often preserve collections of stored seeds in special facilities referred to as seed banks. Many gardens maintain special collections of preserved plants, known as herbaria, used to identify and classify unknown plants. Laboratories for the scientific study of plants and classrooms are also common.

Although landscape gardens have been known for as long as 4,000 years, gardens intended for scientific study have a more recent origin. Kindled by the need for herbal medicines in the sixteenth century, gardens affiliated with Italian medical schools were founded in Pisa about 1543, and Padua in 1545. The usefulness of these medicinal gardens was soon evident, and similar gardens were established in Copenhagen, Denmark (1600), London, England (1606), Paris, France (1635), Berlin, Germany (1679), and elsewhere. The early European gardens concentrated mainly on species with known medical significance. The plant collections were put to use to make and test medicines and to train students in their application.

In the eighteenth and nineteenth centuries, gardens evolved from traditional herbal collections to facilities with broader interests. Some gardens, notably the Royal Botanic Gardens at Kew, near London, played a major role in spreading the cultivation of commercially important plants such as coffee (*Coffea arabica*), rubber (*Hevea* spp.), banana (*Musa paradisiaca*), and tea (*Thea sinensis*) from their places of origin to other areas with an appropriate climate. Other gardens focused on new varieties of horticultural plants. The Leiden garden in Holland, for instance, was instrumental in stimulating the development of the extensive worldwide Dutch bulb commerce. Many other gardens have had an important place in the scientific study of plant diversity as well as the introduction and assessment of plants for agriculture, horticulture, forestry, and medicine.

The total number of botanical gardens in the world can only be estimated, but not all plant collections qualify for the designation because they are deemed to lack serious scientific purpose. A recent estimate places the number of botanical gardens and arboreta at 1,400. About 300 of those are in the United States. Most existing gardens are located in the North Temperate Zone, but there are important gardens on all continents except Antarctica. Although the tropics are home to the vast majority of all plant species, until recently, relatively few gardens were located there. A recognition of the need for further study of the diverse tropical flora has led to the establishment of many new gardens. An estimated 230 gardens are now established in the tropics.

In recent years botanical gardens throughout the world have united to address increasing threats to the planet's flora. The problem is particularly acute in the tropics, where as many as 60,000 species, nearly one-fourth of the world's total, risk extinction by the year 2050. Botanical gardens have organized to produce, adopt and implement a Botanic Gardens Conservation Strategy to help deal with the dilemma.

See also Conservation; Critical habitat; Ecosystem; Endangered species; Forest decline; Organic gardening and farming.

Resources

BOOKS

Aitken, Richard. *Botanical Riches: Stories of Botanical Exploration*. Aldershot, UK: Lund Humphries, 2007.

Lack, Hans Walter. *Alexander Von Humboldt: The Botanicals of America*. New York: Prestels, 2009.

OTHER

Montreal Botanical Garden. "Scientific Activities." http://www2.ville.montreal.qc.ca/jardin/en/act_scien/act_scien.htm (accessed November 11, 2010).

Douglas C. Pratt

Boulding, Kenneth E.

1910–1993
English economist, social scientist, writer, and peace activist

Kenneth Boulding was a highly respected economist, educator, author, Quaker, and pacifist. In an essay in *Frontiers in Social Thought: Essays in Honor of Kenneth E. Boulding* (1976), Cynthia Earl Kerman described Boulding as a person who grew up in the poverty-stricken inner city of Liverpool, broke through the class system to achieve an excellent education, had both scientific and literary leanings, became a well-known American economist, then snapped the bonds of economics to extend his thinking into wide-ranging fields—a person who was a religious mystic and a poet as well as a social scientist.

Kenneth Boulding. *(Photograph by Ken Abbott. University of Colorado at Boulder)*

A major recurring theme in Boulding's work was the need—and the quest—for an integrated social science, even a unified science. He did not see the disciplines of human knowledge as distinct entities, but rather a unified whole characterized by what he described as a diversity of methodologies of learning and testing. For example, Boulding was a firm advocate of adopting an ecological approach to economics, asserting that ecology and economics are not independent fields of study. He identified five basic similarities between the two disciplines: (1) both are concerned not only with individuals, but individuals as members of species; (2) both have an important concept of dynamic equilibrium; (3) a system of exchange among various individuals and species is essential in both ecological and economic systems; (4) both involve some sort of development—succession in ecology and population growth and capital accumulation in economics; (5) both involve distortion of the equilibrium of systems by humans in their own favor.

Boulding's views have been influential in many fields, and he has helped environmentalists reassess and redefine their role in the larger context of science and economics.

Resources

BOOKS

Boulding, Kenneth E. "Economics As an Ecological Science." In *Economics as a Science*. New York: McGraw-Hill, 1970.

Boulding, Kenneth E. *Collected Papers*. Boulder: Colorado Associated University Press, 1971.

Gerald L. Young

Boundary Waters Canoe Area

The Boundary Waters Canoe Area (BWCA), a federally designated wilderness area in northern Minnesota, includes approximately one million acres (410 thousand ha) stretching some two hundred miles (322 km) along the United States-Canadian border. The BWCA contains more than 1,200 miles (1,932 km) of canoe routes and portages. The second largest expanse in the National Wilderness Preservation System (NWPS), the BWCA is administered by the United States Forest Service. Constituting about one-third of the Superior National Forest (established in 1909), the BWCA was set apart as wilderness by an act of Congress in 1958. The 1964 Wilderness Act allowed limited logging in some parts of the BWCA and the use of motorboats on 60 percent of the water area. Under pressure from environmental groups—and over objections by developers, logging interests, and many local residents—Congress finally passed the BWCA Wilderness Act of 1978, which outlawed all logging and limited motorboats to 33 percent of the water surface area (dropping to 24 percent by 1999), and added 45,000 acres (18,450 ha), bringing the total area to 1,075,000 acres (440,750 ha).

Many area residents and resort owners continue to resent and resist efforts to reduce the areas open to motorized watercraft and snowmobile traffic. They have pressed unsuccessfully for federal legislation to that effect. At the urging of Senator Paul Wellstone (D-MN) (1944–2002), a mediation panel was convened in 1996 to consider the future of the BWCA. Environmentalists, resort owners, local residents, and representatives of other groups met for several months to try to reconcile competing interests in the area. Unable to reach an agreement and arrive at a compromise, the panel disbanded in 1997. The fierce and continuing political quarrels over the future of the BWCA contrast markedly with the silence and serenity of this land of sky-blue waters and green forests.

On July 4, 1999, a severe windstorm, referred to as the Boundary Waters–Canadian Derecho or the 1999 Blowdown, damaged nearly 400,000 acres (about 162,000 ha) of forests within and adjacent to the retardant plastic products. Brominated flame-retardants are used in televisions, stereos, computers, and electrical wiring to reduce fire hazard when these common electronic appliances generate excessive heat. Bromine-containing fire-retardant chemicals are also used in carpeting, draperies, and furniture foam padding. While bromine compounds make products more fire-resistant, they do not make them fireproof. Rather, they reduce the likelihood that a plastic item will ignite and delay the spread of fire. As bromine-treated plastic products burn, they release brominated hydrocarbons that threaten the ozone layer not unlike chorofluorocarbons (CFCs). For this reason, research is now directed at finding alternatives to bromine flame-retardant chemicals. For example, promising new fire-resistant compounds use silicon.

Aside from its use in dyes, pesticides, water treatment, pharmaceuticals, and fire retardants, bromine compounds are also used in photographic film and print paper emulsions, hydraulic fluids, refrigeration fluids, inks, and hair products. As useful as bromine is, however, concern for the ozone layer has resulted in heightened vigilance concerning the overuse of bromine-containing chemicals.

Resources

OTHER

Centers for Disease Control and Prevention (CDC). "Bromine." http://emergency.cdc.gov/agent/bromine/ (accessed November 10, 2010).

Terry Watkins

Bronchial constriction *see* **Bronchitis.**

Bronchitis

Bronchitis is an inflammation of the thin mucous lining of the bronchi, the passages that carry air from the trachea to the lungs. Acute bronchitis usually follows a cold or other viral respiratory infection, and features a lingering, dry cough that lasts a few days or weeks. Chronic bronchitis is longer lasting, reoccurs, repeatedly, and includes a persistent productive cough.

Chronic bronchitis is characterized by a daily cough that produces sputum for at least three months each year for two consecutive years, when no other disease can account for these symptoms. The diagnosis of chronic bronchitis is made by this history, rather than by any abnormalities found on a chest x-ray or through a pulmonary function test.

When a person inhales, air, smoke, germs, allergens, and pollutants pass from the nose and mouth into a large central duct called the trachea. The trachea branches into smaller ducts, the bronchi and bronchioles, which lead to the alveoli. These are the tiny, balloon-like air sacs, composed of capillaries, supported by connecting tissue, and enclosed in a thin membrane. Bronchitis can permanently damage the alveoli.

Chronic bronchitis is usually caused by cigarette smoke or exposure to other irritants or air pollutants. The lungs respond to the irritation in one of two ways. They may become permanently inflamed with fluid, which swells the tissue that lines the airways, narrowing them and making them resist airflow. Or, the mucus cells of the bronchial tree may produce excessive mucus.

The first sign of excessive mucus production is usually a morning cough. As smoking or exposure to air pollutants continues, the irritation increases and is complicated by infection, as excess mucus provides food for bacteria growth. The mucus changes from clear to yellow, and the infection becomes deep enough to cause actual destruction of the bronchial wall. Scar tissue replaces the fine cells, or cilia, lining the bronchial tree, and some bronchioles are completely destroyed. Paralysis of the cilia permits mucus to accumulate in smaller airways, and air can no longer rush out of these airways fast enough to create a powerful cough.

With each pulmonary infection, excess mucus creeps into the alveoli, and on its way, blocking portions of the bronchial tree. Little or no gas exchange occurs in the alveoli, and the ventilation-blood flow imbalance significantly reduces oxygen levels in the blood and raises carbon dioxide levels. Chronic bronchitis eventually results in airway or air sac damage; the air sacs become permanently hyperinflated because mucus obstructing the bronchioles prevents the air sacs from fully emptying.

Chronic bronchitis usually goes hand-in-hand with the development of emphysema, another chronic lung disease. These progressive diseases cannot be cured, but can be treated. Treatment includes avoiding the inhalation of harmful substances such as polluted air or cigarette smoke.

Resources

BOOKS

Fothergill, J. M. *Chronic Bronchitis*. Charleston, SC: BiblioLife, 2010.

Sethi, Sanjay. *Respiratory Infections.* New York: Informa Healthcare, 2010.

PERIODICALS

Wenzel, R. P. "Clinical Practice: Acute Bronchitis." *New England Journal of Medicine* 355, no. 20 (2006): 2125.

OTHER

National Heart, Lung, and Blood Institute. "Bronchitis." http://www.nhlbi.nih.gov/health/dci/Diseases/brnchi/brnchi_whatis.html (accessed November 6, 2010).
National Institutes of Health (NIH). "Bronchitis." http://health.nih.gov/topic/Bronchitis (accessed November 6, 2010).

Linda Rehkopf

Brower, David R.

1912–2000
American environmentalist and conservationist

David R. Brower, the founder of both Friends of the Earth and the Earth Island Institute, is considered one of the most radical and effective environmentalists in U.S. history.

Joining the Sierra Club in 1933, Brower became a member of its Board of Directors in 1941 and then its first executive director, serving from 1952 to 1969. In this position, Brower helped transform the group from a regional to a national force, seeing the club's membership expand from 2,000 to 77,000 and playing a key role in the formation of the Sierra Club Foundation. Under Brower's leadership the Sierra Club, among other achievements, successfully opposed the Bureau

David R. Brower. *(© Roger Ressmeyer/Corbis)*

of Reclamation's plans to build dams in Dinosaur National Monument in Utah and Colorado as well as in Arizona's Grand Canyon, but lost the fight to preserve Utah's Glen Canyon. The loss of Glen Canyon became a kind of turning point for Brower, indicating to him the need to take uncompromising and sometimes militant stands in defense of the natural environment. This militancy occasionally caused friction both between the groups he led and the private corporations and governmental agencies with which they interact and also within the increasingly broad-based groups themselves. In 1969 Brower was asked to resign as executive director of the Sierra Club's Board of Directors, which disagreed with Brower's opposition to a nuclear reactor in California's Diablo Canyon, among other differences. Eventually reelected to the Sierra Club's Board in 1983 and 1986, Brower became an honorary vice-president of the club and was the recipient, in 1977, of the John Muir Award, the organization's highest honor.

After leaving the Sierra Club in 1969, Brower founded Friends of the Earth with the intention of creating an environmental organization that would be more international in scope and concern and more political in its orientation than the Sierra Club. Friends of the Earth, which now is operating in some fifty countries, was intended to pursue a more global vision of environmentalism and to take more controversial stands on issues—including opposition to nuclear weapons—than could the larger, generally more conservative organization. But in the early 1980s, Brower again had a falling out with his associates over policy, eventually resigning from Friends of the Earth in 1986 to devote more of his time and energy to the Earth Island Institute, a San Francisco-based organization he founded in 1982.

Over the years, Brower played a key role in preserving wilderness in the United States, helping to create national park's and national seashores in Kings Canyon, the North Cascades, the Redwoods, Cape Cod, Fire Island, and Point Reyes. He was also instrumental in protecting primeval forests in the Olympic National Park and wilderness on San Gorgonio Mountain in California and in establishing the National Wilderness Preservation System and the Outdoor Recreation Resources Review, which resulted in the Land and Water Conservation Fund.

In his youth, Brower was one of this country's foremost rock climbers, leading the historic first ascent of New Mexico's Shiprock in 1939 and making seventy other first ascents in Yosemite National Park and the High Sierra as well as joining expeditions to the Himalayas and the Canadian Rockies. A proficient skier and guide as well as a mountaineer, Brower served with the United States Mountain Troops from 1942 to 1945, training soldiers to scale cliffs and navigate in Alpine areas and serving as a combat-intelligence officer in Italy. For his service, Brower was awarded both the Combat Infantryman's Badge and the Bronze Star, and rose in rank from private to captain before he left active duty. As a civilian, Brower employed many of the same talents and abilities to show people what he had fought so long and so hard to preserve: He initiated the knapsack, river, and wilderness threshold trips for the Sierra Club's Wilderness Outings Program, and between 1939 and 1956 led some 4,000 people into remote wilderness.

Excluding his military service, Brower was an editor at the University of California Press from 1941 to 1952. Appointed to the *Sierra Club Bulletin*'s Editorial Board in 1935, Brower eventually became the *Bulletin*'s editor, serving in this capacity for eight years. He had been involved with the publication of more than fifty environmentally oriented books each for the Sierra Club and Friends of the Earth, several of which earned him prestigious publishing industry awards. He wrote a two-volume autobiography, *For Earth's Sake* and *Work in Progress*. Brower also made several Sierra Club films, including a documentary of raft trips on the Yampa and Green Rivers designed to show people the stark beauty of Dinosaur National Monument, which at the time was threatened with flooding by a proposed dam.

Brower was the recipient of numerous awards and honorary degrees and served on several boards and councils, including the Foundation on Economic Trends, the Council on National Strategy, the Council on Economic Priorities, the North Cascades Conservation Council, the Fate and Hope of the Earth Conferences, Zero Population Growth, the Committee on National Security, and Earth Day. He had twice been nominated for the Nobel Peace Prize. During his life, Brower promoted environmental causes around the globe, giving dozens of lectures in seventeen different countries and organizing several international conferences. In 1990, Brower's life was the subject of a PBS video documentary titled *For Earth's Sake*. He also was featured in the TV documentary *Green for Life*, which focused on the 1992 Earth Summit in Rio de Janeiro.

Before his death on November 5, 2000, Brower continued actively promoting environmental causes. He devoted much of his time to his duties at the Earth

Island Institute and promoting the activities of the International Green Circle. In 1990 and 1991, he led Green Circle delegations to Siberia's Lake Baikal to aid in its protection and restoration. Brower also lectured to companies and schools throughout the United States on Planetary Conservation Preservation and Restoration (CPR). His topics included land conservation, the economics of sustainability, and the meaning of wilderness to science. Brower's book, *Let the Mountains Talk, Let the Rivers Run*, includes a credo for the earth, which reflects what Brower had hoped to accomplish with his lectures and publications: "We urge that all people now determine that an untrammeled wilderness shall remain here to testify that this generation had love for the next."

Resources

BOOKS

Brower, Douglas. *For Earth's Sake*. Layton, UT: Gibbs Smith, 1990.
Brower, Douglas. *Work in Progress*. Layton, UT: Gibbs Smith, 1991.

Lawrence J. Biskowski

Browner, Carol

1955–
American Director of the White House Office of Energy and Climate Change Policy; former administrator of the Environmental Protection Agency

Carol Browner headed the Environmental Protection Agency (EPA) under the Clinton administration. She served from 1993 until 2001, making her the longest-serving director the agency had ever had. As of 2010, she was director of the White House Office of Energy and Climate Change Policy in the Obama administration.

Browner was born in Florida on December 17, 1955. Her father taught English and her mother social science at Miami Dade Community College. Browner grew up hiking in the Everglades, where her lifelong love for the natural world began. She was educated at the University of Florida in Gainesville, receiving both a B.A. in English and her law degree there.

Her political career began in 1980 as an aide in the Florida House of Representatives. Browner moved to Washington, DC, a few years later to join the national

White House Energy and Climate Director Carol Browner during a press conference at the White House. *(Brooks Kraft)*

office of Citizen Action, a grassroots organization that lobbies for a variety of issues, including the environment. She left Citizen Action to work with Florida Senator Lawton Chiles, and in 1989 she joined Senator Al Gore's staff as a senior legislative aide. From 1991 to 1993, Browner headed the Department of Environmental Regulation in Florida, the third-largest environmental agency in the country. She streamlined the process the department used to review permits for expanding manufacturing plants and developing wetlands, reducing the amount of money and time that process was costing businesses as well as the department. Activists had argued that this kind of streamlining interfered with the ability of government to supervise industries and assess their impact on the environment. But in Florida's business community, Browner built a reputation as a formidable negotiator on behalf of the environment. When the Walt Disney Company filed for state and federal permits to fill in 400 acres of wetlands, she negotiated an agreement which allowed the company to proceed with its

development plans in return for a commitment to buy and restore an 8,500-acre ranch near Orlando.

She was also the chief negotiator in the settlement of a lawsuit the government had brought against Florida for environmental damage done to Everglades National Park. The result was the largest ecological restoration project ever attempted in the United States, a plan to purify and restore the natural flow of water to the Everglades with the cost shared by the state and federal governments, as well as Florida's sugar farmers. These actions earned Browner a reputation as an environmentalist who could reconcile environmental protection and economic development; then as now, she believes that the stewardship of the environment requires accommodations with industry.

As director of the EPA, she was determined to protect the environment and public health while not alienating business interests. By many accounts, she was a remarkably successful administrator. She took the job at a time of relatively high environmental fervor, with Vice President Al Gore, who had particularly championed the environment, and Democrats in control of both houses of Congress.

But a conservative backlash led by Representative Newt Gingrich brought Republican control to the House and Senate in 1994. The political climate for environmental reform then became much more embattled. Many conservatives wished to downsize government and cut back the regulatory power of the EPA and other federal agencies. The EPA was shut down twice during her tenure, a temporary victim of congressional budgetary squabbles. As well, appropriations bills for EPA programs were frequently hit with amendments that countermanded the agency's ability to carry out some policies.

Despite the hostility of Congress, Browner had several legislative victories. In 1996 Browner led a campaign to have Congress reauthorize the Safe Drinking Water Act. That same year she spearheaded the Food Quality Protection Act, which modernized standards that govern pesticide use. This landmark legislation was one of the first environmental laws to specifically protect children's health. The law required scientists to determine what levels of pesticides were safe for children. Browner was also successful in getting so-called Superfund sites cleaned up. These were sites that were listed as being particularly polluted. Only twelve Superfund sites had been cleaned up when Browner took over the EPA in 1993. By the time she left the agency, over 700 sites had been cleaned up or were in the process of being cleaned up. After she left the EPA in 2001,

Browner continued to work for the environment by becoming a board member of the Audubon Society.

When Barack Obama was elected as U.S. President, an early decision prior to assuming office was the selection of Browner as an overseer of energy and climate policy. The position was approved and, as of 2010, Browner is Director of the White House Office of Energy and Climate Change Policy.

During 2010, Browner was one of the key public spokespersons following the sinking of the oil rig Deepwater Horizon in the Gulf of Mexico and resulting massive oil spill.

Douglas Smith

Brown, Lester R.

1934–

American founder, president and senior researcher, Earth Policy Institute

Lester Brown is a highly respected and influential authority on global environmental issues. He founded the Worldwatch Institute in 1974 and served as its president until 2000. In 2001 he launched a new initiative, the Earth Policy Institute. Brown is an award-winning author of many books and articles on environmentally

Lester R. Brown. (© Wally McNamee/Corbis)

sustainable economic development and on environmental, agricultural, and economic problems and trends.

Brown was born in Bridgeton, New Jersey. His love and appreciation of nature as a sustainable source of food was developed during high school and college. After earning a degree in agricultural science from Rutgers University in 1955, he spent six months in rural India studying and working on agricultural projects. In 1959, he joined the U.S. Department of Agriculture's Foreign Agricultural Service as an international agricultural analyst. After receiving an M.S. in agricultural economics from the University of Maryland and a master's degree in public administration from Harvard, he went to work for Orville Freeman, the Secretary of Agriculture, as an advisor on foreign agricultural policy in 1964. In 1969, Brown helped establish the Overseas Development Council and in 1974, with the support of the Rockefeller Fund, he founded the Worldwatch Institute to analyze world conditions and problems such as famine, overpopulation, and scarcity of natural resources.

In 1984, Brown launched Worldwatch's annual *State of the World* report, a comprehensive and authoritative account of worldwide environmental and agricultural trends and problems. Eventually published in over thirty languages, *State of the World* is considered one of the most influential and widely read reports on public policy issues. Other Worldwatch publications initiated and overseen by Brown included *Worldwatch*, a bimonthly magazine, the *Environmental Alert* book series, and the annual *Vital Signs: The Trends That Are Shaping Our Future*. Brown has written or coauthored over a dozen books and some two dozen Worldwatch papers on various economic, agricultural, and environmental topics. Among his many awards, Brown has received a "genius award" from the MacArthur Foundation, as well as the United Nation's 1987 environmental prize. *The Washington Post* has described him as "one of the world's most influential thinkers."

Brown has long warned that unless the United States and other nations adopt policies that are ecologically and agriculturally sustainable, the world faces a disaster of unprecedented proportions. The *State of the World* reports have tracked the impact of human activity on the environment, listing things such as the percentage of bird species that were endangered, the number of days China's Yellow River was too depleted to irrigate fields in its lower reaches, the number of females being educated worldwide, and the number of cigarettes smoked per person. The reports made the ecological dimension of global economics clear and concrete. Often these reports were dire. The 1998 report, for example, discussed a concept termed demographic fatigue. This referred to places where population was falling, not because of a birthrate held in check by family planning but because many people were dying through famine, drought, and infectious disease. At the time, conventional economics paid little heed to the environmental cost of development. Brown struggled to unseat that belief. He also wrote about a way out of the doom his work often foresaw. His 2001 *Eco-Economy: Building an Economy for the Earth* argued that creation of a new, ecologically aware economy, with emphasis on renewable energy, tax reform, redesign of cities and transportation, better agricultural methods and global cooperation, could alleviate much of the world's ills. Through the Earth Policy Institute's *Earth Policy Alerts*, Brown continued to add to the themes of his book. The mission of Brown's new organization was to reach policy makers and the public with information about building an environmentally sustainable economy.

Brown has been a prodigious author. As of 2010, he has written or cowritten fifty books. His latest book, published in 2009, is *Plan B 4.0 - Mobilizing to Save Civilization*, which was written as a final wake-up call for humanity.

Resources

BOOKS

Brown, Lester. *Eco-Economy: Building an Economy for the Earth*. New York: Norton, 2001.

ORGANIZATIONS

Earth Policy Institute, 1350 Connecticut Ave. NW, Washington, DC, USA, 20036, (202) 496-9290, (202) 496-9325, epi@earth-policy.org, http://www.earth-policy.org

Lewis Regenstein

Brown pelican

The brown pelican (*Pelecanus occidentalis*) is a large water bird of the family Pelicanidae that is found along both coasts of the United States, chiefly in saltwater habitats. It weighs up to 8 pounds (3.5 kg) and has a wingspan of up to 7 feet (2 m). This pelican has a light brown body and a white head and neck often tinged with yellow. Its distinctive, long, flat bill and large throat pouch are adaptations for catching its primary food, schools of mid-water fishes. The brown pelican hunts while flying a dozen or more feet above the surface of the water, dropping or diving straight down into the water, and using its expandable pouch as a scoop or net to engulf its catch.

Both east and west coast populations, which are considered to be different subspecies, showed various levels of decline over the later part of the twentieth century. It is estimated that there were 50,000 pairs of nesting brown pelicans along the Gulf coast of Texas and Louisiana in the early part of the twentieth century, but by the early 1960s, most of the Texas and all of the Louisiana populations were depleted. The main reason for the drastic decline was the use of organic pesticides, including DDT and endrin. These pesticides poisoned pelicans directly and also caused thinning of their eggshells. This eggshell thinning led to reproductive failure, because the egg were crushed during incubation. Louisiana has the distinction of being the only state to have its state bird become extinct within its borders. In 1970 the brown pelican was listed as endangered throughout its U.S. range.

During the late 1960s and early 1970s, brown pelicans from Florida were reintroduced to Louisiana, but many of these birds were doomed. Throughout the 1970s these transplanted birds were poisoned at their nesting sites at the outflow of the Mississippi River by endrin, which was used extensively upriver. In 1972 the use of DDT was banned in the United States, and the use of endrin was sharply curtailed. Continued reintroduction of the brown pelican from Florida to Louisiana subsequently met with greater success, and the Louisiana population had grown to more than 1,000 pairs by 1989. Texas, Louisiana, and California populations were listed as endangered until 2007; the Alabama and Florida populations of the brown pelican were removed earlier from the federal list due to increases in fledgling success. As of 2010, there were about 15,000 nesting pairs and 20,000 young in Alabama and Florida. The Texas population had about 4,000 active nests and Louisiana had about 16,000 pairs. Scientists monitored the bird's populations in the aftermath of Hurricane Katrina in 2005 and the aftermath of the Gulf of Mexico 2010 oil spill.

Other problems that face the brown pelican include habitat loss, encroachment by humans, and disturbance by humans. Disturbances have included mass visitation of nesting colonies. This practice has been stopped on federally owned lands and access to nesting colonies is restricted. Other human impacts on brown pelican populations have had a more malicious intent. On the California coast in the 1980s there were cases of pelicans' bills being broken purposefully, so that these birds could not feed and would ultimately starve to death. It is thought that disgruntled commercial fishermen faced with dwindling catches were responsible for at least some of these attacks. The brown pelican was a scapegoat for conditions that were due to weather, pollution, or, most likely, overfishing.

Recovery for the brown pelican has been slow, but progress is being made on both coasts. The banning of DDT in the early 1970s was probably the turning point for this species, and the delisting of the Alabama and Florida populations is a hopeful sign for the future.

Resources

BOOKS

Brown Pelican (Pelecanus occidentalis). Arlington, VA: U.S. Dept. of the Interior, U.S. Fish and Wildlife Service, Endangered Species Program, 2008.

Miller, Brian K., and William R. Fontenot. *Birds of the Gulf Coast*. Baton Rouge: Louisiana State University Press, 2001.

PERIODICALS

Eggert, L. M. F., P. G. R. Jodice, and K. M. OReilly. "Stress Response of Brown Pelican Nestlings to Ectoparasite Infestation." *General and Comparative Endocrinology* 166, no. 1 (2010): 33–38.

Sachs, E. B., and P. G. R. Jodice. "Behavior of Parent and Nestling Brown Pelicans During Early Brood Rearing." *Waterbirds* 32, no. 2 (2009): 276–281.

OTHER

National Wildlife Federation. "Oil Spill Puts Pelicans at Risk." http://www.nwf.org/News-and-Magazines/ National-Wildlife/Birds/Archives/2010/Pelicans-Oil. aspx (accessed November 9, 2010).

U.S. Fish and Wildlife Service. "Brown pelican (*Pelecanus occidentalis*)." http://ecos.fws.gov/speciesProfile/ profile/speciesProfile.action?spcode = B02L (accessed November 9, 2010).

Eugene C. Beckham

Brown tree snake

The brown tree snake *(Boiga irregularis)* has caused major ecological and economic damage in Guam, the largest of the Mariana Islands. The snake is native to New Guinea and northern Australia. It has also been introduced to some Pacific islands in addition to Guam.

Brown tree snakes in their natural habitat range from 3–6 ft (0.9–1.8 m) in length. Some snakes in Guam are more than 10 ft (3 m) long. The snake's head is bigger than its neck, and its coloring varies with its habitat. In Guam, the snake's brown-and-olive-green pattern blends in with foliage.

The tree snake was accidentally brought to Guam by cargo ships during the years between the end of World War II (1945) through 1952. On Guam, there were no population controls such as predators that eat snakes. As a result, the snake population boomed. During the late 1990s, there were close to 13,000 snakes per square mile in some areas.

The snake's diet includes birds, and the United States Geographical Survey (USGS) said that the brown tree snake "virtually wiped out" twelve of Guam's native forest birds (three of which are now extinct). Furthermore, snakes crawling on electrical lines may cause power outages.

Liz Swain

Brundtland, Gro Harlem

1939–
Norwegian doctor, former Prime Minister of Norway, Director-General of the World Health Organization

Dr. Gro Harlem Brundtland is a Norweigian who is a physician, politican, and diplomat. Beginning in July 1998, she served for five years as Director-General of the World Health Organization (WHO). As of 2010, she is a special envoy on climate change for the secretary-general of the United Nations.

Brundtland had been instrumental in promoting political awareness of the importance of environmental issues. In her view, the world shares one economy and one environment. With her appointment to the head of WHO, Brundtland continued to work on global strategies to combat ill health and disease.

Brundtland began her political career as Oslo's parliamentary representative in 1977. She became the leader of the Norwegian Labor Party in 1981, when she first became prime minister. At 42, she was the youngest person ever to lead the country and the first woman to do so. She regained the position in 1986 and held it until 1989; in 1990 she was again elected prime minister. Aside from her involvement in the environmental realm, Brundtland promoted equal rights and a larger role in government for women. In her second cabinet, eight of the eighteen positions were filled by women; in her 1990 government, nine of nineteen ministers were women.

Brundtland earned a degree in medicine from the University of Oslo in 1963 and a master's degree in

Gro Harlem Brundtland, United Nations special envoy on climate change, addresses the delegates during the opening of the World Climate Conference-3 in Geneva, Switzerland, 2009. *(© DENIS BALIBOUSE/Reuters/Corbis.)*

public health from Harvard in 1965. She served as a medical officer in the Norwegian Directorate of Health and as medical director of the Oslo Board of Health. In 1974 she was appointed minister of the environment, a position she held for four years. This appointment came at a time when environmental issues, especially pollution, were becoming increasingly important, not only locally but nationally. She gained international attention and in 1983 was selected to chair the United Nation's World Commission on Environment and Development. The commission published *Our Common Future* in 1987, calling for sustainable development and intergenerational responsibility as guiding principles for economic growth. The report stated that present economic development depletes both nonrenewable and potentially renewable resources that must be conserved for future generations. The commission strongly warned against environmental degradation and urged nations to reverse this trend. The report led to the organization of the so-called Earth Summit in Rio de Janeiro in 1992, an international meeting led by the United Nation's Conference on Environment and Development.

Bruntland resigned her position as prime minister in October 1996. This was prompted by a variety of factors, including the death of her son and the possibility of appointment to lead the United Nations.

Instead she was picked to head the World Health Organization, an assembly of almost 200 member nations concerned with international health issues. Brundtland immediately reorganized the leadership structure of WHO, vowed to bring more women into the group, and called for greater financial disclosure from WHO executives. She began dual campaigns intended to reduce the global cases of malaria and tuberculosis. She also launched an unprecedented campaign to combat tobacco use worldwide. Brundtland noted that disease due to tobacco was growing enormously. Tobacco-related illnesses already caused more deaths worldwide than AIDS and tuberculosis combined. Much of the increase in smoking was in the developing world, particularly China. In 2000, WHO organized the Framework Convention on Tobacco Control to come up with a world wide treaty governing tobacco sale, advertising, taxation, and labeling. Brundtland broadened the mission of WHO by attempting the treaty. She was credited with making WHO a more active group, and with making health issues a major component of global economic strategy for organizations such as the United Nations and the so-called G8 group of developed countries. In 2002 Brundtland announced more new goals for WHO, including renewed work on diet and nutrition. She has recognized the need for private sector involvement and serves as a consultant to PepsiCo.

She is also one of the founding members of The Elders, a group of world leaders that convened in 2007 to address pressing world issues, and is a member of the Club of Madrid, an independent think-tank that promotes democracy and global change. In 2007, UN Secretary-General Ban Ki-moon named Brundtland a special envoy for climate change.

Resources

BOOKS

Bugge, Hans Christian, and Christina Voigt. *Sustainable Development in International and National Law: What Did the Brundtland Report Do to Legal Thinking and Legal Development, and Where Can We Go from Here?* Avosetta series, 8. Groningen, Netherlands: Europa Law, 2008.

Cowie, Jonathan. *Climate Change: Biological and Human Aspects.* Cambridge, UK: Cambridge University Press, 2007.

William G. Ambrose
Paul E. Renaud

Brundtland Report *see* **Our Common Future (Brundtland Report).**

Btu

Abbreviation for "British Thermal Unit," the amount of energy needed to raise the temperature of one pound of water by one degree Fahrenheit. One Btu is equivalent to 1,054 joules or 252 calories. To gain an impression of the size of a Btu, the combustion of a barrel of oil yields about 5.6×10^6 joule. A multiple of the Btu, the quad, is commonly used in discussions of national and international energy issues. The term quad is an abbreviation for one quadrillion, or 10^{15}, Btu.

Budyko, Mikhail I.

1920–2001
Belarusian geophysicist, climatologist

Professor Mikhail Ivanovich Budyko is regarded as the founder of physical climatology. Born in Gomel in the former Soviet Union, now Belarus, Budyko earned his master of sciences degree in 1942 from the Division of Physics of the Leningrad Polytechnic Institute. As a researcher at the Leningrad Geophysical Observatory, he received his doctorate in physical and mathematical sciences in 1951. Budyko served as deputy director of the Geophysical Observatory until 1954, as director until 1972, and as head of the Division for Physical Climatology at the observatory from 1972 until 1975. In that year he was appointed director of the Division for Climate Change Research at the State Hydrological Institute in St. Petersburg.

During the 1950s, Budyko pioneered studies on global climate. He calculated the energy or heat balance—the amount of the Sun's radiation that is absorbed by the Earth versus the amount reflected back into space—for various regions of the Earth's surface and compared these with observational data. He found that the heat balance influenced various phenomena including the weather. Budyko's groundbreaking book, *Heat Balance of the Earth's Surface*, published in 1956, transformed climatology from a qualitative into a quantitative physical science. These new physical methods based on heat balance were quickly adopted by climatologists around the world. Budyko directed the compilation of an atlas illustrating the components of the Earth's heat balance. Published in 1963, it remains an important reference work for global climate research.

During the 1960s, scientists were puzzled by geological findings indicating that glaciers once covered much of the planet, even the tropics. Budyko examined a phenomenon called planetary albedo, a quantifiable term that describes how much a given geological feature reflects sunlight back into space. Snow and ice reflect heat and have a high albedo. Dark seawater, which absorbs heat, has a low albedo. Land formations are intermediate, varying with type and heat-absorbing vegetation. As snow and ice-cover increase with a global temperature drop, more heat is reflected back, ensuring that the planet becomes colder. This phenomenon is called ice-albedo feedback. Budyko found an underlying instability in ice-albedo feedback, called the snowball Earth or white Earth solution: If a global temperature drop caused ice to extend to within 30 degrees of the equator, the feedback would be unstoppable and the Earth would quickly freeze over. Although Budyko did not believe that this had ever happened, he postulated that a loss of atmospheric carbon dioxide, for example if severe weathering of silicate rocks sucked up the carbon dioxide, coupled with a sun that was 6 percent dimmer than today, could have resulted in widespread glaciation.

Budyko became increasingly interested in the relationships between global climate and organisms and human activities. In *Climate and Life*, published in 1971, he argued that mass extinctions were caused by climatic changes, particularly those resulting from volcanic activity or meteorite collisions with the Earth. These would send clouds of particles into the stratosphere, blocking sunlight and lowering global temperatures. In the early 1980s Budyko warned that nuclear war could have a similar effect, precipitating a "nuclear winter" and threatening humans with extinction.

By studying the composition of the atmosphere during various geological eras, Budyko confirmed that increases in atmospheric carbon dioxide, such as those caused by volcanic activity, were major factors in earlier periods of global warming. In 1972, when many scientists were predicting climate cooling, Budyko announced that fossil fuel consumption was raising the concentration of atmospheric carbon dioxide, which, in turn, was raising average global temperatures. He predicted that the average air temperature, which had been rising since the first half of the twentieth century, might rise another 5°F (3°C) over the next 100 years. Budyko then examined the potential effects of global warming on rivers, lakes, and groundwater, on twenty-first-century food production, on the geographical distribution of vegetation, and on energy consumption.

The author and editor of numerous articles and books, in 1964 Budyko became a corresponding member of the Division of Earth Sciences of the Academy of Sciences of the Union of Soviet Socialist Republics. In 1992, he was appointed academician in the Division of Oceanology, Atmosphere Physics, and Geography of the Russian Academy of Sciences. His many awards include the Lenin National Prize in 1958, the Gold Medal of the World Meteorological Organization in 1987, the A. P. Vinogradov Prize of the Russian Academy of Sciences in 1989, and the A. A. Grigoryev Prize of the Academy of Sciences in 1995. Budyko was awarded the Robert E. Horton Medal of the American Geophysical Union in 1994, for outstanding contribution to geophysical aspects of hydrology. In 1998 Dr. Budyko won the Blue Planet Prize of the Asahi Glass Foundation of Japan.

Budyko died on December 10, 2001, in Saint Petersburg, Russia.

Resources

BOOKS

Dow, Kirstin, and Thomas Downing. *The Atlas of Climate Change: Mapping the World's Greatest Challenge*, 2nd ed. Berkeley: University of California Press, 2007.

PERIODICALS

Donnadieu, Yannick, et al. "A 'Snowball Earth' Climate Triggered by Continental Break-up Through Changes in Runoff." *Nature* 428 (2004): 303–06.
Hansen, James, and Larissa Nazarenko. "Soot Climate Forcing Via Snow and Ice Albedos." *Proceedings of the National Academy of Sciences* 101 (2004): 423–28.
Kerr, Richard A. "Cosmic Dust Supports a Snowball Earth." *Science Annaler* 308 (2005): 181.
Sturm, Matthew, and Tom Douglas "Changing Snow and Shrub Conditions Affect Albedo with Global Implications." *Journal of Geophysical Research* 110 (September 2005): G01004.

ORGANIZATIONS

State Hydrological Institute, 23 Second Line VO, St. Petersburg, Russia, 199053

Margaret Alic

Buffer

A term in chemistry that refers to the capacity of a system to resist chemical change, although there are buffers against physical change such as temperature and humidity. Most often it is used in reference to the ability to resist change in pH, which is a scale used to indicate the acidity or alkalinity of a solution.

The buffering capacity of a solution refers to its ability to neutralize added acids. A system that is strongly pH buffered will undergo less change in pH with the addition of an acid or a base than a less well-buffered system. Alkalinity refers to the concentration of bicarbonate (HCO_3^-) and carbonate (CO_3^{--}) ions in solution, and the ability of these ions in solution to neutralize introduced acids. A well-buffered lake contains higher concentrations of bicarbonate ions that react with added acid. This kind of lake resists change in pH better than a poorly buffered lake. A highly buffered soil contains an abundance of ion exchange sites on clay minerals and organic matter that react with added acid to inhibit pH reduction.

Acid rain occurs as a result of sulfur dioxide (SO_2) and nitrogen oxide (NO_x) emissions from industrial processes, electric power generation, and motor vehicle exhaust. These gases react with water and oxygen and are converted to acids after release into the atmosphere. The acidic particles are deposited via precipitation (in the form of rain, fog, clouds, snow, or sleet) onto soil, plants, and into bodies of water. The acids can also incorporate into dust and smoke in the air, resulting in dry deposition. Depending on soil and aquatic buffering capacities, this deposition of acidic precipitation can alter the pH of soil and bodies of water such as lakes, rivers, and ponds. This change in pH can affect the growth and health of aquatic organisms and plants by stressing or weakening them. Acid rain affects plants by dissolving and washing away essential nutrients and minerals and enabling the release of harmful or toxic substances in the soil, such as aluminum (Al). Agricultural crops are generally unaffected by acid rain due to the application of nutrient-rich fertilizers and limestone, which is highly alkaline and increases the buffering capacity of the soil. The aluminum released from the soil can be deposited via runoff into nearby bodies of water and is toxic to aquatic organisms. Acid rain can be lethal to aquatic organisms, resulting in reduced aquatic biodiversity.

See also Acid and base.

Resources

BOOKS

Brimblecombe, Peter. *Acid Rain: Deposition to Recovery*. Acid Rain: Deposition to Recovery. Dordrecht, Netherlands: Springer, 2007.
Petheram, Louise. *Acid Rain (Our Planet in Peril)*. Mankato, MI: Capstone Press, 2006.

OTHER

National Geographic Society. "Acid Rain." http:// environment.nationalgeographic.com/environment/

global-warming/acid-rain-overview.html (accessed August 18, 2010).

United States Environmental Protection Agency (EPA). "Air: Air Pollution Effects: Acid Rain." http://www.epa.gov/ebtpages/airairpollutionefacidrain.html (accessed August 18, 2010).

Bulk density

Bulk density is the dry weight of soil per unit bulk volume. The bulk volume consists of mineral and organic materials, water, and air. Bulk density is affected by external pressures such as compaction from the weight of tractors and harvesting equipment and mechanical pressures from cultivation machines, and by internal pressures such as swelling and shrinking due to water-content changes, freezing and thawing, and by plant roots. The bulk density of cultivated mineral soils ranges from 1 to 1.6 milligrams/meter^3. A more porous soil (high percentage of pores or high porosity) has a lower bulk density and a higher water-holding capacity. For instance, organic soils are more porous than mineral soils and thus have a lower bulk density. Bulk density is a general property of powders and granules, but is often applied to soils.

Resources

BOOKS

Chesworth, Ward. *Encyclopedia of Soil Science*. Dordrecht, Netherlands: Springer, 2008.

Burden of proof

The current regulatory system, following traditional legal processes, generally assumes that chemicals are innocent, or not harmful, until proven guilty. Thus, the burden to show proof that a chemical is harmful to human and/or ecosystem health falls on those who regulate or are affected by these chemicals. As evidence increases that many of the more than 70,000 chemicals in the marketplace today—and the 10,000 more introduced each year—are causing health effects in various species, including humans, new regulations are being proposed that would reverse this burden to the manufacturer, importer or user of the chemical and its by-products. They would then have to prove before its production and distribution that the chemical will not be harmful to human health and the environment.

Bureau of Land Management

The Bureau of Land Management (BLM), an agency within the U. S. Department of the Interior, was created by executive reorganization in June 1946. The new agency was a merger of the Grazing Service and the General Land Office (GLO). The GLO was established in the Treasury Department by Congress in 1812 and charged with the administration of the public lands. The agency was transferred to the Department of the Interior when it was established in 1849. Throughout the 1800s and early 1900s, the GLO played the central role in administering the disposal of public lands under a multitude of different laws. But as the nation began to move from disposal of public lands to retention of them, the services of the GLO became less needed, which helped to pave the way for the creation of the BLM. The Grazing Service was created in 1934 (as the Division of Grazing) to administer the Taylor Grazing Act.

In the 1960s, the BLM began to advocate for an organic act that would give it firmer institutional footing, would declare that the federal government planned to retain the BLM lands, and would grant the agency statutory authority to professionally manage these lands (like the Forest Service). Each of these goals was achieved with the passage of the Federal Land Policy and Management Act (FLPMA) in 1976. The agency was directed to manage these lands and undertake long-term planning for the use of the lands, guided by the principle of multiple use.

The BLM manages 253 million acres (102 million ha) of land, primarily in the western states. This land is of three types: Alaskan lands (78 million acres; 31.6 million ha), which is virtually unmanaged; the Oregon and California lands (31.4 million acres; 12.7 million ha), prime timber land in western Oregon that reverted back to the government in the early 1900s due to land grant violations; and the remaining land (143.6 million acres; 58.1 million ha), approximately 70 percent of which is in grazing districts. As a multiple-use agency, the BLM manages these lands for a number of uses: fish and wildlife, forage, minerals, recreation, and timber.

Additionally, FLPMA directed that the BLM review all of its lands for potential wilderness designation, a process that is now well underway. (BLM lands were not covered by the Wilderness Act of 1964). In addition to these general land management responsibilities, the BLM also issues leases for mineral development on all public lands. FLPMA also directed that all mineral claims under the 1872 Mining Law be recorded with the BLM.

The BLM is headed by a director, appointed by the president, and confirmed by the Senate. The chain of command runs from the director in Washington to state directors in twelve western states (all but Hawaii and Washington), to district managers, who administer grazing or other districts, to resource area managers, who administer parts of the districts.

The BLM has often been compared unfavorably to the Forest Service. It has received less funding and less staff than its sibling agency, has been less professional, and has been characterized as captured by livestock and mining interests. Recent studies suggest that the administrative capacity of the BLM has improved.

Resources

ORGANIZATIONS

U.S. Bureau of Land Management, 1849 C Street NW, Room 5665, Washington, DC, USA, 20240, (202) 208-3801, (202) 208-5242, http://www.blm.gov.

Christopher McGrory Klyza

Bureau of Oceans and International Environmental and Scientific Affairs (OES)

The Bureau of OES was established in 1974 under Section 9 of the Department of State Appropriations Act. The OES promotes coordinates U.S. foreign policy on a number of environmental and science issues, including climate change, biodiversity, sustainable development, clean water access, infectious disease, and ocean and polar issues. OES contains various offices that address these issues including: the Oceans and Fisheries Directorate; Office of Marine Conservation; Office of Ocean and Polar Affairs; Office of Environmental Policy; Office of Ecology and Natural Resource Conservation; Office of Global Climate Change; Office of International Health and Biodefense; Office of Space and Advanced Technology; Office of Science and Technology Cooperation. Each of OES's offices coordinates national policy and works with related U.S. agencies and U.S embassies around the world issues under the purview of the OES. The Assistant Secretary of State for Oceans and International Environmental and Scientific Affairs heads the OES.

Resources

ORGANIZATIONS

U.S. Department of State, 2201 C St. NW, Washington, DC, USA, 20520, (202) 647-4000, http://www.state. gov/g/oes

Bureau of Reclamation

The U.S. Bureau of Reclamation was established in 1902 and is part of the U.S. Department of the Interior. It is primarily responsible for the planning and development of dams, power plants, and water transfer projects, such as Grand Coulee Dam on the Columbia River, the Central Arizona Project, and Hoover Dam on the Colorado River. This latter dam, completed in 1935 between Arizona and Nevada, is the highest arch dam in the Western Hemisphere and is part of the Boulder Canyon Project, the first great multipurpose water development project, providing irrigation, electric power, and flood control. It also created Lake Mead, which is supervised by the National Park Service to manage boating, swimming, and camping facilities on the 115-mi-long (185-km-long) reservoir formed by the dam.

The dams on the Colorado River are intended to reduce the impact of the destructive cycle of floods and droughts which makes settlement and farming precarious and to provide electricity and recreational areas; however, the deep canyons and free-flowing rivers with their attendant ecosystems are substantially altered. Along the Columbia River, efforts are made to provide "fish ladders" adjacent to dams to enable salmon and other species to bypass the dams and spawn up river; however, these efforts have not been as successful as desired and many native species are now endangered.

Problems faced by the bureau relate to creating a balance between its mandate to provide hydropower, water control for irrigation, and by-product recreation areas, and the conflicting need to preserve existing ecosystems. For example, at the Glen Canyon Dam on the Colorado River, controls on water releases are being imposed while studies are completed on the best manner of protecting the environment downstream in the Grand Canyon National Park and the Lake Mead National Recreation Area.

Resources

BOOKS

O'Neill, Karen. *Rivers by Design: State Power and the Origins of U.S. Flood Control.* Raleigh, NC: Duke University Press, 2006.

ORGANIZATIONS

Bureau of Reclamation, 1849 C St. NW, Washington DC, USA, 20240-0001, http://www.usbr.gov

Malcolm T. Hepworth

that humans can contract the deadly Ebola virus from infected chimpanzees and gorillas, and that HIV was first passed to humans from chimpanzees.

The Bushmeat Crisis Task Force (BCTF) is working with logging companies in the Congo Basin and West Africa to prevent illegal hunting and the use of company roads for the bush meat trade. The BCTF also works with governments and local communities to develop investment and foreign-aid policies that promote sustainable development and wildlife protection. Other organizations try to turn hunters into conservationists and educate bush meat consumers. TRAFFIC, which monitors wildlife trade for the World Wildlife Fund, and the World Conservation Union are urging that wildlife ownership be transferred from ineffectual African governments to landowners and local communities who have a vested interest in sustaining wildlife populations.

Resources

BOOKS

Davies, Glyn, and David Brown. *Bushmeat and Livelihoods: Wildlife Management and Poverty Reduction.* Oxford, UK: Blackwell, 2007.

Guynup, Sharon. *State of the Wild 2006: A Global Portrait of Wildlife, Wildlands, and Oceans.* Washington, DC: Island Press, 2005.

The Bushmeat Trade. London: Parliamentary Office of Science and Technology, 2005.

PERIODICALS

Bennett, Elizabeth, L., et al. "Hunting for Consensus: Reconciling Bushmeat Harvest, Conservation, and Development Policy in West and Central Africa." *Conservation Biology* 21, no. 3, (2006): 884–87.

Wolfe, Nathan D., et al. "Bushmeat Hunting, Deforestation, and Prediction of Zoonotic Disease." *Emerging Infectious Diseases* 11, no. 12 (2005): 1822–27.

OTHER

Bushmeat Crisis Task Force. http://www.bushmeat.org/ (accessed October 30, 2010).

Jane Goodall Institute. "Chimpanzees and Bushmeat 101." *Jane Goodall Institute.* http://www.Janegoodall.org/Chimpanzees-And-Bushmeat-101 (accessed October 30, 2010).

"Illegal Bushmeat Trade Rife in Europe, Research Finds." *Science Daily,* June 18, 2010. http://www.sciencedaily.com/releases/2010/06/100617210641.htm (accessed October 30, 2010).

ORGANIZATIONS

Bushmeat Crisis Task Force, 8403 Colesville Road, Suite 710, Silver Spring, MD, USA, 20910-3314, (301) 562-0888, info@bushmeat.org, http://www.bushmeat. org

The Bushmeat Project, The Biosynergy Institute, P.O. Box 488, Hermosa Beach, CA, USA, 90254, bushmeat@biosynergy.org, http://www.bushmeat.net

The Bushmeat Research Programme, Institute of Zoology, Zoological Society of London, Regents Park, London, UK, NW1 4RY, 44-207- 449-6601, 44-207- 586-2870, enquiries@ioz.ac.uk, http:// www.zoo.cam.ac.uk/ioz/projects/bushmeat.htm

TRAFFIC East/Southern Africa - Kenya, Ngong Race Course, PO Box 68200, Ngong Road, Nairobi, Kenya, (254) 2 577943, (254) 2 577943, traffic@iconnect.co.ke, http://www.traffic.org

World Wildlife Fund, 1250 24th Street, N.W., P.O. Box 97180, Washington, DC, USA, 20090-7180, 202-293-9211, (800) CALL-WWF, http://www.panda.org

Margaret Alic

BWCA *see* **Boundary Waters Canoe Area.**

Bycatch

The use of certain kinds of commercial fishing technologies can result in large bycatches—incidental catches of unwanted fish, sea turtles, seabirds, and marine mammals. Because the bycatch animals have little or no economic value, they are usually jettisoned, generally dead, back into the ocean. This non-selectivity of commercial fishing is an especially important problem when trawls, seines, and drift nets are used. The bycatch consists of unwanted species of fish and other animals, but it can also include large amounts of under-sized, immature individuals of commercially important species of fish.

The global amount of bycatch has been estimated in recent years at about 30 million tons (27 million tonnes), or more than one-fourth of the overall catch of the world's fisheries. In waters of the United States, the amount of unintentional bycatch of marine life is about 3 million tons per year (2.7 million tonnes per year). However, the bycatch rates vary greatly among fisheries. In the fishery for cod and other groundfish species in the North Sea, the discarded non market biomass averages about 42 percent of the total catch, and it is 44–72 percent in the Mediterranean fishery. Discard rates are up to 80 percent of the catch weight for trawl fisheries for shrimp.

Some fishing practices result in large bycatches of sea turtles, marine mammals, and seabirds. During the fishing year of 1988–1989, for example, the use of pelagic drift nets, each as long as 56 miles (90 km), may have killed as many as 0.3-1.0 million dolphins, porpoises, and other cetaceans. During one 24-day monitoring period, a typical set of a drift net of 19 km/day in the Caroline Islands of the south Pacific entangled ninety-seven dolphins, eleven larger cetaceans, and ten sea turtles. It is thought that bycatch-related mortality is causing population declines in thirteen out of the forty-four species of marine mammals that are suffering high death rates from human activities. Wildlife experts estimate that hundreds of thousands of seabirds have been drowned each year by entanglement in pelagic drift nets.

In 1991, the United Nations passed a resolution that established a moratorium on the use of drift nets longer than 1.6 mi (2 km), and most fishing nations have met this guideline. However, there is still some continued use of large-scale drift nets. Moreover, the shorter nets that are still legal are continuing to cause extensive and severe bycatch mortality.

Sea turtles, many of which are federally listed as endangered, appear to be particularly vulnerable to being caught and drowned in the large, funnel-shaped trawl nets used to catch shrimp. Scientists have, however, designed simple, selective, turtle excluder devices (TEDs) that can be installed on the nets to allow these animals to escape if caught. The use of TEDs is required in the United States and many other countries, but not by all of the fishing nations.

Purse seining for tuna has also caused an enormous mortality of certain species of dolphins and porpoises. This method of fishing is thought to have killed more than 200,000 small cetaceans per year since the 1960s, but perhaps about one-half that number since the early 1990s due to improved methods of deployment used in some regions. Purse seining is thought to have severely depleted some populations of marine mammals.

The use of long-lines also results in enormous bycatches of various species of large fishes, such as tuna, swordfish, and sharks, and it also kills many seabirds. Long-lines consist of a fishing line up to 80 mi (130 km) long and baited with thousands of hooks. A study in the Southern Ocean reported that more than 44,000 albatrosses of various species are killed annually by long-line fishing for tuna.

In addition, great lengths of fishing nets are lost each year during storms and other accidents. Because the synthetic materials used to manufacture the nets are extremely resistant to degradation, these so-called "ghost nets" continue to catch and kill fish and other marine mammals for many years.

In 2007, Congress passed laws creating the Bycatch Reduction Engineering Program (BREP), an organization charged with designing engineering solutions to minimize bycatch in U.S. waters.

Resources

BOOKS

Eayrs, Steve. *A Guide to Bycatch Reduction in Tropical Shrimp-Trawl Fisheries.* Rome: Food and Agriculture Organization of the United Nations, 2007.

PERIODICALS

Barcott, B. "What's the Catch? Of the Many Threats Facing the World's Oceans, One of the Most Serious Is Bycatch–All the Unwanted Species That Are Scooped Up in Fishermen's Indiscriminate Nets." *On Earth* 32, no. 2 (2010): 28–39.

Huang, H. W., and K. M. Liu. "Bycatch and Discards by Taiwanese Large-Scale Tuna Longline Fleets in the Indian Ocean." *Fisheries Research* 106, no. 3 (2010): 261–70.

Jenkins, L. D. "The Evolution of a Trading Zone: a Case Study of the Turtle Excluder Device." *Studies in History and Philosophy of Science* 41, no. 1 (2010): 75–85.

Murphy, Dale D. "The Tuna-Dolphin Wars." *Journal of World Trade* 40, no. 4 (2006): 597.

Piovano, S.; S. Clo; and C. Giacoma. "Reducing Longline Bycatch: the Larger the Hook, the Fewer the Stingrays." *Biological Conservation* 143, no. 1 (2010): 261–64.

OTHER

NOAA Fisheries Service. "Bycatch Reduction Engineering Program." http://www.nmfs.noaa.gov/bycatch.htm (accessed November 9, 2010).

NOAA Fisheries Service. "What is Bycatch?." http://www.nmfs.noaa.gov/by_catch/bycatch_whatis.htm (accessed November 9, 2010).

ORGANIZATIONS

Worldwatch Institute, 1776 Massachusetts Ave., NW, Washington, DC, USA, 20036–1904, (202) 452–1999, (202) 296-7365, worldwatch@worldwatch.org, http://www.worldwatch.org/

Bill Freedman

Bycatch reduction devices

Seabirds, seals, whales, sea turtles, dolphins, and non-targeted fish can be unintentionally caught and killed or maimed by modern fish and shrimp catching methods. This phenomenon is called bycatch or the unintended capture or mortality of living marine resources as a result of fishing. It is managed under such laws as the Migratory Bird Treaty Act, the Endangered Species Act of 1973, the Marine Mammals Protection Act of 1972 (amended in 1994), and, most recently, the Magnuson-Stevens Fishery Conservation and Management Act of 1996. The 1995 United Nations Code of Conduct for Responsible Fisheries, to which the United States is a signatory, also emphasizes the importance of bycatch reduction. Bycatch occurs because most fishing methods are not perfectly "selective," (i.e., they do not catch and retain only the desired size, sex, quality, and quantity of target species). It also occurs because fishermen often have incentive to catch more fish than they will keep.

According to the United Nations' Food and Agriculture Organization, worldwide commercial fishing operations discard millions of tons of fish, approximately one-fourth of the world catch, because they were the wrong type, sex, or size. Some bycatch are protected marine mammals, turtles, or seabirds.

To address this problem, numerous research programs have been established to develop BRDs and other means to reduce bycatch. Much of this research, and the nation's bycatch reduction activities overall, is centered in the Department of Commerce's National Marine Fisheries Service (NMFS), which leads and coordinates the United States' collaborative efforts to reduce bycatch. In March 1997, NMFS proposed a draft long-term strategy, *Managing the Nation's Bycatch*, that seeks to provide structure to the service's diverse bycatch-related research and management programs. These include gear research, technology transfer workshops, and the exploration of new management techniques. NMFS's Bycatch Plan was intended to as a guide for its own programs and for its "cooperators" in bycatch reduction, including eight regional fishery management councils, states, three interstate fisheries commissions, the fishing industry, the conservation community, and other parties. In pursuing its mandate of conserving and managing marine resources, NMFS relies on the direction for bycatch established by the 104th Congress under the new National Standard 9 of the Magnuson-Stevens Act, which states: "Conservation and management measures shall, to the extent practicable, (A) minimize bycatch and (B) to the extent bycatch cannot be avoided, minimize the mortality of such bycatch."

But although the national bycatch standard applies across all regions, bycatch issues are not uniform for all fisheries. Indeed bycatch is not always a problem and can sometimes be beneficial, (e.g., when bycatch species are kept and used as if they had been targeted species). But where bycatch is a problem, the exact nature of the problem and potential solutions will differ depending on the region and fishery. For instance, in the U.S. Gulf of Mexico shrimp fishery, which contributes about 70 percent of the annual U.S. domestic shrimp production, bycatch of juvenile red snapper (*Lutjanus blackfordi*) by shrimp trawlers reduces red snapper stocks for fishermen who target those fish. According to NMFS, "In the absence of bycatch reduction, red snapper catches will continue to be a fraction of maximum economic or biological yield levels." To address this problem, the Gulf of Mexico Fishery Management Council prepared an amendment to the shrimp fishery management plan for the Gulf of Mexico requiring shrimpers to use BRDs in their nets. But BRD effectiveness differs, and devices often lose significant amounts of shrimp in reducing bycatch. For instance, one device called a "30-mesh fisheye" reduced overall shrimp catches, and revenues, by 3 percent, an issue that must be addressed in any analysis of the costs and benefits of requiring BRDs, according to NMFS. In Southern New England, the yellowtail flounder (*Pleuronectes ferrugineus*) has been important to the New England groundfish fisheries for several decades. But the stock has been depleted to a record low because, from 1988 to 1994, most of the catch was discarded by trawlers. Reasons for treating the catch as bycatch were that most of the fish were either too small for marketing or were smaller than the legal size limit. Among the solutions to this complex problem are an increase in the mesh size of nets so smaller fish would not be caught and a redesign of nets to facilitate the escape of undersized yellowtail flounder and other bycatch species.

Although fish bycatch exceeds that of other marine animals, it is by no means the only significant bycatch problem. Sea turtle bycatch has received growing attention in recent years, most notably under the Endangered Species Act Amendments of 1988, which mandated a study of sea turtle conservation and the causes and significance of their mortality, including mortality caused by commercial trawlers. That study, conducted by the National Research Council (NRC), found that shrimp trawls accounted for more deaths of sea turtle juveniles, subadults, and breeders in coastal waters than all other human activities combined.

To address the turtle bycatch problem, NMFS, numerous Sea Grant programs, and the shrimping industry conducted research that led to the development of several types of net installation devices that were called "turtle excluder devices" (TEDs) or "trawler efficiency devices." In 1983, the only TED approved by NMFS was one developed by the service itself. But in the face of industry concerns about using TEDs, the University of Georgia and NMFS tested devices developed by the shrimping industry, resulting in NMFS certification of new TED designs. Each design was intended to divert turtles out of shrimp nets, excluding the turtles from the catch without reducing the shrimp intake. Over a decade of development, these devices have been made lighter and today at least six kinds of TEDs have NMFS's approval. Early in the development of TEDs, NMFS tried to obtain voluntary use of the devices, but shrimpers considered them an expensive, time-consuming nuisance and feared they could reduce the size of shrimp catches. But NMFS, and environmental groups, countered that the best TEDs reduced turtle bycatch by up to 97 percent with slight or no loss of shrimp. By 1985, NMFS faced threats of lawsuits to shut down the shrimping industry because trawlers were not using TEDs. In response, NMFS convened mediation meetings that included environmentalists and shrimpers. The meetings led to an agreement to pursue a "negotiated rulemaking" to phase in mandatory TED use, but negotiations fell apart after state and federal legislators, under intense pressure, tried to delay implementation of TED rules. After intense controversy, NMFS published regulations in 1987 on the use of TEDs by shrimp trawlers.

Dolphins are another marine animal that has suffered significant mortality levels as a result of bycatch associated with the eastern tropical Pacific Ocean tuna fishery. Several species of tuna are often found together with the most economically important tuna species, the yellowfin (*Thunnus albacares*). As a result, tuna fishermen have used a fishing technique called dolphin fishing, in which they set their nets around herds of dolphins to capture the tuna that are always close by. The spotted dolphin (*Stenella attenuata*) is most frequently associated with tuna. Spinner dolphin (*Stenella longirostris*) and the common dolphin (*Delphinus delphis*) also travel with tuna. According to one estimate, between 1960 and 1972 the U.S. fleet in the eastern tropical Pacific Ocean fishery killed more than 100,000 dolphins a year. That number dropped to an estimated 20,000 after 1972, when the Marine Mammals Protection Act was passed. U.S. fishing boats are now required to use techniques that allow dolphins to escape from tuna nets before the catch is hauled in. These include having fishermen jump into the ocean to hold the lip of the net below the water surface so the dolphin can jump out.

Resources

BOOKS

Clover, Charles. *The End of the Line: How Overfishing Is Changing the World and What We Eat*. Berkeley: University of California Press, 2008.

OTHER

United States Department of the Interior, United States Geological Survey (USGS). "Overfishing." http://www.usgs.gov/science/science.php?term=852 (accessed November 10, 2010).

United States Environmental Protection Agency (EPA). "Industry: Industries: Fishing Industry." http://www.epa.gov/ebtpages/induindustriesfishing industry.html (accessed November 10, 2010).

David Clarke

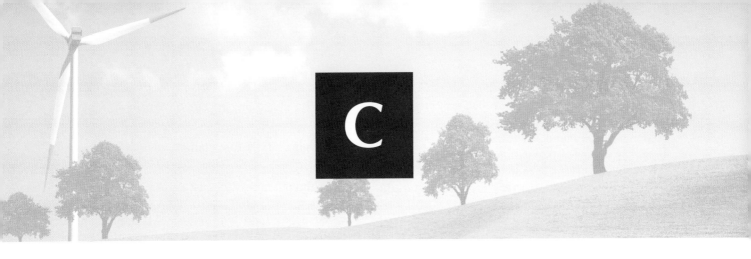

C

Cadmium

A metallic element that occurs most commonly in nature as the sulfide, CdS. Cadmium has many important industrial applications. It is used to electroplate other metals, in the production of paints and plastics, and in nickel-cadmium batteries. The metal also escapes into the environment during the burning of coal and tobacco. Cadmium is ubiquitous in the environment, with detectable amounts present in nearly all water, air, and food samples. In high doses, cadmium is toxic. In lower doses, it may cause kidney disease, disorders of the circulatory system, weakening of bones, and, possibly, cancer.

See also Itai-itai disease.

CAFE *see* Corporate Average Fuel Economy Standards.

Cairo Conference

The 1994 International Conference on Population and Development occurs at a defining moment in the history of international cooperation, the preamble to the Program of Action for the United Nations International Conference on Population and Development (UN ICPD) stated. This historic conference, the fifth such world meeting regarding population issues, was held in Cairo, Egypt September 5–13, 1994, at a time when there was growing, worldwide recognition of the interdependence between global population, development and environment. The Cairo Conference considered itself as building upon the foundation already laid in the 1974 World Population Conference in Bucharest and the 1984 International Conference on Population in Mexico City.

Representatives from 180 countries throughout the world gathered to discuss these problems. In order to represent particular world societal blocs, the UN ensured representation from four communities of nations: developing countries, Muslim states, the industrialized West, and Catholic countries. The makeup of attending blocs was intentional, as controlling the world's population involves many hotly debated issues that bring about sharp differences of opinion on religious and cultural beliefs. Conference attendees and those that would follow up on the conference's recommendations in their host nations would have to address issues like marriage, divorce, abortion, contraception, and homosexuality.

The call to address population growth came at a time when some groups said that the world was already pushing its food supply to natural limits. For example, marine biologists stated that ocean fisheries could not sustain catches upwards of 100 million tons of fish per year. In many countries, underground water supplies are reportedly strained and farming has been exploited to its capacity. The world population in 1994 was estimated at 5.6 billion people, and projected to continue to increase at the 1994 level of over 80 million people per year. Most frightening of all were the United Nations population projections for the twenty years to follow. These estimates ranged between 7.9 billion to 11.9 billion people inhabiting the earth by 2014.

In his opening address on September 5, UN Secretary-General Boutros Boutros-Ghali said, The efficacy of the economic order of the planet on which we live depended in great measure on the conference's outcome. Leaders of several countries spoke of the many issues surrounding the problem and causing social and ethical dilemmas in setting population policy. Then Vice-President Albert Gore of the United States spoke of a holistic approach that also comprehensively addresses world poverty. Over the course of six days, more than 200 speakers shared their concerns about population and strategies to curb growth. Because the conference was mandated to address the issues and how they impacted population, the debate branched out to a number of economic, reproductive, trade, environmental, family structure, healthcare, and education issues.

The following areas of need were spelled out in the UN ICPD Programme of Action:

• Rapid urbanization and movement within and across borders are additional population problems the Cairo conference attempted to address. At that time, more than 93 percent of the 93 million people added to the population each year lived in developing countries. UN data just prior to the conference also showed that 43 percent of the world population resided in urban areas, up from 38 percent in 1975.

• The conference's delegates agreed on common objectives for controlling the world's population and even took on a more ambitious goal of population growth limit than anticipated by some. The conference objectives are targeted for completion by the year 2015. The common strategy limits family size and helps empower women to participate in reproductive decisions.

• The Cairo Conference goals also addressed some mortality issues, and set an objective of reducing infant mortality worldwide by 45 percent. In 1994, infant mortality averaged sixty-two per 1,000 live births. The conference's objective aimed to lower that number to only twelve per 1,000 live births.

• A goal was also outlined to lower maternal mortality to thirty per 100,000 women. These objectives would be furthered by a pledge to offer prenatal care to all pregnant women.

• The conference also included an objective addressing education. It outlined that all children of school age be enabled to complete primary education.

• Contraception availability became another tool to limit population growth. The delegates settled on a goal of making contraception accessible to 70 percent of the world's population by the year 2015. However, they also added universal access to family planning as a final objective.

• In order to meet family planning goals, some leaders in the field suggested a shift in how leaders looked at their policies. Prior to the conference, they tended to focus on organized family planning programs. The Cairo conference objectives shifted the focus to a broadened policy, concerned with issues like gender equality, education, and empowerment of women to better deal with reproductive choices.

• The aging of the populations in more developed countries was also discussed.

Although unprecedented agreement was reached over the few days in Cairo, it didn't happen without heated debate and controversy. Most participants supported the notion that population policies must be based on individual rights and freedom of choice, as well as the rights of women and cultures. Since abortion was considered such a sensitive issue, many delegates agreed that family planning offered women alternative recourse to abortion. The Cairo Conference ended with both hope and concern for the future. If the conference objectives were reached, the delegates projected that world population would rise from 5.7 billion in the mid-1990s to 7.5 billion by the year 2015, and then begin to stabilize. However, if the objectives were not reached, conference attendees concluded that the population would continue to rise to the perilous numbers discussed in the conferences preamble.

Following the Cairo Conference, the United States began implementing the conference objectives. The U.S. Agency for International Development (USAID) held a series of meetings between September 1994 and January 1995 to draw upon the expertise and ideas of American leaders in the field and to begin encouraging U.S. participation in population control goals. Among the USAID's own objectives for the United States were new initiatives and expansion of programs in place that emphasized family planning and health, as well as education and empowerment of women in America. For example, USAID suggested expanding access to reproductive health information and services and improved prevention of HIV/AIDS and other sexually transmitted diseases (STDs). It also promoted research on male methods of contraception and new barrier methods to protect from both pregnancy and STDs. The agency focused on improving women's equality and rights beginning with review of existing U.S. programs. A major focus would also include improving women's economic equity in the United States.

Such initiatives require considerable financial resources. The U.S. Congress appropriated $527 million to family planning and reproductive health in fiscal year 1995 alone. A number of additional programs received substantial funding to support the economic, social, and educational programs needed to complete the United States' objectives related to the Cairo conference. Included in this overall program was establishment of the President's Council on Sustainable Development, a combination public/private group charged with making recommendations on U.S. policies and programs that will help balance population growth with consumption of natural resources. The United States also worked cooperatively with Japan prior to and following the conference on a common agenda to coordinate population and health assistance goals. The program also was designed to help strengthen relations between Japan and the United States.

Five years after the Cairo Conference, a follow-up conference was held in New York City from June 30 to July 2, 1999. In preparation for the conference, a forum was held at The Hague, Netherlands. At that forum, delegates of several countries that had expressed concerns about the Cairo Conference objectives five years earlier began trying to renegotiate the program of action. By the time the Cairo Plus Five Conference convened in 1999, final recommendations had not been completed, as some member delegates, including representatives of the Vatican in Rome, attempted to alter the original Cairo program of action. Money was yet another issue in focus at the 1999 meeting. United Nations Population Fund (UNPFA) Chief Nafis Sadik estimated that $5.7 billion per year was needed to meet the Cairo Conference aims, an amount she described as peanuts. The actual average amount available to the Programme of Action was far less, $2.2 billion.

The real measure of success lies in whether or not world population stabilizes. There is some evidence to suggest that while still growing, the number of human beings inhabiting this planet is not growing at the nightmare rates projected at the beginning of the Cairo Conference. As of 2010, the U.S. Census Bureau and the United Nations Department of Economic and Social Affairs estimate that the world population in 2015 will be around 7.2 or 7.3 billion people.

Resources

BOOKS

Newbold, K. Bruce. *Six Billion Plus: World Population in the Twenty-First Century*. Lanham, MD: Rowman & Littlefield Publishers, 2007.

OTHER

United Nations Population Fund (UNFPA). "Homepage." United Nations Population Fund (UNFPA) (accessed October 14, 2010).

ORGANIZATIONS

United Nations Population Fund (UNFPA), 605 Third Avenue, New York, NY, USA, 10158, (212) 297-5000, (212) 370-0201, hq@unfpa.org, http://www.unfpa.org

Joan M. Schonbeck

Calcareous soil

A calcareous soil is soil that has calcium carbonate ($CaCO_3$) in abundance. If a calcareous soil has hydrochloric acid added to it, the soil will effervesce and give off carbon dioxide and form bubbles because of the chemical reaction. Calcareous soils are most often formed from limestone or in dry environments where low rainfall prevents the soils from being leached of carbonates. Calcareous soils frequently cause nutrient deficiencies for many plants.

Caldicott, Helen M.

1938–
Australian physician and activist

Dr. Helen Caldicott is a pediatrician, mother, antinuclear activist, and environmental activist. Born Helen Broinowski in Melbourne, Australia on August 7, 1938, she is known as a gifted orator and a tireless public speaker and educator. She traces her activism to age fourteen when she read Nevil Shute's *On the Beach*, a chilling novel about nuclear holocaust. In 1961 she graduated from the University of Adelaide Medical School with bachelor of medicine and bachelor of surgery degrees, which are the equivalent of an American M.D. She married Dr. William Caldicott in 1962, and returned to Adelaide, Australia to go into general medical practice. In 1966 she, her husband, and their three children moved to Boston, Massachusetts, where she held a fellowship at Harvard Medical School. Returning to Australia in 1969, she served first as a resident in pediatrics and then as an intern in pediatrics at Queen Elizabeth Hospital. There, she set up a clinic for cystic fibrosis, a genetic disease in children.

In the early 1970s, Caldicott led a successful campaign in Australia to ban atmospheric nuclear testing by the French in the South Pacific. Her success in inspiring a popular movement to stop the French testing has been attributed to her willingness to reach out to the Australian people through letters and television and radio appearances, in which she explained the dangers of radioactive fallout. Next, she led a successful campaign to ban the exportation of uranium by Australia. During that campaign she met strong resistance from Australia's government, which had responded to the 1974 international oil embargo by offering to sell uranium on the world market. (Uranium is the raw material for nuclear technology.) Caldicott chose to go directly to mine workers, explaining the effects of radiation on their bodies and their genes and talking about the effects of nuclear war on them and their children. As a result, the Australian Council of Trade Unions passed a resolution not to mine, sell, or transport uranium. A ban was instituted from 1975 to 1982, when Australia gave in to international pressure to resume the exportation.

In 1977 Dr. Caldicott and her husband immigrated to the United States, accepting appointments at the Children's Hospital Medical Center and teaching appointments at Harvard Medical School in Boston, Massachusetts. She was a co-founder of Physicians for Social Responsibility (PSR), and she was its president at the time of the March 28, 1979 nuclear accident at the Three Mile Island Nuclear Reactor in Pennsylvania. At that time, PSR was a small group of concerned medical specialists. Following the accident, the organization grew rapidly in membership, financial support, and influence. As a result of her television appearances and statements to the media following the Three Mile Island accident, Caldicott became a symbol of the movement to ban all nuclear power and oppose nuclear weapons in any form. Ironically, she resigned as president of PSR in 1983, when the organization had grown to over 20,000 members. At that time, she began to be viewed as an extreme radical in an organization that had become more moderate as it came to represent a wide, diversified membership.

She also founded Women's Action for Nuclear Disarmament (WAND). WAND has been an effective group lobbying Congress against nuclear weapons.

Throughout her career, Caldicott has considered her family to be her first priority. She has three children and she emphasizes the importance of building and maintaining a strong marriage, believing that good interpersonal relationships are essential before a socially-minded person can work effectively for broad social change.

Caldicott has developed videotapes and films and has written over one hundred articles which have appeared in major newspapers and magazines throughout the world. She has written four books. Her first, *Nuclear Madness: What You Can Do* (1978), is considered important reading in the antinuclear movement. Her second, entitled *Missile Envy*, was published in 1986. In *If You Love This Planet: A Plan to Heal the Earth* (1992), Caldicott discusses the race to save the planet from environmental damage resulting from excess energy consumption, pollution, ozone layer depletion, and global warming. She urges citizens of the United States to follow the example set by the Australians, who have adopted policies and laws designed to move their society toward greater corporate and institutional responsibility. She urges the various nations of the world to strive for a what she described as a new legal world order by moving toward a sort of transnational control of the world's natural resources. One of her books, *A Desperate Passion* (1996), is an autobiography in which she reflects upon crucial events that have influenced her and talks about people who have inspired her in her life and work. She also wrote *The New Nuclear Danger: George Bush's Military Industrial Complex*, which was published in 2002. A revised and updated edition of *If You Love This Planet: A Plan to Heal the Earth* was published in 2009.

Since mid-2008, she has hosted a weekly radio program on public radio in the U.S., Australia and Canada.

Caldicott is the recipient of 20 honorary degrees, many awards and prizes including the SANE Peace Prize, the Ghandi Peace Prize, the John-Roger Foundation's Integrity Award (which she shared with Bishop Desmond Tutu), the Norman Cousins Award for Peace-making, the Margaret Mead Award, and the inaugural 2006 Australian Peace Prize.

Resources

BOOKS

Caldicott, H. *If You Love This Planet: A Plan to Heal the Earth*. New York: W. W. Norton, 1992.

Paulette L. Stenzel

Caldwell, Lynton K.

1913–2006

American scholar and environmentalist

Lynton Caldwell was a key figure in the development of environmental policy in the United States. A longtime advocate for adding an environmental amendment to the Constitution, Caldwell insisted that the federal government has a duty to protect the environment in a way that is similar to the defense of civil rights or freedom of speech.

Caldwell was born November 21, 1913 in Montezuma, Iowa. He received his bachelor of arts from the University of Chicago in 1935 and completed a master of arts at Harvard University in 1938. The same year, Caldwell accepted an assistant professorship in government at Indiana University in Bloomington. In 1943, he attained his doctorate from the University of Chicago and began publishing academic works the following year. The subjects of his early writings were not environmental; he published a study of administrative theory in 1944 and a study of New York state government in 1954. By 1964, however, Caldwell had shifted his emphasis, and he began to receive wide recognition for his work on environmental policy. In that year, he was presented with the William E. Mosher Award from the American Society for Public Administration for his article "Environment: A New Focus for Public Policy."

Caldwell's most important accomplishment was his prominent role in the drafting of the National Environmental Policy Act (NEPA) in 1969. As a consultant for the Senate Committee on Interior and Insular Affairs in 1968, he prepared *A Draft Resolution on a National Policy for the Environment*. His special report examined the constitutional basis for a national environmental policy and proposed a statement of intent and purpose for Congress. Many of the concepts first introduced in this draft resolution were later incorporated into the act. As consultant to that committee, Caldwell played a continuing role in the shaping of the NEPA, and he was involved in the development of the environmental impact statement.

Caldwell strongly defended the NEPA, as well as the regulatory agency it created, claiming that they represent the first comprehensive commitment of any modern state toward the responsible custody of its environment. Although the act has influenced policy decisions at every level of government, the enforcement of its provisions have been limited. Caldwell argued that this is because environmental regulations have no clear grounding in the law. Statutes alone are often unable to withstand the pressure of economic interests. He proposed an amendment to the Constitution as the best practical solution to this problem and he maintains that without such an amendment, environmental issues will continue to be marginalized in the political arena.

In addition to advising the Senate during the creation of the NEPA, Caldwell did extensive work on international environmental policy. He advised the Central Treaty Organization and served on special assignments in countries including Colombia, India, the Philippines, and Thailand. Until his death on August 15, 2006, in Bloomington, Indiana, Caldwell served as the Arthur F. Bentley Professor of Political Science emeritus and professor of public and environmental affairs at Indiana University in Bloomington, Indiana.

Resources

BOOKS

Caldwell, Lynton Keith. *Environment: A Challenge for Modern Society*. Garden City, N.Y.: Published for the American Museum of Natural History by the Natural History Press, 1970.
Caldwell, Lynton Keith. *In Defense of Earth: International Protection of the Biosphere*. Bloomington: Indiana University Press, 1972.
Caldwell, Lynton Keith, Lynton R. Hayes, and Isabel M. MacWhirter. *Citizens and the Environment: Case Studies in Popular Action*. Bloomington: Indiana University Press, 1976.

Douglas Smith

California condor

With a wingspan of over nine feet (about three meters), the California condor (*Gymnogyps californianus*) has the largest wingspan of any bird in North America. The condor is a scavenger. It is a member of the New World vulture family (Cathartidae), and is closely related to the Andean condor (*Vultur gryphus*) found in South America.

The California condor is an endangered species (protected under the Endangered Species Act) barely avoiding extinction. While the condor population may

California condor *(Gymnogyps californianus)*. *(©iStockPhoto/ Windzepher.)*

In the 1940s, the National Audubon Society initiated a census of the condor population and recorded approximately sixty birds. In the early 1960s, the population was estimated at forty-two birds. A 1966 survey found fifty-one condors, an increase that may have been due to variability among the sampling systems used. By the end of 1986 there were twenty-four condors alive, twenty-one of which were in captivity.

Direct and indirect human stressors are directly responsible for the decline in the condor population during the past three centuries. California condors have been shot by hunters and poisoned by bait set out to kill coyotes. Their food has been contaminated with pesticides such as dichlorodiphenyltrichloroethane (DDT) or lead (Pb) from lead shot found in animal carcasses shot and lost by hunters. The condor's rarity has made their eggs a valuable commodity for unscrupulous collectors. Original food sources, such as mammoths and giant camels of the Pleistocene era, have disappeared, but as cattle ranching and sheep farming developed, carcasses of these animals have become suitable substitutes. Habitat destruction and general harassment reduced the condor's range and population to a point where intervention was necessary to halt the bird's rapid decline to extinction.

Most other large birds and mammals in North America that lived during the late Pleistocene period have become extinct. However, there is a small chance that with intervention and restoration projects, the California condor can be saved from extinction. As a result, after much debate, the decision was made to remove all California condors from the wild and initiate a captive breeding and reintroduction program. The United States Fish and Wildlife Service (FWS), together with the California Fish and Game Commission, captured the last wild condor in April 1987.

The captive breeding program has thus far increased the condor population dramatically. Condors nest only once every two years and typically lay only one egg. By removing the egg from the nest for laboratory incubation, the female can be tricked into laying a replacement egg. This method helps accelerate the population increase.

In January 1992, a pair of California condors was released into the wild. In October the male was found dead of kidney failure after apparently drinking from a pool of antifreeze in a parking lot near their sanctuary. Six additional condors were released at the end of 1992. In 1994, the total condor population consisted of nine wild condors and sixty-six birds in captivity. In 1995 another pair of captive birds was released into the

have been in decline before the arrival of Europeans in North America, increased settlement of its habitat, hunting, and indirect human influences (e.g., pollution, pesticides, construction hazards such as power lines) have almost eliminated the bird.

The condor was once found in much of North America. Fossil records indicate that it lived from Florida to New York in the East and British Columbia to Mexico in the West. The condor's range became more restricted even before European settlers arrived in North America. By the time Genoese explorer Christopher Columbus (c.1451–1506) arrived in North America (mid-fifteenth century), the condor had already started its retreat west. At the beginning of the twentieth century, it could still be found in southern California and Baja California (northern Mexico), but by 1950 its range had been reduced to a strip of about 150 miles (241 kilometers) at the southern end of the San Joaquin Valley in California.

wild, and with the release of more birds in following years, breeding pairs began to establish nest sites. In 2001, an egg that was laid by captive condors in the Los Angeles Zoo was placed in a condor nest in the wild, and the chick hatched successfully, only to die a few days later. The pair of condors then laid their own egg and hatched the first wild chick in April 2002. This was the first chick hatched in the wild in eighteen years. As of April 2002, there were sixty-three free-living condors in California and Arizona and eighteen more were awaiting release from field pens where they were being acclimated to their natural habitat. The captive population totaled 104 birds.

The International Union for the Conservation of Nature (IUCN) assessment of the California condor listed the bird as Critically Endangered in 2008. In 2003, there were 223 individuals with 138 in captivity and eighty-five birds reintroduced into northern Arizona and California. By 2006, 130 wild birds were located in release areas. At least forty-four of these wild birds were six years of age (the age at which breeding is possible) or older. A program in Baja California is attempting to increase the wild population to twenty pairs. In 2007, the first California condor born in seventy-five years was hatched in Mexico. One of the Baja birds was observed in California in 2007. This sighting is encouraging to the goal of the Baja birds ranging and breeding with U.S. condors.

The California Condor Recovery Plan is a consortium of private groups and public agencies. The goal of the recovery plan is to have 150 condors in each of two separate populations with at least fifteen breeding pairs in each group. However, there are significant obstacles to the recovery of wild populations of California condors, including continued pressure from development that causes accidents such as collisions with power lines, and poisoning by crude oil from drilling operations that the birds mistake for pools of water. Although the capture and release of this species is working and the population of captive birds is growing, the survival of the California condor in the wild is still in doubt. Measures must be taken to reduce risks to wild birds from man-made hazards.

Resources

BOOKS

Mee, Allan, and L. S. Hall. *California Condors in the 21st Century.* Cambridge, MA: Nuttall Ornithological Club, 2007.

Moir, John. *Return of the Condor: The Race to Save Our Largest Bird from Extinction.* New York: Lyons, 2006.

Osborn, Sophie A. H. *Condors in Canyon Country: The Return of the California Condor to the Grand Canyon Region.* Grand Canyon, AZ: Grand Canyon Association, 2007.

Snyder, Noel F. R., and Helen Snyder. *Introduction to the California Condor.* California Natural History Guides, 81. Berkeley: University of California Press, 2005.

OTHER

International Union for Conservation of Nature and Natural Resources (IUCN). "The IUCN Red List of Endangered Species: Gymnogyps Californianus (California Condor)." http://www.iucnredlist.org/details/144771/0 (Accessed August 4, 2010).

Eugene C. Beckham
Marie H. Bundy

Callicott, John Baird

1941–
American environmental philosopher

John Baird Callicott is a founder and seminal thinker in the modern field of environmental philosophy. He is best known as the leading contemporary exponent of Aldo Leopold's land ethic. Callicott has not only interpreted Leopold's original works, but he has applied the reasoning of the land ethic to modern resource issues such as wilderness designation and biodiversity protection.

Callicott's 1987 edited volume, *Companion to A Sand County Almanac,* is the first interpretive and critical discussion of Leopold's classic work. His 1989 collection of essays, *In Defense of the Land Ethic,* explores the intellectual foundations and development of Leopold's ecological and philosophical insights and their ultimate union in his later works. In 1991 Callicott, with Susan L. Flader, introduced to the public the best of Leopold's remaining unpublished and uncollected literary and philosophical legacy in a collection entitled *The River of the Mother of God and Other Essays* by Aldo Leopold.

Since his contribution to the inaugural issue of the journal *Environmental Ethics* in 1979, Callicott's articles and essays have appeared not only in professional philosophical journals and a variety of scientific and technical periodicals, but in a number of lay publications as well. He has contributed chapters to more than twenty books and is internationally known as an author and speaker. Born in Memphis, Tennessee, Callicott completed his Ph.D. in philosophy at Syracuse University in 1971, and has since held visiting professorships at a number of American universities. In 1971, while teaching at the

at the DNA level, disrupting its ability to reproduce. Surgery is the treatment of choice when it has been determined that the tumor is intact and has not metastasized beyond the limits of surgical excision. Surgery is also indicated for benign tumors that could progress into malignant tumors. Premalignant and in situ tumors of epithelial tissues, such as skin, mouth, and cervix, can be removed.

Chemotherapy and radiation treatments are the most commonly used therapies for cancer. Unfortunately, both methods produce unpleasant side effects; they often suppress the immune system, making it difficult for the body to destroy the remaining cancer even after the treatment has been successful. In this regard, immunotherapy holds great promise as an alternative treatment, because it makes use of the unique properties of the immune system.

Immunotherapies for the treatment of cancer are generally referred to as biological response modifiers (BRMs). BRMs are defined as mammalian gene products, agents, and clinical protocols that affect biologic responses in host-tumor interactions. Immunotherapies have a direct cytotoxic effect on cancer cells, initiation or augmentation of the host's tumor-immune rejection response, and modification of cancer cell susceptibility to the lytic or tumor static effects of the immune system. As with other cancer therapies immunotherapies are not without their own side effects. Most common are flu-like symptoms, skin rashes, and vascular-leak syndrome. At their worst, these symptoms are usually less severe than those of current chemotherapy and radiation treatments.

See also Hazardous material; Hazardous waste; Leukemia; Radiation sickness.

Resources

BOOKS

Colditz, Graham A. *Encyclopedia of Cancer and Society*. Los Angeles: Sage Publications, 2007.

King, Roger, J.B., and Mike W. Robins. *Cancer Biology*. Harlow, UK, and New York: Pearson/Prentice Hall, 2006.

PERIODICALS

Meister, Kathleen. *America's War on "Carcinogens": Reassessing the Use of Animal Tests to Predict Human Cancer Risk*. Washington: American Council on Science, 2005.

OTHER

Centers for Disease Control and Prevention (CDC). "Cancer." http://www.cdc.gov/cancer/ (accessed October 2, 2010).

World Health Organization (WHO). "International Agency for Research on Cancer (IARC)." WHO Programs and Projects. http://www.iarc.fr (accessed October 2, 2010).

World Health Organization (WHO). "Radon and cancer." http://www.who.int/entity/mediacentre/factsheets/fs291/en/index.html (accessed October 2, 2010).

Brian R. Barthel

Captive propagation and reintroduction

Captive propagation is the deliberate breeding of wild animals in captivity in order to increase their numbers. Reintroduction is the deliberate release of these species into their native habitat. The Mongolian wild horse, Pere David's deer, and the American bison would probably have become extinct without captive

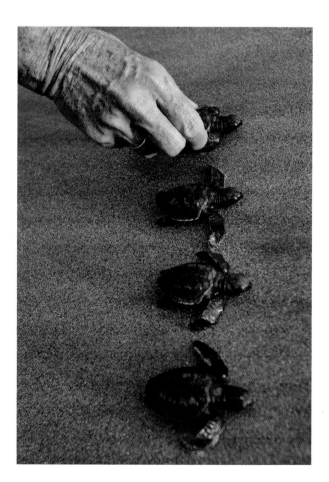

Residents and foreign tourists joined in the release of dozens of sea turtle babies on a beach in Bali, Indonesia. *(© Made Nagi/epa/Corbis.)*

propagation. Nearly all cases of captive propagation and reintroduction involve threatened or endangered species. Zoos are increasingly involved in captive propagation, sometimes using new technologies. One of these, allows a relatively common species of antelope to act as a surrogate mother and give birth to a rare species.

Once suitable sites are selected, a reintroduction can take one of three forms. *Reestablishment reintroductions* take place in areas where the species once occurred but is now entirely absent. Recent examples include the red wolf, the black-footed ferret, and the peregrine falcon east of the Mississippi River. Biologists use *augmentation reintroduction* to release captive-born wild animals into areas in which the species still occurs but only in low numbers. These new animals can help increase the size of the population and enhance genetic diversity. Examples include a small Brazilian monkey called the golden lion tamarin and the peregrine falcon in the western United States. A third type, *experimental reintroduction*, acts as a test case to acquire essential information for use on larger-scale permanent reintroductions. The red wolf was first released as an experimental reintroduction. A 1982 amendment to the Endangered Species Act facilitates experimental reintroductions, offering specific exemptions from the Act's protection, allowing managers greater flexibility should reintroduced animals cause unexpected problems.

Yet captive propagation and reintroduction programs have their drawbacks, the chief one being their high cost. Capture from the wild, food, veterinary care, facility use and maintenance all contribute significant costs to maintaining an animal in captivity. Other costs are incurred locating suitable reintroduction sites, preparing animals for release, and monitoring the results. Some conservationists have argued that the money would be better spent acquiring and protecting habitat in which remnant populations already live.

There are also other risks associated with captive propagation programs such as disease, but perhaps the greatest biological concern is that captive populations of endangered species might lose learned or genetic traits essential to their survival in the wild. Animals fed from birth, for example, might never pick up food-gathering or prey-hunting skills from their parents as they would in the wild. Consequently, when reintroduced such animals may lack the skill to feed themselves effectively. Furthermore, captive breeding of animals over a number of generations could affect their evolution. Animals that thrive in captivity might have a selective advantage over their

"wilder" cohorts in a zoo, but might be disadvantaged upon reintroduction by the very traits that aided them while in captivity.

Despite these shortcomings, the use of captive propagation and reintroduction will continue to increase in the decades to come. Biologists learned a painful lesson about the fragility of endangered species in 1986 when a sudden outbreak of canine distemper decimated the only known group of black-footed ferrets. The last few ferrets were taken into captivity where they successfully bred. Even as new ferret populations become established through reintroduction, some ferrets will remain as captive breeders for insurance against future catastrophes. Biologists are also steadily improving their methods for successful reintroduction. They have learned how to select the combinations of sexes and ages that offer the best chance of success and have developed systematic ways to choose the best reintroduction sites.

Captive propagation and reintroduction will never become the principal means of restoring threatened and endangered species, but it has been proven effective and will continue to act as insurance against sudden or catastrophic losses in the wild.

See also Biodiversity; Extinction; Wildlife management; Wildlife rehabilitation.

Resources

BOOKS

Hayward, Matt. *Reintroduction of Top-Order Predators.* Hoboken, NJ: Wiley-Blackwell, 2009.

James H. Shaw

Carbon

The seventeenth most abundant element on Earth, carbon occurs in at least six different allotropic forms, the best known of which are diamond and graphite. It is a major component of all biochemical compounds that occur in living organisms: carbohydrates, proteins, lipids, and nucleic acids. Carbon-rich rocks and minerals such as limestone, gypsum, and marble often are created by accumulated bodies of aquatic organisms. Plants, animals, and microorganisms cycle carbon through the environment, converting it from simple compounds like carbon dioxide and methane to more complex compounds like sugars and starches, and then, by the action of decomposers, back again to simpler compounds. One of the most important fossil fuels, coal, is composed chiefly of carbon.

Carbon cycle

The series of chemical, physical, geological, and biological changes by which carbon moves through Earth's air, land, water, and living organisms is called the carbon cycle.

Carbon makes up no more than 0.27 percent of the mass of all elements in the universe and only 0.0018 percent by weight of the elements in Earth's crust. Yet, its importance to living organisms is far out of proportion to these figures. In contrast to its relative scarcity in the environment, it makes up 19.4 percent by weight of the human body. Along with hydrogen, carbon is the only element to appear in every organic molecule in every living organism on Earth.

In the atmosphere, carbon exists almost entirely as gaseous carbon dioxide. The best estimates are that the earth's atmosphere contains 740 billion tons of this

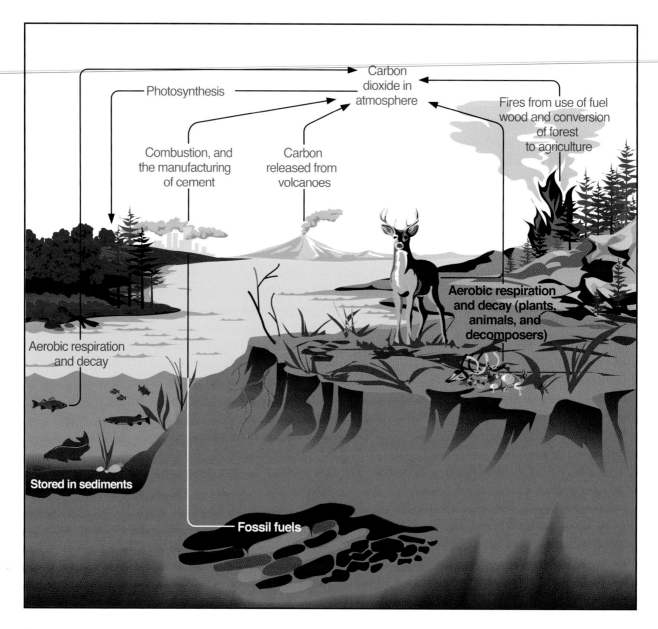

Carbon cycle. *(Reproduced by permission of Gale, a part of Cengage Learning.)*

gas. Its global concentration is about 350 parts per million (ppm), or 0.035 percent by volume. That makes carbon dioxide the fourth most abundant gas in the atmosphere after nitrogen, oxygen and argon. Some carbon is also released as carbon monoxide to the atmosphere by natural and human mechanisms. This gas reacts readily with oxygen in the atmosphere, however, converting it to carbon dioxide.

Carbon returns to the hydrosphere when carbon dioxide dissolves in the oceans, as well as in lakes and other bodies of water. The solubility of carbon dioxide in water is not especially high, 88 milliliters of gas in 100 milliliters of water. Still, the earth's oceans are a vast reservoir of approximately 36,000 billion tons of carbon. About 93 billion tons of carbon flows from the atmosphere into the hydrosphere each year.

Carbon moves out of the oceans in two ways. Some escapes as carbon dioxide from water solutions and returns to the atmosphere. That amount is estimated to be very nearly equal (90 billion tons) to the amount entering the oceans each year. A smaller quantity of carbon dioxide (about 40 billion tons) is incorporated into aquatic plants.

On land, green plants remove carbon dioxide from the air through the process of photosynthesis—a complex series of chemical reactions in which carbon dioxide is eventually converted to starch, cellulose, and other carbohydrates. About 100 billion tons of carbon are transferred to green plants each year, with about 560 billion tons of the element being stored in land plants alone.

The carbon in green plants is eventually converted into a large variety of organic (carbon-containing) compounds. When green plants are eaten by animals, carbohydrates and other organic compounds are used as raw materials for the manufacture of thousands of new organic substances. The total collection of complex organic compounds stored in all kinds of living organisms represents the reservoir of carbon in Earth's biosphere.

The cycling of carbon through the biosphere involves three major kinds of organisms. Producers are organisms with the ability to manufacture organic compounds such as sugars and starches from inorganic raw materials such as carbon dioxide and water. Green plants are the primary example of producing organisms. Consumers are organisms that obtain their carbon (that is, their food) from producers: all animals are consumers. Finally, decomposers are organisms such as bacteria and fungi that feed on the remains of dead plants and animals. They convert carbon compounds in these organisms to carbon

dioxide and other products. The carbon dioxide is then returned to the atmosphere to continue its path through the carbon cycle.

Land plants return carbon dioxide to the atmosphere during the process of respiration. In addition, animals that eat green plants exhale carbon dioxide, contributing to the 50 billion tons of carbon released to the atmosphere by all forms of living organisms each year. Respiration and decomposition both represent, in the most general sense, a reverse of the process of photosynthesis. Complex organic compounds are oxidized with the release of carbon dioxide and water—the raw materials from which they were originally produced.

At some point, land and aquatic plants and animals die and decompose. When they do so, some carbon (about 50 billion tons) returns to the atmosphere as carbon dioxide. The rest remains buried (up to 1,500 billion tons) or on the ocean bottoms (about 3,000 billion tons). Several hundred million years ago, conditions of burial were such that organisms decayed to form products consisting almost entirely of carbon and hydrocarbons. Those materials exist today as pockets of the fossil fuels—coal, oil, and natural gas. Estimates of the carbon stored in fossil fuels range from 5,000 to 10,000 billion tons.

The processes that make up the carbon cycle have been occurring for millions of years, and for most of this time, the systems involved have been in equilibrium. The total amount of carbon dioxide entering the atmosphere from all sources has been approximately equal to the total amount dissolved in the oceans and removed by photosynthesis. However, a hundred years ago changes in human society began to unbalance the carbon cycle. The Industrial Revolution initiated an era in which the burning of fossil fuels became widespread. In a short amount of time, large amounts of carbon previously stored in the earth as coal, oil, and natural gas were burned up, releasing vast quantities of carbon dioxide into the atmosphere.

Between 1900 and 1992, measured concentrations of carbon dioxide in the atmosphere increased from about 296 ppm to over 350 ppm. Scientists estimate that fossil fuel combustion now released about five billion tons of carbon dioxide into the atmosphere each year. In an equilibrium situation, that additional five billion tons would be absorbed by the oceans or used by green plants in photosynthesis. Yet this appears not to be happening: measurements indicate that about 60 percent of the carbon dioxide generated by fossil fuel combustion remains in the atmosphere.

The problem is made even more complex because of deforestation. As large tracts of forest are cut down and burned, two effects result: Carbon dioxide from forest fires is added to that from other sources, and the loss of trees decreases the worldwide rate of photosynthesis. Overall, it appears that these two factors have resulted in an additional one to two billion tons of carbon dioxide in the atmosphere each year.

No one can be certain about the environmental effects of this disruption of equilibria in the carbon cycle, but most believe that the additional carbon dioxide will augment the earth's natural greenhouse effect, resulting in long-term global warming and climate change. It is clear we do not fully understand the way oceans, clouds, and other factors affect climate, but improved knowledge allows better predictions.

See also Atmosphere; Carbon; Greenhouse effect.

Resources

BOOKS

Burroughs, William James. *Climate Change: A Multidisciplinary Approach.* Cambridge: Cambridge University Press, 2007.

Cowie, Jonathan. *Climate Change: Biological and Human Aspects.* Cambridge: Cambridge University Press, 2007.

McCaffrey, Paul. *Global Climate Change.* Minneapolis: H. W. Wilson, 2006.

David E. Newton

Catation exchange *see* **Ion exchange; Radiocarbon dating.**

Carbon dioxide

The fourth most abundant gas in the earth's atmosphere, carbon dioxide occurs in an abundance of about 350 parts per million. The gas is released by volcanoes and during respiration, combustion, and decay. Plants convert carbon dioxide into carbohydrates by the process of photosynthesis. Carbon dioxide normally poses no health hazard to humans. An important factor in maintaining the earth's climate, molecules of carbon dioxide capture infra-red wavelength light (heat) radiated from the earth's surface, raising the planet's temperature to a level at which life can be sustained, a phenomenon known as the greenhouse effect. According to a large consensus of data and studies, excessive levels of carbon dioxide resulting from anthropogenic (human) activities now contributing to global warming and climate change.

Carbon emissions trading

Carbon emissions trading (CET) is a practice allowing countries—and corporations—to trade their harmful carbon emissions for credit to meet their designated carbon emission limits. The concept was first approved in 1992 as part of the United Nations Framework Convention on Climate Change (UNFCCC). Emissions trading is an incentive system and profit-driven system for the purpose of controlling or reducing certain polluting emissions, especially anthropogenic (human activity related) greenhouse gas emissions contributing to global warming. In 1997, most countries also signed the Kyoto Protocol to the UNFCCC. The protocol requires all signatories to monitor and report their greenhouse emissions; developed or industrialized countries are also required to cap and trade emissions of six greenhouses gases, including carbon dioxide, or CO_2.

In most emissions trading schemes, parties buy and sell units of credit that entitle them to emit a certain amount of a given kind of pollution. The total number of credits in circulation for that pollutant is fixed, so the total emissions of the pollutant—the "cap"—can be kept to a predetermined level. Accordingly, the term "cap and trade" is often used to describe emissions trading. Parties that reduce their pollution emissions receive credits that they can then trade with other parties to the agreement. Thus, cuts in emissions can be made where they are most cost-effective, with total emissions not exceeding the cap. Currently, the largest emissions trading systems are carbon trading schemes, which are being implemented by the European Union (EU) and globally under the rules of the Kyoto Protocol.

The carbon trading system established by the Kyoto Protocol, like many other carbon markets, is actually a blend of two kinds of carbon trading. The first is basic emissions trading, the cap-and-trade mechanism described earlier; the second is trade in project-based credits. The system for acquiring project-based credits under Kyoto is called the Clean Development Mechanism. Polluters trade credits directly with each other. Thus, under the Kyoto agreement designed to expire in 2010, industrialized countries can meet their carbon-reduction goals either by reducing their own emissions, by trading emissions, or by acquiring carbon credits earned by funding reductions elsewhere.

As of 2010, the largest emissions market is the European Union Trading Scheme, which began operating in 2005 and includes the twenty-seven states of the EU. This scheme covers emissions from power plants as well as other energy-intensive industries totaling about 46 percent of EU CO_2 emissions.

The United States has several active emissions trading markets (e.g., the U.S. Clean Air Act Amendments of 1990 set up a national cap-and-trade program for pollution rights for sulfur dioxide) and there are proposals to adopt new national scale emissions trading programs. As of 2010, the Chicago Climate Exchange (CCX) operates North America's only cap and trade system for trading emissions related to six major greenhouse gases.

Resources

BOOKS

Lohmann, Larry, et al. *Carbon Trading—A Critical Conversation on Climate Change, Privatisation and Power*. Uppsala, Sweden: Dag Hammarskjöld Foundation, 2006.

Murray, Barrie. *Power Markets and Economics: Energy Costs, Trading, Emissions*. Chichester, U.K.: Wiley, 2009.

Tanenbaum, William A., and Mitchell Zuklie. *Green Technology Law and Business, 2009: Strategies for Finance, Carbon Trading, IT, and Carbon Neutral Policies*. Corporate law and practice course handbook series, no. B–1718. New York, N.Y.: Practicing Law Institute, 2009.

Tietenberg, Thomas H. *Emissions Trading: Principles and Practice*. Washington, DC: Resources for the Future, 2006.

PERIODICALS

Betz, Regina, and Misato Sato. "Emissions Trading: Lessons Learnt from the 1st Phase of the EU ETS and Prospects for the 2nd Phase." *Climate Policy* 6 (2006): 351–359.

Chameides, William, and Michael Oppenheimer. "Carbon Trading Over Taxes." *Science* 315 (2007): 1670.

Hepburn, Cameron. "Carbon Trading: A Review of the Kyoto Mechanisms." *Annual Review of Environment and Resources* 32 (2007): 375–393.

Jackson, Robert B., et al. "Trading Water for Carbon with Biological Carbon Sequestration." *Science* 310 (2005): 1944–1947.

Kanter, James. "Carbon Trading: Where Greed Is Green." *International Herald Tribune* June 20, 2007.

OTHER

Asia-Pacific Emissions Trading Forum. "Emissions Trading." http://www.aetf.emcc.net.au/HTML/whatis.html (accessed October 21, 2010).

International Emissions Trading Association. "Kyoto Protocol." http://www.ieta.org/ieta/www/pages/index.php?IdSitePage = 24 (accessed October 21, 2010).

United States Environmental Protection Agency (EPA). "Air: Air Quality: Emissions Trading." http://www.epa.gov/ebtpages/airairqualityemissionstrading.html (accessed October 21, 2010).

ORGANIZATIONS

U.S. Environmental Protection Agency, 1200 Pennsylvania Avenue, N.W., Washington, D.C., United States, 20460, (202) 260–2090, http://www.epa.gov

Jane E. Spear

Carbon monoxide

A colorless, odorless, tasteless gas that is produced in only very small amounts by natural processes. By far the most important source of the gas is the incomplete combustion of coal, oil, and natural gas. In terms of volume, carbon monoxide is the most important single component of air pollution. Environmental scientists rank it behind sulfur oxides, particulate matter, nitrogen oxides, and volatile organic compounds, however, in terms of its relative hazard to human health. In low doses, carbon monoxide causes headaches, nausea, fatigue, and impairment of judgment. In larger amounts, it causes unconsciousness and death.

Carbon offsets (CO_2-emission offsets)

Many human activities result in large emissions of carbon dioxide (CO_2) and other so-called greenhouse gases into the atmosphere. Especially important in this regard are the use of fossil fuels such as coal, oil, and natural gas to generate electricity, to heat spaces, and as a fuel for vehicles. In addition, the disturbance of forests results in large emissions of CO_2 into the atmosphere. This can be caused by burning and converting forests into agricultural or urbanized land uses, and also by the harvesting of timber.

During the past several centuries, human activities have resulted in large emissions of CO_2 into the atmosphere, causing substantial increases in the concentrations of that gas. Prior to 1850, the atmospheric concentration of CO_2 was about 280 parts per million (ppm), while in 1997 it was about 360 ppm. Other greenhouse gases (these are also known as radiatively active gases) have also increased in concentration during that same period: methane (CH_4) from about 0.7 ppm to 1.7 ppm, nitrous oxide (N_2O) from 0.285 ppm to 0.304 ppm, and chlorofluorocarbons (CFCs) from zero to 0.7 parts per billion (ppb). As of September 2010, the CO$_2$ concentration has grown to 386.80 ppm, according to the Earth Systems Research Laboratory (ESRL) at the U.S. National Oceanic and Atmospheric Administration (NOAA).

The scientific consensus among climatologists and environmental scientists is that the increased concentrations of radiatively active gases are causing an increase in the intensity of Earth's greenhouse effect. The resulting climatic warming could result in important stresses for both natural ecosystems and those that humans depend on for food and other purposes (agriculture, forestry, and fisheries). Overall, CO$_2$ is estimated to account for about 60 percent of the potential enhancement of the greenhouse effect, while CH$_4$ accounts for 15 percent, CFCs for 12 percent, ozone (O$_3$) for 8 percent, and N$_2$O for 5 percent.

Because an intensification of Earth's greenhouse effect is considered to represent a potentially important environmental problem, planning and other actions are being undertaken to reduce the emissions of radiatively active gases. The most important strategy for reducing CO$_2$ emissions is to lessen the use of fossil fuels. This will mostly be accomplished by reducing energy needs through a variety of conservation measures, and by switching to non-fossil fuel sources. Another important means of decreasing CO$_2$ emissions is to prevent or slow the rates of deforestation, particularly in tropical countries. This strategy would help to maintain organic carbon within ecosystems, by avoiding the emissions of CO$_2$ through burning and decomposition that occur when forests are disturbed or converted into other land uses.

Unfortunately, fossil-fuel use and deforestation are economically important activities in the modern world. This circumstance makes it extremely difficult for society to achieve rapid, large reductions in the emissions of CO$_2$. An additional tactic that can contribute to net reductions of CO$_2$ emissions involves so-called CO$_2$ offsets. This involves the management of ecosystems to increase both the rate at which they are fixing CO$_2$ into plant biomass and the total quantity stored as organic carbon. This biological fixation can offset some of the emissions of CO$_2$ and other greenhouse gases through other activities.

Offsetting CO$_2$ emissions by planting trees

As plants grow, their rate of uptake of atmospheric CO$_2$ through photosynthesis exceeds their release of that gas by respiration. The net effect of these two physiological processes is a reduction of CO$_2$ in the atmosphere. The biological fixation of atmospheric CO$_2$ by growing plants can be considered to offset emissions of CO$_2$ occurring elsewhere—for example, as a result of deforestation or the combustion of fossil fuels.

The best way to offset CO$_2$ emissions in this way is to manage ecosystems to increase the biomass of trees in places where their density and productivity are suboptimal. The carbon-storage benefits would be especially great if a forest is established onto poorer-quality farmlands that are no longer profitable to manage for agriculture (this process is known as afforestation, or conversion into a forest). However, substantial increases in carbon storage can also be obtained whenever the abundance and productivity of trees is increased in "low-carbon ecosystems," including urban and residential areas.

Over the longer term, it is much better to increase the amounts of organic carbon that are stored in terrestrial ecosystems, especially in forests, than to just enhance the rate of CO$_2$ fixation by plants. The distinction between the amount stored and the rate of fixation is important. Fertile ecosystems, such as marshes and most agroecosystems, achieve high rates of net productivity, but they usually store little biomass and therefore over the longer term cannot sequester much atmospheric CO$_2$. A possibly better example involves second-growth forests and plantations, which have higher rates of net productivity than do older-growth forests. Averaged over the entire cycle of harvest and regeneration, however, these more-productive forests store smaller quantities of organic carbon than do older-growth forests, particularly in trees and large-dimension woody debris.

Both the greenhouse effect and emissions of CO$_2$ (and other radiatively active gases) are global in their scale and effects. For this reason, projects to gain CO$_2$ offsets can potentially be undertaken anywhere on the planet, but tallied as carbon credits for specific utilities or industrial sectors elsewhere. For example, a fossil-fueled electrical utility in the United States might choose to develop an afforestation offset in a less-developed, tropical country. This might allow the utility to realize significant economic advantages, mostly because the costs of labor and land would be less and the trees would grow quickly due to a relatively benign climate and long growing season. This strategy is known as a joint-implementation project, and such projects are already underway. These involve United States or European electrical utilities supporting afforestation in tropical countries as a means of gaining carbon credits, along with other environmental benefits associated with planting trees. Although this is a valid approach to obtaining carbon offsets, it can be controversial because some people would prefer to see industries develop forest-carbon offsets within the same country where the CO_2 is being emitted.

Afforestation in rural areas

An estimated 5–8 billion acres (2–3 billion ha) of deforested and degraded agricultural lands may be available worldwide to be afforested. This change in land use would allow enormous quantities of organic carbon to be stored, while also achieving other environmental and economic benefits. In North America, millions of acres of former agricultural land have reverted to forest since about the 1930s, particularly in the eastern states and provinces. There are still extensive areas of economically marginal agricultural lands that could be afforested in parts of North America where the climate and soils are suitable for supporting forests. Between 2000 and 2005, afforestation efforts in Asia resulted in an increase of 2,470,000 acres (1,000,000 hectares) of forested area. Now, in 2011, the country of Bangladesh has become an role model for afforestation in Asia since its launch in the late 1990s.

Agricultural lands typically maintain about one-tenth or less of the plant biomass of forests, while agricultural soils typically contain 60–80 percent as much organic carbon as forest soils. Because agricultural sites contain a relatively small amount of organic carbon, reforestation of those lands has a great potential for providing CO_2 offsets.

It is also possible to increase the amounts of carbon stored in existing forests. This can be done by allowing forests to develop into an old-growth condition, in which carbon storage is relatively great because the trees are typically big, and there are large amounts of dead biomass present in the surface litter, dead standing trees, and dead logs lying on the forest floor. Once the old-growth condition is reached, however, the ecosystem has little capability for accumulating "additional" carbon. Nevertheless, old-growth forests provide an important ecological service by tying up so much carbon in their living and dead biomass. In this sense, maintaining old-growth forests represents a strategy of CO_2-emissions deferral, because if those "high-carbon" ecosystems were disturbed by timber harvesting or conversion into another kind of land use, a result would be an enormous emission of CO_2 into the atmosphere.

Carbon-offset projects were initiated in the early 1990s in various parts of the world, most of which involved afforestation of rural lands. The typical costs of the afforestation projects, at that time, were $1–10 per ton of carbon fixed. These are only the costs associated with planting and initial tending of the growing trees; there might also be additional expenses for land acquisition and stand management and protection.

Of course, even while rural afforestation provides large CO_2-emission offsets, other important benefits are also provided. In some cases, the forests might be used to provide economic benefits through the harvesting of timber (although the resulting disturbance would lessen the carbon-storage capability). Even if trees are not harvested from the CO_2-offset forests, it would be possible to hunt animals such as deer, and to engage in other sorts of economically valuable outdoor recreation. Increasing the area of forests also provides many non-economic benefits, such as providing additional habitat for native species, and enhancing ecological services related to clean water and air, erosion control, and climate moderation.

Urban forests

Urban forests consist of trees growing in the vicinity of homes and other buildings, in areas where the dominant character of land use is urban or suburban. Urban forests may originate as trees that are spared when a forested area is developed for residential land use, or they may develop from saplings that are planted after homes are constructed. Urban forests in older residential neighborhoods generally have a relatively high density and extensive canopy cover of trees. These characters are less well developed in younger neighborhoods and where land use involves larger buildings used by institutions, business, or industry.

There are about 70 million acres (28 million ha) of urban land in the United States. The United States' urban forests support an average density of 20 trees/acre (52 trees/ha), and have a canopy cover of 28 percent. In 1994, urban areas of the United States were estimated to contain about 225 million tree-planting opportunities, in which suboptimal tree densities could be subjected to fill-planting. At that time, about 75 percent of all people in the United States lived in urban settings. That percent is predicted to go up to 85 percent by the year 2025, according to the U.S. Forest Service. Additional people in these urban areas mean more concentrations of CO_2, which can be offset by the increased use of urban forests.

Urban forests achieve carbon offsets in two major ways. First, as urban trees grow they sequester atmospheric CO_2 into their increasing biomass. The average carbon storage in urban trees in the United States is about 13 tons per acre (33 tonnes per ha). On a national basis that amounts to approximately 0.9 billion tons of organic carbon, and an annual rate of uptake of approximately six million tons.

In addition, urban trees can offset some of seasonal use of energy for cooling and heating the interior spaces of buildings. Large, well-positioned trees provide a substantial cooling influence through shading. Trees also cool the ambient air by evaporating water from their foliage (a process known as transpiration). Trees also decrease wind speeds near buildings. This results in decreased heating needs during winter, because less indoor warmth is lost by the infiltration of outdoor air into buildings. Over most of North America larger energy offsets associated with urban trees are due to decreased costs of cooling than with decreased heating costs. In both cases, however, much of the energy conserved represents decreased CO_2 emissions through the combustion of fossil fuels.

It is considerably more expensive to obtain CO_2-offset credits using urban trees than with rural trees. This difference is mostly due to urban trees being much larger than rural trees when planted, while also having larger maintenance expenses. The typical costs of rural CO_2-offset projects are much less when compared with urban trees.

Studies have shown that there is plenty of carbon savings associated with planting large number of trees in urban areas within the United States. Much of the total CO_2 offsets were associated with indirect savings of energy for cooling and heating buildings, with much less from carbon sequestration into the growing biomass of the trees. The estimated costs of the carbon offsets were found to decrease considerably as the trees grew larger.

Additional Information

The largest driver of global climate change is carbon dioxide (CO_2) in the atmosphere, which is directly increased by the burning of fossil fuels. In October 2007, scientists announced that since 2000, CO_2 levels had increased faster than even the most pessimistic forecasts of the late 1990s had predicted. Growth in atmospheric CO_2 was only 1.1 percent per year for 1990–1999, but accelerated sharply to more than 3 percent per year for 2000–2004. Climate scientists attribute most of the increased emissions to increases in human population and industrial activity. The amount of CO_2 emitted per unit of economic activity (defined as energy intensity) remained level or increased globally. The increase in CO_2 was due also—to a lesser extent—to decreased absorption of CO_2 by the oceans. In 2004, developing nations accounted for 41 percent of total emissions, but 74 percent of total emissions growth.

Resources

BOOKS

Bishop, Amanda. *How to Reduce Your Carbon Footprint.* St. Catharines, ON: Crabtree, 2008.
Wigley, T. M. L., and D. S. Schimel. *The Carbon Cycle.* Cambridge: Cambridge University Press, 2005.

PERIODICALS

Bin, Shui, and Hadi Dowlatabadi. "Consumer Lifestyle Approach to U.S. Energy Use and the Related CO_2 Emissions." *Energy Policy* 33 (2005): 197–208.
Higgins, Paul A. T., and Millicent Higgins. "A Healthy Reduction in Oil Consumption and Carbon Emissions." *Energy Policy* 33 (2005): 1–4.
Hope, Chris W. "The Social Cost of Carbon: What Does It Actually Depend On?" *Climate Policy* 6 (2006): 566–572.
Revkin, Andrew C. "Climate Panel Reaches Consensus on the Need to Reduce Harmful Emissions." *The New York Times* (May 4, 2007).
Stauffer, Hoff. "New Sources Will Drive Global Emissions." *Energy Policy* 35 (2007): 5433–5435.

OTHER

CO2Now.org, Atmosphere Monthly. "What the World Needs to Watch." http://co2now.org/ (accessed October 25, 2010).
Financial Express. "Bangladesh's Social Afforestation Example for Asian Countries." http://www.thefinancialexpress-bd.com/more.php?news_id = 106475 &date = 2010-07-18 (accessed October 25, 2010).

Bill Freedman

Carbon tax

To limit and control the amount of carbon dioxide (CO_2, sometimes also written as CO2) added to the atmosphere, special (environmental) taxes, called carbon taxes, have been proposed and in some cases adopted, on fuels containing carbon. Fuels such as coal, gasoline, heating oil, and natural gas, release energy by combining the carbon they contain with oxygen in the air, to produce carbon dioxide. Increased use of carbon-containing fossil fuels in modern times has greatly increased the rate at which carbon dioxide is entering the atmosphere. Measurable increases in atmospheric levels of the gas have been detected. Since carbon dioxide is the principle component of so-called greenhouse gases, changes in its concentration in Earth's atmosphere are a concern. Increases in the level of carbon dioxide, and other greenhouse gases such as methane, nitrous oxide, ozone, and artifically-made chlorofluorocarbons can be expected to result in warmer temperatures on the surface of Earth, and wide-ranging changes in the global climate. Greenhouse gases permit radiation from the Sun to reach Earth's surface but prevent the infrared or heat component of sunlight from re-irradiating into space.

The wide spread concern over the possible climatic effects of increases in heat-trapping gases in the atmosphere was exemplified by a 2007 report of the Intergovernmental Panel on Climate Change (IPCC). It concluded that unless greenhouse gas emissions were curtailed, average global temperatures would rise 2.0–11.5°F (1.1–6.4°C) by the year 2100. The IPCC report recommended a 50 to 85 percent reduction of greenhouse gas emissions over 2000 levels by 2050. The IPCC recognized that the rapidly expanding economies of many developing countries make similar reductions less likely. The IPCC working group on climate change mitigation recommended a carbon tax to stabilize CO_2 emissions. The working group estimated that the carbon tax would have to be between $5 and $65 per ton of CO_2 in 2030 and between $15 and $130 per ton of CO_2 by 2050.

While limiting the buildup of heat-trapping gases in the atmosphere has been a goal of natural scientists for some time, many economists have now joined the effort. In 1997, 2,000 prominent economists, including six Nobel Laureates, signed the "Economists' Statement on Climate Change" stating that policies to slow global warming are needed and are economically viable. The economists agreed with a review conducted by a distinguished international panel of scientists under the auspices of the IPCC that, "The balance of evidence suggests a discernible human influence on global climate." They further stated that, "As economists, we believe that global climate change carries with it significant environmental, economic, social, and geopolitical risks, and that preventive steps are justified."

They state that economic studies have found that potential policies to reduce greenhouse-gas emissions can be designed so that benefits outweigh costs, and living standards would not be harmed. They claim that United States productivity may actually improve as a result. The statement goes on to claim that, "the most efficient approach to slowing climate change is through market-based policies" rather than limits or regulations. The economists also suggest that, "A cooperative approach among nations is required, such as an international emissions trading agreement." They recommend, "market mechanisms such as carbon taxes or the auction of emissions permits." Their statement goes on to suggest that, "Revenues generated from such policies can effectively be used to reduce budget deficits or lower existing taxes." New taxes are never popular, but are more apt to be accepted when the revenue is used to replace other taxes or to aid the environment.

As in any public policy debate, views of even well-qualified experts are often sharply divided. Some industry representatives and scientists are unwilling to accept the premise that greenhouse gas emissions represent a serious threat. Industrial spokespersons continue to claim that efforts to reduce greenhouse gas emissions through taxation or other economic means are not cost effective, and would devastate the economy, cost jobs, and reduce living standards. Individual countries worry about the impacts on their competitiveness if other countries do not adopt similar measures. The potential impact on low-income groups must also be considered. Some are concerned that the taxes have to be high to be effective. The amount of year to year variation in climate and temperature throughout the globe, provide ample room for differences of opinion.

International attention was focused on the need to reduce greenhouse gases at the United Nations sponsored Earth Summit held in Rio de Janeiro (Brazil) in 1992. The conference was held to attempt to reconcile worldwide economic development with the need to preserve the environment. Representatives of 178 nations attended, including 117 heads of state, making it the largest gathering of world leaders in history. Documents and treaties were signed committing most of the world's nations to the economic development in ways that were compatible with a healthy environment. A binding treaty called the United Nations Framework Convention on Climate Change (UNFCCC) was adopted, requiring nations to reduce emissions of greenhouse gases. The treaty failed to set binding targets for emission reductions, however. Agreement was hampered by discord between industrialized nations of Western Europe, and North America and developing nations in Africa, Latin America, the Middle East, and Asia. Developing countries were concerned that environmental restrictions would hamper economic growth unless they received increased aid from developed nations to enable them to grow in an environmentally sound way.

In 1997, the Kyoto Protocol was adopted as an amendment to the UNFCCC. The Kyoto Protocol sought to reduce greenhouse gas emissions through flexible mechanisms, such as emissions trading and investment in clean energy initiatives in developing nations. The Kyoto Protocol, which entered into force in 2005, did not adopt a carbon tax. As of November

shown to be carcinogenic in mice. DES is now linked to vaginal and cervical cancers in women born between 1950 and 1966 whose mothers took DES during their pregnancies.

In 1971, DES was banned for use in cattle by the Food and Drug Administration (FDA), but the federal courts reversed the ban, contending that DES posed no danger since it was not directly added to foods but was administered only to cattle. When the FDA subsequently showed that measurable quantities remained present in slaughtered cattle, the courts reinstated the ban. But the issue of using growth additives in meat production remains unresolved today. Environmentalists are still concerned that known carcinogenic chemicals used to "beef up" cattle are being consumed by humans in various meat products, though no direct links have yet been established. In addition, various food additives, such as coal tar dyes used for artificial coloring and food preservatives, have produced cancer in laboratory animals. As yet there is no evidence indicating that human cancer rates are rising because of these substances in food.

Air pollution has been extensively investigated as a possible carcinogen and it is known that people living in cities larger than 50,000 run a 33 percent higher risk of developing lung cancer than people who live in other areas. The reasons behind this phenomenon, referred to as the "urban factor," have never been conclusively determined. Areas with populations exceeding 50,000 tend to have more industry, and air pollutants can have a profound effect in regions such as New Jersey where they are highly concentrated.

Occupational exposure to carcinogenic substances accounts for an estimated 2–8 percent of diagnosed cancers in the United States. Until passage of the Toxic Substances Control Act in 1976, which gave the federal government the power to require testing of potentially hazardous substances before they go on the market, hundreds of new chemicals with unknown side effects came into industrial use each year. Substances such as asbestos are estimated to cause 30–40 percent of all deaths among workers who have been exposed to it. Vinyl chloride, a basic ingredient in the production of plastics, was found in 1974 to induce a rare form of liver cancer among exposed workers. Anaesthetic gases used in operating rooms have been traced as the reason nurse anesthetists develop leukemia and lymphoma at three times the normal rate with an associated higher rate of miscarriage and birth defects among their children. Benzene, an industrial chemical long known as a bone-marrow poison, has been shown to induce leukemia as well. A major step forward in the regulation of these potential cancer causing agents is the implementation by the Occupational Safety and Health Administrations (OSHA) of the Hazard Communication Standard in 1983, intended to provide employees in manufacturing industries access to information concerning hazardous chemicals encountered in the workplace.

In 2008, researchers reported that nanotubes—microscopic, needle-like tubes, usually composed of carbon and used in many new materials because of their strength and conductivity—can have carcinogenic affects on lung tissue similar to those of asbestos. The researchers, whose results were published in the journal *Nature Nanotechnology*, emphasized that nanotubes are not dangerous if sealed inside plastics or other materials; like asbestos fibers, they must be inhaled before they can do harm. Nanotubes are therefore mostly a threat to those handling them in manufacturing, not to the general public. Being forewarned of likely health hazards from nanotubes, manufacturers can take steps to protect workers. Nanoparticles (particles smaller than 100 nanometers across) have also been found to be a cancer hazard, causing DNA damage similar to that caused by ionizing radiation.

There is continuing concern about over-exposure to ultraviolet radiation and the subsequent formation of skin cancers such as basal-cell and squamous-cell skin cancers, along with the highly lethal melanoma skin cancer which currently kills about 9,000 Americans per year.

See also Love Canal; Ozone layer depletion; Radiation exposure; Radiation sickness; Radon; Toxic substance.

Resources

Bignold, Leon P. *Cancer: Cell Structures, Carcinogens and Genomic Instability*. New York: Birkhauser, 2005.
IARC Working Group on the Evaluation of Carcinogenic Risks to Humans. *Combined Estrogen-Progestogen Contraceptives and Combined Estrogen-Progestogen Menopausal Therapy*. Lyon, France: International Agency for Research on Cancer, 2007.

PERIODICALS

Meister, Kathleen *America's War on "Carcinogens": Reassessing the Use of Animal Tests to Predict Human Cancer Risk*. Washington: American Council on Science, 2005.

OTHER

United States Environmental Protection Agency (EPA). "Pollutants/Toxics: Carcinogens." http://www.epa.gov/ebtpages/pollcarcinogens.html (accessed November 9, 2010).

Brian R. Barthel

Carrying capacity

Carrying capacity is a general concept based on the idea that every ecosystem has a limit for use that cannot be exceeded without damaging the system. Whatever the specified use of an area might be, whether for grazing, wildlife habitat, recreation, or economic development, there is a threshold that cannot be breached, except temporarily, without degrading the ability of the environment to support that use. Examinations of carrying capacity attempt to determine, with varying degrees of accuracy, where this threshold lies and what the consequences of exceeding it might be.

The concept of carrying capacity was pioneered early this century in studies of range management and wildlife management. Range surveys of what was then called "grazing capacity" were carried out on the Kaibab Plateau in Arizona as early as 1911, and this term was used in most of the early bulletins issued by the U.S. Department of Agriculture on the subject. In his 1923 classic, *Range and Pasture Management*, Sampson defined grazing capacity as "the number of stock of one or more classes which the area will support in good condition during the time that the forage is palatable and accessible, without decreasing the forage production in subsequent seasons." Sampson was quick to point out that the "grazing capacity equation has not been worked out on any range unit with mathematical precision." In fact, because of the number of variables involved, especially variables stemming from human actions, he did not accept that the "grazing-capacity factor will ever be worked out to a high degree of scientific accuracy." Sampson also pointed out that "grazing the pasture to its very maximum year after year can produce only one result—a sharp decline in its carrying capacity," and he criticized the stocking of lands at their maximum instead of their optimum capacity. Similar discussions of carrying capacity can be found in books about wildlife management from the same period, particularly *Game Management* by Aldo Leopold, published in 1933.

Practitioners of applied ecology have calculated the number of animal-unit months that any given land area can carry over any given period of time. But there have been some controversies over the application of the concept of carrying capacity. The concept is commonly employed without considering the factor of time, neglecting the fact that carrying capacity refers to land use that is sustainable. Another common mistake is to confuse or ignore the implicit distinctions between maximum, minimum, and optimum capacity.

In discussions of land use and environmental impact, some officials have drawn graphs with curves showing maximum use of an area and claimed that these figures represent carrying capacity. Such representations are misleading because they assume a perfectly controlled population, one without fluctuation, which is not likely. In addition, the maximum allowable population can almost never be the carrying capacity of an area, because such a number can almost never be sustained under all possible conditions. A population in balance with the environment will usually fluctuate around a mean, higher or lower, depending on seasonal habitat conditions, including factors critical to the support of that particular species or community.

The concept of carrying capacity has important ramifications for human ecology and population growth. Many of the essential systems on which humans depend for sustenance are showing signs of stress, yet demands on these systems are constantly increasing. William R. Catton has formulated an important axiom for carrying capacity: "For any use of any environment there is a use intensity that cannot be exceeded without reducing that environment's suitability for that use." He then defined carrying capacity for humans on the basis of this axiom: "The maximum human population equipped with a given assortment of technologies and a given pattern of organization that a particular environment can support indefinitely."

The concept of carrying capacity is the foundation for recent interest in sustainable development, an environmental approach which identifies thresholds for economic growth and increases in human population. Sustainable development calculates the carrying capacity of the environment based on the size of the population, the standard of living desired, the overall quality of life, the quantity and type of artifacts created, and the demand on energy and other resources. With his calculations on sustainable development in Paraguay, Herman Daly has illustrated that it is possible to work out rough estimates of carrying capacity for some human populations in certain areas. He based his figures on the ecological differences between the country's two major regions, as well as on differences among types of settlers, and differences between developed good land and undeveloped marginal lands.

If ecological as well as economic and social factors are taken into consideration, then any given environment has an identifiable tolerance for human use and development, even if that number is not now known. For this reason, many environmentalists argue that carrying capacity should always be the basis for what has been called demographic accounting.

Resources

BOOKS

Allan, Catherine, and George Henry Stankey. *Adaptive Environmental Management: A Practitioner's Guide.* New York: Springer, 2009

Cunningham, W. P., and A. Cunningham. *Environmental Science: A Global Concern.* New York: McGraw-Hill International Edition, 2008.

Enger, Eldon, and Bradley Smith. *Environmental Science: A Study of Interrelationships.* New York: McGraw-Hill, 2006.

Molles, Manuel C. *Ecology: Concepts and Applications.* New York: McGraw Hill Science/Engineering/Math, 2009.

Wright, Richard T. *Environmental Science: Toward a Sustainable Future.* Upper Saddle River, N.J.: Pearson, 2008.

Gerald L. Young
Douglas Smith

Carson, Rachel L.

1907–1964
American ecologist, marine biologist, and writer

Rachel Louise Carson was a university-trained biologist, a longtime United States government employee, and a best-selling author of such books as *Edge of the*

Rachel Carson. *(Corbis-Bettmann.)*

Sea, The Sea Around Us (a National Book Award winner), and *Silent Spring.*

Her book on the dangers of misusing pesticides, *Silent Spring*, has become a classic of environmental literature and resulted in her recognition as the fountainhead of modern environmentalism. *Silent Spring* was reissued in a twenty-fifth anniversary edition in 1987, and remains standard reading for anyone concerned about environmental issues.

Carson grew up in the Pennsylvania countryside and reportedly developed an early interest in nature from her mother and from exploring the woods and fields around her home. She was first an English major in college, but a required course in biology rekindled that early interest in the subject and she graduated in 1928 from Pennsylvania College for Women with a degree in zoology and went on to earn a master's degree at Johns Hopkins University. After the publication of *Silent Spring*, she was often criticized for being a "popular science writer" rather than a trained biologist, making it obvious that her critics were unaware of her university work, including a master's thesis entitled "The Development of the Pronephros During the Embryonic and Early Larval Life of the Catfish (*Ictalurus punctatus*)."

Summer work also included biological studies at Woods Hole Marine Biological Laboratory in Massachusetts, where she became more interested in the life of the sea. After doing a stint as a part-time scriptwriter for the Bureau of Fisheries, she was hired full-time as a junior aquatic biologist. When she resigned from the United States Fish and Wildlife Service in 1952 to devote her time to her writing, she was biologist and chief editor there. First, as a biologist and writer with the Bureau and then as a free-lance writer and biologist, she successfully combined professionally the two great loves of her life, biology and writing.

Often described as "a book about death which exalts life," *Silent Spring* is the work on which Carson's position as the modern catalyst of a renewed environmental movement rests. The book begins with a shocking fable of one composite town's "silent spring" after pesticides have decimated insects and the birds that feed upon them. The main part of the book is a massive documentation of the effects of organic pesticides on all kinds of life, including birds and humans. The final sections are quite restrained, drawing a hopeful picture of the future, if feasible alternatives to the use of pesticides—such as biological controls—are used in conjunction with and as a partial replacement of chemical sprays.

Carson was quite conservative throughout the book, being careful to limit examples to those that could be verified and defended. In fact, there was very little new in the book; it was all available earlier in a variety of scientific publications. But her science background allowed her to judge the credibility of the facts she uncovered and provided sufficient knowledge to synthesize a large amount of data. Her literary skills made that data accessible to the general public.

Silent Spring was not a polemic against all use of pesticides but a reasoned argument that potential hazards be carefully and sufficiently considered before any such chemical was approved for use. Many people date modern concern with environmental issues from her argument in this book that "future generations are unlikely to condone our lack of prudent concern for the integrity of the natural world that supports all life." It is not an accident that her book is dedicated to Albert Schweitzer, because she wrote it from a shared philosophy of reverence for life.

Carson provided an early outline of the potential of using biological controls in place of chemicals, or in concert with smaller doses of chemicals, an approach now called integrated pest management. She worried that too many specialists were concerned only about the effectiveness of chemicals in destroying pests and "the overall picture" was being lost, in fact not valued or even sought. She pointed out the false safety of assuming that products considered individually were safe, when in concert, or synergistically, they could lead to human health problems.

Her holistic approach was one of the real, and unusual, strengths of the book. Prior to the publication of *Silent Spring*, she even refused to appear on a National Audubon Society panel on pesticides because such an appearance could provide a forum for only part of the picture and she wanted her material to first appear "as a whole." She did allow it to be partially serialized in *The New Yorker*, but articles in that magazine are long and detailed.

The book was criticized early and often, and often viciously and unfairly. One chemical company, reacting to that pre-publication serialization, tried to get Houghton Mifflin not to publish the book, citing Carson as one of the "sinister influences" trying to reduce the use of agricultural chemicals so that United States food supplies would dwindle to the level of a developing nation. The chemical industry apparently united against Carson, distributing critical reviews and threatening to withdraw magazine advertisements from journals deemed friendly to her. Words and phrases used in the attacks included "ignorant," "biased," "sensational," "unfounded," "distorted," "not written by a scientist," "littered with crass assumptions and gross misinterpretations," to name but a few.

Some balanced reviews were also published, most noteworthy one by Cornell University ecologist LaMont Cole in *Scientific American*. Cole identified errors in her book, but finished by saying "errors of fact are so infrequent, trivial and irrelevant to the main theme that it would be ungallant to dwell on them," and went on to suggest that the book be widely read in the hopes that it "may help us toward a much needed reappraisal of current policies and practices." That was the spirit in which Carson wrote *Silent Spring* and reappraisals and new policies were indeed the result of the myriad of reassessments and studies spawned by its publication. To its credit, it did not take the science community long to recognize her credibility; the President's Science Advisory Committee issued a 1963 report that the journal *Science* suggested "adds up to a fairly thorough-going vindication of Rachel Carson's *Silent Spring* thesis."

While it is important to recognize the importance of *Silent Spring* as a landmark in the environmental movement, one should not neglect the significance of her other work, especially her three books on oceans and marine life and the impact of her writing on people's awareness of one of Earth's great natural ecosystems.

Under the Sea Wind (1941) was Carson's attempt "to make the sea and its life as vivid a reality [for her readers] as it has become for me." And readers are given vivid narratives about the shore, including vegetation and birds, on the open sea, especially by tracing the movements of the mackerel, and on the sea bottom, again by focusing on an example, this time the eel. *The Sea Around Us* (1951) continues Carson's treatment of marine biology, adding an account of the history and development of the sea and its physical features such as islands and tides. She also includes human perceptions of and relationships with the sea. *The Edge of the Sea* (1955) was written as a popular guide to beaches and sea shores, but focusing on rocky shores, sand beaches, and coral and mangrove coasts, it complemented the physical descriptions in *The Sea Around Us* with biological data.

Carson was a careful and thorough scientist, an inspiring author, and a pioneering environmentalist. Her groundbreaking book, and the controversy it generated, was the catalyst for much more serious and detailed looks at environmental issues, including increased governmental investigation that led to creation of the Environmental Protection Agency (EPA). Her work will remain a hallmark in the increasing awareness modern people are gaining of how humans interact with and impact the environment in which they live and on which they depend.

Resources

BOOKS

Carson, Rachel. *Silent Spring*. Boston: Houghton Mifflin, 1962.

Gerald R. Young

Cash crop

A crop that is produced for the purpose of exporting or selling rather than for consumption by the grower. In many Third World countries, cash crops often replace the production of basic food staples such as rice, wheat, or corn in order to generate foreign exchange. For example, in Guatemala, much of the land is devoted to the production of bananas and citrus fruits (97 percent of the citrus crop is exported), which means that majority of the basic food products needed by the native people are imported from other countries. Often these foods are expensive and difficult for many poor people to obtain. Cash crop agriculture also forces many subsistence and tenant farmers to give up their land in order to make room for industrialized farming.

Catalytic converter

Catalytic converters are devices which employ a catalyst to facilitate a chemical reaction. A catalyst is a substance that changes the rate of a chemical reaction, but whose own composition is unchanged by that reaction. For air pollution control purposes, such reactions involve the reduction of nitric oxide (NO) to molecular oxygen and nitrogen or oxidation of hydrocarbons and carbon monoxide (CO) to carbon dioxide (CO_2) and water. Using the catalyst, the activation energy of the desired chemical reaction is lowered. Therefore, exothermic chemical conversion will be favored at a lower temperature.

Traditional catalysts have normally been metallic, although nonmetallic materials, such as ceramics, have

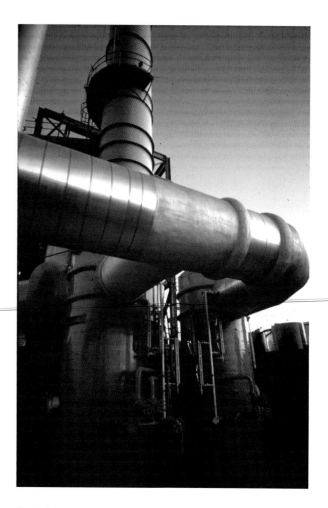

A catalytic converter at a large petrochemical plant. *(Photograph by Tom Carroll. Phototake.)*

been coming into use in recent years. Metals used as catalysts may include noble metals, such as platinum (Pt), or base metals, including nickel (Ni) and copper (Cu). Some catalysts are more effective in oxidation, others are more effective in reduction. Some metals are effective in both kinds of reactions. The catalyst material is normally coated on a porous, inert support structure of varying design. Examples include honeycomb ceramic structures with long channels and pellet beds. The goal is to channel exhaust over a large surface area of catalyst without an unacceptable pressure drop.

In some cases, reduction and oxidation catalysts are combined to control oxides of nitric oxide, carbon monoxide, and hydrocarbon emissions in exhaust from internal combustion engines. The reduction and oxidation processes can be conducted sequentially or simultaneously. Dual catalysts are used in sequential reduction-oxidation. In this case, the exhaust gas from a rich-burn engine initially enters the reducing catalyst to reduce nitric oxide. Subsequently, as the exhaust enters an oxidation catalyst, it is diluted with air to provide oxygen for oxidation. Alternatively, three-way catalysts can be used for simultaneous reduction and oxidation. Engines with three-way catalytic converters are efficient at reducing nitric oxide, carbon monoxide, and hydrocarbon emissions when there is a tight regulation of the air to fuel ratio. Higher levels of oxygen than required result in favored oxidation of carbon monoxide and hydrocarbons and decreased reduction of nitric oxide, while the converse occurs when high levels of fuel are present. Catalytic converters can present a problem in cold conditions or in stop-start driving as they need to be hot to work most effectively.

Reducing catalysts can be made more efficient using a reducing agent, such as ammonia (NH_3). This method of control, referred to as selective catalytic reduction, has been employed successfully on large turbines. In this case, a reducing agent is introduced upstream of a reducing catalyst, allowing for greater rates of nitric oxide reduction.

Regulations in the U.S. and some other countries such as Japan ensure that motor vehicles have catalytic converters installed for reduction of greenhouse gas emissions. The Environmental Protection Agency (EPA) required the installation of three-way catalytic converters on all new passenger cars and light-duty trucks as of 2004, according to an amendment of the Clean Air Act. Three-way catalytic converters are able to convert nitric oxide, hydrocarbons, and carbon monoxide into oxygen, carbon dioxide, and water. Although the emissions of some harmful gases are reduced by three-way catalytic converters, other gases that contribute to climate change are formed such as carbon dioxide and nitrous oxide (N_2). Research and engineering initiatives are being employed to address this issue.

Resources

BOOKS

Bauner, David, and Staffan Laestadius. *Towards a Sustainable Automotive Industry: Experiences from the Development of Emission Control Systems.* Stockholm: Royal Institute of Technology, KTH, Dept. of Industrial Economics and Management, Research Unit on Industrial Dynamics, 2007.

Erjavec, Jack. *Automotive Technology: A Systems Approach.* Clifton Park, NY: Thomson/Delmar Learning, 2005.

National Research Council (U.S.). *Environment and Energy.* Washington, D.C.: Transportation Research Board, 2008.

OTHER

Manufacturers of Emission Controls Association (MECA). "The U.S. Environmental Protection Agency's Motor Vehicle Compliance Program." http://www.meca.org/page.ww?name=The+U.S.+Environmental+Protection+Agency%27s+Motor+Vehicle+Compliance+Program&section=Resources (accessed September 3, 2010).

United States Environmental Protection Agency (EPA). "Nitrous Oxide." http://www.epa.gov/nitrousoxide/sources.html (accessed September 3, 2010).

Robert B. Giorgis Jr.

Cation *see* **Ion.**

Catskill Watershed Protection Plan

New York City has long been proud of its excellent municipal drinking water. Approximately 90 percent of that water comes from the Catskill/Delaware Watershed, which covers about 1,900 square miles (nearly 5,000 square kilometers) of rugged, densely forested land north of the city and west of the Hudson River. Stored in six hard-rock reservoirs and transported through enormous underground tunnels, the city water is outstanding for a large urban area. Yielding 1.3 billion gal (450,000 cubic meters) per day, and serving more than nine million people, this is the largest surface water storage and supply complex in the world. As the metropolitan agglomeration has expended, however, people have moved into the area around the Catskill Forest Preserve, and water quality is not as high as it was a century ago.

When the 1986 U.S. Safe Drinking Water Act mandated filtration of all public surface water systems, the city was faced with building an $8 billion water treatment plant that would cost up to $500 million per year to operate. In 1989, however, the Environmental Protection Agency (EPA) ruled that the city could avoid filtration if it could meet certain minimum standards for microbial contaminants such as bacteria, viruses, and protozoan parasites. In an attempt to limit pathogens and nutrients contaminating surface water and to avoid the enormous cost of filtration, the city proposed land-use regulations for the five counties (Green, Ulster, Sullivan, Schoharie, and Delaware) in the Catskill/Delaware watershed from which it draws most of its water.

With a population of 50,000 people, the private land within the 200 square mi (520 square km) watershed is mostly devoted to forestry and small dairy farms, neither of which are highly profitable. Among the changes the city called for was elimination of storm water runoff from barnyards, feedlots, or grazing areas into watersheds. In addition, farmers would be required to reduce erosion and surface runoff from crop fields and logging operations. Property owners objected strenuously to what they regarded as onerous burdens that would put many of them out of business. They also bristled at having the huge megalopolis impose rules on them. It looked like a long and bitter battle would be fought through the courts and the state legislature.

To avoid confrontation, a joint urban/rural task force was set up to see if a compromise could be reached, and to propose alternative solutions to protect both the water supply and the long-term viability of agriculture in the region. The task force agreed that agriculture is the "preferred land use" on private land, and that agriculture has "significant present and future environmental benefits." In addition, the task force proposed a voluntary, locally developed and administered program of "whole farm planning and best management approaches" very similar to ecosystem-based, adaptive management.

This grass-roots program, financed mainly by the city, but administered by local farmers themselves, attempts to educate landowners, and provides alternative marketing opportunities that help protect the watershed. Economic incentives are offered to encourage farmers and foresters to protect the water supply. Collecting feedlot and barnyard runoff in infiltration ponds together with solid conservation practices such as terracing, contour plowing, strip farming, leaving crop residue on fields, ground cover on waterways, and cultivation of perennial crops such as orchards and sugarbush have significantly improved watershed water quality.

In addition to saving billions of dollars, this innovative program has helped create good will between the city and its neighbors. It has shown that upstream cleanup, prevention, and protection are cheaper and more effective than treating water after it's dirty. Farmers have learned they can be part of the solution, not just part of the problem. This experiment serves as an excellent example of how watershed planning through cooperation is effective when local people are given a voice and encouraged to participate. The success of the Catskill watershed management plans led the EPA to grant a five-year Filtration Avoidance Determination in 2002. The EPA granted New York City a ten-year Filtration Avoidance Determination in 2007.

Resources

OTHER

Catskill Watershed Corporation. "About the CWC." http://www.cwconline.org (accessed November 9, 2010).

ORGANIZATIONS

Catskill Watershed Corporation, PO Box 569 Main St., Margaretville, NY, USA, 12455, (845) 586-1400, (845) 586-1401, (877) 928-7433, invest@cwconline.org, http://www.cwconline.org.

William P. Cunningham

Center for Environmental Philosophy

The Center for Environmental Philosophy was established in 1989 as an organization dedicated to furthering research, publication, and education in the area of environmental philosophy and ethics. Based at the University of North Texas since 1990, the primary activities of the center are the publication of the journal *Environmental Ethics*, the reprinting of significant books on environmental ethics under its own imprint, the sponsorship of various workshops and conferences dedicated to the furthering of research and training in environmental ethics, and the promotion of graduate education, postdoctoral research, and professional development in the field of environmental ethics.

The Center is best known for its journal, which virtually established the field of environmental ethics and remains perhaps the leading forum for serious philosophical work in environmental philosophy. Inspired in part by Aldo Leopold's contention in "A Sand County Almanac" that the roots of most ecological problems were philosophical, Eugene C. Hargrove founded *Environmental Ethics* in 1978. The journal was originally concerned primarily with whether the attribution of rights to animals and to nature itself could be coherently defended as a philosophical doctrine. While remaining true to its central preoccupation with ethics, the journal's interests have more recently broadened to include significant essays on such topics as deep ecology, ecofeminism, social ecology, economics, and public policy. Under Hargrove's editorial leadership, the journal has brought environmental ethics to increasing acceptance as a serious field by mainstream academic philosophers. *Environmental Ethics* is widely read by researchers concerned with the environment in the fields of biological science, economics, and policy science. It also is developing a small but growing following among environmental professionals such as conservation biologists.

The Center also helps to sponsor, in conjunction with the Department of Philosophy and Religion Studies at the University of North Texas, programs in which graduate students may take courses and specialize in the field of environmental ethics.

Resources

BOOKS

Des Jardins, Joseph R. *Environmental Ethics: An Introduction to Environmental Philosophy*. Belmont, CA: Wadsworth Publishing, 2005.

Kuipers, Theo A. F. *General Philosophy of Science: Focal Issues*. Handbook of the philosophy of science. Amsterdam: Elsevier/North Holland, 2007.

ORGANIZATIONS

Center for Environmental Philosophy, University of North Texas, 1704 W. Mulberry St., Suite 370, Denton, TX, USA, 76201, (940) 565-2727, (940) 565-4439, cep@unt.edu, http://www.cep.unt.edu

Lawrence J. Biskowski

Center for Marine Conservation *see* **Ocean Conservatory, The.**

Center for Respect of Life and Environment

Formed in 1986, the Center for Respect of Life and Environment (CRLE) is a nonprofit group based in Washington, D.C., that works to promote humane and environmental ethics, particularly within the academic and religious communities with an emphasis on the links between ecology, spirituality, and sustainability.

CRLE describes itself as being committed to "encourage the well-being of life and living systems—plant, animal, and human relationships..." The work of the Center is "to awaken the public's ecological sensibilities, and to transform lifestyles, institutional practices, and social policies to support the community of life..." In order to accomplish these goals, CRLE sponsors conferences of professionals and experts in various fields and puts out a variety of publications, including a quarterly journal, *Earth Ethics*.

Through workshops and conferences, CRLE's Higher Education Project brings together educators from various institutions to discuss greening policies. The Center's Greening of Academia program works with colleges and universities to make the academic curricula, food services, and other campus policies "ecologically sound, socially just, and humane." CRLE also publishes a *Green Guide to Higher Education*, which describes the availability and ecological orientation of courses at various institutions of higher learning.

The Center's Religion in the Ecological Age program focuses on "ecospirituality," stressing that today's religious leaders and institutions must take the responsibility for addressing the issues of environmental justice, the human population explosion, overconsumption of natural resources, and other environmental problems that threaten the well- being of the natural environment and of the humans and wildlife dependent on it. CRLE sponsors conferences and publications on such issues, including a series of international conferences in Assisi, Italy (home of St. Francis of Assisi, the thirteenth century lover of animals and patron saint of nature).

Fundamental to the Center's mission was the creation of an Earth Charter. The Earth Charter Commission approved the Charter in 2000. The Charter will "prescribe new norms for state and inter-state behavior needed to maintain livelihoods and life on our shared planet." CRLE says that the main purpose of the Earth Charter is to "create a 'soft law' document that sets forth the fundamental principles of this emerging new ethics, principles that include respect for human rights, peace, economic equity, environmental protection and sustainable living...It is hoped that the Charter will become a universal code of conduct for states and people..."

Other activities of the Global Earth Ethic project include sponsoring conferences and publications focusing on ethics as they relate to agriculture, development, the environment, and "the appropriate use of animals in research, education, and agriculture..." This effort to raise consciousness includes a three-year project focused on establishing a new principle for agriculture explained in the "Soul of Agriculture: A Production Ethic for the 21st Century."

CRLE's Sustainable Livelihoods in Sustainable Communities program works with United Nations agencies and other regional and international organizations to promote environmentally sustainable developmental and agricultural practices through publications, conferences and meetings, with particular emphasis on indigenous peoples and rural communities.

CRLE's quarterly journal, *Earth Ethics*, offers book reviews and a calendar of upcoming events, and it provides a forum for scholarly and provocative feature articles discussing and debating sustainability and other environmental topics as they affect the fields of religion, agriculture, education, business, and the arts, often "challenging current economic and developmental practices."

The Center is affiliated with and supported by the Humane Society of the United States, the nation's largest animal protection organization.

Resources

BOOKS

Aronson, James, and Jelte Van Andel. *Restoration Ecology: The New Frontier*. Malden, MA: Blackwell Publishing, 2005.
Babe, Robert E. *Culture of Ecology: Reconciling Economics and Environment*. Toronto, Canada: University of Toronto Press, 2006.
Freeman, Jennifer. *Ecology*. New York: Collins, 2007.
Slobodkin, Lawrence B. *A Citizen's Guide to Ecology*. New York, N.Y.: Oxford University Press, 2003.

ORGANIZATIONS

Center for Respect of Life and Environment, 2100 L Street, NW, Washington, D.C., USA, 20037, (202) 778-6133, (202) 778-6138, info@crle.org, http://www.center1.com

Lewis G. Regenstein

Center for Rural Affairs

The Center for Rural Affairs (CRA) is a nonprofit organization dedicated to the social, economic and environmental health of rural communities. Founded in 1973, the Center for Rural Affairs includes among its participants farmers, ranchers, business people, and educators concerned with the decline of the family farm.

CRA works to provoke public thought on issues and government policies that affect rural Americans, especially in the Midwest and Plains regions of the country. It sponsors research, education, advocacy, organizing and service projects aimed to improve the life of rural dwellers. CRA's sustainable agriculture policy is designed to analyze, propose, and advocate public policies that reward environmental stewardship, promote family farming, and foster responsible technology. CRA assists beginning farmers with design and implement on-site research that helps to make these farms environmentally sound and economically viable.

CRA's conservation and education programs address the environmental problems caused by agricultural practices in the North Central United States.

Through a rural enterprise assistance program, CRA teaches rural communities to support self-employment, and it provides business assistance and revolving loan funds for the self-employed. It also provides professional farm management and brokerage service to landowners who are willing to rent or sell land to beginning farmers. CRA promotes fair competition in the agriculture marketplace by working to prevent monopolies, encouraging enforcement of laws restricting corporate farming in the United States, and advocating for the role of United States farmers in international markets.

Publications offered by CRA include the *Center for Rural Affairs Newsletter*, a monthly report on policy issues and research findings; the *Rural Enterprise Reporter*, which provides information about developing small local enterprises; and a variety of special reports on topics such as small farm technology and business strategy.

See also Environmental health; Sustainable agriculture.

Resources

ORGANIZATIONS

Center for Rural Affairs, 145 Main St., P.O. Box 136, Lyons, NE, USA, 68038, (402) 687-2100, (402) 687-2200, info@cfra.org, http://www.cfra.org

Linda Rehkopf

Center for Science in the Public Interest

The Center for Science in the Public Interest (CSPI) was founded in 1971 by Michael Jacobson. It is a consumer advocacy organization principally concerned with nutrition and food safety, and its membership consists of scientists, nutrition educators, journalists, and lawyers.

CSPI has campaigned on a variety of health and nutrition issues, particularly nutritional problems on a national level. It is the purpose of the group to address "deceptive marketing practices, dangerous food additives or contaminants, conflicts of interests in the academic community, and flawed science propagated by industries concerned solely with profits." It monitors current research on nutrition and food safety, as well as the federal agencies responsible for these areas.

CSPI maintains an office for legal affairs and special projects. It has initiated legal actions to restrict food contaminants and to ban food additives that are either unsafe or poorly tested. The special projects the group has sponsored include: Americans for Safe Food, the Nutrition Project, and the Alcohol Policies Project. The center publishes educational materials on food and nutrition, and it works to influence policy decisions affecting health and the national diet.

CSPI has made a significant impact on food marketing, and they have successfully contested food labeling practices in many sectors of the industry. They were instrumental in forcing fast-food companies to disclose ingredients, and they have recently pressed the Food and Drug Administration to improve regulations for companies which make and distribute fruit juice. Many brands do not reveal the actual percentages of the different juices used to make them, and certain combinations of juices are often misleadingly labeled as cherry juice or kiwi juice, for instance, when they may be little more than a mixture of apple and grape juice. The organization has also taken action against deceptive food advertising, particularly advertising for children's products. In 2007, the FDA awarded CSPI that agency's highest honor, the Harvey W. Wiley Special Citation.

CSPI is funded mainly by foundation grants and subscriptions to its award-winning *Nutrition Action Healthletter*. The newsletter, with approximately 900,000 subscribers in the United States and Canada, is one of the largest-circulation health-oriented publications. The newsletter is chartered to increase public understanding of food safety and nutrition issues. It frequently examines the consequences of legislation and regulation at the state and federal level; it has explored the controversy over organic and chemical farming methods, and it has studied how agribusiness has changed the way Americans eat. CSPI also distributes posters, videos, and computer software, and it offers a directory of mail-order sources for organically-grown food.

See also Food additives.

Resources

ORGANIZATIONS

Center for Science in the Public Interest, 1875 Connecticut Ave., NW, Suite 300, Washington, D.C., USA, 20009, (202) 332-9110, (202) 265-4954, cspi@cspinet.org, http://www.cspinet.org

Douglas Smith

CERCLA *see* **Comprehensive Environmental Response, Compensation, and Liability Act (CERCLA).**

CERES Principles *see* **Valdez Principles.**

Centers for Disease Control and Prevention

The Centers for Disease Control and Prevention (CDC) is the Atlanta, Georgia-based agency of the Public Health Service that has led efforts to prevent diseases such as malaria, polio, smallpox, tuberculosis, and acquired immunodeficiency syndrome (AIDS). As the nation's prevention agency, the CDC's responsibilities have expanded, and it now addresses contemporary threats to health such as injury, environmental and occupational hazards, biological terrorism, behavioral risks, and chronic diseases.

Divisions within the CDC use surveillance, epidemiologic and laboratory studies, and community interventions to investigate and prevent public health threats.

The Center for Chronic Disease Prevention and Health Promotion designs programs to reduce death and disability from chronic diseases—cardiovascular, kidney, liver and lung diseases, and cancer and diabetes.

The Center for Environmental Health and Injury Control assists public health officials at the scene of natural or artificial disasters such as volcano eruptions, forest fires, hazardous chemical spills, and nuclear accidents. Scientists study the effects of chemicals and pesticides, reactor accidents, and health threats from radon, among others. The National Institute for Occupational Safety and Health helps identify chemical and physical hazards that lead to occupational diseases.

Preventing and controlling infectious diseases has been a goal of the CDC since its inception in 1946. The Center for Infectious Diseases investigates outbreaks of infectious disease locally and internationally. The Center

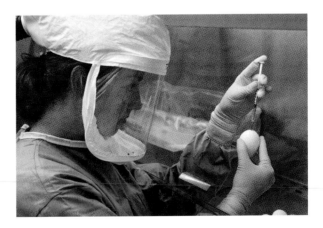

Microbiologist injecting a solution of H5N1 avian influenza viruses into a chicken egg. A chicken egg is used because a large number of viruses are needed for research, and viruses can only replicate within host cells. *(CDC / Photo Researchers, Inc.)*

Chelate

A chemical compound in which one atom is enclosed within a larger cluster of surrounding atoms. The term comes from the Greek word *chela*, meaning claw. Chelating agents—compounds that can form chelates with other atoms—have a wide variety of environmental applications. For example, the compound ethylenediaminetetraacetic acid (EDTA) is used to remove lead from the blood. EDTA molecules surround and bind to lead atoms, and the chelate is then excreted in the urine. EDTA can also be used to soften hard water by chelating the calcium and magnesium ions that cause hardness.

Chelyabinsk, Russia

Chelyabinsk is the name of a province (oblast, or administrative division) in Russia, and also of its capital city in west-central Russia. The oblast covers an area of about 34,000 miles^2 (88,060 km^2) and has a population of about 3.6 million. The city of Chelyabinsk lies on the Miass River on the eastern side of the Ural Mountains. It is about 125 miles (200 km) southeast of the city of Yekaterinburg. Its population in 2002 was about 1.1 million.

Chelyabinsk is best known today as the home of Mayak, a 77-mile^2 (200-km^2) nuclear fuel reprocessing complex where nuclear weapons were built for the former Soviet Union. Mayak is about 45 miles (72 km) northwest from Chelyabinsk. Because of intentional policy decisions and accidental releases of radioactive materials, Mayak (which is Russian for "lighthouse" or "beacon") has been called the most polluted spot on Earth.

The 1994 film *Chelyabinsk: The Most Contaminated Spot on the Planet*, produced and directed by Slawomir Grünberg (of Log In Productions, New York, United States), tells the story of Mayak.

Virtually nothing was known about Mayak by the outside world, the Russian people, or even the residents of Chelyabinsk themselves until 1991. Then, under the new philosophy of *glasnost* (Russian for "openness"), Soviet president Mikhaíl Gorbachev (1931–) released a report on the complex. It listed 937 official cases of chronic radiation sickness among Chelyabinsk residents. Medical authorities believe that the actual number is many times larger.

The report also documented the secrecy with which the Soviet government shrouded its environmental problems at Mayak. Physicians were not even allowed to discuss the cause or nature of the radiation sickness. Instead, they had to refer to it as the "ABC disease."

Chelyabinsk's medical problems were apparently the result of three major "incidents" involving the release of radiation at Mayak. The first dated from the late 1940s to the mid-1950s, when radioactive waste from nuclear weapons research and development was dumped directly into the nearby Techa River. People downstream from Mayak were exposed to radiation levels that were approximately fifty-seven times greater than those at the better-known Chernobyl nuclear power plant accident in 1986. The Gorbachev report admitted that 28,000 people received radiation doses of "medical consequence." Astonishingly, almost no one was evacuated from the area.

The second incident occurred in 1957, when a nuclear waste dump at Mayak exploded with a force equivalent to a five-to-10 kiloton atomic bomb. This incident is often referred to as the Kyshtym disaster because it was named after the nearest known town from the site. The site had been constructed in 1953 as an alternative to simply disgorging radioactive wastes into the Techa. When the automatic cooling system failed, materials in the dump were heated to a temperature of 662°F (350°C). In the resulting explosion, 20 million curies of radiation were released, exposing 270,000 people to dangerous levels of radioactivity. Neither the Soviet Union nor the United States government, which had detected the accident, revealed the devastation at Mayak.

The third incident happened in 1967. In their search for ways to dispose of radioactive waste, officials at Mayak decided in 1951 to use Lake Karachay as a repository. They realized that dumping into the Techa was not a satisfactory solution, and they hoped that Karachay—which has no natural outlet—would be a better choice.

Unfortunately, radioactive materials began leaching into the region's water supply almost immediately. Radiation was eventually detected as far as 2 miles (3 km) away. The 1967 disaster occurred when an unusually dry summer diminished the lake significantly. A layer of radioactive material, deposited on the newly exposed shoreline, was spread by strong winds that blew across the area. This released radiation equivalent to the amount contained in the first atomic bomb explosion over Hiroshima (Japan) in 1945.

The fiftieth anniversary of the 1967 disaster was publicized with a local rally on September 29, 2007. According to a Radio Free Europe article, the environmental group Greenpeace Russia joined with the residents of Chelyabinsk to highlight the troubling situation that these Russian citizens were still living in the contaminated area. The rally also was intended to publically criticize a plan by the Russian government to continue the importation and reprocessing of foreign radioactive waste at Mayak.

Resources

BOOKS

Takada, Jun. *Nuclear Hazards in the World: Field Studies on Affected Populations and Environments.* Tokyo: Kodansha, 2005.

PERIODICALS

Cochran, T. B., and R. S. Norris. "A First Look at the Soviet Bomb Complex." *Bulletin of the Atomic Scientists.* Vol. 47 (May 1991): 25-31.

Hertsgaard, M. "From Here to Chelyabinsk." *Mother Jones.* Vol. 17 (January-February 1992): 51-55+.

Perea, J. "Soviet Plutonium Plant 'Killed Thousands.'" *New Scientist.* Vol. 134 (20 June 1992): 10.

Wald, M. L. "High Radiation Doses Seen for Soviet Arms Workers." *New York Times* (16 August 1990): A3.

OTHER

Log In Productions. "Chelyabinsk: The Most Contaminated Spot on the Planet." http://www.logtv.com/films/chelyabinsk/ (accessed November 2, 2010).

Radio Free Europe, Radio Liberty. "Russia: Living—And Dying—In The Shadow Of Mayak." http://www.rferl.org/content/article/1063825.html (accessed November 2, 2010).

Radio Free Europe, Radio Liberty. "Russia Marks 50th Anniversary Of Mayak Nuclear Blast." http://www.rferl.org/content/article/1078826.html (accessed November 2, 2010).

David E. Newton

Chemical bond

A chemical bond is any force of attraction between two atoms strong enough to hold the atoms together for some period of time. At least five primary types of chemical bonds are known, ranging from very strong to very weak. They are covalent, ionic, metallic, and hydrogen bonds, and London forces.

In all cases, a chemical bond ultimately involves forces of attraction between the positively-charged nucleus of one atom and the negatively-charged electron of a second atom. Understanding the nature of chemical bonds has practical significance since the type of bonding found in a substance explains to a large extent the macroscopic properties of that substance.

An ionic bond is one in which one atom completely loses one or more electrons to a second atom. The first atom becomes a positively charged ion and the second, a negatively charged ion. The two ions are attracted to each other because of their opposite electrical charges.

In a covalent bond, two atoms share one or more pairs of electrons. For example, a hydrogen atom and a fluorine atom each donate a single electron to form a shared pair that constitutes a covalent bond between the two atoms. Both electrons in the shared pair orbit the nuclei of both atoms.

In most cases, covalent and ionic bonding occur in such a way as to satisfy the Law of Octaves. Essentially that law states that the most stable configuration for an atom is one in which the outer energy level of the atom contains eight electrons or, in the case of smaller atoms, two electrons.

Ionic and covalent bonds might appear to represent two distinct limits of electron exchange between atoms, one in which electrons are totally gained and lost (ionic bonding) and one in which electrons are shared (covalent bonding). In fact, most chemical bonds fall somewhere between these two extreme cases. In the hydrogen-fluorine example mentioned above, the fluorine nucleus is much larger than the hydrogen nucleus and, therefore, exerts a greater pull on the shared electron pair. The electrons spend more time in the vicinity of the fluorine nucleus and less time in the vicinity of the hydrogen nucleus. For this reason, the fluorine end of the bond is more negative than the hydrogen end, and the bond is said to be a polar covalent bond. A nonpolar covalent bond is possible only between two atoms with equal attraction for electrons as, for example, between two atoms of the same element.

Metallic bonds are very different from ionic and covalent bonds in that they involve large numbers of atoms. The outer electrons of these atoms feel very little attraction to any one nucleus and are able, therefore, to move freely throughout the metal.

Hydrogen bonds are very weak forces of attraction between atoms with partial positive and negative charges. Hydrogen bonds are especially important in living organisms since they can be broken and reformed easily during biochemical changes.

London forces are the weakest of chemical bonds. They are forces of attraction between two uncharged molecules. The force appears to arise from the temporary shift of electrical charges within each molecule.

Resources

BOOKS

Fliszar, Sandor. *Atomic Charges, Bond Properties, and Molecular Energies.* Hoboken, N.J.: Wiley, 2009.

Manning, Phillip. *Chemical Bonds.* New York: Chelsea House Publishing, 2009.

Pauling, Linus. *The Nature of the Chemical Bond and the Structure of Molecules and Crystals: An Introduction to Modern Structural Chemistry.* 3rd ed. Ithaca, NY: Cornell University Press, 1960.

David E. Newton

Chemical oxygen demand

Chemical oxygen demand (COD) is a measure of the ability of chemical reactions to oxidize matter in an aqueous system. The results are expressed in terms of oxygen so that they can be compared directly to the results of biochemical oxygen demand (BOD) testing. The test is performed by adding the oxidizing solution to a sample, boiling the mixture on a refluxing apparatus for two hours and then titrating the amount of dichromate remaining after the refluxing period. The titration procedure involves adding ferrous ammonium sulfate (FAS), at a known normality, to reduce the remaining dichromate. The amount of dichromate reduced during the test—the initial amount minus the amount remaining at the end—is then expressed in terms of oxygen. The test has nothing to do with oxygen initially present or used. It is a measure of the demand of a solution or suspension for a strong oxidant. The oxidant will react with most organic materials and certain inorganic materials under the conditions of the test. For example, Fe^{2+} and Mn^{2+} will be oxidized for Fe^{3+} and Mn^{4+}, respectively, during the test.

Generally, the COD is larger than the BOD exerted over a five-day period (BOD_5), but there are exceptions in which microbes of the BOD test can oxidize materials that the COD reagents cannot. For a raw, domestic wastewater, the COD/BOD_5 ratio is in the area of 1.5–3.0/1.0. Higher ratios would indicate the presence of toxic, non- biodegradable or less readily biodegradable materials.

The COD test is commonly used because it is a relatively short-term, precise test with few interferences. However, the spent solutions generated by the test are hazardous. The liquids are acidic, and contain chromium, silver, mercury, and perhaps other toxic materials in the sample tested. For this reason laboratories are doing fewer or smaller COD tests in which smaller amounts of the same reagents are used.

Resources

BOOKS

Bitton, Gabriel. *Wastewater Microbiology.* Hoboken, NJ: Wiley-Liss and John Wiley & Sons, 2005.

Fierro, Pedro, and Evan K. Nyer. *The Water Encyclopedia: Hydrologic Data and Internet Resources.* Boca Raton: CRC/Taylor & Francis, 2007.

Greenaway, Theresa. *The Water Cycle.* London: Hodder Wayland, 2006.

Van Oss, Carel J. *The Properties of Water and Their Role in Colloidal and Biological Systems.* Interface science and technology, v. 16. Amsterdam: Elsevier/Academic Press, 2008.

Wachinski, Anthony M. *Ion Exchange Treatment for Drinking Water.* Denver: American Waterworks Association, 2004.

Gregory D. Boardman

Chemical spills

Chemical spills are any accidental releases of synthetic chemicals that pose a risk to the environment.

Spills occur at any of the steps between the production of a chemical and its use. A railroad tank car may spring a leak; a pipe in a manufacturing plant may break; or an underground storage tank may corrode allowing its contents to escape into groundwater. These spills are often classified into four general categories: the release of a substance into a body of water; the release of a liquid on land; the release of a solid on land; and the release of a gas into the atmosphere. The purpose of this method of classification is to provide the basis for a systematic approach to the control of any type of chemical spill.

Some of the most famous chemical spills in history illustrate these general categories. For example, seven cars of a train carrying the pesticide metam sodium fell off the tracks near Dunsmuir, California, in August 1991, breaking open and releasing the chemicals into the Sacramento River. Plant and aquatic life for 43 miles (70 km) downriver died as a result of the accident. The pesticide eventually formed a band 225 feet (70 m) wide across Lake Shasta before it could be contained.

In 1983, the Environmental Protection Agency (EPA) purchased the whole town of Times Beach, Missouri, and relocated more than 2,200 residents because the land was so badly contaminated with highly toxic dioxins. The concentration of these compounds, a by-

Emergency workers practicing decontamination during biohazard training drill. *(© Enigma / Alamy.)*

product of the production of herbicides, was more than a thousand times the maximum recommended level.

In December 1984, a cloud of poisonous gas escaped from a Union Carbide chemical plant in Bhopal, India. The plant produced the pesticide Sevin from a number of chemicals, many of which were toxic. The gas that accidentally escaped probably contained a highly toxic mixture of phosgene, methyl isocyanate (MIC), chlorine, carbon monoxide, and hydrogen cyanide, as well as other hazardous gases. The cloud spread over an area of more than 15 square miles (40 km^2), exposing more than 200,000 people to its dangers. Official estimates of death ranged from approximately 2200 to 3800 people, but other experts and nongovernmental agencies put the death toll, inclusive of subsequent deaths from injuries or disease developed as a result of exposure, at more than 10,000 people.

Chemists have now developed a sophisticated approach to the treatment of chemical spills, which involves one or more of five major steps: containment, physical treatment, chemical treatment, biological treatment, and disposal or destruction. Soil sealants, which can be used to prevent a liquid from sinking into the ground, are an example of containment. One of the most common methods of physical treatment is activated charcoal, because it has the ability to adsorb toxic substances on its surface, thus removing them from the environment. Chemical treatment is possible because many hazardous materials in a spill can be treated by adding some other chemical that will neutralize them, and biological treatment usually involves microorganisms that will attack and degrade a toxic chemical. Open burning, deep-well injection, and burial in a landfill are all methods of ultimate disposal.

Resources

BOOKS

Comyns, Alan E. *Encyclopedic Dictionary of Named Processes in Chemical Technology.* Boca Raton: CRC Press, 2007.

Mackay, Donald, and Donald Mackay. *Handbook of Physical-Chemical Properties and Environmental Fate for Organic Chemicals.* Boca Raton, FL: CRC/Taylor & Francis, 2006.

Wang, Zhendi, and Scott Stout. *Oil Spill Environmental Forensics: Fingerprinting and Source Identification.* New York: Academic, 2006.

OTHER

United States Environmental Protection Agency (EPA). "Cleanup: Storage Tanks: Leaks and Spills." http://www.epa.gov/ebtpages/cleastoragetanksleaksandspills.html (accessed September 4, 2010).

United States Environmental Protection Agency (EPA). "Industry: Storage Tanks: Spill Prevention and Protection." http://www.epa.gov/ebtpages/industorage tanksspillpreventionandprotection.html (accessed September 4, 2010).

United States Environmental Protection Agency (EPA). "Pesticides: Pesticide Effects: Pesticide Spills." http://www.epa.gov/ebtpages/pestpesticideeffecpesticidespills.html (accessed September 4, 2010).

David E. Newton

Chemicals

The general public often construes the word "chemical" to mean a harmful synthetic substance. In fact, however, the term applies to any element or compound, either natural or synthetic. The thousands of compounds that make up the human body are all chemicals, as are the products of scientific research. A more accurate description, however, can be found in the dictionary. Thus, aspirin is a chemical by this definition, since it is the product of a series of chemical reactions.

The story of chemicals began with the rise of human society. Indeed, early stages of human history, such as the Iron, Copper, and Bronze Ages reflect humans' ability to produce important new materials. In the first two eras, people learned how to purify and use pure metals. In the third case, they discovered how to combine two to make an alloy with distinctive properties.

The history of ancient civilizations is filled with examples of men and women adapting natural resources for their own uses. Egyptians of the eighteenth dynasty (1700–1500 B.C.), for example, knew how to use cobalt compounds to glaze pottery and glass. They had also developed techniques for making and using a variety of dyes.

Over the next 3,000 years, humans expanded and improved their abilities to manipulate natural chemicals. Then, in the 1850s, a remarkable breakthrough occurred. A discovery by young British scientist William Henry Perkin led to the birth of the synthetic chemicals industry.

Perkin's great discovery came about almost by accident, an occurrence that was to become common in the synthetics industry. As an 18-year-old student at England's Royal College of Chemistry, Perkin was looking for an artificial compound that could be used as a quinine substitute. Quinine, the only drug available for the treatment of malaria, was itself in short supply.

Following his teacher's lead, Perkin carried out a number of experiments with compounds extracted from coal tar, the black, sticky sludge obtained when coal is heated in insufficient air. Eventually, he produced a black powder which, when dissolved in alcohol, created a beautiful purple liquid. Struck by the colorful solution, Perkin tried dyeing clothes with it.

His efforts were eventually successful. He went on to mass produce the synthetic dye—mauve, as it was named—and to create an entirely new industry. The years that followed are sometimes referred to as The Mauve Decade because of the many new synthetic products inspired by Perkin's achievement. Some of the great chemists of that era have been memorialized in the names of the products they developed or the companies they established: Adolf von Baeyer (Bayer aspirin), Leo Baekeland (Baekelite plastic), Eleuthère Irénée du Pont (DuPont Chemical), George Eastman (Eastman 910 adhesive and the Eastman Kodak Company), and Charles Goodyear (Goodyear Rubber).

Chemists soon learned that from the gooey, ugly by-products of coal tar, a whole host of new products could be made. Among these products were dyes, medicines, fibers, flavorings, plastics, explosives, and detergents. They found that the other fossil fuels—petroleum and natural gas—could also produce synthetic chemicals.

Today, synthetic chemicals permeate our lives. They are at least as much a part of the environment, if not more, than are natural chemicals. They make life healthier, safer, and more enjoyable. People concerned about the abundance of "chemicals" in our environment should remember that everyone benefits from anti-cancer drugs, pain-killing anesthetics, long-lasting fibers, vivid dyes, sturdy synthetic rubber tires, and dozens of other products. The world would be a much poorer place without them.

Unfortunately, the production, use, and disposal of synthetic chemicals can create problems because they may be persistent and/or hazardous. Persistent means that a substance remains in the environment for a long time: dozens, hundreds, or thousands of years in many cases. Natural products such as wood and paper degrade naturally as they are consumed by microorganisms. Synthetic chemicals, however, have not been around long enough for such microorganisms to evolve.

This leads to the familiar problem of solid waste disposal. Plastics used for bottles, wrappings, containers, and hundreds of other purposes do not decay. As a result, landfills become crowded and communities need new places to dump their trash.

Persistence is even more of a problem if a chemical is hazardous. Some chemicals are a problem, for example, because they are flammable. More commonly, however, a hazardous chemical will adversely affect the health of a plant or animal. It may be (1) toxic, (2) carcinogenic, (3) teratogenic, or (4) mutagenic.

Toxic chemicals cause people, animals, or plants to become ill, develop a disease, or die. DDT, chlordane, heptachlor, and aldrin are familiar, toxic pesticides. Carcinogens cause cancer; teratogens produce birth defects. Mutagens, perhaps the most sinister of all, inflict genetic damage.

Determining these effects can often be very difficult. Scientists can usually determine if a chemical will harm or kill a person. But how does one determine if a chemical causes cancer twenty years after exposure, is responsible for birth defects, or produces genetic disorders? After all, any number of factors may have been responsible for each of these health problems.

As a result, labeling any specific chemical as carcinogenic, teratogenic, or mutagenic can be difficult. Still, environmental scientists have prepared a list of synthetic chemicals determined to fall into these categories. Among them are vinyl chloride, trichloroethylene, tetrachloroethylene, the nitrosamines, and chlordane and heptachlor.

Another class of chemicals are hazardous because they may contribute to the greenhouse effect and ozone layer depletion. The single most important chemical in determining the earth's annual average temperature is a naturally-occurring compound, carbon dioxide. Its increased production is partially responsible for a gradual increase in the planet's annual average temperature.

But synthetic compounds may also play a role in global warming. Chlorofluorocarbons (CFCs) are widely used in industry because of their many desirable properties, one of which is their chemical stability. This very property means, however, that when released into the atmosphere, they remain there for many years. Since they capture heat radiated from the earth in much the way carbon dioxide does, they are probably important contributors to global warming.

These same chemicals, highly unreactive on earth, decompose easily in the upper atmosphere. When they do so, they react with the ozone in the stratosphere, converting it to ordinary oxygen. This may have serious consequences, since stratospheric ozone shields the earth from harmful ultraviolet radiation.

There are two ways to deal with potentially hazardous chemicals in the environment. One is to take political or legal action to reduce the production, limit the use, and/or control the disposal of such products. A treaty negotiated and signed in Montreal by more than forty nations in 1987, for example, calls for a gradual ban on CFC production. If the treaty is honored, these chemicals will eventually be phased out of use.

A second approach is to solve the problem scientifically. Synthetic chemicals are a product of scientific research, and science can often solve the problems these chemicals produce. For example, scientists are exploring the possibility of replacing CFCs with related compounds called fluorocarbons (FCs) or hydrochloroflurocarbons (HCFCs). Both are commercially appealing, but they have fewer harmful effects on the environment.

Resources

BOOKS

Armour, M. A. *Hazardous Laboratory Chemicals Disposal Guide*. Boca Raton, Fla: Lewis Publishers, 2003.

Beard, James M. *Environmental Chemistry in Society*. Boca Raton, FL: CRC Press/Taylor & Francis, 2009.

Brady, James E., and Fred Senese. *Chemistry: The Study of Matter and Its Changes*. 5th ed. New York: Wiley, 2008.

Corwin, Charles H. *Introductory Chemistry: Concepts & Connections*. 5th ed. Upper Saddle River, NJ: Prentice-Hall, 2007.

Daintith, John. *A Dictionary of Chemistry*. Oxford paperback reference. Oxford: Oxford University Press, 2008.

Duke, Catherine V. A., and C. D. Williams. *Chemistry for Environmental and Earth Sciences*. Boca Raton: CRC Press, 2008.

Hajian, Harry G., and Robert L. Pecsok. *Working Safely in the Chemistry Laboratory*. Washington, DC: American Chemical Society, 1994.

Hites, R. A. *Elements of Environmental Chemistry*. Hoboken, N.J.: Wiley-Interscience, 2007.

Lewis, Grace Ross. *1001 Chemicals in Everyday Products*. New York: Wiley, 1999.

Lide, David R. *CRC Handbook of Chemistry and Physics [A Ready-Reference Book of Chemical and Physical Data]*. 90th ed. Boca Raton, Fla: CRC, 2009.

Mackay, Donald, and Donald Mackay. *Handbook of Physical-Chemical Properties and Environmental Fate for Organic Chemicals*. Boca Raton, FL: CRC/Taylor & Francis, 2006.

workers volunteered to entomb the reactor ruins with a massive concrete sarcophagus. Bus drivers risked further exposure by making repeated trips into contaminated areas in order to evacuate villagers. Over 600,000 people were involved in the decontamination and clean up of Chernobyl. The health effects on them from their exposure are not completely known. The Chernobyl accident focused international attention on the risks associated with operating a nuclear reactor for the generation of power. Public apprehension has forced some governments to review their own safety procedures and to compare the operation of their nuclear reactors with Chernobyl's. In a review of the Chernobyl accident by the Atomic Energy Authority of the United Kingdom, an effort was made to contrast the design of the Chernobyl reactor and management procedures with those in practice in the United States and the United Kingdom.

Three design drawbacks were noted of the Chernobyl nuclear power plant:

- The reactor was intrinsically unstable below 20% power and never should have been operated in that mode. (U.S. and UK reactors do not have this design flaw.)
- The shut-down operating system was inadequate and contributed to the accident rather than terminating it. (U.S. and UK control systems differ significantly.)
- There were no controls to prevent the staff from operating the reactor in the unstable region or preventing the disabling of existing safeguards.

In addition, the Chernobyl management had no effective watchdog agency to inspect procedures and order closure of the facility. Also in years prior to the accident there was a lack of information given the public of prior nuclear accidents, typical of the press censorship and news management occurring in the period before *glasnost*. The operators were not adequately trained nor were they themselves fully aware of prior nuclear power accidents or near accidents which would have made them more sensitive to the dangers of a runaway reactor system.

Unfortunately in the former Soviet block nations there are several nuclear reactors that are potentially as hazardous as Chernobyl but which must continue operation to maintain power requirements; however, the operational procedures are under constant review to avoid another accident.

Following the Chernobyl disaster, the Soviet Union and, later, Ukraine continued to operate the remaining three reactors at the Chernobyl power station. The Soviet government oversaw the cleanup of Chernobyl, including construction of a concrete containment dome over the damaged reactor. The government also promised to pay lifelong health and other expenses for the so-called liquidators who worked clean up and construction duty after the accident. However, after the collapse of the Soviet Union in 1991, maintenance of the Chernobyl site fell to the Ukrainian government. International aid funds were established to help Ukraine pay for construction of another containment unit as the current concrete structure is already failing. International funds were also distributed to Belarus, Russia, and Ukraine to meet the health needs of those affected by the disaster.

The Ukrainian government closed the final reactor in 2000. In 2010, the Ukrainian government launched a program to clear the Chernobyl site by 2065.

Resources

OTHER

United Nations Development Programme. "United Nations and Chernobyl." http://chernobyl.undp.org/english/history.html (accessed October 15, 2010).

World Health Organization (WHO). "Chernobyl accident: an overview of the health effects." http://www.who.int/entity/mediacentre/factsheets/fs303/en/index.html (accessed October 15, 2010).

Malcolm T. Hepworth

Chesapeake Bay

The Chesapeake Bay is the largest estuary (186 mi [300 km] long) in the United States. The Bay was formed 1500 years ago by the retreat of glaciers and the subsequent sea level rise that inundated the lower Susquehanna River valley. The Bay has a drainage basin of 64,076 square miles (166,000 sq km) covering six states and running through Pennsylvania, Maryland, the District of Columbia, and Virginia before entering the Atlantic Ocean. While 150 rivers enter the Bay, a mere eight account for 90 percent of the freshwater input, with the Susquehanna alone contributing nearly half. Chesapeake Bay is a complex system, composed of numerous habitats and environmental gradients.

Chesapeake Bay's abundant natural resources attracted native Americans, first settling on its shores. The first European record of the Bay was in 1572, and the area surrounding Chesapeake Bay was rapidly colonized by Europeans. In many ways, the United States grew up around the Chesapeake Bay. The

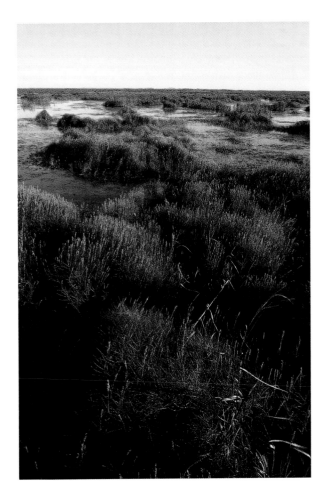

Chesapeake Bay, the largest estuary in the United States, photographed on Deal Island, Maryland. *(Stephen J. Krasemann / Photo Researchers, Inc.)*

colonists harvested the Bay's resources and used its waterways for transportation. Today 10 million people live in the Chesapeake Bay's drainage basin, and many of their activities affect the environmental quality of the Bay as did the activities of their ancestors.

The rivers emptying into the Bay were also used by the colonists to dispose of raw sewage. By the middle 1800s some of the rivers feeding the Bay were polluted: the Potomac was recorded as emitting a lingering stench. The first sewer was constructed in Washington, DC, and it pumped untreated waste into the Bay. It was recognized in 1893 that the diseases suffered by humans consuming shellfish from the Bay were directly related to the discharge of raw sewage into the Bay. Despite this recognition, efforts in 1897 by the mayor of Baltimore to oppose the construction of a sewage system that discharged sewage into the Bay in favor of a "land filtration technique" failed. Ultimately, a secondary treatment system discharging into the Bay was constructed. In the mid–

1970s, a $27 million government-funded study of the Bay's condition concluded that the deteriorating quality of the Chesapeake Bay was a consequence of human impacts. But it was not until the early 1980s that an Environmental Protection Agency (EPA) report on the Chesapeake focused interest on saving the Bay, and $500 million was spent on cleanup and construction of sewage treatment plants.

While the Chesapeake Bay is used primarily as a transportation corridor, its natural resources rank a close second in importance to humans. The most commercially-important fisheries in the Bay are the native American oyster (*Crassostrea virginica*), blue crab (*Callinectes sapidus*), American shad (*Alosa sapidissima*), and striped bass (*Marone saxatilis*). Fisherman first began to notice a decline in fish populations in the 1940s and 1950s, and since then abundances have declined even further. Since 1900, the oyster catch has declined 70 percent, shad 85 percent, and striped bass 90 percent. In the late 1970s, the EPA began to study the declining oyster and striped bass populations and concluded that their decline was due to a combination of over-harvesting and pollution.

Work by the EPA and other federal and state agencies has identified six areas of environmental concern for the Bay: (1) excess nutrient input from both sewage treatment plants discharging into the Bay and runoff from agricultural land; (2) low oxygen levels as a result of increased biochemical oxygen demand, which increases dramatically with loading of organic material; (3) loss of submerged aquatic vegetation due to an increase in turbidity; (4) presence of chemical toxins; (5) loss to development of wetlands surrounding the Bay that serve as nurseries for juvenile fish and shellfish and as buffers for runoff of nutrients and toxic chemicals; and (6) increasing acidity of water (measured by pH) in streams that feed the Bay. These streams are also nursery areas for larval fish that may be adversely affected by decreasing pH.

The increasing growth of phytoplankton—free-floating photosynthetic organisms—in the Bay is generally considered to be a significant contributor to the decline in environmental quality of the Chesapeake Bay. The number of algal blooms has increased dramatically since the 1950s and is attributed to the high levels of the plant nutrients nitrogen and phosphorus that are discharged into the Bay. Some scientists also attribute this increase in algal bloom incidence to climate change. In the 1980s and 1990s, it was estimated that discharge from sewage treatment plants and agricultural runoff accounted for 65 percent of the nitrogen and 22 percent of the phosphorus found in the Bay. The accumulation of nutrients such as phosphates and nitrates in an ecosystem is termed eutrophication. Acid rain, formed from discharges from industrial plants in Canada and the northeast United States, contributed 25 percent of the nitrogen found in the Bay. Excess nutrients encourage phytoplankton growth, and as the large number of phytoplankton die and settle to the bottom, their decomposition robs the water of oxygen needed by fish and other aquatic organisms. When oxygen levels fall too low, these organisms die or flee from the regions of low oxygen. Decomposition of dead organic matter further reduces the concentration of oxygen. During the late twentieth and early twenty-first centuries, Finfish and shellfish kills became increasingly common in the Bay.

Phytoplankton blooms and the increase in suspended sediments resulting from shoreline development and poor agricultural practices have increased turbidity and led to a decline in submerged aquatic vegetation (SAV) such as eelgrass. SAV is extremely important in the prevention of erosion of bottom sediment and as critical habitat for nursery grounds of commercially important fish and shellfish.

Some researchers contend that one particular type of algae, *Pfiesteria piscicida*, was the cause of a rush of harmful algal blooms (HABs) in the Chesapeake Bay waters in the 1990s. The toxin produced by these dinoflagellate algae have adverse effects on humans and are lethal to fish.

Chemicals introduced into the Bay from several sites may have contributed to the decline in the Bay's fish and bird populations. For example, in 1975, the pesticideKepone was leaked or dumped into the James River, poisoning fish and shellfish. Harvests of some species are still restricted in the area of this spill. Chlorine biocides used in wastewater treatment plants and power plants, which discharge into the Bay, are known human carcinogens and can be toxic to aquatic organisms. Polycyclic aromatic hydrocarbons (PAH) have caused dermal lesions in fish populations in the Elizabeth River. PAHs also affect shellfish populations. In the 1990s, public concern focused on tributyl tin (TBT) that was used in anti-fouling paint on recreational and commercial boats. TBTs belong to a family of chemicals known as organotins, which are toxic to shellfish and crustaceans. The diversity of chemical pollutants found in the Bay is exemplified by the results of research that identified 100 inorganic and organic contaminants in striped bass caught in the Bay

Work by private and governmental agencies has reversed the declining environmental quality of the Chesapeake Bay. In 1983 Maryland, Virginia, Pennsylvania, the District of Columbia, the Chesapeake Bay Commission, and the EPA signed the Chesapeake Bay Agreement, which outlined procedures to correct many of the Bay's ecological problems, particularly those caused by nutrient enrichment. Since 1985, increasing compliance with discharge permits, prohibition of the sale of phosphate-based detergents, and the upgrading of wastewater plants has resulted in a significant reductions in the discharge of nitrogen and phosphorous from point sources. Controls on agriculture and urban development reduced the amount of both nitrogen and phosphorus entering the Bay from nonpoint sources. The amount of toxins entering the Bay has also been reduced. Tributyl tin has been banned for use in anti-fouling paints on non-military vessels, and pesticide runoff has been reduced by using alternate strategies for pest control. At the same time, some of the Bay's critical habitats are recovering: man-made oyster reefs are being created to expand suitable habitat for oysters; and rivers are being cleared of obstacles such as dams and spillways to provide access to spawning areas by migratory fish.

Conflict between commercial and environmental interests have encumbered some of the restoration efforts. Research on the life-history of crabs and oysters shows that limiting the size of crabs that can be sold and the numbers of oysters that can be harvested will help these fisheries rebound, but regulations to limit crab and oyster catches have met with strong resistance from watermen who are struggling to survive economically in a declining fishery.

Oysters are designated as a keystone species in the Chesapeake Bay, and thus play an important role in the ecosystem, naturally filtering the water in the Bay. The planned introduction of a non-native Asian oyster (*Crassostrea ariakensis*) by the Virginia Seafood Council raised hopes that a new oyster fishery could be built around the harvest of this disease-resistant, fast-growing species. The native population of oysters is affected by two diseases caused by two different protozoan species, *Haplosporidium nelsoni* (Multinucleated Sphere X; MSX) and *Perkinsus marinus* (known as Dermo). In May 2002, the United States Fish and Wildlife Service called for a moratorium on the introduction of the Asian oyster, until researchers could determine whether its introduction would endanger native species, or bring new and deadly disease organisms into the Chesapeake Bay ecosystem. In 2006, the National Oceanic and Atmospheric Administration (NOAA) Chesapeake Bay Office (NCBO) supplied $4 million to restore the native oyster population and $2 million for funding oyster disease research.

In the Chesapeake 2000 (C2K) agreement, the jurisdictions of Maryland, Pennsylvania, Virginia, and the District of Columbia, together with the Chesapeake Bay

Commission and the Federal Government, committed to "correct the nutrient-and sediment-related problems in the Chesapeake Bay and its tidal tributaries sufficiently to remove the Bay and the tidal portions of its tributaries from the list of impaired waters under the Clean Water Act." This initiative set goals for improved water quality and other markers of ecosystem health in the Chesapeake Bay by 2010. The Chesapeake Bay Program gives periodic updates of its progress towards the goals set in the Chesapeake 2000 agreement. Its 2009 assessment report lists a variety of goals that were set for Bay restoration. For instance, regarding ongoing efforts to reduce pollution entering the Bay, the assessment reported that "Bay Program partners have implemented 62 percent of needed efforts to reduce nitrogen, phosphorus and sediment pollution, which is a 3 percent increase from 2008." Note that in the assessment, not only is the overarching goal stated (62 percent pollution reduction), but year-over-year comparisons are made, demonstrating a small (3 percent), but nonetheless steady, reduction of overall pollution. The Chesapeake Bay Commission is a consortium of stakeholders, jurisdictions, and federal agencies, which demonstrates that citizens, government, and industry can work cooperatively. The Chesapeake Bay program is a national model for efforts to restore other degraded ecosystems.

Resources

BOOKS

Chesapeake Bay Foundation. *Bad Waters Dead Zones, Algal Blooms, and Fish Kills in the Chesapeake Bay Region in 2007*. Annapolis, MD: Chesapeake Bay Foundation, 2007.

Chesapeake Bay Foundation. *Land and the Chesapeake Bay*. Annapolis, MD: The Foundation, 2007.

Miller, Joanne. *Chesapeake Bay*. 2nd ed. Emeryville, CA: Avalon Travel, 2008.

OTHER

National Oceanic and Atmospheric Administration (NOAA) Chesapeake Bay Office. "Science, Service, and Stewardship." http://chesapeakebay.noaa.gov/ (accessed November 10, 2010).

William G Ambrose Jr.
Paul E Renaud
Marie H. Bundy

▌ Child survival revolution

Every year in the developing countries of the world, some 11 million children under the age of five die of common infectious diseases. Most of these children could be saved by simple, inexpensive, preventative medicine. Many public health officials argue that it is as immoral and unethical to allow children to die of easily preventable diseases as it would be to allow them to starve to death or to be murdered. In 1986, the United Nations Children's Find (UNICEF) announced a worldwide campaign to prevent unnecessary child deaths. Called the "child survival revolution," this campaign is based on four principles, designated by the acronym GOBI.

"G" stands for growth monitoring. A healthy child is considered a growing child. Underweight children are much more susceptible to infectious diseases, retardation, and other medical problems than children who are better nourished. Regular growth monitoring is the first step in health maintenance.

"O" stands for oral rehydration therapy (ORT). About one-third of all deaths under five years of age are caused by diarrheal diseases. A simple solution of salts, glucose, or rice powder and boiled water given orally is almost miraculously effective in preventing death from dehydration shock in these diseases. The cost of treatment is only a few cents per child. The British medical journal *Lancet*, called ORT "the most important medical advance of the century."

"B" stands for breastfeeding. Babies who are breastfed receive natural immunity to diseases from antibodies in their mothers' milk, but infant formula companies persuaded mothers in many developing countries that bottle-feeding is more modern and healthful than breastfeeding. Unfortunately, these mothers usually do not have access to clean water to combine with the formula and they cannot afford enough expensive synthetic formula to nourish their babies adequately. Consequently, the mortality among bottle-fed babies is much higher than among breastfed babies in developing countries.

"I" is for universal immunization against the six largest, preventable, communicable diseases of the world: measles, tetanus, tuberculosis, polio, diphtheria, and whooping cough. In 1975, less than 10 percent of the developing world's children had been immunized. By 1990, this number had risen to over 50 percent. Although the goal of full immunization for all children has not yet been reached, many lives are being saved every year. In some countries, yellow fever, typhoid, meningitis, cholera, and other diseases also urgently need attention.

Burkina Faso provides an excellent example of how a successful immunization campaign can be carried out. Although this West African nation is one of the poorest in the world (annual gross national product per capita of only $140), and its roads, health care clinics, communication, and educational facilities are either nonexistent or woefully inadequate, a highly successful "vaccination commando" operation was undertaken in 1985. In a single three-week period, one million children were immunized against three major diseases (measles, yellow fever, and meningitis) with only a single injection. This represents 60 percent of all children under age 14 in the country. The cost was less than $1 per child.

In addition to being an issue of humanity and compassion, reducing child mortality may be one of the best ways to stabilize world population growth. There has never been a reduction in birth rates that was not preceded by a reduction in infant mortality. When parents are confident that their children will survive, they tend to have only the number of children they actually want, rather than "compensating" for likely deaths by extra births. In Bangladesh, where ORT was discovered, a children's health campaign in the slums of Dacca has reduced infant mortality rates 21 percent since 1983. In that same period, the use of birth control increased 45 percent and birth rates decreased 21 percent.

Sri Lanka, China, Costa Rica, Thailand, and the Republic of Korea have reduced child deaths to a level comparable to those in many highly developed countries. This child survival revolution has been followed by low birth rates and stabilizing populations. The United Nations Children's Fund estimates that if all developing countries had been able to achieve similar birth and death rates, there would have been nine million fewer child deaths in 1987, and nearly 22 million fewer births.

Despite the successes of the child survival revolution, a decline in medical infrastructure, drought and resultant famine, and conflict in areas of the developing world led to a continued persistence in child mortality. In 2007, a new effort called the Partnership for Maternal, Newborn, and Child Health, consisting of UNICEF, the World Health Organization, the World Bank, and the United States Agency for International Development (USAID) was launched. Several independent foundations and non-government agencies (NGOs) have joined the Partnership, which coordinates efforts to focus on the UN's Millennium Development Goal number four, reducing child mortality. The Partnership aims to lower childhood mortality rates to thirty-one out of 1000 children by 2015, a reduction of two thirds of child mortality rates in 1990. The partnership focuses on delivering a set of key interventions in partnership with governments including insecticide treated bed nets, oral rehydration and nutrition supplementation, vaccinations, medications against common childhood diseases, clean drinking water,

vitamin supplementation, and building a continuum of care for mothers and children.

See also Demographic transition.

Resources

BOOKS

A Fair Chance at Life: Why Equity Matters for Child Mortality: A Save the Children Report for the 2010 Summit on the Millennium Development Goals. London: International Save the Children Alliance, 2010.

Boone, Peter, and Zhaoguo Zhan. *Lowering Child Mortality in Poor Countries: The Power of Knowledgeable Parents.* London: Centre for Economic Performance, 2006.

Franz, Jennifer S. *Child Mortality, Poverty and Environment in Developing Countries.* St. Andrews: University of St. Andrews, 2006.

Multani, Sukhvinder K. *Infant and Child Mortality: Issues and Initiatives.* Hyderabad, India: ICFAI University Press, 2009.

OTHER

The Partnership for Maternal, Newborn, and Child Health. http://www.who.int/pmnch/en/ (accessed November 7, 2010).

UNICEF. "The State of the World's Children 2010." http://www.unicef.org/sowc/ (accessed November 7, 2010).

William P. Cunningham

Chimpanzees

Common chimpanzees (*Pan troglodytes*) are widespread in the forested parts of West, Central, and East Africa. Pygmy chimpanzees, or bonobos (*P. paniscus*), are restricted to the swampy lowland forests of the Zaire basin. Despite their names, common chimpanzees are no longer common, and pygmy chimpanzees are no smaller than the other species.

Chimpanzees are partly arboreal and partly ground-dwelling creatures. They feed in fruit trees by day, nest in other trees at night, and can move rapidly through treetops. On the ground, chimpanzees usually walk on all fours (knuckle walking), because their arms are longer than their legs. Their hands have fully opposable thumbs and, although lacking a precision grip, can manipulate objects dexterously. Chimpanzees make and use a variety of tools; they shape and strip "fishing sticks" from twigs to poke into termite mounds, and they chew the ends of shoots to fashion fly whisks. They also throw sticks and stones as offensive weapons and hunt and kill young monkeys.

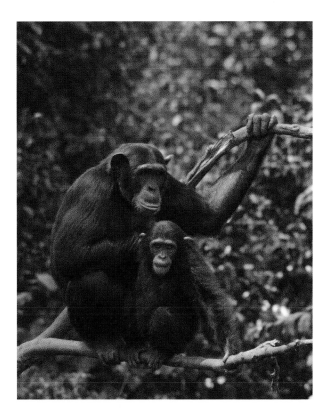

Chimps (*Pan troglodytes*). *(©I Candy/Image from BigStockPhoto.com.)*

These apes live in small nomadic groups of three to six animals (common chimpanzee) or six to fifteen animals (pygmy chimpanzee), which make up a larger community (30–80 individuals) that occupies a territory. Adult males cooperate in defending their territory against predators. Chimpanzee society consists of fairly promiscuous mixed-sex groups. Female common chimpanzees are sexually receptive for only a brief period in mid-month (estrous), while female pygmy chimpanzees are sexually receptive for most of the month. Ovulating females capable of fertilization have swollen pink hind quarters and copulate with most of the males in the group. Female chimpanzees give birth to a single infant after a gestation period of about eight months.

British ethologist Jane Goodall (1934–) has studied common chimpanzees for almost 30 years in the Gombe Stream National Park of Tanzania. She found that chimpanzee personalities are as variable as those of humans, that chimpanzees form alliances, have friendships, have personal dislikes, and run feuds. Chimpanzees also have a cultural tradition, that is, they pass learned behavior and skills from generation to generation. Chimpanzees have been taught complex sign language (the chimpanzee larynx won't allow speech) through which abstract ideas have been conveyed. These studies show that chimpanzees can develop a large vocabulary and that they can manipulate this vocabulary to frame new thoughts.

Humans share 98.4 percent of their genes with chimpanzees, so only 1.6 percent of human DNA is responsible for all the differences between the two species. The DNA of gorillas differs 2.3 percent from chimpanzees, which means that the closest relatives of chimpanzees are humans, not gorillas. Further studies of these close relatives would undoubtedly help to better understand the origins of human social behavior and human evolution. Despite this special status, both species of chimpanzees are threatened by the destruction of their forest habitat, by hunting for bushmeat, and by capture for research. The International Union for Conservation of Nature and Natural Resources (IUCN) considers chimpanzees an endangered species and estimates there are from 175,000–300,000 chimpanzees remaining in the wild.

Resources

BOOKS

Boesch, Christophe. *The Real Chimpanzee: Sex Strategies in the Forest.* Cambridge: Cambridge University Press, 2009.
Cohen, Jon. *Almost Chimpanzee.* New York: Times, 2010.
Goodall, Jane, and Dale Peterson. *Africa in My Blood: An Autobiography in Letters The Early Years.* Boston: Houghton Mifflin, 2000.
Lonsdorf, Elizabeth, Stephen R. Ross, Tetsuro Matsuzawa, and Jane Goodall. *The Mind of the Chimpanzee: Ecological and Experimental Perspectives.* Chicago: University of Chicago Press, 2010.
Redmond, Ian. *The Primate Family Tree: The Amazing Diversity of Our Closest Relatives.* Buffalo, N.Y: Firefly Books, 2008.

OTHER

Great Ape Trust. "Chimpanzees." http://www.greatapetrust.org/great-apes/chimpanzees (accessed November 9, 2010).
Jane Goodall Institute. "Chimpanzees." http://www.janegoodall.org/chimpanzees (accessed November 9, 2010).
National Geographic. "Chimpanzees." http://animals.nationalgeographic.com/mammals/chimpanzee (accessed November 9, 2010).

Neil Cumberlidge

Chipko Andolan movement

India has a long history of non-violent, passive resistance in social movements rooted in its Hindu concept of *ahimsa*, or "no harm." During the British occupation of India in the early twentieth century, Indian leader Mohandas K. Gandhi began to employ a method of resistance against the British that he called *satyagraha* (meaning "force of truth"). Synthesized from his knowledge of Henry David Thoreau, Leo Tolstoy, Christianity, and Hinduism, Gandhi's concept of satyagraha involves the absolute refusal to cooperate with a perceived wrong and the use of nonviolent tactics in combination with complete honesty to confront, and ultimately convert, evil.

During the occupation, the rights of peasants to gather products, including forest materials, was severely curtailed. New land ownership systems imposed by the British transformed what had been communal village resources into the private property of newly created landlords. Furthermore, policies that encouraged commercial exploitation of forests were put into place. Trees were felled on a large scale to build ships for the British Royal Navy or to provide ties for the expanding railway network in India, severely depleting forest resources on which traditional cultures had long depended.

In response to British rule with its forest destruction and impoverishment of native people, a series of non-violent movements utilizing satyagraha spread throughout India. The British and local aristocracy suppressed these protests brutally, massacring unarmed villagers by the thousands, and jailing Gandhi a number of times, but Gandhi and his allies remained steadfast in their resistance. The British, forced to comprehend the horror of their actions and unable to scapegoat the nonviolent Indians, at last withdrew from India.

After India gained independence, two of Gandhi's disciples, Mira Behn and Sarala Behn, moved to the foothills of the Himalayas to establish *ashramas* (spiritual retreats) dedicated to raising women's status and rights. Their project was dedicated to four major goals: (1) organizing local women, (2) campaigning against alcohol consumption, (3) fighting for forest protection, and (4) setting up small, local, forest-based industries.

During the 1970s, commercial loggers began large- scale tree felling in the Garhwal region in the state of Uttar Pradesh in northern India. Landslides and floods resulted from stripping the forest cover from the hills. The firewood on which local people depended was destroyed, threatening the way of life of the traditional forest culture.

In April 1973, village women from the Gopeshwar region who had been educated and empowered by the principles of non-violence devised by the Behns began to confront loggers directly, wrapping their arms around trees to protect them. The outpouring of support sparked by their actions was dubbed the *Chipko*

Andolan movement (literally, "movement to hug trees"). This crusade to save the forests eventually prevented logging on 4,633 square miles (12,000 km^2) of sensitive watersheds in the Alakanada basin. Today, the Chipko Andolan movement has grown to more than four thousand groups working to save India's forests. Their slogan is: "What do the forests bear? Soil, water, and pure air."

The successes of this movement, both in empowering local women and in saving the forests on which they depend, are inspiring models for grassroots green movements around the world.

See also Gandhi, Mohandas Karamchand.

Resources

BOOKS

Bankoti, T.S. *Chipko Movement.* S.l.: Global Vision Pub, 2008.
Guha, Ramachandra. *The Unquiet Woods: Ecological Change and Peasant Resistance in the Himalaya.* Ranikhet: Permanent Black, 2010.

William P. Cunningham
Jeffrey Muhr

Chisel plow *see* **Conservation tillage.**

Chisso Chemical Company *see* **Minamata disease.**

Chlordane

Chlordane and a closely related compound, heptachlor, belong to a group of chlorine-based pesticides known as cyclodienes. They were among the first major chemicals to attract national attention and controversy, mainly because of their devastating effects on wildlife and domestic animals. By the 1970s, they had become two of the most popular pesticides for home and agricultural uses (especially for termite control), despite links between these chemicals and the poisoning of birds and other wildlife, pets and farm animals, as well as links to leukemia and other cancers in humans.

In 1975, environmentalists finally persuaded Environmental Protection Agency (EPA) to issue an immediate temporary ban on most uses of chlordane and heptachlor based on an "imminent hazard of cancer in man." In 1978, the EPA agreed to phase out most remaining uses of chlordane and heptachlor.

In the United States, except for use in termite pesticides, Chlordane was banned from use in 1983. In 1988 the EPA banned all use of Chlordane.

Chlorinated hydrocarbons

Chlorinated hydrocarbons are compounds made of carbon, hydrogen, and chlorine atoms. These compounds can be aliphatic, meaning they do not contain benzene, or aromatic, meaning they do. The chlorine functional group gives these compounds a certain character; for instance, the aromatic organochlorine compounds are resistant to microbial degradation; the aliphatic chlorinated solvents have certain anesthetic properties (e.g., chloroform); some are known for their antiseptic properties (e.g., hexachloraphene). The presence of chlorine imparts toxicity to many organochlorine compounds (e.g., chlorinated pesticides).

Chlorinated hydrocarbons have many uses, including chlorinated solvents, organochlorine pesticides, and industrial compounds. Common chlorinated solvents are dichloromethane (methylene chloride), chloroform, carbon tetrachloride, trichloroethane, trichloroethylene, tetrachloroethane, tetrachloroethylene. These compounds are used in drycleaning solvents, degreasing agents for machinery and vehicles, paint thinners and removers, laboratory solvents, and in manufacturing processes, such as coffee decaffeination. These solvents are hazardous to human health and exposures are regulated in the workplace. Some are being phased out for their toxicity to humans and the environment, as molecules have the potential to react with and destroy stratospheric ozone.

The organochlorine pesticides include several subgroups, including the cyclodiene insecticides (e.g., chlordane, heptachlor, dieldrin), the DDT family of compounds and its analogs, and the hexachlorocyclohexanes (often incorrectly referred to as BHCs, or benzene hexachlorides). These insecticides were developed and marketed extensively after World War II, but due to their toxicity, persistence, widespread environmental contamination, and adverse ecological impacts, most were banned or restricted for use in the United States in the 1970s and 80s. These insecticides generally have low water solubilities, a high affinity for organic matter, readily bioaccumulate in plants and animals, particularly aquatic organisms, and have long environmental half-lives compared to the currently-used insecticides.

There are many chlorinated industrial products and reagent materials. Examples include vinyl chloride, which is used to make PVC (polyvinyl chloride) plastics; chlorinated benzenes, including hexachlorobenzene; PCB (polychlorinated biphenyl), used extensively in electrical transformers and capacitors; chlorinated phenols, including pentachlorophenol (PCP); chlorinated naphthalenes; and chlorinated diphenylethers. They represent a diversity of applications, and are valued for their low reactivity and high insulating properties.

There are also chlorinated byproducts of environmental concern, particularly the polychlorinated dibenzo-p-dioxins (PCDDs) and the polychlorinated dibenzofurans (PCDFs). These families of compounds are products of incomplete combustion of organochlorine-containing materials, and are primarily found in the fly ash of municipal solid waste incinerators. The most toxic component of PCDDs, 2,3,7,8-tetrachlorodibenzo-p-dioxin (2,3,7,8- TCDD), was a trace contaminant in the production of the herbicide 2,4,5-T and is found in trace amounts in 2,4,5- trichlorophenol and technical grade pentachlorophenol. PCDDs and PCDFs can also be formed in the chlorine bleaching process of pulp and paper mills, and have been found in their effluent and in trace amounts in some paper products.

Resources

BOOKS

Comyns, Alan E. *Encyclopedic Dictionary of Named Processes in Chemical Technology*. Boca Raton: CRC Press, 2007.

OTHER

Centers for Disease Control and Prevention (CDC). "Chlorine." http://emergency.cdc.gov/agent/chlorine/index.asp (accessed September 4, 2010).
United States Environmental Protection Agency (EPA). "Pollutants/Toxics: Chemicals: Polychlorinated Biphenyls (PCBs)." http://www.epa.gov/ebtpages/pollchemicalspolychlorinatedbiphenylspcbs.html (accessed September 4, 2010).
United States Environmental Protection Agency (EPA). "Pollutants/Toxics: Soil Contaminants: Polychlorinated Biphenyls (PCBs)." http://www.epa.gov/ebtpages/pollsoilcontaminanpolychlorinatedbiphenylspcbs.html (accessed September 4, 2010).

Deborah L. Swackhammer

Chlorination

Chlorination refers to the application of chlorine for the purposes of oxidation. The forms of chlorine used for chlorination include: chlorine gas, hypochlorous acid (HOCl), hypochlorite ion (OCl), and chloramines or combined chlorine (Mono-, di-, and tri-chloramines). The first three forms of chlorine are known as free chlorine.

Chlorine (Cl) has three valences under normal environmental conditions, -1, 0 and +1. Environmental scientists often refer only to the chlorine forms having 0 and +1 valences as chlorine; they refer to the -1 form as chloride. Chlorine with a valence of 0 (Cl_2) and chlorine with a valence of +1 (HOCl) both have the ability to oxidize materials, whereas chlorine at a -1 valence, chloride, is already at its lowest oxidation state and has no oxidizing power.

The functions of chlorination are to disinfect water or wastewater, decolorize waters or fabrics, sanitize and clean surfaces, remove iron and manganese, and reduce odors. The fundamental principle of each application is that due to its oxidizing potential, chlorine is able to effect many types of chemical reactions. Chlorine can cause alterations in DNA, cell-membrane porosity, enzyme configurations, and other biochemicals; the oxidative process can also lead to the death of a cell or virus. Chemical bonds, such as those in certain dyes, can be oxidized, causing a change in the color of a substance. Textile companies sometimes use chlorine to decolorize fabrics or process waters. In some cases, odors can be reduced or eliminated through oxidation. However, the odor of certain compounds, such as some phenolics, is aggravated through a reaction with chlorine. Certain soluble metals can be made insoluble through oxidation by chlorine (soluble Fe^{2+} is oxidized to insoluble Fe^{3+}), making the metal easier to remove through sedimentation or filtration.

Chlorine is commercially available in three forms; it can also be generated on-site. For treating small quantities of water, calcium hypochlorite ($Ca(OCl)_2$), commonly referred to as high test hypochlorite (HTH) because one mole of HTH provides two OCl ions, is sometimes used. For large applications, chlorine gas (Cl_2) is the most wide used source of chlorine. It reacts readily with water to form various chlorine species and is generally the least expensive source. There are, however, risks associated with the handling and transport of chlorine gas, and these have convinced some to use sodium hypochlorite (NaOCl) instead. Sodium hypochlorite is more expensive than chlorine gas, but less expensive than calcium hypochlorite. Some utilities and industries have generated chlorine on-site for many years, using electrolysis to oxidize chloride ions to chlorine. The process is practical in remote areas where brine, a source of chloride ions, is readily available.

Chlorine has been used in the United States since the early 1900s for disinfection. It is still commonly used to disinfect wastewater and drinking water, but the rules guiding its use are gradually changing. Until recently, chlorine was added to wastewater effluents from treatment plants without great concern over its effects on the environment. The environmental impact was thought to be insignificant since chlorine was being used in such low concentrations. However, evidence has accumulated showing serious environmental consequences from the discharge of even low levels of various forms of chlorine and chlorine compounds, and many plants now dechlorinate their wastewater after allowing the chlorine to react with the wastewater for 30–60 minutes.

The use of chlorine to disinfect drinking water is undergoing a similar review. Since the 1970s, it has been suspected that chlorine and some by-products of chlorination are carcinogenic. Papers published in 1974 indicated that halogenated methanes are formed during chlorination. During the mid-1970s the Environmental Protection Agency (EPA) conducted two surveys of the drinking-water supply in the United States, the National Organics Reconnaissance Survey and the National Organics Monitoring Survey, to determine the extent to which trihalomethanes (THMs) (chloroform, bromodichloromethane, dibromochloromethane, bromoform) and other halogenated organic compounds were present. The studies indicated that drinking water is the primary route by which humans are exposed to THMs and that THMs are the most commonly detected synthetic organic chemicals in United States' drinking water.

Chloroform was the THM found in the highest concentrations during the surveys. The risks associated with drinking water containing high levels of chloroform are not clear. It is known that 0.2 quarts (200 ml) of chloroform is usually fatal to humans, but the highest concentrations in the drinking water surveyed fell far below (311 ug/l) this lethal dose. The potential carcinogenic effects of chloroform are more difficult to evaluate. It does not cause *Salmonella typhimurium* in the Ames test to mutate, but it does cause mutations in yeast and has been found to cause tumors in rats and mice. However, the ability of chloroform to cause cancer in humans is still questionable, and the EPA has classified it and other THMs as probable human carcinogens. Based on these data, the maximum contaminant level for THMs in drinking water is now 100 ug/l. This is an enforceable standard and requires the monitoring and reporting of THM concentrations in drinking water.

There are several ways to test for chlorine, but among the more common methods are iodometric, DPD (N,N-Diethyl-p-phenylenediamine) and amperometric. DPD and amperometric methods are generally used in the water and wastewater treatment industry. DPD is a dye which is oxidized by the presence of chlorine, creating a reddish color. The intensity of the color can then be measured and related to chlorine level; the DPD solution can be titrated with a reducing agent (ferrous ammonium sulfate) until the reddish color dissipates. In the amperometric titration method, an oxidant sets up a current in a solution which is measured by the amperometric titrator. A reducing agent (phenylarsine oxide) is then added slowly until no current can be measured by the titrator. The amount of titrant added is commonly related to the amount of chlorine present.

To minimize the problem of chlorinated byproducts, many cities in the United States, including Denver, Portland, St. Louis, Boston, Indianapolis, Minneapolis, and Dallas, use chloramination rather than simple chlorination. Chlorine is still required for chloramination, but ammonia is added before or at the same time to form chloramines. Chloramines do not react with organic precursors to form halogenated by-products including THMs. The problem in using chloramines is that they are not as effective as the free chlorine forms at killing pathogens.

Questions still remain about whether the levels of chlorine currently used are dangerous to human health. The level of chlorine in most water supplies is approximately 1 mg/l, and there is evidence that the chlorinated by- products formed are not hazardous to humans at these levels. There are some risks, nevertheless, and perhaps the most important question is whether these outweigh the benefits of using chlorine. The final issue concerns the short-term and long-term effects of discharging chlorine into the environment. Dechlorination would be yet another treatment step, requiring the commitment of additional resources. At the present time, the general consensus is that chlorine is more beneficial than harmful. However, it is important to note that a great deal of research is now underway to explore the benefits of using alternative disinfectants such as ozone, chlorine dioxide, and ultraviolet light. Each alternative poses some problems of its own, so despite the current availability of a great deal of research data, the selection of an alternative is difficult.

Resources

OTHER

Centers for Disease Control and Prevention (CDC). "Chlorine." http://emergency.cdc.gov/agent/chlorine/index.asp (accessed November 9, 2010).

Gregory D. Boardman

Chlorine

Chlorine is an element of atomic number 17 and atomic mass of 35.45 atomic mass units. It belongs to Group 17 of the periodic table and is thus a halogen. Halogens are a highly reactive group of elements that, in addition to chlorine, include fluorine, iodine, bromine and another element that does not occur in nature, astatine, but is produced artificially by bombarding bismuth with alpha particles. Halogens are extremely reactive because they have an unpaired electron in their outermost electron shell. Due to its highly reactive nature, chlorine is usually not found in a pure form in nature, but is rather typically bound to other elements such as sodium, calcium, or potassium. In its pure form, chlorine exists as a diatomic molecule, meaning a molecule containing two of the same atoms. This form of chlorine is a yellow-green gas at room temperature. Chlorine gas is more dense than air, condenses to form a liquid at $-29°F(-34°C)$, and freezes into a solid at $-153°F(-103°C)$. Because of its reactivity, desirable properties, and abundance, chlorine is an exceptionally useful element. Since it readily combines with other elements and molecules, chlorine is a main component and vital reactant in the manufacture of thousands of useful products. In addition, almost all municipal water treatment systems in the United States depend on chlorine chemicals to provide clean and safe drinking water. The Chlorine Chemistry Division (part of the American Chemistry Council) estimates that in 2009, chlorine was used in the production of 93 percent of all "life-saving" pharmaceuticals, and in 86 percent of all crop protection chemicals (pesticides and herbicides). Moreover, chlorine chemicals are powerful bleaching agents used in paper processing and inexpensive but highly effective disinfectants.

Chlorine also has an important impact on the economy. The Chlorine Chemistry Division reports that the chlorine industry contributes some $46 billion annually to the North American economy. The automobile industry relies heavily on chlorine because cars contain many components that use chlorine in their manufacture. Three major categories of products produced in the United States—PVC, pickled steel, and paint—account for a large portion of industrial chlorine use. Polyvinylchloride (PVC) is a chlorinated hydrocarbon polymer that is used to make dashboards, air bags, wire covers, and sidings. Chlorine is also used in the manufacture of car exterior paints. Automobile coatings typically use titanium dioxide, which requires chlorine for synthesis. In addition, chlorine is used in the production of pickled steel for automobile frames and undercarriages. The pickling process provides an

applications—their stability, for example—appeared to make them environmentally benign.

However, by the mid–1970s, the error in that view became apparent. Scientists began to find that CFCs in the stratosphere were decomposed by sunlight. One product of that decomposition, atomic chlorine, reacts with ozone (O_3) to form ordinary oxygen (O_2). The apparently harmless CFCs turned out, instead, to be a major factor in the loss of ozone from the stratosphere.

By the time this discovery was made in the early 1970s, levels of CFCs in the stratosphere were escalating rapidly. The concentration of these compounds climbed from 0.8 part per billion in 1950 and 1.0 part per billion in 1970, to 3.5 parts per billion in 1987.

A turning point in the CFC story came in the mid-1980s when scientists found that a large "hole" in the ozone layer (more precisely, a severe thinning of the ozone layer) was opening up over the Antarctic each year. This discovery spurred world leaders to act on the problem of CFC production. In 1987, about forty nations met in Montreal to draft a treaty to greatly reduce the production and use of CFCs worldwide. The agreement that resulted became known as the Montreal Protocol.

While this action was encouraging, the improvements come rather slowly. CFC compounds remain in the atmosphere for long periods of time (about 77 years for CFC-11 and 139 years for CFC–12), so they will continue to pose a threat to the ozone layer for many decades to come. However, measurements have demonstrated that the ozone hole is recovering in the ways predicted from the phase-out of CFC's. Measurements in 2010, for instance, showed that the annually-appearing ozone "hole" over the Antarctic was one the smallest of the decade, indicating a general recovery of the ozone layer there. Atmospheric scientists predict that complete recovery of ozone depletion should come about by 2050. This improvement in atmospheric ozone levels suggests that the Protocol's intent will likely be met, and that concerns over the illegal trade and manufacture of CFC's seems less of a worry than initially thought by environmentalists and scientists.

See also Carbon; Chlorine; Hydrogen.

Resources

BOOKS

Goodstein, Eban S. *Economics and the Environment.* Hoboken, NJ: John Wiley & Sons, 2008.

Manahan, Stanley E. *Environmental Chemistry.* 9th ed. Boca Raton, FL: CRC Press, 2009.

OTHER

United States Environmental Protection Agency (EPA). "Pollutants/Toxics: Air Pollutants: Chlorofluorocarbons (CFCs)." http://www.epa.gov/ebtpages/pollair pollutantschlorofluorocarbonscfcs.html (accessed November 9, 2010).

David E. Newton

Cholera

Cholera is one of the most contagious diseases transmitted by water. Although up to 75 percent of people infected with the bacteria that cause cholera develop few symptoms or a milder form of the disease, severe cholera can kill its victim within hours. Cholera is marked by diarrhea, and in severe cases, copious

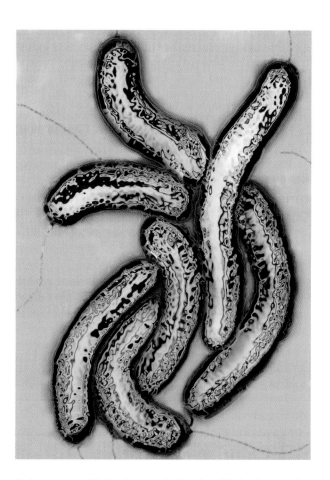

Color enhanced light micrograph showing *Vibrio cholerae*, the bacterium that causes cholera in humans. *(James Cavallini / Photo Researchers, Inc.)*

"rice water" diarrhea that results in dehydration, often followed by shock and death if not treated. The World Health Organization estimates that there are more than three million cases of cholera every year resulting in up to 120,000 deaths. More than 95 percent of fatalities from cholera occur in the Indian sub-continent and Africa.

Cholera is caused by the bacillus *Vibrio cholerae*, a member of the family Vibrionaceae, which are described as Gram-negative, non-sporulating rods that are slightly curved, motile, and have a fermentative metabolism.

The natural habitat of *V. cholerae* is human feces, but some studies have indicated that natural waters may also be a habitat of the organism. Fecal contamination of water is the most common means by which *V. cholerae* is spread, however, food, insects, soiled clothing, or person-to-person contact may also transmit sufficient numbers of the pathogens to cause cholera.

The ability of *V. cholerae* to survive in water is dependent upon the temperature and water type. *V. cholerae* reportedly survive longer at low temperatures, and in seawater, sterilized water, and nutrient rich waters. Also, the particular strain of *V. cholerae* affects the survival of the organism in water, as some strains or types are hardier than others. Most methods to isolate *V. cholerae* from water include concentration of the sample by filtration and exposure to high pH and selective media. Identification of pathogenic strains of *V. cholerae* is dependent upon agglutination tests. Final confirmation of the strain or type must be done in a specialized laboratory.

Persons infected with *V. cholerae* produce 10^7 to 10^9 organisms per milliliter in the stool at the height of the disease, but the number of excreted organisms drops off quickly as the disease progresses. Asymptomatic carriers of *V. cholerae* excrete 10^2 to 10^5 organisms per gram of feces. The mild form of the illness lasts for five to seven days. Hydration therapy and electrolyte replacement is the treatment of choice for less severe cases of cholera. Antibiotics can be used to shorten the course of the disease, but are not generally recommended as they contribute to the emergence of multiple antibiotic resistant strains of cholera in many areas. Vaccines exist to prevent cholera, however, they do not prevent the acquisition of the bacteria in the gastrointestinal tract, do not diminish symptoms in persons already infected, and are effective for less than a year. Proper water treatment should eliminate *V. cholerae* from drinking water, however, the most effective control of this pathogen is dependent upon adequate sanitation and hygiene.

Resources

BOOKS

Hempel, Sandra. *The Strange Case of the Broad Street Pump: John Snow and the Mystery of Cholera.* Berkeley: University of California Press, 2006.

OTHER

Centers for Disease Control and Prevention (CDC). "Cholera." http://www.cdc.gov/cholera/ (accessed November 6, 2010).
World Health Organization (WHO). "Cholera." http://www.who.int/topics/cholera/en (accessed November 6, 2010).
World Health Organization (WHO). "Global Task Force on Cholera Control." WHO Programs and Projects. http://www.who.int/entity/cholera/en (accessed November 6, 2010).

E. K. Black
Gordon R. Finch

Cholinesterase inhibitor

Insecticides kill their target insect species in a variety of ways. Two of the most commonly used classes of insecticide are the organophosphates (nerve gases) and the carbamates. These compounds act quickly (in a matter of hours), are lethal at low doses (parts per billion), degrade rapidly (in hours to days) and leave few toxic residues in the environment. Organophosphates kill insects by inducing loss of control of the peripheral nervous system, leading to uncontrollable spasms followed by paralysis and, ultimately, death. This is often accomplished by a biochemical process called cholinesterase inhibition.

Most animals' nervous systems are composed of individual nerve cells called neurons. Between any two adjacent neurons there is always a gap, called the synaptic cleft; the neurons do not actually touch each other. When an animal senses something—for example, pain—the sensation is transmitted chemically from one neuron to another until the impulse reaches the brain or central nervous system. The first neuron (pre-synaptic neuron) releases a substance, known as a transmitter, into the synaptic cleft. One of the most common chemical transmitters is called acetylcholine. Acetylcholine then diffuses across the gap and binds with receptor sites on the second neuron (post-synaptic neuron). Reactions within the target neuron triggered by occupied receptors result in further transmission of the signal. As soon as the impulse has been transmitted, the acetylcholine in the gap is immediately destroyed by an enzyme called cholinesterase; the destruction of the acetylcholine is an absolutely essential part of the nervous process. If the acetylcholine is not destroyed, it continues to stimulate indefinitely the transmission of impulses from one neuron to the next, leading to loss of all control over the peripheral nervous system. When control is lost, the nervous system is first overstimulated and then paralyzed until the animal dies. Thus, organophosphate insecticides bind to the cholinesterase enzyme, preventing the cholinesterase from destroying the acetylcholine and inducing the death of the insect.

Some trade names for organophosphate insecticides are malathion and parathion. Carbamates include aminocarb and carbaryl. The carbamates produce the same effect of cholinesterase inhibition as the organophosphates, but the chemical reaction of the carbamates is more easily reversible. The potency or power of these compounds is usually measured in terms of the quantity of the pesticide (or inhibitor) that will produce a 50 percent loss of cholinesterase activity. Since acetylcholine transmission of nervous impulses is common to most vertebrates as well as insects, there is a great potential for harm to non-target species from the use of cholinesterase-inhibiting insecticides. Therefore the use of these insecticides is highly regulated and controlled. Access to treated areas and contact with the compounds is prohibited until the time period necessary for the breakdown of the compounds to non-toxic end products has elapsed.

Usha Vedagiri

Chromatography

Chromatography is the process of separating mixtures of chemicals into individual components as a means of identification or purification. It derives from the Greek words *chroma*, meaning color, and *graphy*, meaning writing. The word was coined in 1906 by the Russian chemist Mikhail Tsvett who used a column to separate plant pigments. Currently chromatography is applied to many types of separations far beyond those of just color separations. Common chromatographic applications include gas-liquid chromatography (GC), liquid-solid chromatography (LC), thin layer chromatography (TLC), ion exchange chromatography, and gel permeation chromatography (GPC). All of these methods are invaluable in

analytical environmental chemistry, particularly GC, LC, and GPC.

The basic principle of chromatography is that different compounds have different retentions when passed through a given medium. In a chromatographic system, one has a mobile phase and a stationary phase. The mixture to be separated is introduced in the mobile phase and passed through the stationary phase. The compounds are selectively retained by the stationary phase and move at different rates which allows the compounds to be separated.

In gas chromatography, the mobile phase is a gas and the stationary phase is a liquid fixed to a solid support. Liquid samples are first vaporized in the injection port and carried to the chromatographic column by an inert gas which serves as the mobile phase. The column contains the liquid stationary phase, and the compounds are separated based on their different vapor pressures and their different affinities for the stationary phase. Thus different types of separations can be optimized by choosing different stationary phases, and by altering the temperature of the column. As the compounds elute from the end of the column, they are detected by one of a number of methods that have specificity for different chemical classes.

Liquid chromatography consists of a liquid mobile phase and a solid stationary phase. There are two general types of liquid chromatography: column chromatography and high pressure liquid chromatography (HPLC). In column chromatography, the mixture is eluted through the column containing stationary packing material by passing successive volumes of solvents or solvent mixtures through the column. Separations result as a function of both chemical-solvent interactions as well as chemical-stationary phase interactions. Often this technique is used in a preparative manner to remove interferences from environmental sample extracts. HPLC refers to specific instruments designed to perform liquid chromatography under very high pressures to obtain a much greater degree of resolution. The column outflow is passed through a detector and can be collected for further processing if desired. Detection is typically by ultraviolet light or fluorescence.

A variation of column chromatography is gel permeation chromatography (GPC), which separates chemicals based on size exclusion. The column is packed with porous spheres, which allow certain size chemicals to penetrate the spheres and excludes larger sizes. As the sample mixture traverses the column, larger molecules move more quickly and elute first while smaller molecules require longer elution times.

An example of this application in environmental analyses is the removal of lipids (large molecules) from fish tissue extracts being analyzed for pesticides (small molecules).

Resources

BOOKS

Ettre, Leslie S., and John V. Hinshaw. *Chapters in the Evolution of Chromatography.* London: Imperial College Press, 2008.

McNair, Harold Monroe, and James M. Miller. *Basic Gas Chromatography.* Hoboken, N.J.: John Wiley & Sons, 2009.

Miller, James M. *Chromatography: Concepts and Contrasts.* New York: Wiley-Interscience, 2004.

Niessen, W. M. A. *Liquid Chromatography—Mass Spectrometry.* Boca Raton: CRC/Taylor & Francis, 2006.

Deborah L. Swackhammer

Chronic effects

Chronic effects occur over a long period of time. The length of time termed "long" is dependent upon the life cycle of the organism being tested. For some aquatic species a chronic effect might be seen over the course of a month. For animals such as rats and dogs, chronic would refer to a period of several weeks to years.

Chronic effects can be either caused by chronic or acute exposures. Acute exposure to some metals and many carcinogens can result in chronic effects. With certain toxicants, such as cyanide, it is difficult, if not impossible, to cause a chronic effect. However, at a higher dosage, cyanide readily causes acute effects. Examples of chronic effects include pulmonary tuberculosis and, in many cases, leadpoisoning. In each disease the effects are long-term and cause damage to tissues; acute effects generally result in little tissue reaction. Thus, acute and chronic effects are frequently unrelated, and yet it is often necessary to predict chronic toxicity based on acute data. Acute data are more plentiful and easier to obtain. To illustrate the possible differences between acute and chronic effects, consider the examples of halogenated solvents, arsenic, and lead.

In halogenated solvents, acute exposure can cause excitability and dizziness, while chronic exposure will result in liver damage. Chronic effects of arsenic poisoning are in blood formation and liver and nerve damage. Acute poisoning affects the gastro-intestinal tract. Lead also effects blood formation in chronic

exposure, and damages the gastro-intestinal tract in acute exposure. Other chronic effects of exposure to lead include changes in the nervous system and muscles. In some situations, given the proper combination of dose level and frequency, those exposed will experience both acute and chronic effects.

There are chemicals that are essential and beneficial for the functions and structure of the body. The chronic effect is therefore better health, although people generally do not refer to chronic effects as being positive. However, vitamin D, fluoride, and sodium chloride are just a few examples of agents that are essential and/or beneficial when administered at the proper dosage. Too much of any of the three or too little, however, could cause acute and/or chronic toxic effects.

In aquatic toxicology, chronic toxicity tests are used to estimate the effect and no-effect concentrations of a chemical that is continuously applied over the reproductive life cycle of an organism; for example, the time needed for growth, development, sexual maturity, and reproduction. The range in chemical concentrations used in the chronic tests is determined from acute tests. Criterion for effects might include the number and percent of embryos that develop and hatch, the survival and growth of larvae and juveniles, etc.

The maximum acceptable toxicant concentration (MATC) is defined through chronic testing. The MATC is a hypothetical concentration between the highest concentration of chemical that caused no observed effect (NOEC) and the lowest observed effect concentration (LOEC). Therefore,

$$LOEC > MATC > NOEC$$

Furthermore, the MATC has been used to relate chronic toxicity to acute toxicity through an application factor (AF). AF is defined below:

$$AF = \frac{MATC}{LD_{50}}$$

The AF for one aquatic species might then be used to predict the chronic toxicity for another species, given the acute toxicity data for that species.

The major limitations of chronic toxicity testing are the availability of suitable test species and the length of time needed for a test. In animal testing, mice, rats, rabbits, guinea pigs, and/or dogs are generally used; mice and rats being the most common. With respect to aquatic studies, the most commonly used vertebrates are the fathead (fresh water) and sheepshead (saltwater) minnows. The most commonly used invertebrates are freshwater water fleas (*Daphnia*) and the saltwater mysid shrimp (*Mysidopsis*).

See also Aquatic chemistry; Detoxification; Dose response; Heavy metals and heavy metal poisoning; LD50; Plant pathology; Toxic substance.

Gregory D. Boardman

Cigarette smoke

Cigarette smoke contains more than 4,000 identified compounds. Many are known irritants and carcinogens. Since the first Surgeon General's Report on smoking and health in 1964, evidence linking the use of tobacco to illness, injury, and death has continued to mount. Many thousands of studies have documented the adverse health consequences of any type of tobacco, including cigarettes, cigars, and smokeless tobacco.

Specific airborne contaminants from cigarette smoke include respirable particles, nicotine, polycyclic aromatic hydrocarbons, arsenic, DDT, formaldehyde, hydrogen cyanide, methane, carbon monoxide, acrolein, and nitrogen dioxide. Each one of these compounds impacts some part of the body. Irritating gases like ammonia, hydrogen sulfide and formaldehyde affect the eyes, nose and throat. Others, like nicotine, impact the central nervous system. Carbon monoxide reduces the oxygen-carrying capacity of the blood, starving the body of energy. Carcinogenic agents come into prolonged contact with vital organs and with the delicate linings of the nose, mouth, throat, lungs and airways.

Cigarette smoke is one of the six major sources of indoor air pollution, along with combustion by-products, microorganisms and allergens, formaldehyde and other organic compounds, asbestos fibers, and radon and its airborne decay products. The carbon monoxide concentration in cigarette smoke is more than 600 times the level considered safe in industrial plants, and a smoker's blood typically has four to fifteen times more carbon monoxide in it than that of a nonsmoker. Airborne particle concentrations in a home with several heavy smokers can exceed ambient air quality standards.

Sidestream, or second-hand, smoke actually has higher concentrations of some toxins than the mainstream smoke the smoker inhales. Second-hand smoke carries more than thirty known carcinogens. According to a study by the Centers for Disease Control and Prevention (CDC) released in 1996, nearly nine out of 10 nonsmoking Americans are exposed to environmental tobacco smoke as measured by the levels of cotinine in their blood. The presence of cotinine, a chemical the body metabolizes from nicotine, is documentation of exposure to cigarette smoke. On the basis of health hazards of second-hand smoke, the Environmental Protection Agency has classified second-hand smoke as a Group A carcinogen, known to cause cancer in humans.

Cigarettes probably represent the single greatest source of radiation exposure to smokers in the United States today. Two naturally occurring radioactive materials, lead-210 and polonium-210, are present in tobacco. Both of these long-lived decay products of radon are deposited and retained on the large, sticky leaves of tobacco plants. When the tobacco is made into cigarettes and the smoker lights up, the radon decay products are volatilized and enter the lungs. The resulting dose to small segments of the bronchial epithelium of the lungs of about 50 million smokers in the United States is about 160 mSv per year. (One Sv = 100 rem of radiation.) The dose to the whole body is about 13 mSv, more than ten times the long-term dose rate limit for members of the public.

The CDC reports that as of 2010, more than 400,000 Americans continue to die each year from tobacco-

Cigarettes probably represent the single greatest source of radiation exposure to smokers in the United States today. *(Konstantin Sutyagin/Shutterstock.com.)*

related disease. One in every five deaths in the United States is smoking related, making smoking the largest preventable cause of illness and premature death in the United States. Death is caused primarily by heart disease, lung cancer, and chronic obstructive lung diseases such as emphysema or chronic bronchitis. In addition, the use of tobacco has been linked to cancers of the larynx, mouth and esophagus, and as a contributory factor in the development of cancers of the bladder, kidney, pancreas, and cervix. Cigarette smoke aggravates asthma, triggers allergies, and causes changes in bodily tissues that can leave smokers and nonsmokers prone to illness, especially heart disease.

About 180,000 Americans will die prematurely of coronary heart disease every year due to smoking. The risk of a stroke or heart attack is greatly increased by nicotine, which impacts the platelets that enable the blood to clot. Nicotine causes the surface of the platelets to become stickier, thereby increasing the platelets' ability to aggregate. Thus, a blood clot or thrombus forms more easily. A thrombus in an artery of the heart results in a heart attack; in an artery of the brain it results in a stroke.

Epidemiological studies reveal a direct correlation between the extent of maternal smoking and various illnesses in children. Also, studies show significantly lower heights and weights in six- to eleven-year olds whose mothers smoke. A pregnant woman who smokes faces increased risks of miscarriage, premature birth, stillbirth, infants with low birth weight, and infants with physical and mental impairments. Cigarette smoking also impairs fertility in women and men, contributes to earlier menopause, and increases a woman's risk of osteoporosis.

Cigarette smoke contains benzene which, when combined with the radioactive toxins, can cause leukemia. Although smoking does not cause the disease, smoking may boost a person's risk of getting leukemia by 30 percent.

A long-time smoker increases his risk of lung cancer by 1,000 times. In 2005, according to the CDC, about 150,000 people died of lung cancer directly attributed to cigarette smoke. More than 3,000 people each year develop lung cancer from second-hand smoke. Between 1960 and 2000, deaths from lung cancer among women have increased by more than 400 percent—exceeding breast cancer deaths.

The addiction to nicotine in cigarette smoke, a chemical and behavioral addiction as powerful as that of heroin, is well documented. The immediate effect of smoking a cigarette can range from tachycardia (an abnormally fast heartbeat) to arrhythmia (an irregular heartbeat). Deep inhalations of smoke lower the pressure in a smoker's chest and pulmonary blood vessels, which increases the amount of blood flow to the heart. This increased blood flow is experienced as a relaxed feeling. Seconds later, nicotine enters the liver and causes that organ to release sugar, which leads to a "sugar high." The pancreas then releases insulin to return the blood sugar level to normal, but it makes the smoker irritable and hungry, stimulating a desire to smoke and recover the relaxed, high feeling.

Nicotine also stimulates the nervous system to release adrenaline, which speeds up the heart and respiratory rates, making the smoker feel more tense. Lighting the next cigarette perpetuates the cycle. The greater the number of behaviors linked to the habit, the stronger the habit is and the more difficult to break. Quitting involves combating the physical need and the psychological need, and complete physical withdrawal can take up to two weeks.

From an economic point of view, the Department of Health and Human Services estimates that smoking costs the United States $50 billion in health expenses. That figure is most likely conservative because the medical costs attributable to burn care from smoking-related fires, perinatal care for low birth weight infants of mothers who smoke, and treatment of disease caused by second-hand smoke were not included in the calculation.

In January 2010, independent studies on smoking bans in Europe and the United States showed that the limits on tobacco use in offices and other public buildings and places resulted in reduced rates of heart attacks. On both national and local levels, restrictions on smoking in both Europe and North America corresponded to immediate declines in numbers of reported heart attacks. In 2008, the American Lung Association estimated that 45 million, or 20.6 percent of American adults age 28 or over regularly smoked cigarettes. Overall, this represents a continuing downward trend, as the annual prevalence of smoking in America declined more than 50 percent between 1965 and 2008.

Resources

BOOKS

Balkin, Karen. *Tobacco and Smoking*. San Diego: Greenhaven Press, 2005.

Green, Robert J. *Emphysema and Chronic Obstructive Pulmonary Disease*. San Diego, CA: Aventine Press, 2005.

Schwalbe, Michael. *Smoke Damage: Voices from the Front Lines of America's Tobacco Wars*. Madison, WI: Borderland Books, 2011.

PERIODICALS

Friedrich, M. J. "Preventing Emphysema." *JAMA The Journal of the American Medical Association* 301, no. 5 (2009): 477.

OTHER

Centers for Disease Control and Prevention (CDC). "Smoking and Tobacco Use." http://www.cdc.gov/tobacco (accessed November 7, 2010).

National Institutes of Health (NIH). "Smoking." http://health.nih.gov/topic/Smoking (accessed November 7, 2010).

Linda Rehkopf

CITES *see* **Convention on International Trade in Endangered Species of Wild Fauna and Flora (1975).**

Citizen's Clearinghouse for Hazardous Waste *see* **Gibbs, Lois.**

Citizen science

Citizen science is a term used for individual volunteers, or groups of volunteers, whot do not necessarily have specific scientific training or experiences in a particular area but still help to make observations, measurements, analyzes, and computations, and provide other such help so scientists can more easily or quickly accomplish goals and objectives. Becoming a citizen scientist, according to the Cornell University Laboratory of Ornithology (specializing in the study of birds), can be as simple as glancing periodically at a backyard bird feeder, or as complicated as getting out in the field, collecting data about the relationship between an environment's characteristics and the success of bird-nesting in that milieu (surroundings). Put simply, citizen science is the practice of involving individual citizens, through their voluntary efforts, in the work of environmental science. Such voluntary assistance encompasses a wide variety of environmentally related issues and projects, including the counting of various species of bird, monitoring rainfall amounts, or observing and surveying the habits of threatened species of wildlife. The goal of all environmental science, whether performed by private citizens or environmental scientists, is the same: to provide for the health of the planet, its natural resources, and all living beings.

John Fein, an internationally-renowned science educator, and associate professor at Griffith University (Brisbane, Queensland, Australia), described this participatory process as one that can "bridge the gap between science and the community and between scientific research and policy, decision-making and planning." "Bridging these gaps," he went on to note, "involves a process of social learning through sound environmental research, full public participation, the adoption of adaptive management practices and the development of the democratic values, skills and institutions for an active civil society."

In the United States, citizen science began in the 1800s, and became more formalized in 1886, when the National Audubon Society was created. According to its preamble, the society's aim was to: promote the conservation of wildlife and the natural environment, and educate man regarding his relationship with, and place within, the natural environment as an ecological system. Public awareness of environmental concerns increased with the first Earth Day in 1970. The significance of Earth Day has served as a reminder to the world community that it is not only the scientists who are responsible for the health of the environment. The general public, each individual, has been given the challenge to take measures large or small that might add up to an enormous improvement in the ecological scheme of life.

Today there are so many citizen science projects initiated by various environmental organizations that it would be impossible to list all of them. However, some of the better known include:

- Christmas Bird Count: Beginning in 1900, it is the oldest citizen science project in existence, according to the National Audubon Society. It occurs on a daily basis across the United States between December 14th and January 5th each year. Until the recent era of the home computer, it had been the practice for groups to go on birding outings into nature, in order to do the count.

- The Great Backyard Bird Count: Occurring for three days commencing on Valentine's Day, it involves volunteers counting the birds visiting in their backyards or in nearby parks, and entering the data into their computers. This count is then analyzed and becomes available information to the bird counters through tables, maps, and in other forms. The American ornithologist and director of bird conservation within the National Audubon Society, Gregory Butcher, works in coordinated effort with Dr. John W. Fitzpatrick (1951–), director of the Cornell Lab of Ornithology (Cornell University, Ithaca, New York), in order to calculate the results.

- Project Feeder Watch: A winter-long survey of birds that visit feeders in backyards, nature centers, community areas, and other locales that takes place November through March.

- The Birdhouse Network: The installation of bird houses with volunteers monitoring bird activity throughout the breeding season, collecting data on location, habitat characteristics, nestings, and the number of eggs produced.
- Project Pigeon Watch: Counting the number of each different color, recording the colors of courting birds, and helping scientists determine the mystery of why there are so many different colors of pigeons.
- House Finch Disease Survey: Monitoring of backyard feeders, reporting presence or absence of House Finch eye disease.
- Birds in Forested Landscapes: Studying sites established in forests of varying sizes with volunteers counting the birds during at least two visits (using recordings of vocalizations), searching for markers that breeding was successful, and recording landscape characteristics of the site.
- Golden-winged Warbler Atlas Project: Surveying and conducting point counts at known and potential breeding sites of this bird, using both professionals and volunteers.
- Citizen Science in the Schoolyard: Providing projects in elementary and middle schools that educate children in various aspects of bird-watching and counting.
- BirdSource: An interactive online database operated in conjunction with the Audubon Society that collects information from numerous projects.
- Adirondack Cooperative Loon Program: An ongoing project with volunteers through the Natural History Museum of the Adirondacks, with observation of the common loon as well as the annual count that occurs every July.
- The Mid-Atlantic Integrated Assessment (MAIA): A research, monitoring, and assessment initiative under the auspices of the Environmental Protection Agency, in order to provide high-quality scientific information on the condition of natural resources of the Mid-Atlantic region of the east coast, including the Delaware and Chesapeake Bays, Albemarle-Pamlico Sound, and the Delmarva Coastal Bays, utilizing professional researchers and volunteers.
- Smithsonian Neighborhood Nestwatch: Utilizing backyard bird counters during breeding season to assist the Smithsonian Environmental Research Center in gathering scientific data on various birds.
- Citizen Collaborative for Watershed Sustainability: A project conducted within the Southeast Minnesota Blufflands region, and coordinating with other

farming regions across the state, in order to conserve the watersheds of southeastern Minnesota.

- Connecticut Department of Environmental Protection: A citizen-based project to monitor streams and rivers, determining the chemical, physical, and biological health of the water.

Citizen science projects such as these have provided information used in decision-making resulting in the purchasing of lands that host certain threatened species during breeding time and the development of bird population management guidelines. Further, data obtained from these surveys is often published in scientific and educational journals.

Citizens respond

The explosion of popularity and participation in citizen science has been a welcome development. Both the United Nations (UN) and individual national governments have become involved, sponsoring events such as the 2010 UN Global Compact Leaders Summit in New York City, from June 24 to 25, 2010, which involves sustainability around the world.

For citizens unaware of environmental issues, it appears that more information and publicity is necessary. In an article for the *Environmental News Network* (*ENN*) in October 2001, Erica Gies states, "As a person who cares about environmental issues, I often find myself preaching to the choir. But when I have an opportunity to converse with people who are either uninformed or inclined to disagree, it can be difficult to communicate my passion effectively without alienating them, especially if they are people I have a long history with, such as family." To attain that goal of increasing awareness, Gies suggests using art, literature and the media. There are a variety of nonfiction books and essays, poetry, art, and music featuring environmental or nature themes, as well as the films focusing on environmental issues featured each spring at an Environmental Film Festival held in Washington, D.C. But, even a comic strip can give an ecological focus. Gies quotes *Dilbert* creator, Scott Adams (1957–) from his book, *The Dilbert Future* irreverently noting that, "The children are our future. And that is why, ultimately, we're screwed unless we do something about it. If you haven't noticed, the children who are our future are good looking but they aren't all that bright. As dense as they might be, they will eventually notice that adults have spent all the money, spread disease, and turned the planet into a smoky, filthy ball of death."

Japanese-Canadian environmental activist David Suzuki (1936–) also offered his insights into the challenges facing the environment in an article for *ENN* in June 2002. Enumerating the biggest challenges for the environment in the next century, Suzuki noted, "I'm beginning to think one of the biggest challenges is overcoming the fact that people are tired of all the depressing news about the environment." Citizen science was offering a solution for positive approaches and answers to these issues, giving the public a reason to be hopeful and cherish their involvement in seeing things change. One such person, Robert H. Boyle, of Cold Spring, New York, and author of the 1969 book, *The Hudson River: A Natural and Unnatural History* continued his struggle for thirty years after the publication of his book alerting people to the decay of the Hudson River. Boyle's book was the waterway's equivalent of *Silent Spring* (American biologist Rachel Carson's [1907–1964] book that alerted the public to the danger of pesticides and was responsible for much of the environmental movement of the 1960s). Boyle's book was instrumental in beginning the process of cleaning up not only the Hudson River, but all other polluted waterways across the United States. In 2002, Boyle continued to fight major industrial polluters and the government in courts in his effort to preserve the natural bounty of the Hudson River. In 2010, Boyle, along with the environmental non-profit group Riverkeeper (a part of the larger group Waterkeeper Alliance), continues efforts to protect the Hudson River and its tributaries.

With growing concern regarding global warming, deforestation, and biodiversity loss, more citizens are becoming involved and interested in environmental issues as well as improvement and protection. One citizen science program, Global Learning and Observations to Benefit the Environment (GLOBE), was established in 1995 and is sponsored by the U.S. government. GLOBE is a worldwide program involving primary- and secondary-school students and teachers in science education. There are over 50,000 teachers and over 20,000 schools worldwide in which participants record measurements of air, soil, water, and vegetation in their area for use by scientists. Scientists use this mass of information to study and understand aspects of the environment with the prospect of improving the environment in which we live.

Resources

BOOKS

American Birds The 104th Christmas Bird Count, 2003–2004. New York: National Audubon Society, 2004.

Janke, Rhonda, Rebecca Moscou, and G. Morgan Powell. *Citizen Science: Soil and Water Testing for Enhanced Natural Resource Stewardship.* Manhattan, KS: Agricultural Experiment Station and Cooperative Extension Service, Kansas State University, 2005.

Michigan Audubon Society. *A Century of Conservation and Citizen Science.* Lansing, MI: Michigan Audubon Society, 2004.

OTHER

Cornell Lab of Ornithology. "Citizen Science." http://www.birds.cornell.edu/NetCommunity/Page.aspx?pid = 708 (accessed October 12, 2010).

Environmental Film Festival. "Environmental Film Festiva." http://www.dcenvironmentalfilmfest.org (accessed October 12, 2010).

Global Learning and Observations to Benefit the Environment (GLOBE). "The GLOBE Program." http://www.globe.gov (accessed October 12, 2010).

National Audubon Society. "Christmas Bird Count." http://birds.audubon.org/christmas-bird-count (accessed October 12, 2010).

United Nations. "UN Global Compact Leaders Summit." http://www.leaderssummit2010.org (accessed October 12, 2010).

ORGANIZATIONS

Cornell Lab of Ornithology, 159 Sapsucker Woods Road, Ithaca, NY, USA, 14850, (800) 843-2473, http://www.birds.cornell.edu

The Natural History Museum of the Adirondacks, 45 Museum Drive, Tupper Lake, NY, USA, 12986, (518) 359-7800, (518) 359-3253, http://www.wildcenter.org

Smithsonian Environmental Research Center, 647 Contees Wharf Road, Edgewater, MD, USA, 21037-0028, (443) 482-2200, (443) 482-2380, http://www.serc.si.edu

Joan M. Schonbeck

Citizens for a Better Environment

Citizens for a Better Environment (CBE) is an organization that works to reduce exposure to toxic substances in land, water, and air. Founded in 1971, CBE has 30,000 members and operates with a $1.8 million budget. The organization also maintains regional offices in Minnesota and in Wisconsin.

CBE staff and members focus on research, public information, and advocacy to reduce toxic substances. They also meet with policy-makers on state, regional, and national levels. A staff of scientists, researchers, and policy analysts evaluate specific problems brought to the attention of CBE, testify at legislative and regulatory agency hearings, and file lawsuits in state and federal courts. The organization also conducts public education programs, and provides technical assistance to local residents and organizations that attempt to halt toxic chemical exposures.

The Chicago office won a Supreme Court decision against the construction of an incinerator in a low-income neighborhood in the Chicago area and is researching the issues on volume-based garbage for suburban Chicago areas. In Minnesota, the staff has developed a "Good Neighbor" program of agreements between community groups, environmental activists, businesses, and industries along the Mississippi River to reduce pollution of the river. In Wisconsin, transportation issues under study include selling ride-sharing credits to meet Clean Air Act standards. Selling and buying of credits between and among individuals, businesses, and polluting industries has become a lucrative way for polluters to continue their practices. CBE is attempting to close the legislative loopholes that allows this practice to continue.

The Chicago office maintains a library of books, reports, and articles on environmental pollution issues. Publications include CBE's *Environmental Review*, a quarterly journal on the public health effects of pollution; it includes updates of CBE activities and research projects.

Resources

ORGANIZATIONS

Citizens for a Better Environment, 152 W. Wisconsin Ave., Suite 510, Milwaukee, WI, USA, 53203, (414) 271-7280, (866) 256-5988, (414) 271-5904, cbewi@igc.apc.org, http://www.wsn.org/cbe/livablecommunities.html

Linda Rehkopf

Clay minerals

Clay minerals contribute to the physical and chemical properties of most soils and sediments. At high concentrations they cause soils to have a sticky consistency when wet. Individual particles of clay minerals are very small with diameters less than two micrometers. Because they are so finely divided, clay minerals have a very high surface area per unit weight, ranging form 5 to 800 square meters per gram. They are much more reactive than coarser materials in soils and sediments such as silt and sand and clay minerals account for much of the reactivity of soils and sediments with respect to adsorption and ion exchange.

Mineralogists restrict the definition of clay minerals to those aluminosilicates (minerals predominantly composed of aluminum, silicon, and oxygen) which in nature have particle sizes two micrometers or less in diameter. These minerals have platy structures made up of sheets of silica, composed of silicon and oxygen, and alumina, which is usually composed of aluminum and oxygen, but often has iron and magnesium replacing some or all of the aluminum.

Clay minerals can be classified by the stacking of these sheets. The one to one clay minerals have alternating silica and alumina sheets; these are the least reactive of the clay minerals, and kaolinite is the most common example. The two to one minerals have layers made up of an alumina sheet sandwiched between two silica sheets. These layers have structural defects that result in negative charges, and they are stacked upon each other with interlayer cations between the layers to neutralize the negative layer charges. Common two to one clays are illite and smectite.

In smectites, often called montmorillonite, the interlayer ions can undergo cation exchange. Smectites have the greatest ion exchange capacity of the clay minerals and are the most plastic. In illite, the layer charge is higher than for smectite, but the cation exchange capacity is lower because most of the interlayer ions are potassium ions that are trapped between the layers and are not exchangeable.

Some iron minerals also can be found in the clay-sized fraction of soils and sediments. These minerals have a low capacity for ion exchange but are very important in some adsorption reactions. Gibbsite, an aluminum hydroxide mineral, is also found in the clay-sized fraction of some soils and sediments. This mineral has a reactivity similar to the iron minerals.

Paul R. Bloom

Clay-hard pan

A compacted subsurface soil layer. Hard pans are frequently found in soils that have undergone significant amounts of weathering. Clay will accumulate below the surface and cause the subsoil to be dense, making it difficult for roots and water to penetrate. Soils with clay pans are more susceptible to water erosion. Clay pans can be broken by cultivation, but over time they re-form.

Clean Air Act (1963, 1970, 1990)

The 1970 Clean Air Act and major amendments to the act in 1977 and 1990 serve as the backbone of efforts to control air pollution in the United States. This law established one of the most complex regulatory programs in the country. Efforts to control air pollution in the United States date back to 1881, when Chicago and Cincinnati passed laws to control industrial smoke and soot. Other municipalities followed suit and the momentum continued to build. In 1952, Oregon became the first state to adopt a significant program to control air pollution, and three years later, the federal government became involved for the first time, when the Air Pollution Control Act was passed. This law provided funds to assist the states in their air pollution control activities.

In 1963, the first Clean Air Act was passed. The act provided permanent federal aid for research, support for the development of state pollution control agencies, and federal involvement in cross-boundary air pollution cases. An amendment to the act in 1965 directed the Department of Health, Education and Welfare (HEW) to establish federal emission standards for motor vehicles. (At this time, HEW administered air pollution laws. The Environmental Protection Agency [EPA] was not created until 1970.) This represented a significant move by the federal government from a supportive to an active role in setting air pollution policy. The 1967 Air Quality Act provided additional funding to the states, required them to establish Air Quality Control Regions, and directed HEW to obtain and make available information on the health effects of air pollutants and to identify pollution control techniques.

The Clean Air Act of 1970 marked a dramatic change in air pollution policy in the United States. Following enactment of this law, the federal government, not the states, would be the focal point for air pollution policy. This act established the framework that continues to be the foundation for air pollution control policy today. The impetus for this change was the belief that the state-based approach was not working and increased pressure from a developing environmental consciousness across the country. Public sentiment was growing so significantly that environmental issues demanded the attention of high-ranking officials. In fact, the leading policy entrepreneurs on the issue were President Richard Nixon and Senator Edmund Muskie of Maine.

The regulatory framework of The Clean Air Act featured four key components. First, National Ambient Air Quality Standards (NAAQSs) were established for six major pollutants: carbon monoxide, lead (in 1977), nitrogen dioxide, ground-level ozone (a key component of smog), particulate matter, and sulfur dioxide. For each of these pollutants, sometimes referred to as criteria pollutants, primary and secondary standards were set. The primary standards were designed to protect human health; the secondary standards were based on protecting crops, forests, and buildings if the primary standards were not capable of doing so. The Act stipulated that these standards must apply to the entire country and be set by the EPA, based on the best available scientific information. The costs of attaining these standards were not among the factors considered. The EPA was also directed to set standards for less common toxic air pollutants.

Second, New Source Performance Standards (NSPSs) would be set by the EPA. These standards would determine how much air pollution would be allowed by new plants in various industrial sectors. The standards were to be based on the best available control technology (BACT) and best available retrofit technology (BART) available for the control of pollutants at sources such as power plants, steel factories, and chemical plants.

Third, mobile source emission standards were established to control automobile emissions. These standards were specified in the statute (rather than left to the EPA), and schedules for meeting them were also written into the law. It was thought that such an approach was crucial to ensure success with the powerful auto industry. The pollutants regulated were carbon monoxide, hydrocarbons, and nitrogen oxides, with goals of reducing the first two pollutants by 90 percent and nitrogen oxides by 82 percent by 1975.

The final component of the air protection act dealt with the implementation of the new air quality standards. Each state would be encouraged to devise a state implementation plan (SIP), specifying how the state would meet the national standards. These plans had to be approved by the EPA; if a state did not have an approved SIP, the EPA would administer the Clean Air Act in that state. However, since the federal government was in charge of establishing pollution standards for new mobile and stationary sources, even the states with an SIP had limited flexibility. The main focal point for the states was the control of existing stationary sources, and if necessary, mobile sources. The states had to set limits in their SIPs that allowed them to achieve the NAAQSs by a statutorily determined deadline. One problem with this approach was the construction of tall smokestacks, which helped move pollution out of a particular airshed but did not reduce overall pollution levels. The states were also charged with monitoring and enforcing the Clean Air Act.

The 1977 amendments to the Clean Air Act dealt with three main issues: nonattainment, auto emissions, and the prevention of air quality deterioration in areas where the air was already relatively clean. The first two issues were resolved primarily by delaying deadlines and increasing penalties. Largely in response to a court decision in favor of environmentalists (Sierra Club v. Ruckelshaus, 1972), the 1977 amendments included a program for the prevention of significant deterioration (PSD) of air that was already clean. This program would prevent polluting the air up to the national levels in areas where the air was cleaner than the standards. In Class I areas, areas with near pristine air quality, no new significant air pollution would be allowed. Class I areas are airsheds over large national parks and wilderness areas. In Class II areas, a moderate degree of air quality deterioration would be allowed. And finally, in Class III areas, air deterioration up to the national secondary standards would be allowed. Most of the country that had air cleaner than the NAAQSs was classified as Class II. Related to the prevention of significant deterioration was a provision to protect and enhance visibility in national parks and wilderness areas even if the air pollution was not a threat to human health. The impetus for this section of the bill was the growing visibility problem in parks, especially in the Southwest.

Throughout the 1980s, efforts to further amend the Clean Air Act were stymied. President Ronald Reagan opposed any strengthening of the Act, which he argued would hurt the economy. In Congress, the controversy over acid rain between members from the Midwest and the Northeast further contributed to the stalemate. Gridlock on the issue broke with the election of George Bush, who supported amendments to the Act, and the rise of Senator George Mitchell of Maine to Senate Majority Leader. Over the next two years, the issues were hammered out between environmentalists and industry and between different regions of the country. Major players in Congress were Representatives John Dingell of Michigan and Henry Waxman of California and Senators Robert Byrd of West Virginia and Mitchell.

Major amendments to the Clean Air Act were passed in the fall of 1990. These amendments addressed four major topics: (1) acid rain, (2) toxic air pollutants, (3) nonattainment areas, and (4) ozone layer depletion. To address acid rain, the amendments mandated a 10 million ton reduction in annual sulfur dioxide emissions (a 40 percent reduction based on the 1980 levels) and a two million ton annual reduction in nitrogen oxides to

be completed in a two-phase program by the year 2010. Most of this reduction will come from old utility power plants. The law also creates marketable pollution allowances, so that a utility that reduces emissions more than required can sell those pollution rights to another source. Economists argue that, to increase efficiency, such an approach should become more widespread for all pollution control.

Due to the failure of the toxic air pollutant provisions of the 1970 Clean Air Act, new, more stringent provisions were adopted requiring regulations for all major sources of 189 varieties of toxic air pollution within ten years. Areas of the country still in nonattainment for criteria pollutants were given from three to twenty years to meet these standards. These areas were also required to impose tighter controls to meet the standards. To help these areas and other parts of the country, the Act required stiffer motor vehicle emissions standards and cleaner gasoline. Finally, three chemical families that contribute to the destruction of the stratospheric ozone layer (chlorofluorocarbons (CFCs), hydrochlorofluorocarbons (HCFCs), and methyl chloroform) were to be phased out of production and use.

In 1997, the EPA issued revised national ambient air quality standards (NAAQS), setting stricter standards for ozone and particulate matter. The American Trucking Association and other state and industry groups legally challenged the new standards on the grounds that the EPA did not have the authority under the Act to make such changes. On February 27, 2001, the U.S. Supreme Court unanimously upheld the constitutionality of the Clean Air Act as interpreted by the EPA, and all remaining legal challenges to other aspects of the standards change were rejected by a Washington DC District Court ruling in early 2002.

The Clean Air Act has met with mixed success. The national average pollutant levels for the criteria pollutants have decreased. Nevertheless, many localities have not achieved these standards and are in perpetual nonattainment. Not surprisingly, major urban areas are those most frequently in nonattainment. The pollutant for which standards are most often exceeded is ozone, or smog. This is due in part to increases in nitrogen oxides (NOx), which disperse ozone. NOx emissions increased by approximately 20 percent between 1970 and 2000. As a result, some parts of the country have had worsening ozone levels. According to the EPA, the average ozone levels in twenty-nine national parks increased by over 4 percent between 1990 and 2000.

The greatest successes of air pollution control have come with lead, which between 1981 and 2000 was reduced by 93 percent (largely due to the phasing-out of leaded gasoline), and particulates, which were reduced by 47 percent in the same period. Overall particulate emissions were down 88 percent since 1970. Carbon monoxide has dropped by 25 percent and volatile organic compounds and sulfur dioxides have declined by over 40 percent each between 1970 and 2000. However, air quality analysis is complex, and it is important to note that some changes may be due to shifts in the economy, changes in weather patterns, or other such variables rather than directly attributable to the Clean Air Act.

In February 2002, President Bush introduced the Clear Skies Act of 2003, an initiative that, if fully adopted, would make significant changes to the Clean Air Act. Among them would be a weakening or elimination of new source review regulations and BART rules, and a new cap-and-trade plan that would allow power plants that produced excessive toxic emissions to buy credits from other plants who had levels under the standards. The Bush administration hailed the initiative as a less expensive way to accelerate air pollution clean-up and a more economy-friendly alternative to the Kyoto Protocol for greenhouse gas reductions, while environmental groups and other critics called it a roll-back of the Clean Air Act progress.

In 2007, the Supreme Court of the United States ruled in *Massachusetts v. EPA* that the EPA has the power to regulated carbon dioxide (CO_2) and other greenhouse gas emissions under the Clean Air Act. The court stated that the Clean Air Act terms require the EPA to administer greenhouse gas regulation, unless the EPA could prove that greenhouses gases do not contribute to climate change, or could offer a reasonable explanation as to why the EPA should not regulate greenhouse gases. In April 2009, the EPA declared that CO_2 and other greenhouse gases were a threat to the public's health and welfare. President Obama urged Congress and the Senate to pass comprehensive climate legislation that included regulation of greenhouse gas emissions. In 2009 and 2010, proposed legislation failed to garner enough support. The Obama administration announced that the EPA would move to enact regulations in the absence of Congressional action.

See also Best Available Control Technology.

Resources

BOOKS

Collin, Robert W. *The Environmental Protection Agency: Cleaning Up America's Act*. Westport, CT: Greenwood Press, 2005.

DuPuis, E. Melanie. *Smoke and Mirrors: The Politics and Culture of Air Pollution*. New York: New York University Press, 2004.
Ellerman, A. Denny. *Markets for Clean Air: The U.S. Acid Rain Program*. New York, NY: Cambridge University Press, 2005.
Wooley, David, and Elizabeth Morss. *Clean Air Act Handbook*. Deerfield, IL: Clark Boardman Callaghan, October 2009.

OTHER

United States Environmental Protection Agency (EPA). "Air: Air Pollution Legal Aspects: Legislation." http://www.epa.gov/ebtpages/airairpollutionlelegislation.html (accessed September 2, 2010).

Christopher McGrory Klyza
Paula Anne Ford-Martin

Clean coal technology

Coal is the world's most popular fuel. Globally, nearly 7 billion tons of coal was used in 2006, with use predicted to increase to 10 billion tons by 2030. As of 2010, about 40 percent of the electricity generated world-wide is based on coal, with more than half of the electricity produced in the United States coming from coal-fired power plants. The demand for coal is expected to triple by the middle of the next century, making it more widely used than petroleum or natural gas.

Coal is a relatively dirty fuel. When burned, it releases particulates and pollutants such as carbon monoxide (CO), carbon dioxide (CO_2), nitrogen (NO_x) and sulfur (SO_x) oxides into the atmosphere. These pollutants have an adverse effect on the environment in the form of acid rain (i.e., precipitation that is acidic) and, in the case of CO_2, as a greenhouse gas that traps heat that would normally dissipate into space. Both processes are undesirable. For example, the greenhouse effect is the basis of the warming of the global atmosphere that began in the mid-nineteenth century with the Industrial Revolution and the associated increase in coal-fired technology, a warming that affects the global climate. Other health hazards posed by coal-burning emissions are the emission of trace amounts of mercury (Hg) and the effects of the emitted particles on the human respiratory system.

Since the late 1980s, there has been an increasing amount of research on clean coal technologies, methods by which the combustion of coal releases fewer pollutants to the atmosphere. As early as 1970, the

Bituminous coal cleaning plant in central Pennsylvania. Coal from surface and underground mines in Pennsylvania and throughout the Appalachian Mountains is crushed and washed at cleaning plants to remove ash and sulfur. *(Theodore Clutter / Photo Researchers, Inc.)*

United States Congress acknowledged the need for such technologies in the Clean Air Act of that year. One provision of that Act required the installation of flue gas desulfurization (FGD) systems (referred to as scrubbers) at all new coal-fired plants. In the late 1980s and early 1990s, the U.S. Department of Energy (DOE) teamed up with state and industry organizations to test a number of technologies for reducing pollutants emitted from burning coal. Over twenty of these methods were successful in commercial applications.

The Clean Coal Power Initiative (CCPI) was established in 2002 as a ten-year program focusing on the development and testing of clean coal technologies to address the concerns of coal-burning emissions affecting public health and the environment. In early 2003, eight projects were chosen for testing. Five of these were either discontinued or withdrawn. Of the remaining three, one has been completed, and two are in the operational phase. In late 2004, four projects were chosen for the testing phase. With one project withdrawn, two of the projects are in the developmental phase and focus on integrated gasification

combined cycle (IGCC); a process in which coal is converted to a gas (termed syngas) that can be burned in a conventional power plant. After the conversion, the impurities can be removed from the gas in subsequent steps prior to combustion. One project is in the operational phase and is based on power plant optimization and control of multiple pollutants. As of 2010, two IGCC power plants are operational and generating power in the U.S., with more becoming operational by 2012. The approach is still hindered by its high cost, relative to other means of power generation. The third phase of testing is choosing projects focused on carbon capture and sequestration (CCS) or potential downstream uses of CO_2. The ability of IGCC to capture CO_2 may make it an attractive option to reduce CO_2 emissions to the atmosphere.

Many different technologies have been tested on at least an experimental basis for the cleaning of coal. Some of these technologies are used on coal before it is even burned. Chemical, physical, and biological methods have all been developed for pre-combustion cleaning. For example, pyrite (FeS_2) is often found in

Clean coal technology

conjunction with coal when it is mined. When the coal is burned, pyrite is also oxidized, releasing sulfur dioxide (SO_2) to the atmosphere. Yet, pyrite can be removed from coal by rather simple, straightforward physical means because of differences in the densities of the two substances.

Biological methods for removing sulfur (S) from coal are also being explored. The bacterium *Thiobacillus ferrooxidans* has the ability to change the surface properties of pyrite particles, making it easier to separate them from the coal itself. The bacterium may also be able to extract sulfur that is chemically bound to carbon in the coal.

A number of technologies have been designed to modify existing power plants to reduce the release of pollutants produced during the combustion of coal. In an attempt to improve on traditional wet scrubbers, researchers are now exploring the use of dry injection as one of these technologies. In this approach, dry compounds of calcium (Ca), sodium (Na), or some other element are sprayed directly into the furnace or into the ducts downstream of the furnace. These compounds react with non-metallic oxide pollutants, such as sulfur dioxide and nitrogen dioxide (NO_2), forming solids that can be removed from the system. A variety of technologies are being developed especially for the release of oxides of nitrogen. Since the amount of this pollutant formed is very much dependent on combustion temperature, methods of burning coal at lower temperatures are also being explored.

Some entirely new technologies are also being developed for installation in power plants. Fluidized bed combustion and improved coal pulverization are two of these. In the first of these processes, coal and limestone are injected into a stream of upward-flowing air, improving the degree of oxidation during combustion. The second process involves improving on a technique that has long been used in power plants, reducing coal to very fine particles before it is fed into the furnace.

Resources

BOOKS

Courtney, R. *Clean Coal Technology*. Concord: Paul And Company, 2004.

Douwe, Klaes G. *Clean Coal*. Boston: Nova Science Pub, 2009.

Fraser, Shannon. *Potential Exports of U.S. Clean Coal Technology Through 2030*. Washington, D.C.: U.S. Department of Commerce, International Trade Administration, 2007.

Leonard, Barry. *Clean Coal Technology: Mercury Control Demonstration Projects*. Darby, PA: DIANE Publishing Co., 2008.

United States. *Clean Coal Technology Programs Program Update 2007*. Washington, D.C.: U.S. Department of Energy, Assistant Secretary for Fossil Energy, 2008.

U.S. Clean Coal Technology Demonstration Program. *Clean Coal Technology Demonstration Program Project Profiles*. Washington, D.C.: U.S. Department of Energy, Assistant Secretary for Fossil Energy, 2004.

OTHER

U.S. Department of Energy. "Clean Coal Technology & the Clean Coal Power Initiative." http://fossil.energy.gov/programs/powersystems/cleancoal/ (accessed August 29, 2010).

David E. Newton

Clean Water Act (1972, 1977, 1987)

Federal involvement in protecting the nation's waters began with the Water Pollution Control Act of 1948, the first statute to provide state and local governments with the funding to address water pollution. During the 1950s and 1960s, awareness grew that more action was needed and federal funding to state and local governments was increased. In the Water Quality Act of 1965 water quality standards, to be developed by the newly created Federal Water Pollution Control Administration, became an important part of federal water pollution control efforts.

Despite these advances, it was not until the Water Pollution Control Amendments of 1972 that the federal government assumed the dominant role in defining and directing water pollution control programs. This law was the outcome of a battle between Congress and President Richard M. Nixon. In 1970, Nixon responded to public outcry over pollution problems by resurrecting the Refuse Act of 1899, which authorized the U.S. Army Corps of Engineers to issue discharge permits. Congress felt its prerogative to set national policy had been challenged. It debated for nearly eighteen months to resolve differences between the House and Senate versions of a new law, and on October 18, 1972, Congress overrode a presidential veto and passed the Water Pollution Control Amendments.

Section 101 of the new law set forth its fundamental goals and policies, which continue to this day "to restore and maintain the chemical, physical, and biological integrity of the Nation's waters." This section also set forth the national goal of eliminating discharges of pollution into navigable waters by 1985

The Milwaukee River, free of foam, flows over the Waubeka Dam. The Clean Water Act resulted in cutting the amount of phosphorus that enters Lake Michigan and Lake Superior by 85 percent. *(TOM LYNN/Newscom.)*

and an interim goal of achieving water quality levels to protect fish, shellfish, and wildlife. As national policy, the discharge of toxic pollutants in toxic amounts was now prohibited; federal financial assistance was to be given for constructing publicly owned waste treatment works; area-wide pollution control planning was to be instituted in states; research and development programs were to be established for technologies to eliminate pollution; and nonpoint source pollution—runoff from urban and rural areas—was to be controlled. Although the federal government set these goals, states were given the main responsibility for meeting them, and the goals were to be pursued through a permitting program in the new national pollutant discharge elimination system (NPDES).

Federal grants for constructing publicly owned treatment works (POTWs) had totalled $1.25 billion in 1971, and they were increased dramatically by the new law. The act authorized five billion dollars in fiscal year 1973, six billion for fiscal year 1974, and seven billion for fiscal year 1975, all of which would be automatically available for use without requiring Congressional appropriation action each year. But along

with these funds, the Act conferred the responsibility to achieve a strict standard of secondary treatment by July 1, 1977. Secondary treatment of sewage consists of a biological process that relies on naturally occurring bacteria and other micro-organisms to break down organic material in sewage. The Environmental Protection Agency (EPA) was mandated to publish guidelines on secondary treatment within sixty days after passage of the law. POTWs also had to meet a July 1, 1983, deadline for a stricter level of treatment described in the legislation as "best practicable wastewater treatment." In addition, pretreatment programs were to be established to control industrial discharges that would either harm the treatment system or, having passed through it, pollute receiving waters.

The act also gave polluting industries two new deadlines. By July 1, 1977, they were required to meet limits on the pollution in their discharged effluent using Best Practicable Technology (BPT), as defined by EPA. The conventional pollutants to be controlled included organic waste, sediment, acid, bacteria and viruses, nutrients, oil and grease, and heat. Stricter state water quality standards would also have to be met by that date. The second deadline was July 1, 1983, when industrial dischargers had to install Best Available Control Technology (BAT), to advance the goal of eliminating all pollutant discharges by 1985. These BPT and BAT requirements were intended to be "technology forcing," as envisioned by the Senate, which wanted the new water law to restore water quality and protect ecological systems.

On top of these requirements for conventional pollutants, the law mandated the EPA to publish a list of toxic pollutants, followed six months later by proposed effluent standards for each substance listed. The EPA could require zero discharge if that was deemed necessary. The zero discharge provisions were the focus of great controversy when Congress began oversight hearings to assess implementation of the law. Leaders in the House considered the goal a target and not a legally binding requirement. But Senate leaders argued that the goal was literal and that its purpose was to ensure rivers and streams ceased being regarded as components of the waste treatment process. In some cases, the EPA has relied on the Senate's views in developing effluent limits, but the controversy over what zero discharge means continues to this day.

The law also established provisions authorizing the Army Corps of Engineers to issue permits for discharging dredge or fill material into navigable waters at specified disposal sites. In recent years this program, a key component of federal efforts to protect rapidly diminishing wetlands, has become one of the most explosive issues in the Clean Water Act. Farmers and developers are demanding that the federal government cease "taking" their private property through wetlands regulations, and recent sessions of Congress have been besieged with demands for revisions to this section of the law.

In 1977, Congress completed its first major revisions of the Water Pollution Control Amendments (which was renamed the "Clean Water Act"), responding to the fact that by July 1, 1977 only 30 percent of major municipalities were complying with secondary treatment requirements. Moreover, a National Commission on Water Quality had issued a report which recommended that zero discharge be redefined to stress conservation and reuse and that the 1983 BAT requirements be postponed for five to ten years. The 1977 Clean Water Act endorsed the goals of the 1972 law, but granted states broader authority to run their construction grants programs. The Act also provided deadline extensions and required EPA to expand the lists of pollutants it was to regulate.

In 1981, Congress found it necessary to change the construction grants program; thousands of projects had been started, with $26.6 billion in federal funds, but only 2,223 projects worth $2.8 billion had been completed. The Construction Grant Amendments of 1981 restricted the types of projects that could use grant money and reduced the amount of time it took for an application to go through the grants program.

The Water Quality Act of 1987 phased out the grants program by fiscal year 1990, while phasing in a state revolving loan fund program through fiscal year 1994, and thereafter ending federal assistance for wastewater treatment. The 1987 act also laid greater emphasis on toxic substances; it required, for instance, that the EPA identify and set standards for toxic pollutants in sewage sludge, and it phased in requirements for stormwater permits. The 1987 law also established a new toxics program requiring states to identify "toxic hot spots"—waters that would not meet water quality standards even after technology controls have been established—and mandated additional controls for those bodies of water.

These mandates greatly increased the number of NPDES permits that the EPA and state governments issued, stretching budgets of both to the limit. Moreover, states have billions of dollars worth of wastewater treatment needs that remain unfunded, contributing to continued violations of water quality standards. The new permit requirements for stormwater, together with sewer overflow, sludge, and other permit requirements, as well as POTW construction needs, led to a growing demand for more state flexibility in implementing the clean water laws and for continued federal support for wastewater treatment. State and local governments insist that they cannot do everything the law requires; they argue that they must be allowed to assess and prioritize their particular problems.

Yet despite these demands for less prescriptive federal mandates, on May 15, 1991, a bipartisan group of senators introduced the Water Pollution Prevention and Control Act to expand the federal program. The proposal was eventually set aside after intense debate in both the House and Senate over controversial wetlands issues. Amendments to the Clean Water Act proposed in both 1994 and 1995 also failed to make it to a vote.

In 1998, President Clinton introduced a Clean Water Action Plan, which primarily focused on improving compliance, increasing funding, and accelerating completion dates on existing programs authorized under the Clean Water Act. The plan contained over 100 action items that involved interagency cooperation of EPA, U.S. Department of Agriculture (USDA), Army Corps of Engineers, Department of the Interior, Department of Energy, Tennessee Valley Authority, Department of Transportation, and the Department of Justice. Key goals were to control nonpoint pollution, provide financial incentives for conservation and stewardship of private lands, restore wetlands, and expand the public's right to know on water pollution issues.

New rules were proposed in 1999 to strengthen the requirements for states to set limits for and monitor the Total Maximum Daily Load (TMDL) of pollution in their waterways. The TMDL rule has been controversial, primarily because states and local authorities lack the funds and resources to carry out this large-scale project. In addition, agricultural and forestry interests, who previously were not regulated under the Clean Water Act, would be affected by TMDL rules.

In May 2002 the EPA announced a new rule changing the definition of "fill material" under the Clean Water Act to allow dirt, rocks, and other displaced material from mountaintop removal (MTR) coal mining operations to be deposited into rivers as waste under permit from the Army Corps of Engineers, a practice that was previously unlawful under the Clean Water Act. Shortly thereafter, a group of congressional representatives introduced new legislation to overturn this rule, and a federal district court judge in West Virginia ruled that such amendments to the Clean Water Act could only be made by Congress, not by an EPA rule change. In 2009, the U.S. Court of Appeals for the Fourth Circuit upheld fill material permits in West Virginia. In response, the EPA, under U.S. President Barack Obama's administration, objected to additional fill material permits and requested that the Army Corps of Engineers require additional study or mitigation on such projects. On March 4, 2009, legislation was proposed in Congress that would redefine the term "fill material" to exclude MTR debris and prohibit valley fills. As of September 2010, the House of Representatives had not voted on the proposed changes.

As the Federal Water Pollution Control Act passed its thirtieth anniversary, the EPA state and federal regulators are increasingly recognizing the need for a more comprehensive approach to water pollution problems than the current system, which focuses predominantly on POTWs and industrial facilities. Nonpoint source pollution, caused when rain washes pollution from farmlands and urban areas, is the largest remaining source of water quality impairment, yet the problem has not received a fraction of the regulatory attention addressed to industrial and municipal discharges. Despite this fact, the EPA claims that the Clean Water Act is responsible for a one billion ton decrease in annual soilrunoff from farming, and an associated reduction in phosphorus and nitrogen levels in water sources. EPA also asserts that wetland loss, while still a problem, has slowed significantly—from 1972 levels of 460,000 acres per year to current losses of approximately 90,000 acres annually as of 2009. Over half of this current wetland loss

occurred in the eastern coastal United States from Louisiana to North Carolina. However, critics question these figures, citing abandoned mitigation projects and reclaimed wetlands that bear little resemblance to the habitat they are supposed to replicate.

See also Agricultural pollution; Environmental policy; Industrial waste treatment; Sewage treatment; Storm runoff; Urban runoff.

Resources

BOOKS

Aga, Diana S. *Fate of Pharmaceuticals in the Environment and in Water Treatment Systems.* Boca Raton: CRC Press, 2008.

Calabrese, Edward J., Paul T. Kostecki, and James Gragun. *Contaminated Soils, Sediments and Water: Science in the Real World.* New York: Springer, 2004.

Midkiff, Ken, and Robert F. Kennedy Jr. *Not A Drop To Drink: America's Water Crisis (and What You Can Do).* Novato, Calif.: New World Library, 2007.

Tvedt, Terje, and Eva Jakobsson. *A History of Water.* London: I.B. Tauris, 2004.

PERIODICALS

Vuorinen, H. S., P. S. Juuti, and T. S. Katko. "History of Water and Health from Ancient Civilizations to Modern Times." *Water Science and Technology: Water Supply* 7, 1 (2007): 49–57.

OTHER

California Environmental Protection Agency. "Federal Clean Water Act." http://www.waterboards.ca.gov/laws_regulations/docs/fedwaterpollutioncontrolact.pdf (accessed September 2, 2010).

ORGANIZATIONS

America's Clean Water Foundation, 750 First Street NE, Suite 1030, Washington, DC, USA, 20002, (202) 898-0908, (202) 898-0977, http://yearofcleanwater.org

David Clarke
Paula Anne Ford-Martin

Clear-cutting

Clear-cutting is the removal of all the trees from a section of forest. This forest management technique has been used in a variety of forests around the world. For many years, it was considered the most economical and environmentally sound way of harvesting timber. Since the 1960s and 1970s, the practice has been increasingly called into question as more has been learned about the ecological benefits of old growth timber especially in

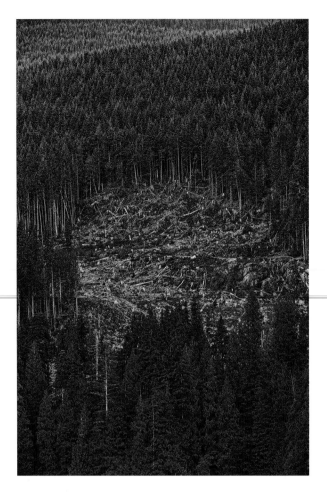

Clear-cutting along the Campell River, British Colombia, Canada. *(© Mira / Alamy)*

itself. Wood in temperate climates decays slowly, resulting in a large number of fallen logs—making even walking difficult, much less dragging cut trees to areas where they can be hauled to mills. Tall trees with narrow branch structures produce a lot of stems in a small area, therefore, the total biomass (or total input of living matter) in these temperate forests is typically four times that of the densest tropical rain forest. Some redwood forest groves in California have been measured with twenty times the biomass of similar sites in the tropics. Those supporting the harvest of these trees and clear-cutting point out that if these trees are not used, it will put increased pressure on other timber producers throughout the world to supply this wood. This could have high environmental costs globally, since it could take 10–30 acres (4–12 ha) of taiga forest in northern Canada, Alaska, or Siberia to produce the wood obtainable from one acre in the state of Washington.

Clear-cutting makes harvesting these trees very lucrative. The number of trees and downed logs makes it difficult to cut only some of the trees, and falling trees can damage the survivors; in addition, they can be damaged when dragged across the ground to be loaded onto trucks. This is made more expensive and time consuming by working around the standing trees. In areas that have been selectively logged, the trees left standing are may be knocked down by winds or may drop branches, presenting a considerable safety risk to loggers working in the area.

Old growth forests leave so much woody debris and half-decayed logs on the ground that it can be difficult to walk through a harvested patch to plant seedlings. This is why an accepted part of the clear-cut practice in the past has been to burn the leftover slash, after which seedlings are planted. Brush that grows on the site is treated with herbicide to allow the seedlings time to grow to the point at which they can compete with other vegetation.

Supporters of clear-cutting contend that it has unique advantages: burning consumes fuel that would otherwise be available to feed forest fires; logging roads built to haul the trees out open the forests to recreational uses and provide firebreaks; clear-cutting leaves no snags or trees to interfere with reseeding the forest, and allows use of helicopters and planes to spray herbicides—the brush that grows up naturally in the sunlight and provides browse for big-game species such as deer and elk.

Detractors point to a number of very negative environmental and economic consequences that need to be considered: particularly where clear cuts have been done near urban areas, plumes of smoke produced by burning slash have polluted cities; animal browse is choked out after a few years by the new seedlings, creating a darkened forest floor that lacks almost any

the Pacific Northwest of the United States and the global ecological value of tropical rain forests. As well, clear-cut land is much more susceptible to soil erosion than land were some trees are left, and whose roots provide some stability to the soil.

Foresters and loggers point out that there are practical economic and safety reasons for the practice of clear-cutting. For example, in the Pacific Northwest, these reasons revolve around the volume of wood that is present in the old growth forests. This volume can actually be an obstacle at times to harvesting, so that the most inexpensive way to remove the trees is to cut everything.

During the post-World War II housing boom in the 1950s, clear-cutting overtook selective cutting as the preferred method for harvesting in the Pacific Northwest. Since that time, worldwide demand for lumber has continued to rise. Practical arguments that the practice should continue despite its ecological implications are tied directly to the volume and character of the forest vegetation or wildlife for at least three decades; herbicides applied to control vegetation contribute to the degradation of surface water quality; mining the forest causes declines in species diversity and loss of habitats (declaration of the northern spotted owl as an endangered species and the efforts to preserve its habitat is an example of the potential loss of diversity); new microclimates appear that promote less desirable species than the trees they replace; so much live and rotting wood is harvested or burned that the soil fertility is reduced, affecting the potential for future tree growth.

Critics also point out that erosion and flooding increases from clear-cut areas have significant economic impact downstream in the watersheds. Studies have shown that since the clear-cut practice increased following World War II, landslides have increased to six times the previous rate. This has resulted in increased sediment delivery to rivers and streams where it has a detrimental impact on the stream fishery.

Loss of critical habitat for endangered species such as the spotted owl and the impact of sediment on the salmon fishery have resulted in government efforts to set aside old growth forest wilderness to preserve the unique forest ecosystem. The practice of clear-cutting with its pluses and minuses, however, continues not only in the Pacific Northwest, but in other forests in the United States and in other areas of the world.

Clear-cutting and rain forests

Another area where clear-cutting has become the focus of ecological debate is in the tropical rain forests. Many of the same economic pressures that make clear-cutting a lucrative practice in the United States make it equally attractive in the rain forest, but there are significant environmental consequences.

Nearly one-half of the earth's rain forests were lumbered by the 1990s. Besides the demand for building material or fuel, these forests are also subject to significant clearing to later be worked to produce food in new agricultural areas and to facilitate the exploration for oil and minerals. If this loss continues at present estimated rates, the rain forests will be totally harvested by the year 2040.

Beginning in the 1990s, agencies and governments began to monitor the pace of lumbering in rain forests, especially the Amazon, using orbiting satellites. The optics of the satellite-borne cameras had improved so much that accurate determinations of the area of clear-cut could be determined. The graphic images of forest decimation have helped mobilize public opinion against the wholesale clear-cutting of the Amazon rain forest, which helped spur the Brazilian government to take action.

As of 2011, the pace of the loss of the Amazon rain forest has slowed. Still, since the 1970s, an estimated 230,000 square miles of forest has been lost.

Rain forest activists continually work to remind the world of the importance of the forests. Locally, for instance, the presence or absence of the rain forest can change the climate and the local water budget. For example, times of drought are more severe and when the rains come, flooding is increased. As well, rain forests can have a major impact on the global climate. When they are cut and the slash burned, significant amounts of carbon dioxide are released into the atmosphere, which may contribute to the overall greenhouse effect and general warming of the earth.

The biological diversity of these forest ecosystems is vast. Rain forests contain about one-half the known plant and animal species in the world and very few of these species have ever been studied by scientists for their potential benefits. (More that 7,000 medicines have already been derived from tropical plants. It is uncertain how many more such uses are yet to be found.)

Most rain forests occur in less-developed countries where it is difficult to meet the expanding needs of rapidly increasing populations for food and shelter. They need the economic benefits that can be derived from the rain forest for their very survival. Any attempt at stopping the clear-cutting practice must provide for their needs to be successful. Until a better practice is developed, clear-cutting will remain an environmental issue on a global scale for the next several decades.

Resources

BOOKS

London, Mark, and Brian Kelly. *The Last Forest: The Amazon in the Age of Globalization*. New York: Random House, 2007.

Marent, Thomas. *Rainforest*. New York: DK Publishing, 2010.

McLeish, Ewan. *Rain Forest Destruction: What If We Do Nothing?* Milwaukee: World Almanac Library, 2007.

James L. Anderson

Clements, Frederic E.

1874–1945
American ecologist

For Frederic Clements, trained in botany as a plant physiologist, ecology became a leader in the new science of plant ecology.

Clements was born in Lincoln, Nebraska, and earned all of his degrees in botany from the University of Nebraska, attaining his Ph.D. under Charles Bessey in 1898. As a student, he participated in Bessey's famous Botanical Seminar and helped carry out an ambitious survey of the vegetation of Nebraska, publishing the results—co-authored with a classmate—in an internationally recognized volume titled *The Phytogeography of Nebraska*, out in print the same year (1898) that he received his doctorate. He then accepted a faculty position at the university in Lincoln.

Clements married Edith Schwartz in 1899, described (in *Ecology* in 1945) as a wife and help-mate, who became his life-long field assistant and also a collaborator on research and books on flowers, particularly those of the Rocky Mountains. Clements rose through the ranks as a teacher and researcher at Nebraska and then, in 1907, he was appointed as Professor and Head of the department of botany at the University of Minnesota. He stayed there until 1917, when he moved to the Carnegie Institution in Washington, D.C., where he focused full-time on research for the rest of his career. Retired from Carnegie in 1941, he continued a year-round workload in research, spending summers on Pikes Peak at an alpine laboratory and winters in Santa Barbara at a coastal laboratory. He died in Santa Barbara on July 26, 1945.

Publication of Research Methods in Ecology in 1905 marked his promotion to full professor at the University of Nebraska, but more importantly it marked his turn from taxonomic and phytogeographical work to ecology. It has been called "the earliest how-to book in ecology," and "a manifesto for the emerging field." More broadly, Arthur Tansley, a leading plant ecologist of the time in Great Britain, though critical of some of Clements' views, described him as by far the greatest individual creator of the modern science of vegetation. Henry Gleason, though an early and severe critic of Clements' ideas, also recognized him as an original ecologist, one inspired by Europeans but developing his ideas entirely de novo from his own mind.

Clements deplored the chaotic state of ecology and assumed the task of remedying it. He did bring rigor, standardization, and an early quantitative approach to research processes in plant ecology, especially through his development of sampling procedures at the turn of the century.

Clements is often described as rigid and dogmatic, which seems at odds with his importance in emphasizing change in natural systems; that emphasis on what he called dynamic ecology became his research trademark.

Clements also stressed the importance of process and function. Clements anticipated ecosystem ecology through his concern for changes through time of plant associations, in correspondence to changes in the physical sites where the communities were found. His rudimentary conception of such systems carried his ecological questions beyond the traditional confines of plant ecology to what McIntosh described as the larger system transcending plants, or even living organisms to link to the abiotic environment, especially climate. McIntosh also credits him with a definition of paleoecology that antedated major developments that would studies of vegetation and pollen analysis (or palynology). Clements authored a book, *Methods and Principles of Paleo-Ecology* (1924), appropriate to the new topic. Clements even played a role, if not a major one relative to figures like Forbes, toward understanding what ecologists came to call eutrophic changes in lakes, an area of major current consideration in the environmental science of water bodies.

One of the stimulants for botanical, and then plant ecological, research for Clements and other students of Bessey's at Nebraska was their location in a major agricultural area dependent on plants for crops. Problems of agriculture remained one of Clements' interests for all of his life. During his long career, Clements remained involved in various ways of applying ecology to problems of land use such as grazing and soil erosion, even the planning of shelter belts. In large part because of the early work by Bessey and Clements and their colleagues and collaborators, the Midwest became a center for the development of grassland ecology and range management. Although considered misguided by some, pleas for wilderness set-asides still argue today that land-use policy should leave the climax undisturbed or preserved areas returned to a perceived climax condition. Clements believed that homesteaders in Nebraska were wrong to destroy the sod covering the sandhills of Nebraska and that the prairies should be grazed and not tilled. Farmers objected to the implications of climax theory, because they feared threats to their livelihoods from calls for cautious use of marginal lands. Even some scientific attempts to discredit the idea of the climax were based on the desire to undermine its importance to the conservation movement.

Clements even anticipated a century of sporadic connections between biological ecology and human ecology in the social sciences, arguing that sociology represented the ecology of a particular species of animal and, therefore, had a similar close association with plant ecology. That connection was lost in professional ecology in the early 1940s and has still not been reestablished, even at the beginning of the twenty-first century. Though Clements' name is still recognized today as one of ecology's foundation thinkers, and though he gained considerable respect and influence among the ecologists of his day, his work from the beginning was controversial. One reason was that ecologists were skeptical that approaches from plant physiology could be transposed directly to ecology. Another reason, and the idea with which Clements is still associated three-quarters of a century later, was his conviction that the successional process was the same as the development of an organism and that the communities emerging as the end-points of succession were in fact super-organisms. Clements believed that succession was a dynamic process of progressive development of the plant formation, that it was controlled absolutely by climate, developed in an absolutely predictable way (and in the same way in all similar climates) and then was absolutely stable for long periods of time.

Clements' fondness for words, and especially for coining his own nonce terms for about every conceivable nuance of his work also got him in trouble with his colleagues in ecology. This fondness for coining new words to describe his work, such as therium to describe a successional stage caused by animals, added to the fuel for critics who wished to find fault with the substance of his ideas. Unfortunately, Clements also gave further cause to those, then and now, wishing to discount all of his ideas by retaining a belief in Lamarckian evolution, and by doing experiments during retirement at his alpine research station, in which he even claimed he been able to convert several Linnean species into each other, histologically as well as morphologically.

BOOKS

Clements, Edith S. *Adventures in Ecology: Half a Million Miles From Mud to Macadam.* New York: Pageant Press, 1960.

Gerald L. Young

Climate

Climate is the general, cumulative pattern of regional or global weather patterns. The most apparent aspects of climate are trends in air temperature and humidity, wind, and precipitation. These observable phenomena occur as the atmosphere surrounding the earth continually redistributes, via wind and evaporating and condensing water vapor, the energy that the earth receives from the sun.

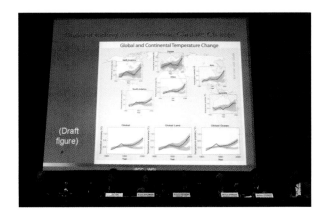

A view of the Intergovernmental Panel on Climate Change final conference. *(AP Photo/Remy de la Mauviniere.)*

Although the climate remains fairly stable on the human time scale of decades or centuries, it fluctuates continuously over thousands or millions of years. A great number of variables simultaneously act and react to create stability or fluctuation in this very complex system. Some of these variables are atmospheric composition, rates of solar energy input, albedo (the earth's reflectivity), and terrestrial geography. Extensive research helps explain and predict the behavior of individual climate variables, but the way these variables control and respond to each other remains poorly understood. Climate behavior is often described as chaos, consisting of changes and movements so complex that patterns cannot be perceived in them, even though patterns may exist. Nevertheless, studies indicate that human activity is disturbing larger climate trends, notably by causing global warming. This prospect raises serious concern because rapid anthropogenic (human-caused) climate change could severely stress ecosystems and species around the world.

Solar energy and climate

Solar energy is the driving force in the earth's climate. Incoming radiation from the sun warms the atmosphere and raises air temperature, warms the earth's surface, and evaporates water, which then becomes humidity, rain, and snow. The earth's surface reflects or re-emits energy back into the atmosphere, further warming the air. Warming air expands and rises, creating air movement patterns in the atmosphere that reach over several degrees of latitude. In these convection cells, low pressure zones develop under rising air, and high pressure zones develop where that air returns downward toward the earth's surface. Such differences in atmospheric pressure force air masses to move, from high to low pressure regions. Movement of air masses creates wind on the earth's surface. When these air masses carry evaporated water, they may create precipitation when they move to cooler regions.

The sun's energy comes to the earth as a collection of long and short radiation wavelengths. The shortest wavelengths are microwaves and infrared waves. Infrared radiation is felt as heat. A small range of medium wavelength radiation makes up the spectrum of visible light. Longer wavelengths include ultraviolet (UV) radiation and radio waves. These longer wavelengths cannot be sensed, but UV radiation can cause damage as organic tissues (such as skin) absorb them. Overexposure to UV rays can cause sunburn and can produce genetic damage that, in some, can prelude the development of skin cancer. The difference in wavelengths is important because long and short wavelengths react differently when they encounter the earth and its atmosphere.

Solar energy approaching the earth encounters filters, reflectors, and absorbers in the form of atmospheric gases, clouds, and the earth's surface. Atmospheric gases filter incoming energy, selectively blocking some wavelengths and allowing other wavelengths to pass through. Blocked wavelengths are either absorbed and heat the air or scattered and reflected back into space. Clouds, composed of atmospheric water vapor, likewise reflect or absorb energy but allow some wavelengths to pass through. Some energy reaching the earth's surface is reflected; a great deal is absorbed in heating the ground, evaporating water, and conducting photosynthesis. Most energy that the Earth absorbs is re-emitted in the form of short, infrared wavelengths, which are sensed as heat. Some of this heat energy circulates in the atmosphere for a time, but eventually it all escapes. If this heat did not escape, the earth would overheat and become uninhabitable.

Variables in the climate system

Climate responds to conditions of the earth's energy filters, reflectors, and absorbers. As long as the atmosphere's filtering effect remains constant, the earth's reflective and absorptive capacities do not change, and the amount of incoming energy does not vary, climate conditions should stay constant. Usually some or all of these elements fluctuate. The earth's reflectivity changes as the shapes, surface features, and locations of continents change. The atmosphere's composition changes from time to time, so that different wavelengths are reflected or pass through. The amount of energy the earth receives also shifts over time.

During the course of a decade the rate of solar energy input varies by a few watts per square meter. Changes in energy input can be much greater over several millennia. Energy intensity also varies with the shape of the earth's orbit around the sun. In a period of 100 million years the earth's elliptical orbit becomes longer and narrower, bringing the earth closer to the sun at certain times of year, then rounder again, putting the earth at a more uniform distance from the sun. When the earth receives relatively intense energy, heating and evaporation increase. Extreme heating can set up exaggerated convection currents in the atmosphere, with extreme low pressure areas receiving intensified rains and high pressure areas experiencing extreme drought.

The earth's albedo depends upon surface conditions. Extensive dark forests absorb a great deal of energy in heating, evaporation of water, and photosynthesis. Light, colored surfaces, such as desert or snow, tend to absorb less energy and reflect more. If highly reflective continents are large or are located near the equator, where energy input is great, then they could reflect a great deal of energy back into the atmosphere and contribute to atmospheric heating. However, if those continents are heavily vegetated, their reflective capacity might be lowered.

Other features of terrestrial geography that can influence climate conditions are mountains and glaciers. Both rise and fall over time and can be high enough to interrupt wind and precipitation patterns. For instance, the growth of the Rocky Mountains probably disturbed the path of upper atmospheric winds known as the jet stream. In southern Asia, the Himalayas block humid air masses flowing from the south. Intense precipitation results on the windward side of these mountains, while the downwind side remains one of the driest areas on Earth.

Atmospheric composition is a climate variable that began to receive increased attention during the 1980s. Each type of gas molecule in the atmosphere absorbs a particular range of energy wavelengths. As the mix of gases changes, the range of wavelengths passing through the filter shifts. For instance, the gas ozone (O_3) selectively blocks long wave UV radiation. A drop in upper atmospheric ozone levels discovered in the late 1980s caused alarm because harmful UV rays were no longer being intercepted as effectively before they reach the earth's surface. Water vapor and solid particulates (dust) in the upper atmosphere also block incoming energy. Atmospheric dust associated with ancient meteor impacts is widely thought responsible for climatic cooling that may have killed the earth's dinosaurs 65 million years ago. Climate cooling could occur today if bombs from a nuclear war threw high levels of dust into the atmosphere. With enough radiation blockage, global temperatures could fall by several degrees, a scenario known as nuclear winter.

A human impact on climate that is more likely than nuclear winter is global warming caused by increased levels of carbon dioxide (CO_2) and other greenhouse gases in the upper atmosphere. Most solar energy enters the atmospheric system as long wavelengths and is reflected back into space in the form of short wavelength (heat) energy. Carbon dioxide blocks these short, warm wavelengths as they leave the earth's surface. Unable to escape, this heat energy remains in the atmosphere and keeps the earth warm enough for life to continue. The burning of fossil fuels and biomass (e.g., slash-and-burn agriculture) have raised atmospheric carbon dioxide levels. Rising CO_2 levels trap excessive amounts of heat and could raise global air temperatures to dangerous levels. This scenario is commonly known as the greenhouse effect. Extreme amounts of trapped heat have the potential to disturb precipitation patterns and overheat ecosystems, killing plant and animal species. Polar ice caps could melt, raising global ocean levels and threatening human settlements.

National Oceanic and Atmospheric Administration (NOAA) and National Aeronautics and Space Administration (NASA) data has shown that the average temperature of the Earth has risen by 1.2 to 1.4° Fahrenheit (-17.1– -17°C) in the past one hundred years. Since 1850, the eight warmest years recorded have occurred since 1998, with 2005 being the warmest year. The factors affecting climate change may also alter weather patterns, inducing droughts and floods, and contribute to sea level changes. Climate models based on accumulating amounts of greenhouse gases predict that temperatures at the end of the twenty-first century may be 3.2 to 7.2° Fahrenheit (-16 to -13.8°C) above 1990 temperatures.

Increased anthropogenic production of other gases such as methane (CH_4) also contributes to atmospheric warming, but carbon dioxide has been a focus of concern because it is emitted in much greater volume and persists in the atmosphere for a longer time.

Researchers are investigating how seriously human activity may be affecting the large and turbulent patterns of climate. Sometimes a very subtle event can have magnified repercussions in larger wind, precipitation, and pressure systems, disturbing major climate patterns

for decades. In many cases the climate appears to have a self-stabilizing capacity—an ability to initiate internal reactions to a destabilizing event that return it to equilibrium. For example, extreme greenhouse heating should cause increased evaporation of water. Resulting clouds could block incoming sunlight, producing an overall cooling effect to counteract heating.

The earth's climate is so complex that human alterations to the atmosphere (such as those caused by carbon dioxide emission) amount to an experiment having an unknown—and possibly life-threatening—outcome.

As of September 2010, the Intergovernmental Panel on Climate Change (IPCC) is progressing toward the issuing of the fifth assessment report on global climate change. Their 2007 report was a landmark document in which the weight of evidence conclusively established the role of human activity in global warming and global climate change.

See also Atmosphere; Greenhouse effect; Photosynthesis.

Resources

BOOKS

Aguado, Edward and James E. Burt. *Understanding Weather and Climate*. Upper Saddle River, NJ: Pearson/Prentice Hall, 2009.

Anderson, Bruce, and Alan H. Strahler. *Visualizing Weather and Climate*. Hoboken, N.J.: Wiley, 2008.

Cotton, William R., and Roger A. Pielke. *Human Impacts on Weather and Climate*. Cambridge: Cambridge University Press, 2007.

McCaffrey, Paul. *Global Climate Change*. Minneapolis, Minn.: H. W. Wilson, 2006.

Pooley, Eric. *The Climate War: True Believers, Power Brokers, and the Fight to Save the Earth*. New York: Hyperion, 2010.

OTHER

United States Environmental Protection Agency (EPA). "Climate Change: Basic Information." http://www.epa.gov/climatechange/basicinfo.html (accessed September 2, 2010).

Mary Ann Cunningham

Climate change controversies

At a press conference at Paris in February, 2007, Intergovernmental Panel on Climate Change (IPCC) scientists offered landmark consensus that strong

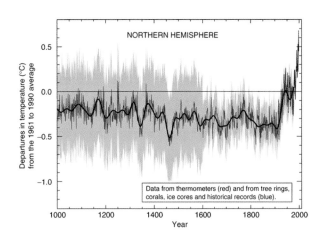

Graph showing Michael Mann's tree ring research. *(Center for Disease Control)*

scientific evidence supports the assertion that "most of the observed increase" in global warming is due to observed increases in greenhouse gases contributed by human (anthropogenic) activity.

The IPCC was formed in 1988 as a joint effort of the World Meteorological Organization (WMO) and the United Nations Environment Programme (UNEP) to study and report on potential global climate change. The IPCC working group does not conduct original research and does not gather the data that form the basis of its reports. Instead, the group studies scientific data published in peer reviewed scientific journals (publications that allow leading scientists in various areas of science to scrutinize the work of their colleagues) and other technical sources in order to attempt to formulate an objective consensus on the science related to climate change and the impact of human activity on climate change.

Working Group I of the IPCC approved and released portions of a report titled, *Climate Change 2007: The Physical Science Basis, Summary for Policymakers.* It was the group's contribution to the broader *Fourth Assessment Report of the Intergovernmental Panel on Climate Change.* The report findings instantly fueled global debate on how to respond to the report's ominous predictions regarding climate change.

The IPCC panel described progress in understanding both the natural and human impacts on climate change, including data related to observed climate change, the methods involved in the scientific study of climate processes, and estimates of projected future climate change. The report expanded upon prior IPCC reports and incorporated new data and findings discovered since the last IPCC report in 2001.

The IPCC working group members also assert that there is a much higher confidence in their predictions regarding future climate change. Because of the increasing focus on the scientific study of climate change, the latest report findings are based upon much more comprehensive data also subjected to much more sophisticated analyses than was possible with prior reports. Accordingly, the 2007 report's findings are issued with increased certainty regarding assessments and predictions.

For example, older IPCC reports stated that satellite measurements of global warming did not agree with surface measurements. The 2007 report, however, asserts that this discrepancy has been reconciled and that "warming of the climate system is unequivocal."

Prior IPCC reports asserted that human activity was only "likely" to be driving global warming. New report says there is "very high confidence" (more than 90 percent probability) that human activity is driving global warming.

Revised assessments and predictions

New data gathered since the last report show that losses from the ice sheets of Greenland and Antarctica have *"very likely* contributed to sea level rise." The new report also contains (for the first time) detailed regional predictions, such as that the western United States will be 9°Fahrenheit (5°C) warmer by 2100.

While the 2001 IPCC report did not address the question of how sensitive the climate is to the amount of CO_2 in the atmosphere: the 2007 report does assert that for each doubling of CO_2, global average temperature will warm 5.4°Fahrenheit (3°C).

The latest report also revises prior predictions of the rise in global average temperature by 2100 from 2.5–10.4°Fahrenheit (1.4–5.8°C) to 2–11.5°Fahrenheit (1.1–6.4°C).

The 2001 IPCC report predicted 3.5–34.6 inches (9–88 cm) rise in sea levels by 2100, but the current report predicts a rise of 7.1–23.2 inches (18–59 cm) in sea levels.

Although predictions of the maximum possible increase in sea levels have actually lowered, scientists and politicians agree that the increase in the minimum expected rise is much more dire because even at the minimum levels there will be profound social and economic ramifications for global humanity.

Early challenges

As dire as the latest sea level rise prediction might be, some scientists challenged the estimates as too optimistic.

For example, the IPCC working group did not consider data gathered after Dec. 2005, but according reports published in the scientific journal *nature*, "Greenland is losing ice at an ever-increasing rate." Scientists argue that this 2006 data, and subsequent studies, further raise estimates of expected increases on sea levels. In addition, a paper published in *Science* the day before the IPCC's press conference reported that sea levels have risen since 2001 at the maximum or uppermost rate predicted in the 2001 IPCC report.

Scientific and political controversy

Despite the high scientific confidence in the conclusions drawn in IPCC reports, errors in data and methodlogy were ultimately discovered, further fueling scientific and political controversy regarding global warming and climate change. Scientific controversy has existed because Earth's climate system is complex, which leads to uncertainties about how it has behaved in the distant past, how it works today, and how it will respond in the future. Public controversy often thrives on the existence of scientific uncertainty, but does not require it. Controversy persists even on points where scientific agreement has been reached. Also, even persons or nations who are convinced that climate change is real and human-caused can, and do, disagree about who should make what changes in order to minimize climate change's future impact and who should pay for them.

There are, then, three basic kinds of climate-change controversy: (1) real scientific disagreement over the details of climate changes past, present, and future; (2) denial of the existence or human origin of present-day climate change; and (3) disagreement over what actions to take against climate change, when to take them, how much they will cost, and how to pay for them.

Background

Many assertions about climate change do not correspond to the best and most complete scientific studies. A number of these arguments are stated below, each with one of the brief counter-arguments normally offered in reply by climate experts.

(1) *The world isn't getting warmer at all.* Millions of direct temperature measurements and satellite observations agree that the world is indeed warming.

(2) *The world did get warmer in the 1990s, but since 1998 it has been getting cooler.* Global warming must by its nature have random ups and downs. The year 1998 was actually an aberration, much hotter than those the years just before or after it. According to

the U.S. National Oceanic and Atmospheric Administration, since 2001 the world has experienced the warmest decade on record. Although it is almost certain that global surface temperatures will continue to climb in the near future. Future spikes and dips in the temperature record are inevitable.

(3) *Climate change is natural.* Natural climate change does occur, but evidence from many sources shows that humans have almost certainly caused the global warming that has been observed in recent years.

(4) *The Sun, cosmic rays, or both are causing global warming, not human activities.* Scientists have carefully examined the possibility that greater energy output from the Sun, or reduced cloud cover on Earth caused by a decrease in cosmic rays from deep space, may have been responsible for recent warming. However, both these effects are much too weak to account for the warming that has occurred. Only mathematical models that take human activities into account have been able to explain the warming that has already been measured. Human beings are changing the global climate.

(5) *Computer models are unreliable—garbage in, garbage out. Plus, they don't even take water vapor into account.* The mathematical models used to simulate past and future climate change are based on physics, are improved by constant testing against physical measurements, and always take water vapor into account. Models created by many independent groups of scientists agree that global warming will continue.

(6) *Climate change isn't anything to worry about anyway. Increased carbon dioxide in the atmosphere will make plants grow faster, soaking up carbon dioxide and keeping warming from getting out of hand.* Scientists gathered under the auspices of the United Nations as the Intergovernmental Panel on Climate Change, have described in detail, mostly recently in 2007, how climate change threatens human beings and is likely to cause the extinction of thousands of plant and animal species over the next century or so. The overall effects of climate change are already negative and are likely to become worse. Also, studies have so far been unable to detect any carbon-dioxide fertilization effect on plants in open-air ecosystems. This effect would, in any case, be too weak to change the overall climate picture.

(7) *Climate predictions must be less reliable than often errant weather prediction.* Weather and climate are different, differing in both duration and extent. Weather is the hour-to-hour, day-to-day behavior of the atmosphere in a given place; climate is weather averaged over years. Long-term, average weather is

more predictable than hourly or daily weather. For example, it is safe to predict in January that average temperatures will be warmer in July (in the Northern Hemisphere), even though the weather on any single day in July cannot be predicted. Climate, like the seasons, can be predicted much more confidently than the weather.

In addition to the scientific arguments, the sources of climate controversy are complex and there are economic costs to both combating climate change that must be weighed against economic losses caused by climate change.

Public impact

Efforts to confuse the public about climate controversies have been partly successful. In February 2008, the outgoing chair of the American Association for the Advancement of Science, John Holdren, told an interviewer for the *Guardian*, a British newspaper, that "deniers of the reality of the climate change problem have been more effective in the United States than they have been in Europe . . . Climate change deniers. . . have received attention in this country out of all proportion to their numbers, their qualifications, or the quality of their arguments. And it has slowed down the whole discussion in the United States."

As of November 2010, the United States had not passed any national legislation to control overall emissions of greenhouse gases. Nor had it yet agreed to any international treaty, such as the Kyoto Protocol, that would set mandatory targets for emissions reductions. The United States and China are the two largest greenhouse-gas emitters, each contributing about 20 percent of the global total.

Climategate

In November 2009, selected excerpts of private emails, data, and computer programs belonging to climate scientists working at the University of East Anglia's (UEA) Climatic Research Unit (CRU), located in Norwich, England, were published online. The CRU is regarded as one of the world's most influential climate science research institutions. The material, made available by hackers, sparked a controversy widely dubbed "Climategate."

In a frenzy of online activity prior to the December 2009 global climate summit in Copenhagen, Denmark, critics of global warming science published excerpts of the CRU material that they claimed supported their assertions that climate scientists manipulated or deleted data in order to support evidence of global warming. The assertions made it into media reports, at times

drawing more media attention than the efforts in Copenhagen to reach a new international treaty intended to limit greenhouse gas emissions and thereby mitigate the most dire predictions regarding climate change.

While many openly disavowed some of the views and hash language contained in some CRU emails (some of which was aimed at climate change skeptics) the vast majority of climate scientists and international agencies stood by the work of the CRU and its major conclusions of climate science as expressed by the IPCC reports, that "warming of global climate is now unequivocal, that is, essentially certain" and that the warming is most likely driven by human (anthropogenic) activity since the industrial revolution. CRU scientists strongly asserted that the allegations against them were false and that CRU data and conclusions had been subject to expert review for decades. CRU defenders argued that while minor errors might be found to exist, such errors are a normal part of the self-correcting scientific method and did not impact the conclusions drawn from CRU data and work, especially work used to prepare IPCC reports.

In response to complaints and concerns surrounding transparency and failure to release data, CRU scientists pointed out that about 95 percent of the raw station data used by CRU is also accessible via other sources.

While expressing varying degrees of dissatisfaction with the content of some of the CRU emails and calling for stricter oversight of climate research, the overwhelming majority of climate scientists, the IPCC, and all major national academies and institutes of science continue to endorse the conclusions of the data and the work produced at CRU.

A March 2010 report by the British House of Commons Science and Technology Committee expressed concern about the CRU scientist conduct but confidence in their conclusions. The committee specifically expressed confidence that words and phrases seized upon by critics in the hacked emails, such as "trick" or "hiding the decline," were "colloquial terms used in private e-mails and the balance of evidence is that they were not part of a systematic attempt to mislead." No data was hidden and the words "trick" and the phrase "hide the decline" merely related to attempts (including attempts by other scientists) to construct graphs and other statistical illustrations that would represent the most reliable data. The report further asserted that the word trick was "richly misinterpreted and quoted out of context." The Committee went on to support the conclusion that there was an attempt by climate scientists to subvert the scientific peer review process.

Later in 2010, a review group consisting of members of the academies of science from the United States, the Netherlands, the United Kingdom, and nearly 100 other countries ultimately concluded that the IPPC reports were sound and that the fundamental conclusions drawn from undisputed evidence concerning climate change were not altered by the minor errors in IPCC reports. However, to avoid future errors, the group advised that the IPCC should make predictions only when it has the highest levels of confidence in evidence and should avoid direct climate policy advocacy.

Resources

Books

Bolin, Bert. *A History of the Science and Politics of Climate Change: The Role of the Intergovernmental Panel on Climate Change.* New York: Cambridge University Press, 2008.

Cowie, Jonathan. *Climate Change: Biological and Human Aspects.* Cambridge, UK: Cambridge University Press, 2007.

Dessler, Andrew Emory, and Edward Parson. *The Science and Politics of Global Climate Change: A Guide to the Debate.* Cambridge, UK: Cambridge University Press, 2006.

DiMento, Joseph, and Pamela M. Doughman. *Climate Change: What It Means for Us, Our Children, and Our Grandchildren.* Boston: MIT Press, 2007.

Giddens, Anthony. *Politics of Climate Change.* Cambridge, UK: Polity Press, 2009.

Hulme, Mike. *Why We Disagree About Climate Change: Understanding Controversy, Inaction and Opportunity.* New York: Cambridge University Press, 2009.

Metz, B., et al. *Climate Change 2007: Mitigation of Climate Change: Contribution of Working Group III to the Fourth Assessment Report of the Intergovernmental Panel on Climate Change.* Cambridge, UK and New York: Cambridge University Press, 2007.

Page, Edward. *Climate Change, Justice and Future Generations.* Cheltenham, UK: Edward Elgar, 2006.

Solomon, S., et al. "IPCC, 2007: Summary for Policymakers." In *Climate Change 2007: The Physical Science Basis. Contribution of Working Group I to the Fourth Assessment Report of the Intergovernmental Panel on Climate Change*, edited by S. Solomon, et al. Cambridge, UK and New York: Cambridge University Press, 2007.

PERIODICALS

Adam, David. "Royal Society Tells Exxon: Stop Funding Climate Change Denial." *The Guardian* (September 20, 2006).

Begley, Sharon. "Global Warming Deniers: A Well-Funded Machine." *Newsweek* (March 18, 2007).

Oreskes, Naomi. "The Scientific Consensus on Climate Change." *Science* 306 (December 3, 2004): 1686.

OTHER

Intergovernmental Panel on Climate Change (IPCC). "IPCC Assessment Reports." http://www.ipcc.ch/ (accessed November 8, 2010).

Royal Society (UK). "Climate Controversies: A Simple Guide." http://royalsociety.org/page.asp?id=6229 (accessed November 8, 2010).

United Nations Framework Convention on Climate Change (UNFCCC). "United Nations Framework Convention on Climate Change (UNFCCC)." http://unfccc.int/2860.php (accessed November 8, 2010).

Larry Gilman
K. Lee Lerner

Climax (ecological)

Referring to a community of plants and animals that is relatively stable in its species composition and biomass, ecological climax is the apparent termination of directional succession—the replacement of one community by another. That the termination is only apparent means that the climax may be altered by periodic disturbances such as drought or stochastic disturbances such as volcanic eruptions. It may also change extremely slowly owing to the gradual immigration and emigration—at widely differing rates—of individual species, for instance following the retreat of ice sheets during the postglacial period. Often the climax is a shifting mosaic of different stages of succession in a more or less steady state overall, as in many climax communities that are subject to frequent fires. Species that occur in climax communities are mostly good competitors and tolerant of the effects (e.g., shade, root competition) of the species around them, in contrast to the opportunistic colonists of early successional communities. The latter are often particularly adapted for wide dispersal and abundant reproduction, leading to success in newly opened habitats where competition is not severe.

In a climax community, productivity is in approximate balance with decomposition. Biogeochemical cycling of inorganic nutrients is also in balance, so that the stock of nitrogen, phosphorus, calcium, etc., is in a more or less steady state.

Frederic E. Clements was the person largely responsible in the early twentieth century for developing the theory of the climax community. Clements regarded climate as the predominant determining factor, though he did recognize that other factors—for instance, fire—could prevent the establishment of the theoretical "climatic climax." Later ecologists placed more stress on interactions among several determining factors, including climate, soil parent material, topography, fire, and the flora and fauna able to colonize a given site.

Resources

BOOKS

Begon, Michael, Colin A. Townsend, and John L. Harper. *Ecology: From Individuals to Ecosystems*. 4th ed. New York: Wiley-Blackwell, 2006.

Eville Gorham

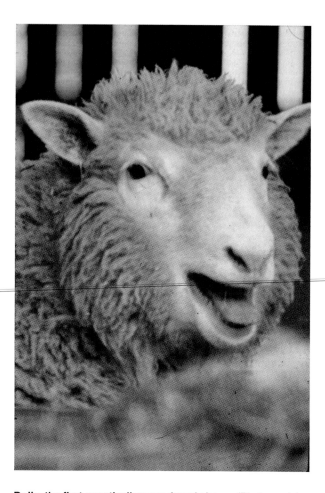

Dolly, the first genetically reproduced sheep. *(Photograph by Jeff Mitchell/Archive Photos)*

Clod

A compact, coherent mass of soil varying in size from 0.39–9.75 inches (10–250 mm). Clods are produced by operations like plowing, cultivation, and digging, especially on soils with atypical moisture levels. They are usually formed by compression, or by breaking off from a larger unit. Tractor attachments like disks, spike-tooth harrows, and rollers are used to break and pulverize clods during seedbed preparation.

Cloning

A clone is an organism or group of organisms or cells descended from a single ancestor and having genetic material (DNA, deoxyribonucleic acid) identical to that of the ancestor. Cloning hit the news headlines in 1997 when scientists in Scotland announced they had successfully cloned a sheep, named Dolly, in 1996. Although several other animal species had been cloned in the previous twenty years, it was Dolly that caught the public's attention. Suddenly, the possibility that humans might soon be cloned jumped from the pages of science fiction stories into the mainstream press. Dolly was the first adult mammal ever cloned.

As of 2010, no human being has yet been cloned but tthical debates over cloning are closely related to debates about other forms of biotechnology that involve the manipulation of DNA.

A clone is a single cell, a group of cells, or an organism produced in a laboratory without sexual reproduction. In effect, the clone is an exact genetic copy of the original source, much like identical twins. There are two types of cloning. Blastomere separation, also called "twinning," named after the naturally occurring process that creates identical twins, involves splitting a developing embryo soon after the egg is fertilized by sperm. The result is identical twins with DNA from both parents. The second cloning type, called nuclear transfer, is what scientists used to create Dolly. In cloning Dolly, scientists transferred genetic material from an adult female sheep to an egg in which the nucleus containing its genetic material had been removed.

Simple methods of cloning plants, such as grafting and stem cutting, have been used for more than 2,000 years. The modern era of laboratory cloning began in 1958 when the English-American plant physiologist Frederick C. Steward cloned carrot plants from mature single cells placed in a nutrient culture containing hormones, chemicals that play various and significant roles in the body.

The first cloning of animal cells occurred in 1964. In the first step of the experiment, biologist John B. Gurdon destroyed with ultraviolet light the genetic information stored in a group of unfertilized toad eggs. He then removed the nuclei (the part of an animal cell that contains the genes) from intestinal cells of toad tadpoles and injected them into those eggs. When the eggs were incubated (placed in an environment that promotes growth and development), Gurdon found that 1–2 percent of the eggs developed into fertile, adult toads.

The first successful cloning of mammals was achieved nearly twenty years later. Scientists in both Switzerland and the United States successfully cloned mice using a method similar to that of Gurdon. However, the Swiss and American methods required one extra step. After the nuclei were taken from the embryos of one type of mouse, they were transferred into the embryos of another type of mouse. The second type of mouse served as a substitute mother that went through the birthing process to create the cloned mice. The cloning of cattle livestock was achieved in 1988 when embryos from cows were transplanted to unfertilized cow eggs whose own nuclei had been removed.

Since Dolly, the pace and scope of cloning mammals has greatly intensified. In February 2002, scientists at Texas A&M University announced they had cloned a cat, the first cloning of a common domestic pet. Named "CC" (for carbon copy or copycat), the cat is an exact genetic duplicate of a two–year–old calico cat. Scientists cloned CC in December 2001 using the nuclear transfer method. In April 2002, a team of French scientists announced they had cloned rabbits using the nuclear transfer process. Out of hundreds of embryos used in the experiment, six rabbits were produced, four that developed normally and two that died. Two of the cloned rabbits mated naturally and produced separate litters of seven and eight babies

The first human embryos were cloned in 1993 using the blastomere technique that placed individual embryonic cells (blastomeres) in a nutrient culture where the cells then divided into forty-eight new embryos. These experiments were conducted as part of some studies on in vitro (out of the body) fertilization aimed at developing fertilized eggs in test tubes that could then be implanted into the wombs of women having difficulty becoming pregnant. However, these fertilized eggs did not develop to a stage that was suitable for transplantation into a human uterus.

The cloning of cells promises to produce many benefits in farming, medicine, and basic research. In farming, the goal is to clone plants that contain specific traits that make them superior to naturally occurring plants. For example, field tests have been conducted using clones of plants whose genes have been altered in the laboratory by genetic engineering to produce resistance to insects, viruses, and bacteria. New strains of plants resulting from the cloning of specific traits have led to fruits and vegetables with improved nutritional qualities, longer shelf lives, and new strains of plants that can grow in poor soil or even underwater.

A cloning technique known as twinning could induce livestock to give birth to twins or even triplets, thus reducing the amount of feed needed to produce meat. Cloning also holds promise for saving certain rare breeds of animals from extinction, such as the giant panda.

In medicine, gene cloning has been used to produce vaccines and hormones. Cloning techniques have already led to the inexpensive production of the hormone insulin for treating diabetes and of growth hormones for children who do not produce enough hormones for normal growth. The use of monoclonal antibodies in disease treatment and research involves combining two different kinds of cells (such as mouse and human cancer cells) to produce large quantities of specific antibodies. These antibodies are produced by the immune system to fight off disease. When injected into the blood stream, the cloned antibodies seek out and attack disease-causing cells anywhere in the body.

Despite the benefits of cloning and its many promising avenues of research, certain moral, religious, and ethical questions concerning the possible abuse of cloning have been raised. At the heart of these questions is the idea of humans tampering with life in a way that could harm society, either morally or in a real physical sense. Some people object to cloning because it allows scientists to "act like God" in manipulating living organisms.

The cloning of Dolly and the fact that some scientists are attempting to clone humans raised the debate over this practice to an entirely new level. A person could choose to make two or ten or 100 copies of himself or herself by the same techniques used with Dolly. This realization has stirred an active debate about the morality of cloning humans. Some people see benefits from the practice, such as providing a way for parents to produce a new child to replace one dying of a terminal disease. Other people worry about humans taking into their own hands the future of the human race.

Another controversial aspect of cloning deals not with the future but the past. Could Abraham Lincoln or Albert Einstein be recreated using DNA from a bone, hair, or tissue sample? If so, perplexing questions arise about whether this is morally or ethically acceptable? While it might ultimately be possible to create genetic duplicates, the clone would not have the same personality as the original Lincoln. This is because Lincoln, like all people, was greatly shaped from birth by his environment and personal experiences in addition to his genetic coding. And although CC, the cloned calico cat, is a genetic duplicate of her mother, she nevertheless has a different color pattern of fur than her mother. This is because environmental factors strongly influenced her development in the womb.

Since the movie "Jurassic Park" was released in 1993, there has been considerable public discussion about the possibility of cloning dinosaurs and other prehistoric or extinct species. In 1999, the Australian Museum in Sydney, Australia, announced scientists were attempting to clone a thylacine (a meat–eating marsupial related to kangaroos and opossums). The species has been extinct since 1932 but the museum has the body of a baby thylacine that has been preserved for 136 years. The problem is that today's cloning techniques are possible only with living tissue. The project was abandoned in 2005 after the DNA proved too degraded to construct a DNA library.

See also Genetic engineering.

Resources

BOOKS

Brown, Terry. *Gene Cloning and DNA Analysis: An Introduction.* 5th ed. Oxford, UK: Blackwell Science, 2006.

Caplan, Arthur and Glenn McGee, eds. *The Human Cloning Debate.* Berkeley: Berkeley Hills Books, 2006.

Harris, John. *On Cloning.* London and New York: Routledge, 2004.

Van Laar, Relinde. *Cloning and Stem Cell Research Legal Documents: An Overview of Research and Studies in Stem.* Tilburg: Wolf Legal Publishers, 2009.

Wilmut, Ian, and Roger Highfield. *After Dolly: The Promise and Perils of Cloning.* New York: W. W. Norton, 2007.

PERIODICALS

Dinnyes, A., and A. Szmolenszky, "Animal Cloning by Nuclear Transfer: State-of-the-Art and Future Perspectives." *Acta Biochimica Polonica* 52, no. 3 (March 2005): 585–588.

OTHER

The Human Genome Project. "Human Genome Project: Cloning." http://www.ornl.gov/sci/techresources/ Human_Genome/elsi/cloning.shtml (accessed October 20, 2010).

Ken R. Wells

Cloud chemistry

One of the exciting new fields of chemical research in the past half century involves chemical changes that take place in the atmosphere. Scientists have learned that a number of reactions are taking place in the atmosphere at all times. For example, oxygen (O_2) molecules in the upper stratosphere absorb solar energy and are converted to ozone (O_3). This ozone forms a layer that protects life on Earth by filtering out the harmful ultraviolet radiation in sunlight. Chloro-fluorocarbons and other chlorinated solvents (e.g., carbon tetrachloride and methyl chloroform) generated by human activities also trigger chemical reactions in the upper atmosphere including the break up of ozone into the two- atom form of oxygen. This reaction depletes the earth's protective ozone layer.

Clouds are often an important locus for atmospheric chemical reactions. They provide an abundant supply of water molecules that act as the solvent required for many reactions. An example is the reaction between carbon dioxide and water, resulting in the formation of carbonic acid. The abundance of both carbon dioxide and water in the atmosphere means that natural rain will frequently be somewhat acidic. Although conditions vary from time to time and place to place, the pH of natural, unpolluted rain is normally about 5.6 (the pH of pure water is 7.0). Other naturally occurring components of the atmosphere also react with water in clouds. In regions of volcanic activity, for example, sulfur dioxide released by outgassing and eruptions is oxidized to sulfur trioxide, which then reacts with water to form sulfuric acid or it dissolves and oxidizes within the droplet.

The water of which clouds are composed also acts as solvent for a number of other chemical species blown into the atmosphere from the earth's surface. Among the most common ions found in solution in clouds are sodium (Na^+), magnesium (Mg^{2+}), chloride (Cl^-), and sulfate (SO_4^{2-}) from sea spray; potassium (K^+), calcium (Ca^{2+}), and carbonate (CO_3^{2-}) from soil dust; and ammonium (NH_4^+) from organic decay.

The nature of cloud chemistry is often changed as a result of human activities. Perhaps the best known and most thoroughly studied example of this involves acid rain. When fossil fuels are burned, sulfur dioxide and nitrogen oxides (among other products) are released into the atmosphere. Prevailing winds often carry these products for hundreds or thousands of miles from their original source. Once deposited in the atmosphere, these oxides tend to be absorbed by water molecules and undergo a series of reactions by which they are converted to acids. Once formed in clouds by these reactions, sulfuric and nitric acids remain in solution in water droplets and are carried to earth as fog, rain, snow, or other forms of precipitation. In recent years it has become clear that clouds have a complex radical and photo-chemistry. The production and loss of hydrogen peroxide, and the related processes with the HO2 and OH radical, seem to lie at the heart of this chemistry in rain droplets. It can lead to the production of oxidized compounds such as formic acid or the nitration of organic solutes.

See also Atmosphere.

Resources

BOOKS

Hewitt, C. N., and Andrea V. Jackson. *Atmospheric Science for Environmental Scientists*. Chichester, U.K.: Wiley-Blackwell, 2009.

Manahan, Stanley E. *Environmental Chemistry*. Boca Raton, FL: Lewis, 2004.

Seinfeld, John H., and Spyros N. Pandis. *Atmospheric Chemistry and Physics: From Air Pollution to Climate Change*. Hoboken, N.J.: J. Wiley, 2006.

David E. Newton

Club of Rome

In April of 1968, thirty people, including scientists, educators, economists, humanists, industrialists, and government officials, met at the Academia dei Lincei in Rome. The meeting was called by Dr. Aurelio Peccei, an Italian industrialist and economist. The purpose of this meeting was to discuss "the present and future predicament of man." The "Club of Rome" was born from this meeting as an informal organization that has been described as an "invisible college." Its purpose, as described by Donella Meadows, is to foster understanding of the varied but interdependent components—economic, political, natural and social—that make up the global system in which we all live; to bring that new understanding to the attention of policy-makers and the public worldwide; and in this way to promote new policy initiatives and action. The original list of members is listed in the preface to Meadows's book entitled *The Limits to Growth*, in which the basic findings of the group are eloquently explained.

This text is a modern-day equivalent to the hypothesis of Thomas Malthus, who postulated that since increases in food supply cannot keep pace with geometric increases in human population, there would therefore be a time of famine with a stabilization of the human population. This eighteenth century prediction has, to a great extent, been delayed by the "green revolution" in which agricultural production has been radically increased by the use of fertilizers and development of special genetic strains of agricultural products. The high cost of agricultural chemicals, however, which are generally tied to the price of oil, has severely limited the capability of developing nations to purchase them.

The development of the Club of Rome's studies is most potently presented by Meadows in the form of graphs which plot on a time axis the supply of arable land needed at several production levels (present, double present, quadruple present, etc.) to feed the world's population based upon growth models.

She states that 7.9 billion acres (3.2 billion ha) of land are potentially suitable for agriculture on the earth; half of that land, the richest and most accessible half, is under cultivation today. She further states that the remaining land will require immense capital inputs to reach, clear, irrigate, or fertilize before it is ready to produce food. Such a conversion would likely produce severe consequences for the environment.

The Club of Rome's studies were not limited to food supply but also considered industrial output per capita, pollution per capita, and general resources available per capita. The key issue is that the denominator, per capita, keeps increasing with time, requiring ever more frugal and careful use of the resources; however, no matter how carefully the resources are husbanded, the inevitable result of uncontrolled population growth is a catastrophe which can only be delayed. Therefore stabilizing the rate of world population growth must be a continuing priority.

As a follow-up to the Club of Rome's original meeting, a global model for growth was developed by Jay Forrester of the Massachusetts Institute of Technology. This model is capable of update with insertion of information on population, agricultural production, natural resources, industrial production, and pollution. Meadows's report *The Limits to Growth* represents a readable summary of the results of this modeling. The Club of Rome continues to run meetings and act as a think tank. It has drawn criticism for being too elitist and over pessimistic in its conclusions about the future. It celebrated its fortieth anniversary at a conference on June 15, 2008.

A new branch of the Club of Rome is the tt30, a group of people between the ages of 25 and 35 who form a "think tank." This group is primarily concerned with problems of today, future issues, and how to deal with them.

Resources

BOOKS

Bartenstein, Martin, and Karin Feiler. *Sustainability Creates New Prosperity: Basis for a New World Order, New Economics and Environmental Protection; Review by Members of the Club of Rome and International Experts.* 2004.

Grachev, Andrei, and Carla Marchese. *From Global Warning to Global Policy: International Conference Organized by the World Political Forum and the Club of Rome, Turin, Italy, 28-29 March 2008.* Ricerche. Venezia: Marsilio, 2009.

Kapica, Sergej P. *Global Population Blow-Up and After The Demographic Revolution and Information Society.* Hamburg: Global Marshall Plan Initiative, 2006.

Kuklinski, Antoni, and B. Skuza. *Turning Points in the Transformation of the Global Scene.* Warsaw: Oficyna Wydawnicza Rewasz, 2006.

Sachs, Wolfgang. *Resource Justice and World Citizenship.* Erasmus-lezing, 2004. Zeist: Club of Rome, Erasmus Liga, 2004.

ORGANIZATIONS

The Club of Rome, Rissener Landstr 193, Hamburg, Germany, 22559, +49 40 81960714, +49 40 81960715, mail@clubofrome.org, http://www.clubofrome.org

Malcolm T. Hepworth

C:N ratio

Organic materials are composed of a mixture of carbohydrates, lignins, tannins, fats, oils, waxes, resins, proteins, minerals, and other assorted compounds. With the exception of the mineral fraction, the organic compounds are composed of varying ratios of carbon and nitrogen. This is commonly abbreviated to the C:N ratio. Carbohydrates are composed of carbon, hydrogen, and oxygen and are relatively easily decomposed to carbon dioxide and water, plus a small amount of other by-products. Protein-like materials are the prime source of nitrogen compounds as well as sources of carbon, hydrogen, and oxygen and are important to the development of the C:N ratio and the eventual decomposition rate of the organic materials.

The aerobic heterotrophic bacteria are primarily responsible for the decay of the large amount of organic compounds generated on the earth's surface. These organisms typically have a C:N ratio of about 8:1. When organic residues are attacked by the bacteria under appropriate habitat conditions, some of the carbon and nitrogen are assimilated into the new and rapidly increasing microbial population, and copious amounts of carbon dioxide are released to the atmosphere. The numbers of bacteria are highly controlled by the C:N ratio of the organic substrate.

As a rule, when organic residues of less than 30:1 ratio are added to a soil, there is very little noticeable decrease in the amount of mineral nitrogen available for higher plant forms. However as the C:N ratio begins to rise to values of greater than 30:1, there may be competition for the mineral nitrogen forms. Bacteria are lower in the food chain/web and become the immediate beneficiary of available sources of mineral nitrogen, while the higher species may suffer a lack of mineral nitrogen. Ultimately, when the carbon source is depleted, the organic nitrogen is released from the decaying microbes as mineral nitrogen.

The variation in the carbon content of organic material is reflected in the constituency of the compound. Carbohydrates usually contain less than 45 percent carbon, while lignin may contain more than 60 percent carbon. The C:N ratio of plant material may well reflect the kind and stage of growth of the plant. A young plant typically contains more carbohydrates and less lignin, while an older plant of the same species will contain more lignin and less carbohydrate. Ligneous tissue such as found in trees may have a C:N ratio of up to 1000:1.

The relative importance of the C:N ratio addresses two concerns: one, the rate of the organic matter decay to the low C:N ratio of humus, (approximately 10:1), and secondly the immediate availability of mineral nitrogen (NH_4^+) to meet the demand of higher plant needs. The addition of mineral nitrogen to organic residues is a common practice to enhance the rate of decay and to reduce the potential for nitrogen deficiency developing in higher plants where copious amounts of organic residue which has a C:N ratio of greater than 30:1 have been added to the soil.

Composting of organic residues permits the breakdown of the residues to occur without competition the of higher plants for the mineral nitrogen and also reduces the C:N ratio of the resulting mass to a C:N value of less than 20:1. When this material is added to a soil, there is little concern about the potential for nitrogen competition between the micro-organisms and the higher plants.

Resources

BOOKS

Burroughs, William James. *Climate Change: A Multidisciplinary Approach*. Cambridge: Cambridge University Press, 2007.

Cowie, Jonathan. *Climate Change: Biological and Human Aspects*. Cambridge: Cambridge University Press, 2007.

Elliott, David. *Sustainable Energy: Opportunities and Limitations*. Basingstoke: Palgrave Macmillan, 2010.

Seifried, Dieter, and Walter Witzel. *Renewable Energy: The Facts*. London: Earthscan Publications, 2010.

Royce Lambert

Coal

Consisting of altered remains of plants, coal is a widely used fossil fuel. Generally, the older the coal, the higher the carbon content and heating value. Anthracite coal ranks highest in carbon content, then bituminous coal, subbituminous coal, and lignite (as determined by the American Society for Testing Materials). Over 80 percent of the world's vast reserves occur in the former Soviet Union, the United States, and China. Though globally abundant, it is associated with many environmental problems, including acid-drainage, degraded land, sulfur oxide emissions, acid rain, and heavy carbon dioxide emissions. However, clean coal-burning technologies, including Liquefied or gasified forms, are now available.

Coal Production From 1950–2000

Date	Anthricite	Bituminous	Lignite	Subbituminous
1950	44.08	516.31	0.00	0.00
1960	18.82	415.51	0.00	0.00
1970	9.73	578.47	8.04	16.42
1980	6.06	628.77	47.16	147.72
1990	3.51	693.21	88.09	244.27
2000	4.51	548.47	88.74	433.78

Coal production. *(Reproduced by permission of Gale, a part of Cengage Learning)*

Anthracite, or "hard" coal, differs from the less altered bituminous coal by having more than 86 percent carbon and less than 14 percent volatile matter. It was formerly the fuel of choice for heat purposes because of high Btu (British Thermal Unit) values, minimally 14,500, and low ash content. In the United States, production has dropped from 100 million tons in 1917 to about seven million tons as anthracite has been replaced by oil, natural gas, and electric heat. Predominantly in eastern Pennsylvania's Ridge and Valley Province, anthracite seams have a wavelike pattern, complicating extraction. High water tables and low demand are the main impediments to expansion.

Bituminous coal, or "soft" coal, is much more abundant and easier to mine than anthracite but has lower carbon content and Btu values and higher volatility. Historically dominant, it energized the Industrial Revolution, fueling steam engines in factories, locomotives, and ships. Major coal regions became steel centers because two tons of coal were needed to produce each ton of iron ore. This is the only coal suitable for making coke, needed in iron smelting processes. Major deposits include the Appalachian Mountains and the Central Plains from Indiana through Oklahoma.

Subbituminous coal ranges in Btu values from 10,500 (11,500 if agglomerating) down to 8,300. Huge deposits exist in Wyoming, Montana, and North Dakota with seams 70 ft (21.4 m) thick. Though distant from major utility markets, it is used extensively for electrical power generation and is preferred because of its abundance, low sulfur content, and good grinding qualities. The latter makes it more useful than the higher grade, but harder, bituminous coal because modern plants spray the coal into combustion chambers in powder form. Demand for this coal skyrocketed following the 1973 OPEC oil embargo and subsequent restrictions on natural gas use in new plants.

Lignite, or "brown" coal, is the most abundant, youngest, and least mature of the coals, with some plant texture still visible. Its Btu values generally range below 8,300. Although over 70 percent of the deposits are found in North America, mainly in the Rocky Mountain region, there is little production there. It is used extensively in many eastern European countries for heating and steam production. Russian scientists have successfully burned lignite *in situ*, tapping the resultant coal gas for industrial heating. If concerns over global warming are satisfied, future liquefying and gasifying technologies could make lignite a prized resource.

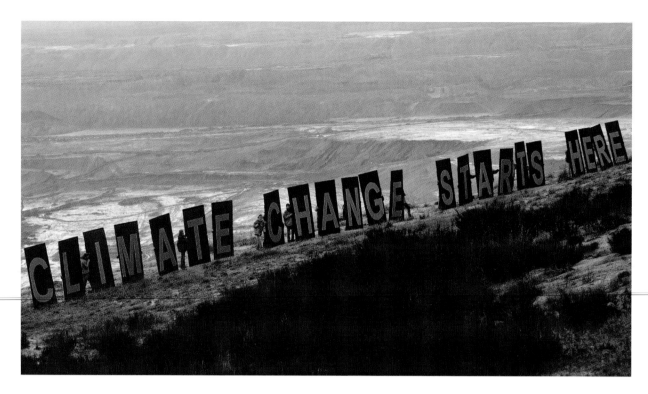

Greenpeace activists pose with huge letters forming the slogan "Climate change starts here" on the edge of a coal mine to commemorate the second anniversary of the validity of the Kyoto protocol. *(AP Photo/Petr David Josek)*

Resources

BOOKS

Andrews, Thomas G. *Killing for Coal: America's Deadliest Labor War*. Cambridge, MA: Harvard University Press, 2010.

Goodell, Jeff. *Big Coal: The Dirty Secret Behind America's Energy Future*. Seattle: Mariner Books, 2007.

Shnayerson, Michael. *Coal River*. New York: Farrar, Straus and Giroux, 2008.

Nathan H. Meleen

Coalition for Environmentally Responsible Economies (CERES) *see* **Valdez Principles.**

Coal bed methane

Coal bed methane (CBM) is a natural gas contained in coal seams. Methane gas is formed during coalification, the transformation of plant material into coal. The gas released during mining is called coal mine methane (CMM). Methane gas is also located in aquifers and in wells penetrating or overlaying CBM deposits.

Methane is a potent greenhouse gas and anthropogenic contributions accelerate global warming.

Liz Swain

Coal gasification

The term coal gasification refers to any process by which coal is converted into some gaseous form that can then be burned as a fuel. Coal gasification technology was relatively well known before World War II, but it fell out of favor after the war because of the low cost of oil and natural gas. Beginning in the 1970s, utilities showed renewed interest in coal gasification technologies as a way of meeting more stringent environmental requirements.

Traditionally, the use of fossil fuels in power plants and industrial processes has been fairly straight-forward. The fuel—coal, oil, or natural gas—is burned in a furnace and the heat produced is used to run a turbine or operate some industrial process. The problem is that such direct use of fuels results in the massive release of oxides of carbon, sulfur, and nitrogen, of unburned

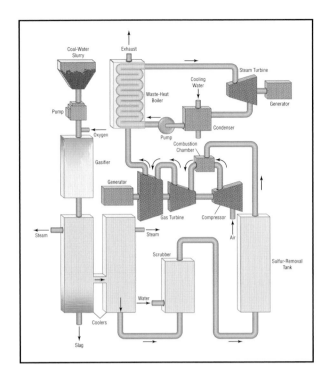

A schematic of coal gasification by heating a coal-and-water slurry in a low-oxygen environment. *(Reproduced by permission of Gale, a part of Cengage Learning)*

oxygen and steam. In the boiler, a complex set of chemical reactions occur, some of which are exothermic (heat releasing) and some of which are endothermic (heat-absorbing).

Below is an example of an exothermic reaction:

$$2C + \frac{3}{2} O_2 \rightarrow CO_2 + CO$$

The carbon monoxide produced in this reaction may then go on to react with hydrogen released from the coal to produce a second exothermic reaction

$$CO + 3H_2 \rightarrow CH_4 + H_2O$$

The energy released by one or both of the reactions is then available to initiate a third reaction that is endothermic

$$C + H_2O \rightarrow > CO + H_2$$

Finally, the mixture of gases resulting from reactions such as these, a mixture consisting most importantly of carbon monoxide, methane, and hydrogen, is used as fuel in a boiler that produces steam to run a turbine and a generator.

In practice, the whole series of exothermic and endothermic reactions are allowed to occur within the same vessel, so that coal, air or oxygen, and steam enter through one inlet in the boiler, coal enters at a second inlet, and the gaseous fuel is removed through an outlet pipe.

One of the popular designs for a coal gasification reaction vessel is the Lurgi pressure gasifier. In the Lurgi gasifier, coal enters through the top of a large cylindrical tank. Steam and oxygen are pumped in from the bottom of the tank. Coal is burned in the upper portion of the tank at relatively low temperatures, initiating the exothermic reactions described above. As unburned coal flows downward in the tank, heat released by these exothermic reactions raises the temperature in the tank and brings about the endothermic reaction in which carbon monoxide and hydrogen are produced. These gases are then drawn off from the top of the Lurgi gasifier.

The exact composition of the gases produced is determined by the materials introduced into the tank and the temperature and pressure at which the boiler is maintained. One possible product, chemical synthesis gas, consists of carbon monoxide and hydrogen. It is used primarily by the chemical industry in the production of other chemicals such as ammonia and methyl alcohol. A second possible product is medium-Btu gas, made up of hydrogen and carbon monoxide. Medium-Btu gas is used as a general purpose fuel for

hydrocarbons, of particulate matter, and of other pollutants. In a more environmentally- conscious world, such reactions are no longer acceptable.

This problem became much more severe with the shift from oil to coal as the fuel of choice in power generating and industrial plants. Coal is "dirtier" than both oil and natural gas and its use, therefore, creates more serious and more extensive environmental problems.

The first response of utilities and industries to new air pollution standards was to develop methods of capturing pollutants after combustion has occurred. Flue gas desulfurization systems, called scrubbers, were one approach strongly favored by the United States government. But such systems are very expensive, and utilities and industries rapidly began to explore alternative approaches in which coal is cleansed of material that produce pollutants when burned. One of the most promising of these clean-coal technologies is coal gasification.

A variety of methods are available for achieving coal gasification, but they all have certain features in common. In the first stages, coal is prepared for the reactor by crushing and drying it and then pre-treating it to prevent caking. The pulverized coal is then fed into a boiler where it reacts with a hot stream of air or utilities and industrial plants. A third possible product is substitute natural gas, consisting essentially of methane. Substitute natural gas is generally used as just that, a substitute for natural gas.

Coal gasification makes possible the removal of pollutants before the gaseous products are burned by a utility or industrial plant. Any ash produced during gasification, for example, remains within the boiler, where it settles to the bottom of the tank, is collected, and then removed. Sulfur dioxide and carbon dioxide are both removed in a much smaller, less expensive version of the scrubbers used in smokestacks.

Perhaps the most successful test of coal gasification technology has been going on at the Cool Water Integrated Gasification Combined Cycle plant near Barstow, California. The plant has been operating since June 1984 and is now capable of generating 100 megawatts of electricity. The four basic elements of the plant are a gasifier in which combustible gases are produced, a particulate and sulfur removal system, a combustion turbine in which the synthetic gas is burned, and a stem turbine run by heat from the combustion turbine and the gasifier.

The Cool Water plant has been an unqualified environmental success. It has easily met federal and state standards for effluent sulfur dioxide, oxides of nitrogen, and particulates, and its solid wastes have been found to be non-hazardous by the California Department of Health.

Waste products from the removal system also have commercial value. Sulfur obtained from the reduction of sulfur dioxide is 99.9 percent pure and has been selling for about $100 a ton. Studies are also being made to determine the possible use of slag for road construction and other building purposes.

Coal gasification appears to be a promising energy technology for the twenty-first century. One of the intriguing possibilities is to use sewage or hazardous wastes in the primary boiler. In the latter case, hazardous elements could be fixed in the slag drawn off from the bottom of the boiler, preventing their contaminating the final gaseous product.

The major impediment in the introduction of coal gasification technologies on a widespread basis is their cost. At the present time, a plant operated with synthetic gas from a coal gasification unit is about three times as expensive as a comparable plant using natural gas. Further research is obviously needed to make this new technology economically competitive with more traditional technologies.

Another problem is that most coal gasification technologies require very large quantities of water. This can be an especially difficult problem since gasification plants should be built near mines to reduce shipping costs. But most mines are located in Western states, where water supplies are usually very limited.

Finally, coal gasification is an inherently less efficient process than the direct combustion of coal. In most approaches, between 30 percent and 40 percent of the heat energy stored in coal is lost during its conversion to synthetic gas. Such conversions would probably be considered totally unacceptable except for the favorable environmental trade-offs they provide.

One possible solution to the problem described above is to carry out the gasification process directly in underground coal mines. In this process, coal would be loosened by explosives and then burned directly in the mine. The low-grade synthetic gas produced by this method could then be piped out of the ground, upgraded and used as a fuel. Underground gasification is an attractive alternative for many reasons. By some estimates, up to 80 percent of all coal reserves cannot be recovered through conventional mining techniques. They are either too deep underground or dispersed too thinly in the earth. The development of methods for gasification in coal seams would, therefore, greatly increase the amount of this fossil fuel available for our use.

A great deal of research is now being done to make coal gasification a more efficient process. A promising breakthrough involves the use of potassium hydroxide or potassium carbonate as a catalyst in the primary reactor vessel. The presence of a catalyst reduces the temperature at which gasification occurs and reduces, therefore, the cost of the operation.

Governments in the United States and Europe, energy research institutes, and major energy corporations are actively involved in research on coal gasification technologies. The Electric Power Research Institute, Institute of Gas Technology, U.S. Department of Energy, Texaco, Shell, Westinghouse, and Exxon are all studying modifications in the basic coal gasification system to find ways of using a wide range of raw materials, to improve efficiency at various stages in the gasification process, and, in general, to reduce the cost of plant construction.

See also Air pollution control; Air quality; Alternative energy sources; Coal washing; Flue-gas scrubbing; Renewable energy; Strip mining; Surface mining.

Resources

BOOKS

Brune, Michael. *Coming Clean: Breaking America's Addiction to Oil and Coal.* San Francisco: Sierra Club Books, 2008.

Miller, Bruce G. *Coal Energy Systems.* Amsterdam and Boston, MA: Elsevier Academic Press, 2005.

PERIODICALS

Associated Press. "Coal Use Grows Despite Warming Worries." *New York Times* (October 28, 2007).

OTHER

National Geographic Society. "High Cost of Cheap Coal: The Coal Paradox ." http://environment.nationalgeogr aphic.com/environment/global-warming/high-cost-coal.html (accessed August 30, 2010).

U.S. Government; science.gov. "Coal." http://www.science. gov/browse/w_121E1.htm (accessed August 30, 2010).

World Energy Council. "Survey of Energy Resources 2007: Coal: Clean Coal Technologies." http://www. worldenergy.org/publications/survey_of_energy_ resources_2007/coal/631.asp (accessed August 30, 2010).

David E. Newton

Coal mining *see* **Black lung disease; Strip mining; Surface mining.**

Coal washing

Coal that comes from a mine is a complex mixture of materials with a large variety of physical properties. In addition to the coal itself, pieces of rock, sand, and various minerals are contained in the mixture. Thus, before coal can be sold to consumers, it must be cleaned. The cleaning process consists of a number of steps that results in a product that is specifically suited to the needs of particular consumers. Among the earliest of these steps is crushing and sizing, two processes that reduce the coal to a form required by the consumer.

The next step in coal preparation is a washing or cleaning step. This step is necessary not only to meet consumer requirements, but also to ensure that its combustion will conform to environmental standards.

Coal washing is accomplished by one of two major processes, by *density separation* or by *froth flotation*. Both processes depend on the fact that the particles of which a coal sample are made have different densities. When water, for example, is added to the sample, particles sink to various depths depending on their densities. The various components of the sample can thus be separated from each other.

In some cases, a liquid other than water may be used to achieve this separation. In a heavy medium bath, for example, a mineral such as magnetite or feldspar in finely divided form may be mixed with water, forming a liquid medium whose density is significantly greater than that of pure water.

A number of devices and systems have been developed for extracting the various components of coal once they have been separated with a water or heavy medium treatment. One of the oldest of these devices is the *jig*. In a jig, the column of water is maintained in a constant up-and-down movement by means of a flow of air. Clean coal particles are carried to the top of the jig by this motion, while heavier refuse particles sink to the bottom.

Another method of extraction, the *cyclone*, consists of a tank in which the working fluid (water or a heavy medium) is kept in a constant circular motion. The tank is constructed so that lighter clean coal particles are thrown out of one side, while heavier refuse particles are ejected through the bottom.

Shaking tables are another extraction method. As the table shakes back and forth, particles are separated by size, producing clean coal at one end and waste products at the other.

In cylindrical separators, a coal mixture is fed into a spinning column of air that throws the heavier waste particles outward. They coat the inner wall of the cylinder and fall to the bottom, where they are drawn off. The clean coal particles remain in the center of the air column and are drawn off at the top of the cylinder.

Froth flotation processes depend on the production of tiny air bubbles to which coal particles adhere. The amount of absorption onto a bubble depends not only on a particle's density, but also on certain surface characteristics. Separation of clean coal from waste materials can be achieved in froth flotation by varying factors, such as pH of the solution, time of treatment, particle size and shape, rate of aeration, solution density, and bubble size.

Resources

BOOKS

Andrews, Thomas G. *Killing for Coal: America's Deadliest Labor War.* Cambridge, MA: Harvard University Press, 2010.

Goodell, Jeff. *Big Coal: The Dirty Secret Behind America's Energy Future.* Seattle: Mariner Books, 2007.

Shnayerson, Michael. *Coal River.* New York: Farrar, Straus and Giroux, 2008.

David E. Newton

Coase theorem

An economic theorem used in discussions of external costs in environment-related situations. The standard welfare economic view states that in order to make the market efficient, external costs—such as pollution produced by a company in making a product—should be internalized by the company in the form of taxes or fees for producing the pollution. Coase theorem, in contrast, states that the responsibility for the pollution should fall on both the producer and recipient of the pollution. For example, people who are harmed by the pollution can pay companies not to pollute, thereby protecting themselves from any potential harm.

Ronald Coase, the economist who proposed the theorem, further states that government should intervene when the bargaining process or transaction costs between the two parties is high. The government's role, therefore, is not to address external costs that harm bystanders but to help individuals organize for their protection.

Coastal Society, The

The Coastal Society (TCS), founded in 1975, is an international, nonprofit organization which serves as a forum for individuals concerned with problems related to coastal areas. Its members, drawn from university settings, government, and private industry, agree that the conservation of coastal resources demands serious attention and high priority.

TCS has four main goals: (1) to foster cooperation and communication among agencies, groups, and private citizens; (2) to promote conservation and intelligent use of coastal resources; (3) to strengthen the education and appreciation of coastal resources; and (4) to help government, industry, and individuals successfully balance development and protection along the world's coastlines. Through these goals, TCS hopes to educate the public and private sectors on the importance of effective coastal management programs and clear policy and law regarding the coasts.

Since its inception, TCS has sponsored numerous conferences and workshops. Individuals from various disciplines are invited to discuss different coastal problems. Past conferences have covered such topics as "Energy Across the Coastal Zone," "Resource Allocation Issues in the Coastal Environment," "The Present and Future of Coasts," and "Gambling with the Shore." Workshops are sponsored in conjunction with government agencies, universities, professional groups, and private organizations. Conference proceedings are subsequently published. TCS also publishes a quarterly magazine, *TCS Bulletin*, which features articles and news covering TCS affairs and the broader spectrum of coastal issues.

TCS representatives present congressional testimony on coastal management, conservation, and water quality. Recently the organization drafted a policy statement and it plans to take public positions on proposed policies affecting coastal issues.

Resources

ORGANIZATIONS

The Coastal Society, P. O. Box 3590, Williamsburg, VA, USA, 22313-5408, (757) 565-0999, (757) 565-0922, coastalsoc@aol.com, http://www.thecoastalsociety.org

Cathy M. Falk

Coastal Zone Management Act (1972)

The Coastal Zone Management Act (CZMA) of 1972 established a federal program to help states in planning and managing the development and protection of coastal areas through the creation of a Coastal Zone Management Program (CZMP). The CZMA is primarily a planning act, rather than an environmental protection or regulatory act. Under its provisions, states can receive grants from the federal government to develop and implement coastal zone programs as long as the programs meet with federal approval. State participation in the program is voluntary, and the authority is focused in state governments. In 2002, 99.9% of the national shoreline and coastal waters were managed by state CZMPs.

In the 1960s, public concern began to focus on dredging and filling, industrial siting, offshore oil development, and second home developments in the coastal zone. The coastal zone law was developed in the context of increased development of marine and coastal areas, need for more coordinated and consistent governmental efforts, an increase in general environmental consciousness and public recreation demands, and a focus on land-use control nationally. In 1969, a report by the Commission on Marine Sciences, Engineering, and Resources (the Stratton Commission) recommended a federal grant program to the states to help them deal with coastal zone management. The Commission found that coastal areas were of prime national interest, but development was taking place without proper consideration of environmental and resource values.

During congressional debate over coastal zone legislation, support came primarily from marine scientists and affected state government officials. The opposition emanated from development and real estate interests and industry, who were also concerned with national land use bills. The major difference between House and Senate versions of the legislation that passed was which department would administer the program. At the executive level there was no debate: the Office of Coastal Zone Management (now the Office of Ocean and Coastal Resource Management, or OCRM), part of the National Ocean Service of NOAA, was placed in charge of the program. But Congressional opinion varied about the administrative oversight of the Act. The House favored the U.S. Department of the Interior (DOI); the Senate, the National Oceanic and Atmospheric Administration (NOAA), part of the Department of Commerce (DOC). The Senate position was adopted in conference.

The congressional committees and executive branch agencies involved in coastal zone management also greatly varied. The Senate Commerce Committee was selected to have jurisdiction over the legislation. In 2010, the Senate Committee on Environment and Public Works exercised jurisdiction over the CZMA. The House originally designated the Merchant Marine and Fisheries Committee as its legislative arm, but it was dissolved in the mid–1990s. In 2010, the House Committee on Resources had jurisdiction over the CZMA.

The CZMA declared that "there is a national interest in the effective management, beneficial use, protection, and development of the coastal zone." The purpose of the law is to further the "wise use of land and water resources for the coastal zone giving full consideration to ecological, cultural, historic and aesthetic values as well as to needs for economic development." The program is primarily a grant program, and the original 1972 Act authorized the spending of $186 million through 1977.

Under CZMA, the Secretary of Commerce was authorized to make grants to the states with coastal areas, including the Great Lakes, to help them develop the coastal zone management programs required by federal standards. The grants would pay for up to two-thirds of a state's program and could be received for no more than three years. In addition to these planning grants, the federal government could also make

grants to the states for administering approved coastal zone plans. Again, the grants could not exceed two-thirds of the cost of the state program. With federal approval, the states could forward federal grant money to local governments or regional entities to carry out the act.

The federal government also has oversight responsibilities, to make sure that the states are following the approved plan and administering it properly. The key components of a state plan are to: (1) identify the boundaries of the coastal zone; (2) define the permissible uses in the coastal zone that have a significant effect; (3) inventory and designate areas of particular concern; (4) develop guidelines to prioritize use in particular areas; (5) develop a process for protection of beaches and public access to them; (6) develop a process for energy facility siting; and (7) develop a process to control shoreline erosion. The states have discretion in these stages. For instance, some states have opted for coastal zones very close to the water, others have drawn boundaries further inland. The states determine what uses are to be allowed in coastal zones. Developments in the coastal area must demonstrate coastal dependence.

Coastal zone management plans deal primarily with private lands, though the management of federal lands and federal activities within the coastal zone is required by the consistency provision of the CZMA which requires state approval and must be consistent with state legislation. Indeed, this was intended as a major incentive for states to participate in the process. Although federal agencies with management responsibility in coastal zones have input into the plans, this state-federal coordination proved to be a problem in the 1980s and 1990s, especially regarding offshore oil development, due to differing interpretations of the consistency section of the CZMA.

At first, states were slow to develop plans and have them approved by federal authorities. This was due primarily to the political complexity of the interests involved in the process. The first three states to have their coastal zone management plans approved were California, Oregon, and Washington, which had their final plans approved by 1978. Both California and Washington had passed state legislation on coastal zone management prior to the federal law, California by referendum in 1972 and Washington in 1971. The California program is the most ambitious and comprehensive in the country. Its 1972 act established six regional coastal commissions with permit authority and a state coastal zone agency, which coordinated the program and oversaw the development of a state coastal plan. The California legislature passed a permanent coastal zone act in 1976 based on this plan. The permanent program stemmed from local plans reviewed by the regional commissions and state agency. Any development altering density or intensity of land use requires a permit from the local government, and sensitive coastal resource areas receive additional protection. As of 2009, of the thirty-five eligible states and territories, thirty-four had approved coastal zone plans, with Illinois choosing not to participate in the program.

Three major issues arose during the state planning processes. Identifying areas of crucial environmental concern was a controversy that pitted environmentalists against developers in many states. In general, developers have proved more successful than environmentalists. A second issue is general development. States that have the most advanced coastal programs, such as California, use a permit system for development within the coastal zone. Environmental concerns and cumulative effects are often considered in these permit decisions. These permit programs often lead developers to alter plans before or during the application process. Such programs have generally served to improve development in coastal zones, and to protect these areas from major abuses.

The final issue, the siting of large scale facilities, especially energy facilities, has proven to be continually controversial as states and localities seek control over siting through their coastal zone plans, while energy companies appeal to the federal government regarding the national need for such facilities. In a number of court cases, the courts ruled that the states did have the power to block energy projects that were not consistent with their approved coastal management plans. This controversy spilled over into offshore oil development in waters in the outer continental shelf (OCS), which were under federal jurisdiction. These waters were often included in state coastal zone plans, many of which sought to prevent offshore oil development. In this case, the courts found in the 1984 ruling (*Secretary of the Interior vs. California*) that such development could proceed over state objections.

Major amendments to the CZMA were passed in 1976, 1980, 1990, 1996, 1998, and 2004. In 1976, the Coastal Energy Impact Fund was created to sponsor grants and loans to state and local governments for managing the problems of energy development. Other changes included an increase in the federal funding level from two-thirds to 80 percent of planning and administration, an increase in planning grant eligibility from three to four years, and the addition of planning requirements for energy facilities, shoreline erosion, and beach access.

The 1980 amendments re-authorized the program through 1985 and established new grant programs for revitalizing urban waterfronts and helping coastal cities deal with the effects of energy developments. The amendments also expanded the policies and objectives of the CZMA to include the protection of natural resources, the encouragement of states to protect coastal resources of national significance, and the reduction of state-federal conflicts in coastal zone policy.

Amendments to the CZMA in 1990 were included in the budget reconciliation bill. Most importantly, the amendments overturned the 1984 decision of *Secretary of the Interior vs. California*, giving states an increased voice regarding federal actions off their coasts. The law, which was strongly opposed by the Departments of Defense and Interior, gives the states the power to try to block or change federal actions affecting the coastal zones if these actions are inconsistent with adopted plans. The amendments also initiated a nonpoint source coastal water pollution grant and planning program, repealed the coastal energy impact program, and reauthorized the CZMA through 1995.

In June 2001, a federal district court judge ruled that the Department of the Interior (DOI) must ensure that any oil and gas leases it grants on the outer continental shelf off the coast of California be consistent with the State of California Coastal Management Program (CCMP). The decision requires the Minerals Management Service of the DOI to provide proof that 36 federally-owned oil and gas drilling leases comply with CCMP guidelines. The case is the first to uphold state rights in federal oil leasing activities granted in the 1990 CZMA amendments.

The 1996 reauthorization of the CZMA extended the Act through September 30, 1999. However, several environmental issues—including debate over funding for nonpoint pollution programs and lobbying by the oil and gas industry to give the states less control over federal projects such as offshore drilling leases—have delayed its further reauthorization. The continuous resistance by industry to curtail the rights of states or the rights of the Federal government to restrict oil and gas exploration and expansion has been in the forefront of the battle to reauthorize the CZMA. New admendments and even new legislation have been proposed that limit states from interfering in industry decisions in certain areas or limit the Federal government in others. It is clear from this contradictory legislation, that industry wants the consistency provision to only work when it is in the industry's interests. The 1998 amendments to the CZMA, the Harmful Algal Bloom and Hypoxia Research and

Control Act, established a task force to prevent and control environmentally and economically devastating algal blooms and hypoxia (lack of oxygen).

See also Environmental law; Environmental policy; International Joint Commission; Marine pollution; National lakeshore; Water pollution.

Resources

BOOKS

Coastal Services Center. *Local Strategies for Addressing Climate Change.* [Charleston, S.C.]: National Oceanic and Atmospheric Administration, Coastal Services Center, 2009.

Dronkers, Job J. *Dynamics of Coastal Systems.* Hackensack, NJ: World Scientific, 2005.

Evans, Edward, and Edmund C. Penning-Rowsell. *Future Flooding and Coastal Erosion Risks.* London: Thomas Telford, 2007.

Valiela, Ivan. *Global Coastal Change.* Malden, MA: Blackwell Pub, 2006.

ORGANIZATIONS

Office of Ocean and Coastal Resource Management, N/ORM, NOAA, Office of Ocean and Coastal Resource Management, N/ORM; 1305 East-West Highway, Silver Spring, MD, USA, 20910, (301) 713-3155, (301) 713-4012, http://coastalmanagement.noaa.gov/programs/czm.html

Christopher McGrory Klyza
Paula A Ford-Martin

Co-composting

As a form of waste management, composting is the process whereby organic waste matter is microbiologically degraded under aerobic conditions to achieve significant volume reduction while also producing a stable, usable end product. Co-composting refers to composting two or more waste types in the same vessel or process, thus providing cost and space savings. The most common type of co-composting practiced by counties and townships in the United States involves mixing sewage sludge and municipal solid waste to speed the process and increase the usefulness of the end product. The processing and ultimate use or disposal of composting end products are regulated by federal and state environmental agencies.

Coevolution

Species are said to "coevolve" when their respective levels of fitness depend not only on their own genetic structure and adaptations but also the development of another species as well. The gene pool of one species creates selection pressure for another species. Although the changes are generally reciprocal, they may also be unilateral and still be considered coevolutionary.

The process of coevolution arises from interactions that establish structure in communities. A variety of different types of interactions can occur–symbiotic, where neither member suffers, or parasitic, predatory, and competitive relationships, where one member of a species pair suffers.

Coevolution can result from mutually positive selection pressure. For example, certain plants have in an evolutionary sense created positive situations for insects by providing valuable food sources for them. In return the insects provide a means to distribute pollen that is more efficient than the distribution of pollen by wind. Unfortunately, the plant and the insect species could evolve into a position of total dependency through increased specialization, thus enhancing the risk of extinction if either species declines.

Coevolution can also arise from negative pressures. Prey species will continually adapt defensive or evasive systems to avoid predation. Predators respond by developing mechanisms to surmount these defenses. However, these species pairs are "familiar" with one another, and neither of the strategies is perfect. Some prey are always more vulnerable, and some predators are less efficient due to the nature of variability in natural populations. Therefore the likelihood of extinction from this association is limited.

Several factors influence the likelihood and strength of coevolved relationships. Coevolution is more likely to take place in pairs of species where high levels of co-occurrence are present. It is also common in cases where selective pressure is strong, influencing important functions such as reproduction or mortality. The type of relationship—be it mutualistic, predator-prey, or competitor—also influences coevolution. Species that have intimate relationships, such as that of a specialist predator or a host-specific parasite, interact actively and thus are more likely to influence each other's selection. Species that do not directly encounter each other but interact through competition for resources are less likely candidates to coevolve, but the strength of the competition may influence the situation.

The result of coevolved relationships is structure in communities. Coevolution and symbiosis create

mistakes. It was thought that the energy increase they detected came from pockets of heat in the fluid, caused by the liquid not being stirred enough. Although a number of scientists around the world continued to research it, within ten years of the Utah experiments the discovery of cold fusion seemed to be discredited.

Pons and Fleischmann suffered a serious loss of reputation after the cold fusion debacle, ultimately losing a libel suit they brought against an Italian newspaper that had labeled them frauds. The Japanese government continued to support cold fusion research through much of the 1990s, but finally abandoned all its financial support in 1998. It seemed that only a few hundred scientists worldwide were working on cold fusion in 2001, judging from attendance at a semi-annual conference on the topic. In 2002, a researcher at the Oak Ridge National Laboratory in Oak Ridge, Tennessee, claimed to have detected something that might have been cold fusion using a process called acoustic cavitation. This produced high pressure in a liquid by means of sound waves. But with the stark example of Pons and Fleischmann, any new claim to cold fusion will have to be carefully substantiated by a number of scientists before it can overcome existing skepticism.

Resources

BOOKS

Chen, Francis F. *Introduction to Plasma Physics and Controlled Fusion*. New York: Springer, 2006.

Miyamoto, Kenro. *Plasma Physics and Controlled Nuclear Fusion*. Berlin, Germany, and New York: Springer, 2005.

Park, Robert. *Voodoo Science*. Oxford: Oxford University Press, 2000.

Taubes, Gary. *Bad Science: The Short and Weird Times of Cold Fusion*. New York: Random House, 1993.

Angela Woodward

Coliform bacteria

Coliform bacteria live in the nutrient-rich environment of animal intestines. Many species fall into this group, but the most common species in mammals is *Escherichia coli*, usually abbreviated *E. coli*. A typical human can easily have several trillion of these tiny individual bacterial cells inhabiting his or her digestive tract. On a purely numerical basis, a human may have more bacterial than mammalian cells in his or her body. Each person is actually a community or ecosystem of diverse species living in a state of cooperation, competition, or coexistence.

The bacterial flora of one's gut provides many benefits. They help break down and absorb food, they synthesize and secrete vitamins such as B_{12} and K on which mammals depend, and they displace or help keep under control pathogens that are ingested along with food and liquids. When the pathogens gain control, disagreeable or even potentially lethal diseases can result. A wide variety of diarrheas, dysenteries, and other gastrointestinal diseases afflict people who have inadequate sanitation. Many tourists suffer diarrhea when they come into contact with improperly sanitized water or food. Some of these diseases, such as cholera or food poisoning caused by *Salmonella*, *Shigella*, or *Lysteria* species, can be fatal. Because identifying specific pathogens in water or food is difficult, time-consuming, and expensive, public health officials usually conduct general tests to detect the presence and concentration of coliform organisms. The presence of any of these species, whether pathogenic or not, indicates that fecal contamination has occurred and that pathogens are likely present.

Colorado River

One of the major rivers of the western United States, the Colorado River flows for some 1,500 miles (2,415 km) from Colorado to northwestern Mexico. Dropping over 2 miles (3.2 km) in elevation over its course, the Colorado emptied into the Gulf of California until human management reduced its water flow. Over millions of years the swift waters of the Colorado have carved some of the world's deepest and most impressive gorges, including the Grand Canyon.

The Colorado River basin supports an unusual ecosystem. Isolated from other drainage systems, the Colorado has produced a unique assemblage of fishes. Of the 32 species of native fishes found in the Colorado drainage, 21–66 percent, are endemic species—that arose in the area and are found nowhere else.

Major projects carried out since the 1920s have profoundly altered the Colorado. When seven western states signed the Colorado River Compact in 1922, the Colorado became the first basin in which "multiple use" of water was initiated. Today the river is used to provide hydroelectric power, irrigation, drinking water, and recreation; over twenty dams have been erected along its length. The river, in fact, no longer drains into the Gulf of Colorado—it simply disappears near the Mexican towns of Tijuana and Mexicali. Hundreds of square miles of land have been submerged by the formation of reservoirs, and the temperature and clarity of the river's water have been profoundly changed by the action of the dams.

Alteration of the Colorado's habitat has threatened many of its fragile fishes, and a number are now listed as endangered species. The Colorado squawfish serves as an example of how river development can affect native wildlife. With the reservoirs formed by the impoundments on the Colorado River also came the introduction of game fishes in the 1940s. One particular species, the Channel catfish, became a prey item for the native squawfish, and many squawfish were found dead, having suffocated due to catfish lodged in their throats with their spines stiffly locked in place. Other portions of the squawfish population have succumbed to diseases introduced by these non-native fishes.

Major projects along the Colorado include the Hoover Dam and its reservoir, Lake Mead, as well as the controversial Glen Canyon Dam at the Arizona-Utah border, which has a reservoir extending into Utah for over 100 miles (161 km).

Resources

BOOKS

Blakey, Ronald C., and Wayne Ranney. *Ancient Landscapes of the Colorado Plateau* . Grand Canyon, AZ: Grand Canyon Association, 2008.
Fradkin, P. L. *A River No More: The Colorado River and the West*. New York: Knopf, 1981.

Eugene C. Beckham
Jeffrey Muhr

Combined sewer overflows

In many older coastal cities, especially in the northeastern United States, storm sewers in the street that collect stormwater runoff from rainfall are connected to municipal sewage treatment plants that process household sewage and industrial wastewater. Under normal, relatively dry conditions runoff and municipal waste go to a sewage treatment plant where they are treated. However, when it rains, in some cases less than an inch, the capacity of a sewage treatment plant can be exceeded; the system is overloaded. The mixed urban stormwater runoff and raw municipal sewage is released to nearby creeks, rivers, bays, estuaries or other coastal waters, completely untreated. This is a combined sewer overflow (CSO) event.

Combined sewer overflow events are not rare. In Boston Harbor, for example, there are eighty-eight pipes or outfalls that discharge combined stormwater runoff and sewage. It has been estimated that CSO events occur approximately sixty times per year, discharging billions of gallons of untreated runoff and wastewater to Boston Harbor.

Materials released during these CSO events can result in serious water quality problems that can be detrimental to both humans and wildlife. Toxic chemicals from households and industries are released during CSO events. In addition, toxic chemicals found in rainwater runoff, such as oil and antifreeze that have dripped onto roads from cars, will wash into coastal waters during these events.

Harmful bacteria and pathogens in the water are another major problem that can result after a CSO event. Some of these bacteria (coliform), live naturally in the intestinal tracts of humans and other warm blooded animals. After heavy rainfalls, scientists have measured increased levels of coliform bacteria in coastal waters near CSO outfalls. These bacteria, which indicate that there are other bacteria and pathogens that can make people sick if they swim in the water or eat contaminated shellfish, come from both animal and human wastes washed in from the streets. The bacteria are not removed or killed because the waters have not been treated in a sewage treatment plant. Because levels of these indicator bacteria are often high after CSO events, many productive shellfish beds are closed to protect human health. This can be a serious economic hardship to the fishing industry.

Combined sewer overflow events also result in increased quantities of trash and floatable debris entering coastal waters. When people litter, the trash is washed into storm sewers with rainwater. Since sewage treatment plants cannot handle the volume of water during these rainfall events, this trash is discharged along with the stormwater and sewage directly into open waters. This floatable debris is unsightly, and can be dangerous to marine animals and birds, which eat it and choke or become entangled within it. This often results in death.

Raw sewage, animal wastes, and runoff from lawns and other fertilized areas contain very high levels of nitrogen and phosphorus, which are nutrients used by marine and aquatic plants for growth. Therefore, CSO events are major contributors of extra nutrients and organic matter to nearshore waters. These nutrients act as fertilizers for many marine and aquatic algae and plants, promoting extreme growth called blooms. When they eventually die, the bacteria decomposing the plants and algae use up vast quantities of oxygen. This results in a condition known as hypoxia or low dissolved oxygen (DO). If DO levels are too low, marine animals will not have enough oxygen to survive and they will either die or move out of the area. Hypoxia has been the cause of some major fish kills, and can result in permanent changes in the ecological community if it is persistent.

There are a number of options available to reduce the frequency and impacts of CSO events. Upgrading sewage treatment plants to handle greater flow or constructing new facilities are two of the best, although most costly options. Another possibility is to separate storm sewers and municipal sewage treatment plants. While this would not prevent discharges of stormwater runoff during rainfall events, untreated household and industrial wastewater (i.e., raw sewage) would not be released. In addition, the resulting stormwater could be minimally treated by screening out trash and disinfecting it to kill bacteria. Use of wetlands to filter this stormwater has also been considered as an effective alternative and is currently being used in some areas. Another option is to build large storage facilities (often underground) to hold materials that would normally be discharged during CSO events. When dry conditions return, the combined runoff and wastewater are pumped to a nearby sewage treatment plant where they are properly treated. This option is being used in several areas, including some locations in New York City. At a minimum, screening of CSO discharges would reduce the quantity of floatable debris in nearshore waters, even if it did not solve all of the other problems associated with CSOs. Of course water conservation is another control that reduces that volume of water treated by sewage treatment plants, and therefore the volume that would be discharged during a CSO event.

Resources

OTHER

United States Environmental Protection Agency (EPA). "Water: Storm Water: Combined Sewer Overflows (CSOs)." http://www.epa.gov/ebtpages/watestormwater combinedseweroverflowscsos.html (accessed November 11, 2010).

Max Strieb

Combustion

The process of burning fuels. Traditionally biomass was used as fuel, but now fossil fuels are the major source of energy for human activities. Combustion is essentially an oxidation process that yields heat and light. Most fuels are carbon and hydrogen which use oxygen in the air as an oxidant. More exotic fuels are used in some combustion processes, particularly in rockets where metals such as aluminum or beryllium or hydrazine (a nitrogen containing compound) are well known as effective fuels. As rockets operate beyond the atmosphere they carry their own oxidants, which may also be quite exotic.

Combustion involves a mixture of fuel and air, which is thermodynamically unstable. The fuel is then converted to stable products, usually water and carbon dioxide, with the release of a large amount of energy as heat. At normal temperatures fuels such as coal and oil are quite stable and have to be ignited by raising the temperature. Combustion is said to be spontaneous when the ignition appears to take place without obvious reasons. Large piles of organic material, such as hay, can undergo slow oxidation, perhaps biologically mediated, and increase in temperature. If the amount of material is very large and the heat cannot escape, the whole pile can suddenly burst into flame. Will-o'-the-wisps or jack-o'-lanterns (known scientifically as *ignis fatuus*) are sometimes observed over swamps where methane is likely to be produced. The reason these small pockets of gas ignite is not certain, but it has been suggested that small traces of gases such as phosphine that react rapidly with air could ignite the methane.

Typical solid fuels like coal and wood begin to burn with a bright turbulent flame. This forms as volatile materials are driven off and ignited. These vapors burn so rapidly that oxygen can be depleted, creating a smoky flame. After a time the volatile substances in the fuel are depleted. At this point a glowing coal is evident and combustion takes place without a significant flame. Combustion on the surface of the glowing coal is controlled by the diffusion of oxygen towards the hot surface. If the piece of fuel is too small, such as a spark from a fire, it is likely to lose temperature rapidly and combustion will stop. By contrast a bed of coals can maintain combustion because of heat storage and the exchange of radiative heat between the pieces. The most intense combustion takes place between the crevices of a bed of coal. In these regions oxygen may be in limited supply which leads to the production of carbon monoxide. This is subsequently oxidized to carbon at the surface of the bed of coals with a faint blue flame. The production of toxic carbon monoxide from indoor fires can occasionally represent a hazard if subsequent oxidation to carbon dioxide is not complete.

Liquid fuels usually need to be evaporated before they burn effectively. This means that it is possible to see liquid combustion and gaseous combustion as similar processes. Combustion can readily be initiated with a flame or spark. Simply heating a fuel-air mixture can cause it to ignite, but temperatures have to be high before reactions occur. A much better way is to initiate combustion with a small number of molecular fragments of radicals. These can initiate chain reactions at much lower temperatures than molecular reactions. In a propane-air flame at about 2000° K, hydrogen and oxygen atoms and hydroxyl radicals account for about 0.3% of a gas mixture. It is these radicals that support combustion. They react with molecules and split them up into more radicals. These radicals can rapidly enter into the exothermic (heat releasing) oxidative processes that lie at the heart of combustion. The reactions also give rise to further radicals that support continued combustion. Under some situations the radicals reaction branch, such that the reaction of each radical produces two new radicals. These can enter further reactions, producing yet further increases in the number of reactions and very soon the system explodes. However the production of radicals can be terminated in a number of ways such as contact with a solid surface. In some systems, such as the internal combustion engine, an explosion is desired, but in others, such as a gas cooker flame, maintaining a stable combustion process is desirable.

In terms of air pollution the reaction of oxygen and nitrogen atoms with molecules in air leads to the formation of the pollutant nitric oxide through a set of reactions known as the Zeldovich cycle. It is this process that makes combustion such an important contributor of nitrogen oxides to the atmosphere.

Resources

BOOKS

Annamalai, Kalyan, and Ishwar Kanwar Puri. *Combustion Science and Engineering.* CRC series in computational mechanics and applied analysis. Boca Raton: CRC Press/Taylor & Francis, 2007.

Cox, Michael, Henk Nugteren, and Maria Janssen-Jurkovicova. *Combustion Residues: Current, Novel and Renewable Applications.* Chichester, England: John Wiley & Sons, 2008.

Jarosinski, Jozef, and Bernard Veyssiere. *Combustion Phenomena: Selected Mechanisms of Flame Formation, Propagation, and Extinction.* Boca Raton: CRC Press, 2009.

Kuan-yun Kuo, Kenneth. *Principles of Combustion.* New York: Wiley Interscience, 2005.

Law, Chun K. *Combustion Physics.* Cambridge, UK: Cambridge University Press, 2006.

PERIODICALS

Quadrelli, Roberta. "The Energy-Climate Challenge: Recent Trends in CO$_2$ Emissions from Fuel Combustion." *Energy Policy* 35 (2007): 5938-2952.

OTHER

United States Environmental Protection Agency (EPA). "Industry: Industrial Processes: Combustion." http://www.epa.gov/ebtpages/induindustcombustion.html (accessed November 10, 2010).

Peter Brimblecombe

Cometabolism

The partial breakdown of a (usually) synthetic compound by microbiological action. Synthetic chemicals are widely used in industry, agriculture, and in the home; many resist complete enzymatic degradation and become persistent environmental pollutants. In cometabolism, the exotic molecule is only partly modified by decomposers (bacteria or fungi), since they are unable to utilize it either as a source of energy, as a source of nutrient elements, or because it is toxic. Cometabolism probably accounts for long-term changes in DDT, dieldrin, and related chlorinated hydrocarbon insecticides in the soil. The products of this partial transformation, like the original exotic chemical, usually accumulate in the environment.

Commensalism

A type of symbiotic relationship. Many organisms depend on intimate physical relationship with organisms of other species, a relationship called symbiosis. In a symbiotic relationship there is a host and a symbiote. The symbiote always derives some benefit from the relationship. In a commensal relationship, the host organism is neither harmed nor benefitted. The relationship that exists between the clown fish living among the tentacles of sea anemones is one example of commensalism. The host sea anemones can exist without their symbiotes, but the fish cannot exist as successfully without the protective cover of the anemone's stinging tentacles.

Commercial fishing

Because fish have long been considered an important source of food, the fisheries were the first renewable resource to receive public attention in the United

Commercial fishermen haul in a net full of pink salmon out of Chatham Straight, Southeast Alaska. *(© Alaska Stock LLC / Alamy)*

sonar and spotting planes, account for the remaining 10 percent of fishers. As a result, wild fish populations have been decimated. About 28 percent of fish stocks have been overfished, and some stocks are nearing extinction.

In recent decades, the size of the industrial fishing fleet grew at twice the rate of the worldwide catch. The expansion in fishing may be coming to an end, however, as environmental, biological, and economic problems beset the fishing industry. As fish harvests decline, the numbers of jobs also decline. Governments have attempted to prop up the failing fisheries industry: in 1994, fishers worldwide spent $124 billion to catch fish valued at $70 billion, and the shortfall was covered by government subsidies. In recent decades, fishery imports have been one of the top five sources of the United States' trade deficit.

The commercial fisheries industry has contributed to its own problems by overfishing certain species to the point where those species' populations are too low to reproduce at a rate sufficient to replace the portion of their numbers lost to harvesting. Cod (*Gadus* species) and haddock (*Melanogrammus aeglefinus*) in the Atlantic Ocean, red snapper in the Gulf of Mexico, and salmon and tuna in the Pacific Ocean have all fallen victim to overfishing. The case of the Peruvian anchovy (*Engraulis ringens*) represents a specific example of how several factors may work together to contribute to species decline. Fishing for anchovies began off the coast of Peru in the early 1950s, and, by the late 1960s, as their fishing fleet had grown exponentially, the catch of Peruvian anchovies made up about 20 percent of the world's annual commercial fish harvest. The Peruvian fishermen were already overfishing the anchovies when meteorological conditions contributed to the problem. In 1972, a strong El Nin;ato struck. This phenomenon is a natural but unpredictable warming of the normally cool waters that flow along Peru's coast. The entire food web of the region was altered as a result, and the Peruvian anchovy population plummeted, leading to the demise of Peru's anchovy fishing industry. Peru has made some economic recovery since then by harvesting other species.

Many of the world's major fishing areas have already been fished beyond their natural limits. Different approaches to the problem of overfishing are under consideration to help prevent the collapse of the world's fisheries. Georges Bank, once one of the most fertile fishing grounds in the North Atlantic, is now closed and is considered commercially extinct. This area underwent strict controls for scallop fishing in 1996, which proved to be a viable remedy for that species in that

States. The National Marine Fisheries Service (NMFS) has existed since 1970, but the original Office of Commissioner of Fish and Fisheries was created over one hundred years ago, signed into law in 1871 by President Ulysses S. Grant (1822–1885). This office was charged with the study of "the decrease of the food fishes of the seacoasts and lakes of the United States, and to suggest remedial measures." From the beginning, the federal fishery agency has been granted broad powers to study aquatic resources ranging from coastal shallow waters to offshore deep-water habitats.

Worldwide, humans get an average of 16 percent of their dietary animal protein from fish and shellfish. In the developing nations of Africa and Asia, fish can account for over 50 percent of human animal protein consumption. With human populations ever increasing, the demand for and marketing of seafood has steadily increased, rising over the last half of the twentieth century to a peak in 1994 of about 100 million tons (91 billion kg) per year. The current annual marine fish catch has fallen to slightly over 90 million tons (about 81.9 billion kilograms). The per capita world fish catch has been steadily declining since 1970 as human population growth outdistances fish harvests. Scientists have projected that by 2020, the per capita consumption of ocean fish will be half of what it was in 1988.

To meet the demand for fish, the commercial fishing industry has expanded as well. According to the United Nations Food and Agriculture Organization (FAO), there were about 30 million fishers in the world as of 2004. About 90 percent of these fishers were categorized as small-scale operations. Industrial fishing crews, manning vessels that deploy highly innovative methods ranging from enormous nets to locale. The scallop population recovered within five years, reaching levels in excess of the original population, and parts of the bay could be re-opened for scallop fishing. But other species in Georges Bank continue to decline. Rapid and direct replenishment is not possible for slow-growing species that take years to reach maturity. For example, the black sea bass (*Stereolepis gigas*), has a life span comparable to that of humans and adults and typically grow to 500 pounds (227 kg). The success of a 1982 ban on fishing the black sea bass off the coast of California became evident early this century when significant numbers of these young fish, already weighing as much as 200 pounds (91 kg), appeared off the shores of Santa Barbara. Yet, full replenishment of the population remains years away.

Environmental problems also plague commercial fishing. Near-shore pollution has altered ecosystems, taking a heavy toll on all populations of fish and shellfish, not only those valued commercially. The collective actions of commercial fishermen also create some major environmental problems. The world's commercial fishermen annually catch and then discard about 20 billion pounds (9 billion kg) of non-target species of sea life. About one half, and in some cases, as much as 90 percent of a catch may be discarded. In addition to fish and shellfish, each year about one million seabirds are caught and killed in fishermen's nets. On average more than 6,000 seals and sea lions, about 20,000 dolphins and other aquatic mammals, and thousands of sea turtles meet the same fate. It is estimated that the amount of fish discarded annually is about 25 percent of the reported catch, or approximately 22 million tons (about 20 million metric tons) per year. Ecologically, two major problems arise from this massive disposal of organisms. One is the disruption of predator-prey ratios, and the other is the addition of a tremendous overload of organic waste to be dealt with in this ecosystem. Amendments to the Marine Mammal Protection Act in 1994 initiated further regulations to protect marine mammals from being inadvertently captured by commercial fishing with a seven-year goal to reduce these captures to "zero mortality and serious injury."

In 2001, a $1.6 billion gas pipeline that was proposed to be routed through neighboring waters from Nova Scotia to New Jersey, to be implemented as early as 2005, posed a new environmental threat to the Georges Bank area. Environmentalists, meanwhile, have lobbied the United States government to establish a marine habitat protection designation similar to wilderness areas and natural parks on land, to provide for the preservation of reefs, marine life, and underwater vegetation. Currently, less than 1 percent of water resources worldwide have the protection of formal legislation to prevent exploitation.

Habitat destruction is serious environmental concern. Fish and other aquatic wildlife rely on the existence of high quality habitat for their survival, and loss of habitat is one of the most pressing environmental threats to shorelines, wetlands, and other aquatic habitats. Approaches to the protection of essential fish habitat include efforts to strengthen and vigorously enforce the Clean Water Act and other protective legislation for aquatic habitats, to develop and implement restoration plans for target regions, to make improved policy decisions based on technical knowledge about shoreline habitats, and to better educate the public on the importance of protecting and restoring habitat. A relatively new approach to habitat recovery is the habitat conservation plan (HCP), in which a multi-species ecosystem approach to habitat management is preferred over a reactive species-by-species plan. Strategies for fish recovery are complex, and, instead of numbers of fish of a given species, the HCP uses quality of habitat to measure the success of restoration and conservation efforts. Long-term situations such as the restoration of black sea bass serve to re-emphasize the importance of resisting the temptation to manage overfishing of single species while failing to address the survival of the ecosystem as a whole.

The Magnuson-Stevens Fishery Conservation and Management Act was passed in 1976 to regulate fisheries resources and fishing activities in Federal waters, those waters extending to the 200 mile (322 km) limit. The act recognizes that commercial fishing contributes to the food supply and is a major source of employment, contributing significantly to the economy of the Nation. However, it also recognizes that overfishing and habitat loss has led to the decline of certain species of fish to the point where their survival is threatened, resulting in a diminished capacity to support existing fishing levels. Further, international fishery agreements have not been effective in ending or preventing overfishing. Fishery resources are limited but renewable and can be conserved and maintained to continue to provide good yields. Also, the act supports the development of underused fisheries, such as bottom-dwelling fish near Alaska.

Another resource to sustain increases in seafood consumption is aquaculture, where commercial food-fish species are grown on fish farms. It is estimated that the amount of farm-raised fish has doubled in the past decade and that about 36 percent of the fish consumed worldwide was raised in captivity as of 2006. Aquaculture shows promise for rescuing the fish industry by providing seafood supply in response to declining wild populations. However, aquaculture may have adverse effects on aquatic ecosystems resulting from heavy nutrient loading and chemicals deposited into waters to promote growth and reduce disease. Large quantities of farmed fish can also deposit significant amounts of waste that may disrupt the ecosystem. Farming fish can reduce some of the fishing pressure on wild populations; however, the farmed fish may compete with natural populations, thus potentially causing a loss of biodiversity.

In the United States, as well as other nations, the commercial fisheries industry faces potential collapse. In addition to overfishing pressures, climate change resulting in warmer temperatures and altered weather patterns may impact wild and farmed fish populations. Severe restrictions and tight controls imposed by the international community may be the only means of salvaging even a portion of this valuable industry. It will be necessary for partnerships to be forged between scientists, fisherman, and the regulatory community to develop and implement measures toward maintaining a sustainable fishery.

Resources

BOOKS

National Research Council (U.S.). *Cooperative Research in the National Marine Fisheries Service.* Washington, D.C.: National Academies Press, 2004.

Woodby, Doug. *Commercial Fisheries of Alaska.* Anchorage: Alaska Dept. of Fish and Game, Division of Sport Fish, Research and Technical Services, 2005.

Workshop on the Challenges and Opportunities of Fisheries Globalisation. *Globalisation and Fisheries: Proceedings of an OECD-FAO Workshop.* Paris, France: OECD, 2007.

OTHER

Food and Agriculture Organization of the United Nations (FAO). "World Fisheries." http://www.st.nmfs.noaa.gov/st1/fus/fus07/04_world2007.pdf (accessed October 13, 2010).

National Oceanic and Atmospheric Administration (NOAA). "NOAA Fisheries Office of Protected Resources: Fisheries Interactions/Protected Species Bycatch." http://www.nmfs.noaa.gov/pr/interactions/ (accessed October 13, 2010).

Eugene C. Beckham

Commingled recyclables *see* **Recycling.**

Commission for Environmental Cooperation

The Commission for Environmental Cooperation (CEC) is a trilateral international commission established by Canada, Mexico, and the United States in 1994 to address transboundary environmental concerns in North America. The original impetus behind the CEC was the perception of inadequacies in the environmental provisions of the North American Free Trade Agreement (NAFTA). A supplementary treaty, the North American Agreement for Environmental Cooperation (NAAEC) was negotiated to remedy these inadequacies, and it is from the NAAEC that the CEC derives its formal mandate.

The general goals set forth by the NAAEC are to protect, conserve, and improve the environment for the benefit of present and future generations. More specifically, the three NAFTA signatories agreed to a core set of actions and principles with regard to environmental concerns related to trade policy. These actions and principles include regular reporting on the state of the environment, effective and consistent enforcement of environmental law, facilitation of access to environmental information, the ongoing improvement of environmental laws and regulations, and promotion of the use of tax incentives and various other economic instruments to achieve environmental goals.

The CEC is to function as a forum for the NAFTA partners to identify and articulate mutual interests and priorities, and to develop strategies for the pursuit or implementation of these interests and priorities. The NAAEC further specifies the following priorities: identification of appropriate limits for specific pollutants; the protection of endangered and threatened species; the protection and conservation of wild flora and fauna and their habitat; the development of new approaches to environmental compliance and enforcement; strategies for addressing environmental issues that have impacts across international borders; the support of training and education in the environmental field; and promotion of greater public awareness of North American environmental issues. Central to the CEC's mission is the facilitation of dialogue among the NAFTA partners in order to prevent and solve trade and environmental disputes.

The governing body of the CEC is a Council of Ministers that consists of the environment ministers (or equivalent) from each country. The executive arm of the Commission is a Secretariat located in Montreal, consisting of a staff of over twenty members and headed by an Executive Director. The Secretariat also maintains a liaison office in Mexico City. The staff is drawn from all three countries and provides technical and administrative support to the Council of Ministers and to committees and groups established by the Council.

Technical and scientific advice is also provided to the Council of Ministers by a Joint Public Advisory Committee consisting of five members from each country appointed by the respective governments. This Committee may, on its own initiative, advise the Council on any matter within the scope of the NAAEC, including the annual program and budget. As a reflection of the CEC's professed commitment to participation by citizens throughout North America, the Committee is intended to represent a wide cross-section of knowledgeable citizens committed to environmental concerns who are willing to volunteer their time in the public interest. The CEC also accepts direct input from any citizen or non-governmental organization who believes that a NAFTA partner is failing to enforce effectively an existing environmental law.

The NAAEC also contains provisions for dispute resolution in cases in which a NAFTA signatory alleges that another NAFTA partner has persistently failed to enforce an existing environmental law, causing specific environmental damage or trade disadvantages to the claimant. These provisions may be invoked when a lack of effective enforcement materially affects goods or services being traded between the NAFTA countries. If the dispute is not resolved through bilateral consultation, the complaining party may then request a special session of the CEC's Council of Ministers. If the Council is likewise unable to resolve the dispute, provisions exist for choosing an Arbitral Panel. Failure to implement the recommendations of the Arbitral Panel subjects the offending party to a monetary enforcement assessment. Failure to pay this assessment may lead to suspension of free trade benefits.

The Council is also instructed to develop recommendations on access to courts (and rights and remedies before courts and administrative agencies) for persons in one country's territory who have suffered or are likely to suffer damage or injury caused by pollution originating in the territory of one of the other countries. In 2002, the Council published a five year study which stated that 3.4 million tonnes of toxins were produced in North America.

The CEC has been subject to some of the same criticisms leveled at the environmental provisions of NAFTA, particularly that it serves as a kind of environmental window- dressing for a trade agreement

that is generally harmful to the environment. The CEC's mandate for conflict resolution is primarily oriented towards consistent enforcement of existing environmental law in the three countries. This law is by no means uniform. By upholding the principles of free trade, and providing penalties for infringements of free trade, NAFTA establishes an environment in which private companies have an economic incentive, other considerations being equal, to locate production where environmental laws are weakest and the costs of compliance are therefore lowest. Countries with stricter environmental regulations face penalties for attempting to protect domestic industries from such comparative disadvantages.

Resources

OTHER

Public Citizen. "North American Free Trade Agreement (NAFTA)." http://www.citizen.org/trade/nafta/ (accessed October 16, 2010).
United States Environmental Protection Agency (EPA). "International Cooperation: Treaties and Agreements: North American Free Trade Agreement (NAFTA)." http://www.epa.gov/ebtpages/intetreatinorthamerican freetradeagre.html (accessed October 16, 2010).

ORGANIZATIONS

Commission for Environmental Cooperation, 393, rue St-Jacques Ouest, Bureau 200, MontreéalQueébec, Canada, H2Y 1N9, (514) 350-4300, (514) 350-4314, info@cec.org, http://www.cec.org.

Lawrence J. Biskowski

Commoner, Barry

1917–
American biologist, environmental scientist, author, and social activist

Born to Russian immigrant parents, Commoner earned a doctorate in biology from Harvard in 1941. As a biologist, he is known for his work with free radicals—chemicals like chlorofluorocarbons, which are suspected culprits in ozone layer depletion. Commoner led a fairly academic life at first, with research posts at various universities, but rose to some prominence in the late 1950s, when he and others protested atmospheric testing of nuclear weapons. He earned a national reputation in the 1960s with books, articles, and speeches on a wide range of environmental concerns, including pollution, alternative energy sources, and population. He wrote *Making Peace with the*

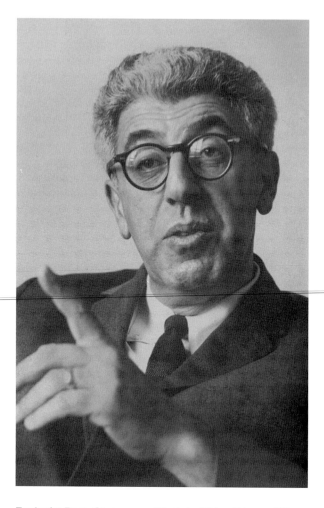

Ecologist Barry Commoner. *(Photo by Michael Mauney//Time Life Pictures/Getty Images)*

Planet, which was published in 1990. Commoner's other works include *Science and Survival* (1967), *The Closing Circle* (1971), *Energy and Human Welfare* (1975), *The Poverty of Power* (1976), and *The Politics of Energy* (1979).

Commoner believes that post-World War II industrial methods, with their reliance on nonrenewable fossil fuels are the root cause of modern environmental pollution. He has been particularly harsh on the petrochemical industry, which he believes is destroying the biosphere.

Almost as distressing as environmental pollution is our inability to clean it up. Commoner rejects attempts at environmental regulation as pointless. Far better, he says, to not produce the toxin in the first place.

Commoner offers radical, sweeping solutions for social and ecological ills. The most urgent of these is a

renewable energy source, primarily photovoltaic cells powered by solar energy. These would not only decentralize electric utilities (another target of Commoner's), but would use sunlight to fuel almost any energy need, including smaller, lighter, battery-powered cars. To ease the transition from fossil fuels to solar power, he proposes methane, cogeneration (which produces electricity from waste heat), and an organic agriculture system that would "produce enough ethanol to replace about 20 percent of the national demand for gasoline without reducing the overall supply of food or significantly affecting its price."

Commoner makes few compromises, and his environmental zeal has made him a crusader for social causes as well. Eliminating Third World debt, he argues, would improve life in impoverished countries and end the spiral of economic desperation that drives countries to overpopulation.

In 1980, Commoner made a bid for the U.S. presidency on the Citizen's Party ticket, a short-lived political attempt to combine environmental and Socialist agendas. From 1981 until 2000 he has been the director of the Center for the Biology of Natural Systems at Queens College in New York City. Since then, he has been a senior scientist at the same institution.

Resources

BOOKS

Commoner, B. *The Closing Circle*. New York: Knopf, 1971.
Commoner, B. *Making Peace With the Planet*. New York: New Press, 1992.

PERIODICALS

Commoner, B. "Ending the War Against Earth." *The Nation* 250 (30 April 1990): 589–90.
Commoner, B. "The Failure of the Environmental Effort." *Current History* 91 (April 1992): 176–81.

Muthena Naseri
Amy Strumolo

Communicable diseases

A communicable disease is any disease that can be transmitted from one organism to another. Agents that cause communicable diseases, called pathogens, are easily spread by direct or indirect contact. These pathogens include viruses, bacteria, fungi, and parasites. Some pathogens make toxins that harm the body's organs. Others actually destroy cells. Some

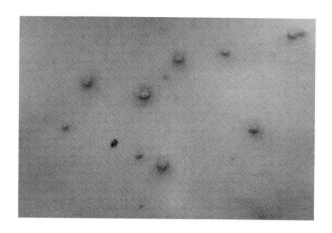
Chickenpox on the back of the child. *(Kasiap/Shutterstock.com)*

can impair the body's natural immune system, and opportunistic organisms set up secondary infections that cause serious illness or death. Once the pathogens have multiplied inside the body, signs of illness may or may not appear. The human body is adept at destroying most pathogens, but the pathogens may still multiply and spread.

The pathogens responsible for some communicable diseases have been known since the mid–1800s, although they have existed for a much longer period of time. European explorers brought highly contagious diseases such as smallpox, measles, typhus and scarlet fever to the New World, to which Native Americans had never been exposed. These new diseases killed 50–90 percent of the native population. Native populations in many areas of the Caribbean were totally eliminated.

In some areas of the world, contaminated soil or water incubate the pathogens of communicable diseases, and contact with those agents will cause diseases to spread. In the 1800s, when Robert Koch discovered the anthrax bacillus (*Bacillus anthracis*), cholera bacillus (*Vibrio cholerae*), and tubercle bacillus (*Mycobacterium tuberculosis*), his work ushered in a new era of public sanitation by showing how water-borne epidemics, such as cholera and typhoid, could be controlled by water filtration.

Malaria, another communicable disease, was responsible for the decline of many ancient civilizations; for centuries, it devitalized vast populations. With the discoveries in the late 1800s of protozoan malarial parasites in human blood, and the discovery of its carrier, the *Anopheles* mosquito, malaria could be combatted by systematic destruction of the mosquitos and their breeding grounds, by the use of barriers between mosquitos and humans such as window screens and mosquito netting, and by drug therapy to kill the parasites in the human host.

Discoveries of the causes of epidemics and transmissible diseases led to the expansion of the fields of sanitation and public health. Draining of marshes, control of the water supply, widespread vaccinations, and quarantine measures improved human health. But, despite the development of advanced detection techniques and control measures to fight pathogens and their spread, communicable diseases still take their toll on human populations. For example, until the 1980s tuberculosis had been declining in the United States due in large part to the availability of effective antibiotic therapy. However, since 1985, the number of tuberculosis cases has risen steadily due to such factors as the emergence of drug-resistant strains of the tubercle bacillus, the increasing incidence of HIV infection which lowers human resistance to many diseases, poverty, and immigration.

Epidemiologists track communicable diseases throughout the world, and their work has helped to eradicate smallpox, one of the world's most deadly communicable diseases. A successful global vaccination campaign wiped out smallpox in the 1980s, and today, the virus exists only in tightly-controlled laboratories in Moscow and in Atlanta at the Centers for Disease Control and Prevention. Scientists are currently debating whether to destroy the viruses or preserve them for study.

Communicable diseases continue to be a major public health problem in developing countries. In the small West African nation of Guinea-Bissau, a cholera epidemic hit in 1990. Epidemiologists traced its outbreak to contaminated shellfish and managed to control the epidemic, but not before it claimed hundreds of lives in that country. Some victims had eaten the contaminated shellfish, others were infected by washing the bodies of cholera victims and then preparing funeral feasts without properly washing their hands. Proper disposal of victims' bodies, coupled with a campaign to encourage proper hygiene, helped stop the epidemic from spreading.

Communicable diseases can be prevented either by eliminating the pathogenic organism from the environment (as by killing pathogens or parasites existing in a water supply) or by placing a barrier in the path of its transmission from one organism to another (as by vaccination or by isolating individuals already infected). But identifying and isolating the causal agent and developing weapons to fight it is time consuming, and, as with the AIDS virus, thousands of people continue to become infected and many die because educational warnings about ways to avoid infection frequently go unheeded.

Acquired immunodeficiency syndrome (AIDS) is caused by the human immunodeficiency virus (HIV). Spread by contact with bodily fluids of an HIV-infected person, the virus weakens and eventually destroys the body's immune system. Researchers assert that the virus originated in monkeys and was first transmitted to humans about 100 years ago.

New diseases deadlier than AIDS exist, including several variations of hemorrhagic fever and Avian Flu (H5N1 influence). Virologists point out that viruses, like human populations, constantly change. Rapidly increasing human populations provide fertile breeding grounds for microbes, including viruses and bacteria. Pathogens can literally travel the globe in a matter of hours.

For example, the completion in 1990 of a major road through the Amazon rain forest in Brazil led to outbreaks of malaria in the region. In 1985, used tires imported to Texas from eastern Asia transported larvae of the Asian tiger mosquito, a dangerous carrier of serious tropical communicable diseases. Deforestation and agricultural changes can unleash epidemics of communicable diseases. Outbreaks of Rift Valley fever followed the construction of the Aswan High Dam, most likely because breeding grounds were created for mosquitoes which spread the disease. In Brazil, the introduction of cacao farming coincided with epidemics of Oropouche fever, a disease linked to a biting insect that thrives on discarded cacao hulls.

In March 2009, a novel virus that was ultimately named 2009 H1N1 influenza virus (alternately, Type A / H1N1) resulted in cases of influenza (flu) in Mexico and the United States. A subsequently confirmed case of H1N1 flu first produced symptoms in a patient in the United States as early as 28 March 2009. The novel virus was detected as early as 16 April 2009. From an outbreak in Mexico, the virus quickly spread globally to become an official global pandemic.

Another important factor in the spread of communicable diseases is the speed and frequency of modern travel. A recent example of the perils of such fluidity in travel was exposed during a 2003 outbreak of severe acute respiratory syndrome (SARS).

Continued rapid transportation of humans around the world is likely to accelerate the movement of communicable diseases. Poverty, lack of adequate sanitation and nutrition, and the crowding of people into megacities in the developing countries of the world only exacerbate the situation. The need for study and control of these disease is likely to grow in the future.

Resources

BOOKS

Battin, M. Pabst. *The Patient As Victim and Vector: Ethics and Infectious Disease*. New York: Oxford University Press, 2009.

Dworkin, Mark S. *Outbreak Investigations Around the World: Case Studies in Infectious Disease Field Epidemiology*. Sudbury, Mass: Jones and Bartlett Publishers, 2009.

Kimbell, Ann Marie. *Risky Trade: Infectious Disease in the Era of Global Trade*. Aldershot, UK: Ashgate Publishing, 2006.

Palladino, Michael A., and Stuart Hill. *Emerging Infectious Diseases*. New York: Benjamin Cummings, 2005.

Woodall, Jack. "ProMed Mail." In *Infectious Diseases: In Context,*. Edited by Brenda Wilmoth Lerner and K. Lee Lerner. Detroit: Gale, 2007.

OTHER

Centers for Disease Control and Prevention (CDC). "Emerging Infectious Diseases." http://www.cdc.gov/ncidod/diseases/eid/index.htm (accessed November 6, 2010).

National Geographic Society. "Infectious Disease Quiz." http://science.nationalgeographic.com/science/health-and-human-body/human-diseases/infectious-disease-quiz.html (accessed November 6, 2010).

World Health Organization (WHO). "Report on Infectious Diseases." WHO Programs and Projects. http://www.who.int/infectious-disease-report/ (accessed November 6, 2010).

Linda Rehkopf

Community ecology

In biological ecology, a community is a set of interacting/non-interacting populations of the same or different species found in an area. Community ecology considers a community to be part of an ecosystem.

Community ecology considers, plant, insect, primate, forest, even herbaceous plant communities, and focuses on the living part of ecosystems, mostly on communities of interacting populations. While a community can include species that are together based on their common physical location that do not necessarily interact, community ecology generally emphasizes the wide diversity of species interactions that exist within the area.

Community ecologists investigate interactions under numerous labels and categories, including on-going studies of traditional topics such as predation, competition, and trophic exchanges, as three examples. Borrowing concepts from other disciplines such as physics, ecologists are also beginning to look closely at the linkages and assemblages that emerge from strong versus weak interactions, or from positive compared to negative interactions. Researchers continue to investigate the relative importance of intra-species and inter-species interactions in terms of their importance to community composition and rates of succession. One recent study, for example, analyzed the importance of interspecific interactions in the structuring of the geographical distribution and abundance of a tropical stream fish community.

The degree of pattern or randomness of community structure has long been an issue in community ecology. Natural communities are immensely complex and it is difficult to simplify this complexity down to useable, predictive models. Researchers continue to investigate the extent to which species' interactions can result in a unit organized enough to be considered a coherent community. The mechanics of community assembly depend heavily on invasions, rates of succession, and on changes in the physical environment, as well as a diversity of co-evolutionary patterns.

A debate still continues among community ecologists about the importance of complexity, the role of species diversity and richness in the maintenance of a community. Community structure patterns might be dependent, in part, on factors as seemingly trivial as seed weight. Accuracy in estimations of the number of species in a given community is difficult, including the large number of microorganisms as yet unidentified and unnamed.

Disturbances and perturbations are important factors in the composition and character of communities. The most significant source of disturbance and change in natural communities is human activity. Humans can also have positive effects through deliberate attempts to offset destructive impacts.

One other type of human impact, however, is receiving a lot of attention: evidence is mounting that climate change is increasingly impacting natural communities.

The information gained in community ecology studies is becoming increasingly important to achieving conservation objectives and establishing guidelines to management of the natural systems on which all humans depend. For example, clearer understanding of linkages established in communities through trophic exchanges can help predict the impacts of concentrations of toxins and pollutants. Better understanding of organismic interactions in a community context can help in comprehending the processes that lead to extinction of species, information critical to attempts to slow the loss of biological diversity. Research into community dynamics can result in better decisions about establishing preserves and refuges, and can create sustainable harvesting strategies.

Resources

BOOKS

Ballesta, Laurent, Pierre Deschamp, and Jean-Michel Cousteau. *Planet Ocean: Voyage to the heart of the Marine Realm*. Washington, DC: National Geographic, 2007.

Chivian, Eric, and Aaron Berstein. *Sustaining Life: How Human Health Depends on Biodiversity*. New York: Oxford University Press, USA, 2008.

Molles, Manuel C. *Ecology: Concepts and Applications*. New York: McGraw Hill Science/Engineering/Math, 2009.

Morin, Peter Jay. *Community Ecology*. New York: Wiley-Blackwell, 2010.

Gerald L. Young

Community right-to-know *see* **Emergency Planning and Community Right-to-Know Act (1986).**

Compaction

Compaction is the mechanical pounding of soil and weathered rock into a dense mass with sufficient bearing strength or impermeability to withstand a load. It is primarily used in construction to provide ground suitable for bearing the weight of any given structure. With the advent of huge earth-moving equipment we are now able to literally move mountains. However, such disturbances loosen and expand the soil. Thus soil must be compacted to provide an adequate breathing surface after it has been disturbed. Inadequate compaction during construction results in design failure or reduced service life of a structure. Compaction, however, is detrimental to crop production because it makes root growth and movement difficult, and deprives the soil of access to life-sustaining oxygen.

With proper compaction we can build enduring roadways, airports, dams, building foundation pads, or clay liners for secure landfills. Because enormous volumes of ground material are involved, it is far less expensive to use on-site or nearby resources wherever feasible. Proper engineering can overcome material deficiencies in most cases, if rigid quality control is maintained.

Successful compaction requires a combination of proper moisture conditioning, the right placement of material, and sufficient pounding with proper equipment. Moisture is important because dry materials seem very hard, but they may settle or become permeable when wet. Because of all the variables involved in the compaction process, standardized laboratory and field testing is essential.

The American Society of State Highway Transportation Officials (ASHTO) and the American Society of Testing Materials (ASTM) have developed specific test standards. The laboratory test involves pounding a representative sample with a drop-hammer in a cylindrical mold. Four uniform samples are tested, varying only in moisture content. The sample is trimmed and weighed, and portions oven- dried to determine moisture content.

The results are then graphed. The resultant curve normally has the shape of an open hairpin, with the high point representing the maximum density at the optimum moisture content for the compactive effort used. This curve reflects the fact that dry soils resist compaction, and overly-moistened soils allow the mechanical energy to dissipate. Field densities must normally meet 95 percent or higher of this lab result.

Because soils are notoriously diverse, several different "curves" may be needed; varied materials require the field engineer to exercise considerable judgment to determine the proper standard. Thus the field engineer and the earth- moving crew must work closely together to establish the procedures for obtaining the required densities.

In the past, density testing has required laboriously digging a hole, and comparing the volume with the weight of the material removed. Nuclear density gages have greatly accelerated this process and allow much more frequent testing if needed or desired. To take a reading, a stake is driven into the ground to form a hole into which a sealed nuclear probe can be lowered.

Resources

BOOKS

Blanco, Humberto, and Rattan Lal. *Principles of Soil Conservation and Management*. New York: Springer, 2008.
Chesworth, Ward. *Encyclopedia of Soil Science*. Dordrecht, Netherlands: Springer, 2008.
Coyne, Mark S. *Fundamental Soil Science*. Clifton Park, NY: Thomson Delmar Learning, 2006.
Gardner, Timothy. *Soil*. Greensboro, NC: Morgan Reynolds Publishing, 2009.

OTHER

United States Department of the Interior, United States Geological Survey (USGS). "Soil Chemistry." http://www.usgs.gov/science/science.php?term = 1078 (accessed November 10, 2010).

Nathan H. Meleen

Competition

Competition is the interaction between two organisms when both are trying to gain access to the same limited resource. When both organisms are members of the same species, such interaction is said to be "intraspecific competition." When the organisms are from different species, the interaction is "interspecific competition."

Intraspecific competition arises because two members of the same species have nearly identical needs for food, water, sunlight, nesting space, and other aspects of the environment. As long as these resources are available in abundance, every member of the community can survive without competition. When those resources are in limited supply, however, competition is inevitable. For example, a single nesting pair of bald eagles requires a minimum of 620 acres (250 ha) that they can claim as their own territory. If two pairs of eagles try to survive on 620 acres, competition will develop, and the stronger or more aggressive pair will drive out the other pair.

Intraspecific competition is also a factor in controlling plant growth. When a mature plant drops seeds, the seedlings that develop are in competition with the parent plant for water, sunlight, and other resources. When abundant space is available and the size of the community is small, a relatively large number of seedlings can survive and grow. When population density increases, competition becomes more severe and more seedlings die off.

Competition becomes an important limiting factor, therefore, as the size of a community grows. Those individuals in the community that are better adapted to gain food, water, nesting space, or some other limited resource are more likely to survive and reproduce. Intraspecific competition is thus an important factor in natural selection.

Interspecific competition occurs when members of two different species compete for the same limited resource(s). For example, two species of birds might both prefer the same type of insect as a food source and will be forced to compete for it if it is in limited supply.

Laboratory studies show that interspecific competition can result in the extinction of the species less well adapted for a particular resource. However, this result is seldom, if ever, observed in nature, at least among animals. The reason is that individuals can adapt to take advantage of slight differences in resource supplies. In the Galapagos Islands, for example, 13 similar species of finches have evolved from a single parent species. Each species has adapted to take advantage of some particular niche in the environment. As similar as they are, the finches do not compete with each other to any specific extent.

Interspecific competition among plants is a different matter. Since plants are unable to move on their own, they are less able to take advantage of subtle differences in an environment. Situations in which one species of plant takes over an area, causing the extinction of competitors, are well known.

One mechanism that plants use in this battle with each other is the release of toxic chemicals, known as allelochemicals. These chemicals suppress the growth of plants in other—and, sometimes, the same—species. Naturally occurring antibiotics are examples of such allelochemicals.

Resources

OTHER

United States Department of the Interior, United States Geological Survey (USGS). "Ecological Competition." http://www.usgs.gov/science/science.php?term = 308 (accessed November 11, 2010).

David E. Newton

Competitive exclusion

Competitive exclusion is the interaction between two or more species that compete for limited resources. It is an ecological principle involving competitors with similar requirements for habitat or resources; they utilize a similar niche. The result of the competition is that one or more of the species is ultimately eliminated by the species that is most efficient at

utilizing the limiting resource, a driving force of evolution. The competitive exclusion principle or "Gause's principle" states that where resources are limiting, two or more species that have the same requirements for the limiting resources cannot co-exist. The co-existing species must therefore adopt strategies that allow resources to be partitioned so that the competing species utilize the resources differently in different parts of the habitat, at different times, or in different parts of the life cycle.

Marie H. Bundy

Composting

Composting is a fermentation process, the break down of organic material aided by an array of microorganisms, earthworms, and other insects in the presence of air and moisture. This process yields compost (residual organic material often referred to as humus), ammonia, carbon dioxide, sulphur compounds, volatile organic acids, water vapor, and heat. Typically, the amount of compost produced is 40–60 percent of the volume of the original waste.

For the numerous organisms that contribute to the composting process to grow and function, they must have access to and synthesize components such as carbon, nitrogen, oxygen, hydrogen, inorganic salts, sulphur, phosphorus, and trace amounts of micronutrients. The key to initiating and maintaining the composting process is a carbon-to-nitrogen (C:N) ratio between 25:1 and 30:1. When C:N ratio is in excess of 30:1, the decomposition process is suppressed due to inadequate nitrogen limiting the evolution of bacteria essential to break the strong carbon bonds. A C:N ratio of less than 25:1 will produce rapid localized decomposition with excess nitrogen given off as ammonia, which is a source of offensive odors.

Attaining such a balance of ratio and range is possible because all organic material has a fixed C:N ratio in its tissue. For example, food waste has a C:N ratio of 15:1, sewage sludge has a C:N ratio of 16:1, grass clippings have a C:N ratio of 19:1, leaves have a C:N ratio of 60:1, paper has a C:N ratio of 200:1, and wood has a C:N ratio of 700:1. When these (and other) materials are mixed in the right proportions, they provide optimum C:N ratios for composting. Typically, nitrogen is the limiting component that is encountered in waste materials and, when insufficient nitrogen is present, the composting mixture can be

(computers, computer peripherals, etc.) to developing countries. The ban is aimed at helping to alleviate the negative effects to the environment and workers' health that is so often associated with third-world electronics recycling. A coalition of industry, environmental, and government groups called the National Electronics Product Stewardship Initiative began meeting in 2001 to come up with national guidelines for computer disposal, which were published in a 2004 resolution. The high cost of computer recycling is expected to decline somewhat as the volume of recycled machines rises. Because of the vast numbers of computers in the United States that could potentially be discarded in the coming years, it is imperative that recycling and safe disposal programs be enhanced and expanded.

Resources

BOOKS

The Download on Disposing of Your Old Computer. Washington, D.C.: Federal Trade Commission, Bureau of Consumer Protection, Division of Consumer and Business Education.

OTHER

Environmental Protection Agency. "eCycling." http://www.epa.gov/osw/conserve/materials/ecycling/ (accessed November 9, 2010).

Intel. "Old Computer Disposal." http://www.intel.com/learn/practical-advice/before-you-buy/old-computer-disposal (accessed November 9, 2010).

ORGANIZATIONS

Electronic Industries Alliance, 2500 Wilson Boulevard, Arlington, VA, USA, 22201, (703) 907-7500, http://www.eia.org

National Electronics Product Stewardship Initiative, http://eerc.ra.utk.edu/clean/nepsi/

Angela Woodward

Condensation nuclei

A physical mass that serves as a platform for condensation or crystallization. When air is cooled below its dew point, the water vapor it contains tends to condense as droplets of water or tiny ice crystals. Condensation may not occur, however, in the absence of tiny particles on which the water or ice can form. These particles are known as condensation nuclei. The most common types of condensation nuclei are crystals of salt, particulate matter formed by the combustion of fossil fuels, and dust blown up from Earth's surface. In the process of cloud-seeding, scientists add tiny crystals of dry ice or silver iodide as condensation nuclei to the atmosphere to promote cloud formation and precipitation.

Condor *see* **California condor.**

Congenital malformiations *see* **Birth defects.**

Congo River and basin

The Congo River (also known as the Zaire River) is the third longest river in the world, and the second longest in Africa (after the Nile River in northeastern Africa). Its river basin, one of the most humid in Africa, is also the largest on that continent, covering over 12 percent of the total land area.

History

The equatorial region of Africa has been inhabited since approximately the middle Stone Age. Late Stone Age cultures flourished in the southern savannas after about 10,000 B.C. and remained functional until the arrival of Bantu-speaking peoples during the first millennium B.C. In a series of migrations taking place from about 1,000 B.C. to the mid-first millennium A.D., many Bantu-speakers dispersed from an area west of the Ubangi-Congo River swamp across the forests and savannas of the region known as the modern-day Democratic Republic of the Congo.

In the precolonial era, this region (modern-day Democratic Republic of the Congo) was dominated by three kingdoms: Kongo (late 1300s), the Loango (at its height in the 1600s), and Tio. Portugese navigator Diogo Cam was the first European to sail up the mouth of the Congo in 1482. After meeting with the rulers of the Kingdom of Kongo, Cam negotiated intercontinental trade and commerce agreements— including the slave trade—between Portugal and the region. And a long history of colonialism began.

Over the centuries, the Congo River has inspired both mystery and legend, from the explorations of Henry Morton Stanley and David Livingstone in the 1870s, to Joseph Conrad, whose novel, *Heart of Darkness* transformed the river into an eternal symbol of the "dark continent" of Africa.

Characteristics

The Congo River is approximately 2,720 miles (4,375 km) long, and its drainage basin consists of about 1.3 million square miles (3.6 million km^2). The basin encompasses nearly the entire Democratic Republic of the Congo (capital: Kinshasa), Republic of Congo (capital: Brazzaville), Central African Republic, eastern Zambia, northern Angola, and parts of Cameroon and Tanzania. The river headwaters emerge at the junction of the Lualaba (the Congo's largest tributary) and Luvua rivers. The flow is generally to the northeast first, then west, and finally south to its outlet into the Atlantic Ocean at Banana, Republic of Congo.

The Congo basin comprises one of the most distinct land depressions between the Sahara desert to its north, and the Atlantic Ocean to its south and west. The river's tributaries flow down slopes varying from 900 to 1,500 feet (274–457 m) into the central depression forming the basin. This depression extends for more than 1,200 miles (1931 km) from the north to the south, from the Congo Lake Chad watershed, to the plateaus of Angola. From the east to west of the depression is another 1,200 miles (1931 km)—from the Nile-Congo watershed to the Atlantic Ocean. The width of the Congo River ranges from 3.5 miles (5.75 km) to 7 miles (about 11.3 km); and its banks contain natural levees formed by silt deposits. During floods, however, these levees overflow, widening the river.

With an average annual rainfall of 1,500 mm of rain (about 60 in), about three-quarters returns to the atmosphere by evapotranspiration; the rest is discharged into the Atlantic. The river is divided into three main regions: the upper Congo, with numerous tributaries, lakes, waterfalls, and rapids; the middle Congo; and, the lower Congo. The middle Congo is characterized by its seven waterfalls, collectively referred to as Boyoma (formerly Stanley) Falls. It is below these falls that navigation on the river becomes possible. The river has approximately 10,000 mi (about 16,000 km) of waterways, creating one of the main transportation routes in Central Africa.

Economic and environmental impact

Due to its size and other key elements, the Congo River and its basin are crucial to the ecological balance of an entire continent. Although the Congo water discharge levels were unstable throughout the second half of the twentieth century—the hydrologic balance of the river has provided some relief from the drought that has afflicted the river basin. This relief occurs even with dramatic fluctuations of rainfall throughout the various terrain through which the river passes.

Researchers have suggested that soil geology plays a key role in maintaining the river's discharge

stability despite fluctuations in rainfall. The sandy soils of the Kouyou region, for example, have a stabilizing effect in their ability to store or disperse water.

In 1999, the World Commission on Water for the twenty-first century, based in Paris and supported by the World Bank and the United Nations, found that the Congo was one of the world's cleanest rivers—in part due to the lack of industrial development along its shores until that time. However, the situation is changing.

The rapidly increasing human population threatens to compromise the integrity of Congo basin ecosystems. Major threats to the large tropical rainforests and savannas, as well as to wildlife, come from the exploitation of natural resources. Uncontrolled hunting and fishing, deforestation (which causes sedimentation and erosion near logging operations) for timber sale or agricultural purposes, unplanned urban expansion (which increases the potential for an increase in untreated sewage and other sources of pollution that could harm nearby freshwater systems), and unrestrained extraction of oil and minerals are some of the major economic and environmental issues confronting the region. And these issues are expected to have a global impact as well.

Wildlife

According to the World Wildlife Fund, the Congo River and its basin, also known as the "Congo River and Flooded Forests ecoregion," is home to the most diverse and distinctive group of animals adapted to a large-river environment in all of tropical Africa.

The Congo river had no outlet to the ocean during the Pliocene Age (5.4–2.4 million years ago) but was instead a large lake. Eventually, the water broke through the rim of the lake, emerging as a river that passed over rocks through a series of rapids, then entered the Atlantic. Except for the beginning and end of its course, the river is uniformly elevated.

With more than 700 fish species, 500 of which are endemic to the river, the Congo basin ranks second only to the Amazon in its diversity of species. Nearly 80 percent of fish species found in the Congo basin exist nowhere else in the world. The various species live both in the river and its attendant habitats—swamps, nearby lakes, and headwater streams. They feed in a variety of ways: scouring the mud at the river's bottom; eating scales off of live fish; and eating smaller fish. Certain fish have even adapted to the river's muddy waters. For example, some have reduced eye size, or no eyes at all, yet easily maneuver through the swift current. The Congo's freshwater fish are a crucial protein source for Central Africa's population;

yet the potential for over-fishing near the urban areas along its banks threatens the available supply.

There are also a wide variety of aquatic mammals—such as unusual species of otters, shrews, and monkeys—that are indigenous to the river basin. Rainforests cover over 60 percent of the Democratic Republic of the Congo, and represents nearly 6 percent of the world's remaining forested area, and 50 percent of Africa's remaining forests. Many of the world's endangered species live near the river, including gorillas.

Resources

OTHER

National Geographic Society. "Congo." http://travel.national geographic.com/places/countries/country_congo.html (accessed November 11, 2010).

ORGANIZATIONS

World Wildlife Fund, 1250 24th St. N.W, P.O. Box 97180., Washington, DC, USA, 20090-7180, 202-293-9211, 1-800-CALL- WWF, http://www.worldwildlife.org

Jane E. Spear

Coniferous forests

Coniferous forests contain trees with cones and generally evergreen needle or scale-shaped leaves. Important genera in the northern hemisphere include pines (*Pinus*), spruces (*Picea*), firs (*Abies*), redwoods (*Sequoia*), Douglas firs (*Pseudotsuga*), and larches (*Larix*). Different genera dominate the conifer forests of the southern hemisphere. Conifer forests occupy regions with cool-moist to very cold winters and cool to hot summers. Many conifer forests originated as plantations of species from other continents. Among conifer formations in North America are the slow-growing circumpolar taiga (boreal), the subalpine-montane, the southern pine, and the Pacific Coast temperate rain forest. Softwoods, another name for conifers, are used for lumber, panels, and paper.

Conservation

Conservation is the philosophy or policy that natural resources should be used cautiously and rationally so that they will remain available for future generations. Widespread and organized conservation

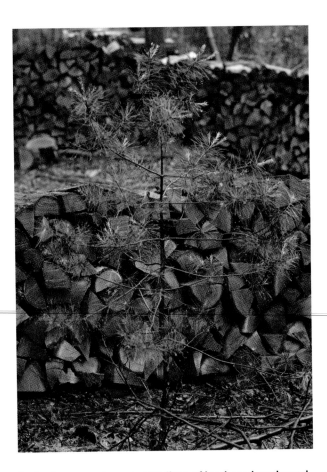

A pine tree sprouts up next to lines of hardwood cord wood, which was harvested as part of a conservation project, on conservation land in Weston, Mass., 2010. *(AP Photo/Charles Krupa)*

movements, dedicated to preventing uncontrolled and irresponsible exploitation of forests, lands, wildlife, and water resources, first developed in the United States in the last decades of the nineteenth century. This was a time at which accelerating settlement and resource depletion made conservationist policies appealing both to a large portion of the public and to government leaders. Since then, international conservationist efforts, including work of the United Nations, have been responsible for monitoring natural resource use, setting up nature preserves, and controlling environmental destruction on both public and private lands around the world.

The name most often associated with the United States' early conservation movement is that of Gifford Pinchot (1865–1946), the first head of the U.S. Forest Service. A populist who fervently believed that the best use of nature was to improve the life of the common citizen, Pinchot brought scientific management methods to the Forest Service. Pinchot also brought a strongly utilitarian philosophy, which continues to prevail in the Forest Service. Beginning as an advisor to former U.S. President Theodore Roosevelt (1858–1919), himself an ardent conservationist, Pinchot had extensive influence in Washington and helped to steer conservation policies from the turn of the century to the 1940s. Pinchot had a number of important predecessors, however, in the development of American conservation. Among these was George Perkins Marsh (1801–1882), a Vermont forester and geographer whose 1864 publication, *Man and Nature*, is widely held as the wellspring of American environmental thought. Also influential was the work of American geologist John Wesley Powell (1834–1902), Clarence King (1842–1901), and other explorers and surveyors who, after the Civil War, set out across the continent to assess and catalog the country's physical and biological resources and their potential for development and settlement.

Conservation, as conceived by Pinchot, Powell, and Roosevelt was about using, not setting aside, natural resources. In their emphasis on wise resource use, these early conservationists were philosophically divided from the early preservationists, who argued that parts of the American wilderness should be preserved for their aesthetic value and for the survival of wildlife, not simply as a storehouse of useful commodities. Preservationists, led by the eloquent writer and champion of Yosemite Valley, Scottish-born American John Muir (1838–1914), bitterly opposed the idea that the best vision for the nation's forests was that of an agricultural crop, developed to produce only useful species and products. Pinchot, however, insisted that "The object of [conservationist] forest policy is not to preserve the forests because they are beautiful. or because they are refuges for the wild creatures of the wilderness. but the making of prosperous homes. Every other consideration is secondary." Because of its more moderate and politically palatable stance, conservation became, by the turn of the century, the more popular position. By 1905 conservation had become a blanket term for nearly all defense of the environment; the earlier distinction was lost until it began to re-emerge in the 1960s as environmentalists began once again to object to conservation's anthropocentric (human- centered) emphasis. More recently deep ecologists and bioregionalists have likewise departed from mainstream conservation, arguing that other species have intrinsic rights to exist outside of human interests.

Several factors led conservationist ideas to develop and spread when they did. By the end of the nineteenth century European settlement had reached across the entire North American continent. The census of 1890 declared the American frontier closed, a blow to the American myth of the virgin continent. Even more important, loggers, miners, settlers, and livestock herders were laying waste to the nation's forests, grasslands, and mountains from New York to California. The accelerating, and often highly wasteful, commercial exploitation of natural resources went almost completely unchecked as political corruption and the economic power of timber and lumber barons made regulation impossible. At the same time, the disappearance of American wildlife was starkly obvious. Within a generation the legendary flocks of passenger pigeons (*Ectopistes migratorius*) disappeared entirely, many of them shot for pig feed while they roosted. Millions of bison were slaughtered by market hunters for their skins and tongues or by sportsmen shooting from passing trains. Natural landmarks were equally threatened—Niagara Falls nearly lost its water to hydropower development, and California's Sequoia groves and Yosemite Valley were threatened by logging and grazing.

At the same time, post-Civil War scientific surveys were crossing the continent, identifying wildlife and forest resources. As a consequence of this data gathering, evidence became available to document the depletion of the continent's resources, which had long been assumed inexhaustible. Travellers and writers, including John Muir, Theodore Roosevelt, and Gifford Pinchot, had the opportunity to witness the alarming destruction and to raise public awareness and concern. Meanwhile an increasing proportion of the population had come to live in cities. These urbanites worked in occupations not directly dependent upon resource exploitation, and they were sympathetic to the idea of preserving public lands for recreational interests. From the beginning this urban population provided much of the support for the conservation movement.

As a scientific, humanistic, and progressive policy, conservation has led to a great variety of projects. The development of a professionally trained forest service to maintain national forests has limited the uncontrolled tree mining practiced by logging and railroad companies of the nineteenth century. Conservation-minded presidents and administrators have set aside millions of acres of public land for national forests, parks, and other uses for the benefit of the public. A corps of professionally trained game managers and wildlife managers has developed to maintain game birds, fish, and mammals for public recreation on federal lands. For much of its history, federal game conservation has involved extensive predator elimination programs, however several decades of protest have led to more ecological approaches to game

Threats to biological diversity

Biodiversity at local, regional, continental, and global scales is critically threatened by human activities. The damages that are being caused to Earth's species and ecosystems are so severe that they are referred to by ecologists as a biodiversity crisis. Many permanent losses of biodiversity have already been caused by human influences, including the extinctions of numerous species and the losses of distinctive, natural communities. Unless there are substantial changes in the ways that humans affect ecosystems, there will be enormously greater losses of biodiversity in the near future.

Earth's natural biodiversity has always been subjected to extinction (that is, the permanent loss of species and other groups) The fossil record shows that species, families, and even entire phyla have appeared and disappeared on Earth. For example, many invertebrate phyla proliferated during an evolutionary radiation at the beginning of the Cambrian era about 570 million years ago, but most of these are now extinct.

Many of the natural extinctions occurred simultaneously, apparently as a result of an unpredictable catastrophe. For instance, about 65 million years ago a mass extinction occurred that resulted in the loss of the last of the dinosaurs and as many as 76 percent of the then-existing species. That catastrophe is believed to have been caused by a meteorite impacting Earth. In other cases, natural extinctions have been caused by more gradual environmental changes, for example in climate or in the intensity of disease or predation.

More recently, however, humans have been responsible for almost all of the extinctions that are occurring. In fact, species are now being lost so quickly that the changes represent a modern mass extinction. Well-known examples of extinctions caused by humans include the dodo (*Raphus cucullatus*), passenger pigeon (*Ectopistes migratorius*), and great auk (*Pinguinus impennis*). Numerous other species have been taken to the brink of extinction, including the plains bison (*Bison bison bison*), whooping crane (*Grus americana*), ivory-billed woodpecker (*Campephilus principalis*), and right whale (*Eubalaena* genus). These losses have been caused by over-hunting and the disturbance and conversion of natural habitats.

In addition to these famous cases involving large animals, an even more ruinous damage to Earth's biodiversity is being caused by extensive losses of tropical ecosystems, particularly the conversion of tropical rain forests into agricultural habitats. Because tropical ecosystems are particularly rich in numbers of species, loss of natural tropical habitat causes extinctions of numerous species. Many of those species occurred nowhere else but in particular tropical locales. Species that are only found in a specific area are referred to as endemic species.

The mission of conservation biology is to understand the causes and consequences of the modern crisis of extinction and degradation of Earth's biodiversity, and then to apply scientific principles to preventing or repairing the damages. This is largely done by conserving populations and by protecting natural areas.

Conservation at the population level

In some cases, endangered species can be enhanced by special programs that increase their breeding success and enhance the survival of their populations. Usually, a variety of actions is undertaken, along with the preservation of appropriate habitat, under a scheme that is known as a population recovery plan. Components of a population recovery plan may include such actions as (1) the careful monitoring of wild populations and the threats that they face; (2) research into the specific habitat needs of the endangered species; (3) the establishment of a captive-breeding program and the release of surplus individuals into the wild; (4) research into genetic variation within the species; and (5) other studies of basic biology and ecology that are considered necessary for preservation of the species, particularly in its natural habitats. Unfortunately, population recovery plans have only been developed for a small fraction of endangered species, and most of these have been prepared for species that occur in relatively wealthy countries.

One example involves the whooping crane (*Grus americana*), an endangered species in North America. Because of excessive hunting and critical habitat loss, this species declined in abundance to the point where as few as only 15 individuals were alive in 1941. Since then, however, the wild population of whooping cranes has been vigorously protected in the United States and Canada, and their critical breeding, migratory, and wintering habitats have been preserved. In addition, the basic biology and behaviour of whooping cranes have been studied, and some wild birds have been taken into captivity and used in breeding programs to increase the total population of the species. Some of the captive-bred animals have been released to the wild, and whooping crane eggs have also been introduced into the nests of the closely related sandhill crane (*Grus canadensis*), which serve as foster parents. These applications of conservation biology have allowed the critically endangered population of whooping cranes to increase to more than 150 individuals in the mid-1980s, and to about 600 birds in 2009, of which about a quarter were in captivity. Because of these actions, there is now guarded optimism for the survival of this endangered species.

Protected areas

Protected areas such as parks and ecological reserves are necessary for the conservation of biodiversity in wild, natural ecosystems. A protected area is defined by the International Union for Conservation of Nature (IUCN) as "an area of land and/or sea especially dedicated to the protection and maintenance of biological diversity, and of natural and associated cultural resources, and managed through legal or other effective means." Most protected areas are established for the preservation of natural values, particularly the known habitats of endangered species, threatened ecological communities, or representative examples of widespread communities. However, many protected areas (particularly parks) are also used for human activities, as long as they do not severely threaten the ecological values that are being conserved. These uses can include ecotourism, and in some cases fishing, hunting, and even timber harvesting. In the most recent United Nations List of Protected Areas in 2009, there were over 102,000 protected areas globally, with a total area of about 7.3 million square miles (18.8 million square kilometers). This area constitutes about 11.6 percent of the world's land surface compared with 6.3 percent of aquatic areas being protected. More areas are being added to the total protected area over time, and more concentrated efforts are focusing on protecting marine areas.

Ideally, a national system of protected areas would provide for the longer-term conservation of all native species and their natural communities, including terrestrial, freshwater, and marine ecosystems. So far, however, no country has implemented a comprehensive system of ecological reserves to fully protect the natural biodiversity of the region. Moreover, many existing reserves are relatively small and are threatened by environmental changes and other disturbances, such as illegal hunting of animals and plants and sometimes intensive tourism.

Ecological knowledge has allowed conservation biologists to make important contributions to the optimized design of networks of protected areas. Important considerations include: (1) the need to protect areas that provide adequate representation of all types of natural ecosystems; (2) the need to preserve all endangered ecosystems and the habitats of threatened species; (3) the requirement of redundancy, so that if one example of an endangered ecosystem becomes lost through an unavoidable natural disturbance (such as a hurricane or wildfire), the type will continue to survive in another protected area; (4) the need to decide whether or not the network of protected areas should be linked by corridors, a matter of some controversy among ecologists.

Conservation biology has also made important contributions towards the spatial design of individual protected areas. Important considerations include (1) the need to make protected areas as large as possible, which will help to allow species and ecosystems to better cope with disturbances and environmental changes; (2) a preference for smaller reserves to have a minimal amount of edge, which helps to avoid damages that can be caused by certain predators and invasive species; (3) the need to take an ecosystem approach which ensures that the reserve and its surrounding area will be managed in an integrated manner.

Although conservation biology is a relatively young field, important progress is being made towards development of the effective ecological and biological tools necessary to preserve biodiversity. Climate change (global warming) induced by human activities such as habitat fragmentation and deforestation, coupled with increased greenhouse gas emissions, is having a negative impact on biodiversity as global temperatures increase. Increasingly, efforts are being made to reduce and potentially reverse these effects.

Resources

BOOKS

Carroll, Scott P., and Charles W. Fox. *Conservation Biology Evolution in Action.* Oxford: Oxford University Press, 2008.

Davis, Frederick Rowe. *The Man Who Saved Sea Turtles: Archie Carr and the Origins of Conservation Biology.* Oxford: Oxford University Press, 2007.

Groom, Martha J., et al. *Principles of Conservation Biology.* 3rd ed. Washington, DC: Sinauer Associates, 2005.

Lovejoy, Thomas E., and Lee Jay Hannah. *Climate Change and Biodiversity.* New Haven: Yale University Press, 2005.

Macdonald, David W., and Katrina Service. *Key Topics in Conservation Biology.* Malden, MA: Blackwell Pub, 2007.

PERIODICALS

McKenzie, Donald, et al. "Climatic Change, Wildfire, and Conservation." *Conservation Biology* 18 (August 2004): 890–902.

Bill Freedman

Conservation design *see* **Urban sprawl.**

Conservation easements

A conservation easement is a covenant, restriction, or condition in a deed, will, or other legal document that allows the owner to maintain ownership and control of real property, but restricts the use of that property so the land is conserved in its natural state, or, in the case of a historic conservation easement, so that it provides a historic benefit. The uses allowed by the easement can include recreation, agriculture, cultural uses, and establishment of wildlife habitat. The federal government allows tax deductions for conservation easements that provide a certified value to the public, such as protecting ecologically valuable natural habitat or, in the case of an easement based on the historical conservation of the property, that contribute to the historic character of the district in which the property is located.

Marie H. Bundy

Conservation International

Conservation International (CI) is a non-profit, private organization dedicated to saving the world's endangered rain forests and the plants and animals that rely on these habitats for survival. CI is basically a scientific organization, a fact which distinguishes it from other conservation groups. Its staff includes leading scientists in the fields of botany, ornithology, herpetology, marine biology, entomology, and zoology.

Founded in 1987 when it split off from the Nature Conservancy, CI now has over 55,000 members. The group, headed by Peter A. Seligmann, has gathered accolades since its inception. In 1991 *Outside Magazine* gave CI an A- (one of the two highest grades received) in its yearly report card rating fourteen leading environmental groups.

The high praise is well founded. CI tends to successfully implement its many projects and goals. Many CI programs focus on building local capacity for conservation in developing countries through financial and technical support of local communities, private organizations, and government agencies. Their "ecosystem conservation" approach balances conservation goals with local economic needs. CI also funds and provides technical support to local communities, private organizations, and government agencies to help build sustainable economies while protecting rain forest ecosystems.

Four broad themes underlie all CI projects: (1) a focus on entire ecosystems; (2) integration of economic interests with ecological interests; (3) creation of a base of scientific knowledge necessary to make conservation-minded decisions; and (4) an effort to make it possible for conservation to be understood and implemented at the local level.

CI has offices in more than thirty countries, including Botswana, Brazil, Colombia, Indonesia, Mexico, New Guinea, and the Philippines, and projects in many more. In 2000, CI expanded into Cambodia and China, among others.

Among CI's many successful projects is the Rapid Assessment Program (RAP), which enlists the world's top field scientists to identify wilderness areas in need of urgent conservation attention. RAP teams have completed around sixty projects around the world and have identified many new species in this way. CI has also helped establish important biosphere reserves in rain forest countries. These efforts successfully demonstrate CI's ecosystem conservation approach, and prove that the economic needs of local communities can be reconciled with conservation needs. No harvesting or hunting is allowed in the reserves, but buffer zones, which include villages and towns, are located just outside the core areas.

CI strongly supports many educational programs. In 1988, it signed a long-term assistance agreement with Stanford University which involves exchange and training of Costa Rican students and resource managers. In 1989 CI began a program with the University of San Carlos which provides financial and technical support to the research activities in northern Guatemala of the university's Center for Conservation Studies.

In 2002, CI has already established several new ways to educate the people and help the environment, such as Centers for Biodiversity Conservation, the Global Conservation Fund, and a joint venture with Ford Motor Company called the Center for Environmental Leadership in Business. CI also focuses on activities in the major wilderness areas identified as the most endangered. The organization also has expanded its conservation efforts to new ecosystems, including marine, desert, and temperate rain forest regions.

Resources

ORGANIZATIONS

Conservation International, 2011 Crystal Drive, Suite 500, Arlington, VA, USA, 22202, (703) 341-2400, (800) 429-5660, inquiry@conservation.org, http://www. conservation.org

Cathy M. Falk

Conservation Reserve Program

The Conservation Reserve Program (CRP) is a voluntary program for agricultural landowners, that encourages farmers to plant long-term resource-conserving vegetative ground cover to improve soil and water, and create more suitable habitat for fish and wildlife. Ground cover options include grasses, legumes, shrubs, and tree plantings. The program is authorized by the federal Food Security Act of 1985, as amended, and is implemented through the Commodity Credit Corporation (CCC). It aims to promote good land stewardship and improve rural aesthetics.

The CRP offers annual rental payments, incentive payments, and cost-share assistance to establish approved cover on eligible cropland. The CCC provides assistance of as much as 50 percent of the landowner's cost in establishing an approved conservation program. Contracts remain in effect for between 10 and 15 years. Annual rental payments are based on the agriculture rental value of the land used in the program. The program provides needed income support for farmers, and helps to curb production of surplus commodities.

Eligibility for participation in CRP extends to individuals, partnerships, associations, Indian tribal ventures corporations, estates, trusts, other business enterprises or legal entities. States, political subdivisions of states, or agencies thereof owning or operating croplands, may also apply.

The CCC via the Farm Service Agency (FSA) manages the CRP. The Natural Resources Conservation Service (NRCS) and the Cooperative State Research and Education Extension Service provide support. State forestry agencies and local soil and water conservation districts also provide assistance.

To be eligible for the CRP, cropland should have been planted or considered planted to an agricultural commodity in two of the five most recent crop years. Eligibility encompasses highly erodible acreage, cropped wetlands, and land surrounding non-cropped wetlands. The cropland must be owned or operated for at least 12 months before the close of the sign-up period. Exceptions can be made for land that was acquired by will or succession, or if the FSA determines that ownership was not acquired for the purpose of placing the land in the conservation reserve.

Initially, erosion reduction was the sole criterion for acceptance in the CRP, and in 1986–87, 22 million acres (8.9 million ha) were enrolled for this purpose. CRP proved to be effective. According NRCS statistics, average erosion on enrolled acres declined by about 90 percent. It was estimated that the program reduced overall erosion nationwide by more than 22 percent even though less than 10 percent of the nation's cropland was enrolled.

An Environmental Benefits Index (EBI) is used to prioritize applications for the CRP. EBI factors include: Wildlife habitat benefits, water quality benefits from reduced erosion, runoff, and leaching, and air quality benefits from reduced wind erosion. The NRCS collects data for each of the factors and, based on its analysis, applications are ranked. Selections are made from that ranking.

The CCC bases rental rates on the productivity of soils within a county, and the average rent for the past three years of local agricultural land. The maximum CRP rental rate for each applicant is calculated in advance of enrollment. Applicants may accept that rate, or may offer a lower rental rate to increase the likelihood that their project will be funded.

The CCC encourages restoration of wetlands by providing a 25 percent incentive payment of the costs involved to establish approved cover. This is in addition to the normal 50 percent cost share for non-wetlands. Eligible acreage devoted to special conservation practices, such as riparian buffers, filter strips, grassed waterways, shelter belts, living snow fences, contour grass strips, salt tolerant vegetation, and shallow water areas for wildlife, may be enrolled at any time and is not subject to competitive bidding.

When CRP contracts expire, participants must continue to follow approved conservation plans. They must comply with wetland, endangered species and other federal, state, and local environmental laws, and they must respect any continuing conservation easement on the property.

The United States Department of Agriculture (USDA) provides information and technical assistance to CRP participants who wish to return CRP land to row-crop producing status as their contracts expire. This is to ensure that the land is developed in a sound, productive, and sustainable manner, preventing excessive erosion, protecting water quality, and employing other measures that enhance soil moisture-retaining ability.

In states such as North Dakota, the landscape has been changed dramatically since the introduction of the CRP. In the 1970s and early 1980s, fields with steep hills and areas of light soil were often cultivated from fencerow-to-fencerow. The result often was severe erosion and permanent loss of soil fertility. In some areas, winters were dominated by "snirtstorms" when a combination of dirt and snow blew across the landscape depositing a dark coating on countryside downwind. In contrast, today's travelers find these landscapes covered with green vegetation in summer and white snow in winter.

The CRP has particularly benefited migratory birds in states such as North Dakota, South Dakota, and Montana. The perennial vegetation on marginal farmland has provided refuge for migrating birds, and added breeding habitat for others. North Dakota farmers have enrolled about 10 percent of the state's cropland in the CRP.

The fiscal year 2000 federal agricultural appropriations bill authorized a pilot project of harvesting of biomass from CRP land to be used for energy production. Six projects were authorized, no more than one of which could be in any state. Vegetation could not be harvested more often than once every two years, and no commercial use could be made of the harvested biomass other than energy production. Annual CRP rental payments are reduced by twenty five percent during the year the acreage is harvested. Land that is devoted to field windbreaks, waterways, shallow water ways for wildlife, contour grass strips, shelter belts, living snow fences, permanent vegetation to reduce salinity, salt tolerant vegetative cover, filter strips, riparian buffers, wetland restoration, and cross-wind trap strips is not eligible for this program. By 2002, contracts had been approved in Iowa, Illinois, Oklahoma, Minnesota, New York, and Pennsylvania.

In June of 2001, the USDA announced a six-state pilot program as part of the CRP to restore up to 500,000 acres (202,000 ha) of farmable wetlands and associated buffers. The Farmable Wetlands Pilot Program is intended to help producers improve the hydrology and vegetation of eligible land in Iowa, Minnesota, Montana, Nebraska, North Dakota, and South Dakota. Restoring wetlands in these states should reduce downstream flood damage, improve surface and groundwater quality, and recharge groundwater supplies. Essential habitat for migratory birds and many other wildlife species, including threatened and endangered species will be created and enhanced. Recreational activities such as hiking and bird watching will also be improved.

In 1985, the year in which CRP originated, spring surveys by the U.S Fish and Wildlife Service estimated waterfowl breeding populations at 25.6 million ducks. A fall flight of 54.5 million was predicted. Several species of ducks including mallards, pintails, and blue-winged teal appeared to be fading away. Numbers were at or near their lowest ebb in thirty years.

Between 1986 and 1990, farmers enrolled 8.2 million acres (3.3 million ha) of cropland in CRP within an area known as the prairie pothole region. This large glaciated area of the north central United States and southern Canada is where up to 70 percent of North America's ducks are hatched. Through CRP, nearly 13,000 square miles (34,000 km^2) was converted to superior nesting habitat through CRP.

In the early 1990s, increased precipitation filled the prairie potholes and many waterfowl ended their spring migration on CRP land, rather than continuing migration to their usual breeding grounds in Canada. Nesting densities increased many fold. Potholes surrounded by CRP grass provided more secure habitat for nests, and hatchlings were no longer easy targets for predators. Nesting success tripled, from 10 to 30 percent, and waterfowl mortality no longer exceeded annual additions. By 1995, ten years after the start of CRP, 36.9 million ducks were included in the annual spring survey numbers, a 40 percent increase in ten years.

Waterfowl are not the only bird species to benefit from CRP. In one study, breeding birds were counted in about 400 fields in eastern Montana, North and South Dakota, and western Minnesota. These states have nearly 30 percent of all land included in the CRP. Fields were planted mostly to mixtures of native and introduced grasses and legumes. For most of the seventy-three different species counted, numbers were far higher in CRP fields than in cropland. Differences were greatest for several grassland species whose numbers had been markedly declining in recent surveys. Two species, lark buntings (*Calamospiza melanocorys*) and grasshopper sparrows (*Ammodramus savannarum*), were ten and sixteen times more common in CRP environment than in cropland. The investigators concluded that restoration of suitable habitat in the form of introduced grasses and legumes can have an enormous beneficial effect on populations of grassland birds.

When CRP was due to expire in 1995, restoration of prairie pothole breeding grounds was in jeopardy. United States farm policy was undergoing major change and it appeared that the CRP would not be continued. Ducks Unlimited and other wildlife organizations lobbied heavily, and funding of one billion dollars was included in the 1996 Farm Bill to continue the program for another seven years. Moreover, the guidelines for the continuing program were geared more directly to the preservation of wetlands and waterfowl conservation. The prairie pothole region was designated a national conservation priority area, and during the March 1997 sign-up more acres in the prairie pothole region were enrolled in CRP than were due to expire.

In its first ten years, the CRP cost nearly $2 billion per year. Opponents have argued that this is too expensive, while proponents maintain that the costs are offset by its conservation and environmental benefits. Estimates of the annual value of benefits range from slightly less than $1 billion to more than $1.5 billion. The CRP, however, produces numerous environmental benefits, including wetland restoration, soil conservation, fertilizer reduction, and carbon sequestration through planting. By 2008, the CRP was responsible for the restoration of approximately 2 million acres (809,000 hectares) of wetlands per year and prevented the erosion of approximately 440 million tons (400 million metric tons) of soil. In 2007, the CRP also prevented the application of over 115 million tons (104 million metric tons) of phosphorus-based fertilizers and was responsible for the sequestration of over 50 million tons (46 metric tons) of carbon dioxide.

Resources

OTHER

Akcakaya, H. Resit. *Species Conservation and Management: Case Studies.* New York: Oxford University Press, 2004.

Frankham, Richard, Jonathan D. Ballou, David A. Briscoe, and Karina H. McInnes. *A Primer of Conservation Genetics.* Cambridge, UK: Cambridge University Press, 2004.

Freyfogle, Eric T. *Why Conservation Is Failing and How It Can Regain Ground.* New Haven: Yale University Press, 2006.

Douglas C. Pratt

Conservation tillage

Conservation tillage is any tilth sequence that reduces loss of soil or water in farmland. It is often a form of non-inversion tillage that retains significant amounts of plant residues on the surface. Definitions vary as to the percentage of soil surface that must be covered with plant residues at crop planting time to qualify as conservation tillage. Many agencies and organizations observe a 30 percent plantable surface area guideline.Other forms of conservation tillage include ridge tillage, rough plowing, and tillage that incorporates plant residues in the top few inches of soil.

A number of implements for primary tillage are used to retain all or a part of the residues from the previous crop on the soil surface. These include machines that fracture the soil, such as chisel plows, combination

chisel plows, disk harrows, field cultivators, undercutters, and strip tillage machines. In a no-till system, the soil is not disturbed before planting. Most tillage systems that employ the moldboard plow are not considered conservation tillage because the moldboard plow leaves only a small amount of residue on the soil surface (usually less than 10 percent).

When compared with conventional tillage (moldboard plow with no residue on the surface), various benefits from conservation tillage have been reported. Chief among the benefits are reduced wind and water erosion and improved water conservation. Erosion reductions from 50 to 90 percent as compared with conventional tillage are common. Conservation tillage often relies on herbicides to help control weeds and may require little or no post-planting cultivation for control of weeds in row crops.

Resources

BOOKS

Blanco, Humberto, and Rattan Lal. *Principles of Soil Conservation and Management.* New York: Springer, 2008.

Brady, Nyle C., and Ray R. Weil. *Nature and Properties of Soils,* 14th ed. Upper Saddle River, NJ: Prentice Hall, 2007.

Chesworth, Ward. *Encyclopedia of Soil Science.* Dordrecht, Netherlands: Springer, 2008.

Hatfield, Jerry L, ed. *The Farmer's Decision: Balancing Economic Agriculture Production with Environmental Quality.* Ankeny, IA: Soil and Water Conservation Society, 2005.

McMahon, M., A.M. Kofranek, and V.E. Rubatzky. *Hartmann's Plant Science: Growth, Development and Utilization of Cultivated Plants,* 4th ed. Englewood Cliffs, NJ: Prentice-Hall, 2006.

Morgan, R. C. P. *Soil Erosion and Conservation,* 3rd ed. New York: Wiley-Blackwell, 2005.

Plaster, Edward. *Soil Science and Management,* 5th ed. Clifton Park, NY: Delmar Cengage Learning, 2008.

William E. Larson

Consultative Group on International Agricultural Research

The Consultative Group on International Agricultural Research (CGIAR) was founded in 1971 to improve food production in developing countries. Research into agricultural productivity and the management of natural resources are the two goals of this organization, and it is dedicated to making the scientific advances of industrialized nations available to poorer countries. The CGIAR emphasizes the importance of developing sustainable increases in agricultural yields and creating technologies that can be used by farmers with limited financial resources.

Membership consists of governments, private foundations, and international and regional organizations. The goals of CGIAR are carried out by a network of International Agricultural Research Centers (IARCs). There are currently fifteen such centers throughout the world, all but two of them in developing countries, and they are each legally distinct entities, over which the CGIAR has no direct authority. The group has no constitution or by-laws, and decisions are reached by consensus after consultations with its members, either informally or at their semiannual meetings. The function of the CGIAR is to assist and advise the IARCs, and to this end it maintains a Technical Advisory Committee (TAC) of scientists who review ongoing research programs at each center.

Each IARC has its own board of trustees as well as its own management, and they formulate individual research programs. The research centers pursue different goals, addressing problems in a particular sector of agriculture, such as livestock production or agricultural challenges in specific parts of the world, such as crop production in the semi-arid regions of Africa and Asia. Some centers conduct research into integrated plant protection, and others into forestry, while some are more concerned with policy issues, such as food distribution and the international food trade. One of the priorities of the CGIAR is the conservation of seed and plant material, known as germplasm, and the development of policies and programs to ensure that these resources are available and fully utilized in developing countries. The International Board for Plant Genetic Resources is devoted exclusively to this goal. Besides research, the basic function of the IARCs is educational, and in the past two decades over 45,000 scientists have been trained in the CGIAR system.

The central challenge facing the CGIAR is world population growth and the need to increase agricultural production over the coming decades while preserving natural resources. The group was one of the main contributors to the so-called "Green Revolution." It helped develop new high-yielding varieties of cereals and introduced them into countries previously unable to grow the food; some of these countries now have agricultural surpluses. In 2001, CGIAR in South Africa worked on developing two new types of maize, which have a 30–50 percent larger crop than what is currently being produced by the smaller farmers.

The CGIAR is working to increase production even further, narrowing the gap between actual and potential yields, while continuing its efforts to limit soil erosion, desertification, and other kinds of environmental degradation.

The World Bank, the Food and Agriculture Organization (FAO), and the United Nations Development Program (UNDP) are among the original sponsors of the CGIAR. The organization has its headquarters at the offices of the World Bank, which also funds central staffing positions. Combined funding has grown from $15 million in 1971 to over $550 million in 2008, the latest year for which financial records are available. In 2010, CGIAR completed a two-year restructuring program to streamline governance and increase accountability.

Resources

ORGANIZATIONS

CGIAR Fund Office, The World Bank, MSN G6-601, 1818 H Street NW, Washington, D.C., USA, 20433, (202) 473-8951, (202) 473-8110, cgiarfund@cgiar.org, http://www.cgiar.org

Douglas Smith

Container deposit legislation

Container deposit legislation requires payment of a deposit on the sale of most or all beverage containers and may require that a certain percentage of beverage containers be allocated for refillable containers. The legislation shifts the costs of collecting and processing beverage containers from local governments and taxpayers to manufacturers, retailers and consumers.

While laws vary from state to state, even city to city, container deposit legislation generally provides a monetary incentive for returning beverage cans and bottles for recycling. Distributors and bottlers are required to collect a deposit from the retailer on each can and bottle sold. The retailer collects the deposit from consumers, reimbursing the consumer when the container is returned to the store. The retailer then collects the deposit from the distributor or bottler, completing the cycle. Consumers who choose not to return their cans and bottles lose their deposit, which usually becomes the property of the distributors and bottlers, though in some states, unredeemed deposits are collected by the state.

Oregon implemented the first deposit law or "bottle bill" in 1972. In the 1970s and 1980s, industry opponents fought container deposit laws on the grounds that they would result in a loss of jobs, an increase in prices, and a reduction in sales. Now opponents denounce the legislation as being detrimental to curbside recycling programs. But over the past four decades, container deposit legislation has proven effective not only in controlling litter and conserving natural resources but in reducing the waste stream as well.

Recovery rates for beverage containers covered under the deposit system depend on the amount of deposit and the size of the container. The overall recovery rate for beverage containers ranges from 75 to 93 percent. The reduction in container litter after implementation of the deposit law ranges from 42 to 86 percent, and reduction in total volume ranges from 30 to 64 percent. Although beverage containers make up just over 5 percent by weight of all municipal solid waste generated in the United States, they account for nearly 10 percent of all waste recovered, according to the Environmental Protection Agency (EPA). While the cans and bottles that are recycled into new containers or new products ease the burden on the environment, recycling is a second-best solution.

As recently as 1960, 95 percent of all soft drinks and 53 percent of all packaged beer was sold in refillable glass bottles. Those bottles required a deposit and were returned for reuse 20 or more times. But the centralization of the beverage industry, the increased mobility of consumers, and the desire for convenience resulted in the virtual disappearance of the reusable beverage container. Today, refillables make approximately 5 percent by volume of packaged soft drinks and approximately 5 percent by volume of packaged beer, according to the National Soft Drink Association and the Beer Institute.

Reuse is a more environmentally responsible waste management option, and is superior to recycling in the waste reduction hierarchy established by the EPA. While the container industry has been unwilling to promote refillable bottles, new interest in container deposit legislation may move industries and governments to adopt reuse as part of their waste management practices.

Industry-funded studies have found that a refillable glass bottle used as few as eight times consumes less energy than any other container, including recycled containers. A study conducted for the National Association for Plastic Container Recovery found that the 16-ounce (1 pt) refillable bottle produces the least amount of waterborne waste and fewest atmospheric emissions of all container types.

To date, eleven states have enacted beverage container deposit systems, designed to collect and process beverage bottles and cans. A deposit of 5–10 cents per can or bottle is an economic incentive to return the container. The states that have some form of container deposit legislation and accompanying deposit system include Oregon, New York, Connecticut, Maine, Iowa, Hawaii, Vermont, Michigan, Massachusetts, Delaware, and California.

Despite the fact that opinion polls show the public supports bottle bills by a wide margin, for decades, the beverage and packaging industries have successfully blocked the passage of bottle bills in nearly forty states. Even the most successful container deposit programs have come under attack and are threatened with repeal.

The United States has a long way to go to catch up to progressive countries such as Sweden, which does not allow aluminum cans to be manufactured or sold without industry assurances of a 75 percent recycling rate. Concerned that voluntary recycling would not meet these standards, the beverage, packaging and retail industries in Sweden have devised a deposit-refund system to collect used aluminum cans. Consumers in Sweden return their aluminum cans at a rate that has not been achieved in any other country. The 75 percent recycling rate was achieved in 1987, and industry experts expect the rate to exceed 86 percent in 1993. Most North American deposit systems rely on the distributor or bottler, but the deposit in Sweden originates with the can manufacturer or drink importer delivering cans within the country. Also, retail participation is voluntary: retailers collect the deposit but are not required to redeem the containers, though most do.

Containers and packaging are the single largest component of the waste stream and they offer the greatest potential for reduction, reuse, and recycling, according to the EPA. Where individuals, industries, and governments will not voluntarily comply with recycling programs, container deposit legislation has decreased the amount of recyclables entering the waste stream. According to a 2008 report by the Aluminum Association, states with container deposit laws have a recycling rate of 74 percent compared to 38 percent in states without container deposit laws.

Resources

BOOKS

Pichtel, John. *Waste Management Practices*. Boca Raton: CRC, 2005.

OTHER

United States Environmental Protection Agency (EPA). "Pollution Prevention: Recycling: Glass." http://www.epa.gov/ebtpages/pollrecyclingglass.html (accessed October 16, 2010).

United States Environmental Protection Agency (EPA). "Pollution Prevention: Recycling: Plastics." http://www.epa.gov/ebtpages/pollrecyclingplastics.html (accessed October 16, 2010).

Linda Rehkopf

Containment structures *see* **Nuclear power.**

Contaminated soil

Soils that contain pollutants at concentrations above background levels that pose a potential health or ecological risk are considered contaminated. Soils can be contaminated by many human actions including the discharge of solids and liquid pollutants at the soil surface; pesticide application; subsurface releases from leaks in buried tanks, pipes, and landfills; and deposition of atmospheric contaminants such as dusts and particles containing lead (Pb). Common contaminants include volatile hydrocarbons—such as benzene (C_6H_6), toluene ($C_6H_5CH_3$), ethylene (C_2H_4), and xylene (BTEX compounds)—found in fuels; heavy paraffins and chlorinated organic compounds such as polychlorinated biphenyls (PCBs) and pentachlorophenol (PCP); inorganic compounds such as lead, cadmium (Cd), arsenic (As), and mercury (Hg); and radionuclides such as tritium ($_3H$). Often, soil is contaminated with a mixture of pollutants. The nature of soil, the contaminant's chemical and physical characteristics, and environmental factors such as climate and hydrology interact to determine the accumulation, mobility, toxicity, and overall significance of the contaminant in any specific instance.

Fate of soil contaminants

Contaminants in soils may be present in solid, liquid, and gaseous phases. When liquids are released,

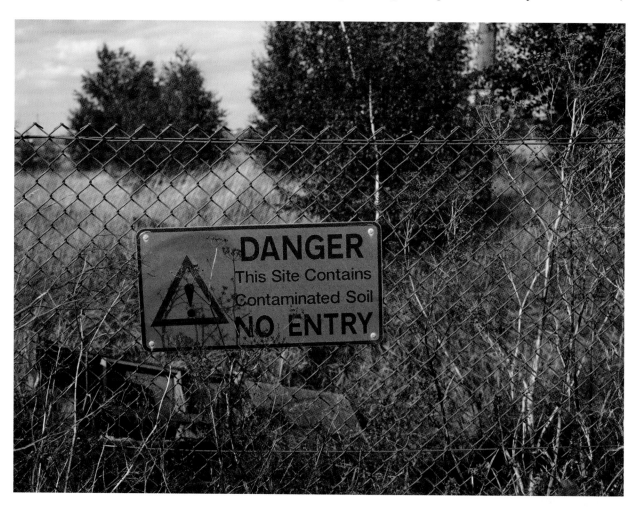

Sign warning public not to enter fenced off land that has been contaminated by toxic chemicals, Rainham Marshes, Essex, United Kingdom, 2005. *(Robert Brook / Photo Researchers, Inc.)*

they move downward through the soil. Some may fill pore spaces as liquids, some may partition or sorb onto mineral soil surfaces, some may dissolve into water in the soft pores, and some may volatilize. For most hydrocarbons, a multiphase system is common. When contaminants reach the water table (where the voids between soil particles are completely filled with water), contaminant behavior depends on its density. Light BTEX-type compounds float on the water table while dense chlorinated compounds may sink. While many hydrocarbon compounds are not very soluble, even low levels of dissolved contaminants may produce unsafe or unacceptable groundwater quality. Other contaminants such as inorganic salts (e.g., nitrates, NO_3^-) may be highly soluble and move rapidly through the environment. Metals demonstrate a range of behaviors. Some may be chemically bound and thus quite immobile; some may dissolve and be transported by groundwater and infiltration.

Contaminated pore water, termed leachate, may be transported in the groundwater flow. Groundwater travels both horizontally and vertically. A portion of the contaminant, termed residual, is often left behind as the flow passes, thus contaminating soils after the major contaminant plume has passed. If the groundwater velocity is fast, for example, hundreds of feet per year, the zone of contamination may spread quickly, potentially contaminating extraction wells or other resources. Over years or decades, especially in sandy or porous soils, groundwater contaminants and leachate may be transported over distances of miles, producing a situation that is very difficult and expensive to remedy. In these cases, immediate action is needed to contain and clean up the contamination. In cases where soils are largely comprised of fine-grained silts and clays, contaminants may spread very slowly. Such examples show the importance of site-specific factors in evaluating the significance of soil contamination as well as the selection of a cleanup strategy.

Superfund and other legislation

Prior to the 1970s, inappropriate land disposal practices, such as dumping untreated liquids in lagoons and landfills, were common and widespread. The presence and potential impacts of soil contamination were brought to public attention after well-publicized incidents at Love Canal, New York and the Valley of the Drums. Congress responded by passing the Comprehensive Environmental Response, Compensation, and Liability Act in 1980 (commonly known as CERCLA or Superfund) to provide funds with which to cleanup the worst sites. After five years of much litigation but little action, Congress updated

this law in 1986 with the Superfund Amendments and Reauthorization Act.

At present, over 1,200 sites across the United States have been selected as National Priorities List (NPL) sites and are eligible under CERCLA for federal assistance for cleanup. These sites are placed on the NPL list following Hazard Ranking System (HRS) screening and upon consideration of public comments. The HRS utilizes information from initial site inspection and assessment to assign a numerical value that represents the degree of potential threat to public health and the environment caused by the contaminated site. The site assessments focus on potential release of hazardous contaminants, characteristics of the wastes, and people or vulnerable ecosystems that would be affected by the contaminants. Possible modes of affecting public health and the environment are considered such as ground water, surface water, soil exposure, and air. Currently, there are also 67 proposed NPL sites. These Superfund sites tend to be the nation's largest and worst sites in terms of the possibility for adverse human and environmental impacts and the most expensive ones to clean up.

While not as well recognized, numerous other sites have serious soil contamination problems. The U.S. Office of Technology Assessment and Environmental Protection Agency (EPA) estimate that about 20 thousand abandoned waste sites and 600 thousand other sites of land contamination exist in the United States. These estimates exclude soils contaminated with lead in older city areas, the accumulation of fertilizers, pesticides, and insecticides in agricultural lands, and other classes of potentially significant soil contamination. Federal and state programs address only some of these sites. Currently, CERCLA is the EPA's largest program, with expenditures of about $1.2 billion annually. However, this is only a fraction of the cost to government and industry that will be needed to clean up all waste sites. Typical estimates of funds required to mitigate the worst 9 thousand waste sites reach at least $500 billion with a fifty-year anticipated time period for cleanup. There are about 300 NPL sites that are listed as deleted, meaning that the sites have been cleaned up or remediated, and thus no longer pose a threat to human health or the environment. The problem of contaminated soil is significant not only in the United States but in all industrialized countries.

U.S. laws such as CERCLA and the Resource Conservation and Recovery Act (RCRA) attempt to prohibit practices that have led to extensive soil contamination in the past. These laws restrict disposal practices; mandate financial liability to recover cleanup costs (as well as personal injury and property damage) and criminal liability to discourage willful misconduct or negligence; require record keeping to track waste; and provide incentives to reduce waste generation and improve waste management.

Soil cleanups

The cleanup or remediation of contaminated soils takes two major approaches: source control and containment, or soil and residual treatment and management. Typical containment approaches include caps and covers over the waste in order to limit infiltration of rain and snow melt and thus decrease the leachate from the contaminated soils. Horizontal transport in near-surface soils and groundwater may be controlled by vertical slurry walls. Clay, cement, or synthetic membranes may be used to encapsulate soil contaminants. Contaminated water and leachate may be hydraulically isolated and managed with groundwater pump- and-treat systems, sometimes coupled with the injection of clean water to control the spread of contaminants. Such containment approaches only reduce the mobility of the contaminant, and the barriers used to isolate the waste must be maintained indefinitely.

The second approach treats the soil to reduce the toxicity and volume of contaminants. These approaches may be broken down into extractive and *in situ* methods. Extractive options involve the removal of contaminated soil, generally for treatment and disposal in an appropriate landfill or for incineration where contaminants are broken down by thermal oxidation. *In situ* processes treat the soil in-place. *In situ* options include thermal, biological, and separation/extraction technologies. Thermal technologies include thermal desorption and in-place vitrification (glassification). Biological treatment includes biodegradation by soil fungi and bacteria that ultimately converts contaminants into carbon dioxide (CO_2) and water. This process is termed mineralization. Biodegradation may produce long-lived toxic intermediate products. Separation technologies include soil vapor extraction for volatile organic compounds that removes a fraction of the contaminants by enhancing volatilization by an induced subsurface air flow; stabilization or chemical fixation that uses additives to bind organics and heavy metals, thus eliminating contaminated leachate; soil washing and flushing using dispersants, solvents, or other means to solubilize certain contaminants such as PCBs and enhance their removal; and finally, groundwater pump and treat schemes that purge the contaminated soils with clean water in a flushing action. The contaminated groundwater may then be treated by air stripping, steam stripping, carbonabsorption, precipitation and flocculation and contaminant removal by ion exchange. A number of these approaches—*in situ* soil vitrification and enhanced bioremediation using engineered microorganisms—are experimental.

The number of potential remediation options is large and expanding due to an active research program that is driven by the need for low cost and more effective solutions. The selection of an appropriate cleanup strategy for contaminated soils requires a thorough characterization of the site and an analysis of the cost-effectiveness of suitable containment and treatment options. A site-specific analysis is required since the distribution and treatment of wastes may be complicated by variation in geology, hydrology, waste characteristics, and other factors at the site. Often, a demonstration of the effectiveness of an innovative or experimental approach may be required by governmental authorities prior to full-scale implementation. In general, large sites use a combination of remediation options. Pump and treat and vapor extraction are the most popular technologies.

Cleanup costs and cleanup standards

The cleanup of contaminated soils can involve significant expense and risk. In general, *in situ* containment is cheaper than soil treatment, at least in the short term. While the Superfund law establishes a preference for permanent remedies, many cleanups that have been funded under this law have used both containment and treatment options. In general, *in situ* treatment methods such as groundwater pump and treat are less expensive than extractive approaches such as soil incineration. *In situ* options, however, may not achieve cleanup goals. Like other processes, costs increase with higher removal levels. Excavation and incineration of contaminated soil can cost $1,500 per ton, leading to total costs of many millions of dollars at large sites. Superfund cleanups have averaged about $26 million. A substantial fraction of these costs are for site investigations. In contrast, small fuel spills at gasoline stations may be mitigated using vapor extraction at costs under $50 thousand.

Unlike air and water, which have specific federal laws and regulations detailing maximum allowable levels of contaminants, no levels have been set for contaminants in soils. Instead, the EPA and states use several approaches to set specific acceptable contaminant levels. For Superfund sites, cleanup standards must exceed applicable or relevant and appropriate requirements (ARARs) under federal environmental and public health laws. More generally, cleanup standards may be based on achieving background levels, that is, the concentrations found in similar, nearby, and unpolluted soils. Second, soil contaminant levels may be acceptable if the soil does not produce leachate with concentration levels above drinking water standards. Such determinations are often based on a test termed the Toxics Characteristic Leaching Procedure, which mildly acidifies and agitates the soil. According to the federal Safe Drinking Water Act, contaminant levels in the leachate below the maximum contaminant levels (MCLs) are acceptable. Third, soil contaminant levels may be set in a determination of health risks based on typical or worst case exposures. Exposures can include inhalation of soils as dust, ingestion of soil (generally by children), and direct skin contact with soil. In part, these various approaches are used based on the complexity of contaminant mobility and toxicity in soils and the difficulty of pinning down acceptable and safe levels.

Resources

BOOKS

Calabrese, Edward J., Paul T. Kostecki, and James Gragun. *Contaminated Soils, Sediments and Water: Science in the Real World.* New York: Springer, 2004.

OTHER

United States Environmental Protection Agency (EPA). "National Priorities List (NPL) Site Totals by Status and Milestone." http://www.epa.gov/superfund/sites/query/queryhtm/npltotal.htm (accessed September 4, 2009).
United States Environmental Protection Agency (EPA). "Superfund: Introduction to the Hazard Ranking System (HRS)." http://www.epa.gov/superfund/programs/npl_hrs/hrsint.htm (accessed September 4, 2009).

Stuart Batterman

Contour plowing

Plowing the soil along the contours of the land. For example, rather than plowing up-and-down the hill, the cultivation takes place around the hill. Contour plowing reduces water erosion.

Contraceptives *see* **Family planning; Male contraceptives.**

Convention on International Trade in Endangered Species of Wild Fauna and Flora (1975)

The Convention on International Trade in Endangered Species of Wild Fauna and Flora (CITES), was signed in 1973 and came into force in 1975. The aim of the treaty is to prevent international trade in listed endangered or threatened animal and plant species and products made from them. A full-time, paid secretariat to administer the treaty was initially funded by the United Nations Environmental Program, but has since been funded by the parties to the treaty. By 2009, 175 nations had become party to CITES, including most of the major wildlife trading nations, making CITES the most widely accepted wildlife conservation agreement in the world. The parties to the treaty meet every two years to evaluate and amend the treaty if necessary.

The species covered by the treaty are listed in three appendices, each of which requires different trade restrictions. Appendix I applies to "all species threatened with extinction," such as African and Asian elephants (*Loxodonta africana* and *Elephas maximus*, the hyacinth macaw (*Anodorhynchus hyacinthinus*), and Queen Alexandria's birdwing butterfly (*Ornithoptera alexandrae*). Commercial trade is generally prohibited for the approximately 900 listed species. Appendix II applies to "all species which although not necessarily now threatened with extinction may become so unless trade in specimens of such species is subject to strict regulation," such as the polar bear, giant clams, and Pacific Coast mahogany. Trade in these species requires an export permit from the country of origin. Currently, over 4,300 animals and 28,000 plants (mainly orchids) are listed in Appendix II. Appendix III is designed to help individual nations control the trade of any species. Any species may be listed in Appendix III for any nation. Once listed, any export of this species from the listing country requires an export permit. Usually a species listed in Appendix III is protected within that nation's borders. These trade restrictions apply only to signatory nations, and only to trade between countries, not to practices within countries.

CITES relies on signatory nations to pass domestic laws to carry out the principles included in the treaty. In the United States, the Endangered Species Act and the Lacey Act include domestic requirements to implement CITES, and the Fish and Wildlife Service is the chief enforcement agency. In other nations, if and when such legislation is passed, effective implementation of these domestic laws is required. This is perhaps the most problematic aspect of CITES, since most signatory nations have poor administrative capacity, even if they have strong desire for enforcement. This is especially a problem in poorer nations of the world, where monies for regulation of trade in endangered species is forced down the agenda by more pressing social and economic issues.

CITES can be regarded as moderately successful. Since the implementation of CITES, only one listed species—Spix's macaw—has become extinct in the wild. There has been steady progress toward compliance with the treaty, though tremendous enforcement problems still exist. Due to the international scope and size of world wildlife trade, enforcement of CITES is estimated to be only 60–65 percent effective worldwide. International trade in endangered species is still big business; the profits from such trade can be huge. In 1988, the world market for exotic species was estimated to be $5 billion, $1.5 billion of which was estimated to be illegal. By 2009, Interpol estimated that the trade of illegal exotic species was worth $20 billion per year. Among the most lucrative products traded are reptile skins, fur coats, and ingredients for traditional drugs. If violations are discovered, the goods are confiscated and penalties and fines for the violators are established by each country. Occasionally, sanctions are imposed on a nation if it flagrantly violates the convention.

Additional information

On May 14, 2008, the U.S. Department of the Interior (DOI) secretary, Dirk Kempthorne, announced that, under the Endangered Species Act (ESA), the polar bear will now be listed as a "threatened" species. ("Threatened" means that a species runs a definite risk of becoming endangered; "endangered" status indicates that the species is at great risk of becoming extinct.)

Scientists estimate that there are 20,000 to 25,000 polar bears living in the Arctic, but that within 50 years, as much as two-thirds of the population could be lost due to loss of ice mass in the Arctic Sea, which is attributed to global warming. Controversy has surrounded the decision to place the polar bears on this list, as the polar bear is the first species to be listed as one whose decline is directly correlated to global warming. In his announcement, Kempthorne asserted that it would be inappropriate to use the ESA or the

protection of the bear to attempt to influence the regulation of greenhouse gases.

A ruling on the polar bear status had been anticipated in January 2008, but on January 7, the U.S. Department of Fish and Wildlife announced that it would not be able to provide its recommendation on listing by the January 9th deadline. The service requested additional time to adequately review and incorporate data on the matter from the U.S. Geological Survey (USGS); they anticipated making the recommendation within a month's time. In April, the U.S. District Court for the Northern District of California ordered the DOI/USFWS to make the final announcement by May 15, 2008.

In February 2008, the DOI sold gas and oil rights covering nearly 30 million acres of Chukchi Sea. This area is just off the Alaskan coast and is home to about 20 percent of the polar bear population. Some environmental groups feel that the delay in announcement was deliberate on the part of the administration, though Kempthorne pointed out that the decision was based solely on the research of the Department's scientists.

Resources

BOOKS

Askins, Robert. *Saving Biological Diversity: Balancing Protection of Endangered Species and Ecosystems.* [New York]: Springer, 2008.

Goodall, Jane, Thane Maynard, and Gail E. Hudson. *Hope for Animals and Their World: How Endangered Species Are Being Rescued from the Brink.* New York: Grand Central Pub, 2009.

Reeve, Rosalind. *Policing International Trade in Endangered Species: The Cites Treaty and Compliance.* London: Royal Institute of International Affairs, 2004.

OTHER

United Nations Environment Programme (UNEP). "Secretariat of the Convention on International Trade in Endangered Species of Wild Fauna and Flora." http://www.cites.org/ (accessed October 14, 2010).

ORGANIZATIONS

CITES Secretariat, International Environment House, Chemin des Ane;aamones, Cháftelaine Geneva, Switzerland, CH–1219, (+4122) 917- 8139/40, (+4122) 797-3417, cites@unep.ch, http:// www.cites.org

Christopher McGrory Klyza

Convention on Long-Range Transboundary Air Pollution (1979)

Held in Geneva in 1979 under the auspices of the United Nations, the goal of the Convention on Long-Range Transboundary Air Pollution was to reduce air pollution and acid rain, particularly in Europe and North America. The accord went into effect in March 1983 and was signed by the United States, Canada, and several European countries. All of the signatories agreed to cooperate in researching and monitoring air pollution and to exchange information on developing technologies for air pollution control. In 2010, fifty-one nations were party to the convention. This convention established the Cooperative Program for Monitoring and Evaluating of the Long-Range Transmission of Air Pollutants in Europe, which was first funded in 1984. The countries that signed the treaty also agreed to reduce their sulfur emissions 30 percent by 1993. All of the countries were able to meet this goal, with many countries reducing more than 50–60 percent of their emissions.

See also Acid rain; Air pollution; Air pollution control.

Resources

BOOKS

Sliggers, Johan, and Willem Kakebeeke, eds. *Clearing the Air: 25 Years of the Convention on Long-range Transboundary Air Pollution.* New York: United Nations, 2005.

OTHER

United Nations Environment Programme (UNEP). "Secretariat for the The Basel Convention on the Control of Transboundary Movements of Hazardous Wastes and their Disposal." http://www.basel.int/ (accessed October 15, 2010).

United States Environmental Protection Agency (EPA). "Air: Air Pollution: Transboundary Pollution." http://www.epa.gov/ebtpages/airairpollutiontrans boundarypollution.html (accessed October 15, 2010).

United States Environmental Protection Agency (EPA). "International Cooperation: Border Issues: Transboundary Pollution." http://www.epa.gov/ebtpages/inteborderissuestransboundarypollution.html (accessed October 15, 2010).

Douglas Smith

Convention on the Conservation of Migratory Species of Wild Animals (1979)

The first attempt at a global approach to wildlife management, the Convention on the Conservation of Migratory Species of Wild Animals was held in Bonn, Germany in 1979. The purpose of the convention was to reach an agreement on the management of wild animals that migrate "cyclically and predictably" across international boundaries. Egypt, Italy, the United Kingdom, Denmark, and Sweden, among other nations, signed an accord on this issue; the treaty went into effect in 1983 and has over 100 member nations. Several notable nations, however, including Canada, China, Russia, and the United States, are not parties to the convention.

The challenge facing the convention was how to assist nations that did not have wildlife management programs while not disrupting those that had already established them. The United States and Canada did not believe that the agreement reached in Bonn met this challenge. Representatives from both countries argued that the definition of a migratory animal was too broad; it would embrace nearly every game bird or animal in North America, including rabbits, deer, and bear. But both countries were particularly concerned that the agreement did not sufficiently honor national sovereignty, and they believed it threatened the effectiveness of the federal-state and federal-provincial systems that were already in place.

The agreement also came into conflict with other laws, particularly laws governing national jurisdictions and territorial boundaries at sea, but it is still considered an important advance in the process of developing international environmental agreements.

See also Environmental law; Wildlife management.

Resources

BOOKS

Alderton, David. *The Illustrated Encyclopedia of Birds of the World.* London: Lorenz Books, 2005.

OTHER

Environmental Law & Policy Center. "Environmental Law & Policy Center." http://www.elpc.org/ (Accessed October 16, 2010).

United Nations Environment Programme (UNEP). "Environmental Law and Conventions (DELC)." http://www.unep.org/dec/ (Accessed October 16, 2010).

Douglas Smith

Convention on the Law of the Sea (1982)

During the age of exploration in the seventeenth century, the Dutch lawyer Hugo Grotius formulated the legal principle that the ocean should be free from exclusive rule by any one nation (*Mare liberum*), which became the basis of the law of the sea. During the eighteenth century, each coastal nation was granted sovereignty over an offshore margin of three nautical miles to provide for better defense against pirates and other intruders. As undersea exploration advanced in the twentieth century, nations became increasingly interested in the potential resources of the oceans. By the 1970s, many were determined to exploit resources such as petroleum on the continental shelf and manganese nodules rich with metals on the deep ocean floor, but uncertainties about the legal status of these resources inhibited investment.

In 1974, the United Nations responded to these concerns and convened a Conference on the Law of the Sea, at which the nations of the world negotiated a consensus dividing the ocean among them. The participants drafted a proposed constitution for the world's oceans, including in it a number of provisions that radically altered their legal status. First, the convention extended the territorial sea from three to 22 nautical miles, giving nations the same rights and responsibilities within this zone that they possess over land. Negotiations were necessary to ensure that ships have a right to navigate these coastal zones when going from one ocean to another.

The convention also acknowledged a 200 nautical mile Exclusive Economic Zone (EEZ), in which nations could regulate fisheries as well as resource exploration and exploitation. By the late 1970s, over 100 nations had already claimed territorial seas extending from 12 to 200 miles. The United States, for example, claims a 200 mile zone. Today, EEZs cover about one third of the ocean, completely dividing the North Sea, the Mediterranean Sea, the Gulf of Mexico, and the Caribbean among coastal states. They also encompass an area that yields more than 90 percent of the ocean's fish catch, and they are considered a potentially effective means to control overfishing because they provide governments clear responsibility for their own fisheries.

The most heated debates during the convention, however, did not concern the EEZs, but involved provisions concerning seabed mining. Many developing nations insisted that deep-sea metals were the common property and sources of sharable wealth. After fierce disagreements, a compromise was reached in 1980 that would establish a new International Seabed Authority (ISA). Under the plan, any national or private enterprise could take half the seabed mining sites and the ISA would take the other half. In exchange for mining rights, industrialized nations would underwrite the ISA and sell it the necessary technology. In December 1982, 117 nations signed the Convention on the Law of the Sea (UNCLOS) at Montego Bay, Jamaica. U.S. President Ronald Reagan refused to sign, citing the seabed-mining provisions as a major reason for United States opposition. Two other key nations, the United Kingdom and Germany, refused, and 21 nations abstained.

UNCLOS came into force in 1994 after Guyana became the 60th nation to ratify the treaty. In 2010, 158 countries and the European Community have joined in the Convention.

See also Coastal Zone Management Act (1972); Commercial fishing; Oil drilling.

Resources

BOOKS

Russell, Denise. *Who Rules the Waves?: Piracy, Overfishing and Mining the Oceans.* London: Pluto Press, 2010.
Stow, Dorrick A.V. *Oceans: An Illustrated Reference.* Chicago, IL: University of Chicago Press, 2006.

OTHER

Environmental Law & Policy Center. "Environmental Law & Policy Center." http://www.elpc.org/ (Accessed October 16, 2010).
United Nations Environment Programme (UNEP). "Environmental Law and Conventions (DELC)." http://www.unep.org/dec/ (Accessed October 16, 2010).

David Clarke

Convention on the Prevention of Marine Pollution by Dumping of Waste and Other Matter (1972)

The 1972 International Convention on the Prevention of Marine Pollution by Dumping of Waste and Other Matter, commonly called the London Dumping Convention, entered into force on August 30, 1975. The London Dumping Convention covers "ocean dumping" defined as any deliberate disposal of wastes or other matter from ships, aircraft, platforms, or other human-made structures at sea. The discharge of sewage effluent and other material through pipes, wastes from other land-based sources, the operational or incidental disposal of material from vessels, aircraft, and platforms (such as fresh or salt water), or wastes from seabed mining are not covered under this convention.

The framework of the London Dumping Convention consists of three annexes. Annex I contains a blacklist of materials that cannot be dumped at sea. Organohalogen compounds, mercury, cadmium, oil, high-level radioactive wastes, warfare chemicals, and persistent plastics and other synthetic materials that may float in such a manner as to interfere with fishing or navigation are examples of substances on the blacklist. The prohibition does not apply to acid and alkaline substances that are rapidly rendered harmless by physical, chemical, or biological processes in the sea. Annex I does not apply to wastes such as sewage sludge or dredge material that contain blacklist substances in trace amounts.

Annex II of the convention comprises a grey list of materials considered less harmful than the substances on the blacklist. At-sea dumping of grey list wastes requires a special permit from contracting states (countries participating in the convention). These materials include wastes containing arsenic, lead, copper, zinc, organosilicon compounds, cyanides, pesticides and radioactive matter not covered in Annex I, and containers, scrap metals, and other bulky debris that may present a serious obstacle to fishing or navigation. Grey list material may be dumped as long as special care is taken with regard to ocean dumping sites, monitoring, and methods of dumping to ensure the least detrimental impact on the environment. A general permit from the appropriate agencies of the contracting states to the Convention is required for ocean dumping of waste not on either list.

Annex III includes criteria that countries must consider before issuing an ocean dumping permit. These criteria require consideration of the effects dumping activities can have on marine life, amenities, and other uses of the ocean, and they encompass factors related to disposal operations, waste characteristics, attributes of the site, and availability of land-based alternatives.

The International Maritime Organization serves as Secretariat for the London Dumping Convention, undertaking administrative responsibilities and ensuring cooperation among the contracting parties. As of 2001, there were 78 contracting parties to the Convention, including the United States.

The London Dumping Convention covers ocean dumping in all marine waters except internal waters of the contracting states, which are required to regulate ocean dumping consistently with the convention's provisions. However, they are free to impose stricter rules on their own activities than those required by the convention. The London Dumping Convention was developed at the same time as the Marine Protection, Research and Sanctuaries Act of 1972 (Public Law 92-532), a law enacted by the United States. The U.S. congress amended this act in 1974 to conform with the London Dumping Convention.

Most nations using the ocean for purposes of dumping waste are developed countries. The United States completely ended its dumping of sewage sludge following the passage of the Ocean Dumping Ban Act (1988), which prohibits ocean dumping of all sewage sludge and industrial waste (Public Law 100-688). Britain and the North Sea countries also intend to end ocean dumping of sewage sludge. During the Thirteenth Consultative Meeting of the London Dumping Convention in 1990, the contracting states agreed to terminate all industrial ocean dumping by the end of 1995.

While the volume of sewage sludge and industrial waste dumped at sea is decreasing, ocean dumping of dredged material is increasing. Incineration at sea requires a special permit and is regulated according to criteria contained in an addendum to Annex I of the convention.

In 1996, the London Dumping Convention was replaced by the 1996 Protocol to the Convention on the Prevention of Marine Pollution by Dumping of Wastes and Other Matter, 1972. Among the changes is the "reverse list." The Protocol only allows dumping material that is listed in Annex I, it is referred to as the "reverse list" since it completely reverses the original Annex I. Incineration at sea is now completely banned, unless there is an emergency, as is the exportation of waste to other countries for ocean dumping. The 1996 Protocol entered into force in March 2004.

See also Marine pollution; Seabed disposal.

Resources

BOOKS

Russell, Denise. *Who Rules the Waves?: Piracy, Overfishing and Mining the Oceans.* London: Pluto Press, 2010.

Stow, Dorrick A.V. *Oceans: An Illustrated Reference.* Chicago, IL: University of Chicago Press, 2006.

OTHER

Environmental Law & Policy Center. "Environmental Law & Policy Center." http://www.elpc.org/ (accessed October 16, 2010).

United Nations Environment Programme (UNEP). "Environmental Law and Conventions (DELC)." http://www.unep.org/dec/ (accessed October 16, 2010).

Marci L. Bortman

Convention on Wetlands of International Importance (1971)

Also called the Ramsar Convention or Wetlands Convention, the Convention on Wetlands of International Importance is an international agreement adopted in 1971 at a conference held in Ramsar, Iran. One of the principal concerns of the agreement was the protection of migratory waterfowl, but the treaty is generally committed, like much wetlands legislation in the United States, to restricting the loss of wetlands in general because of their ecological functions as well as their economic, scientific, and recreational value. The accord went into effect in 1975, establishing a network of wetlands, primarily across Europe and North Africa.

In 2009, there were 159 Contracting Parties, each of whom was required to set aside at least one wetland reserve. Over 1,850 national wetland sites have been established totaling approximately 715,450 square miles (1,853,000 km^2). The convention has secured protection for wetlands around the world, but many environmentalists believe it has the same weakness as many international conventions on the environment. There is no effective mechanism for enforcement. Population growth continues to increase political and economic pressures to develop wetland areas around the world, and there are no provisions in the agreement strong enough to prevent nations from removing protected status from designated wetlands.

Resources

BOOKS

Dugan, Patrick. *Guide to Wetlands*. Buffalo, N.Y.: Firefly Books, 2005.

Lockwood, C., and Rhea Gary. *Marsh Mission: Capturing the Vanishing Wetlands*. Baton Rouge: Louisiana State University Press, 2005.

Tiner, Ralph W. *In Search of Swampland: A Wetland Sourcebook And Field Guide*. Newark: Rutgers University Press, 2005.

ORGANIZATIONS

The Ramsar Convention Bureau, Rue Mauverney 28, Gland, Switzerland, CH–1196, +41 22 999 0170, +41 22 999 0169, ramsar@ramsar.org, http://www.ramsar.org/index.html

Douglas Smith

Conventional pollutant

Conventional water pollutants fall into five categories. These pollutants can be classified as: biological oxygen demanding, suspended solids, pH, fecal coliform, and oil and grease. The presence of these pollutants is commonly determined by measuring biochemical oxygen demand (BOD), total suspended solids (TSS), pH levels, the amount of fecal coliform bacteria, and the quantity of oil and grease in an aquatic ecosystem.

BOD is the quantity of oxygen required by microorganisms to stabilize five-day incubated oxidizable organic matter at 68°F (20°C). Hence, BOD is a measure of the biodegradable organic carbon and at times, the oxidizable nitrogen (N). BOD is the sum of the oxygen used in organic matter synthesis and in the endogenous respiration of microbial cells. Some industrial wastes are difficult to oxidize, and bacterial seed is necessary. In certain cases, an increase in BOD is observed with an increase in dilution. It is hence necessary to determine the detection limits for BOD.

Suspended solids interfere with the transmission of light. Their presence also affects recreational use and aesthetic enjoyment. Suspended solids make fish vulnerable to diseases, reduce their growth rate, prevent successful development of fish eggs and larvae, and reduce the amount of available food. The Environmental Protection Agency (EPA) restricts suspended matter to not more than ten percent of the reasonably established amount for aquatic life. This set maximum allows sufficient sunlight to penetrate and sustain photosynthesis. Suspended solids also cause damage to invertebrates and fill up gravel spawning beds.

The acidity or alkalinity of water is indicated by pH. A pH of seven is neutral. A pH value lower than seven indicates an acidic environment and a pH greater than seven indicates an alkaline environment. Most aquatic life is sensitive to changes in pH. The pH of surface waters is specified to protect aquatic life and prevent or control unwanted chemical reactions such as metal ion dissolution in acidic waters. An increase in toxicity of many substances is often observed with changes in pH. For example, an alkaline environment shifts the ammonium ion (NH_4^+) to a more poisonous form of un-ionized ammonia (NH_3). EPA criteria for pH are 6.5–9.0 for freshwater life, 6.5–8.5 for marine organisms and 5–9 for domestic consumption.

Fecal coliform bacteria are useful indicators for detecting pathogenic or disease-causing bacteria. However, this relationship is not absolute because these bacteria can originate from the intestines of humans and other warm-blooded animals. Prior knowledge of river basins and the possible sources of these bacteria is necessary for a survey to be effective. The strictest EPA criteria for coliforms apply to shellfish, since they are often eaten without being cooked.

Common sources of oil and grease are petroleum derivatives and fats from vegetable oil and meat processing. Both surface and domestic waters should be free from floating oil and grease. Limits for oil and grease are based on lethal concentration or LC_{50} values. LC_{50} is defined as the concentration at which 50 percent of an aquatic species population perishes. EPA criterion is for a ninety-six-hour exposure, and during this period the concentration of individual petrochemicals should not exceed 0.01 of the LC_{50} median. Oil and grease contaminants vary in physical, chemical, and toxicological properties besides originating from different sources.

The lethal concentration values for pollutants may be altered by temperature as well as dissolved oxygen levels. Increases in temperature result in a boost in metabolic rates, potentially increasing the rate of uptake of a pollutant into an organism. Higher temperatures can also inflict added stress onto an organism, weakening the resistance of an organism to a given pollutant. This results in increased vulnerability of organisms to lower concentrations of pollutants, and thus lower LC_{50} values. Increases in temperature also result in a reduction in the amount of dissolved oxygen in a body of water. Studies have shown that decreased dissolved oxygen may increase the toxicity of pollutants, effectively lowering the LC_{50} value for a pollutant. Because the Earth's temperature is rising as a result of climate change, organisms may be more susceptible and vulnerable to lower levels of pollutants in aquatic ecosystems.

Control and prevention of pollution discharges are regulated by the Clean Water Act (CWA). Under the CWA, "the best conventional pollutant control technology" must be employed to treat water that will be discharged from point sources into waterways as effluent. The National Pollution Discharge Elimination System (NPDES) is in place to regulate these discharges and requires that a permit be obtained for a point source discharge.

See also Industrial waste treatment; Sewage treatment; Wastewater; Water pollution; Water quality.

Resources

BOOKS

Viessman, Warren. *Water Supply and Pollution Control.* Upper Saddle River, NJ: Pearson Prentice Hall, 2009.

Welch, E. B., and Jean M. Jacoby. *Pollutant Effects in Freshwater: Applied Limnology.* London: Spon Press, 2004.

James W. Patterson

The Copenhagen Accord

Near the end of the December 2009 conference on climate change held in Copenhagen, Denmark—a summit attended by more than 100 heads of state—contentious negotiations produced the Copenhagen Accord, an international agreement aimed at mitigating global warming and the impacts of climate change.

The Copenhagen Accord

The fifteenth Conference of the Parties (COP) to the United Nations Framework Convention on Climate Change (UNFCCC) took place at Copenhagen, Denmark, in December 2009. Instead of an anticipated comprehensive and legally binding international agreement to replace the Kyoto Protocols that expire in 2012, the global summit of 193 nations and more than 100 world leaders produced the Copenhagen Accord. The Accord creates an international reporting and verification scheme aimed at meeting voluntary and individual targets related to a range of climate change issues. The Accord began as an agreement negotiated by the United States and the BASIC bloc of nations (China India, Brazil, and South Africa), representing the leading greenhouse gas emitters in the world.

Controlling global warming

While recognizing the need to limit temperature rises to less than 2°Celsius (3.6°Fahrenheit), the limit beyond which scientific evidence asserts that climate change consequences become especially dire, the Copenhagen Accord fails to set firm emissions targets. The language of the Accord does not actually set a formal limit for acceptable average global temperature rise, but merely "recognizes the scientific view" that global temperature increases should be held to less than 2°Celsius.

Critics of the Accord immediately argued that the majority of accepted scientific studies indicate that the current pledges of voluntary emissions reductions will be insufficient to limit temperature rise to 2°Celsius. During the conference, representatives of small-island nations expressed hopes that negotiations might lead to stronger cuts in emissions that could limit temperature increase to 1.5°Celsius (2.6°Fahrenheit) in order to further mitigate the extent and impacts of sea level rise.

The Accord also fails to set a target year for peak emissions or specify a target date to conclude a legally binding treaty. The Accord calls only for a review of its provisions and implementation by 2015, more than a year after the next set of formal reports by the Inter-governmental Panel on Climate Change (IPCC) is scheduled to be published.

Emissions limits versus carbon-intensity targets

The Accord does, however, for the first time include pledges by China, India, and other emerging industrial powers to accept some limits on emissions, along with an agreement to submit to international emissions reporting. Countries are asked to specify by February 2010 their individual targets for 2020 carbon emissions. The Accord does not, however, specify penalties for countries failing to set targets or for fail-ure to meet the target ultimately set.

Separate from the Accord negotiations, China and India set carbon-intensity targets (levels of emissions related to measures of economic development such as gross domestic product) rather than specific limits based on the quantity of emissions. Setting carbon-intensity targets allows developing nations to continue to increase the total quantity of their carbon emissions as their industrializing economies continue to expand.

Financial aid to poor and vulnerable nations

The Accord also promises $30 billion in aid to developing nations over the next three years to help them cope and adapt to climate change. Although mechanisms for monitoring and delivery of aid remain to be negotiated, the amount of aid is scheduled to escalate to $100 billion by 2020. A green climate fund will also support technology transfers aimed at low-ering GHG emissions by increasing energy efficiency in developing nations.

Accord status and initial pledges

The Accord was not endorsed by all 193 nations at the talks. Lacking the unanimous vote needed for for-mal approval, conference administrators were only able to officially "note" or recognize the agreement.

In December 2009, the World Meteorological Organization (WMO) released data showing that the decade 2000 to 2009 was the warmest in modern recorded history. WMO officials refuted other reports claiming to show that global warming had leveled off

since 1998, the hottest year in history. Some leveling off or cooling was expected after the surge to 1998 peak levels, but overall, the first decade of the twenty-first century proved warmer than the 1990s and 1980s. Although there are minor differences in rankings of hottest years, the WMO data closely conforms to inde-pendent studies made by the National Oceanic and Atmospheric Administration (NOAA) and the National Aeronautics and Space Administration (NASA).

Despite the WMO data, the initial voluntary national pledges made in response to the Copenhagen Accord failed to achieve a common baseline for cuts by the year 2020. For example, the United States pledged to cut carbon emission by 17 percent of 2005 levels while the European Union pledged a 20 to 30 percent reduction from 1990 levels. Moreover, climate and energy experts argue that, as of November 2010, the United States has failed to pass energy legislation that would allow it to honor its current commitments to the Accord. China and India continued to relate emissions policy to economic activity. Instead of spe-cific limits, China pledged to reduce its carbon inten-sity by 40 to 45 percent compared with 2005 levels and India pledged to reduce carbon intensity by 20 to 25 percent of 2005 levels. Setting carbon-intensity targets allows developing nations to continue to increase the total quantity of their carbon emissions as their indus-trializing economies continue to expand.

Resources

Books

Giddens, Anthony. *Politics of Climate Change*. Cambridge, UK: Polity Press, 2009.

Hulme, Mike. *Why We Disagree About Climate Change: Understanding Controversy, Inaction and Opportunity*. New York: Cambridge University Press, 2009.

Metz, B., et al. *Climate Change 2007: Mitigation of Climate Change : Contribution of Working Group III to the Fourth Assessment Report of the Intergovernmental Panel on Climate Change*. Cambridge, UK and New York: Cambridge University Press, 2007.

Page, Edward. *Climate Change, Justice and Future Generations*. Cheltenham, UK: Edward Elgar, 2006.

Solomon, S., et al. "IPCC, 2007: Summary for Policy-makers." In: *Climate Change 2007: The Physical Science Basis. Contribution of Working Group I to the Fourth Assessment Report of the Intergovernmental Panel on Climate Change*, edited by S. Solomon, et al. Cambridge, UK and New York: Cambridge University Press, 2007.

OTHER

Intergovernmental Panel on Climate Change (IPCC). "IPCC Assessment Reports." http://www.ipcc.ch/ (accessed November 12, 2010).

United Nations Framework Convention on Climate Change (UNFCCC). "United Nations Framework Convention on Climate Change (UNFCCC)." http://unfccc.int/2860.php (accessed November 12, 2010).

United Nations, Secretariat of the United Nations Framework Convention on Climate Change (UNFCCC). "Copenhagen, COP 15/CMP 5." http://unfccc.int/files/press/backgrounders/application/pdf/fact_sheet_copenhagen_cop_15_cmp_5.pdf (accessed November 12, 2010).

K. Lee Lerner

Copper

Atomic symbol Cu, a metallic element with an atomic number of 29 (29 protons in the nucleus), and an average atomic weight of 63.546. Copper is a micronutrient which is needed in many proteins and enzymes. Copper has been and is frequently used in piping systems which convey potable water. Corrosive waters will leach copper from the pipelines, thereby exposing consumers to copper and possibly creating bluish-green stains on household fixtures and clothes. The staining of household fixtures becomes a nuisance when copper levels reach 2–3 mg/l. The drinking water standard for copper is 1 mg/l. It is a secondary standard based on the potential problems of staining and taste. Copper can be toxic to aquatic life, and copper sulfate is often used to control the growth of algae in surface waters.

Copper mining *see* **Ducktown, Tennessee; Sudbury, Ontario.**

Coral bleaching

Coral bleaching is the whitening of coral colonies due to the loss of the symbiotic algae, zooxanthellae, from the tissues of coral polyps. It is mostly caused by stress. The host coral polyp provides the algae with a protected environment and a supply of carbon dioxide for its photosynthetic processes. The golden-brown algae serve as a major source of nutrition and color for the coral. The loss of the algae exposes the translucent calcium carbonate skeletons of the coral colony, and the corals look "bleached." Corals may recover from short-term bleaching (less than a month), but prolonged bleaching causes irreversible damage and mortality, for without the algae, the corals starve and die.

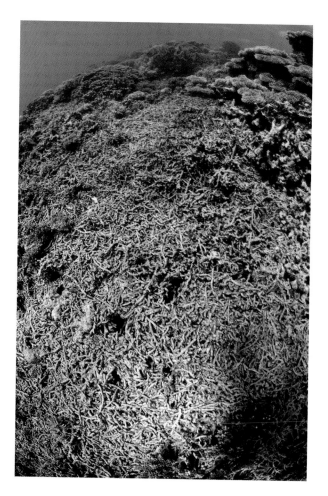

Rising ocean temperatures killed much of this coral reef, leaving only the "bleached" coral skeleton. Coral thrives in a narrow temperature range. *(Kike Calvo via AP Images)*

However, even a sublethal stress may result in increased susceptibility of corals to infections, with resulting significant mortality. Populations of sea urchins, parrot fish, and worms erode and weaken dead reef skeletons, and the reef can be destroyed by storm surges.

The means by which corals expel the zooxanthellae are not yet known. In laboratory experiments the zooxanthellae are released into the gut of the polyp and then expelled through the mouth; however this method has not been observed in the environment. Another hypothesis is that the stressed corals provide the algae with fewer nutrients, which results in the algae leaving the corals. Another possibility is that the algae may produce oxides under stress, which adversely affect the algae.

Bleaching may be caused by a number of stresses or environmental changes, including disease, excess shade, increased levels of ultraviolet radiation, sedimentation, pollution, salinity changes, exposure to air by low tides or low sea level, and increased temperatures. Coral bleaching is most often associated with increased sea surface temperatures, as corals tolerate only a narrow temperature range of between about 77 to 84°Fahrenheit (25 to 29°C).

Historically bleaching was observed on a small scale, such as in overheated tide pools. However, in the early to mid 1980s, coral reefs around the world began to experience large scale bleaching, with a bleaching event occurring somewhere in the world almost every year. In 1998, coral reefs around the world suffered the most extensive and severe bleaching and subsequent mortality in recorded history, in the same year that tropical sea surface temperatures were the highest in recorded history. Coral bleaching was reported in at least 60 countries and island nations in the Pacific Ocean, Indian Ocean, Red Sea, Persian Gulf, and the Caribbean. Only the Central Pacific region was spared. About 16 percent of the world's reefs were lost in a period of only nine months. Previous bleaching events had only affected reefs to a depth of less than 49 feet (15 m), but in 1998, the bleaching extended as deep as 164 feet (50 m). The reason sea temperatures throughout the world were so warm in 1998 remains controversial and uncertain. Three theories have been developed—natural climate variability, El Niño and other climatic variations, and global warming. By the end of 2000, 27 percent of the world's reefs had been lost, with the largest single cause being the coral bleaching event of 1998. More than a dozen significant bleaching events occurred between 2002 and 2009. As of 2010, less than half the reefs lost during 1998 bleaching events have substantially recovered, the others adding to losses due to sediment and nutrient pollution and overexploitation.

Many of the bleached coral reef ecosystems may require decades to recover. Human populations dependent on the reefs will lose fisheries, shoreline protection, and tourism opportunities. Trends of the past century indicate that coral bleaching events may become more frequent and severe if the climate continues to warm.

Resources

BOOKS

Aronson, R. B., ed. *Geological Approaches to Coral Reef Ecology*. Berlin: Springer, 2006.
Cote, I. M., and J. D. Reynold, eds. *Coral Reef Conservation*. Cambridge: Cambridge University Press, 2006.

PERIODICALS

Dean, Cornelia. "Coral Is Dying. Can It Be Reborn?" *New York Times* (May 1, 2007): F:1.

Dean, Cornelia. "Coral Reefs and What Ruins Them." *New York Times* (February 26, 2008).
Fountain, Henry. "More Acidic Ocean Hurts Reef Algae as Well as Corals." *New York Times* (January 8, 2008).

OTHER

National Geographic Society. "Coral Reefs: Canaries of the Sea ." http://environment.nationalgeographic.com/environment/habitats/canaries-of-sea.html (accessed August 29, 2010).
National Oceanic and Atmospheric Administration (NOAA). "25 Things You Can Do to Save Coral Reefs." http://www.publicaffairs.noaa.gov/25list.html (accessed August 29, 2010).
Reef Relief. "Reef Relief: Protect Living Coral." http://www.reefrelief.org/ (accessed August 29, 2010).
U.S. Government; science.gov. "Coral Reefs." http://www.science.gov/browse/w_131B2.htm (accessed August 29, 2010).
United Nations System-Wide EarthWatch. "Coral reefs under pressure." http://earthwatch.unep.net/emerging issues/oceans/coralreefs.php (accessed August 29, 2010).
United Nations System-Wide EarthWatch. "Widespread Coral Bleaching." http://earthwatch.unep.net/emerg ingissues/oceans/coralbleaching.php (accessed August 29, 2010).
United States Environmental Protection Agency (EPA). "Ecosystems: Aquatic Ecosystems: Coral Reefs." http://www.epa.gov/ebtpages/ecosaquaticecosystcoral reefs.html (accessed August 29, 2010).
University of California Museum of Paleontology. "Marine Ecosystems: Corals and coral reefs." http://www.ucmp.berkeley.edu/cnidaria/cnidarialh.html (accessed August 29, 2010).

Judith L. Sims

Coral reef

Coral reefs represent some of the oldest and most complex communities of plants and animals on Earth. About 200–400 million years old, they cover about 231,660 square miles (600,000 km²) worldwide. (The most popular reefs range from 5,000–10,000 years old.) The primary structure of a coral reef is a calcareous skeleton formed by marine invertebrate organisms known as *cnidarians*, which are relatives of sea anemones. Corals are found in most of the oceans of the world, in deep as well as shallow seas and temperate as well as tropical waters. But corals are most abundant and diverse in relatively shallow tropical waters, where they have adapted to the constant temperatures provided by these waters. The reef-forming corals, or

A thriving coral reef. *(Pborowka/Shutterstock.com)*

hermatypic corals, have their highest diversity in the Indian and Pacific Oceans, where over 700 species are found. By contrast, the Atlantic Ocean provides the habitat for less than forty species. Other physical constraints needed for the success of these invertebrate communities are clear water, a firm substrate, high salinity, and sunlight. Clear water and sunlight are required for the symbiotic unicellular plants that live in the surface tissues of the coral polyps. This intimate plant-animal association benefits both participants. Corals obtain oxygen directly from the plants and avoid having to excrete nitrogenous and phosphate waste products because these are absorbed directly as nutrients by the plants. Respiration by the coral additionally provides carbon dioxide to these plants to be used in the photosynthetic process.

The skeletons of hermatypic coral play a major role in the formation of coral reefs, but contributions to reef structure, in the form of calcium carbonate, come from a variety of other oceanic species. Among these are red algae, green algae, foraminifers, mollusk shells, sea urchins, and the exoskeletons of many other reef-dwelling invertebrates. This limestone infrastructure provides the stability needed, not only to support and protect the delicate tissues of the coral polyps themselves, but also to withstand the constant wave action generated in the shallow, near-shore waters of the marine ecosystem.

There are essentially three types of coral reefs. These categories are fringing reefs, barrier reefs, and atolls. Fringing reefs form borders along the shoreline. Some of the reefs found in the Hawaiian Islands are fringing reefs. Barrier reefs also parallel the shoreline but are found further offshore and are separated from the coast by a lagoon. The best example of this type of reef is the Great Barrier Reef off the coast of Australia. Because the coral colonies form an interwoven network of organisms from one end of the reef to the other, this is the largest individual biological feature on earth. The Great Barrier Reef borders about 1,250 mi (2011 km) of Australia's northeast coast. The second largest continuous barrier reef is located in the Caribbean Sea off the coast of Belize, east of the Yucatan Peninsula. The third type of reef, the atoll, is typically a ring-shaped reef, from which several small, low islands may project above the surface of the ocean. The ring structure is present because it represents the remains of a fringing reef that formed around an oceanic volcano. As the volcano eroded or collapsed, the outwardly-growing reef is ultimately all that remains as a circle of coral. Possibly the most infamous atoll is the Bikini Atoll, which was the site of the United States' hydrogen bomb tests during the 1940s and 1950s.

Besides the physical structure of the coral and the reef itself, the most significant thing about these structures is the tremendous diversity of marine life that exists in, on, and around coral reefs. These highly productive marine ecosystems may contain over 3,000 species of fish, shellfish, and other invertebrates. About 33 percent of all of the fishes of the world live and depend on coral reefs. This tremendous diversity provides for a huge commercial fishery in countries such as the Philippines and Indonesia. With the advent and availability of SCUBA gear to the general public in this half of this century, the diversity of life exhibited on coral reefs has been a great lure for tourists to these ecosystems throughout the world.

Even with their calcium carbonate skeleton and exquisite beauty, coral reefs are being degraded and destroyed daily, not only by natural events such as constant wave action and storm surges, but, more importantly, by the actions of man. As of 2010, of the 109 countries that have coral reef formations within their territorial waters, more than 98 have reported loss of corals over the last decade. Corals and their plant symbionts live within a narrow range of physical and chemical conditions that allow coral reefs to thrive. Recent studies have shown that coral reefs around the world are increasingly stressed or on the verge of collapse due to rising marine acidity levels, rising sea temperatures, and over-fishing.

There have been several attempts to save and salvage threatened coral reefs. In 1995, participants from 44 countries representing governments, nongovernmental organizations, international development agencies, and the private sector gathered to launch the International Coral Reef Initiative. In 1997, some 1,400 participants declared the year as the International Year of the Reef, a period they hoped would heighten awareness and further reef salvage activities worldwide.

Many coral reefs are between 5,000 and 10,000 years old, and some have been building on the same site for over a million years. However, recent environmental pressures, including water pollutants from industry continue to degrade and destroy coral. Silt, which washes into the sea from erosion of clearcut forests miles inland, cloud the water or smother the coral, thus prohibiting the photosynthetic process from taking place. Oil spills and other toxic or hazardous chemicals that find their way into the marine ecosystem through man's actions are killing off the coral and/or the organisms associated with the reefs. Mining of coral for building materials takes a massive toll on these communities. Removal of coral to supply the ever increasing demand within the aquarium trade is destroying the reefs as well. The tremendous interest in and appeal of marine aquaria has added another problem to this dilemma. In the race to provide the aquarium market with a great variety of beautiful, brilliantly-colored, and often quite rare, marine fishes, unscrupulous collectors, who are selling their catches illegally merely for the short term monetary gain, spray the coral heads with poison solutions (including cyanide) to stun the fishes, causing them to abandon the reefs. This efficient means of collecting reef fishes leaves the coral head enveloped in a cloud of poison, which ultimately kills that entire section of the reef.

An unusual phenomenon has developed within the past decade with regard to coral reefs and pollution. In the Florida Keys in the early 1980s divers began reporting that the coral, sea whips, sea fans, and sponges of the reefs, around which they had been swimming, had turned white. They also reported that the waters felt unusually warm. The same phenomenon occurred in the Virgin Islands in the late 1980s. As much as 50 percent of the reef was dying due to this bleaching effect. Scientists are still studying these occurrences; however, many argue that it is a manifestation of global warming, and that rising sea temperatures induce bleaching.

Tourism and recreation are inadvertently degrading coral reefs throughout the world as well. Coral is being destroyed by the propellers of recreational boats as well as divers who unintentionally step on coral heads, thus breaking them to pieces, and degrading the very structure of the ecosystem they came to see. Many of the reefs undergoing this degradation are sections that have been set aside for protection. Even with almost a quarter of a million square miles of coral reefs in the world, and about 300 protected regions in 65 countries, ever increasing levels of near-shore pollution, coupled with other acts of man, are destroying these extremely complex communities of marine organisms.

In February 2010, a report issued from a United Nations-backed conference on biodiversity reported that coral reefs around the world are increasingly stressed or on the verge of collapse due to rising marine acidity levels, rising sea temperatures, and over-fishing. Higher acidity levels are corrosive to aragonite, a foundational material of coral reef structure. In addition to environmental and ecosystem loss, the destruction of coral poses a serious threat to local and regional economies in areas of the world that depend on tourism.

RESOURCES

BOOKS

Aronson, R. B., ed. *Geological Approaches to Coral Reef Ecology*. Berlin: Springer, 2006.

Cote, I. M., and J. D. Reynold, eds. *Coral Reef Conservation*. Cambridge: Cambridge University Press, 2006.

PERIODICALS

Dean, Cornelia. "Coral Is Dying. Can It Be Reborn?" *New York Times* (May 1, 2007): F:1.

Dean, Cornelia. "Coral Reefs and What Ruins Them." *New York Times* (February 26, 2008).

Fountain, Henry. "More Acidic Ocean Hurts Reef Algae as Well as Corals." *New York Times* (January 8, 2008).

OTHER

National Geographic Society. "Coral Reefs: Canaries of the Sea ." http://environment.nationalgeographic.com/environment/habitats/canaries-of-sea.html (accessed September 4, 2010).

U.S. Government; science.gov. "Coral Reefs." http://www.science.gov/browse/w_131B2.htm (accessed September 4, 2010).

United Nations Environment Programme (UNEP). "Joint Secretariat of the International Coral Reef Initiative (ICRI)- Rotating Secretariat." http://www.icriforum.org/ (accessed September 4, 2010).

United Nations System-Wide EarthWatch. "Coral reefs under pressure." http://earthwatch.unep.net/emerging issues/oceans/coralreefs.php (accessed September 4, 2010).

United Nations System-Wide EarthWatch. "Widespread Coral Bleaching." http://earthwatch.unep.net/emerging issues/oceans/coralbleaching.php (accessed September 4, 2010).

United States Environmental Protection Agency (EPA). "Ecosystems: Aquatic Ecosystems: Coral Reefs." http://www.epa.gov/ebtpages/ecosaquaticecosystcoral reefs.html (accessed September 4, 2010).

University of California Museum of Paleontology. "Marine Ecosystems: Corals and coral reefs." http://www.ucmp. berkeley.edu/cnidaria/cnidarialh.html (accessed September 4, 2010).

Eugene C. Beckham

Corporate Average Fuel Economy standards

In 1975, as part of the response to the Arab oil embargo of 1973–1974, the U.S. Congress and President Richard Nixon passed the Energy Policy and Conservation Act. One of a number of provisions in this legislation intended to decrease fuel consumption was the establishment of fuel economy standards for passenger cars and light trucks sold in the United States. These Corporate Average Fuel Economy (CAFE) standards require that each manufacturer's entire production of cars or trucks sold in the United States meet a minimum average fuel economy level. Domestic and import cars and trucks are segregated into separate fleets, and each must meet the standards individually. Any manufacturer whose fleet(s) fails to meet the standard is subject to a fine, based on a CAFE shortfall times the total number of vehicles in the particular fleet. Manufacturers are allowed to generate CAFE credits for overachievement in a single year, which may be carried forward or backward up to three years to cover other shortfalls in mileage. This helps to smooth the effects of model introduction cycles or market shifts.

CAFE standards took effect in 1978 for passenger cars and 1979 for light trucks. Truck standards originally covered vehicles under 6,000 pounds (2,724 kg) gross vehicle weight (GVW), but in 1980, they were expanded to trucks up to 8,500 pounds (3,632 kg). Car standards were set at 18 miles per gallon (mpg) for 1978, and increased annually to 27.5 mpg for 1985 and beyond. Manufacturers responded with significant vehicle downsizing, application of new powertrain technology, and reductions in aerodynamic drag and tire friction. These actions helped improve average fuel economy of U.S.-built passenger cars from 13 mpg in 1974 to 25 mpg by 1982. The fall of gasoline prices following an embargo in 1980–1981 encouraged consumers to purchase larger vehicles. This mix shift slowed manufacturers' CAFE improvement by the mid–1980s, despite continued improvements in individual model fuel economy, since CAFE is based on all vehicles sold. The Secretary of Transportation has the authority to reduce the car CAFE standards from 27.5 mpg to 26 mpg. This authority was used in 1986 when it became clear that market factors would prevent CAFE compliance by a large number of manufacturers.

A separate requirement to encourage new car fuel efficiency was created through the Energy Tax Act of 1978. This act created a "Gas Guzzler" tax on passenger cars whose individual fuel economy value fell below a certain threshold, starting in 1980. The tax is progressively higher at lower fuel economy levels. This act has since been amended both to increase the threshold level and to double the tax. The Gas Guzzler tax is independent of CAFE standards.

Environmental groups have long favored increasing the CAFE standards to reduce oil consumption and pollution. At the same time, the automobile industry has fought hard to avoid raising mileage standards, claiming increased regulation would reduce competitiveness and cost jobs. After the Persian Gulf War in 1991, some members of Congress proposed raising CAFE standards to 45 mpg for cars and 35 mpg for light trucks by 2001. Proponents of the plan claimed it would save 2.8 million barrels of oil per day, or more oil than is imported from Persian Gulf countries (about 2.4 million barrels per day in 2000). Strongly opposed by the automobile industry, the efficiency increase failed in Congress.

Cheap oil prices in the 1990s decreased Americans' concerns about fuel efficiency, and sales of larger cars boomed. Since 1988, nearly all technology gains in automobile efficiency have been offset by increased weight and power in new vehicles. The popular Sport Utility Vehicle (SUV), decried by environmentalists, became the new model of gas guzzler. Due to the "SUV loophole" in the CAFE laws, the SUV was classified as a light truck and was exempted from higher mileage standards and fines for inefficiency, although opponents of the vehicles claim that they are used more like cars and should be classified as such. By law, SUVs can emit up to five times more nitrogen oxides than cars, and have a lower fleet mileage standard of 20.7 mpg. By 2001, SUVs and light trucks accounted for nearly half of all new automobiles sold. Because of the high sales of large vehicles, United States automobile efficiency reached its lowest point in 21 years in 2001, despite new technology.

Another battle between environmentalists and the automobile industry occurred in March 2002, when Congress considered raising CAFE standards by 50 percent by the year 2015. Environmental and other groups claimed that increased efficiency would reduce America's dependence on foreign oil, would reduce pollution and greenhouse gases in the environment, and eliminate the need to drill for oil in sensitive regions such as the Arctic National Wildlife Refuge (ANWR) and offshore areas. The bill was strongly opposed by the automobile industry, and did not pass.

In 2007, light trucks and SUVs lost their CAFE exemption and CAFE standards set a new goal of 15 kilometers per liter (35 miles per gallon) by 2020.

In 2009, CAFE standards increased the goal of fuel efficiency of automobiles sold in the United States to 15.09 kilometers per liter (6.62 liters per 100 kilometers, or 35.5 miles per gallon) and accelerated the 2007 deadline to 2016.

Resources

PERIODICALS

Clayton, Mark. "States are Closer to Trimming Autos' CO2 Emissions." *Christian Science Monitor* September 14, 2007.
Marris, Emma. "Car Emissions Are EPA's Problem." *Nature* 446 (April 5, 2007): 589.
Revkin, Andrew C. "US Predicting Steady Increase for Emissions." *The New York Times* (March 3, 2007).
Schiermeier, Quirin. "China Struggles to Square Growth and Emissions." *Nature* 446 (2007): 954–956.

OTHER

United States Environmental Protection Agency (EPA). "Air: Mobile Sources: Vehicle Emissions." http://www.epa.gov/ebtpages/airmobilesourcesvehicleemissions.html (accessed September 4, 2010).

Brian Geraghty
Douglas Dupler

Corrosion and material degradation

Corrosion or degradation involves deterioration of material when exposed to an environment resulting in the loss of that material, the most common case being the corrosion of metals and steel by water. The changes brought about by corrosion include weight loss or gain, material loss, or changes in physical and mechanical properties.

Metal corrosion involves oxidation-reduction reactions in which the metal is lost by dissolution at the anode (oxidation). The electrons travel to the cathode where the reduction takes place, while positive ions move through a conducting solution or electrolyte to form positive and negative poles, called the cathode and the anode, respectively. Current flows between the cathode and anode. Thus the process of corrosion is basically electrochemical.

For corrosion to occur, certain conditions must be present. These are: (1) a potential difference between the cathode and the anode to drive the reaction; (2) an anodic reaction; (3) an equivalent cathodic reaction; (4) an electrolyte for the internal circuit; (5) an external circuit where electrons can travel. Sometimes, polarization of the anodic and the cathodic reactions must be taken into consideration. Polarization is a change in equilibrium electromagnetic field of a cell due to current flow. It has been reported that polarization may retard corrosion, as in the accumulation of unreacted hydrogen on the cathode.

In the corrosion of iron in water, the reactions differ according to whether or not oxygen is present. The common reactions that take place in a deaerated medium are essentially an oxidation reaction releasing ferrous ion into solution at the anode and a reduction reaction releasing hydrogen gas at the cathode. In the presence of oxygen, a complementary cathode reaction involves oxygen being reduced to water.

Degradation of concrete, on the other hand, depends on the composition of cement and the aggressive action of the water in contact with it. Some forms of corrosion may be visibly apparent, but some are not. Surface corrosion, corrosion at discrete areas, and anodic attack in a two-metal corrosion may be readily observed. A less identifiable form, erosion-corrosion, is caused by flow patterns that cause abrasion and wear or sweep away protective films and accelerate corrosion. Another form of corrosion which involves the selective removal of an alloy constituent requires another means of examination. Cracking, a form of corrosion which is caused by the simultaneous effects of tensile stress and a specific corrosive medium, could be verified by microscopy.

Some measures adopted to prevent corrosion in metals are cathodic protection, use of inhibitors, coating, and the formation of a passivating film. Protection of concrete, on the other hand, can be achieved by coating, avoiding corrosive pH of the water with which the concrete is in contact, avoiding excessive concentrations of ammonia, and avoiding deaeration in pipes.

See also Hazardous waste site remediation; Seabed disposal; Waste management.

Resources

BOOKS

Ahmad, Zaki. *Principles of Corrosion Engineering and Corrosion Control.* Boston, MA: Elsevier/BH, 2006.

James W. Patterson

▌Cost-benefit analysis

Environmentalists might believe that total elimination of risk that comes with pollution and other forms of environmental degradation is possible and even desirable, but economists argue that the benefits of risk elimination have to be balanced against the costs. Measuring risk is itself very complicated. Risk analysis in the case of pollution, for instance, involves determining the conditions of exposure, the adverse effects, the levels of exposure, the level of the effects, and the overall contamination. Long latency periods, the need to draw implications from laboratory studies of animal species, and the impact of background contamination complicate these efforts. Under these conditions, simple cause and effect statements are out of the question.

The most that can be said in health risk assessment is that exposure to a particular pollutant *is likely* to cause a particular disease. Risk has to be stated in terms of probabilities, not certainties, and has to be distinguished from safety, which is a societal judgment about how much risk society is willing to bear. When assessing the feasibility of technological systems, different types of risks—from mining, radiation, industrial accidents, or climate impacts, for example—have to be compared. This type of comparison further complicates the judgments that have to be made.

Reducing risk involves asking to what extent the proposed methods of reduction are likely to be effective, and how much these proposed methods will cost. In theory, decision-making could be left to the individual. Society could provide people with information and each person could then decide whether to purchase a product or service, depending upon the environmental and resource consequences. However, relying upon individual judgments in the market may not adequately reflect society's preference for an amenity such as air quality, if that amenity is a public good with no owner and no price attached to it. Thus, social and political judgments are needed.

However much science reduces uncertainty, gaps in knowledge remain. Scientific limitations open the door for political and bureaucratic biases that may not be rational. In some instances, politicians have framed legislation in ways that seriously hinder if not entirely prohibit the consideration of costs (as in the Delaney Clause and the Clean Air Act). In other instances, of which the President's Regulatory Review Council is a good example, they have explicitly required a consideration of cost factors.

There are different ways that cost factors can be considered. Analysts can carry out cost effectiveness analyses, in which they attempt to figure out how to achieve a given goal with limited resources, or they

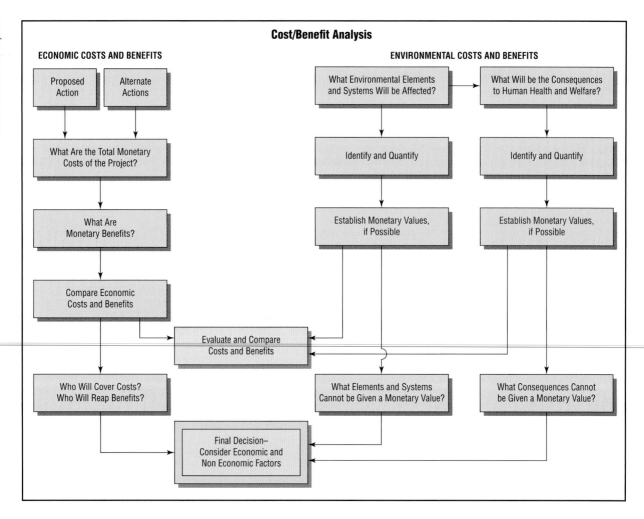

Cost/Benefit Analysis

ECONOMIC COSTS AND BENEFITS

Proposed Action

Alternate Actions

What Are the Total Monetary Costs of the Project?

What Are Monetary Benefits?

Compare Economic Costs and Benefits

Who Will Cover Costs? Who Will Reap Benefits?

Final Decision– Consider Economic and Non Economic Factors

Evaluate and Compare Costs and Benefits

ENVIRONMENTAL COSTS AND BENEFITS

What Environmental Elements and Systems Will be Affected?

What Will be the Consequences to Human Health and Welfare?

Identify and Quantify

Identify and Quantify

Establish Monetary Values, if Possible

Establish Monetary Values, if Possible

What Elements and Systems Cannot be Given a Monetary Value?

What Consequences Cannot be Given a Monetary Value?

A cost-benefit analysis. *(Reproduced by permission of Gale, a part of Cengage Learning.)*

can carry out more formal risk-benefit and cost-benefit analyses in which they have to quantify both the benefits of risk reduction and the costs.

According to the Organization for Economic Cooperation and Development (OECD), the total economic value (TEV) is an "all-encompassing measure of the economic value of any environmental asset." Thus, a project or action that may degrade or damage an environmental asset must take into account the TEV of the asset as a cost. Conversely, any action that benefits an environmental asset must be reflected in the TEV, resulting in a recalculation of the TEV for that asset. Ecosystem valuation is a complicated matter due to the complex network of interactions that encompass a particular ecosystem. Services and products provided by an ecosystem, such as pharmaceutical resources, have to be accounted for in determining the value of an ecosystem and must be acknowledged when calculating the altered TEV in response to any action or policy that may result in a change in the ecosystem.

Economists admit that formal, quantitative approaches to balancing costs and benefits do not eliminate the need for qualitative judgments. Cost-benefit analysis initially was developed for water projects where the issues, while complicated, were not of the same kind as society now faces. For example, it is difficult to assess the value of a magnificent vista obscured by air pollution, to determine the significance of the loss to society if a given genetic strain of grass or animal species becomes extinct, and to assess the lost opportunity costs of spending vast amounts of money on air pollution that could have been spent on productivity enhancement and global competitiveness.

The most recalcitrant question concerns the value of human life. Cost-benefit analysis requires quantifying the value of a human life in dollars, so that specific health risks can be entered into the calculations against the cost of reducing such risks. Many different methods of arriving at an appropriate figure have been undertaken, all with wide-ranging and highly disputed results. Although the question does not lend itself to a logical answer, society must nevertheless decide how much it is willing to pay to save a given number of lives (or how much specific polluters should pay for endangering them).

Equity issues (both interpersonal and intergenerational) cannot be ignored when carrying out cost-benefit analysis. The costs of air pollution reduction may have to be borne disproportionately by the poor in the form of higher gasoline and automobile prices. The costs of water pollution reduction, on the other hand, may be borne to a greater extent by the rich because these costs are financed through public spending. Regions dependent on dirty coal may find it in their interests to unite with environmentalists in seeking pollution control technology. The pollution control technology saves coal mining jobs in West Virginia and the Midwest, where the coal is dirty, but draws away resources from the coal mining industry in the West where large quantities of clean-burning coal are located.

Intergenerational equity also plays a role. Future generations have no current representatives in the market system or political process. The interests concerning future generations ultimately amounts to a philosophical discussion about altruism. Current generations must decide to what extent they should hold back on their own consumption for the sake of posterity. British philosopher Jeremy Bentham's (1748–1832) utilitarian ideal of "achieving the greatest good for the greatest number" could be modified to read "achieving sufficient per capita product for the greatest number over time." Cost-benefit analysis arguments can also invoke the Precautionary Principle which has become a key element for policy decisions concerning environmental protection and management. It may be applied to situations where there are reasonable grounds for concern that an activity could cause harm, but where there is uncertainty about the probability of the risk and the degree of harm.

See also Environmental economics; Intergenerational justice; Pollution control; Risk analysis.

Resources

BOOKS

Baer, Paul, and Clive L. Spash. *Cost-Benefit Analysis of Climate Change: Stern Revisited.* Canberra, Australia: CSIRO Sustainable Ecosystems, 2008.

Pearce, David W., Giles Atkinson, and Susana Mourato. *Cost-Benefit Analysis and the Environment: Recent Developments.* Paris: OECD, 2006.

Revesz, Richard L., and Michael A. Livermore. *Retaking Rationality: How Cost-Benefit Analysis Can Better Protect the Environment and Our Health.* Oxford: Oxford University Press, 2008.

Tisdell, C. A. *Economics of Environmental Conservation.* Cheltenham, UK: Edward Elgar Publishing, 2005.

OTHER

Organization for Economic Co-operation and Development. "Cost-Benefit Analysis and the Environment: Recent Developments Executive Summary." http://www.oecd.org/dataoecd/37/53/36190261.pdf (accessed September 8, 2010).

Alfred A. Marcus

Costle, Douglas M.

1939–
American former director of Environmental Protection Agency

An educator and an administrator, Douglas M. Costle helped design the Environmental Protection Agency (EPA) under Richard Nixon and was appointed to the head of that agency by Jimmy Carter. Costle was born in Long Beach California on July 27, 1939, and spent most of his teenage years in Seattle. He received his B.A. from Harvard in 1961 and his law degree from the University of Chicago in 1964. His career as a trial attorney in the civil rights division of the Justice Department began in Washington in 1965. Later he became a staff attorney for the Economic Development Administration at the Department of Commerce.

In 1969, Costle was appointed to the position of senior staff associate of the President's Advisory Council on Executive Organization, and in this post was instrumental in the formation of the EPA. Although Costle lobbied to be appointed as assistant administrator of the new agency, his strong affiliations with the Democratic party seem to have hindered his bid. Instead he continued as a consultant to the agency for two years and an adviser to the President's Council on Environmental Quality.

The road that led Costle back to the EPA took him to Connecticut in 1972, where he became first deputy commissioner and then commissioner of the Department of Environmental Protection in that state. He proved himself an able and efficient administrator there, admired by many for his ability to work with industry on behalf of the environment. His most important accomplishment was the development of a structure often called the Connecticut Plan, where fines for industrial pollution were calculated on the basis of the costs that business would have incurred if it had complied with environmental regulations.

President Carter appointed Costle head of EPA in 1977 as a compromise candidate during a period of bitter feuding over the direction of the agency (then at the center of a debate over the economic effects of regulation). But many environmentalists believed Costle's record proved he compromised too willingly with business, and they openly questioned his political strength to support environment protection in the face of fierce political and industrial opposition.

By May of his first year in office he was able to secure funding for 600 additional staff positions in the EPA, and under him much was done to provide a rationale for the regulations he had inherited and base them wherever possible on scientific data. Among other decisions Costle made while head of the EPA, he recommended a delay on the imposition of new auto emissions standards, allowed the construction of the nuclear plant in Seabrook, New Hampshire to continue despite protests, and oversaw the formation of an agreement with U.S. Steel on the reduction of air and water pollution.

Throughout his tenure Costle remained a strong proponent of the view that the federal government's responsibility to the environment was not incompatible with the obligations it had to the economy. He often argued that environmental regulation actually assisted economic development. Although conflicts with lobbying groups and hostile litigation, as well as increased controversy over the inflationary effects of environmental regulation, complicated his stewardship of the EPA, Costle continued to believe in what he termed a gradual and quiet victory for environmental protection.

Costle went on to become chairman of the U.S. Federal Regulatory Council until 1981, and was dean of the Vermont Law School from 1987 until his retirement in 1991. In 1994 he unsuccessfully ran for the U.S. Senate for the state of Vermont. He has since retired from public life.

Douglas Smith

Council on Environmental Quality

Until it was abolished by the Clinton Administration in 1993, the President's Council on Environmental Quality (CEQ) was the White House Office that advised the President and coordinated executive branch policy on the environment. The CEQ was established by the National Environmental Policy Act (NEPA) of 1969 to "formulate and recommend national policies" to promote the improvement of the quality of the natural environment. The Environmental Quality Improvement Act of 1970 amended the NEPA with additional provisions.

The CEQ had three basic responsibilities: to serve as advisor to the President on environmental policy matters; to coordinate the positions of the various departments and agencies of government on environmental issues; and to carry out the provisions of the NEPA. The latter responsibility included working with federal agencies on complying with the law and issuing the required regulations for assessing the environmental impacts of federal actions. The NEPA requires that all agencies of the federal government issue "a detailed statement" on "the environmental impact" of "proposals for legislation and other major federal actions significantly affecting the quality of the human environment." This seemingly innocuous provision has been used often by environmental groups to legally challenge federal projects that might damage the environment, on the grounds that the required Environmental Impact Statement was inadequate or had not been issued.

The CEQ also prepared and issued the annual Environmental Quality Report; administered the President's Commission on Environmental Quality, an advisory panel involved in voluntary initiatives to protect the environment; and supervised the President's Environment and Conservation Challenge Awards, which honored individuals and organizations who achieved significant environmental accomplishments. Under the Nixon and Carter administrations, the CEQ had a significant impact on the formulation and implementation of environmental policy. But its role was greatly diminished under the Reagan and Bush administrations, which paid much less attention to environmental considerations.

Perhaps the CEQ's best-known and most influential accomplishment was its landmark work, *The Global 2000 Report to the President*, prepared with the U.S. Department of State and other federal agencies, released in July 1980. This pioneering study was the first report by the U.S. government—or any other—projecting long term environmental, population, and resource trends in an integrated way.

Specifically, the report projected that the world of the future would not be a pleasant place to live for much of humanity, predicting that "if present trends continue, the world in 2000 will be more crowded, more polluted, less stable ecologically and more vulnerable to disruption than the world we live in now. Serious stresses involving population, resources, and environment are clearly visible ahead. The world's people will be poorer in many ways than they are today." The CEQ's *Eleventh Annual Report, Environmental Quality—1980*, further warned that "we can no longer assume, as we could in the past, that the earth will heal and renew itself indefinitely. Human numbers and human works are catching up with the earth's ability to recover. The quality of human existence in the future will rest on careful stewardship and husbandry of the earth's resources."

In the following dozen years, the CEQ was much more reluctant to speak out about the ecological crisis. Early in his presidency, Bill Clinton (1946–) abolished the CEQ and created the White House Office on Environmental Policy to coordinate the provisions of NEPA, environmental policy, and actions of his administration. President Clinton said this new body would "have broader influence and a more effective and focused mandate to coordinate policy" than had the former CEQ.

The White House Office on Environmental Policy is still also referred to as the CEQ.

In 2002 and 2003, CEQ staff member Phillip Cooney (1959–), who was previously a lobbyist for the American Petroleum Institute, altered information regarding climate change to downplay the relationship between greenhouse gas emissions and global warming. These edits were made to introduce uncertainty concerning climate change and its effects and indicated that further research was necessary. Cooney subsequently resigned from the CEQ and accepted a position with Exxon Mobil, an international oil and gas company.

Under the Obama administration, the CEQ is focusing on the development of clean energy such as alternative or renewable energy sources to combat climate change and to protect and restore the environment.

See also Environmental policy.

Resources

PERIODICALS

Lusetich, Robert. "Climate Science was Doctored." *The Australian.* March 21, 2007.

OTHER

U.S. Government; science.gov. "Environment and Environmental Quality." http://www.science.gov/browse/w_123.htm (accessed October 17, 2010).

White House. "The Advanced Energy Initiative." http://www.whitehouse.gov/infocus/energy/ (accessed October 17, 2010).

White House. "White House Climate Change Policy." http://www.whitehouse.gov/ceq/global-change.html (accessed October 17, 2010).

ORGANIZATIONS

Council on Environmental Quality, 722 Jackson Place, NW, Washington, D.C., USA, 20503, (202) 395-5750, (202) 456-6546, http://www.whitehouse.gov/ceq

Lewis G. Regenstein

Cousteau, Jacques-Yves

1910–1997

French oceanographer, inventor, photographer, explorer, and environmentalist

When most people think of marine biology, the person that immediately comes to mind is Jacques-Yves Cousteau. Whether through invention, research, conservation, or education, Cousteau has brought the ocean world closer to scientists and the public, and it is the interest, awareness, and appreciation fostered through

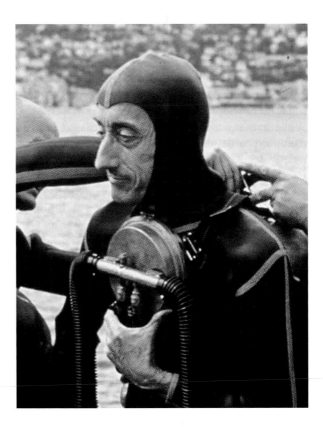

Oceanographer Jacques-Yves Cousteau adjusts his diving gear in 1965 in Paris France. *(AFP/Getty Images)*

the contributions of rivers to ocean vitality, assessing the health of marine and freshwater habitats, and exploring the global connections between major components of the biosphere such as tropical forests, rivers, the atmosphere, seas, oceans, and humankind.

The Society has also developed educational computer programs for young people that explore the consequences of various actions on the environment. It also supports the development of new technologies that will help provide solutions to environmental challenges.

Resources

BOOKS

Dinwiddie, Robert, Louise Thomas, and Fabien Cousteau. *Ocean*. New York: DK Adult, 2006.
Petrie, Kristin. *Jacques Cousteau*. Edina, MN: Abdo Publishing, 2004.

ORGANIZATIONS

The Cousteau Society, 732 Eden Way North, Suite E , Chesapeake, VA, USA, 23320, (757) 523-9335 , cousteau@cousteausociety.org, http://www. cousteausociety.org

Linda Rehkopf

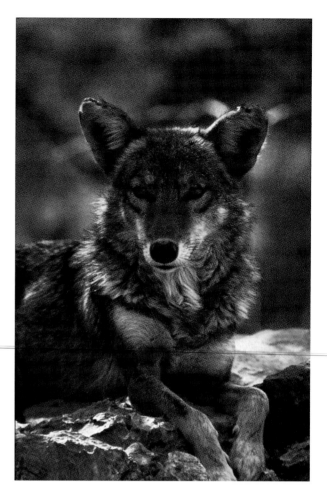

A coyote. (© U. S. Fish & Wildlife Service.)

Coyote

The name coyote comes from the Aztec name *Coyoti*. The Latin name for the coyote is *Canis latrans* which means barking dog. A member of the family Canidae, the coyote is also called God's Dog, brush wolf, prairie wolf, and the Songdog.

Coyotes are especially important to biologists and others studying wildlife and the environment both because of their ability to control their own population and because of their effect upon other wildlife populations, especially that of the white-tailed deer.

The coyote is roughly the size of a small German Shepherd, weighing from 20–35 pounds (9–14 kg) and 4–5 feet (1.2–1.5 m) long, although some can grow as large as 50 pounds (23 kg) in the northern and northeastern part of their range. The coat of a coyote can be gray or brownish gray, depending on the season and habitat, with a bushy tail that is tipped with black. Coyote fur is long and soft and grows heavier in the winter to protect it from the cold. Its fur also grows lighter in color in the winter and darker in the summer to help it blend in with its natural surroundings.

The early European settlers originally found the coyote only in Central and western North America. However, despite many attempts by humans to eliminate them, coyotes today range from Panama to Northern Alaska, and from California to Newfoundland. They have been seen in every state in the continental United States, but as they are predominately nocturnal, are too elusive to count accurately. Coyotes have even been found in isolated regions such as Cape Cod, as well as in various urban areas such as Chicago, Los Angeles, and New York City.

This ongoing expansion of the coyote's range began in the late 1800s, soon after the United States government began programs of killing them in order to protect domestic livestock, especially the sheep that were grazing on public lands. However, unlike the overhunting of other species, efforts to kill off the coyotes have generally failed, and coyotes have continued to thrive.

Several factors account for this expansion. Farming and the clearing of the unbroken forests that once covered much of North America have created new and more suitable habitats. The eradication of the wolf in much of North America has removed one of the coyote's major predators and has led to less competition for many prey animals. Unlike the wolf, the coyote has more easily adapted to changing habitats, and has adjusted to living close to human populations. In addition, the introduction of sheep and other domestic animals has given the coyote new food sources.

Coyotes are opportunistic carnivores with a variable diet depending on the season or habitat. Although they prefer rodents, rabbits, and hares, they are definitely not choosy. They will emulate the larger wolf and hunt deer, elk, and sheep in packs. They will also eat carrion and have been known to eat grasshoppers, beetles, snakes, lizards, frogs, rats, domestic cats, porcupines, turtles, and even watermelons and wild blueberries.

Coyotes are one of the most vocal animals of the wild. Through a series of howls, yelps and barks, coyotes communicate to other coyotes in the area, with others in their pack, and with their young.

Studies of the coyote's breeding patterns have shown that they are able to control their own population. To do this, the coyote relies upon various breeding strategies. If there are too many coyotes for the food supply, they will continue to mate and have young, but only enough pups will be born and will survive to replace those that have died. However, if they are not being overhunted and if there is plenty of food, there will be more pups born in each litter and the number that will survive into adulthood will increase.

Because of this innate ability to vary their own populations, coyotes have thrived and plaed a central role in maintaining the overall ecological balance. Extensive studies of the coyote have shown the crucial interaction between them and many other species, both predator and prey. When the jackrabbit population decreases, as it does periodically in the Curlew Valley of Idaho and Utah for example, the coyote population decreases also, usually within the following year. Then, when the jackrabbit population increases again, coyotes become more numerous also.

Coyotes mate for life and the adults usually breed between January and March. The female maintains the den that she selects from an old badger or woodchuck hole, or a natural cavity in the ground. The female carries her young for over two months, giving birth in April or May. A litter may have from two to twelve pups.

Both parents play an active role in raising the pups. At three weeks old, the pups are allowed to leave the den to play and by twelve weeks old they are taught to hunt. The family stays together through the summer, but by fall, the pups will leave to find their own territories. The survival rate for these pups is low. Between 50 and 70 percent die before they reach adulthood. Eighty percent of those deaths are due to man.

Coyotes limit the populations of various smaller predators, by preying upon them directly or by competing with them for prey. This in turn indirectly affects various prey communities, especially those of birds. When there are no coyotes to keep them in check, the smaller predators expand and are able to kill off a larger segment of the various bird populations.

The coyote and the deer populations interact in an important way as well. Whenever the coyote population decreases, the deer population, particularly that of the white-tailed deer, increases, leading to unhealthy herds and, according to some biologists, to an increase in Lyme disease-bearing ticks in some regions. Coyotes are thus a major factor in maintaining healthy deer populations and in preventing the spread of a disease that is dangerous and potentially fatal to humans.

Coyotes affect humans more directly as well, especially through their impact upon domestic animals. Since Europeans first came to North America, coyotes have preyed upon sheep and other livestock, costing sheep growers, farmers, and ranchers untold millions of dollars. In some areas, livestock may make up 14 percent of a coyote's diet. This has prompted extensive hunting to control the number of coyotes that ranchers and farmers view as pests. Some ranchers keep guard dogs to protect their livestock. New technology, using inaudible (to humans) ultrasound, also allows ranchers to detect or repel coyotes.

Humans cause 90 percent of all adult coyote deaths by hunting, trapping, poisoning, and automobile accidents.

Resources

BOOKS

DeStefano, Stephen. *Coyote at the Kitchen Door: Living with Wildlife in Suburbia.* Cambridge, Mass: Harvard University Press, 2010.

Dobie, J. F. *The Voice of the Coyote.* Lincoln: University of Nebraska Press, 2006.

Martin, Ruth. *The Coyote.* McAllen, TX: Rio Grande Valley Nature Coalition, 2005.

Reid, Catherine. *Coyote: Seeking the Hunter in Our Midst.* Boston: Houghton Mifflin, 2004.

OTHER

Humane Society of the United States. "Coyote."http://www.humanesociety.org/animals/coyotes/ (accessed November 9, 2010).

National Geographic. "Coyote." http://animals.national geographic.com/animals/mammals/coyote.html (accessed November 9, 2010).

Douglas Dupler

Crane (bird) *see* **Whooping crane.**

Creutzfeldt jacob disease *see* **Mad cow disease.**

systems as well as kidney function, reproduction, and development.

See also Air Quality Control Region; Air quality criteria; Primary standards; Secondary standards.

Resources

OTHER

United States Environmental Protection Agency (EPA). "Air Trends." http://www.epa.gov/airtrends/ (accessed October 2, 2010).

United States Environmental Protection Agency (EPA). "Six Common Air Pollutants." http://www.epa.gov/air/urbanair/ (accessed October 2, 2010).

United States Environmental Protection Agency (EPA). "Technology Transfer Network National Ambient Air Quality Standards (NAAQS)." http://www.epa.gov/ttn/naaqs/ (accessed October 2, 2010).

Criteria pollutant

The U.S. Environmental Protection Agency (EPA) is required by the Clean Air Act to set standards for six common air pollutants. The National Ambient Air Quality Standards (NAAQS) are established to prevent harm to public health and the environment. The six pollutants, termed criteria pollutants, are air pollutants which have been studied such that allowable levels of exposure can be defined to protect human health and meet the related EPA standards. The criteria pollutants regulated by these standards are: ground-level (tropospheric) ozone (O_3), carbon monoxide (CO), particulate matter (e.g., organic chemicals, acids, metals, soil, dust), nitrogen oxides (NO_x), sulfur dioxide (SO_2), and lead (Pb).

There are primary and secondary types of standards for the criteria pollutants. Primary NAAQS are levels of air quality with a margin of safety adequate to protect public health and reflect sensitive groups such as the elderly, asthmatics, and children. Secondary NAAQS are levels of air quality which are necessary to protect the public welfare from any known or anticipated adverse effects of a pollutant. The secondary standards are in place to prevent low visibility and to prevent damage to crops, animals, plants, and buildings.

Due to initiatives for reduction of the six criteria pollutants, the levels of each pollutant have declined to a large extent. With the exception of lead, each of the criteria pollutants have been linked to global climate change. Ozone, particulate matter, nitrogen oxides, and sulfur dioxide have been shown to adversely affect the human respiratory system, while carbon monoxide affects the nervous and cardiovascular systems. Lead affects the nervous, immune, and cardiovascular

Critical habitat

As institutionalized in the U. S. Endangered Species Act of 1973, critical habitat is considered the area necessary to the survival of a species, and, in the case of endangered and threatened species, essential to their recovery. An animal's habitat includes not only the area where it lives, but also its breeding and feeding grounds, seasonal ranges, and migration routes. Critical habitat usually refers to the area that is essential for a minimal viable population to survive and reproduce. The Endangered Species Act is intended to conserve "the ecosystems upon which endangered species and threatened species depend." Thus, the Secretary of the Interior is required to identify and designate critical habitats for species that are listed as endangered or threatened under this law. In some cases, areas may be excluded from such designations if the economic, social, or other costs exceed the conservation benefits.

The listing of imperiled species and the designation of their critical habitats have become politically sensitive, since these actions can profoundly affect the development and exploitation of areas so designated, and can, under some circumstances, limit such activities as gas and oil drilling, timber cutting, dam building, mineral exploration and mining. For this and other reasons, the Department of the Interior often has been reluctant to list certain species, and has excluded species from the protected lists in order not to inconvenience certain commercial interests.

Section 7 of the Endangered Species Act requires all federal agencies and departments to ensure that the activities they carry out, fund, or authorize do not jeopardize the continued existence of listed species or adversely modify or destroy their critical habitat. This provision has proven especially significant, since federal agencies such as the Forest Service, Bureau of Land Management, and Fish and Wildlife Service control vast areas of land that constitute habitat for many listed species and on which a variety of commercial activities, such as logging or mining, are undertaken with federal permits.

However, Section 7 of the Endangered Species Act has been implemented in such a way as to generally not affect economic development. The U. S. Fish and Wildlife Service (and, in the case of marine species, the National Marine Fisheries Service) is directed to consult with other federal agencies and review the effects of their actions on listed species. According to a study by the National Wildlife Federation, over 99% of the more than 120,000 reviews or consultations conducted between 1979 and 1991 found that no jeopardy to a listed species was involved. In some cases, "economic and technologically feasible" alternatives and modifications, in the words of the act, were suggested that allowed the federal activities to proceed. In only 34 cases were projects cancelled because of threats to listed species. In rare situations, where the conflict between a project and the Endangered Species Act are absolutely irreconcilable, an agency can apply for an exemption from a seven-member Endangered Species Committee.

The earliest major conflict over critical habitat under the Act was the famous 1979 fight over construction of the $116 million Tellico Dam in Tennessee, which would have flooded and destroyed several hundred family farms as well as what was then the only known habitat of a species of minnow, the snail darter (*Percina tanasi*). (Since then, snail darters have been found in other areas.) Congress exempted this project from the provisions of the Endangered Species Act, and the dam was built as planned, although many consider it a political boondoggle and a huge waste of taxpayers' money.

More recently, efforts by environmentalists to save the remnants of old-growth forest in the Pacific Northwest to preserve habitat for the northern spotted owl (*Strix occidentalis caurina*) created tremendous controversy. Thousands of acres of federally-owned forests in Oregon, Washington, and California were placed off-limits to logging, costing jobs in the timber industry in those states. However, conservationists pointed out, if the federal government allowed the last of the ancient forests to be logged, timber jobs would disappear anyway, along with these unique ecosystems and several species dependent upon them. In mid-1993, Interior Secretary announced a compromise decision that allows logging of some ancient forests to continue, but

also greatly decreases the areas open to this activity. As natural areas and wildlife habitat continue to be destroyed and degraded, conflicts and controversy over saving critical habitats for listed endangered and threatened species can be expected to continue.

Resources

BOOKS

Darby, Stephen E., and David Sear. *River Restoration: Managing the Uncertainty in Restoring Physical Habitat*. Chichester, England: Wiley, 2008.

Dawson, John. *The Nature of Plants: Habitats, Challenges, and Adaptations*. Portland, OR: Timber Press, 2005.

Seattle (Wash.). *Environmentally Critical Areas: Wetlands and Fish & Wildlife Habitat Conservation Areas*. Seattle, Wash: The Dept, 2009.

OTHER

United States Department of the Interior, United States Geological Survey (USGS). "Habitats." http://www.usgs.gov/science/science.php?term=525 (accessed November 9, 2010).

Lewis G. Regenstein

Crocodiles

The largest of the living reptiles, crocodiles inhabit shallow coastal bodies of water in tropical areas throughout the world, and they are often seen floating log-like in the water with only their eyes and nostrils showing. Crocodiles have long been hunted for their hides, and almost all species of crocodilians are now

American Crocodile (*Crocodylus Acutus*) sleeping on the hot sand in the St. Augustine, Florida, Alligator Farm Zoological Park. *(©vonora/Image from BigStockPhoto.com.)*

considered to be in danger of extinction. Members of the crocodile family, called crocodilians (Crocodylidae), are similar in appearance and include crocodiles, alligators, caimans, and gavials. A crocodile can usually be distinguished from an alligator by its pointed snout (an alligator's is rounded), and by the visible fourth tooth on either side of its snout that protrudes when the jaw is shut.

Crocodiles prey on fish, turtles, birds, crabs, small mammals, and any other animals they can catch, including dogs and occasional humans. They hide at the shore of rivers and water holes and grab an animal as it comes to drink, seizing a leg or muzzle, dragging the prey underwater, and holding it there until it drowns. When seizing larger animals, a crocodile will thrash and spin rapidly in the water and tear its prey to pieces. After eating its fill, a crocodile may crawl ashore to warm itself and digest its food, basking in the sun in its classic "grinning" pose, with its jaws wide open, often allowing a sandpiper or plover to pick and clean its teeth by scavenging meat and parasites from between them.

The important role that crocodiles play in the balance of nature is not fully known or appreciated, but, like all major predators, their place in the ecological chain is a crucial one. They eat many poisonous water snakes, and during times of drought, they dig water holes, thus providing water, food, and habitat for fish, birds, and other creatures. When crocodiles were eliminated from lakes and rivers in parts of Africa and Australia, many of the food fish also declined or disappeared. It is thought that this may have occurred because crocodiles feed on predatory and scavenging species of fish that are not eaten by local people, and when left unchecked, these fish multiplied out of control and crowded out or consumed many of the food fish.

Crocodiles reproduce by laying eggs and burying them in the sand or hiding them in nests concealed in vegetation. Recent studies of the Nile and American crocodiles show that some of these reptiles can be attentive parents. According to these studies, the mother crocodile carefully watches over the nest until it is time for the eggs to hatch. Then she digs the eggs out and gently removes the young from the shells. After gathering the newborns together, she puts them in her mouth and carries them to the water and releases them, watching over them for some time. American crocodiles are very shy and reclusive, and disturbance during this critical period can disrupt the reproductive process and prevent successful hatchings.

In recent decades, crocodiles and other crocodilians have been intensively hunted for their scaly hides, which are used to make shoes, belts, handbags, wallets, and other fashion products. As a result, they have disappeared or have become rare in most of their former habitats. In 2010, seven species of crocodile were designated as critically endangered; nearly all crocodiles face some degree of threat. These species are found in Africa, the Caribbean, Central and South America, the Middle East, the Philippines, Australia, some Pacific Islands, southeast Asia, the Malay Peninsula, Sri Lanka, and Iran. They are endangered primarily due to overexploitation and habitat loss.

The American crocodile (*Crocodylus acutus*) occurs all along the Caribbean coast, including the shores of Central America, Colombia, Cuba, Hispaniola, Jamaica, Mexico, extreme south Florida, and on the Pacific coast, from Peru north to southern Mexico. The United States population of the American crocodile consists of some 500–1,2000 individuals. This species breeds only in the southern part of Everglades National Park, mainly Florida Bay, and perhaps on nearby Key Largo, and at Florida Power and Light Company's Turkey Point plant, located south of Miami. The population was thought to be extremely vulnerable and declining, mainly due to human disturbance, habitat loss (from urbanization, especially real estate development), and direct killing such as on highways and in fishing nets. Predation of hatchlings in Florida Bay mainly by raccoons may also have been a factor in the species' decline. However, conservation measures have had some notable impact; the American crocodile was taken off of the endangered list in 2007.

Resources

BOOKS

Fougeirol, Luc. *Crocodiles*. New York: Abrams, 2009.

OTHER

U.S. Government; science.gov. "Reptiles and Amphibians." http://www.science.gov/browse/w_115A10.htm (accessed November 6, 2010).

Lewis G. Regenstein

Cross-Florida Barge Canal

The subject of long and acrimonious debate, this attempt to build a canal across the Florida peninsula began in the 1930s and finally deauthorized in 1991. Although it receives little attention today, the Cross-Florida Barge Canal stands as a landmark, because it was one of the early cases in which the Army Corps of Engineers, whose primary mission has traditionally been to re-design and alter natural waterways, yielded to environmental pressure. The canal's stated purpose, aside from bringing public works funding to the state, was to shorten the shipping distances from the East Coast to the Gulf of Mexico by bypassing the long water route around the tip of Florida. Rerouting barge traffic would also bring commerce into Florida, directing trade and trans-shipment operations through Floridian hands. An additional supporting argument that persisted into the 1980s was that the existing sea route brought American commerce dangerously close to threatening Cuban naval forces.

Construction on the canal began in 1964, on a route running from the St. Johns River near Jacksonville west to the Gulf of Mexico at Yankeetown, Florida. Canal project plans included three dams, five locks, and 110 miles (177 km) of channel 150 feet (46 m) wide and 12 feet (3. 6 m) deep. Twenty-five miles (40 km) of this waterway, along with three locks and three dams, were complete by 1971 when President Richard Nixon, citing economic inefficiency and unacceptable environmental risks, stopped the project by executive order.

From start to finish, the canal's proponents defended the project on economic grounds. The Cross-Florida Canal was proposed as a depression-era job development program. After completion, commerce and recreational fishing would boost the state economy. The Army Corps, well-funded and actively remodelling nature in the 1950s and 1960s, took on the project, vastly overestimating economic benefits and essentially dismissing environmental liabilities with the argument that even modest economic gain justified any habitat or water loss. After work had begun, further studies concluded that most of the canal's minimal benefits would go to non-Floridian agencies and that environmental dangers were greater than first anticipated. Outcry over environmental costs eventually led to a reappraisal of economic benefits, and the state government rallied behind efforts to halt the canal.

Environmental risks were grave. Although Florida has more wetlands than any other state except Alaska, many of the peninsula's natural wetland and riparian habitats had already been lost to development, drainage, and channelization. Along the canal route these habitats sheltered a rich community of migratory and resident birds, crustaceans, fish, and mammals. Fifteen endangered species, including the red-cockaded woodpecker (*Picoides borealis*) and the Florida manatee, stood to loose habitat to channelized rivers and barge traffic. Specialized spring-dwelling mussels and shrimp that depend on reliable and pure water supplies in this porous limestone country were also threatened.

Most serious of all dangers was that to the Floridan aquifer, located in northern Florida but delivering water to cities and wetlands far to the south. Like most of Florida, the reach between Jacksonville and Yankeetown consists of extremely porous limestone full of sinkholes, springs, and underground channels. The local water table is high, often within a few feet of the ground surface, and currents within the aquifer can carry water hundreds of meters or more in a single day. Because the canal route was to cut through 28 miles (45 km) of the Floridan aquifer's recharge zone, the area in which water enters the aquifer, any pollutants escaping from barges would disperse through the aquifer with alarming speed. Even a small fuel leak could contaminate millions of gallons of drinking-quality water. In addition, a canal would expose the aquifer to extensive urban and agricultural pollution from the surrounding region.

Water loss presented another serious worry. A channel sliced through the aquifer would allow water to drain out into the sea, instead of remaining underground. Evaporation losses from impounded lakes were expected to reach or exceed 40 million gallons of fresh water every day. With water losses at such a rate, water tables would fall, and salt water intrusions into the fresh water aquifer would be highly probable. In 1985, 95 percent of all Floridians depended on groundwater for home and industrial use. The state could ill afford the losses associated with the canal.

Florida water management districts joined environmentalists in opposing the canal. By the mid–1980s the state government, eager to reclaim idle land easements for development, sale, and extension of the Ocala National Forest, put its weight against the Corps and a few local development agencies that had been resisting deauthorization for almost twenty years. In 1990, the United States Congress voted to divide and sell the land, effectively eliminating all possibility of completing the canal. In 1991, the governor of Florida agreed to the terms of the federal deauthorization plan.

Resources

BOOKS

Noll, Steven, and David Tegeder. *Ditch of Dreams: The Cross Florida Barge Canal and the Struggle for Florida's Future*. Gainesville: University Press of Florida, 2009.

Davis, Jack E., and Raymond Arsenault. *Paradise Lost?: The Environmental History of Florida*. Gainesville: University Press of Florida, 2005.

Mary Ann Cunningham

CRP *see* **Conservation Reserve Program.**

and chemical disinfectants such as chlorine that are used in municipal drinking water systems and swimming pools (swallowing a small amount of water while swimming in a chlorinated pool can cause cryptosporidiosis). The mechanism that protects oocysts from chlorination has not yet been positively identified—the oocyst membrane may be protective, or an oocyst may pump toxins from its cell before the toxins can cause harm.

Oocysts are present in most surface bodies of water in the United States, many of which are sources of public drinking water. They become more prevalent during periods of runoff (generally from March-June during spring rains in North America) or when wastewater treatment plants become overloaded or break down. Properly drilled and maintained groundwater-wells, with intact well casings, proper seals, and above-ground caps, are not likely to contain *Cryptosporidium* because of natural filtration through soil and aquifer materials.

The detection of *Cryptosporidium* oocysts is unreliable, for recovery and enumeration of oocysts from water samples is difficult. Concentration techniques for oocysts in water samples are poor, and detection methods often measure algae and other debris in addition to oocysts. The volume of water required to concentrate oocysts for detection can range from 26–264 gal (100–1,000 l). Determination of whether oocysts are infective and viable or a member of the species that causes disease is not easy to accomplish. The development of more accurate, rapid and improved assays for oocysts is required, for present tests are time-consuming, highly subjective, and dependent on the skills of the analyst.

In addition, the number of oocysts (the effective dose) required to cause cryptosporidiosis has not yet been well- defined and requires more investigation. Studies to date have suggested that the 50 percent infectious dose may be around 132 oocysts, and in some cases, as few as 30 oocysts (infections have also occurred with the ingestion of a single oocyst). Human susceptibility to *Cryptosporidium* likely varies between individuals and between various *Cryptosporidium* strains.

Therefore protection of drinking water supplies from contamination by *Cryptosporidium* requires multiple approaches. Filtration of drinking water supplies is the only reliable conventional treatment method. Water in a treatment plant is mixed with coagulants that aid in the settling of particles in water; removal can be enhanced by using sand filtration. Ozone disinfection can kill *Cryptosporidium* but ozone does not leave a residual in the distribution system as chlorine does, which provides protection of treated water to the point of use but does not neessarily kill *Cryptosporidium* anyway.

Watershed protection to prevent contamination from entering water sources is also important in protection of drinking water supplies. Regulation of septic systems and best management practices can be used to control runoff of human and animal wastes.

An individual can also take steps to ensure that drinking water is safe. Boiling water (bringing water to a rolling boil for at least one minute) is the best way to kill *Cryptosporidium*. After boiling, the water should be stored in the refrigerator in a clean bottle or pitcher with lid; care should be taken to avoid touching the inside of the bottle or lid to prevent re-contamination. Point-of-use filters, either attached to a faucet, or the pour-through type, can also be used to remove *Cryptosporidium* from water. Only filters with an absolute, rather than a nominal, pore size of one micron or smaller should be used to remove oocysts. Reverse osmosis filters are also effective. Lists of filters and reverse osmosis filters that will remove *Cryptosporidium* oocysts can be obtained from NSF International, an independent non- profit testing agency.

The use of bottled water is not necessarily safer than tap water, as water from a surface water source has the same risks of containing oocysts as tap water from that source, unless it has been treated with appropriate treatment technologies, such as distillation, pasteurization, reverse osmosis, or filtration with an absolute one micron rating, before bottling. Bottled water from deep groundwater wells has a low likelihood of being contaminated with oocysts, so the labels on water bottles should be examined before use to determine water source and treatment methods.

Food can also be a source of *Cryptosporidium*. The parasite may be present in uncooked or unwashed fruits and vegetables that are grown in areas where manure was used or animals were grazed or in beverages or ice prepared with contaminated water. Pasteurization of dairy products will kill oocysts. Bottled and canned drinks, such as soda and beer, are usually heated and/or filtered sufficiently to kill or remove *Cryptosporidium* oocysts. Care should be taken to wash hands thoroughly with soap and water before eating, preparing or serving food. Fruits and vegetables that will be eaten raw should be washed or peeled before being eaten. When traveling to areas with poor sanitation, extra care should be taken in the selection of food and drink.

Resources

OTHER

Centers for Disease Control and Prevention (CDC).
"*Cryptosporidium* Infection [Cryptosporidiosis]."
http://www.cdc.gov/crypto/ (accessed November 9,
2010).

United States Environmental Protection Agency (EPA).
"Pollutants/Toxics: Microorganisms: Cryptosporidium." http://www.epa.gov/ebtpages/pollmicroocry
ptosporidium.html (accessed November 9, 2010).

Judith L. Sims

CSOs *see* **Combined sewer overflows.**

Cubatão, Brazil

Once called the "valley of death" and the "most polluted place on earth," Cubatão, Brazil, is a symbol both of severe environmental degradation and how people can work together to clean up their environment. A determined effort to reduce pollution and restore the badly contaminated air and water in the past decades has had promising results. While not ideal by any means, Cubatão is no longer among the worst places in the world to live.

Cubatão is located in the state of São Paulo, near the Atlantic coastal city of Santos, just at the base of the high plateau on which São Paulo—Brazil's largest city—sprawls. Thirty years ago, Cubatão was an agreeable, well-situated town. Overlooking Santos Bay with forest-covered mountain slopes rising on three sides around it, Cubatão was well removed from the frantic hustle and bustle of São Paulo on the hills above. Several pleasant little rivers ran through the valley and down to the sea. When the rivers were dammed to generate electricity in the early 1970s, however, the situation changed.

Cheap energy and the good location between São Paulo and the port of Santos attracted industry to Cubatão. An oil refinery, a steel mill, a fertilizer plant, and several chemical factories crowded into the valley, while workers and job-seekers scrambled to build huts on the hillsides and the swampy lowlands between the factories. With almost no pollution control enforcement, industrial smokestacks belched clouds of dust and toxic effluents into the air while raw sewage and chemical waste poisoned the river. By 1981, the city had 80,000 inhabitants and accounted for 3 percent of Brazil's industrial output. It was called the most polluted place in the world. More than 1,000 tons of toxic gases were released into the air every day. The steaming rivers seethed with multi-hued chemical slicks, foamy suds, and debris. No birds flew in the air above, and the hills were covered with dying trees and the scars of erosion where rains washed dirt down into the valley.

Sulfur dioxide, which damages lungs, eats away building materials, and kills vegetation, was six times higher than World Health Organization guidelines. After a few hours exposure to sunlight and water vapor, sulfur oxides turn into sulfuric acid, a powerful and dangerous corrosive agent. Winter air inversions would trap the noxious gases in the valley for days on end. One quarter of all emergency medical calls were related to respiratory ailments. Miscarriages, stillbirths, and deformities rose dramatically. The town was practically uninhabitable.

The situation changed dramatically in the mid-1980s, however. Restoration of democracy allowed citizens to organize to bring about change. Governor Franco Montoro was elected on promises to do something about pollution. Between 1983 and 1987, the government worked with industry to enforce pollution laws and to share the costs of clean-up. Backed by a World Bank loan of $100 million, the state and private industry invested more than $200 million for pollution control. By 1988, 250 out of 320 pollution sources were reduced or eliminated. Ammonia releases were lowered by 97 percent, particulate emissions were reduced 92 percent, and sulfur dioxide releases were cut 84 percent. Ozone-producing hydrocarbons and volatile organic compounds dropped nearly 80 percent. The air was breathable again. Vegetation began to return to the hillsides around the valley, and birds were seen once more.

Water quality also improved. Dumping of trash and industrial wastes was cut from some 64 metric tons per day to less than 6 tons. Some 780,000 tons of sediment were dredged out of the river bottoms to remove toxic contaminants and to improve water flow. Fish returned to the rivers after a 20-year absence. Reforestation projects are replanting native trees on hillsides where mudslides threatened the town. The government of Brazil now points to Cubatão with pride as an illustration of its concern for environmental protection. This is a heartening example of what can be done to protect the environment, given knowledge, commitment, and cooperation. As of 2010, the effort to improve the environment around Cubatão had cost over $1.2 billion.

Resources

BOOKS

Hill, Marquita K. *Understanding Environmental Pollution.*
Cambridge: Cambridge University Press. 2004.
Pepper, Ian, Charles Gerba, and Mark Brusseau.
Environmental and Pollution Science. New York:
Academic Press, 2006.

William P. Cunningham

Cultivation *see* **Agricultural pollution.**

Cultural eutrophication

One of the most important types of water pollution,
cultural eutrophication describes human-generated
fertilization of water bodies. Cultural denotes human
involvement, and eutrophication means truly nourished,
from the Greek word *eutrophic.* Key factors in cultural
eutrophication are nitrates and phosphates, and the
main sources are treated sewage and runoff from farms
and urban areas. The concept of cultural eutrophication
is based on anthropocentric values, where clear water
with minimal visible organisms is much preferred over
water rich in green algae and other microorganisms.

Nitrates and phosphates are the most common limiting factors for organism growth, especially in aquatic
ecosystems. Most fertilizers are a combination of nitrogen, phosphorus, and potassium. Nitrates are key components of the amino acids, peptides, and proteins
needed by all living organisms. Phosphates are crucial
in energy transfer reactions within cells. Natural sources
of nitrates (and ammonia) are more readily available
than phosphates, so the latter is often cited as the crucial
limiting factor in plant growth. Nitrates are supplied in
limited quantities by decaying plant material and nitrogen-fixing bacteria, but phosphates must come from
animal bones, organic matter, or from the breakdown
of phosphate-bearing rocks. Consequently, the introduction and widespread use of phosphate detergents,

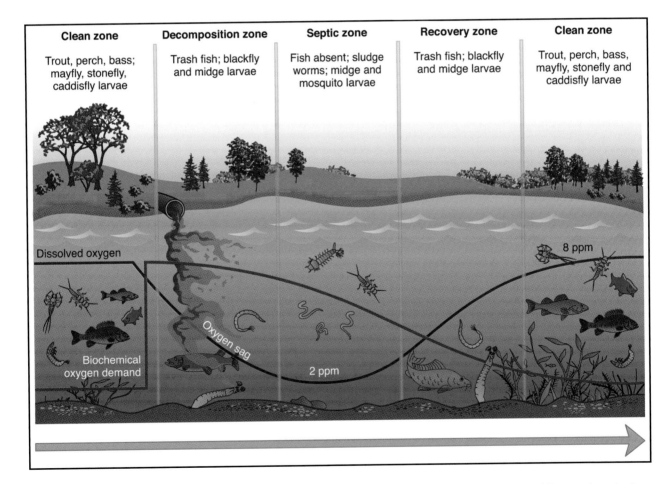

Oxygen reduction downstream of a source of organic pollution. *(Reproduced by permission of Gale, a part of Cengage Learning.)*

combined with excess fertilizer in runoff, has produced a near ecological disaster in some waters.

In ecosystems, there is a continuous cycling of matter, with green algae and plants making food from chemicals dissolved in water via photosynthesis; this provides the food base needed by herbivores and carnivores. Dead plant material and animals are then decomposed by aerobic (oxygen using) and anaerobicdecomposers into the simple elements they came from. Natural water bodies are usually well-suited for handling this matter cycling; however, human impacts often inject large amounts of additional nutrients into the system, changing them from oligotrophic (poorly nourished) to eutrophic water bodies. Once present within a relatively closed body of water, such as a lake or estuary, these extra nutrients may cycle numerous times before leaving the system.

Green algae and trashy fish may thrive in eutrophic water, but most people are offended by what they perceive as "scum." Nutrient-poor water is usually clear and possesses a rich supply of oxygen, but as the nutrient load increases, oxygen levels drop and turbidity rises. For example, when sewage is dumped into a body of water, the sewage fertilizes the algae, and as they multiply and die the aerobic decomposers multiply in turn. The increased demand for oxygen by the decomposers outstrips the system's ability to provide it. As a result, dissolved oxygen levels may fall or sag even below 2.0 parts per million, the threshold below which even trashy fish cannot survive (trout need at least 8.0 ppm to survive). Even though the green algae are producing oxygen as a byproduct of photosynthesis, even more oxygen is consumed by decomposers breaking down the dead algae and other organisms.

As water flows downstream the waste is slowly broken down, so there is less for the decomposers to feed on. Biological oxygen demand slowly falls, while the dissolved oxygen levels rise until the river is finally back to normal levels. Most likely, the nutrients recycled by decomposers are either diluted, turned into biomass by trees and consumer organisms, or tied up in bottom sediments. Thus a river can naturally cleanse itself of organic waste if given sufficient time. Problems arise, however, when discharges are too large and frequent for the river to handle; under extreme conditions it becomes "dead" and is suited only for low-order, often anaerobic, organisms. Municipalities using river water locate their intakes upstream and their sewage treatment plants and storm drains downstream. If communities were required to do

the reverse, the quality of river water would dramatically improve.

The major sources of nitrates and phosphates in aquatic systems are treated sewage effluent; excess fertilizer from farms and urban landscapes; and animal wastes from feedlots, pastures, and city streets. In some areas with pristine water, tertiary sewage treatment has been added; chemicals are used to reduce nitrate and phosphate levels prior to discharge.

Runoff from nonpoint sources is a far more difficult problem because they are harder to control and remediate than runoff from point sources. Point sources can be diverted into treatment plants, but the only feasible way to reduce nonpoint sources is by input reduction or by on-site control. Less fertilizer more frequently applied, especially on urban lawns, for example, would help reduce runoff. Feedlots are a major concern; runoff needs to be collected and treated in the same manner as human sewage; this may also apply to street runoff.

Phosphate detergents are superior cleaning agents than soap, but the resultant wastewater is loaded with this key limiting factor. Phosphate levels in detergents have since been reduced, but its impact is so powerful that abatement may require tertiary treatment.

A success story in the battle to overcome the effects of cultural eutrophication, the Thames River in England was devoid of aquatic life for centuries. Now a massive cleanup effort is restoring the river to vitality. Many fish have returned, most notably the pollution-sensitive salmon, which had not been seen in London for 300 years.

Resources

BOOKS

Hill, Marquita K. *Understanding Environmental Pollution.* Cambridge: Cambridge University Press. 2004.

Maczulak, Anne E. *Pollution: Treating Environmental Toxins.* New York: Facts on File, 2010.

Pepper, Ian, Charles Gerba, and Mark Brusseau. *Environmental and Pollution Science.* New York: Academic Press, 2006.

OTHER

U.S. Government; science.gov. "Pollution Prevention." http://www.science.gov/browse/w_123I.htm (accessed November 10, 2010).

United States Environmental Protection Agency (EPA). "Pollution Prevention." http://www.epa.gov/ebtpages/pollutionprevention.html (accessed November 10, 2010).

Nathan H. Meleen

Cuyahoga River

This 103-mile (166-km) long tributary of Lake Erie is a classic industrial river, with, however, one monumental distinction: it caught fire on numerous occasions. The first fire on the Cuyahoga River occurred in 1868. A 1952 fire on the river caused over $1 million in damages. A major fire in 1959 burned for eight days; fireboats merely spread the blaze. Typical of the times, a November 1959 article in *Fortune* seemed to glorify the industrial pollution here, with words and a portfolio of drawings reminiscent of Charles Dickens. Inspired by feelings of space and excitement, the artist stated: "It is a great expanse, with a smoky cast over everything, smudged with orange dust from the ore–an overall brown color." Clevelanders consoled themselves that the foul water at least symbolized prosperous times.

The second fire occurred on June 22, 1969, as several miles of river along the lowland industrial section called "the Flats" ignited, fed by bunker oil, trash, and tree limbs trapped by a bridge 6 mi (9.7 km) upstream. This fire, along with the Santa Barbara oil well blowout, provided graphic television images that were projected around the world. Jack A. Seamonds's June 18, 1984 article in *U.S. News & World Report* credited the fire with lighting "the fuse that put the bang in the nationwide campaign to clean up the environment." Pamela Brodie described it in the September-October 1983 issue of *Sierra* as "The most infamous episode of water pollution in the U.S . . . It inspired a song . . . and the Clean Water Act."

Although the fire badly hurt Cleveland's image, steps were already underway to correct the problem. The city government had declared the river a fire hazard and won voter approval in the autumn of 1968 for major funding to correct sewage problems. After the fire, Cleveland's three image-conscious steel companies voluntarily quit dumping cyanide-laced water and two installed cooling towers. As a result of just these actions alone, the river once again began to freeze in the winter. The city of Akron banned phosphate detergents, and eventually won the lawsuits brought against it by soap companies. These and other efforts, plus completion in 1983 of the major sewage treatment project, brought about far more encouraging reports.

At one point, over 155 tons of waste were dumped into the Cuyahoga each day. But by 1978, *Business Week* could report substantial improvement: phosphates were cut in half; chemicals were down 20–40%; and oil spills went from 300 per year to 25 in 1977. For the first time, local residents saw ducks on the river. By 1984, point source pollution had been largely eliminated. The waterfront was rediscovered, with restaurants and trendy stores where once fires had burned

The water is still light brown, churned up by deepwater shipping, and Lake Erie is still polluted. Nonetheless, the Cuyahoga seems to have been largely redeemed, and thus helped revitalize the city of Cleveland. The river became one of 14 American Heritage rivers in 1998.

See also Great Lakes; Water quality.

Resources

BOOKS

Pepper, Ian, Charles Gerba, and Mark Brusseau. *Environmental and Pollution Science*. New York: Academic Press, 2006.

OTHER

United States Environmental Protection Agency (EPA). "Water: Water Pollution Control." http://www.epa. gov/ebtpages/watewaterpollutioncontrol.html (accessed November 10, 2010).

United States Environmental Protection Agency (EPA). "Water: Water Pollution Control: Remediation." http://www.epa.gov/ebtpages/watewaterpollutionre mediation.html (accessed November 10, 2010).

United States Environmental Protection Agency (EPA). "Water: Water Pollution Effects." http://www.epa.gov/ ebtpages/watewaterpollutioneffects.html (accessed November 10, 2010).

Nathan H. Meleen

Cyclodienes *see* **Chlorinated hydrocarbons.**

Cyclone *see* **Tornado and cyclone.**

Cyclone collector

A device that removes solids from an effluent stream (solid, liquid, or mixed phas solid/gas) using gravitational settling. The input stream enters a vertical tapered cylinder tangentially and spiral fins or grooves cause the stream to swirl. The centrifugal force created by this swirling action causes particles to settle out of the input stream. These solids are concentrated at the walls and move to the bottom of the tapered cone. The solid-free fluid or gas is removed along the center line of the collector via a vortex finder. A cyclone collector is a compact device capable of processing large volumes of effluent; however, pressure energy is expended in its operation. It removes particles from the effluent stream less effectively than bag filters or electrostatic precipitation, but is simple and can be relatively effective in the collection of dusts. Removal of particulates from air using cyclone collectors is inexpensive and low-maintenance. These collectors are more efficient at extracting larger particles and are usually used with other devices for removal of smaller particles. Cyclone collectors remove and isolate the particulates, but do not destroy them, necessitating a method of disposal.

See also Baghouse.

Siddiqui, Iqtidar H. *Dams and Reservoirs: Planning and Engineering*. Oxford, UK: Oxford University Press, 2009.

United Nations Environment Programme-Dams and Development Project (UNEP-DDP). *Dams and Development: Relevant Practices for Improved Decision-making: A Compendium of Relevant Practices for Improved Decision-making on Dams and Their Alternatives*. Nairobi, Kenya: UNEP-DDP Secretariat, 2007.

ORGANIZATIONS

American Rivers, 1025 Vermont Ave., NW, Suite 720, Washington, DC, USA, 20005, (202) 347-7550, (202) 347-9240, amrivers@amrivers.org, htt://www. amrivers.org/damremoval/default.htm

Paula Anne Ford-Martin

Danube River *see* **Eastern European Pollution.**

Dams (environmental effects)

Most dams are built to control flood hazards, to store water for irrigation or other uses, or to produce electricity. Along with these benefits come environmental costs including riparian habitat loss, water loss through evaporation and seepage, erosion, declining water quality, and air pollution. Other negative consequences of dams include changes in groundwater flow and the displacement of human populations.

Riparian, or streamside, habitats suffer both above and below dams. Valuable ecological zones that support specialized plants, riparian environments, and nearby shallows provide food and breeding grounds for birds, fish, and many other animals. Upstream of a dam, impounded water drowns riparian communities. Because reservoirs can fill hundreds of miles of river channel, and because many rivers have a long sequence of dams and reservoirs, habitat drowning can destroy a great deal of river biodiversity. Downstream, shoreline environments dry up because of water diversions (for irrigation or urban use) or because of evaporation and seepage losses in the reservoir. In addition, dams interrupt the annual floods that occur naturally on nearly all rivers. Seasonal flooding fertilizes and waters floodplains and clears or redistributes debris in river channels. These beneficial effects of flooding cease once a dam is installed on a river. Sixty percent of the world's 227 largest rivers are significantly fragmented, therefore resulting in damage to the river ecosystems.

As a result of covering streamside vegetation with water in the construction of the reservoir, its decomposition results in the emission of greenhouse gases such as methane gas (CH_4). When flooding a large area, the amount of decay can contribute to large emissions of greenhouse gases, thus leading to climate change. However, if the dam is being used for the generation of hydroelectricity to replace other sources of power, there would potentially be a decrease in greenhouse gas emissions to offset those levels generated from decomposition.

Dams and reservoirs alter sediment deposition in rivers. Most rivers carry large amounts of suspended silt and sand, and dams trap sediments normally deposited downstream. Below the dam, erosion reshapes river channels once sediment deposition ceases, thus altering the riverside habitat. If erosion becomes extreme, bridges, walls, levees, and even river deltas can be threatened. Meanwhile, sediment piling up in the still waters of the reservoir behind the dam decreases water storage capacity. An increasingly shallow reservoir also becomes gradually warmer. Oxygen content decreases as the water temperature rises; fish populations fall, and proliferating algae and aquatic plants can begin to block the dam's water intakes. In arid regions a higher percentage of river water evaporates as the reservoir becomes shallower. Evaporating water leaves behind salts, which further decrease water quality in the reservoir and river.

Water losses from evaporation can be extreme: Lakes Powell and Mead on the Colorado River lose about 3 billion cubic feet (1 billion cubic meters) of water to evaporation each year; Egypt's Lake Nasser, on the Nile River, loses about 45 billion cubic feet (15 billion cubic meters). Water losses also result from seepage into bedrock. As river water enters groundwater, water tables usually rise around a reservoir. In arid regions increased groundwater can increase local fertility (sometimes endangering delicate dryland plant species), but in moister regions excessive groundwater can cause swamping. Evaporation from exposed groundwater can leave higher salt concentrations in the soil. The most catastrophic results of reservoir seepage into groundwater occur when saturated rock loses its strength. In such events valley walls can collapse, causing dam failure and disastrous flooding downstream.

Perhaps the most significant environmental effect of dams results from the displacement of human populations. The World Commission on Dams has estimated that 40 to 80 million people have been displaced because of dam construction. Because people normally settle along rivers, where water for drinking, irrigation, power, and transport are readily available, reservoir flooding can displace huge populations. The Three

A torrent of water pours through an opening in the 162-year-old Edwards Dam on the Kennebec River in Augusta, Maine, 1999, after heavy equipment began removing the dam. Environmentalists hailed the dam's removal. *(AP Photo/Robert F. Bukaty)*

Gorges Dam on China's Chang Jiang (Yangtze River) displaced more than 1 million people and flooded some of western China's best agricultural land.

In August 2010, Brazilian officials announced the impending start of construction on an $11 billion hydroelectric dam across the Amazon. Upon completion, the Belo Monte dam will be the third largest in the world. The reservoir behind the dam will displace thousands of people from cities and towns along the Xingu River basin, which will be flooded by the dam. The Belo Monte dam project is opposed by a broad array of environmental and indigenous rights groups, but advocates of the project argue that it is needed to produce electricity for expanding cities such as São Paulo, Brazil. About 200 square miles (520 square km) of land will be flooded by the dam. To mitigate the environmental impacts, the government has pledged $280 million toward a sustainable development plan for the area. In response, some indigenous tribes, including the Xikrin-Kayapó and the Parakanã tribes, have withdrawn their formal opposition to the project. Critics of the project counter that there have been insufficient long-term environmental impact studies, especially of potential downstream impacts.

Where dams are needed for power, they can have a positive effect in offsetting environmental costs associated with other power sources. Hydropower is cleaner and safer than nuclear power. Water turbines are also cleaner than coal-fired generators. Furthermore, both nuclear and coal power require extensive mining, with environmental costs far more severe than those of even a large dam.

Resources

BOOKS

Brooks, K B, and W. Cronon. *Public Power, Private Dams: The Hells Canyon High Dam Controversy*. Seattle: University of Washington Press, 2006.

Gupta, Vishwa Jit. *Hydroelectric Projects, Power, Dams, and Environment*. Delhi, India: S.S. Publishers, 2008.

Hayes, Walter P., and Michael C. Barnes. *Dams: Impacts, Stability, and Design*. New York: Nova Science Publishers, 2009.

International Symposium on Dams in the Societies of the Twenty-first Century, and Luis Berga. *Dams and Reservoirs, Societies and Environment in the Twenty-first Century*. London: Taylor & Francis, 2006.

Leslie, Jacques. *Deep Water: The Epic Struggle Over Dams, Displaced People, and the Environment*. New York: Farrar, Straus, and Giroux, 2005.

OTHER

National Atlas of the United States. "Major Dams of the United States." http://nationalatlas.gov/mld/dams00x.html (accessed August 15, 2010).

Mary Ann Cunningham

Darling, Jay Norwood "Ding"

1876–1962
American environmental cartoonist

Most editorial page cartoonists focus on the tough job of meeting daily deadlines, satisfied that their message will influence public opinion through their drawings. Jay Norwood "Ding" Darling, however, drew Pulitzer prize-winning cartoons, but was also immersed in conservation action, emerging as one of the great innovators in the conservation movement of the first half of the twentieth century.

Norwood, Michigan, was Darling's namesake birthplace, but he grew up in Elkhart, Indiana, and attended high school in Sioux City, Iowa, at a time when the area was relatively undeveloped. Wandering the prairies of nineteenth-century Iowa, Nebraska, and South Dakota instilled in him a lifelong love of the outdoors and wildlife.

After an uneven beginning, Darling graduated from Beloit College in Wisconsin with a degree in biology. (At one college, during one semester, biology was the only course he passed.) And he was expelled from Beloit for a year for caricaturing individuals in the college faculty and administration. Building on an early interest in sketching (and the cartooning skills developed in college), Darling went on to a half-century of drawing political cartoons, including what many view as some of the most memorable conservation and environmental cartoons of the twentieth century. His first job (1900) was as a reporter (and sometimes caricaturist) in Sioux City; but six years later, he was hired by the *Des Moines Register and Leader*, where his primary task was to produce a cartoon each day for the editorial page. He retired from that same paper in 1949. Darling did spend two years as an editorial cartoonist for the *New York Globe*, but returned to Iowa at the first real opportunity.

The only other significant period away from his Des Moines newspaper position illustrates his action mode as a conservationist. Darling was active in political life generally, and this involvement often reflected his love of the outdoors and his dismay at what he felt the United States was doing to destroy its natural resource base. He helped organize the Izaak Walton League in Des Moines, was active in the landscaping of local parks, and worked to establish the first cooperative wildlife research unit at Iowa State College in Ames, a unit that served as a model for the establishment of similar units in other states.

Perhaps his greatest impact as an activist resulted from his appointment, by President Franklin Roosevelt, as head of the U.S. Bureau of Biological Survey, predecessor of the U.S. Fish and Wildlife Service. He was only in the position for two years, but was effective in the job, "raiding" (in Roosevelt's words) the U.S. Treasury for scarce Depression-era funds for waterfowl habitat restoration, and initiating the duck stamp program, which over the years has funded the acquisition of several million acres added to the National Wildlife Refuge system. He also used his drawing skills to design the first duck stamp, as well as the flying goose that has become the signpost and symbol of the national wildlife refuge system.

Darling was also one of the founders, and then first president, of the National Wildlife Federation in 1938. He later criticized the organization, and the proliferation of conservation organizations in general, because he first envisioned the Federation as an umbrella for conservation efforts and thought the emergence of too many groups diluted the focus on solving conservation and environmental problems. Until the end of his life, he tried, and failed, to organize a conservation clearing house that would refocus the conservation effort under one heading.

He put into all his efforts lessons of interdependence learned early from biology classes at Beloit College: "Land, water, and vegetation are [interdependent] with one another. Without these primary elements in natural balance, we can have neither fish nor game, wild flowers nor trees, labor nor capital, nor sustaining habitat for humans."

Resources

BOOKS

Groom, Martha J., et al. *Principles of Conservation Biology*. 3rd ed. Washington, DC: Sinauer Associates, 2005.

Lendt, David L. *Ding: The Life of Jay Norwood Darling*. Ames: Iowa State University Press, 1979.

PERIODICALS

Dudley, Joseph P. "Jay Norwood 'Ding' Darling: A Retrospective." *Conservation Biology* 7, *1* (March 1993): 200–203.

Gerald L. Young

Darwin, Charles Robert

1809–1882
English naturalist

Charles Darwin, an English biologist known for his theory of evolution, was born at Shrewsbury, England, on February 12, 1809. He studied, traveled, and published his famous *On the Origin of Species* (1859).

Darwin's father was an affluent physician and his mother the daughter of the potter Josiah Wedgwood. Charles married Emma Wedgwood, his first cousin, in

Charles Darwin. *(Photo Researchers Inc.)*

1839. Due to his family's wealth, Darwin was made singularly free to pursue his interest in science.

Darwin entered Edinburgh to study medicine, but, as he described in his autobiography, lectures were "intolerably dull," human anatomy "disgusted" him, and he experienced "nausea" on seeing surgery. He subsequently entered Christ's College, Cambridge, to prepare for Holy Orders in the Church of England. While at Cambridge, Darwin became intensely interested in geology and botany, and, because of his knowledge in these sciences, he was asked to join the voyage of His Majesty's Ship (HMS) *Beagle*. Darwin's experiences during the circumnavigational trek of the *Beagle* were of seminal importance in his later views on evolution.

Darwin's *On the Origin of Species* is a monumental catalog of evidence that evolution occurs, together with the description of a mechanism that explains such evolution. This "abstract" of his notions on evolution was hurried to publication because of a letter Darwin received from Alfred Russel Wallace expressing similar views. Darwin's evidence for evolution was drawn from comparative anatomy, embryology, distribution of species, and the fossil record. He believed that species were not immutable but evolved into other species. His theory of evolution by natural selection is based on the premise that species have a great reproductive capacity. The production of individuals in excess of the number that can survive creates a struggle for survival. Variation between individuals within a species was well-documented. The struggle for survival, coupled with variation, led Darwin to postulate that those individuals with favorable variations would have an enhanced survival potential and hence would leave more progeny and this process would lead to new species. This notion is sometimes referred to as "survival of the fittest." While the theory of evolution by natural selection was revolutionary for its day, essentially all biologists in the twenty-first century accept Darwinian evolution as fact.

The first edition of *Origin* had a printing of 1,250 copies. It sold out the first day. Darwin was an extraordinarily productive author for someone who considered himself to be a slow writer. Among his other books are *Structure and Distribution of Coral Reef* (1842), *Geological Observations on Volcanic Islands* (1844), *On the Various Contrivances by which British and Foreign Orchids Are Fertilized by Insects* (1862), *Insectivorous Plants* (1875), and *On the Formation of Vegetable Mould through the Action of Worms* (1881). The last book, of interest to ecologists and gardeners, was published only six months prior to Darwin's death.

Darwin died at age 73 and is buried next to Sir Isaac Newton at Westminster Abbey in London.

Resources

BOOKS

Darwin, Charles, and David Quammen. *On the Origin of Species.* New York: Sterling, 2008.

Darwin, Charles. *The Voyage of the Beagle.* New York: Bantam Books, 1958.

Ruse, Michael, and Robert J. Richards. *The Cambridge Companion to the "Origin of Species."* Cambridge, UK: Cambridge University Press, 2009.

Robert G. McKinnell

DDT *see* **Dichlorodiphenyl-trichloroethane.**

Dead zones

The term *dead zone* refers to those areas in aquatic environments where there is a reduction in the amount of dissolved oxygen in the water. The condition is more appropriately called hypoxia or hypoxic waters or zones. Since scientists began measuring the extent of the marine dead zones in the 1960s, the number of dead zones around the world has doubled every ten years. By 2010 more than 150 permanent or reoccurring marine dead zones exist in locations around the world. Dead zones may vary in size from slightly less than one-half square mile (about 1 square km) to 28,000 square miles (70,000 square km) and may be transient (appearing or disappearing with seasons or over a course of years).

Ocean color changes from winter to summer in the Gulf of Mexico. Reds and oranges represent high concentrations of phytoplankton and river sediment. *(NASA)*

Hypoxia in marine environments is determined when the dissolved oxygen content is less than 2 to 3 milligrams per liter. Five to 8 milligrams per liter of dissolved oxygen is generally accepted as the normal level for most marine life to survive and reproduce. Dead zones can not only reduce the numbers of marine animals, but they can also change the nature of the ecosystem within the hypoxic zone.

The main cause of oxygen depletion in aquatic environments is eutrophication. This process is a chain of events that begins with runoff rich in nitrogen and phosphorus that makes its way into rivers that eventually discharge into estuaries and river deltas. This nutrient-laden water, combined with sunlight, stimulates plant growth, specifically algae, seaweed, and phytoplankton. When these plants die and fall to the ocean floor, they are consumed by bacteria that use large amounts of oxygen, thereby depleting the environment of oxygen.

Dead zones can lead to significant shifts in species balances throughout an ecosystem. Species of aquatic life that can leave these zones—such as fish and shrimp—do so. Bottom-growing plants, shellfish, and others that cannot leave will die, creating an area devoid of aquatic life. This is the reason the term *dead zone* has been aptly used (especially by the media) to describe these areas.

Eutrophication of estuaries and enclosed coastal seas has often been a natural phenomenon when offshore winds and water currents force deep nutrient-laden waters to rise to the surface, stimulating algae bloom. The timing and duration of these conditions varies from year to year and within seasons. Climatic conditions and catastrophic weather events can influence the rapidity with which hypoxia can occur. In the past, these natural hypoxic areas were limited, and the marine environment could recover quickly.

Nutrient availability, temperature, energy supply (i.e., soluble carbon for most microorganisms or light and carbon dioxide for plants), and oxygen status all affect the growth and sustainability of aquatic plants and animals. One condition that perpetuates hypoxia is elevated temperatures because the ability of water to hold oxygen (i.e., water solubility) decreases with increasing temperature. Situations that promote consumption of oxygen such as plant and animal respiration and decomposition can also lead to hypoxic conditions in water. Therefore, situations that stimulate plant (i.e., phytoplankton, benthic algae, and macroalgae) growth in water can lead indirectly to hypoxic conditions. Algal growth can be accelerated with elevated levels of carbon dioxide and certain nutrients (especially nitrogen and phosphorus), provided adequate sunlight is available. Low sediment loads also support increased algal growth because the water is less turbid allowing more light to penetrate to the bottom.

In the last two decades of the twentieth century, the increased incidence of hypoxia and the expanded size of dead zones have been the result of increased nutrients coming from human sources. Runoff from residential and agricultural activities are loaded with fertilizers, animal wastes, and sewage, which have specific nutrients that can stimulate plant growth. In the United States, nutrient runoff has become a major concern for the entire interior watersheds of the Mississippi River basin, which drains into the Gulf of Mexico.

Hypoxic waters occur near the mouths of many large rivers around the world and in coastal estuaries and enclosed coastal seas. In fact, over forty hypoxic zones have been identified throughout the world. Robert J. Diaz from the Virginia Institute of Marine Science has studied the global patterns of hypoxia and has concluded that the extent of these zones has increased over the past several decades. Hypoxic zones are occurring in the Baltic Sea, Kattegat, Skagerrak Dutch Wadden Sea, Chesapeake Bay, Long Island Sound, and northern Adriatic Sea, as well as the extensive one that occurs at the mouth of the Mississippi River in the Gulf of Mexico.

It is interesting to note that one of the largest hypoxic zones documented occurred in conjunction with the increase in ocean temperatures associated with El Niño. This weather event occurs periodically off the west coast of North and South America, and influences not only the winter weather in North America, but also affects the anchovy catch off Peru in the Pacific. This, in turn, affects the worldwide price for protein meal and has a direct impact on soybean farmers (because this is another major source of protein for this product).

The large hypoxic zone in the northern Gulf of Mexico occurs where the Mississippi and Atchafalya rivers enter the ocean. The zone was first mapped in 1985, and had doubled in size by 2010. The Gulf of Mexico dead zone fluctuates in size depending on the amount of river flow entering the Gulf, and on the patterns of coastal winds that mix the oxygen-poor bottom waters of the Gulf with Mississippi River water. By 1999, the zone's maximum size exceeded 8,006 square miles (20,720 square km) in area, making it one of the largest in the world. This zone fluctuates seasonally, as do others. It can form as early as

February and last until October. The most widespread and persistent conditions exist from mid-May to mid-September.

Recent research has shown that increased nitrogen concentrations in the river water, which act like fertilizer, stimulate massive phytoplankton blooms in the Gulf. Bacteria decomposing the dead phytoplankton consume nearly all of the available oxygen. This, combined with a seasonal layering of the freshwater from the river and saltwater in the Gulf, results in the zone of low-dissolved oxygen. Within the zone there are very small fish and shellfish populations.

The effect of periodic events on the transport of nutrients to the Gulf, derived primarily from fertilizer in areas of intensive agriculture in the upper Mississippi River basin, have been implicated as the most likely cause. The largest amount of nutrients is delivered each year after the spring thaw when streams fill and concentrations of nutrients, such as nitrogen, are highest. In addition, during extreme high-flow events, such as those that occurred during the floods of 1993, very high amounts of nutrients are transported to the Gulf. Levels were 100 percent higher in that year than in other years.

The nature of the hypoxia problem in the Gulf is complicated by the fact that some nutrient load from the Mississippi River is vital to maintain the productivity of the Gulf fisheries, but in levels considerably lower than are now entering the marine system. Approximately 40 percent of the U.S. fisheries landings comes from this area, including a large amount of the shrimp harvest. In addition, the area also supports a valuable sport fishing industry. The concern is that the hypoxic zone has been increasing in size since the 1960s because of human activities in the Mississippi watershed that have increased nitrogen loads to the Mississippi River. The impact of an expanding Gulf hypoxia include large algal blooms that affect other aquatic organisms, altered ecosystems with changes in plant and fish populations (i.e., lower biodiversity), reduced economic productivity in both commercial and recreational fisheries, and both direct and indirect impact on fisheries such as direct mortality and altered migration patterns that may lead to declines in populations.

Studies were conducted during the 1990s on the sources of increased nutrient concentrations within the Mississippi River. A significant amount of nutrients delivered to the Gulf come from the Upper Mississippi and Ohio River watersheds. The amount of dissolved nitrogen and phosphorus in the waters of the Mississippi has more than doubled since 1960. The principal areas contributing nutrients are streams draining the corn-belt states, particularly Iowa, Illinois, Indiana, Ohio, and southern Minnesota. About 60 percent of the nitrate transported by the Mississippi River comes from a land area that occupies less than 20 percent of the basin. These watersheds are predominantly agricultural and contain some of the most productive farmland in the world. This area produces approximately 60 percent of the nation's corn. The U.S. Geological Survey has estimated that 56 percent of the nitrogen entering the Gulf hypoxic zone originates from fertilizer. Potential agricultural sources within these regions include runoff from cropland, animal grazing areas, animal waste facilities, and input from agricultural drainage systems. The contributions to nutrient input from sources such as atmospheric deposition, coastal upwelling, and industrial sources within the lower Mississippi watershed are being evaluated also. It is unclear what effect the damming and channelization of the river for navigation has on nutrient delivery. The dead zone area in the Gulf will continue to be monitored to determine whether it continues to expand.

In the meantime, efforts to reduce nutrient loading are being undertaken in agricultural areas across the watershed. Several strategies to reduce nutrient loading have been drafted. They are a reduction in nitrogen-based fertilizers and runoff from feedlots, planting alternative crops that do not require large amounts of fertilizers, removing nitrogen and phosphorus from wastewater, and restoring wetlands so that they can act as reservoirs and filters for nutrients. Depending on whether the zone continues to expand or decreases as nutrient levels diminish will determine whether it remains a significant environmental problem in the Gulf of Mexico. Knowledge gained from the study of changing nutrient loads in the Mississippi River will be useful in addressing similar problems in other parts of the world.

In July 2007, researchers with the United Nations Environment Programme (UNEP) reported increased areas and depth for multiple transient dead zones in the Gulf of Mexico.

Resources

BOOKS

Garrison, Tom. *Oceanography: An Invitation to Marine Science.* 5th ed. Stamford, CT: Thompson/Brooks Cole, 2004.

Karleskint, George, Richard Turner, and James Small. *Introduction to Marine Biology.* 2nd ed. Belmont, CA: Brooks Cole, 2006.

Leigh, G. J. *The World's Greatest Fix: A History of Nitrogen and Agriculture.* Oxford, UK: Oxford University Press, 2004.

Okaichi, Tomotoshi, ed. *Red Tides.* Dordrecht, Netherlands: Kluwer, 2004.

OTHER

Centers for Disease Control and Prevention (CDC). "Marine Toxins." http://www.cdc.gov/ncidod/dbmd/diseaseinfo/ marinetoxins_g.htm (accessed September 4, 2010).

United States Department of the Interior, United States Geological Survey (USGS). "Marine Ecosystems." http://www.usgs.gov/science/science.php?term = 704 (accessed September 4, 2010).

United States Department of the Interior, United States Geological Survey (USGS). "Runoff." http://www.usgs. gov/science/science.php?term = 1375 (accessed September 4, 2010).

James L Anderson
Marie H. Bundy

Debt for nature swap

Debt for nature swaps are designed to relieve developing countries of two devastating problems: spiraling debt burdens and environmental degradation. In a debt for nature swap, developing country debt held by a private bank is sold at a substantial discount on the secondary debt market to an environmental nongovernmental organization (NGO). The NGO cancels the debt if the debtor country agrees to implement a particular environmental protection or conservation project. The arrangement benefits all parties involved in the transaction. The debtor country decreases a debt burden that may cripple its ability to make internal investments and generate economic growth. Debt for nature swaps may also be seen as a good alternative to defaulting on loans, which hurts the country's chances of receiving necessary loans in the future. In addition, the country enjoys the benefits of curbing environmental degradation. The creditor (bank) decreases its holdings of potentially bad debt, which may have to be written off at a loss. The NGO experiences global environmental improvement.

Debt for nature swaps were first suggested by Thomas Lovejoy in 1984. Swaps have taken place between Bolivia, Costa Rica, Ecuador, and NGOs in the United States. The first debt for nature swap was implemented in Bolivia in 1987. Conservation International, an American NGO, purchased $650,000 of Bolivia's foreign debt from a private bank in the United States at a discounted price of $100,000. The NGO then swapped the face value of the debt with the Bolivian government for "conservation payments-in-kind," which involved a conservation program in a 3.7 million acre (1.5 million ha) tropical forest region implemented by the government and a local NGO.

Despite the benefits associated with debt for nature swaps, implementation has been minimal so far. Less than 2 percent of all debt for equity swaps have been debt for nature swaps. A lack of incentives on the part of the debtor or the creditor and the lack of well-developed supporting institutional infrastructure can hinder progress in arranging debt for nature swaps.

If a debtor country is unable to repay foreign debt, it has the option of defaulting on the loans or agreeing to a debt for nature swap. The country has an incentive to agree to a debt for nature swap if defaulting is not a viable option and if the benefits of decreasing debt through a swap outweigh the costs of implementing a particular environmental protection project. The cost of the environmental protection programs can be substantial if the developing country does not have the appropriate institutional infrastructure in place. The program will require the input of professional public administrators and environmental experts. Without institutions to support these individuals, the developing countries may find it impossible to carry out the programs they promise to undertake in exchange for cancellation of the debt. If, in addition, the debtor country is highly capital-constrained, then it might not give high priority to the benefits of an environmental investment.

Whether the creditor has an incentive to sell a debt on the secondary debt market to a NGO depends on the creditor's estimate of the likelihood of receiving payment from the developing country; on the proportion of potentially bad credit the creditor is holding; and on its own financial situation. If the NGOs are willing to pay the price demanded by private banks for developing country debt and swap it for environmental protection projects in the debtor countries, they will have the incentive to pursue debt for nature swaps.

Benefits that may be taken into account by the NGOs are those commonly associated with environmental protection. Many developing countries hold the world's richest tropical rain forests, and the global community will benefit greatly from the preservation of these forests. Tropical forests hold a great deal of carbon dioxide, which is released into the atmosphere and contributes to the greenhouse effect when the forests are destroyed. Another benefit is known as "option value," the value of retaining the option of future use of plant or animal resources that might otherwise become extinct. Although it is not know at present of what use, if any, these species might be, there is a value associated with preserving them for unknown future use. Examples of future uses might be pharmaceutical remedies, scientific understanding, or ecotourism. In addition, NGOs may attach "existence value" to environmental amenities. Existence value refers to the value placed on just knowing that natural environments exist and are being preserved. Many NGOs assert that preservation is important so that future generations can enjoy the environment. This value is known as "bequest value." Finally, the NGO may be interested in decreasing hunger and poverty in developing countries, and both the reduction of external debt and the slowing of the depletion of natural resources in developing countries is perceived as a benefit for this purpose.

To make a swap attractive, however, the NGO must be assured that the environmental project will be carried out after the debt has been canceled. Without adequate enforcement and assistance, a country might promise to implement an environmental project without being able or willing to follow through. Again, the solution to this problem lies in the development of institutions that are committed to monitoring and giving assistance in the implementation of the programs. Such institutions might encourage long-term relationships between the debtor and NGO to facilitate a structure by which debt is canceled piecemeal on the condition that the debtor continues to comply with the agreement.

A complicating factor that may affect an NGO's cost-benefit analysis of debt for nature swaps in the future is that, as the number of swaps and environmental protection projects increases, the value to be derived from any additional projects will decrease, because of diminishing marginal returns.

As described above, the benefits associated with debt for nature swaps both for the debtor countries and NGOs hinge on the presence of supporting institutions in the developing countries. It is particularly important to promote the establishment of appropriate, professionally managed public agencies with adequate resources to hire and maintain environmental experts and managers. These institutions should be responsible for planning and implementing the programs.

It should be noted that, although large debt burdens and environmental degradation are both serious problems faced by many developing countries, there is no direct linkage between them. Nevertheless, debt for nature swaps are an intriguing remedy that seems to address both problems simultaneously. As the quantity and magnitude of swaps so far have been relatively small, it is impossible to say how successful a remedy it may be on a larger scale. The future of debt for nature swaps may depend on the development of incentives and institutions to support the fulfillment of the agreements.

Resources

OTHER

Share the World's Resources (STWR). "Share the World's Resources International Monetary Fund World Debt Page." http://www.stwr.org/aid-debt-development/cancelling-third-world-debt.html (accessed October 13, 2010).

Barbara J. Kanninen

Deciduous forest

Deciduous forests are made up of trees that lose their leaves seasonally and are leafless for part of each year. The tropical deciduous forest is green during the rainy season and bare during the annual drought. The temperate deciduous forest is green during the wet, warm summers and leafless during the cold winters with the leaves turning yellow and red before falling. Temperate deciduous forests once covered large portions of Europe, the eastern United States, Japan, and eastern China. Species diversity is highest in Asia and lowest in Europe.

Decline spiral

A decline spiral is the destruction of a species, ecosystem, or biosphere in a continuing downward trend, leading to ecosystem disruption and impoverishment. The term is sometimes used to describe the loss of biodiversity, when a catastrophic event has led to a sharp decline in the number of organisms in a biological community.

In areas where the habitat is highly fragmented, either because of human intervention or natural disaster, the loss of species is markedly accelerated. Loss of species diversity often initiates a downward spiral, as the weakening of even one plant or animal in an ecosystem, especially a keystone species, can lead to the malfunctioning of the biological community as a whole.

Biodiversity exists at several levels within the same community; it can include ecosystem diversity, species diversity, and genetic diversity. Ecosystem diversity refers to the different types of landscapes that are home to living organisms. Species diversity refers to the different types of species in an ecosystem. Genetic diversity refers to the range of characteristics in the deoxyribonucleic acid (DNA) of the plants and animals of a species. A catastrophic event that affects any aspect of the diversity in an ecosystem can start a decline spiral.

Any major catastrophe that results in a decline in biospheric quality and diversity, known as an ecocatastrophe, may initiate a decline spiral. Herbicides and pesticides used in agriculture, as well as other forms of pollution; increased use of nuclear power; and exponential population growth are all possible contributing factors to an ecocatastrophe. The Lapp reindeer herds were decimated by fallout from the nuclear accident at Chernobyl, Ukraine, in 1986. Similarly, the oil spill from the *Exxon Valdez* in 1989 led to a decline spiral in the Gulf of Alaska ecosystem. More than twenty years after the spill, in 2010 biologists could still measure and attribute species declines to the spill.

The force that begins a decline spiral can also be indirect, as when acid rain, air pollution, water pollution, or climate change kills off many plants or animals in an ecosystem. Diversity can also be threatened by the introduction of nonnative or exotic species, especially when these species have no natural predators and are more aggressive than the native species. In these circumstances, native species can enter into a decline spiral that will impact other native species in the ecosystem.

Restoration ecology is a relatively new discipline that attempts to recreate or revive lost or severely damaged ecosystems. It is a hands-on approach by scientists and amateurs alike designed to reverse the damaging trends that can lead toward decline spirals. Habitat rebuilding for endangered species is an example of restoration ecology. For example, the Illinois chapter of The Nature Conservancy has reconstructed an oak-and-grassland savanna in Northbrook, Illinois, and a prairie in the 100-acre (40-ha) ring formed by the underground Fermi National Accelerator Laboratory at Batavia, Illinois. Because it is easier to reintroduce the flora than the fauna into an ecosystem, practitioners of restoration ecology concentrate on plants first. When the plant mix is right, insects, birds, and small animals return on their own to the ecosystem.

Resources

BOOKS

Begon, Michael; Colin A. Townsend; and John L. Harper. *Ecology: From Individuals to Ecosystems*. 4th ed. New York: Wiley-Blackwell, 2006.

Davis, Barbara J. *Biomes and Ecosystems*. Strongsville, OH: Gareth Stevens Publishing, 2007.

Monechi, Simonetta; R. Coccioni; and Michael R. Rampino. *Large Ecosystem Perturbations: Causes and Consequences*. Boulder, CO: Geological Society of America, 2007.

PERIODICALS

Halpern, B. S., et al. "A Global Map of Human Impact on Marine Ecosystems." *Science*. 319 (2008): 948–952.

Raffaelli, David. "How Extinction Patterns Affect Ecosystems." *Science*. 306 (2004): 1141-1142.

Scholze, Marko, et al. "A Climate Change Risk Analysis for World Ecosystems." *Proceedings of the National Academy of Sciences*. 103 (August 29, 2006): 13, 116–13, 120.

OTHER

United Nations Environment Programme (UNEP). "World Conservation Monitoring Centre Biodiversity of Ecosystems." http://www.unep-wcmc.org/climate/impacts.aspx (accessed November 10, 2010).

United States Environmental Protection Agency (EPA). "Ecosystems." http://www.epa.gov/ebtpages/ecosystems.html (accessed November 10, 2010).

Linda Rehkopf

Decomposers

Decomposers (also called saprophages, meaning "corpse eating") are organisms that perform the critical task of biochemical decomposition in nature. They include bacteria, fungi, and detritivores that break down dead organic matter, releasing nutrients back into the ecosystem. Fungi are the dominant decomposers of plant material, and bacteria primarily break down animal matter. Decomposers secrete enzymes into plant and animal material to break down the organic compounds, starting with compounds such as sugars, which are easily broken down, and ending with more resistant compounds such as cellulose and lignin. Rates of decomposition are faster at higher values of moisture and temperature. Decomposers thus perform a unique and important function in the recycling process in nature.

Decomposition

Decomposition is the chemical and biochemical breakdown of a complex substance into its constituent compounds and elements, releasing energy, and often with the formation of new, simpler substances. Organic decomposition takes place mostly in or on the soil under aerobic conditions. Dead plant and animal materials are consumed by a myriad of organisms, from mice and moles, to worms and beetles, to fungi and bacteria.

Enzymes produced by these organisms attack the decaying material, releasing water, carbon dioxide, nutrients, humus, and heat.

Decomposition is a major process in nutrient cycling, including the carbon and nitrogen cycles. The liberated carbon dioxide can be absorbed by photosynthetic organisms, including green plants, and made into new tissue in the photosynthesis process, or it can be used as a carbon source by autotrophic organisms.

Decomposition also acts on inorganic substances in a process called weathering. Minerals broken free from rocks by physical disintegration can chemically decompose by reacting with water and other chemicals to release elements, including potassium, calcium, magnesium, and iron. These and other elements can be taken up by plants and microorganisms, or they can remain in the soil system to react with other constituents, forming clays.

Deep ecology

The term *deep ecology* was coined by the Norwegian environmental philosopher Arne Naess in 1973. Naess drew a distinction between "shallow" and "deep" ecology. The former perspective stresses the desirability of conserving natural resources, reducing levels of air and water pollution, and other policies primarily for promoting the health and welfare of human beings. Deep ecologists maintain that shallow ecology simply accepts, uncritically and without reflection, the homocentric, or human-centered, view that humans are, or ought to be, if not the masters of nature, then at least the managers of nature for human ends or purposes. Defenders of deep ecology, by contrast, claim that shallow environmentalism is defective in placing human interests above those of animals and ecosystems. Human beings, like all lower creatures, exist within complex webs of interaction and interdependency. If people insist on conquering, dominating, or merely managing nature for their own benefit or amusement, if people fail to recognize and appreciate the complex webs that hold and sustain them, they will degrade and eventually destroy the natural environment that sustains all life.

However, deep ecologists say if people are to protect the environment for all species, now and in the future, they must challenge and change long-held basic beliefs and attitudes about different species' place in nature. For example, people must recognize that animals, plants, and the ecosystems that sustain them

have intrinsic value—that is, are valuable in and of themselves—quite apart from any use or instrumental value they might have for human beings. The genetic diversity found in insects and plants in tropical rain forests is to be protected not (only or merely) because it might one day yield a drug for curing cancer, but also and more importantly because such biodiversity is valuable in its own right. Likewise, rivers and lakes should contain clean water not just because humans need uncontaminated water for swimming and drinking, but also because fish do. Like Indian spiritual leader Mahatma Gandhi, to whom they often refer, deep ecologists teach respect for all forms of life and the conditions that sustain them.

Critics complain that deep ecologists do not sufficiently respect human life and the conditions that promote prosperity and other human interests. Some go so far as to claim that they believe in the moral equivalence of human and all other life-forms. Thus, say the critics, deep ecologists would assign equal value to the life of a disease-bearing mosquito and the child it is about to bite. No human has the right to swat or spray an insect, to kill pests or predators, and so on. But in fact this is a caricature of the stance taken by deep ecology. All creatures, including humans, have the right to protect themselves from harm, even if that means depriving a mosquito of a meal or even eliminating it altogether. Competition within and among species is normal, natural, and inevitable. Bats eat mosquitoes; bigger fish eat smaller fish; humans eat big fish; and so on. But for one species to dominate or destroy all others is neither natural nor sustainable. Yet human beings have, through technology, an ever-increasing power to destroy entire ecosystems and the life that they sustain. Deep ecologists hold that this power has corrupted human beings and has led them to think—quite mistakenly—that human purposes are paramount and that human interests take precedence over those viewed as lower or lesser species. Human beings cannot exist independently from, but only interdependently with, nature's myriad species. Once people recognize the depth and degree of this interdependence, deep ecologists say, they will learn humility and respect. The human species' proper place is not on top, but within nature and with nature's creatures and the conditions that nurture all.

Some cultures and religions have long taught these lessons. Zen Buddhism, Native American religions, and other nature-centered systems of belief have counseled humility toward, and respect for, nature and nonhuman creatures. But the dominant Western reaction is to dismiss these teachings as primitive or mystical. Deep ecologists, by contrast, contend that considerable wisdom is to be found in these native and non-Western perspectives.

Deep ecology is a philosophical perspective within the environmental movement and not a movement in itself. This perspective does, however, inform and influence the actions of some environmentalists. Organizations such as Earth First! and the Sea Shepherd Conservation Society are highly critical of more moderate environmental groups that are prepared to compromise with loggers, developers, dam builders, strip miners, and oil companies.

See also Ecosophy; Environmental ethics; Foreman, Dave; Green politics; Greens; Strip mining.

Resources

BOOKS

Des Jardins, Joseph R. *Environmental Ethics: An Introduction to Environmental Philosophy*. Belmont, CA: Wadsworth Publishing, 2005.
Kuipers, Theo A. F. *General Philosophy of Science: Focal Issues*. Amsterdam: Elsevier/North Holland, 2007.

Terence Ball

Deep-well injection

Deep-well injection refers to the introduction of liquid far below the surface, at depths of 10,000 feet (3,048 m) and greater. Injection of liquid wastes into subsurface geologic formations is a technology that has been widely adopted as a waste-disposal practice. The practice entails drilling a well to a permeable, saline-bearing geologic formation that is confined above and below with impermeable layers known as confining beds. When the injection zones lie below drinking water sources at depths typically between 2,000 to 5,000 feet (610-1,525 m), they are referred to as Class I disposal wells. The liquid hazardous waste is injected at a pressure that is sufficient to replace the native fluid and yet not so high that the integrity of the well and confining beds is at risk. Injection pressure is a limiting factor because excessive pressure can cause hydraulic fracturing of the injection zone and confining strata, and the intake rate of most injection wells is less than 400 gallons (1,500 liters) per minute.

Deep-well injection of liquid waste is one of the least expensive methods of waste management because little waste treatment occurs prior to injection. Suspended solids must be removed from wastewater prior to injection to prevent them from plugging the pores and reducing permeability of the injection zone. Physical and chemical characteristics of the wastewater

must be considered in evaluating its suitability for disposal by injection.

The principal means of monitoring the wastewater injection process is recording the flow rate, the pressure at the site of injection and the annulus (the space between the innermost layer of protective tubing that surrounds the injection tube and the tube itself), and the physical and chemical characteristics of the waste. Not all experts agree that these monitoring practices are adequate and monitoring is still controversial. The major issues concern the placement of monitoring wells and whether they increase the risk that wastewater will migrate out of the injection zone if they are improperly constructed.

Deep-well injection of wastes began as early as the 1950s, and it was then accepted as a means of alleviating surface water pollution. In 2010, most injection wells in the United States are located along the Gulf Coast and near the Great Lakes, and their biggest users are the petrochemical, pharmaceutical, and steel mill industries.

As with all injection wells, there is a concern that the waste will migrate from the injection zone to the overlying aquifers. If this occurs, then the groundwater could be irretrievably contaminated. As well, alterations in underground strata can trigger an earthquake. In the United States and Canada, approximately thirty earthquakes have been traced to deep-well injection of fluids. Experts assert that ensuring deep-well injection sites are in stable geologic zones is a prudent step.

See also Aquifer restoration; Groundwater monitoring; Groundwater pollution; Hazardous waste site remediation; Water quality.

Resources

BOOKS

Basic Environmental Technology: Water Supply, Waste Management, and Pollution Control. 5th ed. New York: Prentice Hall, 2007.
Engineering the Risks of Hazardous Wastes. Oxford, UK: Butterworth-Heineman, 2003.
Underground Injection Science and Technology, Vol. 52, (Developments in Water Science). New York: Elsevier, 2006.

OTHER

United States Department of the Interior, United States Geological Survey (USGS). "Well Drilling." http://www.usgs.gov/science/science.php?term=1325 (accessed August 1, 2010).

Milovan S. Beljin

Defenders of Wildlife

Defenders of Wildlife was founded in 1947 in Washington, D.C., Superseding older groups such as Defenders of Furbearers and the Anti-steel-Trap League, the organization was established to protect wild animals and the habitats that support them. As of 2010, their goals include the preservation of biodiversity and the defense of species as diverse as gray wolves (*Canis lupus*), Florida panthers (*Felis concolor coryi*), and grizzly bears (*Ursus arctos*), as well as the western yellow-billed cuckoo (*Coccyzus americanus*), the desert tortoise (*Gopherus agassizii*), and Kemp's Ridley sea turtle (*Lepidochelys kempii*).

Defenders of Wildlife employs a wide variety of methods to accomplish their goals, from research and education to lobbying and litigation. They have achieved a ban on livestock grazing on 10,000 acres (4,050 ha) of tortoise habitat in Nevada and lobbied for restrictions on the international wildlife trade to protect endangered species in other countries. In 1988 they successfully lobbied Congress for funding to expand wildlife refuges throughout the country. Ten million dollars was appropriated for the Lower Rio Grande National Wildlife Refuge in Texas, two million dollars to purchase land for a new preserve along the Sacramento River in California, and one million dollars for additions to the Rachel Carson National Wildlife Refuge in Maine. In 2010, they campaigned against offshore drilling near environmentally sensitive areas after the Deepwater Horizon oil spill in the Gulf of Mexico threatened marine, coastal, and marsh life.

The organization has been at the forefront of placing preservation on an economic foundation. Defenders maintains a speaker's bureau, and they support a number of educational programs for children designed to nourish and expand their interest in wildlife. But the group also participates in more direct action on behalf of the environment. They coordinate grassroots campaigns through their Defenders Activist Network. They work with the Environmental Protection Agency on a hotline called the Poison Patrol, which receives calls on the use of pesticides that damage wildlife, and they belong to the *Entanglement Network Coalition*, which works to prevent the loss of animal life through entanglement in nets and plastic refuse.

Restoring wolves to their natural habitats has been one of the top priorities of Defenders of Wildlife since 1985. That year, they sponsored an exhibit in Yellowstone National Park and at Boise, Idaho, called Wolves and Humans, which received over 250,000 visitors and won the Natural Resources Council of America Award of Achievement for Education. They have helped reintroduce red and gray wolves back into the northern Rockies. In order to assist farmers and the owners of livestock herds that graze in these areas, Defenders has raised funds to compensate them for the loss of land. They are also working to reduce and eventually eliminate the poisoning of predators, both by lobbying for stricter legislation and by encouraging western farmers to participate in their guard dog program for livestock.

In order to conserve land, the Defenders have also launched their own coffee line called Java Forest. The coffee beans are grown under the forest canopy or on farms that recreate a natural habitat. This reduces the large amount of land that is used for hybrid coffee beans. Twenty-five percent of each purchase is being returned to the Defenders to be used in other programs.

Defenders of Wildlife and its related organizations have approximately 1 million members. In addition to wildlife viewing guides for different states, their publications include a bimonthly magazine for members called *Defenders* and *In Defense of Wildlife: Preserving Communities and Corridors*.

Resources

BOOKS

Sinclair, Anthony R. E. *Wildlife Ecology, Conservation, and Mangement*. Malden, MA; Oxford, UK: Blackwell Publishing, 2006.

Urbigkit, Cat. *Yellowstone Wolves: A Chronicle of the Animal, the People, and the Politics*. Blacksburg, VA: McDonald and Woodward, 2008.

OTHER

United States Department of the Interior, United States Geological Survey (USGS). "Wildlife." http://www.usgs.gov/science/science.php?term=1331 (accessed October 13, 2010).

ORGANIZATIONS

Defenders of Wildlife, 1130 17th Street NW, Washington, DC, USA, 20036, (800) 385-9712, idefenders@mail.defenders.org, http://www.defenders.org.

Douglas Smith

Defoliation

Several factors can cause a plant to lose its leaves and become defoliated. Defoliation is a natural and regular occurrence in the case of deciduous trees and shrubs that drop their leaves each year with the approach of winter. This process is aided by an abscission layer that develops at the base of the leaf petiole, weakens the attachment to the plant, and eventually causes the leaf to drop. Severe drought may also cause leaves to wilt, dry, and drop from a plant. The result of severe dehydration is usually lethal for herbaceous plants, although some woody species may survive an episode of drying. Heavy infestation by leaf-eating insects can lead to partial or complete defoliation. The gypsy moth (*Porthetria dispar*) is an important defoliator that attacks many trees, defoliating, and weakening, or killing them. Parasitic wasps that feed on the larvae can help to control gypsy moth outbreaks. Insecticide sprays have also been used to kill the larvae. Spider mites, any of the plant-feeding mites of the family Tetranychidae (subclass Acari), feed on houseplants and the foliage and fruit of orchard trees. Heavy infestation can lead to serious or complete defoliation. Spider mites are controlled with pesticides, although growing resistance to chemical control agents has made this more difficult, and alternative control measures are under investigation.

Along with natural causes of defoliation, chemicals can cause plants to drop their leaves. The best-known and most widely used chemical defoliators are 2,4,5 trichlorophenoxyacetic acid (2,4,5-T) and 2,4 dichlorophenoxyacetic acid (2,4-D). Both chemicals are especially toxic to broadleaf plants. The herbicide 2,4-D is widely used in lawn care products to rid lawns of dandelions, clover, and other broadleaf plants that interfere with robust turf development. At appropriate application rates, it selectively kills broadleaf herbaceous plants and has little effect on narrow-leaf grasses. The uses of 2,4,5-T are similar, although it has been more widely used against woody species.

A mixture of 2,4-D and 2,4,5-T in a product called Agent Orange has been extensively used to control the growth and spread of woody trees and shrubs in sites earmarked for industrial or commercial development. Agent Orange saw extensive use by American forces in Southeast Asia during the Vietnam War, where it was used initially to clear for power lines, roads, railroads, and other lines of communication. Eventually, as the war continued, it was used to spray enemy hiding places, and U.S. military base perimeters to prevent surprise attacks. Food crops, especially rice, were also targets for Agent Orange to deprive enemy forces of food. Although Agent Orange and other formulations containing the chlorinated phenoxy acetic acid derivatives are generally lethal to herbaceous plants, woody deciduous plants may survive one or more treatments, depending on the species treated, concentrations used, spacing of applications, and weather. In Vietnam it was found that mangrove forests in the Mekong delta were especially sensitive, and often killed by a single treatment. A member of the dioxin family of chemicals, 2,3,7,8-tetrachlorodibenzo-p-dioxin (TCDD) has been found to be an accidental, but common, contaminant of 2,4,5-T and Agent Orange. Dioxins are very resistant to attack by microbes in the environment and are apt to persist in soils for a very long time. Although few disorders have been definitively proven to be caused by dioxins, their effects on laboratory animals have caused some scientists to rank them among the most poisonous substances known. The U.S. government banned some 2,4,5-T containing products in 1979 because of uncertainties regarding its safety, but its use continues in other products.

Resources

BOOKS

Suzuki, David T., and Wayne Grady. *Tree: A Life Story.* Vancouver: Greystone Books, 2004.

Tudge, Colin. *The Tree: A Natural History of What Trees Are, How They Live, and Why They Matter.* New York: Crown Publishers, 2006.

OTHER

World Health Organization (WHO). "Dioxins and their effects on human health." http://www.who.int/entity/mediacentre/factsheets/fs225/en/index.html (accessed November 6, 2010).

Douglas C. Pratt

Deforestation

Deforestation is the complete removal of a forest ecosystem and conversion of the land to another type of landscape. It differs from clear-cutting, which entails complete removal of all standing trees but leaves the soil in a condition to regrow a new forest if seeds are available.

Humans destroy forests for many reasons. American Indians burned forests to convert them to grasslands that supported big game animals. Early settlers cut and burned forest to convert them to croplands. Between 1600 and 1909, European settlement decreased forest cover in the United States by 30 percent. Since that time, total forest acreage in the United States has actually increased. In Germany about two-thirds of the forest was lost through settlement.

The Food and Agriculture Organization (FAO) defines forest as land with more than 10 percent tree

Deforestation of temperate rain forest. *(Photoshot Holdings Ltd./ Alamy)*

cover, natural understory vegetation, nature animals, natural soils, and no agriculture. Analysis of deforestation is difficult because data are unreliable and the definitions for *forest* and *deforestation* keep changing; for example, clear-cuttings that reforest within five years have been considered deforested in some studies but not in others.

The world's forest area totals just over 10 billion acres (4 billion ha) or 31 percent of the total land area. Worldwide, the net loss of forests amounted to 13 million acres (5.2 million ha) per year from 2000 to 2010, totaling an amount of land roughly equivalent in size to the Central American country of Costa Rica. Most deforestation occurred in South America with a predominant amount of the continent's deforestation occurring in the Amazon basin. Forest management, conservation, and preservation efforts have helped global deforestation decline from a peak in the 1990s of 20.75 million acres (8.3 million ha) annually.

The major direct causes of tropical deforestation are the expansion of shifting agriculture, livestock production, and fuelwood harvest in drier regions. Forest conversion to permanent cropland, infrastructure, urban areas, and commercial fisheries also occurs. Although not necessarily resulting in deforestation, timber harvest, grazing, and fires can severely degrade the forest. The environmental costs of deforestation can include species extinction, erosion, flooding, reduced land productivity, desertification, and climate change and increased atmospheric carbon dioxide. As more habitat is destroyed, more species are facing extinctions. Deforestation of watersheds causes erosion, flooding, and siltation. Upstream land loses fertile topsoil and downstream crops are flooded, hydroelectric reservoirs are filled with silt, and fisheries are destroyed. In drier areas, deforestation contributes to desertification.

Deforestation can alter local and regional climates because evaporation of water from leaves makes up as much as two-thirds of the rain that falls in some forest. Without trees to hold back surface runoff and block wind, available moisture is quickly drained away and winds dry the soil, sometimes resulting in desert-like conditions. Another potential effect on climate is the large-scale release into the atmosphere of carbon dioxide stored as organic carbon in forests and forest soils. Tropical deforestation releases 1.65 billion tons (1.5 billion metric tons) of carbon each year into the atmosphere.

As a result of misguided deforestation in the moist and dry tropics, the rural poor are deprived of construction materials, fuel, food, and cash crops harvested from the forest. Species extinctions, siltation, and flooding expand these problems to national and international levels. Despite these human and environmental costs, wasteful deforestation continues. Actions in the early twenty-first century to halt and reverse deforestation focus on creating economic and social incentives to reduce wasteful land conversion by providing for more sustainable ways to satisfy human needs. Other efforts are the reforestation of deforested areas and the establishment and maintenance of biodiversity preserves.

Resources

PERIODICALS

Bala, G., et al. "Combined Climate and Carbon-Cycle Effects of Large-Scale Deforestation." *Proceedings of the National Academy of Sciences* 104 (April 17, 2007): 6550–6555.

Forneri, Claudio, et al. "Keeping the Forest for the Climate's Sake: Avoiding Deforestation in Developing Countries Under the UNFCCC." *Climate Policy* 6 (2006): 275–294.

OTHER

National Geographic Society. "Deforestation." http://environment.nationalgeographic.com/environment/global-warming/deforestation-overview.html (accessed October 2, 2010).

United States Department of the Interior, United States Geological Survey (USGS). "Deforestation." http://www.usgs.gov/science/science.php?term=1354 (accessed October 2, 2010).

Edward Sucoff

Delaney Clause

The Delaney Clause is a part of the Federal Food, Drug, and Cosmetic Act of 1958, Section 409, and it prohibits the addition to food of any substance that will cause cancer in animals or humans. The clause states "no additive will be deemed to be safe if it is found to induce cancer when ingested by man or animal, or if it is found, after tests which are appropriate for the evaluation of the safety of food additives, to induce cancer in man or animals." The clause addresses the safety of food intended for human consumption and few, if any, reasonable individuals would argue with its intent.

There is, however, an emerging scientific controversy over its application, and many now question the merits of the clause as it is written. For example, safrole occurs naturally as a constituent in sassafras tea and spices and thus permissibly under the Delaney Clause, but it is illegal and banned as an additive to natural root beer because it has been proven a carcinogen in animal tests. Coffee is regularly consumed by many individuals, yet more than 70 percent of the tested chemicals that occur naturally in coffee have been shown to be carcinogenic in one or more tests. Naturally occurring carcinogens are found in other foods including lettuce, apples, pears, orange juice, and peanut butter. It is important to note here that the National Cancer Institute recommends the consumption of fruits and vegetables as part of a regimen to reduce cancer risk. This is because it is widely believed that the positive effects of fruits and vegetables far outweigh the potential hazard of trace quantities of naturally occurring carcinogens.

It has been estimated that about 10,000 natural pesticides of plants are consumed in the human diet. These natural pesticides protect the plants from disease and predation by other organisms. Only a few of these natural plant pesticides (less than sixty) have been adequately tested for carcinogenic potential and, of these, about one-half of them tested positive. It has been argued that such naturally occurring chemicals are less hazardous and thus differ in their cancer-causing potential from synthetic chemicals. But this does not appear to be the case; although the mechanisms for chemical carcinogenesis are poorly understood, there seems to be no fundamental difference in how natural and synthetic carcinogens are metabolized in the body.

The Delaney Clause addresses only the issue of additives to the food supply. It is noteworthy that salt, sugar, corn syrup, citric acid, and baking soda comprise 98 percent of the additives listed, while chemical additives, which many fear, constitute only a small fraction. It should also be noted that there are other significant safety issues pertaining to the food supply, including pathogens that cause botulism, hepatitis, and salmonella food poisoning. Of similar concern to health are traces of environmental pollutants, such as mercury in fish, and cooking-induced production of carcinogens, such as benzopyrene in beef cooked over an open flame. Excess fat in the diet is also thought to be a significant health hazard.

Scientists who are rethinking the significance of the "zero risk" requirement of the Delaney Clause do not believe society should be unconcerned about chemicals added to food. They simply believe the clause is no longer consistent with more modern scientific knowledge, and they argue that chemicals added in trace quantities,

Delaney Clause

for worthwhile reasons, should be considered from a different perspective.

The United States Environmental Protection Agency attempted to ease restrictions on several pesticides which posed almost no discernable risk to humans. However, the revised regulation was overturned by a federal Court of Appeals in 1992. In 1996, an amendment to the Food Quality Protection Act removed pesticide use from Delaney Clause regulation.

See also Agricultural chemicals; Agricultural pollution; Drinking-water supply; Food and Drug Administration.

Resources

BOOKS

Davies, Lee, et al. *Addressing Foodborne Threats to Health Policies, Practices, and Global Coordination: Workshop Summary.* Washington, DC: National Academies Press, 2006.

Entis, Phyllis. *Food Safety: Old Habits, New Perspectives.* Washington, DC: ASM Press, 2007.

OTHER

U.S. Government; science.gov. "Food Safety." http://www.science.gov/browse/w_105F.htm (accessed October 13, 2010).

Robert G. McKinnell

Demographic transition

Developed by demographer Frank Notestein in 1945, this concept describes the typical pattern of falling death- and birthrates in response to better living conditions associated with economic development. This idea is important, for it offers the hope that developing countries will follow the same pathway to population stability as have industrialized countries. In response to the Industrial Revolution, for example, Europe experienced a population explosion during the nineteenth century. Emigration helped alleviate overpopulation, but European couples clearly decided on their own to limit family size.

Notestein identified three phases of demographic transition: preindustrial, developing, and modern industrialized societies. Many authors add a fourth phase, postindustrial. In phase one, birthrates and death rates are both high with stable populations. As development provides a better food supply and sanitation, death rates begin to plummet, marking the onset of phase two. However, birthrates remain high, as families follow the pattern of preceding generations. The gap between high birthrates and falling death rates produces a population explosion, sometimes doubling in less than twenty-five years.

After one or two generations of large, surviving families, birthrates begin to taper off; and as the population ages, death rates rise. Finally a new balance is established, phase three, with low birthrates and death rates. The population is now much larger yet stable. The experience of some European countries, especially in Central Europe and Russia, suggests a fourth phase where populations actually decline. This may be a response to past hardships and oppressive political systems there, however.

Historically, birthrates have always been high. With few exceptions, population explosions are linked to declining death rates, not rising birthrates. Infants and young children are especially vulnerable; sanitation and proper food are vital. Infant survival is seen by some as a threat because of the built-in momentum for population growth. However, history reveals that there has been no decline in birthrates, which has not been preceded by a drop in infant mortality. In a burgeoning world this makes infant survival a matter of top priority. To this end, in 1986 the United Nations adopted a program with the acronym GOBI: Growth monitoring, Oral rehydration therapy (to combat killer diarrhea), Breast feeding, and Immunization against major communicable diseases.

See also Child survival revolution.

Resources

BOOKS

Newbold, K. Bruce. *Six Billion Plus: World Population in the Twenty-First Century.* Lanham, MD: Rowman & Littlefield Publishers, 2007.

Poston, Dudley L., and Michael Micklin. *Handbook of Population.* New York: Springer, 2005.

Weeks, John R. *Population: An Introduction to Concepts and Issues.* Belmont, CA: Wadsworth Publishing, 2004.

OTHER

United Nations System-Wide EarthWatch. "Demography." http://earthwatch.unep.net/demography/index.php (accessed November 4, 2010).

Nathan H. Meleen

Dendroica kirtlandii see **Kirtland's warbler.**

Denitrification

A stage in the nitrogen cycle in which nitrates in the soil or in dead organic matter are converted into nitrite, nitrous oxide, ammonia, or (primarily) elemental nitrogen. The process is made possible by certain types of bacteria, known as denitrifying bacteria. Denitrification is a reduction reaction and occurs, therefore, in the absence of oxygen. For example, flooded soil is likely to experience significant denitrification because of low atmospheric oxygen exposure. Although denitrification is an important process for the decay of dead organisms, it can also be responsible for the loss of natural and synthetic fertilizers from the soil.

Deoxyribose nucleic acid

Deoxyribose nucleic acid (DNA) molecules contain genetic information that is the blueprint for life. DNA is made up of long chains of subunits called nucleotides, which are nitrogenous bases attached to ribose sugar molecules. Two of these chains intertwine in the famous double helix structure discovered in 1953 by James Watson (1928–) and Francis Crick (1916–2004).

The genetic information contained in DNA molecules is in a code spelled out by the linear sequence of nucleotides in each chain. Each group of three nucleotides makes up a codon, a unit resembling a letter in the alphabet. A string of codons effectively spells a word of the genetic message.

This message is expressed when enzymes (cellular proteins) synthesize new proteins using a copy of a short segment of DNA as a template. Each nucleotide codon specifies which amino acid subunit is inserted as the protein is formed, thus determining the structure and function of the proteins. Because the chains are very long, a single DNA strand can contain enough information to direct the synthesis of hundreds of different proteins. Since these proteins make up the cell structure and the machinery (enzymes) by which cells carry out the processes of life, such as synthesizing more molecules including more copies of the DNA itself, DNA can be said to be self-replicating. When cells divide, each of the new cells receives a duplicate set of DNA molecules giving them the necessary information to live and reproduce.

See also Ribonucleic acid.

Resources

BOOKS

Watson, James D., and John Tooze. *The DNA Story: A Documentary History of Gene Cloning.* San Francisco: W. H. Freeman and Company, 1981.

OTHER

British Broadcasting Corporation (BBC) News. "Watson and Crick (1928–)." http://www.bbc. co.uk/history/historic_figures/watson_and_crick.shtml (accessed November 8, 2010).
Chemical Heritage Foundation. "James Watson, Francis Crick, Maurice Wilkins, and Rosalind Franklin." http://www.chemheritage.org/discover/chemistry-in-history/themes/biomolecules/dna/watson-crick-wilkins-franklin.aspx (accessed November 8, 2010).

Department of Agriculture *see* **U.S. Department of Agriculture.**

Department of Energy *see* **U.S. Department of Energy.**

Department of Health and Human Services *see* **U.S. Department of Health and Human Services.**

Department of the Interior *see* **U.S. Department of the Interior.**

Desalinization

Desalinization, also known as desalination, is the process of separating seawater or brackish water from their dissolved salts. The average salt content of the ocean water is about 3.4 percent (normally expressed as 34 parts per thousand). The range of salt content varies from 18 parts per thousand in the North Sea and near the mouths of large rivers to a high of 44 parts per thousand in locked bodies of water such as the Red Sea, where evaporation is very high. The desalination process is accomplished commercially by either distillation or reverse osmosis (RO).

Distillation of seawater is accomplished by boiling water and condensing the vapor. The components of the distillation system consist of a boiler and a condenser with a source of cooling water. Reverse osmosis is accomplished by forcing filtered seawater or brackish water through a reverse osmosis membrane. In a reverse osmosis process, approximately 45 percent of the pressurized seawater goes through membranes and becomes freshwater. The remaining brine (concentrated saltwater) is returned to the sea.

In 1980, the United Nations declared the period between 1981 and 1990 as the "International Drinking Water Supply and Sanitation Decade." The objective was to provide safe drinking water and sanitation to developing nations. Despite some progress in India, Indonesia, and a few other countries, the percentage of the world population with access to safe drinking water has not changed much since that declaration. In the period between 1990 and 2000, the amount of people with access has only increased by 5 percent.

The World Health Organization (WHO) estimates that only two in five people in the less-developed countries (LDCs) have access to safe drinking water. The WHO also estimates that at least 25 million people of the LDCs die each year because of polluted water and from water-borne diseases such as cholera, polio, dysentery, and typhoid. Whether by distillation or by reverse osmosis, desalination of water can transform water that is unusable because of its salinity into valuable freshwater. This could be an important water source in many drought-prone areas.

Desalination plants, distribution, and functions

As of 2010, there are approximately 7,500 desalination plants worldwide. Collectively they produce less than 0.1 percent of the world's freshwater supply. This supply is equal to about 3.5 billion gallons (13 million l) per day. The cost and the feasibility of producing desalinated water depends upon the cost of energy, labor, and relative costs of desalinated water to that of imported freshwater. It is estimated that in the United States, commercial desalinated water produced from seawater by reverse osmosis costs about $3 per 1,000 gallons (3,785 l). As of 2010, this price is four to five times the average price paid by urban consumers for drinking water and over 100 times the price paid by farmers for irrigation water. As of 2010, the energy requirement is approximately 3 kilowatt hours of electricity per 1 gallon (3.8 l) of freshwater extracted from seawater. Currently, using desalinated water for agriculture is cost-prohibitive.

About two-thirds of the desalination water is produced in Saudi Arabia, Kuwait, and North Africa. Several small-scale reverse osmosis plants are now operating in the United States, including California (Santa Barbara, Catalina Island, and soon in Ventura and other coastal communities). Generally, desalination plants are used to supplement the existing freshwater supply in areas adjacent to oceans and seas such as southern California, the Persian Gulf region, and other dry coastal areas. Among the advantages of desalinized water are a dependable water supply regardless of rainfall patterns, elimination of water rights disputes, and the preservation of the freshwater supply, all of which are essential for existing natural ecosystems.

Reverse osmosis

Reverse osmosis involves forcing water under pressure through a filtration membrane that has pores small enough to allow water molecules to pass through but that exclude slightly larger dissolved salt molecules. The basic parts of a reverse osmosis system include onshore and offshore components. The onshore components consist of a water pump, an electrical power source, pretreatment filtration (to remove seaweed and debris), reverse osmosis units connected in series, solid waste disposal equipment, and freshwater pumps. The offshore components consist of an underwater intake pipeline, approximately 1,093 yards (995 m) from shore, and a second pipeline for brine discharge.

Small reverse osmosis units for home use with a few gallons-per-day capacity are available. These units use a disposable reverse osmosis membrane. Their main drawback is that they waste four to five times the volume of water they purify.

Producing potable water from seawater is an energy-intensive, costly process. The high cost of producing desalinized water limits its use to domestic consumption. In areas such as the Persian Gulf and Saudi Arabia, where energy is plentiful at a low cost, desalinized water is a viable option for drinking water and very limited greenhouse agriculture. The notion of using desalinized water for wider agricultural purposes is neither practical nor economical at today's energy prices and with the available technology.

See also Salinization of soils.

Resources

BOOKS

Hoffman, Steve. *Planet Water: Investing in the World's Most Valuable Resource.* New York: Wiley, 2009.
Pearce, Fred. *When the Rivers Run Dry: Water — The Defining Crisis of the Twenty-first Century.* Boston: Beacon Press, 2007.
Solomon, Steven. *Water: The Epic Struggle for Wealth, Power, and Civilization.* New York: Harper, 2010.

Muthena Naseri

Desert

Deserts are areas that receive less than 10 inches (25 cm) of rain per year. These arid lands cover 25 percent of the earth's land surface. Deserts usually have seasonal high temperatures, low sporadic rainfall, and a high evaporation rate. They are often found in areas of high pressure—such as subtropical highs or leeward sides of mountains—associated with descending divergent air masses that are common between 30° north and 30° south latitude. Northern and southern Africa, the Arabian Peninsula, southern Asia, Australia, southern South America, and the North American Southwest lie in these subtropical zones.

There are three kinds of deserts—hot (such as the Sahara), temperate (such as the Mojave), and cold (such as the Gobi). The area of global desert is increasing yearly, as marginal lands become degraded by human misuse and climate change resulting in desertification.

Desert tortoise

The desert tortoise (*Gopherus agassizii*) is a large, herbivorous, terrestrial turtle of the family Testudinidae. It is found in both the southwestern United States and in northwestern Mexico. It is the official reptile in the states of California and Nevada. No other turtle in North America shares the extreme conditions of the habitats occupied by the desert tortoise. It inhabits desert oases, washes, rocky hillsides, and canyon bottoms with sandy or gravelly soil under hot, arid conditions.

Desert tortoises dig into dry, gravelly soil under bushes in arroyo banks or at the base of cliffs to construct a burrow, which is their home. Climatic conditions dictate daily activity patterns of these tortoises, and they can relieve the problems of high body temperature and evaporative water loss by retreating into their burrows. Since many desert tortoises live in areas devoid of water, except for infrequent rains, they must rely on their food for their water.

The active period for the desert tortoise is from March through September, after which they enter a hibernation period. Nesting and egg-laying activities extend from May through July. Desert tortoises lay an average of five moisture-proof eggs, an adaptation that helps retain water in its harsh environment. These tortoises reach sexual maturity between fifteen and twenty years, and they have a life span of up to eighty years.

The desert tortoise (*Gopherus agassizii*) is very senstive to human disturbances. *(Janusz Wrobel/Alamy)*

The desert tortoise is very sensitive to human disturbances, and this has led to the decimation of many of its populations throughout the desert Southwest. The Beaver Dam Slope population of southwestern Utah has been studied over several decades and shows some of the general tendencies of the overall population. In the 1930s and 1940s the desert tortoise population in this area exhibited densities of about 160 adults per square mile (2.6 sqaure km). By the 1970s this density had fallen to less than 130 adults per square mile, and more recent studies indicate the level is now about sixty adults per square mile. In southeastern California at least one population reaches densities of 200 adults per square mile, but overall tendencies show that populations are drastically declining. More recent, estimates indicate that there are about 100,000 individual desert tortoises existing in the Mojave and Sonoran deserts.

Desert tortoise populations are listed as threatened in Arizona, California, Nevada, and Utah. Numerous factors are contributing to its decline and vulnerability. Habitat loss through human encroachment and development, overcollecting for the pet trade, and vandalism—including shooting tortoises for target practice and flipping them over onto their backs, causing them to die from exposure—have decimated populations. Other factors contributing to their decline are grazing livestock, which trample them or their burrows, and mining operations, which also causes respiratory infections among the desert tortoises. Numerous desert tortoises have been killed or maimed by off-road vehicles, which also collapse the tortoises' burrows. Concern is mounting as conservation efforts seem to be having little effect throughout much of the desert tortoise's range.

Resources

BOOKS

Chambers, Paul. *A Sheltered Life: The Unexpected History of the Giant Tortoise.* Oxford, UK: Oxford University Press, 2006.

Eugene C. Beckham

Desertification

More than two-thirds of the world's land surface is affected by desertification, with arid areas covering about 40 percent of the earth's surface. About 2 billion people live in arid or semiarid desert lands in 100 countries on five continents. In these drier parts of the world, deserts are increasing rapidly from a combination of natural processes and human activities, a process known as desertification or land degradation. An annual rainfall of less than 10 inches (25 cm) will produce a desert anywhere in the world. In the semiarid areas along the desert margins, where the annual rainfall is around 16 inches (40 cm), the ecosystem is inherently fragile with seasonal rains supporting the temporary growth of plants. Recent changes in the climate of these regions have meant that the rains are now unreliable and the lands that were once semiarid are now becoming desert. The process of desertification is precipitated by prolonged droughts, causing the top layers of the soil to dry out and blow away. The eroded soils become unstable and compacted and do not readily allow for seeding. This means that desertified areas do not regenerate by themselves, but remain bare and continue to erode. Desertification of grazing lands or croplands is accompanied, therefore, by a sharp drop in the productivity of the land.

Natural desertification is greatly accelerated by human activities that leave soils vulnerable to erosion by wind and water. The drier grasslands with too little rain to support cultivated crops have traditionally been used for grazing livestock. When semiarid land is overgrazed (by keeping too many animals on too little land), plants that could survive moderate grazing are uprooted and destroyed altogether. Since plant roots no longer bind the soil together, the exposed soil dries out and is blown away as dust. The destruction and removal of the topsoil means that soil productivity drops drastically. The obvious solution to desertification caused by overgrazing is to limit grazing to what the land can sustain, a concept that is easy to espouse but difficult to practice.

In the Sahel zone along the southern edge of the Sahara desert, settled agriculture and overgrazing livestock on the fragile scrublands have led to widespread soil erosion. Nomadic pastoralists, who have traditionally followed their herds and flocks in search of new pastures, are now prevented by national borders from reaching their chosen grazing grounds. Instead of migrating, the nomads have been encouraged to settle permanently and this has led to their herds overgrazing available pastures.

Other human factors leading to desertification include over-cultivation, deforestation, salting of the soil through irrigation, and the plowing of marginal land. These destructive practices are intensified in developing countries by rapid population growth, high population density, poverty, and poor land management. The consequences of desertification in some

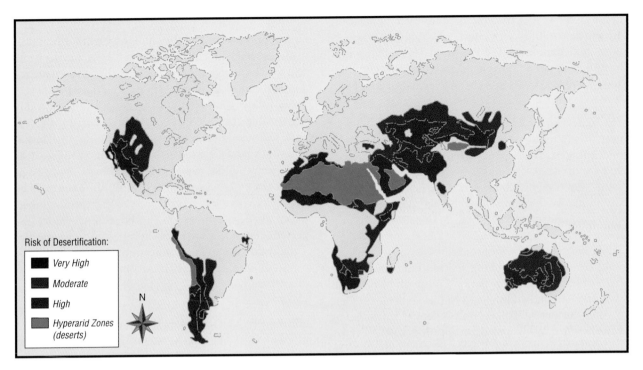

Map of the world's deserts and areas at risk of desertification. *(Reproduced by permission of Gale, a part of Cengage Learning)*

Risk of Desertification:
- Very High
- Moderate
- High
- Hyperarid Zones (deserts)

N

countries mean intensified drought and famine and lowered standards of living. It is estimated that desertification worldwide has claimed an area the size of Brazil (2 billion acres [810 million ha]) in the past fifty years. Each year new deserts consume an area the size of Belgium (15 million acres [6 million ha]), most of which is in the African Sahel.

In marginal areas throughout the world, traditional farming practices can lead to desertification. Plowing turns the top layer of the soil upside down, burying and killing weeds but exposing bare soil to erosion. In arid areas the exposed soil dries out rapidly and is easily lost through wind erosion.

The processes of erosion and soil formation vary with climate and with the composition of the parent material. In balanced ecosystems, soil lost to erosion is replaced by new soil created by natural processes. On average, new soil is formed at a rate of about 5 tons (4.5 metric tons) an acre (0.4 ha) per year, which is equivalent to a layer of soil about 0.2 inches (0.5 cm) thick. This means that soils can sustain an erosion rate of up to 5 tons per acre per year and still remain in balance. However, much of the world's crop and forested land is not within this balance (with erosion running at two to ten times the tolerable rate). In the United States about 6 tons (5.4 metric tons) of topsoil are lost for every ton of grains produced.

Forests are cut down for many different reasons. Some are cleared for agriculture, some for construction, some for paper products, and some to meet cooking and heating needs. Unfortunately, deforestation results in more than just the loss of trees, for soil is eroded, nutrients are lost to the ecosystem, and the water cycle is disrupted. The roots of the trees serve to bind the soil together and to hold water in the ecosystem, while the leaves of the trees break the force of the rain and allow it to soak into the topsoil. The result is that surface runoff from a forested hillside is one-half as much as from a grass-covered slope. Additionally, water soaking into the ground (rather than running off) leads to the natural recharge of groundwater and other water resources. Water and soil runoff from deforested hillsides cause flooding and siltation of agricultural and aquatic ecosystems in adjacent lowlands. Forests are also more efficient at reabsorbing and recycling the nutrients released from decaying detritus than are grasslands. Clearing forests therefore exposes the soil to both erosion and nutrient loss and alters the recharge of water reserves in the ecosystem.

Industrialized countries experienced a period of intense deforestation during the Industrial Revolution and even the early twenty-first century much of the land has low productivity. Fortunately, most of these countries are now reforesting faster than they are deforesting. This is possible because their population growth is low, their agricultural production per acre is high, and their need for fuelwood for cooking and heating is optional, since fossil fuels and electricity are widely available. All of these factors release the pressure to further deforest the land.

In developing countries, by contrast, high population growth rates and widespread poverty put pressure on the forests. Trees are needed for firewood, charcoal, and export, and the land is needed for farmland. In some developing countries deforestation exceeds replanting by five times. In Ethiopia, population and economic pressures pushed people to deforest and cultivate hillsides and marginally dry lands, and more than 1 billion tons (800 million metric tons) of topsoil per year are now lost, resulting in recurrent famines. In parts of India and Africa there is so little wood available that dried animal dung is used to fuel cooking fires, an act that further robs the soil of potential nutrients. In Brazil, soil erosion and desertification have resulted from the conversion of forests to cattle ranches. In China, about one-third of its agricultural land has been lost to erosion, and the story is similar for many other countries.

The answer to erosion from deforestation is reforestation, better forest conservation, and better forest management to increase productivity. Planting trees on hillsides is particularly effective. Reforesting desertified areas first requires mulching the soil to hold moisture and the protection of the seedlings for several years until natural processes can regenerate the soil. Using these methods, Israel has achieved spectacular success in bringing desertlands (a product of past desertification) back to agriculture.

Desertification and its agents, deforestation and erosion, have been powerful shapers of human history. Agriculture had its roots in the once fertile crescent of the Middle East and in the Mediterranean lands. However, deforestation, overgrazing, and poor agricultural practices have turned once-productive pastureland and farmland into the near deserts of 2010. Some people assert that deforestation and desertification may have even contributed to the collapse of the Greek and Roman empires. Similar fates may have befallen the Harappan civilization in India's Indus Valley, the Mayan civilization in Central America, and the Anasazi civilization of Chaco Canyon, New Mexico, in what is today desertland.

Current changes in climate can also affect desertification, but the issue is complex. Arid and semiarid regions have decreased precipitation, which results in the degradation of land, thus reducing vegetation and promoting erosion. Changes in vegetation cover and land properties impact the climate by increasing evaporation and diminishing water storage. In some regions climate change is likely to alter the frequency and severity of drought. However, it does not necessarily follow that drought and desiccation are solely responsible for the desertification occurring in dryland regions. Human contributions to desertification regarding land degradation are significant factors as well. Nevertheless, in policy terms, climate change and desertification are often perceived differently. The impact of climate change is viewed more on a global-scale, while desertification can be seen as widely dispersed effects of local environmental degradation. Thus, addressing the connections between the two processes despite the complexity can aid management of arid regions.

In 1977, the United Nations Convention on Desertification (UNCOD) established the Plan of Action to Combat Desertification (PACD). By 1991, the United Nations Environment Program (UNEP) determined that the rate of land degradation and desertification was escalating. The United Nations Convention to Combat Desertification (UNCCD) was established in 1994 to address the growing concern of land degradation leading to desertification and promoting sustainable development and management practices for desertification prevention. More than 179 countries were involved in this convention as of 2002. The convention has resulted in the development of drought early-warning systems and preventing the risks of drought. Local and national policies have been formed to address the effects of drought and the risks of desertification. Development of farm management practices for improving and maintaining soil fertility and retention of water has been accomplished to ensure crop productivity.

In the developing world, particularly in Africa, where poverty and political unrest are common, progress toward mitigation of desertification and its devastating social costs has been slow. Many of the world's poorest people, who live in countries with the weakest and most corrupt governments, rely on unsustainable agriculture and nomadic grazing to subsist. The World Bank estimates that an estimated 485 million Africans—equivalent to 65 percent of the African population—are affected by land degradation and desertification. Nearly 11 percent of the entire continent is affected by the problem. In Ghana and Mali, the World Bank estimates that the gross income loss due to desertification is equivalent to more than 5 percent of the gross domestic product (GDP). In Ethiopia and Malawi, the loss is about 10 percent of GDP.

Resources

BOOKS

Adapting to Drought: Farmers, Famines, and Desertification in West Africa. Cambridge, UK: Cambridge University Press, 2009.

Ezcurra, Exequiel. *Global Deserts Outlook*. Nairobi, Kenya: United Nations Environment Programme, 2006.

Geist, Helmut. *The Causes and Progression of Desertification (Ashgate Studies in Environmental Policy and Practice)*. Hants, UK: Ashgate Publishing, 2004.

Johnson, Pierre-Marc; Karel Mayrand; and Marc Paquin. *Governing Global Desertification: Linking Environmental Degradation, Poverty, and Participation*. Aldershot, UK: Ashgate, 2006.

United Nations General Assembly (sixty-first session: 2006–2007). 2nd Committee. *Sustainable Development: Implementation of the United Nations Convention to Combat Desertification in Those Countries Experiencing Serious Drought and/or Desertification, Particularly in Africa: Report of the 2nd Committee: General Assembly, sixty-first Session*. New York: United Nations, 2006.

OTHER

United Nations System-Wide EarthWatch. "Desertification." http://earthwatch.unep.net/desertification/index.php (accessed October 15, 2010).

United States Department of the Interior, United States Geological Survey (USGS). "Desertification." http://www.usgs.gov/science/science.php?term = 246 (accessed October 15, 2010).

United States Environmental Protection Agency (EPA). "Air: Mobile Sources: Vehicle Emissions." http://www.epa.gov/ebtpages/airmobilesourcesvehicleemissions.html (accessed October 15, 2010).

Neil Cumberlidge

Detergents

Detergents are a group of organic compounds that serve as cleansing agents based on their surface-active properties. A detergent molecule is termed surface-active because a portion of the molecule is hydrophilic and a portion is hydrophobic. It will therefore collect at the interface of water and another medium such as a gas bubble. Bubbles are stabilized by the surface-active molecules so that when the bubbles rise to the top of the bulk liquid, they maintain their integrity and form a foam. Surface-active agents can agglomerate by virtue of their hydrophilic or hydrophobic nature to form *micelles* that can dissolve, trap, or envelop soil particles, oil, and grease. Surface-active agents are sometimes called *surfactants*. Synthetic detergents are sometimes referred to as *syndets*.

The components of detergents can have an adverse effect on the environment. Many of the ingredients of detergents present low toxicity, but are still harmful with regard to ecosystem health, especially in aquatic ecosystems. Surfactants can biodegrade, but are toxic to aquatic organisms and persistent in the environment. Builders are inorganic phosphates that can accumulate in bodies of water due to runoff or from wastewater. Increased concentration of phosphates in water bodies can induce eutrophication, which occurs when nutrient-rich waters lead to low levels of dissolved oxygen, impairing the health of aquatic organisms. The excess available nutrients lead to enhanced primary productivity, thus causing algal blooms which block sunlight and affect oxygen levels. Builders in household detergents account for as much as 50 percent of the phosphorous (P) in municipal wastewater. Other components are bleaches, solvents, optical brighteners, and colorants.

The United States Environmental Protection Agency (EPA) established a Safer Detergents Stewardship Initiative (SDSI) to acknowledge companies that voluntarily develop detergents containing ingredients that have low environmental impact. Safer surfactants that biodegrade into non polluting substances are promoted to reduce the impact on aquatic ecosystems. Detergents with lower levels of phosphates or use of alternatives to phosphate builders would also help reduce the environmental impact of detergents.

Resources

BOOKS

Showell, Michael S. *Handbook of Detergents. Part D, Formulation*, Vol. 28. Boca Raton, FL: Taylor & Francis, 2006.

OTHER

Water Information System for Europe. "Phosphates and Alternative Detergent Builders." http://ec.europa.eu/environment/water/pollution/phosphates/index_en.htm (accessed September 11, 2010).

United States Environmental Protection Agency (EPA). "Key Characteristics of Laundry Detergent Ingredients." http://www.epa.gov/dfe/pubs/laundry/techfact/keychar.htm (accessed September 11, 2010).

United States Environmental Protection Agency (EPA). "Safer Detergents Stewardship Initiative (SDSI)." http://www.epa.gov/dfe/pubs/projects/formulat/sdsi.htm (accessed September 11, 2010).

Detoxification

When many toxic substances are introduced into the environment, they do not remain in their original form, but are transformed to other products by a variety of biological and nonbiological processes. The chemicals and their transformation products are degraded, converted progressively to smaller molecules, and eventually utilized in various natural cycles, such as the carbon cycle. Toxic metals are not degraded but are interconverted between available and nonavailable forms. Some organic compounds, such as polychlorinated biphenyl (PCBs), are degraded over a period of many years to less toxic compounds, while compounds such as the organophosphate insecticides may break down in only a few hours.

The chemical transformations that occur may either increase (referred to as intoxication, or activation if the parent compound was nontoxic) or decrease (referred to as detoxification) the toxicity of the original compound. For example, elemental mercury, which has low toxicity, can be converted to methylmercury, a very hazardous chemical, through methylation. Parathion is a fairly nontoxic insecticide until it is converted in a living system or by photochemical reactions to paraoxon, an extremely toxic chemical. However, parathion can also be degraded to less toxic products by the process of hydrolysis.

In microbial degradation, the ultimate fate of the toxic chemicals may be mineralization (that is, the conversion of an organic compound to inorganic products), which results characteristically in detoxification; however, intermediates in the degradation sequence, which may be toxic or have unknown toxicity, may persist for a period of time, or even indefinitely. Likewise, since degradation pathways may contain many steps, detoxification may occur early in the degradation pathway, before mineralization has occurred. Detoxification may also be accomplished biologically through cometabolism, the metabolism by microorganisms of a compound that cannot be used as a nutrient; cometabolism does not result in mineralization, and organic transformation products remain. Studies have also shown that the structure and toxicity of many organic compounds can be altered by plants.

In modifying a toxic chemical, detoxifying processes destroy its actual or potential harmful influence on one or more susceptible animal, plant, or microbial species. Detoxification may be measured with the use of bioassays. A bioassay involves the determination of the relative toxicity of a substance by comparing its effect on a test organism with the conditions of a control. The scale or degree of response may include the rate of growth or decrease of a population, colony, or individual; a behavioral, physiological, or reproductive response; or a response measuring mortality.

Bioassays can be used for environmental samples through time to determine detoxification of chemicals.

Both acute and chronic bioassays are used to assess detoxification. In an acute bioassay, a severe and rapid response to the toxic chemical is observed within a short period of time (for example, within four days for fish and other aquatic organisms and within twenty-four hours to two weeks for mammalian species). Detoxification of a chemical may be detected if there is a decrease in the observed toxicity of a test solution over the time of the acute test, indicating removal of the toxic chemical by degradation or other processes. Similarly, an increase in toxicity could indicate the formation of a more toxic transformation product.

Chronic bioassays are more likely to provide information on the rates of degradation, transformation, and detoxification of toxic compounds. Partial or complete life cycle bioassays may be used, with measurements of growth, reproduction, maturation, spawning, hatching, survival, behavior, and bioaccumulation.

Detoxification of chemicals should also be measured by toxicity testing that involves changes in different organisms and interactions among organisms, especially if the chemicals are persistent and stable and may accumulate and magnify in the food chain. Model ecosystems can be used to simulate processes and assess detoxification in a terrestrial-aquatic ecosystem. A typical ecosystem could include soil organisms, lake-bottom fauna, a plant, an insect, a snail, an alga, a crustacean, and a fish species maintained under controlled conditions for a period of time.

Most major types of reactions that result in transformation and detoxification of toxic chemicals can be accomplished either by biological (enzymatic) or by nonbiological (nonenzymatic) mechanisms. Although significant changes in structure and properties of organic compounds may result from nonbiological processes, the biological mechanism is the major and often the only mechanism by which organic compounds are converted to inorganic products. Microorganisms are capable of degrading and detoxifying a wide variety of organic compounds; presumably, every organic molecule can be destroyed by one or more types of microorganisms (referred to as the principle of microbial infallibility) However, since some organic compounds do accumulate in the environment, there must be factors such as unfavorable environmental conditions that prevent the complete degradation and detoxification of these persistent compounds. There are many examples where certain microorganisms have been identified as capable of detoxifying specific organic compounds. In some cases, these microorganisms can be isolated, cultured, and inoculated into contaminated environments in order to detoxify the compounds of concern.

The major types of transformation reactions include oxidation, ring scission, photodecomposition, combustion, reduction, dehydrohalogenation, hydrolysis, hydration, conjugation, and chelation. Conjugation is the only reaction mediated by enzymes alone, while chelation is strictly nonenzymatic. Primary changes in organic compounds are usually accomplished by oxidative, hydrolytic, or reductive reactions.

Oxidation reactions are reactions in which energy is used in the incorporation of molecular oxygen into the toxic molecule. In most mammalian systems, a monooxygenase system is involved. One atom of molecular oxygen is added to the toxic chemical, which usually results in a decrease in toxicity and an increase in water solubility, as well as provides a reaction group that can be used in further transformation processes such as conjugation. Microorganisms use a dioxygenase system, in which oxidation is accomplished by adding both atoms of molecular oxygen to the double bond present in various aromatic (containing benzene-like rings) hydrocarbons.

Ring scission, or opening, of aromatic ring compounds also can occur through oxidation. Though aromatic ring compounds are usually stable in the environment, some microorganisms are able to open aromatic rings by oxidation. After the aromatic rings are opened, the compounds may be further degraded by other organisms or processes. The number, type, and position of substituted molecules on the aromatic ring may protect the ring from enzymatic attack and may retard scission.

Photodecomposition can also result in the detoxification of toxic chemicals in the atmosphere, in water, and on the surface of solid materials such as plant leaves and soil particles. The reaction is usually enhanced in the presence of water; photodecomposition is also important in the detoxification of evaporated compounds. The ultraviolet radiation in sunlight is responsible for most photodecomposition processes. In photooxidation, for example, photons of light provide the necessary energy to mediate the reactions with oxygen to accomplish oxidation.

Combustion of toxic chemicals involves the oxidation of compounds accompanied by a release of energy. Often combustion does not completely result in the degradation of chemicals, and may result in the production of very toxic combustion products. However, if operating conditions are properly controlled, combustion can result in the detoxification of toxic chemicals.

Under anaerobic conditions, toxic compounds may be detoxified enzymatically by reduction. An example of a reductive detoxifying process is the removal of halogens from halogenated compounds. Dehydrohalogenation is another anaerobic process that also results in the removal of halogens from compounds.

Hydrolysis is an important detoxification mechanism in which water is added to the molecular structure of the compound. The reaction can occur either enzymatically or nonenzymatically. Hydration of toxic compounds occurs when water is added enzymatically to the molecular structure of the compound.

Conjugation reactions involve the combination of foreign toxic compounds with endogenous, or internal, compounds to form conjugates that are water-soluble and can be eliminated from the biological organism. However, the toxic compound may still be available for uptake by other organisms in the environment. Endogenous compounds used in the conjugation process include sugars, amino acid residues, phosphates, and sulfur compounds.

Many metals can be detoxified by forming complexes with organic compounds by sharing electrons through the process of chelation. These complexes may be insoluble or nonavailable in the environment; thus the toxicant cannot affect the organism. Sorption of toxic compounds to solids in the environment, such as soil particles, as well as incorporation into humus, may also result in detoxification of the compounds.

Generally, the complete detoxification of a toxic compound is dependent on a number of different chemical reactions, both biological and nonbiological, proceeding simultaneously, and involving the original compound as well as the transformation products formed.

See also Biogeochemistry; Biomagnification; Chemical bond; Environmental stress; Incineration; Oxidation reduction reactions; Persistent compound; Water hyacinth.

Resources

BOOKS

Auyero, Javier, and Débora Alejandra Swistun. *Flammable: Environmental Suffering in an Argentine Shantytown.* New York: Oxford University Press, 2009.

Clapp, Jennifer. *Toxic Exports: The Transfer of Hazardous Wastes from Rich to Poor Countries*. Ithaca, NY: Cornell University Press, 2010.

Deverell, William, and Greg Hise. *Land of Sunshine: An Environmental History of Metropolitan Los Angeles*. Pittsburgh, PA: University of Pittsburgh Press, 2006.

Meuser, Helmut. *Contaminated Urban Soils*. New York: Springer, 2010.

Judith Sims

Detritivores

Detritivores are organisms within an ecosystem that feed on dead and decaying plant and animal material and waste (called detritus); detritivores represent more than one-half of the living biomass. Protozoa, polychaetes, nematodes, Fiddler crabs, and filter-feeders are a few examples of detritivores that live in the salt marsh ecosystem. (Fiddler crabs, for example, scoop up grains of sand and consume the small particles of decaying organic material between the grains.) While microbes would eventually decompose most material, detritivores speed up the process by comminuting and partly digesting the dead, organic material. This allows the microbes to get at such material more readily. The continuing decomposition process is vital to the existence of an ecosystem—essential in the maintenance of nutrient cycles, as well as the natural renewal of soil fertility.

Detritus

Detritus is dead and decaying matter including the wastes of organisms. It is composed of organic material resulting from the fragmentation and decomposition of plants and animals after they die. Detritus is decomposed by bacterial activity, which can help cycle nutrients back into the food chain. In aquatic environments, detritus may make up a substantial percentage of the particulate organic carbon (POC) that is suspended in the water column. Animals that consume detritus are called detritivores. Although detritus is available in large quantities in most ecosystems, it is usually not a very high-quality food and may be lacking in essential nitrogen or carbon compounds. Detritivores generally must expend a larger amount of energy to assimilate carbon and nutrients from detritus than from sources of food based on living plant or animal material. Some detritivores harbor beneficial bacteria or fungi in their intestinal systems to aid in the digestion of compounds that are difficult to degrade.

Marie H. Bundy

Development, sustainable *see* **Sustainable development.**

Dew point

An expression of humidity defined as the temperature to which air must be cooled to cause condensation of its water vapor content (dew formation) without the addition or subtraction of water vapor or changing pressure. At this point the air is saturated and relative humidity becomes 100 percent. When the dew point temperature is below freezing, it is also referred to as the frost point. The dew point is a conservative expression of humidity because it changes very little across a wide range of temperature and pressure, unlike relative humidity, which changes with both. Dew points, however, are affected by water vapor content in the air. High dew points indicate large amounts of water vapor in the air and low dew points indicate small amounts. Scientists measure dew points in several ways: with a dew point hygrometer; from known temperature and relative humidity values; or from the difference between dry and wet bulb temperatures using tables. They use this measurement to predict fog, frost, dew, and overnight minimum temperature.

Diapers *see* **Disposable diapers.**

Diazinon

An organophosphate pesticide. Malathion and parathion are other well-known organophosphate pesticides. The organophosphates inhibit the action of the enzyme cholinesterase, a critical component of the chain by which messages are passed from one nerve cell to the next. They are highly effective against a wide range of pests. However, since they tend to affect the human nervous system in the same way they affect insets, they tend to be dangerous to humans, wildlife, and the environment. One of the most widely used lawn chemicals, diazinon, has been implicated in killing songbirds, pet dogs and cats, and causing near fatal poisonings in humans. The Environmental Protection Agency (EPA) banned its use on golf courses and sod farms after numerous reports of its killing ducks, geese, and other birds. Diazinon was banned from outdoor products at the end of 2004.

See also Cholinesterase inhibitor.

Dichlorodiphenyl-trichloroethane

Dichlorodiphenyl-trichloroethane (DDT) can be degraded to several stable breakdown products, such as dichlorodiphenyldichloroethylene (DDE) and dichloro-diphenyldichloroethane (DDD). Usually DDT refers to the sum of all the DDT-related components.

DDT was first developed for use as an insecticide in Switzerland in 1939, and it was first used on a large scale on the Allied troops in World War II. Commercial, non military use began in the United States in 1945. The discovery of its insecticidal properties was considered to be one of the great moments in public health disease control, as it was found to be effective on the carriers of many leading causes of death throughout the world including malaria, dysentery, dengue fever, yellow fever, filariasis, encephalitis, typhus, cholera, and scabies. It could be sprayed to control mosquitoes and flies or applied directly in powder form to control lice and ticks. It was considered the "atomic bomb" of pesticides, as it benefited public health by direct control of more than fifty diseases and enhanced the world's food supply by agricultural pest control. It had eliminated mosquito transmission of malaria in the United States by 1953. In the first ten years of its use, it was estimated to have saved 5 million lives and prevented 100 million illnesses worldwide.

Use of DDT declined in the mid-1960s due to increased resistance of different species of mosquitos and flies and other pests, and increasing concerns regarding the potential harm ecosystems and human health. Although the potential hazard from dermal absorption is small when the compound is in dry or powdered form, if the compound is in oil or an organic solvent it is readily absorbed through the skin and represents a considerable hazard.

Primarily DDT affects the central nervous system, causing dizziness, hyperexcitability, nausea, headaches, tremors, and seizures from acute exposure. Death can result from respiratory failure. It also is a liver toxin, activating microsomal enzyme systems and causing liver tumors. DDE is of similar toxicity. It became the focus of much public debate in the United States after the publication of *Silent Spring* by Rachel Carson, who effectively dramatized the harm to birds, wildlife, and possibly humans from the widespread use of DDT. Extensive spraying programs to eradicate Dutch elm disease and the gypsy moth (*Porthetria dispar*) also caused widespread songbird mortality. Its accumulation in the food chain also led to chronic exposures to certain wildlife populations. Fish-eating birds were subject to reproductive failure due to eggshell thinning and sterility. DDT is resistant to breakdown and is transported long distances, making it ubiquitous in the world environment today. It was banned in the United States in 1972 after extensive government hearings, but is still in use in other parts of the world, mostly in developing countries, where it continues to be useful in the control of carrier-borne diseases.

See also Acute effects; Chronic effects; Detoxification.

Resources

BOOKS

Hamilton, Denis, and Stephen Crossley. *Pesticide Residues in Food and Drinking Water: Human Exposure and Risks.* New York: Wiley, 2004.

Nett, Mary T. *The Fate of Nutrients and Pesticides in the Urban Environment.* Washington, DC: American Chemical Society, 2008.

Roberts, Rodger. *The Green Killing Fields: The Need for DDT to Defeat Malaria and Reemerging Diseases.* Washington, DC: AEI Press, 2007.

OTHER

National Institutes of Health (NIH). "Pesticides." http://health.nih.gov/topic/Pesticides (accessed November 11, 2010).

U.S. Government; science.gov. "Pesticides, Insecticides, and Herbicides." http://www.science.gov/browse/w_123G. htm (accessed November 11, 2010).

U.S. Environmental Protection Agency (EPA). "Pesticides." http://www.epa.gov/ebtpages/pesticides. html (accessed November 11, 2010).

World Health Organization (WHO). "WHO Pesticides Evaluation Scheme (WHOPES)." WHO Programs and Projects. http://www.who.int/entity/whopes/en (accessed November 11, 2010).

Deborah L. Swackhammer

Dieback

Dieback refers to a rapid decrease in numbers experienced by a population of organisms that has temporarily exceeded, or *overshot*, its carrying capacity. Organisms at low trophic levels such as rodents or deer, as well as weed species of plants, experience dieback most often. Without pressure from predators or other limiting factors, such "opportunistic" species reproduce rapidly, consume food sources to depletion, and then experience a population crash due chiefly to starvation (though reproductive failure can also play a part in dieback). The presence of predators—for example, foxes in a meadow inhabited by small rodents—often results in a stabilizing effect on population numbers.

Digester *see* **Wastewater.**

Die-off

Die-offs describe sudden reductions in species population levels. The most well-known is probably the die-off of dinosaurs, but die-offs continue today in many regions of the world, across a broad range of ecosystems.

Frogs and their kin are mysteriously vanishing in some areas, and scientists suspect that factors including viruses, bacterial infections, fungal infections, and human alteration of ecosystems are variably responsible. Frogs are good indicators of environmental change because of the permeability of their skin. They are extremely susceptible to toxic substances on land and in water. Decreasing rainfall in some areas may be a factor in the die-offs, as is habitat loss due to wetlands drainage. Other scientists are investigating whether increased ultraviolet radiation due to ozone depletion is killing toads in the Cascade Mountain Range.

Other factors—such as acid rain, heavy metal and pesticide contamination of ponds and other surface water, and human predation in some areas—may be causing the amphibians to die off.

Scientists also have recorded widespread declines in the numbers of migratory songbirds in the Western Hemisphere and of wild mushrooms in Europe. The decline in these so-called indicator organisms is a sign of declining health of the overall ecosystems.

The decline of songbirds is attributed to the loss or fragmentation of habitat, particularly forests, throughout the songbirds' range from Canada to Central America and northern South America. Fungi populations in Europe are dying off, and scientists think it is more than a problem of overharvesting. The health of forests in Europe is closely linked to the fungi populations, which point to the ecological decline of the forests. Some scientists argue that air pollution is also playing a role in their decline.

Environmental and wildlife experts argue that, as a consequence, of the international failure to achieve a 2010 target to slow the rate of extinctions, population die-offs leading to extinction pose an increasingly broad environmental and economic peril.

Resources

BOOKS

Chivian, Eric, and Aaron Bernstein. *Sustaining Life: How Human Health Depends on Biodiversity*. Oxford, UK: Oxford University Press, 2008.

Erwin, Douglas H. *Extinction: How Life on Earth Nearly Ended 250 Million Years Ago*. Princeton, NJ: Princeton University Press, 2006.

Fastovsky, David E., et al. *The Evolution and Extinction of the Dinosaurs*. Cambridge, UK: Cambridge University Press, 2005.

McGavin, George C. *Endangered: Wildlife on the Brink of Extinction*. Tonawanda, NY: Firefly Books, 2006.

Ninan, K. N. *Conserving and Valuing Ecosystem Services and Biodiversity: Economic, Institutional, and Social Challenges*. London: Earthscan, 2009.

PERIODICALS

Higgins, Paul A. T. "Biodiversity Loss Under Existing Land Use and Climate Change: An Illustration Using Northern South America." *Global Ecology and Biogeography* 16 (2007): 197–204.

Pounds, J. Alan, et al. "Widespread Amphibian Extinctions from Epidemic Disease Driven by Global Warming." *Nature* 439 (2006): 161–167.

Raffaelli, David. "How Extinction Patterns Affect Ecosystems." *Science* 306 (2004): 1141-1142.

OTHER

National Geographic Society. "Dinosaur Extinction." http://science.nationalgeographic.com/science/prehistoric-world/dinosaur-extinction.html (accessed October 21, 2010).

National Geographic Society. "Mass Extinctions." http://science.nationalgeographic.com/science/prehistoric-world/mass-extinction.html (accessed October 21, 2010).

The Nature Conservancy. "Nature Conservancy Biodiversity Page." http://www.nature.org/initiatives/climate change/strategies/art21202.html (accessed October 21, 2010).

Linda Rehkopf

Dillard, Annie

1945–

American writer

Often compared to the American naturalist Henry David Thoreau, Dillard—a novelist, memoir writer, essayist, poet, and author of books about the natural world—is best-known for her acute observation of the land, the seasons, the changing weather, and the wildlife within her intensely seen environment. Though born in Pittsburgh, Pennsylvania, on April 30, 1945, Dillard's vision of nature's violence and beauty was most fully developed living in Virginia, where she received her BA, 1967, and MA, 1968, from Hollins College. She also lived in the Pacific Northwest from 1975 to 1979 as scholar-in-residence at the University of Western Washington, in Bellingham, and is adjunct professor of English and writer-in-residence at Wesleyan University, in Middletown, Connecticut. As of 2010, she remains married to her third husband, writer Bob Richardson. They live in Hillsborough, North Carolina, and Wythe County, Virginia. Since 1973, she has also been a columnist for *The Living Wilderness*, the magazine of the Wilderness Society, the leading organization advocating expansion of the nation's wilderness.

In 1975, Dillard won the Pulitzer prize for general nonfiction for her first book of prose, *Pilgrim at Tinker Creek* (1974). The book detailed a mystical excursion into the natural world, based on her life in the Roanoke valley, Virginia, where she had lived since 1965. Her vision of "power and beauty, grace tangled in a rapture of violence," of a world in which "carnivorous animals devour their prey alive," is also an intense celebration of the things seen as she wanders the Blue Ridge mountainside and the Roanoke creek banks, observing muskrat, deer, red-winged blackbirds, and the multitude of "free surprises" her environment unexpectedly, and fleetingly, displays. *Seeing* acquires a mystical primacy in Dillard's work.

The urgency of seeing is also conveyed in Dillard's book of essays, *Teaching a Stone to Talk* (1982), in which she writes: "At a certain point you say to the woods, to the sea, to the mountains, the world, Now I am ready. Now I will stop and be wholly attentive." Dillard suggests that, for the natural phenomena that humans do not use or eat, their only task is to witness. But in witnessing, she sees cruelty and suffering, making her question at a religious level what mystery lies at the heart of the created universe, of which she writes: "The world has signed a pact with the devil. The terms are clear: if you want to live, you have to die."

Unlike some natural historians or writers of environmental books, Dillard is not associated with a specific program for curbing the destructiveness of human civilization. Rather, what has been described as her loving attentiveness to the phenomenal world— "nature seen so clear and hard that the eyes tear," as one reviewer commented—allies her with the broader movement of writers whose works teach some other relationship to nature than exploitation. In *Holy the Firm* (1977), her seventy-six-page journal of several days spent in Northern Puget Sound, Dillard records such events as the burning of a seven-year-old girl in a plane crash and a moth's immolation in a candle flame to rehearse her theme of life's harshness but, at the same time, to note, "A hundred times through the fields and along the deep roads I've cried Holy." In the end, Dillard is a sojourner, a pilgrim, wandering the world, ecstatically attentive to nature's bloodiness and its beauty.

The popularity of Dillard's writing can be judged by the frequency with which her work has been reprinted. As well as excerpts included in multiauthor collections, the four-volume *Annie Dillard Library* appeared in 1989, followed by *Three by Annie Dillard* (1990) and *The Annie Dillard Reader* (1994). During these years, she also served as the coeditor of two volumes of prose— *The Best American Essays* (1988), with Robert Atwan, and *Modern American Memoirs* (1995), with Cort Conley—and crafted *Mornings Like This: Found Poems* (1995), a collection of excerpts from other writers' prose, which she reformatted into verse.

Though a minor work, *Mornings Like This* could be said to encapsulate all of the qualities that have made Dillard's work consistently popular among readers: clever and playful, it displays her wide learning and eclectic tastes, her interest in the intersection of nature and science with history and art, and her desire to create beauty and unity out of the lost and neglected fragments of human experience.

In June 2007, Dillard's latest novel as of this writing, *The Maytrees*, was published by HarperCollins.

Resources

BOOKS

Dillard, Annie. *An American Childhood*. New York: Harper and Row, 1987.

Dillard, Annie. *Pilgrim at Tinker Creek*. New York: Bantam Books, 1975.

Dillard, Annie. *Teaching a Stone to Talk: Expeditions and Encounters*. New York: Harper & Row Publishers, 1982.

David Clarke

Dioxin

Dioxin is strictly a single deoxygenated ring (molecular formula $C_4H_4O_2$), but commonly refers to a family of more complex polychlorinated chemicals such as polychlorinated biphenyls (PCBs), polychlorinated dibenzo dioxins (PCDDs), and polychlorinated dibenzo furans (PCDFs). Production of PCBs ceased in 1977, but PCDDs and PCDFs are by-products of combustion and industrial processes. Dioxin is manufactured in several chemical processes, including the production of herbicides and the chlorinated bleaching of pulp and paper. The issue of dioxin's toxicity is one of the most hotly debated in the scientific community, involving the federal government, industry, the press, and the general public. The 2001 Stockholm Convention on Persistent Organic Pollutants (POPs) listed PCBs, PCDDs, and PCDFs in its dirty dozen list and called for an immediate ban on production and use of these substances.

One PCDD is a by-product of certain manufacturing processes or products, 2,3,7,8-tetrachlorodibenzo-p-dioxin or (TCDD). It develops during the manufacture of two herbicides known as 2,4,5-T and 2,4-D, which are the two components found in Agent Orange, a defoliant used widely in the Vietnam War. While dioxin commonly refers to a particular compound known as TCDD, there are actually seventy-five different dioxins. TCDD has been studied in some manner since the 1940s, and the most recent information indicates that it is capable of interfering with a number of physiological systems. Its toxicity has been compared to plutonium and it has proven lethal to a variety of research animals, including guinea pigs, monkeys, rats, and rabbits.

TCDD is also the chemical that some scientists at the Environmental Protection Agency (EPA) and a broad spectrum of researchers have referred to as the most carcinogenic ever studied. During the late 1980s and early 1990s, it was linked to an increased risk of rare forms of cancer in humans—especially soft tissue sarcoma and non-Hodgkins lymphoma—at very high doses. Some have labeled dioxin as "the most toxic chemical known to man." According to American biologist Barry Commoner (1917–), Director of the Center for the Biology of Natural Systems at Washington University, the chemical is "so potent a killer that just 3 ounces of it placed in New York City's water supply could wipe out the populace."

Exposure can come from a number of sources. Dioxin can be airborne and inhaled in drifting incinerator ash, aerial spraying, or the hand-spray application of weed killers. Dioxin can be absorbed through the skin as people walk through a recently sprayed area such as a backyard or a golf course. Water runoff and leaching from agricultural lands treated with pesticides can pollute lakes, rivers, and underground aquifers. Thus, dioxin can be ingested in contaminated water, in fish, and in beef that has grazed on sprayed lands. Residues in plants consumed by animals and humans add to the contaminated food chain. Research has shown that nursing children now receive trace amounts of dioxin in their mother's milk.

Because it bioaccumulates (substances accumulating in living tissues) in the environment, TCDD continues to be found in the soil and waterways in microscopic quantities over twenty-five years after its first application. Dioxin is part of a growing family of chemicals known as organochlorines—a class of chemicals in which chlorine (Cl) is bonded with carbon. These chlorinated substances are created to make a number of products such as polyvinyl chloride, solvents, and refrigerants, as well as pesticides. Hundreds or thousands of organochlorines are produced as by-products when chlorine is used in the bleaching of pulp and paper or the disinfection of wastewater and when chlorinated chemicals are manufactured or incinerated. The by-products of these processes are toxic, persistent, and hormonally active. TCDD is also part of current manufacturing processes, such as the manufacture of the wood preservative, pentachlorophenol (C_6HCl_5O).

If the exposure to dioxin is intense, there can be an immediate response. Tears and watery nasal discharge have been reported, as have intense weakness, giddiness, vomiting, diarrhea, headaches, burning of the skin, and rapid heartbeat. Usually, a weakness persists and a severe skin eruption known as chloracne develops after a period of time. The body excretes very little dioxin, and the chemical can accumulate in the body fat after exposure. Minute quantities may be found in the body years after modest exposure. Since TCDD's half-life has been estimated at as much as ten to twelve years in the soil, it is possible that some TCDD—suggested to be as much as 7 parts per trillion (ppt)—is harbored in the bodies of most Americans.

The development of medical problems may appear shortly after exposure, or they may appear ten, twelve, or twenty years later. If the exposure is large, the symptoms develop more quickly, but there is a greater latency period for smaller exposures. This fact explains why humans exposed to TCDD may appear healthy for years before finally showing what many consider to be typical dioxin-exposure symptoms, such as cancer or immune system dysfunction. There is also a relationship between toxicology and individual susceptibility.

Certain people are more susceptible to the effects of dioxin exposure than others. Once a person has become susceptible to the chemical, he or she tends to develop cross-reactions to other materials that would not normally trigger any response.

Government publications and research funded by the chemical industry have questioned the relationship between dioxin exposure and many of these symptoms. But a growing number of private physicians treating people exposed to dioxins have become increasingly certain about patterns or clusters of symptoms. These studies have reported a higher incidence of cancer at sites of industrial accidents, including increases in rates of stomach cancer, lung cancer, soft-tissue sarcomas, and malignant lymphomas. Some reports have indicated that soft-tissue sarcomas in dioxin-exposed workers have increased by a factor of forty, and there have also been indications of psychological and personality changes and an excess of coronary disease.

Many theories about the medical effects of dioxin exposure are based on the case histories of the thousands of American military personnel exposed to Agent Orange during the Vietnam War. Agent Orange, a chemical defoliant, was used despite the fact that certain chemical companies and select members of the military knew about its toxic properties. Thousands of American ground troops were directly sprayed with the chemical. Those in the spraying planes inhaled the chemical directly when some of the herbicides were blown back by the wind into the open doors of their planes. Others were exposed to accidental applications from the sky, when planes in trouble had to evacuate their loads during emergency procedures.

Despite what many consider to be the obvious dangers of TCDD, industries continue to produce residues and market products contaminated with the chemical. White bleached paper goods contain quantities of TCDD because no agency has required the paper industry to change its bleaching process. In 1998, the EPA issued an effluent guideline—expected to reduce dioxin discharges by at least 95 percent—as part of the air and water regulations for the paper and pulp industry. However, women use dioxin-tainted, bleached tampons, and infants wear bleached, dioxin-tainted paper diapers. Some scientists have estimated that every person in the United States carries a body burden of dioxin that may already be unacceptable.

Many believe that the EPA has done less to regulate dioxin than it has done for almost any other toxic substance. Environmentalists and other activists have argued that any other chemical creating equivalent clusters of problems within specific groups of similarly

exposed victims would be considered an epidemic. Industry experts have often downplayed the problems of dioxin. A spokesman for Dow Chemical has stated that "outside of chloracne, no medical evidence exists to link up dioxin exposure to any medical problems." The federal government and federal agencies have also been accused of protecting their own interests. During congressional hearings in 1989 and 1990, the Centers for Disease Control was found to falsify epidemiology studies on Vietnam veterans.

In April 1991, the EPA initiated a series of studies intended to revise its estimate of dioxin's toxicity. The agency believed there was new scientific evidence worth considering. Several industries, particularly the paper industry, had also pressured the agency to initiate the studies, in the hope that public fears about dioxin toxicity could be allayed. But the first draft of the revised studies, issued in the summer of 1992, indicated more rather than fewer problems with dioxin. It appears to be the most damaging to animals exposed while still in the uterus. It also seems to affect behavior and learning ability, which suggests that it may be a neurotoxin. These studies have also noted the possibility of extensive effects on the immune system. The EPA has labeled dioxin as "a likely human carcinogen" that is "anticipated to increase the risk of cancer at background levels of exposure." The EPA's National Center for Environmental Assessment composed a report outlining the human health effects of dioxin exposure. The EPA dioxin reassessment is expected by 2011.

Other studies have established that dioxin functions like a steroid hormone. Steroid hormones are powerful chemicals that enter cells, bind to a receptor or protein, form a complex that then attaches to the cell's chromosomes, turning on and off chemical switches that may then affect distant parts of the body. It is not unusual for very small amounts of a steroid hormone to have major effects on the body. Based on these reasons, dioxin is considered to be an endocrine disruptor, capable of mimicking endogenous hormones and affecting development. Newer studies conducted on wildlife around the Great Lakes have shown that dioxin has the capacity to feminize male chicks and rats and masculinize female chicks and rats. In male animals, testicle size is reduced as is sperm count.

It is likely that dioxin will remain a subject of considerable controversy, both in the public realm and in the scientific community for some time to come. However, even those scientists who question dioxin's long-term toxic effect on humans, agree that the chemical is highly toxic to experimental animals. Dioxin researcher Nancy I. Kerkvliet of Oregon State University in Corvallis characterizes the situation in these terms, "The fact that you can't clearly show the effects in humans in no way lessens the fact that dioxin is an extremely potent chemical in animals—potent in terms of immunotoxicity, potent in terms of promoting cancer."

See also 2,4,5-T; Agent Orange; Bioaccumulation; Hazardous waste; Kepone; Organochloride; Pesticide residue; Pulp and paper mills; Seveso, Italy; Times Beach.

Resources

BOOKS

Allen, Robert. *The Dioxin War: Truth and Lies about a Perfect Poison*. London: Pluto Press, 2004.

CPHR Symposium in Health Research and Policy, and Heather Purnell. *Dioxin: Exposures, Health Effects and Public Health Policy: Proceedings of the Fifth Annual CPHR Symposium in Health Research and Policy, Wellington, 7th September 2005*. Wellington, New Zealand: Centre for Public Health Research, Massey University, Wellington Campus, 2006.

Pennsylvania Department of Environmental Protection. *What You Should Know about Dioxin*. Harrisburg, PA: Author 2004.

U.S. National Research Council. *Health Risks from Dioxin and Related Compounds: Evaluation of the EPA Reassessment*. Washington, DC: National Academies Press, 2006.

OTHER

United States Environmental Protection Agency (EPA). "EPA's Science Plan for Activities Related to Dioxins in the Environment." http://cfpub.epa.gov/ncea/cfm/recordisplay.cfm?deid=209690 (accessed August 6, 2010).

United States Environmental Protection Agency (EPA). "Pollutants/Toxics: Chemicals: Dioxins." http://www.epa.gov/ebtpages/pollchemicalsdioxins.html (accessed August 6, 2010).

World Health Organization (WHO). "Dioxins and their effects on human health." http://www.who.int/entity/mediacentre/factsheets/fs225/en/index.html (accessed August 6, 2010).

Liane Clorfene Casten

Discharge

Discharge is a term generally used to describe the release of a gas, liquid, or solid to a treatment facility or the environment. For example, wastewater may be discharged to a sewer or into a stream, and gas may be discharged into the atmosphere.

Disposable diapers

Disposable diapers are single-use diapers that are designed to soak up urine and retain feces, in contrast to cotton diapers that are washed and used repeatedly. The superabsorbent polymer-based material can retain several hours' output of urine, again in contrast to cotton diapers that can become soaked after a single urination.

Although there were experiments with disposable diapers by several companies in the 1940s and 1950s, disposable diapers were first successfully marketed by Procter & Gamble in 1961. The acknowledged inventor of the disposable diaper is American architect Marie Donovan (1917–1988); the apparel was termed *boaters* in 1950. In the United Kingdom, they are referred to as nappies.

First used as an occasional convenient substitute for cloth diapers, the popularity of disposable diapers skyrocketed in the 1960s and has remained the standard ever since. Indeed, as of 2010, disposable diapers are the primary diapering method for an estimated 95 percent of American parents. As a result, 3.7 million tons (3.4 million metric tons) of disposable diapers were discarded in 2007, a point decried by environmentalists. The average child will require about 5,000 diapers before being toilet trained and no longer require diaper aid; this represents approximately 1 ton (0.9 metric tons) of diapers per child. This waste accounts for 1.5 percent of total municipal solid waste. For comparison, only 350,000 tons (318,000 metric tons) of disposable diapers were discarded in 1970. Wood pulp is used in the construction of disposable diapers, which results in the use of an estimated 250,000 trees annually.

Proponents of reusable diapers argue that this accounts for only 2 to 3 percent of America's solid waste. However, statistics can reflect the study design and even the biases of the study funding agency (or private corporation), and so it is difficult to gauge the merits of disposable and reusable diapers. In October 2008, the United Kingdom Environment Agency published a life-cycle assessment of disposable and reusable nappies examining the environmental impacts. It suggests that washable nappies can save 40 percent carbon emissions over disposables.

Disposable diapers and their packaging create more solid waste than reusables, and because they are used only once, consume more raw materials—petrochemicals and wood pulp—in their manufacture. And although disposable diapers should be emptied into the toilet before the diapers are thrown away, many people skip this step, which puts feces that may be contaminated with pathogens into landfills and incinerators. There is no indication, however, that this practice has resulted in any increase in health problems. Cloth diapers affect the environment as well. They are made of cotton, which is watered with irrigation systems and treated with synthetic fertilizers and pesticides. They are laundered and dried up to seventy-eight (commercial) or 180 (home) times, consuming more water and energy than disposables. In fact, home laundering is less energy-efficient than commercial because it is done on a smaller scale. Diaper services make deliveries in trucks, which expends another measure of energy and generates more pollution. Human waste from cotton diapers is treated in sewer systems. Some disposable diapers are advertised as biodegradable and claim to pose less of a solid-waste problem than regular disposables. Their waterproof cover contains a cornstarch derivative that decomposes into water and carbon dioxide (CO_2) when exposed to water and air. Unfortunately, modern landfills are airtight and little, if any, degradation occurs. Biodegradable diapers, therefore, are not significantly different from other disposables.

Research studies show that the superabsorbent material used in some disposable diapers can release compounds that are reactive with the air. These volatile organic compounds include toluene, ethylbenzene, dipentene, and xylene.

Resources

PERIODICALS

Time. "LIVING—Cheap, Green, and Often Unclean: Cloth Diapers Make a Comeback." 57 (2008).

Pollock L. "Nappies and the Environment: The Debate Continues." *RCM Midwives: The Official Journal of the Royal College of Midwives*, 8 7 (2005): 2921-2923.

OTHER

Massachusetts Institute of Technology (MIT). "Inventor of the Week Archive: Marion Donovan (1917–1998): Disposable Diaper." http://web.mit.edu/invent/iow/donovan.html (accessed August 8, 2010).

National Geographic's Green Guide. "Diapers Buying Guide: Environmental Issues and Baby's Heath." http://www.thegreenguide.com/buying-guide/diapers/environmental_impact (accessed July 28, 2010).

United States Environmental Protection Agency (EPA). "Municipal Solid Waste in the United States: 2007 Facts and Figures." http://www.epa.gov/osw//nonhaz/municipal/pubs/msw07-rpt.pdf (accessed July 28, 2010).

Teresa C. Donkin

Dissolved oxygen

Dissolved oxygen (DO) refers to the amount of oxygen dissolved in water and is particularly important in limnology (aquatic ecology). Oxygen comprises approximately 21 percent of the total gas in the atmosphere; however, it is much less available in water. The amount of oxygen water can hold depends upon temperature (more oxygen can be dissolved in colder water), pressure (more oxygen can be dissolved in water at greater pressure), and salinity (more oxygen can be dissolved in water of lower salinity). Many lakes and ponds have anoxic (oxygen-deficient) bottom layers in the summer because of decomposition processes depleting the oxygen. The amount of dissolved oxygen often determines the number and types of organisms living in that body of water. For example, fish such as trout are sensitive to low DO levels (less than 8 parts per million) and cannot survive in warm, slow-moving streams or rivers. Decay of organic material in water caused by either chemical processes or microbial action on untreated sewage or dead vegetation can severely reduce dissolved oxygen concentration. This is a common cause of fish kills, especially in summer months when warm water oxygen levels can be further depleted by algae blooms and other factors.

Dissolved solids

Dissolved solids are minerals in solution, typically measured in parts per million (ppm) using an electrical conductance meter calibrated to oven-dried samples. In humid regions, dissolved solids are often the dominant form of sediment transport. Solution features, such as caverns and sinkholes, are common in limestone regions.

Water that contains excessive amounts of dissolved solids is unfit for drinking. Drinking-water standards typically allow a maximum of 250 ppm, the threshold for tasting sodium chloride; by comparison, ocean water ranges from 33,000 to 37,000 ppm. Phosphates and nitrates in solution are the major cause of eutrophication (nutrient enrichment resulting in excessive growth of algae). Dissolved solids buffer acid precipitation; lakes with low levels are especially vulnerable. High levels typically occur in runoff from newly disturbed landscapes, such as strip mines and road construction.

Diversity *see* **Biodiversity.**

DNA *see* **Deoxyribose nucleic acid.**

Dodo

One of the best-known extinct species, the dodo (*Raphus cucullatus*), a flightless bird native to the Indian Ocean island of Mauritius, disappeared around 1680. A member of the dove or pigeon family, and about the size of a large turkey, the dodo was a grayish white bird with a huge black-and-red beak, short legs, and small wings. The dodo did not have natural enemies until humans discovered the island in the early sixteenth century.

The dodo became extinct due to hunting by European sailors who collected the birds for food and to predation of eggs and chicks by introduced dogs, cats, pigs, monkeys, and rats. The Portuguese are credited with discovering Mauritius, where they found a tropical paradise with a unique collection of strange and colorful birds unafraid of humans: parrots and parakeets, pink and blue pigeons, owls, swallows, thrushes, hawks, sparrows, crows, and dodos. Unwary of predators, the birds would walk right up to human visitors, making themselves easy prey for sailors hungry for food and sport.

The Dutch followed the Portuguese and made the island a Dutch possession in 1598 after which Mauritius became a regular stopover for ships traversing the Indian Ocean. The dodos were subjected to regular slaughter by sailors, but the species managed to breed and survive on the remote areas of the island.

When the island became a Dutch colony in 1644, the colonists engaged in a seemingly conscious attempt to eradicate the birds, despite the fact that they were not pests or obstructive to human living. But they were easy to kill. The few dodos in inaccessible areas that could not be found by the colonists were eliminated

by the animals introduced by the settlers. By 1680, the last remnant survivors of the species were "as dead as a dodo."

Interestingly, while the dodo tree (*Calvaria major*) was once common on Mauritius, the tree seemed to stop reproducing after the dodo disappeared, and the only remaining specimens are about 300 years old. Apparently, a symbiotic relationship existed between the birds and the plants. The fruit of this tree was an important food source for the dodo. When the bird ate the fruit, the hard casing of the seed was crushed, allowing it to germinate when expelled by the dodo.

Three other related species of giant, flightless doves were also wiped out on nearby islands. The white dodo (*Victoriornis imperialis*) inhabited Reunion, 100 miles (161 km) southwest of Mauritius, and seems to have survived up to around 1770. The Reunion solitaire (*Ornithoptera solitarius*) was favored by humans for eating and was hunted to extinction by about 1700. The Rodriguez solitaire (*Pezophaps solitarius*), found on the island of Rodriguez 300 miles (483 km) east of Mauritius, was also widely hunted for food and disappeared by about 1780.

Resources

BOOKS

Barrow, Mark V. *Nature's Ghosts: Confronting Extinction from the Age of Jefferson to the Age of Ecology*. Chicago: University of Chicago Press, 2009.

Lewis G. Regenstein

Study of a dodo. Royal Albert Memorial Museum, Exeter, Devon, United Kingdom. *(F Hart/The Bridgeman Art Library/ Getty Images)*

Dolphins

There are thirty-two species of dolphins, members of the cetacean family Delphinidae, that are distributed in all of the oceans of the world. These marine mammals are usually found in relatively shallow waters of coastal zones, but some may be found in the open ocean. Dolphins are a relatively modern group; they evolved about 10 million years ago during the late Miocene. The Delphinidae represents the most diverse group, as well as the most abundant, of all cetaceans. Among the delphinids are the bottlenose dolphins (*Tursiops truncatus*), best-known for their performances in oceanaria; the spinner dolphin (*Stenella longirostris*), which have had their numbers decline due to tuna fishermen's nets; and the orca or killer whale (*Orcinus orca*), the largest of the dolphins. Dolphins

Bottlenosed doplhins (*Tursiops truncatus*) inhabit many of the world's temperate oceans. *(Walter G. Arce/ Shutterstock.com)*

are distinguished from their close relatives, the porpoises, by the presence of a beak.

Dolphins are intelligent, social creatures, and social structure is variously exhibited in dolphins. Inshore species usually form small herds of two to twelve individuals. Dolphins of more open waters have herds composed of up to 1,000 or more individuals. Dolphins communicate by means of echolocation, ranging from a series of clicks to ultrasonic sounds, which may also be used to stun their prey. By acting cooperatively, dolphins can locate and herd their food using this ability. Aggregations of dolphins also have a negative aspect, however. Mass strandings of dolphins, a behavior in which whole herds beach themselves and die en masse is a well-known phenomenon but little understood by biologists. Theories for this seemingly suicidal behavior include nematode parasite infections of the inner ears, which upsets their balance, orientation, or echolocation abilities; simple disorientation due to unfamiliar waters; or even perhaps magnetic disturbances.

Because of their tendency to congregate in large herds, particularly in feeding areas, dolphins have become vulnerable to large nets of commercial fishermen. Gill nets, laid down to catch oceanic salmon and capelin, also catch numerous nontarget species, including dolphins and inshore species of porpoises. In the eastern Pacific Ocean, especially during the 1960s and 1970s, dolphins have been trapped and drowned in the purse seines of the tuna fishing fleets. This industry was responsible for the deaths of an average of 113,000 dolphins annually and, in 1974 alone, killed over one-half million dolphins in their nets. Tuna fishermen have recently adopted special nets and different fishing procedures to protect the dolphins. A panel of netting with a finer mesh, the Medina panel, is part of the net farthest from the fishing vessel. Inflatable powerboats herd the tuna as the net is pulled under and around the school of fish. As the net is pulled toward the vessel, many dolphins are able to escape by jumping over the floats of the medina panel, but others are assisted by hand from the inflatable boats or by divers. The finer mesh prevents the dolphins from getting tangled in the net, unlike the large mesh which previously snared the dolphins as they sought escape. Consumer pressure and tuna boycotts were major factors behind this shift in techniques on the part of the tuna upper Mississippi Valley and in the Great Lakes and winter on the Gulf Coast.

The first double-crested cormorants in the Great Lakes region were sighted on the western shore of Lake Superior in 1913. By the 1940s, the population around the Great Lakes had increased to about 1,000 nesting pairs. Subsequently, the cormorant population began to decline. By 1973 a survey found only 100 pairs in the region.

In the early 1970s, cormorants were not the only waterbird whose population was declining. Around this time, the United States Congress enacted several laws to protect cormorants and other waterfowl. For example, the pesticide dichlorodiphenyl-trichloroethane (DDT) was banned in 1972. This chemical had entered lakes and rivers in runoff water and was implicated in reducing the birthrate of fish-eating birds, including the bald eagle. Congress also amended the Migratory Bird Treaty Act, first passed in 1917, to make it illegal to harm or kill cormorants and other migrating waterfowl.

The double-crested cormorant populations around the Great Lakes began to increase in the 1980s, and the birds expanded their range eastward to Lake Erie and Lake Ontario. The return of the double-crested cormorant seemed to some conservationists to signal that the Great Lakes, once seriously polluted, were recovering. However, one reason the cormorant population may have increased around the Great Lakes is that overfishing during this time seriously depleted the number of large fish in the lakes. This led to an increase in the number of smaller fish such has smelt (*Osmerus mordax*) and alewife (*Alosa pseudoharengus*). Cormorants eat these small fish. An increase in their food supply may have contributed to an increase in the cormorant population.

Another ecological problem, the zebra mussel, may also have helped the cormorant. The zebra mussel is an exotic (nonnative) invasive mussel that competes with native mussels for food resources. It is a voracious feeder and can clean lakes of green plankton, leaving the water particularly clear. Clear water may have helped the cormorant, which hunts fish by sight, find food more easily.

By the 1990s, some local cormorant populations had grown to unprecedented proportions, and at the same time sport fish populations had declined. As a result, some Great Lakes fishermen, refusing to consider the role of overfishing in the decline of fish populations, blamed double-crested cormorants for steep drops in populations of fish such as smallmouth bass (*Micropterus dolomieui*), rock bass (*Ambloplites rupestris*), and brown bullheads (*Ameiurus nebulosus*). An

adult cormorant weighs about 4 pounds (1.8 kg), and eats about 1 pound (0.45 kg) of fish a day. Some areas hosted flocks of thousands of cormorants. A 1991 study of cormorants in Lake Ontario estimated that the birds had consumed about 5 million pounds (2.25 million kg) of fish. In addition, cormorants tend to eat smaller fish; and if enough fish are eaten before they can reproduce, future generations of fish are threatened.

Although fishermen have blamed the cormorant for declining fish populations, conservationists have insisted on scientific studies to determine if these birds are indeed causing a decline in fish populations. A 1998 study of cormorants on Galloo Island in Lake Ontario found that the smallmouth bass, a popular sport fish, made up just 1.5 percent of the cormorant's diet. Yet because the cormorants ate small bass that had not yet grown to reproductive maturity, the bird was considered linked to the smallmouth's decline.

Another 2000 study of the Beaver Islands area in Lake Michigan concluded it likely that the large cormorant population was a factor in the declining numbers of smallmouth bass and other fish. Yet the biologist who led the study was unable to conclude that cormorants were entirely responsible for the decline in the fish population. However, wildlife officials took action to manage this cormorant population.

Cormorants also have made themselves unpopular because they nest in large colonies, where thick layers of their droppings can kill off underlying vegetation and leave the area denuded of all but a few trees. Cormorant colonies have endangered some sensitive woodland habitats and contributed to soil erosion by killing shoreline plants.

Because federal law protects double-crested cormorants, people cannot legally harass or kill these birds. In 1999, nine men were convicted of slaughtering 2,000 cormorants on Little Galloo Island in Lake Ontario. These men had illegally tried to reduce the cormorant colony by shooting adult birds. Soon after the incident, however, the New York Department of Environmental Conservation enacted a plan to reduce the Galloo colony from 7,500 to 1,500 birds over five years by spraying vegetable oil on cormorant eggs. The oil-coated eggs do not hatch. This method of thinning the flock was considered less disruptive to other wildlife and more humane than killing adult birds.

The New York program generated controversy, as not all scientists who had studied the birds believed that double-crested cormorants were responsible for the decline of the fish population, and some conservationists feared similar programs would be enacted against other fish-eating birds. Ironically, success in protecting the double-crested cormorant in the 1970s resulted in controversy two decades later. Even though the population of double-crested cormorants substantially increased, the bird is still under federal protection. Any action to manage the cormorant population must be carefully developed, implemented, and evaluated.

Resources

BOOKS

Lear, Linda J. *Rachel Carson: Witness for Nature.* Boston: Mariner Books, 2009.

Leslie, Scott. *Sea and Coastal Birds of North America.* New York: Key Porter Books, 2008.

O'Brien, Michael; Richard Crossley; and Kevin Karlson. *The Shorebird Guide.* New York: Houghton Mifflin Harcourt, 2006.

Angela Woodward

▌Douglas, Marjory Stoneman

1890–1998
American environmentalist and writer

A newspaper reporter, writer, and environmentalist renowned for her crusade to preserve Florida's Everglades. Marjory Stoneman Douglas was part of the committee that first advocated formation of the Everglades National Park in 1927, and she remained an active advocate of the area's preservation until her death. Born in 1890 in Minneapolis, Minnesota, and raised in Taunton, Massachusetts, Marjory Stoneman Douglas graduated from Wellesley College in 1912. After a brief marriage to Kenneth Douglas, a newspaper editor, she moved south to Florida in 1915 to join her father. She soon began to work as a reporter, columnist, and editor for the *Miami Herald,* founded by her father, Frank Bryant Stoneman. During World War I, she left Miami to become the first female enlistee for the Naval Reserves, then joined the Red Cross, for which she worked and traveled in Europe during and after the war. Returning to Coconut Grove in 1920, Douglas remained there for over seventy-five years.

Douglas was involved very early in efforts to preserve the Everglades from agricultural and residential development, as was her father before her. Her book, *The Everglades: River of Grass* was one of the most important statements publicizing the ecological importance and uniqueness of the area, as well as its plight in the face of drainage, filling, and water diversion. Her framing of the region as a "river of grass" effectively instilled in the general public an idea of the interconnectedness of the land, water, plants, and animals of the area, helping to raise public awareness of the urgency of preserving the entire ecosystem, not just isolated components of it.

Douglas did not start out as a full-time environmental advocate. She wrote *The Everglades: River of Grass* because she loved the history and the natural history of the area, and she was a longtime supporter of preserving the ecosystem, but she did not become deeply involved in the movement to save the Everglades until the 1970s. Friends of Douglas's in the National Audubon Society, confronted in 1969 by proposals to build an airport in the Everglades, enlisted her aid and she helped organize the Friends of the Everglades, an organization that continues to defend the Park and related south Florida ecosystems.

Douglas proved to be an eloquent and forceful speaker and writer in the cause to save the Everglades from development; and, since the 1970s, the Everglades became her single cause for celebrity. However, she also wrote poetry, short stories, histories, natural histories, and novels, nearly all based in Florida. Initially, her writing was popular, at least in part, because Florida and its history were little-known, and rarely written about when she began her literary career. Among her other publications are *Road to the Sun* (1951), *Hurricane* (1958), *Florida: The Long Frontier* (1967), and *Nine Florida Stories* (1990). She also wrote an autobiography, co-authored by J. Rothchild, *Marjory Stoneman Douglas: Voice of the River.*

In 1990, Douglas was honored on her 100th birthday with book signings, interviews, and banquets; and in 1992, she was back in action. That year, Douglas spoke out against President George Bush's proposal to modify the definition of *wetlands,* a move that critics pointed out could open the door to future development. President Bill Clinton in 1993 called to wish her a happy birthday as she turned 103 and, a few months later, awarded her the Presidential Medal of Freedom. In 1994, Florida state lawmakers passed the Everglades Forever Act; and, also in the 1990s, the federal government committed hundreds of millions of dollars to restore and protect the area. In 1996, Florida voters passed an amendment to their state's constitution that makes Everglades polluters, particularly sugar farmers, pay for cleanup costs; in addition, more plans to save the wetlands were expected. However, voters did not pass a law to tax sugar at a penny a pound to assist with the effort; sugar producers had successfully argued that the ruling would cost many jobs.

Contractors move sand pumped in through a dredge pipe in Gulf Shores, Alabama, May 2010. The sand is being piled alongside Little Lagoon Pass so it can be moved into the pass to keep oil from creeping into the lagoon. *(AP Photo/Michelle Rolls-Thomas)*

prior to the disposal of the dredged material. For example, disposal of sediment from a dredging project in Narragansett Bay, Rhode Island, changed the bottom topography and sediment type; and this change in benthic habitat led to a subsequent decline in the clam and finfish fishery at the site, and an increase in the lobster fishery. If the dredged material is similar to the sediment on which it is dumped, the area may be recolonized by the same species that were present prior to any dumping.

If dredged material is dumped in an area that has less than 197 feet (60 m) of water, most of the material will rapidly descend to the bottom as a high-density mass. A radial gradation of large-to-fine grained sediment usually occurs from the impact area of the deposition outward. Fine-grained material spreads outward from the disposal site, in some cases up to 328 feet (100 m), in the form of a fluid mud. It can range in thickness up to 3.9 inches (9.9 cm). From 1-5 percent of the sediment remains suspended in the water as a plume; this sediment plume

is transient in nature and eventually dissipates by dispersion and gravitational settling. The long-term fate of dredged material dumped in the marine environment depends on the location of the dumping site, its physical characteristics such as bottom topography and currents, and the nature of the sediment. Deep-ocean dumping of dredged material results in wider dispersal of the sediment in the water column. The deposition of the dredged material becomes more widely distributed over the ocean bottom than in nearshore areas.

Dredging contaminated sediment poses a much more severe problem for disposal. Disposing of contaminated dredged material in the marine environment can result in long-term degradation to the ecosystem. Sublethal effects, biomagnification of pollutants, and genetic disorders of organisms are some examples of possible long-term effects from toxic pollutants in contaminated dredged material entering the food chain. However, attributing effects from placement of contaminated dredged material at a marine site to a specific cause can be very difficult if other sources of contaminants are present.

Dredged material must be tested to determine contamination levels and the best method of disposal. These tests include bulk chemical analysis, the elutriate test, selective chemical leaching, and bioassays. Bulk chemical analysis involves measurements of volatile solids, chemical oxygen demand, oil and grease, nitrogen, mercury, lead, and zinc. But this chemical analysis does not necessarily provide an adequate assessment of the potential environmental impact on bottom-dwelling organisms from disposal of the dredged material. The elutriate test is designed to measure the potential release of chemical contaminants from suspended sediment caused by dredging and disposal activities. However, the test does not take into account some chemical factors governing sediment-water interactions, such as complexation, sorption, redox, and acid-base reactions.

Selective chemical leaching divides the total concentration of an element in a sediment into identified phases. This test is better than the bulk chemical analysis for providing information that will predict the impact of contaminants on the environment after the disposal of dredged material. Bioassay tests commonly use sensitive aquatic organisms to directly measure the effects of contaminants in dredged material as well as other waste materials. Different concentrations of wastes are measured by determining the waste dilution that results in 50 percent mortality of the test organisms. Permissible concentrations of contaminants can be identified using bioassay tests.

If dredged material is considered contaminated, special management and long-term maintenance are required to isolate it from the rest of the environment. Special management techniques can include capping dredged material disposed in water with an uncontaminated layer of sediment, a technique which is recommended in relatively quiescent, shallow water environments. Other management strategies to dispose of contaminated dredged material include the use of upland containment areas and containment islands. The use of submarine burrow pits has also been examined as a possible means to contain contaminated dredged material.

There is more than one law in the United States governing dredging and disposal operations. The General Survey Act of 1824 delegates responsibility to the Army Corps of Engineers (ACOE) for the improvement and maintenance of harbors and navigation. The ACOE is required to issue permits for any work in navigable waters, according to the Rivers and Harbors Act of 1899. The Marine Protection, Research, and Sanctuaries Act (MPRSA) of 1972 requires the ACOE to evaluate the transportation and ocean dumping of dredged material based on criteria developed by the Environmental Protection Agency (EPA), and to issue permits for approved nonfederal dredging projects. Designating ocean disposal sites for dredged material is the responsibility of the EPA. The discharge of dredged material through a pipeline is controlled by the Federal Water Pollution Control Act, as amended by the Clean Water Act (1977). This act requires the ACOE to regulate ocean discharges of dredged material and evaluate projects based on criteria developed by the EPA in consultation with the ACOE. Other Federal agencies such as the U.S. Fish and Wildlife Service and the National Marine Fisheries Service can provide comments and recommendations on any project, but the EPA has the power to veto the use of proposed disposal sites.

See also Agricultural pollution; Contaminated soil; Hazardous waste; LD_{50}; Runoff; Sedimentation; Synergism; Toxic substance; Urban runoff.

Resources

BOOKS

Denny, Mark W. *How the Ocean Works: An Introduction to Oceanography.* Princeton, NJ: Princeton University Press, 2008.
Earle, Sylvia, and Bill McKibben. *The World Is Blue: How Our Fate and the Ocean's Are One.* Washington, DC: National Geographic, 2010.
Garrison, Tom. *Oceanography: An Invitation to Marine Science.* New York: Brooks Cole, 2009.

Marci L. Bortman

Dreissena polymorpha *see* **Zebra mussel.**

Drift nets

Drift nets are used in large-scale commercial fishing operations. Miles-long in length, nets are suspended from floats at various depths and set adrift in open oceans to capture fish or squid. Drift nets are constructed of a light, plastic monofilament that resists rotting. These nets are generally of a type known as gill nets, because fish usually become entangled in them by their bony gill plates. The fishing industry has found these nets to be cost-effective, but their use has become increasingly controversial. They pose a severe threat to many forms of marine life, and they have long been the

Silky shark (*Carcharhinus falciformis*) in drift gill net, Sea of Cortez, Mexico. *(©Mark Conlin/Alamy)*

object of protests and direct action from a range of environmental groups.

Drift nets are not selective; there is no way to use them to target a particular species of fish. Drift nets can catch almost everything in their path, and there are few protections for species that were never intended to be caught. Although some nets can be quite efficient in capturing only certain species, the bycatch from drift nets can include not only noncommercial fish, but sea turtles, seabirds, seals and sea lions, sharks, porpoises, dolphins, and large whales. Nets that are set adrift from fishing vessels in the open ocean and never recovered pose an even more severe hazard to the marine environment. Lost nets can drift and kill animals for long periods of time, becoming what environmentalists have called "ghost nets."

Drift nets are favored by fishing industries in many countries because of their economic advantages. The equipment itself is relatively inexpensive; it is also less labor-intensive than other alternatives, and it supplies larger yields because of its ability to capture fish over such broad areas. Drift-net fisheries can vary considerably, according to the target species and the type of fishing environment. In coastal areas, short nets can be set and recovered in an hour. The nets do not drift very far in this time and the environmental damage can be limited. But in the open ocean, where the target species may be widely dispersed, nets in excess of 31 miles (50 km) in length may be set and allowed to drift for twenty-four hours before they are recovered and stripped. The primary targets for drift netting include squid in the northern Pacific, salmon in the northeastern Pacific, tuna in the southern Pacific and eastern Atlantic, and swordfish in the Mediterranean.

Because of their cost-effectiveness, the Food and Agricultural Organization (FAO) of the United Nations actively promoted the use of drift nets during the early 1980s. Earthwatch, Earth Island Institute, and other environmental groups instituted drift-net monitoring during this period and founded public education programs to pressure drift-net fishing nations. The Sea Shepherd Conservation Society and other direct action groups have actually intervened with drift-net fishing operations on the high seas. Organizations such as these led international awareness about the dangers of drift nets; and their efforts have affected national and international policy. In December of 1991, the United Nations reversed its earlier endorsement of drift-net fishing and adopted General Assembly Regulations 44-225, 45-197, and 46-215 that totally banned high seas drift nets. The regulations applied to all international waters and went into effect on December 31, 1992. An agreement between the United States and the South and Central American countries of Costa Rica, Ecuador, Mexico, Nicaragua, Panama, and Venezuela for a moratorium on drift nets was also signed. In 1998, the European Union banned all drift nets in its jurisdiction, with the exception of the Baltic Sea. The ban took effect in 2002. The regulatory body for the Baltic Sea is the International Baltic Sea Fishery Commission, which sets more liberal restrictions on nets and catch limits than the European Union.

According to the United States National Marine Fisheries Service, enforcement of the drift net ban in the north Pacific has been successful, and the U.S. salmon fishery in these waters is no longer impacted by illegal drift nets. Enforcement of this ban in other waters has so far proved difficult, however, as drift net fishing is done in open oceans, far from national jurisdictions. In reaching enforceable international agreements about drift net fishing, the primary problem has been the large investment some nations have in this technology. Japan had 457 fishing vessels using drift nets in 1990, and Taiwan and Korea approximately 140 vessels each. France, Italy, and other nations own smaller fleets.

Japan and many of these nations are primarily concerned with protecting their investment, despite worldwide protest. The United States and Canada have both expressed concern about ecological integrity and the rate of unintended catch in drift nets, particularly the bycatch of North American salmon. In the 1990s, bilateral negotiations were pursued with Korea and Taiwan to control their drift net fleets. The International North Pacific Fisheries Commission has provided a forum for the United States, Japan, and Canada to examine and discuss the economic advantages and environmental costs of drift net fishing. A special committee has analyzed bycatch from drift nets used in the northern Pacific. Three avenues are currently being considered to control ecological damage: the use of subsurface nets, research into the construction of biodegradable nets, and alternative gear types for capturing the same species. The Irish Sea Fisheries Board has tested trawl nets as an alternative to drift nets, and found that they are effective in catching tuna. However, environmental groups have raised the concern that trawl nets also capture significant numbers of marine mammals and have called for restrictions on their use after the Irish Sea Fisheries Board study found that more than 140 whales and dolphins were killed by trawlers during the two-year study.

Resources

BOOKS

Clover, Charles. *The End of the Line: How Overfishing Is Changing the World and What We Eat.* Berkeley: University of California Press, 2008.

PERIODICALS

Haedrich, Richard L. "Deep Trouble: Fishermen Have Been Casting Their Nets into the Deep Sea After Exhausting Shallow-water Stocks. But Adaptations to Deepwater Living Make the Fishes There Particularly Vulnerable to Overfishing—and Many Are Now Endangered." *Natural History* 116, *8* (October 2007): 28.

OTHER

United States Environmental Protection Agency (EPA). "Industry: Industries: Fishing Industry." http://www.epa.gov/ebtpages/induindustriesfishingindustry.html (accessed November 6, 2010).

Douglas Smith
Marie H. Bundy

Drinking-water supply

The Safe Drinking Water Act, passed in 1974, required the United States Environmental Protection Agency (EPA) to develop guidelines for the treatment and monitoring of public water systems. In 1986, amendments to the act accelerated the regulation of contaminants, banning the future use of lead pipe, and requiring surface water from most sources to be filtered and disinfected. The amendments also have provisions for greater groundwater protection. Despite the improvement these regulations represent, the Act only covers

public and private systems that serve a minimum of twenty-five people at least sixty days a year. As of 2010, there are more than 160,000 such water systems in the United States. Millions obtain their drinking water from privately owned wells that are not covered under the act. The act also does not pertain to bottled water.

Drinking water comes from two primary sources: surface water and groundwater. Surface water comes from a river or lake, and groundwater, which is pumped from underground sources, generally needs less treatment. Contaminants can originate either from the water source or from the treatment process.

The most common contaminants found in the public water supply are lead, nitrate, and radon—all of which pose substantial health threats. Studies indicate that substances such as chlorine and fluoride which are added to water during the treatment process may also have adverse effects on human health. Over 700 different contaminants have been found in water supplies in the United States, yet the EPA only sets standards for approximately ninety contaminants.

Chlorinated water was first used in the United States in 1908 as a means of reducing diseases in Chicago stockyards. Chlorine or a chlorine-based compound, which kills some disease-causing microbes, is now used to disinfect over 95 percent of the public water supply in the United States. While effective, it is well-known that chlorine reacts with organic compounds to form disinfection by-products (DBPs), which may increase the risk of certain kinds of cancer.

The effectiveness of fluoridated water in reducing dental cavities was first noted in communities with a naturally occurring source of fluoride in their drinking water, and controlled studies of communities where fluoride was added to the water confirmed the results. As of 2010, approximately 70 percent of Americans on public water systems receive fluoridated water. The EPA limit for fluoride in water is 4 parts per million (ppm), and most cities add only 1 ppm to their water. Fluoridated water also has adverse effects, and these may include immune system suppression, tooth discoloration, undesirable bone growth, enzyme inhibition, and carcinogenesis.

The EPA has set the acceptable level of lead in drinking water at 15 parts per billion (ppb), yet according to tests the agency has done, drinking water in almost 20 percent of cities in the United States exceeds that limit. Between 10 to 25 percent of a child's lead intake comes from drinking water, and the EPA cautions that the percentage could be much higher if the water contains high levels of lead. Depending on exposure, lead poisoning can cause permanent learning

When filled, each cask weighed 120 tons (108 metric tons). Casks were placed vertically on concrete pads that were 3-feet (1-m) thick. Each pad would hold twenty-eight casks.

By 2001, the NRC had approved various dry casks designs. The container is usually steel. After it is filled, the container is either bolted or welded shut. The metal casks are then put inside larger concrete casks to ensure radiation shielding. Some systems involve placing the steel cask vertically in a concrete vault. In other systems, the container is placed horizontally in the concrete vault.

Discussions about dry cask safety in the twenty-first century have centered on the Yucca Mountain proposal. In May of 2002, the United States House of Representatives voted to approve the plan. That vote was an override of Nevada Governor Kenny Guinn's veto of the plan to send waste to Yucca. Opponents of the plan like the Nuclear Information and Resource Service maintained that casks of waste could not be transported safely by train or truck. In 2002, the facility was expected to cost $58 billion. It was scheduled to open in 2010 and hold a maximum of 77,000 tons (69,300 metric tons) of waste, but in 2009 the Obama Administration announced it was withdrawing support to use the Yucca Mountain site.

See also Nuclear Regulatory Commission; Radioactive waste.

Resources

BOOKS

Committee on Radioactive Waste Management. *Managing Our Radioactive Waste Safely: CoRWM's Recommendations to Government.* London: Committee on Radioactive Waste Management, 2006.

European Community Conference, and R. Simon. *Radioactive Waste Management and Disposal.* Cambridge, UK: Cambridge University Press, 2009.

Rahman, A. A. *Decommissioning and Radioactive Waste Management.* Dunbeath, UK: Whittles, 2008.

ORGANIZATIONS

Nuclear Information and Resource Service., 1424 16th Street NW, #404, Washington, DC, USA, 20036, (202) 328-0002, (202) 462-2183, nirsnet@nirs.org, http://www.nirs.org.

United States Nuclear Regulatory Commission., One White Flint North, 11555 Rockville Pike, Rockville, MD, USA, 20852-2738, (301) 415-7000, (800) 368-5642, opa@nrc.gov, http://www.nrc.gov

Liz Swain

Dry cleaning

Dry cleaning is a process of cleaning clothes and fabrics with hydrophobic solutions—solutions that do not contain water. The practice has been traced back to France where, around 1825, turpentine was used in the cleaning process. The process is reputed to have been based on an accident where camphene, a fuel for oil lamps, was accidentally spilled on a gown and found to clean it. Because of this, dry cleaning was referred to as "French cleaning" even into the second half of the twentieth century.

By the late 1800s, naphtha, gasoline, benzene, and benzol—the most common solvent—were being used for dry cleaning. Fire hazards associated with using gasoline for dry cleaning prompted the United States Department of Commerce in March 1928 to issue a standard for dry cleaning specifying that a dry cleaning solvent derived from petroleum must have a minimum flash point (the temperature at which it combusts) of $100°F$ ($38°C$). This was known as the Stoddard solvent.

The first chlorinated solvent used in dry cleaning was carbon tetrachloride. It continued to be used until the 1950s when its toxicity and corrosiveness were determined to be hazardous. By the 1930s, the use of trichloroethylene had become common. In the 1990s the chemical was still being used in industrial cleaning plants and on a limited basis in Europe. This chemical's incompatibility with acetate dyes used in the United States brought about the end of its use in the United States. Tetrachloroethylene replaced other dry cleaning solvents almost completely by the 1940s and 1950s. In 1990 about 53 percent of worldwide demand for tetrachloroethylene was for dry cleaning, and approximately 75 percent of all dry cleaners used it. However, in Japan petroleum-based solvents continued in use through the 1990s. By the late 1990s, perchloroethylene (PCE) replaced tetrachloroethylene as the predominant cleaning solvent.

Surface and groundwater sources can be contaminated by spills of dry-cleaning solvents. In the early 1980s, an upsurge in illnesses in and near the town of New Minas, Nova Scotia, Canada, was traced to groundwater that had been contaminated by a local dry cleaning business. Many similar incidents have been reported throughout North America.

When the United States Environmental Protection Agency (EPA) issued national regulations to control air emissions of perc from dry cleaners in September 1993, environmental groups and consumers began to pay closer attention to the possible negative impact this chemical could have on human health. In July 2001, the American Council on Science and Health issued a report concluding that perc was not hazardous to humans at the levels most commonly used in dry cleaning.

However, the claim that PCE is a carcinogen (cancer-causing substance) has received prominent public and governmental attention. Concern has been expressed that environmental exposures to PCE in outdoor or indoor air and in drinking water can cause cancer in humans.

Results of some epidemiological studies of dry cleaning and chemical workers exposed to PCE have been interpreted to suggest a relationship between occupational exposure and various types of cancer. Careful examination of the way in which these studies were conducted reveals serious problems including uncertainties about the amount of PCE to which people were exposed, failure to take into account exposure to other chemicals at the same time, and failure to take into account known confounders. Because of these deficiencies, these studies do not support a link between PCE and cancer or other adverse effects in humans.

Nonetheless, in 2006, the EPA strengthened the regulations for dry cleaning use of perc in the United States. While alternatives to perc exist, such as liquid carbon dioxide, as of 2010, an estimated 28,000 dry cleaning establishments across the United States continue to use perc.

An organization called the State Coalition for Remediation of Dry Cleaners, as of 2010, had chapters in thirteen states seeking to end the use of perc by dry cleaners located in urban areas by 2020.

Resources

BOOKS

Hoffman, Steve. *Planet Water: Investing in the World's Most Valuable Resource.* New York: Wiley, 2009.
Jorgensen, Erik. *Ecotoxicology.* New York: Academic Press, 2010.
Newman, Michael C. *Fundamentals of Ecotoxicology.* 3rd ed. Boca Raton, FL: CRC Press, 2009.

Jane E. Spear

Dry deposition

A process that removes airborne materials from the atmosphere and deposits them on a surface. Dry deposition includes the settling or falling-out of particles due to the influence of gravity. It also includes the deposition of gas-phase compounds and particles too small to be affected by gravity. These materials may be deposited on surfaces due to their solubility with the surface or due to other physical and chemical attractions. Airborne contaminants are removed by both wet deposition, such as rainfall scavenging, and by dry deposition. The sum of wet and dry deposition is called total deposition. Deposition processes are the most important way contaminants such as acidic sulfur compounds are removed from the atmosphere; they are also important because deposition processes transfer contaminants to aquatic and terrestrial ecosystems. Cross-media transfers, such as transfers from air to water, can have adverse environmental impacts. For example, dry deposition of sulfur and nitrogen compounds can acidify poorly buffered lakes.

See also Acid rain; Nitrogen cycle; Sulfur cycle.

Dryland farming

Dryland farming is the practice of cultivating crops without irrigation (rainfed agriculture). In the United States, the term usually refers to crop production in low-rainfall areas without irrigation, using moisture-conserving techniques such as mulches and fallowing. Nonirrigated farming is practiced in the Great Plains, intermountain, and Pacific regions of the country, or areas west of the 23.5 inches (60 cm) annual precipitation line, where native vegetation was short prairie grass. In some parts of the world dryland farming means all rainfed agriculture.

In the western United States, dryland farming has often resulted in severe or moderate wind erosion. Alternating seasons of fallow and planting has left the land susceptible to both wind and water erosion. High demand for a crop sometimes resulted in cultivating lands not suitable for long-term farming, degrading the soil measurably.

Conservation tillage, leaving all or most of the previous crop residues on the surface, decreases erosion and conserves water. Methods used are stubble mulch, mulch, and ecofallow. In the wetter parts of the Great Plains, fallowing land has given over to annual cropping, or four-year rotations with three years of planting followed by a year of fallowing.

See also Arable land; Desertification; Erosion; Soil; Tilth.

Resources

BOOKS

Peterson, Gary A.; Paul W. Unger; and William A. Payne, eds. *Dryland Agriculture*. Madison, WI: American Society of Agronomy, 2006.

William E. Larson

Dubos, René Jules

1901–1982

French American microbiologist, ecologist, and writer

Dubos, a French-born microbiologist, spent most of his career as a researcher and teacher at Rockefeller University in New York state. His pioneering work in microbiology, such as isolating the anti-bacterial substance *gramicidin* from a soil organism and showing the feasibility of obtaining germ-fighting drugs from microbes, led to the development of antibiotics.

Nevertheless, most people know Dubos as a writer. Dubos's books centered on how humans relate to their surroundings, books informed by what he

René J. Dubos, Pulitzer Prize-winning microbiologist and environmentalist, poses on the grounds of Rockefeller University in New York, 1973. Dubos said that conservation, not production of more energy, is the answer to the energy crisis. *(RON FREHM)*

described as "the main intellectual attitude that has governed all aspects of my professional life...to study things, from microbes to man, not *per se* but in their complex relationships." That pervasive intellectual stance, carried throughout his research and writing, reflected what *Saturday Review* called "one of the best-formed and best-integrated minds in contemporary civilization."

A related theme was Dubos's conviction that "the total environment" played a role in human disease. By total environment, he meant "the sum of the facts which are not only physical and social conditions but emotional conditions as well." Though not a medical doctor, he became an expert on disease, especially tuberculosis, and headed Rockefeller's clinical department on that disease for several years.

"Despairing optimism" also pervaded Dubos's human-environment writings, his own title for a column he wrote for *The American Scholar*, beginning in 1970. *Time* magazine even labeled him the "prophet of optimism." "My life philosophy is based upon a faith in the immense resiliency of nature," he once commented.

Dubos held a lifelong belief that a constantly changing environment meant organisms, including humans, had to adapt constantly to keep up, survive, and prosper. But he worried that humans were too good at adapting, resulting in both his optimism and his despair: "Life in the technologized environment seems to prove that [humans] can become adapted to starless skies, treeless avenues, shapeless buildings, tasteless bread, joyless celebrations, spiritless pleasures—to a life without reverence for the past, love for the present, or poetical anticipations of the future." He stated that "the belief that we can manage the earth may be the ultimate expression of human conceit," but insisted that nature is not always right and even that humankind often improves on nature. As Thomas Berry suggested, "Dubos sought to reconcile the existing technological order and the planet's survival through the resilience of nature and changes in human consciousness."

Beginning in 1980, Dubos dedicated most of his professional energy to the Dubos Center for Human Environment, a nonprofit research and education center dedicated in his honor, which sought to inform and educate about the environment and environmental problems and values.

Dubos died on February 20, 1982.

Resources

BOOKS

Ward, B., and R. Dubos. *Only One Earth: The Care and Maintenance of a Small Planet.* New York: Norton, 1972.

Gerald L. Young

Ducks Unlimited

Ducks Unlimited (DU) is an international (United States, Canada, Mexico, New Zealand, and Australia), membership organization founded during the Depression years in the United States by a group of sportsmen interested in waterfowl conservation. DU was incorporated in early 1937, and DU (Canada) was established later that spring. The organization was established to preserve and maintain waterfowl populations through habitat protection and development, primarily to provide game for sport hunting. During the "dust bowl" of the 1930s, the founding members of DU recognized that most of the continental waterfowl populations were maintained by breeding habitat in the wetlands of Canada's southern prairies in Saskatchewan, Manitoba, and Alberta. The organizers established DU Canada and used their resources to protect the Canadian prairie breeding grounds. Cross-border funding has since been a fundamental component of DU's operation, although in recent years funds also have been directed to the northern American prairie states. In 1974 Ducks Unlimited de Mexico was established to restore and maintain wetlands south of the U.S.-Mexican border where many waterfowl spend the winter months.

Throughout most of its existence, DU has funded habitat restoration projects and worked with landowners to provide water management benefits on farmlands. But, from its inception, DU has been subject to criticism. Early opponents characterized it as an American intrusion into Canada to secure hunting areas. More recently, critics have suggested that DU defines waterfowl habitat too narrowly, excluding upland areas where many ducks and geese nest. The group plans to broaden its focus to encompass preservation of these upland breeding and nesting areas.

Aerial view of a wetlands habitat in Eastern Prince Edward Island, Canada. (© *All Canada Photos / SuperStock*)

Since many of these areas are found on private land, DU also plans to expand its cooperative programs with farmers and ranchers. Most commonly, however, DU is criticized for placing the interests of waterfowl hunters above wildlife management concerns. The organization does allow duck hunting on its preserves.

Following the fundamental principle of "users pay," duck hunters still provide the majority of DU's funding. For that reason DU has not addressed some issues that have a serious effect on continental waterfowl populations. The combination of illegal hunting and liberal bag limits is blamed by some for the continued decline in waterfowl numbers. DU has not addressed this issue, preferring to leave management issues to government agencies in the United States and Canada, while focusing on habitat preservation and restoration. Critics of DU suggest that the organization will not act on population matters and risk offending the hunters who provide its financial support.

In North America DU has expanded its scope and activities to address ecological and land use problems through the work of the North American Waterfowl Management Plan (NAWMP) and the Prairie conservation of agriculture, resources, and environment (CARE) program. The wetlands conservation and other habitat projects addressed in these and similar programs, not only benefit game species, but other endangered species of plants and animals as well. The NAWMP (an agreement between the United States and Canada) alone protects over 5.5 million acres (2.2 million ha) of waterfowl habitat. In 2002, the North American Wetlands Conservation Act (NAWCA) granted the DU one million dollars to be put toward a new wetlands in Ohio. The NAWCA has also contributed to DU's "rescue the duck factory" campaign that was launched in 2008. The aim is to protect 300,000 acres (120,000 ha) of grasslands in the Dakotas as wildfowl habitat. As of 2010, DU was nearly halfway to this goal.

Resources

BOOKS

Akcakaya, H. Resit. *Species Conservation and Management: Case Studies*. New York: Oxford University Press, 2004.

Alderfer, Jonathan K. *National Geographic Complete Birds of North America*. Washington, DC: National Geographic, 2006.

Alderton, David. *The Illustrated Encyclopedia of Birds of the World*. London: Lorenz Books, 2005.

Dickson, Barney; Jonathan Hutton; and W. M. Adams. *Recreational Hunting, Conservation, and Rural Livelihoods: Science and Practice*. Chichester, UK: Blackwell, 2009.

Ladle, Richard J. *Biodiversity and Conservation: Critical Concepts in the Environment*. London: Routledge, 2009.

ORGANIZATIONS

Ducks Unlimited, Inc., One Waterfowl Way, Memphis, TN, USA, 38120, (901) 758-3825, (800) 45DUCKS, http://www.ducks.org

David A. Duffus

Ducktown, Tennessee

Tucked in a valley of the Cherokee National Forest, on the border of Tennessee, North Carolina, and Georgia, Ducktown once reflected the beauty of the surrounding Appalachian Mountains. Instead, Ducktown and the valley known as the Copper Basin now form the only desert east of the Mississippi. Mined for its rich copper lode since the 1850s, it had become a vast stretch of lifeless, red-clay hills. It was an early and stark lesson in the devastation that acid rain and soil erosion can wreak on a landscape, one of the few man-made landmarks visible to the astronauts who landed on the moon.

Prospectors came to the basin during a gold rush in 1843, but the closest thing to gold they discovered was copper, and most went home. By 1850, entrepreneurs realized the value of the ore, and a new rush began to mine the area. Within five years, thirty companies had dug beneath the topsoil and made the basin the country's leading producer of copper.

The only way to separate copper from the zinc, iron, and sulfur present in the copper basin rock was to roast the ore at extremely high temperatures. Mining companies built giant open pits in the ground for this purpose, some as wide as 600 feet (183 m) and as deep as a ten-story building. Fuel for these fires came from the surrounding forests. The forests must have seemed a limitless resource, but it was not long before every tree, branch, and stump for 50 square miles (130 km^2) had been torn up and burned. The fires in the pits emitted great billows of sulfur dioxide gas—so thick people could get lost in the clouds even at high noon—and this gas mixed with water and oxygen in the air to form sulfuric acid, which is the main component in acid rain. Saturated by acidic moisture and choked by the remaining sulfur dioxide gas and dust, the undergrowth died and the soil became poisonous to new plants. Wildlife fled the shelterless hillsides. Without root systems, virtually all the soil washed into the Ocoee River, smothering aquatic life. Open-range grazing of cattle, allowed in Tennessee until 1946, denuded the land of what little greenery remained.

Soon after the turn of the century, Georgia filed suit to stop the air pollution which was drifting out of this corner of Tennessee. In 1907, the Supreme Court, in a decision written by Justice Oliver Wendell Holmes, ruled in Georgia's favor, and the sulfur clouds ceased in the Copper Basin. It was one of the first environmental-rights decisions in the United States. That same year, the Tennessee Copper Company designed a way to capture the sulfur fumes, and sulfuric acid, rather than copper, became the area's main product. It remains so as of 2010.

Ducktown was the first mining settlement in the area. Since the 1930s, the Tennessee Copper Company, the Tennessee Valley Authority, and the Soil Conservation Service have worked to restore the land, planting hundreds of loblolly pine and black locust trees. Their efforts have met with little success, but new reforestation techniques such as slow-release fertilizer have helped many new plantings survive. Scientists hope to use the techniques practiced here on other deforested areas of the world. Many of the townspeople lobbied to preserve a piece of the scar, both for its unique beauty and as a symbol of human impact on the environmental lesson.

See also Mine spoil waste; Smelter; Sudbury, Ontario; Surface mining; Trail Smelter arbitration.

Resources

BOOKS

Visgilio, Gerald R. *Acid in the Environment: Lessons Learned and Future Prospects.* New York: Springer, 2007.

PERIODICALS

Barnhardt, W. "The Death of Ducktown." *Discover* 8 (October 1987): 34-36.

OTHER

United States Environmental Protection Agency (EPA). "Industry: Industrial Processes: Mining." http://www.epa.gov/ebtpages/induindustmining.html (accessed October 13, 2010).

L. Carol Ritchie

Dunes and dune erosion

Dunes are small hills, mounds, or ridges of wind-blown soil material, usually sand, that are formed in both coastal and inland areas. The formation of coastal or inland dunes requires a source of loose sandy

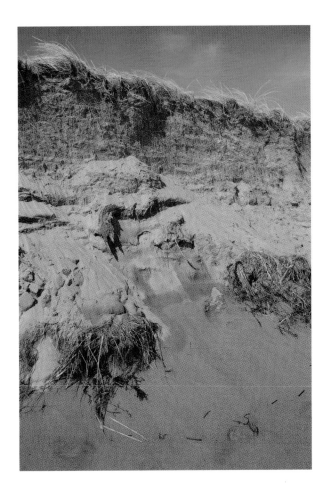

Partially eroded sand dune. *(© iStockPhoto.com/ visionofmaine)*

material and dry periods during which the sand can be picked up and transported by the wind. Dunes exist independently of any fixed surface feature and can move or drift from one location to another over time. They are the result of natural erosion processes and are natural features of the landscape in many coastal areas and deserts, yet they also can be symptoms of land degradation. Inland dunes are either an expression of aridity or can be indicators of desertification—the result of long-term land degradation in dryland areas.

Coastal dunes are the result of marine erosion in which sand is deposited on the shore by wave action. During low tide, the beach sand dries and is dislodged and transported by the wind, usually over relatively short distances. Depending on the local topography and direction of the prevailing winds, a variety of shapes and forms can develop—from sand ridges to parabolic mounds. The upper few centimeters of coastal dunes generally contain chlorides from salt spray and wind-blown salt. As a result, attempts to stabilize coastal dunes with vegetation are often limited to salt-tolerant plants.

The occurrence of beaches and dunes together have important implications for coastal areas. A beach absorbs the energy of waves and acts as a buffer between the sea and the dunes behind it. Low-lying coastlines are best defended against high tides by consolidated sand dunes. In such cases, maintaining a wide, high beach that is backed by stable dunes is desirable.

Engineering structures along coastal areas and the mouths of rivers can affect the formation and erosion of beaches and coastal dunes. In some instances it is desirable to build and widen beaches to protect coastal areas. This can require the construction of structures that trap littoral drift, rock mounds to check wave action, and seawalls that protect areas behind the beach from heavy wave action. Where serious erosion has occurred, artificial replacement of beach sands may be necessary. Such methods are expensive and require considerable engineering effort as well as the use of heavy equipment.

The weathering of rocks, mainly sandstone, is the origin of material for *inland dunes*. However, whether or not sand dunes form depends on the vegetative cover condition and use of the land. In contrast to coastal dunes, which are often considered to be beneficial to coastal areas, inland dunes can be indicators of land degradation where the protective cover of vegetation has been removed as a result of inappropriate cultivation, overgrazing, construction activities, and so forth. When vegetative cover is absent, soil is highly susceptible to both water and wind erosion. The two work together in dry lands to create sources of soil that can be picked up and transported either downwind or downstream. The flow of water moves and exposes sand grains and supplies fresh material that results in deposits of sand in floodplains and ephemeral drainage systems. Before dunes can develop in such areas, there must be long dry periods between periodic or episodic sediment-laden flows of water. Wind erosion occurs where such sand deposits from water erosion are exposed to the energy of wind, or in areas that are devoid of vegetative cover.

Where sand is the principle size soil particle and where high wind velocities are common, sand particles are moved by a process called saltation and creep. Sand dunes form under such conditions and are shaped by wind patterns over the landscape. Complex patterns can be formed—the result of interactions of wind, sand, the ground surface topography, and any vegetation or other physical barriers that exist. These patterns can be sword-like ridges, called longitudinal dunes, crescentic accumulations or barchans, turret-shaped mounds, shallow sheets of sand, or large seas of transverse dunes. The typical pattern is one of a gradual long slope on the windward side of the dune, dropping off sharply on the leeward side.

Exposed sand dunes can move up to 11 yards (10 m) annually in the direction of the prevailing wind. Such dunes encroach upon areas, covering farmlands, pasture lands, irrigation canals, urban areas, railroads, and highways. Blowing sand can mechanically injure and kill vegetation in its path and can eventually bury croplands or rangelands. If left unchecked, the drifting sand will expand and lead to serious economic and environmental losses.

Worldwide, dryland areas are those most susceptible to wind erosion. For example, 22 percent of Africa north of the equator is severely affected by wind erosion as is over 35 percent of the land area in the Near East. As a result, inland dunes represent a significant landscape component in many desert regions. Although dunes can be symptoms of land-use problems, in some areas they are part of a natural dryland landscape that are considered to be features of beauty and interest. Sand dunes have become popular recreational areas in parts of the United States, including the Great Sand Dune National Monument in southern Colorado with its 229-yard (210-m) high dunes that cover a 158 square mile (254.4-km^2) area, and the Indiana Dunes State Park along the shore of Lake Michigan.

When dune formation and encroachment represent significant environmental and economic problems, sand dune stabilization and control should be undertaken. Dune stabilization may initially require one or more of the following: applications of water, oil, bitumens emulsions, or chemical stabilizers to improve the cohesiveness of surface sands; the reshaping of the landscape such as construction of foredunes that are upwind of the dunes, and armoring of the surface using techniques such as hydroseeding, jute mats, mulching, and asphalt; and constructing fences to reduce wind velocity near the ground surface. Although sand dune stabilization is the necessary first step in controlling this process, the establishment of a vegetative cover is a necessary condition to achieve long-term control of sand dune formation and erosion. Furthermore, stabilization and revegetation must be followed with appropriate land management that deals with the causes of dune formation in the first place. Where dune erosion has not progressed to a seriously degraded state, dunes can become reclaimed through natural regeneration simply by protecting the area against livestock grazing, all-terrain vehicles, and foot traffic.

Vegetation stabilizes dunes by decreasing wind speed near the ground and by increasing the cohesiveness of sandy material by the addition of organic colloids and the binding action of roots. Plants trap the finer wind-blown soil particles, which helps improve soil texture; and they also improve the microclimate of the site, reducing soil surface temperatures. Upwind barriers or windbreak plantings of vegetation, often trees or other woody perennials, can be effective in improving the success of revegetating sand dunes. They reduce wind velocities, help prevent exposure of plant roots from the drifting sand, and protect plantings from the abrasive action of blowing sand. Areas that are susceptible to sand dune encroachment can likewise be protected by using fences or windbreak plantings that reduce wind velocities near the ground surface. Because of the severity of sand dune environments, it can be difficult to find plant species that can be established and survive. In addition, any plantings must be protected against exploitation, for example, from grazing or fuelwood harvesting.

The expansion of sand dunes resulting from desertification not only represents environmental problems, but it also represents serious losses of productive land and a financial hardship for farmers and others who depend upon the land for their livelihood. Such problems are particularly acute in many of the poorer dryland countries of the world and deserve the attention of governments, international agencies, and nongovernmental organizations who need to direct their efforts toward the causes of soil erosion and dune formation.

The erosion of sand dunes, especially those located along the eastern coastline of the United States, is accelerating as a the global climate changes in response to the warming of the atmosphere. Increased severe storm activity and consequent wave action is producing more seawater-delivered destructive force. The situation is not likely to improve, given the anticipated continuation of global climate change for at least the next century.

Resources

BOOKS

Albert, Dennis A. *Borne of the Wind: Michigan Sand Dunes.* Ann Arbor: University of Michigan Press, 2006.

Pilkey, Orrin H.; Tracy Monegan Rice; and William J. Neal. *How to Read a North Carolina Beach: Bubble Holes, Barking Sands, and Rippled Runnels.* Winston-Salem: University of North Carolina Press, 2006.

Valiela, Ivan. *Global Coastal Change.* New York: Wiley-Blackwell, 2006.

Kenneth N. Brooks

Dust Bowl

Dust Bowl is a term coined by a reporter for the *Washington* (DC) *Evening Star* to describe the effects of severe wind erosion in the Great Plains during the

Dust Bowl-era photo of a cloud of topsoil parched by drought and picked up by winds, moving down a road near Boise City, Oklahoma, 1935. *(AP Photo)*

1930s, caused by severe drought and lack of conservation practices.

For a time after World War I, agriculture prospered in the Great Plains. Land was rather indiscriminantly plowed and planted with cereals and row crops. In the 1930s, the total cultivated land in the United States increased, reaching 530 million acres (215 million ha), its highest level ever. Cereal crops, especially wheat, were most prevalent in the Great Plains. Summer fallow (cultivating the land, but only planting every other season) was practiced on much of the land. Moisture, stored in the soil during the fallow (uncropped) period, was used by the crop the following year. In a process called dust mulch, the soil was frequently clean tilled to leave no crop residues on the surface, control weeds, and, it was thought at the time, preserve moisture from evaporation. Frequent cultivation and lack of crop canopy and residues optimized conditions for wind erosion during the droughts and high winds of the 1930s.

During the process of wind erosion, the finer particles (silt and clay) are removed from the topsoil, leaving coarser-textured sandy soil. The fine particles carry with them higher concentrations of organic matter and plant nutrients, leaving the remaining soil impoverished and with a lower water storage capacity. Wind erosion of the Dust Bowl reduced the productivity of affected lands, often to the point that they could not be farmed profitably.

While damage was particularly severe in Texas, Oklahoma, Colorado, and Kansas, erosion occurred in all of the Great Plains states, from Texas to North

many environmentalists believe is urgently needed to clarify, synthesize, and further develop international sustainable-development law.

The United Nations Conference on Environment and Development (UNCED), or Earth Summit held in Rio de Janeiro, Brazil, in 1992, did take up the challenge of drafting the Earth Charter. A number of governments prepared recommendations. Many nongovernmental organizations, including groups representing the major faiths, became actively involved. Although the resulting Rio Declaration on Environment and Development is a valuable document, it falls short of the aspirations that many groups have had for the Earth Charter.

The Earth Charter Project, 1994–2000

A new Earth Charter initiative began in 1994 under the leadership of Maurice Strong, the former secretary general of UNCED and chairman of the newly formed Earth Council, and Mikhail Gorbachev, acting in his capacity as chairman of Green Cross International. The Earth Council was created to complete the unfinished business of UNCED and to promote implementation of Agenda 21, the Earth Summit's action plan. Jim MacNeill, former secretary general of the WCED and Prime Minister Ruud Lubbers of The Netherlands were instrumental in facilitating the organization of the new Earth Charter project. Ambassador Mohamed Sahnoun of Algeria served as the executive director of the project during its initial phase, and its first international workshop was held at the Peace Palace in The Hague in May 1995. Representatives from thirty countries and more than seventy different organizations participated in the workshop. Following this event, the secretariat for the Earth Charter project was established at the Earth Council in San José Costa Rica.

A worldwide Earth Charter consultation process was organized by the Earth Council in connection with the Rio+5 review in 1996 and 1997. The Rio+5 review, which culminated with a special session of the United Nations General Assembly in June 1997, sought to assess progress toward sustainable development since the Rio Earth Summit and to develop new partnerships and plans for implementation of Agenda 21. The Earth Charter consultation process engaged men and women from all sectors of society and all cultures in contributing to the Earth Charter's development. A special program was created to contact and involve the world's religions, interfaith organizations, and leading religious and ethical thinkers. The Earth Council also organized a special indigenous people's network.

Early in 1997 an Earth Charter Commission was formed to oversee the project. The initial twenty-three members were geographically and economically diverse. The Commission issued a Benchmark Draft Earth Charter in March 1997 at the conclusion of the Rio+5 Forum in Rio de Janeiro. The Forum was organized by the Earth Council as part of its independent Rio+5 review, and it brought together more than 500 representatives from civil society and national councils of sustainable development. The Benchmark Draft reflected the many and diverse contributions received through the consultation process and from the Rio+5 Forum. The Commission extended the Earth Charter consultation until early 1998, and the Benchmark Draft was circulated widely as a document in progress.

The Earth Charter concept

A consensus developed that the Earth Charter should be: a statement of fundamental principles of enduring significance that are widely shared by people of all races, cultures, and religions; a relatively brief and concise document composed in a language that is inspiring, clear, and meaningful in all tongues; the articulation of a spiritual vision that reflects universal spiritual values, including but not limited to ethical values; a call to action that adds significant new dimensions of value to what has been expressed in earlier relevant documents; a people's charter that serves as a universal code of conduct for ordinary citizens, educators, business executives, scientists, religious leaders, nongovernmental organizations, and national councils of sustainable development; and a declaration of principles that can serve as a "soft law" document when adopted by the UN General Assembly. The Earth Charter was designed to focus on fundamental principles with the understanding that the IUCN Covenant and other treaties will set forth the more specific practical implications of these principles.

The Earth Charter draws upon a variety of resources, including ecology and other contemporary sciences, the world's religious and philosophical traditions, the growing literature on global ethics and the ethics of environment and development, the practical experience of people living sustainably, as well as relevant intergovernmental and nongovernmental declarations and treaties. At the heart of the new global ethics and the Earth Charter is an expanded sense of community and moral responsibility that embraces all people, future generations, and the larger community of life on Earth. Among the values affirmed by the Benchmark Draft are respect for Earth and all life; protection and restoration of the health of Earth's ecosystems; respect for human rights, including the right to an environment adequate for human well-being; eradication of poverty; nonviolent problem solving and peace; the equitable sharing of resources; democratic participation in decision making; accountability and transparency in administration; universal education for sustainable living; and a sense of shared responsibility for the well-being of the Earth community.

Resources

BOOKS

Brown, Oli. *The Environment and Our Security: How Our Understanding of the Links Has Changed*. Winnipeg, Manitoba, Canada: International Institute for Sustainable Development, 2005.

OTHER

United Nations Environment Programme (UNEP). "UNEP System-wide Earthwatch Coordination Office–Geneva, Switzerland." http://earthwatch.unep.net/ (accessed October 13, 2010).
United Nations System-Wide EarthWatch. "Development." http://earthwatch.unep.net/development/index.php (accessed October 13, 2010).

ORGANIZATIONS

The Earth Charter Initiative, The Earth Council, P.O. Box 319-6100, San Jose, Costa Rica, (+506) 205-1600, (+506) 249-3500, info@earthcharter.org, http://www.earthcharter.org

Steven C. Rockefeller

Earth Day

The first Earth Day, April 22, 1970, attracted over twenty million participants in the United States. It launched the modern environmental movement and spurred the passage of several important environmental laws. It was the largest demonstration in history. People from all walks of life took part in marches, teach-ins, rallies, and speeches across the country. Congress adjourned so that politicians could attend hometown events, and cars were banned from New York's Fifth Avenue.

The event had a major impact on the nation. Following Earth Day, conservation organizations saw their memberships double and triple. Within months the Environmental Protection Agency (EPA) was created; Congress also revised the Clean Air Act, the Clean Water Act, and other environmental laws.

A windmill and inflatable globe serve as visual props in encouraging ecology at a 1990 Earth Day celebration on the lawn of the Capitol Building, Washington, DC. *(© Todd Gipstein/CORBIS)*

The concept for Earth Day began with United States Senator Gaylord Nelson, a Wisconsin Democrat, who in 1969 proposed a series of environmental teach-ins on college campuses across the nation. Hoping to satisfy a course requirement at Harvard by organizing a teach-in there, law student Denis Hayes flew to Washington, DC, to interview Nelson. The senator persuaded Hayes to drop out of Harvard and organize the nationwide series of events that were only a few months away. According to Hayes, Wednesday, April 22 was chosen because it was a weekday and would not compete with weekend activities. It also came before students would start "cramming" for finals, but after the winter thaw in the north.

Twenty years later, Earth Day anniversary celebrations attracted even greater participation. An estimated 200 million people in more than 140 nations were involved in events ranging from a concert and rally of over a million people in New York's Central Park, to a festival in Los Angeles that attracted 30,000, to a rally of 350,000 at the National Mall in Washington, DC.

Earth Day 1990 activities included planting trees; cleaning up roads, highways, and beaches; building bird houses; presenting ecology teach-ins; and recycling cans and bottles. A convoy of garbage trucks drove through the streets of Portland, Oregon, to dramatize the lack of landfill space. Elsewhere, children wore gas masks to protect air pollution, others marched in parades wearing costumes made from recycled materials, and some even released ladybugs into the air to demonstrate alternatives to harmful pesticides. The gas-guzzling car that was buried in San Jose, California, during the first Earth Day was dug up and recycled.

Abroad, Berliners planted 10,000 trees along the East-West border. In Myanmar there were protests against the killing of elephants. Brazilians demonstrated against the destruction of their tropical rain forests. In Japan people demonstrated against disposable chopsticks, and 10,000 people attended a concert on an island built on reclaimed land in Tokyo Bay.

Denis Hayes, with help from hundreds of volunteers, also organized the 1990 Earth Day. This time, the event was well organized and funded; it was widely-supported by both environmentalists and the business community. The United Auto Workers Union sent Earth Day booklets to all of its members, the National Education Association sent information to almost every teacher in the country, and the Methodist Church mailed Earth Day sermons to more than 30,000 ministers.

The sophisticated advertising and public relations campaign, licensing of its logo, and sale of souvenirs provoked criticism that Earth Day had become too commercial. Even oil, chemical, and nuclear firms joined in and proclaimed their love for nature. But Hayes defended the approach as necessary to maximize interest and participation in the event, to broaden its appeal, and to launch a decade of environmental activism that would force world leaders to address the many threats to the planet. He also pointed out that, although foundations, corporations, and individuals had donated $3.5 million, organizers turned down over $4 million from companies that were thought to be harming the environment.

The thirty-year anniversary of the event in 2000 was also organized by Hayes. Unfortunately, it did not produce the large numbers of the prior anniversary celebration. The movement had reached more than 5,000 environmental groups who helped organize local rallies, and hundreds of thousands of people met in Washington to hear political, environmental, and celebrity speakers.

Hayes believes that the long-term success of Earth Day in securing a safe future for the planet depends on getting as many people as possible involved in environmentalism. The fortieth annual celebration of Earth Day was held in 2010. More than one billion people worldwide took part in their local Earth-Day-related events.

Resources

BOOKS

1001 Easy Ways for Earth-Wise Living: Natural and Eco-Friendly Ideas That Can Make a Real Difference to Your Life. Ultimo, NSW, New Zealand: Reader's Digest, 2006.

OTHER

National Geographic Society. "Earth Day Tips: Rx for the Planet and Your Health." http://environment.national geographic.com/environment/habitats/earth-day-tips. html (accessed October 14, 2010).

Lewis G. Regenstein

Earth First!

Earth First! is a radical and often controversial environmental group founded in 1979 in response to what Dave Foreman and other founders believed to be the increasing co-optation of the environmental movement. For Earth First! members, too much of the environmental movement has become lethargic, compromising, and corporate in its orientation. To avoid a similar fate, Earth First! members have restricted their use of traditional fund-raising techniques and have sought a non-hierarchical organization with neither a professional staff nor formal leadership.

A movement established by and for self-acknowledged environmental hardliners, Earth First!'s general stance is reflected in its slogan, "No compromise in the defense of Mother Earth." Its policy positions are based upon principles of deep ecology and in particular on the group's belief in the intrinsic value of all natural things. Its goals include preserving all remaining wilderness, ending environmental degradation of all kinds, eliminating major dams, establishing large-scale ecological preserves, slowing and eventually reversing human population growth, and reducing excessive and environmentally harmful consumption.

Combining biocentrism with a strong commitment to activism, Earth First! does not restrict itself to lobbying, lawsuits, and letter-writing, but also employs direct action, civil disobedience, "guerrilla theater," and other confrontational tactics, and in fact is probably

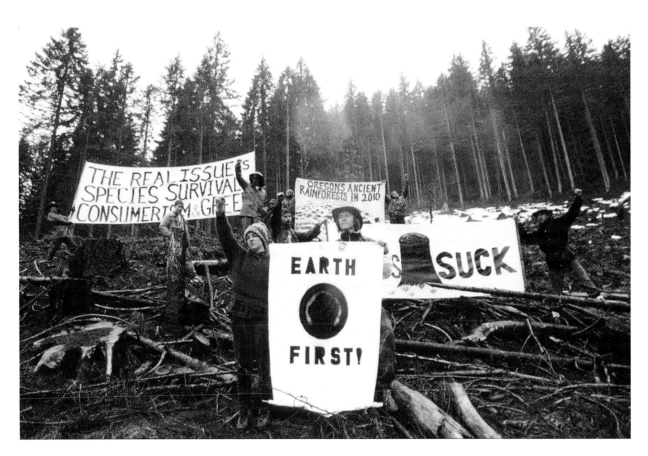

Members of the environmental group Earth First hold protest banners on May 1, 1990, in Oregon. Logging of old-growth or "ancient" forests in the Pacific Northwest has triggered high-stake public struggles over policy and management practices. *(Photo by Stephen Ferry/Liaison/Getty Images.)*

best known for its various clashes with the logging industry, particularly in the Pacific Northwest. Earth First! members and sympathizers have been associated with controversial tactics including the chopping down of billboards and monkey-wrenching, which includes pouring sand in bulldozer gas tanks, spiking trees, sabotaging drilling equipment, and so forth. In 1992 the more radical Earth Liberation Front formed from former members of Earth First! after the organization moved away from potentially violent or destructive criminal acts as a means of direct action. Earth First! asserts that acts of civil disobedience are their preferred method of direct action.

Earth First! encourages people to respect species and wilderness, to refrain from having children, to recycle, to live simpler, less destructive lives, and to engage in civil disobedience to thwart environmental destruction. During the summer of 1990, the group sponsored its most noted event, "Redwood Summer." Activists from around the United States gathered in the Northwest to protest large- scale logging operations, to

call attention to environmental concerns, to educate and establish dialogues with loggers and the local public, and to engage in civil disobedience.

Earth First! also sponsors a Biodiversity Project for protecting and restoring natural ecosystems. Its Ranching Task Force educates the public about the consequences of overgrazing in the American West. The Grizzly Bear Task Force focuses on the preservation of the grizzly bear in the Rockies and the reintroduction of the species to its historical range throughout North America. Earth First!'s wider Predator Project seeks the restoration of all native predators to their respective habitats and ecological roles. Other Earth First! projects seek to defend redwoods and other native forests, encourage direct action against the fur and construction industries, intervene in government-sponsored wolf-control programs in the United States and Canada, protest government and business decisions that have environmentally destructive consequences for tropical rain forests, campaign against genetically modified foods, protest the export of e-waste to developing nations, and advocate for alternative energy development.

Resources

BOOKS

1001 Easy Ways for Earth-Wise Living: Natural and Eco-Friendly Ideas That Can Make a Real Difference to Your Life. Ultimo, NSW, New Zealand: Reader's Digest, 2006.

Crane, Jeff, and Michael Egan. *Natural Protest: Essays on the History of American Environmentalism.* New York: Routledge, 2009.

Lytle, Mark H. *The Gentle Subversive: Rachel Carson, Silent Spring, and the Rise of the Environmental Movement.* New York: Oxford University Press, 2007.

Lawrence J. Biskowski

Earth Island Institute

The Earth Island Institute (EII) was founded by David Brower in 1982 as a nonprofit organization dedicated to developing innovative projects for the conservation, preservation, and restoration of the global environment. In its earliest years, the Institute worked primarily with a volunteer staff and concentrated on projects like the first Conference on the Fate of the Earth, publication of *Earth Island Journal*, and the production of films about the plight of indigenous peoples. In 1985 and again in 1987, EII expanded its facilities and scope, opening office space and providing support for a number of allied groups and projects. Its membership now numbers approximately 35,000.

EII conducts research on, and develops critical analyses of, a number of contemporary issues. With sponsored projects ranging from saving sea turtles to encouraging land restoration in Central America, EII does not restrict its scope to traditionally "environmental" goals but rather pursues what it sees as ecologically-related concerns such as human rights, economic development of the Third World, economic conversion from military to peaceful production, and inner-city poverty, among others. But much of its mission is to be an environmental educator and facilitator. In that role EII sponsors or participates in numerous programs designed to provide information, exchange viewpoints and strategies, and coordinate efforts of various groups. EII even produces

music videos as part of its environmental education efforts.

EII is perhaps best known for its efforts to halt the use of drift nets by tuna boats, a practice that is often fatal to large numbers of dolphins. After an EII biologist signed on as a crew member aboard a Latin American tuna boat and documented the slaughter of dolphins in drift nets, EII brought a lawsuit to compel more rigorous enforcement of existing laws banning tuna caught on boats using such nets. EII also joined with other environmental groups in urging a consumer boycott of canned tuna. These efforts were successful in persuading the three largest tuna canners to pledge not to purchase tuna caught in drift nets. The monitoring of tuna fishing practices is an ongoing EII project.

EII also sponsors a wide variety of other projects, some of which have now spun off to become independent organizations. For instance, Baikal Watch works for the permanent protection of biologically unique Lake Baikal, Russia. Energy Action is another project, supporting the youth and student clean-energy movement in North America. EII founded the Sea Turtle Restoration Project, now an independent organization, which investigates threats to the world's endangered sea turtles, organizes and educates United States citizens to protect the turtles, and works with Central American sea turtle restoration projects. The Rain Forest Action Network campaigns for the forests and their inhabitants through education and non-violent direct action. Meanwhile, The Urban Habitat Program develops multicultural environmental leadership and organizes efforts to restore urban neighborhoods. Both were founded by IEE and are now independent.

EII administers a number of funds designed to support creative approaches to environmental conservation, preservation, and restoration to support activists exploring the use of citizen suit provisions of various statutes to enforce environmental laws and to help develop a Green political movement in the United States. EII also sponsors several international conferences, exchange programs, and publication projects in support of various environmental causes.

Resources

ORGANIZATIONS

Earth Island Institute, 2150 Allston Way, Suite 460, Berkeley, CA, USA, 94704-1375, (510) 859-9100, (510) 859-9091, http://www.earthisland.org

Lawrence J. Biskowski

Earth Liberation Front

Earth Liberation Front (ELF) is a grassroots environmental group that the Federal Bureau of Investigation (FBI) labeled "a serious terrorism threat." Founded in the United Kingdom in 1992, ELF is dedicated to using economic sabotage and vandalism to prevent perceived exploitation of the environment. Since 1996 ELF and the Animal Liberation Front (ALF) have committed thousands of acts of vandalism that resulted in tens of millions of dollars in damage.

James Jarboe, the FBI section chief for Domestic Terrorism, told Congress that it was hard to track down ELF and ALF members because the two groups have little organized structure. According to the ELF link on the ALF Web site, there is no designated leadership nor formal membership. ELF and ALF claim responsibility for their activities by e-mail, fax, and other communications usually sent to the media.

Media information about ELF attacks is provided by the North American Earth Liberation Front Press Office, which was relaunched in October 2008. The press office receives anonymous communiqués from ELF and further distributes them. The press office claims that it does not know the identities of any ELF members and merely disseminates information received anonymously from ELF activists.

ELF describes itself as an international underground organization dedicated to stopping "the continued destruction of the natural environment." Members join anonymous cells that may consist of one person or more. Members of one cell do not know the identity of members in other cells, a structure that prevents activists in one cell from being compromised should members in another cell become disaffected. People act on their own and carry out actions following anonymous ELF postings.

ELF postings include guidelines for taking action. One guideline is to inflict "economic damage on people who profit from the destruction and exploitation of the natural environment." Another guideline is to reveal and educate the public about the "atrocities" committed against the environment. The third

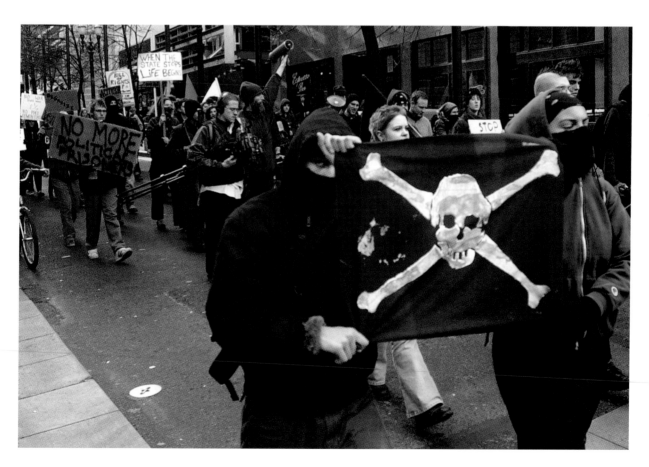

Hooded and masked protesters, some proclaiming to be supporters of the Earth Liberation Front, march in Portland, Oregon, in 2002. *(AP Photo/Don Ryan.)*

guideline is to take all needed precautions against harming any animal, human and non-human.

ELF is an outgrowth of Earth First!, a group formed during the 1980s to promote environmental causes. Earth First! held protests and civil disobedience events, according to the FBI. In 1984 Earth First! members began a campaign of "tree spiking" to prevent loggers from cutting down trees. Members inserted metal or ceramic spikes into trees. The spikes damaged saws when loggers tried to cut down trees.

When Earth First! members in Brighton, England, disagreed with proposals to make their group more mainstream, radical members of the group founded the Earth Liberation Front in 1992. The following year ELF aligned itself with ALF.

The Animal Liberation Front was started in Great Britain during the mid-1970s. The loosely organized movement had the goal of ending animal abuse and exploitation of animals. An American ALF branch was started in the late 1970s. According to the FBI, people became members by participating in "direct action" activities against companies or people using animals for research or economic gain. ALF activists targeted animal research laboratories, fur companies, mink farms, and restaurants.

ELF and ALF declared mutual solidarity in a 1993 announcement. The following year the San Francisco branch of Earth First! recommended that the group move away from ELF and its unlawful activities. ELF calls its activities "monkeywrenching," a term that refers to actions such as tree spiking, arson, sabotage of logging equipment, and property destruction.

In a forty-three-page document titled "Year End Report for 2001," ELF and ALF claimed responsibility for sixty-seven illegal acts that year. ELF claimed sole credit for setting a 2002 fire that caused $5.4 million in damage to a University of Washington horticulture building. The group also took sole credit for a 1998 fire set at a Vail, Colorado, ski resort. Damage totaled $12 million for the arson that destroyed four ski lifts, a restaurant, a picnic facility, and a utility building, according to the FBI.

ELF issued a statement after the arson saying that the fire was set to protect the lynx, which was being reintroduced to the Rocky Mountains. "Vail, Inc. is already the largest ski operation in North America and now wants to expand even further. This action is just a warning. We will be back if this greedy corporation continues to trespass into wild and unroaded areas," the statement said.

In 2002 ELF also claimed credit for a fire that caused $800,000 in damage at the University of Minnesota Microbial Plant and Genomics Center. ELF targeted the genetic crop laboratory because of its efforts to "control and exploit" nature. In August 2003 ELF activists set fire to a condominium project in San Diego, California, causing $50 million in damages. In March 2009, ELF arsonists burned four suburban houses near Seattle, Washington, causing $7 million in damages.

"Eco-terrorism" is the term used by the FBI to define illegal activities related to ecology and the environment. These activities involve the "use of criminal violence against innocent victims or property by an environmentally-oriented group."

Although ELF members are difficult to track, several arrests have been made. In February 2001 two teen-age boys pleaded guilty to setting fires at a home construction site in Long Island, New York. In December of that year, a man was also charged with spiking 150 trees in Indiana state forests. In his Congressional testimony, Jarboe said that cooperation among law enforcement agencies was essential to responding efficiently to eco-terrorism. In December 2005 and January 2006, an FBI investigation, Operation Backfire, resulted in the indictment and arrest of eighteen ecoterrorists. The North American Earth Liberation Front Press Office reported that eleven of these individuals claimed to be ELF activists.

Resources

BOOKS

Liddick, Donald R. *Eco Terrorism: Radical Environmental and Animal Liberation Movements*. Westport CT: Praeger Publishers, 2006.

ORGANIZATIONS

Federal Bureau of Investigation, 935 Pennsylvania Ave., Washington, DC, USA, (202) 324-3000, http://www.fbi.gov

North American Earth Liberation Front Press Office, James Leslie Pickering, P.O. Box 14098, Portland, OR, USA, 97293, (503) 804-4965, elfpress@tao.ca, http://www.animalliberation.net/library/facts/elf

Liz Swain

Earth Pledge Foundation

Created in 1991 by attorney Theodore W. Kheel, the Earth Pledge Foundation (EPF) is concerned with the impact of technology on society. Recognizing the often delicate balance between economic growth and environmental protection, EPF encourages the implementation of sustainable practices, especially in community development, tourism, cuisine, and architecture.

As a result of the United Nations Earth Summit (Rio de Janeiro, 1992), the UN pledged its commitment to the principles of sustainable development—to foster "development meeting the needs of the present without compromising the ability of future generations to meet their own needs."

Created for the summit in support of the principles, the Earth Pledge was prominently displayed throughout the event. Heads of state, ambassadors, delegates, and prominent dignitaries from around the world stood in line to sign their names on a large Earth Pledge board. Since the Summit, millions have taken the Earth Pledge: "Recognizing that people's actions towards nature and each other are the source of growing damage to the environment and to resources needed to meet human needs and ensure survival, I pledge to act to the best of my ability to help make the Earth a secure and hospitable home for present and future generations."

In early 1996 Earth Pledge created the Business Coalition for Sustainable Cities (BCSC) to influence the development of cities as centers of commerce, employment, recreation, and settlement. Chaired by William L. Lurie, former president of The Business Roundtable, the BCSC provides business leaders a forum to address issues of major importance to our cities in ways that ensure economic viability, while promoting respect for the environment. One event sponsored by the BCSC was a seven-course dinner prepared by twelve of the nation's most environmentally-conscious chefs to show that restaurants, one of the largest industries and employers, can practice the principles of sustainable cuisine. The BCSC hosted the event with the theme that good food can be well-prepared without adversely impacting health, culture, or environment. This theme was elaborated on in 2000 when the Sustainable Cuisine Project was established to develop and teach cooking classes.

One of the most significant developments from Earth Pledge was the creation of the web site, farm to table. This web site highlights local farmers of the New York region and give consumers a direct link to fresh food news. More recently, Earth Pledge set up the Farmers Education Fund to support farmers in transitioning to methods of planting, growing, and harvesting that have lower environmental impact.

Earth Pledge has also formed a number of alliances to further their goals with groups such as The Foundation for Prevention and Resolution of Conflict (PERC, founded by Theodore Kheel), the United Nations Environmental Programme (UNEP), Earth-Kind International and the New England Aquarium.

As a joint effort with the New England Aquarium, EPF sponsors a marine awareness project that educates people on the importance of coastlines and aquatic resources as well as the sustainable development of the world's cities. The project emphasizes that many countries have water shortages due to inefficient use of their water supply, degradation of their water by pollution and unsustainable usage of groundwater resources.

In late 1995 Earth Pledge co-sponsored the first Caribbean Conference on Sustainable Tourism with the UN Department for Policy Coordination and Sustainable Development, EarthKind International, and UNEP. The conference brought together officials from government and business to discuss strategies for developing a healthy tourist economy, sound infrastructure, environmental protection, and community participation.

The foundation has constructed an environmentally sensitive building, Foundation House, to display their solutions for improving air quality and energy efficiency. Sustainable features include heating, cooling, and lighting systems that minimize consumption of fossil fuels, increased ventilation and use of natural daylight, an auditorium for conferences with Internet access and a computer lab for training. The Foundation House houses exhibits, including one on enhancing efficiency in the workplace for the benefit of the staff.

Earth Pledge continues to develop smaller organizations and promote companies that participate in sustainable practices.

Resources

ORGANIZATIONS

Earth Pledge Foundation, 122 East 38th Street, New York, NY, USA, 10016, (212) 725-6611, (212) 725-6774, http:// www.earthpledge.org

Nicole Beatty

Earth Summit see United Nations Earth Summit (1992).

Earthquake

Major earthquakes typically strike both populated and unpopulated areas of the world every year, killing hundreds, injuring thousands, and causing hundreds of millions of dollars in damage. Despite millions of dollars and decades of research, seismologists (scientists who study earthquakes) are still unable to predict precisely when and where an earthquake will happen.

An earthquake is a geological event in which rock masses below the surface of the earth suddenly shift, releasing energy and sending out strong vibrations to the surface. Most earthquakes are caused by movement along a fault line, which is a fracture in the earth's crust. Thousands of earthquakes happen each day around the world, but most are too small to be felt.

Earth is covered by a crust of rock that is broken into numerous plates, sometimes referred to as tectonic plates. The plates float on a layer of molten (liquid) rock within the earth called the mantel. This molten rock moves and flows, and this movement is thought to cause the shifting of the plates. When plates move, they either slide past, bump into, overrun, or pull away from each other. The movement of plates is called plate tectonics. Boundaries between plates are called faults.

Earthquakes can occur when any of the four types of movement take place along a fault. Earthquakes along the San Andreas and Hayward faults in California occur because two plates slide past one another. Earthquakes also occur if one plate overruns another. When this happens one plate is pushed under the other plate, as on the western coast of South America, the northwest coast of North America, and in Japan. If plates collide but neither is pushed downward, as they do crossing Europe and Asia from Spain to Vietnam, earthquakes result as the plates are pushed into each other and are forced upward, creating high mountain ranges. Many faults at the floor of the ocean are between two plates moving apart. Many earthquakes with centers at the floor of the ocean are caused by this kind of movement.

The relative size of earthquakes is determined by the Richter Scale, which measures the energy an earthquake releases. Each whole number increase in value on the Richter scale indicates a ten-fold increase in the energy released and a thirty–fold increase in ground motion. An earthquake measuring 8 on the Richter scale is ten times more powerful, therefore, than an earthquake with a Richter magnitude of 7. Another

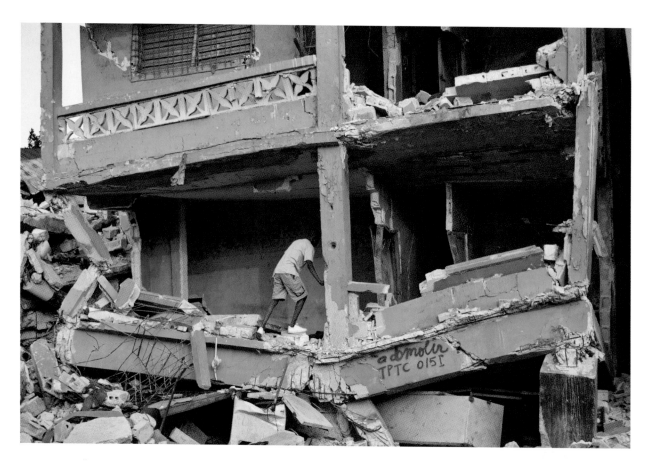

Man crawling through a building in downtown Port-au-Prince, Haiti, that was demolished from the 7.0 earthquake that hit Haiti on January 12, 2010. *(Julie Dermansky)*

scale, called the Mercalli Scale, uses observations of damage (such as fallen chimneys) or people's assessments of effects (such as mild or severe ground shaking) to describe the intensity of a quake. The Richter Scale is open-ended, while the Mercalli scale ranges from 1–12.

Catastrophic earthquakes happened just as often in the past as they do today. Earthquakes shattered stone-walled cities in the ancient world, sometimes hastening the ends of civilizations. Earthquakes destroyed Knossos, Chattusas, and Mycenae, ancient cities in Europe located in tectonically active mountain ranges. Scribes have documented earthquakes in the chronicles of ancient countries.

Many faults are located in California because two large plates are sliding past each other there. Of the fifteen largest recorded earthquakes ever to hit the continental United States, eight have occurred in California, according to the United States Geological Survey (USGS). The San Francisco earthquake of 1906 is perhaps the most famous. It struck on April 4, 1906, killing an estimated 3,000 people, injuring thousands more, and causing $524 million in property loss. Many of the casualties and much of the damage resulted from the ensuing fires. This earthquake registered a 7.7 magnitude on the Richter Scale and 11 on the Mercalli Scale. Four other devastating earthquakes have occurred in California since the 1906 San Francisco quake, with all four occurring in the twentieth century: 1933 in Long Beach, 1971 in the San Fernando Valley, 1989 in the San Francisco Bay area, and 1994 in Los Angeles.

The Long Beach earthquake struck on March 10, 1933, killing 120, injuring hundreds, and causing more than $50 million in property damage. It led to the passage of the state's Field Act, which established strict building code standards designed to make structures better able to withstand strong earthquakes.

Centered about 30 miles (48 km) north of downtown Los Angeles, the San Fernando earthquake killed 65, injured more than 2,000, and caused an estimated $505 million in property damage. The quake hit on February 9, 1971, and registered 6.5 on the Richter Scale and 11 on the Mercalli Scale. Most of the deaths occurred when the Veterans Administration Hospital in San Fernando collapsed.

The Loma Prieta earthquake occurred on October 18, 1989, in the Santa Cruz Mountains about 62 miles (100 km) south of San Francisco. It killed 63, injured 3,757, and caused an estimated $6 billion in property damage, mostly in San Francisco, Oakland, and Santa Cruz. The earthquake was a 6.9 on the Richter Scale and 9 on the Mercalli Scale.

The Northridge earthquake that struck Los Angeles on January 17, 1994, killed 72, injured 11,800, and caused an estimated $40 billion in damage. It registered 6.7 on the Richter Scale and 9 on the Mercalli Scale. It was centered about 30 miles (48 km) northwest of downtown Los Angeles.

Many earthquakes occurred in California in the early twenty-first century, but none (as of late 2010) created anywhere near the deaths and property damage as the five major earthquakes between 1906 and 1994. However, the 2010 Baja California earthquake (also called the 2010 Easter earthquake) was the strongest earthquake to hit southern California in almost two decades. Initially estimating its magnitude as 6.9 on the Richter Scale, the United States Geological Survey (USGS) later upgraded the 2010 Easter earthquake to 7.2. Though the epicenter was south of the U.S. border in the Baja region of Mexico, the earthquake was sufficiently massive that it was felt throughout much of the southwestern United States. The National Aeronautics and Space Administration (NASA) reported that the quake actually moved the Mexican city of Calexico, which is situated at the U.S.-Mexican border, southwards some 2.5 feet (0.8 m). Although four deaths and some property damage were attributed to the earthquake, many more deaths and much greater property damage would have occurred had the 2010 Easter earthquake struck further north in the more populated areas of California.

In the past 100 years, Alaska has had many more severe earthquakes than California. However, they have occurred in mostly sparsely populated areas, so deaths, injuries, and property damage have been light. Of the fifteen strongest earthquakes ever recorded in the fifty states, ten have been in Alaska, with the strongest registering a 9.2 (the second strongest ever recorded in the world) on the Richter Scale and 12 on the Mercalli Scale. It struck the Anchorage area on March 28, 1964, killing 125 (most from a tsunami caused by the earthquake), injuring hundreds, and causing $311 million in property damage.

The strongest earthquake ever recorded in the world registered 9.5 on the Richter scale and 12 on the Mercalli Scale. It occurred on May 22, 1960, and was centered off the coast of Chile. It killed 2,000, injured 3,000, and caused $675 million in property damage. A resulting tsunami caused death, injuries, and significant property damage in Hawaii, Japan, and the West Coast of the United States.

Every major earthquake raises the question of whether scientists will ever be able to predict exactly when and where one will strike. Today, scientists can only make broad predictions. For example, scientists believe there is at least a 50 percent chance that a devastating earthquake will strike somewhere along the San Andreas fault within the next 100 years. A more precise prediction is not yet possible. However, scientists in the United States and Japan are working on ways to make predictions more specific.

Ultra-sensitive instruments placed across faults at the surface can measure the slow, almost imperceptible movement of fault blocks. This measurement records the great amount of potential energy stored at the fault boundary. In some areas small earthquakes called foreshocks that precede a larger earthquake may help seismologists predict the larger earthquake. In other areas where seismologists believe earthquakes should be occurring but are not, this discrepancy between what is expected and what is observed may be used to predict an inevitable large-scale earthquake.

Other instruments measure additional fault-zone phenomena that seem to be related to earthquakes. The rate at which radon gas issues from rocks near faults has been observed to change before an earthquake. The properties of the rocks themselves (such as their ability to conduct electricity) also changed as the tectonic forces exerted on them slowly alters the rocks of the fault zone between earthquakes. Unusual animal behavior has been reported before many earthquakes, and research into this phenomenon is a legitimate area of scientific inquiry, even though no definite answers have been found.

Techniques of studying earthquakes from space are also being explored. Scientists have found that ground displacements cause waves in the air that travel into the ionosphere and disturb electron densities. By using the network of satellites and ground stations that are part of the Global Positioning System (GPS), and data about the ionosphere that is being collected, scientists may better understand the energy released from earthquakes. This may help scientists to predict them.

According to the USGS, several million earthquakes occur each year, with most being small in size and magnitude. On average globally, about 140 earthquakes above magnitude 6.0 are expected each year. About 700 earthquakes each year are strong enough and located close enough to a populated area to cause reportable damage.

In January 12, 2010, a 7.0 magnitude earthquake in Haiti killed an estimated 250,000 to 300,000 people, and left around one million people homeless. In February 2010, a powerful magnitude 8.8 quake in central Chile killed at least 450 people. In April 2010 the death toll quickly surpassed 2,000 people following a 6.9

magnitude earthquake near Jiegu, a town located in the remote Tibetan Qinghai province of China.

The 2010 earthquakes followed other recent major quakes. In 2009 a 6.3 magnitude earthquake damaged historical sites in L'Aquila, Italy, while killing 300 people and leaving tens of thousands homeless. The 7.8 magnitude Sichuan earthquake in May 2008 killed 87,000 people. In October 2005 an 7.6 magnitude earthquake struck in north Pakistan, killing 73,000 people.

As in the case of the December 2004 Indian Ocean tsunami that killed an estimated 220,000 to 250,000 people, undersea earthquakes can result in large and destructive waves known as tsunamis.

Resources

BOOKS

Hough, Susan Elizabeth. *Earthshaking Science: What We Know (and Don't Know) about Earthquakes*. Princeton, NJ: Princeton University Press, 2004.

Kusky, Timothy. *Earthquakes: Plate Tectonics and Earthquake Hazards*. New York: Facts On File, 2008.

OTHER

National Geographic Society. "Earthquake Safety Tips." http://environment.nationalgeographic.com/environment/natural-disasters/earthquake-safety-tips.html (accessed November 7, 2010).

National Geographic Society. "Earthquakes." http://environment.nationalgeographic.com/environment/natural-disasters/earthquake-profile.html (accessed November 7, 2010).

ORGANIZATIONS

National Earthquake Information Center, P.O. Box 25046, DFC, MS 967, Denver, CO, USA, 80225, (303) 273-8500, (303) 273-8450, sedas@neis.cr.usgs.gov, http://www.neic.usgs.gov

Ken R. Wells

Earthwatch

Earthwatch is a non-profit institution that provides paying volunteers to help scientists around the world conduct field research on environmental and cultural projects. It is one of the world's largest private sponsors of field research expeditions. Its mission is "to improve human understanding of the planet, the diversity of its inhabitants, and the processes which affect the quality of life on earth" by working "to sustain the world's environment, monitor global change, conserve endangered habitats and species, explore the vast heritage of our peoples, and foster world health and international cooperation."

The group carries out its work by recruiting volunteers to serve in an environmental EarthCorps and to work with research scientists on important environmental issues. The volunteers, who pay to join expeditions (usually one to three weeks in duration) to the far corners of the globe, gain valuable experience and knowledge on situations that affect the earth and human welfare.

Earthwatch has sponsored thousands of projects around the world. Research teams address such topics as tropical rain forest ecology and conservation; marine studies (ocean ecology); geosciences (climatology, geology, oceanography, glaciology, volcanology, paleontology); life sciences (wildlife management, biology, botany, ichthyology, herpetology, mammalogy, nithology, primatology, zoology); social sciences (agriculture, economic anthropology, development studies, nutrition, public health); and art and archaeology (architecture, archaeoastronomy, ethnomusicology, folklore).

No special skills are needed to be part of an expedition, and special teen expeditions are available. Scholarships for students and teachers are also available. Earthwatch's affiliate, The Center for Field Research, receives several hundred grant applications and proposals every year from scientists and scholars who need volunteers to assist them on study expeditions.

Earthwatch maintains an active online site with multiple offices located around the world.

Resources

ORGANIZATIONS

Earthwatch, 3 Clock Tower Place, Suite 100, Box 75, Maynard, MA, USA, 01754, (978) 461-0081, (978) 461-2332, (800) 776-0188, info@earthwatch.org, http://www.earthwatch.org

Lewis G. Regenstein

Eastern European pollution

Between 1987 and 1992 the disintegration of Communist governments of Eastern Europe allowed the people and press of countries from the Baltic to the Black Sea to recount tales of life-threatening pollution and disastrous environmental conditions in which they lived. Villages in Czechoslovakia were black and barren

Exterior of the Nowa Huta steelworks, near Cracow, Poland. *(Simon Fraser/Photo Researchers, Inc.)*

because of acid rain, smoke, and coal dust from nearby factories. Drinking water from Estonia to Bulgaria was tainted with toxic chemicals and untreated sewage. Polish garden vegetables were inedible because of high lead and cadmium levels in the soil. Chronic health problems were endemic to much of the region, and none of the region's new governments had the necessary cash to alleviate their environmental liabilities.

The air, soil, and water pollution exposed by new environmental organizations and by a newly vocal press had its roots in Soviet-led efforts to modernize and industrialize Eastern Europe after 1945. (Often the term "Central Europe" is used to refer to Poland, Czech Republic, Slovakia, Hungary, Yugoslavia, and Bulgaria, and "Eastern Europe" to refer to the Baltic states, Belarus, and Ukraine. For the sake of simplicity, the latter term is used for all these states.) Following Stalinist theory that modernization meant industry, especially heavy industries such as coal mining, steel production, and chemical manufacturing, Eastern European leaders invested heavily in industrial growth. Factories were often built in resource-poor areas, as in traditionally agricultural Hungary and Romania, and they rarely had efficient or clean technology. Production quotas generally took precedence over health and environmental considerations, and billowing smokestacks were considered symbols of national progress. Emission controls on smokestacks and waste effluent pipes were, and are, rare. Soft, brown lignite coal, cheap and locally available, was the main fuel source. Lignite contains up to 5 percent sulfur and produces high levels of sulfur dioxide, nitrogen oxides, particulates, and other pollutants that contaminate air and soil in population centers, where many factories and power plants were built. The region's water quality also suffers, with careless disposal of toxic industrial wastes, untreated urban waste, and runoff from chemical-intensive agriculture.

By the 1980s the effects of heavy industrialization began to show. Dependence on lignite coal led to sulfur dioxide levels in Czechoslovakia and Poland eight times greater than those of Western Europe. The industrial triangle of Bohemia and Silesia had Europe's highest concentrations of ground-level ozone, which harms human health and crops. Acid rain, a result of industrial air pollution, had destroyed or damaged half of the forests in the former East Germany and the Czech Republic. Cities were threatened by outdated factory equipment and aging chemical storage containers and pipelines, which leaked chlorine, aldehydes, and other noxious gases. People in cities and villages experienced alarming numbers of birth defects and short life expectancies. Economic losses, from health care expenses, lost labor, and production inefficiency further handicapped hard-pressed Eastern European governments.

Popular protests against environmental conditions crystallized many of the movements that overturned Eastern and Central European governments. In Latvia, expose;aas on petrochemical poisoning and on environmental consequences of a hydroelectric project on Daugava River sparked the Latvian Popular Front's successful fight for independence. Massive campaigns against a proposed dam on the Danube River helped ignite Hungary's political opposition in 1989. In the same year, Bulgaria's Ecoglasnost group held Sofia's first non-government rally since 1945. The Polish Ecological Club, the first independent environmental organization in Eastern Europe, assisted the Solidarity movement in overturning the Polish government in the mid-1980s.

Citizens of these countries rallied around environmental issues because they had first-hand experience with the consequences of pollution. In Espenhain, of former East Germany, 80 percent of children developed chronic bronchitis or heart ailments before they were eight years old. Studies showed that up to 30 percent of Latvian children born in 1988 may have suffered from birth defects, and both children and adults showed unusually high rates of cancer, leukemia, skin diseases, bronchitis, and asthma. Czech children in industrial regions had acute respiratory diseases, weakened immune systems, retarded bone development, and high concentrations of lead and cadmium in their hair. In the industrial regions of Bulgaria, skin diseases were seven times more common than in cleaner areas, and cases of rickets and liver diseases were four times as common. Much of the air and soil contamination that produced these symptoms remains today and continues to generate health problems. In the area surrounding a lead smelting facility in Copsa Mica, Romania, life expectancy remains nine years below the national average and 96 percent of children have bronchitis or other respiratory diseases. A 2006 report by the Czech government noted that air pollution affects the health of 60 percent of Czech citizens.

Water pollution is at least as threatening as air and soil pollution. Many cities and factories in the region have no facilities for treating wastewater and sewage. Existing treatment facilities are usually inadequate or ineffective. Toxic waste dumps containing old and rusting barrels of hazardous materials are often unmonitored or unidentified. Chemical leaching from poorly monitored waste sites threatens both surface water and groundwater, and water clean enough to drink has become a rare commodity. In Poland untreated sewage, mine drainage, and factory effluents make much of Polands surface water and groundwater unsafe for drinking. Cleanup of Poland's water supplies was slow. More than a decade after the fall of Communism, at least half of Polish rivers were too polluted, by government assessment, even for industrial use.

Few pollution problems are geographically restricted to the country in which they were generated. Shared rivers and aquifers and regional weather patterns carry both air-borne and water-borne pollutants from one country to another. The Chernobyl nuclear reactor disaster, which spread radioactive gases and particulates from Belarus across northern Europe and the Baltic Sea to northern Norway and Sweden is one infamous example of trans-border pollution, but other examples are common. Toxic wastes flowing into the Baltic Sea from Poland's Vistula River continue to endanger fisheries and shoreline habitats in Sweden, Germany, and Finland.

The Danube River is a particularly critical case. Accumulating and concentrating urban and industrial waste from Vienna to the Black Sea, this river supports industrial complexes of Austria, Czech Republic, Hungary, Croatia, Serbia, Bulgaria, and Romania. Before the Danube leaves Budapest, it is considered unsafe for swimming. Like other rivers, the Danube flows through a series of industrial cities and mining regions, uniting the pollution problems of several countries. Each city and farm along the way uses the contaminated water and contributes some pollutants of its own. Also like other rivers, the Danube carries its toxic load into the sea, endangering the marine environment.

Western countries from Sweden to the United States have their share of pollution and environmental disasters. The Rhine and the Elbe have disastrous chemical spills like those on the Danube and the Vistula. Like recent communist regimes, most western business leaders would prefer to disregard environmental and human health considerations in their pursuit of production quotas. Yet several factors set apart environmental conditions in Eastern Europe. Aside from its aged and outdated equipment and infrastructure, Eastern Europe is handicapped by its compressed geography, intense urbanization near factories, a long-standing lack of information and accurate records on environmental and health conditions, and severe shortages of clean-up funds.

Eastern Europe's dense settlement crowds all the industrial regions of the Baltic states, Poland, the Czech and Slovak republics, and Hungary into an area considerably smaller than Texas but with a much higher population. This industrial zone lies adjacent to crowded manufacturing regions of Western Europe. In this compact region, people farm the same fields and live on the same mountains that are stripped for mineral extraction. Cities and farms rely on aquifers and rivers that receive factory effluent and pesticide runoff immediately upstream. Furthermore, post-1945 industrialization gathered large labor forces into factory towns more quickly than adequate infrastructure could be built. Expanding urban populations had little protection from the unfiltered pollutants of nearby furnaces. Although many Eastern Europeans were eyewitnesses to environmental transgressions, little public discussion about the problem was possible. Official media disliked publicizing health risks or the destruction of forests, rivers, and lakes. Those statistics that existed were often unreliable. Air- and water-quality data were collected and reported by industrial and government officials, who could not afford bad test results.

Now that environmental conditions are being exposed, cleanup efforts remain hampered by a shortage of funding. Poland's long-term environmental restoration may cost $260 billion. Efforts to cut just sulfur dioxide emissions to Western standards would cost Poland about $2.4 billion a year. Hungary plans to spend nearly $6 billion on sewage treatment by 2013. Cleanup of oil and other toxic chemicals in the port of Ventspils, Latvia, is expected to cost $1.5 billion. Air, soil, and water remediation in former East Germany received a boost from their western neighbors, but the bill is expected to run between $40 and $150 billion.

Ironically, East European leaders see little choice for raising this money aside from expanded industrial production. Meanwhile, business leaders urge production expansion for other capital needs. Some Western investment in cleanup work has begun, especially on the part of such countries as Sweden and Germany, which share rivers and seas with polluting neighbors. Already in 1989 Sweden had begun work on water-quality monitoring stations along Poland's Vistula River, which carries pollutants into the Baltic Sea. Capital necessary to purchase mitigation equipment, improve factory conditions, rebuild rusty infrastructure, and train environmental experts will probably be severely limited for decades to come, however.

Meanwhile, western investors are flocking to Eastern and Central Europe to build or rebuild business ventures for their own gain. The region is seen as one of quick growth and great potential. Manufacturers in heavy and light industries, automobiles, power plants, and home appliances are coming from Western Europe, North America, and Asia. From textile manufacturing to agribusiness, outside investors hope to reshape Eastern economies. Many Western companies are improving and updating equipment and adding pollution control devices. In a climate of uncertain regulation and rushed economic growth, however, it is not clear if the region's new governments will be able or willing to enforce environmental safeguards or if the new investors will take advantage of weak regulations and poor enforcement as did their predecessors. Yet, there have been encouraging signs through the first decade of the twenty-first century. Air-pollution data from the Czech Republic, Hungary, and Poland show ambient sulfur dioxide levels have fallen.

Membership of the European Union may be the instrument that Eastern Europe needs to help clean up its pollution. Ten new countries, including the Czech Republic, Slovakia, and Poland, joined the EU in 2004 and were followed by Romania and Bulgaria in 2007. To become EU members, these countries had to commit to integrating more than 200 directives and regulations on environmental matters, including pollution, into national legislation. The EU has supported environmental improvements, including pollution clean up, in these countries with considerable technical, legal, and financial assistance.

Resources

BOOKS

Hill, Marquita K. *Understanding Environmental Pollution.* Cambridge: Cambridge University Press. 2004.

Mary Ann Cunningham

Ebola

Ebola is a highly deadly viral disease that is hemorrhagic; that is, it involves the destruction of blood cells. As the disease progresses, the walls of blood vessels break down and blood gushes from every tissue and organ. The disease is caused by the Ebola virus, named after the river in Zaire (now the Democratic Republic of Congo) where the first known outbreak occurred in 1976.

The disease is extremely contagious and exceptionally lethal. Where a 10 percent mortality rate is considered high for most infectious diseases, Ebola can kill up to 90 percent of its victims, usually within only a few days after exposure. Direct contact with contaminated blood or bodily fluids seem to transmit the disease. Health personnel and caregivers are often the most likely to be infected. Even after a patient has died, preparing the body for a funeral can be deadly for families members.

The Ebola virus is one of two members of a family of RNA viruses called the Filoviridae. The other Filovirus causes Marburg fever, an equally contagious and lethal hemorrhagic disease, named after a German town where it was first contracted by laboratory workers who handled imported monkeys infected with the virus. Together with members of three other families (arenaviruses, bunyanviruses, and flaviviruses), these viruses cause a group of deadly, episodic diseases including Lassa fever, Rift Valley fever, Bolivian fever, and Hanta or Four-Corners fever (named after the region of the southwestern United States where it was first reported).

There are five known species of the Ebola virus. They are Bundibugyo, Sudan, Zaire, Côte d'Ivoire, and Reston. These first three species are the cause of the

Workers wearing protective clothing lower the corpse of five-year-old Adamou, who died of Ebola in Mekambo, northeast Gabon in 2001. Adamou, who only had one name, was the fifth member of his family to catch the highly lethal disease during the Ebola outbreak. *(AP Photo/Christine Nesbitt)*

majority of the outbreaks that have occurred prior to 2010, and which produce the pronounced mortality rate.

The viruses associated with most of these emergent, hemorrhagic fevers are zoonotic. That means a reservoir of pathogens, or disease-causing agents, naturally resides in an animal host or arthropod vector (insect that transmits disease). As of 2010 the specific host or vector for Ebola has not been definitively confirmed. The fruit bat appears to be a reservoir for the virus. But, whether fruit bats are the natural host is unclear. Monkeys and other primates can contract related diseases and appear to be other reservoirs. People who initially become infected with Ebola often have been involved in killing, butchering, and eating gorillas, chimps, or other primates. Why the viruses remain in their hosts for many years without causing much more trouble than a common cold, but then erupt sporadically and unpredictably into terrible human epidemics, is a perplexing issue in environmental health.

The geographical origin for Ebola is unknown, but all recorded outbreaks have occurred in or around Central Africa, or in animals or people from this area. Ebola appears every few years in Africa. Confirmed cases have occurred in the Democratic Republic of the Congo, Gabon, Sudan, Uganda, and the Ivory Coast. No case of the disease in humans has ever been reported in the United States, but a variant called Ebola-Reston virus killed a number of monkeys in a research facility in Reston, Virginia. There probably are isolated cases in remote areas that go unnoticed. In fact, the disease may have been occurring in secluded villages deep in the jungle for a long time without outside attention.

As of 2010, the worst documented epidemic of Ebola in humans occurred in 1995, in Kikwit, Zaire (now the Democratic Republic of Congo). Although many more people died in Kikwit than in any other outbreak, in many ways, the medical and social effects of the epidemic there was typical of what happens elsewhere. The first Kikwit victim was a thirty-six-year-old laboratory technician named Kimfumu, who checked into a medical clinic complaining of a severe headache, stomach pains, fever, dizziness, weakness, and exhaustion. Surgeons did an exploratory operation to try to find the cause of his illness. They found the patient's gastrointestinal tract was necrotic (containing dying tissue) and putrefying. He bled uncontrollably, and within hours was dead. By the next day the five medical workers who had cared for Kimfumu, including an Italian nun who had assisted in the operation, showed similar symptoms, including high fevers, fatigue, bloody diarrhea, rashes, red and itchy eyes, vomiting, and bleeding from every body orifice, or opening. Less than forty-eight hours later, they, too, were dead, and the disease spread throughout the city of 600,000.

As panicked residents fled into the bush, government officials responded to calls for help by closing off all travel—including humanitarian aid—into or out of Kikwit, about 250 miles (400 km) from Kinshasa, the national capital. Fearful neighboring villages felled trees across the roads to seal off the pestilent city. No one dared enter houses where dead corpses rotted in the intense tropical heat. Boats plying the adjacent Kwilu River refused to stop to take on or discharge passengers or cargo. Food and clean water became scarce. Hospitals could hardly function as medicines and medical personal became scarce. Within a few weeks about 400 people in Kikwit had contracted the disease, and at least 350 were dead. Eventually, the epidemic dissipated and disappeared.

This pattern of appearance has since been shown to be the norm for the very virulent types of Ebola. After an incubation period that can last up to three weeks, symptoms develop and progress very rapidly.

As of 2010 there is no vaccine or other antiviral drug available for Ebola to prevent or halt an infection, but research and experimental tests continue.

Several factors seem to be contributing to the appearance and spread of highly contagious diseases such as Ebola and Marburg fevers. With 6 billion people now inhabiting the planet, human densities are much higher, enabling germs to spread farther and faster than ever before. Expanding populations push people into remote areas where they encounter new pathogens and parasites. Environmental change is occurring on a larger scale: cutting forests, creating unhealthy urban surroundings, and causing global climate change, among other things. Elimination of predators and changes in habitats favor disease-carrying organisms such as mice, rats, cockroaches, and mosquitoes.

Another important factor in the spread of many diseases is the speed and frequency of modern travel. Millions of people go every day from one place to another by airplane, boat, train, or automobile. Very few places on earth are more than twenty-four hours by jet plane from any other place.

Tracking down the source of Ebola has been hampered by the sudden appearance and resolution of outbreaks, especially because the conventional response occurs after an outbreak has developed.

Resources

BOOKS

Hewlett, Barry S., and Bonnie L. Hewlett. *Ebola, Culture, and Politics: The Anthropology of an Emerging Disease.* Case studies on contemporary social issues. Belmont, CA: Thomson Higher Education, 2007.

Klenk, H-D., and Heinz Feldmann. *Ebola and Marburg Viruses: Molecular and Cellular Biology.* Wymondham, U.K.: Horizon Bioscience, 2004.

Lupoli John Carlo. *Bioterrorism and the Ebola Virus.* Valhalla, NY: New York Medical College, 2004.

Smith, Tara C. *Ebola (Deadly Diseases and Epidemics).* New York: Chelsea House, 2005.

PERIODICALS

Leroy, E. M., et al. "Fruits Bats as Reservoirs of Ebola Virus." *Nature* 438 (December 1, 2005): 575–576.

Ecojustice

The concept of ecojustice has at least two different usages among environmentalists. The first refers to a general set of attitudes about justice and the environment, at the center of which is dissatisfaction with traditional theories of justice. With few exceptions (notably a degree of concern about excessive cruelty to animals), anthropocentric (human-centered) and egocentric Western moral and ethical systems have been unconcerned with individual plants and animals, species, oceans, wilderness areas, and other parts of the biosphere, except as they may be used by humans. In general, that which is non-human is viewed mainly as raw material for human uses, largely or completely without moral standing.

Relying upon holistic principles of biocentrism and deep ecology, the ecojustice alternative suggests that the value of non-human life-forms is independent of the usefulness of the non-human world for human purposes. Antecedents of this view can be found in sources as diverse as Eastern philosophy, Aldo Leopold's "land ethic," Albert Schweitzer's "reverence for life," and Martin Heidegger's injunction to "let beings be." The central idea of ecojustice is that ethical and moral reflection on justice should be expanded to encompass nature itself and its constituent parts and that human beings have an obligation to consider the inherent value of other living things whenever these living things are affected by human actions.

Some advocates of ecojustice base standards of just treatment on the evident capacity of many life-forms to experience pain. Others assert the equal inherent worth of all individual life-forms. More typically, environmental ethicists assert that all life-forms have at least some inherent worth, although perhaps not the same worth, and thus deserve moral consideration. The practical goals associated with ecojustice include fostering stability and diversity within and between self-sustaining ecosystems, harmony and balance in nature and within competitive biological systems, and sustainable development.

Ecojustice can also refer simply to the linking of environmental concerns with various social justice issues. The advocate of ecojustice typically strives to understand how the logic of a given economic system results in certain groups or classes of people bearing the brunt of environmental degradation. This entails, for example, concern with the frequent location of polluting industries and hazardous waste dumps near the economically disadvantaged (i.e., those with the least mobility and fewest resources to resist).

Ecojustice also involves the fostering of sustainable development in less-developed areas of the globe. It aims to prevent economic development that exports polluting industries and other environmental problems from wealthier nations to developing nations, especially if the industry will primarily benefit or serve consumers in the wealthier nations. An additional point of concern is the allocation of costs and benefits in environmental reclamation and preservation—for example, the preservation of Amazonian rain forests affects the global environment and may benefit the whole world, but the costs of this preservation fall disproportionately upon Brazil and the other countries of the region. An advocate of ecojustice would be concerned that the various costs and benefits of development be apportioned fairly.

See also Biodiversity; Environmental ethics; Environmental racism; Environmentalism; Holistic approach.

Resources

BOOKS

Curry, Patrick. *Ecological Ethics: An Introduction.* Cambridge, UK: Polity Press, 2006.

Ebbesson, Jonas, and Phoebe N. Okowa. *Environmental Law and Justice in Context.* Cambridge, UK: Cambridge University Press, 2009.

Page, Edward. *Climate Change, Justice and Future Generations.* Cheltenham, UK: Edward Elgar, 2006.

Pojman, Louis P., ed. *Environmental Ethics: Readings in Theory and Application.* Belmont, CA: Wadsworth Publishing, 2004.

OTHER

United States Environmental Protection Agency (EPA). "Environmental Management: Environmental Justice." http://www.epa.gov/ebtpages/envienvironmentaljustice. html (accessed October 14, 2010).

Lawrence J. Biskowski

Ecological consumers

Organisms that feed either directly or indirectly on producers, plants that convert solar energy into complex organic molecules. Primary consumers are animals that eat plants directly. They are also called herbivores. Secondary consumers are animals that eat other animals. They are also called carnivores. Consumers that consume both plants and animals are omnivores. Parasites are a type of consumer that lives in or on the plant or animal on which it feeds.

Detrivores (detritus feeders and decomposers) constitute a specialized class of consumers that feed on dead plants and animals.

See also Biotic community.

Ecological economics

Although ecology and economics share the common root eco- (from Greek *Oikos* or household), these disciplines have tended to be at odds with each other in recent years over issues such as the feasibility of continued economic growth and the value of natural resources and environmental services.

Economics deals with resource allocation or trade-offs between competing wants and needs. Economists focus on consumers and the products that meet the needs of those consumers. Furthermore, they are concerned with the time frame and manner in which these goods are produced. In mainstream, neoclassical economics, economists are usually concerned with the cost to obtain the things human consumers desire and the benefits that will be derived from those products.

According to classical economists, the costs of goods and services are determined by the interaction of supply and demand in the marketplace. If the supply of a particular commodity or service is high but the demand is low, the price will be low. If the commodity is scarce but everyone wants it, the price will be high. But high prices also encourage invention of new technology and substitutes that can satisfy the same demands. The cyclic relationship of scarce resources and development of new technology or new materials, in this view, allows for unlimited growth. And continued economic growth is seen as the best, perhaps the only, solution to poverty and environmental degradation.

Ecologists, however, view the world differently than economists. From their studies of the interactions between organisms and their environment, ecologists see the world as a dynamic, but finite system that can support only a limited number of humans with their demands for goods and services. Many ecological processes and the nonrenewable natural resources on which the economy is based have no readily available substitutes. Further, much of the natural world is being degraded or depleted at unsustainable rates. Ecologists criticize the narrow focus of conventional economics and its faith in unceasing growth, market valuation, and endless substitutability. Ecologists warn that unless humans changes their patterns of production and consumption to ways that protect natural resources and ecological systems, they will soon deplete these resources.

Ecological economics

Ecological or environmental economics is a relatively new field that introduces ecological understanding into the economic discourse. It takes a transdisciplinary, holistic, contextual, value-sensitive approach to economic planning and resource allocation. This view recognizes the dependence on the natural world and the irreplaceable life-support services it renders. Rather than expressing values solely in market prices, ecological economics pays attention to intangible values, nonmarketed resources, and the needs and rights of future generations and other species. Issues of equitable distribution of access to resources and the goods and services they provide need to be solved, in this perspective, by means other than incessant growth.

Whereas neoclassical economics sees our environment as simply a supply of materials, services, and waste sinks, ecological economics regards human activities as embedded in a global system that places limits on what humans can and cannot do. Uncertainty and dynamic change are inherent characteristics of this complex natural system. Damage caused by human activities may trigger sudden and irreversible changes. The precautionary principle suggests that society should leave a margin for error in the use of resources and plan for adaptive management policies.

Natural capital

Conventional economists see wealth generated by human capital (human knowledge, experience, and enterprise) working with manufactured capital (buildings, machines, and infrastructure) to transform raw materials into useful goods and services. In this view economic growth and efficiency are best accomplished, by increasing the throughput of raw materials extracted from nature. Until they are transformed by human activities, natural resources are regarded as having little value. In contrast, ecological economists see natural resources as a form of capital equally important with human-made capital. In addition to raw materials such as minerals, fuels, fresh water, food, and fibers, nature provides valuable services on which we depend. Natural systems assimilate our wastes and regulate the earth's energy balance, global climate, material recycling, the chemical composition of the atmosphere and oceans, and the maintenance of biodiversity. Nature also provides aesthetic, spiritual, cultural, scientific, and educational opportunities that are rarely given a monetary value but are, nevertheless, of great significance.

Ecological economists argue that the value of natural capital should be taken into account rather than treated as a set of unimportant externalities. Our goal, in this view, should be to increase our efficiency in natural resource use and to reduce its throughput. Harvest rates for renewable resources (those like organisms that regrow or those like fresh water that are replenished by natural processes) should not exceed regeneration rates. Waste emissions should not exceed the ability of nature to assimilate or recycle those wastes. Nonrenewable resources (such as minerals) may be exploited by humans, but only at rates equal to the creation of renewable substitutes.

Accounting for natural capital

Where neoclassical economics seeks to maximize present value of resources, ecological economics calls for recognition of the real value of those resources in calculating economic progress. A market economist, for example, once argued that the most rational management policy for whales was to harvest all the remaining ones immediately and to invest the proceeds in some profitable business. Whales reproduce too slowly, he claimed, and are too dispersed to make much money in the long run by allowing them to remain wild. Ecologists reject this limited view of whales as only economic units of production, observing many other values in these wild, beautiful, sentient creatures. Furthermore whales may play important roles in marine ecology that people don't yet fully understand.

Ecologists are similarly critical of Gross National Product (GNP) as a measure of national progress or well-being. GNP measures only the monetary value of goods and services produced in a national economy. It doesn't distinguish between economic activities that are beneficial or harmful. People who develop cancer from smoking, for instance, contribute to the GNP by running up large hospital bills. The pain and suffering they experience doesn't appear on the balance sheets. When calculating GNP in conventional economics, a subtraction is made, for capital depreciation in the form of wear and tear on machines, vehicles, and buildings used in production, but no account is made for natural resources used up or ecosystems damaged by that same economic activity.

Robert Repetto of the World Resources Institute estimates that soil erosion in Indonesia reduces the value of crop production about 40 percent per year. If natural capital were taken into account, total Indonesian GNP would be reduced by at least 20 percent

annually. Similarly, Costa Rica experienced impressive increases in timber, beef, and banana production between 1970 and 1990. But decreased natural capital during this period represented by soil erosion, forest destruction, biodiversity losses, and accelerated water runoff add up to at least $4 billion, or about 25 percent, of annual GNP. Ecological economists call for a new System of National Accounts that recognizes the contribution of natural capital to economic activity.

Valuation of natural capital

Ecological economics requires new tools and new approaches to represent nature in GNP. Some categories in which natural capital might fit include:

- use values: the price paid to use or consume a resource
- option value: preserving options for the future
- existence value: those things people like to know still exist even though they may never use or even see them
- aesthetic value: things we appreciate for their beauty
- cultural value: things important for cultural identity
- scientific and educational value: information or experience-rich aspects of nature.

Measuring the value of natural resources and ecological services not represented in market systems is difficult. Ecological economists often have to resort to shadow pricing or other indirect valuation methods for natural resources. For instance, assigning a value to represent the worth of a day of canoeing on a wild river is complicated. It is possible to measure opportunity costs such as the amount paid to get to the river or to rent a canoe. The direct out-of-pocket costs might represent only a small portion, however, of what it is really worth to participants. Another approach is contingent valuation, in which potential resource users are asked such as how much they would be willing to pay for the experience or the sale price they would be willing to accept to forego the opportunity. These approaches are controversial and problematic because people may report what they think they ought to pay rather than what they would really pay for these activities.

Carrying capacity and sustainable development

Carrying capacity is the maximum number of organisms of a particular species that a given area can sustainably support. Whereas neoclassical economists believe that technology can overcome any obstacle and that human ingenuity frees the world from any constraints on population or economic growth, ecological economists argue that nature places limits on humans just as it does on any other species.

One of the ultimate limits society faces is energy. Because of the limits of the second law of thermodynamics, whenever work is done, some energy is converted to a lower quality, less useful form and ultimately is emitted as waste heat. This means that a constant input of external energy is needed. Many fossil fuel supplies are nearing exhaustion, and continued use of these sources by current technology carries untenable environmental costs. Vast amounts of solar energy reach the earth, and this solar energy already drives the generation of all renewable resources and ecological services. By some calculations, humans now control or directly consume about 40 percent of all the solar energy reaching the earth. It is uncertain how much more natural resources people can monopolize for their own purposes without seriously jeopardizing the integrity of natural systems for which there is no substitute. If an infinite supply of clean, renewable energy were available, there would need to be a solution for removal of heat without causing harm to our environment.

Ecological economics urges society to restrain growth of both human populations and the production of goods and services to conserve natural resources and to protect remaining natural areas and biodiversity. This does not necessarily mean that the billion people in the world who live in absolute poverty and cannot, on their own, meet the basic needs for food, shelter, clothing, education, and medical care are condemned to remain in that state. Ecological economics calls for more efficient use of resources and more equitable distribution of the benefits among those now living as well as between current generations and future ones.

A mechanism for attaining this goal is sustainable development, that is, a real improvement in the overall welfare of all people on a long-term basis. In the words of the World Commission on Economy and Development, sustainable development means "meeting the needs of the present without compromising the ability of future generations to meet their own needs." This development requires increased reliance on renewable resources in harmony with ecological systems in ways that do not deplete or degrade natural capital. It does not necessarily mean that all growth must cease. Many human attributes, such as knowledge, kindness, compassion, cooperation, and creativity, can expand infinitely without damaging the environment. Employing these attributes to develop sustainable strategies may provide solutions for the current problems concerning continued growth and

depletion of resources. Climate change is also an issue of significant concern and is affected by population growth and resource use. Although ecological economics offers a sensible framework for approaches to resource use that can harmonize with ecological systems over the long term, it remains to be seen whether society will be wise enough to adopt this framework before it is too late.

An organization concerned with the ecological effects of economy on the environment is the International Society for Ecological Economics (ISEE), formed in 1989, which promotes sustainable development, thus unifying ecological and economic principles. The society is an umbrella organization for regional societies, holding conferences and publishing a journal featuring research in the field to share strategies and ideas for a sustainable world.

Resources

BOOKS

Common, Michael S., and Sigrid Stagl. *Ecological Economics: An Introduction.* Cambridge, UK: Cambridge University Press, 2005.

Daly, Herman E. *Ecological Economics and Sustainable Development: Selected Essays of Herman Daly.* Cheltenham, UK: Edward Elgar, 2007.

Daly, Herman E., and Joshua C. Farley. *Ecological Economics: Principles and Applications.* Washington, D.C.: Island Press, 2004.

Lawn, Philip A. *Sustainable Development Indicators in Ecological Economics.* Cheltenham, UK: Edward Elgar Publishing, 2006.

Patterson, M. G., and Bruce C. Glavovic. *Ecological Economics of the Oceans and Coasts.* Cheltenham, UK: Edward Elgar Publishing, 2008.

Zografos, Christos, and Richard B. Howarth. *Deliberative Ecological Economics.* New Delhi: Oxford University Press, 2008.

OTHER

International Society for Ecological Economics (ISEE). "About the ISEE." http://www.ecoeco.org/about.php (accessed September 9, 2010).

William P. Cunningham

Ecological integrity

Ecological (or biological) integrity is a measure of how intact or complete an ecosystem is. Ecological integrity is a relatively new and somewhat controversial notion, however, which means that it cannot be defined exactly. Human activities cause many changes in environmental conditions, and these can benefit some species, communities, and ecological processes, while causing damages to others. The notion of ecological integrity is used to distinguish between ecological responses that represent improvements and those that are degradations.

Challenges to ecological integrity

Ecological integrity is affected by changes in the intensity of environmental stressors. Environmental stressors can be defined as physical, chemical, and biological constraints on the productivity of species and the processes of ecosystem development. Many environmental stressors are associated with the activities of humans, but some are also natural factors. Environmental stressors can exert their influence on a local scale, or they may be regional or even global in their effects. Stressors represent environmental challenges to ecological integrity.

Environmental stressors are extremely complex, but they can be categorized in the following ways:

(1) Physical stressors are associated with brief, but intense, exposures to kinetic energy. Because of its acute, episodic nature, this represents a type of disturbance. Examples include volcanic eruptions, windstorms, and explosions; (2) Wildfire is another kind of disturbance, characterized by the combustion of much of the biomass of an ecosystem, and often the deaths of the dominant plants; (3) Pollution occurs when chemicals are present in concentrations high enough to affect organisms and thereby cause ecological changes. Toxic pollution may be caused by such gases as sulfur dioxide (SO_2) and ozone (O_3), metals such as mercury (Hg) and lead (Pb), and pesticides. Nutrients such as phosphate (PO_4^{3-}) and nitrate (NO_3^-) can affect ecological processes such as productivity, resulting in a type of pollution known as eutrophication; (4) Thermal stress occurs when releases of heat to the environment cause ecological changes, as occurs near natural hot-water vents in the ocean, or where there are industrial discharges of warmed water; (5) Radiation stress is associated with excessive exposures to ionizing energy. This is an important stressor on mountaintops because of intense exposures to ultraviolet radiation, and in places where there are uncontrolled exposures to radioactive wastes; (6) Climatic stressors are associated with excessive or insufficient regimes of temperature, moisture, solar radiation, and combinations of these. Tundra and deserts are climatically stressed ecosystems, whereas tropical rain forests occur in places where the climatic regime is relatively benign; (7)

Biological stressors are associated with the complex interactions that occur among organisms of the same or different species. Biological stresses result from competition, herbivory, predation, parasitism, and disease. The harvesting and management of species and ecosystems by humans can be viewed as a type of biological stress.

All species and ecosystems have a limited capability for tolerating changes in the intensity of environmental stressors. Ecologists refer to this attribute as resistance. When the limits of tolerance to environmental stress are exceeded, however, substantial ecological changes results.

Large changes in the intensity of environmental stress result in various kinds of ecological responses. For example, when an ecosystem is disrupted by an intense disturbance, there will be substantial mortality of some species and other damages. This is followed by recovery of the ecosystem through the process of succession. In contrast, a longer-term intensification of environmental stress, possibly caused by chronic pollution or climate change, will result in longer lasting ecological adjustments. Relatively vulnerable species decrease in abundance or are eliminated from sites that are stressed over the longer term, and their modified niches will be assumed by more tolerant species. Other common responses to an intensification of environmental stress include a simplification of species richness and decreased rates of productivity, decomposition, and nutrient cycling. These changes represent a longer-term change in the character of the ecosystem.

Components of ecological integrity

Many studies have been made of the ecological responses to both disturbance and to longer-term changes in the intensity of environmental stressors. Such studies have, for instance, examined the ecological effects of air or water pollution, of the harvesting of species or ecosystems, and the conversion of natural ecosystems into managed agroecosystems. The commonly observed patterns of change in stressed ecosystems have been used to develop indicators of ecological integrity, which are useful in determining whether this condition is improving or being degraded over time. It has been suggested that greater ecological integrity is displayed by systems that, in a relative sense (1) *are resilient and resistant to changes in the intensity of environmental stress.* Ecological resistance refers to the capacity of organisms, populations, or communities to tolerate increases in stress without exhibiting significant responses. Once thresholds of tolerance are exceeded, ecological changes occur rapidly. Resilience refers to the ability to recover from disturbance; (2) *are biodiverse*. Biodiversity is defined as the total richness of biological variation, including genetic variation within populations and species, the numbers of species in communities, and the patterns and dynamics of these over large areas; (3) *are complex in structure and function*. The complexity of the structural and functional attributes of ecosystems is limited by natural environmental stresses associated with climate, soil, chemistry, and other factors, and also by stressors associated with human activities. As the overall intensity of stress increases or decreases, structural and functional complexity respond accordingly. Under any particular environmental regime, older ecosystems will generally be more complex than younger ecosystems; (4) *have large species present*. The largest species in any ecosystem appropriate relatively large amounts of resources, occupy a great deal of space, and require large areas to sustain their populations. In addition, large species tend to be long-lived, and consequently they integrate the effects of stressors over an extended time. As a result, ecosystems that are affected by intense environmental stressors can only support a few or no large species. In contrast, mature ecosystems occurring in a relatively benign environmental regime are dominated by large, long-lived species; (5) *have higher-order predators present*. Top predators are sustained by a broad base of ecological productivity, and consequently they can only occur in relatively extensive and/or productive ecosystems; (6) *have controlled nutrient cycling*. Ecosystems that have recently been disturbed lose some of their biological capability for controlling the cycling of nutrients, and they may lose large amounts of nutrients dissolved or suspended in stream water. Systems that are not susceptible to loss of their nutrient capital are considered to have greater ecological integrity; (7) *are efficient in energy use and transfer*. Large increases in environmental stress commonly result in community-level respiration exceeding productivity, resulting in a decrease in the standing crop of biomass in the system. Ecosystems that are not losing their capital of biomass are considered to have greater integrity than those in which biomass is decreasing over time; (8) *have an intrinsic capability for maintaining natural ecological values*. Ecosystems that can naturally maintain their species, communities, and other important characteristics, without being managed by humans, have greater ecological integrity. If, for example, a population of a rare species can only be maintained by management of its habitat by humans, or by a program of captive-breeding and release, then its population, and the ecosystem of which it is a component, are lacking in ecological integrity; (9) *are*

components of a natural community. Ecosystems that are dominated by non-native, introduced species are considered to have less ecological integrity than ecosystems composed of indigenous species.

Indicators (8) and (9) are related to naturalness and the roles of humans in ecosystems, both of which are philosophically controversial topics. However, most ecologists would consider that self-organizing, unmanaged ecosystems composed of native species have greater ecological integrity than those that are strongly influenced by humans. Examples of strongly human-dominated systems include agroecosystems, forestry plantations, and urban and suburban areas. None of these ecosystems can maintain their character in the absence of management by humans, including large inputs of energy and nutrients.

Indicators of ecological integrity

Indicators of ecological integrity vary greatly in their intent and complexity. For instance, certain metabolic indicators have been used to monitor the responses by individuals and populations to toxic stressors, as when bioassays are made of enzyme systems that respond vigorously to exposures to dichlorodiphenyl-trichloroethane (DDT), pentachlorophenols (PCBs), and other chlorinated hydrocarbons. Other simple indicators include the populations of endangered species; these are relevant to the viability of those species as well as the integrity of the ecosystem of which they are a component. There are also indicators of ecological integrity at the level of landscape, and even global indicators relevant to climate change, such as depletion of stratospheric ozone and deforestation.

Relatively simple indicators can sometimes be used to monitor the ecological integrity of extensive and complex ecosystems. For example, the viability of populations of spotted owls (*Strix occidentalis*) is considered to be an indicator of the integrity of the old-growth forests in which this endangered species breeds in the western United States. These forests are commercially valuable, and if plans to harvest and manage them are judged to threaten the viability of a population of spotted owls, this would represent an important challenge to the integrity of the old-growth forest ecosystem.

Ecologists are also beginning to develop composite indicators of ecological integrity. These are designed as summations of various indicators and are analogous to such economic indices as the Dow-Jones Index of stock markets, the Consumer Price Index, and the gross domestic product of an entire economy. Composite economic indicators of this sort are relatively simple to design, because all of the input data are measured in a common way (e.g., in dollars). In ecology, however, no common currency exists among the many indicators of ecological integrity. Consequently it is difficult to develop composite indicators that ecologists will agree upon.

Still, some research groups have developed composite indicators of ecological integrity that have been used successfully in a number of places and environmental contexts. For instance, the ecologist James Karr and his co-workers developed composite indicators of the ecological integrity of aquatic ecosystems, which are being used in modified form in many places in North America.

In spite of all of the difficulties, ecologists are making substantial progress in the development of indicators of ecological integrity. This is an important activity, because society needs objective information about complex changes that are occurring in environmental quality, including degradations of indigenous species and ecosystems. Without such information, actions may not be taken to prevent or repair unacceptable damages that may be occurring.

In terms of sustainable methods of maintaining ecological integrity, ecologists can approach ecosystem or environmental management in three ways. The normative view assumes that humans and the ecosystem are separate entities focusing on controlling human practices to preserve ecosystem health. The ecosystem-pluralistic view considers humans as a part of the ecosystem and assumes that all involved are equal stakeholders in the process of managing the ecosystem. The third approach is the transpersonal-collaborative, which takes into account human and ecosystem relationships with regard to cultural and social implications of these interactions. All three of these approaches have their merits, but ultimately the solutions for maintaining ecological integrity are complicated and require compromise and changing practices to preserve ecosystem health and monitor resource use to reduce climate change and biodiversity losses.

Increasingly, it is being recognized that human economies can only be sustained over the longer term by ecosystems with integrity. Ecosystems with integrity are capable of supplying continuous flows of such renewable resources as timber, fish, agricultural products, and clean air and water. Ecosystems with integrity are also needed to sustain populations of native species and their natural ecosystems, which must be maintained even while humans are exploiting the resources of the biosphere.

Resources

BOOKS

Askins, Robert. *Saving Biological Diversity: Balancing Protection of Endangered Species and Ecosystems.* New York: Springer, 2008.

Begon, M., C.R. Townsend, and J.L. Harper. *Ecology: From Individuals to Ecosystems.* Malden, MA: Blackwell, 2005.

Brown, Valerie A., et al. *Sustainability and Health: Supporting Global Ecological Integrity in Public Health.* Crows Nest, NSW, New Zealand: Allen & Unwin, 2005.

Egan, Dave, and Evelyn A. Howell, eds. *The Historical Ecology Handbook: A Restorationist's Guide to Reference Ecosystems.* 2nd ed. Washington, DC: Island Press, 2005.

Jarvis, Devra I., Christine Padoch, and H. D. Cooper. *Managing Biodiversity in Agricultural Ecosystems.* New York: Columbia University Press, 2007.

Moran, Emilio F., and Elinor Ostrom, eds. *Seeing the Forest and the Trees: Human-environment Interactions in Forest Ecosystems.* Cambridge, MA: MIT Press, 2005.

Perry, David A., Ram Oren, and Stephen C. Hart. *Forest Ecosystems.* Baltimore: Johns Hopkins University Press, 2008.

Waltner-Toews, David, James Kay, and Nina-Marie E. Lister. *The Ecosystem Approach: Complexity, Uncertainty, and Managing for Sustainability.* Complexity in ecological systems series. New York: Columbia University Press, 2008.

Westra, Laura, Klaus Bosselmann, and Richard Westra. *Reconciling Human Existence with Ecological Integrity: Science, Ethics, Economics and Law.* London: Earthscan, 2008.

OTHER

United Nations Environment Programme (UNEP). "World Conservation Monitoring Centre Biodiversity of Ecosystems." http://www.unep-wcmc.org/climate/impacts.aspx (accessed September 7, 2010).

United States Environmental Protection Agency (EPA). "Ecosystems." http://www.epa.gov/ebtpages/ecosystems.html (accessed September 7, 2010).

United States Environmental Protection Agency (EPA). "Ecosystems: Species: Endangered Species." http://www.epa.gov/ebtpages/ecosspeciesendangeredspecies.html (accessed September 7, 2010).

Bill Freedman

Ecological productivity

One of the most important properties of an ecosystem is its productivity, which is a measure of the rate of incorporation of energy by plants per unit area per unit time. In terrestrial ecosystems ecologists usually estimate plant production as the total annual growth—the increase in plant biomass over a year. Since productivity reflects plant growth, it is often used loosely as a measure of the organic fertility of a given area.

The flow of energy through an ecosystem starts with the fixation of sunlight by green plants during photosynthesis. Photosynthesis supplies both the energy (in the form of chemical bonds) and the organic molecules (glucose) that plants use to make other products in a process known as biosynthesis. During biosynthesis, glucose molecules are rearranged and joined together to become complex carbohydrates (such as cellulose and starch) and lipids (such as fats and plant oils). These products are also combined with nitrogen, phosphorus, sulfur, and magnesium to produce the proteins, nucleic acids, and pigments required by the plant. The many products of biosynthesis are transported to the leaves, flowers, and roots, where they are stored to be used later.

Ecologists measure the results of photosynthesis as increases in plant biomass over a given time. To do this more accurately, ecologists distinguish two measures of assimilated light energy: gross primary production (GPP), which is the total light energy fixed during photosynthesis, and net primary production (NPP), which is the chemical energy that accumulates in the plant over time.

Some of this chemical energy is lost during plant respiration (R) when it is used for maintenance, reproduction, and biosynthesis. The proportion of GPP that is left after respiration is counted as net production (NPP). In an ecosystem it is the energy stored in plants from net production that is passed up the food chain/web when the plants are eaten. This energy is available to consumers either directly as plant tissue or indirectly through animal tissue.

One measure of ecological productivity in an ecosystem is the production efficiency. This is the rate of accumulation of biomass by plants, and it is calculated as the ratio of net primary production to gross primary production. Production efficiency varies among plant types and among ecosystems. Grassland ecosystems which are dominated by non-woody plants are the most efficient at 60-85 percent, since grasses and annuals do not maintain a high supporting biomass. On the other end of the efficiency scale are forest ecosystems; they are dominated by trees, and large old trees spend most of their gross production in maintenance. For example, eastern deciduous forests have a production efficiency of about 42 percent.

Ecological productivity in terrestrial ecosystems is influenced by physical factors such as temperature and rainfall. Productivity is also affected by air and water currents, nutrient availability, land forms, light intensity, altitude, and depth. The most productive ecosystems are tropical rain forests, coral reefs, salt marshes, and estuaries; the least productive are deserts, tundra, and the open sea.

See also Ecological consumers; Habitat; Restoration ecology.

Neil Cumberlidge

Ecological risk assessment

Ecological risk assessment (ERA) is a procedure for evaluating the likelihood that adverse ecological effects are occurring, or may occur, in ecosystems as a result of one or more human activities. These activities may include the alteration and destruction of wetlands or other habitats; the introduction of herbicides, pesticides, or other toxic materials into the environment; oil spills; or the cleanup of contaminated hazardous waste sites. Ecological risk assessments consider many aspects of an ecosystem, including biotic plants and animals and abiotic water, soils, and other elements. Ecosystems can be as small as a pond or stream or as large as thousands of square miles of land or lengthy coastlines in which communities exist.

Although closely related to human health risk assessment, ecological risk assessment is not only a newer discipline but also uses different procedures, terminology, and concepts. Both human health and ecological risk assessment provide frameworks for collecting information to define a risk and to help make risk management or regulatory decisions, but human health risk assessment follows four basic steps that were defined in 1983 by the National Research Council: hazard assessment, dose response assessment, exposure assessment, and risk characterization. In contrast, ecological risk assessment relies on the Guidelines for Ecological Risk Assessment published by the Environmental Protection Agency (EPA) in 1998. The National Contingency Plan (NCP), which is the plan for responding to oil spills and the release of hazardous substances, requires the characterization of "current and potential threats to human health and the environment" and states that "environmental evaluations shall be performed to assess threats to the environment, especially sensitive habitats and critical habitats of species protected under the Endangered Species Act."

The guidelines define three phases: problem formulation, analysis, and risk characterization. In some cases a screening step conducted prior to the assessment involves a site visit to investigate the area and potentially affected species. The toxicity of the stressor is evaluated, and estimates of exposure and risk calculation are accomplished. This initial step provides the information necessary to decide whether a threat needs to be evaluated by the guidelines. The problem formulation phase involves defining the value of the ecological resource and developing a conceptual model to understand and illustrate exposure pathways. In the analysis phase, a sampling plan is devised and data is collected to determine the resource's exposure to the stressor and the stressor's effects on the ecological resource. Risk characterization and management involves the calculation of risk based on the integration of the exposure- and stressor-response profiles.

The problems that human health risk assessments seek to address are clearly defined: cancer, birth defects, mortality, and similar concerns. The problems that ecological risk assessments attempt to understand and address, however, are less straightforward. For instance, a major challenge ecological risk assessors face is distinguishing natural changes in an ecosystem from changes caused by human activities and defining what changes are unacceptable. As a result, the initial problem formulation step of an ecological risk assessment requires extensive discussions between risk assessors and risk managers to define ecological significance, a key concept in ecological risk assessment. Since it is not immediately clear whether an ecological change is positive or negative—unlike cancer or birth defects, which are known to be adverse—judgments must be made early in the assessment about whether a change is significant and whether it will alter a socially valued ecological condition. For example, Lake Erie was declared dead in the 1960s as a result of phosphorous (P) loadings from cities and farms. In fact, there were more fish in the lake after it was dead than before; however, these fish were carp, suckers, and catfish, not the walleyed pike, yellow perch, and other fish that had made Lake Erie one of the most highly valued freshwater sport-fishing lakes in the United States. More recently, with pollution inputs greatly reduced, the lake has recovered much of its former productivity. Choosing one ecological condition over the other is a social value fundamental to ecological risk assessment problem formulation. Once judgments have been made about

what values to protect, analysts can examine the stressors that ecosystems are exposed to, and can characterize the ecological effects likely to result from such stressors.

Since 1989 the EPA has held workshops on ecological significance and other technical issues pertaining to ecological risk assessment and has published the results of its workshops in a series of reports and case studies. Overall, the direction of ecological protection priorities has moved away from earlier concerns with narrow goals (e.g., use of commercially valuable natural resources) toward broader interest in protecting natural areas and ecosystems, such as National Parks and Scenic Rivers, for both present and future generations to enjoy.

Resources

BOOKS

Paustenbach, Dennis J. *Human and Ecological Risk Assessment: Theory and Practice.* Hoboken, N.J.: Wiley, 2008.

Suter, Glenn W. *Ecological Risk Assessment.* Boca Raton: CRC Press/Taylor & Francis, 2007.

OTHER

United States Environmental Protection Agency (EPA). "Environmental Management: Risk Assessment: Ecological Risk Assessment." http://www.epa.gov/ebtpages/enviriskassessmentecologicalriskassessment.html (accessed October 2, 2010).

David Clarke

Ecological Society of America

The Ecological Society of America (ESA), representing more than 10,000 ecological researchers worldwide, was founded in 1915 as a non-profit, scientific organization. Members include ecologists from academia, government agencies, industry, and non-governmental organizations (NGOs). In pursuing its goal of promoting "the responsible application of ecological principles to the solution of environmental problems," the society publishes reports, membership research, and three scientific journals a year, and it provides expert testimony to Congress. In addition ESA holds a conference every summer attended by more than 3,000 scientists and students at which members present the latest ecological research. The Society's three journals are: *Ecology* (eight issues per year), *Ecological Monographs* (four issues per year), and *Ecological Applications* (four issues per year). ESA also publishes a bimonthly member newsletter.

A milestone in the Society's development was its 1991 proposal for a Sustainable Biosphere Initiative (SBI), which was published in a 1991 issue of *Ecology* as a "call-to- arms for ecologists." Based on research priorities identified in the proposal, ESA chartered the SBI Project Office in the same year to focus on global change, biodiversity, and sustainable ecosystems. The SBI marked a commitment by the Society to more actively convey its members' findings to the public and to policy makers, and, as such, included research, education, and environmental decision-making components. The three-pronged SBI proposal grew out of a "period of introspection" during which ESA led its members to examine "the whole realm of ecological activities" in the face of decreasing funds for research, an urgent need to set priorities, and "the need to ameliorate the rapidly deteriorating state of the environment and to enhance its capacity to sustain the needs of the world's population."

Since the SBI Project Office was chartered, it has focused on linking the ecological scientific community to other scientists and decision makers through a multi-disciplinary twelve-member steering committee and five-member staff. For instance, in 1995 the SBI began a series of semi-annual meetings with federal government officials to discuss "overlapping areas of interest and possible collaborative opportunities." In addition, SBI hosts various symposia and panel discussions of key ecological topics.

In 1993 ESA chartered a Special Committee on the Scientific Basis of Ecosystem Management to establish the scientific grounds for discussing the increasingly prominent ecosystem approach to addressing land and natural resource management problems. The committee published its findings in the August 1996 issue of *Ecological Applications*. Articles discussed the emerging consensus on essential elements of ecosystem management, including its holistic nature—incorporating the biological and physical elements of an ecosystem and their interrelationships—and the concept of sustainability as the "essential element and precondition" of ecosystem management.

The ESA also maintains a blog, EcoTone, that presents the public with information about ESA activities as well as information and updates on various environmental issues.

ESA's headquarters in Washington, DC, consistent with the SBI's goal of broader public education, includes a Public Affairs Office. Its Publications Office is in Ithaca, New York.

Resources

BOOKS

Curry, Patrick. *Ecological Ethics: An Introduction.* Cambridge, UK: Polity Press, 2006.
Ebbesson, Jonas, and Phoebe N. Okowa. *Environmental Law and Justice in Context.* Cambridge, UK: Cambridge University Press, 2009.

OTHER

United States Environmental Protection Agency (EPA). "Ecosystems: Ecological Monitoring." http://www. epa.gov/ebtpages/ecosecologicalmonitoring.html (accessed October 14, 2010).
United States Environmental Protection Agency (EPA). "Environmental Management: Risk Assessment: Ecological Risk Assessment." http://www.epa.gov/ebt-pages/enviriskassessmentecologicalriskassessment. html (accessed October 14, 2010).

ORGANIZATIONS

Ecological Society of America, 1990 M Street, NW, Suite 700, Washington, DC, USA, 20036, (202) 833-8773, (202) 833-8775, esahq@esa.org, http://www.esa.org

David Clarke

Ecological succession *see* **Succession.**

Ecology

The word ecology was first used in published works in 1870 by the German zoologist Ernst Haeckel (1834–1919) from the Greek words *oikos* (house) and *logos* (logic or knowledge) to describe the scientific study of the relationships among organisms and their environment. The environment consists of both abiotic (e.g., water availability and temperature) and biotic (e.g., other organisms of the same species or different species) factors. Biologists began referring to themselves as ecologists at the end of the nineteenth century and shortly thereafter the first ecological societies and journals appeared. Since that time ecology has become a major branch of biological science. The contextual, historical understanding of organisms as well as the systems basis of ecology set it apart from the reductionist, experimental approach prevalent in many other areas of science.

This broad ecological view is gaining significance today as modern resource-intensive lifestyles consume much of nature's supplies. Although intuitive ecology has always been a part of some cultures, current environmental crises make a systematic, scientific understanding of ecological principles especially important.

For many ecologists the basic structural units of ecological organization are species and populations. A biological species consists of all the organisms potentially able to interbreed under natural conditions and to produce fertile offspring. A population consists of all the members of a single species occupying a common geographical area at the same time. An ecological community is composed of a number of populations that live and interact in a specific region. In ecology there is an organizational hierarchy from the most inclusive—the biosphere—to the most specific—each individual organism. The levels of organization are (from broadest to most narrow): biosphere, region, landscape, ecosystem, community, interactions, population, and individual organism. The biosphere contains all ecosystems in the world, whereas each region holds only a subset of ecosystem types. Ecologists study organisms and their relationships with their environments with respect to these levels of organization.

This population-community view of ecology is grounded in natural history—the study of where and how organisms live—and the Darwinian theory of natural selection and evolution. Proponents of this approach generally view ecological systems primarily as networks of interacting organisms. Abiotic forces such as weather, soils, and topography are often regarded as external factors that influence, but are apart from, the central living core of the system.

In the past three decades, the emphasis on species, populations, and communities in ecology has been replaced by a more quantitative, thermodynamic analysis of the processes through which energy flows and the cycling of nutrients and toxins are carried out in ecosystems. This process-functional approach is concerned more with the ecosystem as a whole than the particular species or populations that make it up. In this perspective both the living organisms and the abiotic physical components of the environment are equal members of the system.

The feeding relationships among different species in a community are a key to understanding ecosystem function. Who eats whom, where, how, and when determine how energy and materials move through the system. Predator-prey relationships also influence natural selection, evolution, and species adaptation to a particular set of environmental conditions. Ecosystems are open systems, insofar as energy and materials flow through them. Nutrients, however, are often recycled extremely efficiently so that the annual losses to sediments or through surface water runoff are relatively small in many mature ecosystems. In undisturbed tropical rain forests, for instance, nearly 100 percent of leaves and detritus are decomposed and recycled within a few days after they fall to the forest floor.

Because of thermodynamic losses every time energy is exchanged between organisms or converted from one form to another, an external energy source is an indispensable component of every ecological system. Green plants capture solar energy through photosynthesis and convert it into energy-rich organic compounds that are the basis for all other life in the community. This energy capture is referred to as primary productivity. These green plants form the producers, which are the first trophic (or feeding) level of most communities.

Secondary productivity is represented by consumers and decomposers, making up the other components of the food chain. Herbivores (animals that consume plants) make up the next trophic level, whereas carnivores (animals that consume other animals) add to the complexity and diversity of the community. Detritivores (such as beetles and earthworms) and decomposers (generally bacteria and fungi) convert dead organisms or waste products to inorganic chemicals. The nutrient recycling they perform is essential to the continuation of life. Together, all these interacting organisms form a food chain/web through which energy flows and nutrients and toxins are recycled. Due to intrinsic inefficiencies in transferring material and energy between organisms, the energy content in successive trophic levels is usually represented as a pyramid in which primary producers form the base and the top consumers occupy the apex.

This introduces the problem of persistent contaminants in the food chain. Because they tend not to be broken and metabolized in each step of the food chain in the way that other compounds are, persistent contaminants such as pesticides and heavy metals tend to accumulate in top carnivores, often reaching toxic levels many times higher than original environmental concentrations. This biomagnification is an important issue in pollution control policies. In many lakes and rivers, for instance, game fish have accumulated dangerously high levels of mercury (Hg) and chlorinated hydrocarbons that present a health threat to humans and other fish-eating species.

Diversity, in ecological terms, is a measure of the number of different species in a community, whereas abundance is the total number of individuals. Tropical rain forests, although they occupy only about five percent of the earth's land area, are thought to contain somewhere around half of all terrestrial plant and animals species, whereas coral reefs and estuaries are generally the most productive and diverse aquatic communities. Community complexity refers to the number of species at each trophic level as well as the total number of trophic levels and ecological niches in a community.

Structure describes the patterns of organization, both spatial and functional, in a community. In a tropical rain forest, for instance, distinctly different groups of organisms live on the surface, at mid-levels in the trees, and in the canopy, giving the forest vertical structure. A patchy mosaic of tree species, each of which may have a unique community of associated animals and smaller plants living in its branches, gives the forest horizontal structure as well.

For every physical factor in the environment there are both maximum and minimum tolerable limits beyond which a given species cannot survive. The factor closest to the tolerance limit for a particular species at a particular time is the critical factor that will determine the abundance and distribution of that species in that ecosystem. Natural selection is the process by which environmental pressures—including biotic factors such as predation, competition, and disease, as well as physical factors such as temperature, moisture, soil type, and space—affect survival and reproduction of organisms. Over a very long time, given a large enough number of organisms, natural selection works on the randomly occurring variation in a population to allow evolution of species and adaptation of the population to a particular set of environmental conditions.

Habitat describes the place or set of environmental conditions in which an organism lives; niche describes the role an organism plays. A yard and garden, for instance, may provide habitat for a family of cottontail rabbits. Their niche is being primary consumers (eating vegetables and herbs).

Organisms interact within communities in many ways. Symbiosis is the intimate living together of two species; commensalism describes a relationship in which one species benefits while the other is neither helped nor harmed. Lichens, the thin crusty organisms often seen on exposed rocks, are an obligate symbiotic association of a fungus and an alga. Neither can survive without the other, and both components benefit from the other. Some orchids and bromeliads (air plants), conversely, live commensally on the branches of tropical trees. The orchid benefits by having a place to live, but the tree is neither helped nor hurt by the presence of the orchid.

Predation—feeding on another organism—can involve pathogens, parasites, and herbivores as well as carnivorous predators. Competition is another kind of antagonistic relationship in which organisms vie for space, food, or other resources. Predation, competition, and natural selection often lead to niche specialization and resource partitioning that reduce competition between species. The principle of competitive exclusion states that no two species will remain in direct competition for very long in the same habitat because natural selection and adaptation will cause organisms to specialize in when, where, or how they live to minimize conflict over resources. This can contribute to the evolution of a given species into new forms over time.

It is also possible, on the other hand, for species to co-evolve, meaning that each changes gradually in response to the other to form an intimate and often highly dependent relationship either as predator and prey or for mutual aid. Because individuals of a particular species may be widely dispersed in tropical forests, many plants have become dependent on insects, birds, or mammals to carry pollen from one flower to another. Some amazing examples of coevolution and mutual dependence have resulted.

Ecological succession, the process of ecosystem development, describes the changes through which whole communities progress as different species colonize an area and change its environment. A typical successional series starts with pioneer species such as grasses or fireweed that colonize bare ground after a disturbance. Organic material from these pioneers helps build soil and hold moisture, allowing shrubs and then tree seedlings to become established. Gradual changes in shade, temperature, nutrient availability, wind protection, and living space favor different animal communities as one type of plant replaces its predecessors. Primary succession starts with a previously unoccupied site. Secondary succession occurs on a site that has been disturbed by external forces such as fires, storms, or humans. In many cases, succession proceeds until a mature climax community is established. Introduction of new species by natural processes, such as opening a land bridge, or by human intervention can upset the natural relationships in a community and cause catastrophic changes for indigenous species.

Biomes consist of broad regional groups of related communities. Their distribution is determined primarily by climate, topography, and soils. Often similar niches are occupied by different but similar species (termed ecological equivalents) in geographically separated biomes. Some of the major biomes of the world are deserts, grasslands, wetlands, forests of various types, and tundra.

The relationship between diversity and stability in ecosystems is a controversial topic in ecology.

F.E. Clements (1874–1945), an early American bio-geographer, championed the concept of climax communities: stable, predictable associations toward which ecological systems tend to progress if allowed to follow natural tendencies. Deciduous, broad-leaved forests are climax communities in moist, temperate regions of the eastern United States according to Clements, whereas grasslands are characteristic of the dryer western plains. In this view, homeostasis (a dynamic steady-state equilibrium), complexity, and stability are endpoints in ecological succession. Ecological processes, if allowed to operate without external interference, tend to create a natural balance between organisms and their environment.

H.A. Gleason (1882–1975), another American pioneer biogeographer and contemporary of Clements, argued that ecological systems are much more dynamic and variable than the climax theory proposes. Gleason saw communities as temporary or even accidental combinations of continually changing biota rather than predictable associations. Ecosystems may or may not be stable, balanced, and efficient; change, in this view, is thought to be more characteristic than constancy. Diversity may or may not be associated with stability. Some communities such as salt marshes that have only a few plant species may be highly resilient and stable while species-rich communities, such as coral reefs, may be highly sensitive to disturbance.

Although many ecologists now tend to agree with the process-functional view of Gleason rather than the population-community view of Clements, some retain a belief in the balance of nature and the tendency for undisturbed ecosystems to reach an ideal state if left undisturbed. The efficacy and ethics of human intervention in natural systems may be interpreted very differently in these divergent understandings of ecology. Those who see stability and constancy in nature often call for policies that maintain historic conditions and associations. Those who see greater variability and individuality in communities may favor more activist management and be willing to accept change as inevitable.

In spite of some uncertainty, however, about how to explain ecological processes and the communities they create, humans have learned a great deal about the world around them through scientific ecological studies in the past century. This important field of study remains a crucial component in the ability to manage resources sustainably and to avoid or repair environmental damage caused by human actions. Deforestation, biodiversity loss, pollution, and climate change are causing dramatic changes in the world's ecosystems. Ecological research may aid in identifying problems caused by the expanding world population

and in developing sustainable solutions for the management and maintenance of the biosphere and its resources.

Resources

BOOKS

Babe, Robert E. *Culture of Ecology: Reconciling Economics and Environment.* Toronto, Canada: University of Toronto Press, 2006.

Curry, Patrick. *Ecological Ethics: An Introduction.* Cambridge, UK: Polity Press, 2006.

Freeman, Jennifer. *Ecology.* New York: Harper Collins, 2007.

Fulbright, Timothy E., and David G. Hewitt. *Wildlife Science: Linking Ecological Theory and Management Applications.* Boca Raton, FL: CRC Press, 2008.

Grant, William, and Todd Swannack. *Ecological Modeling.* Malden, MA: Blackwell Publishing, 2007.

Gurevitch, Jessica, Samuel M. Scheiner, and Gordon A. Fox. *The Ecology of Plants.* Sunderland, MA: Sinauer Associates, 2006.

Herrel, Anthony, Thomas Speck, and Nicolas P. Rowe, eds. *Ecology and Biomechanics: A Mechanical Approach to the Ecology of Animals and Plants.* Boca Raton, FL: CRC/Taylor & Francis, 2006.

Karban, Richard, and Mikaela Huntzinger. *How to Do Ecology: A Concise Handbook.* Princeton, NJ: Princeton University Press, 2006.

Kingsland, Sharon E. *The Evolution of American Ecology, 1890–2000.* Baltimore: Johns Hopkins University Press, 2005.

Mayhew, Peter J. *Discovering Evolutionary Ecology: Bringing Together Ecology and Evolution.* Oxford, UK, and New York: Oxford University Press, 2006.

McArthur, J. Vaun. *Microbial Ecology: An Evolutionary Approach.* Amsterdam, The Netherlands, and Boston, MA: Elsevier/Academic Press, 2006.

Merchant, Carolyn. *Ecology.* Key concepts in critical theory. Amherst, N: Humanity Books, 2008.

Molles, Manuel C. *Ecology: Concepts and Applications.* Dubuque, IA: McGraw-Hill, 2009.

Moran, Emilio F. *People and Nature: An Introduction to Human Ecological Relations.* Blackwell primers in anthropology. Malden, MA: Blackwell, 2006.

Pickett, Steward T., Jurek Kolasa, and Clive G. Jones. *Ecological Understanding: The Nature of Theory and the Theory of Nature.* Amsterdam: Elsevier/Academic Press, 2007.

Schultz, Jurgen. *The Ecozones of the World: The Ecological Divisions of the Geosphere.* Berlin: Springer, 2005.

Slobodkin, Lawrence B. *A Citizen's Guide to Ecology.* New York, NY: Oxford University Press, 2003.

Speight, Martin R., Mark D. Hunter, and Allan D. Watt. *Ecology of Insects: Concepts and Applications.* Oxford: Wiley-Blackwell, 2008.

Spellerberg, Ian F. and Martin W. Holdgate. *Monitoring Ecological Change.* Malden, MA: Blackwell Science, 2005.

Wilkinson, David M. *Fundamental Processes in Ecology: An Earth Systems Approach.* Oxford: Oxford University Press, 2006.

OTHER

U.S. Government, science.gov. "Ecology." http://www.science.gov/browse/w_123C.htm (accessed September 23, 2010).

United States Department of the Interior, United States Geological Survey (USGS). "Ecological Processes." http://www.usgs.gov/science/science.php?term=310 (accessed September 23, 2010).

United States Environmental Protection Agency (EPA). "Ecosystems: Ecological Monitoring." http://www.epa.gov/ebtpages/ecosecologicalmonitoring.html (accessed September 23, 2010).

William P. Cunningham

Ecology, deep *see* **Deep ecology.**

Ecology, human *see* **Human ecology.**

Ecology, restoration *see* **Restoration ecology.**

Ecology, social *see* **Social ecology.**

EcoNet

EcoNet is a computer network that focused on environmental topics and, through the Institute for Global Communications, lists links to the international community. Several thousand organizations and individuals have accounts on the network. EcoNet's electronic conferences contain press releases, reports, and electronic discussions on hundreds of topics, ranging from clean air to pesticides. As of 2010, EcoNet remains a branch of ICG Internet, as are PeacNet, WomensNet, and AntiRacismNet. ICG actively supports Progressive causes and remains active in fund-raising to relieve suffering related to the 2010 Earthquake in Haiti.

Resources

ORGANIZATIONS

Institute for Global Communications, P.O. Box 29904, San Francisco, CA, USA, (941) 29-0904, support@igc, apc.org, http://www.igc.org/

Economic growth and the environment

The issue of economic growth and the environment primarily concerns the kinds of pressures that economic growth, at the national and international level, places on the environment over time. The relationship between ecology and the economy has become increasingly significant as humans gradually understand the impact that economic decisions have on the sustainability and quality of the planet.

Economic growth is commonly defined as increases in total output from new resources or better use of existing resources; it is measured by increased real incomes per capita. All economic growth involves transforming the natural world, and it may effect environmental quality in one of three ways. Environmental quality can increase with growth. Increased incomes, for example, provide the resources for public services, such as sanitation and rural electricity. With these services widely available, individuals need to worry less about day-to-day survival and can devote more resources to conservation. Secondly, environmental quality can initially worsen but then improve as the growth rate rises. In the cases of air pollution, water pollution, deforestation, and encroachment, there is little incentive for any individual to invest in maintaining the quality of the environment. These problems can only improve when countries deliberately introduce long-range policies to ensure that additional resources are devoted to dealing with them. Thirdly, environmental quality can decrease when the rate of growth increases. In the cases of emissions generated by the disposal of municipal solid waste, for example, abatement is relatively expensive and the costs associated with the emissions and wastes are not perceived as high because they are often borne by someone else.

The World Bank estimates that, under present productivity trends and given projected population increases, the output of developing countries would be about five times higher by the year 2030 than it was in 2002. The output of industrial countries would rise more slowly, but it would still triple over the same period. If environmental pollution were to rise at the same pace, severe environmental hardships would occur. Tens of millions of people would become sick or die from environmental causes, and the planet would be significantly and irreparably harmed.

The earth's natural resources place limits on economic growth. These limits vary with the extent of resource substitution, technical progress, and structural changes. For example, in the late 1960s many feared that the world's supply of useful metals would run out. Yet, by 2010 there was a glut of useful metals, and prices had fallen dramatically. The demand for other natural resources such as water, however, often exceeds supply. In arid regions such as the Middle East and in non-arid regions such as northern China, aquifers have been depleted and rivers so extensively drained that not only irrigation and agriculture are threatened but also the local ecosystems.

Some resources such as water, forests, and clean air are under attack, whereas others such as metals, minerals, and energy are not threatened. This is because the scarcity of metals and similar resources is reflected in market prices. Here, the forces of resource substitution, technical progress, and structural change have a strong influence. But resources such as water are characterized by open access, and there are, therefore, no incentives to conserve. Many believe that effective policies designed to sustain the environment are most necessary because society must take account of the value of natural resources, and governments must create incentives to protect the environment. Economic and political institutions have failed to provide these necessary incentives for four separate yet interrelated reasons: (1) short time horizons; (2) failures in property rights; (3) concentration of economic and political power; and (4) immeasurability and institutional uncertainty.

Although economists and environmentalists disagree on the definition of sustainability, the essence of the idea is that current decisions should not impair the prospects for maintaining or improving future living standards. The economic systems of the world should be managed so that societies live on the dividends of the natural resources, always maintaining and improving the asset base.

Promoting growth, alleviating poverty, and protecting the environment may be mutually supportive objectives in the long run, but they are not always compatible in the short run. Poverty is a major cause of environmental degradation, and economic growth is thus necessary to improve the environment. Yet, ill-managed economic growth can also destroy the environment and further jeopardize the lives of the poor. In many poor, but still forested, countries, timber is a good short-run source of foreign exchange. When demand for Indonesia's traditional commodity export—petroleum—fell and its foreign exchange income slowed, Indonesia began depleting its hardwood forests at non-sustainable rates to earn export income.

In developed countries, competition may shorten time horizons. Competitive forces in agricultural markets, for example, induce farmers to take short-term perspectives for financial survival. Farmers must maintain cash flow to satisfy bankers and make a sufficient return on their land investment. They therefore adopt high-yield crops, monoculture farming, increased fertilizer and pesticide use, salinizing irrigation methods, and more intensive tillage practices which cause erosion.

"The Tragedy of the Commons" is the classic example of property rights failure. When access to a grazing area, or commons is unlimited, each herdsman knows that grass not eaten today will not be there tomorrow. As a rational economic being, each herdsman seeks to maximize his gain and adds more animals to his herd. No herdsman has an incentive to prevent his livestock from grazing the area. Degradation follows as does the loss of a common resource. In a society without clearly defined property rights, those who pursue their own interests ruin the public good.

In many nations, political upheaval may void property rights virtually overnight, so an individual with a concession to harvest trees is motivated to harvest as many and as quickly as possible. The government-granted timber-cutting concession may belong to someone else tomorrow. The same is true of some developed countries. For example, in Louisiana mineral rights revert to the state when wetlands become open water and there has been no mineral development on the property. Thus, the cheapest methods of avoiding loss of mineral revenues has been to hurry the development of oil and gas in areas that might revert to open water, thereby hastening erosion and saltwater intrusion, or putting up levies around the property to maintain it as private property, thus interfering with normal estuarine processes.

Global or transnational problems such as ozone layer depletion or acid rain produce a similar problem. Countries have little incentive to reduce damage to the global environment unilaterally when doing so will not reduce the damaging behavior of others or when reduced fossil fuel use would leave that country at a competitive disadvantage. International agreements are thus needed to impose order on the world's nations that would be analogous to property rights.

Concentration of wealth within the industrialized countries allows for the exploitation and destruction of ecosystems in less developed countries (LDC) through, for example, timber harvests and mineral extraction. The concentration of wealth inside a less developed country skews public policy toward benefiting the wealthy and politically powerful, often at the expense of the ecosystem on which the poor depend. Local sustainability is dependent upon the goals of those who have power—goals which may or may not be in line with a healthy, sustainable ecosystem. Furthermore, when an exploiting party has substitute ecosystems available, it can exploit one and then move to the next. In such a situation, the benefits of sustainability are low, and companies or people that exploit natural resources have shorter time horizons than those with local interests. This is also an example of how the high discount rates in developed countries are imposed on the management of developing countries' assets.

Environmental policy-making is always more complicated than merely measuring the effects of a proposed policy on the environment. Due to scientific uncertainty about biophysical and geological relations and a general inability to measure a policy's effect on the environment, economic rather than ecological effects are more often relied upon to make policy. Policy-makers and institutions do not often grasp the direct and indirect effects of policies on ecological sustainability nor do they know how their actions will affect other areas not under their control.

Many contemporary economists and environmentalists argue that the value of the environment should nonetheless be factored into the economic policy decision-making process. The goal is not necessarily to put monetary values on environmental resources; it is rather to determine how much environmental quality is being given up in the name of economic growth, and how much growth is being given up in the name of the environment. A danger always exists that too much income growth may be lost in the future by failing to clarify and minimize tradeoffs or to take advantage of policies that are good for both economic growth and the environment.

See also Energy policy; Environmental economics; Environmental policy; Environmentally responsible investing; Exponential growth; Sustainable agriculture; Sustainable biosphere; Sustainable development.

Resources

BOOKS

Elliot, Jennifer A. *An Introduction to Sustainable Development*. New York: Viking, 2005.

Rogers, Peter, et al. *An Introduction to Sustainable Development*. Cambridge, MA: Harvard Division of Continuing Education, 2006.

OTHER

International Institute for Sustainable Development (IISD). "International Institute for Sustainable Development (IISD)." http://www.iisd.org/ (accessed October 2, 2010).

United States Environmental Protection Agency (EPA). "Environmental Management: Smart Growth." http://www.epa.gov/ebtpages/envismartgrowth.html (accessed October 2, 2010).

Kevin Wolf

Ecopsychology *see* **Roszak, Theodore.**

Ecosophy

A philosophical approach to the environment that emphasizes the importance of action and individual beliefs. Often referred to as "ecological wisdom," it is associated with other environmental ethics, including deep ecology and bioregionalism.

Ecosophy originated with the Norwegian philosopher Arne Dekke Eide Næss (1912–2009). Naess described a structured form of inquiry he called *ecophilosophy*, which examines nature and humankind's relationship to it. He defined it as a discipline, like philosophy itself, which is based on analytical thinking, reasoned argument, and carefully examined assumptions. Naess distinguished ecosophy from ecophilosophy; it is not a discipline in the same sense but what he called a "personal philosophy," which guides people's conduct toward the environment. He defined ecosophy as a set of beliefs about nature and other people which varies from one individual to another. Everyone, in other words, has his or her own ecosophy, and, though these personal philosophies may share important elements, they are based on norms and assumptions that are particular to each person.

Naess proposed his own ecophilosophy as a model for individual ecosophies, emphasizing the intrinsic value of nature and the importance of cultural and natural diversity. Other discussions of ecosophy concentrate on similar issues. Many environmental philosophers argue that all life has a value that is independent of human perspectives and human uses, and that it is not to be tampered with except for the sake of survival. Human population growth threatens the integrity of other life systems; they argue that population must be reduced substantially and that radical changes in human values and activities are

Ecosystem health (and ecological integrity) is an indicator of the well-being and natural condition of ecosystems and their functions. These indicators are influenced by natural changes in environmental conditions and are related to such factors as climate change and disturbances, such as wildfire, windstorms, and diseases. Increasingly, however, ecosystems are being affected by environmental stressors associated with human activities that cause pollution and disturbance, which result in many changes in environmental conditions. Some species, communities, and ecological processes benefit from those environmental changes, but others suffer great damages.

The notion of ecosystem health is intended to help distinguish between ecosystem-level changes that represent improvements and those that are degradations. In this sense, ecosystem level refers to responses occurring in ecological communities, landscapes, or seascapes. Effects on individual organisms or populations do not represent an ecosystem-level response to changes in environmental conditions.

The notion of health

The notion of ecosystem health is analogous to that of medical health. In the medical sense, health is a term used to refer to the vitality or well-being of individual organisms. Medical health is a composite attribute, because it is characterized by a diversity of inter-related characteristics and conditions. These include blood pressure and chemistry, wounds and injuries, rational mental function, and many other relevant variables. Health is, in effect, a summation of all of these characters related to vitality and well-being. In contrast, a diagnosis of unhealthiness would focus on abnormal values for only one or for several variables within the diverse congregation of health-related attributes. For example, individuals might be judged as unhealthy with a broken leg, a high fever, unusually high blood pressure, or non-normal behavioral traits, even though all their other traits are normal.

To compare human and ecosystem health is, however, imperfect in some important respects. Health is a relative concept. It depends on what is considered normal at a particular stage of development. The aches and pains that are considered normal in a human at age eighty would be a serious concern in a twenty-year old. It is much more difficult, however, to define what to expect in an ecosystem. Ecosystems don't have a prescribed lifespan and generally don't die but rather change into some other form. Because of these problems, some ecologists prefer the notions of ecological or biological integrity rather than ecosystem health.

It should also be pointed out that many ecologists like none of these notions (that is, ecosystem health, ecological integrity, or biological integrity), because by their very nature, these concepts are imprecise and difficult to define. For these reasons, scientists have had difficulty in agreeing upon the specific variables that should be included when designing composite indicators of health and integrity in ecological contexts.

Ecosystem health

Ecosystem health is a summation of conditions occurring in communities, watersheds, landscapes, or seascapes. Ecosystem health conditions are higher-level components of ecosystems, in contrast with individual organisms and their populations.

Although ecosystem health cannot be defined precisely, ecologists have identified a number of specific components that are important in this concept. These include the following indicators: (1) an ability of the system to resist changes in environmental conditions without displaying a large response (this is also known as resistance or tolerance); (2) an ability to recover when the intensity of environmental stress is decreased (this is known as resilience); (3) relatively high degrees of biodiversity; (4) complexity in the structure and function of the system; (5) the presence of large species and top predators; (6) controlled nutrient cycling and a stable or increasing content of biomass in the system; and (7) domination of the system by native species and natural communities that can maintain themselves without management by humans. Higher values for any of these specific elements imply a greater degree of ecosystem health, while decreasing or lower values imply changes that reflect a less healthy condition.

Ecologists are also working to develop composite indicators (or multivariate summations) that would integrate the most important attributes of ecosystem health into a single value. Indicators of this type are similar in structure to composite economic indicators such as the Dow-Jones Stock Market Index and the Consumer Price Index. Because they allow complex situations to be presented in a simple and direct fashion, composite indicators are extremely useful for communicating ecosystem health to the broader public.

Resources

BOOKS

Begon, Michael, Colin A. Townsend, and John L. Harper. *Ecology: From Individuals to Ecosystems,* 4th ed. New York: Wiley-Blackwell, 2006.

Chapin III, F. Stuart, Harold A. Mooney, and Melissa C. Chapin. *Principles of Terrestrial Ecosystem Ecology.* New York: Springer, 2004.

Ninan, K. N. *Conserving and Valuing Ecosystem Services and Biodiversity: Economic, Institutional and Social Challenges.* London: Earthscan, 2009.

Bill Freedman

Ecosystem management

Ecosystem management (EM) is a concept that has germinated since the last decades of the twentieth century and continues to increase in popularity across the United States and Canada. It is a concept that eludes one concise definition, however, because it embodies different meanings in different contexts and for different people and organizations. This can be witnessed by the multiple variations on its title (e.g., ecosystem-based management or collaborative ecosystem management). The definitions that have been given for EM, though varied, fall into two distinct groups. One group emphasizes long-term ecosystem integrity, while the other group emphasizes an intention to address all concerns equally, be they economic, ecological, political or social, by actively engaging and incorporating the

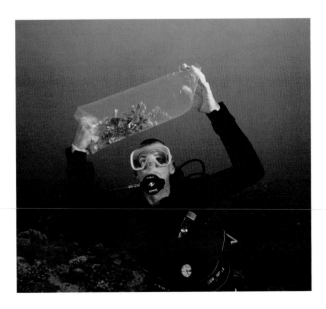

Scientist Jean Turquet collecting a sample of Halimeda seaweed for analysis in a laboratory to establish which species of microalgae are present on its surface. *(Alexis Rosenfeld/Photo Researchers, Inc.)*

multitude of stakeholders (literally, those who hold a stake in the issue) into the decision-making process.

One usable though incomplete definition of EM is provided by R. Edward Grumbine, former director of the Sierra Institute (undergraduate program in wilderness and cultural field studies): "Ecosystem management integrates scientific knowledge of ecological relationships within a complex sociopolitical and values framework toward the general goal of protecting native ecosystem integrity over the long term." Ultimately, EM is a new way to make decisions about how we humans should live with each other and with the environment that supports us. And, it is best defined not only by articulating an ideal description of its contents, as Grumbine has done, but also through a rigorous analysis of actual EM examples.

EM—A new management style

Between the years 1992 and 1994, each of the four predominant federal land management agencies in the United States the National Park Service, the Bureau of Land Management, the Forest Service and the Fish and Wildlife Service implemented EM as their operative management paradigm. Combined, these four agencies control 97 percent of the 650 million federally-owned acres (267 million ha) in the United States, or roughly 30 percent of the United States' entire land area. EM has become the primary management style for these agencies because it became apparent in the 1980s and 1990s that the traditional resource-management style did not work. It was largely ineffective in addressing the loss and fragmentation of wild areas, the increasing number of threatened or endangered species, and the increased occurrence of environmental disputes. This ineffectiveness has been attributed to the traditional management style's main focus being on species with economic value, its exclusion of the public from the decision-making process, and its reliance on outdated ecological beliefs. This explicit acknowledgment that the traditional management style is inadequate has coalesced within state and federal agencies, academia, and environmental organizations, and has been bolstered by advances in other relevant fields, such as ecology and conflict management.

Break the traditional management style into individual component shows that each ineffective attribute has a new or altered counterpart in EM. One of the best ways to describe what EM actually entails is to explicate this juxtaposition between traditional and new management styles.

EM, as its name makes clear, concentrates on managing at the scale of an ecosystem. Alternately, traditional resource management has focused only on one or a handful of species, especially those species that have a utilitarian, or more specifically economic, value. For example, the U.S. Forest Service has traditionally managed the national forests so as to produce a sustained yield of timber. This management style is often harmful to species other than timber and can have negative effects on the entire ecosystem. In EM, all significant biotic and abiotic components of the ecosystem, as well as aspects such as economic factors, are, ideally, reviewed, and the important ecological data incorporated into the decision-making process. A review of a forest ecosystem may include an analysis of habitat for significant song birds, a description of the requirements needed to maintain a healthy black bear (*Ursus americanus*) population, *and* a discussion of acceptable levels of timber production.

A major problem associated with using an ecosystem to define a management area is that boundaries of jurisdictional authority, or political boundaries, rarely follow ecological ones. This implies that by following political boundaries alone, ecological components may be left out of the management plan, so one may be forced to manage only part of an ecosystem, that part which is within one's political jurisdiction. For example, the Greater Yellowstone Ecosystem goes far beyond the boundaries of Yellowstone National Park. Therefore, a large scale EM project for Yellowstone would require crossing several political boundaries (e.g., national park lands and national forest lands), which is a difficult task because it entails several political jurisdictions and political entities (e.g., state and federal agencies, and county governments).

EM projects address this obstacle by forming decision teams that include, among others, representatives from all of the relevant jurisdictions. These decision-making bodies can either act as advisory committees without decision-making authority, or they can attempt to become vested with the power to make decisions. This collaborative process involves all of the stakeholders, whether that stakeholder is a logging company interested in timber production or a private citizen concerned with water quality. Such collaboration diverges from the traditional resource-management method, which made most decisions without public awareness, asking for and receiving little public input. These agencies traditionally shied away from actively engaging the public because it is less complicated and faster to make decisions on one's own than to ask for input from many sources. There has often been an antagonistic and distrustful relationship between state and federal agencies and the public, and there has been little institutional support (i.e., within the structure of

the agency itself) encouraging the manager in the field to invite the public into the decision-making process.

EM's more collaborative and inclusive decision-making style ideally fosters a wiser and more effective decision. This happens because as the decision team works toward consensus, personal relationships are established, some trust may form between parties, and, ultimately, people are more likely to support a decision or plan they help create. EM attempts to transcend the traditional antagonistic relationship between agency personnel and the public. Because 70 percent of the United States is privately owned, many environmental issues arise on private land—land that is only partially affected by federal and state natural resource legislation. EM allows groups to deal with these issues on private lands by establishing a dialogue between private and public decision makers. Finally, even though the EM decision-making style takes longer to conduct, time is saved in the end, because the decision achieved is more agreeable to all interested parties. Having all parties agree to a particular management plan decreases the number of potential lawsuits that can arise and delay the plan's implementation.

Nonequilibrium ecology and EM

A change in the dominant theories in ecology has encouraged this switch to EM and an ecosystem-level focus.

The idea that environments achieve a climax state of homeostasis has been a significant theory in ecology since the early 1900s. This view, now discredited by most scholars, was most vigorously articulated by American ecologist F.E. Clements (1874–1945) and holds that all ecosystems have a particular end point to which they each progress, and that ecosystems are closed systems. Disturbances such as fires or floods are considered only temporary setbacks on the ecosystem's ultimate progression to a final state. This theory offers a certain level of predictable stability, the type of stability desired within traditional resource management. For example, if a forest is in its climax state, that condition can be maintained by eliminating disturbances such as forest fires, and a predictable level of harvestable timber can be extracted (hence, this theory contributed to the creation of Smokey the Bear, an icon for forest fire prevention, and the national policy of stopping forest fires on public land).

This teleological view of nature has ebbed and waned in importance, but has lost favor especially within the past two decades. Ecologists, among others, have realized that from certain temporal and spatial points of view ecosystems may seem to be in equilibrium, but in the long term all ecosystems are in a state of nonequilibrium. That is, ecosystems always change. They change because their ecological structure and function is often regulated by dynamic external forces such as storms or droughts and because they are comprised of varied habitat types that change and affect one another.

This acknowledgment of a changing ecosystem means that predictable stability does not really exist and that an adaptive management style is needed to meet the changing requirements of a dynamic ecosystem. With all of the interactions and factors involved in an ecosystem's operation, it becomes necessary to be adaptive and flexible when determining methods of ecosystem management and to focus less on predictions and control of ecosystems. EM is adaptive. After an EM decision team has formulated and implemented a management plan, the particular ecosystem is monitored. The team watches significant biotic and abiotic factors to see if and how they are altered by the management practices. If logging produces changes which effect the fish in one of the ecosystem's streams, the decision team could adapt to the new data and decide to relocate the logging.

The ability to adaptively manage is a crucial aspect of EM, one that is not incorporated in traditional resource management, because the traditional management style emphasizes one or a few species and believes that ecosystems are mostly stable. In traditional ecosystem management, one only needs to view how a particular species fares to determine the necessary management practices; little monitoring is conducted and previous management practices are rarely altered. In EM, management practices are constantly reviewed and adjusted as a result of ongoing data gathering and the goals articulated by the decision team.

The future of EM

The future of EM in the United States and Canada looks stable, and other countries such as France are beginning to use EM. There are many impediments, though, to the successful implementation of EM. Institutions, such as the United States' federal land management agencies, are often hesitant to change, and, when they do change, it happens very slowly; there are still many legal questions surrounding the legitimacy of implementing EM on federal, state, and private lands; and, even though attempts are made to review the entire ecosystem using EM examples, ecologists still lack significant understanding of how even the most basic ecosystems operate. Even given these impediments, EM may be the primary land management style of the United States and Canada well into the future. Adaptive EM methods will enable users to learn more about ecosystems in the process of monitoring and management and thus have more detailed insight and involvement in the inner workings of an ecosystem. The continuous involvement required for adaptive management strategies is becoming more important in the face of climate change and imminent natural resource depletion.

Resources

BOOKS

Adams, Jonathan S. *The Future of the Wild: Radical Conservation for a Crowded World*. Boston: Beacon Press, 2006.

Great Britain. *Ecosystem Management Indicators*. Bristol: Environment Agency, 2008.

Doyle, Mary, and Cynthia A. Drew. *Large-Scale Ecosystem Restoration: Five Case Studies from the United States*. Washington, DC: Island Press, 2008.

Nagle, John Copeland, and J. B. Ruhl. *The Law of Biodiversity and Ecosystem Management*. New York, NY: Foundation Press, 2006.

Norton, Bryan G. *Sustainability: A Philosophy of Adaptive Ecosystem Management*. Chicago: University of Chicago Press, 2005.

Sen, Simantee. *Ecosystem Management: Issues and Trends*. Hyderabad, India: Icfai University Press, 2008.

OTHER

United Nations Environment Programme (UNEP). "World Conservation Monitoring Centre Biodiversity of Ecosystems." http://www.unep-wcmc.org/climate/impacts.aspx (accessed September 9, 2010).

United States Environmental Protection Agency (EPA). "Ecosystems: Ecological Monitoring: Ecological Assessment." http://www.epa.gov/ebtpages/ecosecological moniecologicalassessment.html (accessed September 9, 2010).

United States Environmental Protection Agency (EPA). "Ecosystems: Ecological Monitoring: Environmental Indicators." http://www.epa.gov/ebtpages/ecosecological monienvironmentalindicators.html (accessed September 9, 2010).

Paul Phifer

Ecotage *see* **Ecoterrorism; Monkey-wrenching.**

Ecoterrorism

In the wake of the terrorist attacks on the World Trade Center in New York City on September 11, 2001, the line between radical environmental protest (sometimes called ecoanarchism) and terrorism became blurred by strong emotions on all sides. Environmentalists in America have long held passionate beliefs about protecting the environment and saving threatened species from extinction, as well as treating animals humanely and protesting destructive business practices. One of the first and greatest environmentalists, Henry David Thoreau (1817–1862), wrote about the doctrine of "civil disobedience," or using active protest as a political tool. The author Edward Abbey (1927–1989) became a folk hero among environmentalists when he wrote the novel, *The Monkey Wrench Gang*, in 1975. In that book, a group of militant environmentalists practiced monkey-wrenching, or sabotaging machinery in desperate attempts to stop logging and mining. Monkey-wrenching, in its destruction of private property, goes beyond civil disobedience and is unlawful. The American public has tended to view monkey-wrenchers as idealistic youth fighting for the environment and has not strongly condemned them for their actions. However, the U.S. government views monkey- wrenching as domestic terrorism, and since September 11, 2001, law enforcement activity concerning environmental groups has been significantly increased.

Ecoanarchism is the philosophy of certain environmental or conservation groups that pursue their goals through radical political action. The name reflects both their relation to older anarchist revolutionary groups and their distrust of official organizations. Nuclear issues, social responsibility, animal rights, and grassroots democracy are among the concerns of ecoanarchists. Ecoanarchists tend to view mainstream political and environmental organizations as too passive, and those who maintain them as compromising or corrupt. Ecoanarchists may resort to direct confrontation, direct action, civil disobedience, and guerrilla tactics to fight for survival of wild places. Monkey-wrenchers perform sit-ins in front of bulldozers, they disable machinery in various ways including pouring sand in a bulldozers gas tanks, they ram whaling ships, and they spike trees by driving metal bars into them to discourage logging. Ecoanarchists may practice ecotage, which is sabotage for environmental ends, often of machines that alter the landscape. In the early 2000s, radical environmentalists committed arson on thirty-five sport utility vehicles (SUVs) at a car dealership in Eugene, Oregon, to protest the gas-guzzling vehicles and set fire to buildings at the University of Washington to protest genetic engineering.

Ecoanarchists do not necessarily view the destruction of machinery as out-of-bounds, but the U.S. government views it as terrorism. For instance, Earth First!, a radical environmental group whose motto is "No Compromise in Defense of Mother Earth," eschews violence against humans but does approve of monkey-

wrenching. The Federal Bureau of Investigation (FBI) defines terrorism as "the unlawful use, or threatened use, of violence by a group or individual...committed against persons or property to intimidate or coerce a government, the civilian population or any segment thereof, in furtherance of political or social objectives." The FBI reports that an estimated 1,100 criminal acts of ecoterrorism occurred in the United States between 1996 and 2010, with damages estimated at $110 million. Two groups associated with these acts of sabotage are the militant Earth Liberation Front (ELF) and Animal Liberation Front (ALF) groups. In 1998 ELF took credit for arson at Vail Ski Resort, which resulted in damages of $12 million. The act was a protest over the resort's expansion on mountain ecosystems.

Most environmental groups are more peaceful and have voiced concern over radical environmentalists, as well as over law enforcement officials who view all environmental protesters as terrorists. For instance, Greenpeace activists have been known to steer boats in the path of whaling ships and throw paint on nuclear vessels as protest, and have been jailed for doing so, although no people were targeted or injured. People for the Ethical Treatment of Animals (PETA) members have thrown pies in the faces of business executives whom they found guilty of inhumane treatment of animals, considering the act a form of civil disobedience and not violence.

The issues of ecoterrorism and ecoanarchism become more heated as environmental degradation worsens. Environmentalists become more desperate to protect rapidly disappearing endangered areas or species, whereas industry continually seeks new resources to replace those being used up. Thrown in the middle are law enforcement officials, who must protect against violence and destruction of property, and also uphold citizens' basic rights to protest.

The most famous ecoterrorist has been Theodore Kaczynski, also known as the Unabomber, who was convicted of murder in the mail-bombing of the president of the California Forestry Association. On the other end of the spectrum of environmental protest is Julia Butterfly Hill, whom ecoterrorists and ecoanarchists would do well to emulate. Hill is an activist who lived in a California redwood tree for two years to prevent it from being cut down by the Pacific Lumber Company. Hill practiced a nonviolent form of civil disobedience and endured what she perceived as violent actions from loggers and timber company. Her peaceful protest brought national attention to the issue of logging in ancient growth forests, and a compromise was eventually reached between environmentalists and the timber company. Unfortunately, the tree in which Hill sat, which she named Luna, was damaged by angry loggers.

Resources

BOOKS

Abbey, Edward. *The Monkey Wrench Gang.* Salt Lake City: Dream Garden Press, 1990.

Hill, Julia Butterfly. *The Legacy of Luna: The Story of a Tree, a Woman, and the Struggle to Save the Redwoods.* San Francisco: Harper, 2000.

Liddick, Donald R. *Eco Terrorism: Radical Environmental and Animal Liberation Movements.* Westport CT: Praeger Publishers, 2006.

Thoreau, Henry David. *Civil Disobedience, Solitude and Life Without Principle.* Amherst, NY: Prometheus Books, 1998.

Douglas Dupler

Ecotone

The boundary between adjacent ecosystems is known as an ecotone. For example, the intermediary zone between a grassland and a forest constitutes an ecotone that has characteristics of both ecosystems. The transition between the two ecosystems may be abrupt or, more commonly, gradual. Because of the overlap between ecosystems, an ecotone usually contains a larger variety of species than is to be found in either of the separate ecosystems and often includes species unique to the ecotone. This effect is known as the edge effect. Ecotones may be stable or variable. Over time a forest may invade a grassland. Changes in precipitation are one important factor in the movement of ecotones.

Ecotourism

Ecotourism is ecology-based tourism, focused primarily on natural or cultural resources such as scenic areas, coral reefs, caves, fossil sites, archeological or historical sites, and wildlife, particularly rare and endangered species.

The successful marketing of ecotourism depends on destinations that have biodiversity, unique geologic features, and interesting cultural histories, as well as an

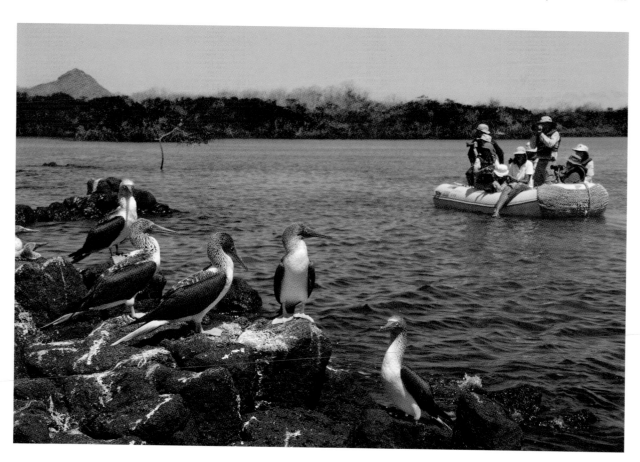

Tourists photographing blue footed boobies from dingy in the Galapagos islands *(© David Hosking / Alamy)*

adequate infrastructure. In the United States, national parks are perhaps the most popular destinations for ecotourism, particularly Yellowstone National Park, the Grand Canyon, the Great Smoky Mountains, and Yosemite National Park. In 2007 there were 272 million recreational visits to the national parks. Some of the leading ecotourist destinations outside the United States include the Galapagos Islands in Ecuador; the wildlife parks of Kenya, Tanzania, and South Africa; the mountains of Nepal; and the national parks and forest reserves of Costa Rica.

Tourism is one of the world's leading industries, on a par with the oil, food, and automotive industries, with international travel alone producing more than $852 billion in export earnings in 2009. There were 800 million international tourists in 2009, and 235 million tourism jobs. Adventure tourism, which includes ecotourism, accounts for 10 percent of this market. In developing countries tourism can comprises as much as one-third of trade in goods and services, and much of this is ecotourism. Wildlife-based tourism in Kenya, for example, generated $752 million in 2008, a fall from the previous year's high of $917 million because of political unrest in the country at the start of 2008.

Ecotourism is not a new phenomenon. In the late 1800s railroads and steamship companies were instrumental in the establishment of the first national parks in the United States, recognizing even then the demand for experiences in nature and profiting from transporting tourists to destinations such as Yellowstone and Yosemite. However, ecotourism has recently taken on increased significance worldwide.

There has been a tremendous increase in demand for such experiences, with adventure tourism increasing at a rate of 30 percent annually. But there is another reason for the increased significance of ecotourism. It is a key strategy in efforts to protect cultural and natural resources, especially in developing countries, because resource-based tourism provides an economic incentive to protect resources. For example, rather than converting tropical rain forests to farms, which may be short-lived, income can be earned by providing goods and services to tourists visiting the rain forests.

Although ecotourism has the potential to produce a viable economic alternative to exploitation of the environment, it can also threaten it. Water pollution, litter, disruption of wildlife, trampling of vegetation, and mistreatment of local people are some of the negative impacts of poorly planned and operated ecotourism. To distinguish themselves from destructive tour companies, many reputable tour organizations have adopted environmental codes of ethics which explicitly

state policies for avoiding or minimizing environmental impacts. In planning destinations and operating tours, successful firms are also sensitive to the needs and desires of the local people, for without native support efforts in ecotourism often fail.

Ecotourism can provide rewarding experiences and produce economic benefits that encourage conservation. The challenge upon which the future of ecotourism depends is the ability to carry out tours that the clients find rewarding, without degrading the natural or cultural resources upon which they is based.

See also Earthwatch; National Park Service.

Resources

BOOKS

Fennell, David A. *Ecotourism*. London: Routledge, 2007.
Honey, Mary. *Ecotourism and Sustainable Development: Who Owns Paradise?*. Washington, DC: Island Press, 2008.
Patterson, Carol. *The Business of Ecotourism*. New York: Trafford Publishing, 2007.

Ted T. Cable

Ecotoxicology

Ecotoxicology is a field of science that studies the effects of toxic substances on ecosystems. It analyzes environmental damage from pollution and predicts the consequences of proposed human actions in both the short-term (days, weeks, and months) and the long-term (years and decades). With more than 150,000 chemicals in commercial use as of 2010 and thousands more being introduced each year, the scale of the task is daunting.

Ecotoxicologists use a variety of methods to measure the impact of harmful substances on people, plants, and animals. Toxicity tests measure the response of biological systems to a substance to determine if it is poisonous. A test can study, for example, how well fish live, grow, and reproduce in various concentrations of industrial effluent. Another test determines what concentration of metal contaminants in soil damages plants' ability to convert sunlight into food. Another test measures how various concentrations of pesticides in agricultural runoff affect sediment and nutrient absorption in wetlands. Analyses of the chemical fate of a pollutant (i.e., where a pollutant goes once it is released into the environment) can be combined with toxicity-test information to predict environmental response to pollution. Because toxicity is the interaction between a living system and a substance, only living plants, animals, and systems can be used in these experiments.

Although not used routinely in the field, ecotoxicology research also involves the detection of changes to the genetic material of organisms exposed to various toxic compounds. Some compounds are able to alter deoxyribonucleic acid (DNA) by causing deletions of a portion of the DNA or by changing the composition of the DNA building blocks (nucleotides). Because DNA is an information code for the production of various proteins and other molecules needed by the organism, these changes (mutations) can affect organism function. Some of the changes can be harmful or even lethal.

The genetic approach to ecotoxicology tends to be confined to the research lab, where the possible consequences of various compounds are explored. As with other genetic tests, however, the aim is to develop tests that are applicable in real-life situations.

More practically, another tool used in ecotoxicology is the field survey. The purpose of a field survey is to clarify and detail the ecological conditions, including pollution levels, in both healthy and damaged natural systems. Surveys often focus on the number and variety of plants and animals supported by the ecosystem, but they can also characterize other valuable attributes, such as crop yield, commercial fishing, timber harvest, or aesthetics. Information from a number of field surveys can be combined for an overview of the relationship between pollution levels and the ecological condition.

At first glance, it might seem that ecotoxicology might only involve measuring the concentration of a toxic substance and predicting what the consequences of the toxin levels could be. However, this approach by itself is insufficient, because chemical analysis alone cannot predict environmental consequences in most cases. Unfortunately, interactions between toxicants and the components of ecosystems are not clear; in addition, because the interactions in the components of an ecosystem are complex and may not all be known, ecotoxicologists have not yet developed simulation models that allow them to make predictions based on chemical concentration alone. For example:

• An ecosystem's response to toxic materials is greatly influenced by environmental conditions. The concentration of zinc that will kill fish or larvae in a relatively soft water may not be lethal in harder waters. As well, different aquatic species can exhibit different sensitivities to a given compound. Many relationships between environmental conditions and toxicity have not been established.

• Most pollutants are a complex mixture of chemicals, not just a single substance. In addition, some chemicals are more harmful when combined with other toxicants, a phenomenon termed synergy.

• Some chemicals are toxic at concentrations and levels too small to be measured. Yet they may still have an adverse effect, in particular in creating a genetic mutation or in affecting the physiology of the exposed organism.

• An organism's response to toxic materials can be influenced by other organisms in the community. For example, a fish exposed to pollution may be unable to escape from its predators.

Ecotoxicologists rely on information from all the approaches available to them. Field surveys prove that some important characteristic of the ecosystem has been damaged, chemical measurements confirm the presence of a toxicant, and toxicity tests link a particular toxicant to a certain type of damage. Finally, genetic monitoring can indicate the possibility that the pollutant is capable of causing mutations.

The scope of environmental protection has broadened considerably over the years, and the types of ecotoxicological information required have also changed. Toxicity testing began as an interest in the effects of various substances on human health. Gradually this concern extended to the other organisms that were most important to humans—domestic animals and crop plants—and finally spread to other organisms that are less apparent or universal in their importance. In addition, the toxicant must not eliminate or taint the plants and/or animals upon which they feed nor can it destroy their habitat. Indirect effects of toxicants, which can also be devastating, are difficult to predict. A chemical that is not toxic to an organism, but instead destroys the grasses in which it lays eggs or hides from predators, will be indirectly responsible for the death of that organism.

Because each one of the millions of species on this planet cannot be tested before a chemical is used, ecotoxicologists test a few representative species to characterize toxicity. If a pollutant is found in rivers, testing might be done on an alga, an insect that eats algae, a fish that eats insects, and a fish that eats other fish. Other representative species are chosen by their habitat: on the bottom of rivers, midstream, or in the soil. Regardless of the sampling scheme, however, thousands of organisms that will be affected by a pollutant will not be tested.

Despite the unknowns of ecosystems, predictions need to be made about the consequences of ecotoxicants. Statistical techniques can predict the response of organisms in general from information on a few randomly selected organisms. Another approach tests the well-being of higher levels of biological organization. Since natural communities and ecosystems are composed of a large number of interacting species, the response of the whole reflects the responses of its many constituents. The health of a large and complex ecosystem cannot be measured in the same way as the health of a single species, however. Different attributes are important. For example, examining a single species, such as cattle or trout, might require measuring respiration, reproduction, behavior, growth, or tissue damage. The condition of an ecosystem might be determined by measuring production, nutrient distribution, or colonization.

The consequences of environmental degradation can be severe and long-lasting. As one example, monitoring studies have determined that the damage to portions of the Alaskan shoreline caused by the release of oil from the Exxon Valdez into Prince William Sound beginning in 1989 persisted to at least 2007. Another example occurred, on April 20, 2010 when an explosion on the Deepwater Horizon oil-drilling platform in the Gulf of Mexico triggered the rupture of a deep-water oil pipeline. Upwards of 60,000 barrels of oil (2,520,000 U.S. gallons) per day gushed into the Gulf waters for eighty-seven days. Although some short-tem consequences were well-documented, as of July 2010, the ultimate environmental fate of the massive amount of oil that remained underwater was unknown.

Resources

BOOKS

Carson, R. *Silent Spring.* Boston: Houghton Mifflin, 1962.

Jorgensen, Erik. *Ecotoxicology* New York: Academic Press, 2010.

Newman, Michael C. *Fundamentals of Ecotoxicology. 3rd ed.* Boca Raton, FL: CRC Press, 2009.

Smart, Robert C., and Ernest Hodgson. *Molecular and Biochemical Toxicology.* Hoboken, N.J.: John Wiley & Sons, 2008.

Walker, C.H. *Principles of Ecotoxicology.* Boca Raton, FL, and London: CRC, 2006.

Wilson, E. O. *Biodiversity.* Washington, DC: National Academy Press, 1988.

John Cairns Jr.

Ecotype

A recognizable geographic variety, population, or ecological race of a widespread species equivalent to a taxonomic subspecies. Typically, ecotypes are restricted to one habitat and are recognized by distinctive characteristics resulting from adaptations to local selective pressures and isolation. For example, a population or ecotype of species found at the foot of a mountain may differ in size, color, or physiology from another ecotype living at higher altitudes, thus reflecting a sharp change in local selective pressures. Members of an ecotype are capable of interbreeding with other ecotypes within the same species, usually without loss of fertility or vigor.

Ectopistes migratorius see **Passenger pigeon.**

Edaphic

Refers to the relationship of soil characteristics to attributes expressed in plants and animals. Soil characteristics influence living things, particularly plants. For example, soils with a low pH will more likely have plants growing on them that are adapted to this level of soil acidity. Animals living in the area will most likely eat these acid-loving plants. In general, the more extreme a soil characteristic, the smaller the population able to adapt to the soil environment.

Edaphology

The study of soil, including its ecosystem role, value, and management as a medium for plant growth and as a habitat for animals. This branch of soil science covers physical, chemical, and biological properties, including soil fertility, acidity, water relations, gas and energy exchanges, microbial ecology, and organic decay.

Eelgrass

Eelgrass is the common name for a genus of perennial grass-like flowering plants referred to as *Zostera*. *Zostera* is from the Greek word *zoster*, meaning belt, which describes the dark green, long, narrow, ribbon shape of the leaves that range in size from 20 to 50 centimeters in length, but can grow up to 2 meters. Eelgrass grows underwater in estuaries and in shallow coastal areas. Eelgrass is a member of a group of land plants that migrated into the sea in relatively recent geologic times and is not a seaweed.

Eelgrass grows by the spreading of rhizomes and by seed germination. Eelgrass flowers are hidden behind a transparent leaf sheath. Long filamentous pollen is released into the water, where it is spread by waves and currents. Both leaves and rhizomes contain lacunae, which are air spaces that provide buoyancy.

Eelgrass grows rapidly in shallow waters and is highly productive, thus providing habitats and food for many marine organisms in its stems, roots, leaves, and rhizomes. Eelgrass communities (meadows) provide many important ecological functions, including:

- anchoring of sediments with the spreading of rhizomes, which prevents erosion and provides stability

- decreasing the impact of waves and currents, resulting in a calm environment where organic materials and sediments can be deposited

- providing food, breeding areas, shelter, and protective nurseries for marine organisms of commercial, recreational, and ecological importance

- concentrating nutrients from seawater that are then available for use in the food chain

- serving as food for water fowl and other animal,s such as snails and sea urchins

- as detritus (decaying plant matter), providing nutrition to organisms within the eelgrass community, in adjoining marshes, and in offshore sinks at depths up to 30,000 feet.

Eelgrass growth can be adversely impacted by human activities, including dredging, logging, shoreline or over-water construction, power plants, oil spills, pollution, and species invasion.

Resources

OTHER

United States Environmental Protection Agency (EPA). "Water: Aquatic Ecosystems: Estuaries." http://www.epa.gov/ebtpages/wateaquatiestuaries.html (accessed October 9, 2010).

Judith L. Sims

Effluent

For geologists, effluent refers to a wide range of substances and situations, from lava outflow to watercourse origins and transformations. However, effluent is now most commonly used by environmentalists in reference to the Clean Water Act of 1977. In this act, effluent is a discharge from a point source, and the legislation specifies allowable quantities of pollutants. These discharges are regulated under Section 402 of the act, and these standards must be met before these types of industrial and municipal wastes can be released into surface waters.

See also Industrial waste treatment; Thermal pollution; Water pollution; Water quality.

Effluent tax

Effluent tax refers to the fee paid by a company to discharge to a sewer. As originally proposed, the fee would have been paid for the privilege of discharging into the environment. However, there is presently no fee structure that allows a company, municipality, or person to contaminate the environment above the levels set by water-quality criteria and effluent permits, unless fines levied by a regulatory agency could be deemed to be such fees.

Fees are now charged on the basis of simply being connected to a sewer and the types and levels of materials discharged to the sewer. For example, a municipality might charge all sewer customers the same rate for domestic sewage discharges below a certain flowrate. Customers discharging wastewater at a higher strength and/or flowrate would be assessed an incremental fee proportional to the increased amount of contaminants and/or flow. This charge is often referred to as a sewer charge, fee, or surcharge.

There are cases in which wastewater is collected and tested to ensure that it meets certain criteria (e.g., required level of oil or a toxic metal, etc.) before it is discharged into a sewer. If the criteria are exceeded, the wastewater would require pretreatment. Holding the water before discharge is generally only practical when flowrates are low. In other situations it may not be possible to discharge to a sewer because the treatment facility is unable to treat the waste; thus, many hazardous materials must be managed in a different manner.

Effluent taxes force dischargers to integrate environmental concerns into their economic plans and operational procedures. The fees cause firms and agencies to rethink water conservation policies, waste minimization, processing techniques and additives, and pollution control strategies. Effluent taxes, as originally proposed, might be instituted as an alternative to stringent regulation, but the sentiment of the current public and regulatory agencies is to block any significant degradation of the environment. It appears to be too risky to allow pollution on a fee basis, even when the fees are high. Thus, in the foreseeable future, effluents to the environment will continue to be controlled by means of effluent limits, water quality criteria, and fines. However, the taxing of effluents to a sewer is a viable means of challenging industries to enhance pollution control measures and stabilizing the performance of downstream treatment facilities.

Resources

OTHER

United States Environmental Protection Agency (EPA). "Water: Wastewater: Effluent Guidelines." http://www.epa.gov/ebtpages/watewastewatereffluentguidelines.html (accessed October 14, 2010).

Gregory D. Boardman

Eggshell thinning *see* **Dichlorodiphenyltrichloroethane.**

E_H

A measure of the oxidation/reduction potential (redox potential) of a chemical species or entity (natural water, sediment, soil, etc). E_H can be measured with a potentiometer (e.g., a pH meter adjusted to read in volts) using an inert platinum electrode and a reference electrode (calomel or silver/silver chloride). E_H is often reported in volts or millivolts, and is referenced to the potential for the oxidation of hydrogen gas to hydrogen ions (H^+). This electron transfer reaction is assigned a potential of zero on the relative potential scale. The oxidation and reduction reactions in natural systems are pH dependent and the interpretation of E_H values requires a knowledge of the pH. At pH 7 the E_H in water in equilibrium with the oxygen in air is +0.76v. The lowest possible potential is - 0.4v when oxygen and other electron acceptors are depleted and methane, carbon dioxide, and hydrogen gas are produced by the decay of organic matter.

See also Electron acceptor and donor.

Ehrlich, Paul R.

1932–
American population biologist and ecologist

Paul Ehrlich is an American population biologist who is a member of the department of Biological Sciences at Stanford University. An entomologist by training, he is best known for his warnings of the potential consequences of population growth.

Born in Philadelphia, Paul R. Ehrlich had a typical childhood during which he cultivated an early interest in entomology and zoology by investigating the fields and woods around his home. As he entered his teen years, Ehrlich grew to be an avid reader. He was particularly influenced by ecologist William Vogt's book, *Road to Survival* (1948), in which the author outlined the potential global consequences of imbalance between the growing world population and level of food supplies available. This concept is one Ehrlich has discussed and examined throughout his career. After high school, Ehrlich attended the University of Pennsylvania where he earned his undergraduate degree in zoology in 1953. He received his master's degree from University of Kansas two years later and continued at the university to receive his doctorate in 1957. His degrees led to postgraduate work on various aspects of entomological projects, including observing flies on the Bering Sea, the behavioral characteristics of parasitic mites, and (his favorite) the population control of butterfly caterpillars with ants rather than pesticides. Other related field projects have taken him to Africa, Alaska, Australia, the South Pacific and Southeast Asia, Latin America, and Antarctica. His travels enabled him to learn firsthand the ordeals endured by those in overpopulated regions.

In 1954 he married Anne Fitzhugh Howland, biological research associate, with whom he wrote the best-selling book, *The Population Bomb* (1968). In the book, the Ehrlichs focus on a variety of factors contributing to overpopulation and, in turn, world hunger. It is evident throughout the book that the words and warnings of *The Survival Game* continued to exert a strong influence on Ehrlich. The authors warned that birth and death rates worldwide need to be brought into line before nature intervenes and causes (through ozone layer depletion, global warming, and soil exhaustion, among other environmental degradation) massive deaths among humans. Human reproduction, especially in highly developed countries like the United States, should be discouraged through levying taxes on diapers, baby food, and other related items; compulsory sterilization among the populations of certain countries should be enacted. The authors' feelings on *compulsory* sterilization have relaxed somewhat since 1968. Ehrlich himself underwent a vasectomy after the birth of the couple's first and only child.

In 1968 Ehrlich founded Zero Population Growth, Inc., an organization established to create and rally support for balanced population levels and the environment. He has been a faculty member at Stanford University since 1959 and as of 2010 is a full professor of Population Studies in the Biological Sciences Department. In addition, Ehrlich has been a news correspondent for NBC since 1989. In 1993 he was awarded the Crafoord Prize in Population Biology and the Conservation of Biological Diversity from the Royal Swedish Academy of Sciences and received the World Ecology Medal from the International Center for Tropical Ecology. In 1994 Ehrlich was given the United Nations Environment Programme Sasakawa Environment Prize.

Among Ehrlich's published works are *The Population Bomb* (1968); *The Cold and the Dark: The World After Nuclear War* (1984); *The Population Explosion* (1990); *Healing the Planet* (1991); *Betrayal of Science and Reason* (1996), which was written with his wife; and *Human Natures: Genes, Cultures, and the Human Prospect*, which was published in 2000. In 2001 Ehrlich was the recipient of the Distinguished Scientist Award. His latest book, co-authored with his wife and published in 2008, is *The Dominant Animal: Human Evolution and the Environment*.

Resources

BOOKS

Ehrlich, Paul R. *The Population Bomb.* New York: Ballantine Books, 1968.

Ehrlich, Paul R. *Healing the Planet: Strategies for Solving the Environmental Crisis.* Reading, MA: Addison Wesley, 1992.

Ehrlich, Paul R. *Human Natures: Genes, Cultures, and the Human Prospect.* Washington, DC: Island Press/ Shearwater Books, 2000.

Ehrlich, Paul R., and Anne. H. Ehrlich. *The Population Explosion.* New York: Simon & Schuster, 1990.

Ehrlich, Paul R., Anne. H. Ehrlich, and John P. Holdren. *Eco-Science: Population, Resources, and Environment.* San Francisco: W. H. Freeman, 1970.

Kimberley A. Peterson

Eichhornia crassipes see **Water hyacinth.**

El Niño

El Niño, also referred to as El Niño/La Niña-Southern Oscillation (ENSO), is the most powerful weather event on Earth, disrupting weather patterns across half of Earth's surface. El Niño is caused by temperature changes in the surface waters in the tropical Pacific Ocean. Generally, when these waters of the eastern Pacific Ocean become warmer (the warm phase) it is called El Niño, but when they cool down (the cold phase), it is called La Niña. In addition, both phases run parallel to air pressure differences in the Pacific Ocean: El Niño is present with high atmospheric pressure in the west Pacific, whereas La Niña occurs when low atmospheric pressure is found in the same waters.

The three-to-seven-year cycle of the ENSO brings lingering rain to some areas and severe drought to others. El Niño develops when currents in the Pacific Ocean shift, bringing warm water eastward from Australia toward Peru and Ecuador. Heat rising off warmer water shifts patterns of atmospheric pressure, interrupting the high-altitude wind currents of the jet stream and causing climate changes.

El Niño, which means "the boy" in Spanish but is sometimes also referred to as the "Christ Child" or "the child," tends to appear in December (at around Christmas). The phenomenon was first noted by Peruvian fishermen in the 1700s, who saw a warming of normally cold Peruvian coastal waters and a simultaneous disappearance of anchovy schools that provided their livelihood.

A recent El Niño began to develop in 1989, but significant warming of the Pacific did not begin until late in 1991, reaching its peak in early 1992 and lingering until 1995–the longest-running El Niño on record. Typically, El Niño results in unusual weather and short-term climate changes that cause losses in crops and commercial fishing. El Niño contributed to North America's mild 1992 winter, torrential flooding in southern California, and severe droughts in southeastern Africa. Wild animals in central and southern

Africa died by the thousands, and 20 million people were plagued by famine. The dried prairie of Alberta, Canada, failed to produce wheat, and Latin America received record flooding. Droughts were felt in the Philippines, Sri Lanka, and Australia, while Turkey experienced heavy snowfall. The South Pacific saw unusual numbers of cyclones during the winter of 1992. El Niño's influence also seems to have suppressed some of the cooling effects of Mount Pinatubo's 1991 explosion.

Another El Niño peaked early in 2010. However, like many of them, this El Niño did not do what scientists thought it would. In fact, Dr. William Patzert, an oceanographer for the Jet Propulsion Laboratory in California, called this El Niño "quirky" (cited from NASA). Patzert made this comment because heavy rains were expected but, instead, the West Coast of the United States alternated between months of warm temperatures and dry weather and months of cold temperatures and wet weather.

Scientists mapping the sea floor of the South Pacific near Easter Island found one of the greatest concentration of active volcanoes on Earth. The discovery has intensified debate over whether undersea volcanic activity could change water temperatures enough to affect weather patterns in the Pacific. Some scientists speculate that periods of extreme volcanic activity underwater could trigger El Niño.

El Niño ends when the warm water is diverted toward to the North and South Poles, emptying the moving reservoir of stored energy. During this phase the cycle is called La Niña (which means "the girl" in Spanish). Before El Niño can develop again, the western Pacific must refill with warm water, which takes at least two years.

Scientific research has explored the connection between El Niño and climate change. Although current research suggests that this disruption in the ocean-atmosphere system is not caused by climate change, the frequency and intensity of ENSO has increased in the past few decades along with increases in global ocean temperatures. Research has shown the warmer sea temperatures can enhance the occurrence of ENSO. Researchers are continuing studies to uncover the relationship between El Niño and climate change.

See also Atmosphere; Desertification.

Resources

BOOKS

Clarke, Allan J. *An Introduction to the Dynamics of El Niño and the Southern Oscillation.* London: Academic, 2008.

Fagan, Brian M. *Floods, Famines, and Emperors: El Niño and the Fate of Civilizations.* New York: Basic Books, 2009.

Long, John A., and David S. Wells. *Ocean Circulation and El Niño: New Research.* Hauppage, NY: Nova Science Publishers, 2009.

Philander, S. George. *Our Affair with El Niño: How We Transformed an Enchanting Peruvian Current into a Global Climate Hazard.* Princeton, NJ: Princeton University Press, 2004.

Rosenzweig, Cynthia, and Daniel Hillel. *Climate Variability and the Global Harvest: Impacts of El Niño and Other Oscillations on Agroecosystems.* New York: Oxford University Press, 2008.

Sandweiss, Daniel H., and Jeffrey Quilter. *El Niño, Catastrophism, and Culture Change in Ancient America.* Washington, DC: Dumbarton Oaks Research Library, 2008.

PERIODICALS

Turney, Chris S., et al. "Millennial and Orbital Variations of El Niño/Southern Oscillation and High-Latitude Climate in the Last Glacial Period." *Nature* 428 (2004): 306–310.

OTHER

National Oceanic and Atmospheric Administration. "NOAA's El Niño Page." http://www.elnino.noaa.gov/ (accessed November 2, 2010).

Jet Propulsion Laboratory, National Aeronautics and Space Administration. "A Quirky El Niño." http://science.nasa.gov/science-news/science-at-nasa/2003/14mar_elnino2002/ (accessed November 2, 2010).

Linda Rehkopf

EIS *see* **Environmental Impact Statement.**

Electric automobiles *see* **Transportation.**

▎Electric utilities

Energy utilities do not produce energy, but convert it from one form to another. Electric utilities are involved in the conversion of energy to electricity. The conversion can involve petroleum, natural gas, coal, nuclear power, and water (e.g., dams).

The major participants in the electric power industry in the United States were, as of 2010, about 159 investor-owned utilities that generated nearly 80 percent of the power and supply about three-quarters of the customers. When publicly owned cooperative and federal government utilities are added, the list grows to almost 3,300 utilities.

In the period after World War II, the electric utilities industry focused primarily on expansion.

A transformer station for the Eon electric utility plant in Grosskrotzenburg, Germany. *(© imagebroker.net / SuperStock)*

During this period, demand increased at a rate of more than 7 percent per year, with construction of new generation facilities and power lines needed to meet the growing demand. Public utility commissions lowered prices, which stimulated additional demand. As long as prices continued to fall, demand continued to rise, and additional construction was necessary. New construction also occurred because the rate of return for the industry was regulated, and the only way for it to increase profits was to expand its rate base by building new plants and equipment.

This period of industry growth came to an end in the 1970s, primarily as a result of the energy crisis. Economic growth slowed, and fuel prices escalated. The financial condition of the industry was further affected as capital costs for nuclear power and coal power plants increased.

During the 1970s, many people also came to believe that coal and nuclear power plants were a threat to the environment. They argued that new options had to be developed and that conservation was important. The federal government implemented new environmental

and safety regulations that further increased utility costs. Changes in government regulations also affected the electric utilities industry. In 1978 the federal government deregulated interstate power sales and required utilities to purchase alternative power, such as solar energy, from qualifying facilities at fully avoided costs.

By the end of the 1970s, the era of expansion of the electric utilities had passed. Since the beginning of the 1980s, the industry has adopted many different strategies, with different segments following different courses based on divergent perceptions of the future. Almost all utilities have tried to negotiate long-term contracts, which would lower their fuel-procurement costs, and attempts have also been made to limit the costs of construction, maintenance, and administration. Many utilities redesigned their rate structures to promote use when excess capacity was available and discourage use when it was not. Multiple rate structures for different classes of customers were also implemented for this purpose.

See also Economic growth and the environment; Energy and the environment; Energy conservation;

Since the beginnings of widespread cell-phone use in the 1980s, there has been concern that electromagnetic fields (radio signals) emitted by cell phones might cause brain cancer or other harm. In the 1990s, a number of studies found no tendency for heavy cell-phone users to get more brain cancers than other people. However, critics pointed to the short-term nature of these studies, with an average period of cell-phone use of only three years. In 2007 and 2008, the debate over cell phones and brain cancer revived when several scientific studies from Israel and Europe tracking ten year cell-phone use were published. The studies found a correlation between certain rare head cancers and cell-phone use, with the cancers happening more often on the side of the head where the patients habitually held their cell phones. Many experts disputed the results, but others urged caution, especially with regard to children. Because children have thinner skulls than adults, more radio energy from a cell phone held to a child's ear is absorbed into the brain. Cell-phone use continues to increase among children under age twelve; by 2010 there were more than 20 million pre-teen cell-phone users in the United States alone. Another cell-phone health concern was the possible effect of cell-phone radio signals on brain function: A 2004 study published by the Institute of Electrical and Electronics Engineers (IEEE) found that radio waves from mobile telephones could modify both the waking and sleeping electrical activity of the human brain, with unknown consequences. Using a headset, which allows the phone's radio transmitter to be kept away from the head, is one way of reducing any possible risk from cell phone emissions.

A major 2005 medical study found that distraction (unrelated to electromagnetic fields) while driving and talking on a cell phone makes a driver four times more likely to be hospitalized by a traffic accident and that headset use does not alter this statistic.

Resources

BOOKS

Cohen-Tannoudji, Claude. *Atoms in Electromagnetic Fields.* 2nd ed. Hackensack, NJ: World Scientific, 2005.

Funk, Richard H. W., Thomas Monsees, and Nurdan Özkucur. *Electromagnetic Effects: From Cell Biology to Medicine.* Amsterdam: Elsevier, 2009.

OTHER

National Institutes of Health (NIH). "Electromagnetic Fields." http://health.nih.gov/topic/Electromagnetic Fields (accessed October 13, 2010).

World Health Organization (WHO). "Electromagnetic fields (EMF)." WHO Programs and Projects. http://www.who.int/entity/peh-emf/en (accessed October 13, 2010).

Lewis G. Regenstein

Electron acceptor and donor

Electron acceptors are ions or molecules that act as oxidizing agents in chemical reactions. Electron donors are ions or molecules that donate electrons and are reducing agents. In the combustion reaction of gaseous hydrogen and oxygen to produce water (H_2O), two hydrogen atoms donate their electrons to an oxygen atom. In this reaction the oxygen is reduced to an oxidation state of -2, and each hydrogen is oxidized to +1. Oxygen is an oxidizing agent (electron acceptor), and hydrogen is a reducing agent (electron donor). In aerobic (with oxygen) biological respiration, oxygen is the electron acceptor accepting electrons from organic carbon molecules, and as a result oxygen is reduced to -2 oxidation state in H_2O, and organic carbon is oxidized to +4 in CO_2. In flooded soils, after oxygen is reduced by aerobic respiration, nitrate and sulfate, as well as iron and manganese oxides can act as electron acceptors for microbial respiration. Other common electron acceptors include peroxide and hypochlorite (household bleach), which are bleaching agents because they can oxidize organic molecules. Other common electron donors include antioxidants, such as sulfite.

Electrostatic precipitation

A technique for removing particulate pollutants from waste gases prior to their exhaustion to a stack. A system of thin wires and parallel metal plates are charged by a high-voltage direct current (DC) with the wires negatively charged and the plates positively charged. As waste gases containing fine particulate pollutants (i.e., smoke particles, fly ash, etc.) are passed through this system, electrical charges are transferred from the wire to the particulates in the gases. The charged particulates are attracted to the plates within the device, where they are then dislodged during short intervals when the DC current is interrupted. (Stack gases can be shunted to a second parallel device during this period). They fall to a collection bin below the plates. Under optimum conditions, electrostatic precipitation is 99 percent efficient in removing particulates from waste gases.

Elemental analysis

Chemists have developed a number of methods by which they can determine the elements present in a material as well as the amount and percentage composition of each element. Elemental analysis includes use of nuclear magnetic resonance (NMR), flame spectroscopy, mass spectrometry, and tests designed to detect specific elements. As technology improves, elemental analysis becomes more accurate. Sensitivity is also enhanced, and concentrations of a few parts per million of an element or less can often be detected even under field conditions. Elemental analysis is valuable in environmental work to determine the presence of a contaminant or pollutant.

Elephants

The elephant is a large mammal with a long trunk and tusks. The trunk is an elongated nose used for feeding, drinking, bathing, blowing dust, and testing the air. The tusks are upper incisor teeth composed entirely of dentine (ivory) used for defense, levering trees, and scraping for water. Elephants are long-lived (50–70 years) and reach maturity at twelve years. They reproduce slowly (one calf every two to three years) due to a twenty-one-month gestation period and an equally long weaning period. A newborn elephant stands 3 feet (1 m) at the shoulder and weighs 200 lb (90 kg). The Elephantidae includes two living species and various extinct relatives.

Asian elephants (*Elephas maximus*) grow to 10 feet (3 m) high and weigh 4 tons. The trunk ends in a single lip, the forehead is high and domed, the back convex, and the ears small. Asian elephants are commonly trained as work animals. They range from India to southeast Asia. There are four subspecies, the most abundant of which is the Indian elephant (*E. m. bengalensis*) with a wild population of about 40,000. Another 20,000 are estimated to be domesticated in Asia. The Sri Lankan (*E. m.*

Elephants at a watering hole. *(Villiers Steyn/Shutterstock.com)*

maximus), Malayan (*E. m. hirsutus*), and Sumatran elephants (*E. m. sumatranus*) are all endangered subspecies.

In Africa, adult bush elephants (*Loxodonta africana oxyotis*) are the world's largest land mammals, growing 11 feet (3.3 m) tall and weighing 6 tons. The trunk ends in a double lip, the forehead slopes, the back is hollow, and the ears are large and triangular. African elephants are also endangered and have never been successfully trained to work. The rare round-eared African forest elephant (*L. a. cyclotis*) is smaller than the bush elephant and inhabits dense tropical rain forests.

Elephants were once abundant throughout Africa and Asia, but they are now threatened or endangered nearly everywhere because of widespread ivory poaching. In 1970 there were about 4.5 million elephants in Africa, but by 2007 there were only 500,000–600,000. Protection from poachers and the 1990 ban on the international trade in ivory (which caused a drop in the price of ivory) temporarily slowed the slaughter of African bush elephants. However, the relatively untouched forest elephants are now coming under increasing pressure due to human encroachment. Organized poaching to increase the coffers of rival African tribal leaders and warlords has also reduced forest elephant populations in West Africa to less than 3,000.

Elephants are keystone species in their ecosystems, and their elimination could have serious consequences for other wildlife. Wandering elephants disperse fruit seeds in their dung, and the seeds of some plants must pass through elephants to germinate. Elephants are also "bulldozer herbivores," habitually trampling plants and uprooting small trees. In African forests elephants create open spaces that allow the growth of vegetation favored by gorillas and forest antelope. In woodland savanna elephants convert wooded land into grasslands, thus favoring grazing animals. However, large populations of elephants confined to reserves can also destroy most of the vegetation in a region. Culling exploding elephant populations in reserves has been practiced in the past to protect the vegetation for other animals that depend on it.

Resources

BOOKS

Ammann, Karl, and Dale Peterson. *Elephant Reflections.* Berkeley: University of California Press, 2009.

Anthony, Lawrence, and Graham Spence. *The Elephant Whisperer: My Life with the Herd in the African Wild.* New York: Thomas Dunne Books/St. Martin's Press, 2009.

O'Connell, Caitlin. *The Elephant's Secret Sense: The Hidden Life of the Wild Herds of Africa.* New York: Free Press, 2007.

Owens, Mark, and Delia Owens. *Secrets of the Savanna: Twenty-three Years in the African Wilderness Unraveling the Mysteries of Elephants and People.* Boston: Houghton Mifflin, 2006.

Wylie, Dan. *Elephant.* London: Reaktion Books, 2008.

BOOKS

International Elephant Foundation. http://www.elephant conservation.org/ (accessed November 9, 2010).

Save the Elephants. http://www.savetheelephants.org/ (accessed November 9, 2010).

Neil Cumberlidge

ELI *see* **Environmental Law Institute.**

Elton, Charles S.

1900–1991
English ecologist

Charles Elton was a British ecologist and one of the leaders in that field in the twentieth century.

Elton was born March 29, 1900, in Manchester, England. His interest in what he later called "scientific natural history" was sparked early in his life by his older brother Geoffrey. By the age of nineteen, Charles Elton was already investigating species relationships in ponds, streams, and sand-dunes around Liverpool.

His formal education in ecology was shaped by an undergraduate education at Oxford University, and by his participation in three scientific expeditions to Spitsbergen in the Arctic (in 1921, 1923, and 1924), the first one as an assistant to Julian Huxley. Even though an undergraduate, he was allowed to begin an ecological survey of animal life in Spitsbergen, a survey completed on the third trip. These experiences and contacts led him to a position as biological consultant to the Hudson's Bay Company, which he used to conduct a long-term study of the fluctuations of fur-bearing mammals, drawing on company records dating back to 1736.

During this period, Elton became a member of the Oxford University faculty (in 1923), and was eventually elected a senior research fellow of Corpus Christi College. His whole academic career was spent at Oxford, from which he retired in 1967.

He applied the skills and insights gained through the Spitsbergen and Hudson Bay studies to work on the fluctuations of mice and voles in Great Britain. To advance and coordinate this work, he started the Bureau of Animal Populations at Oxford. This institution (and Elton's leadership of it) played a vital role in the shaping of research in animal ecology and in the training and education of numerous ecologists in the early twentieth century.

Elton published a number of books, but four proved to be of particular significance in ecology. In 1929 he published his first book, *Animal Ecology*, a volume now considered a classic, and its author was one of the pioneers in the field of ecology, especially animal ecology. In the preface to a 1966 reissue, Elton suggested that the book "must be read as a pioneering attempt to see . . . the outlines of the subject at a period when our knowledge [of] terrestrial communities was of the roughest, and considerable effort was required to shake off the conventional thinking of an earlier zoology and enter upon a new mental world of populations, inter-relations, movements and communities—a world [of] direct study of natural processes." Topics in that book remain major topics in ecology today: the centrality of trophic relationships, the significance of niche as a functional concept, ecological succession, the dynamics of dispersal, and the relationships critical to the fluctuation of animal populations, including interactions with habitat and physical environment.

His year of work on small mammals in Spitzbergen, for the Hudson's Bay Company, and in British localities accessible to Oxford, culminated in the publication, in 1942, of *Voles, Mice and Lemmings*. This work, which is still in print, "brought together . . . his own work and a collection of observations from all over the world and from ancient history onward." Elton begins the book by establishing a context of "vole and mouse plagues" through history. The second section is on population fluctuations in north-west Europe, voles and mice in Britain, but also lemmings in Scandinavia. The other two sections focus on wildlife cycles in northern Labrador, including chapters on fox and marten, voles, foxes, the lemmings again, and caribou herds. In all this work, the emphasis is on the dynamics of change, the constant interactions and subsequent fluctuations of these various populations, and the often stringent environments they populate.

Elton's 1958 book, *The Ecology of Invasions by Animals and Plants*, focused on a problem that is of even more concern in the 2000s—the arrival and impact of exotic species introduced from other places, sometimes naturally, increasingly through the actions of humans. As always, Elton is careful to set the historical context by showing how biological "invaders" have been moving around the globe for a long time, but he also emphasizes that "we are living in a period of the world's history when the mingling of thousands of kinds of organisms from different parts of the world is setting up terrific dislocations in nature."

The Pattern of Animal Communities, published in 1966, emerged from years of surveying species, populations, communities, and habitats in the Wytham Woods, not far from Oxford. In this book, his primary intent was to describe and classify the diverse habitats available to terrestrial animals, and most of the chapters of the book are given to specific habitats for specialized organisms. Though not generally considered a theoretical ecologist, his early thinking did help to shape the field. In this book, late in his career, he summarized that thinking in a chapter titled "The Whole Pattern," in which he presents a set of fifteen "new concepts of the structure of natural systems," which he stated as a "series of propositions," though some reviewers labeled them "principles" of ecology.

Always the pragmatist, Elton devoted considerable time to the practical, applied aspects of ecology. Nowhere is this better demonstrated than in the work he turned to early in World War II. One of his original purposes in establishing the Bureau of Animal Populations was to better understand the role of disease in the fluctuations of animal numbers. At the beginning of the war, he turned the research focus of the bureau to controlling rodent pests, especially to help in reducing human disease and to reduce crop losses to rodents.

Elton was an early conservationist, stating in the preface to his 1927 text that "ecology is a branch of zoology which is perhaps more able to offer immediate practical help to mankind than any of the others [particularly important] in the present rather parous state of civilization." Elton strongly advocated the preservation of biological diversity and pressed hard for the prevention of extinctions, which he emphasizes in his chapter on "The Reasons for Conservation" in the *Invasions* book. But he also expanded his conception of conservation to mean finding some "general basis for understanding what it is best to do" and "looking for some wise principle of co-existence between man and nature, even if it has to be a modified kind of man and a modified kind of nature." He even took the unusual step (unusual for professional ecologists in his time and still unusual today) of going into the broadcast booth to popularize the importance of ecology in achieving those goals though environmental management as applied ecology.

Elton's service to ecology as a learned discipline was enormous. In the early twentieth century, ecology was still in its formative years, so Elton's ideas and

contributions came at a critical time. He took the infant field of animal ecology to maturity, building it to a status equal to that of the more established plant ecology. His research bureau at Oxford fostered innovative research and survey methods, and provided early intellectual nurture and professional development to ecologists who went on to become major contributors to the field, including the American ecologist Eugene Odum. As its first editor, Elton was "in a very real sense the creator" of the *Journal of Animal Ecology*, serving in the position for almost twenty years. He was one of the founders of the British Ecological Society. Elton's books and ideas continue to influence ecologists of the twenty-first century.

Resources

BOOKS

Elton, Charles S. *Animal Ecology*. London: Sidgwick & Jackson, 1927.

Elton, Charles S. *The Ecology of Invasions by Animals and Plants*. London: Methuen, 1958.

Elton, Charles S. *The Pattern of Animal Communities*. London: Methuen, 1966.

Elton, Charles S. *Voles, Mice and Lemmings: Problems in Population Dynamics*. Oxford: The Clarendon Press, 1942.

Gerald L. Young

Emergency Planning and Community Right-to-Know Act (1986)

The Emergency Planning and Community Right-to-Know Act (EPCRA), also known as Title III, is a statute enacted by Congress in 1986 as a part of the Superfund Amendments and Reauthorization Act (SARA). It was passed in response to public concerns raised by the accidental release of poisonous gas from a Union Carbide plant in Bhopal, India, which killed more than 2,000 people.

EPCRA has two distinct yet complementary sets of provisions. First, it requires communities to establish plans for dealing with emergencies created by chemical leaks or spills and defines the general structure these plans must assume. Second, it extends to communities the same kind of right-to-know provisions that were guaranteed to employees earlier in the 1980s. Overall, EPCRA is an important step away from crisis-by-crisis environmental enforcement toward a proactive or preventative approach. This approach depends on government monitoring of potential environmental hazards, which is being accomplished by using computerized files of data submitted by businesses.

Under the provisions of EPCRA, the governors of every state were required to establish a State Emergency Response Commission by 1988. Each commission in turn had to establish various emergency planning districts and to appoint a local committee for each district. The committees were to prepare plans for potential chemical emergencies in their communities, which included the identities of facilities, the procedures to be followed in the event of a chemical release, and the names of community emergency coordinators as well as a facility coordinator from each business subject to EPCRA.

A facility is subject to EPCRA if it has a substance in a quantity equal to or greater than the threshold specified on a list of about 400 extremely hazardous substances published by the Environmental Protection Agency (EPA). Also, after public notice and comment, either the state governor or the State Emergency Response Commission may designate facilities to be covered outside of these guidelines. Each business must provide facility notification information to the state commission and must designate a facility coordinator to work with the local planning committee.

EPCRA requires these facilities to report immediately to the Community Coordinator of its local emergency committee any accidental releases of hazardous material. There are two classifications for such hazardous substances. The substance must either be on the EPA's extremely hazardous substance list or be defined under the Comprehensive Environmental Response Compensation and Liability Act (CERCLA). In addition to the initial emergency notice, follow-up notices and information are required.

EPCRA's second major set of provisions is designed to establish and implement a community right-to-know program. Information about the presence of chemicals at facilities within the community is collected from businesses and made available to public officials and the general public. Businesses must submit two sets of annual reports: the Hazardous Chemical Inventory and Toxic Release Inventory (TRIs), also known as Chemical Release Forms.

For the Hazardous Chemical Inventory, each business in the community must prepare or obtain a Material Safety Data Sheet for each chemical on its premises meeting the threshold quantity. This information is then submitted to the Local Emergency Planning Committee, the local fire department, and the State Emergency Response Commission. These data sheets are identical to those required under the Occupational Safety and Health Act's worker right-to-know provisions. For each chemical reported in the Hazardous Chemical Inventory, a Chemical Inventory Report must be filed each year.

The second set of annual reports required as a part of the community right-to-know program is the Toxic Release Inventory (TRI), which must be filed annually. Releases reported on this form include even those made legally with permits issued by the EPA and its state counterparts. Releases made into air, land, and water during the preceding twelve months are summarized in this inventory. The form must be filed by companies having ten or more employees if that company manufactures, stores, imports, or otherwise uses designated toxic chemicals at or above threshold levels.

The information submitted pursuant to both the emergency planning and the right-to-know provisions of EPCRA is available to the general public through the Local Emergency Planning Committees. In addition, health professionals may obtain access to specific chemical identities to treat exposed individuals or protect potentially exposed individuals. Even if that information is claimed by the business to be a trade secret.

Congress provided stiff penalties for noncompliance. Civil penalties of tens of thousands of dollars may be assessed against a business failing to comply with reporting requirements, and citizens have the right to sue companies that fail to report. Further enforcement by the government may include criminal prosecution and imprisonment.

In June of 1996, in response to community pressure, the U.S. Congress took its first major vote on Community Right to Know since 1986. In the 1996 vote, the House of Representatives removed provisions from an EPA budget appropriation that would have made substantial cuts in funds allocated to compiling of Toxics Release Inventories (TRI). In addition, Congress passed an EPA proposal that added seven additional industries to the number of industries that must report under TRI, thus bringing the total number of industries required to report to twenty-seven. From 1996 to 2009, some changes were made to TRI requirements. Several new classes of industries, including coal and metal mining, were required to report to the TRI. Reporting thresholds were lowered to include companies that manufactured 25,000 pounds or handled more than 10,000 pounds of certain toxic chemicals or hazardous compounds annually. Companies that recycle, dispose, or release more than 500 pounds of TRI-listed chemicals into the environment have stricter, more comprehensive reporting requirements. Reporting and archiving of TRI data also expanded to online databases, such as TRI Explorer, that allow for easier access to information.

Studies have revealed that EPCRA has had far-reaching effects on companies and that industrial practices and attitudes toward chemical risk management are changing. Some firms have implemented new waste reduction programs or adapted previous programs. Others have reduced the potential for accidental releases of hazardous chemicals by developing safety audit procedures, reducing their chemical inventories, and using less hazardous chemicals in their operations.

As information included in reports, such as the annual Toxics Release Inventory, has been disseminated throughout the community, businesses have found they must be concerned with risk communication. Various industry groups throughout the United States have begun making the information required by EPCRA readily available and helping citizens to interpret that information. For example, the Chemical Manufacturers Association has conducted workshops for its members on communicating EPCRA information to the community and on how to communicate about risk in general. Similar seminars are now made available to businesses and their employees through trade associations, universities, and other providers of continuing education.

See also Chemical spills; Environmental monitoring; Environmental Monitoring and Assessment Program; Hazardous Materials Transportation Act (1975); Toxic substance; Toxic Substances Control Act (1976); Toxics use reduction legislation.

Resources

BOOKS

Ryan, Jeffrey R. *Pandemic Influenza: Emergency Planning and Community Preparedness.* Boca Raton, FL: CRC Press, 2009.

OTHER

Centers for Disease Control and Prevention (CDC). "Emergency Preparedness & Response." http://emergency.cdc.gov (accessed October 15, 2010).
United States Environmental Protection Agency (EPA). "Emergency Planning and Community Right-to-Know Act." http://epa.gov/oecaagct/lcra.html (accessed October 16, 2010).

Paulette L. Stenzel

Emergent diseases (human)

Diseases are referred to as emergent diseases if they are novel (first-appearing) diseases. Diseases that are reintroduced into a population after long periods of control or presumed elimination can also be termed emergent diseases.

Although many diseases such as measles, pneumonia, and pertussis (whooping cough) have likely been present as human infections for millennia, at least thirty new infectious diseases have appeared since 1980. In addition, other infectious diseases (e.g. tuberculosis) have reappeared and become widespread, due to the acquisition of genes that have made the microorganisms more capable of causing infections (commonly, this is due to increased resistance of bacteria to antibiotics).

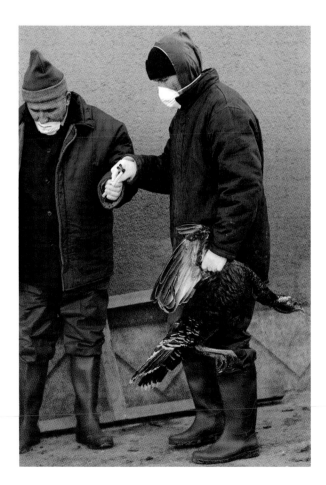

Veterinary officials hold a turkey before testing it for avian flu in the Danube Delta village of Murighiol, Romania, 300 kilometers east of Bucharest, in 2007 after an outbreak of the deadly H5N1 avian flu virus occurred in a farm in the village. *(AP Photo/Vadim Ghirda)*

Emergent ecological diseases

Emergent ecological diseases are a relatively recent phenomena involving extensive damage to natural communities and ecosystems. In some cases the specific causes of the ecological damage are known, but in others they are not yet understood.

Examples of relatively well-understood ecological diseases mostly involve cases in which introduced, non-native disease causing organisms (pathogens) are the source of the damage. One of many examples is the introduced chestnut blight fungus (*Endothia parasitica*), which has virtually eliminated the once extremely abundant American chestnut (*Castanea dentata*) from the hardwood forests of eastern North America. A similar ongoing pandemic involves the Dutch elm disease fungus (*Ceratocystis ulmi*), which is removing the white elm (*Ulmus americana*) and other native elms from North America.

Some introduced insects are also causing important forest damage, including the effects of the balsam wooly adelgid (*Adelges picea*) on Fraser fir (*Abies fraseri*) in the Appalachian Mountains.

Other cases of ecological diseases involve widespread damages that are well-documented, but the causes are not yet understood. One example is known as birch decline and, which occurred across great regions of the northeastern United States and eastern Canada from the 1930s to the 1950s. The disease killed yellow birch (*Betula alleghaniensis*), paper birch (*B. papyrifera*), and grey birch (*B. populifolia*) over a huge area. The specific cause of this extensive forest damage was never determined, but it could have involved the effects of freezing ground conditions during winters with little snow cover.

Rather similar forest declines and diebacks have affected red spruce (*Picea rubens*) and sugar maple (*Acer saccharum*) in the same broad region of eastern North America during the 1970s to 1990s. Although the causes of these forest damages are not yet fully understood, it is thought that air pollution or acidifying atmospheric deposition may have played a key role. In western Europe extensive declines of Norway spruce (*Picea abies*) and beech (*Fagus sylvatica*) are also thought to somehow be related to exposure to air pollution and acidification. In comparison the damage caused by ozone to forests dominated by ponderosa pine (*Pinus ponderosa*) in California is a relatively well-understood emergent ecological disease.

In the marine realm, widespread damage to diverse species of corals has been documented in various locales globally. The phenomenon is known as coral bleaching and involves the corals expelling their symbiotic algae (known as zooxanthellae), often resulting in death of the coral. Coral bleaching is thought to possibly be related to climate warming, although it can be caused by both unusually high or low water temperatures, changes in salinity, and other environmental stresses.

Another unexplained case of an ecological disease appears to be afflicting species of amphibians in many parts of the world. The amphibian declines involve severe population collapses and have even caused the extinction of some species. The specific causes are not yet known, but they likely involve introduced microbial pathogens or possibly increased exposure to solar ultraviolet radiation, climate change, or some other factor.

The release of a report on July 29, 2010, from the United States National Oceanographic and Atmospheric Administration (NOAA) provided convincing evidence of global climate change that is affecting terrestrial and aquatic environments. With the changing global climate, further emergence of ecological diseases will likely be a certainty.

Bacterial blights, outbreaks of bacterial growth, can severely impact agricultural production and result in substantial economic loss. The impacts can range from crop loss to blights that diminish crop quality or economic value.

For example, during the 2009 and 2010 growing season, onion crops in Argentina suffered a blight that resulted in onions with excess water retention, which accelerated spoilage during storage and transport. The blight substantially reduced exports to traditional European markets.

Bacterial blights can spread internationally and subside and erupt in outbreaks over several years. Pathovars (bacterial strains that infects only certain species of plants) can spread in seed or be spread by wind and water. Blights are most common in tropical and subtropical areas. Heavy rains may provide the environmental trigger for an outbreak.

Blights can be caused by subtle changes in bacteria to create bacterial pathovars. A pathovar is a pathological (disease-causing) variation of the normally nonpathological bacteria. A pathovar may refer to one specific bacterial strain or a set of closely related strains with similar characterisics. In addition to genus and species names, pathovars are often designated and differentiated by a suffix. For example, the pathovar responsible for an onion blight in Japan in 2000 was deginated, *Xanthomonas axonopodis pv. allii*.

Pathovars may also infect specific crops. For example, there are variations of *Xanthomonas axonopodis* that infect different crops. *Xanthomonas axonopodis* pv. *citri* causes citrus crop blights while *Xanthomonas axonopodis* pv. vesicatoria causes a bacterial blight in tomato crops.

Growers generally respond by trying to develop varieties of plants resistant to specific pathovars.

Resources

BOOKS

Callahan, Joan R. *Emerging Biological Threats: A Reference Guide.* Santa Barbara, CA: Greenwood Press, 2010.
Lashley, Felissa R., and Jerry D. Durham. *Emerging Infectious Diseases: Trends and Issues.* New York: Springer. 2007.

OTHER

Centers for Disease Control and Prevention (CDC). "EID Journal (Emerging Infectious Diseases Journal)." http://www.cdc.gov/ncidod/EID/index.htm (accessed November 8, 2010).
Centers for Disease Control and Prevention (CDC). "Emerging Infectious Diseases." http://www.cdc.gov/ncidod/diseases/eid/index.htm (accessed November 8, 2010).
World Health Organization (WHO). "Emerging Diseases." http://www.who.int/topics/emerging_diseases/en (accessed November 8, 2010).
World Health Organization (WHO). "Foodborne Diseases, Emerging." http://www.who.int/entity/mediacentre/factsheets/fs124/en/index.html (accessed November 8, 2010).

Bill Freedman

EMF *see* **Electromagnetic field.**

Emission

Release of material into the environment either by natural or anthropogenic (human-caused) activity. This term is often used to describe gaseous or particulate contributions to air pollution. Definitions of pollution are complicated by the fact that many of the materials that damage or degrade Earth's atmosphere result from natural processes. Volcanoes emit ash, acid mists, hydrogen sulfide, and other toxic gases. Natural forest fires release smoke, soot, carcinogenic hydrocarbons, dioxins, and other toxic chemicals as well as large amounts of carbon dioxide. Many experts consider natural or background pollution levels when estimating the significance of anthropogenic emissions.

Another key consideration in pollutant analysis is the regenerative capacity of the environment to remove or neutralize excess contaminants.

Emissions trading *see* **Trade in pollution permits.**

Emission standards

Emission standards are agreed-upon requirements, usually by governments or other regulatory groups, that set limits on the amount of pollutants that can be released into the environment. Many such standards are targeted specifically to regulating pollutants coming from motorized vehicles, such as cars and trucks. Other standards regulate pollutants generated from industry and power plants. Many of these environmental pollutants include carbon dioxide, carbon monoxide, sulfur oxides, nitrogen oxides, and others.

Federal, state, and local emission standards regulate flue gas stack and automobile exhaust emissions. These laws limit the quantity, rate, and concentration of such emissions. Emission standards can also regulate the opacity of plumes of smoke and dust from point and area emission sources. They can also assess the type and quality of fuel and the way the fuel is burned. With the exception of plume opacity, such standards are normally applied to the specific type of source for a given pollutant. Federal standards include New Source Performance Standards (NSPS) and National Emission Standards for Hazardous Air Pollutants (NESHAPS), both issued and enforced by the U.S. Environmental Protection Agency (EPA). Emission standards may include prohibitory rules that restrict existing and new source emission to specific concentration levels, mass rates, plume opacity, and emissions relative to process throughput emission rates. They may also require the most practical or best available technology in case of new emission in pristine areas.

New sources and modifications to existing sources can be subject to new source permitting procedures that require technology-forcing standards such as Best Available Control Technology (BACT) and Lowest Achievable Emission Rate (LAER). However, these standards are designed to consider the changing technological and economic feasibility of evenmore stringent emission controls. As a result, such requirements are not stable and are determined through a process involving the governing air pollution authority's discretionary judgments of appropriateness.

See also Point source.

Resources

BOOKS

Committee on the Significance of International Transport of Air Pollutants, Board on Atmospheric Sciences and Climate, Division on Earth and Life Studies, National Research Council of the National Academies. *Global Sources of Local Pollution: An Assessment of Long-range Transport of Key Air Pollutants to and from the United States.* Washington DC: National Academies Press, 2010.

Murray, Barrie. *Power Markets and Economics: Energy Costs, Trading, Emissions.* Chichester, U.K.: Wiley, 2009.

Rabe, Barry G., ed. *Greenhouse Governance: Addressing Climate Change in America.* Washington DC: Brookings Institution Press, 2010.

Tietenberg, Thomas H., and Thomas H. Tietenberg. *Emissions Trading: Principles and Practice.* Washington, DC: Resources for the Future, 2006.

OTHER

Environmental Protection Agency. "Environmental Protection Agency." http://www.epa.gov/ (accessed October 12, 2010).

Environmental Protection Agency. "Selected New Source Performance Standards." http://www.epa.gov/ttnatw01/nsps/nspstbl.html (accessed October 12, 2010).

Emphysema

Emphysema, also known as chronic obstructive pulmonary disease or COPD, is an abnormal, permanent enlargement of the airways responsible for gas exchange in the lungs. Primary emphysema is commonly linked to a genetic deficiency of the enzyme α_1-antitrypsin, a major component of α_1-globulin, a plasma protein. Under normal conditions α_1-antitrypsin inhibits the activity of many enzymes that break down proteins. This results in the increased likelihood of developing emphysema as a result of proteolysis (breakdown) of the lung tissues.

Emphysema begins with destruction of the alveolar septa, the thin walls that separate the tiny alveoli in the lungs. This results in "air hunger" characterized by labored or difficult breathing, sometimes accompanied by pain. Although emphysema is genetically linked to deficiency in certain enzymes, the onset and severity of asthmatic symptoms has been definitively linked to irritants and pollutants in the environment. A significantly greater proportion of emphysema symptoms has been observed in smokers, populations clustered around industrial complexes, and coalminers.

Emphysema or COPD also often co-exists with chronic bronchitis. About 12 million adults in the United States live with COPD, and it is also responsible for more than 100,000 deaths in the United States per year. Smoking is the primary risk factor for emphysema, although regular exposure to second-hand smoke or air pollutants, along with repeated childhood respiratory infections also predispose a person to developing COPD.

See also Asthma; Cigarette smoke; Respiratory diseases.

Resources

BOOKS

Green, Robert J. *Emphysema and Chronic Obstructive Pulmonary Disease.* San Diego, CA: Aventine Press, 2005.

PERIODICALS

Friedrich, M. J. "Preventing Emphysema." *JAMA: The Journal of the American Medical Association* 301, no. 5 (February 4, 2009): 477.

Katz, Patricia P., et al. "Disability in Valued Life Activities Among Individuals with COPD and Other Respiratory Conditions." *Journal of Cardiopulmonary Rehabilitation and Prevention* 30, no. 2 (March/April 2010): 126–136.

Wenzel, R., "Clinical Practice - Acute Bronchitis." *The New England Journal of Medicine* 355, no. 20 (November 16, 2006): 2125.

OTHER

American Lung Association. "COPD." http://www.lungusa.org/lung-disease/copd/ (accessed November 6, 2010).

Endangered species

An endangered species under United States law (the Endangered Species Act [1973]) is a creature "in danger of extinction throughout all or a significant portion of its range." A threatened species is one that is likely to become endangered in the foreseeable future.

For most people the endangered species problem involves the plight of such well-known animals as eagles, tigers, whales, chimpanzees, elephants, wolves, and whooping cranes (*Grus americana*). However, literally millions of lesser-known or unknown species are endangered or becoming so, and the loss of these life forms could have even more profound effects on humans than that of large mammals with whom we more readily identify and sympathize.

Most experts on species extinction, such as American biologist Edward O. Wilson (1929–) of Harvard and British environmentalist Norman Myers (1934–), estimate current and projected annual extinctions at anywhere from 15 thousand to 50 thousand species, or fifty to 150 per day, mainly invertebrates such as insects in tropical rain forests. At this rate, 5–10 percent of the world's species, perhaps more, could be lost in the next decade and a similar percentage in coming decades.

The single most important common threat to wildlife worldwide is the loss of habitat, particularly the destruction of biologically-rich tropical rain forests. Additional factors have included commercial exploitation, the introduction of non-native species, pollution, hunting, and trapping. Climate change— the warming of the Earth—may also contribute to the loss of biodiversity. Thus, the planet is rapidly losing a most precious heritage, the diversity of living species that inhabit the earth. Within one generation, the world will witness the threatened extinction of between one-fifth and one-half of all species on the planet.

Species of wildlife are becoming extinct at a rate that defies comprehension and threatens our own future. These losses are depriving this generation and future ones of much of the world's beauty and diversity, as well as irreplaceable sources of food, drugs, medicines, and natural processes that are or could prove extremely valuable, or even necessary, to the well-being of our society.

The early 2000s rate of extinction exceeds that of all of the mass extinction in geologic history, including the disappearance of the dinosaurs 65 million years ago. It is impossible to know how many species of plants and animals are actually being lot, or even how many species exist, because many have never been discovered or identified. Facts show that countless unique life forms that will never again exist are rapidly being exterminated.

Most of these species extinctions will occur—and are occurring—in tropical rain forests, which are the richest biological areas on earth and are being cut down at a rate of 1–2 acres (0.4–0.8 hectares) a second. Although tropical forests cover only about 5–7 percent of the world's land surface, they are thought to contain more than half of the species on earth.

More bird species live in one Peruvian preserve than in the entire United States. More species of fish exist in one Brazilian river than in all the rivers of the United States. And a single square mile in lowland Peru, Amazonian Ecuador, or Brazil may contain more than 1500 species of butterflies, more than twice as many as are found in all of the United States and Canada. Half an acre of Peruvian rain forest may contain more than 40 thousand species of insects.

Erik Eckholm in *Disappearing Species: The Social Challenge* notes that when a plant species is wiped out, some ten to thirty dependent species, such as insects and even other plants, may also be jeopardized. An example of the complex relationship that has evolved between many tropical species is the forty different kinds of fig trees native to Central America, each of which has a specific insect pollinator. Other insects, including pollinators for other plants, depend on certain of these fig trees for food.

Thus, the extinction of one species can set off a chain reaction, the ultimate effects of which cannot be foreseen. As Eckholm puts it, "Crushed by the march of civilization, one species can take many others with it, and the ecological repercussion and arrangements that follow may well endanger people." The loss of so many unrecorded, unstudied species will deprive the world not only of beautiful and interesting life forms, but also of much-needed sources of medicines, drugs, and food that could be of critical value to humanity. Every day, the world could be losing plants that might provide cures for cancer or AIDS or could become food staples as important as rice, wheat, or corn. No one will ever know the value or importance of the untold thousands of species vanishing each year.

In 2010 the U.S. Department of the Interior's list of endangered species included 1,541 total species, of which 768 are vertebrates (mammals, birds, reptiles, amphibians, fish), 171 are invertebrates (snails, clams, crustaceans, insects, and arachnids), and 602 are plants. There are 354 threatened species listed, including 172 vertebrates, 34 invertebrates, and 148 plants.

Under the Endangered Species Act, the Department of the Interior Fish and Wildlife Service (FWS) is given general responsibility for listing and protecting endangered wildlife, except for marine species (such as whales and seals), which are the responsibilities of the Department of Commerce's National Oceanic and Atmospheric Administration (NOAA) Fisheries Division. Recovery plans to rebuild the populations and improve the overall health of the species are initiated by these organizations for threatened and endangered species. Methods of recovery include restoring or acquiring habitat, removing invasive species or introduced predators, monitoring and surveying individual populations, and breeding species in captivity for future release into the wild. Recovery efforts have focused on bald eagles (*Haliaeetus leucocephalus*), peregrine falcons (*Falco peregrinus*), red wolves (*Canis rufus*), whooping cranes, and the Aleutian Canada goose (*Branta hutchinsii leucopareia*).

In addition, the United States is subject to the provisions of the Convention on International Trade in Endangered Species of Wild Flora and Fauna (CITES), which regulates global commerce in rare species. But in many cases, the government has not been enthusiastic about administering and enforcing the laws and regulations protecting endangered wildlife. Conservationists have for years criticized the Department of the Interior for its slowness and even refusal to list hundreds of endangered species that, without government protection, were becoming extinct. Indeed, the department admits that some three dozen species have become extinct while undergoing review for listing.

In December 1992 the department settled a lawsuit brought by animal protection groups by agreeing to expedite the listing process for some 1,300 species and to take a more comprehensive "multispecies, ecosystem approach" to protecting wildlife and their habitats. In October 1992 at the national conference of the Humane Society of the United States held in Boulder, Colorado, Secretary of the Interior Bruce Babbitt (1938–) in his keynote address lauded the Endangered Species Act as "an extraordinary achievement" and emphasized the importance of preserving endangered species and biological diversity, noting: "The extinction of a species is a permanent loss for the entire world. It is millions of years of growth and development put out forever."

Related information

According to the International Union for the Conservation of Nature (IUCN), 38 percent of the world's species are threatened based on assessments as of 2008. In May 2008 the Zoological Society of London and World Wide Fund for Nature (WWF) released results of a study indicating a dramatic decline in global wildlife populations during the last four decades. The data showed that between 1970 and 2005, the populations of land-based species fell by 25 percent, of marine species by 28 percent, and of freshwater species by 29 percent. Biodiversity loss was also reported in the form of extinction of about 1 percent of the world's species each year. The two groups, in agreement with most biologists, stated that one of the "great extinction episodes" in Earth's history is now under way. Species whose populations shrink are more likely to be endangered than species with large populations.

The losses were attributed to pollution, urbanization, overfishing, and hunting, but both groups emphasized that climate change would play an increasing role in species decline over coming decades. The Zoological Foundation and WWF based their conclusion upon population estimates and other data related to 1,400 species tracked as part of its joint Living Planet index. Data were obtained from publications in scientific journals.

See also Biodiversity.

Resources

BOOKS

Askins, Robert. *Saving Biological Diversity: Balancing Protection of Endangered Species and Ecosystems.* New York Springer, 2008.

Goodall, Jane, Thane Maynard, and Gail E. Hudson. *Hope for Animals and Their World: How Endangered Species Are Being Rescued from the Brink.* New York: Grand Central, 2009.

Mackay, Richard. *The Atlas of Endangered Species.* Berkeley: University of California Press, 2008.

Reeve, Rosalind. *Policing International Trade in Endangered Species: The CITES Treaty and Compliance.* London: Royal Institute of International Affairs, 2004.

Vega, Evelyn T. *Endangered Species Act Update and Impact.* New York: Nova Science Publishers, 2008.

PERIODICALS

Clayton, Mark. "New Tool to Fight Global Warming: Endangered Species Act?" *Christian Science Monitor* (September 7, 2007).

OTHER

International Union for Conservation of Nature and Natural Resources (IUCN). "The IUCN Red List of Endangered Species." http://www.iucnredlist.org (accessed August 4, 2010).

International Union for Conservation of Nature and Natural Resources (IUCN). "Numbers of Threatened Species by Major Groups of Organisms." http://www.iucnredlist.org/documents/2008RL_stats_table_1_v1223294385.pdf (accessed August 4, 2010).

United Nations Environment Programme (UNEP). "Secretariat of the Convention on International Trade in Endangered Species of Wild Fauna and Flora." http://www.cites.org/ (accessed August 4, 2010).

United States Environmental Protection Agency (EPA). "Ecosystems: Species: Endangered Species." http://www.epa.gov/ebtpages/ecosspeciesendangeredspecies.html (accessed August 4, 2010).

U.S. Fish & Wildlife Service. "Endangered Species Recovery Program." http://www.fws.gov/endangered/factsheets/recovery.pdf (accessed August 4, 2010).

Lewis G. Regenstein

Endangered Species Act (1973)

The Endangered Species Act (ESA) is a law designed to save species from extinction. What began as an informal effort to protect several hundred North American vertebrate species in the 1960s has expanded into a program that could involve hundreds of thousands of plant and animal species throughout the world. As of 2010 approximately 1,959 species were listed as endangered or threatened, with approximately 1,372 in the United States and approximately 587 in other countries. The law has become increasingly controversial as it has been viewed by commercial interests as a major impediment to economic development. This issue came to a head in the Pacific Northwest, where the northern spotted owl was listed as threatened. This action had significant effects on the regional forest products industry. The ESA was due to be reauthorized in 1992, but this was postponed due to that year's election. Although it expired on October 1, 1992, Congress allotted enough funds to keep the ESA active.

Government action to protect endangered species began in 1964, with the formation of the Committee on Rare and Endangered Wildlife Species within the Bureau of Sport Fisheries and Wildlife (now the Fish and Wildlife Service [FWS]) in the U.S. Department of the Interior. In 1966 this committee issued a list of eighty-three native species (all vertebrates) that it considered endangered. Two years later the first act designed to protect species in danger of extinction, the Endangered Species Preservation Act of 1966, was passed. The Secretary of the Interior was to publish a list, after consulting the states, of native vertebrates that were endangered. This law directed federal agencies to protect endangered species when it was "practicable and consistent with the primary purposes" of these agencies. The taking of listed endangered species was prohibited only within the national wildlife refuge system; that is, species could be killed almost anywhere in the United States. Finally, the law authorized the acquisition of critical habitat for these endangered species.

In 1969 the Endangered Species Conservation Act was passed, which included several significant amendments to the 1966 act. Species could now be listed if they were threatened with worldwide extinction. This substantially broadened the scope of species to be covered, but it also limited the listing of specific populations that might be endangered in some parts of the United States but not in danger elsewhere (e.g., grizzly bears, bald eagles, timber wolves, all of which flourish in Canada and Alaska). The 1969 law stated that mollusks and crustaceans could now be included on the list, further broadening of the scope of the law. Finally, trade in illegally taken endangered species was prohibited. This substantially increased the protection offered such species, compared to the 1966 law.

The Endangered Species Act of 1973 built upon and strengthened the previous laws. The impetus for the law was a call by President Nixon in his state of the union message for further protection of endangered species and the concern in Congress that the previous acts were not working well enough. The goal of the ESA was to protect all endangered species through the use of "all methods and procedures necessary to bring any endangered or threatened species to the point at which the measures provided pursuant to [the] Act are no longer necessary." In other words the goal was to bring endangered species to full recovery. This goal, like others included in environmental legislation at the time, was unrealistic. The ESA also expanded the number of species that could be considered for listing to all animals (except those considered pests) and plants. It stipulated that the listing of such species should be based on the best scientific data available. Additionally, it included a provision that allowed groups or individuals to petition the government to list or de-list a species. If the petition contained reasonable support, the agency had to respond to it.

The law created two levels of concern: endangered and threatened. An endangered species was "in danger of extinction throughout all or a significant portion of its range." A threatened species was "likely to become an endangered species within the foreseeable future throughout all or a significant portion of its range." Also, the species did not have to face worldwide extinction before it could be listed. No taking of any kind was allowed for endangered species; limited taking could be allowed for threatened species. Thus, the distinction between "endangered" and "threatened" species allowed for some flexibility in the program.

The 1973 Act divided jurisdiction of the program between the FWS and the National Marine Fisheries Service (NMFS), an agency of the National Oceanic and Atmospheric Administration in the Department of Commerce. The NMFS would have responsibility for species that were primarily marine; responsibility for marine mammals (whales, dolphins, etc.) was shared by the two agencies. The law also provided for the establishment of cooperative agreements between the federal government and the states on endangered species protection. This did not prove very successful due to a lack of funds to entice the states to participate and due to frequent conflict between the states (favoring development and hunting) and the FWS.

has lost most of its earlier geographic range. A familiar autochthonous endemic species is the Australian koala, which evolved in its current environment and continues to occur nowhere else. A well-known example of allochthonous endemism is the California coast redwood (*Sequoia sempervirens*), which millions of years ago ranged across North America and Eurasia, but today exists only in isolated patches near the coast of northern California. Another simpler term for allochthonous endemics is relict, meaning something that is left behind.

In addition to geographic relicts, plants or animals that have greatly restricted ranges, there are what is known as taxonomic relicts. These are species or genera that are sole survivors of once-diverse families or orders. Elephants are taxonomic relicts: Millions of years ago the family Elephantidae had twenty-five different species (including woolly mammoths) in five genera. Today only two species remain, one living in Africa (*Loxodonta africana*) and the other in Asia (*Elephas maximus*). Horses are another familiar species whose family once had many more branches. Ten million years ago North America alone had at least 10 genera of horses. Today only a few Eurasian and African species remain, including the zebra and the ass. Common horses, all members of the species *Equus caballus*, returned to the New World only with the arrival of Spanish conquistadors.

Taxonomic relicts are often simultaneously geographic relicts. The ginkgo tree, for example was one of many related species that ranged across Asia 100 million years ago. Today the family Ginkgoales contains only one genus, *Ginkgo*, with a single species, *Ginkgo biloba*, that occurs naturally in only a small portion of eastern China. Similarly the coelacanth, a rare fish found only in deep waters of the Indian Ocean near Madagascar, is the sole remnant of a large and widespread group that flourished hundreds of millions of years ago.

Where living things become relict endemics, some sort of environmental change is usually involved. The redwood, the elephant, the ginkgo, and the coelacanth all originated in the Mesozoic era, 245–65 million years ago, when the earth was much warmer and wetter than it is today. All of these species managed to survive catastrophic environmental change that occurred at the end of the Cretaceous period, changes that eliminated dinosaurs and many other terrestrial and aquatic animals and plants. The end of the Cretaceous was only one of many periods of dramatic change; more recently two million years of cold ice ages and warmer interglacial periods in the Pleistocene substantially altered the distribution of the world's

plants and animals. Species that survive such events to become relicts do so by adapting to new conditions or by retreating to isolated refuges where habitable environmental conditions remain.

When endemics evolve in place, isolation is a contributing factor. A species or genus that finds itself on a remote island can evolve to take advantage of local food sources or environmental conditions, or its characteristics may simply drift away from those of related species because of a lack of contact and interbreeding. Darwin's Galapagos finches, for instance, are isolated on small islands, and on each island a unique species of finch has evolved. Each finch is now endemic to the island on which it evolved. Expanses of water isolated these evolving finch species, but other sharp environmental gradients can contribute to endemism, as well. The humid southern tip of Africa, an area known as the Cape region, has one of the richest plant communities in the world. A full 90 percent of the Cape's 18,500 plant species occur nowhere else. Separated from similar habitat for millions of years by an expanse of dry grasslands and desert, local families and genera have divided and specialized to exploit unique local niches. Endemic speciation, or the evolution of locally unique species, has also been important in Australia, where 32 percent of genera and 75 percent of species are endemic. Because of its long isolation, Australia even has family-level endemism, with forty families and sub-families found only on Australia and a few nearby islands.

Especially high rates of endemism are found on long-isolated islands, such as St. Helena, New Caledonia, and the Hawaiian chain. St. Helena, a volcanic island near the middle of the Atlantic, has only sixty native plant species, but fifty of these exist nowhere else. Because of the island's distance from any other landmass, few plants have managed to reach or colonize St. Helena. Speciation among those that have reached the remote island has since increased the number of local species. Similarly Hawaii and its neighboring volcanic islands, colonized millions of years ago by a relatively small number of plants and animals, now have a wealth of locally-evolved species, genera, and sub-families. Today's 1,200–1,300 native Hawaiian plants derive from about 270 successful colonists; 300–400 arthropods that survived the journey to these remote islands have produced more than 6,000 descendent species today. Ninety-five percent of the archipelago's native species are endemic, including all ground birds. New Caledonia, an island midway between Australia and Fiji, consists partly of continental rock, suggesting that at one time the island was attached to a larger landmass and its resident species had contact with those of the mainland. Nevertheless, because of long isolation 95 percent of native animals and plants are endemic to New Caledonia.

Ancient, deep lakes are like islands because they can retain a stable and isolated habitat for millions of years. Siberia's Lake Baikal and East Africa's Lake Tanganyika are two notable examples. Lake Tanganyika occupies a portion of the African Rift Valley, 0.9 miles (1.5 km) deep and perhaps 6 million years old. Fifty percent of the lake's snail species are endemic, and most of its fish are only distantly related to the fish of nearby Lake Nyasa. Siberia's Lake Baikal, another rift valley lake, is 25 million years old and 1 mile (1.6 km) deep. Eighty-four percent of the lake's 2,700 plants and animals are endemic, including the nerpa, the world's only freshwater seal.

Because endemic animals and plants by definition have limited geographic ranges, they can be especially vulnerable to human invasion and habitat destruction. Island species are particularly at risk because islands commonly lack large predators, so many island endemics evolved without defenses against predation. Cats, dogs, and other carnivores introduced by sailors have decimated many island endemics. The flora and fauna of Hawaii, exceptionally rich before Polynesians arrived with pigs, rats, and agriculture, were severely depleted because their range was limited, and they had nowhere to retreat as human settlement advanced. Tropical rain forests, with extraordinary species diversity and high rates of endemism, are also vulnerable to human invasion. Many of the species eliminated daily in Amazonian rain forests are locally endemic, so that their entire range can be eliminated in a short time.

Mary Ann Cunningham

Endocrine disruptors

In recent years, scientists have proposed that chemicals released into the environment may be disrupting the endocrine system of humans and wildlife. The endocrine system is a network of glands and hormones that regulates many of the body's functions, such as growth, development, behavior, and maturation. The endocrine glands include the pituitary, thyroid, adrenal, thymus, pancreas, and the male and female gonads (testes and ovaries). These glands secrete regulated amounts of hormones into the bloodstream, where they act as chemical messengers as they are carried throughout the body to control and regulate many the body's functions. The hormones bind to

specific cell sites called receptors. By binding to the receptors, the hormones trigger various responses in the tissues that contain the receptors.

An endocrine disruptor is an external agent that interferes in some way with the role of the hormones in the body. The agent might disrupt the endocrine system by affecting any of the stages of hormone production and activity, such as preventing the synthesis of a hormone, directly binding to hormone receptors, or interfering with the breakdown of a natural hormone. Disruption in endocrine function during highly sensitive prenatal periods is especially critical, as small changes in endocrine functions may delay consequences that only become evident later in adult life or in a subsequent generation. Adverse effects that might be a result of endocrine disruption include the development of cancers, reproductive and developmental effects, neurological effects (effects on behavior, learning and memory, sensory function, and psychomotor development), and immunological effects (immunosuppression, with resulting disease susceptibility).

Exposure to suspected endocrine disruptors may occur through direct contact with the chemicals or through ingestion of contaminated water, food, or air. These disruptors can enter air or water from chemical and manufacturing processes, and through incineration of products. Industrial workers may be exposed in work settings. Documented examples of health effects on humans exposed to endocrine disrupting chemicals include shortened penises in the sons of women exposed to dioxin-contaminated rice oil in China and reduced sperm count in workers exposed to high doses of Kepone in a Virginia pesticide factory. Diethylstilbestrol (DES), a synthetic estrogen, was used in the 1950s and 1960s by pregnant women to prevent miscarriages. Unfortunately it did not prevent miscarriages, but the teenage daughters of women who had taken DES suffered high rates of vaginal cancers, birth defects of the uterus and ovaries, and immune system suppression. These health effects were traced to their mothers' use of DES.

A variety of chemicals, including some pesticides, have been shown to result in endocrine disruption in animal laboratory studies. However, except for the incidences of endocrine disruption due to chemical exposures in the workplace and to the use of DES, causal relationships between exposure to specific environmental agents and adverse health effects in humans due to endocrine disruption have not yet been firmly established.

There is more evidence that the endocrine systems of fish and wildlife have been affected by chemical contamination in their habitats. Groups of animals that have been affected by endocrine disruption include snails, oysters, fish, alligators and other reptiles, and birds, including gulls and eagles. Whether effects on individuals of a particular species have an impact on populations of that organism is difficult to prove. Scientists also do not know if endocrine disruption is confined to specific areas or is more widespread. In addition, proving that a specific chemical causes a particular endocrine effect is difficult, as animals are exposed to a variety of chemicals and non-chemical stressors. However, some persistent organic chemicals such as DDT (dichlorodiphenyltrichloroethane), PCBs (polychlorinated biphenyls), dioxin, and some pesticides have been shown to act as endocrine disruptors in the environment. Adverse effects that may be caused by endocrine disrupting mechanisms include abnormal thyroid function and development in fish and birds; decreased fertility in shellfish, fish, birds, and mammals; decreased hatching success in fish, birds, and reptiles; demasculinization and feminization of fish, birds, reptiles, and mammals; defeminization and masculinization of gastropods, fish, and birds; and alteration of immune and behavioral function in birds and mammals. Many potential endocrine disrupting chemicals are persistent and bioaccumulate in fatty tissues of organisms and increase in concentration as they move up through the food web. Because of this persistence and mobility, they can accumulate and harm organisms far from their original source.

More information is needed to define the ecological and human health risks of endocrine disrupting chemicals. Epidemiological investigations, exposure assessments, and laboratory testing studies for a wide variety of both naturally occurring and synthetic chemicals are tools that are being used to determine whether these chemicals as environmental contaminants have the potential to disrupt hormonally mediated processes in humans and animals.

Resources

BOOKS

Greenstein, Ben. *The Endocrine System at a Glance*. Malden, MA: Blackwell Publishing, 2006.

Romeo, June Hart. *Advances in Endocrine Disorders*. Nursing clinics of North America, v. 42, no. 1. Philadelphia, PA: Saunders/Elsevier, 2007.

Rushton, Lynette. *The Endocrine System*. Philadelphia, PA: Chelsea House Publishers, 2004.

OTHER

National Institutes of Health (NIH). "Endocrine Diseases (General)." http://health.nih.gov/topic/Endocrine DiseasesGeneral (accessed November 11, 2010).

United States Environmental Protection Agency (EPA). "Pollutants/Toxics: Chemicals: Endocrine Disruptors." http://www.epa.gov/ebtpages/pollchemicalsendocrine disruptors.html (accessed November 11, 2010).

Judith L. Sims

Energy and the environment

Energy in the form of moving forces (e.g., wind, tides, water flow) and in materials such as natural gas and oil can be captured directly (e.g., wind turbine, water turbines) or extracted by processing of the material. The obtained energy is invaluable as a source of heat, electricity, and fuel.

However, the acquisition of energy is a prime factor in environmental quality. Extracting, processing, shipping, and the combustion of coal, oil, and natural gas are the largest sources of air pollutants, thermal and chemical pollution of surface waters, accumulation of mine tailings and toxic ash, and land degradation caused by surface mining in the United States.

On the other hand, a cheap, inexhaustible source of energy would allow humans to eliminate or repair much of the environmental damage done already and to improve the quality of the environment in many ways. Often, the main barrier to reclaiming degraded land, cleaning up polluted water, destroying wastes, restoring damaged ecosystems, or remediating other environmental problems is that solutions are expensive—and much of that expense is energy costs.

Our ability to use external energy to do useful work is one of the main characteristics that distinguishes humans from other animals. Clearly, technological advances based on this ability have made our lives much more comfortable and convenient than that of our early ancestors. However, a large part of our current environmental crisis is that our ability to modify our environment has outpaced our capacity to use energy and technology wisely.

In the United States, fossil fuels supply about 85 percent of the commercial energy. This situation cannot continue indefinitely, because fossil fuel supplies are finite. They will eventually be exhausted. Americans now get more than 60 percent of their oil from foreign sources at great economic and political costs. At current rates of use, the known and economically extractable world supplies of oil and natural gas will probably last only a century or so. Reserves of coal are much larger, but coal is one of the dirtiest fuels, contributing particulate material and greenhouse gases that fuel global warming. In addition, the burning of coal is the largest single source in the United States of sulfur dioxide and nitrogen oxides (which cause respiratory health problems, ecosystem damage, and acid precipitation). Paradoxically, coal-burning power plants also release radioactivity, since radioactive minerals such as uranium and thorium are often present in low concentrations in coal deposits.

Nuclear power was once thought to be an attractive alternative to fossil fuels. Nuclear power was promoted in the 1960s as the energy source for the future. Nuclear power production accidents, such as the explosion and fire at Chernobyl in the former Soviet Union (present day Ukraine) in 1986, problems with releases of radioactive materials in mining and processing of fuels, and the inability to find a safe, acceptable permanent storage solution for nuclear waste have made nuclear power seem much less attractive in recent years. The majority of the citizens of most European and North American countries now regard nuclear power as unacceptable.

The U.S. government once projected that 1,500 nuclear plants would be built. In 2002 only 105 plants were in operation and no new construction has been undertaken since 1975. Many of these aging plants are now reaching the end of their useful life. There will be enormous costs and technical difficulties in dismantling them and disposing of the radioactive debris. However, as other energy resources become less available or increasingly expensive, nuclear power is likely to return to the agenda.

Damming rivers to create hydroelectric power from spinning water turbines has the attraction of providing a low-cost, renewable, air pollution-free energy source. Only a few locations remain in the United States, where large hydroelectric projects are feasible. Many more sites are available in Canada, Brazil, India, and other countries. The social and ecological effects of building large dams, flooding valuable river valleys, and eliminating free-flowing rivers are such that opposition is mounting to this energy source.

An example of the ecological and human damage done by large hydroelectric projects is seen in the James Bay region of Eastern Quebec. A series of huge dams and artificial lakes have flooded thousands of square miles of forest. Migration routes of caribou have been disrupted, the habitat for game on which indigenous people depended has been destroyed, and

decaying vegetation has acidified waters, releasing mercury from the bedrock and raising mercury concentrations in fish to toxic levels. The hunting and gathering traditions of local Cree and Inuit people has been affected significantly.

There are several sustainable, environmentally benign energy sources that could be developed. Among these are wind power, biomass (burning renewable energy crops such as fast-growing trees or shrubs), small-scale hydropower (low head or run-of-the- river turbines), passive-solar space heating, active-solar water heaters, photovoltaic energy (direct conversion of sunlight to electricity), and ocean tidal or wave power. There may be unwanted environmental consequences of some of these sources as well, but they seem much better in aggregate than current energy sources. A disadvantage is that most of these alternative energy sources are diffuse and not always available when or where the energy is needed.

Ways to store and ship energy generated from these sources are needed. There have been many suggestions that a breakthrough in battery technology could be on the horizon. Other possibilities include converting biomass into methane or methanol fuels, or using electricity to generate hydrogen gas through electrolysis of water. These fuels would be easy to store, transport, and use with current technology without greatly altering existing systems. Estimates indicate that some combination of these sustainable energy sources could supply all of the United States' energy needs by utilizing only a small fraction (perhaps less than one percent) of United States land area. If means are available to move this energy efficiently, these energy farms could be in remote locations with little other value.

See also Acid rain; Air pollution; Greenhouse effect; Photovoltaic cell; Solar energy; Thermal pollution; Wind energy.

Resources

BOOKS

Elliott, David. *Sustainable Energy: Opportunities and Limitations.* Basingstoke, UK: Palgrave Macmillan, 2010.

Kruger, Paul. *Alternative Energy Resources: The Quest for Sustainable Energy.* Hoboken, NJ: John Wiley, 2006.

Seifried, Dieter, and Walter Witzel. *Renewable Energy: The Facts.* London: Earthscan Publications, 2010.

Wolfson, Richard. *Energy, Environment, and Climate.* New York: W.W. Norton & Company, 2008.

William P. Cunningham

Energy conservation

Energy conservation is the term used for efforts to reduce energy consumption through more efficient energy use and decreased energy consumption from conventional (non-renewable) energy sources.

The concept of energy conservation was largely unfamiliar to America—and to much of the rest of the world—prior to 1973. Certainly, some thinkers prior to that date thought about, wrote about, and advocated a more judicious use of the world's energy supplies. But in a practical sense, it seemed that the world's supply of petroleum, such as coal, oil, and natural gas, was virtually unlimited.

In 1973, however, the Organization of the Petroleum Exporting Countries (OPEC)—which in 2010 was made up of the following twelve countries: Algeria, Angola, Ecuador, Iran, Iraq, Kuwait, Libya, Nigeria, Qatar, Saudi Arabia, the United Arab Emirates, and

Display of a plug-in energy monitor showing the electricity consumption of a kettle (rated at 240 volts, 2200 watts) to be 2208 watts. Monitoring power consumption can help reduce energy consumption and reduce the impact on the environment. (Sheila Terry/Photo Researchers, Inc.)

Venezuela (but included Indonesia in 1973)—placed an arbitrary limit on the amount of petroleum that non-producing nations could buy from them. Although the OPEC embargo lasted only a short time, the nations of the world were suddenly forced to consider the possibility that they might have to survive on a reduced and ultimately finite supply of fossil fuels.

In the United States the OPEC embargo set off a flurry of administrative and legislative activity, designed to ensure a dependable supply of energy for the nation's further needs. Out of this activity came acts such as the Energy Policy and Conservation Act of 1976, the Energy Conservation and Production Act of 1976, and the National Energy Act of 1978.

An important feature of the nation's and the world's new outlook on energy was the realization of how much energy is wasted in transportation, residential and commercial buildings, and industry. When energy supplies appeared to be without limit, waste was a matter of small concern. However, when energy shortages began to be a possibility, conservation of energy sources assumed a high priority.

Energy conservation is certainly one of the most attainable goals the federal government can set for the United States. Almost every way people use energy results in enormous waste. Only about 20 percent of the energy content of gasoline, for example, is actually put to productive work in an automobile. Each time humans make use of electricity, waste is produced; for example, coal is burned to heat water to drive a turbine to operate a generator to make electricity. More than 90 percent of the energy generated in the electrical process is wasted.

Fortunately, a vast array of conservation techniques are available in each of the major categories of energy use: transportation, residential and commercial buildings, and industry. In the area of transportation, conservation efforts focus on the nation's use of the private automobile for most personal travel. Certainly, the private automobile is an enormously wasteful method for moving people from one place to another. It is hardly surprising, therefore, that conservationists have long argued for the development of alternative means of transportation: bicycles, motorcycles, mopeds, carpools and vanpools, dial-a-rides, and various forms of mass transit. The amount of energy needed to move a single individual on average on a bus is about one-third of the amount for a private car. One need only compare the relative energy cost per passenger for eight people traveling in a commuter vanpool to the cost for a single individual in a private automobile to see the advantages of some form of mass transit.

For a number of reasons, however, mass transit systems in the United States are not very popular. Whereas the number of new cars sold continues to rise year after year, investment in and use of heavy and light rail systems, trolley systems, subways, and various types of automobile pools remain modest.

Many authorities believe that the best hope for energy conservation in the field of transportation is to make private automobiles more efficient or to increase the tax on their use. Some experts argue that technology already exists for the construction of 100-mile-per-gallon (42.5 kilometer/liter) automobiles if industry would make use of the technology. They also argue for additional research on electric cars as an energy-saving and pollution-reducing alternative to internal combustion vehicles. Several automobile manufacturers have produced gasoline/electric hybrid cars. The U.S. government offers tax credits toward purchase of a hybrid vehicle as an incentive for buying a more environmentally friendly automobile. These vehicles use significantly less fuel, reducing the amount of fossil fuel emissions into the environment. The Obama administration set goals for improving fuel efficiency for both cars and trucks, calling for a raise in the average fuel economy to thirty-five miles a gallon by the year 2020.

Increasing the cost of using private automobiles has also been explored. One approach is to raise the tax on gasoline to a point where commuters begin to consider mass transit as an economical alternative. Increased parking fees and more aggressive traffic enforcement have also been tried. Such approaches often fail—or are never attempted—because public officials are reluctant to anger voters.

Other methods that have been suggested for increasing energy efficiency in automobiles include the design and construction of smaller, lighter cars, extending the useful life of a car, improving the design of cars and tires, and encouraging the design of more efficient cars through federal grants or tax credits.

A number of techniques are well known and could be used, however, to conserve energy in large buildings or even small two-room cottages. A thorough insulation of floors, walls, and ceilings, for example, can save up to 80 percent of the cost of heating and cooling a building.

In addition, buildings can be designed and constructed to take advantage of natural heating and cooling factors in the environment. A home in Canada, for example, should be oriented with windows tilting toward the south to take advantage of the sun's heating rays. A home in Mexico might have quite a different orientation to benefit from natural heating and cooling.

One of the most extreme examples of environmentally friendly buildings are those that have been constructed at least partially underground. The earthen walls of these buildings provide a natural cooling effect in the summer and provide excellent insulation during the winter.

The kind, number, and placement of trees around a building can also contribute to energy efficiency. Trees that lose their leaves in the winter will allow sunlight to heat a building during the coldest months, but will shield the building from the sun during the hot summer months.

Energy can also be conserved by modifying appliances used within a building. Prior to 1973, consumers became enamored with all kinds of electrical devices, from electric toothbrushes to electric shoeshine machines to trash compactors. As convenient as these appliances may be, they waste energy and are not necessarily considered a basic human need.

Even items as simple as light bulbs can become a factor in energy conservation programs. Fluorescent light bulbs use at least 75 percent less energy than do incandescent bulbs, and they often last twenty times longer. Although many commercial buildings now use fluorescent lighting exclusively, it still tends to be relatively less popular in private homes. Many U.S. homeowners are replacing their incandescent light bulbs with compact fluorescent light bulbs (CFLs). CFLs can be used in most existing light fixtures, use less energy (75 percent less energy on average), and have a longer life (six to ten times longer than incandescent bulbs); however, they are more expensive than incandescent bulbs. Over the life of a CFL, the savings generated in energy costs more than compensate for the upfront cost of the bulb.

As the largest single user of energy in American society, industry is a prime candidate for conservation measures. Always sensitive to possible money-saving changes, industry has begun to develop and implement energy savings devices and procedures. One such idea is cogeneration, the use of waste heat from an industrial process for use in the generation of electricity.

Another approach is the expanded use of recycling by industry. In many cases, reusing a material requires less energy than producing it from raw materials. Finally, researchers are continually testing new designs for equipment that will allow that equipment to operate using less energy.

Governments and utilities have two primary methods by which they can encourage energy conservation. One approach is to penalize individuals and companies that use too much energy. For example, an industry that uses large amounts of electricity might be charged at a higher rate per kilowatt hour than one that uses less electricity, a policy just the opposite of that is now in practice in most places.

A more positive approach is to encourage energy conservation by techniques such as tax credits. Those who insulate their homes with more energy-efficient insulation might, for example, be given cash bonuses by the local utility or a tax deduction by state or federal government.

In recent years another issue of energy conservation has come to the forefront—its environmental advantages. Obviously, the less coal, oil, and natural gas that humans use, the fewer pollutants are released into the environment. Thus, a practice that is energy-efficient, conserving energy, can also provide environmental benefits. Those concerned with global warming and climate change have been especially active in this area. Proponents point out that reducing human use of fossil fuels will both reduce the consumption of energy and the release of carbon dioxide (CO_2)—a major greenhouse gas—into the atmosphere. Climate change can be combated, they point out, by taking the steps to use less energy.

The general public concern about energy waste engendered by the 1973 OPEC oil embargo eventually dissolved into complacency. Some of the sensitivity to energy conservation created by that event has not been lost. Many people have switched to more energy-efficient forms of transportation, think more carefully about leaving house lights on all night, have switched from incandescent light bulbs to CFLs, and consider energy efficiency when buying major appliances. Energy Star specifications are determined for each type of appliance to identify the most energy efficient products and are set by the U.S. Environmental Protection Agency (EPA) and the Department of Energy (DOE).

But some of the more aggressive efforts to conserve energy have become stalled. Higher taxes on gasoline, for example, still are certain to raise an enormous uproar among the populace. And energy-saving construction steps that might be mandated by law still remain optional and are frequently ignored.

For some time, renewed confidence in an endless supply of fossil fuels meant many people were no longer convinced that energy conservation was very important. However, now in the twenty-first century, government attitude, public perception, and changes in the price of energy have meant a greater amount of conservation has become embedded in policy and regulation.

See also Energy efficiency; Energy policy; Fossil fuels.

Resources

BOOKS

Diwan, Parag, and Prasoom Dwivedi. *Energy Conservation.* New Delhi: Pentagon Energy Press, 2009.

International Code Council. *International Energy Conservation Code, 2006.* Country Club Hills, IL: International Code Council, 2006.

Krupp, Fred, and Miriam Horn. *Earth, the Sequel: The Race to Reinvent Energy and Stop Global Warming.* New York: W. W. Norton & Co, 2008.

PERIODICAL

Hargreaves, Steve, CNN.com. "Obama Acts on Fuel Efficiency, Global Warming."(January 26, 2009), http://www.cnn.com/2009/BUSINESS/01/26/obama.green/ (accessed October 12, 2010).

OTHER

U.S Department of Energy. "Energy Star Product Specifications." http://www.energystar.gov/index.cfm?c = prod_development.prod_development_index (accessed October 12, 2010).

David E. Newton

Energy crops *see* **Biomass.**

Energy efficiency

The utilization of energy for human purposes is a defining characteristic of industrial society. The conversion of energy from one form to another and the efficient production of mechanical work for heat energy has been studied and improved for centuries. The science of thermodynamics deals with the relationship between heat and work and is based on two fundamental laws of nature, the first and second laws of thermodynamics. The utilization of energy and the conservation of critical, nonrenewable energy resources are controlled by these laws and the technological improvements in the design of energy systems.

The First Law of Thermodynamics states the principle of conservation of energy: energy can be neither created nor destroyed by ordinary chemical or physical means, but it can be converted from one form to another. Stated another way, *in a closed system, the total amount of energy is constant.* An interesting example of energy conversion is the incandescent light bulb. In the incandescent light bulb, electrical energy is used to heat a wire (the bulb filament) until it is hot enough to glow. The bulb works satisfactorily except that the great majority (95 percent) of the electrical energy supplied to the bulb is converted to heat rather than light. The incandescent bulb is not very efficient as a source of light. In contrast a fluorescent bulb uses electrical energy to excite atoms in a gas, causing them to give off light in the process at least four times more efficiently than the incandescent bulb. Both light sources, however, conform to the First Law in that no energy is lost and the total amount of heat and light energy produced is equal to the amount of electrical energy flowing to the bulb.

The Second Law of Thermodynamics states that whenever heat is used to do work, some heat is lost to the surrounding environment. The complete conversion of heat into work is not possible. This is not the result of inefficient engineering design or implementation but, rather, a fundamental and theoretical thermodynamic limitation. The maximum, theoretically possible efficiency for converting heat into work depends solely on the operating temperatures of the heat engine and is given by the equations: $E = 1 - T_2/T_1$. T_1 is the absolute temperature at which heat energy is supplied, and T_2 is the absolute temperature at which heat energy is exhausted.

The maximum possible thermodynamic efficiency of a four-cycle internal combustion engine is about 54 percent; for a diesel engine, the limit is about 56 percent; and for a steam engine, the limit is about 32 percent. The actual efficiency of real engines, which suffer from mechanical inefficiencies and parasitic losses (eg. friction, drag, etc.) is significantly lower than these levels. Although thermodynamic principles limit maximum efficiency, substantial improvements in energy utilization can be obtained through further development of existing equipment such as power plants, refrigerators, and automobiles and the development of new energy sources such as solar and geothermal.

Experts have estimated the efficiency of other common energy systems. The most efficient of these appear to be electric power generating plants (33% efficient) and steel plants (23% efficient). Among the least efficient systems are those for heating water (1.5–3%), for heating homes and buildings (2.5–9%), and refrigeration and air-conditioning systems (4–5%). It has been estimated that about 85 percent of the energy available in the United States is lost due to inefficiency.

The predominance of low-efficiency systems reflects the fact that such systems were invented and developed when energy costs were low and there was little customer demand for energy efficiency. It made more sense then to build appliances that were inexpensive rather than efficient, because the cost to operate them was so low. Since the 1973 oil embargo by OPEC, that philosophy has been carefully re-examined. Experts began to point out that more expensive appliances could be designed and built if they were also more efficient. The additional cost to the manufacturer, industry, and homeowner could usually be recovered within a few years because of the savings in fuel costs.

The concept of energy efficiency suggested a new way of looking at energy systems, the examination of the total lifetime energy use and cost of the system. Consider the common light bulb. The total cost of using a light bulb includes both its initial price and the cost of operating it throughout its lifetime. When energy was cheap, this second factor was small. There was little motivation to make a bulb that was more efficient when the life-cycle savings for its operation was minimal.

But as the cost of energy rises, that argument no longer holds true. An inefficient light bulb costs more and more to operate as the cost of electricity rises. Eventually, it made sense to invent and produce more efficient light bulbs. Even if these bulbs cost more to buy, they pay back that cost in long-term operating savings.

Thus, consumers might balk at spending $25 for a fluorescent light bulb unless they know that the bulb will last ten times as long as an incandescent bulb that costs $3.75. Similar arguments can and have been used to justify the higher initial cost of energy-saving refrigerators, solar-heating systems, household insulation, improved internal combustion engines, and other energy-efficient systems and appliances.

Governmental agencies, utilities, and industries are gradually beginning to appreciate the importance of increasing energy efficiency. The 1990 amendments to the Clean Air Act encourage industries and utilities to adopt more efficient equipment and procedures. Certain leaders in the energy field, such as Pacific Gas and Electric and Southern California Edison have already implemented significant energy-efficiency programs.

Resources

BOOKS

Deffeyes, Kenneth S. *Hubbert's Peak: The Impending World Oil Shortage.* Princeton, NJ: Princeton University Press, 2008.

McLean-Conner, Penni. *Energy Efficiency: Principles and Practices.* Tulsa, OK: PennWell Publishing, 2009.

OTHER

Alliance To Save Energy. "Creating an Energy-efficient World." http://www.ase.org/ (accessed November 11, 2010).

American Council for an Energy-efficient Economy. "American Council for an Energy-Efficient Economy." http://www.aceee.org/ (accessed November 11, 2010).

Energy Efficient Building Association. "Energy Efficient Building Association." http://www.eeba.org/ (accessed November 11, 2010).

David E. Newton
Richard A. Jeryan

Energy flow

Understanding energy flow is vital to many environmental issues. One can describe the way ecosystems function by saying that matter cycles and energy flows. This is based on the laws of conservation of matter and energy and the second law of thermodynamics, or the law of energy degradation.

Energy flow is strictly one way, such as from higher to lower or from hotter to colder. Objects cool only by loss of heat. All cooling units, such as refrigerators and air conditioners, are based on this principle: they are essentially heat pumps, absorbing heat in one place and expelling it to another.

This heat flow is explained by the laws of radiation, as seen in fire and the color wheel. All objects emit radiation, or heat loss, but the hotter the object the greater the amount of radiation, and the shorter and more energetic the wavelength. As energy intensities rise and wavelengths shorten, the radiation changes from infrared to red, then orange, yellow, green, blue, violet, and ultraviolet. A blue flame, for example, is desired for gas appliances. A well-developed wood fire is normally yellow, but as the fire dies out and cools, the color gradually changes to orange, then red, then black. Black coals may still be very hot, giving off invisible, infrared radiation. These varying wavelengths are the main differences seen in the electromagnetic spectrum.

All chemical reactions and radioactivity emit heat as a by-product. Because this heat radiates out from the source, the basis of the second law of thermodynamics, one can never achieve 100 percent energy efficiency. There will always be a heat-loss tax. One can slow down the rate of heat loss through insulating devices, but never stop it. As the insulators absorb heat, their temperatures rise, and they in turn lose heat.

There are three main applications of energy flow to environmental concerns. First, only 10 percent of the food passed on up the food chain/web is retained as body mass; 90 percent flows to the atmosphere as heat. In terms of caloric efficiency, more calories are obtained by eating plant food than meat. Since fats are more likely to be part of the 10 percent retained as body mass, pesticides dissolved in fat are subject to bioaccumulation and biomagnification. This explains the high levels of DDT in birds of prey like the peregrine falcon (*Falco peregrinus*) and the brown pelican (*Pelecanus occidentalis*).

Second, the percentage of waste heat is an indicator of energy efficiency. In light bulbs 5 percent produces light and 95 percent heat, just the opposite of the highly efficient firefly. Electrical generation from fossil fuels or nuclear power produces vast amounts of waste heat.

Third, control of heat flow is a key to comfortable indoor air and solving global warming. Well-insulated buildings retard heat flow, reducing energy use. Atmospheric greenhouse gases, such as anthropogenic carbon dioxide and methane, retard heat flow to space, which theoretically should cause global temperatures to rise. Policies that reduce these greenhouse gases allow a more natural flow of heat back to space.

See also Greenhouse effect.

Resources

BOOKS

Dean, Andrew, and David G. Haase. *Energy.* New York: Chelsea House Publications, 2007.

Indonesian Petroleum Association. *Managing Resources and Delivering Energy in a Challenging Environment: Proceedings of the Thirty-Third Annual Convention, Jakarta, 5-7 May 2009.* Jakarta, Indonesia: Indonesian Petroleum Association, 2009.

Kemp, William H. *Smart Power: An Urban Guide to Renewable Energy and Efficiency.* Kingston, Canada: Aztext Press, 2006.

Nathan H. Meleen

Energy Information Administration *see* **U.S. Department of Energy.**

Energy path, hard vs. soft

What will energy use patterns in the year 2100 look like? Such long-term predictions are difficult, risky, and perhaps impossible. Could an American citizen in 1860 have predicted what the pattern of 2010's energy use would be?

Dramatic changes in the ways people use energy may be in store over the next century. Most importantly, the world's supplies of non-renewable energy—especially, coal, oil, and natural gas—will continue to decrease. Critics have been warning for decades that time was running out for the fossil fuels and that society could not count on using them as prolifically as it had in the past.

For at least three decades, experts have debated the best way to structure energy use patterns in the future. The two most common themes have been described (originally by physicist Amory Lovins) as the "hard path" and the "soft path."

Proponents of the hard path argue essentially that society should continue to operate in the future as it has in the past, except more efficiently. They point out that 1960s and 1970s predictions that oil supplies would be depleted by the end of the twentieth century have been proved wrong. If anything the known reserves of fossil fuels may actually have increased as economic incentives encouraged further exploration.

The energy future, the hard-pathers say, should focus on further incentives to develop conventional energy sources, such as fossil fuels and nuclear power. Such incentives might include tax breaks and subsidies for coal, uranium, and petroleum companies. When supplies of fossil fuels become depleted, the emphasis should shift to a greater reliance on nuclear power.

An important feature of the hard energy path is the development of huge, centralized coal-fired and nuclear-powered plants for generating electricity. One characteristic of most hard energy proposals, in fact, is the emphasis on very large, expensive, centralized systems. For example, one would normally think of solar energy as a part of the soft energy path. But one proposal developed by the National Aeronautics and Space Administration (NASA) calls for a gigantic solar power station to be orbited around the earth. The station could then transmit power via microwaves to centrally located transmission stations at various points on the earth's surface.

Those who favor a soft energy path have a completely different scenario in mind. Fossil fuels and nuclear power must diminish as sources of energy as soon as possible, they say. In their place, alternative sources of power such as hydropower, geothermal energy, wind energy, and photovoltaic cells must be developed.

In addition, the soft-pathers say, people should encourage conservation to extend coal, oil, and natural gas supplies as long as possible. Also because electricity is one of the most wasteful of all forms of energy, its use should be curtailed.

Most importantly, soft-path proponents maintain energy systems of the future should be designed for small-scale use. The development of more efficient solar cells, for example, would make it possible for individual facilities to generate a significant portion of the energy they need.

Underlying the debate between hard- and soft-pathers is a fundamental question as to how society should operate. On the one hand are those who favor the control of resources in the hands of a relatively small number of large corporations. On the other hand are those who prefer to have that control decentralized to individual communities, neighborhoods, and families. The choice made between these two competing philosophies will probably determine which energy path the United States and the world will ultimately follow.

See also Alternative energy sources; Energy and the environment; Renewable energy.

Resources

BOOKS

Darley, Julian. *Standard High Noon for Natural Gas: The New Energy Crisis*. White River Junction, VT: Chelsea Green Publishing Company, 2004.

Fanchi, John R., and John R. Fanchi. *Energy in the 21st Century*. Hackensack, NJ: World Scientific, 2005.

Friedman, Lauri S., ed. *Energy Alternatives: An Opposing Viewpoints Guide*. Detroit, MI: Greenhaven Press/ Thomson Gale, 2006.

Kishore, V. V. N. *Renewable Energy Engineering and Technology: Principles and Practice*. London: Earthscan, 2009.

Nersesian, Roy L. *Energy for the 21st Century: A Comprehensive Guide to Conventional and Alternative Sources*. Armonk, New York: M.E. Sharpe, 2007.

David E. Newton

Energy policy

Energy policies are the actions governments take to affect the demand for energy as well as the supply of it. These actions include the ways in which governments cope with energy supply disruptions and their efforts to influence energy consumption and economic growth.

The energy policies of the United States government have often worked at cross purposes, both stimulating and suppressing demand. Taxes are perhaps the most important kind of energy policy, and energy taxes are much lower in the United States than in other countries. This is partially responsible for the fact that energy consumption per capita is higher than elsewhere, and there is less incentive to invest in conservation or alternative technologies. Following the 1973 Arab oil embargo, the federal government instituted price controls, which kept energy prices lower than they would otherwise have been, thereby stimulating consumption. Yet the government also instituted policies at the same time, such as fuel-economy standards for automobiles, which were designed to increase conservation and lower energy use. Thus, policies in the period after the embargo were contradictory: what one set of policies encouraged, the other discouraged.

The United States government has a long history of different types of interference in energy markets. The Natural Gas Act of 1938 gave the Federal Power Commission the right to control prices and limit new pipelines from entering the market. In 1954 the Supreme Court extended price controls to field production. Before 1970 the Texas Railroad Commission effectively controlled oil output in the United States through prorationing regulations that provided multiple owners with the rights to underground pools. The federal government provided tax breaks in the form of intangible drilling expenses and gave the oil companies a depletion allowance. A program was also in place from 1959 to 1973 that limited oil imports and protected domestic producers from cheap foreign oil. The ostensible purpose of this policy was maintaining national security, but it contributed to the depletion of national reserves.

After the oil embargo, Congress passed the Emergency Petroleum Allocation Act giving the federal

U.S. President Barack Obama meets with a bipartisan group of governors from across the country in the State Dining Room to discuss energy policy in Washington. *(RON SACHS/UPI /Landov)*

government the right to allocate fuel in a time of shortage. In 1974 President Gerald Ford announced Project Independence, which was designed to eliminate dependence on foreign imports. Congress passed the Federal Non-Nuclear Research and Development Act in 1974 to focus government efforts on non-nuclear research. Finally, in 1977 Congress approved the cabinet-level creation of the U.S. Department of Energy (DOE), which had a series of direct and indirect policy approaches at its disposal, designed to encourage and coerce both the energy industry as well as the commercial and residential sectors of the country to make changes. After Ronald Reagan became president, many DOE programs were abolished, though DOE continued to exist, and the net impact has probably been to increase economic uncertainty.

In 2009 President Barack Obama declared "a new era of energy exploration." Obama's energy plan called for massive government investment in clean, renewable, and sustainable energy initiatives. Obama's energy plan also called for greater energy efficiency. Between 1990 and 2009, the U.S. fuel economy standard for cars remained 27.5 miles per gallon (mpg) (11.7 km/l). In 2009 the standard for light trucks was 23.0 mpg (8.6 km/l). Under the fuel efficiency standards announced by Obama in 2009, by 2016 cars and light trucks had to achieve approximately 39 mpg (16.6 km/l) and 30 mpg (12.8 km/l), respectively.

Energy policy issues have always been very political in nature. Different segments of the energy industry have often been differently affected by policy changes, and various groups have long proposed divergent solutions. The energy crisis, however, intensified these conflicts. Advocates of strong government action called for policies which would alter consumption habits, reducing dependence on foreign oil and the nation's vulnerability to an oil embargo. They have been opposed by proponents of free markets, some of whom considered the government itself responsible for the crisis. Few issues were subject to such intensive scrutiny and fundamental conflicts over values as energy policies were during this period. Interest groups representing causes from energy conservation to nuclear power mobilized. Business interests also expanded their lobbying efforts.

An influential advocate of the period was Amory Lovins, who helped create the renewable energy movement. His book, *Soft Energy Paths: Toward A Durable Peace* (1977), argued that energy problems existed because large corporations and government bureaucracies had imposed expensive centralized technologies such as nuclear power on society. Lovins argued that the solution was in small scale, dispersed, technologies. He believed that the "hard path" imposed by corporations

and the government led to an authoritarian, militaristic society while the "soft path" of small-scale dispersed technologies would result in a diverse, peaceful, self-reliant society.

Because coal was so abundant, many in the 1970s considered it a solution to American dependence on foreign oil, but this expectation has proved to be mistaken. During the 1960s the industry had been controlled by an alliance between management and the union, but this alliance disintegrated by the time of the energy crisis, and wildcat strikes hurt productivity. Productivity also declined because of the need to address safety problems following passage of the 1969 Coal Mine Health and Safety Act. Environmental issues also hurt the industry following passage of the National Environmental Policy Act of 1969, the Clean Air Act of 1970, the Clean Water Act of 1972, and the 1977 Surface Mining Control and Reclamation Act. Worker productivity in the mines dropped sharply from 19 tons per worker day to 14 tons, and this decreased the advantage coal had over other fuels. The 1974 Energy Supply and Environmental Coordination Act and the 1978 Fuel Use Act, which required utilities to switch to coal, had little effect on how coal was used because so few new plants were being built.

Other energy-consuming nations responded to the energy crises of 1973–1974 and 1979–1980 with policies that were different from the United States. Japan and France, although via different routes, made substantial progress in decreasing their dependence on Mideast oil. Great Britain was the only major industrialized nation to become completely self-sufficient in energy production, but this fact did not greatly aid its ailing economy. When energy prices declined and then stabilized in the 1980s, many consuming nations eliminated the conservation incentives they had put in place.

Japan is one of the most heavily petroleum-dependent industrialized nations. To pay for a high level of energy and raw material imports, Japan must export the goods that it produces. When energy prices increased after 1973, it was forced to expand exports. The rate of economic growth in Japan began to decline. Annual growth in GNP averaged nearly 10 percent from 1963–1973, but from 1973–1983 it was just under 4 percent, although the association between economic growth and energy consumption has weakened.

The Energy Rationalization Law of 1979 was the basis for Japan's energy conservation efforts, providing for the financing of conservation projects and a system of tax incentives. It has been estimated that more than 5 percent of total Japanese national investment in 1980 was for energy-saving equipment. In the cement, steel, and chemical industries more than 60 percent of total investment was for energy conservation, Japanese society shifted from petroleum to a reliance on other forms of energy including nuclear power and liquefied natural gas.

In France, energy resources at the time of the oil embargo were extremely limited. France possessed some natural gas, coal, and hydropower, but together these sources constituted only 0.7 percent of the world's total energy production. By 1973 French dependence on foreign energy had grown to 76.2 percent: oil made up 67 percent of the total energy used in France, up from 25 percent in 1960.

France had long been aware of its dependence on foreign energy and had taken steps to overcome it. Political instability in the Mideast and North Africa had led the government to take a leading role in the development of civilian nuclear power after World War II. In 1945 Charles de Gaulle set up the French Atomic Energy Commission to develop military and peaceful uses for nuclear power. The nuclear program proceeded at a very slow pace until the 1973 embargo, after which there was rapid growth in France's reliance on nuclear power. By 1990 more than fifty reactors had been constructed and more than 70 percent of France's energy came from nuclear power. France now exports electricity to nearly all its neighbors, and its rates are about the lowest in Europe. Starting in 1976 the French government also subsidized 3,100 conservation projects at a cost of more than 8.4 billion francs, and these subsidies were particularly effective in encouraging energy conservation.

Concerned about oil supplies during World War I, the British government had taken a majority interest in British Petroleum and tried to play a leading role in the search for new oil. After the World War II, the government nationalized the coal, gas, and electricity industries, creating, for ideological reasons as well as for postwar reconstruction, the National Coal Board, British Gas Corporation, and Central Electricity Generating Board. After the discovery of oil reserves in the 1970s in the North Sea, the government established the British National Oil Company. This government corporation produced about 7 percent of North Sea oil and ultimately handled about 60 percent of the oil produced there.

All the energy sectors in the United Kingdom were thus either partially or completely nationalized. Government relations with the nationalized industries often were difficult, because the two sides had different interests. The government intervened to pursue macroeconomic objectives such as price restraint, and it attempted to stimulate investment at times of unemployment. The electric and gas industries had substantial operating profits and they could finance their capital requirements from their revenues, but profits in the coal industry were poor, the work force was unionized, and opposition to the closure of uneconomic mines was great. Decision-making was highly politicized in this nationalized industry, and the government had difficulty addressing the problems there. It was estimated that 90 percent of mining losses came from thirty of the 190 pits in Great Britain, but only since 1984 to 1985 was there rapid mine closure and enhanced productivity. New power-plant construction was also poorly managed, and comparable coal-fired power stations cost twice as much in Great Britain as in France or Italy.

The Conservative Party proposed that the nationalized energy industries be privatized. However, with the exception of coal, these energy industries had natural monopoly characteristics: economies of scale and the need to prevent duplicate investment in fixed infrastructure. The Conservative Party called for regulation after privatization to deal with the natural monopoly characteristics of these industries, and it took many steps toward privatization. In only one area, however, did it carry its program to completion, abolishing the British National Oil Company and transferring its assets to private companies.

See also Alternative energy sources; Corporate Average Fuel Economy Standards; Economic growth and the environment; Electric utilities; Energy and the environment; Energy efficiency; Energy path, hard vs. soft; Lovins, Amory Bloch.

Resources

BOOKS

U.S. Congress. "Energy Policy Act of 2005." Sec. 1342, Sec. 30B. Washington, DC: U.S. Congress, 2005.

PERIODICALS

Aune, Margrethe. "Energy Comes Home." *Energy Policy* 35 (2007): 5475-5465.

Dias, Rubens A., et al. "The Limits of Human Development and the Use of Energy and Natural Resources." *Energy Policy* 34 (2006): 1026–1031.

Quadrelli, Roberta. "The Energy-Climate Challenge: Recent Trends in CO_2 Emissions from Fuel Combustion." *Energy Policy* 35 (2007): 5938–2952.

OTHER

Renewable Energy Policy Project. "Renewable Energy Policy Homepage." http://www.repp.org/ (accessed November 8, 2010).

U.S. Government; science.gov. "Energy Policy and Studies." http://www.science.gov/browse/w_121D.htm (accessed November 8, 2010).

Alfred A. Marcus

Energy recovery

A fundamental fact about energy use in modern society is that huge quantities are lost or wasted in almost every field and application. For example, the series of processes by which nuclear energy is used to heat a home with electricity results in a loss of about 85 percent of all the energy originally stored in the uranium used in the nuclear reactor. Industry, utilities, and individuals could use energy far more efficiently if they could find ways to recover and reuse the energy that is being lost or wasted.

One such approach is cogeneration, the use of waste heat for some useful purpose. For example, a factory might be redesigned so that the steam from its operations could be used to run a turbine and generate electricity. The electricity could then be used elsewhere in the factory or sold to power companies. Cogeneration in industry can result in savings of between 10 and 40 percent of energy that would otherwise be wasted.

Cogeneration can work in the opposite direction also. Hot water produced in a utility plant can be sold to industries that can use it for various processes. Proposals have been made to use the wasted heat from electricity plants to grow flowers and vegetables in greenhouses, to heat water for commercial fish and shellfish farms, and to maintain warehouses at constant temperatures. The total energy efficiency resulting from this sharing is much greater than it would be if the utility's water was simply discarded.

Another possible method of recovering energy is by generating or capturing natural gas from biomass. For example, as organic materials decay naturally in a landfill, one of the products released is methane, the primary component of natural gas. Collecting methane from a landfill is a relatively simple procedure. Vertical holes are drilled into the landfill and porous pipes are sunk into the holes. Methane diffuses into the pipes and is drawn off by pumps. The recovery system at the Fresh Kills landfill on Staten Island, New York, for example, produces enough methane to heat 10,000 homes.

Biomass can also be treated in a variety of ways to produce methane and other combustible materials.

Sewage, for example, can be subjected to anaerobic digestion, the primary product of which is methane. Pyrolysis is a process in which organic wastes are heated to high temperatures in the absence of oxygen. The products of this reaction are solid, liquid, and gaseous hydrocarbons whose composition is similar to those of petroleum and natural gas. Perhaps the most known example of this approach is the manufacture of methanol from biomass. When mixed with gasoline, a new fuel, gasohol, is obtained.

Energy can also be recovered from biomass simply by combustion. The waste materials left after sugar is extracted from sugar cane, known as *bagasse*, have long been used as a fuel for the boilers in which the sugar extraction occurs. The burning of garbage has also been used as an energy source in a wide variety of applications such as the heating of homes in Sweden, the generation of electricity to run streetcars and subways in Milan, Italy, and the operation of a desalination plant in Hempstead, Long Island.

The recovery of energy that would otherwise be lost or wasted has a secondary benefit. In many cases, that wasted energy might cause pollution of the environment. For example, the wasted heat from an electric power plant may result in thermal pollution of a nearby waterway. Or the escape of methane into the atmosphere from a landfill could contribute to air pollution. Capture and recovery of the waste energy not only increases the efficiency with which energy is used, but may also reduce some pollution problems.

Resources

BOOKS

Barbaro, Pierluigi, and Claudio Bianchini, eds. *Catalysis for Sustainable Energy Production*. Wiley-VCH, 2009.

Boyle, Godfrey. *Renewable Energy*. New York: Oxford University Press USA, 2004.

Brewer, Dennis C. *Green My Home!: 10 Steps to Lowering Energy Costs and Reducing Your Carbon Footprint*. New York, NY: Kaplan Pub, 2009.

Dean, Andrew, and David G. Haase. *Energy*. New York: Chelsea House Publications, 2007.

Fridell, Ron. *Earth-Friendly Energy*. Saving our living earth. Minneapolis, MN: Lerner Publications, 2009.

Friedman, Lauri S., ed. *Energy Alternatives: An Opposing Viewpoints Guide*. Detroit, MI: Greenhaven Press/ Thomson Gale, 2006.

Gibilisco, Stan. *Alternative Energy Demystified*. New York: McGraw-Hill, 2007.

Kemp, WIlliam H. *Smart Power: An Urban Guide to Renewable Energy and Efficiency* Kingston, Canada: Aztext Press, 2006.

hostile force to be tamed and civilized. Society uses materials to make the things they want, then discard them when they no longer are useful. "Dilution is the solution to pollution" suggests that if people just spread their wastes out in the environment widely enough, no one will notice.

This approach has given the world an abundance of material things, but also has produced massive pollution and environmental degradation. It also is incredibly wasteful. On average, for every truckload of products delivered in the United States, thirty-two truckloads of waste are produced along the way. The automobile is a typical example. Industrial ecologist, Lovins, calculates that for every one hundred gallons (380 liters) of gasoline burned in your car engine, only 1 percent (0 gallons or 3.8 liters) actually moves the passengers inside. All the rest is used to move the vehicle itself. The wastes produced—carbon dioxide (CO_2), nitrogen oxides (NO_x), unburned hydrocarbon, rubber dust, and heat—are spread through the environment where they pollute air, water, and soil. And when the vehicle wears out after only a few years of service, thousands of pounds of metal, rubber, plastic, and glass become part of our rapidly growing waste stream.

This is not the way things work in nature, environmental designers point out. In living systems almost nothing is discarded or unused. The wastes from one organism become the food of another. Industrial processes, to be sustainable over the long term, should be designed on similar principles, designers argue. Rather than following current linear patterns in which manufacturers try to maximize the throughput of materials and minimize labor, products and processes should be designed to be energy efficient and use renewable materials. Industrial processes should create products that are durable and reusable or easily dismantled for repair and remanufacture, and are non-polluting throughout their entire life cycle. Countries should base their economies on renewable solar energy rather than fossil fuels. Rather than measuring conomies progress by how much material is used, they should evaluate productivity by how many people are gainfully and meaningfully employed. They should judge how well they are doing by how many factories have no smokestacks or dangerous effluents. They ought to produce nothing that will require constant vigilance from future generations.

Inspired by how ecological systems work, McDonough proposes three simple principles for designing processes and products:

- Waste equals food. This principle encourages elimination of the concept of waste in industrial design. Every process should be designed so that the products themselves, as well as leftover chemicals, materials, and effluents, can become food for other processes.
- Rely on current solar income. This principle has two benefits: First, it diminishes, and may eventually eliminate, reliance on hydrocarbon fuels. Second, it means designing systems that sip energy rather than gulping it down.
- Respect diversity. Evaluate every design for its impact on plant, animal, and human life, examining the effects of products and processes on identity, independence, and integrity of humans and natural systems. Every project should respect the regional, cultural, and material uniqueness of its particular place.

According to McDonough, the first question about a product should be whether it is really needed. Would it be possible to obtain the same satisfaction, comfort, or utility in another way that has less environmental and social impact? For the well-being of the environment, the things companies design should be restorative and regenerative: they should help reduce the damage done by earlier, wasteful approaches and help nature heal rather than simply adding to existing problems. McDonough invites society to reinvent businesses and institutions to work with nature, and redefine people as consumers, producers, and citizens to promote a new sustainable relationship with the earth. In an eco-efficient economy, McDonough says, products might be divided into three categories:

- Consumables are products such as food, natural fabrics, or paper that are produced from renewable materials and can go back to the soil as compost.
- Service products are durables such as cars, televisions, and refrigerators. These products should be leased to the customer to provide their intended service, but would always belong to the manufacturer. Eventually, they would be returned to the maker, who would be responsible for recycling or remanufacturing.
- Unmarketables are materials such as radioactive isotopes, persistent toxins, and bioaccumulating chemicals. Ideally, no one would make or use these products. But because eliminating their use will take time, McDonough suggests that for now, these materials should belong to the manufacturer and be molecularly tagged with the maker's mark. If they are discarded illegally, the manufacturer would be liable.

Following these principles McDonough Braungart Design Chemistry (MBDC) has created nontoxic, easily recyclable, healthy materials for buildings and for consumer goods. Rather than designing products

Solar panels on houses in a housing development in Richmond, California. *(Proehl Studios/Corbis)*

for a cradle-to-grave life cycle, MBDC aims for a fundamental conceptual shift to Cradle to Cradle® processes, whose materials perpetually circulate in closed systems that create value and are inherently healthy and safe. Among some important examples are carpets designed to be recycled at the end of their useful life, paints and adhesives that are non-toxic and non-allergenic, and clothing that is both healthy for the wearer and that has minimal environmental impact in its production.

Braungart founded the Environmental Protection and Encouragement Agency (EPEA) in 1987 to promote the design of products, processes, and services based on the application of the Cradle to Cradle principles. The EPEA is "an international scientific research and consultancy institute that improves product quality, utility and environmental performance via eco-effectiveness." The design and principles of the EPEA are focused on continual utility of materials in either a recycled or upcycled manner exceeding environmental standards. In addition to the design of products, the EPEA is also involved with policy-making, planning, and administrative processes related to environmental design.

In his architecture firm, McDonough + Partners, these new design models and environmentally friendly materials have been used in a number of innovative building projects. A few notable examples include The Gap, Inc. offices in California and the Environmental Studies building at Oberlin College in Ohio.

Built in 1994 The Gap building in San Bruno, California, is designed to maintain the unique natural features of the site, while providing comfortable, healthy, and flexible office spaces. Intended to promote employee well-being and productivity as well as eco-efficiency, The Gap building has high ceilings, open, airy spaces, a natural ventilation system including operable windows, a full-service fitness center (including a pool), and a landscaped atrium for each office bay that brings the outside in. Skylights in the roof deliver daylight to interior offices and vent warm, stale air. Warm interior tones and natural woods (all wood used in the building was harvested by certified sustainable methods) give a friendly feel. Paints, nature. Nature's limited capacities to absorb wastes also set a limit on the economy's ability to produce. Energy plays a role in this process. Energy inputs make food, forest products, chemicals, petroleum products, metals, and structural materials such as stone, steel, and cement. Energy also supports materials processing by providing electricity, heating, and cooling services, and energy aids in transportation and distribution. According to the law of the conservation of energy, the material inputs and energy that enter the economy cannot be destroyed. Rather they change form, finding their way back to nature in a disorganized state as unwanted and perhaps dangerous by-products.

Environmentalists use the laws of physics (the notion of entropy) to show how society systematically dissipates low entropy, highly concentrated forms of energy, by converting it to high entropy, little concentrated waste that cannot be used again except at very high cost. They project current resource use and environmental degradation into the future to demonstrate that civilization is running out of critical resources. The earth cannot tolerate additional contaminants. Human intervention in the form of technological innovation and capital investment complemented by substantial human ingenuity and creativity is insufficient to prevent this outcome unless drastic steps are taken soon. Nearly every economic benefit has an environmental cost, and the sum total of the costs in an affluent society often exceed the benefits. The notion of carrying capacity is used to show that the earth has a limited ability to tolerate the disposal of contaminants and the depletion of resources.

Economists counter these claims by arguing that limits to growth can be overcome by human ingenuity, that benefits afforded by environmental protection have a cost, and that government programs to clean up the environment are as likely to fail as the market forces that produce pollution. The traditional economic view is that production is a function of labor and capital and, in theory, that resources are not necessary because labor and/or capital are infinitely substitutable for resources. Impending resource scarcity results in price increases that lead to technological substitution of capital, labor, or other resources for those that are in scarce supply. Price increases also create pressures for efficiency-in-use, leading to reduced consumption. Thus, resource scarcity is reflected in the price of a given commodity. As resources become scarce, their prices rise accordingly. Increases in price induce substitution and technological innovation.

People turn to less scarce resources that fulfill the same basic technological and economic needs provided by the resources no longer available in large quantities. To a large extent, the energy crises of the 1970s (the 1973 price shock induced by the Arab oil embargo and 1979 price shock following the Iranian Revolution) were alleviated by these very processes: higher prices leading to the discovery of additional supply and to conservation. By 1985 energy prices in real terms were lower than they were in 1973.

Humans respond to signals about scarcity and degradation. Extrapolating past consumption patterns into the future without considering the human response is likely to be a futile exercise, economists argue. As far back as the end of the eighteenth century, thinkers such as Thomas Malthus have made predictions about the limits to growth, but the lesson of modern history is one of technological innovation and substitution in response to price and other societal signals, not one of calamity brought about by resource exhaustion. In general the prices of natural resources have been declining despite increased production and demand. Prices have fallen because of discoveries of new resources and because of innovations in the extraction and refinement process.

See also Greenhouse effect; Trade in pollution permits; Tragedy of the commons.

Resources

BOOKS

Babe, Robert E. *Culture of Ecology: Reconciling Economics and Environment*. Toronto, Canada: University of Toronto Press, 2006.
Green Economics Conference, and Miriam Kennett. *Proceedings of the Green Economics Conference 2008: Civilisation, the First 10,000 Years, an Audit, Economics in an Age of Uncertainty and Instability*. Reading: Green Economics Institute, 2008.
Tietenberg, Thomas, H. *Environmental and Natural Resource Economics*. Boston, MA: Pearson/Addison Wesley, 2006.

Alfred A. Marcus

Environmental education

Environmental education is fast emerging as one of the most important disciplines in the United States and worldwide. Merging the ideas and philosophy of environmentalism with the structure of formal education systems, it strives to increase awareness of environmental problems as well as to foster the skills and strategies for solving those problems. Environmental issues have traditionally fallen to the state, federal, and international policymakers, scientists, academics, and legal scholars. Environmental education (often referred to simply as EE) shifts the focus to the general population. In other words, it seeks to empower individuals with an understanding of environmental problems and the skills to solve them.

Background

The first seeds of environmental education were planted roughly a century ago and are found in the works of such writers as American environmentalist George Perkins Marsh (1801–1882), Scottish-born American naturalist John Muir (1838–1914), American naturalist Henry David Thoreau (1817–1862), and American ecologist Aldo Leopold (1887–1948). Their writings served to bring the country's attention to the depletion of natural resources and the often detrimental impact of humans on the environment. In the early 1900s three related fields of study arose that eventually merged to form the present-day environmental education.

Nature education expanded the teaching of biology, botany, and other natural sciences out into the natural world, where students learned through direct observation. Conservation education took root in the 1930s, as the importance of long-range, wise-use management of resources intensified. Numerous state and federal agencies were created to tend public lands, and citizen organizations began forming in earnest to protect a favored animal, park, river, or other resource. Both governmental and citizen entities included an educational component to spread their message to the general public. Many states required their schools to adopt conservation education as part of their curriculum. Teacher training programs were developed to meet the increasing demand. The Conservation Education Association formed to consolidate these efforts and help solidify citizen support for natural resource management goals. The third pillar of modern EE is outdoor education, which refers more to the method of teaching than to the subject taught. The idea is to hold classrooms outdoors; the topics are not restricted to environmental issues but include art, music, and other subjects.

With the burgeoning of industrial output and natural resource depletion following World War II, people began to glimpse the potential environmental disasters looming ahead. The environmental movement exploded upon the public agenda in the late 1960s and early 1970s, and the public reacted emotionally and vigorously to isolated environmental crises and events. Yet it soon became clear that the solution would involve nothing short of fundamental changes in values, lifestyles, and individual behavior, and that meant a comprehensive educational approach.

In August 1970 the newly-created Council on Environmental Quality called for a thorough discussion of the role of education with respect to the environment. Two months later Congress passed the Environmental Education Act, which called for EE programs to be incorporated in all public school curricula. Although the act received little funding in the following years, it energized EE proponents and prompted many states to adopt EE plans for their schools. In 1971 the National Association for Environmental Education formed, as did myriad of state and regional groups.

Definition

What EE means depends on one's perspective. Some see it as a teaching method or philosophy to be applied to all subjects, woven into the teaching of political science, history, economics, and so forth. Others see it as a distinct discipline, something to be taught on its own. As defined by federal statute, it is the "education process dealing with people's relationships and their natural and manmade surroundings, and includes the relation of population, pollution, resource allocation and depletion, conservation, transportation, technology, and urban and rural planning to the total human environment."

One of the early leaders of the movement is American environmentalist William Stapp (1930–2001), a former professor at the University of Michigan's School of Natural Resources and the Environment. Stapp's three-pronged definition has formed the basis for much subsequent thought: "Environmental education is aimed at producing a citizenry that is *knowledgeable* concerning the biophysical environment and its associated problems, *aware* of how to help solve these problems, and *motivated* to work toward their solution."

Many environmental educators believe that programs covering kindergarten through twelfth grade are necessary to successfully instill an environmental ethic in students and a comprehensive understanding of environmental issues so that they are prepared to deal with environmental problems in the real world. Further, an emphasis is placed on problem-solving, action, and informed behavioral changes. In its broadest sense, EE is not confined to public schools but includes efforts by governments, interest groups, universities, and news media to raise awareness. Each citizen should understand the environmental issues of his or her own community: land-use planning, traffic congestion, economic development plans, pesticide use, water pollution, and air pollution, and so on.

International level

Concurrently with the emergence of EE in the United States, other nations began pushing for a comprehensive approach to environmental problems within their own borders and on a global scale. In 1972, at the United Nations Educational, Scientific, and Cultural Organization (UNESCO) and United Nations Environment Program (UNEP) Conference on the Human Environment in Stockholm, the need for an international EE effort was clearly recognized and emphasized. Three years later an International Environmental Education Workshop was held in Belgrade, from which emerged an eloquent, urgent mandate for the drastic reordering of national and international development policies. The Belgrade Charter called for an end to the military arms race and a new global ethic in which "no nation should grow or develop at the expense of another nation." It called for the eradication of poverty, hunger, illiteracy, pollution, exploitation, and domination. Central to this impassioned plea for a better world was the need for environmental education of the world's youth. That same year, the UN approved a $2 million budget to facilitate the research, coordination, and development of an international EE program among dozens of nations.

In 1977 the UNESCO/UNEP Conference on Environmental Education established twelve guiding principles of environmental education in the Tbilisi Declaration. These guidelines encouraged hands-on learning focusing on sustainable development, establishing an appreciation for and valuing the environment, problem-solving, decision-making, diversified subjects, and cooperation on the local, national, and international level to improve the environment. The 1992 United Nations Conference on Environment and Development promoted the direction of education toward Education for Sustainable Development (ESD) in Agenda 21, calling for the integration of environment and development topics into all educational disciplines. Both the 2002 Johannesburg World Summit for Sustainable Development (WSSD) and the International Consultation on Education for Sustainable Development in Sweden in 2004 discussed the issue that school curriculums were not incorporating ESD. The United Nations declared 2005–2014 the UN Decade of Education for Sustainable Development with UNESCO as the head organization.

Effectiveness

There has been criticism over the last fifteen years that EE too often fails to educate students and makes little difference in their behavior concerning the environment. Researchers and environmental educators have formulated a basic framework for how to improve EE: (1) Reinforce individuals for positive environmental behavior over an extended period of time. (2) Provide students with positive, informal experiences outdoors to enhance their environmental sensitivity. (3) Focus instruction on the concepts of ownership and empowerment. The first concept means that the learner has some personal interest or investment in the environmental issues being discussed. Perhaps the student can relate more readily to concepts of solid waste disposal if there is a landfill in the neighborhood. Empowerment gives learners the sense that they can make changes and help resolve environmental problems. (4) Design an exercise in which students thoroughly investigate an environmental issue and then develop a plan for citizen action to address the issue, complete with an analysis of the social, cultural, and ecological consequences of the action.

Despite the efforts of environmental educators, the movement has a long way to go. The scope and number of critical environmental problems facing the world today far outweigh the successes of EE. Further, most countries still do not have a comprehensive EE program that prepares their students, as future citizens, to make ecologically sound choices and to participate in cleaning up and caring for the environment. Lastly, educators, including the media, are largely focused on explaining the problems but fall short on explaining or offering possible solutions. The notion of empowerment is often absent.

Recent developments and successes in the United States

Project WILD, based in Boulder, Colorado, is a K–12 supplementary conservation and environmental education program emphasizing wildlife protection, sponsored by fish and wildlife agencies and environmental educators. The project sets up workshops in which teachers learn about wildlife issues. These teachers in turn teach children and help students understand how they can act responsibly on behalf of wildlife and the environment. The program, begun in 1983, has grown tremendously in terms of the number of educators reached and the monetary support from states, which, combined, are spending about $3.6 million annually.

The Global Rivers Environmental Education Network (GREEN), begun at the University of Michigan under the guidance of William Stapp, has likewise been enormously successful. Teachers worldwide take their students down to their local river and show them how to monitor water quality, analyze watershed usage, and identify socioeconomic sources of river degradation. Lastly, and most importantly, the students then present their findings and recommendations to the local officials. These students also exchange information with other GREEN students around the world via computers.

Another promising development is the National Consortium for Environmental Education and Training (NCEET), also based at the University of Michigan. The partnership of academic institutions, non-profit organizations, and corporations, NCEET was established in 1992 with a three-year, $4.8 million grant from the Environmental Protection Agency (EPA). Its main purpose is to dramatically improve the effectiveness of environmental education in the United States. The program has attacked its mission from several angles: to function as a national clearinghouse for K–12 teachers, to make available top-quality EE materials for teachers, to conduct research on effective approaches to EE, to survey and assess the EE needs of all fifty states, to establish a computer network for teachers needing access to information and resources, and to develop a teacher training manual for conducting EE workshops around the country.

See also Marsh, George Perkins.

Resources

BOOKS

Chiranjeev, Avinash, and Anil K. Jamwal. *Environmental Education and Management.* New Delhi: Jnanada Prakashan, 2008.

Conkin, Paul Keith. *The State of the Earth: Environmental Challenges on the Road to 2100.* Lexington: University Press of Kentucky, 2007.

Hill, Jennifer, Alan Terry, and Wendy Woodland. *Sustainable Development: National Aspirations, Local Implementation.* Aldershot, England: Ashgate, 2006.

Johnson, E. A., and Michael Mappin. *Environmental Education and Advocacy: Changing Perspectives of Ecology and Education.* Cambridge: Cambridge University Press, 2005.

Tomar, Archana. *Environmental Education.* Delhi: Kalpaz Publications, 2007.

Cathryn McCue

Environmental enforcement

Environmental enforcement is the set of actions that a government takes to achieve full implementation of environmental requirements (compliance) within the regulated community and to correct or halt situations or activities that endanger the environment or public health. Experience with environmental programs has shown that enforcement is essential to compliance because many people and institutions will not comply with a law unless there are clear consequences for noncompliance. Enforcement by the government usually includes inspections to determine the compliance status of the regulated community and to detect violations; negotiations with individuals or facility managers who are out of compliance to develop mutually agreeable schedules and approaches for achievement of compliance; legal action when necessary to compel compliance and to impose some consequences for violation of the law or for posing a threat to public health and the environment; and compliance promotion, such as educational programs, technical assistance, and subsidies, to encourage voluntary compliance.

Nongovernmental organizations (NGOs) may become involved in enforcement by detecting noncompliance, negotiating with violators, and commenting on governmental enforcement actions. They may also, if the law allows, take legal actions against a violator for noncompliance or against the government for not enforcing environmental requirements. The banking and insurance industries may be indirectly involved with enforcement by requiring assurance of compliance with environmental requirements before they issue a loan or an insurance policy to a facility. Strong social sanctions for noncompliance with environmental requirements can also be effective to ensure compliance. For example, the public may choose to boycott a product if they believe the manufacturer is harming the environment.

Environmental enforcement is based on environmental laws. An environmental law provides the vision, scope, and authority for environmental protection and restoration. Some environmental laws contain requirements while others specify a structure and criteria for establishing requirements, which are then developed separately. Requirements may be general, in which they apply to a group of facilities, or facility-specific.

Examples of environmental enforcement programs include those that govern the ambient environment, performance, technology, work practices, dissemination of information, and product or use bans.

Ambient standards (media quality standard) are goals for the quality of the ambient environment (that is, air and water quality). Ambient standards are usually written in units of concentration, and they are used to plan the levels of emissions that can be accommodated from individual sources while still meeting an area-wide goal. Ambient standards can also be used as triggers (i.e., when a standard is exceeded, monitoring or enforcement efforts are increased). Enforcement of these standards involves relating an ambient measurement to emissions or activities at a specific facility, which can be difficult.

Performance standards, widely used for regulations, permits, and monitoring requirements, limit the amount or rate of particular chemicals or discharges that a facility can release into the environment in a given period of time. These standards allow sources to choose which technologies they will use to meet the standards. Performance standards are often based on output that can be achieved by using the best available control technology. Some standards allow a source with multiple emissions to vary its emissions from each stack as long as the total sum of emissions does not exceed the permitted total. Compliance with emission standards is accomplished by sampling and monitoring, which in some cases may be difficult and/or expensive.

Technology standards require the regulated community to use a particular type of technology (i.e., "best available technology") to control and/or monitor emissions. Technology standards are effective if the equipment specified is known to perform well under the range of conditions experienced by the source. Compliance is measured by whether the equipment is installed and operating properly. However, proper operation over a long period of time is more difficult to monitor. The use of technology standards can inhibit technological innovation.

Practice standards require or prohibit work activities that may have environmental impacts (e.g., prohibition of carrying hazardous liquids in uncovered containers). Regulators can easily inspect for compliance and take action against noncomplying sources, but ongoing compliance is not easy to ensure.

Dissemination of information and product or use bans are also governed by environmental enforcement programs. Information standards require a source of potential pollution (e.g., a manufacturer or facility involved in generating, transporting, storing, treating, and disposing of hazardous wastes) to develop and submit information to the government. For example, a source generating pollution may be required to monitor, maintain records, and report on the level of pollution generated and whether or not the source exceeds performance standards. Information requirements are also used when a potential pollution source is a product such as a new chemical or pesticide. The manufacturer may be required to test and report on the potential of the product to cause harm if released into the environment. Finally, product or use bans are used to prohibit a product (i.e., ban the manufacture, sale, and/or use of a product), or they may prohibit particular uses of a product.

An effective environmental law should include the authority or power necessary for its own enforcement. An effective authority should govern implementation of environmental requirements, inspection, and monitoring of facilities, and legal sanctions for noncompliance. One type of authority that is used is guidance for the implementation of environmental laws by issuance of regulations, permits, licenses, and/or guidance policies. Regulations establish, in greater detail than is specified by law, general requirements that must be met by the regulated community. Some regulations are directly enforced, whereas others provide criteria and procedures for developing facility-specific requirements utilizing permits and licenses to provide the basis of enforcement. Permits are used to control activities related to construction or operation of facilities that generate pollutants. Requirements in permits are based on specific criteria established in laws, regulations, and/or guidance. General permits specify what a class of facilities is required to do, whereas a facility-specific permit specifies requirements for a particular facility, often taking into account the conditions there. Licenses are permits to manufacture, test, sell, or distribute a product that may pose an environmental or public health risk if improperly used. Licenses may be general or facility specific. Written guidance and

A sign enforcing a local noise ordinance. (© rabh images / Alamy)

policies, which are prepared by the regulator, are used to interpret and implement requirements to ensure consistency and fairness. Guidance may be necessary because not all applications of requirements can be anticipated, when regulation is achieved by the use of facility-specific permits or licenses.

Authority is also required to provide for inspection and monitoring of facilities, with legal sanctions for noncompliance. Requirements may either be waived or prepared for facility-specific conditions. The authority will inspect regulated facilities and gain access to their records and equipment to determine if they are in compliance.

Authority is necessary to ensure that the regulated community monitors its own compliance, maintains records of its compliance activities and status, reports this information periodically to the enforcement program, and provides information during inspections.

An effective law should also include the authority to take legal action against non-complying facilities, imposing a range of monetary penalties and other sanctions on facilities that violate the law, as well as criminal sanctions on those facilities or individuals who deliberately violate the law (e.g., facilities that knowingly falsify data). Also, power should be granted to correct situations that pose an immediate and substantial threat to public health and/or the environment.

The range and types of environmental enforcement response mechanisms available depend on the number and types of authorities provided to the enforcement program by environmental and related laws. Enforcement mechanisms may be designed to return violators to compliance, impose a sanction, or remove the economic benefits of noncompliance. Enforcement may require that specific actions be taken to test, monitor, or provide information. Enforcement may also correct environmental damages and modify internal company management problems.

Enforcement response mechanisms include informal responses, such as phone calls, site visits and inspections, warning letters, and notices of violations, which are more formal than warning letters. They provide the facility manager with a description of the violation, what should be done to correct it, and by what date. Informal responses do not penalize but can lead to more severe responses if ignored. The more formal enforcement mechanisms are backed by law and are accompanied by procedural requirements to protect the rights of the individual. Authority to use formal enforcement mechanisms for a specific situation must be provided in the applicable environmental law. Civil administrative orders are legal, independently enforceable orders issued directly by enforcement program officials that define the violation, provide evidence of the violation, and require the recipient to correct the violation within a specified time period. If the recipient violates the order, program managers can take further legal action using additional orders or the court system to force compliance.

Further legal action includes the use of field citations, which are administrative orders issued by inspectors in the field. They require the violator to correct a clear-cut violation and pay a small monetary fine. Field citations are used to handle more routine types of violations that do not pose a major threat to the environment. Legal action may also lead to civil judicial enforcement actions, which are formal lawsuits before the courts. These actions are used to require action to reduce immediate threats to public health or the environment, to enforce administrative orders that have been violated, and to make final decisions regarding orders that have been appealed. Finally, a criminal judicial response is used when a person or facility has knowingly and willfully violated the law or has committed a violation for which society has chosen to impose the most serious legal sanctions available. This response involves criminal sanction, which may include monetary penalties and imprisonment. The criminal response is the most difficult type of enforcement, requiring intensive investigation and case development, but it can also create a significant deterrence.

Environmental enforcement must include processes to balance the rights of individuals with the government's need to act quickly. A notice of violation should be issued before any action is taken so that the finding of violation can be contested, or so that the violation can be corrected before further government action. Appeals should be allowed at several stages in the enforcement process so that the finding of violation, the required remedial action, or the severity of the proposed sanction can be reviewed. There should also be dispute resolution processes for negotiations between program officials and the violator, which may include face-to-face discussions, presentations before a judge or hearing examiner, or use of third party mediators, arbitrators, or facilitators.

Resources

BOOKS

Tacconi, Luca. *Illegal Logging: Law Enforcement, Livelihoods and the Timber Trade*. London: Earthscan Publications, 2007.

OTHER

United States Environmental Protection Agency (EPA). "Compliance and Enforcement." http://www.epa.gov/ebtpages/complianceenforcement.html (accessed October 27, 2010).

Judith Sims

Environmental estrogens

The United States Environmental Protection Agency (EPA) defines an environmental endocrine disruptor—the term the agency uses for environmental estrogens—as "an exogenous agent that interferes with the synthesis, secretion, transport, binding, action, or elimination of natural hormones in the body that are responsible for the maintenance of homeostasis, reproduction, development, and/or behavior." Dr. Theo Colborn, a zoologist and senior scientist with the World Wildlife Fund, and the person most credited with raising national awareness of the issue, describes these chemicals as "hand-me-down poisons" that are passed from mothers to offspring and may be linked to a wide range of adverse effects, including low sperm counts, infertility, genital deformities, breast and prostate cancer, neurological disorders in children such as hyperactivity and attention deficits, and developmental and reproductive disorders in wildlife. Colborn discusses these effects in her book, *Our Stolen Future*, co-authored with Dianne Dumanoski and John Peterson Myers, which asks: "Are we threatening our fertility, intelligence, and survival?" Some other names used for the same class of chemicals are hormone disruptors, estrogen mimics, endocrine-disrupting chemicals, and endocrine modulators.

Although EPA takes the position that it is "aware of and concerned" about data indicating that exposure to environmental endocrine disruptors may cause adverse impacts on human health and the environment, the agency at present does not consider endocrine disruption to be "an adverse endpoint *per se*." Rather, it is "a mode or mechanism of action potentially leading to other outcomes"—such as the health effects Colborn described drawing from extensive research of numerous scientists—but, in EPA's view, the link to human health effects remains an unproven hypothesis. For Colborn and a significant number of other scientists, however, enough is known to support prompt and far-reaching action to reduce exposures to these chemicals and myriad products that are manufactured using them. Foods, plastic packaging, and pesticides are among the sources of exposure Colborn raises concerns about in her book.

Ultimately, the environmental estrogens issue is about whether these chemicals are present in the environment at high enough levels to disrupt the normal functioning of wildlife and human endocrine systems and thereby cause harmful effects. The endocrine system is one of at least three important regulatory systems in humans and other animals (the nervous and immune systems are the other two) and includes such endocrine glands as the pituitary, thyroid, pancreas, adrenal, and the male and female gonads (testes and ovaries). These glands secrete hormones into the bloodstream where they travel in very small concentrations and bind to specific sites called cell receptors in target tissues and organs. The hormones affect development, reproduction, and other bodily functions. The term *endocrine disruptors* includes not only estrogens but also antiandrogens and other agents that act on the endocrine system.

The question of whether environmental endocrine disruptors may be causing effects in humans has arisen over the past decade based on a growing body of evidence about effects in wildlife exposed to dichlorodiphenyl-trichlorethane (DDT), polychlorinated biphenyls (PCBs), and other chemicals. For instance, field studies have proven that tributyltin (TBT), which is used as an antifouling paint on ships, can cause "imposex" in female snails, which are now commonly found with male genitalia, including a penis and vas deferens, the sperm-transporting tube. TBT has also been shown to cause decreased egg production by the periwinkle (*Littorina littorea*). As early as 1985, concerns arose among scientists and the public in the United Kingdom over the effects of synthetic estrogens from birth control pills entering rivers, a concern that was heightened when anglers reported catching fish with both male and female characteristics. Other studies have found Great Lakes salmon to invariably have thyroids that were abnormal in appearance, even when there were no overt goiters. Herring gulls (*Larus argentatus*) throughout the Great Lakes have also been found with enlarged thyroids. In the case of the salmon and gulls, no agent has been determined to be causing these effects. But other studies have linked DDT exposure in the Great Lakes to eggshell thinning and breakage among bald eagles and other birds. In Lake Apopka, Florida, male alligators (*Alligator mississippiensis*) exposed to a mixture of dicofol, DDT, and dichlorodiphenyldichloroethylene (DDE) have been "demasculinized," with phalluses one-half to one-fourth the normal size. Red-eared turtles (*Trachemys scripta*) in the lake have also been demasculinized. One 1988 study reported that four of fifteen female black bears (*Ursus americanus*) and one of four female brown bears (*Ursus arctos*) had, to varying degrees, male sex organs. These and nearly 300 other peer-reviewed studies led the EPA—in conjunction with a multi-agency White House Committee on Environment and Natural Resources—to develop a "framework for planning" and an extensive research agenda to answer questions about the effects of endocrine disruptors. The goal has been to better understand the potential effects of such chemicals on human beings before implementing regulatory actions.

The federal research agenda has been evolving through a series of workshops. As early as 1979, the U.S. National Institute of Environmental Health Sciences (NIEHS) held an "Estrogens in the Environment" conference to evaluate the chemical properties and diverse structures among environmental estrogens. NIEHS held a second conference in 1985 that addressed numerous potential toxicological and biological effects from exposure to these chemicals. NIEHS's third conference, held in 1994, focused on detrimental effects in wildlife. At an April 1995 EPA-sponsored workshop on "Research Needs for the Risk Assessment of Health and Environmental Effects of Endocrine Disruptors,"a number of critical research questions were discussed: What do we know about the carcinogenic effects of endocrine-disrupting agents in humans and wildlife? What are the research needs in this area, including the highest priority research needs? Similar questions were discussed for reproductive effects, neurological effects, immunological effects, and a variety of risk assessment issues. Drawing on the preceding conferences and workshops, in February 1997 EPA issued a *Special Report on Environmental Endocrine Disruption: An Effects Assessment and Analysis* that recommended key research needs to better understand how environmental endocrine disruptors may be causing the variety of specific effects in human beings and wildlife hypothesized by some scientists. For instance, male reproductive research should include tests that evaluate both the quantity and quality of sperm produced. Furthermore, when testing the endocrine-disrupting potential of chemicals, it is important to test for both estrogenic and antiandrogenic activity because new data suggest that it is possible the latter—antiandrogenic activity—not estrogenic activity, is causing male reproductive effects. In the area of ecological research, EPA's special report highlighted the need for research on such issues as what chemicals or class of chemicals can be considered genuine endocrine disruptors and what dose is needed to cause an effect.

Even before environmental estrogens received a place on the federal environmental agenda as a priority concern, Colborn and other scientists met in in Racine, Wisconsin, to discuss their misgivings about the prevalence of estrogenic chemicals in the environment. From that meeting came the landmark "Wingspread Consensus Statement" of twenty-one leading researchers. The statement asserted that the scientists were certain that a large number of human-made chemicals that have been released into the environment, as well as a few natural ones, "have the potential to disrupt the endocrine system of animals, including humans," and that many wildlife populations are already affected by these chemicals. Furthermore, the scientists expressed certainty that the effects may be entirely different in the embryo, fetus, or perinatal organisms than in the adult; that effects are more often manifested in offspring than in exposed parents; that the timing of exposure in the developing organism is crucial; and that, while embryonic development is the critical exposure period, "obvious manifestations may not occur until maturity." Besides these and other "certain" conclusions, the scientists estimated with confidence that "some of the developmental impairments reported in humans today are seen in adult offspring of parents exposed to synthetic hormone disruptors (agonists and antagonists) released in the environment" and that "unless the environmental load of synthetic hormone disruptors is abated and controlled, large scale dysfunction at the population level is possible." The Wingspread Statement included numerous other consensus views on what models predict and the judgment of the group on the need for much greater research and a comprehensive inventory of these chemicals.

The Food Quality Protection Act of 1996 (FQPA) and the Safe Drinking Water Act Amendments of 1996 require EPA to develop a screening program to determine whether pesticides or other substances cause effects in humans similar to effects produced by naturally occurring estrogens and other endocrine effects. The FQPA requires pesticide registrants to test their products for such effects and submit reports, and it requires that registrations be suspended if registrants fail to comply. Besides the EPA screening program, the United Nations Environment Programme is pursuing a multinational effort to manage "persistent organic pollutants," including DDT and PCBs, which, though banned in the United States, are still used elsewhere and can persist in the environment and be transported long-distance. In 1998 the EPA announced the Endocrine Disruptor Screening Program (EDSP) to test a variety of chemicals for their potential to disrupt the human endocrine system. A lengthy, drawn-out process ensued, and a draft list of chemicals was not published until 2007. A so-called final list of the first group of chemicals to be tested using the EDSP-derived set of protocols was published in April 2009. Prioritizing which chemicals to test first has been part of the EPA's EDSP program, since there are many thousands of chemicals that could potentially be tested. The first group of chemicals to be tested consisted of those deemed most likely to cause endocrine problems in people based upon prior evidence.

Resources

BOOKS

IARC Working Group on the Evaluation of Carcinogenic Risks to Humans. *Combined Estrogen-Progestogen Contraceptives and Combined Estrogen-Progestogen Menopausal Therapy*. Lyon, France: International Agency for Research on Cancer, 2007.

David Clarke

Environmental ethics

Ethics is a branch of philosophy that deals with morals and values. Environmental ethics refers to the moral relationships between humans and the natural world. It addresses such questions as, do humans have obligations or responsibilities toward the natural world, and, if so, how are those responsibilities balanced against human needs and interests? Are some interests more important than others?

Efforts to answer such ethical questions have led to the development of a number of schools of ethical thought. One of these is utilitarianism, a philosophy associated with the English eccentric Jeremy Bentham and later modified by his godson John Stuart Mill. In its most basic terms, utilitarianism holds that an action is morally right if it produces the greatest good for the greatest number of people. The early environmentalist Gifford Pinchot was inspired by utilitarian principles and applied them to conservation. Pinchot proposed that the purpose of conservation is to protect natural resources to produce "the greatest good for the greatest number for the longest time." Although utilitarianism is a simple, practical approach to human moral dilemmas, it can also be used to justify reprehensible actions. For example, in the nineteenth century many white Americans believed that the extermination of native peoples and the appropriation of their land

was the right thing to do. However, most would now conclude that the good derived by white Americans from these actions does not justify the genocide and displacement of native peoples.

The tenets of utilitarian philosophy are presented in terms of human values and benefits, a clearly anthropocentric (human-centered) world view. Many philosophers argue that only humans are capable of acting morally and of accepting responsibility for their actions. Not all humans, however, have this capacity to be moral agents. Children, the mentally ill, and others are not regarded as moral agents, but, rather, as moral subjects. However, they still have rights of their own—rights that moral agents have an obligation to respect. In this context, moral agents have intrinsic value independent of the beliefs or interests of others.

Although humans have long recognized the value of non-living objects, such as machines, minerals, or rivers, the value of these objects is seen in terms of money, aesthetics, cultural significance, or other benefit. The important distinction is that these objects are useful or inspiring to some person—they are not ends in themselves but are means to some other end. Philosophers term this instrumental value, because these objects are the instruments for the satisfaction of some other moral agent. This philosophy has also been applied to living things, such as domestic animals. These animals have often been treated as simply the means to some humanly-desired end without any inherent rights or value of their own.

Aldo Leopold, in his famous essay on environmental ethics, "A Sand County Almanac" pointed out that not all humans have been considered to have inherent worth and intrinsic rights. As examples he points to children, women, foreigners, and indigenous peoples—all of whom were once regarded as less than full persons, as objects or the property of an owner who could do with them whatever he wished. Most civilized societies now recognize that all humans have intrinsic rights, and, in fact, these intrinsic rights have also been extended to include such entities as corporations, municipalities, and nations.

Many environmental philosophers argue that the human race must also extend recognition of inherent worth to all other components of the natural world, both living and non-living. In their opinion, the anthropocentric view, which considers components of the natural world to be valuable only as the means to some human end, is the primary cause of environmental degradation. As an alternative, they propose a biocentric view which gives inherent value to all the natural world regardless of its potential for human use.

Paul Taylor outlines four basic tenets of biocentrism in his book, *Respect for Nature*. These are: (1) Humans are members of earth's living community in the same way and on the same terms as all other living things; (2) Humans and other species are interdependent; (3) Each organism is a unique individual pursuing its own good in its own way; (4) Humans are not inherently superior to other living things. These tenets underlie the philosophy developed by Norwegian Arne Naess known as deep ecology.

From this biocentric philosophy Paul Taylor developed three principles of ethical conduct: (1) Do not harm any natural entity that has a good of its own; (2) Do not try to manipulate, control, modify, manage, or interfere with the normal functioning of natural ecosystems, biotic communities, or individual wild organisms; (3) Do not deceive or mislead any animal capable of being deceived or misled. These principles led Professor Taylor to call for an end to hunting, fishing, and trapping; to espouse vegetarianism; and to seek the exclusion of human activities from wilderness areas. However, Professor Taylor did not extend intrinsic rights to non-living natural objects, and he assigned only limited rights to plants and domestic animals. Others argue that all natural objects, living or not, have rights.

Regardless of the appeal that certain environmental philosophies may have in the abstract, it is clear that humans must make use of the natural world if they are to survive. They must eat other organisms and compete with them for all the essentials of life. Humans seek to control or eliminate harmful plants or animals. How is this intervention in the natural world justified? Stewardship is a principle that philosophers use the justify such interference. Stewardship holds that humans have a unique responsibility to care for domestic plants and animals and all other components of the natural world. In this view, humans, their knowledge, and the products of their intellect are an essential part of the natural world, neither external to it nor superfluous. Stewardship calls for humans to respect and cooperate with nature to achieve the greatest good. Because of their superior intellect, humans can improve the world and make it a better place, but only if they see themselves as an integral part of it.

Ethical dilemmas arise when two different courses of action each have valid ethical underpinnings. A classic ethical dilemma occurs when any course of action taken will cause harm, either to oneself or to others. Another sort of dilemma arises when two parties have equally valid, but incompatible, ethical interests. To resolve such competing ethical claims Paul Taylor suggests five guidelines: (1) it is usually permissible for moral agents to defend themselves; (2) basic interests, those interests necessary for survival, take precedence over other interests; (3) when basic interests are in conflict, the least amount of harm should be done to all parties involved; (4) whenever possible, the disadvantages resulting from competing claims should be borne equally by all parties; (5) the greater the harm done to a moral agent, the great is the compensation required.

Ecofeminists do not find that utilitarianism, biocentrism, or stewardship provide adequate direction to solve environmental problems or to guide moral actions. In their view these philosophies come out of a patriarchal system based on domination—of women, children, minorities and nature. As an alternative, ecofeminists suggest a pluralistic, relationship-oriented approach to human interactions with the environment. Ecofeminism is concerned with nurturing, reciprocity, and connectedness rather than with rights, responsibilities, and ownership. It challenges humans to see themselves as related to others and to nature. Out of these connections, then, will flow ethical interactions among individuals and with the natural world.

See also Animal rights; Bioregionalism; Ecojustice; Environmental racism; Environmentalism; Future generations; Humanism; Intergenerational justice; Land stewardship; Speciesism; Callicott, John Baird; Naess, Arne; Rolston, Holmes.

Resources

BOOKS
Curry, Patrick. *Ecological Ethics: An Introduction.* Cambridge, UK: Polity Press, 2006.
Des Jardins, Joseph R. *Environmental Ethics: An Introduction to Environmental Philosophy.* Belmont, CA: Wadsworth Publishing, 2005.

Christine B. Jeryan

Environmental health

Environmental health is concerned with the medical effects of chemicals, pathogenic (disease-causing) organisms, or physical factors in our environment. Because our environment affects nearly every aspect of people's lives in some way or other, environmental health is related to virtually every branch of medical science. The special focus of this discipline, however, tends to be health effects of polluted air and water, contaminated food, and toxic or hazardous materials in the environment. Concern about these issues makes environmental health one of the most compelling reasons to be interested in environmental science.

For a majority of humans, the most immediate environmental health threat has always been pathogenic organisms. Improved sanitation, nutrition, and modern medicine in the industrialized countries have reduced or eliminated many of the communicable diseases that once threatened people. But for those in the less developed countries where nearly 80 percent of the world's population lives, bacteria, viruses, fungi, parasites, worms, flukes, and other infectious agents remain major causes of illness and death. Hundreds of millions of people suffer from major diseases such as malaria, gastrointestinal infections (diarrhea, dysentery, cholera), tuberculosis, influenza, and pneumonia spread through the air, water, or food. Many of these terrible diseases could be eliminated or greatly reduced by a cleaner environment, inexpensive dietary supplements, and better medical care.

For the billion or so richest people in the world—including most of the population of the United States and Canada—diseases related to lifestyle or longevity tend to be much greater threats than more conventional environmental concerns such as dirty water or polluted air. Heart attacks, strokes, cancer, depression and hypertension, traffic accidents, trauma, and AIDS lead as causes of sickness and death in wealthy countries. These problems are becoming increasingly common in the developing world as people live longer, exercise less, eat a richer diet, and use more drugs, tobacco, and alcohol. Epidemiologists predict that by the middle of the next century, diseases of affluence will be leading causes of sickness and death.

Although a relatively minor cause of illness compared to the factors above, toxic or hazardous synthetic chemicals in the environment are becoming an increasing source of concern as industry uses more and more exotic materials to manufacture goods for purchase. There are many of these compounds to worry about. Somewhere around five million different chemical substances are known, about 100,000 are used in commercial quantities, and about 10,000 new ones are discovered or invented each year. Few of these materials have been thoroughly tested for toxicity. Furthermore, the process of predicting what the chances of exposure and potential harm might be from those released into the environment remains highly controversial. Toxins are poisonous, which means that they react specifically with cellular components or interfere with unique physiological processes. A particular

chemical may be toxic to one organism but not another, or dangerous in one type of exposure but not others. Because of this specificity, they may be harmful even in very dilute concentrations. Ricin, for instance, is a protein found in castor beans and one of the most toxic materials known. Three hundred picograms (trillionths of a gram) injected intravenously is enough to kill an average mouse. A single molecule can kill an individual cell. If humans were as sensitive as mice, a few teaspoons of this compound, divided evenly and distributed uniformly could kill everyone in the world. This points illustrates that not all toxins are produced by industry. Many natural products are highly toxic.

Toxins that have chronic (long-lasting) or irreversible effects are of special concern. Among some important examples are neurotoxins (attack nerve cells), mutagens (cause genetic damage), teratogens (result in birth defects), and carcinogens (cause cancer). Many pesticides and metals such as mercury (Hg), lead (Pb), and chromium (Cr) are neurotoxins. Loss of even a few critical neurons can be highly noticeable or may even be lethal, making this category of great importance. Chemicals or physical factors, such as radiation, that damage genetic material can harm not only cells produced in the exposed individual, but also the offspring of those individuals as well.

Among the most feared characteristics of all these chronic environmental health threats are that the initial exposure may be so small or have results so unnoticeable that the victim doesn't even know that anything has happened until years later. Furthermore the results may be catastrophic and irreversible once they do appear. These are among the worst fears and are powerful reasons that many are so apprehensive about environmental contaminants. There may be no absolutely safe exposure—no matter how small—of some chemicals. Because of these fears, people often demand absolute protection from some of the most dreaded contaminants. Unfortunately, this may not be possible. There may be no way to insure that humans are never exposed to any amount of some hazards. It may be that the only recourse is to ask to reduce exposure or mitigate the consequences of that exposure.

In spite of the foregoing discussion of the dangers of chronic effects from minute exposures to certain materials or factors, not all pollutants are equally dangerous; nor is every exposure an unacceptable risk. Fear of unknown and unfamiliar industrial chemicals can lead to hysterical demands for zero exposure to risks. The fact is that most aspects of life are risky.

Furthermore, some materials are extremely toxic, whereas others are only moderately or even slightly so.

This is expressed in the adage of the German physician, Paracelsus (1493–1541), who said in 1540 that "the dose makes the poison." It has become a basic principle of toxicology that nearly everything is toxic at some concentration, but most materials have some lower level at which they present an insignificant risk. Sodium chloride (NaCl; table salt), for instance, is essential for human life in small doses. If people were forced to eat a kilogram all at once, however, it would make them very sick. A similar amount injected all at once into the bloodstream would be lethal.

How a material is delivered—at what rate, through which route of entry, in what form—is often as important as what the material is. The movement, distribution, and fate of materials in the environment are important aspects of environmental health. Solubility is one of the most important characteristics in determining how, when, and where a material will travel through the environment and into people's bodies. Chemicals that are water-soluble move more rapidly and extensively but also are easier to wash off, excrete, or eliminate. Oil or fat-soluble chemicals may not move through the environment as easily as water-soluble materials but may penetrate very efficiently through the skin and into tissues and organs. Fat-soluble substances also may be more likely to be concentrated and stored permanently in fat deposits in the body.

The most common route of entry into the body for many materials is through ingestion and absorption in the gastrointestinal (GI) tract. The GI tract, as well as the urinary system, are the main routes of excretion of dangerous materials. Not surprisingly, those cells and tissues most intimately and continuously in contact with dangerous materials are among the ones most likely to be damaged. Ulcers, infections, lesions, or tumors of the mouth, esophagus, stomach, intestine, colon, kidney, bladder, and associated glands are among the most common manifestations of environmental toxins. Other common routes of entry for toxins are through the respiratory system and the skin. These also are important routes for excreting or discharging unwanted materials.

Some of the most convincing evidence about the toxicity of particular chemicals on humans has come from experiments in which volunteers (students, convicts, or others) were deliberately given measured levels under controlled conditions. Because it is now considered unethical to experiment on living humans, scientists are forced to depend on proxy experiments using computer models, tissue cultures, or laboratory animals.

These proxy tests are difficult to interpret. Experts can't be sure that experimental methods can be extrapolated to how real living humans would react. The most commonly used laboratory animals in toxicity tests are rodents, such as rats and mice. However, different species can react very differently to the same compound. Of some two hundred chemicals shown to be carcinogenic in either rats or mice, for instance, about half caused cancer in one species but not the other. This raises many questions about how to interpret these results, based on uncertainties about the meaning of these results with respect to effects on humans.

It is especially difficult to determine responses to very low levels of particular chemicals, especially when they are not highly toxic. The effects of random events, chance, and unknown complicating factors become troublesome, often resulting in a high level of uncertainty in predicting risk. The case of the sweetener saccharin is a good example of the complexities and uncertainties in risk assessment. Studies in the 1970s suggested a link between saccharin and bladder cancer in male rats. Critics pointed out that humans would have to drink eight hundred cans of soft drink per day to get a dose equivalent to that given to the rats. Furthermore, they argued, most people are not merely large rats.

The Food and Drug Administration (FDA) uses a range of estimates of the probable toxicity of saccharin in humans. At current rates of consumption, the lower estimate predicts that only one person in the United states will get cancer every 1,000 years from saccharin. That is clearly inconsequential considering the advantages of reduced weight, fewer cases of diabetes, and other benefits from this sugar substitute. The upper estimate, however, suggests that 3,640 people will die each year from this same exposure. That is most certainly a risk worthy of concern.

An emerging environmental health concern with a similarly high level of uncertainty but potentially dire consequences is the disruption of endocrine hormone functions by synthetic chemicals. About ten years ago, wildlife biologists began to report puzzling evidence of reproductive failures and abnormal development in certain wild animal populations. Alligators in a lake in central Florida, for instance, were reported to have a 90 percent decline in egg hatching and juvenile survival along with feminization of adult males including abnormally small penises and lack of sperm production. Similar reproductive problems and developmental defects were reported for trout in the Great Lakes, seagulls in California, panthers in Florida, and a number of other species. Even humans may be effected if reports of global reduction of sperm counts and increases of hormone-dependent cancers prove to be true.

Both laboratory and field studies point to a possible role of synthetic chemicals in these problems. More than fifty chemicals, if present in high enough concentrations, are now known to mimic or disrupt the signals conveyed by naturally occurring endocrine hormones that control almost every aspect of development, behavior, immune functions, and metabolism. These chemicals are referred to as endocrine disruptors. Some research studies show that small amounts of these chemicals are just as harmful (possibly more harmful) as larger amounts. Among these chemicals are dioxin, polychlorinated biphenyl, and several persistent pesticides. This new field of research promises to be of great concern in the next few years because it combines dreaded factors of great emotional power, such as undetectable exposure, threat to future generations, unknown or delayed consequences, and involuntary or inequitable distribution of risk.

In spite of the seriousness of the concerns expressed above, the Environmental Protection Agency (EPA) warns that people need to take a balanced view of environmental health. The risks associated with allowable levels of certain organic solvents in drinking water or some pesticides in food are thought to carry a risk of less than one cancer in a million people in a lifetime. Many people are outraged about being exposed to this risk, yet they accept risks thousands of times higher from activities they enjoy, such as smoking, driving a car, or eating an unhealthy diet. According to the EPA, the most important things individuals can do to improve their health are to abstain from smoking, drive safely, eat a balanced diet, exercise reasonably, lower stress, avoid dangerous jobs, lower indoor pollutants, practice safe sex, avoid sun exposure, and prevent household accidents. Many of these factors that can be controlled may be more risky than unknown, uncontrollable environmental hazards.

These potential environmental hazards still generate concern about environmental health. Organizations have been established to develop standards and guidelines for the regulation of exposures to potentially harmful substances and situations in an effort to preserve environmental health. Some of these agencies are the American Conference of Governmental Industrial Hygienists (ACGIH; promotes standards for limiting chemical and physical stresses in the workplace) and two groups that issue guidelines for exposure to ionizing radiation: the International Commission on Radiological Protection (ICRP) and the National Council on Radiation Protection and Measurements (NCRP). Federal organizations such as the FDA, the EPA, the

Occupational Health and Safety Administration (OSHA), and the U.S. Nuclear Regulatory Commission (USNRC) compile the available information to set regulations and standards to limit exposures to physical and chemical stresses in the environment. These standards can be primary or secondary. Primary standards refer to the protection of human health, whereas secondary standards concern the protection of the environment (e.g., regulations regarding air pollution).

Resources

BOOKS

Brown, Phil. *Toxic Exposures: Contested Illnesses and the Environmental Health Movement.* New York: Columbia University Press, 2007.

Friis, Robert H. *Essentials of Environmental Health.* Essential Public Health Series. Sudbury, MA: Jones and Bartlett, 2007.

Frumkin, Howard. *Environmental Health: From Global to Local.* New York: Jossey-Bass, 2005.

Moeller, D. W. *Environmental Health.* Cambridge, MA: Harvard University Press, 2005.

PERIODICALS

Knowlton, Kim, et al. "Assessing Ozone-Related Health Impacts under a Changing Climate." *Environmental Health Perspectives* 112 (2004): 1557–1563.

OTHER

Centers for Disease Control and Prevention (CDC). "Environmental Health." http://www.cdc.gov/Environmental/ (accessed August 9, 2010).

National Institutes of Health (NIH). "Environmental Health." http://health.nih.gov/topic/Environmental Health (accessed August 9, 2010).

World Health Organization (WHO). "Environmental Health." http://www.who.int/topics/environmental_health/en (accessed August 9, 2010).

World Health Organization (WHO). "Quantifying Environmental Health Impacts." WHO Programs and Projects. http://www.who.int/entity/quantifying_ehimpacts/en (accessed August 9, 2010).

William P. Cunningham

Environmental history

Much of human history has been a struggle for food, shelter, and survival in the face of nature's harshness. Three major events or turning points have been the use of fire, the development of agriculture, and the invention of tools and machines. Each of these advances has brought benefits to humans but often at the cost of environmental degradation. Agriculture, for instance, increased food supplies but also caused soil erosion, population explosions, and support of sedentary living and urban life. It was the Industrial Revolution that gave humankind the greater power to conquer and devastate the environment. Polish-born British mathematician Jacob Bronowski (1908–1974) called it an energy revolution, with power as the prime goal. As Bronowski noted, it is an ongoing revolution, with the fate of literally billions of people hanging on the outcome.

The Industrial Revolution, with its initial dependence on the steam engine, iron works, and heavy use of coal, made possible a modern lifestyle with its high consumption of energy and material resources. With it, however, has come devastating levels of air, water, land, and chemical pollution. In essence, environmental history is the story of the growing recognition of humankind's negative impact upon nature and the corresponding public interest in correcting these abuses. American botanist William Cunningham and Barbara Saigo describe four stages of conservation history and environmental activism: 1) pragmatic resource conservation; 2) moral and aesthetic resource preservation; 3) growing concern over the impact of pollution on health and ecosystems; and 4) global environmental citizenship.

Environmental history, like all history, is very much a study of key individuals and events. Included here are British economist Thomas Robert Malthus (1766–1834), American environmentalist George Perkins Marsh (1801–1882), American President Theodore Roosevelt (1858–1919), and American marine biologist Rachel Carson (1907–1964). Writing at the end of the eighteenth century, Malthus was the first to develop a coherent theory of population, arguing that growth in food supply could not keep up with the much larger growth in population. Of cruel necessity, population growth would inevitably be limited by famine, pestilence, disease, or war. Modern supporters are labeled neo-Malthusians and include notable spokespersons American entomologist Paul Erlich (1932–) and American environmentalist Lester Brown (1934–).

In his 1864 book *Man in Nature*, Marsh was the first to attack the American myth of superabundance and inexhaustible resources. Citing many examples from Mediterranean lands and the United States, Marsh described the devastating impact of land abuse through deforestation and soil erosion. American urban planner Lewis Mumford (1895–1990) called this book "the fountainhead of the conservation movement," and American politician Stewart Udall (1920–) described it as the beginning of land wisdom in this country. Marsh's work led to forest preservation and influenced President Theodore Roosevelt and his chief forester, American Gifford Pinchot (1865–1946).

Effective forest and wildlife protection began during Theodore Roosevelt's presidency; his term of office (1901–1909) has been referred to as The Golden Age of Conservation. Roosevelt's administration established the first wildlife refuges and national forests.

At this time key differences emerged between proponents of conservation and preservation. Pinchot's policies were utilitarian, emphasizing the wise use of resources. By contrast, preservationists led by Scottish-born American naturalist John Muir (1838–1914) argued for leaving nature untouched. A key battle was fought over the Hetch Hetchy Reservoir in Yosemite National Park, a proposed water supply for San Francisco, California. Although Muir lost, the Sierra Club (founded in 1882) gained national prominence. Similar battles are were waged in the early 2000s over petroleum extraction in Alaska's Arctic National Wildlife Refuge and mining permits on federal lands, including wilderness areas.

Rachel Carson gained widespread fame through her battle against the indiscriminate use of pesticides. Carson's 1962 book *Silent Spring* has been hailed as the "fountainhead of the modern environmental movement." It has been translated into more than twenty languages and is still a best seller. Carson argued against pesticide abuse and for the right of common citizens to be safe from pesticides in their own homes. Though vigorously opposed by the chemical industry, Carson's views were vindicated by her overpowering reliance on scientific evidence, some given surreptitiously by government scientists. In effect *Silent Spring* was the opening salvo in the battle of ecologists against chemists. Much of the current mistrust of chemicals stems from her work.

Several historical events are relevant to environmental history. The closing of the American frontier at the end of the nineteenth century gave political strength to the Theodore Roosevelt presidency. The 1908 White House Conference on Conservation, organized and chaired by Pinchot, is perhaps the most prestigious and influential meeting ever held in the United States.

During the 1930s the drought in the American Dust Bowl awakened the country to the soil erosion concerns first voiced by Marsh. The establishment of the Soil Conservation Service in 1935 was a direct response to this national tragedy.

and consulting firms complete the assessments. The experience of the United States Army Corps of Engineers in detailing the impacts of projects such as dams and waterways is particularly noteworthy, as the Corps has developed comprehensive methodologies to assess impacts of such major and complex projects. These include evaluation of direct environmental impacts as well as social and economic ramifications.

The content of the assessments generally follows guidelines in the National Environmental Policy Act. Assessments and usually includes the following sections:

- Background information describing the affected population and the environmental setting, including archaeological and historical features, public utilities, cultural and social values, topography, hydrology, geology and soil, climatology, natural resources, and terrestrial and aquatic communities;

- Description of the proposed action detailing its purpose, location, time frame, and relationship to other projects;

- The environmental impacts of proposed action on natural resources, ecological systems, population density, distribution and growth rate, land use, and human health. These impacts should be described in detail and include primary and secondary impacts, beneficial and adverse impacts, short and long term effects, the rate of recovery, and, importantly, measures to reduce or eliminate adverse effects;

- Adverse impacts that cannot be avoided are described in detail, including a description of their magnitude and implications;

- Alternatives to the project are described and evaluated. These must include the "no action" alternative. A comparative analysis of alternatives permits the assessment of environmental benefits, risks, financial benefits and costs, and overall effectiveness;

- The reason for selecting the proposed action is justified as a balance between risks, impacts, costs, and other factors relevant to the project;

- The relationship between short and long term uses and maintenance is described, with the intent of detailing short and long term gains and losses;

- Reversible and irreversible impacts;

- Public participation in the process is described;

- Finally, the EIS includes a discussion of problems and issues raised by interested parties, such as specific federal, state, or local agencies, citizens, and activists.

The environmental impact assessment process provides a wealth of detailed technical information. It has been effective in stopping, altering, or improving some projects. However, serious questions have been raised about the adequacy and fairness of the process. For example, assessments may be too narrow or may not have sufficient depth. The alternatives considered may reflect the judgment of decision-makers who specify objectives, the study design, and the alternatives considered. Difficult and important questions exist regarding the balance of environmental, economic, and other interests. Finally, these issues often take place in a politicized and highly charged atmosphere that may not be amenable to negotiation. Despite these and other limitations, environmental impact assessments help to provide a systematic approach to sharing information that can improve public decision-making.

See also Risk assessment (public health).

Resources

BOOKS

Gilpin, Alan. *Environmental Impact Assessment.* Cambridge, UK: Cambridge University Press, 2006.

Hanna, Kevin S. *Environmental Impact Assessment: Practice and Participation.* Don Mills, Ont., Canada: Oxford University Press, 2005.

Ramachandra, T. V. *Cumulative Environmental Impact Assessment.* New York: Nova Science Publishers, 2006.

OTHER

United States Environmental Protection Agency (EPA). "Environmental Management: Environmental Impact Statement (EIS)." http://www.epa.gov/ebtpages/ envienvironmentalimpactstatement.html (accessed October 20, 2010).

Stuart Batterman

Environmental Impact Statement

The National Environmental Policy Act (1969) made all federal agencies responsible for analyzing any activity of theirs "significantly affecting the quality of the human environment." Environmental Impact Statements (EIS) are the assessments stipulated by this act, and these reports are required for all large projects initiated, financed, or permitted by the federal government. In addition to examining the damage a particular project might have on the environment, federal agencies are also expected to review ways of minimizing or alleviating these adverse effects, a review that can include consideration of the environmental benefits of abandoning the project altogether. The agency compiling an EIS is required to hold public hearings; it is also required to submit a draft to public review, and it is forbidden from proceeding until it releases a final version of the statement.

NEPA has been called "the first comprehensive commitment of any modern state toward the responsible custody of its environment," and the EIS is considered one of the most important mechanisms for its enforcement. It is often difficult to identify environmental damages with remedies that can be pursued in court, but the filing of an EIS and the standards the document must meet are clear and definite requirements for which federal agencies can be held accountable. These requirements have allowed environmental groups to focus legal challenges on the adequacy of the report, contesting the way an EIS was prepared or identifying environmental effects that were not taken into account. The expense and the delays involved in defending against these challenges have often given these groups powerful leverage for convincing a company or an agency to change or omit particular elements of a project. Many environmental organizations have taken advantage of these opportunities by filing thousands of lawsuits based on improper EISs.

Although litigation over impact statements can have a decisive influence on a wide range of decisions in government and business, the legal status of these reports and the legal force of the NEPA itself are not as strong as many environmentalists believe they should be. The act does not require agencies to limit or prevent the potential environmental damage identified in an EIS. The Supreme Court upheld this interpretation in 1989, deciding that agencies are "not constrained by NEPA from deciding that other values outweigh the environmental costs." The government, in other words, is required only to identify and evaluate the adverse impacts of proposed projects; it is not required, at least by NEPA, to do anything about them. Environmentalists have long argued that environmental protection needs a stronger legal grounding than this act provides.

In addition to the controversies over what should be included in these reports and what should be done about the information, there have also been a number of debates over who is required to file them. For example, an EIS is not required of all government agencies; the U.S. Department of Agriculture, for instance, is not required to file such reports on its commodity support programs.

Impact statements have been opposed by business and industrial groups since they were first introduced.

An EIS can be extremely costly to compile, and the process of filing and defending them can take years. Businesses can be left in limbo over projects in which they have already invested large amounts of money, and the uncertainties of the process itself have often stopped development before it has begun. In the debate about these statements, many advocates for business interests have pointed out that environmental regulation accounts for a significant percentage of all federal regulatory expenditures. They argue that impact statements restrict the ability of the United States to compete in international markets by forcing American businesses to spend money on compliance that could be invested in research or capital improvements. Many people believe that impact statements seriously delay many aspects of economic growth, and business leaders have questioned the priorities of many environmental groups, who seem to value conservation over social benefits such as high-levels of employment.

See also Economic growth and the environment; Environmental auditing; Environmental economics; Environmental impact assessment; Environmental Monitoring and Assessment Program; Environmental policy; Life cycle assessment; Risk analysis; Sustainable development.

Resources

BOOKS

Hanna, Kevin S. *Environmental Impact Assessment: Practice and Participation.* Don Mills, Ont., Canada: Oxford University Press, 2005.

OTHER

United States Environmental Protection Agency (EPA). "Environmental Management: Environmental Impact Statement (EIS)." http://www.epa.gov/ebtpages/envienvironmentalimpactstatement.html (accessed October 20, 2010).

Douglas Smith

Environmental labeling *see* **Green Cross; Green seal.**

Environmental law

Environmental law is concerned with protecting the planet and its people from activities that upset the earth and its life-sustaining capabilities, and it is aimed at controlling or regulating human activity toward that end. Until the 1960s most environmental legal issues in the United States involved efforts to protect and conserve natural resources, such as forests and water. Public debate focused on who had the right to develop and manage those resources.

In the succeeding decades lawyers, legislators, and environmental activists increasingly turned their attention to the growing and pervasive problem of pollution. In both instances, environmental law—a term not coined until 1969—evolved mostly from a grassroots movement that forced Congress to pass sweeping legislation, much of which contained provisions for citizen suits. As a result the courts were thrust into a new era of judicial review of the administrative processes and of scientific uncertainty.

Initially environmental law formed around the principles of common law, which is law created by courts and judges that rests upon a foundation of judicial precedents. Environmental law, however, soon moved into the arena of administrative and legislative law, which encompasses most of today's environmental law. The following discussion looks at both areas of law, reviews some of the basic issues involved in environmental law, and outlines some landmark cases.

Three types of common law torts have been applied in environmental law with varying degrees of success. A tort is an intentional or negligent act that causes an injury for which the law provides a civil remedy. Trespass is one type of tort that has been applied in the environmental law setting. Trespass is the physical invasion of one's property, which has been interpreted to include situations such as air pollution, runoff of liquid wastes, or contamination of groundwater.

Closely associated with trespass are the torts of private and public nuisance. *Private nuisance* is interference with the use of one's property. Environmental examples include noise pollution, odors and other air pollution, and water pollution. The operation of a hazardous waste site fits the bill for private nuisance, where the threat of personal discomfort or disease interferes with the enjoyment of one's home. A *public nuisance* adversely affects the safety or health of the public or causes substantial annoyance or inconvenience to the public. In these situations the courts tend to balance the plaintiff's interest against the social and economic need for the defendant's activity.

Lastly, *negligence* involves the failure to exercise the care that a reasonable person would exercise under the circumstances. To prove negligence one must show that the defendant owed a duty to the plaintiff, that the defendant breached that duty, that the plaintiff suffered actual loss or damages, and that there is a reasonable connection between the defendant's conduct and the plaintiff's injury.

These common law remedies have not been very effective in protecting the overall quality of the environment. The lawsuits and resulting decisions were fragmented and site specific as opposed to issue oriented. Further, they relied heavily on a level of hard scientific evidence that is elusive in environmental issues. For instance, a trespass action must be based on a somehow visible or tangible invasion, which is difficult if not impossible to prove in pollution cases. Common law presents other barriers to action. Plaintiffs must prove actual physical injury (so-called aesthetic injuries don't count) and a causal relationship to the plaintiff's activity, which again, is a difficult task in environmental issues.

In the early 1970s, environmental groups, aided by the media, focused public attention on the broad scope of the environmental crisis, and Congress reacted. It passed a host of comprehensive laws, including the Clean Air Act (CAA), the Endangered Species Act (ESA), the National Environmental Policy Act (NEPA), the Resource Conservation and Recovery Act (RCRA), the Toxic Substances Control Act (TSCA), and others. These laws, or statutes, are implemented by federal agencies, who gain their authority through "organic acts" passed by Congress or by executive order.

As environmental problems grew more complicated, legislators and judges increasingly deferred to the agencies' expertise on issues such as the health risk from airborne lead, the threshold at which a species should be considered endangered, or the engineering aspects of a hazardous waste incinerator. Environmental and legal activists then shifted their focus toward administrative law—challenging agency discretion and procedure as opposed to specific regulations—in order to be heard. Hence, most environmental law today falls into the administrative category.

Most environmental statutes provide for administrative appeals by which interest groups may challenge agency decisions through the agency hierarchy. If no solution is reached, the federal Administrative Procedures Act provides that any person aggrieved by an agency decision is entitled to judicial review.

The court must first grant the plaintiff "standing," the right to be a party to legal action against an agency. Under this doctrine, plaintiffs must show they have been injured or harmed in some way. The court must then decide the level of judicial review based on one of three issues—interpretation of applicable statutes, factual basis of agency action, and agency procedure—and apply a different level of scrutiny in each instance.

Generally, courts are faced with five basic questions when reviewing agency action: Is the action or decision constitutional? Did the agency exceed its statutory authority or jurisdiction? Did the agency follow legal procedure? Is the decision supported by substantial evidence in the record? Is the decision arbitrary or capricious? Depending on the answers, the court may uphold the decision, modify it, remand or send it back to the agency to redo, or reverse it.

By far the most important statute that opened the administrative process to judicial review is NEPA. Passed in 1969 the law requires any agencies to prepare an Environmental Impact Statement (EIS) for any major federal actions, including construction projects and issuing permits. Environmental groups have used this law repeatedly to force agencies to consider the environmental consequences of their actions, attacking various procedural aspects of EIS preparation. For example, they often claim that a given agency failed to consider alternative actions to the proposed one, which might reduce environmental impact.

In filing a lawsuit, plaintiffs might seek an injunction against a certain action, say, to stop an industry from dumping toxic waste into a river, or stop work on a public project such as a dam or a timber sale that they claim causes environmental damage. They might seek compensatory damages for a loss of property or for health costs, for instance, and punitive damages, money awards above and beyond repayment of actual losses.

Boomer v. Atlantic Cement Co. (1970) is a classic common law nuisance case. The neighbors of a large cement plant claimed they had incurred property damage from dirt, smoke, and vibrations. They sued for compensatory damages and to enjoin, or stop, the polluting activities, which would have meant shutting down the plant, a mainstay of the local economy. The New York court rejected a long-standing practice and denied the injunction. Further, in an unusual move, the court ordered the company to pay the plaintiffs for present and future economic loss to their properties. A dissenting judge said the rule was a virtual license for the company to continue the nuisance so long as it paid for it.

Sierra Club v. Morton (1972) opened the way for environmental groups to act on behalf of the public interest, and of nature, in the courtroom. The Sierra Club challenged the U.S. Forest Service's approval of Walt Disney Enterprises' plan to build a $35 million complex of motels, restaurants, swimming pools, and ski facilities that would accommodate up to 14,000 visitors daily in Mineral King Valley, a remote, relatively undeveloped national game refuge in the Sierra Nevada Mountains of California. The case posed the now-famous question: Do trees have standing? The Supreme Court held that the Sierra Club was not "injured in fact" by the development and therefore did not have standing. The Sierra Club reworded its petition, gained standing, and stopped the development.

Citizens to Preserve Overton Park v. Volpe (1971) established the so-called "hard look" test to which agencies must adhere even during informal rule making. It opened the way for more intense judicial review of the administrative record to determine if an agency had made a "clear error of judgment." The plaintiffs, local residents and conservationists, sued to stop the U.S. Department of Transportation from approving a six-lane interstate through a public park in Memphis, Tennessee. The court found that Secretary Volpe had not carefully reviewed the facts on record before making his decision and had not examined possible alternative routes around the park. The case was sent back to the agency, and the road was never built.

Tennessee Valley Authority v. Hill (1978) was the first major test of the Endangered Species Act and gained the tiny snail darter fish fame throughout the land. The Supreme Court authorized an injunction against completion of a multi-million dollar dam in Tennessee because it threatened the snail darter, an endangered species. The court balanced the act against the money that had already been spent and ruled that Congress's intent in protecting endangered species was paramount.

Just v. Marinette County (1972) involved wetlands, the public trust doctrine, and private property rights. The plaintiffs claimed that the county's ordinance against filling in wetlands on their land was unconstitutional, and that the restrictions amounted to taking their property without compensation. The county argued it was exercising its normal police powers to protect the health, safety, and welfare of citizens by protecting its water resources through zoning measures. The Wisconsin appellate court ruled in favor of the defendant, holding that the highest and best use of land does not always equate to monetary value, but includes the natural value. The opinion reads, "we think it is not an unreasonable exercise of that [police] power to prevent harm to public rights by limiting the use of private property to its natural uses."

In 2007 the Supreme Court ruled in *Massachusetts v. Environmental Protection Agency* (549 U.S. 497) that carbon dioxide emissions qualified as a pollutant under standards set forth by the Environmental Protection

Agency (EPA) and that the EPA thus was obliged to regulate carbon dioxide as a pollutant. The EPA had previously declined to regulate carbon emissions.

Resources

BOOKS

Ebbesson, Jonas, and Phoebe N. Okowa. *Environmental Law and Justice in Context*. Cambridge, UK: Cambridge University Press, 2009.

McGeoch, Sally. *The Challenge of Green Tape: Growth of Environmental Law and Its Impact on Small and Medium Enterprises Across Australia*. Sidney: New South Wales Business Chamber, 2007.

OTHER

United States Environmental Protection Agency (EPA). "Air: Air Pollution Legal Aspects." http://www.epa.gov/ebtpages/airairpollutionlegalaspects.html (accessed October 22, 2010).

United States Environmental Protection Agency (EPA). "Cleanup: Cleanup Legal Aspects." http://www.epa.gov/ebtpages/cleacleanuplegalaspects.html (accessed October 22, 2010).

United States Environmental Protection Agency (EPA). "Environmental Protection Agency: Legislative and Regulatory Resources." http://www.epa.gov/ebtpages/envilegislativeandregulatoryresour.html (accessed October 22, 2010).

Cathryn McCue

Environmental Law Institute

Environmental Law Institute (ELI) is an independent research and education center involved in developing environmental laws and policies at both national and international levels. The institute was founded in 1969 by the Public Law Education Institute and the Conservation Foundation to conduct and promote research on environmental law. In the ensuing years it has maintained a strong and effective presence in forums ranging from college courses to law conferences. For example, ELI has organized instructional courses at universities for both federal and non-governmental agencies. In addition, it has sponsored conferences in conjunction with such bodies as the American Bar Association, the American Law Institute, and the Smithsonian Institute.

Within the field of environmental law, ELI provides a range of educational programs and services. Through funding and endowments, the institute has established judicial education programs nationwide to teach judges about environmental law issues.

ELI also offers various workshops to the general public. In New Jersey the institute provided a course designed to guide citizens through the state's environmental laws and thus enable them to better develop pollution-prevention programs in their communities. Broader right-to-know guidance has since been provided—in collaboration with the World Wildlife Fund—at the international level.

ELI's endeavors at the federal level include various interactions with the Environmental Protection Agency (EPA). The two groups worked together to develop the National Wetlands Protection Hotline, which answers public inquiries on wetlands protection and regulation, and to assess the dangers of exposure to various pollutants.

Since its inception, ELI has evolved into a formidable force in the field of environmental law. In 1991 it drafted a statute to address the continuing problem of lead poisoning in children. The institute has also worked—in conjunction with federal and private groups, including scientists, bankers, and even realtors—to address health problems attributable to radon gas.

ELI has compiled and produced several publications. Among the leading ELI books are *Law of Environmental Protection*, a two-volume handbook (updated annually) on pollution control law; *Practical Guide to Environmental Management*, a resource book on worker health and safety; and the *Environmental Law Deskbook*, a collection of environmental statues and resources. The institute's principal periodical is *Environmental Law Reporter*, which provides analysis and coverage of topics ranging from courtroom decisions to regulation developments. ELI also publishes *Environmental Forum*—a policy journal intended primarily for individuals in environmental law, policy, and management—and *National Wetlands Newsletter*, which reports on ensuing developments—legal, scientific, regulatory—related to wetlands management.

Resources

BOOKS

Ebbesson, Jonas, and Phoebe N. Okowa. *Environmental Law and Justice in Context*. Cambridge, UK: Cambridge University Press, 2009.

McGeoch, Sally. *The Challenge of Green Tape: Growth of Environmental Law and Its Impact on Small and Medium Enterprises Across Australia*. Sidney: NSW Business Chamber, 2007.

OTHER

Environmental Law & Policy Center. "Environmental Law & Policy Center." http://www.elpc.org/ (Accessed October 22, 2010).

ORGANIZATIONS

Environmental Law Institute, 2000 L Street, NW, Suite 620, Washington, DC, USA, 20036, (202) 939-3800, (202) 939-3868, law@eli.org, http://www.eli.org.

Les Stone

Environmental liability

Environmental liability refers primarily to the civil and criminal responsibility in the environmental issues of hazardous substances that threaten to endanger public health. Compliance with the standards issued by the U.S. Environmental Protection Agency (EPA) became a major issue following the December 11, 1980, enactment by Congress of the original Comprehensive Environmental Response, Compensation, and Liability Act (CERCLA). In 1986 the Superfund Amendments and Reauthorization Act (SARA) provided an amendment to CERCLA. The initial legislation created a tax on chemical and petroleum companies, and gave the federal government authority to handle the releases or threatened releases of the hazardous waste. That tax created $1.6 billion over the first five years of the act, which was put into a trust fund to cover the costs of cleaning up abandoned or uncontrolled hazardous waste sites. When the Superfund was created, changes to the original legislation, as well as additions, were made that reflected the experience gained from the first years of administering the program. It also raised the trust fund to $8.5 billion.

The issue of liability was made more complex following 1986, when regulations increased state involvement and encouraged greater citizen participation in the decisions of site cleanup. In addition to the civil liability of claims due to federal, state, and local governments, a possibility of criminal liability emerged as a matter of particular concern. As those who might be responsible, CERCLA defines four categories of individuals and corporations against whom judgment could be rendered, referred to as potentially responsible parties (PRPs):

- current owners or operators of a specific piece of real estate
- past owners if they owned or operated the property at the time of the hazardous contamination
- generators and possessors of hazardous substances who arranged for disposal, treatment, or transport
- certain transporters of hazardous substances

In acting under EPA-expanded powers, some states have provided exemptions from liability in certain cases. For example, the state of Wisconsin has provided that the person responsible for the discharge of a hazardous substance is the one who is required to report it, investigate it, and clean up the contamination. According to Wisconsin Department of Natural Resources information, the state defines the responsible person as the one who "causes, possesses, or controls" the contamination—the one who owns the property with a contaminant discharge or owns a container that has ruptured. Other Wisconsin exemptions include limiting liability for parties who voluntarily remediate contaminated property; limiting liability for lenders and representatives, such as banks, credit unions, mortgage bankers, and similar financial institutions, or insurance companies, pension funds, or government agencies engaged in secured lending; limiting liability for local government units; and, limiting liability for property affected by off-site discharge.

Courts have also acted in finding persons responsible not specifically listed as PRPs:

- lessees of contaminated property
- lessors of the contaminated property for contamination caused by their lessees
- landlords and lessees for the contamination cause by their sub-lessees
- corporate officers in their personal capacity
- shareholders
- parent corporations liable for their subsidiaries
- trustees and representatives personally liable for contaminated property owned by a trust or estate
- successor corporations
- donees
- lenders who foreclose on and subsequently manage contaminated property

Environmental liability continues into the early twenty-first century to be a matter of grave concern not only with land, but also in maritime issues. After the 2010 Deepwater Horizon oil spill in the Gulf of Mexico, the U.S. government assured citizens that BP, the company who owned the leaking well, would be held finically accountable for all damages done to the coastal environment and economy. BP created a $20 billion fund to settle claims from those impacted by the spill. Additionally the EPA is permitted to assess fines for the amount of pollutants discharged during the several months that the damaged well spewed oil.

Some environmental advocates assert that the chemical dispersants used by to keep some oil from reaching shore should also be considered as pollutants for liability purposes.

Resources

OTHER

United States Environmental Protection Agency (EPA). "Cleanup: Cleanup Legal Aspects: Liability." http://www.epa.gov/ebtpages/cleacleanuplegalasliability.\html (accessed October 24, 2010).

United States Environmental Protection Agency (EPA). "Compliance and Enforcement." http://www.epa.gov/ebtpages/complianceenforcement.html (accessed October 24, 2010).

United States Environmental Protection Agency (EPA). "Compliance and Enforcement: Environmental Liability." http://www.epa.gov/ebtpages/comp environmentalliability.html (accessed October 24, 2010).

Jane E. Spear

Environmental literacy and ecocriticism

Environmental literacy and ecocriticism refer to the work of educators, scholars, and writers to foster a critical understanding about environmental issues. Environmental literacy includes educational materials and programs designed to provide lay citizens and students with a broad understanding of the relationship between humans and the natural world, borrowing from the fields of science, politics, economics, and the arts. Environmental literacy also seeks to develop the knowledge and skills citizens and students may need to identify and resolve environmental crises, individually or as a group. Ecocriticism is a branch of literary studies that offers insights into the underlying philosophies in literature that address the theme of nature and have been catalysts for change in public consciousness concerning the environment.

Americans have long turned to literature and popular culture to develop, discuss, and communicate various ideals about the natural world. This literature has also made people think about the idea of progress: what constitutes advancement in culture, what are the goals of a healthy society, and how nature would be considered and treated by such a society. In contemporary times, the power and visibility of modern media in influencing these debates is also widely recognized. Given this trend, understanding how these forms of communication work and developing them further to broaden public participation, which is a task of environmental literacy and ecocriticism, is vital to the environmental movement.

Educators and ecocritics take diverse approaches to the task of raising consciousness about environmental issues, but they share a collective concern for the global environmental crisis and begin with the understanding that nature and human needs require rebalancing. In that, they become emissaries, as American writer Barry Lopez (1945–) suggests in *Orion* magazine, who have to "reestablish good relations with all the biological components humanity has excluded from its moral universe." For Lopez, as with many generations of nature writers, including American naturalist Henry David Thoreau (1817–1862), Scottish-born American John Muir (1838–1914), American Edward Abbey (1927–1989), American Terry Tempest Williams (1955–), and American Annie Dilliard (1945–), the lessons to be imparted are learned from long experience with and observation of nature. Lopez suggests another pervasive theme, that observing the ever-changing natural world can be a humbling experience, when he writes of "a horizon rather than a boundary for knowing, toward which we are always walking."

The career of Henry David Thoreau was one of the most influential and early models for being a student of the natural world and for the development of an environmental awareness through attentive participation within nature. Thoreau also made a fundamental contribution to American's identification with the ideals of individualism and self-sufficiency. His most important work, *Walden*, was a book developed from his journal written during a two-and-a-half-year experiment of living alone and self-sufficiently in the woods near Concord, Massachusetts. Thoreau describes his process of education as an awakening to a deep sense of his interrelatedness to the natural world and to the sacred power of such awareness. This is contrasted to the human society from which he isolated himself, of whose utilitarianism, materialism, and consumerism he was extremely critical. Thoreau famously writes in *Walden*: "I went to the woods because I wished to live deliberately, to front only the essential facts of life, and see if I could not learn what it had to teach, and not, when I came to die, discover that I had not lived." For Thoreau, living with awareness of the greater natural world became a matter of life and death.

Many educators have also been influenced by two founding policy documents, created by commissions of the United Nations, in the field of environmental literacy. The Belgrade Charter (UNESCO-UNEP, 1976) and the Tbilisi Declaration (UNESCO, 1978) share the goal "to develop a world population that is aware of, and concerned about, the environment and its associated problems." Later governmental bodies such as the Brundtland Commission (Brundtland, 1987), the United Nations Conference on Environment and Development in Rio (UNCED, 1992), and the Thessaloniki Declaration (UNESCO, 1997) have built on these ideas.

One of the main goals of environmental literacy is to provide learners with knowledge and experience to assess the health of an ecological system and to develop solutions to problems. Models for environmental literacy include curriculums that address key ecological concepts, provide hands-on opportunities, foster collaborative learning, and establish an atmosphere that strengthens a learner's belief in responsible living. Environmental literacy in such programs is seen as more than the ability to read or write. As in nature writing, it is also about a sensibility that views the natural world with a sense of wonder and experiences nature through all the senses. The element of direct experience of the natural world is seen as crucial in developing this sensibility. The Edible Schoolyard program in the Berkeley, California, school district, for example, integrates an organic garden project into the curriculum and lunch program, where students become involved in the entire process of farming, while learning to grow and prepare their own food. The program aims to promote participation in and awareness of the workings of the natural world, and also to awaken all the senses to enrich the process of an individual's development.

Public interest in environmental education came to the forefront in the 1970s. Much of the impetus as well as the funding for integrating environmental education into school curriculums comes from non-profit foundations and educators' associations such as the Association for Environmental and Outdoor Education, the Center for Ecoliteracy, and The Institute for Earth Education. In 1990 the U.S. Congress created the National Environmental Education and Training Foundation (NEETF), whose efforts include expanding environmental literacy among adults and providing funding opportunities for school districts to advance their environmental curriculums. The National Environmental Education Act of 1990 directed the Environmental Protection Agency (EPA) to provide national leadership in the environmental literacy arena. To that end, the EPA established several initiatives including the Environmental Education Center as a resource for educators and the Office of Environmental Education, which provides grants, training, fellowships, and youth awards.

The Public Broadcasting System also plays an active role in the promotion of environmental literacy as evidenced by the partnership of the Annenberg Foundation and the Corporation for Public Broadcasting to create and disseminate educational videos for students and teachers.

A common thread woven through these organizations is a definition of environmental learning that goes beyond simple learning to an appreciation of nature. However, appreciation is measured differently by each organization, and segments of the American population differ on which aspects of the environment should be preserved. In 1993 the North American Association for Environmental Education (NAAEE) established the National Project for Excellence in Environmental Education, which developed standards for environmental education. At the end of the 1990s, the George C. Marshall Institute directed an independent commission to study whether the goals of environmental education were being met. The Commission's 1997 report found that curricula and texts vary widely on many environmental concepts, including what constitutes conservation. The Commission later became known as the Environmental Literacy Council, which hosts a wealth of background information and resources for both teachers and students concerning environmental issues on their website. With the inception of an Advanced Placement exam in Environmental Science by the College Board in 1997, many more high schools are offering Environmental Science as a course.

Thus, the main challenges to environmental literacy are the lack of unifying programs that would bring together the many approaches to environmental education, and the fact that there is inconsistent support for these programs from the government and public school system. Observers of environmental literacy movements suggest that the new perspectives that learners gain may often be at odds with the concerns and ethics of mainstream society, issues that writers such as Thoreau grappled with. For instance, consumerism and conservationism may be at opposite ends of the spectrum of how people interact with the natural world and its resources. To be effective, literacy initiatives must address these dilemmas and provide tools to solve them. Environmental literacy is thus about providing new ways of seeing the world, about providing language tools to address these new perceptions, and to provide ethical frameworks through which people can make informed choices on how to act.

Ecocriticism develops the tools of literary criticism to understand how the relationship of humans to nature is addressed in literature, as a subject, character, or as a component of the setting. Ecocritics also highlight the ways in which literature is a vehicle to create environmental consciousness. For critic William Rueckert, the scholar who coined the term *ecocriticism* in 1978, poetry and literature are the "verbal equivalent of fossil fuel, only renewable," through which abundant energy is transferred between nature and the reader.

Ecocritics highlight aspects of nature described in literature, whether frontiers, rivers, regional ecosystems, cities, or garbage, and ask what the purposes of these descriptions are. Their interests have included understanding how historical movements such as the Industrial Revolution have changed the relationship between human society and nature, giving people the false illusion that they can completely control nature, for instance. Ecocriticism also brings together perspectives from various academic disciplines and draws attention to their shared purposes. Studies in ecology and cellular biology, for example, echo the theme of interconnectedness of the individual and the natural world seen in poetry, by demonstrating how the life of all organisms is dependent upon their ongoing interactions with the environment around them.

Although nature writers have expressed their philosophies of nature and reflected on their modes of communication since the nineteenth century, as a self-conscious practice, ecocriticism's history did not begin until the late 1970s. By the 1990s it had gained wide currency. In his 1997 article "Wild Things," published in *Utne Reader*, Gregory McNamee notes that courses in environmental literature are available at colleges across the nation and that "ecocritcism has become something of an academic growth industry." In 1992 the Association for the Study of Literature and Environment (ASLE) was founded with the mission "to promote the exchange of ideas and information about literature and other cultural representations that consider that human relationships with the natural world."

Nature's role in theatre and film are also popular ecocriticism topics for academic study in the form of seminars on, for example, The Nature of Shakespeare, and suggested lists of commercial films for class discussion.

Ecocritics Carl Herndl and Stuart Brown suggest that there are three underlying philosophies in evaluating nature in modern society. The language used by institutions that make government policies usually regards nature as a resource to be managed for greater social welfare. This is described as an ethnocentric perspective, which begins with the idea that one opinion or way of looking at the world is superior to others. Thus, the benefits of environmental issues are always measured against various political and social interests, and not seen as important simply in themselves.

Another viewpoint is the anthropocentric perspective, wherein human perspectives are central in the world and are the ultimate source of meaning. The specialized language of the sciences, which treats nature as an object of study, is an example of this. The researcher is seen as existing outside of or above nature, and science is grounded on the faith that humans can come to know all of nature's secrets.

In contrast, poetry often describes nature in terms of its beauty and emotional and spiritual power. This language sees man as part of the natural world and seeks to harmonize human values and actions with a respect for nature. This is the ecocentric perspective, which means putting nature and ecology as the central viewpoint when considering the various interactions in the world, including human ones. That is, this perspective acknowledges that humans are a part and parcel of nature and ultimately depend upon the ecology's living and complex interactions for survival.

Scholars make the distinction between environmental writing and other kinds of literature that use images of nature in some fashion or another. Environmental writing explores at length ecocentric perspectives. They include discussions about human ethical responsibility toward the natural world, such as in American ecologist Aldo Leopold's (1887–1948) *A Sand County Almanac*, considered one of the best explorations of environmental ethics. Many ecocritics also share a concern for the environment, and one aim of ecocriticism is to raise awareness within the literary world about the environmental movement and nature-centered perspectives in understanding human relationships and cultural practices.

In *Silent Spring*, a major text in the field of environmental literacy and ecocriticism, American biologist Rachel Carson (1907–1964) writes that society faces two choices: to travel as it now does on a superhighway at high speed but ending in disaster, or to walk the less traveled other road which offers the chance to preserve the earth. The challenge of ecocriticism is to spread the word of the other road, and to simultaneously offer constructive criticism to the environmental movement from within.

See also Environmental education; Environmental ethics; Muir, John; Thoreau, Henry David.

Resources

BOOKS

Alex, Rayson K., Nirmaldasan, and Nirmal Selvamony. *Essays in Ecocriticism*. Chennai, India: Osle, 2007.

Bate, Jonathan, and Laurence Coupe, eds. *The Green Studies Reader: From Romanticism to Ecocriticism*. London: Routledge, 2004.

Borlik, Todd Andrew. *Green Pastures: Ecocriticism and Early Modern English Literature*. PhD diss., University of Washington, 2008.

Carson, Rachel. *Silent Spring*. Boston: Houghton Mifflin, 1962.

Egan, Gabriel. *Green Shakespeare: From Ecopolitics to Ecocriticism*. Accents on Shakespeare. London: Routledge, 2006.

Fromm, Harold, and Cheryll Glotfelty. *The Ecocriticism Reader: Landmarks in Literary Ecology*. Athens, GA: University of Georgia Press, 2004.

Garrard, Greg. *Ecocriticism*. London: Routledge, 2004.

Gersdorf, Catrin, and Sylvia Mayer. *Nature in Literary and Cultural Studies: Transatlantic Conversations on Ecocriticism*. Nature, culture and literature, 03. Amsterdam: Rodopi, 2006.

Gifford, Terry. *Reconnecting with John Muir: Essays in Post-Pastoral Practice*. Athens: University of Georgia Press, 2006.

Huggan, Graham, and Helen Tiffin. *Postcolonial Ecocriticism*. New York: Routledge, 2009.

Ingram, Annie Merrill. *Coming into Contact: Explorations in Ecocritical Theory and Practice*. Athens: University of Georgia Press, 2007.

Lytle, Mark H. *The Gentle Subversive: Rachel Carson, Silent Spring, and the Rise of the Environmental Movement*. New narratives in American history. New York: Oxford University Press, 2007.

McKibben, Bill, and Albert Gore. *American Earth: Environmental Writing since Thoreau*. New York: Literary classics of the United States, 2008.

Murphy, Priscilla Coit. *What a Book Can Do: The Publication and Reception of Silent Spring*. Amherst: University of Massachusetts Press, 2007.

Slovic, Scott. *Going Away to Think: Engagement, Retreat, and Ecocritical Responsibility*. Reno: University of Nevada Press, 2008.

Thoreau, Henry David. *Walden*. Sandy: UT: Quiet Vision Publishing, 2008.

Wheeler, Wendy, and Hugh Dunkerley. *Earthographies: Ecocriticism and Culture*. London: Lawrence & Wishart, 2008.

PERIODICALS

Velasquez-Manoff, Moises. "Picking Up Where 'Silent Spring' Left Off." *Christian Science Monitor* (August 14, 2007).

OTHER

Environmental Education and Training Partnership (EETAP). "Environmental Education Resources." http://eetap.org/ (accessed October 24, 2010).

Environmental Literacy Council. "Environmental Literacy Council." http://www.enviroliteracy.org/ (accessed June 19, 2009).

National Environmental Education Advisory Council. "Setting the Standard, Measuring Results, Celebrating Successes: A Report to Congress on the Status of Environmental Education in the United States." http://www.epa.gov/enviroed/pdf/reporttocongress2005.pdf (accessed October 22, 2010).

Douglas Dupler

Environmental mediation and arbitration *see* **Environmental dispute resolution.**

Environmental monitoring

Environmental monitoring detects changes in the health of an ecosystem and indicates whether conditions are improving, stable, or deteriorating. This quality, too large to gauge as a whole, is assessed by measuring indicators, which represent more complex characteristics. The concentration of sulfur dioxide, for example, is an indicator that reflects the presence of other air pollutants. The abundance of a predator indicates the health of the larger environment. Other indicators include metabolism, population, biological and microbiological factors, community, and landscape. All changes are compared to an ideal, pristine ecosystem. The SER (stressor-exposure-response) model, a simple but widely used tool in environmental monitoring, classifies indicators as one of three related types:

- Stressors, which are agents of change associated with physical, chemical, or biological constraints on environmental processes and integrity. Many stressors are caused by humans, such as air pollution, the use of pesticides and other toxic substances, or habitat change caused by forest clearing. Stressors can also be natural processes, such as wildfire, hurricanes, volcanoes, and climate change.

- Exposure indicators, which link a stressor's intensity at any point in time to the cumulative dose received. Concentrations or accumulations of toxic substances are exposure indicators; so are clear-cutting and urbanization.

- Response indicators, which shows how organisms, communities, processes, or ecosystems react when

For example, the key actors in forming policy on clear-cutting in the national forests are the House Subcommittee on Forests, Family Farms and Energy, the U.S. Forest Service (USFS), and the National Forest Products Association, which represents many industries dependent on timber.

For more than a century, conservation and environmental groups worked at the fringes of the traditional "iron triangle." Increasingly, however, these public interest groups, which derived their financial support and sense of mission from an increasing number of citizen members, began gaining more influence. Scientists, whose studies and research play a pivotal role in decision-making, also began to emerge as major players.

The watershed years

Catalyzed by vocal, energetic activists and organizations, the emergence of an "environmental movement" in the late 1960s prompted the government to grant environmental protection a greater priority and visibility. In 1970, the year of the first celebration of Earth Day, the federal government implement landmark environmental legislation, Clean Air Act and the National Environmental Policy Act of 1969. U.S. President Richard Nixon also created the Environmental Protection Agency (EPA), which was given control of many environmental policies previously administered by other agencies. In addition, some of the most serious problems, such as DDT and mercury contamination, began to be addressed between 1969 and 1972. Yet, environmental policies in the 1970s developed largely in an adversarial setting pitting environmental groups on one side and the traditional iron triangles on the other.

The first policies that came out of this era were designed to clean up visible pollution—clouds of industrial soot and dust, detergent-filled streams, and so forth—and employed "end-of-pipe" solutions to target point sources, such as waste-water-discharge pipes, smokestacks, and other easily identifiable emitters.

An initial optimism generated by improvements in air and water quality was dashed by a series of frightening environmental episodes at Times Beach, Missouri; Three Mile Island, Pennsylvania; Love Canal, New York; and other locations. Such incidents (as well as memory of the devastation caused by the recently-banned DDT) shifted the focus of public concern to specific toxic agents. By the early 1980s a fearful public led by environmentalists had steered governmental policy toward tight regulation of individual, invisible toxic substances—dioxin, PCBs and others—by

backing measures limiting emissions to within a few parts per million. Without an overall governmental framework for action, the result has been a multitude of regulations and laws that address specific problems in specific regions that sometimes conflict and often fail to protect the environment in a comprehensive manner. "It's been reactionary, and so we've lost the integration of thought and disciplines that is essential in environmental policy making," says Carol Browner, former administrator of the U.S. EPA.

The 1980 Comprehensive Environmental Response, Compensation and Liability Act (CERCLA), or Superfund toxic waste program grew as much out of the public's perception and fear of toxic waste as it did from crude scientific knowledge of actual health risks. Roughly $2 billion dollars a year was spent cleaning up a handful of the nation's worst toxic sites to near pristine condition. EPA officials now believe the money could have been better spent cleaning up more sites, although to a somewhat lesser degree.

Current trends in environmental policy

Today, governmental bodies and public interest groups are drawing back from "micro management" of individual chemicals, individual species, and individual industries to focus more on the interconnections of environmental systems and problems. This new orientation has been shaped by several (sometimes conflicting) forces, including:

- industrial and public resistance to tight regulations fostered by fears that such laws impact employment and economic prosperity;
- financial limitations that prevent government from carrying out tasks related to specific contaminants, such as cleaning up waste sites or closely monitoring toxic discharges;
- a perception that large-scale, global problems such as the greenhouse effect, ozone layer depletion, habitat destruction, and the like should receive priority;
- the emergence of a preventative orientation on the part of citizen groups that attempts to link economic prosperity with environmental goals. This approach emphasizes recycling, efficiency, and environmental technology and stresses the prevention of problems rather than their remediation after they reach a critical stage. This strategy also marks an attempt by some citizen organizations to a more conciliatory stance with industry and government.

Since the 1990s scientists and policymakers have stressed the need for international cooperation to tackle global environmental issues, including global climate change and sustainable development. The

Kyoto Protocol, which was adopted in 1997 and entered force in 2005, is an international environmental treaty designed to stabilize greenhouse gas emissions. Although the Kyoto Protocol expires at the end of 2012, the treaty highlighted the need for a coordinated international response to reduce greenhouse emissions. Policymakers hope to have a new greenhouse gas emissions reduction treaty in place by the time the Kyoto Protocol expires.

See also Pollution Prevention Act (1990).

Resources

BOOKS

Greenberg, Michael R. *Environmental Policy Analysis and Practice.* New Brunswick, NJ: Rutgers University Press, 2007.

PERIODICALS

Byrne, John, et al. "American Policy Conflict in the Greenhouse: Divergent Trends in Federal, Regional, State, and Local Green Energy and Climate Change Policy." *Energy Policy* 35 (2007): 4555–4573.

Economides, George, and Apostoles Philippopulos. "Should Green Governments Give Priority to Environmental Policies Over Growth-Enhancing Policies?" CESifo working papers, No. 1433 : Category 5, Fiscal policy, macroeconomics and growth. Munich: CES, 2005.

Metz, B., et al. "IPCC, 2007: Summary for Policymakers." In: *Climate Change 2007: Mitigation. Contribution of Working Group III to the Fourth Assessment Report of the Intergovernmental Panel on Climate Change.* Cambridge, UK: Cambridge University Press, 2007.

OTHER

United Nations, Secretariat of the United Nations Framework Convention on Climate Change (UNFCCC). "Introduction to the UNFCCC and its Kyoto Protocol." http://unfccc.int/files/press/backgrounders/application/pdf/unfccc_and_kyoto_protocol.pdf (accessed October 27, 2010).

Cathryn McCue
Kevin Wolf
Jeffrey Muhr

Environmental Protection Agency (EPA)

The Environmental Protection Agency (EPA) was established in July 1970, a landmark year for environmental concerns, having been preceded by the implementation of the National Environmental Policy Act

(NEPA) in January 1970, the passage of the Clean Air Act (CAA) Extension of 1970, and the celebration of the first Earth Day in April 1970. President Richard Nixon and Congress, working together in response to the growing public demand for cleaner air, land, and water, sought to create a new agency of the federal government structured to make a coordinated attack on the pollutants that endanger human health and degrade the environment. The EPA was charged with repairing the damage already done to the environment and with instituting new policies designed to maintain a clean environment.

The EPA's mission is "to protect human health and to safeguard the natural environment." At the time the EPA was formed, at least fifteen programs in five different agencies and cabinet-level departments were handling environmental policy issues. For the EPA to work effectively, it was necessary to consolidate the environmental activities of the federal government into one agency. Air pollution control, solid waste management, radiation control, and the drinking water program were transferred from the U.S. Department of Health, Education and Welfare (currently known as the U.S. Department of Health and Human Services). The water pollution control and pesticides research programs were acquired from the U.S. Department of the Interior. Registration and regulation of pesticides was transferred from the U.S. Department of Agriculture, and the responsibility for setting tolerance levels for pesticides in food was acquired from the Food and Drug Administration. The EPA also took over from the Atomic Energy Commission the responsibility for setting some environmental radiation protection standards and assumed some of the duties of the Federal Radiation Council.

For some environmental programs, the EPA works with other agencies: for example, the United States Coast Guard and the EPA work together on flood control, shoreline protection, and dredging and filling activities. And, since most state governments in the United States have their own environmental protection departments, the EPA delegates the implementation and enforcement of many federal programs to the states.

The EPA's headquarters is in Washington, DC, and there are ten regional offices and field laboratories. The main office develops national environmental policy and programs, oversees the regional offices and laboratories, requests an annual budget from Congress, and conducts research. The regional offices implement national policies, oversee the environmental programs that have been delegated to the states, and review

Environmental Impact Statements for federal actions. The field laboratories conduct research, the data from which are used to develop policies and provide analytical support for monitoring and enforcement of EPA regulations, and for the administration of permit programs.

The administrator of the EPA is appointed by the president, subject to approval by the Senate. The same procedure is used to appoint a deputy administrator, who assists the administrator, and nine assistant administrators, who oversee programs and support functions. Other posts include the chief financial officer, who manages the EPA's budget and funding operations; the inspector general, who is responsible for investigating environmental crimes; and a general counsel, who provides legal support.

In addition to the administrative offices, the EPA is organized into the following program offices: the Office of Air and Radiation; the Office of Chemical Safety and Pollution; the Office of Environmental Information; the Office of International and Tribal Affairs, which includes the American Indian Environment Office; the Office of Research and Development; the Office of Water; and the Office of Solid Waste and Emergency Response; and the Office of Water.

One of the major activities of the EPA is the management of Superfund sites. For many years uncontrolled dumping of hazardous chemical and industrial wastes in abandoned warehouses and landfills continued without concern for the potential impact on public health and the environment. Concern over the extent of the hazardous waste site problem led Congress to establish the Superfund Program in 1980 to locate, investigate, and clean up the worst such sites. The EPA Office of Solid Waste and Emergency Response, oversees management of the program in cooperation with individual states. When a hazardous-waste site is discovered, the EPA is notified. The EPA makes a preliminary assessment of the site and gives a numerical score according to Hazard Ranking System (HRS), which determines whether the site is placed on the National Priorities List (NPL). As of 2009 more than 1,250 sites were listed on the final NPL. The final NPL lists Superfund sites in which the clean-up plan is under construction or ongoing. NPL proposed sites include sites for which the HRS indicates that placement on the final NPL is appropriate. A final NPL site is deleted from the list when it is determined that no further clean up is needed to protect human health or the environment. Finally, under Superfund's redevelopment program, former hazardous-waste sites have been remade into office buildings, parking lots, or even golf courses to be re-integrated as productive parts of the community.

The offices and programs of the EPA recognize a set of main objectives, or core functions. These core functions help define the agency's mission and provide a common focus for all agency activities. The core functions are:

- Pollution Prevention—taking measures to prevent pollution from being created rather than only cleaning up what has already been released, also known as source reduction

- Risk Assessment and Risk Reduction—identifying problems that pose the greatest risk to human health and the environment and taking measures to reduce those risks

- Science, Research and Technology—conducting research that will help in developing environmental policies and promoting innovative technologies to solve environmental problems

- Regulatory Development—developing requirements such as operating procedures for facilities and standards for emissions of pollutants

- Enforcement—assuring compliance with established regulations

- Environmental Education—developing educational materials, serving as an information clearinghouse, and providing grant assistance to local educational institutions

Environmental Protection Agency scientists Peter Kalla (L) and Archie Lee (R) carry samples collected May 3, 2010, on the beach in Biloxi, Mississippi, in anticipation of the oil spill from the BP *Deepwater Horizon* platform disaster. *(STAN HONDA/AFP/Getty Images)*

Many EPA programs are established by legislation enacted by Congress. For example, most activities carried out by the Office of Solid Waste and Emergency Response originated in the Resource Conservation and Recovery Act (RCRA). Among other laws that form the legal basis for the programs of the EPA are the National Environmental Policy Act (NEPA) of 1969, which represents the basic national charter of the EPA, the Clean Air Act (CAA) of 1970, the Occupational Safety and Health Act (OSHA) of 1970, the Endangered Species Act (ESA) of 1973, the Safe Drinking Water Act (SDWA) of 1974, the Toxic Substances Control Act (TSCA) of 1976, the Clean Water Act (CWA) of 1977, the Superfund Amendments and Reauthorization Act (SARA) of 1986, and the Pollution Prevention Act (PPA) of 1990. It is through such legislation that the EPA obtains authority to develop and enforce regulations. Environmental regulations drafted by the agency are subjected to intense review before being finalized. This process includes approval by the President's Office of Management and Budget and input from the private sector and from other government agencies.

Growing public concern about water pollution led to a landmark piece of legislation, the Federal Water Pollution Control Act of 1972, amended in 1977, and commonly known as the Clean Water Act. The Clean Water Act gives the EPA the authority to administer pollution control programs and to set water quality standards for contaminants of surface waters.

In 2007 the Supreme Court of the United States ruled in *Massachusetts v. Environmental Protection Agency* that the EPA may regulate carbon dioxide and other greenhouse gases as pollutants under the Clean Air Act. The EPA, under the administration of U.S. President George W. Bush, had asserted that the EPA did not have the authority under the Clean Air Act to regulate greenhouse gases as pollutants. In 2003 the EPA had asserted that, even if it had the authority to regulate greenhouse gases, it would refuse to do so. The Court ruled that the EPA, not only has the authority to regulate greenhouse gases, but the EPA must regulate greenhouse gases unless "it determines that greenhouse gases do not contribute to climate change, or if it provides some reasonable explanation as to why it cannot or will not exercise its discretion to determine whether they do."

Resources

BOOKS

Collin, Robert W. *The Environmental Protection Agency: Cleaning Up America's Act*. Westport, CT: Greenwood Press, 2005.

STRIVE: An Environmental Protection Agency Program, 2007–2013. Wexford, Ireland: Johnstown Castle Estate: Environmental Protection Agency, 2007.

PERIODICALS

Greenhouse, Linda. "Justices Say the EPA Has Power to Act on Harmful Gases." *The New York Times* (April 3, 2007).

OTHER

U.S. Environmental Protection Agency (EPA). http://www.epa.gov/ (accessed November 9, 2010).

Teresa C. Donkin

Environmental racism

The term environmental racism was coined by Reverend Dr. Benjamin F. Chavis Jr. in a 1987 study conducted by the United Church of Christ that examined the location of hazardous waste dumps and found an "insidious form of racism." Concern had surfaced five years before, when opposition to a polychlorinated biphenyl (PCB) landfill prompted Congress to examine the location of hazardous waste sites in the Southeast, the Environmental Protection Agency (EPA)'s Region IV. They found that three of the four facilities in the area were in communities primarily inhabited by people of color. Subsequent studies, such as Ben Goldman's *The Truth about Where You Live*, have contended that exposure to environmental risks is significantly greater for racial and ethnic minorities than for nonminority populations. However, an EPA study contends that there is not enough data to draw such broad conclusions.

The *National Law Journal* found that official response to environmental problems may be racially biased. According to their study, penalties for environmental crimes were higher in white communities. They also found that the EPA takes 20 percent longer to place a hazardous waste site in a minority community on the Superfund's National Priorities List (NPL). And, once assigned to the NPL, these clean ups are more likely to be delayed.

Advocates also contend that environmentalists and regulators have tried to solve environmental problems without regard for the social impact of the solutions. For example, the Los Angeles Air Quality Management District wanted to require businesses to set up programs that would discourage their employees from driving to work. As initially conceived, employers could have simply charged fees for parking spaces without helping workers set up car pools. The Labor-Community Strategy Center, a local activist group, pointed out that this would have disproportionately affected people who could not afford to pay for parking spaces. As a compromise, regulators now review employers' plans and only approve those that mitigate the any unequal effects on poor and minority populations.

In response to the concern that traditional environmentalism does not recognize the social and economic components of environmental problems and solutions, a national movement for "environmental and economic justice" has spread across the country. Groups such as the Southwest Network for Environmental and Economic Justice attempt to frame the environment as part of the fight against racism and other inequalities.

In addition, the federal government has begun to address the debate over environmental racism. In 1992 the EPA established an Environmental Equity office. In addition, several bills that advocate environmental justice have been introduced. A 2007 study by the University of Colorado at Boulder, examined the sixty-one largest metropolitan areas in the United States to determine patterns in environmental racism. The study indicated that, although African-American and Hispanic residents generally live in more polluted areas of cities than Caucasians, the levels of inequality varied widely from city to city.

See also Comprehensive Environmental Response, Compensation, and Liability Act (CERCLA); Environmental economics; Environmental law; South.

Resources

BOOKS

Curry, Patrick. *Ecological Ethics: An Introduction.* Cambridge, UK: Polity Press, 2006.

Des Jardins, Joseph R. *Environmental Ethics: An Introduction to Environmental Philosophy.* Belmont, CA: Wadsworth Publishing, 2005.

PERIODICALS

Russell, D. "Environmental Racism." *Amicus Journal* 11 (Spring 1989): 22–32.

OTHER

U.S. General Accounting Office. *Siting of Hazardous Waste Landfills and Their Correlation with Racial and Economic Status of Surrounding Communities.* Washington, DC: U.S. Government Printing Office, 1983.

Alair MacLean

Environmental refugees

The term environmental refugee was coined in the late 1980s by the United Nations Environment Programme and refers to people who are forced to leave their community of origin because the land can no longer support them. Environmental factors such as soil erosion, drought, or floods, which are often coupled with poor socioeconomic conditions, are the cause of this loss of stability and security. Many environmental influences may cause such a displacement and include the deterioration of agricultural land, natural or "unnatural" disasters, climate change, the destruction resulting from war, and environmental scarcity.

Environmental scarcity can be supply induced, demand induced, or structural. Supply-induced scarcity refers to the depletion of agricultural resources, as in the erosion of cropland or overgrazing. Demand-induced scarcity occurs when consumption of a resource increases or when population increases, as is occurring in countries such as Philippines, Kenya, and Costa Rica. Structural scarcity results from the unequal social distribution of a resource within a community. The causes of environmental scarcity can occur simultaneously and in combination with each other, as seen in South Africa during the years of apartheid. Approximately 60 percent of the cropland in South Africa is marked by low organic content, and half of the country receives less than 19.5 inches (500 mm) of annual precipitation. When these factors were coupled with the rapid soil erosion, overcrowding, and unequal social distribution of resources, environmental scarcity resulted. Other environmental influences such as climate change and natural disasters can greatly compound the problems related to scarcity. Those countries which are especially vulnerable to these other influences are those which are already experiencing the precursors to scarcity, for example, highly populated countries such as Egypt and Bangladesh. Environmental problems place an added burden on a situation that is already under pressure. When this combination occurs in societies without strong social ties or political and economic stability, many times the people within the population have no choice but to relocate.

Environmental refugees tend to come from rural areas and developing countries—those most vulnerable to the influences of scarcity, climate change, and natural disasters. According to the Centers for Disease Control and Prevention, since the early 1960s most emergencies involving refugees have taken place in

Having just walked for two days from Eritrea with her family, Fatima Hassan, 32, constructs her home using sticks and woven mats that she carried, in a refugee camp in Suola, Ethiopia, 2008. *(AP Photo / Mandatory Credit: John Stanmeyer / VII)*

these less developed countries where resources are inadequate to support the population during times of need. In 1995 CDC directed forty-five relief missions to developing countries such as Angola, Bosnia, Haiti, and Sierra Leone.

The number of displaced people is rising worldwide. Of these, the number forced to migrate because of economic and environmental conditions is growing more rapidly than refugees from political strife. According to Dr. Norman Myers at the University of Oxford, there are 25 million environmental refugees today, compared with 20 million officially recognized refugees migrating due to political, religious, or ethical problems. It has been predicted that by the year 2010, this number could rise to 50 million. The number of migrants seeking environmental refuge is grossly underestimated because many do not actually cross borders but are forced to wander within their own country. As global warming causes major climate changes, these numbers could increase even more. Climate change alone may displace 150 million more people by the middle of the next century. Not only would a global climate change increase the number

of refugees, it could have a negative impact on agricultural production that would seriously limit the amount of food surpluses available to help displaced people.

Although approximately three out of every five refugees are fleeing from environmental hardships, this group of refugees is not legally recognized. According to the 1951 Convention on the Status of Refugees as modified by the 1967 Protocol, a legal refugee is a person who escapes a country and cannot re enter due to fear of persecution for reasons of race, religion, nationality, social affiliation, or political opinion. This definition requires both the element of persecution as well as cross-border migration. Because of these two requirements, states are not legally compelled to recognize environmental refugees; as mentioned above, many of these refugees never leave their own country, and it is unreasonable to expect them to prove fear of persecution. Environmental refugees are often forced to enter a country illegally because they cannot be granted protection or asylum. Many of the Mexican immigrants who enter the United States are escaping the sterile, unproductive land they have been living on. Over 60

Costa Rica's major industries, providing much needed economic development. People from around the world appreciate the beauty and the wonder—the intangible values—of these resources. Local people who would have been hired temporarily to cut down a forest can now be hired for a lifetime to work as park rangers and guides. Some would also argue that these nature preserves have value in themselves without reference to human needs, simply because they are filled with beautiful living birds, insects, plants, and animals.

Much of the dialogue in environmental resource management is about the need to balance the needs for economic growth and prosperity with needs for sustainable resource use. In a limited, finite world, there is a need to close the gap between the rates of consumption and rates of supply. The debate over how to assign value to different environmental resources is a lively one because the way that people think about their environment directly affects how they interact with the world.

Resources

BOOKS

Climate Variability and El Niño: Southern Oscillation: Implications for Natural Coastal Resources and Management. Helgoland marine research, v.62, supp.1. Berlin: Springer, 2008.

Hall, Michelle K., et al. *Exploring Water Resources.* Belmont, CA: Brooks Cole, 2006.

Rogers, Jerry R. *Environmental and Water Resources: Milestones in Engineering History : May 15–19, 2007, Tampa, Florida.* Reston, VA: American Society of Civil Engineers, 2007.

PERIODICALS

Coghlan, Andy. "Earth Suffers as We Gobble Up Resources." *New Scientist* 195 no. 2611 (July 7, 2007): 15.

Dias, Rubens A., et al. "The Limits of Human Development and the Use of Energy and Natural Resources." *Energy Policy* 34 (2006): 1026–1031.

OTHER

United Nations System-Wide EarthWatch. "Global/ Regional Assessments and Resources: World." http:// earthwatch.unep.net//assessment/global/index.php (accessed September 29, 2010).

United States Environmental Protection Agency (EPA). "Environmental Management: Resources Management." http://www.epa.gov/ebtpages/enviresources management.html (accessed September 29, 2010).

John Cunningham

Environmental restoration *see* **Restoration ecology.**

Environmental risk analysis *see* **Risk analysis.**

Environmental science

Environmental science is not constrained within any one discipline; it is a comprehensive field. A considerable amount of environmental research is accomplished in specific departments such as chemistry, physics, civil engineering, or the various biology disciplines. Much of this work is confined to a single field, with no interdisciplinary perspective. These programs produce scientists who build on their specific training to continue work on environmental problems, sometimes in a specific department, sometimes in an interdisciplinary environmental science program.

Many new academic units are interdisciplinary, their members and graduates specifically designated as environmental scientists. Most have been trained in a specific discipline, but they may have degrees from almost any scientific background. In these units the degrees granted—from B.S. to Ph.D.—are in Environmental Science, not in a specific discipline.

Environmental science is not ecology, though that discipline may be included. Ecologists are interested in the interactions between an organism and its surroundings. Most ecological research and training does not focus on environmental problems except as those problems impact the organism of interest. Environmental scientists may or may not include organisms in their field of view: they mostly focus on the environmental problem, which may be purely physical in nature. For example, acid deposition can be studied as a problem of emissions and as a characteristic of the atmosphere without necessarily examining its impact on organisms. An alternate focus might be on the acidification of lakes and the resulting implications for resident

Air quality monitoring, Hadera, israel. (© PhotoStock-Israel / Alamy)

fish. Both studies require expertise from more than one traditional discipline; they are studies in environmental science.

Environmental science issues focus on the impact that humans have on the environment, how people can resolve past actions and change the affect on the environment, and what future effects these actions will have. Environmental science incorporates research from biology, chemistry, ecology, geology, engineering, and physics as well as economics, political science, anthropology, law, history, and sociology to approach environmental problems. Key topics in environmental science include climate change; air, water, and soil pollution; biodiversity; waste management; and conservation of natural resources. Climate change, a global warming of the earth, is being caused by the rapid accumulation of human-generated pollutants, such as carbon dioxide (CO_2), in the atmosphere. A significant amount of current environmental science research is concentrated on the causes, present status, and predictions of climate change and methods of mitigating or eliminating pollutant affects and emissions.

See also Air quality; Environment; Environmental ethics; Water quality.

Resources

BOOKS

Arms, Karen. *Environmental Science*. Orlando: Holt, Rinehart & Winston, 2006.

Berg, Linda, and Mary Catherine Hager. *Visualizing Environmental Science*. New York: Wiley, 2006.

Chiras, Daniel D. *Environmental Science*. Sudbury, MA: Jones and Bartlett, 2006.

Cunningham, W. P. and A. Cunningham. *Environmental Science: A Global Concern*. New York: McGraw-Hill International Edition, 2008.

Enger, Eldon, and Bradley Smith. *Environmental Science: A Study of Interrelationships*. New York: McGraw-Hill, 2006.

Wright, Richard T. *Environmental Science: Toward a Sustainable Future*. Upper Saddle River, NJ: Pearson, 2008.

Gerald L. Young

Environmental stress

In the ecological context, environmental stress can be considered any environmental influence that causes a discernible ecological change, especially in terms of a constraint on ecosystem development. Stressing agents (or stressors) can be exogenous to the ecosystem, as in the cases of long-range transported acidifying substances, toxic gases, or pesticides. Stress can also cause change as a result of an accentuation of some pre-existing site factor beyond a threshold for biological tolerance, for example, thermal loading, nutrient availability, wind, or temperature extremes.

Often implicit within the notion of environmental stress, particularly from the perspective of ecosystem managers, is a judgement about the quality of the ecological change. That is, from the human perspective, whether the effect is good or bad.

Environmental stressors can be divided into several, not necessarily exclusive, classes of causal agencies:

- Physical stress refers to episodic events (or disturbances) associated with intense, but usually brief, loadings of kinetic energy, perhaps caused by a windstorm, volcanic eruption, or an explosion.
- Wildfire is another episodic stress, usually causing a mass mortality of ecosystem dominants such as trees or shrubs and a rapid combustion of much of the biomass of the ecosystem.
- Pollution occurs when certain chemicals are bioavailable in a sufficiently large amount to cause toxicity. Toxic stressors include gaseous air pollutants such as sulfur dioxide and ozone, metals such as lead and mercury, residues of pesticides, and even nutrients that may be beneficial at small rates of supply but damaging at higher rates of loading.
- Nutrient impoverishment implies an inadequate availability of physiologically essential chemicals, which imposes an oligotrophic constraint upon ecosystem development.
- Thermal stress occurs when heat energy is released into an ecosystem, perhaps by aquatic discharges of low-grade heat from power plants and other industrial sources.
- Exploitative stress refers to the selective removal of particular species or size classes. Exploitation by humans includes the harvesting of forests or wild animals, but it can also involve natural herbivory and predation, as with infestations of defoliating insects such as locusts, spruce budworm, or gypsy moth, or irruptions of predators such as crown-of-thorns starfish.
- climatic stress is associated with an insufficient or excessive regime of moisture, solar radiation, or temperature. These can act over the shorter term as weather, or over the longer term as climate.

Within most of these contexts, stress can be exerted either chronically or episodically. For example, the toxic gas sulfur dioxide can be present in a chronically elevated concentration in an urbanized region with a large number of point sources of emission. Alternatively, where the emission of sulfur dioxide is dominated by a single, large point source such as a smelter or power plant, the toxic stress associated with this gas occurs as relatively short-term events of fumigation.

Environmental stress can be caused by natural agencies as well as resulting directly or indirectly from the activities of humans. For example, sulfur dioxide can be emitted from smelters, power plants, and homes, but it can also be emitted in large quantities by volcanoes. Similarly, climate change has always occurred naturally, but it may also be forced by human activities that result in emissions of carbon dioxide, methane, and nitrous oxide into the atmosphere.

Over most of Earth's history, natural stressors have been the dominant constraints on ecological development. Increasingly, however, the direct and indirect consequences of human activities are becoming dominant environmental stressors. This is caused by both the increasing human population and by the progressively increasing intensification of the per-capita effect of humans on the environment.

Resources

BOOKS

Cunningham, William, and Mary Ann Cunningham. *Environmental Science: A Global Concern*, 10th ed. New York: McGraw-Hill, 2007.

Bill Freedman

Environmental Working Group

The Environmental Working Group is a public interest research group that monitors public agencies and public policies on topics relating to environmental and social justice. EWG publicizes its findings in research reports that emphasize both national and local implications of federal laws and activities. These research reports are based on analysis of public databases, often obtained through the Freedom of Information Act. In operation since 1993 EWG is a non-profit organization funded by grants from private foundations. EWG is based in Washington, DC, and Oakland, California, and is associated with the Tides Center of San Francisco. The organization performs its research both independently and in collaboration with other public interest research groups such as the Sierra Club and the Surface Transportation Policy

Project (a nonprofit coalition focusing on social and environmental quality in transportation policy).

EWG specializes in analyzing large computer databases maintained by government agencies, such as the Toxic Release Inventory database maintained by the Environmental Protection Agency to record spills of toxic chemicals into the air or water, or the Regulatory Analysis Management System, maintained by the Army Corps of Engineers for internal tracking of permits granted for filling and draining wetlands. Because many of these data sources are capable of exposing actions or policies embarrassing to public agencies, EWG has often obtained data by using the Freedom of Information Act, which legally enforces release of data belonging to the public domain. Many of the databases researched by EWG have never been thoroughly analyzed before, even by the agency collecting the data. EWG is unusual in working with primary data—going directly to the original database—rather than basing its research on secondary sources, anecdotal information, or interviews.

Research findings are published both in print and electronically on the Internet. Electronic publishing allows immediate and inexpensive distribution of reports that concern issues of current interest. EWG is a prolific source of information, producing extensive, detailed reports, often at a rate of more than one a month. Among the environmental topics on which the EWG has reported are drinking water quality, wetland protection and destruction, and the impacts of agricultural pesticides on both farm workers and consumers. Social justice and policy issues that EWG has researched include campaign finance reform, inequalities and inefficiency in farm subsidy programs, and threats to public health and the environment from medical waste. For each general topic, EWG usually publishes a series of articles ranging from the nature and impact of a federal law to what individuals can do about the current problem. Also included with research reports are state-by-state and county summaries of statistics, which provide details of local implications of the general policy issues.

Resources

BOOKS

Arms, Karen. *Environmental Science*. Orlando, FL: Holt, Rinehart & Winston, 2006.

Burns, Ronald G., and Michael J. Lynch. *Environmental Crime: A Sourcebook*. New York: LFB Scholarly Publishing, 2004.

Crane, Jeff, and Michael Egan. *Natural Protest: Essays on the History of American Environmentalism*. New York: Routledge, 2009.

OTHER

Center for Environmental Citizenship. "Center for Environmental Citizenship." http://www.envirocitizen.org/ (accessed September 29, 2010).

Environmental Law & Policy Center. "Environmental Law & Policy Center." http://www.elpc.org/ (accessed September 29, 2010).

ORGANIZATIONS

Environmental Working Group, 1436 U Street, NW, Suite 100 , Washington, DC, USA, 20009, (202) 667-6982, info@ewg.org, http://www.ewg.org

Mary Ann Cunningham

Environmentalism

Environmentalism is the ethical and political perspective that places the health, harmony, and integrity of the natural environment at the center of human attention and concern. From this perspective human beings are viewed as *part of* nature rather than as overseers. Therefore, to care for the environment is to care about human beings, since they cannot live without the survival of the natural habitat.

Although there are many different views within the very broad and inclusive environmentalist perspective, several common features can be discerned. The first is environmentalism's emphasis on the interdependence of life and the conditions that make life possible. Human beings, like other animals, need clean air to breathe, clean water to drink, and nutritious food to eat. Without these necessities, life would be impossible. Environmentalism views these conditions as being both basic and interconnected. For example, fish contaminated with polychlorinated biphenyl (PCB), mercury, and other toxic substances are not only hazardous to humans but to bears, eagles, gulls, and other predators. Likewise, mighty whales depend on tiny plankton, cows on grass or corn, koala bears on eucalyptus leaves, bees on flowers, and flowers on bees and birds, and so on through all species and ecosystems. All animals, human and nonhuman alike, are interdependent participants in the cycle of birth, life, death, decay, and rebirth.

A second emphasis of environmentalism is on the value of life—not only human life but all life, from the tiniest microorganism to the largest whale. Because the fate of the human species is inextricably tied with all others and because life requires certain conditions to sustain it, environmentalists contend that people have an obligation to respect and care for anything that nurtures and sustains life in its many forms.

Although environmentalists agree on some issues, there are also a number of disagreements about the purposes of environmentalism and about how to best achieve those ends. Some environmentalists emphasize the desirability of conserving natural resources for recreation, sightseeing, hunting, and other human activities, both for present and future generations. Such a utilitarian view has been sharply criticized by Arne Naess and other proponents of deep ecology who claim that the natural environment has its own intrinsic value apart from any aesthetic, recreational, or other value assigned to it by human beings. Bears, for example, have their own intrinsic value or worth, quite apart from that assigned to their existence via shadow pricing or other mechanisms, by bear-watchers, hunters, or other human beings.

Environmentalists also differ on how best to conserve, reserve, and protect the natural environment. Some groups, such as the Sierra Club and the Nature Conservancy, favor gradual, low-key legislative and educational efforts to inform and influence policy makers and the general public about environmental issues. Other more radical environmental groups, such as the Sea Shepherd Conservation Society and Earth First!, favor carrying out direct action by employing ecotage (ecological sabotage), or monkey-wrenching, to stop strip mining, logging, drift net fishing, and other activities that they deem dangerous to animals and ecosystems. Within this environmental spectrum are many other groups, including the World Wildlife Fund, Greenpeace, Earth Island Institute, Clean Water Action, and other organizations which use various techniques to inform, educate, and influence public opinion regarding environmental issues and to lobby policy makers.

Despite these and other differences over means and ends, environmentalists agree that the natural environment, whether valued instrumentally or intrinsically, is valuable and worth preserving for present and future generations.

Resources

BOOKS

Crane, Jeff, and Michael Egan. *Natural Protest: Essays on the History of American Environmentalism*. New York: Routledge, 2009.

Lytle, Mark H. *The Gentle Subversive: Rachel Carson, Silent Spring, and the Rise of the Environmental Movement*. New narratives in American history. New York: Oxford University Press, 2007.

Maher, Neil M. *Nature's New Deal: The Civilian Conservation Corps and the Roots of the American Environmental Movement*. New York: Oxford University Press, 2007.

Olson, Robert L., and David Rejeski. *Environmentalism and the Technologies of Tomorrow: Shaping the Next Industrial Revolution*. Washington, DC: Island Press, 2005.

Terence Ball

Environmentally preferable purchasing

Environmentally preferable purchasing (EPP) invokes the practice of buying products with environmentally-sound qualities—reduced packaging, reusability, energy efficiency, recycled content, and rebuilt or re-manufactured products. It was first addressed officially with Executive Order (EO) 12873 in October 1993, "Federal Acquisition, Recycling and Waste Prevention," but was further enhanced on September 14, 1998, by EO 13101 also signed by President Clinton. Titled "Greening the Government through Waste Prevention, Recycling, and Federal Acquisition," it superseded EO 12873, but retained similar directives for purchasing. The "Final Guidance" of directives was issued through the Environmental Protection Agency in 1995.

What the federal government would adopt as a guideline for its purchases also would mark the beginning of environmentally preferable purchasing for the private sector, and create an entirely new direction for individuals and businesses as well as governments. At the federal level, the EPA's "Final Guidance" applied to all acquisitions, from supplies and services to buildings and systems. Initially the EPA developed five guiding principles for incorporating the plan into the federal government setting.

The five guiding principles are:

- Environment + Price + Performance = Environmentally Preferable Purchasing
- Pollution prevention
- Life cycle perspective/multiple attributes
- Comparison of environmental impacts
- Environmental performance information

In January 2007 President George W. Bush issued EO 13423, "Strengthening Federal Environmental, Energy, and Transportation Management." The order also defines environmentally preferable as "products or services that have a lesser or reduced effect on human health and the environment when compared with competing products or services that serve the same purpose." Furthermore, EO 13423 ordered all federal agencies to use sustainable practices when purchasing products and services.

In the private world of business, environmentally preferable purchasing promises to save money, in addition to meeting EPA regulations and improving employee safety and health. In an age of environmental liability, EPP can make the difference when a question of environmental ethics or damage arises.

For the private consumer, purchasing "green" in the late 1960s and 1970s tended to mean something as simple as recycled paper used in Christmas cards or unbleached natural fibers for clothing. In the twenty-first century, the average American home is affected in countless additional ways—energy-efficient kitchen appliances and personal computers, environmentally-sound household cleaning products, and neighborhood recycling centers. To be certified as "green," characteristics of items, such as recyclability, biodegradability, amount of organic ingredients, and lack of ozone-depleting chemicals, are tested.

Of those everyday uses, the concern over cleaning products for home, industrial, and commercial use has the focus of particular attention. Massachusetts was one of the states that took a lead on the issue of decreasing toxic chemicals with its Massachusetts Toxic Use Reduction Act. With a focus on products that have known carcinogens and ozone-depleting substances, excessive phosphate concentrations, and volatile organic compounds, testing identifies alternative products that are more environmentally acceptable—and safer—for humans and all forms of life as well.

The products approved for purchasing must meet the following mandated criteria:

- contain no ingredients from the Massachusetts Toxic Use Reduction Act list of chemicals
- contain no carcinogens appearing on lists established by the International Agency for Research on Cancer, the National Toxicology Program, or the Occupational Safety and Health Administration; and cannot contain any chemicals defined as Class A, B, or C carcinogens by the EPA
- contain no ozone-depleting ingredients
- must be compliant with the phosphate content levels stipulated in Massachusetts law
- must be compliant with the Volatile Organic Compound (VOC) content levels stipulated in Massachusetts law

The National Association of Counties offers an extensive list of EPP resources links through its Web site. In addition to offices and agencies of the federal government, the states of Minnesota and Massachusetts, and the local governments of King County, Washington, and Santa Monica, California, the list includes such organizations as Buy Green, Earth Systems' Virtual Shopping Center for the Environment (an online database of recycling industry products and services), the Environmental Health Coalition, Green Seal, the National Institute of Government Purchasing, Inc., and the National Pollution Prevention Roundtable. Businesses and business-related companies mentioned include the Chlorine Free Products Association, Pesticide Action Network of North America, Chlorine-Free Paper Consortium, and the Smart Office Resource Center.

Resources

BOOKS

Reeve, Tim, and Barb Everdene. *Applying Total Cost of Ownership to Sustainability Purchasing: Workbook.* Vancouver: Sustainability Purchasing Network, 2006.

OTHER

United States Environmental Protection Agency (EPA). "Compliance And Enforcement: Voluntary Standards: Environmentally Preferable Purchasing." http://www.epa.gov/ebtpages/compvoluntarystand environmentally preferablepurch.html (accessed August 1, 2009).

ORGANIZATIONS

National Association of Counties, 25 Massachusetts Avenue, Washington, DC, USA, 20001, (202) 393-6226, (202) 393-2630, http://www.naco.org.
U.S. Environmental Protection Agency, 1200 Pennsylvania Avenue, NW, Washington, DC, USA, 20460, (202) 260–2090, http://www.epa.gov

Jane E. Spear

Environmentally responsible investing

Environmentally responsible investing is one component of a larger phenomenon known as *socially responsible investing*. The idea is that investors should use their money to support industries whose operations accord with the investors' personal ethics. This concept is not a new one. In the early part of the century, Methodists, Presbyterians, and Baptists shunned companies that promoted sinful activities such as smoking, drinking, and gambling. More recently many investors chose to protest apartheid by divesting from companies with operations in South Africa. Investors today might arrange their investment portfolios to reflect companies' commitments to affirmative action, human rights, animal rights, the environment, or any other issues the investors believe to be important.

The purpose of environmentally responsible investing is to encourage companies to improve their environmental records. The emergence and growth of mutual funds identifying themselves as environmentally oriented funds indicates that environmentally responsible investing is a popular investment area. By 2007 socially responsible investing, which includes environmentally responsible investing as a major component in investment selections, had grown to a $2.7 trillion industry. The naming of these funds can be misleading, however. Some funds have been developed for the purpose of being environmentally responsible; others have been developed for the purpose of reaping the anticipated profits in the environmental services sector as environmentalists in the marketplace and environmental regulations encourage the purchasing of green products and technology. These funds are not necessarily environmentally responsible; some companies in the environmental clean-up industry, for example, have less-than-perfect environmental records.

As the idea of environmentally responsible investing is still new, a generally accepted set of criteria for identifying environmentally responsible companies has not yet emerged.

When grading a company in terms of its behavior toward the environment, one could use an absolute standard. For example, one could exclude all companies that have violated any Environmental Protection Agency (EPA) standards. The problem with such a standard is that some companies that have very good overall environmental records have sometimes failed to meet certain EPA standards. Alternatively, a company could be graded on its efforts to solve environmental problems. Some investors prefer to divest of all companies in heavily polluting industries, such as oil and chemical companies; others might prefer to use a relative approach and examine the environmental records of companies within industry groups. By directly comparing oil companies with other oil companies, for example, one can identify the particular companies committed to improving the environment.

For consistency some investors might choose to divest from all companies that supply or buy from an environmentally irresponsible company. It then becomes an arbitrary decision as to where this process stops. If taken to an extreme, the approach rejects holding United States treasury securities, since public funds are used to support the military, one of the world's largest polluters and a heavy user of nonrenewable energy.

A potential new indicator for identifying environmentally responsible companies has been developed by the Coalition for Environmentally Responsible Economies (CERES); it is a code called the Ceres Principles. The principles are the environmental equivalent of the Sullivan Principles, a code of conduct for American companies operating in South Africa. Formerly called the Valdez Principles, these directives encourage companies to strive for sustainable use of natural resources and the reduction and safe disposal of waste. By signing the principles, companies commit themselves to continually improving their behavior toward the environment over time. So far, however, few companies have signed the code, possibly because it requires companies to appoint environmentalists to their boards of directors.

As there is no generally accepted set of criteria for identifying environmentally responsible companies, investors interested in such an investment strategy must be careful about accepting "environmentally responsible" labels. Investors must determine their own set of screening criteria based on their own personal beliefs about what is appropriate behavior with respect to the environment.

See also Valdez Principles.

Resources

BOOKS

1001 Easy Ways for Earth-Wise Living: Natural and Eco-Friendly Ideas That Can Make a Real Difference to Your Life. Ultimo, NSW, New Zealand: Reader's Digest, 2006.

Kiernan, Matthew J. *Investing in a Sustainable World: Why GREEN Is the New Color of Money on Wall Street.* New York: AMACOM, 2009.

Barbara J. Kanninen

Enzyme

Enzymes are catalysts that speed the rate at which chemical reactions occur. In living systems catalysts are proteins that facilitate reactions by lowering the activation energy or energy required to complete physiological and cellular reactions. As catalysts, enzymes alter the pace of reactions without undergoing specific permanent change. They are crucial to life because without them, the vast majority of biochemical reactions would occur too slowly for organisms to survive.

In general, enzymes catalyze two quite different kinds of reactions. The first type of reaction includes those by which simple compounds are combined with each other to make new tissue from which plants and animals are made. For example, the most common enzyme in nature is probably carboxydismutase, the enzyme in green plants that couples carbon dioxide with an acceptor molecule in one step of the photosynthesis process by which carbohydrates are produced.

Enzymes also catalyze reactions by which more complex compounds are broken down to provide the energy needed by organisms. The principal digestive enzyme in the human mouth, for example, is ptyalin (also known as a - amylase), which begins the digestion of starch.

Enzymes have both beneficial and harmful effects in the environment. For example environmental hazards such as heavy metals, pesticides, and radiation often exert their effects on an organism by disabling one or more of its critical enzymes. As an example, arsenic is poisonous to animals because it forms a compound with the enzyme glutathione. The enzyme is disabled and prevented from carrying out its normal function, the maintenance of healthy red blood cells.

David E. Newton

EPA *see* **Environmental Protection Agency (EPA).**

Ephemeral species

Ephemeral species are plants and animals whose lifespan lasts only a few weeks or months. The most common types of ephemeral species are desert annuals, plants whose seeds remain dormant for months or years but which quickly germinate, grow, and flower when rain does fall. In such cases the amount and frequency of rainfall determine entirely how frequently ephemerals appear and how long they last. Tiny, usually microscopic, insects and other invertebrate animals often appear with these desert annals, feeding on briefly available plants, quickly reproducing, and dying in a few weeks or less. Ephemeral ponds, short-duration desert rain pools, are especially noted for supporting ephemeral species. Here small insects and even amphibians have short lives. The spadefoot toad (*Scaphiopus multiplicatus*), for example, matures and breeds in as little as eight days after a rain, feeding on short-lived brine shrimp, which in turn consume algae and plants that live as long as water or soil moisture lasts. Eggs, or sometimes the larvae of these animals, then remain in the soil until the next moisture event.

Ephemerals play an important role in many plant communities. In some very dry deserts, as in North Africa, ephemeral annuals comprise the majority of living species—although this rich flora can remain hidden for years at a time. Often widespread and abundant after a rain, these plants provide an essential food source for desert animals, including domestic livestock. Because water is usually unavailable in such environments, many desert perennials also behave like ephemeral plants, lying dormant and looking dead for months or years but suddenly growing and setting seed after a rare rain fall.

The frequency of desert ephemeral recurrence depends upon moisture availability. In the Sonoran Desert of California and Arizona, annual precipitation allows ephemeral plants to reappear almost every year. In the drier deserts of Egypt, where rain may not fall for a decade or more, dormant seeds must survive for a much longer time before germination. In addition, seeds have highly sensitive germination triggers. Some annuals that require at least 1 inch (2-3 cm) of precipitation to complete their life cycle will not germinate when less than that has fallen. In such a case seed coatings may be sensitive to soil salinity, which decreases as more rainfall seeps into the ground. Annually-recurring ephemerals often respond to temperature as well. In the Sonoran Desert some rain falls in both summer and winter. Completely different summer and winter floral communities appear in response. Such adaptation to different temporal niches probably helps decrease competition for space and moisture and increase each species' odds of success.

Although they are less conspicuous, ephemeral species also occur outside desert environments. Short-duration food supplies or habitable conditions in some marine environments lead to ephemeral species growth. Ephemerals successfully exploit such unstable environments as volcanoes and steep slopes prone to slippage. More common are spring ephemerals in temperate deciduous forests. For a few weeks between snow melt and closure of the overstory canopy, quick-growing ground plants, including small lilies and violets, sprout and take advantage of available sunshine. Flowering and setting seed before they are shaded out by larger

vegetation, these ephemerals disappear by mid summer. Some persist in the form of underground root systems, but others are true ephemerals, with only seeds remaining until the next spring.

See also Adaptation; Food chain/web; Opportunistic organism.

Resources

BOOKS

Turner, R. M., Janice Emily Bowers, Tony L. Burgess, and James Rodney Hastings. *Sonoran Desert Plants: An Ecological Atlas.* Tucson: University of Arizona Press, 2005.

Mahmoud, Tamer. *Desert Plants of Egypt's Wadi El Gemal National Park.* Cairo: American University in Cairo Press, 2010.

Chambers, Nina, and John A. Hall. *Lessons Learned: Sonoran Desert Ecosystem Initiative.* Tucson, AZ: Nature Conservancy in Arizona, 2005.

Mary Ann Cunningham

Epidemiology

Epidemiology is the study of the origin and cause of disease among populations. Epidemiology is an increasingly important part of public health because rapidly increasing human populations provide a fertile breeding ground for microbes. As the planet becomes more crowded and the distances that separate communities become smaller, infectious diseases that were one limited by geographic isolation now spread globally. In addition, many viruses and bacteria constantly evolve and shift among populations and across species. Epidemiologists use an array of skills to trace the source of disease.

The epidemiologist is concerned with the interactions of organisms and their environments. Environmental factors related to disease may include geographical features, climate, the concentration of pathogens in soil and water, or other factors. Epidemiologists determine the numbers of individuals affected by a disease, the environmental contributions to disease, the causative agent or agents of disease, and the transmission patterns and lethality of disease.

Origins of epidemiology

Epidemiology is commonly thought to be limited to the study of infectious diseases, but that is only one aspect of the medical specialty. The epidemiology of the environment and lifestyles has been studied since Hippocrates's time. More recently, scientists have broadened the worldwide scope of epidemiology to

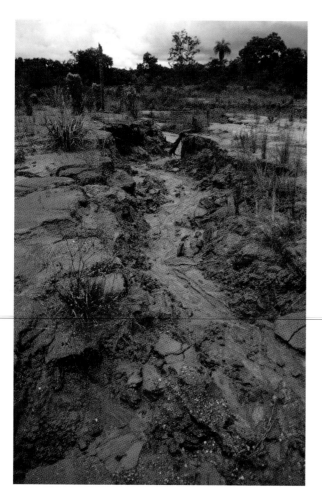

Erosion due to deforestation in the Brazilian rainforest.
(© Peter Arnold, Inc. / Alamy)

The term geologic erosion refers to the normal, natural loss of land caused by geological processes acting over long periods of time, undisturbed by humans. Accelerated erosion is a more rapid erosion process influenced by human, or sometimes animal, activities. Accelerated erosion in North America has only been recorded for the past few centuries, and in research studies, post-settlement erosion rates were found to be eight to 350 times higher than pre-settlement erosion rates.

Soil erosion has been both accelerated and controlled by humans since recorded history. In Asia, the Pacific, Africa, and South America, complex terracing and other erosion control systems on arable land go back thousands of years. Soil erosion and the resultant decreased food supply have been linked to the decline of historic, particularly Mediterranean, civilizations, though the exact relationship with the decline of governments such as the Roman Empire is not clear.

A number of terms have been used to describe different types of erosion, including gully erosion, rill erosion, interrill erosion, sheet erosion, splash erosion, saltation, surface creep, suspension, and siltation. In gully erosion, water accumulates in narrow channels and, over short periods, removes the soil from this narrow area to considerable depths, ranging from 1.5 feet (0.5 m) to as much as 82–98 feet (25–30 m).

Rill erosion refers to a process in which numerous small channels of only a few inches in depth are formed, usually occurring on recently cultivated soils. Interrill erosion is the removal of a fairly uniform layer of soil on a multitude of relatively small areas by rainfall splash and film flow.

Usually interpreted to include rill and interril erosion, sheet erosion is the removal of soil from the land surface by rainfall and surface runoff. Splash erosion, the detachment and airborne movement of small soil particles, is caused by the impact of raindrops on the soil.

Saltation is the bouncing or jumping action of soil and mineral particles caused by wind, water, or gravity. Saltation occurs when soil particles 0.1–0.5 millimeters in diameter are blown to a height of less than 6 inches (15 cm) above the soil surface for relatively short distances. The process includes gravel or stones affected by the energy of flowing water, as well as any soil or mineral particle movement downslope due to gravity.

Surface creep, which usually requires extended observation to be perceptible, is the rolling of dislodged particles 0.5–1.0 millimeters in diameter by wind along the soil surface. Suspension occurs when soil particles less than 0.1 millimeters diameter are blown through the air for relatively long distances, usually at a height of less than 6 inches (15 cm) above the soil surface. In siltation, decreased water speed causes deposits of water-borne sediments, or silt, to build up in stream channels, lakes, reservoirs, or flood plains.

In the water erosion process, the eroded sediment is often higher (enriched) in organic matter, nitrogen, phosphorus, and potassium than in the bulk soil from which it came. The amount of enrichment may be related to the soil, amount of erosion, the time of sampling within a storm, and other factors. Likewise, during a wind erosion event, the eroded particles are often higher in clay, organic matter, and plant nutrients. Frequently, in the Great Plains, the surface soil becomes increasingly more sandy over time as wind erosion continues.

Erosion measurements using the Universal Soil Loss Equation (USLE) and the Wind Erosion Equation (WEE) estimate erosion on a point basis expressed in mass per unit area. If aggregated for a large area (e.g., state or nation), very large numbers are generated and have been used to give misleading conclusions. The estimates of USLE and WEE indicate only the soil moved from a point. They do not indicate how far the sediment moved or where it was deposited. In cultivated fields the sediment may be deposited in other parts of the field with different crop cover or in areas where the land slope is less. It may also be deposited in riparian land along stream channels or in flood plains.

Only a small fraction of the water-eroded sediment leaves the immediate area. For example, in a study of five river watersheds in Minnesota, it was estimated that less than 1–27 percent of the eroded material entered stream channels, depending on the soil and topographic conditions. The deposition of wind-eroded sediment is not well quantified, but much of the sediment is probably deposited in nearby areas more protected from the wind by vegetative cover, stream valleys, road ditches, woodlands, or farmsteads.

Although a number of national and regional erosion estimates for the United States have been made since the 1920s, the methodologies of estimation and interpretations have been different, making accurate time comparisons impossible. The most extensive surveys have been made since the Soil, Water and Related Resources Act was passed in 1977. In these surveys a large number of points were randomly selected, data assembled for the points, and the Universal Soil Loss Equation (USLE) or the Wind Erosion Equation (WEE) used to estimate erosion amounts. Although these equations were the best available at the time, their results are only estimations, and subject to interpretation. Considerable research on improved methods of estimation is underway by the U.S. Department of Agriculture (USDA).

In the cornbelt of the United States, water erosion may cause a 1.7–7.8 percent drop in soil productivity over the next one hundred years, as compared to current levels, depending on the topography and soils of the area. The USDA results, based on estimated erosion amounts for 1977, only included sheet erosion, not losses of plant nutrients. Though the figures may be low for this reason, other surveys have produced similar estimates.

In addition to depleting farmlands, eroded sediment causes off-site damages that, according to one study, may exceed on site loss. The sediment may end up in a domestic water supply, clog stream channels, even degrade wetlands, wildlife habitats, and entire ecosystems.

See also Environmental degradation; Gullied land; Soil eluviation; Soil organic matter; Soil texture.

Resources

BOOKS

Evans, Edward, and Edmund C. Penning-Rowsell. *Future Flooding and Coastal Erosion Risks.* London: Thomas Telford, 2007.

Gifford, Clive. *Weathering and Erosion.* North Mankato, MN: Smart Apple Media, 2005.

Montgomery, David R. *Dirt: The Erosion of Civilizations.* Berkeley: University of California Press, 2007.

Morgan, R. C. P. *Soil Erosion and Conservation,* 3rd ed. New York: Wiley-Blackwell, 2005.

OTHER

National Geographic Society. "Erosion and Weathering." http://science.nationalgeographic.com/science/earth/the-dynamic-earth/weathering-erosion-article.html (accessed November 11, 2010).

United States Department of the Interior, United States Geological Survey (USGS). "Erosion." http://www.usgs.gov/science/science.php?term=353 (accessed November 11, 2010).

United States Environmental Protection Agency (EPA). "Ecosystems: Soils: Erosion." http://www.epa.gov/ebtpages/ecossoilserosion.html (accessed November 11, 2010).

William E. Larson

Escherichia coli

Escherichia coli, or *E. coli* as a common short form, is a bacterium in the family *Enterobacteriaceae* that is found in the intestines of warm-blooded animals, including humans. *E. coli* represent about 0.1 percent of the total bacteria of an adult's intestines (on a Western diet). As part of the normal flora of the human intestinal tract, *E. coli* aids in food digestion by producing vitamin K and B-complex vitamins from undigested materials in the large intestine and suppresses the growth of harmful bacterial species. However, *E. coli* has also been linked to diseases in about every part of the body. Pathogenic strains of *E. coli* have been shown to cause pneumonia, urinary tract infections, wound and blood infections, and meningitis.

Toxin-producing strains of *E. coli* can cause severe gastroenteritis (hemorrhagic colitis), which can include abdominal pain, vomiting, and bloody diarrhea. In most people, the vomiting and diarrhea stop within two to three days. However, about 5–10 percent of the those affected will develop hemolytic-uremic syndrome (HUS), which is a rare condition that affects mostly children under the age of ten but also may affect the elderly as well as persons with other illnesses. About 75 percent of HUS cases in the United States are caused by an enterohemorrhagic (intestinally-related organism that causes hemorrhaging) strain of *E. coli* referred to as *E. coli* O157:H7, while the remaining cases are caused by non-O157 strains. *E. coli*. O157:H7 is found in the intestinal tract of cattle. In the United States the Centers for Disease Control and Prevention estimates that there are about 10,000–20,000 infections and 500 deaths annually that are caused by *E. coli* O157:H7.

E. coli O157:H7 is also a concern globally. In one example, runoff from a cattle farm contaminated the well-water drinking supply of Walkerton, Ontario, Canada in the summer of 2000. In the resulting illness outbreak, about 2,000 people were sickened, seven people died, and several were afflicted with life-long damage to the kidneys and other organs.

E. coli O157:H7, first identified in 1982, and isolated with increasing frequency since then, is found in contaminated foods such as meat, dairy products, and juices. Symptoms of an *E. coli* O157:H7 infection start about seven days after infection with the bacteria. The first symptom is sudden onset of severe abdominal cramps. After a few hours, watery diarrhea begins, causing loss of fluids and electrolytes (dehydration), which causes the person to feel tired and ill. The watery diarrhea lasts for about a day, and then changes to bright red bloody stools, as the infection causes sores to form in the intestines. The bloody diarrhea lasts for two to five days, with as many as ten bowel movements a day. Additional symptoms may include nausea and vomiting, without a fever or with only a mild fever. After about five to ten days, HUS can develop. HUS is characterized by destruction of red blood cells, damage to the lining of blood vessel walls, reduced urine production, and in severe cases, kidney failure. Toxins produced by the bacteria enter the blood stream, where they destroy red blood cells and platelets, which contribute to the clotting of blood. The damaged red blood cells and platelets clog tiny blood vessels in the kidneys, or cause lesions to form in the kidneys, making it difficult for the kidneys to remove wastes and extra fluid from the body, resulting in hypertension, fluid accumulation, and reduced production of urine. The diagnosis of an *E. coli* infection is made through a stool culture.

Treatment of HUS is supportive, with particular attention to management of fluids and electrolytes. Some studies have shown that the use of antibiotics and antimotility agents during an *E. coli* infection may worsen the course of the infection and should be avoided. Ninety percent of children with HUS who receive careful supportive care survive the initial acute stages of the condition, with most having no long-term effects. In about 50 percent of the cases, short-term replacement of kidney function is required in the form of dialysis. However, between 10 and 30 percent of the survivors will have kidney damage that will lead to kidney failure immediately or within several years. These children with kidney failure require on going dialysis to remove wastes and extra fluids from their bodies or may require a kidney transplant.

The most common way an *E. coli* O157:H7 infection is contracted is through the consumption of undercooked ground beef (e.g., eating hamburgers that are still pink inside). Healthy cattle carry *E. coli* within their intestines. During the slaughtering process, the meat can become contaminated with the *E. coli* from the intestines. When contaminated beef is ground up, the *E. coli* bacteria are spread throughout the meat.

In 1992 and 1993, the bacterium was the source of illnesses that afflicted people who had eaten at several fast food outlets in the U. S. Midwest. The cause of the illness and several deaths were traced to contaminated beef that had been processed into hamburger and subsequently undercooked during preparation.

Additional ways to contract an *E. coli* infection include drinking contaminated water and unpasteurized milk and juices, eating contaminated fruits and vegetables, and working with cattle. The infection is also easily transmitted from an infected person to others in settings such as day care centers and nursing homes when improper sanitary practices are used.

Prevention of HUS caused by ingestion of foods contaminated with *E. coli* O157:H7 and other toxin-producing bacteria is accomplished through practicing hygienic food preparation techniques, including adequate hand washing; cooking of meat thoroughly; defrosting meats safely; vigorous washing of fruits and vegetables; and handling leftovers properly. Irradiation of meat has been approved by the United States Food and Drug Administration and the United States Department of Agriculture to decrease bacterial contamination of consumer meat supplies.

Sediments can act as a reservoir for *E. coli*, as the sediments protect the organisms from bacteriophages and microbial toxicants. The *E. coli* can persist in the sediments and contribute to concentrations in the overlying waters for months after the initial contamination.

The presence of *E. coli* in surface waters indicates that there has been fecal contamination of the water body from agricultural and/or urban and residential areas. Although the detected *E. coli* may not necessarily cause disease (not all strains of the bacterium are pathogenic), it is certainly an indicator of the presence of feces, which can harbour other pathogenic types of bacteria. Examples include *Salmonella typhimurium*, *Vibrio cholerae*, and *Shigella*. Furthermore, because most strains of *E. coli* do not survive long once outside the intestinal tract, the detection in water is usually an indicator of recent (hours to a few days) fecal pollution. These attributes have made *E. coli* a popular indicator organism for the health quality of drinking and recreational waters.

To provide safe drinking water, the water is treated with chlorine, ultra-violet light, and/or ozone. Traditionally fecal coliform bacteria have been used as the indicator organisms for monitoring, but the test for these bacteria also detects thermotolerant non-fecal coliform bacteria.

Therefore, the U.S. Environmental Protection Agency (EPA) is recommending that *E. coli* as well as enterococci be used as indicators of fecal contamination of a water body instead of fecal coliform bacteria. The test for *E. coli* does not include non-fecal thermotolerant coliforms.

The U.S. EPA recreational water quality standard is based on a threshold concentration of *E. coli* above which the health risk from waterborne disease is unacceptably high. The recommended standard corresponds to approximately 8 gastrointestinal illnesses per 1000 swimmers. The standard is based on two criteria: 1) a geometric mean of 126 organisms per 100 ml, based on several samples collected during dry weather conditions; or 2) 235 organisms/100 ml sample for any single water sample. During 2002 the U.S. EPA finalized guidance on the use of *E. coli* as the basis for bacterial water quality criteria to protect recreational freshwater bodies.

Resources

BOOKS

Costerton, J. William. *The Biofilm Primer*. New York: Springer, 2010.

Goldsmith, Connie. *Invisible Invaders: Dangerous Infectious Diseases*. Minneapolis: Twenty-First Century Books, 2006.

Gualde, Norbert. *Resistance: The Human Struggle against Infection*. Washington, DC Dana Press, 2006.

Judith L. Sims

Essential fish habitat

Essential Fish Habitat (EFH) is a federal provision to conserve and sustain the habitats that fish need to go through their life cycles. The EFH was a 1996 provision added by the United States Congress to the Magnuson Fishery Conservation and Management Act of 1976. Renamed the Magnuson-Stevens Conservation and Management Act in 1996, the act became the federal law governing marine (sea) fishery management in the United States.

The amended act required that fishery management plans include designations and descriptions of essential fish habitats. The plan is a document describing the strategy to reach management goals in a fishery, an area where fish breed and people catch them. The Magnuson-Stevens Act covers plans for waters located within the United States' exclusive economic zone. The zone extends offshore from the coastland for three to 200 miles.

The designation of EFH was necessary because the continuing loss of aquatic habitat posed a major long-term threat to the viability of commercial and recreational fisheries, Congress said in 1996. The mandate of the EFH concerns waters and water-related surfaces (sediment and structures below the water) that are needed for the spawning, breeding, feeding, or full growth of fish.

The Magnuson-Stevens Act called for identification of EFH by eight regional fishery management councils and the Highly Migratory Species Division of the National Marine Fisheries Service (NMFS), an agency of the Commerce Department's National Oceanic and Atmospheric Administration (NOAA). Under the Magnuson-Stevens Act, NOAA manages more than 700 species.

NOAA fisheries and the councils are required by the act to minimize "to the extent practicable" the adverse effects of fishing on EFH. The act also directed the councils and NOAA to devise plans to conserve and enhance EFH. Those plans are included in the management plans. Also in the plan are what is described as "habitat areas of particular concern," rare habitats or habitats that are ecologically important.

Furthermore, the act required federal agencies to work with NMFS when the agencies plan to authorize, finance, or carry out activities that could adversely affect EFH. This process, called an EFH consultation, is required if the agency plans an activity such as dredging near an essential fishing habitat. Although NMFS does not have veto power over the project, NOAA Fisheries will provide conservation recommendations.

Eight regional fishery management councils were established by the 1976 Magnuson Fishery and Conservation Management Act. That legislation also established the exclusive economic zone and staked the United States' claim to it. The 1976 act also addressed issues such as foreign fishing and how to connect the fishing community to the management process, according to an NOAA report. The councils manage living marine resources in their regions and address issues such as EFH.

The New England Fishery Management Council manages fisheries in federal waters off the coasts of Maine, New Hampshire, Massachusetts, Rhode Island, and Connecticut. New England fish species include Atlantic cod, Atlantic halibut, and white hake.

The Mid-Atlantic Fishery Management Council manages fisheries in federal waters off the mid-Atlantic coast. Council members represent the states of New York, New Jersey, Pennsylvania, Delaware, Maryland, and Virginia. North Carolina is represented on this council and the South Atlantic Council. Fish species found within this region include ocean quahog, Atlantic mackerel, and butterfish.

The South Atlantic Fishery Management Council is responsible for the management of fisheries in the federal waters within a 200-mile area off the coasts of North Carolina, South Carolina, Georgia, and east Florida to Key West. Marine species in this area include cobia, golden crab, and Spanish mackerel.

The Gulf of Mexico Fishery Management Council draws its membership from the Gulf Coast states of Florida, Alabama, Mississippi, Louisiana, and Texas. Marine species in this area include shrimp, red drum, and stone crab.

The Caribbean Fishery Management Council manages fisheries in federal waters off the Commonwealth of Puerto Rico and the U.S. Virgin Islands. The management plan covers coral reefs and species including queen triggerfish and spiny lobster.

The North Pacific Fishery Management Council includes representatives from Alaska and Washington state. Species within this area include salmon, scallops, and king crab.

The Pacific Fishery Management Council draws its members from Washington, Oregon, and California. Species in this region include salmon, northern anchovy, and Pacific bonito.

The Western Pacific Fishery Management Council is concerned with the United States exclusive economic zone that surrounds Hawaii, American Samoa, Guam, the Northern Mariana Islands, and other U.S. possessions in the Pacific. Fishery management encompasses coral and species such as swordfish and striped marlin.

Resources

BOOKS

Glantz, Michael H. *Climate Variability, Climate Change and Fisheries.* Cambridge: Cambridge University Press, 2005.

Longhurst, Alan R. *Mismanagement of Marine Fisheries.* Cambridge: Cambridge University Press, 2010.

Pauly, Daniel. *5 Easy Pieces: The Impact of Fisheris on Marine Ecosystems.* Washington DC: Island Press, 2010.

Liz Swain

Estuary

Estuaries represent one of the most biologically productive aquatic ecosystems on Earth. An estuary is a coastal body of water where chemical and physical conditions modulate in an intermediate range between the freshwater rivers that feed into them and the salt water of the ocean beyond them. It is the point of mixture for these two very different aquatic ecosystems. The freshwater of the rivers mix with the salt water pushed by the incoming tides to provide a brackish water habitat ideally suited to a tremendous diversity of coastal marine life.

Estuaries are nursery grounds for the developing young of commercially important fish and shellfish. The young of any species are less tolerant of physical extremes in their environment than adults. Many species of marine life cannot tolerate the concentrations of salt in ocean water as they develop from egg to sub-adult, and by providing a mixture of fresh and salt water, estuaries give these larval life forms a more moderate environment in which to grow. Because of this, the adults often move directly into estuaries to spawn.

Estuaries are extremely rich in nutrients, and this is another reason for the great diversity of organisms in these ecosystems. The flow of freshwater and the periodic flooding of surrounding marshlands provides an influx of nutrients, as do the daily surges of tidal fluctuations. Constant physical movement in this environment keeps valuable nutrient resources available to all levels within the food chain/web.

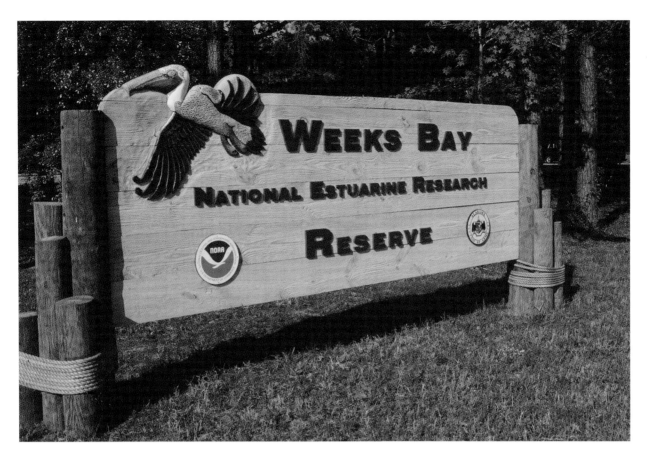

Estuary at Weeks Bay. *(Courtesy of the NOAA)*

Coastal estuaries also provide a major filtration system for waterborne pollutants. This natural water treatment facility helps maintain and protect water quality, and studies have shown that one acre of tidal estuary can be the equivalent of a small waste treatment plant.

Chesapeake Bay on the Atlantic coast is the largest estuary in the United States, draining six states and the District of Columbia. It is the largest producer of oysters in the country; it is the single largest producer of blue crabs in the world, and 90 percent of the striped bass found on the East Coast hatch there. It one of the most productive estuaries in the world, yet its productivity has declined in recent decades due to a huge increase in the number of people in the region. Between the 1940s and the 1990s, the population jumped from about 3.5 million to more than 15 million, bringing with it an increase in pollution and overfishing of the bay. Sewage treatment plants contribute large amounts of phosphates, and agricultural, urban, and suburban discharges deposit nitrates, which in turn contribute to algal blooms and oxygen depletion. Pesticides and industrial toxics also contribute to the bay's problems.

Beginning in the early 1980s, concerted state and federal government initiatives began, with the aims to clean up the Chesapeake Bay and restore its seafood productivity.

In the 1990s fish populations in Chesapeake Bay were decimated by a disease caused by the fungus *Pfiesteria piscicida*. The cause was determined to be runoff from chicken farms in the region (chickens are a source of the organism). The outbreak highlighted the deleterious effects of agricultural runoff on coastal waters. Runoff of fertilizer into estuaries encourages the overgrowth of algae, which can deplete the nutrients in the water to such an extent that the capacity of the water system as a reservoir of life is affected.

As of October 2010, the United States Environmental Protection Agency was still working on revised plans for the long-term restoration of the Chesapeake Bay.

See also Agricultural pollution; Aquatic chemistry; Commercial fishing; Dissolved oxygen; Nitrates and nitrites; Nitrogen cycle; Restoration ecology.

Resources

BOOKS

Cronin, William B. *The Disappearing Islands of the Chesapeake.* Baltimore: The Johns Hopkins University Press, 2006.

Haydamacker, Nelson. *Deckhand: Life on Freighters of the Great Lakes.* Ann Arbor: Michigan State University Press, 2009

Lippson, Alice Jane and Robert L. Lippson. *Life in the Chesapeake Bay.* Baltimore: The Johns Hopkins University Press, 2006.

Schindler, David W. and John R. Vallentyne. *The Algal Bowl: Overfertilization of the World's Freshwaters and Estuaries.* Edmonton, Canada: University of Alberta Press, 2008.

Eugene C. Beckham

Ethanol

Ethanol is an organic compound with the chemical formula C_2H_5OH. Its common names include ethyl alcohol and grain alcohol. The latter term reflects one method by which the compound can be produced: the distillation of corn, sugar cane, wheat, and other grains. Ethanol is the primary component in many alcoholic drinks such as beer, wine, vodka, gin, and whiskey. The use of ethanol in automotive fuels has become common in recent years in the United States and some other countries, especially Brazil. When mixed in a (usually) one to nine ratio with gasoline, it is sold as gasohol. The reduced costs of producing gasohol—thanks largely to government subsidies—make it a viable economic alternative to other automotive fuels.

The production of fuel ethanol from corn and other crops has been criticized both as a net energy consumer (or poor net-energy producer) and as injurious to the world's poor. In the spring of 2008, surging oil prices increased the cost of food production and transport, and increases in biofuel production diverted crops away from growth for consumption. The result of the price and production forces caused the lowest estimated level in world food stocks in more than twenty-five years. Food experts agreed that the push to raise crops for ethanol biofuel for vehicles was part of the problem. Biofuel manufacturers compete directly with food buyers in the market for corn, and high demand for biofuel crops causes growers to switch acreage away from food production. The result is higher food prices, which some the world's poorest simply cannot pay.

See also Renewable energy.

Ethnobotany

The field of ethnobotany is concerned with the relationship between indigenous cultures and plants. Plants play a major and complex role in the lives of indigenous peoples, providing nourishment, shelter, and medicine. Some plants have had such a major effect on traditional cultures that religious ceremonies and cultural beliefs were developed around their use. Ethnobotanists study and document these relationships.

The discovery of many plant-derived foods and medicines first used by indigenous cultures has changed the modern world. On the economic side, the field of ethnobotany determines the traditional uses of plants to find other potential applications for food, medicine, and industry. As an academic discipline, ethnobotany studies the interaction between peoples and plant life to learn more about human culture, history, and

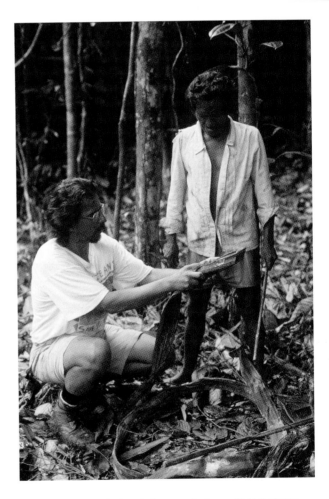

Peter Gorman and a local shaman study medicinal plants near the Javari River in the Amazon rainforest. *(Alison Wright/ Photo Researchers, Inc.)*

development. Ethnobotany draws upon many academic areas including anthropology, archeology, biology, ecology, chemistry, geography, history, medicine, religious studies, economics, linguistics, and sociology to help understand the complex interaction between traditional human cultures and the plants around them.

Early explorers who observed how native peoples used plants and then carried those useful plants back to their own cultures might be considered the first ethnobotanists, although that is not a word they would have used to describe themselves. The plant discoveries these explorers made caused the expansion of trade between many parts of the globe. For example, civilizations were changed by the discovery and subsequent trade of sugar, tea, coffee, and spices, including cinnamon and black pepper.

During his 1492 voyage to the Caribbean, Genoese explorer Christopher Columbus (c.1451–1506) discovered tobacco in Cuba and took it back with him to Europe, along with later discoveries of corn, cotton, and allspice. Other Europeans traveling to the Americas discovered tomatoes, potatoes, cocoa, bananas, pineapples, and other useful plants and medicines. Latex rubber was discovered in South America when European explorers observed native peoples dipping their feet in the rubber compound before walking across hot coals.

The study of plants and their place in culture has existed for centuries. In India the *vedas*, which are long religious poems that have preserved ancient beliefs for thousands of years, contain descriptions of the use and value of certain plants. Descriptions of how certain plants can be used have been found in the ancient Egyptian scrolls. More than 500 years ago the Chinese made records of medicinal plants. In ancient Greece, the philosopher Aristotle (384 BCE–322 BCE) wrote about the uses of plants in early Greek culture, and theorized that each plant had a unique spirit. The Greek surgeon Dioscorides in AD 77 recorded the medicinal and folk use of nearly 600 plants in the Mediterranean region. During the Middle Ages, there existed many accounts of the folk and medicinal use of plants in Europe in books called herbals.

As the study of plants became more scientific, ethnobotany evolved as a field. One of the most important contributors to the field was Carl Linnaeus (1707–1778), a Swedish botanist who developed the system of naming organisms that is still used today. This system of binomial classification gives each organism a Latin genus and species name. It was the first system that enabled scientists speaking different languages to organize and accurately record new plant discoveries. In addition to naming the 5,900 plants known to European botanists, Linnaeus sent students around the world looking for plants that could be useful to European civilization. This was an early form of economic botany, an offshoot of ethnobotany that is interested primarily in practical developments of new uses for plants. Pure ethnobotany is more sociology-based, and is primarily interested in the cultural relationship of plants to the indigenous peoples who use them.

In the nineteenth century, bioprospecting, (the active search for valuable plants in other cultures) multiplied as exploration expanded across the globe. The most famous ethnobotanist of the 1800s was Richard Spruce (1817–1893), an Englishman who spent seventeen years in the Amazon living among the native people and observing and recording their use of plants. Observing a native treatment for fever, Spruce collected cinchona bark (*Cinchona* species) from which the drug quinine was developed. Quinine has saved the lives of millions of people infected with malaria. Spruce also documented the native use of hallucinogenic plants used in religious rituals and the accompanying belief systems associated with these hallucinogenic plants.

Many native cultures on all continents believed that certain psychoactive plants that caused visions or hallucinations gave them access to spiritual and healing powers. Spruce discovered that for some cultures, plants were central to the way human life and the world was perceived. Thus, the field of ethnobotany began expanding from the study of the uses of plants by native peoples, to how plants and cultures interacted on more sociological or philosophical/religious grounds.

An American botanist named John Harshberger (1869–1929) first coined the term ethnobotany in 1895, and the field evolved into an academic discipline in the twentieth century. The University of New Mexico offered the first master's degree in ethnobotany. Their program concentrated on the ethnobotany of the Native Americans of the southwestern United States.

In the twentieth century one of the most famous ethnobotanists and plant explorers was American Harvard professor Richard Evans Schultes (1915–2001). Inspired by the story of Richard Spruce, Schultes lived for twelve years with several indigenous tribes in South America while he observed and documented their use of plants as medicine, poison, and for religious purposes. Schultes discovered many plants that have subsequently been investigated for pharmaceutical use. Schultes' work also provided further insight into the use

A swampy area in the Florida Everglades. *(Tomasz Szymanski/Shutterstock.com)*

years the United States Army Corps of Engineers constructed and enlarged flood control canals.

It was only in the mid-1950s, with the development and implementation of the Central and Southern Florida Project for Flood Control & Other Purposes (C&SF Project), that water control took priority over uncontrolled drainage of the Everglades. The project, completed by 1962, was to provide flood protection, water supply, and environmental benefits over a 1,600 square-mile (41,440-square-kilometer) area. It consists of 1,500 miles (2,415 km) of canals and levees, 125 major water control structures, eighteen major pumping stations, thirteen boat locks, and several hundred smaller structures. Interspersed throughout the Everglades is a series of habitats, each dominated by a few or in some cases a single plant species. Seasonal wetlands and upland pine forests, which once dominated the historic border of the system, have come under the heaviest pressure from urban and agricultural development. In the system's southern part, freshwater wetlands are superseded by muhly grass (*Muhlenbergia filipes*), prairies, upland pine and tropical hardwood forests, and mangrove forests that are influenced by the tides.

Attached algae, also known as periphyton, are an important component of the Everglades food web, providing both organic food matter and habitat for various grazing invertebrates and forage fish that are eaten by wading birds, reptiles, and sport fish. These algae include calcareous and filamentous algae (*Scytonema hoffmani, Schizothrix calcicola*) and diatoms (*Mastogloia smithii v. lacustris*). Sawgrass (*Cladium jamaicense*) constitutes one of the main plants occurring throughout the Everglades, being found in 65–70 percent of the remaining freshwater marsh. In the north the sawgrass grows in deep peat soils and is both dense and tall, reaching up to 10 feet (3 m) in height. In the south it grows in low-nutrient marl soils and is less dense and shorter, averaging 2.5–5 feet (0.75–1.5 m). Sawgrass is adapted to survive both flooding and burning. Stands of pure sawgrass as well as mixed communities are found in the Everglades. The mixed communities can include maidencane (*Panicum hemitomon*), arrowhead (*Sagittaria lancifolia*), water hyssop (*Bacopa caroliniana*), and spikerush (*Eleocharis cellulosa*).

Wet prairies, which together with aquatic sloughs provide habitat during the rainy season for a wide variety of aquatic invertebrates and forage fish, are another important habitat of the Everglades system. They are seasonally inundated wetland communities that require certain standing water for six to ten months. Once common, today more than 1,500 square miles (3,885 square kilometers) of these prairies have been drained or destroyed. The lowest elevations of the Everglades are ponds and sloughs, which have deeper water and longer inundation periods. They occur throughout the system, and in some cases can be formed by alligators in peat soils. Among the types of emergent vegetation commonly found in these areas are white water lily (*Nymphaea odorata*), floating heart (*Nymphoides aquatica*), and spatterdock (*Nuphar luteum*). Common submerged species include bladderwort (*Utricularia*) and the periphyton mat community. Ponds and sloughs serve as important feeding areas and habitat for Everglades wildlife.

At the highest elevations are found communities of isolated trees surrounded by marsh, called tree islands. These provide nesting and roosting sites for colonial birds and habitat for deer and other terrestrial animals during high-water periods. Typical dominant species constituting tree islands are red bay (*Persa borbonia*), swamp bay (*Magnolia virginiana*), dahoon holly (*Ilex cassine*), pond apple (*Annona glabra*), and wax myrtle (*Myrica cerifera*). Beneath the canopy grows a dense shrub layer of cocoplum (*Chrysobalanus icacao*), buttonbush (*Cephalanthus accidentalis*), leather leaf fern (*Acrostichum danaeifolium*), royal fern (*Osmunda regalis*), cinnamon fern (*O. cinnamonea*), chain fern (*Anchistea virginica*), bracken fern (*Pteridium aquilinium*), and lizards tail (*Saururus cernuus*).

In addition to the indigenous plants of the Everglades, numerous exotic and nuisance species have been brought into Florida and have now spread in the wild. Some threaten to invade and displace indigenous species. Brazilian pepper (*Schinus terebinthifolius*), Australian pine (*Casuarina equisetifolia*), and melaleuca (*Melaleuca quinquenervia*) are three of the most serious exotic species that have gained a foothold and are displacing native plants.

There are sixty-nine species on the federal list of endangered species within the Everglades. Of the forty mammal species found in the Everglades, there are two endangered species: the Florida panther (*Felis concolor coryi*) and the West Indian manatee (*Trichechus manatus*). There are three threatened and four endangered species of birds including piping plover (*Charadrius melodus*), the wood stork (*Mycteria americana*), snail kite (*Rostrhamus sociabilis*), and the red-cockaded

woodpecker (*Picoides borealis*). Endangered or threatened reptiles and amphibians include the green sea turtle (*Chelonia mydas*), the eastern indigo snake (*Drymarchon corais couperi*), and the loggerhead sea turtle (*Caretta caretta*).

The alligator (*Alligator mississippiensis*) was once endangered due to excessive alligator hide hunting. In 1972 the state made alligator product sales illegal. Protection allowed the species to recover, and it is now widely distributed in wetlands throughout the state. It is still listed as threatened by the federal government, but in 1988 Florida instituted an annual alligator harvest. According to the non-profit National Parks Conservation Association (NPCA), by 2010 there were over a million alligators in the United States, with most living in Florida and three other southern coastal states (Texas, Louisiana, and Georgia).

Faced with pressures on Everglades habitats and the species within them, as well as the need for water management within the rapidly developing state, in 1987 the Florida legislature passed the Surface Water Improvement and Management Act. The law requires the state's five water management districts to identify areas needing preservation or restoration. The Everglades Protection Area was identified as a priority for preservation and improvement planning. Within the state's protection plan, excess nutrients, in large part from agriculture, have been targeted as a major problem that causes natural periphyton to be replaced by species more tolerant of pollution. In turn, sawgrass and wet prairie communities are overrun by other species, impairing the Everglades' ability to serve as habitat and forage for higher trophic level species.

A federal lawsuit was filed against the South Florida Management District in 1988 for phosphorus (chemical symbol P) pollution, and in 1989 President George H. W. Bush (1924–) authorized the addition of more than 100,000 acres to the Everglades National Park. The law that authorized this addition was Public Law 101-229, otherwise known as the Everglades National Park Protection and Expansion Act of 1989. Included in this legislation was the stipulation that the Army Corps of Engineers improve water flow to the Park. In 1994 the Everglades Forever Act was passed by the Florida state legislature. The act called for construction of experimental marshes termed Stormwater Treatment Areas that were designed to remove phosphorus from water entering the Everglades. In 1997 six more Stormwater Treatment Areas were constructed and phosphorus removal was estimated to be as much as 50 percent, due in part to better management practices that were food sources, especially around the disturbed environments of human settlement.

This competitiveness has been a problem, for example, with house sparrows (*Passer domesticus*). These birds were intentionally introduced from Europe to North America in 1850 to control insect pests. Their aggressive foraging and breeding habits often drive native sparrows, martins, and bluebirds from their nests, and today house sparrows are one of the most common birds in North America. Exotic plants can also become nuisance species when they crowd, shade, or out-propagate their native competitors. They can be extraordinarily effective colonists, spreading quickly and eliminating competition as they become established.

The list of species introduced to the Americas from Europe, Asia, and Africa is immense, as is the list of species that have made the reverse trip from the Americas to Europe, Asia, and Africa. Some notable examples are kudzu (*Pueraria lobata*), the zebra mussel (*Dreissena polymorpha*), Africanized bees *Apis mellifera scutellata*), and Eurasian milfoil (*Myriophyllum spicatum*).

Kudzu is a cultivated legume in Japan. It was intentionally brought to the southern United States for ground cover and erosion control. Fast growing and tenacious, kudzu quickly overwhelms houses, tangles in electric lines, and chokes out native vegetation.

Africanized "killer" bees were accidentally released in Brazil by a beekeeper in 1957. These aggressive insects have no more venom than standard honey bees (also an Old World import), but they attack more quickly and in great numbers. Breeding with resident bees and sometimes traveling with cargo shipments, Africanized bees have spread north from Brazil at a rate of up to 200 miles (322 km) each year and now threaten to invade commercially valuable fruit orchards and domestic bee hives in Texas and California.

The zebra mussel, accidentally introduced to the Great Lakes around 1985, presumably in ballast water dumped by ships arriving from Europe, colonizes any hard surface, including docks, industrial water intake pipes, and the shells of native bivalves. Each female zebra mussel can produce 50,000 eggs a year. Growing in masses with up to 70,000 individuals per square foot, these mussels clog pipes, suffocate native clams, and destroy breeding grounds for other aquatic animals. They are also voracious feeders, competing with fish and native mollusks for plankton and microscopic plants. The economy and environment of the Great Lakes now pay the price of zebra mussel infestations. Area industries spend hundreds of millions of dollars annually unclogging pipes and equipment, and commercial fishermen complain of decreased catches.

Eurasian milfoil is a common aquarium plant that can propagate from seeds or cuttings. A tiny section of stem and leaves accidentally introduced into a lake by a boat or boat trailer can grow into a huge mat covering an entire lake. When these mats have consumed all available nutrients in the lake, they die and rot. The rotting process robs fish and other aquatic animals of oxygen, causing them to die.

Exotic species have brought ecological disasters to every continent, but some of the most extreme cases have occurred on isolated islands where resident species have lost their defensive strategies. For example, rats, cats, dogs, and mongooses introduced by eighteenth-century sailors have devastated populations of ground-breeding birds on Pacific islands. Rare flowers in Hawaii suffer from grazing goats and rooting pigs, both of which were brought to the island for food, but have escaped and established wild populations. Grazing sheep threaten delicate plants on ecologically fragile North Atlantic islands, while rats, cats, and dogs endanger northern seabird breeding colonies. Rabbits introduced into Australia overran parts of the island and wiped out hundreds of acres of grassland.

Humans have always carried plants and animals as they migrated from one region to another with little regard to the effects these introductions might have on their new habitat. Many introduced species seem benign, useful, or pleasing to have around, making it difficult to predict which imports will become nuisance species. When an exotic plant or animal threatens human livelihoods or economic activity, as do kudzu, zebra mussels, and killer bees, people begin to seek ways to control these invaders.

Control efforts include using pesticides and herbicides, and introducing natural predators and parasites from the home range of the exotic plant or animal. For example, beetles that naturally prey on purple loosestrife have been experimentally introduced in American loosestrife populations. This deliberate introduction requires a great deal of care, research, and monitoring, however, to ensure that an even worse problem does not result, as happened with the house sparrow. Such solutions, and the time and money to develop them, are usually elusive and politically controversial, so in many cases effective control methods remain unavailable.

In 1999 President Bill Clinton signed the Executive Order on Invasive Species. This order established the Invasive Species Council to coordinate the activities of federal agencies, such as the Aquatic Nuisance Species Task Force, the Federal Interagency Committee for the Management of Noxious and Exotic Weeds, and the Committee on Environment and Natural Resources.

The Invasive Species Council is responsible for the development of a National Invasive Species Management Plan. This plan is intended to be updated every two years to provide guidance and recommendations about the identification of pathways by which invasive species are introduced, and measures that can be taken for their control.

Non-profit environmental organizations across the globe are leading the effort for control of exotic species. For example, The Nature Conservancy has established Landscape Conservation Networks to address issues of land conservation that include invasive species management. These networks bring in outside experts and land conservation partners to develop innovative and cost effective means of controlling exotic species. The Great Lakes to Information Network, managed by the Great Lakes Commission based in Ann Arbor, Michigan, provides online access to information about environmental issues, including exotics species, in the Great Lakes region. The Florida Exotic Pest Plant Council, founded in 1984, provides funding to organizations that educate the public about the impacts of exotic invasive plants in the State of Florida.

Resources

BOOKS

Coates, Peter A. *American Perceptions of Immigrant and Invasive Species: Strangers on the Land.* Berkeley: University of California Press, 2006.
Terrill, Ceridwen. *Unnatural Landscapes: Tracking Invasive Species.* Tucson: University of Arizona Press, 2007.

Mary Ann Cunningham
Marie H. Bundy

Experimental Lakes Area

The Experimental Lakes Area (ELA) in northwestern Ontario is in a remote landscape characterized by Precambrian bedrock, northern mixed-species forests, and oligotrophic lakes, and bodies of water deficient in plant nutrients. The Canadian Department of Fisheries and Oceans began developing a field-research facility at ELA in the 1960s, and the area has become the focus of a large number of investigations by David W. Schindler of the University of Alberta and others into chemical and biological conditions in these lakes. Schindler is a pre-eminent environmental researcher in Canada and a vocal critic of industry-driven environmental degradation.

Of the limnological investigations conducted at ELA, the best known is a series of whole-lake experiments designed to investigate the ecological effects of perturbation by a variety of environmental stress factors, including eutrophication, acidification, metals, radionuclides, and flooding during the development of reservoirs.

The integrated, whole-lake projects at ELA were initially designed to study the causes and ecological consequences of eutrophication. In one long-term experiment, Lake 227 was fertilized with phosphate and nitrate. This experiment was designed to test whether carbon could limit algal growth during eutrophication, so none was added. Lake 227 responded with a large increase in primary productivity by drawing on the atmosphere for carbon, but it was not possible to determine which of the two added nutrients, phosphate or nitrate, had acted as the primary limiting factor.

Observations from experiments at other lakes in ELA, however, clearly indicated that phosphate is the primary limiting nutrient in these oligotrophic water bodies. Lake 304 was fertilized for two years with phosphorus, nitrogen, and carbon, and it became eutrophic. It recovered its oligotrophic condition again when the phosphorus fertilization was stopped, even though nitrogen and carbon fertilization were continued. Lake 226, an hourglass-shaped lake, was partitioned with a vinyl curtain into two basins, one of which was fertilized with carbon and nitrogen, and the other with phosphorus, carbon, and nitrogen. Only the latter treatment caused an algal bloom. Lake 302 received an injection of all three nutrients directly into its hypolimnion during the summer. Because the lake was thermally stratified at that time, the hypolimnetic nutrients were not available to fertilize plant growth in the epilimnetic euphotic zone, and no algal bloom resulted. Nitrogen additions to Lake 227 were reduced in 1975 and eliminated in 1990. The lake continued with high levels of productivity by fixing nitrogen from the atmosphere.

Research of this sort was instrumental in confirming conclusively the identification of phosphorus as the most generally limiting nutrient to eutrophication of freshwaters. This knowledge allowed the development of waste management systems which reduced eutrophication as an environmental problem by reducing the phosphorus concentration in detergents, removing phosphorus from sewage, and diverting sewage from lakes.

Another well known ELA project was important in gaining a deeper understanding of the ecological consequences of the acidification of lakes. Sulfuric acid was added to Lake 223, and its acidity was increased progressively, from an initial pH near 6.5 to pH 5.0–5.1 after six years. Sulfate and hydrogen ions were also added to the lake in increasing concentrations during this time. Other chemical changes were caused indirectly by acidification: manganese increased by 980 percent, zinc by 550 percent, and aluminum by 155 percent.

As the acidity of Lake 223 increased, the phytoplankton shifted from a community dominated by golden-brown algae to one dominated by chlorophytes and dinoflagellates. Species diversity declined somewhat, but productivity was not adversely affected. A mat of the green alga *Mougeotia* sp. developed near the shore after the pH dropped below 5.6. Because of reduced predation, the density of cladoceran zooplankton was larger by 66 percent at pH 5.4 than at pH 6.6, and copepods were 93 percent more abundant. The nocturnal zooplankton predator *Mysis relicta*, however, was an important extinction. The crayfish *Orconectes virilis* declined because of reproductive failure, inhibition of carapace hardening, and effects of a parasite. The most acid-sensitive fish was the fathead minnow (*Pimephales promelas*), which declined precipitously when the lake pH reached 5.6.

The first of many year-class failures of lake trout (*Salvelinus namaycush*) occurred at pH 5.4, and failure of white sucker (*Catastomus commersoni*) occurred at pH 5.1. One minnow, the pearl dace (*Semotilus margarita*), increased markedly in abundance but then declined when pH reached 5.1. Adult lake trout and white sucker were still abundant, though emaciated, at pH 5.0–5.5, but in the absence of successful reproduction they would have become extinct. Overall, the Lake 223 experiment indicated a general sensitivity of many organisms to the acidification of lake water. However, within the limits of physiological tolerance, the tests showed that there can be a replacement of acid-sensitive species by relatively tolerant ones.

In a similarly designed experiment in Lake 302, nitric acid was shown to be nearly as effective as sulfuric acid in acidifying lakes, thereby alerting the international community to the need to control atmospheric emissions of gaseous nitrogen compounds.

As of 2010, research continues at the 58 lakes that comprise the ELA. The area continues to be free of activities that would affect lake water quality, allowing the area to remain one of the world's best sites for studies of freshwater degradation and preservation.

See also Acid rain; Algicide; Aquatic chemistry; C:N ratio; Cultural eutrophication; Water pollution.

Resources

OTHER

United States Department of the Interior, United States Geological Survey (USGS). "Lakes." http://www.usgs.gov/science/science.php?term=596&type=feature (Accessed October 13, 2010).
United States Environmental Protection Agency (EPA). "Ecosystems: Aquatic Ecosystems: Lakes." http://www.epa.gov/ebtpages/ecosaquatilakes.html (Accessed October 13, 2010).

Bill Freedman

Exponential growth

Exponential growth refers to growth that occurs at a constant or even accelerating rate with time.

A well-known example is the growth of bacteria under ideal conditions of temperature and nutrients. In such a case, the population can double in set time (for example 15 minutes). In this example, one bacterium will grow and divide to produce two bacteria in 15 minutes. In another 15 minutes there would be four bacteria, then eight in another 15 minutes. Even in very short time of a few hours, the number of bacteria can become very large. Over 24 hours, millions of bacteria are produced.

The notion of exponential growth is of particular interest in population biology, because all populations of organisms have the capacity to undergo exponential growth. The biotic potential or maximum rate of reproduction for all living organisms is very high. Put another way, all species theoretically have the capacity to reproduce themselves many, many times over during their lifetimes. In reality, only a few of the offspring of most species survive, due to reproductive failure, limited availability of space and food, diseases, predation, and other mishaps. A few species, such as the lemming, go through cycles of exponential population growth resulting in severe overpopulation. A catastrophic dieback follows, during which the population is reduced enormously, readying it for the next cycle of growth and dieback. Interacting species will experience related fluctuations in population levels. Generally, however, populations are held stable by environmental resistance, unless an environmental disturbance takes place.

Climatic changes and other natural phenomena may cause such habitat disturbances, but more usually they result from human activity. Pollution, predator control, and the introduction of foreign species into habitats that lack competitor or predator species are a few examples among many of human activities that may cause declines in some populations and exponential growth in others.

An altogether different case of exponential population growth is that of humans themselves. The human population has grown at an accelerating rate, starting at a low average rate of 0.002% per year early in its history and reaching a record level of 2.06% in 1970. Since then the rate of increase has dropped below 2%, but human population growth is still alarming. In 2010, the estimated number of humans, according to the United States Census Bureau, is about 6.9 billion people. Some scientists have expressed concern that continued growth of the global population in the absence of substantive changes in global food supply could ultimately cause a marked decrease in the population.

Resources

BOOKS

McKee, Jeffrey K.*Sparing Nature: The Conflict between Human Population Growth and Earth's Biodiversity* Rutgers, NJ: Rutgers University Press, 2005.

Neal, Dick*Introduction to Population Biology*Cambridge: Cambridge University Press, 2004.

Marijke Rijsberman

Externality

Most economists argue that markets ordinarily are the superior means for fulfilling human wants. In a market, deals are ideally struck between consenting adults only when the parties feel they are likely to benefit. Society as a whole is thought to gain from the aggregation of individual deals that take place. The wealth of a society grows by means of what is called the hidden hand of free market mechanisms, which offers spontaneous coordination with a minimum of coercion and explicit central direction. However, the market system is complicated by so-called externalities, which are effects of private market activity not captured in the price system.

Economics distinguishes between positive and negative externalities. A positive externality exists when producers cannot appropriate all the benefits of their activities. An example would be research and development, which yields benefits to society that the producer cannot capture, such as employment in subsidiary industries. Environmental degradation, on the other hand, is a negative externality, or an imposition on society as a whole of costs arising from specific market activities. Historically, the United States have encouraged individuals and corporate entities to make use of natural resources on public lands, such as water, timber, and even the land itself, in order to speed development of the country. Many undesirable by-products of the manufacturing process, in the form of exhaust gases or toxic waste, for instance, were simply released into the environment at no cost to the manufacturer. The agricultural revolution brought new farming techniques that relied heavily on fertilizers, pesticides, and irrigation, all of which affect the environment. Automobile owners did not pay for the air pollution caused by their cars. Virtually all human activity has associated externalities in the environmental arena, which do not necessarily present themselves as costs to participants in these activities. Over time, however, the consequences have become unmistakable in the form of a serious depletion of renewable resources and in pollution of the air, water, and soil. All citizens suffer from such environmental degradation, though not all have benefited to the same degree from the activities that caused them.

In economic analysis, externalities are closely associated with common property and the notion of free riders. Many natural resources have no discrete owner and are therefore particularly vulnerable to abuse. The phenomenon of degradation of common property is known as the Tragedy of the Commons. The costs to society are understood as costs to nonconsenting third parties, whose interests in the environment have been violated by a particular market activity. The consenting parties inflict damage without compensating third parties because without clear property rights there is no entity that stands up for the rights of a violated environment and its collective owners.

Nature's owners are a collectivity that is hard to organize. They are a large and diverse group that cannot easily pursue remedies in the legal system. In attempting to gain compensation for damage and force polluters to pay for their actions in the future the collectivity suffers from the free rider problem. Although everyone has a stake in ensuring, for example, good air quality, individuals will tend to leave it to others to incur the cost of pursuing legal redress. It is not sufficiently in the interest of most members of the group to sue because each has only a small amount to gain. Thus, government intervention is called for to protect the interests of the collectivity, which otherwise would be harmed.

The government has several options in dealing with externalities such as pollution. It may opt for regulation and set standards of what are considered acceptable levels of pollution. It may require reduced lead levels in gasoline and require automakers to manufacture cars with greater fuel economy and reduced emissions, for instance. If manufacturers or social entities such as cities exceed the standards set for them, they will be penalized. With this approach, many polluters have a direct incentive to limit their most harmful activities and develop less environmentally costly technologies. So far, this system has not proved to be very effective. In practice, it has been difficult (or not politically expedient) to enforce the standards and to collect the fines. Supreme Court decisions since the early 1980s have reinterpreted some of the laws to make standards much less stringent. Many companies have found it cheaper to pay the fines than to invest in reducing pollution. Or they evade fines by declaring bankruptcy and reorganizing as a new company.

Economists tend to favor pollution taxes and discharge fees. Since external costs do not enter the calculations a producer makes, the producer manufactures more of the good than is socially beneficial. When polluters have to absorb the costs themselves, to internalize them, they have an incentive to reduce production to acceptable levels or to develop alternative technologies. A relatively new idea has been to give out marketable pollution permits. Under this system, the government sets the maximum levels of pollution it will tolerate and leaves it to the market system to decide who will use the permits. The costs of past pollution (in the form of permanent environmental damage or costly clean-ups) will still be borne disproportionately by society as a whole. The government generally tries to make responsible parties pay for clean-ups, but in many cases it is impossible to determine who the culprit was and in others the parties responsible for the pollution no longer exist.

A special case is posed by externalities that make themselves felt across national boundaries, as is the case with acid rain, ozone layer depletion, and the pollution of rivers that run through more than one country. Countries that suffer from environmental degradation caused in other countries receive none of the benefits and often do not have the leverage to modify the polluting behavior. International conservation efforts must rely on agreements specific countries may or may not follow and on the mediation of the United Nations.

See also Internalizing costs; Trade in pollution permits.

Resources

BOOKS

Harrison, Kathryn and Lisa McIntosh Sundstrom, eds. *Global Commons, Domestic Decisions: The Comparative Politics of Climate Change*. Boston: The MIT Press, 2010.

Miller, Richard W. *God, Creation, and Climate Change: A Catholic Response to the Environmental Crisis*. Maryknoll, NY: Orbis Books, 2010.

Somerville, Richard. *The Unforgiving Air: Understanding Environmental Change, Second Edition*. Washington, DC: American Meteorological Society, 2008.

Alfred A. Marcus
Marijke Rijsberman

External costs *see* **Internalizing costs.**

Extinction

Extinction is the complete disappearance of a species, when all of its members have died or been killed. As a part of natural selection, the extinction of species has been on-going throughout the earth's history. However, with modern human strains on the environment, plants, animals, and invertebrates are becoming extinct at an unprecedented rate of thousands of species per year, especially in tropical rain forests. Many thousands more are threatened and endangered.

Scientists have determined that mass extinctions have occurred periodically in prehistory, coming about every 50 million years or so. The greatest of these came at the end of the Permian period, some 250 million years ago, when up to 96 percent of all species on the earth may have died off. Dinosaurs and many ocean species disappeared during a well-documented mass extinction at the end of the Cretaceous period (about 65 million years ago). It is estimated that of the billions of species that have lived on the earth during the last 3.5 billion years, 99.9 percent are now extinct.

It is thought that most prehistoric extinctions occurred because of climatological changes, loss of food sources, destruction of habitat, massive volcanic eruptions, or asteroids or meteors striking the earth. Extinctions, however, have never been as rapid and massive as they have been in the modern era. During the last two centuries, more than 75 species of mammals and over 50 species of birds have been lost, along with countless other species that had not yet been identified. According to one estimate, since 1600, including species and subspecies, the world has lost at least 100 types of mammals and 258 kinds of birds.

The first extinction in recorded history was the European lion, which disappeared around A.D. 80. In 1534, seamen first began slaughtering the great auk, a large, flightless bird once found on rocky North Atlantic islands, for food and oil. The last two known auks were killed in 1844 by an Icelandic fisherman motivated by rewards offered by scientists and museum collectors for specimens. Humans have also caused the extinction of many species of marine mammals. Steller's sea cow, once found on the Aleutian Islands off Alaska, disappeared by 1768. The sea mink, once abundant along the coast and islands of Maine, was hunted for its fur until about 1880, when none could be found. The Caribbean monk seal, hunted by sailors and fishermen, has not been found since 1962.

The early European settlers of America succeeded in wiping out several species, including the Carolina parakeet and the passenger pigeon. The pigeon was one of most plentiful birds in the world's history, and accounts from the early 1800s describe flocks of the birds blackening the sky for days at a time as they passed overhead. By the 1860s and 1870s tens of millions of them were being killed every year. As a result of this overhunting, the last passenger pigeon, Martha, died in the Cincinnati Zoo in 1914. The pioneers who settled the West were equally destructive, causing the disappearance of 16 separate types of grizzly bear, six of wolves, one type of fox, and one cougar. Since the Pilgrims arrived in North America in 1620, over 500 types of native American animals and plants have disappeared.

In the last decade of the twentieth century, the rate of species loss was unprecedented and accelerating. Up to 50 million species could be extinct by 2050, with a rate of three per day. Most of these species extinctions will occur—and are occurring—in tropical rain forests, the richest biological areas on the earth. Rain forests are being cut down at a rate of one to two acres per second.

In 1988, Harvard professor and biologist Edward O. Wilson estimated the current annual rate of extinction at up to 17,500 species, including many unknown rain forest plants and animals that have never been studied or even seen, by humans. Botanist Peter Raven, director of the Missouri Botanical Garden, calculated that a total of one-quarter the world's species could be gone by 2010. A study by the World Resources Institute pointed out that humans have accelerated the extinction rate to 100 to 1,000 times its natural level.

While it is impossible to predict the magnitude of these losses or the impact they will have on the earth and its future generations, it is clear that the results will be profound, possibly catastrophic.

Species whose populations shrink can be at greater risk of extinction than species with large populations. In May 2008, the Zoological Society of London and World Wide Fund for Nature released results of a study indicating a dramatic decline in global wildlife populations during the last four decades. The data showed that between 1970 and 2005, the populations of land-based species fell by 25 percent, of marine species by 28 percent, and of freshwater species by 29 percent. Biodiversity loss was also reported in the form of extinction of about 1 percent of the world's species each year. The two groups, in agreement with most biologists, stated that one of the "great extinction episodes" in Earth's history is now under way.

The losses were attributed to pollution, urbanization, overfishing, and hunting, but both groups emphasized that climate change would play an increasing role in species decline over coming decades.

See also Biodiversity; Climate; Dodo; Endangered species.

Resources

BOOKS

Costa, Rebecca. *The Watchman's Rattle: Thinking Our Way Out of Extinction.* New York: Vanguard Press, 2010.
Erwin, Douglas H. *Extinction: How Life on earth Nearly Ended 250 Million Years Ago.* Princeton, NJ: Princeton University Press, 2008.
Ward, Peter Douglas. *Under s Green Sky: Global Warming, the Mass Extinctions of the Past, and What They Can tell Us About Our Future.* New York: Collins, 2008.

Lewis G. Regenstein

Exxon Valdez

On March 24, 1989, the 987-foot super tanker *Exxon Valdez* outbound from Port Valdez, Alaska, with a full load of oil from Alaska's Prudhoe Bay passed on the wrong side of a lighted channel marker guarding a shallow stretch of Prince William Sound. The momentum of the large ship carried it onto Bligh Reef and opened a 6 x 20 ft hole in the ship's hull. Through this hole poured 260,000 barrels (11 million gallons) to 750,000 barrels (31.5 million gallons) of crude oil out of the ship's approximately 1.26 million

A clean-up worker uses high pressure, high temperature water to wash crude oil off the rocky shore of Block Island, Alaska, after the 1989 *Exxon Valdez* oil spill. *(AP Photo/John Gaps III)*

barrel (53 million gallon) cargo, making it the largest oil spill in the history of the United States to that date. The *Exxon Valdez* spill was surpassed by the *Deepwater Horizon*, which released approximately 4.9 million barrels (185 million gallons) of oil.

The oil spill resulting from the *Exxon Valdez* accident spread oil along 1,500 miles of pristine shoreline on Prince William Sound and the Kenai Peninsula, covering an area of 460 miles. Oil would eventually reach shores southwest of the spill up to 600 miles away.

The *Exxon Valdez* Oil Spill Trustee Council estimates that 250,000 seabirds, 2,800 sea otters, 300 harbor seals, 250 bald eagles, and 22 killer whales, were killed. These figures may be an underestimate of the animals killed by the oil, because many of the carcasses likely sank or washed out to sea before they could be collected. Most of the birds died from

hypothermia due to the loss of insulation caused by oil-soaked feathers. Many predatory birds, such as bald eagles, died as a result of ingesting contaminated fish and birds. Hypothermia affected sea otters as well, and many of the dead mammals suffered lung damage due to oil fumes. Billions of salmon eggs were also lost to the spill. While a record 43 million pink salmon were caught in Prince William Sound in 1990, the 1993 harvest declined to a record low of three million.

Response to the oil spill was slow and generally ineffective. The Alyeska Oil Spill Team responsible for cleaning up oil spills in the region took more than 24 hours to respond, despite previous assurances that they could mount a response in three hours. Much of the oil containment equipment was missing, broken, or barely operable. By the time oil containment and recovery equipment were in place, 11.1 million gallons (42 million liters) of oil had already spread over a large area. Ultimately, less than 10% of this oil was recovered, the remainder dispersing into the air, water, and sediment of Prince William Sound and adjacent sounds and fjords. Exxon reports spending a total of $2.2 billion to clean up the oil. Much of this money employed 10,000 people to clean up oil-fouled beaches; yet after the first year, only 3% of the soiled beaches had been cleaned.

In response to public concern about the poor response time and uncoordinated initial cleanup efforts following the Valdez spill, the Oil Pollution Act (OPA; part of the Clean Water Act) was signed into law in August 1990. The Act established a Federal trust fund to finance clean-up efforts for up to $1 billion per spill incident.

Ultimately, nature was the most effective surface cleaner of beaches; winter storms removed the majority of oil and by the winter of 1990, less than 6 miles (10 km) of shoreline was considered seriously fouled. Cleanup efforts were declared complete by the U.S. Coast Guard and the State of Alaska in 1992.

On October 9, 1991, a settlement between Exxon and the State of Alaska and the United States government was approved by the U.S. District Court. Under the terms of the agreement, Exxon agreed to pay $900 million in civil penalties over a 10-year period. The civil settlement also provides for a window of time for new claims to be made should unforeseen environmental issues arise. That window is from September 1, 2002 to September 1, 2006.

Exxon was also fined $150 million in a criminal plea agreement, of which $125 million was forgiven in return for the company's cooperation in cleanup and various private settlements. Exxon also paid $100 million in criminal restitution for the environmental damage caused by the spill.

A flood of private suits against Exxon have also deluged the courts in the years since the spill. In 1994, a district court ordered Exxon to pay $287 million in compensatory damages to a group of commercial fishermen and other Alaskan natives who were negatively impacted by the spill. The jury who heard the case also awarded the plaintiffs $5 billion in punitive damages. However, in November 2001 a federal appeals judge overturned the $5 billion punitive award, deeming it excessive and ordering the district court to reevaluate the settlement. The trial judge reduced punitive damages to $4.5 billion, which the court of appeals lowered to $2.5 billion in 2006. Exxon appealed to the Supreme Court of the United States, which, in 2008, vacated the punitive damages judgment and ruled that, under maritime law, punitive damages could not exceed the $507 million in compensatory damages owed by Exxon.

The Captain of the *Exxon Valdez*, Joseph Hazelwood, had admitted to drinking alcohol the night the accident occurred, and had a known history of alcohol abuse. Nevertheless, he was found not guilty of charges that he operated a shipping vessel under the influence of alcohol. He was found guilty of negligent discharge of oil, fined $50,000, and sentenced to 1,000 hours of community service work.

Despite official cleanup efforts by Exxon having ended, the environmental legacy of the *Valdez* spill lives on. The Auke Bay Laboratory of the Alaskan Fisheries Science Center conducted a beach study of Prince William Sound in the summer of 2001. Researchers found that approximately 20 acres of Sound shoreline is still contaminated with oil, the majority of which has collected below the surface of the beaches where it continues to pose a danger to wildlife. Of the 30 species of wildlife affected by the spill, only two—the American bald eagle and the river otter—were considered recovered in 1999. Preliminary 2002 reports from the Exxon Valdez Oil Spill Trustee Council reflect progress is being made in the area of wildlife recovery.

Resources

BOOKS

Bushell, Sharon, and Stan Jones. *The Spill: Personal Stories from the Exxon Valdez Disaster*. Kenmore, WA: Epicenter Press, 2009.

Exxon Valdez Oil Spill Trustee Council. *Comprehensive Plan for Habitat Restoration Projects Pursuant to Reopener for Unknown Injury*. Anchorage: Exxon Valdez Oil Spill Trustee Council, 2006.

Exxon Valdez Oil Spill Trustee Council. *Then and Now - a Message of Hope: 15th Anniversary of the Exxon Valdez Oil Spill*. Anchorage, AK: Exxon Valdez Oil Spill Trustee Council, 2004.

Ott, Riki. *Not One Drop: Betrayal and Courage in the Wake of the Exxon Valdez Oil Spill*. White River Junction, VT: Chelsea Green Pub, 2008.

ORGANIZATIONS

The Exxon Valdez Oil Spill Trustee Council, 441 West Fifth Avenue, Suite 500, Anchorage, AL, USA, 99501, (907) 278-8012, (907) 276-7178, (800) 478-7745, dfg.evos. restoration@alaska.gov, http://www.evostc.state.ak.us.

William G Ambrose
Paul E Renaud
Paula A Ford-Martin

Falcon *see* **Peregrine falcon.**

Falco peregrinus see **Peregrine falcon.**

Fallout *see* **Radioactive fallout.**

Family planning

Family planning generally refers to the use of various birth control methods to limit family size. The roots of family planning are related to the global population and food supply. As of August 2010 the global human population was 6.7 billion. In 2011 the population will pass 7 billion. As it has since the fifteenth century, the population has been continuously increasing, especially in developing countries. Worldwide famine has lessened because of the increased food production resulting from modern agricultural procedures (the Green Revolution), which have greatly increased grain production. Nevertheless, with limited land and resources that can be devoted to food production—and increasing numbers of humans who need both space and food—the global population could outstrip the food availability. Because of this, there is increased interest in family planning.

Family planning can also act in reverse. Jurisdictions such as the Canadian province of Quebec are encouraging more children per family in an effort to increase the population. Additionally, family planning is not limited to birth control but includes procedures designed to overcome difficulties in becoming pregnant. About 15 percent of couples are unable to conceive children after a year of sexual activity without using birth control. Many couples feel an intense desire and need to conceive children. Aid to these couples is thus a reasonable part of family planning.

Birth control procedures have evolved rapidly in this century. Further, utilization of existing procedures is changing with different age groups and different populations. An example of the changing technology includes oral contraception with pills containing hormones. Birth control pills have been marketed in the United States since the 1960s. Since that time there have been many formulations with significant reductions in dosage. There was much greater acceptance of pills by American women under the age of thirty. The intrauterine device (IUD) was much more popular in Sweden than in the United States, whereas sterilization was more common in the United States than in Sweden.

A very common form of birth control is the condom, which is a thin rubber sheath worn by men during sexual intercourse. They are generally readily accessible, cheap, and convenient for those individuals who may not have sexual relations regularly. Sperm cannot penetrate the thin (0.3—0.8 mm thick) latex. Condom use also greatly lessens the likelihood that human immunodeficiency virus (HIV) associated with AIDS or other pathogenic agents of sexually transmitted diseases (STDs) will be passed between individuals during sexual intercourse. Some individuals are opposed to treating healthy bodies with drugs (hormones) for birth control, and, for these individuals, condoms have a special appeal. Natural "skin" (lamb's intestine) condoms are still available for individuals who may be allergic to latex, but this product provides less protection from HIV and other STDs.

Spermicides—surface active agents that inactivate sperm and STD pathogens—can be placed in the vagina in jellies, foam, and suppositories. Condoms used in conjunction with spermicides have a failure rate lower than either method used alone and may provide added protection against some infectious agents.

The vaginal diaphragm, like the condom, is another form of barrier. The diaphragm was in use in World War I and was still used by about one-third of couples by the time of World War II. However, because of the efficacy and ease of use of oral contraceptives, and

A volunteer and staff of Planned Parenthood of South Texas pass out free condoms to spring breakers at J.P. Luby Surf Park Beach in Corpus Christi, Texas. *(AP Photo/Corpus Christi Caller-Times, Todd Yates)*

perhaps because of the protection against disease by condoms, the use of vaginal diaphragms has decreased. The diaphragm, which must be fitted by a physician, is designed to prevent sperm access to the cervix and upper reproductive tract. It is used in conjunction with spermicides. Other similar barriers include the cervical cap and the contraceptive sponge. The cervical cap is smaller than the diaphragm and fits only around that portion of the uterus that protrudes into the vagina. The contraceptive sponge, which contains a spermicide, is inserted into the vagina prior to sexual intercourse and retained for several hours afterwards to insure that no living sperm remain.

Intrauterine devices (IUDs) were popular during the 1960s and 1970s in the United States, but their use has plummeted. However in China, a nation which is rapidly attending to its population problems, about 60 million women use IUDs. The failure rate of IUDs in less-developed countries is reported to be less than that with the pill. The devices may be plastic, copper, or stainless steel. The plastic versions may be impregnated with barium sulfate to permit visualization by x ray and also may slowly release hormones such as progesterone. Ovulation continues with IUD use. Efficacy probably results from a changed uterine environment which kills sperm.

Oral contraception is overwhelmingly associated with use of The Pill, a formulation containing an estrogen and a progestational agent, and current dosage is very low compared with several decades ago. The combination of these two agents is taken daily for three weeks, followed by one week with neither hormone. Frequently a drug-free pill is taken for the last week to maintain the pill-taking habit and thus enhance the efficacy of the regimen. The estrogenic component prevents follicle maturation, and the progestational component prevents ovulation. This method of birth control is best designed for women who have one sexual partner. Those with multiple sexual partners, and so who may be at greater risk of exposure to someone with a STD, may more prudently consider the addition of a barrier method to minimize risk for STDs. The reliability of the pill reduces the need for abortion or surgical sterilization. There may be other salutary health effects. which include less endometrial and ovarian cancer as well as fewer uterine fibroids. Use of oral contraceptives in women over the age of thirty-five who also smoke is thought to increase the risk of heart and vascular disease.

Contraceptive hormones can be administered by routes other than oral. Subdermal implants of progestin-containing tubules have been available since 1990 in the United States. In this device familiarly known as Norplant, six tubules are surgically placed on the inside of the upper arm, and the hormone diffuses through the wall of the tubules to provide long-term contraceptive activity. Another form of progestin-only contraception is by intramuscular injection that must be repeated every three months.

Fears engendered by IUD litigation are thought to have increased the reliance of many American women on surgical sterilization (tubal occlusion). Whatever the reason, more American women rely on the procedure than do their European counterparts. Tubal occlusion involves the mechanical disruption of the oviduct, the tube that leads from the ovary to the uterus, and prevents sperm from

Types of Contraceptives

Effectiveness	Predicted (%)	Actual (%)
Birth control pills	99.9	97
Condoms	98	88
Depo Provera	99.7	99.7
Diaphragm	94	82
IUDS	99.2	97
Norplant	99.7	99.7
Tubal sterilization	99.8	99.6
Spermicides	97	79
Vasectomy	99.9	99.9

Types of contraceptives. *(Reproduced by permission of Gale, a part of Cengage Learning)*

reaching the egg. Inasmuch as the fatality rate for the procedure is lower than that of childbirth, surgical sterilization is now the safest method of birth control. Tubal occlusion is far more common now that it was in the 1960s because of the lower cost and reduced surgical stress. Use of the laparoscope and very small incisions into the abdomen have allowed the procedure to be completed during an office visit.

Male sterilization, another method, involves severing the vas deferens, the tube that carries sperm from the testes to the penis. Sperm comprise only a small portion of the ejaculate volume, and thus ejaculation is little changed after vasectomy. The male hormone is produced by the testes and production of that hormone continues as does erection and orgasm.

Another family-planning option for a woman who has become pregnant is the termination of the pregnancy. Legal abortion has become one of the leading surgical procedures in the United States. Morbidity and mortality associated with pregnancy have been reduced more with legal abortion than with any other event since the introduction of antibiotics to fight puerperal fever.

Other methods of birth control are used by individuals who do not wish to use mechanical barriers, devices, or drugs (hormones). One of the oldest of these methods is withdrawal (*coitus interruptus*), in which the penis is removed from the vagina just before ejaculation. Withdrawal must be exquisitely timed, is probably frustrating to both partners, is not thought to be reliable, and provides no protection against HIV and other STD infections. Another barrier- and drug-free procedure is natural family planning (also known as the rhythm method). Abstinence of sexual intercourse is scheduled for a period of time before and after ovulation. Ovulation is calculated by temperature change, careful record keeping of menstruation (the calendar method), or by vaginal mucous inspection. Natural family planning has appeal for individuals who wish to limit their exposure to drugs, but it provides no protection against HIV and other STDs.

See also Population growth.

Resources

BOOKS

Jütte, Robert .*Contraception: A History*. Boston: Polity, 2008.
May, Elaine T. *America and the Pill: A History of Promise, Peril, and Liberation*. New York: Basic Books, 2010.

Robert G. McKinnell

Famine

Famine is widespread hunger and starvation. A region struck with famine experiences acute shortages of food, massive loss of lives, social disruption, and economic chaos. Images of starving mothers and children with emaciated eyes and swollen bellies during recent crises in Ethiopia and Somalia have brought international attention to the problem of famine. Other well-known famines include the great Irish potato famines of the 1850s that drove millions of immigrants to America, and a Russian famine during Stalin's agricultural revolution that killed 20 million people in the 1930s. The worst recorded famine in recent history occurred in China between 1958 and 1961, when 23–30 million people died as a result of the failed agricultural program, termed the Great Leap Forward.

Even though these tragedies seem to be single, isolated events, famine and chronic hunger continue to be serious problems. Between 18 and 20 million people, three-quarters of them children, die each year of starvation or diseases caused by malnourishment. Environmental problems such as overpopulation, scarce resources, and natural disasters affect people's ability to produce food. Political and economic problems such as unequal distribution of wealth, delayed or insufficient action by local governments, and imbalanced trade relationships between countries affect people's ability to buy food when they cannot produce it.

Perhaps the most common explanation for famine is overpopulation. The world's population, now more than 6.8 billion people, grows by about 250,000 people every day. It seems impossible that the natural world could support such rapid growth. Indeed, the pressures of rapid growth have had a devastating impact on the environment in many places. Land that once fed one family must now feed ten families and resulting over use harms the quality of the land. The world's deserts are rapidly expanding as people destroy fragile topsoil by poor farming techniques, clearing vegetation, and overgrazing.

Although the demands of population growth and industrialization are straining the environment, the world has yet to exceed the limits of growth. Since the 1800s some have predicted that humans, like rabbits living without predators, would foolishly reproduce far beyond the carrying capacity of their environment and then die in masses from lack of food. This argument assumes that the supply of food will remain the same as populations grow, but as populations have grown, people have learned to grow more food. World food production increased two-and-a-half times between 1950 and 1980. After World War II, agriculture specialists initiated the green revolution, developing new crops and farming techniques that radically increased food production per acre. Farmers began to use special hybrid crop strains, chemical fertilizers, pesticides, and advanced irrigation systems. These innovations improved the ability of farmers to produce enough food for the population.

Many famines occur in the aftermath of natural disasters such as floods and droughts. In times of drought, crops cannot grow because they do not have enough water. In times of flood, excess water washes out fields, destroying crops and disrupting farm activity. These disasters have several effects. First, damaged crops cause food shortages, making nutrients difficult to find and making any food available too expensive for many people. Second, reduced food production means less work for those who rely on temporary farm work for their income. Famines usually affect only the poorest five to ten percent of a country's population. This subset of the population is most vulnerable because during a crisis wages for the poorest workers go down as food prices go up.

Famine is a problem of distribution as well as production. Environmental, economic, and political factors together determine the supply and distribution of food in a country. Starvation occurs when people lose their ability to obtain food by growing it or by buying it. Often, poor decisions and organizational problems aggravate environmental factors to cause human suffering. In Bangladesh, floods during the summer of 1974 interfered with rice transplantation, the planting of small rice seedlings in their rice patties. Although the crop was only partly damaged, speculators hoarded rice, and fears of a shortage drove prices beyond the reach of the poorest in Bangladesh. At the same time, disruption of the planting meant lost work for the same people. Even though there was plenty of rice from the previous year's harvest, deaths from starvation rose as the price of rice went up. In December of 1974, when the damaged rice crop was harvested, the country found that its crop had been only partly ruined. Starvation resulted not from a shortage of rice, but from price speculation. The famine could have been avoided completely if the government had responded more quickly, acting to stabilize the rice market and to provide relief for famine victims.

In other cases governments have acted to avoid famine. The Indian state of Maharashtra offset effects of a severe drought in 1972 by hiring the poorest people to work on public projects, such as roads and wells. This relief system provided a service for the country and at the same time it diverted a catastrophe by providing an income for the most vulnerable citizens to compete with the rest of the population for a limited food supply. At the same time, the countries of Mali, Niger, Chad, and Senegal experienced severe famine, even though the average amount of food per person in these countries was the same as in Maharashtra. The difference, it would seem, lies in the actions and intentions of the governments. The Indian government provides a powerful example. Although India lags behind many countries in economic development, education, and health care, the Indians have managed to avert serious famine since 1943, four years before they gained independence from the British.

Responsibility for hunger and famine rests also with the international community. Countries and peoples of the world are increasingly interconnected, continuously exchanging goods and services. They are increasingly dependent on one another for success and for survival. The world economic and political order dramatically favors the wealthiest industrialized countries, Europe and North America. Following patterns established during colonial expansion, Third World nations often produce raw materials, unfinished goods, and basic commodities, such as bananas and coffee that they sell to the First World at low prices. The First World nations then manufacture and refine these products and sell back information and technology, such as machinery and computers for a very high price. As a result the wealthiest nations amass capital and resources and enjoy very high standards of living, while the poorest nations retain huge national debts and struggle to remain stable economically and politically. The word's poorest countries, then, are left vulnerable to all of the conditions that cause famine, economic hardship, political instability, overpopulation, and over-taxed resources.

Furthermore, large colonial powers often left behind unjust political and social hierarchies that are very good at extracting resources and sending them north, but not as good at promoting social justice and human welfare. Many Third World countries are dominated by a small ruling class, who own most of the land, control industry, and run the government. Because the poorest people, who suffer most in famines, have little power to influence government policies and manage the countries economy, their needs are often unheard and unmet. A government that rules without democratic support of its people has less incentive to protect those who would suffer in times of famine. In addition, the poorest often do not benefit from the industry and agriculture that does exist in a developing country. Large corporate farms often force small subsistence farmers off of their land. These farmers must then work for day wages, producing food for export, while local people go without adequate nutrition.

Economic and social arrangements, as well as environmental conditions, are central to the problems of hunger and starvation. Famine is much less likely to occur in countries that are concerned with issues of social justice. Environmental pressures of population growth and human use of natural resources will continue to be issues of great concern. Natural disasters like droughts and floods will continue to occur. The best response to the problem of famine lies in working to better manage environmental resources and crisis situations and to change political and economic structures that cause people to go without food.

Resources

BOOKS

Devereux, Stephen. *The New Famines: Why Famines Persist in an Era of Globalization.* Routledge studies in development economics, 52. London: Routledge, 2007.

Donnelly, James S. *The Great Irish Potato Famine.* Stroud: Sutton, 2007.

Kannan, Seshadri. *Food, Famine and Fertilizers.* New Delhi: A.P.H. 2008.

Nabhan, Gary Paul. *Where Our Food Comes from: Retracing Nikolay Vavilov's Quest to End Famine.* Washington, DC: Island Press/Shearwater Books, 2009.

O'Grada, Cormac. *Famine: A Short History.* Princeton: Princeton University Press, 2006.

John Cunningham

Farming *see* **Agricultural revolution; Conservation tillage; Dryland farming; Feedlots; Organic gardening and farming; Shifting cultivation; Slash and burn agriculture; Strip-farming; Sustainable agriculture.**

Fauna

All animal life, or compilation of animal species, that lives in a particular geographic area during a particular time in history. The type of fauna to be found in any certain region is determined by factors such as plant life, physical environment, topographic barriers, and evolutionary history. Zoologists sometimes divide the

earth into six regions inhabited by distinct faunas: Ethiopian (Africa south of the Sahara, Madagascar, Arabia), Neotropical (South and Central America, part of Mexico, the West Indies), Australian (Australia, New Zealand, New Guinea), Oriental (Asia south of the Himalaya Mountains, India, Sri Lanka, Malay Peninsula, southern China, Borneo, Sumatra, Java, the Philippines), Palearctic (Europe, Asia north of the Himalaya Mountains, Afghanistan, Iran, North Africa), and Nearctic (North America as far south as southern Mexico).

Fecundity

Fecundity comes from the Latin word *fecundus,* meaning fruitful, rich, or abundant. It is the rate at which individual organisms in the population produce offspring. Although the term can apply to plants, it is typically restricted to animals.

There are two aspects of reproduction. One is fertility, which refers to the physiological ability to breed. The second is fecundity, which refers to the ecological ability to produce offspring. Thus, higher fecundity is dependent on advantageous conditions in the environment that favor reproduction (e.g., abundant food, space, water, and mates; limited predation, parasitism, and competition). The intrinsic rate of increase equals the birth rate minus the death rate. It is a population characteristic that takes into account that not all individuals have equal birth rates and death rates. So, it refers to the reproductive capacity in the population made up of individual organisms. Fecundity, on the other hand, is an individual characteristic. It can be further subdivided into the potential and actual fecundity. For example, deer can potentially produce four or more fawns per year, but they typically give birth to only one or two per year. In good years with ample food, they often have only two fawns.

Animals in nature are limited by environmental conditions that control their life history characteristics, such as birth, survivorship, and death. A graph of the number of offspring per female per age class (e.g., year) is a fecundity curve. This can then be used to interpret the individuals of a certain age class who contribute more to the population growth than others. In other words, certain age classes have a greater reproductive output than others. Wildlife managers often use this type of information in deciding which

retained responsibility for interstate regulation of electric utilities and the siting of hydroelectric power plants as well as their operation. It also set rates and charges for the transportation and sale of natural gas and electricity. The five members of the commission were appointed by the president with approval of the Senate; three of the members were the Secretaries of the Interior, Agriculture, and War (later designated as U.S. Department of the Army). During its existence the commission retained its status as an independent regulatory agency.

Feedlot runoff

Feedlots are containment areas used to raise large numbers of animals to an optimum weight within the shortest time span possible. Most feedlots are open air, and so are affected by variable weather conditions. A substantial portion of the feed is not converted into meat, and is excreted. If the excrement is not contained, air, ground, and surface water quality can suffer. The issues of odor and water pollution from such facilities center on the traditional attitudes of producers that farming has always produced odors, and manure is a fertilizer, not a waste from a commercial undertaking.

Legislative and regulatory actions have increased with encroachment of urban population and centers of high sensitivity, such as shopping malls and recreation facilities. Because odor is difficult to measure, control of these facilities is being achieved on the grounds that they must not pose a "nuisance," a principle that is being sustained by the courts.

Odor is influenced by feed, number and species of animal, lot surface and manure removal frequency, wind, humidity, and moisture. These factors, individually and collectively, influence the type of decomposition that will occur. Typically, it is an anaerobic process that produces a sharp pungent odor of ammonia, the nauseating odor of rotten eggs from hydrogen sulfide, and the smell of decaying cabbage or onions from methyl mercaptan.

Odorous compounds seldom reach concentrations that are dangerous to the public. However, levels can become dangerously elevated with reduced ventilation in winter months or during pit cleaning. It is this latter activity, in conjunction with disposal onto the surface of the land, that is most frequently the cause of complaints. Members of the public respond to feedlot odors depending on their individual sensitivity, previous experience, and disposition. It can curtail outdoor activities and require windows to be closed, which means the additional use of air purifiers or air-conditioning systems.

Surface water contamination is the problem most frequently attributed to open feedlot and manure spreading activities. It is due to the dissolving, eroding action of rain striking the manured-covered surface. Duration and intensity of rainfall dictates the concentration of contaminants that will flow into surface waters. Their dilution or retention in ponds, rivers, and streams depends on area hydrology (dry or wet conditions) and topography (rolling or steeply graded landscape). Such factors also influence conditions in those parts of the continent where precipitation is mainly in the form of snow. Large snow drifts form around wind breaks, and in the early spring substantial volumes of snowmelt are generated.

There are many examples of problems due to feedlot runoff. In Walkerton, Ontario, Canada, runoff contaminated well water and lead to an illness outbreak in May of 2000 that sickened thousands and killed six people. In 1998 the bursting of a hog sewage lagoon in North Carolina released 25 million gallons of sewage into a neighboring river, killing millions of fish and closing coastal shellfish grounds.

Odor and water pollution control techniques include simple operational changes, such as increasing the frequency of removing manure, scarifying the surface to promote aerobic conditions, and applying disinfectants and feed-digestion supplements. Other control measures require construction of additional structures or the installation of equipment at feedlots. These measures include installing water sparge-lines, adding impervious surfaces, drains, pits and roofs, and installing extraction fans.

As of 2010 the number of hog farms in the United States continued to drop, although the total number of hogs raised remained about the same. The increasing size of the individual farms increases the sewage load and so the possibility of runoff.

See also Animal waste; Odor control.

Resources

BOOKS

Johnsen, Carolyn. *Raising a Stink: The Struggle over Factory Hog Farms in Nebraska*. Winnipeg, ONT, Canada: Bison Books, 2003.

Mcdowell, Richard W., David J. Houlbrooke, Richard W. Muirhead, and Karin Muller. *Grazed Pastures and Surface Water Quality*. Boston: Nova Science, 2008.

Niman, Nicolette H. *Righteous Porkchop: Finding a Life and Good Food Beyond Factory Farms*. New York: William Morrow, 2009.

OTHER

United States Department of the Interior, United States Geological Survey (USGS). "Runoff." http://www.usgs.gov/science/science.php?term = 1375 (accessed October 13, 2010).

United States Environmental Protection Agency (EPA). "Water: Ground Water: Runoff." http://www.epa.gov/ebtpages/wategroundwaterrunoff.html (accessed October 13, 2010).

George M. Fell

Felis concolor coryi see **Florida panther.**

Fens see **Wetlands.**

Ferret see **Black-footed ferret.**

Fertility see **Biological fertility.**

Feedlots

A feedlot, also referred to as an animal feeding operation (AFO), is an open space where animals are fattened before slaughter. There are approximately 450,000 feedlots in the United States in 2010. Beef cattle usually arrive at the feedlot directly from the ranch or farm where they were raised, while poultry and pigs often remain in an automated feedlot from birth until death. Feed (often grains, alfalfa, and molasses) is provided to the animals so they do not have to forage for their food. This feeding regimen promotes the production of higher quality meat more rapidly. There are no standard parameters for the number of animals per acre in a feedlot, but the density of animals is usually very high. Some feedlots can contain 100,000 cows and steers. Animal rights groups actively campaign against confining animals in feedlots, a practice they consider inhumane, wasteful, and highly polluting.

Feedlots were first introduced in California in the 1940s, but many are now found in the Midwest, closer to grain supplies. Feedlot operations are highly mechanized, and large numbers of animals can be handled with relatively low labor input. Half of the beef produced in the United States is feedlot raised.

Feedlots are a significant source of the pollution flowing into surface waters and groundwater in the United States. As one example, in 1998 a lagoon housing the sewage from a pig feedlot in North Carolina burst, releasing 25 million gallons of sewage into a neighboring river. Millions of fish were killed and, as the sewage flowed out to the coast, shellfish grounds were closed.

At least half a billion tons of animal waste are produced in feedlots each year. Since this waste is concentrated in the feedlot rather than scattered over grazing lands, it overwhelms the soil's ability to absorb and buffer it and creates nitrate-rich, bacteria-laden runoff to pollute streams, rivers, and lakes. Dissolved pollutants can also migrate down through the soil into aquifers, leading to groundwater pollution over wide areas. To protect surface waters, most states require that feedlot runoff be collected. However, protection of groundwater has proved to be a more difficult problem, and successful regulatory and technological controls have not yet been developed. Nevertheless, most feedlots require some type of state permit and have plans in place to deal with the large amount of waste that is generated. The Clean Water Action Plan (CWAP), initiated by the Clinton administration in 1998, included strategies to control runoff from AFOs. Concentrated AFOs (CAFOs) house and feed a large number of animals in a restricted area for forty-five days or more within a year. CAFOs are required under the Clean Water Act to obtain a permit for discharges through the National Pollutant Discharge Elimination System. Proper regulation also considers issues such as noise and air pollution.

Due to the large numbers of animals being raised, the amount of air pollutants contributed by AFOs is an environmental concern. Methane gas (CH_4), carbon dioxide (CO_2), and nitrous oxide (N_2O) are greenhouse gases that are emitted from livestock production. Of the human-related methane emissions, ruminant livestock production accounts for 28 percent of those annual emissions, amounting to 88.2 million tons (80 million metric tons) of methane gas produced each year. Methane is twenty times more efficient at trapping heat in the atmosphere compared with carbon dioxide.

Resources

BOOKS

Imhoff, Daniel. *The CAFO Reader: The Tragedy of Industrial Animal Factories.* Berkeley: University of California Press, 2010.

Kirby, David. *Animal Factory: The Looming Threat of Industrial Pig, Dairy, and Poultry Farms to Humans and the Environment.* New York: St. Martin's Press, 2010.

OTHER

Centers for Disease Control and Prevention (CDC). "Concentrated Animal Feeding Operations

Feedlots

(CAFOs)." http://www.cdc.gov/cafos/about.htm (accessed October 13, 2010).

United States Environmental Protection Agency (EPA). "International Cooperation: Global Climate Change: Methane." http://www.epa.gov/ebtpages/integlobalcli matecmethane.html (accessed October 13, 2010).

Christine B. Jeryan

Fertilizer

Any substance that is applied to land to encourage plant growth and produce higher crop yield. Fertilizers may be made from organic material—such as recycled waste, animal manure, compost, etc.—or chemically manufactured. Most fertilizers contain varying amounts of nitrogen, phosphorus, and potassium, inorganic nutrients that plants need to grow.

Since the 1950s crop production worldwide has increased dramatically because of the use of fertilizers. In combination with the use of pesticides and insecticides, fertilizers have vastly improved the quality and yield of such crops as corn, rice, wheat, and cotton. However overuse and improper use of fertilizers have also damaged the environment and affected the health of humans, animals, and plants.

In the United States it is estimated that as much as 25 percent of fertilizer is carried away as runoff. Fertilizer runoff has contaminated groundwater and polluted bodies of water near and around farmlands. High and unsafe nitrate concentrations in drinking water have been reported in countries that practice intense farming, including the United States. Accumulation of nitrogen and phosphorus in waterways from chemical fertilizers has also contributed to the eutrophication of lakes and ponds. Ammonia, released from the decay of fertilizers, causes minor irritation to the respiratory system.

While few advocate the complete eradication of chemical fertilizers, many environmentalists and scientists urge more efficient ways of using them. For example, some farmers use up to 40 percent more fertilizer than they need. Frugal applications—in small doses and on an as-needed-basis on specific crops—helps reduce fertilizer waste and runoff. The use of organic fertilizers, including animal waste, crop residues, or grass clippings, is also encouraged as an alternative to chemical fertilizers.

Within U.S. waters, oxygen-depleted dead zones are annually identified in the Gulf of Mexico and Chesapeake Bay. The zones are, in part, linked to runoff of nutrient pollution from fertilizers. For example, nitrogen runoff from the length of the Mississippi River ultimately drains into the Gulf of Mexico to create a marine dead zone that fluctuates in size and intensity.

See also Cultural eutrophication; Recycling; Sustainable agriculture; Trace element/micronutrient.

Fibrosis

A medical term that refers to the excessive growth of fibrous tissue in some part of the body. Many types of fibroses are known, including a number that affect the respiratory system. A number of these respiratory fibroses, including such conditions as black lung disease, silicosis, asbestosis, berylliosis, and byssinosis, are linked to specific environmental exposures. A fibrosis develops when a person inhales very tiny solid particles or liquid droplets over many years or decades. Part of the body's reaction to these foreign particles is to enmesh them in fibrous tissue. The disease name usually suggests the agent that causes the disease. Silicosis, for example, is caused by the inhalation of silica, tiny sand-like particles. Occupational sources of silicosis include rock mining, quarrying, stone cutting, and sandblasting. Berylliosis is caused by the inhalation of beryllium particles over a period of time, and byssinosis (from byssos, the Greek word for flax) is found among textile workers who inhale flax, cotton, or hemp fibers.

Field capacity

Field capacity refers to the maximum amount of water potentially held in soil after excess water drains due to gravitational force. For example, sandy soils have a lower field capacity (hold less water) than clay soils. The higher a soil's field capacity (the more water it retains) the greater amount of water available for plants, etc. Field capacity is an important factor in determining or limiting the agricultural use of land.

Filters

Filters are devices that remove particles from the air or fluid that passes through. The size of the particles that are prevented from crossing the filter depends on the size of the openings of the filter (the pore size). Filters with a pore size in the nanometer range are capable of restricting the passage of particles as small as viruses.

The most common type of filters used to filter air or gas are fibrous filters, in which the fibers are of cellulose (paper filters). But, almost any fibrous material, including glass fiber, wool, asbestos, and finely spun polymers can be used. Microscopically, these fibers collect fine particles because fine particles vibrate around their average position due to collision with air molecules (Brownian motion). These vibrations are likely to cause them to collide with the fibers as they pass through the filter. Larger particles are removed because, as the air stream carrying them passes through the filter, some of the particles are intercepted as they contact the fibers. Other particles are in air streams that would cause them to miss the fibers, but when the air stream bends to go around the fibers the momentum of the particles is too much to let them remain with the stream. In effect, centrifugal force causes them to contact the fibers. Still other particles may be attracted to the fibers because the particles and the fiber have opposite electric charges. Finally, particles may simply be larger than the space between fibers and will be sifted out of the air in a process called sieving.

Filters are also formed by a process in which polymers such as cellulose esters are made into a film from a solution in an organic solvent containing water. As the solvent evaporates, a point is reached at which the water separates out as microscopic droplets, in which the polymer is not soluble. The final result is a film of polymer full of microscopic holes where the water droplets once were. Such filters can have pore sizes from a small fraction of a micrometer to a few micrometers (a micrometer equals 0.00004 in). These are called membrane filters.

In circumstances where filter strength is of paramount importance, such as in industrial filters where a large air flow must pass through a relatively small filter area, filters of woven cloth are used, made of materials ranging from cotton to glass fiber and asbestos. The latter are used when very hot gases must be filtered. The woven fabric itself is not a particularly good filter, but it retains enough particles to form a particle cake on the surface, and that soon becomes the filter. When the cake becomes thick enough to slow airflow to an unacceptable degree, the air flow is interrupted briefly, and the filters are shaken to dislodge the filter cake, which falls into bins at the bottom of the filters. Then filtration is resumed, allowing the cloth filters to be used for months before being replaced. A familiar domestic example is the bag of a home vacuum cleaner. Cement plants and some electric power plants use dozens of cloth bags up to several feet in diameter and more than 10 feet (3 meters) in length to remove particles from their waste gases.

Filters are also useful for filtering disease-causing agents from water. The degree of removal of these pathogens can be so good that the water is rendered fit to drink. Some water purification systems utilize filters in a process called reverse osmosis, although the integrity of the finished water is typically guaranteed by the combined use of another water treatment method, such as chlorination or exposure to ultraviolet light.

See also Baghouse; Electrostatic precipitation; Odor control; Particulate.

James P. Lodge Jr.

Filtration

A common technique for separating substances in two physical states. For example, a mixture of solid and liquid can be separated into its components by passing the mixture through a filter paper. Filtration has many environmental applications. In water purification systems, impure water is often passed through a charcoal filter to remove the solid and gaseous contaminants that give water a disagreeable odor, color, or taste. Trickling filters are used to remove solid wastes in plants. Some solid and liquid contaminants in waste industrial gases can be removed by passing them through a filter or series of filters prior to discharge in a smokestack.

Fire *see* **Prescribed burning; Wildfire.**

Fire ants

Fire ants are an example of a species that was accidently introduced into a new habitat and which spread quickly in the new setting. In the United States, fire ants have spread throughout Southern and Southwestern states. In 2009 fire ants caused an estimated $750 million in agricultural losses, both directly due to crop loss and indirectly due to care of animals bitten by the ants. The total cost of battling fire ants in the United States is estimated to be $5 billion annually.

Two distinct species of fire ants (genus *Solenopsis*) from South America have been introduced into the

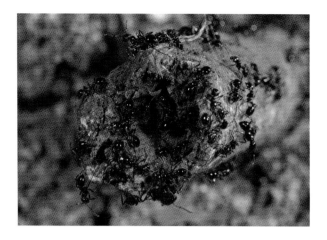

Red fire ants (*Selonopsis invicta*) moving on their mound, Texas. *(Francesco Tomasinelli/Photo Researchers, Inc.)*

United States. The South American black fire ant (*S. richteri*) was first introduced into the United States in 1918. Its close relative, the red fire ant (*S. wagneri*), was introduced in 1940, probably escaping from a South American freighter docked in Mobile, Alabama. Both species became established in the southeastern United States, spreading into nine states from Texas across to Florida and up into the Carolinas. It is estimated that they have infested more than 320 million acres (130 million ha) covering states as well as Puerto Rico.

Successful introduced species are often more aggressive than their native counterparts, and this is definitely true of fire ants. They are very small, averaging 0.2 inches (5 mm) in length, but their aggressive, swarming behavior makes them a threat to livestock and pets as well as humans. These industrious, social insects build their nests in the ground—the location is easily detected by the elevated earthen mounds created from their excavations. The mounds are 18–36 inches (46–91 cm) in diameter and may be up to 36 inches (91 cm) high, although mounds are generally 6–10 inches (15–25 cm) high. Each nest contains as many as 25,000 workers, and there may be more than 100 nests on an acre of land.

If the nest is disturbed, fire ants swarm out of the mound by the thousands and attack with swift ferocity. As with other aspects of ant behavior, a chemical alarm pheromone is released that triggers the sudden onslaught. Each ant in the swarm uses its powerful jaws to bite and latch onto whatever disturbed the nest, while using the stinger on the tip of its abdomen to sting the victim repeatedly. The intruder may receive thousands of stings within a few seconds.

The toxin produced by the fire ant is extremely potent, and it immediately causes an intense burning pain that may continue for several minutes. After the pain subsides, the site of each sting develops a small bump that expands and becomes a tiny, fluid-filled blister. Each blister flattens out several hours later and fills with pus. These swollen pustules may persist for several days before they are absorbed and replaced by scar tissue. Fire ants obviously pose a problem for humans. Some people may become sensitized to fire ant venom, have a generalized systematic reaction, and go into anaphylactic shock. Fire-ant induced deaths have been reported. Because these species prefer open, grassy yards or fields, pets and livestock may fall prey to fire ant attacks as well.

Attempts to eradicate this pest involved the use of several different generalized pesticides, as well as the widespread use of gasoline either to burn the nest and its inhabitants or to kill the ants with strong toxic vapors. Another approach involved the use of specialized crystalline pesticides, which were spread on or around the nest mound. The workers collected them and took them deep into the nest, where they were fed to the queen and other members of the colony, killing the inhabitants from within. A more recent method involves the release of a natural predator of the fire ant, the "phorid" fly. The fly releases an egg into the fire ant. The larva then eats the ant's brain while releasing an enzyme. The enzyme systematically destroys the joints causing the ant's head to fall off. The flies were released in eleven states as of 2001. Although the strategy can lessen the population growth of fire any colonies, as of 2010 fire ants remained a problem. It is likely that they are too numerous and well established to be completely eradicated in North America.

Resources

BOOKS

Lockwood, Julie L., Martha Hoopes, and Michael Marchetti. *Invasive Ecology*. New York: Wiley-Blackwell, 2006.
Terrill, Ceiridwen. *Unnatural Landscapes: Tracking Invasive Species*. Tucson: University of Arizona Press, 2007.
Tschinkel, Walter R. *The Fire Ants*. Boston: Belknap Press of Harvard University Press, 2006.

Eugene C. Beckham

First World

The world's more wealthy, politically powerful, and industrially developed countries are unofficially, but commonly, designated as the First World. In common use, the term generally differentiates the powerful, capitalist states of Western Europe, North America and Japan from the (formerly) communist states (Second World) and from the nonaligned, developing countries (Third World) in world systems theory. In common usage First World refers mainly to a level of economic strength. The level of industrial development of the First World, characterized by an extensive infrastructure, mechanized production, efficient and fast transport networks, and pervasive use of high technology, consumes huge amounts of natural resources and requires an educated and skilled work force. However, such a system is usually highly profitable. Often depending upon raw materials imported from poorer countries (wood, metal ores, petroleum, food, and so on), First World countries efficiently produce goods that less developed countries desire but cannot produce themselves, including computers, airplanes, optical equipment, and military hardware. Generally, high domestic and international demand for such specialized goods keeps First World countries wealthy, allowing them to maintain a high standard of material consumption, education, and health care for their citizens. An increasing number of critics consider the use of the terms First, Second, or Third World as offensive and inarticulate; preferring instead to use terms such as capitalist, wealthy, poor, industrialized, or developing.

Fish and Wildlife Service

The United States Fish & Wildlife Service, which is headquartered in Washington, DC, is charged with conserving, protecting, and enhancing fish, wildlife, and their habitats for the benefit of the American people. As a division of the U.S. Department of the Interior, the Service's primary responsibilities are for the protection of migratory birds, endangered species, freshwater and anadromous (saltwater species that spawn in freshwater rivers and streams) fisheries, and certain marine mammals.

In addition to its Washington, DC, headquarters, the service maintains seven regional offices and field units. Those include national wildlife refuges, national fish hatcheries, research laboratories, and a nationwide network of law enforcement agents.

As of 2010 the service managed 548 refuges that provided habitats for migratory birds, endangered species, and other wildlife, and sixty-six national fish hatcheries. It sets migratory bird hunting regulations

and leads an effort to protect and restore endangered and threatened animals and plants in the United States and other countries.

Service scientists assess the effects of contaminants on wildlife and habitats. Its geographers and cartographers work with other scientists to map wetlands and carry out programs to slow wetland loss, or preserve and enhance these habitats. Restoring fisheries that have been depleted by overfishing, pollution, or other habitat damage is a major program of the service. Efforts are underway to help four important species: lake trout in the upper Great Lakes, striped bass in both the Chesapeake Bay and Gulf Coast, Atlantic salmon in New England, and salmonid species of the Pacific Northwest. A notable effort taking place in the U.S. Pacific Northwest in 2010 is directed at protecting the habitat of a threatened species of fish known as the bull trout.

Fish and Wildlife biologists working with scientists from other federal and state agencies, universities, and private organizations develop recovery plans for endangered and threatened species. Among its successes are the American alligator, no longer considered endangered in some areas and a steadily increasing bald eagle population. The service continually evaluates the condition of species of interest to determine if they are endangered or not. As two examples, in 2010 the Sacramento Splitttail, a fish that inhabits many California freshwaters, was determined not to be endangered, whereas a species of eider duck called Steller's Edier was designated an endangered species.

Internationally the Service cooperates with forty wildlife research and wildlife management programs and provides technical assistance to many other countries. Its 200 special agents and inspectors enforce wildlife laws and treaty obligations. They investigate cases ranging from individual migratory-bird hunting violations to large-scale poaching and commercial trade in protected wildlife.

Resources

BOOKS

Helfman, Gene S. *Fish Conservation: A Guide to Understanding and Restoring Global Aquatic Biodiversity and Fishery Resources.* Washington, DC: Island Press, 2007.

U.S. Fish and Wildlife Service. *Fish and Wildlife News.* Ann Arbor: University of Michigan Library, 2010.

ORGANIZATIONS

U.S. Fish and Wildlife Service, contact@fws.gov, http://www.fws.gov

Linda Rehkopf

panthers are catching armadillos and raccoons for food. Panthers then become underweight and anemic due to poor nutrition.

Development contributes to the Florida panther's decline in other ways too. The panther's range is currently split in half by the east-west highway known as Alligator Alley. During peak seasons more than 30,000 vehicles traverse this stretch of highway daily, and since 1972 dozens of panthers have been killed by cars—the single largest cause of death for these cats in recent decades.

Biology is also working against the Florida panther. Because of the extremely small population size, inbreeding of panthers has yielded increased reproductive failures, due to deformed or infertile sperm. The spread of feline distemper virus also is a concern to wildlife biologists. All these factors have led officials to develop a recovery plan that includes a captive breeding program using a small number of injured animals, as well as a mark and recapture program, using radio collars, to inoculate against disease and track young panthers with hopes of saving this valuable part of the biota of south Florida's Everglades ecosystem.

Resources

BOOKS

Askins, Robert. *Saving Biological Diversity: Balancing Protection of Endangered Species and Ecosystems.* New York: Springer, 2008.
Macdonald, David W., and Andrew J. Loveridge. *Biology and Conservation of Wild Felids.* Oxford: Oxford University Press, 2010.
Mackay, Richard. *The Atlas of Endangered Species.* Berkeley: University of California Press, 2008.

OTHER

United States Environmental Protection Agency (EPA). "Ecosystems: Species: Endangered Species." http://www.epa.gov/ebtpages/ecosspeciesendangeredspecies.html (accessed November 11, 2010).

Eugene C. Beckham

Flotation

An operation in which submerged materials are brought to the surface under buoyant force created by the induction and adherence of small air bubbles. Bubbles are generated through a system called dissolved air flotation (DAF), which is capable of producing clouds of very fine, very small bubbles. A large number of small-sized bubbles is generally most efficient for removing material from water.

This process is commonly used in wastewater treatment and by industries, but not in water treatment. For example, the mining industry uses flotation to concentrate fine ore particles, and flotation has been used to concentrate uranium from sea water. It is commonly used to thicken the sludges and to remove grease and oil at wastewater treatment plants. The textile industry often uses flotation to treat process waters resulting from dyeing operations. Flotation might also be used to remove surfactants. Materials that are denser than water or that dissolve well in water are poor candidates for flotation. Flotation should not be confused with foam separation, a process in which surfactants are added to create a foam that affects the removal or concentration of some other material.

Flu pandemic

The Spanish influenza (flu) outbreak of 1918–1919 killed an estimated 20 to 40 million people worldwide. Although the virus was not especially lethal in terms of number of deaths per total cases (by 1918 standards), the virus infected at least an estimated 500 million people. It was, however, very lethal to otherwise healthy adults ages twenty to forty-four, as opposed to most flu outbreaks, which kill only the very young, the elderly, and people with weakened immune systems. Scientists and public health officials

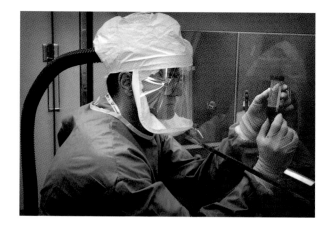

This undated photo provided by the journal *Science* shows Dr. Terrence Tumpey, microbiologist for the National Center for Infectious Diseases, examining specimens of the 1918 Pandemic Influenza Virusa virus that was reconstructed.
(AP Photo/ Cynthia Goldsmith, CDC, Science)

continue to study Spanish flu in the hopes of preventing a similar outbreak.

The Spanish flu virus caused one of the worst infectious disease pandemics ever recorded in modern history. And although the threats of some diseases, such as smallpox, have been contained by vaccination programs, influenza remains a difficult disease. There are worldwide outbreaks of influenza every year (approximately 300,000–500,000 people die of influenza or influenza complications each year—about 36,000 in the United States alone), and the flu typically reaches pandemic proportions (lethally afflicting an unusually high portion of the population) every ten to forty years. Prior to the declaration of a global pandemic of 2009 A H1N1 influenza in June 2009, the last influenza pandemic was the Hong Kong flu of 1968–1969, which caused an estimated one million deaths worldwide and killed approximately 33,000 Americans. The influenza virus is highly mutable, so each year's flu outbreak presents the human body with a slightly different virus. Because of this, people do not build immunity to influenza. Vaccines are successful in protecting people against influenza, but vaccine manufacturers must prepare a new batch each year, based on their best supposition of which particular virus will spread.

Most influenza viruses originate in Asia, and doctors, scientists, and public health officials closely monitor flu cases there to make the appropriate vaccine. The two main organizations tracking influenza are the Centers for Disease Control (CDC) and the World Health Organization (WHO). The CDC and other government agencies have been preparing for a flu pandemic on the level of Spanish flu since the early 1990s.

In April 2009 a new virus with a mixture of swine, avian, and human influenza genes emanated from Mexico to sites around the world. Initially classified as a swine flu, the flu was renamed the 2009 H1N1 flu (or 2009 A H1N1 influenza) because there was no evidence that pigs were involved in the most recent transmission to humans. The virus must have passed through pigs at one time, but as of August 2010, there was no evidence that pigs played a direct part in the 2009 H1N1 outbreak. The 2009 H1N1 flu spread quickly and reached pandemic status within two months, because it was a new and highly transmissible virus to which humans had no natural immunity. By the end of July 2009, more than 55,867 laboratory-confirmed cases of H1N1 influenza, including 700 deaths, were reported to WHO. Officials at WHO ceased collecting data on individual case counts, as

the epidemic was well established in both the Northern and Southern hemispheres. In August 2010 WHO officials officially declared an end to the global pandemic.

The 2009 pandemic H1N1 virus proved to be less lethal than many other flu viruses (it is not as lethal as the H5N1, Spanish, or Hong Kong viruses, for example), and infectious disease experts predicted that the 2009 influenza pandemic would not approach the severity of prior pandemics such as the 1918–1919 Spanish flu. Priorities for responding to the pandemic included manufacturing and delivering sufficient quantities of antiviral drugs and vaccine specific to the 2009 H1N1 virus.

The Spanish flu actually did not originate in Spain, but presumably in Kansas, where the first case was recorded in March 1918, at the army base Camp Funston. It quickly spread across the United States and then to Europe with American soldiers who were fighting in the last months of World War I (1914–1918). Infected ships brought the outbreak to India, New Zealand, and Alaska. Spanish flu killed quickly. People often died within forty-eight hours of first feeling symptoms. The disease afflicted the lungs, and caused the tiny air sacs, called alveoli, to fill with fluid. Victims were soon starved of oxygen and sometimes drowned in the fluid clogging their lungs. Children and old people recovered from the Spanish flu at a much higher rate than young adults. In the United States, the death rate from Spanish flu was several times higher for men ages twenty-five to twenty-nine than for men in their seventies.

Social conditions at the time probably contributed to the remarkable power of the disease. The flu struck just at the end of World War I, when thousands of soldiers were moving from America to Europe and across that continent. In a peaceful time, sick people may have gone home to bed, and thus passed the disease only to their immediate family. But in 1918 men with the virus were packed in already crowded hospitals and troop ships. The unrest and devastation left by the war probably hastened the spread of Spanish flu. So it is possible that if a similarly virulent virus were to arise again soon, especially with modern antiviral medicines, it would not be as deadly.

Researchers are concerned about a return of Spanish flu because little is known about what made it so virulent. The flu virus was not isolated until 1933, and since then there have been several efforts to collect and study the 1918 virus by exhuming graves in Alaska and Norway, where bodies were preserved in permanently frozen ground. In 1997 a Canadian researcher,

In electrostatic precipitators the particles are given a negative charge and then attracted to a positive electrode where they are collected and removed. Cloth or paper filters and spraying water through the exhaust gases can be useful in removing fly ash.

Fly ash is a nuisance at high concentrations because it accumulates as grit on the surfaces of buildings, clothes, cars, and outdoor furnishings. It is a highly visible and very annoying aspect of industrial air pollution. The deposition of fly ash increases cleaning costs incurred by people who live near poorly controlled combustion sources. Fly ash also has health impacts because the finer particles can penetrate into the human lung. If the deposits are especially heavy, fly ash can also inhibit plant growth.

Each year millions of tons of fly ash are produced from coal-powered furnaces and electrical generation facilities. Where fly ash settles or from designated waste dumps, high levels of toxic metals and alkalis can potentially leach into surrounding watershed and waterways.

Fly ash may be used as a low-grade cement in road building because it contains a large amount of calcium oxide, but generally the demand is low. Fly ash is also used as a high performance mineral filler for use in carpet backing.

Resources

BOOKS

French, David, and Jim Smitham. *Fly Ash Characteristics and Feed Coal Properties.* Pullenvale, Qld, Australia: QCAT Technology Transfer Centre, 2007.

Telone, Peter H. *Fly Ash Reuse, Environmental Problems and Related Issues.* Hauppauge, NY, USA: Nova Science Publishers, 2009.

Peter Brimblecombe

Flyway

The route taken by migratory birds and waterfowl when they travel between their breeding grounds and their winter sanctuary. Flyways often follow geographic features such as mountain ranges, rivers, or other bodies of water. Protecting flyways is one of the many responsibilities of wildlife managers. Draining of wetlands, residential and commercial development, and overhunting are some of the factors that threaten flyway sites visited by birds for food and rest during migration. In most cases international agreements are needed to guarantee protection along the entire length of a flyway. In the United States flyway protection is financed to a large extent by funds produced through the Migratory Bird Hunting Stamp Act passed by the U.S. Congress in 1934.

Food additives

Food additives are substances added to food as flavorants, nutrients, preservatives, emulsifiers, or colorants. In addition, foods may contain residues of chemicals used during the production of plant or animal crops, including pesticides, antibiotics, and growth hormones. The use of most food additives is clearly beneficial because it results in improved public health and prevention of spoilage, which enhances the food supply. Nevertheless, there is controversy about the use of many common additives and over the presence of contaminants in food. This is partly because some people are hypersensitive and suffer allergic reactions if they are exposed to certain types of these chemicals. In addition, some people assume that low levels of chronic toxicity and diseases may be caused in the larger population by exposure to some of these substances. Although there is no compelling scientific evidence that this is indeed the case, the possibility of chronic damage caused by food additives and chemical residues is an important social and scientific issue.

The use of food additives in the United States is closely regulated by the government agencies responsible for health, consumer safety, and agriculture (e.g., Food and Drug Administration or FDA). The FDA composes a partial list of tested food additives approved to be "generally recognized as safe" or GRAS, which means that qualified experts have shown the substance to be safe for intended use, or that the substance does not qualify as a food additive under the FDA definition (e.g., pesticides, color additives). Any additives that do not fit the GRAS description are subject to premarket review and approval by the FDA. This type of regulatory system is also in effect in other developed countries in Europe, Canada, and elsewhere. Chemicals cannot be used as additives in those countries unless regulators are convinced that they have been demonstrated to be toxicologically safe, with a wide margin of security. In addition, chemicals added to commercially prepared foods must be listed on the packaging so that consumers can know what is present in the foodstuffs that they choose to eat.

Because of the intrinsic nature of low-level, toxicological risks, especially those associated with diseases

Unhealthy Food Additives

Name	Description	Example products
Aspartame	An artificial sweetener associated with rashes headaches, dizziness, depression, etc.	Diet sodas, sugar substitutes, etc.
Brominated vegetable oil (BVO)	Used as an emulsifier and clouding agent. Its main ingredient, bromate, is a poison.	Sodas, etc.
Butylated hydroxyanisole (BHA)/ butylated hydroxytoluene (BHT)	Prevents rancidity in foods and is added to food packagings. It slows the transfer of nerve impulses and affects sleep, aggressiveness, and weight in test animals.	Cereal and cheese packaging
Citrus red dye #2	Used to color oranges, it is a probable carcinogen. The FDA has recommended it be banned.	Oranges
Monosodium glutamate (MSG)	A flavor enhancer that can cause headaches, heart palpitations, and nausea.	Fast food, processed and packaged food
Nitrites	Used as preservatives, nitrites form cancer-causing compounds in the gastrointestinal tract and have been associated with cancer and birth defects.	Cured meats and wine
Saccharin	An artificial sweetener that may be carcinogenic.	Diet sodas and sugar substitutes
Sulfites	Used as a food preservative, sulfites have been linked to at least four deaths reported to the FDA in the United States.	Dried fruits, shrimp, and frozen potatoes
Tertiary butyhydroquinone (TBHQ)	It is extremely toxic in low doses and has been linked to childhood behavioral problems.	Candy bars, baking sprays, and fast foods
Yellow dye #6	Increases the number of kidney and adrenal gland tumors in lab rats. It has been banned in Norway and Sweden.	Candy and sodas

Unhealthy food additives. *(Reproduced by permission of Gale, a part of Cengage Learning)*

that may take a long time to develop, scientists are never able to demonstrate that trace exposures to any chemical are absolutely safe—there is always a level of risk, however small. Because some people object to these potential, low-level, often involuntary risks, a certain degree of controversy will always be associated with the use of food additives. This is also true of the closely related topic of residues of pesticides, antibiotics, and growth hormones in foods.

Flavorants

Certain chemicals are added to foods to enhance their flavor. This is particularly true of commercially processed or prepared foods, such as canned vegetables and frozen foods and meals. One of the most commonly added flavorants is table salt (or sodium chloride, NaCl), a critical nutrient for humans and other animals. In large amounts, however, sodium chloride can predispose people to developing high blood pressure, a factor that is important in strokes and other circulatory and heart diseases.

Table sugar (or sucrose, $C_{12}H_{22}O_{11}$), manufactured from sugar cane or sugar beets, and fructose ($C_6H_{12}O_6$), or fruit sugar, are commonly used to sweeten prepared foods. Such foods include sugar candies, chocolate products, artificial drinks, sweetened fruit juices, peanut butter, jams, ketchup, and most commercial breads. Sugars are easily assimilated from foods and are a useful form of metabolic energy. In large amounts, however, sugars can lead to weight gain, tooth decay, and hypoglycemia or diabetes in genetically predisposed people. Artificial sweeteners such as saccharine, aspartame, and sucralose avoid the nutritional problems associated with eating too much sugar. Another naturally-derived sweetener is stevia, which was approved by the FDA in 2008. These nonsugar sweeteners may have their own problems, however, and some people consider them to be a low-level health hazard.

Monosodium glutamate (or MSG) is commonly used as a flavor enhancer, particularly in processed meats, prepared soups, and Asian foods. Some people are relatively sensitive to this chemical, developing headaches and other symptoms that are sometimes referred to as Chinese food syndrome. Other flavorants used in processed foods include many kinds of spices, herbs, vanilla, mustard, nuts, peanuts, and wine. Some people are extremely allergic to even minute exposures to peanuts or nuts in food and can rapidly develop a condition known as anaphylactic shock, which is life-threatening unless quickly treated with medicine. This is one of the reasons why any foods containing peanuts or nuts as a flavoring ingredient must be clearly labeled as such.

Many flavorants are natural in origin. Increasingly, however, synthetic flavorants are being discovered and used. For example, vanilla used to be extracted from a particular species of tropical orchid (*Vanilla* species) and was therefore a rather expensive flavorant. However, a synthetic vanilla flavorant can now be manufactured from wood-pulp lignins, and this has made this pleasant flavor much more readily available than it used to be.

Nutrients

Many foods are fortified with minerals, vitamins, and other micronutrients. One such example is table salt, which has iodine (I) added (as potassium iodide, KI) to help prevent goiter in the general population. Goiter used to be relatively common but is now rare, in part because of the widespread use of iodized salt.

Other foods that are commonly fortified with minerals and vitamins include milk and margarine (with vitamins A and D), flour (with thiamine, riboflavin, niacin, iron), and some commercial breads and breakfast cereals (with various vitamins and minerals, particularly in some commercial cereal preparations). Micronutrient additives in these and other commercial foods are carefully formulated to help contribute to a balanced diet in their consumers. Nevertheless, some people believe that it is somehow unnatural and unhealthy to consume foods that have been adulterated in this manner, and they prefer to eat "natural" foods that do not have any vitamins or minerals added to them.

Preservatives

Preservatives are substances added to foods to prevent spoilage caused by bacteria, fungi, yeasts, insects, or other biological agents. Spoilage can lead to a decrease in the nutritional quality of foods, to the growth of food-poisoning microorganisms such as the botulism bacterium, or to the production of deadly chemicals, such as aflatoxin, that can be produced in stored grains and seeds (such as peanuts) by species of fungi.

Salt has long been used to preserve meat and fish, either added directly to the surface or by immersing the food in a briny solution. Nitrates and nitrites (such as sodium nitrate $NaNO_3$, or saltpetre) are also used to preserve meats, especially cured foods, such as sausages, salamis, and hams. These chemicals are especially useful in inhibiting the growth of *Clostridium botulinum*, the bacterium that causes deadly botulism. Vinegar and wood smoke are used for similar purposes. Sulfur dioxide (SO_2), sodium sulfite (Na_2SO_3), and benzoic acid ($C_7H_6O_2$) are often used as preservatives in fruit products, such beverages as wine and beer, and in ketchup, pickles, and spice preparations.

Anti-oxidants are chemicals added to certain foods to prevent a deterioration in their quality or flavor, occurring due to the exposure of fats and oils to atmospheric oxygen. Examples of commonly used antioxidants are ascorbic acid (or vitamin C), butylated hydroxyanisole (or BHA), butylated hydroxytoluene (or BHT), gallates, and ethoxyquin.

Stabilizers and emulsifiers

Stabilizers and emulsifiers are added to prepared foods to maintain suspensions of fats or oils in water matrices (or vice versa), or to prevent the caking of ingredients during storage or preparation. One example of an emulsifying additive is glyceryl monostearate ($C_{21}H_{42}O_4$), often added to stored starch products to maintain their texture. Alginates, which are substances found in the cell walls of algae, are compounds added to commercial ice cream, salad dressing, and other foods to stabilize emulsions of oil- or fat-in-water during storage.

Colorants

Some prepared foods have colors added to improve their aesthetic qualities and thereby to make them more attractive to consumers. This practice is especially common in the preparation of confectionaries such as candies, chocolate bars, ice creams, and similar products, and in fancy cakes and pastries. Similarly, products such as ketchup and strawberry preserves have red dyes added to enhance their color, relishes and tinned peas have green colors added, and dark breads may contain brown colorants. Most margarines have yellow colors added, to make them appear more similar to butter. Artificial drinks and drink-mixes contain food colorants appropriate to their flavor—cherry and raspberry contain red dyes, and so forth. There are nine certified color additives approved by the FDA. Color additives produced from natural sources are exempt from FDA testing.

Various chemicals are used as food colorants, some of them being extracted from plants (e.g., yellow and orange carotenes), whereas many others are synthetic chemicals derived from coal tars and other organic substances. The acute toxicity (i.e., short-term poisoning) and chronic toxicity (i.e., longer-term damage associated with diseases, cancers, and developmental abnormalities) of these colorants are stringently tested on animals in the laboratory, and the substances must be demonstrated to be safe before they are allowed to be used as food additives. Still, some people object to having these chemicals in their food, and choose to consume products that are not adulterated with colorants.

Residues of pesticides, antibiotics, and growth hormones

Insecticides, fungicides, herbicides, and other pesticides are routinely used in modern, industrial agriculture. Some of these chemicals are persistent, because they do not quickly break down in the environment to simpler substances, and/or they do not readily wash off produce. The chemicals in such cases are called residues, and it is not unusual for them to be present on or in foodstuffs in low concentrations. The permissible residue levels allowed in foodstuffs intended for human consumption are closely regulated by government. However, not all foods can be properly inspected, so it is common for people to be routinely exposed to small concentrations of these chemicals in their diet.

In addition, most animals cultivated in intensive agricultural systems, such as feedlots and factory farms, are routinely treated with antibiotics in their feed. This preemptive treatment is done to prevent outbreaks of communicable diseases under densely crowded conditions. Antibiotic use is especially common during the raising of chickens, turkeys, pigs, and cows. Small residues of these chemicals remain in the meat, eggs, milk, or other products of these animals, and are ingested by human consumers. Also, growth hormones are given to beef and dairy cows to increase their productivity. Small residues of these chemicals also occur in products eaten by consumers.

Strictly speaking, residues of pesticides, antibiotics, and growth hormones are not additives because they are not added directly to foodstuffs. Nevertheless, these chemicals are present in foods eaten by people, and many consumers find this to be objectionable. Foods grown under organic conditions are cultivated without the use of synthetic pesticides, antibiotics, or growth hormones, and many people prefer to eat these foods instead of the much more abundantly available foodstuffs that are typically sold in commercial outlets. (Note that the term organic foods is somewhat of a misnomer, because all foods are organic in nature. The phrase organic in this sense is used to refer to foods that do not contain additives and/or residues, etc.)

Irradiation of food

Irradiation is a new technology that can be used to prevent spoilage of foods by sterilizing most or all of the microorganisms and insects that they may contain. This process utilizes gamma radiation, and it is not known to cause any chemical or physical changes in foodstuffs, other than the intended benefit of killing organisms that can cause spoilage. Although this process displaces some of the uses of preservative chemicals as food additives, irradiation itself is somewhat controversial. Even though there is no scientific evidence that food irradiation poses tangible risks to consumers, some people object to the use of this technology and prefer not to consume foodstuffs processed in this manner.

Resources

BOOKS

Emerton, Victoria, and Eugenia Choi. *Essential Guide to Food Additives.* Leatherhead, Surrey: Leatherhead Food International, 2008.

Ettlinger, Steve. *Twinkie, Deconstructed: My Journey to Discover How the Ingredients Found in Processed Foods Are Grown, Mined (Yes, Mined), and Manipulated into What America Eats.* New York: Hudson Street Press, 2007.

Metcalfe, Dean D., Hugh A. Sampson, and Ronald A. Simon. *Food Allergy: Adverse Reactions to Foods and Food Additives.* Malden, Mass: Blackwell Pub, 2008.

Skurray, Geoffrey. *Decoding Food Additives: A Comprehensive Guide to Food Additive Codes and Food Labelling.* South Melbourne: Lothian Books, 2006.

Winter, Ruth. *A Consumer's Dictionary of Food Additives: Descriptions in Plain English of More Than 12,000 Ingredients Both Harmful and Desirable Found in Foods.* New York: Three Rivers Press, 2004.

OTHER

Joint FAO/WHO Expert Committee on Food Additives. *Safety Evaluation of Certain Food Additives and Contaminants.* Geneva: World Health Organization, International Programme on Chemical Safety, 2008. http://site.ebrary.com/lib/berkeley/Doc?id = 10227089

New Zealand Food Safety Authority. Identifying Food Additives. Wellington: New Zealand Food Safety Authority, 2007. http://www.nzfsa.govt.nz/consumers/chemicals-toxins-additives/additives-booklet.pdf

U.S. Food and Drug Administration (FDA). "Guidance for Industry: Frequently Asked Questions about GRAS." http://www.fda.gov/Food/GuidanceComplianceRegulatoryInformation/GuidanceDocuments/FoodIngredientsandPackaging/ucm061846.htm#Q2 (accessed November 6, 2009).

World Health Organization (WHO). "Food Additives." http://www.who.int/topics/food_additives/en (accessed November 6, 2009).

Bill Freedman

Food and Drug Administration

Founded in 1927, the Food and Drug Administration (FDA) is an agency of the Untied States Department of Health and Human Services. One of the nation's oldest consumer protection agencies, the FDA is charged with enforcing the Federal Food, Drug, and Cosmetics Act and other related public health laws. The agency assesses risks to the public posed by foods, drugs, and cosmetics, as well as medical devices, blood, and medications, such as insulin, which are made from living organisms. It also tests food samples for contaminants, sets labeling standards, and monitors the public health effects of drugs given to animals raised for food.

To carry out its mandate of consumer protection, the FDA employs more than 9,300 people who collect domestic and imported product samples for examination by FDA scientists. The FDA has the power to remove from the market those foods, drugs, chemicals, or medical devices it finds unsafe. The FDA often seeks voluntary recall of the product by manufacturers, but the agency can also stop sales and destroy products through court action. About 3,000 products a year are found to be unfit for consumers and are withdrawn from the marketplace based on FDA action. Also, about 30,000 import shipments each year are detained at the port of entry on FDA orders.

FDA scientists analyze samples of products to detect contamination, or review test results submitted by companies seeking agency approval for drugs, vaccines, food additives, dyes, and medical devices. The FDA also operates the National Center for Toxicological Research at Jefferson, Arkansas, which conducts research to investigate the biological effects of widely used chemicals. The FDA also has centers that test medical devices, radiation-emitting products, and radioactive drugs. The Bureau of Radiological Health, now the Center for Devices and Radiological Health, was formed in 1971 to protect against unnecessary human exposure to radiation from electronic products such as microwave ovens.

The FDA is one of several federal organizations that oversees the safety of biotechnology, such as the industrial use of microorganisms to processes waste and water products.

In 1996 when the FDA declared that cigarettes and smokeless tobacco are nicotine-delivery devices, it took responsibility for regulating those products under the authority of the Federal Food, Drug, and Cosmetics Act. With regard to these products, the FDA has issued federal mandates concerning sales to minors, sales from vending machines, and advertising campaigns. In 2000, however, in *FDA v. Brown & Williamson Tobacco Corp.*, the Supreme Court of the United States ruled that the FDA did not have the authority to regulate tobacco. The Court noted that Congress had not explicitly granted the FDA that authority under the Food, Drug, and Cosmetics Act. Furthermore, subsequent tobacco legislation enacted by Congress indicated that they did not intend for the FDA to have the power to regulate tobacco. In 2009, however, Congress passed the Family Smoking Prevention and Tobacco Control Act, which explicitly granted the FDA the power to regulate the tobacco industry.

In 2004 the FDA allowed the production and marketing of the first living organism for use as a prescription medical device. The FDA approved medical maggots for "debriding non-healing necrotic skin and soft tissue wounds." Also in 2004 the FDA approved the use of a species of leaches to drain pooled blood around regrafted amputated appendages.

Resources

BOOKS

Ekins, Sean, and Jinghai J. Xu. *Drug Efficacy, Safety, and Biologics Discovery: Emerging Technologies and Tools.* Hoboken, NJ: John Wiley & Sons, 2009.

Pampel, Fred C. *Threats to Food Safety*. Library in a book. New York: Facts On File, 2006.

Simon, Michele. *Appetite for Profit: How the Food Industry Undermines Our Health and How to Fight Back*. New York: Nation Books, 2006.

United States Food and Drug Administration. *Bad Bug Book: Food-borne Pathogenic Microorganisms and Natural Toxins Handbook*. McLean, VA: International Medical Publishing, 2004.

ORGANIZATIONS

U.S. Food and Drug Administration, 5600 Fishers Lane, Rockville, MD, USA, 20857-0001, (888) INFO-FDA, http://www.fda.gov

Linda Rehkopf

Food chain/web

Food chains and food webs are methods of describing an ecosystem by explaining how energy flows from one species to another.

First proposed by the English zoologist Charles Elton in 1927, food chains and food webs describe the successive transfer of energy from plants to the animals that eat them, and to the animals that eat those animals, and so on. A food chain is a model for this process that assumes that the transfer of energy within the community is relatively simple. A food chain in a grassland ecosystem, for example, might be: Insects eat grass, and mice eat insects, and fox eat mice. But such an outline is not exactly accurate, and many more species of plants and animals are actually involved in the transfer of energy. Rodents often feed on both plants and insects, and some animals, such as predatory birds, feed on several kinds of rodents. This more complex description of the way energy flows through an ecosystem is called a food web. Food webs can be thought of as interconnected or intersecting food chains.

The components of food chains and food webs are producers, consumers, and decomposers. Plants and chemosynthetic bacteria are producers. They are also called primary producers or autotrophs ("self-nourishing") because they produce organic compounds from inorganic chemicals and outside sources of energy. The groups that eat these plants are called primary consumers or herbivores. They have adaptations that allow them to live on a purely vegetative diet which is high in cellulose. They usually have teeth modified for chewing and grinding; ruminants such as deer and cattle have well-developed stomachs, and lagomorphs such as rabbits have caeca which aid their digestion. Animals that eat herbivores are called secondary consumers or primary carnivores, and predators that eat these animals are called tertiary consumers. Decomposers are the final link in the energy flow. They feed on dead organic matter, releasing nutrients back into the ecosystem. Animals that eat dead plant and animal matter are called scavengers, and plants that do the same are known as saprophytes.

The components of food chains and food webs exist at different stages in the transfer of energy through an ecosystem. The position of every group of organisms obtaining their food in the same manner is known as a trophic level. The term comes from a Greek word meaning "nursing," and the implication is that each stage nourishes the next. The first tropic level consists of autotrophs, the second herbivores, the third primary carnivores. At the final trophic level exists what is often called the top predator. Organisms in the same trophic level are not necessarily connected taxonomically; they are connected ecologically by the fact they obtain their energy in the same way. Their trophic level is determined by how many steps it is above the primary producer level. Most organisms occupy only one trophic level; however some may occupy two. Insectivorous plants like the venus flytrap are both primary producers and carnivores. Horseflies are another example: the females bite and draw blood, whereas the males are strictly herbivores.

In 1942 Raymond Lindeman published a paper titled "The Tropic-Dynamic Aspect of Ecology." Although a young man and only recently graduated from Yale University, he revolutionized ecological thinking by describing ecosystems in the terminology of energy transformation. He used data from his studies of Cedar Bog Lake in Minnesota to construct the first energy budget for an entire ecosystem. He measured harvestable net production at three trophic levels, primary producer, herbivore, and carnivore. He did this by measuring gross production minus growth, reproduction, respiration, and excretion. He was able to calculate the assimilation efficiency at each tropic level and the efficiency of energy transfers between each level. Lindeman's calculations are still widely regarded today, and his conclusions are usually generalized by saying that the ecological efficiency of energy transfers between trophic levels averages about 10 percent.

Lindeman's calculations and some basic laws about physics reveal important truths about food chains, food webs, and ecosystems in general. The First Law of Thermodynamics states that energy cannot be created or destroyed; energy input must equal energy output. The Second Law of Thermodynamics states that all physical processes proceed in such a way that the availability of the energy involved decreases. In other words, no transfer of energy is completely efficient. Using the generalized 10 percent figure from Lindeman's study, a hypothetical ecosystem with 1,000 kcal of energy available (net production) at the primary-producer level would mean that only 100 kcal would be available to the herbivores at the second trophic level, 10 kcal to the primary carnivores at the third level, and 1 kcal to the secondary carnivores at the

In an ecosystem, food chains become interconnected to form food webs. *(Reproduced by permission of Gale, a part of Cengage Learning)*

fourth level. Thus, no matter how much energy is assimilated by the autotrophs at the first level of an ecosystem, the eventual number of trophic levels is limited by the laws that govern the transfer of energy. The number of links in most natural food chains is four.

The relationships between trophic levels has sometimes been compared to a pyramid, with a broad base that narrows to an apex. Trophic levels represent successively narrowing sections of the pyramid. These pyramids can be described in terms of the number of organisms at each trophic level. This was first proposed by Charles Elton, who observed that the number of plants usually exceeded the number of herbivores, which in turn exceeded the number of primary carnivores, and so on. Pyramids

of number can be inverted, particularly at the base; an example of this would be the thousands of insects which might feed on a single tree. The pyramid-like relationship between trophic levels can also be expressed in terms of the accumulated weight of all living matter, known as biomass. Although upper-level consumers tend to be large, the population of organisms at lower trophic levels are usually much higher, resulting in a larger combined biomass. Pyramids of biomass are not normally inverted, though they can be under certain conditions. In aquatic ecosystems the biomass of the primary producers may be less than that of the primary consumers because of the rate at which they are being consumed; phytoplankton can be eaten so rapidly that the biomass of zooplankton and other herbivores are greater at any particular time. The relationship between trophic levels can also described in terms of energy, but pyramids of energy cannot be inverted. There will always be more energy at the bottom than there is at the top.

Humans are the top consumer in many ecosystems, and they exert strong and sometimes damaging pressures on food chains. For example, overfishing or overhunting can cause a large drop in the number of animals, resulting in changes in the food-web interrelationships. On the other hand, overprotection of some animals like deer or moose can be just as damaging. Another harmful influence is that of biomagnification. Toxic chemicals such as mercury and DDT released into the environment tend to become more concentrated as they travel up the food chain. Some ecologists have proposed that the stability of ecosystems is associated with the complexity of the internal structure of the food web and that ecosystems with a greater number of interconnections are more stable. Although more studies must be done to test this hypothesis, we do know that food chains in constant environments tend to have a greater number of species and more trophic links, whereas food chains in unstable environments have fewer species and trophic links.

Resources

BOOKS

Tarbox, A.D. *Food Chain*. Mankato, Minn: Creative Education, 2009.

Food Chain. South Brisbane, Australia: Griffith University, 2010.

Alpas, Hami. *Food Chain Security*. Berlin: Springer Verlag, 2010.

Morgan, Kevin, Terry Marsden, and Jonathan Murdoch. *Worlds of Food: Place, Power, and Provenance in the Food Chain*. Oxford Geographical and Environmental Studies. Oxford: Oxford University Press, 2006.

John Korstad
Douglas Smith

Food irradiation

The treatment of food with ionizing radiation has been in practice for nearly a century since the first irradiation process patents were filed in 1905. Regular use of the technology in food processing started in 1963 when the U.S. Food and Drug Administration (FDA) approved the sale of irradiated wheat and wheat flour. Today irradiation treatment is used on a wide variety of food products and is regulated in the United States by the FDA under a Department of Health and Human Services regulation.

Irradiation of food has three main applications: extension of shelf life, elimination of insects, and the destruction of bacteria and other pathogens that cause foodborne illness. This final goal may have the most far-reaching implications for Americans; the U.S. Centers for Disease Control (CDC) estimate that seventy-six million Americans get sick, and five thousand die each year from illnesses caused by foodborne microorganisms, such as *E. coli*, *Salmonella*, the botulism toxin, and other pathogens responsible for food poisoning.

Irradiation technology involves exposing food to ionizing radiation. The radiation is generated from gamma rays emitted by cobalt-60 (^{60}Co) or cesium–137 (^{137}Cs), or from x rays or electron beams. The amount of radiation absorbed during irradiation processing is measured in units termed radiant energy absorbed (RADs). One hundred RADs is equivalent to one Gray (Gy). Depending on the food product being irradiated, treatment can range from 0.05 to 30 kGy. A dosimeter, or film badge, verifies the kGy dose. The ionizing radiation displaces electrons in the food, which slows cell division and kills bacteria and pests.

The irradiation process itself is relatively simple. Food is packed in totes or containers, which are typically placed on a conveyor belt. Beef and other foods that require refrigeration are loaded into insulated containers prior to treatment. The belt transports the food bins through a lead-lined irradiation cell or chamber, where they are exposed to the ionizing radiation that kills the microorganisms. Several trips through the chamber may be required for full irradiation. The length of the treatment depends upon the food being processed and the technology used, but each rotation takes only a few minutes.

The FDA has approved the use of irradiation for wheat and wheat powder, spices, enzyme preparations, vegetables, pork, fruits, poultry, beef, lamb, and goat meat. In 2000 the FDA also approved the use of irradiation to control salmonella in fresh eggs.

Labeling guidelines introduced by the Codex Alimentarius Commission, an international food standards organization sponsored jointly by the United Nations Food and Agricultural Organization (FAO) and the World Health Organization (WHO), requires that all irradiated food products and ingredients be clearly labeled as such for consumers. Codex also created the radura, a voluntary international symbol that represents irradiation. In the United States the food irradiation process is regulated jointly by FDA and the U.S. Department of Agriculture (USDA). Facilities using radioactive sources such as cobalt-60 are also regulated by the Nuclear Regulatory Commission (NRC). The FDA regulates irradiation sources, levels, food types and packaging, as well as required recordkeeping and labeling. Records must be maintained and made available to FDA for one year beyond the shelf-life of the irradiated food to a maximum of three years. These records must describe all aspects of the treatment and foods that have been irradiated must be denoted with the radura symbol and by the statement "treated with radiation" or "treated by irradiation." As of 2002 food irradiation was allowed in some fifty countries and was endorsed by the WHO and many other organizations.

The Farm Security and Rural Investment Act of 2002 (the Farm Bill) passed in May 2002 may relax the food irradiation standards. The Farm Bill calls for the Secretary of Health and Human Services and the FDA to implement a new regulatory program for irradiated foods. The program will allow the food industry to instead label irradiated food as pasteurized as long as they meet appropriate food safety standards. These new guidelines have not yet been implemented. Some people argue that labeling irradiated food as pasteurized is misleading and assert that research has not shown the long-term effects of irradiated foods on health.

Food that has been treated with ionizing energy typically looks and tastes the same as non-irradiated food. Just like a suitcase going through an airport x-ray machine, irradiated food does not come into direct contact with a radiation source and is not radioactive. However, depending on the strength and duration of the irradiation process, some slight changes in appearance and taste have been reported in some foods after treatment. Some of the flavor changes may be attributed to the generation of substances known as radiolytic products in irradiated foods.

When food products are irradiated, the energy displaces electrons in the food and forms compounds termed free radicals. The free radicals react with other molecules to form new stable compounds termed radiolytic products. Benzene (C_6H_6), formaldehyde (CH_2O), and hydrogen peroxide (H_2O_2) are just a few of the radiolytic products that may form during the irradiation process. These substances are only present in minute amounts, however, and the FDA reports that 90 percent of all radiolytic products from irradiation are also found naturally in food.

The chemical change that creates radiolytic products also occurs in other food processing methods, such as canning or cooking. However, about 10 percent of the radiolytic products found in irradiated food are unique to the irradiation process, and little is known about the effects that they may have on human health. It should be noted, however, that the WHO, the American Medical Association (AMA), the American Dietetic Association (ADA), and a host of other professional healthcare organizations endorse the use of irradiation as a food safety measure.

Treating fruit and vegetables with irradiation can also eliminate the need for chemical fumigation after harvesting. Produce shelf life is extended by the reduction and elimination of organisms that cause spoilage. It also slows cell division, thus delaying the ripening process, and in some types of produce irradiation extends the shelf life for up to a week. Advocates of irradiation claim that it is a safe alternative to the use of fumigants, several of which have been banned in the United States.

Nevertheless, irradiation removes some of the nutrients from foods, particularly vitamins A, C, E, and the B- complex vitamins. Whether the extent of this nutrient loss is significant enough to be harmful is debatable. Advocates of irradiation say the loss is insignificant, and standard methods of cooking can destroy these same vitamins. However, research suggests that cooking an irradiated food may further increase the loss of nutrients.

Although research studies have shown that consuming irradiated foods does not cause carcinogenic, mutagenic, or toxic effects, critics of irradiation question the long-term safety of consumption of irradiated food and their associated radiolytic products. The changes in texture, flavor, and odor as well as the reduction in nutritional content in irradiated foods are also issues that critics cite in opposing irradiation. These critics charge that the technology does nothing to address the unsanitary food processing practices and inadequate inspection programs that breed foodborne pathogens. Additionally, a potential type of carcinogen termed 2-alkylcyclobutanone (2-ACB) has been identified in foods as a product of irradiation and is being studied with regard to health effects.

Even if irradiation is 100 percent safe and beneficial, there are numerous environmental concerns. Many opponents of irradiation cite the proliferation of radioactive material and the environmental hazards. The mining and on-site processing of radioactive materials are devastating to regional ecosystems. There are also safety hazards associated with the transportation of radioactive material, production of isotopes, and disposal.

Resources

BOOKS

Sommers, Christopher and Xuetong Fan. *Food Irradiation Research and Technology*. Boston: Blackwell, 2006.

OTHER

Centers for Disease Control and Prevention (CDC). "Food Irradiation." http://www.cdc.gov/ncidod/dbmd/diseaseinfo/foodirradiation.htm (accessed November 6, 2010).
United States Environmental Protection Agency (EPA). "Human Health: Food Safety: Irradiated Food." http://www.epa.gov/ebtpages/humafoodsafetyirradiatedfood.html (accessed November 6, 2010).
United States Environmental Protection Agency (EPA). "Radiation and Radioactivity: Radiation Sources and Uses: Irradiated Food." http://www.epa.gov/ebtpages/radiradiationsourirradiatedfood.html (accessed November 6, 2010).

ORGANIZATIONS

The Food Irradiation Website, http://www.food-irradiation.com
National Food Processors Association, 1350 I Street, NW Suite 300, Washington, DC, USA, 20005, (202) 639-5900, (202) 639-5932, nfpa@nfpa-food.org, http://www.nfpa-food.org
Public Citizen, Critical Mss Energy & Environmental Program, 1600 20th St. NW, Washington, DC, USA, 20009, (202) 588-1000, CMEP@citizen.org, http://www.citizen.org/cmep/foodsafety/food;usirrad

Paula Ford-Martin
Debra Glidden

Food policy

Through a variety of agricultural, economic, and regulatory programs that support or direct policies related to the production and distribution of food, the United States government has a large influence on how agricultural business is conducted. The government's major impact on agriculture is the setting of prices and mandates regarding how land can be used by farmers that participate in government programs. These policies can also have a large impact on the adoption or use of alternative practices and technologies that may be more efficient and sustainable in the world marketplace.

Farm commodity programs have had a large influence in the past on the kinds and amounts of crops grown as well as on the choice of management practices used to grow them. Prices under government commodity programs have often been above world market prices, which meant that many farmers felt compelled to preserve or build their farm commodity program base acres, because acreage determines program eligibility and future income. These programs strongly influenced land-use decisions on about two-thirds of the harvested cropland in the United States.

Price and income support programs for major commodities also influence growers not in the programs. For example, pork producers are not a part of a government program, and in the past they have paid higher feed prices because of high price supports on production of feed grains. At other times, particularly after the Food Security Act of 1985, they benefited from policies resulting in lower food costs. So as the government changes policy in one area, there can be widespread indirect impacts in other areas. For example, the federal dairy termination program, which ran from 1985–1987, was designed to reduce overproduction of milk. Those farmers who sold their milk cows and decided to produce hay for local cash markets caused a steep decline in the prices received by other established hay producers. Farm bills such as the Food Security Act of 1985 are the federal government's main vehicle for legislating farm policy, and are usually amended every few years, or new versions are passed. The 2008 farm bill, known as the Food, Conservation, and Energy Act of 2008 continues subsidies and provides funds for research in ethanol fuels.

Federal policy evolved as a patchwork of individual programs, each created to address individual problems. There was not a coherent strategy to direct programs toward a common set of goals. Many programs such as soil conservation and export programs have had conflicting objectives, but attempts have now been made in the most current farm legislation to address some of these problems.

Government food policy has produced a wide variety of results. The policy has not only affected commodity prices and the level of output, but it has also shaped technological change, encouraged uneconomical capital investments in machinery and facilities, inflated the value of land, subsidized crop production practices that have led to resource degradation (such as soil erosion and surface and groundwater pollution), expanded the interstate highway system, financed irrigation projects, and promoted farm commodity exports. Together with other economic forces, government policy has had a far-reaching structural influence on agriculture, much of it unintended and unanticipated.

Federal commodity programs were put into place beginning in the 1930s, primarily with the Agriculture Adjustment Act of 1938. The purpose of these programs born out of the Depression was primarily to protect crop prices and farmer income, which has been done by a number of programs over the years. A variety of methods have been used including setting prices, making direct payments to farmers, and subsidizing exports. However, by 2000, an increasing number of people felt that these programs impeded movement toward alternative types of agriculture, to the detriment of family farms and society in general.

Two components in particular were highlighted as being problems: base acre requirements and cross-compliance. All crop price and income support programs relied on the concept of an acreage base planted with a given commodity that would produce a predictable yield. Most of this acreage was planted to maximize benefits and was based on a five-year average. Farmers knew that if they reduced their acres for a given crop, they would not only lose the current year's benefits, but would also lose future benefits.

Cross-compliance was instituted in the Food Security Act of 1985. It was designed to control government payments and production by attaching financial penalties to the expansion of program crop base acres. It served as an effective financial barrier to diversification of crops by stipulating that to receive any benefits from an established crop acreage base, farmers must not exceed their acreage base for any other program crop. This had a profound impact on farmers and crop growers because about 70 percent of the United States' cropland acres were enrolled in the programs.

In addition to citing these problem areas, critics of food policy programs argued that many farmers faced economic penalties for adopting beneficial practices, such as crop rotation or strip cropping, practices that reduce soil erosion and improve environmental quality. The economic incentives built into commodity programs, for example, encouraged heavier use of fertilizer, pesticides, and irrigation. These programs also encouraged surplus production, subsidized inefficient use of inputs, and they resulted in increased government expenditures. Critics argued that the rules associated with these programs discouraged farmers from pursuing alternative practices or crops, or technologies that might have proved more effective

in the long term or that were more environmentally friendly.

Critics also contend that much of the research conducted over the past forty years has responded to the needs of farmers operating under a set of economic and policy incentives that encouraged high yields without regard to the long-term environmental impacts. During the late 1980s and early 1990s, several U.S. Department of Agriculture research and education programs were instituted to determine whether current levels of production can be maintained with reduced levels of fertilizers and pesticides, to examine more intensive management practices, to increase understanding of biological principles, and to improve profitability per unit of production with less government support. As the impacts of the alternative production systems on the environment continue to be evaluated, it will be important to have policies in place that will allow the farmer to easily adopt those practices that increase efficiency and reduce impacts. In a farm bill passed in 1996 there are provisions that change these commodity programs. Labeled the "right to farm provisions," these allow farmers to make decisions on what they grow and establish a phased seven-year reduction in price supports.

Food quality and safety are major concerns addressed as a part of federal policy. Programs addressing these concerns are primarily designed to prevent health risks and acute illnesses from chemical and microbial contaminants in food. Supporters say that this has provided the country with the safest food supply in the world. However, critics contend that a number of regulations do not enhance quality or safety and put farmers that use or would adopt alternative agricultural practices at a disadvantage. Several examples can be cited. Critics of government food policy point out that until recently, meat grading standards awarded producers of fatty beef which has been linked to the increased likelihood of heart disease.

The use of pesticides provides another example. The Environmental Protection Agency establishes pesticide residual tolerance levels in food which are monitored for compliance. For some types of risk, cancer in particular, there is a great deal of uncertainty, and critics point out that cosmetic standards that increase prices for fruits and vegetables may encourage higher risks of disease among consumers. Also, certain poultry slaughter and handling practices can result in microbiological contamination. Of particular concern is *Salmonella* food poisoning which has become widespread. After a 2010 *Salmonella* outbreak that resulted in more than 2,400 *Salmonella*-related illnesses in the United States was linked to contaminated eggs, new policies were formulated for the poultry industry regarding egg sorting and poultry food.

Government food policy heavily influences on-farm decision making. In some cases one part of the policy has negative or unintended consequences for another policy or segment of farmers. Look to the future, the struggle will be to provide a coherent, coordinated policy. The recent changes in policy will need to be evaluated from the standpoint of sustainability and environmental impacts.

Resources

BOOKS

Babe, Robert E. *Culture of Ecology: Reconciling Economics and Environment*. Toronto, Ontario, Canada: University of Toronto Press, 2006.
Hatfield, Jerry L, ed. *The Farmer's Decision: Balancing Economic Agriculture Production with Environmental Quality*. Ankeny, Iowa: Soil and Water Conservation Society, 2005.
Peterson, E. Weseley F. *A Billion Dollars a Day: The Economics and Politics of Agricultural Subsidies*. Malden, MA: Wiley-Blackwell, 2009.

OTHER

U.S. Department of Agriculture (USDA). http://www.usda.gov/wps/portal/usda/usdahome (accessed November 6, 2010).
World Trade Organization. "Understanding the WTO: The Agreements: Anti-dumping, Subsidies, Safeguards." http://www.wto.org/english/thewto_e/whatis_e/tif_e/agrm8_e.htm#subsidies (accessed November 6, 2010).

James L. Anderson

Food waste

Waste food from residences, grocery stores, and food services accounts for nearly 7 percent of the municipal solid waste stream. The per-capita amount of food waste in municipal solid waste has been declining since 1960 due to increased use of garbage disposals and increased consumption of processed foods. Food waste ground in garbage disposals goes into sewer systems and thus ends up in wastewater. Waste generated by the food processing industry is considered to be industrial waste, and is not included in municipal solid waste estimates.

Waste from the food processing industry includes: vegetables and fruits unsuitable for canning or freezing; vegetable, fruit, and meat trimmings; and pomace from juice manufacturing. Vegetable and fruit processing waste is sometimes used as animal feed, and waste from meat and seafood processing can be composted. Liquid waste from juice manufacturing can be applied to cropland as a soil amendment. Much of the waste generated by all types of food processing is wastewater due to such processes as washing, peeling, blanching, and cooling. Some food industries recycle wastewaters back into their processes, but there is potential for more of this wastewater to be reused.

Grocery stores generate food waste in the form of lettuce trimmings, excess foliage, unmarketable produce, and meat trimmings. Waste from grocery stores located in rural areas is often used as hog or cattle feed, whereas grocery waste in urban areas is usually ground in garbage disposals. There is potential for more urban grocery store waste to be either used on farms or composted, but lack of storage space, odor, and pest problems prevent most of this waste from being recycled.

Restaurants and institutional cafeterias are the major sources of food service industry waste. In addition to food preparation wastes, they also generate large amounts of cooking oil and grease waste, post-consumer waste (uneaten food), and surplus waste. In some areas small amounts of surplus waste is utilized by feeding programs, but most waste generated by food services goes to landfills or into garbage disposals.

Most food waste is generated by sources other than households. However, a greater percentage of household food waste is disposed of because there is a higher rate of recycling of industrial and commercial food waste. Only a very small segment of households compost or otherwise recycle their food wastes.

Resources

BOOKS

Adley, Catherine C., ed. *Food-borne Pathogens: Methods and Protocols*. Totowa, NJ: Humana Press, 2006.
Bitton, Gabriel. *Wastewater Microbiology*. Hoboken, NJ: Wiley-Liss and John Wiley & Sons, 2005.
Pichtel, John. *Waste Management Practices*. Boca Raton: CRC, 2005.
Williams, Paul T. *Waste Treatment and Disposal*. New York: Wiley, 2005.

OTHER

Centers for Disease Control and Prevention (CDC). "Food Safety." http://www.cdc.gov/foodsafety/ (accessed November 9, 2010).

World Health Organization (WHO). "Food Safety." http://www.who.int/topics/food_safety/en (accessed November 9, 2010).

Teresa C. Donkin

Food-borne diseases

Food-borne diseases are illnesses caused when people consume contaminated food or beverages. Contamination with disease-causing microbes called pathogens is usually due to improper food handling or storage. Other causes of food-borne diseases are toxic chemicals or other harmful substances in food and beverages. More than 250 food-borne illnesses have been described, according to the United States Centers for Disease Control and Prevention (CDC). The CDC estimates that food-borne pathogens cause approximately 76 million illnesses, 5,000 deaths, and 325,000 hospitalizations in the United States each year.

Most food-borne illnesses are infections caused by bacteria, viruses, and parasites such as *Cryptosporidium*. Harmful toxins cause food poisonings. Because there are so many different types of food-borne illnesses, symptoms will vary. However, some early symptoms are similar because the microbe or toxin travels through the gastrointestinal tract. The initial symptoms of food-borne disease are usually nausea, vomiting, abdominal cramps, and diarrhea.

According to CDC the most common food-borne viruses are caused by three bacteria and a group of viruses. *Campylobacter* are bacteria that live in the intestines of healthy birds. It is also found in raw poultry meat. An infection is caused by eating undercooked chicken or food contaminated by juices from raw chicken. The bacterial pathogen causes fever, diarrhea, and abdominal cramps. *Campylobacter* is also among the primary causes of bacteria-related diarrhea illness throughout the world.

Salmonella bacteria are prevalent in the intestines of birds, mammals, and reptiles. The bacteria spread to humans through various foods. *Salmonella* causes the illness *salmonellosis*. Symptoms include fever, diarrhea, and abdominal cramps. This illness can result in a life-threatening infection for a person who is in poor health or has a weakened immune system.

E. coli O157:H7 is a bacterial pathogen that has a reservoir in cattle and similar animals. *E. coli* sometimes causes a serious illness. People become ill after

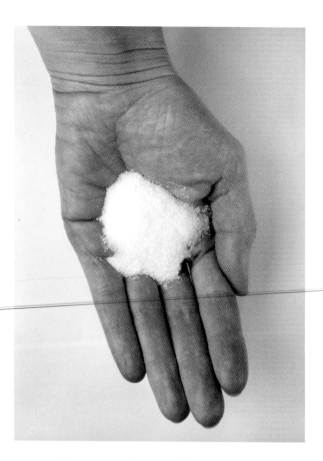

Handful of monsodium glutamate (MSG) crystalline powder. A common additive in food, it has been implicated in a number of conditions and allergic reactions. *(Cordelia Molloy/Photo Researchers, Inc.)*

eating food or drinking water that was contaminated with microscopic amounts of cow feces, according to CDC. A person often experiences severe and bloody diarrhea and painful abdominal cramps. Hemolytic uremic syndrome (HUS) occurs in 3–5 percent of *E. coli* cases. This complication may occur several weeks after the first symptoms. HUS symptoms include temporary anemia, profuse bleeding, and kidney failure.

Food-borne illnesses are also caused by *Calicivirus*, which is also known as the Norwalk-like virus. This group of viruses is thought to spread from one person to another. An infected health service worker preparing a salad or sandwich could contaminate the food. According to CDC, infected fishermen contaminated oysters that they handled. These viruses are characterized by severe gastrointestinal illness. There is more vomiting than diarrhea, and the person usually recovers in two days.

The types and causes of food-borne illnesses have changed through the years. The pasteurization of milk, improved water quality, and safer canning techniques led to a reduction in the number of cases of common food-borne illness such as typhoid fever, tuberculosis, and cholera. Causes of contemporary food-borne illnesses range from parasites living in imported food to food-processing techniques.

Food may be contaminated during processing. For example, the meat contained in one hamburger may come from hundreds of animals, according to CDC. In 2009 more than 500,000 pounds of ground beef contaminated with *E. coli* was recalled from market shelves after a multi-state outbreak of illness. Nineteen people were hospitalized with illness due to *E. coli* O157:H7. Five of these persons developed hemolytic uremic syndrome, and two persons died.

Technology in the form of food irradiation may eliminate the pathogens that cause food-borne disease. Advocates say that radiating food with gamma rays is effective and can be done when the food is packaged. Opponents say the process is dangerous and could produce the free radicals that cause cancer.

CDC ranks "raw foods of animal origin" as the foods most likely to be contaminated. This category includes meat, poultry, raw eggs, raw shellfish, and unpasteurized milk. Furthermore, raw fruit and vegetables could also pose a health risk. Vegetables fertilized by manure can also be contaminated by the fertilizer. CDC said that some outbreaks of food-borne illness were traced to unsanitary processing procedure. Water quality was crucial when washing vegetables, as was chilling the produce after it was harvested.

CDC advises consumers to thoroughly cook meat, poultry, and eggs. Produce should be washed. Meat, dairy products, and leftovers should be chilled promptly. CDC is part of the United States Public Health Service. It researches and monitors health issues. Federal regulation of food safety is the responsibility of agencies such as the Food and Drug Administration, the United States Department of Agriculture, and the National Marine Fisheries Service.

Resources

BOOKS

Morrone, Michele. *Poisons on Our Plates: The Real Food Safety Problem in the United States.* Westport, CT: Praeger, 2008.

Watson, Ronald R., and Victor R. Preedy. *Bioactive Foods in Promoting Health: Fruits and Vegetables.* Amsterdam: Academic Press, 2010.

PERIODICALS

Hoffman, Richard E. "Preventing Foodborne Illness." *Emerging Infectious Diseases* 11, no. 1 (2005): 11–16.

Kuehn, Bridget M. "Surveillance and Coordination Key to Reducing Foodborne Illness." *JAMA: The Journal of the American Medical Association* 294, no. 21 (2005): 2683–2684.

Machado, Antonio E. "Preventing Foodborne Illness in the Field." *Journal of Environmental Health* 72, no. 3 (2009): 56.

Noèel, Harold, et al. "Consumption of Fresh Fruit Juice: How a Healthy Food Practice Caused a National Outbreak of Salmonella Panama Gastroenteritis." *Foodborne Pathogens and Disease* 7, no. 4 (2010): 375–381.

"The Price of Foodborne Illness in the USA." *The Lancet* 375, no. 9718 (2010): 866.

OTHER

"Foodborne Disease." *World Health Organization* http://www.who.int/topics/foodborne_diseases/en/ (accessed November 7, 2010).

"Foodborne Illness." *U.S. Centers for Disease Control and Prevention (CDC).* http://www.cdc.gov/ncidod/dbmd/diseaseinfo/foodborneinfections_g.htm (accessed November 7, 2010).

ORGANIZATIONS

Centers for Disease Control and Prevention, 1600 Clifton Road, Atlanta, GA, USA, 30333, (404) 639-3311, (800) 311-3435, http://www.cdc.gov

Liz Swain

Foot and mouth disease

Foot and mouth disease (FMD), also called hoof and mouth disease, is a highly contagious and economically devastating viral disease of cattle, swine, and other cloven-hoofed (split-toed) ruminants, including sheep, goats, and deer. The disease is highly contagious—nearly 100 percent of exposed animals become infected—and it spreads rapidly through susceptible populations. Although there is no cure for FMD, it is seldom fatal, but it can kill young animals.

The initial symptoms of the disease include fever and blister-like lesions (vesicles). The vesicles rupture into erosions on the tongue, in the mouth, on the teats, and between the hooves. Vesicles that rupture discharge clear or cloudy fluid and leave raw, eroded areas with ragged fragments of loose tissue. Erosions in the mouth result in excessive production of sticky, foamy, stringy saliva, which is a characteristic of FMD. Another characteristic symptom is lameness with reluctance to move. Other possible symptoms and effects of FMD include elevated temperatures in the early stages of the disease for two to three days, spontaneous abortion of fetuses, low conception rates, rapid weight loss, and drop in milk production. FMD lasts for two to three weeks, with most animals recovering within six months. However, it can leave some animals debilitated, thus causing severe losses in the production of meat and milk. Even cows that have recovered seldom produce milk at their original rates. Animals grown for meat do not usually regain lost weight for many months. FMD can also lead to myocarditis, which is an inflammation of the muscular walls of the heart, and death, especially in newborn animals. Infected animals can spread the disease throughout their lives, so the only way to stop an outbreak is to destroy the animals.

The virus that causes the disease survives in lymph nodes and bone marrow at neutral pH. There are at least seven types and many subtypes of the FMD virus. The virus persists in contaminated fodder and in the environment for up to one month, depending on the temperature and pH. FMD thrives in dark, damp places, such as barns, and can be destroyed with heat, sunlight, and disinfectants.

The disease is not likely to affect humans, either directly or indirectly through eating meat from an infected animal, but humans can spread the virus to animals. FMD can remain in human nasal passages for up to twenty-eight hours. FMD viruses can be spread by other animals and materials to susceptible animals. The viruses can also be carried for several miles on the wind if environmental conditions are appropriate for virus survival. Specifically, an outbreak can occur when:

- people wearing contaminated clothes or footwear or using contaminated equipment pass the virus to susceptible animals
- animals carrying the virus are introduced into susceptible herds
- contaminated facilities are used to hold susceptible animals
- contaminated vehicles are used to move susceptible animals
- raw or improperly cooked garbage containing infected meat or animal products is fed to susceptible animals
- susceptible animals are exposed to contaminated hay, feedstuffs, or hides
- susceptible animals drink contaminated water
- a susceptible cow is inseminated by semen from an infected bull

Widespread throughout the world, FMD has been identified in Africa, South America, the Middle East, Asia, and parts of Europe. North America, Central America, Australia, New Zealand, Chile, but some European countries are considered to be free of FMD. The United States has been free of FMD since 1929, when the last of nine outbreaks that occurred during the nineteenth and early twentieth centuries was eradicated.

In 2001 an FMD outbreak was confirmed in the United Kingdom, France, the Netherlands, the Republic of Ireland, Argentina, and Uruguay. Officials in the United Kingdom detected 2,030 cases of FMD, slaughtered almost four million animals, and took seven months to control the outbreak. The economic losses were estimated to be in the billions of pounds, and tourism in the affected countries was adversely affected. The outbreak was detected on February 20, 2001; no new cases were reported after September 30, 2001. On January 15, 2002, the British government declared the FMD outbreak to be over. The outbreak appeared to have started in a pig finishing unit in Northumberland, which was licensed to feed processed waste food. The disease appeared to have spread through two routes: through infected pigs who were sent to a slaughterhouse and through windborne spread to sheep on a nearby farm. These sheep entered the marketing chain and were sold through dealers in markets where they infected other sheep, people, and vehicles, spreading the FMD virus throughout England, Wales, and southern Scotland. As the outbreak continued, cases were detected in other European countries. Infections of foot-and-mouth disease were again found in the United Kingdom in 2007 in Surrey, where the strain seemed to be identical with that used at the nearby research facility.

The Animal and Plant Health Inspection Service (APHIS) of the United States Department of Agriculture (USDA) has developed a ongoing comprehensive prevention program to protect American agriculture from FMD. APHIS continuously monitors for FMD cases worldwide. When FMD outbreaks are identified, APHIS initiates regulations that prohibit importation of live ruminants and swine and many animal products from the affected countries. In the 2001 outbreak in some European Union member countries, APHIS temporarily restricted importation of live ruminants and swine and their products from all European Union member states. APHIS officials are on duty at all United States land and maritime ports-of-entry to ensure that passengers, luggage, cargo, and mail are checked for prohibited agricultural products or other materials that could carry FMD. The USDA

Beagle Brigade, dogs trained to sniff out prohibited meat products and other contraband items, are also on duty at airports to check incoming flights and passengers.

The cooperation of private citizens is a crucial component of the protection program. APHIS prevents travelers entering the United States from bringing any agricultural products that could spread FMD and other harmful agricultural pests and diseases. Therefore passengers must declare all food items and other materials of plant or animal origin that they are carrying. Prohibited agricultural products that are found are confiscated and destroyed. Passengers must also report any visits to farms or livestock facilities. Failure to declare any items may result in delays and fines up to $1,000. Individuals traveling from countries that have been designated as FMD-affected must have shoes disinfected if they have visited farms, ranches, or other high risk areas, such as zoos, circuses, fairs, and other facilities and events where livestock and animals are exhibited.

APHIS recommends that travelers should shower and shampoo prior to and again after returning to the United States from an FMD-affected country. They should also launder or dry clean clothes before returning to the United States. Full-strength vinegar can be used by passengers to disinfect glasses, jewelry, watches, belts, hats, cell phones, hearing aids, camera bags, backpacks, and purses. If travelers had visited a farm or had any contact with livestock on their trip, they should avoid contact with livestock, zoo animals, or wildlife for five days after their return. Although dogs and cats cannot become infected with FMD, their feet, fur, and bedding should be cleaned of excessive dirt or mud. Pet bedding should not contain straw, hay, or other plant materials. The pet should be bathed as soon as it reaches its final destination and be kept away from all livestock for at least five days after entering the United States.

In the United States, animal producers and private veterinarians also monitor domestic livestock for symptoms of FMD. Their surveillance activities are supplemented with the work of 450 specially trained animal disease diagnosticians from federal, state, and military agencies. These diagnosticians are responsible for collecting and analyzing samples from animals suspected of FMD infection. If an outbreak were confirmed, APHIS would quickly try to identify infected and exposed animals, establish and maintain quarantines, destroy all infected and exposed animals using humane euthanization procedures as quickly as possible, and dispose of the carcasses by an approved method such as incineration or burial. After cleaning and disinfecting of facilities where the infected animals were housed, the facility would be left vacant for several weeks. After this period, a few susceptible animals would be placed in the facility and observed for signs of FMD. A large area around the facility would be quarantined, where animal and human movement would be restricted or prohibited. In some cases all susceptible animals within a two-mile radius would be also euthanized and disposed of properly by incineration or burial. In addition, APHIS has developed plans to pay affected producers the fair market value of their animals.

APHIS would consider vaccinating animals against FMD to enhance other eradication activities as well as to prevent spread to disease-free areas. However, vaccinated animals may still become infected and serve as a reservoir for the disease, even though they do not develop the disease symptoms themselves. Also, for continued protective immunity, the vaccines must be given every few months. In late 2010 the APHIS announced plans to field test a FMD vaccination by inoculating 600 cattle. Vaccination is part of the overall strategy the USDA has developed to deal with an outbreak of the disease were it ever to reappear in the United States.

APHIS is working with the U.S. Armed Forces to ensure that military vehicles and equipment are cleaned and disinfected before returning to the United States.

Preventing FMD from infecting and becoming established in an FMD-free area requires constant vigilance and a well-developed, thorough plan to control and eradicate any cases that might occur.

Resources

BOOKS

Barrah, David. *Through My Eyes: The Inside Story of the UK 2001 Foot and Mouth Crisis.* Poole: TimeBox Press, 2005.

Bourne, Debra. *Foot-and-Mouth Disease.* London: Wildlife Information Network, 2007.

Convery, Ian. *Animal Disease and Human Trauma: Emotional Geographies of Disaster.* Basingstoke England: Palgrave Macmillan, 2008.

Doring, Martin, and Brigitte Nerlich. *The Social and Cultural Impact of Foot and Mouth Disease in the UK in 2001 Experiences and Analyses.* Manchester University press, 2009.

Great Britain. *Foot and Mouth Disease: Applying the Lessons: Ninth Report of Session 2005-06.* London: Stationery Office, 2005.

Forel, Francois-Alphonse. *La Faune Profonde des Lacs Suisse*. (Reprint) Manchester, NH: Ayer Company Publishers, 1977.

Forel, Francois-Alphonse. *Handbuch der Seenkunde*. (Reprint) Manchester, NH: Ayer Company Publishers, 1978.

ORGANIZATIONS

F.-A. Forel Institute, 10, route de Suise, Geneva, Switzerland, CP 416, +41 22 379 03 00, +41 22 379 03 29, secretariatforel-terre@unige.ch, http://www.unige.ch/forel

Jane Spear

Foreman, Dave

1947–
American radical environmental activist

Dave Foreman is a self-described radical environmentalist, co-founder of Earth First!, and leading defender of "monkey-wrenching" as a direct-action tactic to slow or stop strip mining, clear-cut logging of old-growth forests, the damming of wild rivers, and other environmentally destructive practices. He no longer participates in monkey-wrenching but has no regrets for past destruction carried out with environmental protection as the aim.

The son of a United States Air Force employee, Foreman traveled widely while growing up. In college he chaired the conservative Young Americans for Freedom and worked in the 1964 presidential election campaign of Senator Barry Goldwater. In the 1970s Foreman was a conservative Republican and moderate environmentalist who worked for the Wilderness Society in Washington, DC.

Foreman came to believe that the petrochemical, logging, and mining interests were "extremists" in their pursuit of profit and that government agencies—the Forest Service, the Bureau of Land Management, the U.S. Department of Agriculture, and others—were "gutless" and unwilling or unable to stand up to wealthy and powerful interests intent upon profiting from the destruction of American wilderness. Well-meaning moderate organizations such as the Sierra Club, Friends of the Earth, and the Wilderness Society were, with few exceptions, powerless to prevent the continuing destruction.

What was needed, Foreman reasoned, was an immoderate and unrespectable band of radical environmentalists like those depicted in Edward Abbey's novel *The Monkey, Wrench Gang* (1975), to take direct action against anyone who destroyed the wilderness in the name of development. With several like-minded friends, Foreman founded Earth First!, whose motto is "No compromise in defense of Mother Earth."

From the beginning Earth First! was unlike any other radical group. It did not issue manifestoes or publish position papers; it had "no officers, no bylaws or constitution, no incorporation, no tax status; just a collection of women and men committed to the Earth." Earth First!, Foreman wrote, "would be big enough to contain street poets and cowboy bar bouncers, agnostics and pagans, vegetarians and raw steak eaters, pacifists and those who think that turning the other cheek is a good way to get a sore face." Its weapons would include "monkey-wrenching," civil disobedience, music, "media stunts [to hold] the villains up to ridicule," and self-deprecating humor: "Radicals frequently verge toward a righteous seriousness. But we felt that if we couldn't laugh at ourselves we would be merely another bunch of dangerous fanatics who should be locked up (like the oil companies). Not only does humor preserve individual and group sanity, it retards hubris, a major cause of environmental rape, and it is also an effective weapon." But besides humor, Foreman called for "fire, passion, courage, and emotionalism. We [environmentalists] have been too reasonable, too calm, too understanding. It's time to get angry, to cry, to let rage flow at what the human cancer is doing to Mother Earth."

In 1987 Foreman published *Ecodefense: A Field Guide to Monkeywrenching*, in which he described in detail the tools and techniques of environmental sabotage or monkey-wrenching. These techniques included "spiking" old-growth redwoods and Douglas firs to prevent loggers from felling them, "munching" logging roads with nails, sabotaging bulldozers and other earth-moving equipment, pulling up surveyors' stakes, and toppling high-voltage power lines. These tactics, Foreman said, were aimed at property, not at people. But critics quickly charged that loggers' lives and jobs were endangered by tree-spiking and other techniques that could turn deadly. Moderate or mainstream environmental organizations joined in the condemnation of the confrontational tactics favored by Foreman and Earth First!

In his autobiography *Confessions of an Eco-Warrior* (1991), Foreman defends monkey-wrenching as an unfortunate tactical necessity that has achieved its primary purpose of attracting the attention of the American people and the media to the destruction of the nation's remaining wilderness. It also attracted the attention of the FBI, whose agents arrested Foreman at his home in 1989 for allegedly financing and encouraging *eco-teurs* (ecological saboteurs) to topple high-voltage power poles. Foreman was put on trial to face felony charges, which he denied. The charges were questioned when it was disclosed that an FBI informant had infiltrated Earth First! with the intention of framing Foreman and discrediting the organization. In a plea bargain, Foreman pleaded guilty to a lesser charge and received a suspended sentence.

Foreman left Earth First! to found and direct The Wildlands Project in Tucson, Arizona. In 2007 he and fellow board directors founded the Rewidling Institute, an environmental conservation think-tank. As of 2010 he continues to lecture and write about the protection of the wilderness.

Resources

BOOKS

Foreman, Dave. *Confessions of an Eco-Warrior*. New York: Harmony Books, 1991.
Foreman, Dave and B. Haywood. *Ecodefense: A Guide to Monkeywrenching*. Tucson: Ned Ludd Books, 1985.

Terence Ball

Forestry Canada *see* **Canadian Forest Service.**

Forests *see* **Coniferous forest; Deciduous forest; Hubbard Brook Experimental Forest; National forest; Old-growth forest; Rain forest; Taiga; Temperate rain forest; Tropical rain forest.**

Forest and Rangeland Renewable Resources Planning Act (1974)

The Forest and Rangeland Renewable Resources Planning Act (RPA) was passed in response to the growing tension between the timber industry and environmentalists in the late 1960s and the early 1970s. These tensions can be traced to increased controversy over and restrictions on timber harvesting on the national forests, primarily due to wilderness designations and study areas and clear-cutting. These environmental restrictions, coupled with a dramatic increase in the price of timber in 1969, made Congress receptive to timber industry demands for a steadier supply of timber. Numerous bills addressing timber supply were introduced and debated in Congress, but none passed due to strong environmental pressure. A task force appointed by President Richard Nixon, the President's Panel on Timber and the Environment, delivered its recommendations in 1973. The panel's recommendations were geared toward dramatically increased harvests from the national forests, and hence were also unacceptable to environmentalists.

One aspect of the various proposals that proved to be acceptable to all interested parties—the timber industry, environmentalists, and the Forest Service—was increased long-range resource planning. Senator Hubert Humphrey of Minnesota drafted a bill creating such a program and helped guide it to passage in Congress. RPA planning is based on a two-stage process, with a document accompanying each stage. The first stage is called the *Assessment*, which is an inventory of the nation's forest and range resources (public and private). The second stage, which is based on the *Assessment*, is referred to as the *Program*. Based on the completed inventory, the Forest Service provides a plan for the use and development of the available resources. The assessment is to be done every ten years. A program based on the assessment will be completed every five years. This planning was to be done by interdisciplinary teams and to incorporate widespread public involvement.

The RPA was quite popular with the Forest Service because the plans generated through the process gave the agency a solid foundation on which to base its budget requests, increasing the likelihood of greater funding. This has proved to be successful, as the Forest Service budget increased dramatically in 1977, and the agency fared much better than other resource agencies in the late twentieth century.

The RPA was amended by the National Forest Management Act of 1976. Based on this law, in addition to the broad national planning mandated in the 1974 law, an assessment and program were required for each unit of the national forest system. This has allowed the Forest Service to use the plans to help shield itself from criticism. Since these plans address all uses of the forests and make budget recommendations for these uses, if Congress does not fund these recommendations, the Forest Service can point to Congress as the culprit. However, the plans have also been more visible targets for interest group criticism.

Overall, the RPA has met with mixed results. The Forest Service has received increased funds, and the planning process has been expanded to each national forest unit, but planning at such a scale is a difficult task. The act has also led to increased controversy and to increased bureaucracy. Perhaps most importantly, planning cannot solve a problem based on conflicting values, commodity use versus forest preservation, which is at the heart of forest management policy.

See also Old-growth forest.

Resources

BOOKS

Wuerthner, George. *Wildfire: A Century of Failed Forest Policy*. Washington, DC: Island Press, 2006.

Christopher McGrory Klyza

Forest decline

Thirty percent of the world's surface area is currently covered with forests. In recent decades there have been observations of widespread declines in vigor and dieback of mature forests in many parts of the world. In many cases pollution may be a factor contributing to forest decline, for example, in regions where air quality is poor because of acidic deposition or contamination with ozone (O_3), sulfur dioxide (SO_2), nitrogen (N) compounds, or metals. However, forest decline also occurs in some places where the air is not polluted, and in these cases it has been suggested that the phenomenon is natural.

Forest decline is characterized by a progressive, often rapid deterioration in the vigor of trees of one or several species, sometimes resulting in mass mortality (or dieback) within stands over a large area. Decline often selectively affects mature individuals and is thought to be triggered by a particular stress or a combination of stressors, such as severe weather, nutrient deficiency, toxic substances in soil, or air pollution. According to this scenario, excessively stressed trees suffer a large decline in vigor. In this weakened condition, trees are relatively vulnerable to lethal attack by insects and microbial pathogens. Such secondary agents may not be so harmful to vigorous individuals, but they can cause the death of severely stressed trees.

The preceding is only a hypothetical etiology of forest dieback. It is important to realize that although the occurrence and characteristics of forest decline can be well documented, the primary environmental variables that trigger the disease-causing decline are not usually known. As a result the etiology of the decline syndrome is often attributed to a vague but unsubstantiated combination of biotic (living factors) and abiotic (non-living such as physical and chemical factors) factors.

The symptoms of decline differ among tree species. Frequently observed effects include: (1) decreased productivity; (2) chlorosis (paling of leaves due to lack of chlorophyll causing reduced photosynthesis), abnormal size or shape, and premature abscission of foliage; (3) a progressive death of branches that begins at the extremities and often causes a stag-headed appearance; (4) root dieback; (5) an increased frequency of secondary attack by fungal pathogens and defoliating or wood-boring insects; and (6) ultimately mortality, often as a stand-level dieback.

One of the best-known cases of an apparently natural forest decline, unrelated to human activities, is the widespread dieback of birches that occurred throughout the northeastern United States and eastern Canada from the 1930s to the 1950s. The most susceptible species were yellow (*Betula alleghaniensis*) and paper birch (*B. papyrifera*), which were affected over a vast area, often with extensive mortality. For example, in 1951 at the time of peak dieback in Maine, an estimated 67 percent of the birch trees had been killed. In spite of considerable research effort, a single primary cause has not been determined for birch dieback. It is known that a heavy mortality of fine roots usually preceded deterioration of the above-ground tree, but the environmental cause(s) of this effect are unknown, although deeply frozen soils caused by a sparse winter snow cover are suspected as being important. No biological agent was identified as a primary predisposing factor, although fungal pathogens and insects were observed to secondarily attack weakened trees and cause their death.

Another apparently natural forest decline is that of ohia (*Metrosideros polymorpha*), an endemic species of tree usually occurring in monospecific (consisting of one species) stands that dominates the native forest of Hawaiian Islands. There are anecdotal accounts of events of widespread mortality of ohia extending back at least a century, but the phenomenon is probably more ancient than this. The most recent widespread decline began in the late 1960s and resulted in about 200 square miles (518 km^2) of forest with symptoms of ohia decline in a 1982 survey of 308 square miles (798 km^2). In most declining stands only the canopy individuals were affected. Understory saplings and seedlings were not in decline, and in fact were released from competitive stresses by dieback of the overstory.

A hypothesis to explain the cause of ohia decline has been advanced by ecologist D. Mueller-Dombois and coworkers, who believe that the stand-level dieback is caused by the phenomenon of cohort senescence. This is a stage of the life history of ohia characterized by a simultaneously decreasing vigor in many individuals, occurring in old-growth stands. The development of senescence in individuals is governed by genetic factors, but the timing of its onset can be influenced by environmental stresses. The decline-susceptible, over-mature, life-history stage follows a more vigorous, younger, mature stage in an even-aged stand of individuals of the same generation (i.e., a cohort) that had initially established following a severe disturbance. In Hawaii, lava flows, events of deposition of volcanic ash, and hurricanes are natural disturbances that initiate succession. Sites disturbed in this way are colonized by a cohort of ohia individuals, which produce an even-aged stand. If there is no intervening catastrophic disturbance, the stand matures, then becomes senescent (aged) and enters a decline and dieback phase. The original stand is then replaced by another ohia forest comprised of an advance regeneration of individuals released from the understory. Therefore, according to the cohort senescence theory, the ohia dieback should be considered to be a characteristic of the natural population dynamics of the species.

Other forest declines are occurring in areas where the air is contaminated by various potentially toxic chemicals, and these cases might be triggered by air pollution. In North America prominent declines have occurred in ponderosa pine (*Pinus ponderosa*), red spruce (*Picea rubens*) and sugar maple (*Acer saccharum*). In western Europe, Norway spruce (*Picea abies*) and beech (*Fagus sylvatica*) have been severely affected.

The primary cause of the decline of ponderosa pine in stands along the western slopes of the mountains of southern California is believed to be the toxic effects of ozone. Ponderosa pine is susceptible to the effects of this gas at the concentrations that are commonly encountered in the declining stands, and the symptoms are indicative of ozone damage.

In the other cases of decline noted above that are putatively related to air pollution, the available evidence is less convincing. The recent forest damage in Europe has been described as a new decline syndrome that may in some way be triggered by stresses associated with air pollution. Although the symptoms appear to be similar, the new decline is believed to be different from diebacks that are known to have occurred historically and are believed to have been

natural. The modern decline syndrome was first noted in fir (*Abies alba*) in Germany in the early 1970s. In the early 1980s a larger-scale decline was apparent in Norway spruce, the most commercially-important species of tree in the region, and in the mid 1980s decline became apparent in beech and oak (*Quercus* spp.).

Decline of this type has been observed in countries throughout Europe, extending at least to western Russia. The decline has been most intensively studied in Germany, which has many severely damaged stands, although a widespread dieback has not yet occurred. Decline symptoms are variable in the German stands, but in general: (1) mature stands older than about sixty years tend to be more severely affected; (2) dominant individuals are relatively vulnerable; and (3) individuals located at or near the edge of the stand are more-severely affected, suggesting that a shielding effect may protect trees in the interior. Interestingly epiphytic lichens (symbiosis between a fungus and a photosynthetic organism) often flourish in badly damaged stands, probably because of a greater availability of light and other resources caused by the diminished cover of tree foliage. In some respects this is a paradoxical observation, because lichens are usually hypersensitive to air pollution, especially toxic gases.

From the information that is available, it appears that the new forest decline in Europe is triggered by a variable combination of environmental stresses. The weakened trees then decline rapidly and may die as a result of attack by secondary agents, such as fungal disease or insect attack. Suggestions of the primary inducing factor include gaseous air pollutants, acidification, toxic metals in soil, nutrient imbalance, and a natural climatic effect, in particular drought. However, there is not yet a consensus as to which of these interacting factors is the primary trigger that induces forest decline in Europe, and it is possible that no single stress will prove to be the primary cause. In fact, there may be several different causes contributing to declines occurring simultaneously in different areas.

The declines of red spruce and sugar maple in eastern North America involve species that are long-lived and shade-tolerant, but shallow-rooted and susceptible to drought. The modern epidemic of decline in sugar maple began in the late 1970s and early 1980s, and has been most prominent in Quebec, Ontario, New York, and parts of New England. During the late 1980s and early 1990s, the decline appeared to reverse, and the health of most stands improved. The symptoms are similar to those described for an earlier dieback and include abnormal coloration, size, shape, and premature abscission of foliage; death of branches from the top of the tree downward; reduced productivity; and death of trees. There is a frequent association with the pathogenic fungus *Armillaria mellea*, but this is believed to be a secondary agent that only attacks weakened trees. Many declining stands had recently been severely defoliated by the forest tent caterpillar (*Malacosoma disstria*), and many stands were tapped each spring for sap to produce maple sugar. Because the declining maple stands are located in a region subject to a high rate of atmospheric deposition of acidifying substances, this has been suggested as a possible predisposing factor, along with soil acidification and mobilization of available aluminum (Al). Natural causes associated with climate, especially drought, have also been suggested. However, little is known about the modern sugar maple decline, apart from the fact that it occurred extensively; no conclusive statements can yet be made about its causation.

The stand-level dieback of red spruce has been most frequent in high-elevation sites of the northeastern United States, especially in upstate New York, New England, and the mid- and southern-Appalachian states. These sites are variously subject to acidic precipitation (mean annual pH about 4.0–4.1), to very acidic fog water (pH as low as 3.2–3.5), to large depositions of sulfur (S) and nitrogen from the atmosphere, and to stresses from metal toxicity in acidic soil.

Declines of red spruce are anecdotally known from the 1870s and 1880s in the same general area where the modern decline is occurring. Up to one-half of the mature red spruce in the Adirondacks of New York was lost during that early episode of dieback, and there was also extensive damage in New England. As with the European forest decline, the past and current episodes appear to have similar symptoms, and it is possible that both occurrences are examples of the same kind of disease.

The hypotheses suggested to explain the modern decline of red spruce are similar to those proposed for European forest decline. These potential causes include acidic deposition, soil acidification, aluminum toxicity, drought, winter injury exacerbated by insufficient hardiness due to nitrogen fertilization, heavy metals in soil, nutrient imbalance, and gaseous air pollution.

Climate change, or global warming, may be a significant factor involved in forest decline. Heat- and drought-related stress leading to tree mortality has been recorded on all forested continents. The global mean temperatures are currently higher than temperatures in the past thirteen hundred years. Warmer temperatures and shifted or altered precipitation patterns are being observed worldwide as limiting the availability of water for trees, thus potentially contributing to worldwide forest diebacks. Many of the recently documented diebacks are occurring on climate boundaries where trees are on the edge of temperature sensitivity and water stress. Although climate change may not be the only cause of tree mortality, stress induced by heat is thought to be the underlying factor for disease- and pest-related diebacks. Additionally, forest diebacks may further contribute to climate change in becoming a source of atmospheric carbon when originally those forests were carbon sinks.

At present, not enough is known about the etiology of the forest declines in Europe and eastern North America to allow an understanding of possible role(s) of air pollution and of natural environmental factors. This lack of information does not necessarily mean that air pollution and climate change are not involved. Rather, it suggests that more research is required before any conclusive statements can be made regarding the causes and effects of the phenomenon of forest decline. To evaluate the health of forests, long-term global monitoring of forest status would contribute to forest maintenance and management by allowing scientists to check for symptoms of decline such as water stress and predict areas susceptible to mortality.

See also Forest management.

Resources

BOOKS

Clatterbuck, Wayne K. *Dieback and Decline of Trees.* Knoxville: University of Tennessee, UT Extension, 2006.

Thomas, Peter, and John R. Packham. *Ecology of Woodlands and Forests: Description, Dynamics and Diversity.* Cambridge: Cambridge University Press, 2007.

United States. *Invasive Species and Forest Health.* APHIS factsheet. Riverdale, MD.: USDA, APHIS, 2006. http://purl.access.gpo.gov/GPO/LPS99528

Washington (State). *Forest Health.* Olympia, WA: The Dept, 2006.

OTHER

Allen, C.D. "Climate-Induced Forest Dieback: An Escalating Global Phenomenon?" The Official Website of the Food and Agriculture Organization of the United Nations. ftp://ftp.fao.org/docrep/fao/011/i0670e/i0670e10.pdf (accessed June 30, 2009).

Bill Freedman

Forest management

The question of how forest resources should be used goes beyond the science of growing and harvesting trees; forest management must solve the problems of balancing economic, aesthetic, and biological value for entire ecosystems. The earliest forest managers in North America were native peoples, who harvested trees for building and burned forests to make room for grazing animals. But many native populations were wiped out by European diseases soon after Europeans arrived. By the mid-nineteenth century, it became apparent to many Americans that overharvesting of timber along with wasteful practices, such as uncontrolled burning of logging waste, was denuding forests and threatening future ecological and economic stability. The Forest Service, established in 1905, began studying ways to preserve forest resources for their economic as well as aesthetic, recreational, and wilderness value.

From the 1600s to 1820s, 370 million acres (150 million hectares) of forests—about 34 percent of the United States' total—were cleared, leaving about 749 million acres (303 million hectares) as of 2005, covering approximately 33 percent of U.S. land area. Only 10 to 15 percent of the forests have never been cut. Many previously harvested areas, however, have been replanted, but the nature of the forests has been altered, often detrimentally. If logging of old-growth forests were to continue at the rate maintained during the 1980s, all remaining unprotected stands would be gone by 2015. Some 33,000 timber-related jobs could also be lost during that time, not just from environmental protection but also from overharvesting, increased mechanization, and increasing reliance on foreign processing of whole logs cut from private lands. Recent federal and court decisions, most notably to protect the northern spotted owl (*Strix occidentalis caurina*) in the United States, have slowed the pace of old-growth harvesting and for now have put more mature forests under protection. But the questions of how to use forest resources is still under fierce debate.

For decades, clear-cutting of tracts has been the standard forestry management practice. Favored by timber companies, clear-cutting takes virtually all material from a tract. But clear-cutting has come under increasing criticism from environmentalists, who point out that the practice replaces mixed-age, biologically diverse forests with single-age, single or few species plantings. Single species planting is termed monoculture, which results in decreased biodiversity and increased disease susceptibility. Clear-cutting also relies heavily on roads to haul out timber, causing root damage, top soil erosion, and siltation of streams.

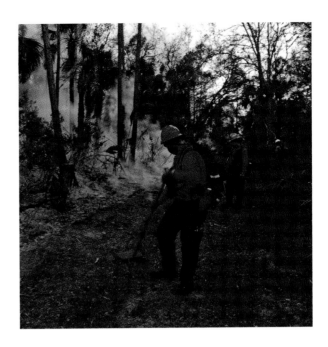

A prescribed fire in south Florida is responsible for fuel reduction and wildlife habitat management. (© Danita Delimont / Alamy)

Industry standards, such as best management practices (BMPs), prevent most erosion and siltation by keeping roads away from stream beds. But BMPs only address water quality. Clear-cutting also removes small trees, snags, boles, and woody debris that are important to invertebrates and fungi.

Rather than focusing on what is removed from a forest, sustainable forest management focuses on what is left behind. In sustainable forestry, tracts are never clear-cut: instead, individual trees are selected and removed to maintain diversity and health of the remaining ecosystem. Such methods avoid artificial replanting, herbicides, insecticides, and fertilizers. However, much debate remains on which trees and how many are chosen for harvesting under sustainable forestry.

In a management style known most commonly as new forestry (also referred to as ecological or natural forestry), select trees on a site are harvested over time. At each harvest some mature trees are retained to maintain a diverse and complex ecosystem. Standing or fallen dead trees and other wood debris are left in forests managed in this way. This method reduces erosion and promotes diversity left behind on a tract. As opposed to clearcutting and other harvesting practices, the methods employed in new forestry more closely resemble natural forest processes. American forest ecologist Jerry Franklin and colleagues coined the term new forestry and developed the method for the natural management of forests. The establishment of new forestry management is a step toward sustainable forest management and retaining forest resources and ecosystems for future generations.

Those who make their living from America's forests, and those who value the biological ecosystems they support, must resolve the debate on how to best preserve our forests. One-third of forest resources now come from Forest Service lands, and the debate is an increasingly public one, involving interests ranging from the Sierra Club and sporting clubs to the Environmental Protection Agency (EPA), the Department of Agriculture (and its largest agency, the Forest Service), and timber companies and their employees. The future of our forests depends on balancing real short-term needs with the high price of long-term forest health.

Forests have an influence on climate change in that they remove carbon dioxide (CO_2) from the air and store carbon, acting as significant carbon sinks in the global ecosystem. In 2008 U.S. forests offset about 10 percent of American carbon emissions. Reducing the amount of deforestation and focusing on sustainable forest management can potentially result in a higher compensation for U.S. carbon emissions.

Resources

BOOKS

Bell, Simon, and Dean Apostol. *Designing Sustainable Forest Landscapes*. London: Taylor & Francis Group, 2008.
Bettinger, Pete. *Forest Management and Planning*. Amsterdam: Academic, 2009.
Chowdhury, J. A. *Towards Better Forest Management*. Dhaka: Oitijjhya, 2006.
Grossberg, Steven P. *Forest Management*. Hauppauge NY: Nova Science Publishers, 2009.
Hasenauer, Hubert. *Sustainable Forest Management*. New York: Berlin: Springer, 2006.
Parkins, J. *Public participation in forest management*. Edmonton: Northern Forestry Centre, 2006.

L. Carol Ritchie

Forest Service

The national forest system in the United States must be considered one of the great success stories of the conservation movement despite the continual controversies that seem to accompany administration of national forest lands by the United States Forest Service.

The roots of the Forest Service began with the appointment in 1876 of Franklin B. Hough as a "forestry agent" in the U.S. Department of Agriculture to gather information about the nation's forests. Ten years later, Bernhard E. Fernow was appointed chief of a fledgling Division of Forestry. Part way through Fernow's tenure, Congress passed the Forest Reserve Act of 1891, which authorized the president to withdraw lands from the public domain to establish federal forest reserves. The public lands were to be administered, however, by the General Land Office in the U.S. Department of the Interior.

Gifford Pinchot succeeded Fernow in 1899 and was Chief Forester when President Theodore Roosevelt approved the transfer of 63 million acres (25 million ha) of forest reserves into the Department of Agriculture in 1905. That same year, the name of the Bureau of Forestry was changed to the United States Forest Service. Two years later the reserves were redesignated national forests.

The Forest Service today is organized into four administrative levels: the office of the Chief Forester in Washington, D.C.; nine regional offices; 155 national forests; and more than 600 ranger districts. The Forest Service also administers twenty national grasslands. In addition, a research function is served by a forest products laboratory in Madison, Wisconsin, and eight other field research stations.

These lands are used for a wide variety of purposes and given official statutory status with the passage of the Multiple Use-Sustained Yield Act of 1960. That act officially listed five uses—timber, water, range, wildlife, and recreation—to be administered on national forest lands. Wilderness was later included.

Despite a professionally trained staff and a sustained spirit of public service, Forest Service administration and management of national forest lands has been controversial from the beginning. The agency has experienced repeated attempts to transfer it to the Department of the Interior (or once to a new Department of Natural Resources); its authority to regulate grazing and timber use, including attempts to transfer national forest lands into private hands, has been frequently challenged, and some of the Service's management policies have been the center of conflict. These policies have included clear-cutting and "subsidized" logging, various recreation uses, preservation of the northern spotted owl, and the cutting of old-growth forests in the Pacific Northwest.

Resources

BOOKS

Williams, Gerald W. *The Forest Service: Fighting for Public Lands.* Understanding our Government. Westport, CT: Greenwood Press, 2007.

Hays, Samuel P. *The American People & the National Forests: The First Century of the U.S. Forest Service.* Pittsburgh: University of Pittsburgh Press, 2009.

ORGANIZATIONS

Forest Service, U. S. Department of Agriculture, 1400 Independence Avenue, SW, Washington, D.C., USA, 20250, info@fs.fed.us, http://www.fs.fed.us

Gerald L. Young

Dian Fossey, shown with a baby gorilla, was an American zoologist who undertook an extensive gorilla study in Rwanda, Africa. *(© Liam White / Alamy)*

Fossey, Dian

1932–1985
American naturalist and primatologist

Dian Fossey is remembered by her fellow scientists as one of the world's foremost authorities on mountain gorillas. But to the millions of wildlife conservationists who came to know Fossey through her articles and book, she will always be remembered as a martyr. Throughout the nearly twenty years she spent studying mountain gorillas in central Africa, the American primatologist tenaciously fought the poachers and bounty hunters who threatened to wipe out the endangered primates. She was brutally murdered at her research center in 1985 by what many believe was a vengeful poacher.

Fossey's dream of living in the wilds of Africa dates back to her lonely childhood in San Francisco. She was born in 1932, the only child of George, an insurance agent, and Kitty, a fashion model, (Kidd) Fossey. The Fosseys divorced when Dian was six years old. A year later, Kitty married a wealthy building contractor named Richard Price. Price was a strict disciplinarian who showed little affection for his stepdaughter. Although Fossey loved animals, she was allowed to have only a goldfish. When it died, she cried for a week.

Fossey began her college education at the University of California at Davis in the preveterinary medicine program. She excelled in writing and botany, but she failed chemistry and physics. After two years, she transferred to San Jose State University, where she earned a bachelor of arts degree in occupational therapy in 1954. While in college, Fossey became a prize-winning equestrian. Her love of horses in 1955 drew her from California to Kentucky, where she directed the occupational therapy department at the Kosair Crippled Children's Hospital in Louisville.

Fossey's interest in Africa's gorillas was aroused through primatologist George Schaller's 1963 book, *The Mountain Gorilla: Ecology and Behavior.* Through Schaller's book, Fossey became acquainted with the largest and rarest of three subspecies of gorillas, *Gorilla beringei beringei.* She learned that these giant apes make their home in the mountainous forests of Rwanda, Democratic Republic of Congo (formerly Zaire), and Uganda. Males grow up to 6 feet (1.8 m) tall and weigh 400 pounds (182 kg) or more. Their arms span up to 8 feet (2.4 m). The smaller females weigh about 200 pounds (91 kg).

Schaller's book inspired Fossey to travel to Africa to see the mountain gorillas in their homeland. Against her family's advice, she took out a three-year bank loan for $8,000 to finance the seven-week safari. While in Africa, Fossey met the celebrated paleoanthropologist Louis Leakey, who had encouraged Jane Goodall in her research of chimpanzees in Tanzania. Leakey was impressed by Fossey's plans to visit the mountain gorillas.

Those plans were nearly destroyed when she shattered her ankle on a fossil dig with Leakey. But just two weeks later, she hobbled on a walking stick up a mountain in the Congo (later Zaire, and now Democratic Republic of the Congo) to her first encounter with the great apes. The sight of six gorillas set the course for her future. "I left Kabara (gorilla site) with reluctance but with never a doubt that I would, somehow, return to learn more about the gorillas of the misted mountains," Fossey wrote in her book, *Gorillas in the Mist*.

Her opportunity came three years later, when Leakey was visiting Louisville on a lecture tour. Fossey urged him to hire her to study the mountain gorillas. He agreed, if she would first undergo a preemptive appendectomy. Six weeks later, he told her the operation was unnecessary; he had only been testing her resolve. But it was too late. Fossey had already had her appendix removed.

The L.S.B. Leakey and the Wilkie Brothers foundations funded her research, along with the National Geographic Society. Fossey began her career in Africa with a brief visit to Jane Goodall in Tanzania to learn the best methods for studying primates and collecting data.

Fossey set up camp early in 1967 at the Kabara meadow in Zaire's Parc National des Virungas, where Schaller had conducted his pioneering research on mountain gorillas a few years earlier. The site was ideal for Fossey's research. Because Zaire's park system protected them against human intrusion, the gorillas showed little fear of Fossey's presence. Unfortunately, civil war in Zaire forced Fossey to abandon the site six months after she arrived.

She established her permanent research site September 24, 1967, on the slopes of the Virunga Mountains in the tiny country of Rwanda. She called it the Karisoke Research Centre, named after the neighboring Karisimbi and Visoke mountains in the Parc National des Volcans. Although Karisoke was just five miles from the first site, Fossey found a marked difference in Rwanda's gorillas. They had been harassed so often by poachers and cattle grazers that they initially rejected all her attempts to make contact.

Theoretically, the great apes were protected from such intrusion within the park. But the government of the impoverished, densely populated country failed to enforce the park rules. Native Batusi herdsmen used the park to trap antelope and buffalo, sometimes inadvertently snaring a gorilla. Most trapped gorillas escaped, but not without seriously mutilated limbs that sometimes led to gangrene and death. Poachers who caught gorillas could earn up to $200,000 for one

by selling the skeleton to a university and the hands to tourists. From the start, Fossey's mission was to protect the endangered gorillas from extinction—indirectly, by researching and writing about them, and directly, by destroying traps and chastising poachers.

Fossey focused her studies on some fifty-one gorillas in four family groups. Each group was dominated by a sexually mature silverback, named for the characteristic gray hair on its back. Younger, bachelor males served as guards for the silverback's harem and their juvenile offspring.

When Fossey began observing the reclusive gorillas, she followed the advice of earlier scientists by concealing herself and watching from a distance. But she soon realized that the only way she would be able to observe their behavior as closely as she wanted was by habituating the gorillas to her presence. She did so by mimicking their sounds and behavior. She learned to imitate their belches that signal contentment, their barks of curiosity, and a dozen other sounds. To convince them she was their kind, Fossey pretended to munch on the foliage that made up their diet. Her tactics worked. One day early in 1970, Fossey made history when a gorilla she called Peanuts reached out and touched her hand. Fossey called it her most rewarding moment with the gorillas.

She endeared laypeople to Peanuts and the other gorillas she studied through her articles in National Geographic magazine. The apes became almost human through her descriptions of their nurturing and playing. Her early articles dispelled the myth that gorillas are vicious. In her 1971 *National Geographic* article she described the giant beasts as ranking among "the gentlest animals, and the shiest." In later articles, Fossey acknowledged a dark side to the gorillas. Six of thirty-eight infants born during a thirteen-year-period were victims of infanticide. She speculated the practice was a silverback's means of perpetuating his own lineage by killing another male's offspring so he could mate with the victim's mother.

Three years into her study, Fossey realized she would need a doctoral degree to continue receiving support for Karisoke. She temporarily left Africa to enroll at Cambridge University, where she earned her Ph.D. in zoology in 1974. In 1977 Fossey suffered a tragedy that would permanently alter her mission at Karisoke. Digit, a young male she had grown to love, was slaughtered by poachers. Walter Cronkite focused national attention on the gorillas' plight when he reported Digit's death on the CBS Evening News. Interest in gorilla conservation surged. Fossey took advantage of that interest by establishing the Digit Fund, a non-profit organization to raise money for anti-poaching patrols and equipment.

Unfortunately, the money wasn't enough to save the gorillas from poachers. Six months later, a silver-back and his mate from one of Fossey's study groups were shot and killed defending their three-year-old son, who had been shot in the shoulder. The juvenile later died from his wounds. It was rumored that the gorilla deaths caused Fossey to suffer a nervous breakdown, although she denied it. What is clear is that the deaths prompted her to step up her fight against the Batusi poachers by terrorizing them and raiding their villages. "She did everything short of murdering those poach-ers," Mary Smith, senior assistant editor at National Geographic, told contributor Cynthia Washam in an interview. A serious calcium deficiency that causes bones to snap and teeth to rot forced Fossey to leave Africa in 1980. She spent her three-year sojourn as a visiting associate professor at Cornell University. Fos-sey completed her book, *Gorillas in the Mist,* during her stint at Cornell. It was published in 1983. Although some scientists criticized the book for its abundance of anecdotes and lack of scientific discussion, lay readers and reviewers received it warmly.

When Fossey returned to Karisoke in 1983, her scientific research was virtually abandoned. Funding had run dry. She was operating Karisoke with her own savings. "In the end, she became more of an animal activist than a scientist," Smith said. "Science kind of went out the window."

On December 27, 1985, Fossey, 54, was found murdered in her bedroom at Karisoke, her skull split diagonally from her forehead to the corner of her mouth. Her murder remains a mystery that has prompted much speculation. Rwandan authorities jointly charged American research assistant Wayne McGuire, who discovered Fossey's body, and Emma-nuel Rwelekana, a Rwandan tracker Fossey had fired several months earlier. McGuire maintains his inno-cence. At the urging of U.S. authorities, he left Rwanda before the charges against him were made public. He was convicted in absentia and sentenced to die before a firing squad if he ever returns to Rwanda.

Farley Mowat, the Canadian author of Fossey's biography, *Woman in the Mists,* believes McGuire was a scapegoat. He had no motive for killing her, Mowat wrote, and the evidence against him appeared contrived. Rwelekana's story will never be known. He was found dead after apparently hanging himself a few weeks after

he was charged with the murder. Smith, and others, believe Fossey's death came at the hands of a vengeful poacher. "I feel she was killed by a poacher," Smith said. "It definitely wasn't any mysterious plot."

Fossey's final resting place is at Karisoke, sur-rounded by the remains of Digit and more than a dozen other gorillas she had buried. Her legacy lives on in the Virungas, as her followers have taken up her battle to protect the endangered mountain gorillas. The Dian Fossey Gorilla Fund, formerly the Digit Fund, finances scientific research at Karisoke and employs camp staff, trackers, and anti-poaching patrols.

The Rwanda government, which for years had ignored Fossey's pleas to protect its mountain gorillas, on September 27, 1990, recognized her scientific achievement with the Ordre National des Grandes Lacs, the highest award it has ever given a foreigner. Gorillas in Rwanda are still threatened by cattle ranch-ers and hunters squeezing in on their habitat. Shortly after Fossey's death, a census of gorillas in the Virunga Mountains counted 320 gorillas. A 2004 census, how-ever, revealed a significant increase to about 380 goril-las. Smith is among those convinced that the number would be much smaller if not for Fossey's eighteen years of dedication to save the great apes. "Her con-servation efforts stand above everything else (she accomplished at Karisoke)," Smith said. "She single-handedly saved the mountain gorillas."

Resources

BOOKS

Fossey, Dian. *Gorillas in the Mist.* Boston: Houghton Mifflin Company, 1983.
Mowat, Farley. *Woman in the Mists.* New York: Warner Books, 1987.

Cynthia Washam

Fossil fuels

In early societies, wood or other biological fuels were the main energy source. Today in many non-industrial societies, they continue to be used widely. Biological fuels may be seen as part of a solar economy where energy is extracted from the sun in a way that makes them renewable. However industrialization requires energy sources at much higher density and these have generally been met through the use of fossil fuels such as coal, gas, or oil. In the twentieth century a

Oil well pump, British Columbia, Canada. (© All Canada Photos / SuperStock)

number of other options, such as nuclear or higher density renewable energy sources (wind power, hydro-electric power, etc.), became available. Nevertheless, into the early part of the twenty-first century, fossil fuels still represented the principal source of energy for most of the industrialized world.

Fossil fuels are types of sedimentary organic materials, often loosely called bitumens, with asphalt, a solid, and petroleum, the liquid form. More correctly bitumens are sedimentary organic materials that are soluble in carbon disulfide. It is this that distinguishes asphalt from coal, which is an organic material largely insoluble in carbon disulfide.

Petroleum can probably be produced from any kind of organism, but the fact that these sedimentary deposits are more frequent in marine sediments has suggested that oils arise from the fats and proteins in material deposited on the sea floor. These fats would be stable enough to survive the initial decay and burial but sufficiently reactive to undergo conversion to petroleum hydrocarbons at low temperature. Petro-leum consists largely of paraffins or simple alkanes, with smaller amounts of napthenes. There are traces of

aromatic compounds such as benzene present at the percent level in most crude oils. Natural gas is an abundant fossil fuel that consists largely of methane and ethane, although traces of higher alkanes are present. In the past, natural gas was regarded very much as a waste product of the petroleum industry and was simply burnt or flared off. Increasingly it is being seen as the favored fuel.

Coal, unlike petroleum, contains only a little hydrogen. Fossil evidence shows that coal is mostly derived from the burial of terrestrial vegetation with its high proportion of lignin and cellulose.

Most sediments contain some organic matter, and this can rise to many percent in shales. Here the organic matter can consist of both coals and bitumens. This organic material, often called sapropel, can be distilled to yield petroleum. Oil shales containing the sapropel kerogen are very good sources of petroleum. Shales are considered to have formed where organic matter was deposited along with fine grain sediments, perhaps in fjords, where restricted circulation keeps the oxygen concentrations low enough to prevent decay of the organic material.

Fossil fuels are mined or pumped from geological reservoirs where they have been stored for long periods of time. The more viscous fuels, such as heavy oils, can be quite difficult to extract and refine, which has meant that the latter half of the twentieth century has seen lighter oils being favored. However, in recent decades natural gas has been popular because it is easy to pipe and has a somewhat less damaging impact on the environment. These fossil fuel reserves, although large, are limited and non-renewable. According to the U.S. Department of Energy (DOE, from the *Oil and Gas Journal,*), as of 2010, the total proved light-to-medium oil reserves in the world were estimated at about 1.354 trillion barrels, of which about a third had already been used. About 56 percent of the world's proved oil reserves are located in the countries of the Middle East. In addition, 80 percent of these reserves are contained within the OPEC countries, along with the non-OPEC countries of Canada and Russia. Natural gas reserves are estimated (again by the DOE, 2010) at the equivalent of 1.177 trillion barrels of oil (6,609 trillion cubic feet) and about a sixth had already been used. Heavy oil and bitumen amount to about 0.6 and 0.34 trillion barrels, most of which remained unutilized. The gas and lighter oil reserves lie predominantly in the eastern hemisphere, which accounts for the enormous petroleum industries of the Middle East. The heavier oil and bitumen reserves lie mostly in the western hemisphere. These are more costly to use and have been for the most part untapped. According to DOE figures as of 2008, of the estimated 909-billion ton coal reserve (which is recoverable), only about 2.5 to 2.6 percent has been used. About 72 percent of the available coal is shared between China, Russia, and the United States.

Petroleum is not burnt in its crude form but must be refined, which is essentially a distillation process that splits the fuel into batches of different volatility. The lighter fuels are used in automobiles, with heavier fuels used as diesel and fuel oils. Modern refining can use chemical techniques in addition to distillation to help make up for changing demands in terms of fuel type and volatility.

The combustion of fuels represents an important source of air pollutants. Although the combustion process itself can lead to the production of pollutants, such as carbon or nitrogen oxides, it has often been the trace impurities in fossil fuels that have been the greatest source of air pollution. Natural gas is a much-favored fuel because it has only traces of impurities such as hydrogen sulfide. Many of these impurities are removed from the gas by relatively simple scrubbing techniques, before it is distributed. Oil is refined, so

although it contains more impurities than natural gas, these become redistributed in the refining process. Sulfur compounds tend to be found only in trace amounts in the light automotive fuels. Thus automobiles are only a minor source of sulfur dioxide in the atmosphere. Diesel oil can have as much as a percent of sulfur, and heavier fuel oils can have even more, so in some situations these can represent important sources of sulfur dioxide in the atmosphere. Oils also dissolve metals from the rocks in which they are present. Some of the organic compounds in oil have a high affinity for metals, most notably nickel and vanadium. Such metals can reach high concentration in oils, and refining will mean that most become concentrated in the heavier fuel oils. Combustion of fuel oil will yield ashes that contain substantial fractions of the trace metals present in the original oil. This means that an element like vanadium is a useful marker of fuel oil combustion.

Coal is often seen as the most polluting fuel because low grade coals can contain large quantities of ash, sulfur, and chlorine. However, it should be emphasized that the quantity of impurities in coal can vary widely, depending on where it is mined. The sulfur present in coal is found both as iron pyrites (inorganic) and bound up with organic matter. The nitrogen in coal is almost entirely organic nitrogen. Coal users are often careful to choose a fuel that meets their requirements in terms of the amount of ash, smoke, or pollution risk it imposes. High rank coals such as anthracite have high carbon content. They are mined in locations such as Pennsylvania (United States) and South Wales (Wales), and they contain little volatile matter and burn almost smokeless. Much of the world's coal reserve is bituminous, which means that it contains from 20 to 25 percent of volatile matter.

The fuel industry is often seen as responsible for pollutants and environmental risks that go beyond those produced by the combustion of its products. Mining and extraction processes result in spoil heaps, huge holes in open cast mining, and the potential for slumping of land (conventional mining). Petroleum refineries are large sources of hydrocarbons, although not usually the largest anthropogenic source of volatile organic compounds in the atmosphere. Refineries also release sulfur, carbon, and nitrogen oxides from the fuel that they burn. Liquid natural gas and oil spills are experienced both in the refining and transport of petroleum.

Being a solid, coal presents somewhat less risk when transported, although wind-blown coal dust can cause localized problems. Coal is sometimes converted to coke or other refined products such as Coalite, a smokeless coal. These derivatives are less polluting, although much concern has been expressed about the pollution damage that occurs near the factories that manufacture them. Despite this, the conversion of coal to less polluting synthetic solid, and liquid and gaseous fuels would appear to offer much opportunity for the future.

One of the principal concerns about the current reliance on fossil fuels relates not so much to their limited supply, but more to the fact that combustion releases such large amounts of carbon dioxide. Human use of fossil fuels over the twentieth century and now into the twenty-first century has increased the concentration of carbon dioxide in the atmosphere. Already there is mounting evidence that this has increased the temperature of Earth's atmosphere through an enhanced greenhouse effect.

Resources

BOOKS

Cheng, Jay. *Biomass to Renewable Energy Processes*. Roca Raton, FL: CRC Press/Taylor and Francis, 2010.
Gevorkian, Peter. *Alternative Energy Systems in Building Design*. New York: McGraw-Hill, 2010.
Hodge, B. K. *Alternative Energy Systems and Applications*. Hoboken, NJ: Wiley, 2010.

OTHER

Department of Energy. "Fossil Fuels" http://www.energy. gov/energysources/fossilfuels.htm (accessed October 12, 2010).
Energy Information Administration, Department of Energy. "International Energy Outlook, 2010." http:// www.eia.doe.gov/oiaf/ieo/pdf/0484(2010).pdf (accessed October 14, 2010).
Energy Information Administration, Department of Energy. "World Proved Reserves of Oil and Natural Gas, Most Recent Estimates." http://www.eia.doe. gov/emeu/international/reserves.html (accessed October 12, 2010).

Peter Brimblecombe

Fossil water

Water that occurs in an aquifer or zone of saturation that is temporarily excluded or shielded from participation in the the dynamic hydrologic cycle. Such water is often prized both for drinking and scientific study because it does not contain elevated levels of chemicals or contaminants common to water exposed to an increasingly industrialized and polluted world.

See also Drinking-water supply; Groundwater; Water quality.

Four Corners

The Hopi culture and religious tradition taught that the Four Corners—where Colorado, Utah, New Mexico and Arizona meet—is the center of the universe and holds all life on Earth in balance. It also has some of the largest deposits of coal, uranium, and oil shale in the world. According to the National Academy of Sciences the Four Corners is a "national sacrifice area." This ancestral home of the Hopi and Dineh (Navajo) people is the center of the most intense energy development in the United States. Traditional grazing and farm land are being swallowed up by uranium mines, coal mines, and power plants.

The Four Corners, sometimes referred to as the "joint-use area," is comprised of 1.8 million acres (729,000 ha) of high desert plateau where Navajo sheep herders have grazed their flocks on idle Hopi land for generations. In 1972 Congress passed Public Law (PL) 93-531, establishing the Navajo/Hopi Relocation Commission, which had the power to enforce livestock reduction and the removal of more than 10,000 traditional Navajo and Hopi, the largest forced relocation within the United States since the Japanese internment during World War II. Elders of both Nations issued a joint statement that officially opposed the relocation: "The traditional Hopi and Dineh (Navajo) realize that the so-called dispute is used as a disguise to remove both people from the JUA (joint use area), and for non-Indians to develop the land and mineral resources...Both the Hopi and Dineh agree that their ancestors lived in harmony, sharing land and prayers for more than four hundred years...and cooperation between us will remain unchanged."

The traditional Navajo and Hopi leaders have been replaced by Bureau of Indian Affairs (BIA) tribal councils. These councils, in association with the U.S. Department of the Interior, Peabody Coal, the Church of Jesus Christ of Later-day Saints, attorneys and public relation firms, created what is commonly known as the "Hopi-Navajo land dispute" to divide the joint-use area, so that the area could be opened up for energy development.

In 1964, 223 Southwest utility companies formed a consortium known as the Western Energy and Supply Transmission Associates (WEST), which includes water and power authorities on the West Coast as well as Four Corners area utility companies. WEST drafted plans for massive coal surface mining operations and six coal-fired, electricity- generating plants on Navajo and Hopi land. By 1966 John S. Boyden, attorney for the Bureau of Indian Affairs Hopi Tribal Council, secured lease arrangements with Peabody Coal to surface mine 58,000 acres (23,490 ha) of Hopi land and contracted WEST to build the power plants. This was done despite objections by the traditional Hopi leaders and the self-sufficient Navajo shepherds. Later that same year Kennecott Copper, owned in part by the Mormon Church, bought Peabody Coal. Peabody supplies the Four Corners' power plant with coal. The plant burned 5 million tons of coal a year, which is the equivalent of ten tons per minute. It emits over 300 tons of fly ash and other particles into the San Juan River Valley every day. Since 1968 the coal mining operations and the power plant have extracted over 60 million gallons (227 million l) of water a year from the Black Mesa water table, which has caused extreme desertification of the area, causing the ground in some areas to sink by up to 12 feet (3.6 m).

One of the worst nuclear accident in American history occurred at Church Rock, New Mexico, on July 26, 1979, when a Kerr-McGee uranium tailings pond spilled over into the Rio Puerco. The spill contaminated drinking water from Church Rock to the Colorado River, over 200 miles (322 km) to the west. The mill tailings dam broke—two months prior to the break cracks in the dam structure were detected, yet repairs were never made—and discharged more than 100 million gallons (379 million l) of highly radioactive water directly into the Rio Puerco River. The main source of potable water for more than 1,700 Navahoes was contaminated. When Kerr-McGee abandoned the Shiprock site in 1980, they left behind 71 acres (29 ha) of "raw" uranium tailings, which retained 85 percent of the original radioactivity of the ore at the mining site. The tailings were at the edge of the San Juan River and have since contaminated communities located downstream.

Resources

BOOKS

Childs, Craig. *House of Rain: Tracking a Vanished Civilization Across the American Southwest*. New York: Little, Brown and Co, 2007.

Bloomfield, Debra. *Four Corners*. Albuquerque: University of New Mexico Press, 2004.

Benson, Sara. *The Four Corners Region*. Woodstock, VT: Countryman Press, 2008.

Debra Glidden

Fox hunting

Fox hunting is the sport of mounted riders chasing a wild fox with a pack of hounds. The sport is also known as riding to the hounds, because the fox is pursued by horseback riders following the hounds that chase the fox. The specially trained hounds pursue the fox by following its scent. The riders are called the "field" and their leader is called the "master of foxhounds." A huntsman manages the pack of hounds.

Foxhunting originated in England, and dates back to the Middle Ages. People hunted foxes because they were predators that killed farm animals, such as chickens and sheep. Rules were established reserving the hunt to royalty, the aristocracy (people given titles by royalty), and landowners. As the British Empire expanded, the English brought fox hunting to the lands they colonized. The first fox hunt in the United States was held in Maryland in 1650, according to the Masters of Foxhounds Association (MFHA), the organization that controls foxhunting in the United States.

Although the objective of most fox hunts is to kill the fox, some hunts do not involve any killing. In a drag hunt, hounds chase the scent of a fox on a trail prepared before the hunt. In the United States a hunt ends successfully when the fox goes into a hole in the ground called the "earth."

In the United Kingdom, a campaign to outlaw fox hunting started in the late twentieth century. Fox hunting was banned in Scotland in 2002 and in England and Wales in 2004.

Organized hunt supporters such as, the Countryside Alliance, said a hunting ban would result in the loss of 14,000 jobs. In addition, advocates said that hunts help to eliminate a rural threat by controlling the fox population. Hunt supporters described foxes as vermin, a category of destructive, disease-bearing animals. Hunting was seen as less cruel than other methods of eliminating foxes.

Opponents called foxhunting a "blood sport" that was cruel to animals. According to the International Fund for Animal Welfare (IAFW), more than 15,000 foxes are killed during the 200 hunts held each year.

Free riders

The IAFW reported that young dogs are trained to hunt by pitting them against fox cubs. The group led efforts to have the British Parliament outlaw fox hunting. Drag hunting was suggested as a more humane alternative.

Another opposition group, the Nottingham Hunt Saboteurs Association, would disrupt "blood sports" through "non-violent direct action." Saboteurs' methods included trying to distract hounds by laying false scent trails, shouting, and blowing horns.

Both supporters and opponents of fox hunting claimed public support for their positions. In 1997 the Labour Party won the general election and proposed a bill to ban hunting with hounds. The following year, the bill passed through some legislative readings. However, time ran out before a final vote was taken.

In July 1999 Prime Minister Tony Blair promised to make fox hunting illegal. That November, Home Secretary Jack Straw called for an inquiry to study the effect of a ban on the rural economy. The Burns Inquiry concluded in June 2000 that a hunting ban would result in the loss of between 6,000 and 8,000 jobs. The inquiry did not find evidence that being chased was painful for foxes. However, the inquiry stated that foxes did not die immediately. This conclusion about a slower, painful death echoed opponents' charges that hunting was a cruel practice.

Two months before the inquiry was released, Straw proposed that lawmakers should have several options to vote on. One choice was a ban on fox hunting; another was to make no changes. A third option was to tighten fox hunting regulations.

In November 2004 Parliament adopted the Hunting Act 2004, which banned fox hunting in England and Wales. The ban went into effect on February 18, 2005. Fox hunting remains legal in Northern Ireland. Hunts now involve chasing artificial trails laid for the hounds. The League Against Cruel Sports, however, alleges that actual fox hunting remains widespread in Great Britain.

Fox hunting in other countries

The Scottish Parliament banned hunting in February of 2002. The ban was to take effect on August 1, 2002. The Countryside Alliance announced plans to legally challenge that ruling.

Fox hunting is legal in the following countries: Ireland, Belgium, Portugal, Italy, and Spain. Hunting with hounds is banned in Switzerland. In the United States, the MFHA was established in 1907. In March of 2002, the MFHA reported that there were 171 organized hunt clubs in North America. Organized fox hunts in the United States rarely involve killing foxes, although non-registered hunts may engage in such behavior.

Resources

BOOKS

Green, Kate, and Trevor Meeks. *Foxhunting*. London: Andre Deutsch, 2010.
Cook, John. *Observations on Fox Hunting*. Gardners Books, 2007.
Orendi, Dagmar. *Fox Hunting*. Gardners Books, 2007.

ORGANIZATIONS

International Fund for Animal Welfare, 290 Summer Street, Yarmouth Port, MA, USA, 02675, (508) 744–2000, (508) 744–2009, (800) 932-4329, info@iafw.org, http://www.iafw.org
Masters of Foxhounds Association, P.O. Box 363, Millwood, VA, USA, 22646, office@mfha.com, http://www.mfha.com

Liz Swain

Free riders

A free rider, in the broad sense of the term, is anyone who enjoys a benefit provided, probably unwittingly, by others. In the narrow sense, a free rider is someone who receives the benefits of a cooperative venture without contributing to the provision of those benefits. A person who does not participate in a cooperative effort to reduce air pollution by driving less, for instance, will still breathe cleaner air—and thus be a free rider—if the effort succeeds.

In this sense free riders are a major concern of the theory of collective action. As developed by economists and social theorists, this theory rests on a distinction between private and public (or collective) goods. A public good differs from a private good because it is indivisible and non-rival. A public good, such as clean air or national defense, is indivisible because it cannot be divided among people the way food or money can. It is non-rival because one person's enjoyment of the good does not diminish anyone else's enjoyment of it. Smith and Jones may be rivals in their desire to win a prize, but they cannot be rivals in their desire to breathe clean air, for Smith's breathing clean air will not deprive Jones of an equal chance to do the same.

Problems arise when a public good requires the cooperation of many people, as in a campaign to reduce pollution or conserve resources. In such cases, individuals have little reason to cooperate, especially when cooperation is burdensome. After all, one person's contribution—using less gasoline or electricity, for example—will make no real difference to the success or failure of the campaign, but it will be a hardship for that person. So the rational course of action is to try to be a free rider who enjoys the benefits of the cooperative effort without bearing its burdens. If everybody tries to be a free rider, however, no one will cooperate, and the public good will not be provided. If people are to prevent this from happening, some way of providing selective or individual incentives must be found, either by rewarding people for cooperating or punishing them for failing to cooperate.

The free rider problem posed by public goods helps to illuminate many social and political difficulties, not the least of which are environmental concerns. It may explain why voluntary campaigns to reduce driving and to cut energy use so often fail, for example. As formulated in Garrett Hardin's Tragedy of the Commons, moreover, collective action theory accounts for the tendency to use common resources—grazing land, fishing banks, perhaps the earth itself—beyond their carrying capacity. The solution, as Hardin puts it, is "mutual coercion, mutually agreed upon" to prevent the overuse and destruction of vital resources. Without such action, the desire to ride free may lead to irreparable ecological damage.

Resources

PERIODICALS

Hardin, G. "The Tragedy of the Commons." *Science* 162 (December 13, 1968): 1243–1248.

Richard K. Dagger

Freon

Freon is the generic name for several chlorofluorocarbons (CFCs) widely used in refrigerators and air conditioners, including systems in houses and cars. Freon—comprised of chlorine (Cl), fluorine (F), and carbon atoms—is a non-toxic gas at room temperature. It is environmentally significant because it is extremely long-lived in the atmosphere, with a typical residence time of seventy years. This long life-span permits CFCs to disperse, ultimately reaching the stratosphere 19 miles (30 kilometers) above the earth's surface. Here, high-energy photons in sunlight break

down Freon, and chlorine atoms liberated during this process participate in other chemical reactions that consume ozone (O_3). The final result is to decrease the stratospheric ozone layer that shields Earth from damaging ultraviolet radiation. Under the 1987 Montreal Protocol, thirty-one industrialized countries agreed to phase out CFC Freon production, banning their production and use as of 1996. As of 2010 more than 190 countries now observe and enforce provisions of the Montreal Protocol. Freon substitutes such as HCFC-22 (chlorodifluoromethane, $CHClF_2$) and HFC–134a contain hydrogen in the molecule, causing them to be more readily broken down and destroyed in the lower atmosphere. However, HCFC-22 still contains chlorine, which contributes to the depletion of the ozone layer, although to a lesser extent than Freon. Amendments to the Montreal Protocol in 1992 initiated the phaseout of HCFCs as part of the U.S. Clean Air Act (CAA). As of 2020, HCFC-22 will no longer be produced, but existing HCFC-22 can be used and recycled for maintenance of existing systems. The Environmental Protection Agency (EPA) lists HFC–134a and a blend of HFCs termed HFC-410a as substitutes for banned refrigerants. Freon, HCFCs, and HFCs are considered greenhouse gases and contribute to climate change.

Resources

OTHER

United Nations Environment Programme (UNEP). "Multilateral Fund Secretariat for the Implementation of the Montreal Protocol." http://www.multilateralfund.org/ (accessed September 2, 2010).
United States Environmental Protection Agency (EPA). "International Cooperation: Treaties and Agreements: Montreal Protocol." http://www.epa.gov/ebtpages/ intetreatiesandagrmontrealprotocol.html (accessed September 2, 2010).
United States Environmental Protection Agency (EPA). "Oone Layer Depletion—Regulatory Programs: What You Should Know about Refrigerents When Purchasing or Repairing a Residential A/C System or Heat Pump." http://www.epa.gov/Ozone/title6/phaseout/ 22phaseout.html (accessed August 10, 2010).

Fresh water ecology

The study of fresh water habitats is termed limnology, coming from the Greek word *limnos*, meaning pool, lake, or swamp. Fresh water habitats are normally divided into two groups: the study of standing bodies of water such as lakes and ponds (termed lentic ecosystems) and the study of rivers, streams, and other moving sources of water (termed lotic ecosystems). Another important area that should be included is fresh water wetlands.

The historical roots of limnology go back to Swiss hydrologist F. A. Forel (1841–1912), who studied Lake Geneva, Switzerland, in the late 1800s and American scientists E. A. Birge (1851–1950) and C. Juday (1871–1944), who studied lakes in Wisconsin in the early 1900s. More recently, the modern father of limnology can arguably be attributed to English-born American zoologist G. Evelyn Hutchinson (1903–1991), who taught at Yale University for more than forty years and died in 1991. Among his prolific writings are four treatises on limnology which offer the most detailed descriptions of lakes that have been published.

Fresh water ecology is an intriguing field because of the great diversity of aquatic habitats. For example, lakes can be formed in different ways: volcanic origin such as Crater Lake, Oregon; tectonic (earth movement) origin such as Lake Tahoe in California/Nevada and Lake Baikal in Siberia; glacially-derived lakes such as the Great Lakes or smaller kettle hole or cirque lakes; oxbow lakes, which form as rivers change their meandering paths; and human-created reservoirs. Aquatic habitats are strongly influenced by the surrounding watershed, and lakes in the same geographic area tend to be of the same origin and have similar water quality.

Lakes are characteristically non-homogeneous. Physical, chemical, and biological factors contribute to both horizontal and vertical zonations. For example, light penetration creates an upper photic (lighted) zone and a deeper aphotic (unlit) zone. Phytoplankton (microscopic algae such as diatoms, desmids, and filamentous algae) inhabit the photic zone and produce oxygen through photosynthesis. This creates an upper productive area termed the trophogenic (productive) zone. The deeper area where respiration prevails is termed the tropholytic (unproductive) zone. Zooplankton (microscopic invertebrates such as cladocerans, copepods, and rotifers) and nekton (free-swimming animals such as fish) inhabit both of these zones. The boundary area where oxygen produced from photosynthesis equals that consumed through respiration is referred to as the compensation depth. Nearshore areas where light penetrates to the bottom and aquatic macrophytes such as cattails and bulrushes grow are termed the littoral zone. This is typically the most productive area, and it is more pronounced in ponds than in lakes. Open water areas are termed the limnetic zone, where most plankton are found. Some species of zooplankton are more concentrated in deeper waters during the day and in greater numbers in the upper waters during the night. One explanation for this vertical migration is that these zooplankton, often large and sometimes pigmented, are avoiding visually feeding planktivorous fish. These zooplankton are thus able to feed on phytoplankton in the trophogenic zone during periods of darkness and then swim to deeper waters during daylight hours. Phytoplankton are also adapted for inhabiting the limnetic zone. Some species are quite small in size (less than 20 microns in diameter and termed nannoplankton), allowing them to be competitive at nutrient uptake due to their high surface-to-volume ratio. Other groups form large colonies, often with spines, lessening the negative impacts of herbivory and sinking. Cyanobacteria (commonly referred to as blue-green algae) produce oils that help them float on or near the water's surface. Some are able to fix atmospheric nitrogen (termed nitrogen fixation), giving them a competitive advantage in low-nitrogen conditions. Other species of cyanobacteria produce toxic chemicals, making them inedible to herbivores. There have even been reports of cattle deaths following ingestion of water with dense growths of these cyanobacteria.

Lakes can be isothermal (uniform temperature from top to bottom) during some times of the year, but during the summer months they are typically thermally stratified with an upper, warm layer termed the epilimnion (upper lake), and a colder, deeper layer termed the hypolimnion (lower lake). These zones are separated by the metalimnion, which is determined by the depths with a temperature change of more than 1.8°F (1°C) per meter depth, termed the thermocline. The summer temperature stratification creates a density gradient that effectively prevents mixing between zones.

Wind is another physical factor that influences aquatic habitats, particularly in lakes with broad surface areas exposed to the main direction of the wind, termed fetch. Strong winds can produce internal standing waves termed seiches that create a rocking motion in the water once the wind dies down. Other types of wind-generated water movements include Ekman spirals (rotation of flow direction with respect distance from a horizontal boundary) and Langmuir circulation (helical wind patterns induce lines of foam parallel to wind direction). Deep or chemically-stratified lakes that never completely mix are termed meromictic. Lakes in tropical regions that mix several times a year are termed polymictic. In regions with severe winters resulting in ice covering the surface of the lake, mixing normally occurs only during the spring and fall when the water is isothermal. These lakes are termed dimictic, and the mixing process is termed overturn. People living downwind of these lakes often notice a rotten egg smell caused by the release of hydrogen sulfide (H_2S). This gas is a product of benthic (bottom-dwelling) bacteria that inhabit the anaerobic muds of productive lakes.

Lakes that receive a low-nutrient input remain fairly unproductive, and are termed oligotrophic (low nourished). These lakes typically have low concentrations of phytoplankton, with diatoms being the main representative. Moderately productive lakes are termed mesotrophic. Eutrophic (well nourished) lakes receive more nutrient input and are therefore more productive. They are typically shallower than oligotrophic lakes and have more accumulated bottom sediments that often experience summer anoxia. These lakes have an abundance of algae, particularly cyanobacteria which are often considered a nuisance because they float on the surface and out-compete the other algae for nutrients and light.

Most lakes naturally age and become more productive over time; however, large, deep lakes may remain oligotrophic. The maturing process is termed eutrophication and it is regulated by the input of nutrients, which are needed by the algae for growth. Definitive limnological studies done in the 1970s concluded that phosphorus (P) is the key limiting nutrient in most lakes. Thus, the accelerated input of this chemical into streams, rivers, and eventually lakes by excess fertilization, sewage input (both human and animal), and erosion is termed cultural eutrophication. Much debate and research has been spent on how to slow down or control this process. One interesting management tool is termed biomanipulation, in which piscivorous (fish-eating) fish are added to lakes to consume planktivorous (plankton-eating) fish. Because the planktivores are visual feeders on the largest prey, this allows higher numbers of large zooplankton to thrive in the water, which consume more phytoplankton, particularly non-toxic cyanobacteria. A more practical approach to controlling cultural eutrophication is by limiting the nutrient loading into our bodies of water. Although this isn't easy, people must consider ways of limiting excessive uses of fertilizers, both at home and on farms, as well as more effectively regulating the release of treated sewage into rivers and lakes, particularly those which are vulnerable to eutrophication. Another lake management tool is to aerate the bottom (hypolimnetic) water so that it remains oxygenated. This keeps iron in the oxidized state (Fe^{+3}), which chemically binds with phosphate (PO_4) and prevents it from being available for algal uptake. Lakes that have anoxic bottom water keep iron in the reduced state (Fe^{+2}), and phosphate is released from the sediment into the water. Fall and spring turnover then returns this limiting nutrient to the photic zone, promoting high algal growth. When these organisms eventually die and sink to the bottom, decomposers use up more oxygen, producing a snowball effect. Thus, productive lakes can become more eutrophic with time, and may eventually develop into hypertrophic (overly productive) systems.

Lotic ecosystems differ from lakes and ponds in that currents are more of a factor and primary production inputs are generally external (allochthonous) instead of internal (autochthonous). Thus, a river or stream is considered a heterotrophic ecosystem along most of its length. Gradients in physical and chemical parameters also tend to be more horizontal than vertical in running water habitats, and organisms living in lotic ecosystems are specially adapted for surviving in these conditions. For example, trout require higher amounts of dissolved oxygen, and are primarily found in relatively cold, fast-moving water with low nutrient input. Carp are able to tolerate warmer, slower, more productive bodies of running water. Darters are fish that quickly dart back and forth behind rocks in the bottom of fast-moving streams and rivers as they feed on aquatic insects. Many of these insects are shredders and detritivores on the organic material like leaves that enter the water. Other groups specialize by scraping algae and bacteria off rocks in the water.

Recently, ecologists have begun to take a greater interest in studying fresh water wetlands. These areas are defined as being inundated or saturated by surface or ground water for most of the year, and therefore have characteristic wetland vegetation. Although some people consider these areas a nuisance and prefer them being drained, ecologists realize that they are valuable habitats for migrating water fowl. They also serve as major adsorptive areas for nutrients, which are particularly useful around sewage lagoons. Such benefits provide a strong argument for preservation of these habitats.

Climate change and pollution are threatening bodies of fresh water and their inhabitants. Global temperatures are rising as a result of accumulating greenhouse gases such as carbon dioxide (CO_2). Precipitation patterns are also being altered as a consequence of climate change and are affecting the availability of water due to induced drought. Rising temperatures can reduce the amount of dissolved oxygen in a body of water below levels necessary for survival of some aquatic organisms. Water limitations, warmer temperatures, and pollutants can negatively affect aquatic organisms, sometimes leading to death. Reducing emissions and input of agricultural chemicals can aid in preserving the fresh water habitats and their organisms.

Resources

BOOKS

Closs, Gerry, Barbara J. Downes, and Andrew J. Boulton. *Freshwater Ecology: A Scientific Introduction*. Malden, MA.: Blackwell, 2004.

Cooke, G. Dennis. *Restoration and Management of Lakes and Reservoirs*. Boca Raton, FL: CRC Press, 2005.

Dodson, Stanley I. *Introduction to Limnology*. New York: McGraw-Hill, 2005.

Graybill, George, and Benjamin Samuel Bloom. *Conservation: Fresh Water Resources*. San Diego, CA: Classroom Complete Press, 2009.

O'Sullivan, Patrick E. *Lake Restoration and Rehabilitation*. The lakes handbook, 2. Malden, MA: Blackwell, 2005.

Vincent, Warwick F., and Johanna Laybourn-Parry. *Polar Lakes and Rivers: Limnology of Arctic and Antarctic Aquatic Ecosystems*. Oxford: Oxford University Press, 2008.

Ziglio, G., Maurizio Siligardi, and Giovanna Flaim. *Biological Monitoring of Rivers: Applications and Perspectives*. Water Quality Measurements Series. Chichester, England: Wiley, 2006.

John Korstad

Friends of the Earth

Friends of the Earth (FOE) is a public interest environmental group committed to the conservation, restoration, and rational use of the environment. Founded by David Brower and other environmentalists in San Francisco in 1969, FOE works on the local, national, and international levels to prevent and reverse environmental degradation, and to promote the wise use of natural resources.

FOE has an international membership of more than two million people. Its particular areas of interest include global climate change, environmental justice, biodiversity, coal mining, coastal and ocean pollution, tropical forests destruction, groundwater contamination, corporate accountability, and food independence and sustainability. Over the years, FOE has published numerous books and reports on various topics of concern to environmentalists.

FOE was originally organized to operate internationally and now has national organizations and more than 5,000 affiliated groups in over seventy-five countries. In several of these, most notably the United Kingdom, FOE is considered to be one of the best-known and most effective public interest group concerned with environmental issues.

The organization has changed its strategies considerably over the years, and not without considerable controversy within its own ranks. Under Brower's leadership, FOE's tactics were media-oriented and often confrontational, sometimes taking the form of direct political protests, boycotts, sit-ins, marches, and demonstrations. Taking a holistic approach to the environment, the group argued that fundamental social change was required for lasting solutions to many environmental problems.

FOE eventually moved away from confrontational tactics and toward a new emphasis on lobbying and legislation, which helped provoke the departure of Brower and some of the group's more radical members. FOE began downplaying several of its more controversial stances (for example, on the control of nuclear weapons) and moved its headquarters from San Francisco to Washington, DC. More recent controversies have concerned FOE's endorsement of so-called green products and its acceptance of corporate financial contributions.

FOE remains committed, however, to most of its original goals, even if it has foresworn its earlier illegal and disruptive tactics. Relying more on the technical competence of its staff and the technical rationality of its arguments than on idealism, FOE has been highly successful in influencing legislation and in creating networks of environmental, consumer, and human rights organizations worldwide. Its publications and educational campaigns have been quite effective in raising public consciousness of many of the issues with which FOE is concerned.

Resources

ORGANIZATIONS

Friends of the Earth, 1100 15th Street NW, 11th Floor, Washington, D.C., USA, 20005, (202) 783-7400, (202) 783-0444, foe@foe.org, http://www.foe.org

Lawrence J. Biskowski

Frogs

Frogs are amphibians belonging to the order anura. The anuran group has nearly 2,700 species throughout the world and includes both frogs and toads. The word *anura* means "without a tail," and the term applies to most adult frogs. The anura are distinguished from tailed amphibians (urodeles) such as salamanders, because the latter retain a tail as an adult.

One of the most studied and best understood frogs is the northern leopard frog, *Rana pipiens*. This species is well- known to most children, to people who love the outdoors, and to scientists. Leopard frogs live throughout much of the United States as well as Canada and northern Mexico. Inhabiting a diverse array of environments, the order anura exhibits an impressive display of anatomical and behavioral variations among its members. Despite such diversity, the leopard frog is often used as a model that represents all members of the group.

Leopard frogs mate in the early spring. The frogs deposit their eggs in jelly-like masses. These soft formless clumps may be seen in temporary ponds where frogs are common. Early embryonic development occurs within the jelly mass, after which the eggs hatch, releasing small swimming tadpoles. The tadpoles feed on algal periphyton, fine organic detritus, and yolk reserves through much of the spring and early summer. Next, metamorphosis begins a few weeks or up to two years after the eggs hatch depending of the species. Metamorphosis is the process whereby amphibious tadpoles lose their gills and tails and develop arms and legs. The process of metamorphosis is complex and involves not only the loss of the tail and the development of limbs, but also a fundamental reorganization of the gut. For example, in the leopard frog, the relatively long intestine of the vegetarian tadpole is reorganized to form the short intestine of the carnivorous adult frog because nutrients are more difficult to extract from plant sources. Additionally, metamorphosis profoundly changes the method of respiration in the leopard frog. As the animal loses it gills, air-breathing lungs form and become functional. A significant portion of respiration and gas exchange will occur through the skin of the adult frog. When metamorphosis of a tadpole is complete, the result is a terrestrial, insect-eating, air-breathing frog.

Frogs are important for biological learning and research. Many students first encounter vertebrate anatomy with the dissection of an adult leopard frog. Consequently, physiologists have used frogs for the study of muscle contraction and the circulation of blood which is easily seen in the webbing between the toes of a frog. Embryologists have used frogs for study because they lay an abundance of eggs. A mature *R. pipiens* female may release 3,000 or more eggs during a single spawning. Frog eggs are relatively large and abundant, which simplifies manipulation and experimentation. Another anuran, the South African clawed frog, *Xenopus laevis*, is useful in research because it can be cultivated to metamorphosis easily and can be bred any time of year. For these reasons frogs emerge as extremely useful laboratory test animals that provide valuable information about human biology.

Frogs and biomedical research

Many biological discoveries have been made or enhanced by using frogs. For example, the role of sperm in development was studied in the late 1700s in frogs. Amphibians in general, including frogs, have significantly more deoxyribonucleic acid (DNA) per cell than do other chordates. Thus, their chromosomes are large and easy to see with a microscope. In addition, induced ovulation in frogs by pituitary injection was developed during the early 1930s. Early endocrinologists studied the role of the hormone thyroxine (also found in human beings) in vertebrate development in 1912. The role of viruses in animal and human cancer is receiving renewed interest—the first herpes virus known to cause a cancer was the frog cancer herpes virus. Furthermore, mammalian and human cloning, a controversial topic, has its foundations in the cloning of frogs. The first vertebrate ever cloned was *Rana pipiens*, in an experiment published by Thomas King and Robert Briggs in 1952. As such, experimentation with frogs has contributed greatly to biomedical research.

Unfortunately, the future of amphibians, like the northern leopard frog, appears to be in jeopardy. Most amphibians in the world are frogs and toads. Since the 1980s scientists have noted a distinct decline in amphibian populations. Worldwide, more than 200 species of amphibians have experienced recent population declines. At least thirty-two documented species extinctions have occurred. Of the 242 native North American amphibian species, the U.S. Nature Conservancy has identified three species that are presumed to be extinct, another three classified as possibly extinct, with an additional 38 percent categorized as vulnerable to extinction. According to the United States Fish and Wildlife Service, there are four frog species listed as endangered, and three listed as threatened. Occasionally, new frog species continue to be identified, including a tiny (10–12 millimeter) micro-species that can fit on the end of a pencil, which was found in Borneo in 2010.

The actual cause of the reductions in frog and amphibian populations remains unclear, but many well-supported hypotheses exist. One speculation is the run-off of chemicals that poison ponds producing defective frogs that cannot survive well. A second possibility is an increase in amphibian disease caused by virulent pathogens. Viral and fungal infection of frogs has led to recent declines in many populations. A third explanation involves parasitic infections of frogs with flatworms, causing decreased survival. Another

Energy Initiative legislation supported the development of this technology and aimed for the production of commercial hydrogen-powered vehicles by the year 2020. The U.S. government supplied over $1 billion to fuel cell development as of 2008. In 2009 the Obama administration decided to stop providing funds for hydrogen fuel cell development arguing that hydrogen fuel cell powered automobiles will not be practical in ten to twenty years. Nevertheless, hydrogen fuel cell uses are increasingly common, and fuel cell buses can be seen in many cities. In 2008 an automobile manufacturer released a hydrogen fuel cell car (Honda FCX Clarity), while the German navy's Type 212 class submarine uses a fuel cell propulsion system, which allows it to be submerged for up to three weeks. Development of and innovations in hydrogen fuel cell technologies will continue, but the U.S. government is primarily interested in technologies that show promise for reducing greenhouse gas emissions in the short-term. As of 2009 Daimler, Honda, Nissan, Hyundai, Ford, General Motors, Kia, Renault, and Toyota had announced intentions to produce hydrogen fuel-cell vehicles in the future.

Resources

BOOKS

Busby, Rebecca L. *Hydrogen and Fuel Cells: A Comprehensive Guide*. Tulsa, OK: PennWell Corporation, 2005.

Li, Xianguo. *Principles of Fuel Cells*. New York: Taylor & Francis, 2006.

Press, Roman J. *Introduction to Hydrogen Technology*. Hoboken, NJ: John Wiley, 2009.

Sorenson, Brent. *Hydrogen and Fuel Cells: Emerging Technologies and Applications*. San Diego, CA: Academic Press, 2005.

PERIODICALS

Kennedy, Donald. "The Hydrogen Solution." *Science* 305 (2004): 917.

Turner, John A. "Sustainable Hydrogen Production." *Science* 305 (2004): 972-974.

Wald, Matthew L. "U.S. Drops Research into Fuel Cells for Cars." *New York Times* (May 7, 2009).

OTHER

Edmunds.com. "9 Major Automakers Sign Letter Agreeing to Develop and Launch Fuel-Cell Vehicles." http://blogs.edmunds.com/greencaradvisor/2009/09/9-major-automakers-sign-letter-agreeing-to-develop-and-launch-fuel-cell-vehicles.html (accessed October 27, 2010).

U.S. Department of Energy. "Energy Secretary Highlights Hydrogen Fuel Initiative In Western New York." http://www.energy.gov/news/archives/3256.htm (accessed October 26, 2010).

U.S. Department of Energy. "Solid State Energy Conversion Alliance (SECA)." http://www.netl.doe.gov/technologies/coalpower/fuelcells/seca/ (accessed October 27, 2010).

U.S. Department of Energy. "Tubular Solid Oxide Fuel Cell Technology." http://www.fossil.energy.gov/programs/powersystems/fuelcells/fuelcells_solidoxide.html (accessed October 27, 2010).

ORGANIZATIONS

Argonne National Laboratory, 9700 S. Cass Avenue, Argonne, IL, USA, 60439, (630) 252–2000, http://www.anl.gov

Fuel Cells 2000, 1100 H Street NW, Suite 800, Washington, D.C., USA, 20005, (202) 785-4222, http://www.fuelcells.org

U.S. DOE Energy Efficiency and Renewable Energy, 1100 H Street NW, Suite 800, http://www.eere.energy.gov/

David E Newton
Marie H. Bundy

Fuel switching

Fuel switching is the substitution of one energy source for another to meet requirements for heat, power, or electrical generation. Generally this term refers to the practices of some industries that can substitute among natural gas, electricity, coal, and LPG within thirty days without modifying their fuel-consuming equipment and that can resume the same level of production following the change. Price is the primary reason for fuel switching; however, additional factors may include environmental regulations, agreements with energy or fuel suppliers, and equipment capabilities.

Fugitive emissions

Contaminants that enter the air without passing through a smokestack and, thus, are often not subject to control by conventional emission control equipment or techniques. Most fugitive emissions are caused by activities involving the production of dust, such as soil erosion and strip mining, building demolition, or the use of volatile compounds. In a steel-making complex, for example, there are several identifiable smokestacks from which emissions come, but there are also numerous sources of fugitive emissions, which escape into the air as a result of processes such as producing coke, for which there is no identifiable smokestack. The control of fugitive emissions is generally much more complicated and costly than the control of smokestack emissions for which known add-on technologies to the smokestack have been developed. Baghouses and other costly mechanisms typically are needed to control fugitive emissions.

Fumigation

Most commonly, fumigation refers to the process of disinfecting a material or an area by using some type of toxic material in gaseous form. The term has a more specialized meaning in environmental science, where it refers to the process by which pollutants are mixed in the atmosphere. Under certain conditions, emitted pollutants rise above a stable layer of air near the ground. These pollutants remain aloft until convective currents develop, often in the morning, at which time the cooler pollutants trade places with air at ground level as it is warmed by the sun and rises. The resulting damage to ecosystems from the pollutants is most obvious around metal smelters.

Fund for Animals

Founded in 1967 by author and humorist Cleveland Amory, the Fund for Animals is one of the most activist of the national animal protection groups. Formed "to speak for those who can't," it has led sometimes militant campaigns against sport hunting, trapping, and wearing furs, as well as the killing of whales, seals, bears, and other creatures. Amory, in particular, has campaigned tirelessly against these activities on television and radio, and in lectures, articles, and books.

In the early 1970s the Fund worked effectively to rally public opinion in favor of passage of the Marine Mammal Protection Act, which was signed into law in October 1972. This act provides strong protection for whales, seals and sea lions, dolphins, sea otters, polar bears, and other ocean mammals. In 1978 the Fund bought a British trawler and renamed it *Sea Shepherd*. Under the direction of its captain, Paul Watson, they used the ship to interfere with the baby seal kill on the ice floes off Canada. Activists sprayed some 1,000 baby harp seals with a harmless dye that destroyed the commercial value of their white coats as fur, and the ensuing publicity helped generate worldwide opposition to the seal kill and a ban on imports into Europe. In 1979 *Sea Shepherd* hunted down and rammed *Sierra*, an outlaw whaling vessel that was illegally killing protected and endangered species of whales. After *Sea Shepherd* was seized by Portuguese authorities, Watson and his crew scuttled the ship to prevent it from being given to the owners of *Sierra* for use as a whaler.

Also in 1979 the Fund used helicopters to airlift from the Grand Canyon almost 600 wild burros that were scheduled to be shot by the National Park Service.

The airlift was so successful and generated so much favorable publicity, that it led to similar rescues of feral animals on public lands that the government wanted removed to prevent damage to vegetation. Burros were also airlifted by the Fund from Death Valley National Monument, as were some 3,000 wild goats on San Clemente Island, off the coast of California, scheduled to be shot by the United States Navy.

Many of the wild horses, burros, goats, and other animals rescued by the Fund end up, at least temporarily before adoption, at Black Beauty Ranch, a 1430-acre (578-ha) sanctuary near Arlington, Texas. The ranch has provided a home for abused race and show horses, a non-performing elephant, and Nim, the famous signing chimpanzee who was saved from a medical laboratory.

Legal action initiated by the Fund has resulted in the addition of almost 200 species to the U.S. Department of the Interior's list of threatened and endangered species, including the grizzly bear, the Mexican wolf, the Asian elephant, and several species of kangaroos. The Fund is also active on the grassroots level, working on measures to restrict hunting and trapping.

Resources

ORGANIZATIONS

The Fund for Animals, 200 West 57th Street, New York, NY, USA, 10019, (212) 246-2096, (212) 246-2633, info@ fundforanimals.org, http://www.fundforanimals.org

Lewis G. Regenstein

Fungi

Fungi are broadly characterized as cells that possess nuclei and rigid cell walls but lack chlorophyll. Fungal spores germinate and grow slender tube like structures called *hyphae*, separated by cell walls called *septae*. The vegetative biomass of most fungi in nature consists of masses of hyphae, or *mycelia*. Most species of fungi inhabit soil, where they are active in the decomposition of organic matter. The biologically most complex fungi periodically form spore-producing fruiting structures, known as mushrooms. Some fungi occur in close associations, known as *mycorrhizae*, with the roots of many species of vascular plants. The plant benefits mostly through an enhancement of nutrient uptake, while the fungus benefits through access to metabolites. Certain fungi are also partners in the symbioses with algae known as lichens.

See also Fungicide.

Fungicide

A fungicide is a chemical that kills fungi, the tiny plant-like organisms that obtain their nourishment from dead or living organic matter. Examples of fungi include mushrooms, toadstools, smuts, molds, rusts, and mildew.

Fungicides can be important in protecting crops from the growth of fungi that cause disease (pathogens). The first known fungicide was the naturally occurring substance sulfur. One of the most effective of all fungicides, Bordeaux mixture, was invented in 1885. Bordeaux mixture is a combination of two inorganic compounds, copper sulfate and lime.

With the growth of the chemical industry during the twentieth century, researchers invented synthetic fungicides. Ferbam, ziram, captan, naban, dithiocarbonate, quinone, and 8-hydroxyquinoline are examples of synthetic fungicides.

Compounds containing mercury and cadmium were popular fungicides. Methylmercury was widely used by farmers in the United States to protect growing plants and treat stored grains.

During the 1970s, however, evidence of a number of adverse effects of mercury- and cadmium-based fungicides began to emerge. The most serious effects were observed among birds and small animals who were exposed to sprays and dusting, or who ate treated grain. A few dramatic incidents of methylmercury poisoning among humans, however, were also recorded. The best known of these was the 1953 disaster at Minamata Bay, Japan, in which residents of the area developed nervous disorders ultimately linked to methylmercury in fish they had The fungicides affected their ability to walk, caused mental disorders, and, in some, caused permanently disability.

The problems with mercury and cadmium compounds spurred the development of less toxic substitutes for the more dangerous fungicides. Dinocap, binapacryl, and benomyl are examples of less toxic compounds.

Another approach has been to use integrated pest management and to develop plants that are resistant to fungi. The latter approach was used with great success during the corn blight disaster of 1970. Researchers worked quickly to develop strains of corn that were resistant to the corn-leaf blight fungus and by 1971 had provided farmers with seeds of the new strain.

Research has also identified natural fungicides, which include tea tree oil, cinnamaldehyde, neem oil, and rosemary oil. The bacterium *Bacillus subtilis* can also function as a fungicide.

A problem with many fungicides is the development of resistance by the target organism, similar to the development of resistance by disease causing bacteria to antibiotics. If the resistance arises because of a genetic change in the target organism, the resistance can be passed on to subsequent generations. As with antibiotics, a particular fungicide may need to be replaced with another as resistance develops.

See also Minamata disease.

Resources

BOOKS

Conklin, Alfred R., and Thomas Stilwell. *World Food: Production and Use*. New York: Wiley Interscience, 2007.

Deppe, Carol. *The Resilient Gardener: Food Production and Self-Reliance in Uncertain Times*. White River Junction, VT: Chlesea Green Publishing, 2010.

Sleper, D.A. *Breeding Food Crops*. New York: Wiley-Blackwell, 2006.

David E. Newton

Farmer spraying carrots with fungicide. (© *IndexStock / SuperStock*)

Furans

Furans are by-products of natural and industrial processes and are considered environmental pollutants. They are chemical substances found in small amounts in the environment, including air, water, and soil, and are also present in some foods. Of particular concern are the polychlorinated dibenzo furans (PCDFs), which although present in small amounts, are persistent and remain in the environment for long periods of time, accumulating in the food chain. The U.S. Environmental Protection Agency's (EPA) Persistent Bioaccumulative and Toxic (PBT) Chemical Program classifies furans as priority PBTs.

Furans belong to a class of organic compounds known as heterocyclic aromatic hydrocarbons. The basic furan structure is a five-membered ring consisting of four atoms of carbon and one oxygen. Various types of furans have additional atoms and rings attached to the basic furan structure. Some furans are used as solvents or as raw materials for synthesizing chemicals.

PCDFs are considered environmental pollutants of significant concern. These furans consist of three-ringed structures, with two rings of six carbon atoms each (benzene rings) attached to the furan. Between one and eight chlorine (Cl) atoms are attached to the rings. There are more than 135 types of PCDFs, whose properties are determined by the number and position of the chlorine atoms. PCDFs are closely related to polychlorinated dioxins (PCDDs) and polychlorinated biphenyls (PCBs). These three types of toxic compounds often occur together and PCDFs are major contaminants of manufactured PCBs. Production of PCBs ceased in 1977, but PCDDs and PCDFs are by-products of combustion and industrial processes. In fact, the term dioxin commonly refers to a subset of these compounds that have similar chemical structures and toxic mechanisms. This subset includes ten of the PCDFs, as well as seven of the PCDDs and twelve of the PCBs. Less frequently, the term dioxin is used to refer to all 210 structurally-related PCDFs and PCDDs, regardless of their toxicities.

PCDFs are present as impurities in various industrial chemicals and are trace byproducts of most types of combustion, including the incineration of chemical, industrial, medical, and municipal waste; the burning of wood, coal, and peat; and automobile emissions. Thus most PCDFs are released into the environment through smokestacks. However the backyard burning of common household trash in barrels has been identified as potentially one of the largest sources of dioxin and furan emissions in the United States. Because of the lower temperatures and inefficient combustion in burn barrels, burning releases more PCDFs than municipal incinerators. Some industrial chemical processes, including chlorine bleaching in pulp and paper mills, also produce PCDFs.

PCDFs that are released into the air can be carried by currents to all parts of the globe. Eventually they fall to earth and are deposited in soil, sediments, and surface water. Although furans are slow to volatilize and have a low solubility in water, they can wash from soils into bodies of water, evaporate, and be re deposited elsewhere. Furans have been detected in soils, surface waters, sediments, plants, and animals throughout the world, even in arctic organisms. They are very resistant to both chemical breakdown and biological degradation by microorganisms.

Most people have low, but detectable, levels of PCDDs and PCDFs in their tissues. Furans enter the food chain from soil, water, and plants. They bioaccumulate (build up in living tissues) at the higher levels of the food chain, particularly in fish and animal fat. The concentrations of PCDDs and PCDFs may be hundreds or thousands of times higher in aquatic organisms than in the surrounding waters. Most humans are exposed to furans through animal fat, milk, eggs, and fish. Some of the highest levels of furans are found in human breast milk. The presence of dioxins and furans in breast milk can lead to the development of soft, discolored molars in young children. Industrial workers can be exposed to furans while handling chemicals or during industrial accidents.

PCDFs bind to aromatic hydrocarbon receptors in cells throughout the body, causing a wide range of deleterious effects, including developmental defects in fetuses, infants, and children. Furans also may adversely affect the reproductive and immune systems. At high exposure levels, furans can cause chloracne, a serious acne-like skin condition. Furan itself, as well as PCDFs, are potential cancer-causing agents.

The switch from leaded to unleaded gasoline, the halting of PCB production in 1977, changes in paper manufacturing processes, and new air and water pollution controls, have reduced the emissions of furans. Between 1987 and 1995 there was an over 75 percent reduction in dioxin and furan release from known industrial sources in the United States due to regulatory actions as well as voluntary efforts by U.S. industry. In 1999 the United States, Canada, and Mexico agreed to cooperate to further reduce the release of dioxins and furans. As a requirement of the Clean Air Act (CAA), the "maximum achieveable control technology" must be employed with regard to dioxin and furan release. Sources of dioxin and furan regulated by the CAA are pulp and paper manufacturing; municipal, medical, and hazardous waste incineration; and specific metals production and refining processes.

In May 2001 American EPA Administrator Christie Whitman (1946–), along with representatives from more than ninety other countries at the Stockholm Convention, signed the global treaty on Persistent Organic Pollutants (POPs). The treaty phases out the manufacture and use of twelve manmade toxic chemicals—the so-called dirty dozen—that persist in the environment for long periods of time without degrading. The goal is to reduce or eliminate the further release of POPs into the environment. The United States opposed a complete ban on furans and dioxins; thus, unlike eight of the other chemicals that were banned outright, the treaty calls for the use of dioxins and furans to be minimized and eliminated where feasible.

Resources

BOOKS

Devotta, Sukumar. *Dioxins and Furans: Unintentional By-products of Chlorine Base Activities.* Dordrecht, Netherlands: Springer, 2007.

Lippmann, Morton. *Environmental Toxicants: Human Exposures and Their Health Effects.* Hoboken, NJ: John Wiley & Sons, 2009.

Mommsen, T. P., and T. W. Moon. *Environmental Toxicology.* Biochemistry and molecular biology of fishes, 6. Amsterdam: Elsevier, 2005.

National Research Council (U.S.). *Human Biomonitoring for Environmental Chemicals.* Washington, DC: National Academies Press, 2006.

OTHER

United States Environmental Protection Agency (EPA). "Pollutants/Toxics: Chemicals: Furans." http://www.epa.gov/ebtpages/pollchemicfurans.html (accessed October 12, 2010).

ORGANIZATIONS

Clean Water Action Council, 1270 Main Street, Suite 120, Green Bay, WI, USA, 54302, (920) 437-7304, (920) 437-7326, CleanWater@cwac.net, http://www.cwac.net/index.html

United Nations Environment Programme: Chemicals, 11–13, chemin des Ane;aamones, 1219 Cháftelaine, Geneva, Switzerland, opereira@ unep.ch, http://www.chem.unep.ch

United States Environmental Protection Agency, 1200 Pennsylvania Avenue, NW, Washington, DC, USA, 20460, (800) 490-9198, public-access@epa.gov, http://www.epa.gov

Margaret Alic

Fusion *see* **Nuclear fusion.**

Future generations

According to demographers, a generation is an age-cohort of people born, living, and dying within a few years of each other. Human generations are roughly defined categories, and the demarcations are not as distinct as they are in many other species. As the Scottish philosopher David Hume noted in the eighteenth century, generations of human beings are not like generations of butterflies, who come into existence, lay their eggs, and die at about the same time, with the next generation hatching thereafter. But distinctions can still be made, and future generations are all age-cohorts of human beings who have not yet been born.

The concept of future generations is central to environmental ethics and environmental policy, because the health and well-being of human beings depends on how people living today care for the natural environment.

Proper stewardship of the environment affects not only the health and well-being of people in the future but also their character and identity. In *The Economy of the Earth*, Mark Sagoff compares environmental damage to the loss of our rich cultural heritage. The loss of all our art and literature would deprive future generations of the benefits other generations have enjoyed and render them nearly illiterate. By the same token, if humankind destroyed all the wildernesses and dammed all the rivers, allowing environmental degradation to proceed at the same pace, it would do more than deprive people of the pleasures others have known. It would make them into what Sagoff calls "environmental illiterates," or "yahoos" who would neither know nor wish to experience the beauties and pleasures of the natural world. "A pack of yahoos," says Sagoff, "will like a junkyard environment" because they will have known nothing better.

The concept of future generations emphasizes both our ethical and aesthetic obligations to our environment. In relations between existing and future generations, however, the present generation holds all the power. While it is possible to affect them, they can do nothing to affect the past. Though, as some environmental philosophies have argued, moral code is in large degree based on reciprocity, the relationship between generations cannot be reciprocal. Adages such as "like for like," and "an eye for an eye," can apply only among contemporaries. Because an adequate environmental ethic would require that moral consideration be extended to include future people, views of justice based on the norm of reciprocity may be inadequate.

A good deal of discussion has gone into what an alternative environmental ethic might look like and on what it might be based. But perhaps the important point to note is that the treatment of generations yet unborn has now become a lively topic of philosophical discussion and political debate.

See also Environmental education; Environmentalism; Intergenerational justice.

Resources

BOOKS

Sagoff, M. *The Economy of the Earth: Philosophy, Law and the Environment*. Cambridge and New York: Cambridge University Press, 1988.

Terence Ball

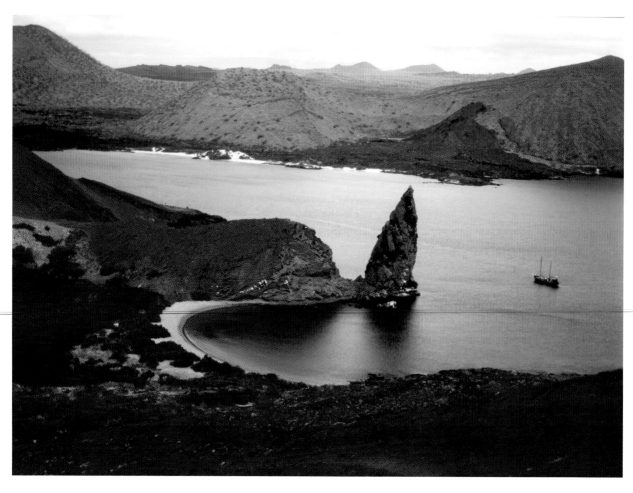

Coastline in the Galápagos Islands. *(Dr. Morley Reed/Shutterstock.com)*

prime example, a group of birds known as Darwin's finches. Charles Darwin discovered and collected specimens of these birds from the Galápagos Islands in 1835 on his five-year voyage around the world aboard the HMS *Beagle*. His cumulative experiences, copious notes, and vast collections ultimately led to the publication of his monumental work, *On the Origin of Species*, in 1859. The Galápagos Islands and their unique assemblage of plants and animals were an instrumental part of the development of Darwin's evolutionary theory.

The Galápagos Islands are located at 90° W longitude and 0° latitude (the equator), about 600 miles (965 km) west of Ecuador. These islands are volcanic in origin and are about 10 million years old. The original colonization of the Galápagos Islands occurred by chance transport over the ocean as indicated by the gaps in the flora and fauna of this archipelago compared to the mainland. Of the hundreds of species of birds along the northwestern South American coast, only seven species colonized the Galápagos

Islands. These evolved into 57 resident species, 26 of which are endemic to the islands, through adaptive radiation. The only native land mammals are a rat and a bat. The land reptiles include iguanas, a single species each of snake, lizard, and gecko, and the Galápagos giant tortoise (*Geochelone elephantopus*). No amphibians and few insects or mollusks are found in the Galápagos. The flora has large gaps as well—no conifers or palms have colonized these islands. Many of the open niches have been filled by the colonizing groups. The tortoises and iguanas are large and have filled niches normally occupied by mammalian herbivores. Several plants, such as the prickly pear cactus, have attained large size and occupy the ecological position of tall trees.

The most widely known and often used example of adaptive radiation is Darwin's finches, a group of fourteen species of birds that arose from a single ancestor in the Galápagos Islands. These birds have specialized on different islands or into niches normally filled by other groups of birds. Some are strictly seed eaters, while others have evolved more warbler-like bills and eat insects, still others eat flowers, fruit, and/or nectar, and others find insects for their diet by digging under the bark of trees, having filled the niche of the woodpecker. Darwin's finches are named in honor of their discoverer, but they are not referred to as Galápagos finches because there is one of their numbers that has colonized Cocos Island, located 425 miles (684 km) north-northeast of the Galápagos.

Because of the Galápagos Islands' unique ecology, scenic beauty and tropical climate, they have become a mecca for tourists and some settlement. These human activities have introduced a host of environmental problems, including introduced species of goats, pigs, rats, dogs, and cats, many of which become feral and damage or destroy nesting bird colonies by preying on the adults, young, or eggs. Several races of giant tortoise have been extirpated or are severely threatened with extinction, primarily due to exploitation for food by humans, destruction of their food resources by goats, or predation of their hatchlings by feral animals. Most of the thirteen recognized races of tortoise have populations numbering only in the hundreds. Three races are tenuously maintaining populations in the thousands, one race has not been seen since 1906, but it is thought to have disappeared due to natural causes, another race has a population of about twenty-five individuals, and the Pinta Island tortoise is represented today by only one individual, "Lonesome George," a captive male at the Charles Darwin Biological Station. For most of these tortoises to survive, an active capture or extermination program of the feral animals will have to continue. One other potential threat to the Galápagos Islands is tourism. Thousands of tourists visit these islands each year and their numbers can exceed the limit deemed sustainable by the Ecuadorian government. These tourists have had, and will continue to have, an impact on the fragile habitats of the Galápagos.

See also Ecotourism; Endemic species.

Resources

BOOKS

Kricher, John C. *Galápagos: A Natural History*. Princeton: Princeton University Press, 2006.

Nicholls, Henry. *Lonesome George: The Life and Loves of a Conservation Icon*. London: Macmillan, 2006.

Pons, Alain, and Christine Baillet. *Galápagos*. London: Evans Mitchell, 2007.

Bassett, Carol Ann. *Galápagos at the Crossroads: Pirates, Biologists, Tourists and Creationists Battle for Darwin's Cradle of Evolution*. Washington, D.C.: National Geographic, 2009.

De Roy, Tui. *Galápagos: Preserving Darwin's Legacy*. Richmond Hill, Ont: Firefly Books, 2009.

Eugene C. Beckham

Galdikas, Birute M.

1948–
Lithuanian/Canadian primatologist

One of the world's leading expert on orangutans, Birute Galdikas has dedicated much of her life to studying the orangutans of Indonesia's Borneo and Sumatra islands. Her work, which has complemented that of such other scientists as Dian Fossey and Jane Goodall, has led to a much greater understanding of the primate world and more effective efforts to protect orangutans from the effects of human infringement. Galdikas has also been credited with providing valuable insights into human culture through her decades of work with primates. She discusses this aspect of her work in her 1995 autobiography, *Reflections of Eden: My Years with the Orangutans of Borneo*.

Galdikas was born on May 10, 1948 in Wiesbaden in what was then West Germany, while her family was en route from their native Lithuania. She was the first of four children. The family moved to Toronto,

Primatologist Birute M. Galdikas was sent by anthropologist Louis Leakey to Asia and Africa to study great apes. *(Irwin Fedriansyah)*

Canada when she was two, and she grew up in that city. As a child, Galdikas was already enamored of the natural world, and spent much of her time in local parks and reading books on jungles and their inhabitants. She was already especially interested in orangutans. The Galdikas family eventually moved to Los Angeles, where Birute attended the local campus of the University of California. She earned a BA there in 1966 and immediately began work on a master's degree in anthropology. Galdikas had already decided to start a long-term study of orangutans in the rain forests of Indonesia, where most of the world's last remaining wild orangutans live.

Galdikas began to realize her dream in 1969 when she approached famed paleoanthropologist Louis Leakey after he gave a lecture at UCLA. Leakey had helped launch the research efforts of Fossey and Goodall, and she asked him to do the same for her. He agreed, and by 1971 had helped her raise enough money to get started. With her first husband, Galdikas traveled to southern Borneo's Tanjung Puting National Park in East Kalimantan to start setting up her research station. Challenges such as huge leeches, extremely toxic plants, perpetual dampness, swarms of insects, and aggressive viruses slowed Galdikas down, but did not ruin her enthusiasm for her new project.

After finally locating the area's elusive orangutan population, Galdikas faced the difficulty of getting the shy animals accustomed enough to her presence that they would permit her to watch them even from a distance.

Once Galdikas accomplished this, she was able to begin documenting some of the traits and habits of the little-studied orangutans. She compiled a detailed list of staples in the animals' diets, discovered that they occasionally eat meat, and recorded their complex behavioral interactions.

Eventually, the animals came to accept Galdikas and her husband so thoroughly that their camp was often overrun by them. Galdikas recalled in a 1980 National Geographic article that she sometimes felt as though she were "surrounded by wild, unruly children in orange suits who had not yet learned their manners." Meanwhile, she applied her findings to her UCLA education, earning both her master's degree and doctorate in 1978.

During her first decade on Borneo, Galdikas founded the Orangutan Project, which has since been funded by such organizations as the National Geographic Society, the World Wildlife Fund, and Earthwatch. The Project not only carries out primate research, but also rehabilitates hundreds of former captive orangutans. She also founded the Los Angeles-based nonprofit Orangutan Foundation International in 1987.

From 1996 to 1998, Galdikas served as a senior adviser to the Indonesian Forestry Ministry on orangutan issues as that government attempted to rectify the mistreatment of the animals and the mismanagement of their dwindling rain forest habitat. As part of these efforts, the Jakarta government also helped Galdikas establish the Orangutan Care Center and Quarantine near Pangkalan Bun, which opened in 1999. This center has since cared for many of the primates injured or displaced by the devastating fires in the Borneo rain forest in 1997–1998.

Divorced in 1979, Galdikas married a native Indonesian man of the Dayak tribe in 1981. She has one son with her first husband and two children with her second. Galdikas and her second husband. Galdikas is a professor at Simon Fraser University in Canada and a professor extraordinaire at Universitas Nasional in Indonesia.

Acolades she has received include the PETA Humanitarium Award (1990), United Nations Global 500 Award (1993), and the Tyler Prize for Environmental Achievement (1997). She is also an Officer of the Order of Canada.

In the late 1990s, questions were raised regarding the legality and treatment of orangutans at her home in Indonesia. Charges were never laid and Galdikas publicly declared the incident to be a malicious attempt by critics to discredit her and her work.

Resources

BOOKS

Galdikas, Birute. *Reflections of Eden: My Years with the Orangutans of Borneo*. New York: Little, Brown, 1995.

Game animal

Birds and mammals commonly hunted for sport. The major groups include upland game birds (quail, pheasant, and partridge), waterfowl (ducks and geese), and big game (deer, antelope, and bears). Game animals are protected to varying degrees throughout most of the world, and hunting levels are regulated through the licensing of hunters as well as by seasons and bag limits. In the United States, state wildlife agencies assume primary responsibility for enforcing hunting regulations, particularly for resident or non-migratory species. The Fish and Wildlife Service shares responsibility with state agencies for regulating harvests of migratory game animals, principally waterfowl.

Game preserves

Game preserves (also known as game reserves, or wildlife refuges) are a type of protected area in which hunting of certain species of animals is not allowed, although other kinds of resource harvesting may be permitted. Game preserves are usually established to conserve populations of larger game species of mammals or waterfowl. The protection from hunting allows the hunted species to maintain relatively large populations within the sanctuary. However, animals may be legally hunted when they move outside of the reserve during their seasonal migrations or when searching for additional habitat.

Game preserves help to ensure that populations of hunted species do not become depleted through excessive harvesting throughout their range. This conservation allows the species to be exploited in a sustainable fashion over the larger landscape.

Although hunting is not allowed, other types of resource extraction may be permitted in game reserves, such as timber harvesting, livestock grazing, some types of cultivated agriculture, mining, and exploration and extraction of fossil fuels. However, these land-uses are managed to ensure that the habitat of game species is not excessively damaged. Some game preserves are managed as true ecological reserves, where no extraction of natural resources is allowed. However, low-intensity types of land-use may be permitted in these more comprehensively protected areas, particularly non-consumptive recreation such as hiking and wildlife viewing.

Game preserves as a tool in conservation

The term "conservation" refers to the sustainable use of natural resources. Conservation is particularly

Zebras and giraffes in the Masai-Mara Game Reserve, Kenya. (© Cusp / SuperStock)

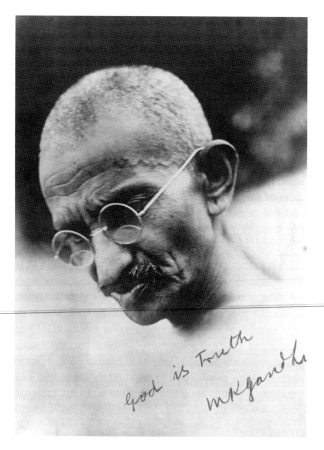

A portrait of Mohandas Gandhi, called Mahatma ("great-souled") by his followers. *(© Hulton-Deutsch Collection/CORBIS)*

would allow India to be both self-sustaining and egalitarian. He did not believe that an independent economy in India could be based on the Western model; he considered a consumer economy of unlimited growth impossible in his country because of the huge population base and the high level of poverty. He argued instead for the development of an economy based on the careful use of indigenous natural resources. His was a philosophy of conservation, and he advocated a lifestyle based on limited consumption, sustainable agriculture, and the utilization of labor resources instead of imported technological development.

Gandhi's plans for India's future were firmly rooted both in moral principles and in a practical recognition of its economic strengths and weaknesses. He believed that the key to an independent national economy and a national sense of identity was not only indigenous resources but indigenous products and industries. Gandhi made a point of wearing only homespun, undyed cotton clothing that had been hand-woven on cottage looms. He anticipated that the practice of wearing homespun cotton cloth would create an industry for a product that had a ready market, for cotton was a resource that was both indigenous and renewable. He recognized that India's major economic strength was its vast labor pool, and the low level of technology needed for this product would encourage the development of an industry that was highly decentralized. It could provide employment without encouraging mass migration from rural to urban areas, thus stabilizing rural economies and national demography. The use of cotton textiles would also prevent dependence on expensive synthetic fabrics that had to be imported from Western nations, consuming scarce foreign exchange. He also believed that synthetic textiles were not suited to India's climate, and that they created an undesirable distinction between the upper classes that could afford them and the vast majority that could not.

The essence of his economic planning was a philosophical commitment to living a simple lifestyle based on need. He believed it was immoral to kill animals for food and advocated vegetarianism; advocated walking and other simple forms of transportation contending that India could not afford a car for every individual; and advocated the integration of ethical, political, and economic principles into individual lifestyles. Although many of his political tactics, particularly his strategy of civil disobedience, have been widely embraced in many countries, his economic philosophies have had a diminishing influence in a modern, independent India, which has been pursuing sophisticated technologies and a place in the global economy. But to some, his work seems increasingly relevant to a world with limited resources and a rapidly growing population.

Resources

BOOKS

Gandhi, and Gandhi. *Hind Swaraj, or, Indian Home Rule.* Memphis, Tenn: General Books, 2010.

Gandhi, Rajmohan. *Gandhi: The Man, His People, and the Empire.* Berkeley: University of California Press, 2008.

Gandhi, and Judith M. Brown. *Mahatma Gandhi: The Essential Writings.* Oxford: Oxford University Press, 2008.

Usha Vedagiri
Douglas Smith

Garbage

In 2007, the United States generated over 254 million tons of municipal solid waste, compared with 195 million tons in 1990, according to Environmental Protection Agency (EPA) estimates. On average, each person generated 4.6 pounds (2.1 kg) of such waste per day in 2007, and the EPA expects that amount to continue increase. That waste includes cans, bottles, newspapers, paper and plastic packages, uneaten food, broken furniture and appliances, old tires, lawn clippings, and other refuse. This waste can be placed in landfills, incinerated, recycled, or in some cases composted.

Landfilling—waste disposed of on land in a series of layers that are compacted and covered, usually with soil—is the main method of waste management in this country, accounting for about 54 percent of the waste. But old landfills are being closed and new ones are hard to site because of community opposition. Landfills once were open dumps, causing unsanitary conditions, methane explosions, and releases of hazardous chemicals into groundwater and air. Old dumps make up 22 percent of the sites on the Superfund National Priorities List. Today, landfills must have liners, gas collection systems, and other controls mandated under Subtitle D of the Resource Conservation and Recovery Act (RCRA).

Incineration has been popular among solid waste managers because it helps to destroy bacteria and toxic chemicals and to reduce the volume of waste. But public opposition, based on fears that toxic metals and other chemical emissions will be released from incinerators, has made the siting of new facilities extremely difficult. In the past, garbage burning was done in open fields, in dumps, or in backyard drums, but the Clean Air Act (1970) banned open burning, leading to new types of incinerators, most of which are designed to generate energy.

Recycling, which consists of collecting materials from waste streams, preparing them for market, and using those materials to manufacture new products, is catching national attention as a desirable waste management method. All states and the District of Columbia have some type of statewide recycling law aimed at promoting greater recycling of glass, paper, metals, plastics, and other materials. Used oil, household batteries, and lead-acid automotive batteries are recyclable waste items of particular concern because of their toxic constituents.

Composting is a waste management approach that relies on heat and microorganisms—mostly bacteria and fungi—to decompose yard wastes and food scraps, turning them into a nutrient-rich mix called humus or compost. This mix can be used as fertilizer. However, as with landfills and incinerators, composting facilities have been difficult to site because of community opposition, in part because of the disagreeable smell generated by some composting practices.

Recently, waste managers have shown interest in source reduction, reducing either the amount of garbage generated in the first place or the toxic ingredients of garbage. Reusable blankets instead of throw-away cardboard packaging for protecting furniture is one example of source reduction. Businesses are regarded as a prime target for source reduction, such as implementing double-sided photocopying to save paper, because the approach offers potentially large cost savings to companies.

Resources

BOOKS

Applegate, J.S., and J.G. Laitos. *Environmental Law: RCRA, CERCLA, and the Management of Hazardous Waste.* New York: Foundation Press, 2006.

Pichtel, John. *Waste Management Practices.* Boca Raton: CRC, 2005.

OTHER

United States Environmental Protection Agency (EPA). "Wastes: Solid Waste–Nonhazardous." http://www.epa.gov/ebtpages/wastsolidwaste.html (accessed November 11, 2010).

United States Environmental Protection Agency (EPA). "Wastes: Solid Waste–Nonhazardous: Composting." http://www.epa.gov/ebtpages/wastsolidwastecomposting.html (accessed November 11, 2010).

United States Environmental Protection Agency (EPA). "Wastes: Solid Waste–Nonhazardous: Household Waste." http://www.epa.gov/ebtpages/wastsolidwastehouseholdwaste.html (accessed November 11, 2010).

United States Environmental Protection Agency (EPA). "Wastes: Solid Waste–Nonhazardous: Municipal Solid Waste." http://www.epa.gov/ebtpages/wastsolidwastemunicipalsolidwaste.html (accessed November 11, 2010).

United States Environmental Protection Agency (EPA). "Wastes: Waste Disposal: Landfills." http://www.epa.gov/ebtpages/wastwastedisposallandfills.html (accessed November 11, 2010).

David Clarke

Garbage Project

The Garbage Project was founded in 1973, shortly after the first Earth Day, by William Rathje, professor of anthropology, and fellow archaeologists at the

University of Arizona. The objective was to apply the techniques and tools of their science to the study of modern civilization by analyzing its garbage.

Using sample analysis and assessing biodegradation, they also hoped to increase their understanding of resource depletion and environmental and landfill-related problems. Because it requires sunlight, moisture, and oxygen, as well as organic material and bacteria, little biodegradation actually takes place in landfills, resulting in perfectly preserved heads of lettuce, forty-year-old hot dogs, and completely legible fifty-year-old newspapers.

In *Rubbish: The Archaeology of Garbage*, published in 1992, Rathje and *Atlantic Monthly* managing editor Cullen Murphy discuss some of the data gleaned from the project. For example, the accumulation of refuse has raised the City of New York 6–30 feet (1.8–9 m) since its founding. In 1992, the largest proportion—40 percent—of landfilled garbage was paper, followed by the leftovers from building construction and demolition. In fact, newspapers alone made up about 13 percent of the total volume of trash in 1992.

Just as interesting as what they found was what they did not find. Contrary to much of public opinion, fast-food packaging made up only one-third of 1 percent of the total volume of trash landfilled between 1980 and 1989, while expanded polystyrene foam accounted for no more than 1 percent. Even disposable diapers averaged out at only 1 percent by weight of the total solid waste contents (1.4 percent by volume). Of all the garbage examined, plastics constituted from 20–24 percent. Surveys of several national landfills revealed that organic materials made up 40–52 percent of the total volume of waste.

The Garbage Project also debunked the idea that the United States is running out of space for landfills. While it is true that many landfills have been shut down, it is also true that many of those were quite small to begin with and that they now pose fewer environmental hazards. It is estimated that one landfill 120 feet (35.4 m) deep and measuring 44 square miles (71 km^2) would adequately handle the needs of the entire nation for the next 100 years (assuming current levels of waste production).

In "A Perverse Law of Garbage," Rathje extrapolated from "Parkinson's Law" to define his Parkinson's Law of Garbage: "Garbage expands so as to fill the receptacles available for its containment." As evidence he cites a Garbage Project study of the recent mechanization of garbage pick-up in some larger cities and the ensuing effects. As users were provided with increasingly larger receptacles (in order to accommodate the mechanized trucks), they continued to fill

them up. Rathje attributes this to the newfound convenience of disposing of that which previously had been consigned to the basement or secondhand store, and concludes that the move to automation may be counterproductive to any attempt to reduce garbage and increase recycling.

Resources

BOOKS

Rathje, W. L., and C. Murphy. *Rubbish!: The Archaeology of Garbage*. New York: Harper Collins, 1992.

PERIODICALS

Rathje, W. L. "A Perverse Law of Garbage." *Garbage* 4, no. 6 (December-January 1993): 22.

Ellen Link

Garbology

The study of garbage, through either archaeological excavation of landfills or analysis of fresh garbage, to determine what the composition of municipal solid waste says about the society that generated it. The term is associated with the Garbage Project of the University of Arizona (Tucson), co-directed by William Rathje and Wilson Hughes, which began studying trash in 1973 and excavating landfills in 1987.

Gardens *see* **Botanical garden; Organic gardening and farming.**

Gasohol

Gasohol is a term used for the mixture of ethyl alcohol (also called ethanol or grain alcohol) with gasoline. Ethanol raises the octane rating of lead-free automobile fuel and significantly decreases the carbon monoxide released from tailpipes. Ethanol raises the vapor pressure of gasoline, and it has been reported to increase the release of "evaporative" volatile hydrocarbons from the fuel system and oxides of nitrogen from the exhaust. These substances are components of urban smog, and thus the role of ethanol in reducing pollution is controversial.

The production of fuel ethanol from corn and other crops has been criticized both as a net energy consumer (or poor net-energy producer) and as injurious to the world's poor. In early 2008, food shortages gripped much of the world, with riots against high prices becoming more common in some poor countries. Food prices worldwide had risen 83 percent since 2005, according to the World Bank. Although there were a number of causes for rising prices, including a drought in Australia that lowered production of rice, the main food of over half the world's population, food experts agreed that the push to raise crops for ethanol biofuel for vehicles was part of the problem. Biofuel manufacturers compete directly with food buyers in the market for corn, and high demand for biofuel crops causes growers to switch acreage away from food production. The result is higher food prices.

Gasoline

Crude oil in its natural state has very few practical uses. However, when it is separated into its component parts by the process of fractionation, or refining, those parts have an almost unlimited number of applications.

In the first sixty years after the process of petroleum refining was invented, the most important fraction produced was kerosene, widely used as a home heating product. The petroleum fraction slightly lighter than kerosene—gasoline—was regarded as a waste product and discarded. Not until the 1920s, when the automobile became popular in the United States, did manufacturers find any significant use for gasoline. From then on, however, the importance of gasoline has increased with automobile use.

The term gasoline refers to a complex mixture of liquid hydrocarbons that condense in a fractionating tower at temperatures between 100° and 400° Fahrenheit (40° and 205° C). The hydrocarbons in this mixture are primarily single- and double-bonded compounds containing five to twelve carbon atoms.

Gasoline that comes directly from a refining tower, known as naphtha or "straight-run" gasoline, was an adequate fuel for the earliest motor vehicles. But as improvements in internal combustion engines

Refilling a car with a gas pump. (zhu difeng/Shutterstock.com)

were made, problems began to arise. The most serious problem was "knocking."

If a fuel burns too rapidly in an internal combustion engine, it generates a shock wave that makes a "knocking" or "pinging" sound. The shock wave will, over time, also cause damage to the engine. The hydrocarbons that make up straight-run gasoline proved to burn too rapidly for automotive engines developed after 1920.

Early in the development of automotive fuels, engineers adopted a standard for the amount of knocking caused by a fuel and, hence, for the fuel's efficiency. That standard was known as "octane number." To establish a fuel's octane number, it is compared with a very poor fuel (n-heptane), assigned an octane number of zero, and a very good fuel (isooctane), assigned an octane number of 100. The octane number of straight-run gasoline is anywhere from 50 to 70.

As engineers made more improvements in automotive engines after the 1920s, chemists tried to keep pace by developing better fuels. One approach they used was to subject straight-run gasoline (as well as other crude oil fractions) to various treatments that changed the shape of hydrocarbon molecules in the gasoline mixture. One such method, called cracking, involves the heating of straight-run gasoline or another petroleum fraction to high temperatures. The process results in a better fuel from newly-formed hydrocarbon molecules.

Another method for improving the quality of gasoline is catalytic reforming. In this case, the cracking reaction takes place over a catalyst such as copper, platinum, rhodium, or other "noble" metal, or a form of clay known as zeolite. Again, hydrocarbon molecules formed in the fraction are better fuels than straight-run gasoline. Gasoline produced by catalytic cracking or reforming has an octane number of at least eighty.

A very different approach to improving gasoline quality is the use of additives, chemicals added to gasoline to improve the fuel's efficiency. Automotive engineers learned more than fifty years ago that adding as little as two grams of tetraethyl lead to one gallon of gasoline raises its octane number by as much as ten points.

Until the 1970s, most gasoline contained tetraethyl lead. Then, concerns began to grow about the release of lead to the environment during the combustion of gasoline. Lead concentrations in urban air had reached a level five to ten times that of rural air. Residents of countries with few automobiles, such as Nepal, had only one-fifth the lead in their bodies as did residents of nations with many automotive vehicles.

The toxic effects of lead on the human body have been known for centuries, and risks posed by leaded gasoline became a major concern. In addition, leaded gasoline became a problem because it damaged a car's catalytic converter, which reduced air pollutants in exhaust.

Finally, in 1973, the Environmental Protection Agency (EPA) acted on the problem and set a timescale for the gradual elimination of leaded fuels. According to this schedule, the amount of lead was to be reduced from 2 to 3 grams per gallon (the 1973 average) to 0.5 g/gal by 1979. Ultimately, the additive was to be totally eliminated from all gasoline.

The elimination of leaded fuels was made possible by the invention of new additives. One of the most popular is methyl tert-butyl ether (MTBE). By 1988, MTBE had become so popular that it was among the forty most widely produced chemicals in the United States. MTBE is highly soluble in water, however, which allows it to spread easily in moist environments and contaminate groundwater and soil. Consequently, as of 2007, twenty-five states had enacted legislation to phase out the use of MTBE.

Yet another approach to improving fuel efficiency is the mixing of gasoline and ethyl or methanol. This product, known as gasohol, has the advantage of high octane rating, lower cost, and reduced emission of pollutants, compared to normal gasoline.

See also Air pollution; Renewable energy.

Resources

OTHER

United States Environmental Protection Agency (EPA). "Pollutants/Toxics: Air Pollutants: Lead." http://www.epa.gov/ebtpages/pollairpollutantslead.html (accessed November 11, 2010).
United States Environmental Protection Agency (EPA). "Pollutants/Toxics: Chemicals: Methyl-T-Butyl-Ether (MTBE)." http://www.epa.gov/ebtpages/pollchemic methyl-t-butyl-ethermtbe.html (accessed November 11, 2010).

David E. Newton

Gasoline tax

Gasoline taxes include federal, state, county, and municipal taxes imposed on gasoline motor vehicle fuel. In the United States, most of the federal tax is used to fund maintenance and improvements in such transportation infrastructures as interstate highways. As of 2010, the federal excise tax for gasoline stood at 18.4 cents per gallon, and state excise taxes ranged from 8 cents in Alaska to 46.6 cents in California. As of 2010, the national average state tax was 29 cents per gallon. In total, the U.S. national average gasoline tax (combining federal and state taxes) was 47.4 cents per gallon.

Oregon became the first state to institute a tax on gasoline in 1919. By the time the federal government established its own 1 cent gas tax in 1932, every state had a gas tax. After several small increases in the 1930s and 1940s, the gas tax was raised to 3 cents to finance the Highway Trust Fund in 1956. The Trust Fund was earmarked to pay for federal interstate construction and other road work. In 1982, the federal gasoline tax was increased to 9 cents to fund road maintenance and mass transit. The tax was hiked again in 1990 to 14.1 cents, and to 18.4 cents in 1993—where it remained as of 2010.

Over this time, gasoline prices increased from about 20 cents per gallon in 1938 to a U.S. national average of 280.6 cents in November 2010. The average national gasoline tax (both federal and a weighted average of state taxes) accounts for approximately 16 percent of the retail price of a gallon of gas.

In some countries, diesel fuels are taxed and priced less than gasoline. Commercial vehicles are major consumers of diesel, and lower taxes avoid undue impacts on trucking and commerce. In the United States, diesel is taxed at a higher rate than gasoline—an average of 52.5 cents per gallon (including 24.4 cents federal tax and the weight average of state taxes).

Although federal gasoline taxes are a manufacturer's excise tax, meaning that the government collects the tax directly from the manufacturer, rate hikes are often passed on to consumers at the pump. In this light, gasoline taxes have been criticized as regressive and thus inequitable, i.e., lower income individuals pay a greater share of their income as tax than higher income individuals. Also, the tax as a share of the pump price has been increasing.

Politics often play a large factor in setting gasoline taxes. In May 1996, Congress attempted to rollback the 4.3 cents tax increase of 1993. The impact of this repeal for a family of four who drive 12,000 miles a year at 20 miles per gallon is a savings of $26, which in the House debates was compared to the cost of a family dinner at McDonald's. On the other hand, a rollback could have bigger consequences for the future upkeep of the country's highways and interstates; a 2000 Congressional report estimated that a repeal of

the federal gasoline tax would mean a $5.2 billion annual loss in revenues for the Highway Trust Fund.

Outside of the United States, both gasoline prices and gas tax rates are typically far higher (e.g., gasoline taxes were 445 U.S. cents per gallon in the United Kingdom as of November 2010). In addition to funding governments, high gasoline taxes form part of a strategy to encourage the use of public transportation, reduce pollution, conserve energy, and improve national security (since most gasoline is imported).

Resources

BOOKS

Davis, Lucas W., and Lutz Kilian. *Estimating the Effect of a Gasoline Tax on Carbon Emissions.* Cambridge, Mass: National Bureau of Economic Research, 2009.

Nersesian, Roy L. *Energy for the 21st Century: A Comprehensive Guide to Conventional and Alternative Sources.* Armonk, N.Y.: M.E. Sharpe, 2007.

Stuart Batterman
Paula Anne Ford-Martin

Gastropods

Gastropods are invertebrate animals that make up the largest class in the phylum Mollusca. Examples of common gastropods include all varieties of snails, abalone, limpets, and land and sea slugs. There are between 60,000 and 80,000 existing species, as well an additional 15,000 separate fossil species. Gastropods first appeared in the fossil record during the early Cambrian period, approximately 550 million years ago.

This diverse group of animals is characterized by a soft body, made up of three main parts: the head, foot, and visceral mass. The head contains a mouth and often sensing tentacles. The lower portion of the body makes up the foot, which allows slow creeping along rocks and other solid surfaces. The visceral mass is the main part of the body, containing most of the internal organs. In addition to these body parts, gastropods possess a mantle, or fold which secretes a hard, calcium carbonate shell. The single, asymmetrical shell of a gastropod is most often spiral shaped, however, it can be flattened or cone-like. This shell is an important source of protection. Predators have a difficult time accessing the soft flesh inside, especially if there are sharp points on the outside, as there are on the shells of some of the more ornate gastropods. There are also some gastropods, such as slugs and sea hares, that do not have shells or have greatly reduced shells. Some of the shelless types that live in the ocean (i.e., nudibranchs or sea slugs) are able to use stinging cells from prey that they have consumed as a means of protection.

In addition to a spiraling of their shells, the soft bodies of most gastropods undergo 180 degrees of twisting, or torsion, during early development, when one side of the visceral mass grows faster than the other. This characteristic distinguishes gastropods from other molluscs. Torsion results in a U-shaped digestive tract, with the anal opening slightly behind the head. The torsion of the soft body and the spiraling of the shell are thought to be unrelated evolutionary events.

Gastropods have evolved to live in a wide variety of habitats. The great majority are marine, living in the world's oceans. Numerous species live in fresh water, while others live entirely on land. Of those that live in water, most are found on the bottom, attached to rocks or other surfaces. There are even a few species of gastropods without shells, including sea butterflies, that are capable of swimming. Living in different habitats has resulted in a wide variety of structural adaptations within the class Gastropoda. For example, those gastropods that live in water use gills to obtain the oxygen necessary for respiration, while their terrestrial relatives have evolved lungs to breathe.

Gastropods are important links in food webs in the habitats in which they live, employing a wide variety of feeding strategies. For example, most gastropods move a rasping row of teeth on a tongue like organ called a radula back and forth to scrape microscopic algae off rocks or the surface of plants. Because the teeth on the radula gradually wear away, new teeth are continuously secreted. Other gastropods have evolved a specialized radula for drilling through shells of animals to get at their soft flesh. For example, the oyster drill, a small east coast gastropod, bores a small hole in the shell of neighboring molluscs such as oysters and clams so that it can consume the soft flesh. In addition, some terrestrial gastropods such as snails use their radula to cut through pieces of leaves for food.

Gastropods are eaten by numerous animals, including various types of fish, birds, and mammals. They are also eaten by humans throughout the world. Abalone, muscular shelled gastropods that cling to rocks, are consumed on the west coast of the United States and in Asia. Fritters and chowder are made from the large, snail-like queen conch on many Caribbean islands. Escargot (snails in a garlicky butter sauce) are a European delicacy.

Resources

BOOKS

Cimino, Guido, and Margherita Gavagnin. *Molluscs.* New York: Springer, 2006.

Robin, Alain. *Encyclopedia of Marine Gastropods.* Paris: Association Française de Conchyliologie, 2008.

Salgeback, Jenny. *Functional Morphology of Gastropods and Bivalves.* Uppsala, Sweden: Acta Universitatis Upsaliensis, 2006.

Max Strieb

Gene bank

The term gene bank refers to any system by which the genetic composition of some population is identified and stored. Many different kinds of gene banks have been established for many different purposes. Perhaps the most numerous gene banks are those that consist of plant seeds, known as germ banks.

The primary purpose for establishing a gene bank is to preserve examples of threatened or endangered species. Each year, untold numbers of plant and animal species become extinct because of natural processes and more commonly, as the result of human activities. Once those species become extinct, their gene pools are lost forever.

Scientists want to retain those gene pools for a number of reasons. For example, agriculture has been undergoing a dramatic revolution in many parts of the world over the past half century. Scientists have been making available to farmers plants that grow larger, yield more fruit, are more disease-resistant, and have other desirable characteristics. These plants have been produced by agricultural research in the United States and other nations. Such plants are very attractive to farmers, and they are also important to governments as a way of meeting the food needs of growing populations, especially in Third World countries.

When farmers switch to these new plants, however, they often abandon older, more traditional crops that may then become extinct. Although the traditional plants may be less productive, they have other desirable characteristics. They may, for example, be able to survive droughts or other extreme environmental conditions that new the new variety of plants cannot.

Placing seeds from traditional plants in a gene bank allows them to be preserved. At some later time, scientists may want to study these plants further and perhaps identify the genes that are responsible for various desirable properties of the plants. The U.S. Department of Agriculture (USDA) has long maintained a seed bank of plants native to the United States. About 200,000 varieties of seeds are stored at the USDA's Station at Fort Collins, Colorado, and another 100,000 varieties are kept at other locations around the country.

Efforts are now underway to establish gene banks for animals, too. Such banks consist of small colonies of the animals themselves. Animal gene banks are desirable as a way of maintaining species whose natural population is very low. Sometimes the purpose of the bank is simply to maintain the species to prevent its becoming extinct. In other cases, species are being preserved because they were once used as farm animals although they have since been replaced by more productive modern hybrid species. The Fayoumi chicken native to Egypt, for example, has now been abandoned by farmers in favor of imported species. The Fayoumi, without some form of protection, is likely to become extinct. Nonetheless, it may well have some characteristics (genes) that are worth preserving.

In recent years, another type of gene bank has become possible. In this kind of gene bank, the actual base sequence of important genes in the human body will be determined, collected, and catalogued. This effort, begun in 1990, was a part of the Human Genome Project effort to map all human genes. The Human Genome Project was completed in 2003 with the mapping of the 20,000 to 25,000 genes comprising the human genome. As of late 2010, research continued on the data gleaned from the Project. For instance, in 2008 a scholarly paper using data collected by the Human Genome Project appeared in the science journal *Nature* titled "Mapping and sequencing of structural variation from eight human genomes."

See also Agricultural revolution; Extinction; Genetic engineering; Population growth.

Resources

BOOKS

Caetano-Anollés, Gustavo, editor. *Evolutionary Genomics and Systems Biology*. Hoboken, NJ: Wiley-Blackwell, 2010.

McElheny, Victor K. *Drawing the Map of Life: Inside the Human Genome Project*. New York City: Basic Books, 2010.

Palladino, Michael A. *Understanding the Human Genome Project*. San Francisco: Pearson/Benjamin Cummings, 2006.

OTHER

Agricultural Research Service, U.S. Department of Agriculture. "National Center for Genetic Resources Preservation." http://www.ars.usda.gov/main/site_main.htm?mode code = 54-02-05-00 (accessed November 10, 2010).

Department of Energy Genome Program. "Human Genome Project Information." http://www.ornl.gov/sci/techresources/Human_Genome/home.shtml (accessed November 10, 2010).

David E. Newton

Gene pool

The term gene pool refers to the sum total of all the genetic information stored within any given population. A gene is a specific portion of a DNA (deoxyribose nucleic acid) molecule, so a gene pool is the sum total of all of the DNA contained within a population of individuals.

The concept of gene pool is important in ecological studies because it reveals changes that may or may not be taking place within a population. In a population living in an ideal environment for its needs, the gene pool is likely to undergo little or no change. If individuals are able to obtain all the food, water, energy, and other resources they need, they experience relatively little stress and there is no pressure to select one or another characteristic.

Changes do occur in gene frequency because of natural factors in the environment. For example, natural radiation exposure causes changes in DNA molecules that are revealed as genetic changes. These natural mutations are one of the factors that make possible continuous changes in the genetic constitution of a population that, in turn, allows for evolution to occur.

Natural populations seldom live in ideal situations, however, and so they experience various kinds of stress that lead to changes in the gene pool. A classic example of this kind of change was reported by J. B. S. Haldane in 1937. Haldane found that a population of moths gradually became darker in color over time as the trees on which they lived also became darker because of pollution from factories. Moths in the population who carried genes for darker color were better able to survive and reproduce than were lighter-colored moths, so the composition of the gene pool changed to relieve stress.

Humans have the ability to make conscious changes in gene pools that no other species has. Sometimes we make those changes in the gene pools of plants or animals to serve our own needs for food or other resource. Hybridization of plants to produce populations that have some desirable quality such as resistance to disease, shorter growing season, or better-tasting fruit. The modern science of genetic engineering is perhaps the most specific and deliberate way of changing in gene pools today.

Humans can also change the gene pool of their own species. For example, individuals with various genetic disorders were at one time doomed to death. Our inability to treat diabetes, sickle-cell anemia, phenylketonuria, and other hereditary conditions meant that the frequency of the genes causing those disorders in the human gene pool was kept under control by natural forces.

Today, many of those same disorders can be treated by medical or genetic techniques. That results in positive benefit for the individuals who are cured, but raises questions about the quality of the human gene pool overall. Instead of having many of those deleterious genes being lost naturally by an individual's death, they are now retained as part of the gene pool. This fact has at times raised questions about the best way in which medical science should deal with genetic disorders.

See also Agricultural revolution; Birth defects; Extinction; Gene bank; Population growth.

Resources

BOOKS

Acquaah, George. *Principles of Plant Genetics and Breeding.* Malden, MA: Blackwell, 2006.

Dobzhansky, T. *Genetics and the Origin of Species.* New York: Columbia University Press, 1951.

Fox, Charles W., and Jason B. Wolf. *Evolutionary Genetics: Concepts and Case Studies.* Oxford: Oxford University Press, 2006.

Hartl, Daniel L., and Andrew G. Clark. *Principles of Population Genetics*, 4th ed. Sunderland, MA: Sinauer Associates, 2006.

Klug, William S., William S. Klug, and Sarah M. Ward. *Concepts of Genetics.* San Francisco: Pearson Benjamin Cummings, 2009.

David E. Newton

Genetic engineering

Genetic engineering is the deliberate manipulation of the genetic material of organisms, typically deoxyribonucleic acid (DNA), which is present in chromosomes in eukaryotes and dispersed throughout the interior of prokaryotes. The DNA manipulation can involve the addition, removal, or alteration of genes, the sequences of DNA that code for a product.

DNA consists of a chain of individual nucleotides, which are composed of a nitrogen-containing base (adenine, cytosine, guanine, or thymine), a five-carbon sugar, and a phosphate group. Two DNA chains are bound together to form a two-stranded helical structure that is known as a double helix, with the adenine residues of one strand linked only with thymine on the other chain, and cytosine on one chain linked only with guanine on the other chain.

The total DNA of an organism is referred to as its genome. In the 1950s, scientists first discovered how the structure of DNA molecules worked and how they stored and transmitted genetic information. Beginning in the 1970s, scientists refined the technology that enabled the deliberate and targeted alteration of DNA. This was the beginning of both genetic engineering and the discipline of molecular biology.

A recent example illustrates the power of genetic engineering. A 2010 paper published in the journal *Science* by a research team headed by J. Craig Venter described the construction of a genome and the insertion of the genetic material into a bacterium from which the native genetic material had been removed. The resulting construct was capable of growth and division.

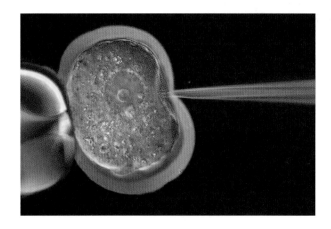

Foreign DNA (deoxyribonucleic acid) being injected into a pronuclear mouse egg (female reproductive cells). This procedure is known as transgenesis. *(Martin Oeggerli/Photo Researchers, Inc.)*

Genetic engineering relies on recombinant DNA technology (the use of select enzymes to cut DNA at specific sites, allowing for insertion or removal of DNA) to manipulate genes. Methods now permit the rapid determination of the nucleotide sequence of pieces of DNA, as well as for identifying particular genes of interest, and for isolating individual genes from complex genomes. This allows genetic engineers to alter genetic materials to produce new substances or create new functions.

The biochemical tools used by genetic engineers or molecular biologists include a series of enzymes called restriction enzymes that can cut and paste genes or regions of genes. As of 2010, more than 500 restriction enzymes have been discovered, allowing a large number of DNA sequences to be recognized manipulated.

Most restriction enzymes are endonucleases—enzymes that cut one or both strands of the double helix of DNA at the specific nucleotide sequences or, for some of the enzymes, at a region removed from the recognition sequence. Every restriction enzyme is given a specific name to identify it uniquely. The first three letters, in italics, indicate the biological source of the enzyme, the first letter being the initial of the genus, the second and third letters being the first two letters of the species name. Thus, restriction enzymes from *Escherichia coli* are termed *Eco*, those from *Haemophilus influenzae* are *Hin*, from *Diplococcus pneumoniae* are *Dpn*, as three examples.

Particular restriction enzymes can be used in a tailor-made fashion to locate and cut almost any sequence of nucleotide bases. Cuts can be made anywhere along the DNA, dividing it into many small fragments or a few longer ones. The cut made by a specific restriction enzyme occurs at a restriction site characterized by a particular sequence of bases recognized by that restriction enzyme. The results are repeatable: Cuts made by the same enzyme on a given sort of DNA will always be the same. Some enzymes recognize sequences as long as six or seven bases; these are used for opening a circular strand of DNA at just one point. Other enzymes have a smaller recognition site, three or four bases long; these produce small fragments that can then be used to determine the sequence of bases along the DNA. For instance, if a scientist has the sequence of a gene and wants to cut the strands, a restriction enzyme can be selected to correspond with the bases at the desired cutting point to cut the strands of DNA at a very specific point.

The cut that each enzyme makes varies from enzyme to enzyme. Some, like *Hin* dII, make a clean, or blunt, cut straight across the double helix, leaving DNA fragments with ends that are flush. Other enzymes such as *Eco* RI make a staggered break, leaving single strands with protruding cohesive ends (sticky ends) that are complementary in base sequence. This is advantageous as it allows two complementary protruding regions to associate, which creates the opportunity to introduce new DNA into existing DNA. Following breakage, DNA ligase, an enzyme that allows formation of the bonds between bases, can be utilized to rejoin the complementary bases from different sources. The result is termed recombinant DNA.

Another important biochemical tool used by genetic engineers is DNA polymerase, an enzyme that normally catalyses the growth of a nucleic acid chain. DNA polymerase is used by genetic engineers to seal the gaps between the two sets of fragments in newly joined chimera molecules of recombinant DNA. DNA polymerase is also used to label DNA fragments, for DNA polymerase labels practically every base, allowing minute quantities of DNA to be studied in detail. If a piece of ribonucleic acid (RNA) of the target gene is the starting point, then the enzyme reverse transcriptase is used to produce a strand of complementary DNA (cDNA).

Genetic engineers usually need large numbers of genetically identical copies of the DNA fragment of interest. One way of doing this is to insert the gene into a suitable gene carrier, termed a cloning vector. Common cloning vectors are bacterial plasmids or viruses such as the bacteriophage lambda, which are viruses capable of inserting their circular genetic material (DNA) into bacterial cells. When the cloning vectors divide, they replicate both themselves and the foreign DNA segment contained within the bacterial cell.

In the plasmid insertion method, restriction enzymes are used to cleave the plasmid double helix so that a stretch of DNA (previously cleaved with the same enzyme) can be inserted into the plasmid. As a result, the sticky ends of the plasmid DNA and the foreign DNA are complementary and base-pair when mixed together. The fragments held together by base pairing are permanently joined by DNA ligase. The host bacterium, with its twenty- to thirty-minute reproductive cycle, is like a manufacturing plant. With repeated doublings of its offspring on a controlled culture medium, millions of clones of the purified DNA fragments can be produced overnight.

Similarly, if viruses (bacteriophages) are used as cloning vectors, the gene of interest is inserted into the phage DNA, and the virus is allowed to enter the host bacterial cell where it multiplies. These cloning vectors can be purchased for scientific research and contain a multiple cloning site that has many different restriction sites for creating recombinant DNA. A single parental lambda phage particle containing recombinant DNA can multiply to several hundred progeny particles inside the bacterial cell (*E. coli*) within roughly twenty minutes.

Cosmids are another type of viral cloning vehicle that attaches foreign DNA to the packaging sites of a virus and introduces the foreign DNA into an infective viral particle. Cosmids allow researchers to insert very long stretches of DNA into host cells where cell multiplication amplifies the amount of DNA available. Large artificial chromosomes of yeast (termed mega yeast artificial chromosomes) are also used as cloning vehicles, since they can store even larger pieces of DNA, thirty-five times more than can be stored conventionally in bacteria.

The polymerase chain reaction (PCR) technique that was developed in 1983 by Kary Mullis, who shared the 1993 Nobel Prize in Chemistry for the accomplishment, is one of the most significant developments in the field of genetic engineering. PCR allows the direct mass production of short segments of DNA, bypassing several steps involved in using bacterial and viruses as cloning vectors.

DNA fragments can be introduced into mammalian cells, but a different method must be used. Here, genes packed in solid calcium phosphate are placed next to a cell membrane that surrounds the fragment and transfers it to the cytoplasm. The gene is delivered to the nucleus during mitosis (when the nuclear membrane has disappeared) and the DNA fragments are incorporated into daughter nuclei, then into daughter cells. A mouse containing human cancer genes (the oncomouse) was generated in this way and patented in 1988.

The potential benefits of recombinant DNA research are enormous. There are more than 6,000 known disorders and diseases that involve the disruption or deletion of a single gene. As of 2010, scientists have identified the genetic basis of a number of these, with the list growing continually. Genetic engineering helps scientists replace a particular missing or defective gene with correct copies of that gene. If that gene then begins functioning normally in an individual, a genetic disorder may be cured. Researchers have already met with great success in finding the single genes that are responsible for common diseases like cystic fibrosis and hemochromatosis.

In August 2001, former U.S. President George W. Bush (1946–) struggled with the debate over allowing federal funding for embryonic stem cell research. He allowed funding for only existing lines of cells, leaving further research in the hands of those who could seek private funding. In early 2009, U.S. President Barack Obama (1961–) signed an order to allow use of federal funding for stem cell research.

See also Gene bank; Gene pool.

Resources

BOOKS

Alberts, Bruce. *Molecular Biology of the Cell*. New York: Garland Science, 2008.

Berry, Roberta M. *The Ethics of Genetic Engineering*. Routledge Annals of Bioethics. London: Routledge, 2007.

Carson, Susan, and Dominique Robertson. *Manipulation and Expression of Recombinant DNA: A Laboratory Manual*. Burlington, MA: Elsevier Academic Press, 2006.

Jackwood, Mark W. *Vaccine Development Using Recombinant DNA Technology*. Ames, Iowa: Counil for Agricultural Science and Technology, 2008.

LeVine, Harry, and Mildred Vasan. *Genetic Engineering*. Santa Barbara, CA: ABC-CLIO, 2006.

Watson, James D., Jan Witkowski, Richard M. Myers, and Amy A. Caudy. *Recombinant DNA Fundamentals of Genes And Genomes*. W H Freeman & Co, 2007.

PERIODICALS

Gibosn, Daniel G., et al. "Creation of a bacterial cell controlled by a chemically synthesized genome" *Science* 329: 52-56, May 20, 2010.

ORGANIZATIONS

Pew Initiative on Food and Biotechnology, 1331 H Street, Suite 900, Washington, DC, USA, 20005, (202) 347-9044, (202) 347-9047, inquiries@pewagbiotech.org, http://pewagbiotech.org

Neil Cumberlidge

Genetic resistance (or genetic tolerance)

Genetic resistance (or genetic tolerance) refers to the ability of certain organisms to endure environmental conditions that are extremely stressful or lethal to non-adapted individuals of the same species. Such tolerance has a genetic basis, and it evolves at the population level in response to intense selection pressures.

Genetic resistance occurs when genetically variable populations contain some individuals that are relatively tolerant of an exposure to some environmental factor, such as the presence of a high concentration of a specific chemical. If the tolerance is genetically based (i.e., due to specific information embodied in the DNA of the organism's chromosomes), some or all of the offspring of these individuals will also be tolerant. Under conditions in which the chemical occurs in concentrations high enough to cause toxicity to non-tolerant individuals, the resistant ones will be relatively successful. As time passes their offspring will become increasingly more prominent in the population. Acquiring genetic resistance is an evolutionary process, involving increased tolerance within a population, for which there is a genetic basis, and occurring in response to selection for resistance to the effects of a toxic chemical. Some of the best examples of genetic resistance involve the tolerance of certain bacteria to antibiotics and of certain pests to pesticides.

Resistance to antibiotics

Antibiotics are chemicals used to treat bacterial infections of humans and domestic animals. Examples of commonly used antibiotics include various kinds of penicillins, streptomycins, and tetracyclines, all of which are metabolic byproducts created by certain microorganisms, especially fungi. There are also many synthetic antibiotics.

Antibiotics are extremely toxic to non-resistant strains of bacteria, and this has been very beneficial in the control of bacterial infections and diseases. However, if even a fraction of a bacterial population has a genetically based tolerance to a specific antibiotic, evolution will quickly result in the development of a population that is resistant to that chemical. Bacterial resistance to antibiotics was first demonstrated for penicillin, but the phenomenon is now quite widespread. This is an important medical problem because some serious pathogens are now resistant to virtually all of the available antibiotics, which means that infections by these bacteria can be extremely difficult to control. Bacterial resistance has recently become the cause of infections by some virulent strains of *Staphylococcus* and other potentially deadly bacteria. Some biologists assert that this problem has been made worse by the failure of many people to finish their course of prescribed antibiotic treatments, which can allow tolerant bacteria to survive and flourish. Also possibly important has been the routine use of antibiotics to prevent diseases in livestock kept under crowded conditions in industrial farming. The small residues of antibiotics in meat, eggs, and milk may be resulting in low-level selection for resistant bacteria in exposed populations of humans and domestic animals.

Resistance to pesticides

The insecticide dichlorodiphenyl-trichloroethane (DDT) was the first pesticide to which insect pests developed resistance. This occurred because the exposure of insect populations to toxic DDT results in intense selection for resistant genotypes. Tolerant populations can evolve because genetically resistant individuals are not killed by the pesticide and therefore survive to reproduce. Almost 500 species of insects and mites have populations that are known to be resistant to at least one insecticide. There are also more than 100 examples of fungicide-resistant plant pathogens and about fifty herbicide-resistant weeds. Insecticide resistance is most frequent among species of flies and their relatives (order Diptera), including more than fifty resistant species of malaria-carrying *Anopheles* mosquitoes. In fact, the progressive evolution of insecticide resistance by *Anopheles* has been an important factor in the recent resurgence of malaria in countries with warm climates. In addition, the protozoan *Plasmodium*, which actually causes malaria, has become resistant to some of the drugs that used to effectively control it.

Crop geneticists have recently managed to breed varieties of some plant species that are resistant to glyphosate, a commonly used agricultural herbicide that is effective against a wide range of weeds, including both monocots and dicots. The development of glyphosate-tolerant varieties of such crops as rapeseed means that this effective herbicide can be used to control difficult weeds in planted fields without causing damage to the crop.

Resources

BOOKS

Birge, Edward A. *Bacterial and Bacteriophage Genetics*. New York: Springer, 2006.

Snustad, D. Peter, and Michael J. Simmons. *Principles of Genetics*. Hoboken, NJ: Wiley, 2006.

Hamilton, Matthew B. *Population Genetics*. Chichester, UK: Wiley-Blackwell, 2009.

Bill Freedman

Genetically engineered organism

The modern science of genetics began in the midnineteenth century with the work of Gregor Mendel, but the nature of the gene itself was not understood

AquaBounty salmon (rear) have an added growth hormone gene from the Chinook salmon to a normal Atlantic salmon (front) that results in a transgenic salmon that grown to market size in about half the time as a normal salmon. *(Photo courtesy AquaBounty/MCT)*

until James Watson (1928–) and Francis Crick (1916–2004) announced their findings in 1953. According to the Watson and Crick model, genetic information is stored in molecules of DNA (deoxyribose nucleic acid) by means of certain patterns of nitrogen base that occur in such molecules. Each set of three such nitrogen bases were codes, they said, for some particular amino acid, and a long series of nitrogen bases were codes for a long series of amino acids or a protein.

Deciphering the genetic code and discovering how it is used in cells has taken many years of work since that of Watson and Crick. The basic features of that process, however, are now well understood. The first step involves the construction of a RNA (ribonucleic acid) molecule in the nucleus of a cell, using the code stored in DNA as a template. The RNA molecule then migrates out of the nucleus to a ribosome in the cell cytoplasm. At the ribosome, the sequence of nitrogen bases stored in RNA act as a map that determines the sequence of amino acids to be used in constructing a new protein.

This knowledge is of critical importance to biologists because of the primary role played by proteins in an organism. In addition to acting as the major building materials of which cells are made, proteins have a number of other crucial functions. All hormones and enzymes, for example, are proteins, and therefore nearly all of the chemical reactions that occur within organisms are mediated by one protein or another.

Our current understanding of the structure and function of DNA makes it at least theoretically possible to alter the biological characteristics of an organism. By changing the kind of nitrogen bases in a DNA molecule, their sequence, or both, a scientist can change the genetic instructions stored in a cell and thus change the kind of protein produced by the cell.

One of the most obvious applications of this knowledge is in the treatment of genetic disorders. A large majority of genetic disorders occur because an organism is unable to correctly manufacture a particular protein molecule. An example is Lesch-Nyhan syndrome. It is a condition characterized by self-mutilation, mental retardation, and cerebral palsy which arises because a person's body is unable to manufacture an enzyme known as hypoxanthine guanine phosphoribosyl transferase (HPRT).

The general principles of the techniques required to make such changes are now well understood. The technique is referred to as genetic engineering or genetic surgery because it involves changes in an organism's gene structure. When used to treat a particular disorder in humans, the procedure is also called human gene therapy. Developing specific experimental techniques for carrying out genetic engineering has proved to be an imposing challenge, yet impressive strides have been made. A common procedure is known as recombinant DNA (rDNA) technology.

The first step in an rDNA procedure is to collect a piece of DNA that carries a desired set of instructions. For a genetic surgery procedure for a person with Lesch-Nyhan syndrome, a researcher would need a piece of DNA that codes for the production of HPRT. That DNA could be removed from the healthy DNA of a person who does not have Lesch-Nyhan syndrome, or the researcher might be able to manufacture it by chemical means in the laboratory.

One of the fundamental tools used in rDNA technology is a closed circular piece of DNA found in bacteria called a plasmid. Plasmids are the vehicle or vector that scientists use for transferring new pieces of DNA into cells. The next step in an rDNA procedure, then, would be to insert the correct DNA into the plasmid vector. Cutting open the plasmid can be accomplished using certain types of enzymes that recognize specific base sequences in a DNA molecule. When these enzymes, called restriction enzymes, encounter the recognized sequence in a DNA molecule, they cleave the molecule. After the plasmid DNA has been cleaved and the correct DNA mixed with it, a second type of enzyme is added. This kind of enzyme inserts the correct DNA into the plasmid and closes it up. The process is known as gene splicing.

In the final step, the altered plasmid vector is introduced into the cell where it is expected to function. In the case of a Lesch-Nyhan patient, the plasmid would be introduced into the cells where it would start producing HPRT from instructions in the correct DNA. Many technical problems remain with rDNA technology, and this last step has caused some of the greatest obstacles. It has proven very difficult to make introduced DNA function. Even when the plasmid vector with its new DNA gets into a cell, it may never actually begin to function.

Any organism whose cells contain DNA altered by this or some other technique is called a genetically engineered organism. The first human patient with a genetic disorder who is treated by human gene therapy will be a genetically engineered organism. The use of genetic engineering on human subjects has gone forward very slowly for a number of reasons. One reason is that humans are very complex organisms. Another reason is that changing the genetic make-up of a human involves more ethical questions and more difficult questions than does the genetic engineering of bacteria, mice, or cows.

Most of the existing examples of genetically engineered organisms, therefore, involve plants, non-human animals, or microorganisms. One of the earliest success stories in genetic engineering involved the altering of DNA in microorganisms to make them capable of producing chemicals they do not normally produce. Recombinant DNA technology can be used, for instance, to insert the DNA segment or gene that codes for insulin production into bacteria. When these bacteria are allowed to grow and reproduce in large fermentation tanks, they produce insulin. The list of chemicals produced by this mechanism now includes somatostatin, alpha interferon, tissue plasminogen activator (tPA), Factor VIII, erythroprotein, and human growth hormone, and this list continues to grow each year.

The introduction of genetically engineered plants and animals has, at times, generated quite a bit of controversy. One example is the genetically engineered maize (otherwise known as corn) referred to by its brand name of MON810. Developed by the Monsanto Corporation, MON810 is corn genetically engineered with the toxin Bt, which is poisonous to certain insects, including certain insect pests that feed on the corn plant. Although originally cleared by the European Union as safe to grow and consume, several European member nations have since banned the cultivation or sale of MON810 within their borders. The governments that have ruled out use or sale of MON810 include that of France, in 2008, and Germany, in 2009.

Resources

BOOKS

Challenges and Risk of Genetically Engineered Organisms. Paris: Organization for Economic Co-operation and Development, 2004.

Hodge, Russ. *Genetic Engineering: Manipulating the Mechanisms of Life.* New York City: Facts On File, 2009.

Newton, David E. *DNA Technology: A Reference Handbook.* Santa Barbara, CA: ABD-CLIO, 2010.

OTHER

U.S. Food and Drug Administration. "Genetically Engineered Animals." http://www.fda.gov/animalveterinary/developmentapprovalprocess/geneticengineering/geneticallyengineeredanimals/default.htm (accessed November 9, 2010).

David E. Newton

Genetically modified organism

A genetically modified organism, or GMO, is an organism whose genetic structure has been altered by incorporating one or more single genes from another

Bathers enjoy the waters of the Blue Lagoon surrounded by steam in the shadow of a geothermal power plant in Grindavik, Iceland. The lagoon is the liquid run-off from one of Iceland's geothermal power plants. *(Robert Rozbora/Shutterstock.com)*

the water is able to boil normally, producing steam. These regions are known as dry steam fields.

Humans have long been aware of geothermal energy. Geysers and fumaroles are obvious indications of water heated by underground rock. The Maoris of New Zealand, for example, have traditionally used hot water from geysers to cook their food. Natural hot spring baths and spas are a common feature of many cultures where geothermal energy is readily available.

The first geothermal well was apparently opened accidentally by a drilling crew in Hungary in 1867. Eventually, hot water from such wells was used to heat homes in some parts of Budapest. Geothermal heat is still an important energy source in some parts of the world. More than 99 percent of the buildings in Reykjavik, the capital of Iceland, are heated with geothermal energy.

The most important application of geothermal energy today is in the generation of electricity. In general, hot steam or super-heated water is pumped to the planet surface where it is used to drive a turbine.

Cool water leaving the generator is then pumped back underground. Some water is lost by evaporation during this process, so the energy that comes from geothermal wells is actually non-renewable. However, most zones of heated water and steam are large enough to allow a geothermal mine to operate for a few hundred years.

A dry steam well is the easiest and least expensive geothermal well to drill. A pipe carries steam directly from the heated underground rock to a turbine. As steam drives the turbine, the turbine drives an electrical generator. The spent steam is then passed through a condenser where much of it is converted to water and returned to the Earth.

Dry steam fields are relatively uncommon. One, near Larderello, Italy, has been used to produce electricity since 1904. The geysers and fumaroles in the region are said to have inspired Dante's *Inferno*. The Larderello plant is a major source of electricity for Italy's electric railway system. Other major dry steam fields are located near Matsukawa, Japan, and at Geysers, California. The first electrical generating plant at the Geysers was installed in 1960. It and companion plants now provide about 5 percent of all the electricity produced in California.

Wet steam fields are more common, but the cost of using them as sources of geothermal energy is greater. The temperature of the water in a wet steam field may be anywhere from 360 to 660°Fahrenheit (from 180 to 250°C). When a pipe is sunk into such a reserve, some water immediately begins to boil, changing into very hot steam. The remaining water is carried out of the reserve with the steam.

At the surface, a separator is used to remove the steam from the hot water. The steam is used to drive a turbine and a generator, as in a dry steam well, before being condensed to a liquid. The water is then mixed with the hot water (now also cooled) before being returned to the earth.

The largest existing geothermal well using wet steam is in Wairakei, New Zealand. Other plants have been built in Russia, Japan, and Mexico. In the United States, pilot plants have been constructed in California and New Mexico. The technology used in these plants is not yet adequate, however, to allow them to compete economically with fossil- fueled power plants.

Hot water (in contrast to steam) from underground reserves can also be used to generate electricity. Plants of this type make use of a binary (two-step) process. Hot water is piped from underground into a heat exchanger at the surface. The heat exchanger contains some low-boiling point liquid (the "working fluid"), such as a freon or isobutane. Heat from the hot water causes the working fluid to evaporate. The vapor then produced is used to drive the turbine and generator. The hot water is further cooled and then returned to the rock reservoir from which it came.

In addition to dry and wet steam fields, a third kind of geothermal reserve exists: pressurized hot water fields located deep under the ocean floors. These reserves contain natural gas mixed with very hot water. Some experts believed that these geopressurized zones are potentially rich energy sources although technology does not currently exists for tapping them.

Another technique for the capture of geothermal energy makes use of a process known as hydrofracturing. In hydrofracturing, water is pumped from the surface into a layer of heated dry rock at pressures of about 7,000 pounds per square inch (500 kg/cm^2). The pressurized water creates cracks over a large area in the rock layer. Then, some material such as sand or plastic beads is also injected into the cracked rock. This material is used to help keep the cracks open.

Subsequently, additional cold water can be pumped into the layer of hot rock, where it is heated just as natural groundwater is heated in a wet or dry steam field. The heated water is then pumped back out of Earth and into a turbine-generator system. After cooling, the water can be re-injected into the ground for another cycle. Since water is continually re-used in this process and Earth's heat is essentially infinite, the hydrofracturing system can be regarded as a renewable source of energy.

Considerable enthusiasm was expressed for the hydrofracturing approach during the 1970s and a few experimental plants were constructed. But, as oil prices dropped and interest in alternative energy sources decreased in the 1980s, these experiments were terminated.

Geothermal energy clearly has some important advantages as a power source. The raw material—heated water and steam—is free and readily available, albeit in only certain limited areas. The technology for extracting hot water and steam is well developed from petroleum-drilling experiences, and its cost is relatively modest. Geothermal mining, in addition, produces almost no air pollution and seems to have little effect on the land where it occurs.

On the other hand, geothermal mining does have its disadvantages. One is that it can be achieved in only limited parts of the world. Another is that it results in the release of gases, such as hydrogen sulfide, sulfur dioxide, and ammonia, that have offensive odors and are mildly irritating. Some environmentalists also object that geothermal mining is visually offensive, especially in some areas that are otherwise aesthetically attractive. Pollution of water by runoff from a geothermal well and the large volume of cooling water needed in such plant are also cited as disadvantages.

At their most optimistic, proponents of geothermal energy claim that up to 15 percent of the power needs for the United States can be met from this source. Lagging interest and research in this area over the past decade have made this goal unreachable. As of the end of the 2000s, no more than 1 percent of the nation's electricity comes from geothermal sources. Only in California is geothermal energy a significant power source. As an example, GeoProducts Corporation, of Moraga, California, has constructed a $60 million geothermal plant near Lassen National Park that generates 30 megawatts of power. As of 2006, the energy generation in this region has expanded so now three geothermal projects are actively producing electrical power. The Honey Lake Geothermal Area is located in three locations within Lassen County, California, and Washoe County, Nevada.

In 2010, the United States leads the world in geothermal electrical production with about 3,086 megawatts (MW) of installed capacity from 77 power plants. As of the end of 2009, 103 new geothermal projects are actively being developed in the United States. The Philippines is second in the world with 1,904 MW.

Even though geothermal generation of electricity has grown about 3 percent annually in the 2000s, until the U.S. government and the general public becomes more concerned about the potential of various types of alternative energy sources, geothermal is likely to remain a minor energy source in the country as a whole.

See also Fossil fuels; Renewable energy; Water pollution.

Resources

OTHER

Department of Energy. "Fossil Fuels." http://www.energy. gov/energysources/fossilfuels.htm (accessed October 12, 2010).

Energy Efficiency and Renewable Energy, Department of Energy. "Honey Lake Geothermal Area." http:// www.energy.gov/energysources/fossilfuels.htm (accessed October 12, 2010).

Pew Center. "Honey Lake Geothermal Area." http:// www.pewclimate.org/technology/factsheet/geothermal (accessed October 14, 2010).

David E. Newton

Giant panda

Today, the giant panda (*Ailuropoda melanoleuca*) is one of the best known and most popular large mammals among the general public. Although its existence was known long ago, having been mentioned in a 2,500-year-old Chinese geography text, Europeans did not learn of the giant panda until its discovery by a French missionary in 1869. The first living giant panda did not reach the Western Hemisphere until 1937. The giant panda, variously classified with the true bears or, often, in a family of its own, once ranged throughout much of China and Burma, but is now restricted to a series of 13 wildlife reserves totaling just over 2,200 square miles (5,700 km^2) in three central and western Chinese provinces. The giant panda population has been decimated

Panda eating bamboo. *(javarman/Shutterstock.com)*

over the past 2,000 years by hunting and habitat destruction. In the years since 1987, they have lost more than 30 percent of their habitat. Giant pandas are one of the rarest mammals in the world, with current estimates of their population size at about 1,800 individuals (150 reside in captivity). Today, human pressure on giant panda populations has diminished, although poaching continues. Giant pandas are protected by tradition and sentiment, as well as by law in the Chinese mountain forest reserves. Despite this progress, however, IUCN—The International Union for Conservation of Nature and Natural Resources—and the U. S. Fish and Wildlife Service consider the giant panda to be endangered. Some of this species' unique requirements and habits help put them in jeopardy.

The anatomy of the giant panda indicates that it is a carnivore, however, its diet consists almost entirely of bamboo, whose cellulose cannot be digested by the panda. Since the giant panda obtains so little nutrient value from the bamboo, it must eat enormous quantities of the plant each day, about 35 pounds (16 kg) of leaves and stems, in order to satisfy its energy requirements. Whenever possible, it feeds solely on the young succulent shoots of bamboo, which, being mostly water, requires it to eat almost 90 pounds (41 kg) per day. This translates into ten to twelve hours per day that pandas spend eating. Giant pandas have been known to supplement their diet with other plants such as horsetail and pine bark, and they will even eat small animals, such as rodents, if they can catch them, but well over 95 percent of their diet consists of the bamboo plant.

Bamboo normally grows by sprouting new shoots from underground rootstocks. At intervals from forty to 100 years, the bamboo plants blossom, produce seeds, then die. New bamboo then grows from the seed. In some regions it may take up to six years for new plants to grow from seed and produce enough food for the giant panda. Undoubtedly, this has produced large shifts in panda population size over the centuries. Within the last quarter century, two bamboo flowerings have caused the starvation of nearly 200 giant pandas, a significant portion of the current population. Although the wildlife reserves contain sufficient bamboo, much of the vast bamboo forests of the past have been destroyed for agriculture, leaving no alternative areas to move to should bamboo blossoming occur in their current range.

Low fecundity and limited success in captive breeding programs in zoos does not bode well for replenishing any significant losses in the wild population. Although there are 150 pandas in captivity, only about 28 percent are breeding. For the time being, the giant panda population appears stable, a positive sign for one of the world's scarcest and most popular animals.

Resources

BOOKS

Croke, Vicki. *The Lady and the Panda: The True Adventures of the First American Explorer to Bring Back China's Most Exotic Animal.* New York: Random House, 2005.

Johnson, Jinny, and Graham Rosewarne. *Giant Panda.* London: Franklin Watts, 2005.

Lindburg, Donald G., and Karen Baragona. *Giant Pandas: Biology and Conservation.* Berkeley: University of California Press, 2004.

OTHER

International Union for Conservation of Nature and Natural Resources. "IUCN Red List of Threatened Species: *Ailuropoda melanoleuca.*"http://www.iucnredlist.org/apps/redlist/details/712/0 (accessed November 9, 2010).

World Wildlife Fund. "Giant Panda." http://wwf.panda.org/what_we_do/endangered_species/giant_panda/ (accessed August 24, 2010).

Eugene C. Beckham

Giardia

Giardia is the genus (and common) name of a protozoan parasite in the phylum Sarcomastigophora. It was first described in 1681 by Antoni van Leeuwenhoek (called "The Father of Microbiology"), who discovered it in his own stool. The most common species is *Giardia intestinalis* (also called *lamblia*), which is a fairly common parasite found in humans. The disease it causes is called giardiasis.

The trophozoite (feeding) stage is easily recognized by its pear-shaped, bilaterally-symmetrical form with two internal nuclei and four pairs of external flagella; the thin-walled cyst (infective) stage is oval. Both stages are found in the upper part of the small intestine in the mucosal lining. The anterior region of the ventral surface of the troph stage is modified into a sucking disc used to attach to the host's abdominal epithelial tissue. Each troph attaches to one epithelial cell. In extreme cases, nearly every cell will be covered, causing severe symptoms. Infection usually occurs through drinking contaminated water. Symptoms include diarrhea, flatulence (gas), abdominal cramps, fatigue, weight loss, anorexia, and/or nausea and may last for more than five days. Diagnosis is usually done by detecting cysts or trophs of this parasite in fecal specimens.

Giardia has a worldwide distribution. It is more common in warm, tropical regions than in cold regions. Hosts include frogs, cats, dogs, beaver, muskrat, horses, and humans. Children as well as adults can be affected, although it is more common in children. It is highly contagious. Normal infection rate in the United States ranges from 1.5 percent to 20 percent.

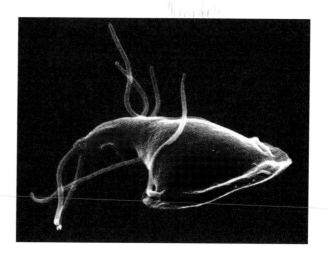

Digitally-colorized scanning electron micrograph (SEM) depicting the dorsal (upper) surface of a *Giardia* protozoan. *(© P-59 Photos / Alamy)*

In one case involving scuba divers from the New York City police and fire fighters, 22–55 percent were found to be infected, presumably after they accidentally drank contaminated water in the local rivers while diving. In another case, an epidemic of giardiasis occurred in Aspen, Colorado, in 1965 during the popular ski season and 120 people were infected. Higher infection rates are common in some areas of the world, including Iran and countries in Sub-Saharan Africa.

Giardia can typically withstand sophisticated forms of sewage treatment, including filtration and chlorination. It is therefore hard to eradicate and may potentially increase in polluted lakes and rivers. For this reason, health officials should make concerted efforts to prevent contaminated feces from infected animals (including humans) from entering lakes used for drinking water.

The most effective treatment for giardiasis is the drug Atabrine (quinacrine hydrochloride). Adult dosage is 0.1 g taken after meals three times each day. Side effects are rare and minimal.

See also Cholera; Coliform bacteria.

Resources

BOOKS

Lujan, Hugo D. *Giardia.* New York: Springer Verlag, 2010.

Ortega-Pierres, Guadalupe. *Giardia and Cryptosporidium: From Molecules to Disease.* Wallingford, UK: CABI, 2009.

Sterling, Charles R., and Rodney D. Adam. *The Pathogenic Enteric Protozoa Giardia, Entamoeba, Cryptosporidium, and Cyclospora.* World Class Parasites, vol. 8. Boston: Kluwer Academic, 2004.

OTHER

Centers for Disease Control and Prevention (CDC). "*Giardia* Infection [Giardiasis]." http://www.cdc.gov/ncidod/dpd/parasites/giardiasis/default.htm (accessed November 12, 2010).

Centers for Disease Control and Prevention (CDC). "Giardiasis." http://www.cdc.gov/ncidod/dpd/parasites/giardiasis/default.htm (accessed November 12, 2010).

John Korstad

Gibbons

Gibbons (genus *Hylobates*, meaning "dweller in the trees") are the smallest members of the ape family which also includes gorillas, chimpanzees, and orangutans.

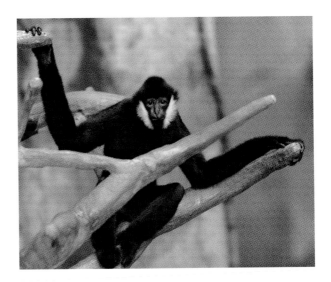

Gibbon in tree. *(Nagel Photography/Shutterstock.com)*

They spend most of their lives at the tops of trees in the jungle, eating leaves and fruit. They are extremely agile, swinging at speeds of 35 miles per hour (56 km/h) with their long arms on branches to move from tree to tree. The trees can even be 50 feet (15 m) apart. They have no tails and are often seen walking upright on tree branches. Gibbons are known for their loud calls and songs, which they use to announce their territory and warn away others. They are devoted parents, raising usually one or two offspring at a time and showing extraordinary affection in caring for them. Conservationists and animal protectionists who have worked with gibbons describe them as extremely intelligent, sensitive, and affectionate.

Gibbons have long been hunted for food, for medical research, and for sale as pets and zoo specimens. A common method of collecting them is to shoot the mother and capture the nursing or clinging infant, if it is still alive. The mortality rate in collecting and transporting gibbons to areas where they can be sold is extremely high. This, coupled with the fact that their jungle habitat is being destroyed at a rate of 32 acres (13 ha) per minute, has resulted in severe depletion of their numbers.

Gibbons are found in southeast Asia, China, and India, and nine species are recognized. All nine species are considered endangered by the U.S. Department of the Interior and are listed in the most endangered category of the Convention on International Trade in Endangered Species of Wild Fauna and Flora (CITES). IUCN—The World Conservation Union considers three species of gibbon to be endangered and two species to be vulnerable. Despite the ban on international trade in gibbons conferred by listing in Appendix I of CITES, illegal trade in gibbons continues on a wide scale in markets throughout Asia.

Resources

BOOKS

Campbell, Christina J. *Primates in Perspective*. New York: Oxford University Press, 2007.

Hart, Donna, and Robert W. Sussman. *Man the Hunted: Primates, Predators, and Human Evolution*. New York: Westview, 2005.

Nystrom, Pia, and Pamela Ashmore. *The Life of Primates*. Upper Saddle River, N.J.: Pearson Prentice Hall, 2008.

Lewis G. Regenstein

Gibbs, Lois M.

1951–
American environmentalist and community organizer

An activist dedicated to protecting communities from hazardous wastes, Lois Gibbs began her political career as a housewife and homeowner near the Love Canal, New York. She was born in Buffalo on June 25, 1951, the daughter of a bricklayer and a full-time

Environmentalist Lois Gibbs speaks to an interviewer at a 1970s rally in West Virginia to raise factory workers' awareness of toxins and the environment. (© *Wally McNamee/ CORBIS*)

homemaker. Gibbs was 21 and a mother when she and her husband bought their house near a buried dump containing hazardous materials from industry and the military, including wastes from the research and manufacture of chemical weapons.

From the time the first articles about Love Canal began appearing in newspapers in 1978, Gibbs has petitioned for state and federal assistance. She began when she discovered the school her son was attending had been built directly on top of the buried canal. Her son had developed epilepsy and there were many similar, unexplained disorders among other children at the school, yet the superintendent was refusing to transfer anyone. The New York State Health Department then held a series of public meetings in which officials appeared more committed to minimizing the community perception of the problem than to solving the problem itself. The governor made promises he was unable to keep, and Gibbs herself was flown to Washington to appear at the White House for what she later decided was little more than political grandstanding. In the book she wrote about her experience, *Love Canal: My Story*, Gibbs describes her frustration and her increasing disillusionment with government, as the threats to the health of both adults and children in the community became more obvious and as it became clearer that no one would be able to move because no one could sell their homes.

While state and federal agencies delayed, the media took an increasing interest in their plight, and Gibbs became more involved in political action. To force federal action, Gibbs and a crowd of supporters took two officers from the Environmental Protection Agency (EPA) hostage. A group of heavily armed FBI agents occupied the building across the street and gave her seven minutes before they stormed the offices of the Homeowners' Association, where the men were being held. With less than two minutes left in the countdown, Gibbs appeared outside and released the hostages in front of a national television audience. By the middle of the next week, the EPA had announced that the Federal Disaster Assistance Administration would fund immediate evacuation for everyone in the area.

But the families who left the Love Canal area still could not sell their homes, and Gibbs fought to force the federal government to purchase them and underwrite low-interest loans. After she accused President Jimmy Carter of inaction on a national talk show, in the midst of an approaching election, he agreed to purchase the homes. But he refused to meet with her to discuss the loans. Carter signed the appropriations bill in a televised ceremony at the Democratic National

Convention in New York City, and Gibbs simply walked onstage in the middle of it and repeated her request for mortgage assistance. The president could do nothing but promise his political support, and the assistance she had been asking for was soon provided.

Gibbs was divorced soon after her family left the Love Canal area. She moved to Washington, D.C. with her two children and founded the Citizen's Clearinghouse for Hazardous Wastes in 1981 (later renamed the Center for Health, Environment and Justice, CHEJ, in 1997). Its purpose is to assist communities in fighting toxic waste problems, particularly plans for toxic waste dumping sites, and the organization has worked with over 7,000 neighborhood and community groups. Gibbs has also published *Dying from Dioxin, A Citizen's Guide to Reclaiming Our Health and Rebuilding Democracy*. She has appeared on many television and radio shows and has been featured in hundreds of newspaper and magazine articles. Gibbs has also been the subject of several documentaries and television movies. She often speaks at conferences and seminars and has been honored with numerous awards, including the prestigious Goldman Environmental Prize in 1991, Heinz Award (1998), and two honorary doctorates. In 2003, she was nominated for the Nobel Peace Prize.

Because of Gibbs' activist work, no commercial sites for hazardous wastes have been opened in the United States since 1978.

As of 2010, Gibbs serves as the Executive Director of CHEJ and continues to be a vocal proponent of environmental justice.

Resources

BOOKS

Gibbs, L. *Love Canal: My Story*. Albany: State University of New York Press, 1982.

Lewis G. Regenstein
Douglas Smith

▌Gill nets

Gill nets are panels of diamond-shaped mesh netting used for catching fish. When fish attempt to swim through the net their gill covers get caught and they cannot back out. Depending on the target species, different mesh sizes are available for use. The top line of the net has a series of floats attached for buoyancy, and the bottom line has lead weights to hold the net vertically in the water column.

Gill nets have been in use for many years. They became popular in commercial fisheries in the nineteenth century, evolving from cotton twine netting to the more modern nylon twine netting and monofilament nylon netting. As with many other aspects of commercial fishing, the use of gill nets has developed from minor utilization to a major environmental issue. Coupled with overfishing, the use of gill nets has caused serious concern throughout the world.

Because gill nets are so efficient at catching fish, they are just as efficient at catching many non-target species, including other fishes, sea turtles, sea mammals, and sea birds. Gill nets have been used extensively in the commercial fishery for salmon and capelin (*Mallotus villosus*). Dolphins, seals, and sea otters (*Enhydra lutris*) get tangled in the nets, as do diving sea birds such as murres, guillemots, auklets, and puffins that rely on capelin as a mainstay in their diet. Sea turtles are also entangled and drown.

The introduction and extensive use of drift nets by fishing fleets based outside of the U.S. is also problematic. Described as "the most indiscriminate killing device used at sea," drift nets are monofilament gill nets up to 40 miles (64 km) in length. Left at sea for several days and then hauled on board a fishing vessel, these drift nets contain vast numbers of dead marine life, besides the target species, that are simply discarded over the side of the boat. The outrage expressed regarding these "curtains of death" led to a 1992 United Nations resolution banning their use in commercial fishing. Commercial fishermen who use other types of nets for catching fish, such as the purse seines used in the tuna fishing industry and the bag trawls used in the shrimping industry, have modified their nets and fishing techniques to attempt to eliminate the killing of dolphins and sea turtles, respectively. Unfortunately, such modifications of gill nets are nearly impossible due to the nets' design and the way these nets are used.

See also Turtle excluder device.

Resources

BOOKS

Clover, Charles. *The End of the Line: How Overfishing Is Changing the World and What We Eat*. Berkeley: University of California Press, 2008.

Russell, Denise. *Who Rules the Waves?: Piracy, Overfishing and Mining the Oceans*. London: Pluto Press, 2010.

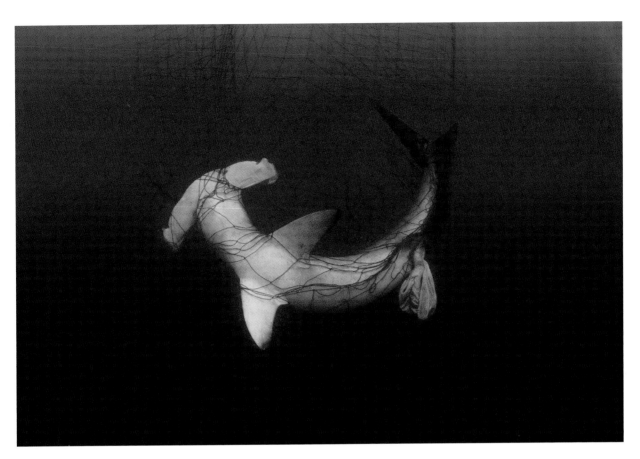

Hammerhead shark (*Sphyrna lewini*) in drift gill net, Sea of Cortez, Mexico. *(Mark Conlin/photolibrary.com)*

OTHER

United States Environmental Protection Agency (EPA). "Industry: Industries: Fishing Industry." http://www.epa.gov/ebtpages/induindustriesfishingindustry.html (accessed October 25, 2010).

Eugene C. Beckham

GIS *see* **Geographic information systems.**

Glaciation

The covering of Earth's surface with glacial ice. The term also includes the alteration of the surface of the earth by glacial erosion or deposition. Due to the passage of time, ice erosion can be almost unidentifiable; the weathering of hard rock surfaces often eliminates minor scratches and other evidence of such glacial activities as the carving of deep valleys. The evidence of deposition, known as depositional imprints, can vary. It may consist of specialized features a few meters above the surrounding terrain, or it may consist of ground materials several meters in thickness covering wide areas of the landscape.

Only 10 percent of Earth's surface is currently covered with glacial ice, but it is estimated that 30 percent has been covered with glacial ice at earlier times. During the last major glacial period, most of Europe and more than half of the North American continent were covered with ice. The glacial ice of modern day is much thinner than it was in the ice age, and the majority of it (over 85 percent) is found in Antarctica. About 11 percent of the remaining glacial ice is in Greenland, and the rest is scattered in high altitudes throughout the world.

Moisture and cold temperatures are the two main factors for the formation of glacial ice. Glacial ice in Antarctica is the result of relatively small quantities of annual snow deposition and low loss of ice because of the cold climate. In the middle and low latitudes where ice loss, known as ablation, is higher, snowfall also tends to be much higher and glaciers overcome ablation by generating large amounts of ice. These types of systems tend to be more active than the glaciers in Antarctica, and the most active of these are often

campsites and have been a major mechanism for the removal of silt from backwater channels used by native fish. Depending on the final results of this study, the Bureau will determine what changes, if any, should be taken in adjusting flow patterns over the dam to provide for maximum environmental benefit downstream along with power output.

Based on the results of the earlier EIS, the Bureau of Reclamation decided to perform another environmental impact statement in 2006. This new EIS, called the Long-Term Experimental Plan Environmental Impact Statement (LTEP EIS), was intended to help protect the natural resources further down steam of Glen Canyon Dam. It was also necessitated on the settlement agreement made through the litigation record of *Center for Biodiversity et al. v. Kempthorne.* Based on the latest news from the Bureau of Reclamation, as of October 21, 2009, the EIS has yet to be completed, being indefinitely suspended.

See also Alternative energy sources; Riparian Land; Wild river.

Resources

BOOKS

Hamill, John F. *Status and trends of resources below Glen Canyon Dam update, 2009.* Reston, VA: U.S. Department of the Interior, U.S. Geological Survey, 2009.

Powell, James Lawrence. *Dead Pool: Lake Powell, Global Warming, and the Future of Water in the West.* Berkeley: University of California Press, 2008.

OTHER

Bureau of Reclamation, U.S. Department of the Interior. "Glen Canyon Dam Construction History." http://www.glencanyonnha.org/ (accessed November 4, 2010).

Bureau of Reclamation, U.S. Department of the Interior. "Glen Canyon Dam: Long-Term Experimental Plan." http://www.usbr.gov/uc/rm/gcdltep/index.html (accessed November 4, 2010).

Glen Canyon Natural History Association. "Home web page." http://www.glencanyonnha.org/ (accessed November 4, 2010).

David E. Newton

Global Environment Monitoring System

A data-gathering project administered by the United Nations Environment Programme. The Global Environment Monitoring System (GEMS) is one aspect of the modern understanding that environmental problems ranging from the greenhouse effect and ozone layer depletion to the preservation of biodiversity are international in scope. The system was inaugurated in 1975, and it monitors weather and climate changes around the world, as well as variations in soils, the health of plant and animal species, and the environmental impact of human activities.

GEMS was not intended to replace any existing systems; it was designed to coordinate the collection of data on the environment, encouraging other systems to supply information it believed was being omitted. In addition to coordinating the gathering of this information, the system also publishes it in an uniform and accessible fashion, where it can be used and evaluated by environmentalists and policy makers.

GEMS operates twenty-five information networks in over 145 countries. These networks monitor air pollution, including the release of greenhouse gases and changes in the ozone layer, and air quality in various urban center; they also gather information on water quality and food contamination in cooperation with the World Health Organization and the Food and Agriculture Organization (FAO) of the United Nations.

Resources

BOOKS

Spellerberg, Ian F., and Martin W. Holdgate. *Monitoring Ecological Change.* Malden, MA: Blackwell Science Inc., 2005.

Global Releaf

Global Releaf, an international citizen action and education program, was initiated in 1988 by the 115-year-old American Forestry Association in response to the worldwide concern over global warming and the greenhouse effect. Campaigning under the slogan "Plant a tree, cool the globe," its over 112,000 members began the effort to reforest the earth one tree at a time.

In 1990, Global Releaf began Global Releaf Forest, an effort to restore damaged habitat on public lands through tree plantings Global Releaf Fund is its urban counterpart. Using each one-dollar donation to plant one tree resulted in the planting of more than four million trees on seventy sites in thirty-three states. By involving local citizens and resource experts in each project, the program ensures that the right species are planted in the right place at the right time. Results include the protection of endangered and threatened animals, restoration of native species, and improvement of recreational opportunities.

Funding for the program has come largely from government agencies, corporations, and non-profit organizations. Chevrolet-Geo celebrated the planting of its millionth tree in October 1996. The Texaco/Global Releaf Urban Tree Initiative, utilizing more than 6,000 Texaco volunteers, has helped local groups plant more than 18,000 large trees and invested over $2.5 million in projects in twelve cities. Outfitter Eddie Bauer began an "Add a Dollar, Plant a Tree" program to fund eight Global Releaf Forest sites in the United States and Canada, planting close to 350,000 trees.

The Global Releaf Fund also helps finance urban and rural reforestation on foreign soil in projects undertaken with its international partners. Engine manufacturer, Briggs & Stratton, for example, has made possible tree plantings both in the United States and in Ecuador, England, Germany, Poland, Romania, Slovakia, South Africa, and Ukraine, while Costa Rica, Gambia, and the Philippines have benefitted from picture-frame manufacturer Larsen-Juhl.

Unfortunately, not enough funding exists to grant all the requests; in 1996, only 40 percent of the proposed projects received financial backing. Forced to pick and choose, the review board favors those projects which aim to protect endangered and threatened species. Burned forests and natural disaster areas—like the Francis Marion National Forest in South Carolina, devastated by 1989's Hurricane Hugo—are also high on the priority list, as are streamside woodlands and landfills.

Looking to the future, Global Releaf 2000 was launched in 1996 with the aim of encouraging the planting of 20 million trees, increasing the canopy in select cities by 20 percent, and expanding the program to include private lands and sanitary landfills. A twenty-city survey done in 1985 by American Forests showed that four trees die for every one planted in United States cities and that the average city tree lives only thirty-two years (just seven years, downtown). With these facts in mind, Global Releaf asks that communities plant twice as many trees as are lost in the next decade. In August of 2001, more than 19 million trees had been planted. By the end of 2010, an estimated 35 million trees had been planted across the United States and internationally. Releaf's goal is to plant 100 million trees by 2020.

Resources

BOOKS

Mansourian, Stephanie, Daniel Vallauri, and Nigel Dudley. *Forest Restoration in Landscapes: Beyond Planting Trees.* New York, NY: Springer, 2005.

Perry, David A., Ram Oren, and Stephen C. Hart. *Forest Ecosystems.* Baltimore: Johns Hopkins University Press, 2008.

OTHER

United Nations System-Wide EarthWatch. "Forest loss." http://earthwatch.unep.net/emergingissues/forests/forestloss.php (accessed September 29, 2010).

ORGANIZATIONS

American Forests, P.O. Box 2000, Washington, D.C., USA, 20013, (202) 737 1944, info@amfor.org, http://www.americanforests.org

Ellen Link

Global warming *see* **Greenhouse effect.**

GOBO *see* **Child survival revolution.**

Goiter

Generally refers to an abnormal enlargement of the thyroid gland. The most common type of goiter, the simple goiter, is caused by a deficiency of iodine in the diet. In an attempt to compensate for this deficiency, the thyroid gland enlarges and may become the size of a large softball in the neck. The general availability of table salt to which potassium iodide has been added ("iodized" salt) has greatly reduced the incidence of simple goiter in many parts of the world. A more serious form of goiter, toxic goiter, is associated with hyperthyroidism. The etiology of this condition is not well understood. A third form of goiter occurs primarily in women and is believed to be caused by changes in hormone production.

Golf courses

The game of golf appears to be derived from ancient stick-and-ball games long played in western Europe. However, the first documented rules of golf were established in 1744, in Edinburgh, Scotland. Golf was first played in the United States in the 1770s, in Charleston, South Carolina. It was not until the 1880s, however, that the game began to become widely popular, and it has increasingly flourished since then. In 2010, there were over 17,000 golf courses in the United States, and many thousands more throughout much of the rest of the world.

Golfers playing golf. (phloen/Shutterstock.com)

Golf is an excellent form of outdoor recreation. There are many health benefits of the game, associated with the relatively mild form of exercise and extensive walking that can be involved. However, the development and management of golf courses also results in environmental damage of various kinds. The damage associated with golf courses can engender intense local controversy, both for existing facilities and when new ones are proposed for development.

The most obvious environmental affect of golf courses is associated with the large amounts of land that they appropriate from other uses. Depending on its design, a typical 18-hole golf course may occupy an area of about 100–200 acres. If the previous use of the land was agricultural, then conversion to a golf course results in a loss of food production. Alternatively, if the land previously supported forest or some other kind of natural ecosystem, then the conversion results in a large, direct loss of habitat for native species of plants and animals.

In fact, some particular golf courses have been extremely controversial because their development caused the destruction of the habitat of endangered species or rare kinds of natural ecosystems. For instance, the Pebble Beach Golf Links course, one of the most famous in the world, was developed in 1919 on the Monterey Peninsula of central California, in natural coastal and forest habitats that harbor numerous rare and endangered species of plants and animals. Several additional golf courses and associated tourist facilities were subsequently developed nearby, all of them also displacing natural ecosystems and destroying the habitat of rare species. Most of those recreational facilities were developed at a time when not much attention was paid to the needs of endangered species. Today, however, the conservation of biodiversity is considered an important issue. It is quite likely that if similar developments were now proposed in such critical habitats, citizen groups would mount intense protests and government regulators would not allow the golf courses to be built.

The most intensively modified areas on golf courses are the fairways, putting greens, aesthetic lawns and gardens, and other highly managed areas. Because these kinds of areas are intrinsic to the design of golf courses, a certain amount of loss of natural habitat is inevitable. To some degree, however, the net amount of habitat loss can be decreased by attempting, to the degree possible, to retain natural community types within the golf course. This can be done particularly effectively in the brushy and forested areas between the holes and their approaches. The habitat quality in these less-intensively managed areas can also be enhanced by providing nesting boxes and brush piles for use by birds and small mammals, and by other management practices known to favor wildlife. Habitat quality is also improved by planting native species of plants wherever it is feasible to do so.

In addition to land appropriation, some of the management practices used on golf courses carry the risk of causing local environmental damage. This is particularly the case of putting greens, which are intensively managed to maintain an extremely even and consistent lawn surface.

For example, to maintain a monoculture of desired species of grasses on putting greens and lawns, intensive management practices must be used. These include frequent mowing, fertilizer application, and the use of a variety of pesticidal chemicals to deal with various pests affecting the turfgrass. This may involve the application of such herbicides as Roundup (glyphosate), 2,4-D, MCPP, or Dicamba to deal with undesirable weeds. Herbicide application is particularly necessary when putting greens and lawns are being first established. Afterward their use can be greatly reduced by only using spot-applications directly onto turf-grass weeds. Similarly, fungicides might be used to combat infestations of turf-grass disease fungi, such as the fusarium blight (*Fusarium culmorum*), take-all patch (*Gaeumannomyces graminis*), and rhizoctonia blight (*Rhizoctonia solani*).

Infestations by turf-damaging insects may also be a problem, which may be dealt with by one or more insecticide applications. Some important insect pests of golf-course turfgrasses include the Japanese beetle (*Popillia japonica*), chafer beetles (*Cyclocephala* spp.), June beetles (*Phyllophaga* spp.), and armyworm beetle (*Pseudaletia unipuncta*). Similarly, rodenticides may be needed to get rid of moles (*Scalopus aquaticus*) and their burrows.

Golf courses can also be major users of water, mostly for the purposes of irrigation in dry climates, or during droughty periods. This can be an important problem in semi-arid regions, such as much of the southwestern United States, where water is a scare and valuable commodity with many competing users. To some degree, water use can be decreased by ensuring that irrigation is only practiced when necessary,

and only in specific places where it is needed, rather than according to a fixed schedule and in a broadcast manner. In some climatic areas, nature-scaping and other low-maintenance practices can be used over extensive areas of golf courses. This can result in intensive irrigation only being practiced in key areas, such as putting greens, and to a lesser degree fairways and horticultural lawns.

Many golf courses have ponds and lakes embedded in their spatial design. If not carefully managed, these waterbodies can become severely polluted by nutrients, pesticides, and eroded materials. However, if care is taken with golf-course management practices, their ponds and lakes can sustain healthy ecosystems and provide refuge habitat for local native plants and animals.

Increasingly, golf-course managers and industry associations are attempting to find ways to support their sport while not causing an unacceptable amount of environmental damage. One of the most important initiatives of this kind is the Audubon Cooperative Sanctuary Program for Golf Courses, run by the Audubon International, a private conservation organization. The program provides environmental education and conservation advice to golf-course managers and designers.

The Audubon Cooperative Sanctuary Program for Golf Courses provides advice to help planners and managers with: (a) environmental planning; (b) wildlife and habitat management; (c) chemical use reduction and safety; (d) water conservation; and (e) outreach and education about environmentally appropriate management practices. If a golf course completes recommended projects in all of the components of the program, it receives recognition as a Certified Audubon Cooperative Sanctuary. This allows the golf course to claim that it is conducting its affairs in a certifiably "green" manner. This results in tangible environmental benefits of various kinds, while being a source of pride of accomplishment for employees and managers, and providing a potential marketing benefit to a clientele of well-informed consumers.

There are many specific examples of environmental benefits that have resulted from golf courses engaged in the Audubon Cooperative Sanctuary Program. For instance, seven golf courses in Arizona and Washington have allowed the installation of 150 artificial nesting burrows for burrowing owls (*Athene cunicularia*), an endangered species, on suitable habitat on their land. Audubon International conducted a survey of cooperating golf courses, and the results were rather impressive. About 78 percent of the respondents reported that they had decreased the total amount of turfgrass area on their property; 73 percent had taken steps to increase the amount of wildlife habitat; 45 percent were engaged in an ecosystem restoration project; 90 percent were attempting to use native plants in their horticulture; and 85 percent had decreased their use of pesticides and 91 percent had switched to lower-toxicity chemicals. Just as important, about half of the respondents believed that there had been an improvement in the playing quality of their golf course and in the satisfaction of both employees and their client golfers. Moreover, none of the respondents believed that any of these values had been degraded as a result of adopting the management practices advised by the Audubon International program.

These are all highly positive indicators. They suggest that the growing and extremely popular sport of golf can, within limits, potentially be practiced in ways that do not cause unacceptable levels of environmental and ecological damage.

Resources

BOOKS

Virginia, and Chesapeake Bay Program (U.S.). *Golfing Green Virginia: Golf Course Environmental Stewardship.* Richmond, VA: Virginia Dept. of Conservation & Recreation, 2004.

OTHER

United States Department of the Interior, United States Geological Survey (USGS). "Water Use." http://www.usgs.gov/science/science.php?term = 1313 (accessed November 8, 2010)

ORGANIZATIONS

United States Golf Association, P.O. Box 708, Far Hills, N.J., USA, 07931, (908) 781–1735, usga.org, http://www.usga.org

Bill Freedman

Good wood

Good wood, or smart wood, is a term certifying that the wood is harvested from a forest operating under environmentally sound and sustainable practices. A "certified wood" label indicates to consumers that the wood they purchase comes from a forest operating within specific guidelines designed to ensure future use of the forest. A well-managed forestry operation takes into account the overall health of the forest and its ecosystems, the use of the forest by indigenous people and cultures, and the economic influences the forest has on local communities. Certification of wood allows the wood to be traced from harvest through processing to the final product (i.e., raw wood or an item made from wood) in an attempt to reduce uncontrollable deforestation, while meeting the demand for wood and wood products by consumers around the world.

Public concern regarding the disappearance of tropical forests initially spurred efforts to reduce the destruction of vast acres of rainforests by identifying environmentally responsible forestry operations and encouraging such practices by paying foresters higher prices. Certification, however, is not limited to tropical forests. All forest types—tropical, temperate, and boreal (those located in northern climes)—from all countries may apply for certification. Plantations (stands of timber that have been planted for the purpose of logging or that have been altered so that they no longer support the ecosystems of a natural forest) may also apply for certification.

Certification of forests and forest owners and managers is not required. Rather, the process is entirely voluntary. Several organizations currently assess forests and forest management operations to determine whether they meet the established guidelines of a well-managed, sustainable forest. The Forest Stewardship Council (FSC), founded in 1993, is an organization of international members with environmental, forestry, and socioeconomic backgrounds that monitors these organizations and verifies that the certification they issue is legitimate.

A set of 10 guiding principles known as Principles and Criteria (P&C) were established by the FSC for certifying organizations to utilize when evaluating forest management operations. The P&C address a wide range of issues, including compliance with local, national, and international laws and treaties; review of the forest operation's management plans; the religious or cultural significance of the forest to the indigenous inhabitants; maintenance of the rights of the indigenous people to use the land; provision of jobs for nearby communities; the presence of threatened or endangered species; control of excessive erosion when building roads into the forest; reduction of the potential for lost soil fertility as a result of harvesting; protection against the invasion of non-native species; pest management that limits the use of certain chemical types and of genetically altered organisms; and protection of forests when deemed necessary (for example, a forest that protects a watershed or that contains threatened and/or endangered species).

Guarding against illegal harvesting is a major hurdle for those forest managers working to operate within the established regulations for certification. Forest devastation occurs not only from harvesting timber for wood sales but when forests are clear cut to make way for cattle crazing or farming, or to provide a fuel source for local inhabitants. Illegal harvesting often occurs in developing countries where enforcement against such activities is limited (for example, the majority of the trees harvested in Indonesia are done so illegally).

Critics argue against the worthiness of managing forests, suggesting that the logging of select trees from a forest should be allowed and that once completed, the remaining forest should be placed off limits to future logging. Nevertheless, certified wood products are in the market place; large wood and wood product suppliers are offering certified wood and wood products to their consumers. In 2001, the Forest Leadership Forum (a group of environmentalists, forest industry representatives, and retailers) met to identify how wood retailers can promote sustainable forests. It is hoped that consumer demand for good wood will drive up the number of forests participating in the certification program, thereby reducing the rate of irresponsible deforestation of the world's forests.

Resources

BOOKS

Ravenel, Ramsey, Ilmi Granoff, and Carrie Magee. *Illegal Logging in the Tropics*. Boca Raton, FL: CRC Press, 2005.
Tacconi, Luca. *Illegal Logging: Law Enforcement, Livelihoods and the Timber Trade*. London: Earthscan Publications, 2007.

ORGANIZATIONS

Forest Stewardship Council United States, 212 Third Avenue N, Suite 504, Minneapolis, MN, USA, 55401, (612) 353-4511, (612) 208-1565, info@fscus.org, http://www.fscus.org

Monica Anderson

▌Goodall, Jane

1934–
English primatologist and ethnologist

Jane Goodall is known worldwide for her studies of the chimpanzees of the Gombe Stream National Park in Tanzania, Africa. She is well respected within the scientific community for her ground-breaking field studies and is credited with the first recorded observation of chimps eating meat and using and making tools. Because of Goodall's discoveries, scientists have been forced to redefine the characteristics once considered as solely human traits. Goodall is now leading efforts to ensure that animals are treated humanely both in their wild habitats and in captivity.

Goodall was born in London, England, on April 3, 1934, to Mortimer Herbert Goodall, a businessperson and motor-racing enthusiast, and the former Margaret Myfanwe Joseph, who wrote novels under the name Vanne Morris Goodall. Along with her sister, Judy, Goodall was reared in London and Bournemouth, England. Her fascination with animal behavior began in early childhood. In her leisure time, she observed native birds and animals, making extensive notes and sketches, and read widely in the literature of zoology and ethnology. From an early age, she dreamed of traveling to Africa to observe exotic animals in their natural habitats.

Goodall attended the Uplands private school, receiving her school certificate in 1950 and a higher certificate in 1952. At age eighteen she left school and found employment as a secretary at Oxford University. In her spare time, she worked at a London-based documentary film company to finance a long-anticipated trip to Africa. At the invitation of a childhood friend, she visited South Kinangop, Kenya. Through other friends, she soon met the famed anthropologist Louis Leakey, then curator of the Coryndon Museum in Nairobi. Leakey hired her as a secretary and invited her to participate in an anthropological dig at the now famous Olduvai Gorge, a site rich in fossilized prehistoric remains of early ancestors of humans. In addition, Goodall was sent to study the vervet monkey, which lives on an island in Lake Victoria.

Leakey believed that a long-term study of the behavior of higher primates would yield important evolutionary information. He had a particular interest in the chimpanzee, the second most intelligent primate. Few studies of chimpanzees had been successful; either the size of the safari frightened the chimps, producing unnatural behaviors, or the observers spent too little time in the field to gain comprehensive knowledge. Leakey believed that Goodall had the proper temperament to endure long-term isolation in the wild. At his prompting, she agreed to attempt such a study. Many experts objected to Leakey's selection of Goodall because she had no formal scientific education and lacked even a general college degree.

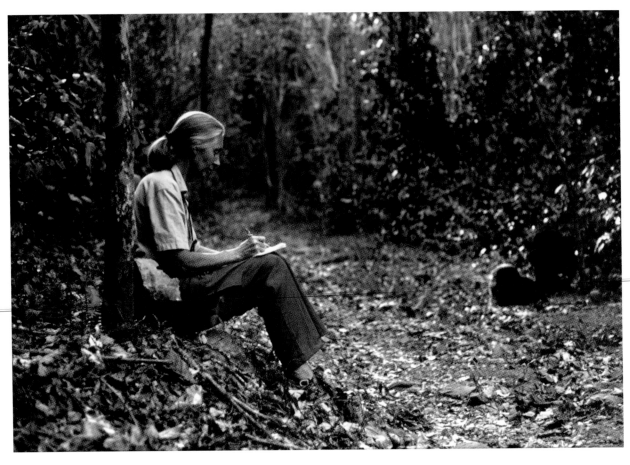

Scientist Jane Goodall studies the behavior of a chimpanzee during her research in Tanzania in 1987. *(Penelope Breese/Liaison/ Getty Images)*

While Leakey searched for financial support for the proposed Gombe Reserve project, Goodall returned to England to work on an animal documentary for Granada Television. On July 16, 1960, accompanied by her mother and an African cook, she returned to Africa and established a camp on the shore of Lake Tanganyika in the Gombe Stream Reserve. Her first attempts to observe closely a group of chimpanzees failed; she could get no nearer than 500 yards (457 m) before the chimps fled. After finding another suitable group of chimpanzees to follow, she established a nonthreatening pattern of observation, appearing at the same time every morning on the high ground near a feeding area along the Kakaombe Stream valley. The chimpanzees soon tolerated her presence and, within a year, allowed her to move as close as 30 feet (9 m) to their feeding area. After two years of seeing her every day, they showed no fear and often came to her in search of bananas.

Goodall used her newfound acceptance to establish what she termed the "banana club," a daily systematic feeding method she used to gain trust and to obtain a more thorough understanding of everyday chimpanzee behavior. Using this method, she became closely acquainted with more than half of the reserve's one hundred or more chimpanzees. She imitated their behaviors, spent time in the trees, and ate their foods. By remaining in almost constant contact with the chimps, she discovered a number of previously unobserved behaviors. She noted that chimps have a complex social system, complete with ritualized behaviors and primitive but discernible communication methods, including a primitive "language" system containing more than twenty individual sounds. She is credited with making the first recorded observations of chimpanzees eating meat and using and making tools. Tool making was previously thought to be an exclusively human trait, used, until her discovery, to distinguish man from animal. She also noted that chimpanzees throw stones as weapons, use touch and embraces to comfort one another, and develop long-term familial bonds. The male plays no active role in family life but is part of the group's social stratification. The chimpanzee "caste" system places the dominant males at the top. The lower castes often act obsequiously in their presence, trying to ingratiate themselves to avoid possible harm. The male's rank is often related to the intensity of his entrance performance at feedings and other gatherings.

Ethologists had long believed that chimps were exclusively vegetarian. Goodall witnessed chimps stalking, killing, and eating large insects, birds, and some bigger animals, including baby baboons and bushbacks (small antelopes). On one occasion, she recorded acts of cannibalism. In another instance, she observed chimps inserting blades of grass or leaves into termite hills to lure worker or soldier termites onto the blade. Sometimes, in true toolmaker fashion, they modified the grass to achieve a better fit. Then they used the grass as a long-handled spoon to eat the termites.

In 1962, Baron Hugo van Lawick, a Dutch wildlife photographer, was sent to Africa by the National Geographic Society to film Goodall at work. Goodall and van Lawick married on March 28, 1964. Their European honeymoon marked one of the rare occasions on which Goodall was absent from Gombe Stream. Her other trips abroad were necessary to fulfill residency requirements at Cambridge University, where she received a Ph.D. in ethology in 1965, becoming only the eighth person in the university's long history who was allowed to pursue a Ph.D. without first earning a baccalaureate degree. Her doctoral thesis, "Behavior of the Free- Ranging Chimpanzee," detailed her first five years of study at the Gombe Reserve.

Van Lawick's film, *Miss Goodall and the Wild Chimpanzees,* was first broadcast on American television on December 22, 1965. The film introduced the shy, attractive, unimposing yet determined Goodall to a wide audience. Goodall, van Lawick (along with their son, Hugo, born in 1967), and the chimpanzees soon became a staple of American and British public television. Through these programs, Goodall challenged scientists to redefine the long-held "differences" between humans and other primates.

Goodall's fieldwork led to the publication of numerous articles and five major books. She was known and respected first in scientific circles and, through the media, became a minor celebrity. *In the Shadow of Man,* her first major text, appeared in 1971. The book, essentially a field study of chimpanzees, effectively bridged the gap between scientific treatise and popular entertainment. Her vivid prose brought the chimps to life, although her tendency to attribute human behaviors and names to chimpanzees struck some critics being as manipulative. Her writings reveal an animal world of social drama, comedy, and tragedy where distinct and varied personalities interact and sometimes clash.

From 1970 to 1975, Goodall held a visiting professorship in psychiatry at Stanford University. In 1973, she was appointed honorary visiting professor of Zoology at the University of Dar es Salaam in Tanzania, a position she still holds. Goodall and van Lawick divorced, and she wed Derek Bryceson, a former member of Parliament, in 1975. Bryceson died in 1980. Goodall's life continued to revolve around Gombe Stream. After attending a 1986 conference in Chicago that focused on the ethical treatment of chimpanzees, however, she began directing her energies more toward educating the public about the wild chimpanzee's endangered habitat and the unethical treatment of chimpanzees used for scientific research.

To preserve the wild chimpanzee's environment, Goodall encourages African nations to develop nature-friendly tourism programs, a measure that makes wildlife into a profitable resource. She actively works with business and local governments to promote ecological responsibility. Her efforts on behalf of captive chimpanzees have taken her around the world on a number of lecture tours. She outlined her position strongly in her 1990 book *Through a Window:* "The more we learn of the true nature of non-human animals, especially those with complex brains and corresponding complex social behaviour, the more ethical concerns are raised regarding their use in the service of man-whether this be in entertainment, as 'pets,' for food, in research laboratories or any of the other uses to which we subject them. This concern is sharpened when the usage in question leads to intense physical or mental suffering-as is so often true with regard to vivisection."

Goodall's stance is that scientists must try harder to find alternatives to the use of animals in research. She has openly declared her opposition to militant animal rights groups who engage in violent or destructive demonstrations. Extremists on both sides of the issue, she believes, polarize thinking and make constructive dialogue nearly impossible. While she is reluctantly resigned to the continuation of animal research, she feels that young scientists must be educated to treat animals more compassionately. "By and large," she has written, "students are taught that it is ethically acceptable to perpetrate, in the name of science, what, from the point of view of animals, would certainly qualify as torture."

Goodall's efforts to educate people about the ethical treatment of animals extends to young children as well. Her 1989 book, *The Chimpanzee Family Book*, was written specifically for children, to convey a new, more humane view of wildlife. The book received the 1989 Unicef/Unesco Children's Book-of-the-Year award, and Goodall used the prize money to have the text translated into Swahili. It has been distributed throughout Tanzania, Uganda, and Burundi to educate children who live in or near areas populated by chimpanzees. A French version has also been distributed in Burundi and Congo.

In recognition of her achievements, Goodall has received numerous honors and awards, including the Gold Medal of Conservation from the San Diego Zoological Society in 1974, the J. Paul Getty Wildlife Conservation Prize in 1984, the Schweitzer Medal of the Animal Welfare Institute in 1987, the National Geographic Society Centennial Award in 1988, and the Kyoto Prize in Basic Sciences in 1990. In 1995, Goodall was presented with a CBE (Commander of the British Empire) from Queen Elizabeth II. She was presented with a DBE (Dame of the British Empire) by Prince Charles in 2003. Many of Goodall's endeavors are conducted under the auspices of the Jane Goodall Institute for Wildlife Research, Education, and Conservation, a nonprofit organization located in Ridgefield, Connecticut. In April 2002, Goodall was chosen to be a United Nations Messenger of Peace. She has received numerous honorary doctorates. In 2008, Goodall received the Leakey Prize in human evolutionary science from the Leakey Foundation.

As of 2010, Goodall continues to lecture and write extensively, and remains devoted to chimpanzee welfare and environmental issues. On behalf of the Jane Goodall Institute, she continues to globe-trot over 300 days each year.

Resources

BOOKS

Goodall, Jane. *Jane Goodall: 50 Years at Gombe*. New York: Stewart, Tabori & Chang, 2010.

Peterson, Dale. *Jane Goodall: The Woman Who Redefined Man*. Boston: Houghton Mifflin Co, 2006.

Goodall, Jane, Thane Maynard, and Gail E. Hudson. *Hope for Animals and Their World: How Endangered Species Are Being Rescued from the Brink*. New York: Grand Central Pub., 2009.

Tom Crawford

Gopherus agassizii see **Desert tortoise.**

Gore Jr., Albert

1948–

American former U.S. representative, senator and vice president of the United States

Albert Gore Jr. is an author, businessman, environmental activist, and was forty-fifth Vice-President of the United States in the administration of Bill Clinton. Following his stint in the White House, Gore has become world-famous for his staunch defense of the environment, particularly for his book and video documentary *An Inconvenient Truth*. Gore has received numerous awards for his efforts, including the Nobel Peace Prize.

Gore was the Democratic Party presidential candidate in the 2000 U.S. election. While receiving the most popular votes, he lost the Electoral College to George W. Bush in a very contentious and controversial election.

Gore was born and raised in Washington, D.C., where his father was a well-known and widely respected representative and later senator from Tennessee. He attended St. Alban's Episcopal School for Boys, where he excelled both academically and athletically. He later went to Harvard, earning a bachelor's degree in government. After graduation he enlisted in the army, serving as a reporter in Vietnam, even though he opposed the war. After completing his tour of duty, in 1974 Gore entered the law school at Vanderbilt University. Following in his father's footsteps, Gore ran for Congress, was elected, and served five terms before running for and winning a Senate seat in 1984. He served in the Senate until 1992, when then-governor and Democratic presidential candidate Bill Clinton selected him as his vice presidential running-mate. After winning the 1992 election, Vice President Gore became the Clinton administration's chief environmental advisor. He was also largely responsible for President Clinton's selection of Carol Browner as head of the Environmental Protection Agency (EPA) and Bruce Babbit as Secretary of the Interior.

A self-described "raging moderate," Gore for more than two decades championed environmental causes and drafted and sponsored environmental legislation in the Senate. He was one of two U.S. senators to attend and take an active part in the 1992 U.N.-sponsored Rio Summit on the environment, and after he became vice president in 1993, he took a leading role in shaping the Clinton administration's environmental agenda. As vice president, Gore's ability to shape that agenda was perhaps somewhat limited. At the close of Clinton's first term, most environmental organizations rated his

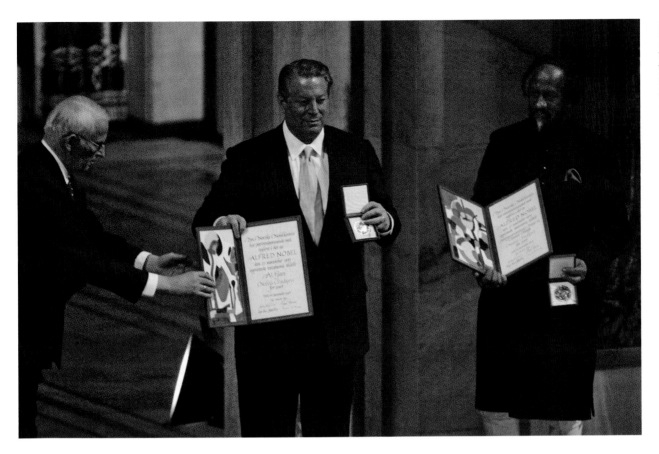

Nobel Peace Prize winners Al Gore, center, and Chairman of the Intergovernmental Panel on Climate Change Dr. Rajendra K. Pachauri, right, receive their medal and diploma from Nobel Committee Chairman Ole at City Hall in Oslo, 2007. *(AP Photo/John McConnico)*

administration's record on environmental matters as mixed, at best. Yet clearly environmental groups had more access to the White House than they had ever had before. In his second term, Gore was thought to be responsible for salvaging the 1997 Kyoto agreement on climate change when he flew to Japan as negotiations were falling apart and personally represented a new American position.

Gore's most notable contribution to the environmental movement might be as an author. With a wealth of statistical and scientific evidence, Gore's 1992 book *Earth in the Balance* makes the case that careless development and growth have damaged the natural environment; but better policies and regulations will supply incentives for more environmentally responsible actions by individuals and corporations. For example, a so-called carbon tax could provide financial incentives for developing new, nonpolluting energy sources such as solar and wind power. Corporations could also be given tax credits for using these new sources. By structuring a system of incentives that

favors the protection and restoration of the natural environment, government at the local, national, and (through the United Nations) international levels can restore the balance between satisfying human needs and protecting the earth's environment. But restoring this balance, Gore maintained in his book, requires more than public policy and legislation; it requires changes in basic beliefs and attitudes toward nature and all living creatures. More specifically, environmental protection and restoration requires a willingness on the part of individuals to accept responsibility for their actions (or inaction). At the individual level, environmental protection means living, working, eating, and recreating responsibly, with an eye to one's effects on the natural and social environment, now and in the future in which our children and their children will survive.

Despite the strong stances Gore articulated in *Earth in the Balance*, his achievements as vice president under Clinton were more moderate. His former staffer Carol Browner headed the EPA, and she consumers that seek out and regularly purchase green products. The research identified an additional 68 percent of American consumers as "Light Greens," or consumers who occasionally purchase green products. Studies indicate that, generally, the most environmentally-concerned consumers were well-to-do with the most discretionary income and the highest educational level. In short, they were trend-setters that advertising and marketing people could not ignore.

Marketers began to commonly use the terms "environmentally friendly," "safe for the environment," "recycled," "degradable," "biodegradable," "compostable," and "recyclable." Cause-related marketing also became popular as companies promised to support moderate environmental organizations such as World Wildlife Fund. While the advertising practices of many companies went uncontested by environmental groups, concerns arose regarding the claims of certain companies. For example, the Mobil Oil Corporation was sued for misleading advertising after claiming that its plastic Hefty garbage bags were recyclable. After suffering much embarrassment, British Petroleum was forced to withdraw its claim that its new brand of unleaded gasoline caused no pollution. Reacting to these and similar findings, ten state Attorney Generals issued a report in 1990 calling for greater accountability in "green" marketing. The Environmental Protection Agency (EPA) and the Federal Trade Commission also devised standards to evaluate the claims made by advertisers.

Often, the issue has been whether one product is really better for the environment than another. For instance, phosphate-free detergents created a controversy when they were introduced in France. Some companies claimed that they were no more environmentally benign than detergents that had phosphates. Rhone-Poulenc, the French producer of the detergents with phosphates, ran ads of dead fish apparently killed by the substances in the detergents which did not have phosphates. Proctor & Gamble launched a campaign which claimed that disposable diapers actually had less negative environmental impacts than reusable diapers. They pointed to the detergents, hot water, and energy used in washing cloth diapers, the energy needed to bring them to consumers, and the pesticides that were in the cotton out of which they were made.

Life-cycle assessments came into vogue as companies argued about the relative environmental merits of various products. Assessments exam the total environmental impact of using the product and how it rates—environmentally—to other similar products. Migros,

the large Swiss retailer, has developed an "eco-balance" or life-cycle program to analyze the impact of its packaging in terms of the resources used and how they are disposed of.

Green labeling programs exist in Germany (Blue Angel), Canada (Environmental Choice), and Japan (Ecomark). They are run by the governments of these countries, but the United States government has not been willing to give this kind of endorsement to commercial products. Instead, various environmental groups have seals of approval which they have applied to selected goods that pass their tests of environmental acceptability.

All of the problems associated with defining and labeling green products has generated concern about a phenomena known as "greenwashing." Greenwashing is the unjustified advertising and marketing of products or policies as environmentally friendly to gain an economic advantage. The term also refers to companies claiming an environmentally sound reason for an action that is undertaken for an economic benefit. For example, many hotel chains promote the reuse of linens and towels as part of their green philosophy without noting that the hotel chain saves millions of dollars through washing less frequently. While less frequent washing saves water and the discharge of phosphates into the environment, greenwashing promotes a environmentally and economically sound policy as being undertaken solely as a green policy.

See also Environmentalism; Green packaging; Green products; Recycling.

Resources

BOOKS

ABC News, and Films for the Humanities & Sciences (Firm). *Going Green Real-World Solutions for the Environment*. Princeton, N.J.: Films for the Humanities & Sciences, 2007.

Barker, Jill. *Baby Green: Caring for Your Baby the Eco-Friendly Way*. London: Gaia,2007.

Berman, Alan. *Green Design: A Healthy Home Handbook*. London: Frances Lincoln, 2008.

TerraChoice Environmental Marketing, Inc. *The "Six Sins of GreenwashingTM": A Study of Environmental Claims in North American Consumer Markets*. [Philadelphia, PA]: TerraChoice Environmental Marketing, Inc, 2007.

Todd, Anne M. *Get Green!* Chicago, Ill: Heinemann Library, 2009.

Alfred A. Marcus

party elected one member of the federal parliament in 2008, who was defeated in a subsequent election. Green politics have been more influential in Germany, where the party formed a coalition government with the Social democratic Party from 1998 to 2005.

See also Environmental Defense; Green advertising and marketing; Green products; Sea Shepherd Conservation Society; Sierra Club; Abbey, Edward; Bookchin, Murray; Brower, David Ross; Foreman, Dave.

Resources

BOOKS

Boston, Tim. *Commonwealth Greens: Comparing Australian and Canadian Green Parties.* Mandurah, WA: Equilibrium Books, 2009.

Conca, Ken, and Geoffrey D. Dabelko. *Green Planet Blues: Four Decades of Global Environmental Politics.* Boulder, CO: Westview Press, 2010.

Giddens, Anthony. *Politics of Climate Change.* Cambridge, MA: Polity Press, 2009.

Kraft, Michael E. *Environmental Policy and Politics.* London: Longman, 2010.

Terence Ball

Green products

Green products refer to marketed products that do not affect the environment as much as their traditional counterparts in terms of their manufacture and/or disposal. Examples are products made from recycled materials or products made in a way that is minimally disruptive, if at all, to the environment.

Green products are now preferred by many consumers and manufacturers. Some companies have thrived by marketing product lines as environmentally correct or green. One of the first international companies to exclusively promote green products was the Body Shop, a cosmetics company that is strongly and explicitly pro-environment with regard to its products. As of 2010, the Body Shop offered over 700 beauty products that are considered to be green products.

Similarly, the American ice cream manufacturer, Ben and Jerry's, has adopted a similar approach to using rain forest products in what it sells. Mercury- and cadmium-free batteries have been marketed by Varta, a German company. Ecover, a small Belgian company, made major sales gains when it began to

market a line of phosphate-free detergents. Loblaw, a Canadian grocery chain, has introduced a line of environmentally-friendly products and has sold more than twice the amount than it had initially projected.

Many factors comprise a green product. The product has to be made with few raw materials, use recycled or renewable materials when available, produce low or no contaminants, and have the smallest negative effect on human health possible. Consideration must also be given to how consumers will use the product and how they will dispose of it when they are finished. To reduce its waste potential, a product must often last a significant amount of time or be reusable or recyclable. These factors help determine a product's environmental footprint, the environmental cost of a product over its lifetime.

When discussing greenhouse gas emissions and climate change impacts of products, green consumers consider the carbon footprint of a product. Carbon footprints refer to the emissions produced in the manufacture, shipping, use, and disposal of a product. A product's carbon footprint is one part of its total environmental footprint. The greenest products are often those with the smallest footprints. Many green consumers advocate choosing local products to reduce carbon emissions associated with freezing, storage, and shipping.

As consumers become more aware of environmental issues, they will likely look to producers and governments to provide more products that will permit them to maintain a life-style that is less harmful to the environment. Therefore, the very nature of products evolve.

Energy-efficient appliances, low-water washing machines, solar hot water heaters, hybrid and alternative fuel cars, increased investment in public transportation, and green building construction are all examples of commonly available green alternatives to common infrastructure. In contrast to old smokestack industries, new technologies and emerging industries should be able to offer products that are less environmentally harmful.

Resources

BOOKS

Asian Productivity Organization. *Green Productivity and Green Supply Chain Manual.* Tokyo: Asian Productivity Organization, 2008.

Berman, Alan. *Green Design: A Healthy Home Handbook.* London: Frances Lincoln, 2008.

Green Computing Cut It Costs and Reduce Your Carbon Footprint on Any Budget. Sybex Inc, 2009.

See also Corporate Average Fuel Economy Standards; Environmental economics; Externality; Internalizing costs; Pollution control.

Resources

BOOKS

Cholakov, Georgi Stefanov, and Bhaskar Nath. *Pollution Control Technologies Volume 1*. Oxford: Eolss Publishers Co Ltd, 2009.
Environmental Protection UK., Environment Agency, and Scottish Environment Protection Agency. *Pollution Control Handbook 2009*. Brighton: Environmental Protection UK, 2009.

PERIODICALS

Chameides, William, and Michael Oppenheimer. "Carbon Trading Over Taxes." *Science* 315 (2007): 1670.
Mankiw, N. Gregory. "One Answer to Global Warming: A New Tax." *The New York Times* (September 16, 2007).

Bill Asenjo

Greenhouse effect

The greenhouse effect is a natural phenomenon through which gases present in the atmosphere absorb heat, which lessens the amount of radiating heat that exists the earth's atmosphere heating the Earth's surface. Natural greenhouse gases include water vapor, carbon dioxide (CO_2), nitrous oxide (N_2O), methane (CH_4), and ozone (O_3), all of which are essential to support life. The enhanced greenhouse effect, which the weight of evidence, such as that presented in a series of reports by the Intergovernmental Panel on Climate Change (IPCC), is directly due to human activities, increases concentrations of these gases in the atmosphere, and leads to pollution of the lower atmosphere contributing to climate change. These gases let in sunlight but tend to insulate Earth against the loss of heat, as do the glass walls of a greenhouse (hence the name). A higher concentration of the greenhouse gases means a warmer climate. For example, the twentieth century was 1° warmer on worldwide average than the nineteenth century—warming at a rate twenty times faster than average.

CO_2 is the predominant greenhouse gas and the gas that most influences atmospheric temperature. From April 1958, when monthly measurements of CO_2 from atop the Mauna Loa volcano began, through July 2010, 2009, the CO_2 concentration in parts per million went from 316 ppm to 390 ppm. During that time, the concentration has always increased, except for 1974 when the concentration remained approximately unchanged.

The examination of ice cores retrieved from the Antarctic has provided a snapshot through time of the concentration of various parameters, including carbon dioxide. Data from several studies has conclusively demonstrated that, prior to the mid-eighteenth century, the level of CO_2 was maintained at a baseline level of approximately 280 ppm. Then, coinciding with the Industrial Revolution, CO_2 levels began their sharp upward rise.

CH_4 is produced when oxygen is not freely available and bacteria have access to organic matter, such as in swamps, bogs, rice paddies and moist soils. CH_4 is also produced in the guts of creatures as varied as termites and cows, in garbage dumps, landfills, emissions from coal mining, natural gas production and distribution, and changing land use. CH_4 concentrations have increased over 100 percent since 1765.

On an equal weight basis, CH_4 is approximately 21 times more potent as an atmospheric warming agent than CO_2. Fortunately, in terms of dealing with atmospheric warming, its' half-life (the time for half of a set amount of the compound to degrade) is about 12 years. In contrast, the half-life of CO_2 is approximately 38 years. Strategies to reduce the amount of CH_4 being added to the atmosphere will be apparent as a change in atmospheric temperature long before the change due to reduced CO_2

N_2O concentrations in the atmosphere have increased due to fertilizer use and chemical production, such as in the manufacture of nylon. N_2O is also dispersed during fossil fuel combustion, biomass burning, and changing land use.

Chlorofluorocarbons (CFCs), also implicated in depletion of the ultraviolet-trapping O_3 layer, act as greenhouse gases. While useful and widely used as refrigerants, their total effect is significant because compared to a molecule of CO_2, each CFC molecule absorbs much more radiation, thereby trapping heat in the atmosphere.

Rain forest destruction (deforestation) also contributes to global warming. When the canopy of leaves is removed through clear-cutting or burning, the sudden warming of the forest floor releases CH_4 and CO_2. The massive increase in the number of dead tree trunks and branches leads to a population explosion of termites, which themselves produce methane. Dead trees can no longer take up CO_2 or convert it to oxygen and are not able to serve as a carbon sink.

The greenhouse effect. *(Reproduced by permission of Gale, a part of Cengage Learning)*

The accelerating melting of the permafrost region at higher northern latitudes as a consequence of the warming atmosphere may also be another source of CH_4. Permafrost is estimated to contained 500 billion tons of CH_4 that has remained locked in the frozen ground since prior to the last ice age. A relatively sudden addition of much of this CH_4 to the atmosphere would more than double the amount of carbon in the atmosphere.

Two factors that appear to mitigate the effect of enhanced greenhouse gases are aerosols and dust. Aerosols are minute solid particles produced by combustion and from natural sources, primarily volcanoes. The particles along with dust are dispersed in the atmosphere and can offset the greenhouse effect by blocking light. For example, a significant cooling trend in the spring and summer of 1992 seemed to correlate with the eruption of Mount Pinatubo in the Philippines. The fall and winter of 1992 were fairly mild on worldwide average. As all the particulate matter from the Mt. Pinatubo eruption settled out of the atmosphere, the surface cooling effects abated and the climate change trend resumed. Other aerosol particles are soot and can be produced as a result of cooking fires and coal burning. These particles absorb heat and promote climate change.

Anthropogenic (human-caused) greenhouse gases now appear to be responsible for increasing the global average temperature. According to current projections, global temperatures may rise as much as 35.6–37.4°Fahrenheit (2–3°C) above the pre-industrial temperatures by the year 2100. To place this change in perspective, the temperature rise that brought the planet out of the most recent ice age was only about 37.4–39.2°Fahrenheit (3–4°C).

The top ten warmest years of average global recorded temperatures were in the last fifteen years of the twentieth century and saw devastating fires in Yellowstone National Park, flooding in Bangladesh, record number of hurricanes and tornadoes, and a deadly heat wave and drought in the southeastern United States. It is probable, based on computer models, that warming will accompany changes in regional weather. A forty-year trend of increased precipitation in Europe and decreased precipitation in the African Sahel (Ethiopia, the Sudan, Somalia) may be an early consequence of global warming due to the greenhouse effect. Longer and more frequent heat waves would result in public health threats as well as inconveniences such as road buckling, electrical brownouts, or blackouts.

Precipitation is likely to increase regionally because as the temperature increases, more evaporation takes place, leading to more precipitation. The average precipitation event is likely to be heavier: wetter monsoons in coastal subtropics; more frequent and heavier winter snows at high altitudes and high latitudes; an earlier snowmelt, and a wetter spring.

Increases in rain or snowfall are not expected to offset the effects of higher temperatures on soil, however. Higher temperatures are expected to dry the soil in North America and southern Europe, among other places, by boosting the rates of evaporation and transpiration through plants. More favorable agricultural conditions in high latitudes could move the center of agriculture farther north into Canada and Siberia and out of the United States.

Other consequences of global warming from the enhanced greenhouse effect include the reduction of sea ice, coastal sea level rises of several feet per century, more frequent and powerful hurricanes, and more frequent and severe forest fires.

The rise of sea level is the most easily predicted consequence. The one-degree increase in temperature over the past century contributed to a 4–8 inch (10–20 cm) rise in mean sea level. This rise could lead to severe and frequent storm damage, flooding and disappearance of wetlands and lowlands, coastal erosion, loss of beaches and low islands, wildlife extinctions, and increased salinity of rivers, bays and aquifers. However, because the global atmosphere operates as a complex system, it is difficult, even with today's sophisticated computer models, to predict the exact nature of the changes we are likely to cause with increased greenhouse gas emissions. Scientists have predicted that low-lying areas and islands, including the Seychelles, the Maldives, the Marshall Islands, and large areas of Bangladesh, Egypt, Florida, Louisiana, and North Carolina will disappear over the next few decades.

The predicted rise in sea level along the east coast of the United States would swamp portions of major cities including New Orleans, Boston, and New York. In New York State, a Sea Level Rise Task Force was created in 2007 to recommend measures to protect the city from damage and to adapt to the changing coastline. Their report is due in January, 2011.

The earth's natural atmospheric cleanser—rain—may wash excess greenhouse gases out of the atmosphere. But, until rates of greenhouse gases slow their rapid increases or actually begin to decrease, the planet will get warmer. In response to climate projections, the United Nations Framework Convention on Climate Change (UNFCC), adopted and signed by 162 countries in 1992 at the Rio Earth Summit, sets country-by-country standards to reduce the emissions of greenhouse gases, particularly CO_2.

Policy-makers in the United States, including former Vice President Albert Gore Jr., proposed stricter requirements for more fuel-efficient cars, environment taxes that penalize heavy polluters and help pay for cleansing the atmosphere, and trading technological advances for rain forest protection in Third World countries.

The greatest controversy over slowing the rate of greenhouse gases injected into the atmosphere concerns methods of reducing emissions. Some scientists advocate increased use of nuclear power to reduce dependence on fossil fuels, but that carries its own controversies. Nuclear power plants are so energy-intensive just to build, the trade-off from current nuclear technology is negligible. As of 2010, thorium-based nuclear power is gaining popularity as a safe and relatively cost-effective nuclear option. Conservation and a switch from a dependence on fossil fuels to dependence on renewable resources such as wind, biofuels, and solar energy, would slow the rate of increase of CO_2 emissions into the atmosphere. Others oppose the use of nuclear power because of concerns over the long-term, safe storage of radioactive waste and the potential for environmental harm from rare catastrophic accidents.

Additional information

As of mid-2010, 2010 has the potential to be the hottest year in recorded weather history. The eight years following 1998 have had the record highest global average temperatures since 1850. By some analyses, 1998 was slightly warmer than 2005 globally. The warmest year on record so far for the continental United States, as of 2008, was actually 1934, which according to an August 2007 update from NASA was warmer than 1998 in the U.S. by a few hundredths of a degree. The temperature of the earth's surface is currently increasing at $0.32°$ Fahrenheit ($-17.6°C$) per decade corresponding to $3.2°$ Fahrenheit ($-16°C$) per century.

Increased extreme weather is a predicted consequence of global climate change. In August 2007, scientists at the World Meteorological Organization, an agency of the United Nations, announced that during the first half of 2007, Earth showed significant increases above long term global averages in both high temperatures and frequency of extreme weather events including heavy rainfall, cyclones, wind storms, and the El Nin;ato-Southern Oscillation (ENSO); an extreme weather event brought about by temperature changes in the surface waters of the tropical Pacific Ocean.

In February 2007, the Earth Challenge Prize was established to provide incentives for research on methods to remove CO_2 from the atmosphere. The prize, to be judged by an international panel that includes climate experts, is intended for the person or group who develops the best and least ecologically harmful method of removing at least one billion tons of carbon per year from Earth's atmosphere. The vast majority of climate and atmospheric scientists have concluded that excess levels of carbon dioxide from human activity are partially responsible for climate change. England's Virgin Airways

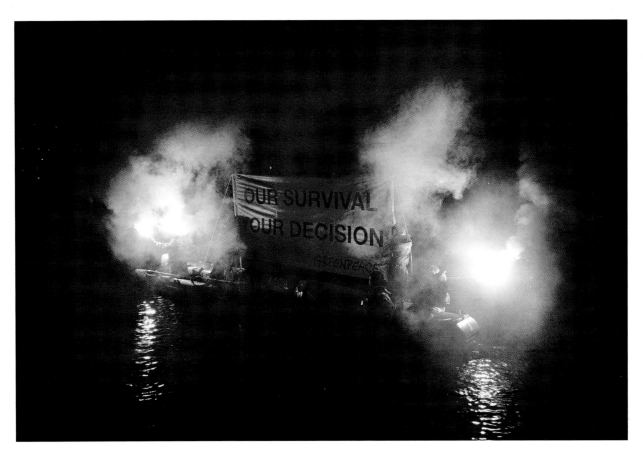

Greenpeace activists display a sign stating 'Our survival - your decision' at the Copenhagen Planetarium in 2009, while a dinner was hosted there by Danish Climate Minister Connie Hedegaard. *(© JENS ASTRUP/epa/Corbis)*

protecting the global environment through non-violent direct action, public education, and legislative lobbying. As of September 2010, Greenpeace is represented in over forty countries with a worldwide membership of nearly 2.9 million and maintains a scientific base in Antarctica.

In 2008, Greenpeace revenue was approximately $500 million USD. These funds are used to address priority issues that include deforestation, commercial whaling, overfishing, global climate change and nuclear power.

Having mounted successful campaigns on a wide variety of environmental issues, Greenpeace is perhaps best known for its direct and often confrontational crusades against nuclear testing and commercial whaling. The group has also garnered wide publicity for protesting various environmental abuses by hanging enormous banners from smokestacks, buildings, bridges, and the scaffolding used in the renovation of the Statue of Liberty.

Greenpeace works to eliminate widespread dependence upon fossil fuels and lobbies for laws and policies encouraging energy efficiency and renewable energy sources. The group is also working to halt the spread of nuclear power and the dumping of radioactive waste as well as to seek a global ban on the manufacture and use of ozone-depleting chemicals such as chlorofluoro-carbons (CFCs).

The organization seeks to protect both habitats and threatened species, including whales, harp seals, dolphins, sea turtles, elephants, and birds of prey. It works to discourage overfishing and other wasteful fishing practices, particularly the killing of dolphins in tuna nets. Greenpeace was instrumental in protecting Antarctica by persuading twenty-three nations to sign an accord banning all mining in Antarctica for at least fifty years. Supporting the principle of biodiversity, the group also works to protect tropical and temperate forests around the world. In 2002, the ships *MV Esperanza* and *Rainbow Warrior* stopped illegally logged timber from Africa and the Amazon from being imported.

Greenpeace is especially concerned with stopping the use of unneeded chlorine in the bleaching of paper and with preventing the dumping of hazardous waste in Third World nations. Particularly concerned in recent years with dioxin, polychlorinated biphenyl (PCB), CFCs, and pesticides, the group regularly investigates, publicizes, and lobbies against chemical pollution. Greenpeace also conducts research on the effects of toxic substances on human beings and the environment and encourages recycling as a means of reducing pollution. In 1998, Greenpeace activists prevented a PVC plant from opening in Convent, Louisiana.

Greenpeace conducts research into the effects of warfare on human beings and the environment and advocates the global elimination of nuclear weapons. More immediately, the group also urges the cessation of all nuclear and chemical weapons testing and is trying to persuade the major powers to agree to a global ban on naval nuclear propulsion.

In an effort to avoid compromising its goals and activities, Greenpeace does not seek corporate or government funding. Nor does it become directly involved in the electoral process in any of the nations in which it is active. The confrontational tactics have provoked angry responses from various governmental authorities, including the bombing and sinking of Greenpeace's flagship vessel *Rainbow Warrior* by agents of the French government in 1985. The *Rainbow Warrior* had been in New Zealand preparing to protest French nuclear testing in the South Pacific when it was sabotaged.

Resources

BOOKS

Connolly, Sean. *Greenpeace*. Global organizations. North Mankato, MN: Smart Apple Media, 2008.
Greenpeace Foundation. *Greenpeace Living Guide*. Toronto: Greenpeace, 2007.
Pash, Chris*The Last Whale*. New York: Freemantle, 2008.

ORGANIZATIONS

Greenpeace, 702 H Street NW, Washington, D.C., USA, 20001, (800) 326-0959, http://www.greenpeaceusa.org

Lawrence J. Biskowski

Greens

The term greens applies to those who engage in green politics. The term originated in Germany, where members of the environmentally-oriented Green Party were quickly dubbed *die Grünen*, or "the Greens." In the United States, greens refers not to a particular political party, but to any individual or group making environmental issues the central focus and main political concern. Thus, the term covers a wide array of political perspectives and organizations, ranging from moderate or mainstream groups such as the Sierra Club and Greenpeace to more militant movements and direct-action organizations such as the Sea Shepherd Conservation Society and Earth First!, as well as ecofeminists, bioregionalists, social ecologists, and deep ecologists.

Greens in the United States and Canada are divided over many issues. Some, for example, are in favor of organizing as interest groups to lobby for environmental legislation, while others reject politics in favor of a more spiritual orientation. Some greens (for example, social ecologists and ecofeminists) see their cause as connected to questions of social justice—the elimination of exploitation, militarism, racism, sexism, and so on—while others (deep ecologists, for instance) seek to separate their cause from such humanistic concerns, favoring a biocentric instead of an anthropocentric orientation. Despite such differences, however, all greens agree that the preservation and protection of the natural environment is a top priority and a precondition for every other human endeavor.

See also Environmental ethics; Environmentalism.

Resources

BOOKS

Boston, Tim. *Commonwealth Greens: Comparing Australian and Canadian Green Parties*. Mandurah, W.A.: Equilibrium Books, 2009.
Conca, Ken, and Geoffrey D. Dabelko. *Green Planet Blues: Environmental Politics from Stockholm to Kyoto*. Boulder, Colo: Westview, 2004.
Dessler, Andrew Emory, and Edward Parson. *The Science and Politics of Global Climate Change: A Guide to the Debate*. Cambridge, UK: Cambridge University Press, 2006.

Greenwashing

Greenwashing, or greensheening, describes deceptive marketing that promotes a product, policy, or company as environmentally beneficial. Greenwashing can include making products appear more environmentally friendly than they actually are by using nature imaging on their labels or including vague affirmations like "natural" for which there may be no legal standard. Giving a harmful product a name

Campaigners dressed as mock clean-up crew called the "Greenwash Guerrillas" gather outside the National Portrait Gallery in London in June 22, 2010, where the BP Portrait Award ceremony is held. *(AP Photo/Akira Suemori)*

that evokes the earth or wholesomeness is also considered greenwashing. Companies engaged in environmentally harmful work that, through marketing, present themselves as environmentally sound also engage in greenwashing. The practice can also be extended to cover policies that may or may not have significant environmental benefit, but which were initially motivated only by economic considerations.

The term greenwashing is a *portmanteau* (combination) of "green," indicating environmental soundness, and "whitewashing" or glossing over flaws or concealing the truth. The term also developed as a play on words, as the laundering of domestic linens is sometimes referred to as "washing the whites." New York environmentalist Jay Westerveld coined the word greenwashing in a 1982 article describing new laundering practices in hotels. Campaigns encouraging hotel guests to reuse their towels and bed linens for several days initially advertised the move as a green policy to save water and lessen detergent

use. However, Westerveld asserted that the campaigns' true objective was to increase profits by reducing hotelier laundering and labor expenses; any environmental benefit was incidental.

There is no legal standard for what is considered greenwashing. However, the environmental marketing group TerraChoice has established an evaluative list of seven "sins of greenwashing." The identified sins represent a logical flaw in the marketing story or labeling of green products. The sin of the hidden trade-off suggests that a product is entirely environmentally responsible by emphasizing one, narrow attribute of greenness without sufficient consideration of the entire environmental impact of the product. The sin of no proof applies to green claims that are not supported by science or independent verification. Similarly, the TerraChoice dubbed the presentation of meaningless certifications (ones not backed up by scientific rigor or independent verification) as worshipping false labels. Greenwashing also includes presenting outdated environmental claims (such as not containing a substance that was long-ago banned) or otherwise truthful claims that are unimportant when comparing the environmental impact of two products. This is labeled the sin of irrelevance. The sin of fibbing covers greenwashing via untruthful statements about a product's environmental benefits. The sin of the lesser of two evils, as in the case of organic pesticides, lulls consumers into thinking that a product is a completely safe, green alternative type of product, when all similar products have significant negative environmental consequences.

TerraChoice considers vague claims that a product is natural or green without defining any specific criteria as the Sin of Vagueness. In its initial survey of over 3,000 green products, TerraChoice found this was the type of greenwashing most commonly committed on product packaging. Common examples of greenwashing using vagueness include claims that products are "natural," without acknowledgement that some natural ingredients can be just as environmentally toxic or harmful as their synthetic counterparts. Claims that products are "chemical free" misrepresent the scientific definition of chemicals, which are omnipresent. Employed in greenwashing, the phrase "chemical-free" plays on public fears of manufactured chemicals that may or may not have negative health effects or cause environmental harm.

Greenwashing is not only applicable to products and labels, but also whole company advertising campaigns and industrial processes. Environmental advocates assert that clean coal is a greenwashed, industrial process that makes consumers focus only on re-reductions in air pollution at the expense of the industry's reliance on damaging processes like strip mining, mountain top removal, slurry impounding, and valley filling. In 2000, the international energy company British Petroleum re-branded itself as BP. It re-designed its corporate logo—one that appears on all of its auto gasoline stations—as a flower design in green and yellow. The company's new slogan, still in use over a decade later, was "Beyond Petroleum." Critics asserted that BP did not change its activities and still engaged in drilling, shipping, and refining practices that ultimately caused environmental damage. BP claimed that the campaign represented the company's research on alternative energy sources. General Electric faced similar accusations of greenwashing when it launched its $90 million "Ecomagnination" advertising campaign in 2005.

Opponents of greenwashing, including media and environmental watchdog groups, rely on media coverage and public education to expose greenwashing practices.

Greenpeace, the international environmental activism group, features greenwashing claims on its blog, Stop Greenwash. The Greenwashing Index, maintained by the University of Oregon and EnvrioMedia Social Marketing, features online examples of greenwashed advertising and products sent in by members of the public.

Laws have been slow to address greenwashing concerns. In the United States, the parameters for natural and organic standards apply predominantly to food, and there is some verification on the inclusion and labeling of post-consumer recycled products. Untrue marketing claims about human health are covered by advertising regulations. In Britain, the Code of Advertising Sales Promotion and Direct Marketing has a specific provision (section 49) addressing false and misleading environmental claims.

Greenwashing is prevalent in the modern marketplace where consumers have expressed a demand for greener products. Manufacturers of green products claim that greenwashing dilutes consumer confidence and creates confusion with legitimately green alternatives. An October 2010 survey found that more than 95 percent of products claiming environmental benefits, or reduced environmental impacts, make excessive claims of greenness. There are, however, signs that consumers are becoming more knowledgeable about what they accept as a green product, and that product advertising is getting better with environmental claims. TerraChoice's 2010 survey of products found that there were 73 percent more allegedly green products in the marketplace than there were just one year before. Evaluating almost 5,300 products under their "deadly sins" criteria, 4.5 percent were free of greenwashing, representing an increase of 3.5 percent since 2007.

Resources

BOOKS

TerraChoice Environmental Marketing, Inc. *The "Six Sins of GreenwashingTM": A Study of Environmental Claims in North American Consumer Markets*. Philadelphia, PA: TerraChoice Environmental Marketing, Inc, 2007.

OTHER

Greenpeace. "Greenpeace." http://www.greenpeace.org/ (accessed November 10, 2010).

World Energy Council. "Survey of Energy Resources 2007: Coal: Clean Coal Technologies." http://www.worldenergy.org/publications/survey_of_energy_resources_2007/coal/631.asp (accessed November 10, 2010).

Adrienne Wilmoth Lerner

Grinevald, Jacques

1946–
French university professor and historian

Philospher and historian Jacques Grinevald is recognized as a key expert on biospheres His studies and publications regarding the concept's initiator, Vladimir Vernadsky who published his work on the subject first in 1926, have comprised the substance of Grinevald's work as a scientific historian and philosopher.

Grinevald was a faculty member of the University of Geneva (Switzerland) from 1980 to 2005, serving as a part-time lecturer. He received a science degree in policies from the University Institute of High International Studies in Geneva in 1970; and his doctorate of third cycle of philosophy, Paris X-Nanterre, in 1979. His early career included positions at the University of Geneva as an assistant in charge of research and teaching to the faculty of law, in addition to duties as the person in charge of press information; part-time lecturer position at the federal polytechnic school of Lausanne, serving as program man technique environment beginning in 1981; and, serving as an invited professor at the Federal Universidade of Rio de Janeiro (Brazil) in 1980 and 1984. His active schedule has taken him all over the world for conferences and seminars. Grinevald retains membership in several professional societies, including International Society for Ecological Economics; European Association for Bioeconomic Studies; and, World Council for the Biosphere.

Grinevald has been published extensively. His writings—among them chapters and articles in various books and journals—include, *The Greening of Europe* in 1990; "The Revolution Carnotienne: thermodynamics, economy and ideology," from the *European Review of Social Sciences*, 1976; and, "There is holistic total concept for deep and ecology: the Biosphere," for *Fundamenta Scientiae*, 1987. He has lectured and written on subjects that include the biosphere, the greenhouse effect, and famous scientists such as Stephen H. Schneider, a native New Yorker, whose research has focused on the greenhouse effect on civilization. He has been a regular contributor to a journal established by a group at the University of Geneva in 1990, *Strategies Energetiques Biosphere et Society* (Energy Strategies, Biosphere and Company) (SEBES). What began as a special volume became a publication devoted to the biosphere.

Writing for the publication *Etat De La Planete* (State of the Planet), Grinevald discussed the key issues of the biosphere. "This concept underlines the fact that the Life exceeds the individuals and is an ecological phenomenon of solidarity on various scales, microbial communities on a planetary scale of the Biosphere. It is the observer which decides scale of observation, so much [more] at the geographical level than at the temporal level. It is our world civilization which discovers the Biosphere as a phenomenon characteristic of the face of the Earth in cosmos. That implies a certain responsibility. The interdisciplinary and holistic concept of Biosphere associates astronomy, geophysics, meteorology, biogeography, evolutionary biology, geology, the geochemistry and, in fact, all science of the ground and the living."

As of 2010, Grinevald is a professor of development studies at The Graduate Institute, Geneva.

Resources

PERIODICALS

Grinevald, Jacques. "Biodiversity and Biosphere." *Etat De La Planete*, No. 1.
Grinevald, Jacques. "On Holistic Aconcept for Deep and Global Ecology: The Biosphere." *Fundamenta Scientiae* (1987): 197–226.

Jane Spear

Grizzly bear

The grizzly bear (*Ursus arctos*), a member of the family Ursidae, is the most widely distributed of all bear species. Although reduced from prehistoric times, its range today extends from Scandinavia to eastern Siberia, Syria to the Himalayan Mountains, and, in North America, from Alaska and northwest Canada into the northwestern portion of the lower forty-eight states. Even though the Russian, Alaskan, and Canadian populations remain fairly large, the grizzly bear population in the northwestern continental United States represents only about 1 percent of its former size of less than 200 years ago. Grizzly bears occupy a variety of habitats, but in North America they seem to prefer open areas including tundra, meadows, and coastlines. Before the arrival of Europeans on the continent, grizzlies were common on the Great Plains. Now they are found primarily in wilderness forests with open areas of moist meadows or grasslands.

Female grizzly bears vary in size from 200–450 pounds (91–204 kg), whereas the much larger males can weigh up to 800 pounds (363 kg). The largest individuals—from the coast of southern Alaska—weigh up to 1,720 pounds (780 kg). Grizzly bears measure from 6.5–9 feet

Male grizzly bears can get up to 800 pounds (363 kg). (*Ursus arctos*). *(Photograph)*

(2–2.75 m) tall when standing erect. To maintain these tremendous body sizes, grizzly bears must eat large amounts of food daily. They are omnivorous and are highly selective feeders. During the six or seven months spent outside their den, grizzly bears will consume up to 35 pounds (16 kg) of food, chiefly vegetation, per day. They are particularly fond of tender, succulent vegetation, tubers, and berries, but also supplement their diet with insect grubs, small rodents, carrion, salmon, trout, young deer, and livestock, when the opportunity presents itself. In Alaska, along the McNeil River in particular, when the salmon are migrating upstream to spawn in July and August, it is not unusual to see congregations of dozens of grizzly bears, along the riverbank or in the river, catching and eating these large fish.

Grizzly bears breed during May or June, but implantation of the fertilized egg is delayed until late fall when the female retreats to her den in a self-made or natural cave, or a hollow tree. Two or three young are born in January, February, or March, and are small (less than 1 lb/0.45 kg) and helpless. They remain in the den for three or four months before emerging, and stay with their mother for one and a half to four years. The age at which a female first reproduces, litter size, and years between litters are determined by nutrition, which induces females to establish foraging territories which exclude other females. These territories range from 10–75 square miles (26–194 km^2). Males tend to have larger ranges extending up to 400 square miles (1,036 km^2) and incorporate the territories of several females. Young females, however, often stay within the range of their mother for some time after leaving her care, and one case was reported of three generations of female grizzly bears living within the same range.

Grizzly bear populations have been decimated over much of their original range. Habitat destruction and hunting are the primary factors involved in their decline. The North American population, particularly in the lower forty-eight states, has been extremely hard hit. Grizzly bears numbered near 100,000 in the lower forty-eight states as little as 180 years ago, but today, fewer than 1,000 remain on less than 2 percent of their original range. This population has been further fragmented into seven small, isolated populations in Washington, Idaho, Montana, Wyoming, and Colorado. This decline and fragmentation makes their potential for survival in these habitats tenuous. The U.S. Fish and Wildlife Service considers the grizzly bear to be threatened in the lower forty-eight states. Related brown bear species and populations are not threatened, with over 100,000 estimated to inhabit Russia alone.

Little has been done to protect this declining species in the lower 48 states. In 1999 it was agreed to begin slowly reintroducing grizzlies into the 1.2 million acre (49,000 ha) area of the Selway-Bitterroot Wilderness on the border of Idaho and Montana. Unfortunately, the project was put on hold as of 2001 due to unfounded fear of the animal. Habitat loss due to timbering, road building, and development in this region is still a major problem and will continue to impact these threatened populations of bears.

Resources

BOOKS

Anderson, Casey. *The Story of Brutus: My Life with Brutus the Bear and the Grizzlies of North America*. New York: Pegasus Books, 2010.

Cole, Jim, and Tim Vandehey. *Blindsided: Surviving a Grizzly Attack and Still Loving the Great Bear*. New York: St. Martin's Press, 2010.

Final Conservation Strategy for the Grizzly Bear in the Yellowstone Area. Missoula, Montana: Interagency Conservation Strategy Team, 2007.

Knibb, David. *Grizzly Wars: The Public Fight Over the Great Bear*. Spokane: Eastern Washington University Press, 2008.

Sartore, Joel. *Face to Face with Grizzlies*. Washington, D.C: National Geographic, 2007.

OTHER

Craighead Institute. http://www.grizzlybear.org (accessed November 9, 2010).

Defenders of Wildlife. "Grizzly Bear." http://www.defenders.org/programs_and_policy/wildlife_conservation/imperiled_species/grizzly_bear/ (accessed November 9, 2010).

International Union for Conservation of Nature and Natural Resources. "IUCN Red List of Threatened Species: *Ursus arctos.*" http://www.iucnredlist.org/apps/redlist/details/41688/0 (accessed November 9, 2010).

Eugene C. Beckham

Groundwater

Groundwater refers to water that is located underground in the spaces between soil particles and bedrock, and even in cracks within rocks.

Almost all groundwater originates as surface water. Some portion of rain hitting the earth runs off into streams and lakes, and another portion soaks into the soil, where it is available for use by plants and subject to evaporation back into the atmosphere. The third portion soaks below the root zone and continues moving downward until it enters the groundwater.

Precipitation is the major source of groundwater. Other sources include the movement of water from lakes or streams and contributions from such activities as excess irrigation and seepage from canals. Water has also been purposely applied to increase the available supply of groundwater. Water-bearing formations called aquifers act as reservoirs for storage and conduits for transmission back to the surface.

The occurrence of groundwater is usually discussed by distinguishing between a zone of saturation and a zone of aeration. In the zone of saturation the pores are entirely filled with water, while the zone of aeration has pores that are at least partially filled by air. Suspended water does occur in this zone. This water is called vadose, and the zone of aeration is also known as the vadose zone. In the zone of aeration, water moves downward due to gravity, but in the zone of saturation it moves in a direction determined by the relative heights of water at different locations.

Water that occurs in the zone of saturation is termed groundwater. This zone can be thought of as a natural storage area of reservoir whose capacity is the total volume of the pores of openings in rocks.

An important exception to the distinction between these zones is the presence of ancient sea water in some sedimentary formations. The pore spaces of materials that have accumulated on an ocean floor, which has then been raised through later geological processes, can sometimes contain salt water. This is called connate water.

Formations or strata within the saturated zone from which water can be obtained are called aquifers. Aquifers must yield water through wells or springs at a rate that can serve as a practical source of water supply. To be considered an aquifer the geological formation must contain pores or open spaces filled with water, and the openings must be large enough to permit water to move through them at a measurable rate. Both the size of pores and the total pore volume depends on the type of material. Individual pores in fine-grained materials such as clay, for example, can be extremely small, but the total volume is large. Conversely, in coarse material such as sand, individual pores may be quite large but total volume is less. The rate of movement from fine-grained materials, such as clay, will be slow due to the small pore size, and it may not yield sufficient water to wells to be considered an aquifer. However, the sand is considered an aquifer even though they yield a smaller volume of water because, they will yield water to a well.

The water table is not stationary but moves up or down depending on surface condition such as excess precipitation, drought, or heavy use. Formations where the top of the saturated zone or water table define the upper limit of the aquifer are called unconfined aquifers. The hydraulic pressure at any level with an aquifer is equal to the depth from the water table, and there is a type known as a water- table aquifer, where a well drilled produces a static water level which stands at the same level as the water table.

A local zone of saturation occurring in an aerated zone separated from the main water table is called a perched water table. These most often occur when there is an impervious strata or significant particle-size change in the zone of aeration which causes the water to accumulate. A confined aquifer is found between impermeable layers. Because of the confining upper layer, the water in the aquifer exists within the pores at pressures greater than the atmosphere. This is termed an artesian condition and gives rise to an artesian well.

Groundwater has always been an important resource, and it will become more so in the future as the need for good quality water increases due to urbanization and agricultural production. It has recently been estimated that 50 percent of the drinking water in the United States comes from groundwater; 75 percent of the nation's cities obtain all or part of their supplies from groundwater, and rural areas are 95 percent dependent upon it. For these reasons, it is widely believed that every precaution should be taken to protect groundwater purity. Once contaminated, groundwater is difficult, expensive, and sometimes impossible to clean up. The most prevalent sources of contamination are waste disposal, the storage, transportation and handling of commercial materials, mining operations, and nonpoint sources such as agricultural activities.

An example of a groundwater resource that is threatened is the Ogalla Aquifer, an approximately 17,000 square mile region of groundwater that underlies eight mid-west states. The aquifer, which supplies the bulk of irrigation water for the cropland in the eight states (which are a major crop source for the country) is being depleted at a rate faster than the addition of water. At this rate of depletion, the aquifer will run dry by about 2030.

See also Agricultural pollution; Aquifer restoration; Contaminated soil; Drinking-water supply; Safe Drinking Water Act (1974); Water quality.

Resources

BOOKS

Barlow, Maude. *Blue Covenant: The Global Water Crisis and the Coming Battle for the Right to Water*. New York: New Press, 2009.

Garte, Seymour. *Where We Stand: A Surprising Look at the Real State of Our Planet*. Vancouver, BC: AMACOM, 2007.

Glennon, Robert J. *Unquenchable: America's Water Crisis and What To Do About It*. Washington, DC: Island Press, 2009.

Pearce, Fred. *When the Waters Run Dry: Water -The Defining Crisis of the Twenty-first century*. Boston: Beacon Press, 2007.

James L. Anderson

Groundwater monitoring

Monitoring groundwater quality and aquifer conditions can detect contamination before it becomes a problem. The appropriate type of monitoring and the design of the system depends upon hydrology, pollution sources, and the population density and climate of the region. There are four basic types of groundwater monitoring systems: ambient monitoring, source monitoring, enforcement monitoring, and research monitoring.

Ambient monitoring involves collection of background water quality data for specific aquifers as a way to detect and evaluate changes in water quality. Source monitoring is performed in an area surrounding a specific, actual, or potential source of contamination such as a landfill or spill site. Enforcement monitoring systems are installed at the direction of regulatory agencies to determine or confirm the origin and concentration gradients of contaminants relative to regulatory compliance. Research monitoring wells are installed for detection and assessment of cause and effect relationships between groundwater quality and specific land use activities.

See also Aquifer restoration; Contaminated soil; Drinking-water supply; Leaching; Water quality standards.

Groundwater pollution

When contaminants in groundwater exceed the levels deemed safe for the use of a specific underground reservoir of water (aquifer), the groundwater is considered polluted. There are three major sources of groundwater pollution: natural sources; waste disposal; and spills, leaks, and nonpoint source activities such as run-off from agricultural land.

All groundwater naturally contains some dissolved salts or minerals. These salts and minerals may be leached from the soil and from the aquifer materials themselves and can result in water that poses problems for human consumption, is considered polluted, or does not meet the secondary standards for water quality. Natural minerals or salts that may result in polluted ground water include chloride, nitrate, fluoride, iron and sulfate.

The large-scale disposal of waste will always carry a risk for serious pollution of the environment. Waste-disposal practices the specifically threaten groundwater range from separate sewage treatment systems for individual residences, such as septic fields, which, as of 2010, are the basis of sewage disposal for about 25 percent of US households, to the storage and disposal of industrial wastes. Many of the problems posed by industrial waste arise from the use of surface storage facilities that rely on evaporation for disposal. These facilities are also known as discharge ponds. In other types, the waste is treated to standards suitable for discharge to surface water. The potential exists in both types of sewage disposal for the movement of contaminants into groundwater. Many of the numerous sanitary landfills in the country are in the same situation. Water moving down and away from these sites into groundwater aquifers carries with it a variety of chemicals leached from the material deposited in the landfills. The liquid that moves out of landfills is called leachate.

Modern landfills are constructed with an outer barrier that is impermeable to water. This should restrict the movement of leachate into the groundwater. However, any imperfection in the barrier can allow percolation of liquid into the soil. Older landfills that were built when barrier construction was mandatory may not have a surrounding impermeable barrier.

Agricultural practices also contribute to groundwater pollution. For example, there have been increases in nitrate concentrations and low-level concentrations of pesticides. For control of groundwater pollution, one of the most important agricultural practices is the management of nitrogen from all sources including fertilizer, nitrogen-fixing plants, and organic waste. Once nitrogen is in the nitrate form it is subject to leaching, so it is important that the amount applied not exceed the crops' ability to use it. At the same time, crops need adequate nitrogen to obtain high yields, and a good balance must be maintained. Low-level pesticide contamination occurs in areas where aquifers are sensitive to surface activity, particularly areas of shallow aquifers beneath rapidly permeable soils, and regions of topography where deep and wide range pollution can occur due to fractures in the bedrock.

Except in cases of deep-well injection waste or substances contained in sanitary landfills, most contaminants move from the land surface to aquifers. The water generally moves through an unsaturated zone, in which biological and chemical processes may act to degrade or change the contaminant. Plant uptake can also reduce some of the pollution. Once in the aquifer, however, the movement of the contaminant with the water will depend the solubility of the compound, and the speed of contamination will depend on how fast water moves through the aquifer. Chemical and biological degradation of contaminants can occur in the aquifer, but usually at a slower rate than it does on the surface due to lower temperatures, less available oxygen, and reduced biological activity. In addition, aquifer contaminants exist in lower concentrations, diluted by the large water volume. Most pollution remains relatively localized in aquifers, since movement of the contaminants usually occurs in plumes that have definite boundaries and do not mix with the rest of the water. This does provide an advantage for isolation and treatment.

The types of chemicals that pollute groundwater are as varied as their sources. They range from such simple inorganic materials as nitrate from fertilizers, septic tanks, and feedlots, chloride from high salt, and heavy metals such as chromium from metal plating processes, to very complex organic chemicals used in manufacturing and household cleaners.

The most efficient way to protect groundwater is to limit activities in recharge areas (areas where the groundwater is replenished). For confined aquifers it may be possible to control activities that can result in pollution, but this is extremely difficult for unconfined aquifers, which are essentially open systems and that are subject to effects from any land activity. In areas of potential salt-water intrusion excess pumping can be regulated, and this can also be done where water is being used for irrigation faster than the recharge rate, so that the water becomes saline. Another important activity for the protection of groundwater is the proper sealing of all wells that are not currently being used.

Classification of aquifers according to their predominant use is another management tool now employed in a number of states. This establishes water-quality goals and standards for each aquifer, and means that aquifers can be regulated according to their major use. This protects the most valuable aquifers, but leaves the problem of predicting future needs. Once an aquifer is contaminated, it is very expensive if not impossible to restore, and this management tool may have serious drawbacks in the future.

In rural areas of the United States, 95 percent of the population draws their drinking water from the groundwater supply. As of 2010, almost half of the U.S. population drinks water that has been obtained from a groundwater source. Nearly 20 billion gallons of groundwater are withdrawn from aquifers every day. With a growing population, continued industrialization, and increasing agricultural reliance on the use of chemicals, many believe it is now more important than ever to protect groundwater. Contamination problems have been encountered in every state, but prevention is far more efficient and effective than restoration after damage has been done. Prevention can be achieved through regional planning and enforcement of state and federal regulations.

See also Agricultural pollution; Aquifer restoration; Contaminated soil; Drinking-water supply; Feedlot runoff; Groundwater monitoring; Hazardous waste site remediation; Heavy metals and heavy metal poisoning; Waste management; Water quality; Water quality standards; Water treatment.

Resources

BOOKS

Glennon, Robert Jerome. *Unquenchable: America's Water Crisis and What to do About It.* Washington, DC: Island Press, 2010.

Pearce, Fred. *When the Rivers Run Dry: Water–The Defining Crisis of the Twenty-first Century.* Boston: Beacon Press, 2007.

Solomon, Steven. *Water: The Epic Struggle for Wealth, Power, and Civilization.* New York: Harper, 2010.

James L. Anderson

Growth curve

A graph in which the number of organisms in a population is plotted against time. Such curves are amazingly similar for populations of almost all organisms

from bacteria to human beings and are considered characteristic of populations.

Growth curves typically have a sigmoid or S-shaped curve. When a few individuals enter a previously unoccupied area, growth is at first slow during the positive acceleration phase. The growth then becomes rapid and increases exponentially, called the logarithmic phase. The growth rate eventually slows down as environmental resistance gradually increases; this phase is called the negative acceleration phase. It finally reaches an equilibrium or saturation level. The final stage of the growth curve is termed the carrying capacity of the environment.

A good example of a species' growth curve is demonstrated by the sheep population in Tasmania. Sheep were introduced into Tasmania in 1800. Careful records of their numbers were kept, and by 1850 the sheep population had reached 1.7 million. The population remained more or less constant at this carrying capacity for nearly a century.

The figures used to plot a growth curve—time and the total number in the population—vary from one species to another, but the shape of the growth curve is similar for all populations. Once a population has become established in a certain region and has reached the equilibrium level, the numbers of individuals will vary from year to year depending on various environmental factors. Comparing these variations for different species living in the same region is helpful to scientists who manage wildlife areas or who track factors that affect populations.

For example, a study of the population variations of the snowshoe hare and the lynx (*Lynx canadensis*) in Canada is a classic example of species interaction and interdependence. The peak of the hare population comes about a year before the peak of the lynx population. Since the lynx feeds on the hare, it is obvious that the lynx cycle is related to the hare cycle. This leads to a decline in the population of hares and secondarily to a decline in the lynx population. This permits the plants to recover from the overharvesting by the hares, and the cycle can begin again.

Growth curves are just one of the characteristics of populations. Other characteristics that are a function of the whole group and not of the individual members include population density, birth rate, death rate, age distribution, biotic potential, and rate of dispersion.

See also Population growth.

Linda Rehkopf

Growth, exponential *see*
Exponential growth.

Growth limiting factors

There are a number of essential conditions which all organisms, both plants and animals, require to grow. These are known as growth factors. Plants, for example, require sunlight, water, and carbon dioxide in order to perform photosynthesis. They require nutrients such as nitrogen, phosphorus, and various trace elements in order to form tissues. The environment in which the plant is growing does not contain a unlimited supply of these growth factors. When one or more of them is present in levels or concentrations low enough to constrain the growth of the plant, it is known as a growth limiting factor. The rate or magnitude of the growth of any organism is controlled by the growth factor that is available in the lowest quantities. This concept is analogous to the saying that a chain is only as strong as its weakest link.

These factors limit population growth. If they did not exist, a population could increase exponentially, limited only by its own intrinsic lifespan. Growth limiting factors are essential to the traditional concept of carrying capacity, which rests on the assumption that the available resources limit the population that can be sustained in that area. Advances in technology have enabled people to increase the carrying capacity in certain areas by manipulating the growth limiting factors. Perhaps the best example of this is the use of fertilizers on farmland.

In the field of population ecology, identifying growth limiting factors is part of establishing the constraints and pressures on populations and predicting growth in various conditions. Algal growth in New York Harbor provides an example of the importance of identifying growth limiting factors. In New York Harbor, several billion gallons of untreated wastewater are released daily, bringing enormous quantities of nutrients and suspended solids into the water. Algae in the harbor take advantage of the nutrient loads and grow more than they would under nutrient-poor conditions. At the same time, however, the suspended solids and silts brought into harbor cause the water to become very turbid, limiting the amount of sunlight that penetrates it. Sunlight is rarely a growth limiting factor for algae; nutrients are usually what limits their growth, but in this case nutrients are in excess supply. This means that if pollution control in the harbor ever results in control of the turbidity in the water, there will probably be a sharp increase in the growth of algae.

Consideration of growth limiting factors is also very important in the field of conservation biology and habitat protection. If the goal is to protect a bird such as the heron, which may feed on fish from a lake and nest in upland trees nearby, limiting factors must be taken into account not only for the growth of the individual but also for the population. Conservation efforts must not be directed only toward ensuring there are enough fish in the lake. Enough trees must also be left uncut and undisturbed for nesting in order to address all of the growth requirements for the population. Regardless of how abundant the fish are, the number of herons will only grow to the extent allowed by the number of available nesting sites.

Environmentalists use growth limiting factors to distinguish between undisturbed ecosystems and unstable or stressed systems. In an ecosystem that has been distressed or disturbed, the nature of growth limiting factors changes, and these changes are often human-induced, as they are in New York Harbor. Though the change in circumstances may not always appear negative in impact, it still represents a shift away from the original balance, and it may have effects on other species or lead to subtle long-term changes in the system. Any cleanup or management strategy must use these new growth limiting factors to identify the nature of the imbalance that has occurred and develop a procedure to restore the system to its original condition.

Growth limiting factors are extensively used in the field of bioremediation, in which microbes are used to clean up environmental contaminants by breakdown and decomposition. Oil spills are a good example. Bacteria that can break down and degrade oils are naturally present in small quantities in soil, but under normal conditions their growth is limited by both the availability of essential nutrients and the availability of oil. In the event of an oil spill on land, the only growth limiting factor for these bacteria is nutrients. Bioremediation scientists can add nitrogen and phosphorus to the soil in these circumstances to stimulate growth, which increases degradation of the oil. Techniques such as these, which use naturally occurring bacterial populations to control contamination, are still in development; they are most useful when the contaminants are present in high concentrations and confined to a limited area.

See also Algal bloom; Decline spiral; Ecological productivity; Exponential growth; Food chain/web; Restoration ecology.

Resources

BOOKS

Bolen, Eric, and William Robinson. *Wildlife Ecology and Management*. New York: Benjamin Cummings, 2008.

Usha Vedagiri
Douglas Smith

Growth, logistic *see* **Logistic growth.**

Growth, population *see* **Population growth.**

Grus americana see **Whooping crane.**

Guano

Guano is manure created by flying animals that is deposited in a central location because of nesting habits. Guano can occur in caves from bats or in nesting grounds where large populations of birds congregate. Guano was frequently used as a source of nitrogen (N) fertilizer prior to the time when nitrogen fertilizer was commercially manufactured from methane (CH_4) in natural gas. Guano was also used as saltpeter for the production of gunpowder. Due to demand for guano as nitrogen fertilizer and for its use in producing gunpowder in the mid–1800s, the U.S. passed the Guano Islands Act in 1856, stating that the U.S. could take possession of any unoccupied, unclaimed island found to house guano deposits.

See also Animal waste.

Resources

BOOKS

Bown, Stephen R. *A Most Damnable Invention: Dynamite, Nitrates, and the Making of the Modern World*. New York: T. Dunne Books, 2005.

Chesworth, Ward. *Encyclopedia of Soil Science*. Dordrecht, Netherlands: Springer, 2008.

Guinea worm eradication

In 1986, the world health community began a campaign to eliminate the guinea worm (*Dracunculus medinensis*) from the entire world. If successful, this will be only the second global disease affecting humans ever completely eradicated (smallpox, which was eradicated in 1977 was first), and the only time that a human parasite will have been totally exterminated worldwide. Known as the fiery serpent, the guinea worm has been a scourge in many tropical countries. Dracunculiasis (pronounced dra-KUNK-you-LIE-uh-sis) or guinea worm disease, starts when people drink stagnant water contaminated with tiny copepod water fleas (called cyclops) containing guinea worm larvae. Inside the human body, the worms grow to as long as 3 feet (1 m). After a year of migrating through the body, a threadlike adult worm emerges slowly through a painful skin blister. Most worms come out of the legs or feet, but they can appear anywhere on the body. The eight to twelve weeks of continuous emergence are accompanied by burning pain, fever, nausea, and vomiting. Many victims bathe in a local pond or stream to soothe their fever and pain. When the female worm senses water, she releases tens of thousands of larvae, starting the cycle once again. Once the worms become established in local ponds, infections among people living nearby are at high risk for further infections.

As the worm emerges from the wound, it is often rolled around small stick and pulled out a few centimeters each day. Sometimes the entire worm can be extracted in a few days, but the process usually takes weeks. Unfortunately, if the worm is removed too fast and breaks off, the part left in the body can die, leading to serious secondary infections. If the worm exits the body through a joint, permanent crippling can occur. There is no cure for guinea worm disease once the larvae are ingested. There is no vaccine, and having been infected once doesn't provide immunity. Many people in affected villages suffer the disease repeatedly year after year. The only way to break the cycle is through behavioral changes. Community health education, providing clean water from wells or by filtering or boiling drinking water, eliminating water fleas by chemical treatment, and teaching infected victims to stay out of drinking supplies are the only solutions to this problem.

Although people rarely die as a direct effect of the parasite, the social and economic burden at both the individual and community level is great. During the weeks that worms are emerging, victims usually are unable to work or carry out family duties. This debilitation often continues for several months after worms are no longer visible. In severe cases, arthritis-like conditions can develop in infected joints, and the person may be permanently crippled.

When the eradication campaign was started in 1986, guinea worms were endemic to 16 countries in sub-Saharan Africa as well as Yemen, India, and Pakistan. Every year, about 3.5 million people were stricken and at least 100 million people were at risk.

With the leadership of former United States President Jimmy Carter, a consortium of agencies, institutions, and organizations—the World Health Organization (WHO), UNICEF, the United Nations Development Program (UNDP), the World Bank, bilateral aid agencies, and the governments of many developed countries—banded together to fight this disease. Although complete success has not yet occurred, encouraging progress has been made. Already the guinea worm infections are down more than 96 percent. Pakistan was the first formerly infested country to be declared completely free of these parasites. More than 80 percent of all remaining cases occur in Sudan, where civil war, poverty, drought, and governmental resistance to outside aid have made treatment difficult. By 2010, there were only about 3200 cases of known guinea worm infection, all occurring in the African nations of Ethiopia, Ghana, Mali, and Sudan.

An encouraging outcome of this crusade is the demonstration that public health education and community organization can be effective, even in some of the poorest and most remote areas. Village-based health workers and volunteers conduct disease surveillance and education programs, allowing funds and supplies to be distributed in an efficient manner. Once people understand how the disease spreads and what they need to do to protect themselves and their families, they do change their behavior. A great advantage of this community health approach is educating villagers about the importance of proper sanitation and clean drinking water is effective not only against dracunculiasis, but also can help eliminate many other water-borne diseases.

Resources

BOOKS

Despommier, Dickson D., et al. *Parasitic Diseases*. 5th ed. New York: Apple Trees Productions, 2005.

OTHER

World Health Organization. "Eradicating Guinea Worm Disease." http://whqlibdoc.who.int/hq/2008/WHO_HTM_NTD_PCT_2008.1_eng.pdf (accessed November 6, 2010).

The Carter Center. "Guinea Worm Disease Eradication: Countdown to Zero."http://www.cartercenter.org/health/guinea_worm/mini_site/index.html (accessed November 6, 2010).

ORGANIZATIONS

The Carter Center, One Copenhill 453 Freedom Parkway, Atlanta, GA, USA, 30307, carterweb@emory.edu, http://www.cartercenter.org

William P. Cunningham

Gulf oil spill

On April 22, 2010, fires from an explosion taking place two days earlier, sank the *Deepwater Horizon* oil rig, located in the Gulf of Mexico about 52 miles (84 kilometers) off the coast of Louisiana. The explosion killed eleven workers and seriously injured seventeen others. The destruction created a massive oil leak from the wellhead located 5,000 feet (1,500 meters) below the surface. Over the next eighty-seven days, the resulting oil spill would become the largest accidental oil spill in history, creating a swath of environmental devastation and death along broad areas of the central and northern Gulf Coast.

By April 24, 2010, the oil slick was visible on surrounding surface waters. Over the following weeks and months, NASA satellite photographs—obtained from the Moderate Resolution Imaging Spectroradiometer (MODIS) mounted on the Aqua satellite and the Advanced Land Imager aboard the Earth Observing-1 (EO-1) satellite—provided evidence of an expanding surface oil slick over the northern Gulf of Mexico.

Scientists using fluorometers mapped and measured clouds and plumes of sub-surface oil, and later layers of subsurface oil mixed with chemical dispersants.

The oil leak was initially estimated at 1,000 barrels of oil (42,000 gallons) per day. Expert estimates of the volume of crude oil spilling into the Gulf each day quickly increased to 5,000 barrels (210,000 gallons). Many experts asserted that it was clear from pictures of the continuing underwater gusher eventually released that a significant amount of oil continued to spewing into the Gulf, especially in light of the fact that the company responsible for the spill, BP (formerly British Petroleum) claimed that a temporary funnel-like cap was collecting more than 15,000 (630,000 gallons) of oil per day. An array of marine and oil industry experts argued that the underwater pictures and surface observations provided clear evidence of an underreporting of the size of the spill.

Estimates of the volume of oil gushing into the Gulf increased steadily throughout the spill, ultimately reaching 5 million barrels (210 million gallons) of oil. The Gulf of Mexico spill (also called the BP oil spill or *Deepwater Horizon* oil spill) surpassed the estimated 3.3 million barrels (approximately 140 million gallons) of oil released during the 1979 Ixtoc I spill to become the worst accidental marine oil spill in history. The Ixtoc 1 spill followed a Petroleos Mexicanos' (PEMEX) rig explosion in the Bay of Campeche (the southern Gulf of Mexico off Mexico's coast). The 2010 Gulf of Mexico spill far surpassed the 11 million

This baby tern stuck in an oil patch on a Grand Isle, Louisiana, beach was rescued by Chris Hernandaz, the Street Superintendent of Grand Isle after the 2010 Gulf of Mexico oil spill. *(Julie Dermansky/Photo Researchers, Inc.)*

gallons of oil spilled into Alaskan waters following the 1989 grounding of the tanker *Exxon Valdez*.

Although there was initial uncertainty in estimating the rate of leakage, within days of the accident it became clear that enough oil would be spilled into the Gulf of Mexico to create a significant—and perhaps unprecedented— ecological disaster.

Undersea oil leak

Using remotely operated submersibles to examine the wreckage, engineers quickly discovered at least three major leaks. Oil was leaking from a ruptured drill pipe near the wellhead and from the crumpled riser pipe that had once connected the *Deepwater Horizon* rig to the well head.

Stopping the spill proved difficult. Automatic shut-off valves on the blowout preventer, a 50-foot (17 yd) stack of valves that sits on top of the wellhead, failed to operate and then failed to respond to remote commands. The blowout preventer's valves are designed to close when there are sudden surges or drops in oil or gas pressure. Such surges or fluctuations in pressure are often the cause of blowout explosions. With the valves damaged or open, the oil spill continued as engineers also attempted to use robot submersibles to close the valves. While continuing efforts to stop the leak at the source, engineers immediately began to assemble oil collection domes. Containment domes are normally used in shallower waters and prior to the *Deepwater Horizon* spill had never been deployed at the depth required to contain the leak. BP officials immediately dispatched drilling equipment and two rigs capable of drilling nearby relief wells to reduce the pressure within the leaking well, thus reducing the rate and amount of oil spill, and ultimately plug the well with heavy drilling mud and concrete. However, engineers cautioned that drilling relief wells would take months.

Efforts to stop the leak were performed under difficult marine conditions. Rough seas hampered initial efforts to close the blowout preventer shut-off valves. The vertical column of pipes from the seafloor was so badly damaged that, akin to kinks in a hose, the twisted remains of the connecting pipe actually acted to slow the oil leak. Engineers had to proceed with caution because while attempting repairs, they ran the risk of opening new leaks or inadvertently increasing the rate of leaks already spewing oil into the Gulf.

At the time the disaster occurred, the platform was located within the Mississippi Canyon Block 252 (Macondo Prospect oil field). BP, Transocean, and Halliburton were the three primary companies involved in operations related to the catastrophe.

The federal government formed the *Deepwater Horizon* Unified Command, which includes BP and Transocean, along with numerous government agencies, to fight the environmental problems associated with the BP Deepwater oil spill. On June 1, the organization commanded 1,400 vessels, 20,000 personnel, and seventeen staging areas. As of that date, the organization has already used over 910,000 gallons (3,445,000 l) of dispersants, recovered over 12.1 million gallons (45.8 million l) of oily water, and deployed over 3.7 million feet (1.13 million m) of booms. Several official investigations were also underway, including the Deepwater Horizon Joint Investigation (by the MMS and Coast Guard) and an investigation by the National Academy of Engineering. In addition, President Barack Obama announced that a bipartisan National Commission on the BP *Deepwater Horizon* Oil Spill and Offshore Drilling to investigate the incident

As of November 2010, investigations into the *Deepwater Horizon* sinking provided tentative conclusions that a series of technical and human failures resulted in the massive Gulf oil spill. Congressional hearings have produced documents and testimony to show the reliability of deepwater technology and the ability to handle "worst-case scenarios" are questionable at best. In addition, lax government inspections and improper relationships between government regulators and oil representatives have also been highlighted during ongoing investigations.

On May 28, 2010, President Obama ordered a suspension of any future drilling deepwater offshore pending a safety review. Obama also invoked a six-month moratorium on the issuance of drilling permits. In addition, the U.S. government named BP the responsible party in the disaster. BP accepted responsibility and agreed to pay all cleanup costs, but added that the accident was not entirely its fault because the rig's owner and operator was Transocean and operations potentially related to the explosion and spill were carried out by other companies.

As of November 2010, the oil spill has cost BP approximately $10 billion, but the final cost could go upwards of $30 billion.

Containment efforts

BLOWOUT PREVENTER. Numerous methods to limit the impact of the Deepwater Horizon spill were attempted throughout May and early June, including using ROVs to manually close the blowout preventer (BOP), a shutoff device at the wellhead. In all, BP sent six of these ROVs to cut the flow of oil at the wellhead. These attempts ultimately failed.

RELIEF WELLS. During this time, BP also began drilling two relief wells to intercept the original well at about 12,800 feet (3,900 m) below the seafloor. These relief wells were to be used in case the other attempts failed. BP hoped that by August these relief wells would be ready to end the oil flow.

CONTAINMENT DOME. Early in May, an attempt was made to place a 98-ton steel and concrete containment dome (called a "top hat") on top of the largest leak. The top hat, which was four feet (1.2 m) in diameter and 5 feet (1.5 m) tall, would attach to a drill pipe that would siphon the oil to a ship waiting on the ocean's surface. However, this procedure failed when the pipe became blocked with gas hydrates (crystalline solids of methane gas and water molecules). In the second week, the insertion of a Riser Insertion Tube Tool (RITT) between the platform pipe and the broken seafloor pipe was attempted. The apparatus allowed some collection of leaking oil.

TOP KILL AND JUNK SHOT. Then, on May 25, the RITT apparatus was removed so a "top kill" technique and a "junk shot" technique could be attempted to permanently close the leak. Heavy drilling fluids were pumped through two lines into the blowout preventer on the seabed. The top kill technique was designed to restrict the flow of oil so that cement could be poured in to permanently seal the leak. After temporarily stopping the flow, BP announced on May 29 that the "top kill" method had failed to permanently stop it. The junk shot technique, which consisted of shooting shredded tire bits, golf balls, knotted rope, and other selected materials into the BOP with the intention of clogging it, also failed.

LMRP CAP. Thereupon, BP began using the Lower Marine Riser Package (LMRP) Cap Containment System. A diamond saw blade began cutting the damaged riser so a custom-built cap could be placed on the newly cut pipe; however, the saw became stuck. With a substituted pair of shears, a successful cut was accomplished on June 3 and a cap attached. Recovery began the next day, with less than one-tenth of the oil captured. On June 8, according to BP, a total of about 15,000 barrels of oil had been collected that day, and about 57,500 barrels over the past four days. Ultimately, this technique also failed to contain the leaking oil.

NEW CAP ASSEMBLY. On July 10, the LMRP cap was removed so that a different cap could be installed. The new cap assembly, which was hopefully a better fit that the older one, consisted of a flange transition spool and a 3 ram stack. Five days later, BP announced that the leak had stopped when the BOP was closed shut with the new cap assembly.

STATIC KILL. With the success of the LMRP containment cap, on August 3, BP began a process called "static kill," or hydrostatic kill. The spill prevention process involves injecting several thousand barrels of cement and mud through the containment cap and into the top of the damaged BP well. The cement plug created from the process is designed to hold back the pressure of the oil, which was estimated at about 7,000 pounds per square inch, or almost 500 times atmospheric pressure. On August 9, BP reported that the static kill procedure was holding and leaking oil was no longer present.

BOTTOM KILL. The final major step is called the "bottom kill" technique. It involved pumping cement and mud from the bottom of the well, similar to the "static kill" technique used earlier. A storm entering the Gulf Coast area delayed the bottom kill procedure for several days. However, in late August, the bottom kill was completed and the flow of oil was permanently stopped from flowing out of the damaged well.

NEW BOP. On September 4, the damaged BOP was removed from the site, and lifted to the Gulf surface, a process that lasted just over one day. A new BOP was then installed in order to prevent any new leaks from occurring.

The two relief wells were completed in September, which allowed engineers to permanently seal the well with drilling mud and cement at levels deep into the reservoir. On September 19, Incident Commander and Retired Coast Guard Admiral Thad Allen announced that the well was dead.

Environmental impact

By June 2010, the surface slick extended over most of the northern Gulf of Mexico. While the bulk of the spill initially remained at sea, oil began washing into ecologically sensitive marshlands in Louisiana. Extending eastward into Florida waters, the surface slick spotted white-sand beaches vital to local tourist-based economies. Fishing bans extended over more than a quarter of the Gulf of Mexico, resulting in crippling economic hardship and apocalyptic predictions for the future of a Gulf seafood industry integral to the regional economy and deeply entwined with the culture of the region. In Alabama, oil flowed into inland waterways and wetland areas. Deaths of marine mammals, fish, birds, and other wildlife began to spike upwards.

In addition to the surface slick, more than 1.1 million gallons of dispersants, much of it sprayed deep underwater as oil gushed from the damaged well, reduced the surface slick at the expense of clouds and plumes of oil suspended in the water column.

Predicting precise landfalls and the degree of damage to specific areas proved difficult. Oil slicks are generally not consistent in thickness; the vast majority of a discernable slick is a thin sheen that can dissipate to create gaps in the slick area. However, within oil slicks are areas of thicker oil that pose substantial threat to wildlife, coastal environments, and the economies of impacted areas. Estimating what portion of a visible spill might dissipate before reaching shore is also complex, and subject to wave action, distance to the shore, and other variables. Wave action can churn the slick, hardening and clumping oil so that "globs" sink to ocean floor. However, such globs of oil can wash up on beaches and contaminate coastal areas for weeks and months following a spill. Ultimately, oil washed up along more than 600 miles (966 km) of coastline.

Mitigation and cleanup efforts

Initial mitigation and cleanup efforts include measures to both contain and directly remove surface oil. Such efforts relied on floating booms and skimmers to contain the slick until it could be pumped into container vessels. Boats and aircraft can also applied massive amounts of oil dispersants.

While still at sea, the slick killed and threatened birds, marine mammals, plankton, and species of fish that lay eggs at the surface.

CONTAINMENT BOOMS. Many miles of floating containment booms were used to restrict where oil could go, such as into mangroves, marshes, and other ecologically sensitive areas. These booms were about 1 to 4 feet (0.3 to 1.2 m) above and below the water line in order to fulfill its purpose.

SKIMMING AND CONTROLLED BURNS. The U.S. Coast Guard used dozens of skimmer ships ("skimmers") to collect ("skim") oil that was on the surface waters of the Gulf of Mexico. These skimmer vessels were used to contain this oil in preparation for controlled fires. This activity was done in an attempt to burn off the spilled oil before it reached land and devastate the environment.

DISPERSANTS. Dispersants are detergent-like chemicals that break up oil slicks. The molecular nature of the dispersants (one part of the molecular structure of dispersants has a polar affinity to water, the other end a non-polar affinity to oil) allows them to surround and coat small droplets of oil. Oil remains on the inside of the oil-dispersant glob in contact with the non-polar parts of the dispersant molecule. On the surface of the glob the polar portions of the dispersant molecules allow the glob to drop out of the spill and mix with water (a polar substance). Ultimately the oil-dispersant globs drop to the sea floor. Over thousands of years, those globules that do not wash up on beaches are consumed by microorganisms.

Chemical dispersants were used to accelerate the way that oil is naturally dispersed in water following oil spills. Such artificially made dispersants used on the Gulf oil spill were primarily Corexit EC9500A and Corexit EC9527A. Although many marine experts experessed disagreement or caution , according to their manufacturer (Nalco), "[COREXIT 9500] is a simple blend of six well-established, safe ingredients that biodegrade, do not bioaccumulate and are commonly found in popular household products. COREXIT products do not contain carcinogens or reproductive toxins. All the ingredients have been extensively studied for many years and have been determined safe and effective by the EPA."

By September 4, 2010, approximately 1.1 million gallons (4.2 million l) of chemical dispersant was applied to the wellhead. There remains much controversy with the use of such dispersants. Chemical dispersants have been used for over fifty years to treat oil spills around the world. However, the medical community has yet to decide the long-term effects of such dispersants on marine life.

Some dispersants have proven toxic to marine organisms. In addition, dispersed oil globules can also be highly toxic. By June, U.S. Environmental Protection Agency (EPA) officials expressed concern about the untested toxicity of the nearly million gallons of dispersants used to reduce the surface slick. EPA officials ordered changes on the types of dispersants used and, at one point, issued a ban on the use of some types of dispersants. Experts contend that it will take years to measure the full impact of the unprecedented use of dispersants.

Challenging scientific questions

Although it may take decades, the Gulf and its ecosystem have an enormous restorative capacity. Microbial life devours an estimated 1,000 barrels of crude oil naturally seeping into Gulf waters each day and microbes flourishing in the warm Gulf waters quickly devoured significant amounts of oil. However, there are differences in scientist's estimates of how much oil was consumed. While composed primarily of hydrocarbons, crude oil contains thousands of other chemical species in trace amounts, and bacterial responses to the particular crude spilled vary. Some bacteria consume selected elements of crude, leaving residues for other bacteria or for slower physical degradation. The rate at which microbes feed on oil is also related to levels of other nutrients present such as nitrogen, phosphorus, and iron.

Several species of prokaryotic microorganisms are responsible devouring the petroleum hydrocarbons emanating from natural seafloor seeps of oil and gas found around the world. The microbe Vibrio parahaemolyticus, common in warm Gulf waters and a rare, but known source of shellfish poisoning is an avid consumer of petroleum hydrocarbons derived from oil and methane. A related species *Vibrio vulnificus*, sometimes found in raw oysters, is far more pathogenic (able to cause disease). The combination of warm water and abundances of hydrocarbon food fueled nearly exponential growth rates in some areas of the Northern Gulf of Mexico. Scientists remain uncertain, however, at the real extent of the enhanced growth, whether pathogenic bacteria respond differently, how long enhanced growth rates will last, and what additional threats the higher bacterial counts pose to human health.

A challenging question for marine scientists is also whether the microbial population explosion in Gulf will create larger hypoxic regions devoid of oxygen and life. Such "dead zone" areas already existed in the Gulf prior to the spill, but experts feared that the spill will expand their the number, area, and depth of such zones.

The abundance of oil may also alter the population balances between microbes and have lasting impacts on their evolutionary development. There are also unanswered question as to how the microbes that normally feed on oil will respond to the partially emulsified oil in large undersea oil clouds and plumes. Microbiologists initially defended the use of dispersants because reducing the droplet size of the spilled oil created a larger surface area upon which microbes could feed. Some microorganisms, including *Alcanivorax borkumensis* naturally feed on oil by producing their own detergent-like surfactant substances to break down oil film into more digestible micro-droplets.

Adding complexity to the analysis are the natural checks on the growth of bacteria. For example, as a consequence of the higher bacterial counts, predatory bacterial viruses and protozoa not normally pathogenic to humans, but normally effective in stabilizing the populations of pathogenic bacteria such as *Vibrio* also increased in numbers.

Following the *Exxon Valdez* spill in 1989, the percentage of petroleum-consuming microbes in contaminated waters soared to ten times normal levels.

HEAVILY IMPACTED SPECIES. During the disaster, damage to the coastal environment of Louisiana visibly worsened. Workers helping to recover wildlife on Louisiana's East Grand Terre Island report that birds had been found "coated in thick, black goo" and brown pelicans "drenched in thick oil, struggling and flailing in the surf." The wetlands system—supporting a complex array of wildlife including seabirds and wading birds, speckled trout, shrimp, whooping cranes, wood storks, songbirds, sea turtles including the endangered Kemp's Ridley turtle, along with untold numbers of lower life forms—were destroyed or damaged to varying degrees by the oil spill.

In addition to helping measure and remediate devastating impacts on the environment and wildlife, scientists also face an array of continuing challenges. In addition to fighting for access to data, they must also take into account the influence of natural factors and preexisting phenomena. For example, media reports of dolphin deaths in the northern Gulf in May 2010 were quickly attributed to the oil spill. However, there was already an observed spike in bottlenose dolphin deaths in the region prior to the oil spill and prior to the oil spill, bottlenose dolphin deaths were already at a seven-year high. Prior to the spill, in March 2010 wildlife officials recorded more than three times the normal number of dead bottlenose dolphins. NOAA officials declared the deaths an "unusual mortality event" and ordered an investigation. Although oil residues are highly toxic to marine mammals and significant deaths and damage were observed pathologists conducting necropsies did not definitively link observed bottlenose dolphin death directly to the oil spill prior to June 2010. Scientists are also investigating alternative causes, including the influence of an abnormally cold winter, possible paralytic shellfish poisoning, or deaths caused by viruses such as the Morbillivirus.

Experts contend that such scientific rigor, at times unyielding to personal perceptions, political correctness, or popular media influences, is critical to determining the full extent of the damage caused by the spill. Incontrovertible scientific evidence strengthens claims made against parties responsible for the spill—making it difficult for them to escape legal accountability—and also strengthens efforts to create effective solutions.

HUMAN HEALTH CONCERNS. Medical professions state that the oil spilled into the Gulf waters and the dispersants used to clean them up will pose short- and long-term health problems to humans involved in the cleanup and to the citizens of the Gulf area. Specifically, hundreds of workers were treated for various medical problems during the cleanup. Symptoms were present, such as vomiting, coughing, chest pain, headaches, dizziness, nausea, and respiratory stress. Experts found that these symptoms of toxic origins are common within humans exposed to oil spills. For example, workers at the *Exxon Valdez* spill were treated for numerous respiratory problems during the clean-up activities, and subsequently showed a higher than normal rate of chronic airway disease.

Additional impacts

Carbon from the oil spill was ultimately traced into the marine food chain. Although the impacts remain uncertain and under study, a lack of consumer confidence in the safety of Gulf seafood severely damaged a Gulf seafood industry already crippled by fishing bans. Gulf fishing and tourism industries also suffered devastating losses.

A number of efforts continue in order to minimize damage and remediate existing damage to bays, estuaries, and wetlands, including the use of protective sand berms.

The U.S. Department of the Interior reported on August 9 that its Fish and Wildlife Service and National Parks Service cleanup crews were continuing shoreline duties at Gulf Islands National Seashore in Mississippi. So far the crews had removed 3,540 pounds (1,605 kg) of oily debris from Horn Island, 1,000 pounds (454 kg) from Petit Bois, 1,925 pounds (873 kg) from Cat Island, and 1,600 pounds (726 kg) from Ship Island.

As of November 2010, at least 11,000 people are still working to clean up the Gulf Coast shore affected by the BP *Deepwater Horizon* oil spill. The highest number of cleanup workers was estimated to be about 48,000 throughout the many months of cleaning of the shoreline. As of October 27, 2010, about 93 miles (150 km) of shoreline still had moderate to heavy oil present, with most of oiled coastline in Louisiana. An additional 483 miles (779 km) of shoreline had light-to-trace amounts of oil present, with about 226 miles (365 km) in Louisiana, 119 miles (192 km) in Florida, 78 miles (126 km) in Mississippi, and 60 miles

(97 km) in Alabama. Clean-up efforts are expected to be completed, with respect to "deep beach cleaning of oil" in Pensacola, Florida, and Orange Beach and Gulf Shores, Alabama, by 2011.

Resources

BOOKS

Exxon Valdez Oil Spill Trustee Council. *Then and Now—A Message of Hope: 15th Anniversary of the Exxon Valdez Oil Spill.* Anchorage, AK: Exxon Valdez Oil Spill Trustee Council, 2004.

Freudenburg, William R., and Robert Gramling. *Blowout in the Gulf: The BP Oil Spill Disaster and the Future of Energy in America.* Cambridge, Mass: MIT Press, 2011.

Robertson, Scott B. *Guidelines for the Scientific Study of Oil Spill Effects.* Thousand Oaks, CA: Robertson Environmental Services, 2004.

Wang, Zhendi, and Scott Stout. *Oil Spill Environmental Forensics: Fingerprinting and Source Identification.* New York: Academic, 2006.

OTHER

U.S. Government: Unified Command's Joint Information Center (JIC). "Gulf of Mexico Oil Spill Response: Deepwater Horizon Reponse" http://www.deepwaterhorizon response.com/go/site/2931/ (accessed November 12, 2010).

U.S. National Commission on the BP Deepwater Horizon Oil Spill and Offshore Drilling. "Media Advisories." http://www.oilspillcommission.gov/news#media-alerts (accessed November 12, 2010).

U.S. Environmental Protection Agency (EPA). "Emergencies: Oil Spills." http://www.epa.gov/ebtpages/ emeroilspills.html (accessed November 12, 2010).

K. Lee P. Lerner

Gulf War syndrome

Approximately 697,000 United States service members were deployed to the Persian Gulf from January to March 1991 as part of a multinational effort to stop Iraq's attack against Kuwait. And while the war itself was short, a long battle has been taking place ever since by veterans, the United States government, and scientists to determine what has caused "Gulf War Syndrome," a mysterious collection of symptoms reported by as many as 70,000 U.S. men and women who served in the war. They are joined by British veterans in their health complaints, and in smaller numbers by Canadians, Czechs, and Slovaks.

Gulf War Syndrome is a complex array of symptoms, including chronic fatigue, rashes, headaches, diarrhea, sleep disorders, joint and muscle pain, digestive problems, memory loss, difficulty concentrating, and depression. A small percentage of veterans have had babies born with twisted limbs, congestive heart failure, and missing organs. The veterans blamed these abnormalities on their service in the Gulf. The U.S. Environmental Protection Agency (EPA) has also found high rates of brain and nervous system cancers among these veterans, up to seven to fourteen times higher than among the general population, depending on the age group. Considering that most soldiers and veterans are younger and in better physical shape than the general population, researchers find such figures unusual.

Collectively, these ailments suggest that neurological processes may have been altered, or immune systems damaged. While no single cause has been identified, various analyses of the Gulf War experience point to low-level exposure of chemical weapons, combined with other environmental and medical factors, as key contributors to the health problems triggered years after exposure.

The war was unique in the levels of physical and emotional stresses created for those who served, as well as for their families. A significant portion of troops were from the reserves, rather than active enlistees. Deployment occurred at unprecedented speed. Most troops were given multiple vaccinations that singularly do not have adverse effects, but their combined effects were not tested before distribution. Detectors often signaled the presence of chemical weapons during the conflict, but were mostly ignored as inaccurate. The soldiers worked long hours in extreme temperatures, lived in crowded and unsanitary conditions where pesticides were used indiscriminately to rid areas of flies, snakes, spiders, and scorpions, and breathed and had dermal exposures to chemicals from the continuous oil fires—burning trash, feces, fuels, and solvents. Blazing sun, blowing sand and biting sandflies further increased the discomfort and stress of military life in the desert. Exposures to the various fumes often exceeded federal standards and World Health Organization (WHO) health guidelines; these alone could have caused "permanent impairment," according to a 1994 National Institutes of Health (NIH) report.

The U.S. military now admits it was inadequately prepared for chemical and biological warfare, which it knew Iraq had previously used. Three of four reserve units, for example, didn't have protective gear. The drug pyridostigmine bromide (PB, 3-dimethylaminocarbonyloxy-N-methylpyridium bromide) was given to almost 400,000 troops before and during the Gulf War to combat the effects of nerve gas, even though it is approved by the Food and Drug Administration (FDA) only for treatment of the neurological disorder myasthenia gravis. The FDA agreed on the condition that commanders inform troops what they were taking and what the potential side effects were. One survey, however, found that sixty-three of seventy-three veterans who had taken the drug did not receive such information. Records were not kept on who took which drugs or vaccines, as required by FDA and Defense Department guidelines.

While the Defense Department and other government agencies have spent more than $80 million to try to identify the cause of veterans' ailments, a privately funded team of toxicologists and epidemiologists may have discovered an explanation for at least some problems experienced by Gulf War veterans. Researchers treated chickens in 1996 with nonlethal doses of three chemicals veterans were exposed to: DEET (N,N-diethyl-m-toluamide) and chlorpyrifos (O,O- diethyl O-3,5,6-trichloropyridinyl phosphorothioate), used topically or sprayed on uniforms as insecticides, and the anti-nerve gas drug pyridostigmine bromide. They found that simultaneous exposure to two or more of the insecticides and drugs damaged the chickens' nervous system, even though none of the chemicals caused problems by itself. The range of symptoms the chickens developed is similar to those the veterans describe. A similar study by the Defense Department found that the chemicals were more toxic to rats when given together than individually. Follow up studies are underway to determine if this also holds for humans.

The researchers hypothesize that multiple chemicals overwhelmed the animals' ability to neutralize them. The enzyme butyrylcholinesterase, which circulates in the blood, breaks down a variety of nitrogen-containing organic compounds, including the three substances tested. But the anti-nerve gas drug, in particular, can monopolize the enzyme, preventing it from dealing with the insecticides. Those chemicals could then sneak into the brain and cause damage they would not produce on their own.

Many veterans believe that, while the drugs and pesticides may have played a role in their ailments, so have chemical weapons. Troops could have been subjected to a higher degree of sustained, low-level exposure of chemical weapons than previously believed, either directly or via air plumes, because 75 percent of Iraq's chemical weapons production capability, along with 21 chemical weapons storage sites, were destroyed by allied air raids.

In addition, U.S. battalions blew up an Iraqi arms dump soon after the war was over, before many troops had left the Gulf. Khamisiyah, an enormous ammunition storage site, covered 20 square miles (50 km^2) with 100 ammunition bunkers and other storage facilities. Two large explosions were set off, one on March 4 and a second on March 10, 1991. Smaller demolition operations continued in the area through most of April 1991.

While the site was not believed to have contained chemical weapons at the time, the Defense Department admitted in June 1996 that the complex had included nerve and mustard gases. The Central Intelligence Agency (CIA) also admitted in April 1997 that it knew in 1986 that thousands of mustard gas weapons had been stored at the Khamisiyah depot, but the agency failed to include it on a list of suspected sites provided to the Defense Department before the 1991 war, which led troops to assume it was safe to blow it up.

Weather data shows that upper-level winds in the gulf were blowing in a southerly direction during and after the bombing. Thus, vapors carried by these winds could have contaminated troops hundreds of miles away. A 1974 report, *Delayed Toxic Effects of Chemical Warfare Agents*, found that chemicals weapons plant workers suffer as many chronic symptoms as those now suffered by Gulf War veterans, including neurological, gastrointestinal and heart problems, loss of memory, and a greater risk of cancer; exposure to these chemicals may also create birth effects in children. A 1995 study by a British medical researcher found many of the same symptoms in Third World people exposed to organophosphate insecticides, like DEET, which are diluted versions of chemical weapons.

While British, Canadian and Slovak veterans have reported similar ailments, albeit in smaller numbers, no French veterans have complained of such illnesses, despite extensive publicity. This is also providing valuable clues to the U.S. veterans' maladies, in several ways. For example, the French did not use many of the vaccines that the British and Americans used, including pyridostigmine bromide. French camps were not sprayed with insecticides as a preventive measure, rather only when needed to control pest populations. When they did spray, they did not use organophosphates. Finally, the French were nowhere near the Khamisiyah munitions depot when the destruction occurred.

In February 1997, a series of study results established the most definitive links between Gulf War syndrome and chemicals to date. The research identifies six "syndromes," or clusters of like symptoms in discrete groups of veterans, and associates each with distinct events during the war. Troops who reported exposure to chemical weapons, for example, are likely to suffer from confusion, balance problems, impotence, and depression. Other sets of symptoms correspond to the use of insect repellants and anti-nerve gas drugs. While not conclusive, these findings will likely spur further research into the effects of low-level exposure to certain chemicals.

Such research was advocated by the presidential advisory committee, a twelve-member panel of veterans, scientists, and health care and policy experts established in 1995. The committee held 18 public meetings between August 1995 and November 1996 to investigate the nature of Gulf War veterans' illnesses, health effects of Gulf War risk factors, and the government's response to Gulf War illnesses. While the committee's final report in January 1997 concluded that no single, clinically recognizable disease can be attributed to Gulf War service, it recommended additional research on the long-term health effects of low-level exposures to chemical weapons, on the synergistic effects of pyridostigmine bromide with other Gulf War risk factors, and on the body's physical response to stress.

While the debate continues, the Veterans Affairs (VA) and Defense Departments are providing free medical help to any veteran who believes he or she is suffering from Gulf War Syndrome. In January 1997, President Clinton proposed new regulations that would extend the time available to veterans to prove their disabilities are related to Gulf War service from two to ten years. He also initiated a presidential review to ensure that in any future deployments the health of servicemen and women and their families is better protected.

Definitive answers as to the causes and treatments for veterans' ailments have yet to be found. What is clear is that the complex biological, chemical, physical, and psychological stresses of the Persian Gulf War appear to have produced a variety of complex adverse health effects. No single disease or syndrome is apparent, but rather multiple illnesses with overlapping symptoms and causes. If what had been considered acceptable trace levels of chemical agents in the war environment are found to be harmful, the U.S. military will have to revamp the way it protects its forces against even those tiny amounts. Tragically, that would mean that not only did "friendly fire" account for nearly 25 percent of the

146 U.S. deaths, but also that allied actions were responsible for the war's most persistent and haunting pain.

In November 2008, the U.S. Congress-mandated Research Advisory Committee on Gulf War Veterans' Illnesses issued a report, "Gulf War Illness and the Health of Gulf War Veterans: Scientific Findings and Recommendations," which stated that "scientific evidence leaves no question that Gulf War illness is a real condition with real causes and serious consequences for affected veterans." The report indicated that exposure to two neurotoxins—pesticides and anti-nerve agent pyridostigmine bromide (PB) pills—were "causally associated with Gulf War illness." The VA announced in February 2010 that their Department was conducting a reexamination of the disability claims of Gulf War veterans who appeared to still be suffering from Gulf War Syndrome.

Resources

PERIODICALS

Barber, Mike. "First Gulf War Still Claims Lives." *Seattle Post-Intelligencer* (January 16, 2006).

OTHER

Research Advisory Committee on Gulf War Veterans' Illnesses, U.S. Department of Veterans Affairs. "Gulf War Illness and the Health of Gulf War Veterans." http://www1.va.gov/rac-gwvi/ (accessed November 8, 2010).

Sally Cole-Misch

Gullied land

Areas where all diagnostic soil horizons have been removed by flowing water, resulting in a network of V-shaped or U-shaped channels. Generally, gullies are so deep that extensive reshaping is necessary for most uses. They cannot be crossed with normal farm machinery. While gullied land can occur on any land, they are often most prevalent on loess, sandy, or other soils with low cohesion.

See also Erosion; Soil profile; Soil texture.

Gymnogyps californianus see **Gypsy moth.**

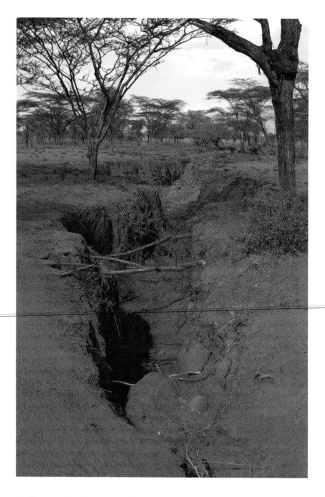

Gully erosion in Australia. *(Photograph by A. B. Joyce. Photo Researchers Inc.)*

Gypsy moth

The gypsy moth (*Portheria dispar*), a native of Europe and parts of Asia, has been causing both ecological and economic damage in the eastern United States and Canada since its introduction in New England in the 1860s.

In 1869, french entomologist Leopold Trouvelet brought live specimens of the insect to Medford, Massachusetts for experimentation with silk production. Several individual specimens escaped and became an established population over the next twenty years. The destructive abilities of the gypsy moth became readily apparent to area residents, who watched large sections of forest be destroyed by the larvae. From the initial infestation in Massachusetts, the gypsy moth spread throughout the northeastern United States and southeastern Canada. In 2000, states located along the leading edge of gypsy moth populations, along with the United States

Department of Agriculture Forest Service, launched a project to stop its spread. In 2008, North Carolina, Virginia, West Virginia, Kentucky, Ohio, Indiana, Illinois, Wisconsin and Minnesota participated in the program that reduces gypsy moth spread by 70 percent.

Gypsy moths have a voracious appetite for leaves, and the primary environmental problem caused by them is the destruction of huge areas of forest. Gypsy moth caterpillars defoliate a number of species of broadleaf trees including birches, larch, and aspen, but prefer the leaves of several species of oaks, though they have also been found to eat some evergreen needles. One caterpillar can consume up to one square foot of leaves per day. In 2001, 84.9 million acres (34.4 million ha) had been defoliated by these insects. The sheer number of gypsy moth caterpillars produced in one generation can create other problems as well. Some areas become so heavily infested that the insects have covered houses and yards, causing psychological difficulties as well as physical.

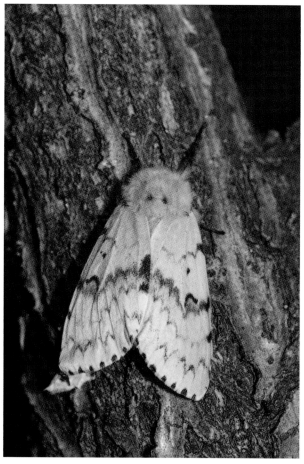

Female adult gypsy moth (*Lymantria dispar*). *(© Marcos Veiga / Alamy)*

There are few natural predators of the gypsy moth in North America, and none that can keep its population under control. Attempts have been made since the 1940s to control the insect with pesticides, including DDT, but these efforts have usually resulted in further contaminating of the environment without controlling the moths or their caterpillars. Numerous attempts have also been made to introduce species from outside the region to combat it, and almost 100 different natural enemies of the gypsy moth have been introduced into the northeast United States. Most of these have met, at best, with limited success. Recent progress has been made in experiments with a Japanese fungus that attacks and kills the gypsy moth. The fungus enters the body of the caterpillar through its pores and begins to destroy the insect from the inside out. It is apparently non-lethal to all other species in the infested areas, and its use has met with limited success in parts of Rhode Island and upstate New York. It remains unknown however, whether it will control the gypsy moth, or at least stem the dramatic population increases and severe infestations.

Resources

BOOKS

Conner, William E. *Tiger Moths and Woolly Bears: Behavior, Ecology, and Evolution of the Arctiidae.* Oxford: Oxford University Press, 2009.

Davies, Hazel, and Carol A. Butler. *Do Butterflies Bite? Fascinating Answers to Questions About Butterflies and Moths.* New Brunswick, NJ: Rutgers University Press, 2008.

Eugene C. Beckham

Haagen-Smit, Arie Jan

1900–1977

Dutch atmospheric chemist

The discoverer of the causes of photochemical smog, Haagen-Smit was one of the founders of atmospheric chemistry, but first made significant contributions to the chemistry of essential oils. Haagen-Smit was born in Utrecht, Holland, in 1900, and graduated from the University of Utrecht. He became head assistant in organic chemistry there and, later, served as a lecturer until 1936. He came to the United States as a lecturer in biological chemistry at Harvard, then became an associate professor at the California Institute of Technology in Pasadena. He retired as Professor of Biochemistry in 1971, having served as executive officer of the department of biochemistry and director of the plant environment laboratory. Haagen-Smit also contributed to the development of techniques for decreasing nitrogen oxide formation during combustion in electric power plants and autos, and to studies of damage to plants by air pollution.

While early workers on the haze and eye irritation that developed in Los Angeles, California, tried to treat it as identical with the smog (smoke + fog) then prevalent in London, England, Haagen-Smit knew at once it was different. In his reading, Haagen-Smit had encountered a 1930s Swiss patent on a process for introducing random oxygen functions into hydrocarbons by mixing the hydrocarbons with nitrogen dioxide and exposing the mixture to ultraviolet light. He thought this mixture would smell much more like a smoggy day in Los Angeles than would some sort of mixture containing sulfur dioxide, a major component of London smog. He followed the procedure and found his supposition was correct. Simple analysis showed the mixture now contained ozone, organic peroxides, and several other compounds. These findings showed that the

sources of the problems in Los Angeles were petroleum refineries, petrochemical industries, and ubiquitous automobile exhaust.

Haagen-Smit was immediately attacked by critics, who set up laboratories and developed instruments to prove him wrong. Instead the research proved him right, except in minor details.

Haagen-Smit had a long and distinguished career. In addition to his work on the chemistry of essential flower oils and famous findings on smog, he also contributed to the chemistry of plant hormones and plant alkaloids and the chemistry of microorganisms. He was a founding editor of the *International Journal of Air Pollution*, now known as *Atmospheric Environment*, one of the leading air pollution research journals. Though he found the work uncongenial, he stayed with it for the first year, then retired to the editorial board, where he served until 1976.

Once it was obvious that he had correctly identified the cause of the Los Angeles smog, he was showered with honors. These included membership in the National Academy of Science, receipt of the Los Angeles County Clean Air Award, the Chambers Award of the Air Pollution Control Association (now the Air and Waste Management Association), the Hodgkins Medal of the Smithsonian Institution, and the National Medal of Science. In his native Netherlands he was made a Laureate of Labor by the Netherlands Chemical Society, and Knight of the Order of Orange Nassau.

Resources

BOOKS

DuPuis, E. Melanie. *Smoke and Mirrors: The Politics and Culture of Air Pollution.* New York: New York University Press, 2004.

Seinfeld, John H., and Spyros N. Pandis. *Atmospheric Chemistry and Physics: From Air Pollution to Climate Change.* Hoboken, NJ: J. Wiley, 2006.

OTHER

United States Environmental Protection Agency (EPA). "Air: Atmosphere: Smog." http://www.epa.gov/ebt pages/airatmospheresmog.html (accessed November 8, 2010).

James P. Lodge Jr.

Habitat

Refers to the type of environment in which an organism or species exists in, as defined by its physical properties (e.g., rainfall, temperature, topographic position, soil texture, soil moisture) and its chemical properties (e.g., soil acidity, concentrations of nutrients and toxins, oxidation reduction status). Some authors include broad biological characteristics in their definitions (e.g., forest versus prairie habitats), when referring to the different types of environments occupied by trees and grasses. Within a given habitat there may be different micro-habitats, such as the hummocks and hollows on bogs or the different soil horizons in forests.

Habitat conservation plans

Protection of the earth's flora and fauna, the myriad of plants and animals that inhabit the planet, is totally dependent upon preserving their habitats. This is because a habitat, or natural environment for a specific variety of plant or animal, provides everything necessary for sustaining life for that species. Habitat conservation is part of a larger picture involving interdependency between all living things. Human life has been sustained since its dawning through the utilization of both plants and animals for food, clothing, shelter, and medicines. It follows then that the destruction of any species' environment, which will result in the eventual destruction of the species itself, adversely affects all other species.

Habitat destruction occurs for a number of reasons. Human industry has usually resulted in pollution that has often destroyed the balance of natural elements in soil, water, and air that are necessary to life. The need for forest products such as lumber has threatened woodlands in more ways than one. A lumbering practice called clear-cutting, in addition to over-harvesting a forest, leaves behind barren ground that results in erosion and threatens species dependent on the vegetation that grows on the forest floor. This wearing away of soil often results in negative changes to nearby streams and thus the water supply to multitudes of living things. The introduction of non-native species into an area can also threaten a habitat, as plants and animals with no natural enemies may thrive abnormally. This in turn will throw off the delicate balance and natural biological controls on population growth that each environment provides for its inhabitants.

As the understanding of these facts became more widespread, the demand for habitat conservation throughout the world increased. In the United States, in 1973, Congress passed the Endangered Species Act to protect both at-risk species and their environments. In order to control activities by private and nonfederal government landowners who might disturb habitats, the Act included a section outlining Habitat Conservation Plans (HCPs).

These HCPs were not implemented without a good deal of controversy. Some environmentalists were critical of the plans, believing that they failed to actually preserve the habitats and species they were designed to protect. Conservationists, landowners, and industrialists all argued that these HCPs were based upon faulty science and invalid and insufficient data. Impartial reviews did indicate that there were flaws. One cited weakness was that a species could theoretically be added to the endangered list but have no modification made in the plan to protect that species' living space.

It is clearly a measure of this controversy that in the law's first twenty years, only fourteen such plans were developed and approved nationwide. During the administration of President Bill Clinton (b. 1946), the federal government revised policies to encourage more numerous and more effective HCP applications. A more responsive to change *no-surprises* policy helped encourage participation in habitat conservation plans. During the 1990s, 259 HCPs were approved and there are now over 330, affecting more than 100 threatened or endangered species.

These policies recognize that natural resources change and environments require continual monitoring and alterations in plans, that those trying to balance environments can do so with adaptive management. Such alterations encouraged participation in HCPs, and by 1997, more than 200 plans covering nine million acres (3.6 million ha) of land had been approved. As of April 2002, the United States Fish and Wildlife Service noted that nearly double that number, 379 plans covering nearly 30 million acres (12.1 million ha) and protecting more than 200 endangered and threatened species have now been approved.

One positive example of the possibilities created by such plans is the work of the Plum Creek Timber Company, a nationwide timberland company whose corporate offices are located in Seattle, Washington. Self-described as "the second largest private timberland owner in the United States, with 7.8 million acres (3.2 million ha) located in the northwestern, southern, and northeastern regions of the country," the company lists many initiatives it has undertaken to preserve the environments under its charge. In the early 1990s, Plum Creek Timber Company developed a plan to provide a *ladder-like* framework in Coquille River tributaries in Oregon to aid fish in reaching upper areas blocked for many years by a culvert. Spawning surveys afterward showed that coho salmon and steelhead trout, for the first time in forty years, were present above the culvert.

HCPs were created to focus attention on the problem of declining wildlife on land not owned and protected by the federal government and to attempt to maintain the biodiversity so necessary for all life. This goal would ideally assure that land is developed in such a way that it serves both the needs of the landowner and of threatened wildlife. However, in reality, it is not always possible to achieve such a goal. Often the more appropriate aim, if habitat conservation is to be successful, will be total protection of a wildlife environment even at the price of banning all development.

Resources

BOOKS

Ladle, Richard J. *Biodiversity and Conservation: Critical Concepts in the Environment.* London: Routledge, 2009.

Park, Chris C. *A Dictionary of Environment and Conservation.* Oxford: Oxford University Press, 2007.

OTHER

Conservation International. "Conservation International." http://www.conservation.org/ (accessed November 6, 2010).

World Conservation Monitoring Center. "World Conservation Monitoring Center." http://www.wcmc.org.uk/ (accessed November 6, 2010).

ORGANIZATIONS

Fish and Wildlife Reference Service, 54300 Grosvenor Lane, Suite 110, Bethesda, MD, USA, 20814, (301) 492-6403, (800) 582-3421, fw9fareferenceservice@fws.gov, http://www.lib.iastate.edu/ collections/ db/usfwrs.html

National Wildlife Federation, 111000 Wildlife Center Drive, Reston, VA, USA, 20190-5362, (800) 822-9919, info@nwf.org, http://www.nwf.org

Joan M. Schonbeck

Habitat fragmentation

The habitat of a living organism, plant, animal, or microbe is a place, or a set of environmental conditions, where the organism lives. Net loss of habitat obviously has serious implications for the survival and well-being of dependent organisms, but the nature of remaining habitat is also very important. One factor affecting the quality of surviving habitat is the size of its remaining pieces. Larger areas tend to be more desirable for most species. Various influences, often a result of human activity, cause habitat areas to be divided into smaller and smaller, widely separated pieces. This process of habitat fragmentation has profound implications for species living there.

Each patch created when larger habitat areas are fragmented results in more edge area where patches interface with the surrounding environment. These smaller patches with a relatively large ratio of edge to interior area have some unique characteristics. They are often distinguished by increased predation when predators are able to hunt or forage along this edge more easily. The decline of songbirds throughout the United States is due in part to the increase of the brown-headed cowbird competing with other birds along habitat edges. The cowbird acts as a parasite by laying its eggs in other birds' nests and leaving them for other birds to hatch and raise. After hatching, the young cowbirds compete with the smaller birds of the nest, almost always killing them.

In the smaller patches formed from fragmentation, habitat areas are less protected from adverse environmental events, and a single storm may destroy the entire area. A disease outbreak may eliminate an entire population of a species. When the number of breeding adults becomes very low, some species can no longer reproduce successfully.

Some songbirds found in the United States are declining in number as their habitat shrinks or disappears. When they migrate south in the winter they find that habitat to be more scarce and fragmented. When they return from the tropics in the spring, they discover that the nesting territory that they used the previous year has disappeared.

Species dispersal is decreased as organisms must travel farther to go from one habitat area to another, increasing their exposure to predation and possibly harmful environmental conditions. Populations become increasingly insular as they become separated from related populations, losing the genetic benefits of a larger interbreeding population.

Road building often divides habitat areas, seriously disrupting migration of some mammals and herptiles (frogs, snakes, and turtles). Large swaths of land used by modern freeways are particularly effective in this regard. In earlier times, railroads built across the Great Plains to connect the west coast of the United States with states east of the Mississippi River, divided bison habitat and hampered their migration from one grazing area to another. This was one of the factors that led to their near extinction.

Habitat fragmentation, usually a result of human activity, is found in all major habitat types around the world. Rain forests, wetlands, grasslands, and hardwood and conifer forests are all subject to various degrees of fragmentation. Globally, rain forests are currently by far the most seriously impacted ecosystem. Because they contain 50 percent or more of the world's species, the resulting number of species extinctions is particularly disturbing. It is estimated that 25 percent of the world's rainforests disappeared during the twentieth century, and another 25 percent were seriously fragmented and degraded. In the last two centuries, nearly all of the prairie grassland once found in the United States has disappeared. Remaining remnants occur in small, scattered, and isolated patches. This has resulted in the extinction or near extinction of many plant and animal species.

Ability to survive habitat fragmentation and other environmental changes varies greatly among species. Most find the stress overwhelming and simply disappear. A few of the common species that have been very successful in adapting to changing conditions include animals such as the opossum, raccoon, gray squirrel, and European starling. Plant examples include dandelions, crab grass, creeping Charlie, and many other weed species. The wetland invader, purple loosestrife, originally imported from Europe to the United States, is rapidly spreading into disrupted habitat previously occupied by native emergent aquatic vegetation such as reeds and cattails. Animal species that once found a comfortable home in cattail stands must move on.

Resources

OTHER

United States Department of the Interior, United States Geological Survey (USGS). "Habitat Alteration." http://www.usgs.gov/science/science.php?term = 522 (accessed November 6, 2010).

United States Department of the Interior, United States Geological Survey (USGS). "Habitats." http://www.usgs.gov/science/science.php?term = 525 (accessed November 6, 2010).

Douglas C. Pratt, Ph.D.

Haeckel, Ernst H.

1834–1919

German naturalist, scientist, biologist, philosopher, and professor

Ernst Haeckel was born in Potsdam, Germany. As a young boy he was interested in nature, particularly botany, and kept a private herbarium, where he noticed that plants varied more than the conventional teachings of his day advocated. Despite these natural interests, he studied medicine—at his father's insistence—at Wurzburg, Vienna, and Berlin, Germany, between 1852 and 1858. After receiving his license, Haeckel practiced medicine for a few years, but his desire to study *pure science* won over, and he enrolled at the University of Jena to study zoology. Following completion of his dissertation, he served as professor of zoology at the university from 1862 to 1909. The remainder of his adult life was devoted to science.

Haeckel was considered a liberal nonconformist of his day. He was a staunch supporter of Charles Darwin, one of his contemporaries. Haeckel was a prolific researcher and writer. He was the first scientist to draw a family tree of animal life, depicting the proposed relationships between various animal groups. Many of his original drawings are still used in current textbooks. One of his books, *The Riddle of the Universe* (1899), exposited many of his theories on evolution. Prominent among these was his theory of recapitulation, which explained his views on evolutionary vestiges in related animals. This theory, known as the *biogenic law*, stated that ontogeny recapitulates phylogeny—the development of the individual (ontogeny) repeats the history of the race (phylogeny). In other words, he argued that when an embryo develops, it passes through the various evolutionary stages that reflect its evolutionary ancestry. Although this theory was widely prevalent in biology for many years, scientists today consider it inaccurate or only partially correct. Some even argue that Haeckel falsified his diagrams to prove his theory.

Ernst Haeckel. *(Corbis-Bettmann)*

In environmental science, Haeckel is perhaps best known for coining the term *ecology* in 1869, which he defined as "the body of knowledge concerning the economy of nature—the investigation of the total relationship of the animal both to its organic and its inorganic environment including, above all, its friendly and inimical relations with those animals and plants with which it directly or indirectly comes into contact—in a word, ecology is the study of all those complex interrelations referred to by Darwin as the conditions for the struggle for existence."

Resources

BOOKS

Richards, Robert J. *The Tragic Sense of Life: Ernst Haeckel and the Struggle over Evolutionary Thought.* Chicago: University of Chicago Press, 2008.

John Korstad

Half-life

A term primarily used to describe the atomic transformations, how radioactive decay processes cause unstable atoms to be transformed into other isotopes and elements, but which can also refer to the biological half-life of substances that are not radioactive.

Specifically, the physical half-life is the time required for half of a given initial quantity to decay into another entity or disappear (via conversion or removal) from a defined system.

See also Radioactivity.

Haliaeetus leucocephalus see **Bald eagle.**

Halons

Halons are chemicals that contain carbon (C), fluorine (F), and bromine (Br). They are used in fire extinguishers and other firefighting equipment. Because of their bromine content, halons can very efficiently destroy molecules of ozone (O_3), contributing to the depletion of ozone in the stratosphere. The ozone layer is located 10 to 28 miles (16–47 km) above the surface of the earth and it protects humans and the environment from damaging solar ultraviolet-B radiation. Halons account for approximately 20 percent of the ozone depletion.

Halons have been used since the 1940s, when they were discovered by United States Army researchers looking for a fire-extinguishing agent to replace carbon tetrachloride (CCl_4). Halons are very effective against most types of fires, are nonconductive, and dissipate without leaving a residue. They are also economical, very stable, and safe for human use.

Halons consist of carbon atom chains with attached hydrogen atoms that are replaced by the halogens fluorine and bromine. Some also contain chlorine (Cl). Halon–1211 ((bromochlorodifluoromethane, CF_2ClBr), halon–1301 (bromotrifluoromethane, CF_3Br), and halon-2402 (dibromotetrafluoroethane, $C_2F_4Br_2$) are the major fire-suppressing halons. Halon–1211 is discharged as a liquid and vaporizes into a cloud within a few feet. Halon–1301 is stored as a liquid but discharges as a gas. Halons suppress fires because they bond with the free radicals and intermediates of the decomposing fuel molecules that fuel the fire. They also lower the temperature of the fire.

Halons may take up to seven years to drift up and distribute throughout the stratosphere, with the highest concentrations over the poles. High-energy ultraviolet radiation breaks their bromine and chlorine bonds, releasing these very reactive components, which in turn break down the ozone molecules and react with free oxygen to interfere with ozone creation. Although chlorine is more abundant, bromine is more than one hundred times more damaging to the ozone. Because of the ozone-depleting actions of halons, the substances also contribute to global warming and climate change.

Halons are categorized as class I ozone-depleting substances, along with chlorofluorocarbons (CFCs) and other substances with ozone-depleting potentials (ODPs) of 0.2 or greater. The ODP is the ratio that refers to the amount of ozone depletion caused by a substance compared to a similar mass of CFC–11—a common refrigerant—which has an ODP of 1.0. Halon–1211 has an ODP of 3.0 and an atmospheric lifetime of sixteen years. It also has a global warming potential (GWP) of 1300 to 1890. The GWP of a chemical is a ratio that indicates the amount of global warming caused by the substance compared with a similar mass of carbon dioxide (CO_2). Halon–1301 has an ODP of 10.0, an average lifetime of sixty-five years, and a GWP of 6900 to 7140. Data from the World Meterological Organization has shown that the ODP values for halon–1301 and halon–1211 are higher than originally reported at 16.0 and 7.1, respectively. Halon-2402 has an ODP of 11.5. Halon–1211 and halon–1301 are the most common halons in the United States. Halon-2402 is widely used in Russia and the developing world. Although total halon production between 1986 and 1991 accounted for only about 2 percent of the total production of class I substances, it accounted for about 23 percent of the ozone depletion caused by class I substances.

Halon production in the United States ended on December 31, 1993, because of their ozone depletion. Under the Montreal Protocol on Ozone Depleting Substances, first negotiated in 1987 and now including more than 191 countries, halons became the first ozone-depleting substances to be phased out in industrialized nations, with production stopped in 1994. Under the Clean Air Act, the United States banned the production and importation of halons as of January 1, 1994. The use of existing halons in fire protection systems continues and recycled halons can be purchased to recharge such systems. It is estimated that about 50 percent of all halons ever produced currently exist in portable fire extinguishers and firefighting equipment. The United States has 40 percent of the world's supply of halon–1301. In 1997 approximately 1,080 tons (977 metric tons) of halon–1211 and 790 tons (717 metric tons) of halon–1301 were released in the United States. In 1998, the U.S. Environmental Protection Agency prohibited the venting of halons during training, testing, repair, or disposal of equipment, and banned the blending of halons, to prevent the accumulation of nonrecyclable stocks.

The European Union has gone beyond the Montreal Protocol, banning the sale and noncritical use of halons after December 31, 2002, and, as of December 31, 2003, mandating the decommissioning of noncritical halon systems.

Alternatives are now available for most halon applications. Existing halon supplies from fire-suppression systems are being recycled for critical uses where no alternative exists.

Resources

BOOKS

Burroughs, William James. *Climate Change: A Multidisciplinary Approach.* Cambridge: Cambridge University Press, 2007.

Cowie, Jonathan. *Climate Change: Biological and Human Aspects.* Cambridge: Cambridge University Press, 2007.

Shulk, Bernard F. *Greenhouse Gases and Their Impact.* New York: Nova Science Publishers, 2007.

OTHER

United States Environmental Protection Agency (EPA). "Class I Ozone-depleting Substances." http://www.epa.gov/Ozone/science/ods/classone.html (accessed August 29, 2010).

United States Environmental Protection Agency (EPA). "Pollutants/Toxics: Chemicals: Halons." http://www.epa.gov/ebtpages/pollchemicalshalons.html (accessed August 29, 2010).

ORGANIZATIONS

Halon Alternatives Research Corporation, Halon Recycling Corporation, 2111 Wilson Boulevard, Eighth Floor, Arlington, VA, USA, 22201, (703) 524-6636, (703) 243-2874, (800) 258–1283, harc@harc.org, http://www.harc.org

Stratospheric Ozone Information Hotline, United States Environmental Protection Agency, 1200 Pennsylvania Avenue, NW, Washington, D.C., USA, 20460, (202) 775-6677, (800) 296–1996, public-access@epa.gov, http://www.epa.gov/ozone

United Nations Environment Programme, Division of Technology, Industry and Economics, Energy and OzonAction Programme, Tour Mirabeau, 39-43 quai Andre;aa Citroe;aun, 73759 Paris Cedex 15, France, (33-1) 44 37 14 50, (33-1) 44 37 14 74, ozonaction@unep.fr, http://www.uneptie.org/ozonaction

Margaret Alic

Hanford Nuclear Reservation

The Hanford Engineering Works was conceived in June 1942 under the direction of Major General Leslie R. Groves (1896–1970), head of the famous Manhattan Project, to produce plutonium and other materials for use in the development of nuclear weapons. By December 1942, a decision was reached to proceed with the construction of three plants—two to be located at the Clinton Engineering Works in Tennessee and a third at the Hanford Engineering Works in Washington.

Hanford was established in the southeastern portion of Washington state between the Yakima Range and the Columbia River, about 15 miles (24 km) northwest of Pasco, Washington. The site occupies approximately 586 square miles (1,517 km^2) of desert with the Columbia River flowing through its northern region. Once a linchpin of U.S. nuclear weapons production during the Cold War era (1947–1991), Hanford has now become the world's largest environmental cleanup project.

Hanford Engineering Works (HEW), known by various other names such as Hanford Works, Hanford Project, or "site W" in classified terms, was originally under the control of the Manhattan District of the Army Corps of Engineers (MED) until the Atomic Energy Commission (now the Department of Energy, DOE) took over in 1947. The actual operation of the site has been managed by a series of contractors since its inception. The first organization granted a contract to run site operations at Hanford was E.I. DuPont de Nemours and Company. In 1946, General Electric took over, and with the aid of several subcontractors, ran construction and operation of the site through 1965. A series of contractors have directed operations at both the main DOE-Richland Operations Office and the DOE-Office of River Protection (ORP), the agency responsible for overseeing hazardous waste tank farm clean up along the Columbia River, since then.

In Battelle Memorial Institute, a nonprofit organization, assumed management of the federal government's DOE research laboratories on the Hanford site. The newly formed Pacific Northwest Laboratory (which became Pacific Northwest National Laboratory [PNNL] in 1995) supports the Hanford site cleanup through the development and testing of new technologies. Battelle still runs the PNNL today.

A number of contractors have directed operations at HEW throughout its history. As of 2010, the prime contractors at the DOE-Richland Operations Office include the following: AdvanceMed Hanford (AMH), CH2M HILL Plateau Remediation Company (CHPRC), Mission Support Alliance (MSA), and Washington Closure Hanford (WCH).

The DOE-ORP, responsible for overseeing hazardous waste tank farm clean-up along the Columbia River, is managed by prime contractors: Advanced Technologies and Laboratories International, Inc. (ATL), Bechtel National, Inc. (BNI), and Washington River Protection Solutions, LLC (WRPS).

From 1943 to 1963, a total of nine plutonium-production reactors and five processing centers were built at Hanford, with the last of the reactors ceasing operations in 1987 (permanent shut-down of the Fast Flux Test Facility [FFTF], a sodium-cooled breeding reactor that produced isotopes for medical and industrial use, was ordered to close in 2001). Plutonium produced at Hanford was used in the world's first atomic explosion at Alamogordo, New Mexico, in 1945. The Hanford site processed an estimated 74 tons (67 metric tons) of plutonium during its years of active operation, accounting for approximately two-thirds of all plutonium produced for U.S. military use.

The plutonium fuel was processed on site at the Plutonium and Uranium Extraction Plant (PUREX). In the processing of plutonium, a substantial quantity of radioactive waste is produced; at Hanford, it amounts to 40 percent of the nation's one billion curies of high-level radioactive waste from weapons production. PUREX ceased regular production in 1988 and was officially closed in 1992. Deactivation of the plant was completed in 1997.

During the HEW's operational lifetime, some high-level radioactive wastes were diverted from the relatively safe underground storage to surface trenches. Lower-level contaminated water was also released into ditches and the nearby Columbia River. The DOE reports that over 450 billion gallons of waste liquid from the Hanford plants was improperly discharged into the soil column during their operational lifetime. These wastes contained cesium-137, technetium-99, plutonium-239 and -240, strontium-90, and cobalt-60. So much low-level waste was dumped that the groundwater under the reservation was observed to rise by as much as 75 feet (23 m).

Today more than 50 million gallons (189.3 million l) of high-level radioactive waste is contained in approximately 177 underground storage tanks, many of which have known leaks. One million gallons (3.8 million l) of this tank waste has already leaked into the surrounding ground and groundwater. An estimated 25 million cubic feet of radioactive solid wastes remains buried in trenches, which are similar to septic-tank drainage fields. Spent nuclear fuel basins have leaked over 15 million gallons (56.8 million l) of radioactive waste. In total, over 1,900 waste sites and 500 contaminated facilities have been identified for clean up at Hanford. The contaminated zone encompasses a variety of ecosystems and the nearby Columbia River. According to a 2007 Associated Press article, the site is considered to be one of the most contaminated nuclear sites in the United States.

In the early 1990s, Congress passed a bill requiring the DOE to create a watchdog list of those Hanford tanks at risk for explosion or other potential release of high-risk radioactive waste. By 1994 the list had fifty-six tanks listed as a high-risk potential. By late 2001, the final twenty-four tanks had been removed from the congressional watch list, which is sometimes referred to as the "Wyden Watch List" (after sponsoring Senator Ron Wyden).

In 1989 the DOE, U.S. Environmental Protection Agency (EPA), and the State of Washington Department of Ecology signed the Hanford Federal Facility Agreement and Consent Order (i.e., the Tri-Party Agreement [TPA]). The TPA outlines cleanup efforts required to achieve compliance with the Comprehensive Environmental Response Compensation and Liability Act (CERCLA; or Superfund) and the Resource Conservation and Recovery Act (RCRA).

Compliance with federal and state authorities under the TPA has not always progressed without problems for Hanford officials. The EPA fined DOE for poor waste management practices in 1999, levying $367,078 in civil penalties that were later reduced to $25,000 and a promise to spend $90,000 on additional clean-up activities. The following year EPA issued an additional $55,000 fine against the DOE for noncompliance with the TPA. However, the five-year review of the project completed by the EPA in 2001, while specifying eighteen "action items" for DOE compliance, also states that cleanup of soil waste sites and burial grounds "are proceeding in a protective and effective manner." In March 2002 DOE was fined $305,000 by the Washington Department of Ecology for failing to start construction of a waste treatment facility as outlined in the TPA. Over the next eight years, little was accomplished concerning the agreement, as negotiations continued fruitlessly among the parties.

Even though the Hanford site poses significant environmental concerns, the facility has generated a number of innovative technological advances in the cleanup and immobilization of radioactive wastes. For instance, one process developed in Hanford's Pacific Northwest Laboratory is *in situ vitrification*, which uses electricity to treat waste and surrounding contaminated soil, melting it into a glass material that is more easily disposed. The William R. Wiley Environmental and Molecular Sciences Laboratory (EMSL), which opened in 1997, houses a wide range of experimental programs aimed at solving nuclear waste treatment issues.

And despite the extent of its contamination, the isolation and security of the Hanford site has made the area home to a large and diverse population of flora and fauna. A 1997 Nature Conservancy study at Hanford found dozens of previously undiscovered and rare plant and animal species in various site habitats. The biodiversity study led to the establishment of the federally-protected 195,000-acre (79,000-ha) Hanford Reach National Monument in June 2000.

On October 26, 2010, a public announcement was made that the U.S. District Court in Spokane (Washington state) had approved a judicial decree that stipulated an enforceable schedule to clean up the Hanford waste, as per the TPA. The key points of the decree include the following: (1) agreed-upon milestones for the construction of the Waste Treatment Plant (WTP), with completion in 2047; (2) retrieval of single-shell tanks at C Farm in 2014, with full retrieval in 2040; (3) treatment of tank waste commencing in 2019, with full operations by 2022; and (4) closure of the double-shell tank farms in 2052.

Resources

BOOKS

Gephart, Roy E. *Hanford: A Conversation about Nuclear Waste and Cleanup.* Columbus, OH: Battelle Press, 2003.

Gerber, Michele S. *On the Home Front: The Cold War Legacy of the Hanford Nuclear Site*, 3rd ed. Lincoln: University of Nebraska Press, 2007.

Taylor, Bryan C. *Nuclear Legacies: Communication, Controversy, and the U.S. Nuclear Weapons Complex.* Lanham, MD: Lexington Books, 2007.

OTHER

Associated Press/SeattlePI.com. "U.S. to Assess the Harm from Hanford." http://www.seattlepi.com/local/310247_hanford04.html (accessed November 5, 2010).

Department of Energy. "Agencies Agree to Integration of Soil, Facility, and Groundwater Cleanup Milestones at Hanford Site." http://www.hanford.gov/c.cfm/media/attachments.cfm/DOE/Press%20--%20Central%20Plateau%20Milestones%20M15.pdf (accessed November 5, 2010).

Department of Energy. "Hanford." http://www.hanford.
gov/ (accessed November 5, 2010).
Washington State Department of Ecology. "Court
Approves Agreement on New Commitments for
Hanford Tank Waste Cleanup." http://www.ecy.wa.
gov/news/2010news/2010-284.html (accessed
November 5, 2010).
Washington State Department of Ecology Nuclear Waste
Program. "Nuclear Waste." http://www.ecy.wa.gov/
programs/nwp/ (accessed November 5, 2010).

Paula A. Ford-Martin

Hardin, Garrett

1915–2003
American environmentalist and writer

Trained as a biologist (University of Chicago undergraduate degree; PhD from Stanford in 1942 in microbial ecology), Garrett Hardin spent most of his career at the University of California at Santa Barbara, where his title was Professor of Human Ecology. He was born in Dallas, Texas, and grew up in various places in the Midwest, spending summers on his grandparents' farm in Missouri.

Very few biologists, short of Charles Darwin, have generated the levels of controversy that Hardin's thinking and writing have. The controversies, which continue today, center on two metaphors of human-ecological relationships: "the tragedy of the commons" and "the lifeboat ethic."

Hardin is widely credited with inventing the idea of the tragedy of the commons, but his work was long preceded by an ancient rhyme about the tragedy that results from stealing the commons from the goose. He did popularize the idea, though, and it made him a force to reckon with in population studies. Seldom is an academic author so identified with one article (though his thoughts on lifeboat ethics have since become almost equal in identification and impact). The idea is very simple: Resources held in common will be exploited by individuals for personal gain in disregard of public impacts; individual profit belongs to individual exploiters while they bear the brunt of only part of the impacts. Much of the controversy centers on Hardin's solutions to the tragedy: first, that private property owners "recognize their responsibility to care for" the land, thus lending at least implicit support to privatization efforts; and, second, the paradoxical idea that because exercise of individual freedom leads to ruin, such freedom cannot be tolerated, thus we must turn to "mutual coercion mutually agreed upon." As Hardin noted, "If everyone would restrain himself, all would be well; but it takes only one less than everyone to ruin a system of voluntary restraint. In a crowded world of less than perfect human beings, mutual ruin is inevitable if there are no controls. This is the tragedy of the commons." Hardin later expanded on his thesis, answering his critics by incorporating the differences between an open access system and a closed one. But the debate continues.

His other widely debated metaphor was of a lifeboat (standing for a nation's land and resources) occupied by rich people in an ocean of poor people (who have fallen out of their own, inadequate lifeboats). If the rich boat is close to its margin of safety, what should its occupants do about the poor people in the ocean? Or in the other boats? If the occupants let in even a few more people, the boat may be swamped, thus creating an ethical dilemma for the rich occupants. How do these occupants of the already full lifeboat justify taking in additional people if it guarantees the collapse for all? Ever since, Hardin has been accused of social Darwinism, of racism, of ignoring the possibility that "the poverty of the poor may be caused in part by the affluence of the rich," and of isolationism (because he argued that "for the foreseeable future survival demands that we govern our actions by the ethics of a [sovereign] lifeboat" rather than "space-ship" ethics that arguably try to care for all equitably).

What Hardin was trying to do in both of these cases was to look unemotionally at population growth and resource use, employing the rationale and language of a scientist viewing human populations in an objective, evolutionary perspective. Biologists know that "natural" populations that overuse their resources, that exceed the carrying capacity of their range, are then adjusted—in numbers or in resource use levels—also naturally and often brutally (from the perspective of many people). The problems remain that the division between the rich and the poor in the world is widening, and that many resources and ecosystems on earth are being stressed, though how close to the breaking point no one really knows. Also, nation-states are arbitrary divisions, and perhaps—in a corollary to the spaceship metaphor—the ultimate lifeboat is the earth and all the inhabitants must exist in it together.

Hardin retired as a professor emeritus of human ecology, a rare title that reflects his attempts to fold the human species into the evolutionary ecology perspective developed in biology. Ultimately what he tried to get across was what he called an *ecolate* view: *ecolacy* asks the question "and then what?" The basic insight of the ecolate citizen is that the world is a complex variety of systems so intricately interconnected that one can seldom be very confident that a proposed intervention in this system of systems will produce the consequences one wants. Hardin's views are rich and varied—the themes presented here are only two of many—but he still summarizes the human condition by what he labels "the ecolate predicament": that all human interventions are doomed to failure if a population exceeds its carrying capacity, whether of a region or of the world. Doing so "will bring everyone down to a level of poverty."

Hardin and his wife, both of whom suffered from fatal, chronic illnesses, took their lives on September 14, 2003.

Resources

BOOKS

Hardin, G. *Living within Limits: Ecology, Economics, and Population Taboos.* New York: Oxford University Press, 1993.
Hardin, G. *The Ostrich Factor: Our Population Myopia.* New York: Oxford University Press, 1999.

PERIODICALS

Hardin, G. "Living on a Lifeboat." *BioScience* 24, no. 10, (October 1974): 561–568.
Hardin, G. "The Tragedy of the Commons." *Science* 162, (13 December 1968): 1243–1248.

Gerald L. Young

Hawaiian Islands

The Hawaiian Islands are made up of a chain of ancient volcanic islands that have formed at irregular intervals over the last ten million years. There are over 100 islands in the chain, eight of which are considered major. Of the eight major islands, Kauai is the oldest, and the island of Hawaii is the youngest, having formed within the past million years. This vast discrepancy in age relative to the other islands on the chain has contributed to the tremendous biodiversity that has evolved there. The other factor responsible for Hawaii's great species diversity is the island chain's remoteness. The nearest continent is 2,400 miles (3,862 km) away, thus limiting the total colonization that could, or has, taken place. The niches available to these colonizing species are very diverse because of geophysical events. For example, on the island of Kauai, the average annual rainfall on the windward side of Mount Waialeale is 460 inches (1,169 cm), whereas on its leeward side, it is only 19 inches (48 cm). The temperature on the islands ranges from 75–90°Fahrenheit (24–32°C) for 300 days each year. The lowest temperature ever recorded in Hawaii was 54°Fahrenheit (12°C). Thus with its tropical climate and unique biotic communities, it is easy to understand why Hawaii has been considered a paradise. But now this paradise is threatened by serious environmental problems caused by humans.

The impact of humans has been felt in the Hawaiian Islands since their first arrival, but perhaps never more so than today. In the last quarter century tourism has replaced sugar cane and pineapple as the islands' main revenue source. Over six million tourists visit Hawaii each year, who collectively spend $11 million a day. With tourism comes development that often destroys natural habitats and strains existing natural resources. Hawaii has more than sixty golf courses, with plans to develop at least 100 more. This would destroy thousands of acres of natural vegetation, require the use of millions of gallons of freshwater for irrigation, and necessitate the use of tons of chemical fertilizers and pesticides in their maintenance. The increased number of tourists, many of whom visit the islands to enjoy its natural beauty, are often destroying the very thing they are there to see. Along the shore of Hanauma Bay, just south of Honolulu, over 90 percent of the coral reef is dead, primarily because people have trampled it in their desire to see and experience this unique piece of nature.

Much of Hawaii's fauna and flora are unique. Over ten thousand species are native to the Hawaiian Islands, having evolved and filled specialized niches there through the process of adaptive radiation from the relatively few original colonizing species. Examples are found in virtually every group of plants and animals in Hawaii. The avian adaptive radiation in the Hawaiian Islands surpasses even the best-known example, Darwin's finches of the Galapagos Islands. From as few as fifteen colonizing species evolved ninety native species of birds in the Hawaiian archipelago. Included in this number are the Hawaiian honeycreepers of the family Drepanididae. This endemic family of birds arose from a single ancestral species, which gave rise to twenty-three species, including twenty-four subspecies, of honeycreepers spread throughout the main islands. Niche availability and reproductive isolation contributed greatly to the spectacular diversity of forms that evolved. These birds developed adaptations such as finch-like bills for crushing seeds, parrot-like bills for foraging on larvae in wood, long decurved bills for taking nectar and insects from specialized flowers, and small forcep-like bills for capturing insects.

Introductions of vast numbers of alien species is taking its toll on native Hawaiian species, a process that began when the first humans settled the islands over 1,500 years ago. Many native species of birds had experienced drastic population declines by the time Europeans first encountered the islands over two hundred years ago. Since then at least twenty-three species of Hawaii's native birdlife have become extinct, and currently over thirty additional avian species are threatened with extinction. Through the process of primary ecological succession, one new species became established in the Hawaiian Islands every seventy thousand years. Introductions of alien species are taking place a million times faster, and are thus eliminating native species at an unprecedented rate. Recent estimates indicate that there are over eight thousand introduced species of plants and animals throughout the Hawaiian Island chain. The original Polynesian settlers brought with them pigs, dogs, chickens, and rats, along with a variety of plants they had cultivated for fiber, food, and medicine. With the Europeans came cattle, horses, sheep, cats, and additional rodents. The mongoose was introduced purposefully to control the rat populations; however, they presented more of a threat to ground-nesting birds. The introduction of rabbits caused the loss of vast quantities of vegetation and ultimately the extinction of three bird species on Laysan Island. Over 150 species of birds, including escaped cage birds, have been introduced to the islands. However, most of these have not established breeding populations.

Although Hawaii represents only 0.2 percent of the United States' land base, almost 75 percent of the total extinctions of birds and plants of the nation have occurred in this state. Hawaii's endemic flora are disappearing rapidly. Introduced mammalian browsers are decimating the native plantlife, which never needed to evolve defense mechanisms against such predators, and other introduced animals are destroying populations of native pollinators. Habitat destruction and opportunistic nonnative vegetation are also working against the endemic Hawaiian plant species. Organizations such as the Hawaii Plant Conservation Center are working to preserve the state's floral diversity by collecting and propagating as many of the rare and endangered species as possible. Their greenhouses now contain over two thousand plants representing almost two-thirds of Hawaii's native species, and their goal is to propagate four hundred of the state's most endangered plants.

Introduced plant and animal species and their assault on native species are by no means the only environmental problems facing the Hawaiian Islands. Because Hawaii's population is growing at a higher rate than the national average, and in part because of its isolation, it is facing many environmental problems on a grander scale and at a more rapid rate than its sister states on the mainland. Energy is one of the primary problems. Much of Hawaii's electricity is produced by burning imported oil, which is extremely expensive. Because of limited reserve capabilities with regard to electrical generation, Hawaii faces the potential for blackouts.

Planners have been, over the past several decades, looking at the feasibility of tapping into Hawaii's seemingly vast geothermal energy resources. Hawaii has the most active volcano in the world, Kilauea, whose underground network of geothermal reserves is the largest in the state. The proposed Hawaii Geothermal Deep Water Cable project would supply the energy for a 500-megawatt power plant, the electrical power of which would be transmitted from the island of Hawaii to Oahu by three undersea cables. This would, of, course, be of economic and environmental benefit by reducing dependence on oil reserves. Opponents of this geothermal power plant point to several problems. They are concerned with the potential for the release of toxic substances, such as hydrogen sulfide, lead, mercury, and chromium, into the environment from wellheads. They also have voiced negative opinions concerning construction of a geothermal plant so near the lava flows and fissures of Hawaii's two active volcanoes. An alternately proposed site would have the facility located in Hawaii's last major inholding of lowland tropical rain forest. To avoid economic disaster from escalating prices for imported oil and the problem of frequent blackouts, Hawaii must reach some compromise on geothermal energy or research the potential for getting its electricity from wind, wave, or solar energy.

Hawaii is also faced with environmental problems of another sort: natural disasters. Volcanoes not only provide the potential for geothermal energy, they also have the potential for massive destruction. Over the past two hundred years, Hawaii's two active volcanoes, Kilauea and Mauna Loa, have covered nearly 200,000 acres (81,000 ha) of land with lava, and geologists expect them to remain active for centuries to come. Severe hurricanes and tsunamis also hold the potential for vast destruction, not only of human property and lives, but of natural areas and the wildlife it holds. Many of the endemic species of Hawaii are threatened or endangered, and their populations are often so low that a single storm could wipe out most or all of its numbers.

There are efforts to reverse the environmental destruction in the Hawaiian Islands. Several organizations comprised of native Hawaiians are working to stem the destruction and loss of the wilderness paradise discovered by their ancestors.

Resources

BOOKS

Schmincke, Hans-Ulrich. *Volcanism.* New York: Springer, 2004.

U.S. Army Corps of Engineers. *Coastal Geology.* Honolulu, Hawaii: University Press of the Pacific, 2004.

Eugene C. Beckham

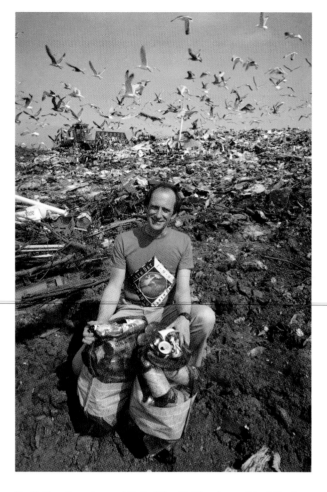

Earth Day founder and Palo Alto, California, resident Denis Hayes at a dump site in Palo Alto, California, 1990, as he displays burlap sacks filled with empty glass bottles and metal cans that local residents use in a weekly recycling program. *(Paul Sakuma)*

Hayes, Denis A.

1944–
American environmental activist and Earth Day organizer

Denis Hayes, an environmental activist and early advocate of solar power, is best known as the organizer of Earth Day and creation of the Earth Day Network. As of 2010, the network comprises over 180 member countries.

As executive director of the first Earth Day in April 1970, Hayes helped launch the modern movement of environmentalism, and has promoted the use of solar energy and other renewable resources. A native of Camas, Washington, Hayes acquired his appreciation of nature exploring and enjoying the mountains, lakes, and beaches of the Pacific Northwest. At age nineteen he dropped out of Clark College in Vancouver, Washington, and spent the next three years traveling the world. He installed church pews in Honolulu, taught swimming and modeled in Tokyo, and hitchhiked through Africa.

After returning to the United States, Hayes enrolled at Stanford as a history major, was elected student body president, and became active in the anti-Vietnam War movement, occupying laboratories that researched military projects. After graduating from Stanford, he went to Harvard Law School, but dropped out in 1970 to help organize the first Earth Day. During Jimmy Carter's presidential administration, Hayes headed the federal Solar Energy Research Institute. He left after the agency's $120 million budget was cut by Ronald Reagan's presidential administration. From 1983 to

1992, after completing his law degree at Stanford, he served there as an adjunct professor of engineering.

In 1990, as the international chairman of Earth Day on its twentieth anniversary, Hayes helped organize participation by over two hundred million people in 141 countries. This event generated extraordinary publicity for and concern about global environmental problems. In 1992 Hayes was named president of the Seattle-based Bullitt Foundation, which works to protect the environment of the Pacific Northwest and to help disadvantaged children. He remains in this position as of 2010. He also serves as chairman of the board of Green Seal, a group that endorses consumer products meeting strict environmental standards and cochairs the group promoting corporate responsibility.

Hayes has received awards and honors from many groups, including the Humane Society of the United

States, the Interfaith Center for Corporate responsibility, the National Wildlife Federation, and the Sierra Club. In 1990, *Life* magazine named him one of the eighteen Americans most likely to have an impact on the twenty-first century.

Hayes has written over one hundred papers and articles, and his book on solar energy, *Rays of Hope: The Transition to a Post-Petroleum World*, has been published in six languages. He has long advocated increased development of solar and other renewable energy sources, which he believes could provide most of the nation's energy supply within a few years.

Resources

BOOKS

Hayes, D. *Rays of Hope: The Transition to a Post-petroleum World*. New York: Norton, 1977.

PERIODICALS

Hayes, D. "Earth Day 1990: Threshold of the Green Decade." *Natural History* 99 (April 1990): 55–60.
Hayes, D. "The Green Decade." *Amicus Journal* 12 (Spring 1990): 10–21.

Lewis G. Regenstein

Hazard ranking system

The hazard ranking system (HRS) is a numerical scoring procedure that the federal Environmental Protection Agency (EPA) uses to place and prioritize waste sites on the National Priorities List. Only these priority sites can be cleaned up through the Superfund Trust Fund program.

The HRS score is based on an evaluation of threats related to the release or potential release of hazardous substances. The HRS assessment of a site ranks public health factors such as threats to drinking water, the food chain, and populations exposed through occupational and ambient environments. Also evaluated are environmental threats such as the effect of substances on air quality, resources, and sensitive ecosystems.

Federal investigators score a site by evaluating four pathways that could be affected by hazardous releases. The pathways are ground water migration, surface water migration, air migration, and soil exposure.

The pathway scores are combined using a root-mean-square equation. This calculation produces the overall score for a site. A high HRS score does not guarantee immediate action because clean-up work

may be going on at other sites. The decision to take action on a site is based on additional research that includes a remedial investigation of what corrective action is needed.

Liz Swain

Hazardous chemicals *see* **Hazardous material; Hazardous waste.**

Hazardous material

Any agent that presents a risk to lifeforms or the environment can be considered a hazardous material. This is a very broad term that encompasses pure compounds and mixtures, raw materials, and other naturally occurring substances, as well as industrial products and wastes. Depending on the nature and the length of exposure, virtually all substances can have toxic effects, ranging from headaches and dizziness to cancer. The challenge facing any legislation is not only to devise regulations for the safe handling of hazardous materials but also to define the term itself.

The U.S. legislation that offers the most detailed and comprehensive definition of hazardous materials is the Resource Conservation and Recovery Act (RCRA), enacted in 1976. RCRA classifies a waste mixture or compound as hazardous if it fails what is called a *characteristic test* or appears on one of a few lists. The lists of hazardous wastes include those from specific and non-specific sources and those that are acutely hazardous and generally hazardous. There are four characteristic tests: ignitability, reactivity, corrosivity, and extraction-procedure toxicity.

A waste fails the ignitable test if it is a liquid with a flash point below 140°Fahrenheit (60° C); a solid that, under standard temperature and pressure, causes fire through friction, absorbing moisture, or spontaneous changes and burns vigorously and persistently; or a compressed gas defined by the Department of Transportation (DOT) as an oxidizer or as being ignitable. Spent solvents, paint removers, epoxy resins, and waste inks are often classified as hazardous under this definition.

A waste fails the corrosivity test if it is aqueous and has a pH of either 2 or less or 12.5 or more, or if it is a liquid that corrodes steel at a rate equal to or more than 0.25 inches (6.35 mm) per year. Examples of corrosive wastes include various acids and bases such as nitric acid, ammonium hydroxide, perchloric acid, sulfuric

The law authorized the Secretary of the Department of Health, Education, and Welfare (HEW) to require warning labels for household substances that were deemed hazardous. These substances were categorized as toxic, corrosive, irritant, strong sensitizer, flammable or combustible, pressure generating, or radioactive. The law does not cover pesticides (which are regulated by the Federal Insecticide, Fungicide, and Rodenticide Act); food, drugs, or cosmetics (which are covered by the Federal Food, Drug, and Cosmetics Act); radioactive materials related to nuclear power; fuels for cooking, heating, or refrigeration; or tobacco products.

A product is defined to be hazardous if it might lead to personal injury or substantial illness, especially if there is a reasonable danger that a child might ingest the substance. When the HEW Secretary declares a substance to be hazardous, a label is required. The label must include a description of the chief hazard and first aid instructions, along with handling and storage instructions. If its chief hazard is flammable, corrosive, or toxic, it must say "danger" on the label; other hazardous substances require either "caution" or "warning." In addition, all labels must include the statement "Keep out of the reach of children."

Major amendments to the Child Protection Act were passed in 1966. These amendments, largely in response to the message on consumer issues by President Lyndon Johnson, expanded federal control over hazardous substances. The Food and Drug Administration (FDA), which administered the law, could now ban substances (after formal hearings) that were deemed too hazardous, even if they had a warning label or if "the degree or nature of the hazard involved in the presence or use of such substance in households is such that the objective of the protection of the public health and safety can be adequately served" only by such a ban. The amendments also extended the scope of the law to pay greater attention to toys and children's articles. This meant the government could require a warning on all household items, rather than just packaged items.

The Child Protection and Toy Safety Act of 1969 further amended the Hazardous Substances Act. Toys could be declared hazardous if they presented electrical, mechanical, or thermal dangers. Also, substances that were hazardous to children, including toys, could be banned automatically. As the titles of these amendments suggest, the Hazardous Substances Act became the primary vehicle to protect children from dangerous substances and toys. Administration of the act has been shifted to the Consumer Product Safety Commission. If the Commission finds a "substantial risk of injury," children's clothes, furniture, and toys can be pulled from the market immediately.

See also Environmental law; Hazardous material; Hazardous Materials Transportation Act (1975); Hazardous waste; Radioactive waste; Radioactivity.

Resources

BOOKS

Lippmann, Morton, ed. *Environmental Toxicants: Human Exposures and Their Health Effects*. Hoboken, NJ: Wiley-Interscience, 2006.
Rapp, Doris. *Our Toxic World: A Wake-up Call*. Buffalo, NY: Environmental Research Foundation, 2004.

Christopher McGrory Klyza

Hazardous waste

Of the thousands of millions of tons of waste generated in the United States annually, approximately 60 million tons (54.4 million metric tons) are classified as hazardous. Hazardous waste is legally defined by the Resource Conservation and Recovery Act (RCRA) of 1976. The RCRA defines hazardous waste as any waste or combination of wastes, which because of its quantity, concentration, or physical, chemical, or infectious

Toxic sludge being dumped into the River Tame, Walsall, West Midlands, United Kingdom. *(Robert Brook/Photo Researchers, Inc.)*

characteristics may: (A) cause, or significantly contribute to, an increase in mortality or an increase in serious irreversible or incapacitating illness; or, (B) pose a substantial present or potential hazard to human health or the environment when improperly treated, stored, transported, disposed of, or otherwise managed.

In the Code of Federal Regulations, the Environmental Protection Agency (EPA) specifies that a solid waste is hazardous if it meets any of these four conditions: (1) It exhibits ignitability corrosivity, reactivity, or EP toxicity; (2) has been listed as a hazardous waste; (3) is a mixture containing a listed hazardous waste and a nonhazardous waste, unless the mixture is specifically excluded or no longer exhibits any of the four characteristics of hazardous waste; (4) is not specifically excluded from regulation as a hazardous waste.

The EPA established two criteria for selecting the characteristics given above. The first criterion is that the characteristic is capable of being defined in terms of physical, chemical, or other properties. The second criterion is that the properties defining the characteristic must be measurable by standardized and available test procedures. For example under the term ignitability (Hazard code label "I"), any one of four criteria can be met: (1) A liquid with a flash point less than 60°Fahrenheit (16°C); (2) if not a liquid, then it is capable under standard temperature and pressure of causing fire through friction, absorption of moisture, or spontaneous chemical changes, and when ignited, burns so vigorously and persistently that it creates a hazard; (3) it may be an ignitable compressed gas; (4) it is an oxidizer.

Similarly under the characteristics of corrosivity, reactivity, and toxicity, there are specifically defined requirements that are spelled out in the Code of Federal Register (CFR). Further examples are given below:

Corrosivity (hazard code "C") has either of the following properties: an aqueous waste with a pH equal to or less than 2.0 or greater than 12.5, or a liquid that will corrode carbon steel at a rate greater than 0.25 inches (0.64 cm) per year.

Reactivity (hazard code "R") has at least one of the following properties: a substance which is normally unstable and undergoes violent physical or chemical change without being detonated; a substance that reacts violently with water (for example, sodium metal); a substance that forms a potentially explosive mixture when mixed with water; a substance that can generate harmful gases, vapors, or fumes when mixed with water; a cyanide- or sulfide-bearing waste that can generate harmful gases, vapors, or fumes when exposed to pH conditions between 2.0 and 12.5; a waste that, when subjected to a strong initiating source or when heated in confinement, will detonate or generate an explosive reaction; a substance that is readily capable of detonation at standard temperature and pressure.

Toxicity (hazard code "E") has the properties such that an aqueous extract contains contamination in excess of that allowed (e.g., arsenic greater than 5 milligrams per liter, barium greater than 0.1 milligrams per liter, cadmium greater than 1 milligrams per liter, chromium greater than 5 milligrams per liter, lead greater than 5 milligrams per liter). Additional codes under toxicity include an "acute hazardous waste" with code "H": a substance that has been found to be fatal to humans in low doses or has been found to be fatal in corresponding human concentrations in laboratory animals. Toxic waste (hazard code "T") designates wastes that have been found through laboratory studies to be a carcinogen, mutagen, or teratogen for humans or other life forms.

Certain wastes are specifically excluded from classification as hazardous wastes under RCRA, including domestic sewage, irrigation return flows, household waste, and nuclear waste. The latter is controlled via other legislation. The impetus for this effort at legislation and classification comes from several notable cases in locations such as Love Canal, New York; Bhopal, India; Stringfellow Acid Pits (Glen Avon, California); and Seveso, Italy; which have brought media and public attention to the need for identification and classification of dangerous substances, their effects on health and the environment, and the importance of having knowledge about the potential risk associated with various wastes.

A notable feature of the legislation is its attempt at defining terms so that professionals in the field and government officials will share the same vocabulary. For example, the difference between toxic and hazardous has been established: Toxic denotes the capacity of a substance to produce injury, and hazardous denotes the *probability* that injury will result from the use of (or contact with) a substance.

The RCRA legislation on hazardous waste is targeted toward larger generators of hazardous waste rather than small operations. The small generator is one who generates less than 2,205 pounds (1,000 kg) per month; accumulates less than 2,205 pounds (1,000 kg); produces wastes that contain no more than 2.2 pounds (1 kg) of acutely hazardous waste; has containers no larger than 5.3 gallons (20 L) or contained in liners less than 22 pounds (10 kg) of weight of acutely hazardous waste; has no greater than 220 pounds (100 kg) of residue or soil contaminated from a spill. The purpose of this exclusion is to enable the system of of cost and performance information generated by their use at many hazardous waste sites.

The last category is that of emerging technologies. These technologies are at a very early stage of development and therefore require additional laboratory and pilot scale testing to demonstrate their technical viability. No cost or performance information is available. An example of an emerging technology is electro-kinetic treatment of soils for metals removal.

Groundwater contaminated by hazardous materials is a widespread concern. Most hazardous waste site remediations use a pump-and-treat approach as a first step. Once the groundwater has been brought to the surface, various treatment alternatives exist, depending upon the constituents present. In situ air sparging of the groundwater using pipes, wells, or curtains is also being developed for removal of volatile constituents. The vapor is either treated above ground with technologies for off-gas emissions, or biologically in the unsaturated or vadose zone above the aquifer. While this approach eliminates the costs and difficulties in treating the relatively large volumes of water (with relatively low contaminant concentrations) generated during pump-and-treat, it does not necessarily speed up remediation.

Contaminated bedrock frequently serves as a source of groundwater or soil recontamination. Constituents with densities greater than water enter the bedrock at fractures, joints, or bedding planes. From these locations, the contamination tends to diffuse in all directions. After many years of accumulation, bedrock contamination may account for the majority of the contamination at a site. Currently, little can be done to remediate contaminated bedrock. Specially designed vapor stripping applications have been proposed when the constituents of concern are volatile. Efforts are ongoing in developing means to enhance the fractures of the bedrock and thereby promote removal. In all cases, the ultimate remediation will be driven by the diffusion of contaminants back out of the rock, a very slow process.

The remediation of buildings contaminated with hazardous waste offers several alternatives. Given the cost of disposal of hazardous wastes, the limited disposal space available, and the volume of demolition debris, it is beneficial to determine the extent of contamination of construction materials. This contamination can then be removed through traditional engineering approaches, such as scraping or sandblasting. It is then only this reduced volume of material that requires treatment or disposal as hazardous waste. The remaining building can be reoccupied or disposed of as nonhazardous waste.

See also Hazardous material; Solidification of hazardous materials; Vapor recovery system.

Resources

BOOKS

Crowley, Kevin D. *Science and Technology for DOE Site Cleanup: Workshop Summary*. Washington, DC: National Academies Press, 2010.

Maczulak, Anne E. *Cleaning up the Environment: Hazardous Waste Technology*. New York: Facts On File, 2009.

Maczulak, Anne E. *Pollution: Treating Environmental Toxins*. New York: Facts On File, 2010.

OTHER

CLU-IN, U.S. Environmental Protection Agency. "About Remediation Technologies." http://www.clu-in.org/remediation/ (accessed November 9, 2010).

Ann N. Clarke
Jeffrey L. Pintenich

Haze

An aerosol in the atmosphere of sufficient concentration and extent to decrease visibility significantly when the relative humidity is below saturation is known as haze. Haze may contain dry particles or droplets or a mixture of both, depending on the precise value of the humidity. In the use of the word, there is a connotation of some degree of permanence. For example, a dust storm is not a haze, but the coarse particles may settle rapidly and leave a haze behind once the velocity drops.

Human activity is responsible for many hazes. Enhanced emission of sulfur dioxide results in the formation of aerosols of sulfuric acid. In the presence of ammonia, which is excreted by most higher animals including humans, such emissions result in aerosols of ammonium sulfate and bisulfate. Organic hazes are part of photochemical smog, such as the smog often associated with Los Angeles, California, and they consist primarily of polyfunctional, highly oxygenated compounds with at least five carbon atoms. Such hazes can also form if air with an enhanced nitrogen oxide content meets air containing the natural terpenes emitted by vegetation.

All hazes, however, are not products of human activity. Natural hazes can result from forest fires, dust storms, and the natural processes that convert gaseous contaminants into particles for subsequent removal by precipitation or deposition to the surface or to vegetation. Still other hazes are of mixed origin, as noted above, and an event such as a dust storm can be enhanced by human-caused devegetation of soil.

Though it may contain particles injurious to health, haze is not of itself a health hazard. It can have a significant economic impact, however, when tourists cannot see scenic views, or if it becomes sufficiently dense to inhibit aircraft operations.

See also Air pollution; Air quality; Air quality criteria; Los Angeles Basin; Mexico City, Mexico.

Resources

BOOKS

DuPuis, E. Melanie. *Smoke and Mirrors: The Politics and Culture of Air Pollution.* New York: New York University Press, 2004.

Kidd, J. S., and R. A. Kidd. *Air Pollution: Problems and Solutions.* New York: Facts on File, 2005.

Seinfeld, John H., and Spyros N. Pandis. *Atmospheric Chemistry and Physics: From Air Pollution to Climate Change.* Hoboken, NJ: J. Wiley, 2006.

PERIODICALS

Casey, Michael. "Indian Ocean Haze Adds to Global Warming." *The Washington Post* (August 2, 2007).

OTHER

Centers for Disease Control and Prevention (CDC). "Air Pollution and Respiratory Health." http://www.cdc.gov/nceh/airpollution/default.htm (accessed October 2, 2010).

United States Environmental Protection Agency (EPA). "Air: Air Pollution Monitoring." http://www.epa.gov/ebtpages/airairpollutionmonitoring.html (accessed October 2, 2010).

United States Environmental Protection Agency (EPA). "Air: Air Pollution: Urban Air Pollution." http://www.epa.gov/ebtpages/airairpourbanairpollution.html (accessed October 2, 2010).

World Health Organization (WHO). "Air pollution." http://www.who.int/topics/air_pollution/en (accessed October 2, 2010).

James P. Lodge Jr.

Heat (stress) index

The heat index (HI) or heat stress index—sometimes called the apparent temperature or comfort index—is a temperature measure that takes into account the relative humidity. Based on human physiology and on clothing science, it measures how a given air temperature feels to the average person at a given relative humidity. The HI temperature is measured in the shade and assumes a wind speed of 5.6 miles per hour (9.0 kph) and normal barometric pressure.

At low relative humidity, the HI is less than or equal to the air temperature. At higher relative humidity, the HI exceeds the air temperature. For example, according to the National Weather Service's (NWS) HI chart, if the air temperature is 70°Fahrenheit (21°C), the HI is 64°Fahrenheit (18°C) at 0 percent relative humidity and 72°Fahrenheit (22°C) at 100 percent relative humidity. At 95°Fahrenheit (35°C) and 55 percent relative humidity, the HI is 110°Fahrenheit (43°C). In very hot weather, humidity can raise the HI to extreme levels; at 115°Fahrenheit (46°C) and 40 percent relative humidity, the HI is 151°Fahrenheit (66°C). This is because humidity affects the body's ability to regulate internal heat through perspiration. The body feels warmer when it is humid because perspiration evaporates more slowly; thus the HI is higher.

The HI is used to predict the risk of physiological heat stress for an average individual. Caution is advised at an HI of 80–90°Fahrenheit (27–32°C): fatigue may result with prolonged exposure and physical activity. An HI of 90–105°Fahrenheit (32–41°C) calls for extreme caution, as it increases the likelihood of sunstroke, muscle cramps, and heat exhaustion. Danger warnings are issued at HIs of 105–130°Fahrenheit (41–54°C), when sunstroke and heat exhaustion are likely and there is a potential for heat stroke. Category IV, extreme danger, occurs at HIs above 130°Fahrenheit (54°C), when heatstroke and sunstroke are imminent.

Individual physiology influences how people are affected by high HIs. Children and older people are more vulnerable. Acclimatization (being used to the climate) can alleviate some of the danger. However sunburn can increase the effective HI by slowing the skin's ability to shed excess heat from blood vessels and through perspiration. Exposure to full sunlight can increase HI values by as much as 15°Fahrenheit (8°C). Winds, especially hot dry winds, also can increase the HI. In general, the NWS issues excessive heat alerts when the daytime HI reaches 105°Fahrenheit (41°C) and the nighttime HI stays above 80°Fahrenheit (27°C) for two consecutive days; however, these values depend somewhat on the region or metropolitan area. In cities, high HIs often mean increased air pollution. The concentration of ozone, the major component of smog, tends to rise at ground level as the HI increases, causing respiratory problems for many people.

At higher temperatures, the air can hold more water vapor; thus humidity and HI values increase as the atmosphere warms. Since the late nineteenth century, the mean annual surface temperature of the earth has risen between 0.5 and 1.0°Fahrenheit (0.3 and 0.6°C). According to the National Aeronautics and Space Administration (NASA), the five-year mean temperature increased about 0.9°Fahrenheit (0.5°C) between 1975 and 1999, the fastest rate of recorded increase. In 1998 global surface temperatures were the warmest since the advent of reliable measurements and the 1990s accounted for seven of the ten warmest years on record. Nighttime temperatures have been increasing twice as fast as daytime temperatures.

Greenhouse gases, including carbon dioxide, methane, nitrous oxide, and chlorofluorocarbons, increase the heat-trapping capabilities of the atmosphere. Evaporation from the ocean surfaces increased during the twentieth century, resulting in higher humidity that enhanced the greenhouse effect. It is projected that during the twenty-first century greenhouse gas concentrations will double or even quadruple from preindustrial levels. Increased urbanization also contributes to global warming, as buildings and roads hold in the heat. Climate simulations predict an average surface air temperature increase of 4.5–7°Fahrenheit (2.5–4°C) by 2100. This will increase the number of extremely hot days and, in temperate climates, double the number of very hot days, for an average increase in summer temperatures of 4–5°Fahrenheit (2–3°C). More heat-related illnesses and deaths will result.

The National Oceanic and Atmospheric Administration projects that the HI could rise substantially in humid regions of the tropics and subtropics. Warm, humid regions of the southeastern United States are expected to experience substantial increases in the summer HI because of increased humidity, even though temperature increases may be smaller than in the continental interior. Predictions for the increase in the summer HI for the southeast United States over the next century range from 8–20°Fahrenheit (4–11°C).

Resources

BOOKS

Barnosky, Anthony D. *Heatstroke: Nature in an Age of Global Warming* . Washington, DC: Island Press/ Shearwater Books, 2009.

Cullen, Heidi. *The Weather of the Future: Heat Waves, Extreme Storms, and Other Scenes from a Climate-Changed Planet* . New York: HarperCollins, 2010.

Monbiot, George. *Heat: How to Stop the Planet from Burning* . Cambridge, MA: South End Press, 2007.

Pittock, A. Barrie. *Climate Change: Turning up the Heat* . London: Earthscan, 2005.

PERIODICALS

Meehl, Gerald A., and Claudia Tebaldi. "More Intense, More Frequent, and Longer Lasting Heat Waves in the Twenty-first Century." *Science* 305, no. 5686 (August 13, 2004): 994–997.

OTHER

Centers for Disease Control and Prevention (CDC). "Extreme Heat [Hyperthermia]." http://emergency. cdc.gov/disasters/extremeheat/index.asp (accessed November 8, 2010).

Centers for Disease Control and Prevention (CDC). "Heat Stress." http://www.cdc.gov/niosh/topics/heatstress/ (accessed November 8, 2010).

National Institutes of Health (NIH). "Heat Illness." http://health.nih.gov/topic/HeatIllness (accessed November 8, 2010).

ORGANIZATIONS

National Weather Service, National Oceanic and Atmospheric Administration, U. S. Department of Commerce, 1325 East West Highway, Silver Spring, USA, 20910, http://www.nws.noaa.gov

Physicians for Social Responsibility, 1875 Connecticut Avenue, NW, Suite 1012, Washington, DC, USA, 20009, (202) 667-4260, (202) 667- 4201, psrnatl@ psr.org, http://www.psr.org

Union of Concerned Scientists, 2 Brattle Square, Cambridge, MA, USA, 02238, (617) 547-5552, ucs@ucsusa.org, http://www.ucsusa.org

Margaret Alic

Heavy metals and heavy metal poisoning

There is no agreed upon chemical definition of the group pf elements commonly described as "heavy metals." Heavy metals are a loosely defined group of elements with variable chemical and physical properties (e.g., atomic weight and position on the periodic table) all of which exhibit some of the characteristics of metals. Many heavy metals are essential for life in some form or lower concentrations yet toxic in higher concentrations.

Whether based on their physical or chemical properties, the distinction between heavy metals and nonmetals is not sharp. For example, arsenic (As), germanium (Ge), selenium (Se), tellurium (Te), and antimony (Sb) possess chemical properties of both metals and nonmetals. Defined as metalloids, they are often loosely classified as heavy metals. The category heavy metal is, therefore, somewhat arbitrary and highly nonspecific because it can refer to approximately eighty of

Local health workers remove earth contaminated by lead from a family compound in the village of Dareta in Gusau, Nigeria, in June 2010. *(AP Photo/Sunday Alamba)*

the 103 elements in the periodic table. The term trace element is commonly used to describe substances that cannot be precisely defined but most frequently occur in the environment in concentrations of a few parts per million (ppm) or less. Only a relatively small number of heavy metals such as cadmium (Cd), copper (Cu), iron (Fe), cobalt (Co), zinc (Zn), mercury (Hg), vanadium (V), lead (Pb), nickel (Ni), chromium (Cr), manganese (Mn), molybdenum (Mo), silver (Ag), and tin (Sn) as well as the metalloids arsenic and selenium are associated with environmental, plant, animal, or human health problems.

Some chemical forms of heavy metals are persistent environmental contaminants. Natural processes such as bedrock and soilweathering, wind and water erosion, volcanic activity, sea salt spray, and forest fires release heavy metals into the environment. While the origins of anthropogenic releases of heavy metals are lost in antiquity, they probably began as our prehistoric ancestors learned to recover metals such as gold, silver, copper, and tin from their ores and to produce bronze. The modern age of heavy metal pollution has its beginning with the Industrial Revolution. The rapid development of industry, intensive agriculture, transportation, and urbanization over the past 150 years, however, has been the precursor of today's environmental contamination problems. Anthropogenic utilization has also increased heavy metal distribution by removing the substances from localized ore deposits and transporting them to other parts of the environment. Heavy metal by-products result from many activities including the following: ore extraction and smelting, fossil fuel combustion, dumping of industrial wastes into landfills, exhausts from leaded gasolines, steel, iron, cement and fertilizer production, or refuse and wood combustion. Heavy metal cycling has also increased through activities such as farming, deforestation, construction, dredging of harbors, and the disposal of municipal sludges and industrial wastes on land.

Thus, anthropogenic processes, especially combustion, have substantially supplemented the natural atmospheric emissions of selected heavy metals or metalloids such as selenium, mercury, arsenic, and antimony. They can be transported as gases or adsorbed on particles. Other metals such as cadmium, lead, and zinc are transported atmospherically only as particles. In either state heavy metals may travel long distances before being deposited on land or water.

The heavy metal contamination of soils is a far more serious problem than either air or water pollution because heavy metals are usually tightly bound by the organic components in the surface layers of the soil and may, depending on conditions, persist for centuries or millennia. Consequently, the soil is an important geochemical sink that accumulates heavy metals rapidly and usually depletes them very slowly by leaching into groundwater aquifers or bioaccumulating into plants. Bioaccumulation is the accumulation of a substance in an organism. However, heavy metals can also be very rapidly translocated through the environment by erosion of the soil particles to which they are adsorbed or bound and redeposited elsewhere on the land or washed into rivers, lakes, or oceans to the sediment.

The cycling, bioavailability, toxicity, transport, and fate of heavy metals are markedly influenced by their physicochemical forms in water, sediments, and soils. Whenever a heavy metal-containing ion or compound is introduced into an aquatic environment, it is subjected to a wide variety of physical, chemical, and biological processes. These include the following: hydrolysis (splitting of water molecules), chelation and complexation (binding of a compound, usually organic, to a metal ion), redox (oxidation-reduction), biomethylation (biochemical addition of methyl group), precipitation (a solid forming in a solution), and adsorption (accumulation of surface molecules) reactions. Often heavy metals experience a change in the chemical form or speciation as a result of these processes and so their distribution, bioavailability, and other interactions in the environment are also affected.

The interactions of heavy metals in aquatic systems are complicated because of the possible changes due to many dissolved and particulate components and non-equilibrium conditions. For example, the speciation of heavy metals is controlled not only by their chemical properties but also by environmental variables such as: (1) pH; (2) redox potential; (3) dissolved oxygen; (4) ionic strength; (5) temperature; (6) salinity; (7) alkalinity; (8) hardness; (9) concentration and nature of inorganic ligands such as carbonate (CO_2^{3-}), bicarbonate (HCO_3^-), sulfate (SO_4^{2-}), sulfides (S^{2-}), chlorides (Cl^-); (10) concentration and nature of dissolved organic chelating agents such as organic acids, humic materials, peptides, and polyamino-carboxylates; (11) the concentration and nature of particulate matter with surface sites available for heavy metal binding; and (12) biological activity.

In addition, various species of bacteria can oxidize arsenate or reduce arsenate (As^{5+}) to arsenite (As^{3+}), or oxidize ferrous iron (Fe^{2+}) to ferric iron (Fe^{3+}), or convert mercuric ion (Hg^{2+}) to elemental mercury or the reverse. Various enzyme systems in living organisms can biomethylate a number of heavy metals. While it had been known for at least sixty years that arsenic and selenium could be biomethylated, microorganisms capable of converting inorganic mercury into monomethyl ($[CH_3Hg]^+$) and dimethylmercury ($[CH_3]_2Hg$) in lake sediments were not discovered until 1967. Since then, numerous heavy metals such as lead, tin, cobalt, antimony, platinum, gold, tellurium, thallium (Tl), and palladium (Pd) have been shown to be biomethylated by bacteria and fungi in the environment.

As environmental factors change the chemical reactivities and speciation of heavy metals, they influence not only the mobilization, transport, and bioavailability, but also the toxicity of heavy metal ions toward biota in both freshwater and marine ecosystems. The factors affecting the toxicity and bioaccumulation of heavy metals by aquatic organisms include the following: (1) the chemical characteristics of the ion; (2) solution conditions, which affect the chemical form (speciation) of the ion; (3) the nature of the response such as acute toxicity, bioaccumulation, various types of chronic effects, and so on; (4) the nature and condition of the aquatic animal such as age or life stage, species, or trophic level in the food chain. The extent to which most of the methylated metals are bioaccumulated or biomagnified is limited by the chemical and biological conditions and how readily the methylated metal is metabolized by an organism. At present, only methylmercury seems to be sufficiently stable to bioaccumulate to levels that can cause adverse effects in aquatic organisms. All other methylated metal ions are produced in very small concentrations and are degraded naturally faster than they are bioaccumulated. Therefore, they do not biomagnify in the food chain. Biomagnification occurs when a bioaccumulated substance is present in higher concentrations in organisms because of persistence in the environment as well as the inability of organisms to degrade or excrete the substance.

The largest proportion of heavy metals in water is associated with suspended particles, which are ultimately deposited in the bottom sediments where concentrations are orders of magnitude higher than those in the overlying or interstitial waters. The heavy metals associated with suspended particulates or bottom sediments are complex mixtures of the following: (1)

weathering and erosion residues such as iron and aluminum oxyhydroxides, clays, and other aluminosilicates; (2) methylated and nonmethylated forms in organic matter such as living organisms, bacteria and algae, detritus, and humus; (3) inorganic hydrous oxides and hydroxides, phosphates, and silicates; and (4) diagenetically produced iron and manganese oxyhydroxides in the upper layer of sediments and sulfides in the deeper, anoxic layers.

In anoxic waters (decreased amounts of oxygen) the precipitation of sulfides may control the heavy metal concentrations in sediments, whereas in oxic waters adsorption, absorption, surface precipitation, and coprecipitation are usually the mechanisms by which heavy metals are removed from the water column. Moreover, physical, chemical, and microbiological processes in the sediments often increase the concentrations of heavy metals in the pore waters, which are released to overlying waters by diffusion or as the result of consolidation and bioturbation, which is the mixing of sediments by bottom-dwelling organisms. Transport by living organisms does not represent a significant mechanism for local movement of heavy metals. However, accumulation by aquatic plants and animals can lead to important biological responses. Even low environmental levels of some heavy metals may produce subtle and chronic effects in animal populations.

Despite these adverse effects, at very low levels, some metals have essential physiological roles as micronutrients. Heavy metals such as chromium, manganese, iron, cobalt, molybdenum, nickel, vanadium, copper, and selenium are required in small amounts to perform important biochemical functions in plant and animal systems. In higher concentrations they can be toxic, but usually some biological regulatory mechanism is available by means of which animals can speed up their excretion or retard their uptake of excessive quantities.

In contrast, nonessential heavy metals are primarily of concern in terrestrial and aquatic systems because they are toxic and persist in living systems. Metal ions commonly bond with sulfhydryl (-SH) and carboxylic acid (CO_2H) groups in amino acids, which are components of proteins (enzymes) or polypeptides. This increases their bioaccumulation and inhibits excretion. For example, heavy metals such as lead, cadmium, and mercury bind strongly with sulfur-containing groups such as -SH and $-SCH_3$ in cysteine and methionine and so inhibit the metabolism of the bound enzymes. In addition, other heavy metals may replace an essential element, decreasing its availability and causing symptoms of deficiency.

Uptake, translocation, and accumulation of potentially toxic heavy metals in plants differ widely depending on soil type, pH, redox potential, moisture, and organic content. Public health officials closely regulate the quantities and effects of heavy metals that move through the agricultural food chain to be consumed by human beings. While heavy metals such as zinc, copper, nickel, lead, arsenic, and cadmium are translocated from the soil to plants and then into the animal food chain, the concentrations in plants are usually very low and generally not considered to be an environmental problem. However, plants grown on soils either naturally enriched or highly contaminated with some heavy metals can bioaccumulate levels high enough to cause toxic effects in the animals or human beings that consume them.

Some heavy metals are toxic or carcinogenic, having adverse affects on the central nervous system of humans. Minimata disease is caused by mercury poisoning and affects the central nervous system, in some cases resulting in death. In Japan, the Chisso Corporation released methylmercury-containing wastewater into Minimata Bay from 1932 to 1968. Mercury bioaccumulated in fish and shellfish harvested and consumed by humans, causing mercury poisoning and neurological symptoms. Approximately three thousand people were diagnosed with Minimata disease. Because of the persistence of mercury in the environment, a clean-up project was initiated in 1974 to dredge the bottom sediment in Minimata Bay for removal of mercury. The project took sixteen years to complete and cost forty-eight billion yen (or $500 million 2009 U.S. dollars).

Contamination of soils caused by land disposal of sewage and industrial effluents and sludges may pose the most significant long-term problem. While cadmium and lead are the greatest hazard, other elements such as copper, molybdenum, nickel, and zinc can also accumulate in plants grown on sludge-treated land. High concentrations can, under certain conditions, cause adverse effects in animals and human beings that consume the plants. For example, when soil contains high concentrations of molybdenum and selenium, these metals can be translocated into edible plant tissue in sufficient quantities to produce toxic effects in ruminant animals. Consequently, the U.S. Environmental Protection Agency has issued regulations that prohibit or tightly regulate the disposal of contaminated municipal and industrial sludges on land to prevent heavy metals, especially cadmium, from entering the food supply in toxic amounts. However, presently, the most serious known human toxicity is not through bioaccumulation from crops but from either

Logic of Capitalism (1985). He has also served on the editorial board of the socialist journal *Dissent*.

In 1974 Heilbroner published *An Inquiry into the Human Prospect*, in which he argued that three so-called external challenges confronted humanity: the population explosion, the threat of war, and, quoting from the book, "the danger...of encroaching on the environment beyond its ability to support the demands made on it." Each of these problems, he maintained, has arisen from the development of scientific technology, which has increased the human life span, multiplied weapons of destruction, and encouraged industrial production that consumes natural resources and pollutes the environment. Heilbroner believed that these challenges confront all economies, and that meeting them will require more than adjustments in economic systems. Societies will have to muster the will to make sacrifices.

Heilbroner argued that persuading people to make these sacrifices may not be possible. Those living in one part of the world are not likely to give up what they have for the sake of those in another part, and people living now are not likely to make sacrifices for future generations. His reluctant conclusion was that coercion is likely to take the place of persuasion. Authoritarian governments may well supplant democracies because "the passage through the gantlet ahead may be possible only under governments capable of rallying obedience far more effectively than would be possible in a democratic setting. If the issue for mankind is survival, such governments may be unavoidable, even necessary."

Heilbroner wrote *An Inquiry into the Human Prospect* in 1972 and 1973, but his position had not changed by the end of the decade. In a revised edition written in 1979, he continued to insist upon the environmental limits to economic growth: "The industrialized capitalist and socialist worlds can probably continue along their present growth paths" for about twenty-five years, at which point "we must expect...a general recognition that the possibilities for expansion are limited, and that social and economic life must be maintained within fixed...material boundaries." Heilbroner published a number of books, including *Twenty-first Century Capitalism* and *Visions of the Future*. He also received the New York Council for the Humanities Scholar of the Year award in 1994.

Until his death on January 4, 2005, at the age of 85 in New York City, Heilbroner was Norman Thomas Professor of Economics, Emeritus, at the New School for Social Research.

Resources

BOOKS

Heilbroner, Robert L. *An Inquiry into the Human Prospect.* Rev. ed. New York: Norton, 1980.

Heilbroner, Robert L. *The Making of an Economic Society.* 6th ed. Englewood Cliffs, NJ: Prentice-Hall, 1980.

Heilbroner, Robert L. *The Nature and Logic of Capitalism.* New York: Norton, 1985.

Heilbroner, Robert L. *Twenty-first Century Capitalism.* Don Mills, Ontario, CA: Anansi, 1992.

Heilbroner, Robert L. *The Worldly Philosophers: The Lives, Times, and Ideas of the Great Economic Thinkers.* 6th ed. New York: Simon & Schuster, 1986.

Richard K. Dagger

Hells Canyon

Hells Canyon is a stretch of canyon on the Snake River between Idaho and Oregon. This canyon, deeper than the Grand Canyon and formed in ancient basalt flows, contains some of the United States' wildest rapids and has provided extensive recreational and scenic boating since the 1920s. The narrow canyon has also provided outstanding dam sites. Hells Canyon became the subject of nationwide controversy between 1967 and 1975, when environmentalists challenged hydroelectric developers over the last stretch of free-flowing water in the Snake River from the border of Wyoming to the Pacific.

Historically Hells Canyon, over 100 miles (161 km) long, filled with rapids, and averaging 6,500 feet (1,983 m) deep, presented a major obstacle to travelers and explorers crossing the mountains and deserts of southern Idaho and eastern Oregon. Nez Perce, Paiute, Cayuse, and other Native American groups of the region had long used the area as a mild wintering ground with good grazing land for their horses. European settlers came for the modest timber and with cattle and sheep to graze. As early as the 1920s travelers were arriving in this scenic area for recreational purposes, with the first river runners navigating the canyon's rapids in 1928. By the end of the Great Depression, the Federal Power Commission was urging regional utility companies to tap the river's hydroelectric potential, and in 1958 the first dam was built in the canyon.

Falling from the mountains in southern Yellowstone National Park through Idaho, and into the Columbia River, the Snake River drops over 7,000 vertical feet (2,135 m) in 1,000 miles (1,609 km) of river. This drop and the narrow gorges the river has carved presented excellent dam opportunities, and by the end of the 1960s there were eighteen major dams along the river's course. By that time the river was also attracting great numbers of whitewater rafters and kayakers, as well as hikers and campers in the adjacent national forests. When a proposal was developed to dam the last free-running section of the canyon, protesters brought a suit to the United States Supreme Court. In 1967 Justice William O. Douglas led the majority in a decision directing the utilities to consider alternatives to the proposed dam.

Hells Canyon became a national environmental issue. Several members of Congress flew to Oregon to raft the river. The Sierra Club and other groups lobbied vigorously. Finally, in 1975 President Gerald Ford signed a bill declaring the remaining stretch of the canyon a National Scenic Waterway, creating a 650,000-acre (260,000-ha) Hells Canyon National Recreation Area, and adding 193,000 acres (77,200 ha) of the area to the National Wilderness Preservation System.

See also Wild and Scenic Rivers Act (1968); Wild river.

Resources

BOOKS

Brooks, Karl Boyd. *Public Power, Private Dams: The Hells Canyon High Dam Controversy*. Seattle: University of Washington Press, 2006.

Mary Ann Cunningham

Henderson, Hazel

1933–
English/American environmental activist and writer

Hazel Henderson is an environmental activist and futurist who has called for an end to current unsustainable industrial modes and urges redress for what she describes as "unequal access to resources which is now so dangerous, both ecologically and socially."

Born in Clevedon, England, Henderson immigrated to the United States after finishing high school; she became a naturalized citizen in 1962. After working for several years as a freelance journalist, she married Carter F. Henderson, former London bureau chief of the *Wall Street Journal* in 1957. Her activism began when she became concerned about air quality in

New York City, where she was living. To raise public awareness, she convinced the Federal Communications Commission (FCC) and television networks to broadcast the air pollution index with the weather report. She persuaded an advertising agency to donate their services to her cause and teamed up with a New York City councilman to cofound Citizens for Clean Air. Her endeavors were rewarded in 1967, when she was commended as Citizen of the Year by the New York Medical Society.

Henderson's career as an advocate for social and environmental reform took flight from there. She argued passionately against the spread of industrialism, which she called "pathological" and decried the use of an economic yardstick to measure quality of life. Indeed, she termed economics "merely politics in disguise" and even "a form of brain damage." Henderson believed that society should be measured by less tangible means, such as political participation, literacy, education, and health. "Per-capita income," she felt, is "a very weak indicator of human well-being."

She became convinced that traditional industrial development wrought little but "ecological devastation, social unrest, and downright hunger...I think of development, instead,...as investing in ecosystems, their restoration and management."

Even the fundamental idea of labor should, Henderson argued, "be replaced by the concept of 'Good Work'—which challenges individuals to grow and develop their faculties; to overcome their ego-centeredness by joining with others in common tasks; to bring forth those goods and services needed for a becoming existence; and to do all this with an ethical concern for the interdependence of all life forms."

To advance her theories, Henderson has published several books, *Creative Alternative Futures: The End of Economics* (1978), *The Politics of the Solar Age: Alternatives to Economics* (1981), *Building a Win-Win World* (1996), *Toward Sustainable Communities: Resources for Citizens and Their Governments* (1998), and *Beyond Globalization: Shaping a Sustainable Global Economy* (1999). She has also contributed to several periodicals, and lectured at colleges and universities. In 1972 she cofounded the Princeton Center for Alternative Futures, of which she is still a director. She is a member of the board of directors for Worldwatch Institute and the Council for Economic Priorities, among other organizations. In 1982 she was appointed a Horace Allbright Professor at the University of California at Berkeley. In 1996 Henderson was awarded the Global Citizen Award.

In 2005 she founded Ethical Markets Media, LLC, dedicated to providing information on green or ethical investing, business, energy, and sustainable development. This led to the establishment of Ethical Markets TV in 2007. As of 2010, Henderson is a producer of the public television series *Ethical Markets* and is a lecturer at the University of California at Santa Barbara.

Resources

BOOKS

Henderson, Hazel. *Beyond Globalization: Shaping a Sustainable Global Economy.* West Hartford, CT: Kumarian Press, 1999.

Henderson, Hazel, and Daisaku Ikeda. *Planetary Citizenship: Your Values, Beliefs, and Actions Can Shape a Sustainable World.* Santa Monica, CA: Middleway Press, 2004.

Henderson, Hazel. *The Politics of the Solar Age: Alternatives to Economics.* Garden City, NY: Anchor Press/Doubleday, 1981.

Henderson, Hazel, and Simran Sethi. *Ethical Markets: Growing the Green Economy.* White River Junction, VT: Chelsea Green, 2006.

Henderson, Hazel. *Building a Win-Win World: Life beyond Global Economic Warfare.* San Francisco: Berrett-Koehler Publishers, 1996.

Amy Strumolo

Herbicide

Herbicides are chemical pesticides that are used to manage vegetation. Usually, herbicides are used to reduce the abundance of weeds, so as to release desired crop plants from competition for space, light, water, and nutrients. This is the context of most herbicide use in agriculture, forestry, and lawn management. Sometimes herbicides are not used to protect crops, but to reduce the quantity or height of vegetation, for example along highways and transmission corridors. The reliance on herbicides to achieve these ends has increased greatly in recent decades, and the practice of chemical weed control appears to have become an entrenched component of the modern technological culture of humans, especially in agroecosystems.

Millions of pounds of pesticides are used each year with herbicides the most widely used form of chemical pesticide used in agriculture. In many areas, the same tracts of land are treated numerous times each year with various pesticides.

Herbicides are either contact or systemic, acting in two different ways. Contact herbicides kill any plant tissue that comes into contact with the chemical. Systemic herbicides are taken up either by the leaves or roots and are distributed throughout the plant. Additionally, herbicides can be applied for preemergence control or postemergence. Preemergence herbicides inhibit the development of weeds before they emerge through the soil, whereas postemergence chemicals act on plants that have emerged from the soil.

A wide range of chemicals is used as herbicides, including the following:

- chlorophenoxy acids, especially 2,4-D and 2,4,5-T, which have an auxin-like (auxin is a naturally occurring plant hormone) growth-regulating property and are selective against broadleaved angiosperm plants

- triazines such as atrazine, simazine, and hexazinone

- chloroaliphatics such as dalapon and trichloroacetate

- the phosphonoalkyl chemical, glyphosate

- inorganics such as various arsenicals, cyanates, and chlorates

A weed is usually considered to be any plant that interferes with the productivity of a human-desired plant, even though in other contexts weed species may have positive ecological and economic values. Weeds exert this effect by competing with the crop for light, water, and nutrients. Studies in Illinois demonstrated an average reduction of yield of corn or maize (*Zea mays*) of 81 percent in unweeded plots, while a 51 percent reduction was reported in Minnesota. Weeds also reduce the yield of small grains, such as wheat (*Triticum aestivum*) and barley (*Hordeum vulgare*), by 25–50 percent.

Because there are several herbicides that are toxic to dicotyledonous weeds but not grasses, herbicides are used most intensively in grain crops of the Gramineae. For example, in North America almost all of the area of maize cultivation is treated with herbicides. In part this is due to the widespread use of no-tillage cultivation, a system that reduces erosion and saves fuel. Because an important purpose of plowing is to reduce the abundance of weeds, the no-tillage system would be impracticable if not accompanied by herbicide use. The most important herbicides used in maize cultivation are atrazine, propachlor, alachlor, 2,4-D, and butylate. Most of the area planted to other agricultural grasses such as wheat, rice (*Oryza sativa*), and barley is also treated with herbicide, mostly with the phenoxy herbicides 2,4-D or MCPA.

Farmer applying herbicide with bell sprayer to a tomato plant. *(© age fotostock / SuperStock)*

The intended ecological effect of any pesticide application is to control a pest species, usually by reducing its abundance to below some economically acceptable threshold. In a few situations, this objective can be attained without important nontarget damage. For example, a judicious spot-application of a herbicide can allow a selective kill of large lawn weeds in a way that minimizes exposure to nontarget plants and animals.

Of course, most situations where herbicides are used are more complex and less well-controlled. Whenever a herbicide is broadcast-sprayed over a field or forest, a wide variety of onsite, nontarget organisms is affected, and sprayed herbicide also drifts from the target area. In addition to drifting, runoff of herbicides as a result of irrigation or precipitation may drain into nearby bodies of water, thus affecting aquatic organisms. These cause ecotoxicological effects directly, through toxicity to nontarget organisms and ecosystems, and indirectly, by changing habitat or the abundance of food species for wildlife. These effects can be illustrated by the use of herbicides in forestry, with glyphosate used as an example.

The most frequent use of herbicides in forestry is for the release of small coniferous plants from the effects of competition with economically undesirable weeds. Usually the silvicultural use of herbicides occurs within the context of an intensive harvesting and management system, which may include clear-cutting, scarification, planting seedlings of a single desired species, spacing, and other practices.

Glyphosate is a commonly used systemic herbicide in forestry and agriculture. The typical spray rate in silviculture is about 2.2–4.9 pounds (1–2.2 kg) active ingredient, and the typical projection is for one to two treatments per forest rotation of forty to one hundred years. In agriculture, many crops have been genetically modified to be resistant to the effects of glyphosate, which enables growers to spray herbicide onto crops to suppress weeds without damaging the crops.

Immediately after an aerial application in forestry, glyphosate residues on foliage are about six times higher than litter on the forest floor, which is physically shielded from spray by overtopping foliage. The persistence of glyphosate residues is relatively short, with typical half-lives of two to four weeks in foliage and the forest floor, and up to eight weeks in soil. The disappearance of residues from foliage is mostly due to translocation and wash-off, but in the forest floor and soil, glyphosate is immobile (and unavailable for root uptake or leaching) because of binding to organic matter and clay, and residue disappearance is due to microbial oxidation. Residues in oversprayed waterbodies tend to be small and short-lived. For example, two hours after a deliberate overspray on Vancouver Island, Canada, residues of glyphosate in stream water rose to high levels, then rapidly dissipated through flushing to only trace amounts ninety-four hours later.

Because glyphosate is soluble in water, there is no propensity for bioaccumulation in organisms in preference to the inorganic environment, or to occur in larger concentrations at higher levels of the food chain. This is in marked contrast to some other pesticides such as DDT, which is soluble in organic solvents but not in water, so it has a strong tendency to bioaccumulate into the fatty tissues of organisms.

As a plant poison, glyphosate acts by inhibiting the pathway by which four essential amino acids are synthesized. Only plants and some microorganisms have this metabolic pathway; animals obtain these amino acids from food. Consequently, glyphosate has a relatively small acute toxicity to animals, and there are large margins of toxicological safety in comparison with environmental exposures that are realistically expected during operational silvicultural sprays.

Acute toxicity of chemicals to mammals is often indexed by the oral dose required to kill 50 percent of a test population, usually of rats (i.e., rat LD_{50}). The LD_{50} value for pure glyphosate is 5,600 milligrams per kilogram, and its silvicultural formulation has a value of 5,400 milligrams per kilogram. Compare these to LD_{50}s for some chemicals that many humans ingest voluntarily: nicotine 50 milligrams per kilogram, caffeine 366, acetylsalicylic acid (ASA) 1,700, sodium chloride 3,750, and ethanol 13,000. The documented risks of longer-term, chronic exposures of mammals to glyphosate are also small, especially considering the doses that might be received during an operational treatment in forestry.

Considering the relatively small acute and chronic toxicities of glyphosate to animals, it is unlikely that wildlife inhabiting sprayed clear-cut areas would be directly affected by a silvicultural application. However, glyphosate causes large habitat changes through species-specific effects on plant productivity, and by changing habitat structure. Therefore, wildlife such as birds and mammals could be secondarily affected through changes in vegetation and the abundance of their prey arthropod foods. These indirect effects of herbicide spraying are within the context of ecotoxicology. Indirect effects can affect the abundance and reproductive success of terrestrial and aquatic wildlife on a sprayed site, irrespective of a lack of direct, toxic effects.

Studies of the effects of habitat changes caused by glyphosate spraying have found relatively small effects on the abundance and species composition of wildlife. Much larger effects on wildlife are associated with other forestry practices, such as clear-cutting and the broadcast spraying of insecticides. For example, in a study of clear-cuts sprayed with glyphosate in Nova Scotia, Canada, only small changes in avian abundance and species composition could be attributed to the herbicide treatment. However, such studies of bird abundance are conducted by enumerating territories, and the results cannot be interpreted in terms of reproductive success. Regrettably, there are not yet any studies of the reproductive success of birds breeding on clear-cuts recently treated with a herbicide. This is an important deficiency in terms of understanding the ecological effects of herbicide spraying in forestry.

An important controversy related to herbicides focused on the military use of herbicides during the Vietnam War. During this conflict, the U.S. Air Force broadcast-sprayed herbicides to deprive their enemy of food production and forest cover. More than 5,600 square miles (14,503 km^2) were sprayed at least once, about one-seventh of the area of South Vietnam. More than 55 million pounds (25 million kg) of 2,4-D, 43 million pounds (21 million kg) of 2,4,5-T, and 3.3 million pounds (1.5 million kg) of picloram were used in this military program. The most frequently used herbicide was a 50-50 formulation of 2,4,5-T and 2,4-D known as Agent Orange. The rate of application was relatively large, averaging about ten times the application rate for silvicultural purposes. About 86 percent of spray missions were targeted against forests, and the remainder against cropland.

As was the military intention, these spray missions caused great ecological damage. Opponents of the practice labeled it ecocide (i.e., the intentional use of antienvironmental actions as a military tactic). The broader ecological effects included severe damage to mangrove and tropical forests, and a great loss of wildlife habitat.

In addition, the Agent Orange used in Vietnam was contaminated by the dioxin isomer known as TCDD, an incidental by-product of the manufacturing process of 2,4,5-T. Using post-Vietnam manufacturing technology, the contamination by TCDD in 2,4,5-T solutions can be kept to a concentration well below the maximum of 0.1 parts per million (ppm) set by the U.S. Environmental Protection Agency (EPA). However, the 2,4,5-T used in Vietnam was grossly contaminated with TCDD, with a concentration as large as 45.0 ppm occurring in Agent Orange, and an average of about 2.0 ppm. Perhaps 243–375 pounds (110–170 kg) of TCDD was sprayed with herbicides onto Vietnam. TCDD is well-known as being extremely toxic, and it can cause birth defects and miscarriage in laboratory mammals, although as is often the case, toxicity to humans is less well understood. There has been great controversy about the effects on soldiers and civilians exposed to TCDD in Vietnam, but epidemiological studies have been equivocal about the damages.

A preferable approach to pesticide use is integrated pest management (IPM). In the context of IPM, pest control is achieved by employing an array of complementary approaches based on monitoring in addition to biological and mechanical controls. A successful IPM program can greatly reduce, but not necessarily eliminate, the reliance on pesticides. With specific relevance to herbicides, more research into organic systems and into procedures that are pest-specific are required for the development of IPM systems. Examples of pest-specific practices are the biological control of certain introduced weeds, for example:

- St. John's wort (*Hypericum perforatum*) is a serious weed of pastures of the southwest United States because it is toxic to cattle, but it was controlled by the introduction in 1943 of two herbivorous leaf beetles;
- the prickly pear cactus (*Opuntia* spp.) became a serious weed of Australian rangelands after it was introduced as an ornamental plant, but it has been controlled by the release of the moth *Cactoblastis cactorum*, whose larvae feed on the cactus.

Unfortunately, effective IPM systems have not yet been developed for most weed problems for which herbicides are now used. With a growing interest in organic farming and reducing the input of potentially toxic chemicals into the environment, there are some natural herbicides that can be used to control weeds such as garlic oil, vinegar, and clove oil. Pelargonic acid, a natural chemical found in soaps, can be used as a contact herbicide that causes the plant cells to break apart. Until there are other alternatives, pest-specific methods to achieve an economically acceptable degree of control of weeds in agriculture and forestry, herbicides will continue to be used for that purpose.

See also Agricultural chemicals.

Resources

BOOKS

Cobb, Andrew. *Herbicides and Plant Physiology*. Malden, MA: Blackwell, 2008.
Dodge, A. D. *Herbicides and Plant Metabolism*. Cambridge, UK: Cambridge University Press, 2008.
Gillman, Jeff. *The Truth about Organic Gardening: Benefits, Drawbacks, and the Bottom Line*. Portland, OR: Timber Press, 2008.
Institute of Medicine (U.S.). *Veterans and Agent Orange: Update 2006*. Washington, DC: National Academies Press, 2007.
LeBaron, Homer M., Janis E. McFarland, and Orvin Burnside. *The Triazine Herbicides: Fifty Years Revolutionizing Agriculture*. Amsterdam: Elsevier, 2008.

OTHER

Centers for Disease Control and Prevention (CDC). "Herbicides." http://www.cdc.gov/exposurereport/ (accessed October 2, 2010).

Bill Freedman

Heritage Conservation and Recreation Service

The Heritage Conservation and Recreation Service (HCRS) was created in 1978 as an agency of the U.S. Department of the Interior (Secretarial Executive Order 3017) to administer the National Heritage Program initiative of President Jimmy Carter. The new agency was an outgrowth of and successor to the former Bureau of Outdoor Recreation. The HCRS resulted from the consolidation of some thirty laws, executive orders, and interagency agreements that provided federal funds to states, cities, and local community organizations to acquire, maintain, and develop historic, natural, and recreation sites. HCRS focused on the identification and protection of the nation's significant natural, cultural, and recreational resources. It classified and established registers for heritage resources, formulated policies and programs for their preservation, and coordinated federal, state, and local resource and recreation policies and actions. In February 1981 HCRS was abolished as an agency and its responsibilities were transferred to the National Park Service.

Hetch Hetchy Reservoir

The Hetch Hetchy Reservoir, located on the Tuolumne River in Yosemite National Park, was built to provide water and hydroelectric power to San Francisco, California. Its creation in the early 1900s led to one of the first conflicts between preservationists and those favoring utilitarian use of natural resources. The controversy spanned the presidencies of Theodore Roosevelt, William Taft, and Woodrow Wilson.

A prolonged conflict between San Francisco and its only water utility, Spring Valley Water Company, drove the city to search for an independent water supply. After surveying several possibilities, the city decided to build a dam and reservoir in the Hetch Hetchy Valley because the river there could supply the most abundant and purest water. This option was also the least expensive because the city planned to use the dam to generate hydroelectric power. It would also provide an abundant supply of irrigation water for area farmers and the recreation potential of a new lake.

The city applied to the U.S. Department of the Interior in 1901 for permission to construct the dam, but the request was not approved until 1908. The department then turned the issue over to Congress to work out an exchange of land between the federal government and the city. Congressional debate spanned several years and produced a number of bills. Part of the controversy involved the Right of Way Act of 1901, which gave Congress power to grant rights of way through government lands; some claimed this was designed specifically for the Hetch Hetchy project.

Opponents of the project likened the valley to Yosemite on a smaller scale. They wanted to preserve its high cliff walls, waterfalls, and diverse plant species. One of the most well-known opponents, California naturalist John Muir, described the Hetch Hetchy Valley as "a grand landscape garden, one of Nature's rarest and most precious mountain temples." Campers and mountain climbers fought to save the campgrounds and trails that would be flooded.

As the argument ensued, often played out in newspapers and other public forums, overwhelming national opinion appeared to favor the preservation of the valley. Despite this public support, a close vote in Congress led to the passage of the Raker Act, allowing the O'Shaughnessy Dam and Hetch Hetchy Reservoir to be constructed. President Wilson signed the bill into law on December 19, 1913.

The Hetch Hetchy Reservoir was completed in 1923 and still supplies water and electric power to San Francisco. In 1987, Secretary of the Interior Donald Hodel created a brief controversy when he suggested tearing down O'Shaughnessy Dam.

See also Economic growth and the environment; Environmental law; Environmental policy.

Resources

BOOKS

Jones, Holway R. *John Muir and the Sierra Club: The Battle for Yosemite.* San Francisco: Sierra Club, 1965.

Nash, Roderick. "Conservation as Anxiety." In *The American Environment: Readings in the History of Conservation.* 2nd ed. Reading, MA: Addison-Wesley Publishing, 1976.

Teresa C. Donkin

Heterotroph

A heterotroph is an organism that derives its nutritional carbon and energy by oxidizing (i.e., decomposing) organic materials. The higher taxonomic animals, fungi, actinomycetes, and most bacteria are heterotrophs. These are the biological consumers that eventually decompose most of the organic matter on the earth. The decomposition products then are available for chemical or biological recycling.

See also Oxidation reduction reactions.

High-grading (mining, forestry)

The practice of high-grading can be traced back to the early days of the California gold rush, when miners would sneak into claims belonging to others and steal the most valuable pieces of ore. The practice of high-grading remains essentially unchanged today. An individual or corporation will enter an area and selectively mine or harvest only the most valuable specimens, before moving on to a new area. High-grading is most prevalent in the mining and timber industries. It is not uncommon to walk into a forest, particularly an old-growth forest, and find the oldest and finest specimens marked for harvesting.

See also Forest management; Strip mining.

High-level radioactive waste

High-level radioactive waste consists primarily of the by-products of nuclear power plants and defense activities. Such wastes are highly radioactive and often decay very slowly. They may release dangerous levels of radiation for hundreds or thousands of years. Most high-level radioactive wastes have to be handled by remote control by workers who are protected by heavy shielding. They present, therefore, a serious health and environmental hazard. No entirely satisfactory method for disposing of high-level wastes has as yet been devised. In the early twenty-first century, the best approach seems to involve immobilizing the wastes in a glass-like material and then burying them deep underground.

See also Low-level radioactive waste; Radioactive decay; Radioactive pollution; Radioactive waste management; Radioactivity.

High-solids reactor

Solid waste disposal is a serious problem in the United States and other developed countries. Solid waste can constitute valuable raw materials for commercial and industrial operations, however, and one of the challenges facing scientists is to develop an economically efficient method for utilizing it.

Although the concept of bacterial waste conversion is simple, achieving an efficient method for putting the technique into practice is difficult. The main problem is that efficiency of conversion requires increasing the ratio of solids to water in the mixture, and this makes mixing more difficult mechanically. The high-solids reactor was designed by scientists at the Solar Energy Research Institute (SERI) to solve this problem. (In 1991, the SERI became known as the National Renewable Energy Laboratory [NREL],

when it was designated a national laboratory for the U.S. Department of Energy.) It consists of a cylindrical tube on a horizontal axis, and an agitator shaft running through the middle of it, which contains a number of Teflon-coated paddles oriented at ninety-degree angles to each other. The pilot reactors operated by SERI had a capacity of 2.6 gallons (10 l).

SERI (now NREL) scientists modeled the high-solids reactor after similar devices used in the plastics industry to mix highly viscous materials. With the reactor, they have been able to process materials with 30 to 35 percent solids content, whereas existing reactors normally handle wastes with 5–8 percent solid content. With higher solid content, NREL reactors have achieved a yield of methane five to eight times greater than that obtained from conventional mixers. Researchers hope to be able to process wastes with solid content ranging anywhere from 0 to 100 percent. They believe that they can eventually achieve 80 percent efficiency in converting biomass to methane.

The most obvious application of the high-solids reactor is the processing of municipal solid wastes. Initial tests were carried out with sludge obtained from sewage treatment plants in Denver (Colorado), Los Angeles (California), and Chicago (Illinois). In all cases, conversion of solids in the sludge to methane was successful, and other applications of the reactor are also being considered. For example, it can be used to leach out uranium from mine wastes: Anaerobic bacteria in the reactor will reduce uranium in the wastes and the uranium will then be absorbed on the bacteria or on ion exchange resins. The use of the reactor to clean contaminated soil is also being considered in the hope that this will provide a desirable alternative to current processes for cleaning soil, which create large volumes of contaminated water.

As of October, 2010, the NREL recently completed its Integrated Biorefinery Research Facility (IBRF). The facility will expand its Alternative Fuels User Facility (AFUF) by adding additional laboratory space, which will house new biomass processing equipment, and an additional storage area, which will hold feedstock milling and general storage materials. This advancement to the NREL facility will allow its researchers to better develop and test bioprocessing technologies for the production of fuels, such as ethanol, and cellulosic biomass chemicals.

See also Biomass fuel; Solid waste incineration; Solid waste recycling and recovery; Solid waste volume reduction; Waste management.

Resources

BOOKS

Collin, Robin Morris. *Encyclopedia of Sustainability*. Santa Barbara, CA: Greenwood Press, 2010.

Mitchell, Ralph, and Ji-Dong Gu. *Environmental Microbiology*. Hoboken, NJ: Wiley-Blackwell, 2010.

Young, Gary C. *Municipal Solid Waste to Energy Conversion Processes: Economic, Technical, and Renewable Comparisons*. Hoboken, NJ: Wiley, 2010.

OTHER

National Renewable Energy Laboratory. "Home Web page." http://www.nrel.gov/ (accessed October 18, 2010).

David E. Newton

High-voltage power lines *see*
Electromagnetic field.

High-yield crops *see* **Consultative Group on International Agricultural Research; Borlaug, Norman E.**

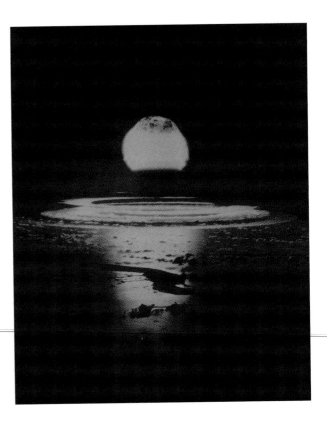

August 6th, 1945: The first of two atomic bombs was dropped on the city of Hiroshima, Japan. *(Science Source/Photo Researchers, Inc.)*

Hiroshima, Japan

Hiroshima is a beautiful modern city located near the southwestern tip of the main Japanese island of Honshu. It had been a military center with the headquarters of the Japanese southern army and a military depot prior to the end of World War II. The city is now a manufacturing center with a major university and medical school. It is most profoundly remembered because it was the first city to be exposed to the devastation of an atomic bomb.

At 8:15 A.M. on the morning of August 6, 1945, a single B-29 bomber flying from Tinian Island in the Marianas Islands released the bomb at 31,060 feet (9,467.1 m) above the city. The target was a T-shaped bridge near the city center. The only surviving building in the city center after the atomic bomb blast was a domed cement building at ground zero, just a few yards from the bridge. An experimental bomb developed by the Manhattan Project had been exploded at Alamogordo, New Mexico, only a few weeks earlier. The Alamogordo bomb had the explosive force of 15,000 tons (13,607 metric tons) of TNT. The Hiroshima uranium-235 bomb, with the explosive power of 20,000 tons (18,143 metric tons) of TNT, was even more powerful than the New Mexico bomb. The immediate effect of the bomb was to destroy by blast, winds, and fire an area

of 4.4 square miles (7 km^2). Two-thirds of the city was destroyed. A portion of Hiroshima was protected from the blast by hills, and this is all that remains of the old city. Destruction of human lives was caused immediately by the blast force of the bomb or by burns or radiation sickness later. The blast killed 70,000 people, and by the end of 1945 between 90000 and 160,000 people had died. Nagasaki, to the south and west of Hiroshima, was bombed on August 9, 1945, with much loss of life. The bombing of these two cities brought World War II to a close. The lessons that Hiroshima and Nagasaki teach are the horrors of war with its random killing of civilian men, women, and children.

The reasons for bombing Hiroshima should be taken in the context of the battle for Okinawa, which occurred only weeks before. America forces suffered twelve thousand dead with thirty-six thousand wounded in the battle for that small island 350 miles (563.5 km) from the mainland of Japan. The Japanese were reported to have lost 100,000 men. The determination of the Japanese to defend their homeland was well known, and it was estimated that the invasion of Japan would cost no less than 500,000

American lives. Japanese casualties were expected to be larger. It was the military judgment of American President Harry Truman that a swift termination of the war would save more lives, both American and Japanese, than it would cost. Whether this rationale for the atomic bombing of Hiroshima was correct (i.e., whether more people would have died if Japan was invaded) will never be known. However, it certainly is a fact that the war came to a swift end after the bombing of the two cities.

The second lesson to be learned from Hiroshima is that radiation exposure is hazardous to human health and radiation damage results in radiation sickness and increased cancer risk. It had been known since the development of X-rays at the turn of the century that radiation has the potential to cause cancer. However, the thousands of survivors at Hiroshima and Nagasaki were to become the largest group ever studied for radiation damage. The Atomic Bomb Casualty Commission, now referred to as the Radiation Effects Research Foundation (RERF), was established to monitor the health effects of radiation exposure and has studied these survivors since the end of World War II. The RERF has reported a ten to fifteen times excess of all types of leukemia among the survivors compared with populations not exposed to the bomb. The leukemia excess peaked four to seven years after exposure but still persists among the survivors. All forms of cancer tended to develop more frequently in heavily irradiated individuals, especially children under the age of ten at the time of exposure. Thyroid cancer was also increased in these children survivors of the bomb.

Robert G. McKinnell

Holistic approach

First formulated by Jan Smuts, holism has been traditionally defined as a philosophical theory that states that the determining factors in nature are wholes that are irreducible to the sum of their parts and that the evolution of the universe is the record of the activity and making of such wholes. More generally, it is the concept that wholes cannot be analyzed into parts or reduced to discrete elements without unexplainable residuals. Holism may also be defined by what it is not: It is not synonymous with organicism; holism does not require an entity to be alive or even a part of living processes. And neither is holism confined to spiritual mysticism, unaccessible to scientific methods or study.

The holistic approach in ecology and environmental science derives from the idea proposed by Harrison Brown that "a precondition for solving [complex] problems is a realization that all of them are interlocked, with the result that they cannot be solved piecemeal." For some scholars holism is the rationale for the very existence of ecology. As David Gates notes, "the very definition of the discipline of ecology implies a holistic study."

The holistic approach has been successfully applied to environmental management. The U.S. Forest Service, for example, has implemented a multilevel approach to management that takes into account the complexity of forest ecosystems, rather than the traditional focus on isolated incidents or problems.

Some people claim belief that a holistic approach to nature and the world will counter the effects of reductionism—excessive individualism, atomization, mechanistic worldview, objectivism, materialism, and anthropocentrism. Advocates of holism claim that its emphasis on connectivity, community, processes, networks, participation, synthesis, systems, and emergent properties will undo the ills of reductionism. Others warn that a balance between reductionism and holism is necessary. American ecologist Eugene Odum mandated that "ecology must combine holism with reductionism if applications are to benefit society." Parts and wholes, at the macro- and micro-level, must be understood. The basic lesson of a combined and complementary parts-whole approach is that every entity is both part *and* whole—an idea reinforced by philosopher Arthur Koestler's concept of a *holon*. A holon is any entity that is both a part of a larger system and itself a system made up of parts. It is essential to recognize that holism can include the study of *any* whole, the entirety of any individual in all its ramifications, without implying any organic analogy other than organisms themselves. A holistic approach alone, especially in its extreme form, is unrealistic, condemning scholars to an unproductive wallowing in an unmanageable complexity. Holism and reductionism are both needed for accessing and understanding an increasingly complex world.

See also Environmental ethics.

Resources

BOOKS

Koopsen, Cyndie, and Caroline Young. *Integrative Health: A Holistic Approach for Health Professionals.* Sudbury, MA: Jones and Bartlett Publishers, 2009.

Gerald L. Young

requirement of residence and cultivation for five years, against payment of $1.25 per acre. With the advent of machinery to mechanize farm labor, 160-acre (65-ha) tracts soon became uneconomical to operate, and Congress modified the original act to allow acquisition of larger tracts. The Homestead Act is still in effect, but good unappropriated land is scarce. Only Alaska still offers opportunities for homesteaders.

The Homestead Act was designed to speed development of the United States (especially westward) and to achieve an equitable distribution of wealth. Poor settlers, who lacked the capital to buy land, were able to start their own farms. Indeed, the Act contributed greatly to the growth and development of the country, particularly in the period between the Civil War and World War I, and it did much to speed settlement west of the Mississippi River. In all, well over a quarter of a billion acres (101 million ha) of land has been distributed under the Homestead Act and its amendments. However, only a small percentage of land granted under the Act between 1862 and 1900 was in fact acquired by homesteaders. According to estimates, between one of every six acres (0.4 of every 2.4 ha) and possibly only one in nine acres (0.4 in 3.6 ha) passed into the hands of family farmers.

The railroad companies and land speculators obtained the bulk of the land, sometimes through gross fraud using dummy entrants. Moreover, the railroads often managed to get the best land, whereas the homesteaders, ignorant of farming conditions on the Plains, often ended up with tracts least suitable to farming. Speculators frequently encouraged settlement on land that was too dry or had no sources of water for domestic use. When the homesteads failed, many settlers sold the land to speculators.

The environmental consequences of the Homestead Act were many and serious. The Act facilitated railroad development, often in excess of transportation needs. In many instances, competing companies built lines to connect the same cities. Railroad development contributed significantly to the destruction of bison herds, which in turn led to the destruction of the way of life of the Plains Indians. Cultivation of the Plains caused wholesale destruction of the vast prairies, so that whole ecological systems virtually disappeared. Overfarming of semiarid lands led to another environmental disaster, whose consequences were fully experienced only in the 1930s. The great Dust Bowl, with its terrifying dust storms, made huge areas of the country unlivable.

Resources

BOOKS

Lause, Mark A. *Young America: Land, Labor, and the Republican Community*. Urbana: University of Illinois Press, 2005.

OTHER

National Park Service. "The Last Homesteader." http://www.nps.gov/home/historyculture/lasthomesteader.htm (accessed October 15, 2010).

William E. Larson
Marijke Rijsberman

Horizon

Layers in the soil develop because of the additions, losses, translocations, and transformations that take place as the soil ages. The soil layers occur as a result of water percolating through the soil and leaching substances downward. The layers are parallel to the soil surface and are called horizons. Horizons will vary from the surface to the subsoil and from one soil to the next because of the different intensities of the above processes. Soils are classified into different groups based on the characteristics of the horizons.

Horseshoe crabs

The horseshoe crab (*Limulus polyphemus*) is the American species of a marine animal that is only a distant relation of crustaceans, such as crabs and lobsters. Horseshoe crabs are more closely related to spiders and scorpions. The crabs have been called living fossils because the genus dates back millions of years, and *Limulus* evolved very little over the years.

Fossils found in British Columbia indicate that the ancestors of horseshoe crabs were in North America about 520 million years ago. During the late twentieth century, the declining horseshoe crab population concerned environmentalists. Horseshoe crabs are a vital food source for dozens of species of birds that migrate from South America to the Arctic Circle. Furthermore, crabs are collected for medical research. After blood is taken from the crabs, they are returned to the ocean.

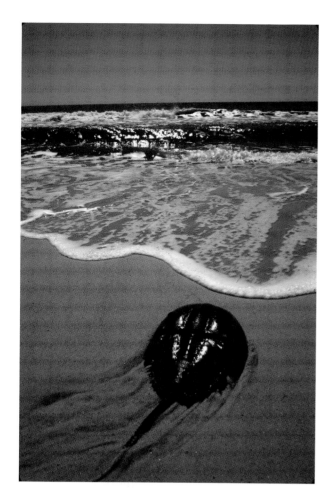

A horseshoe crab (*limulus polyphemus*). *(©John M. Burnley, National Audubon Society Collection. Photo Researchers Inc.)*

American horseshoe crabs live along the Atlantic Ocean coastline. Crab habitat extends south from Maine to the Yucatáan in the Gulf of Mexico. Several other crab species are found in Southeast Asia and Japan.

The American crab is named for its helmet-like shell that is shaped like a horseshoe. *Limulus* has a sharp tail shaped like a spike. The tail helps the crab move through the sand. If the crab tips over, the tail serves as a rudder so the crab can get back on its feet. The horseshoe crab is unique; its blood is blue and contains copper. The blood of other animals is red and contains iron.

Mature female crabs measure up to 24 inches (61 cm) in length. Males are about two-thirds smaller. Horseshoe crabs can live for nineteen years, and they reach sexual maturity in ten years. The crabs come to shore to spawn in late May and early June. They spawn during the phases of the full and new moon. The female digs

nests in the sand and deposits from two hundred to three hundred eggs in each pit. The male crab fertilizes the eggs with sperm, and the egg clutch is covered with sand.

During the spawning season, a female crab can deposit as many as ninety thousand eggs. This spawning process coincides with the migration of shorebirds. Flocks of birds, such as the red knot and the sandpiper, eat their fill of crab eggs before continuing their northbound migration.

Through the years, people found a variety of uses for horseshoe crabs. During the sixteenth century, Native Americans in South Carolina attached the tails to the spears that they used to catch fish. In the nineteenth century, people ground the crabs up for use as fertilizer or food for chickens and hogs.

During the twentieth century, researchers learned much about the human eye by studying the horseshoe crab's compound eye. Furthermore, researchers discovered that the crab's blood contained a special clotting agent that could be used to test the purity of new drugs and intravenous solutions. The agent called *limulus amoebocyte lysate* is obtained by collecting horseshoe crabs during the spawning season. Crabs are bled and then returned to the beach.

Horseshoe crabs are also used as bait. The harvesting of crabs increased sharply during the 1990s when people in the fishing industry used crabs as bait to catch eels and conch. The annual numbers of crabs harvested jumped from the thousands to the millions during the 1990s, according to environmental groups and organizations such as the National Audubon Society.

The declining horseshoe crab population could affect millions of migrating birds.

The Audubon Society and environmental groups have campaigned for state and federal regulations to protect horseshoe crabs. By 2002, coastal states and the Atlantic States Marine Fisheries Commission had set limits on the amount of crabs that could be harvested. The state of Virginia made bait bags mandatory when fishing with horseshoe crab bait. The mesh bag made of hard plastic holds the crab. That made it more difficult for predators to eat the crab so fewer *Limulus* crabs were needed as bait.

Furthermore, the federal government created a 1,500-square-mile (3,885-square-km) refuge for horseshoe crabs in the Delaware Bay. The refuge extends from Ocean City, New Jersey, to north of Ocean City, Maryland. As of March of 2002, harvesting was banned in the refuge. People who took crabs from the area faced a fine of up to $100,000, according to the National Marine Fisheries Service.

As measures such as those were enacted, marine biologists said that it could be several decades before the crab population increased. One reason for slow population growth was that it takes crabs ten years to reach maturity.

Resources

BOOKS

Ballesta, Laurent, Pierre Deschamp, and Jean-Michel Cousteau. *Planet Ocean: Voyage to the Heart of the Marine Realm.* Washington, DC: National Geographic, 2007.

Sargent, William. *Crab Wars: A Tale of Horseshoe Crabs, Bioterrorism, and Human Health.* Lebanon, NH: UPNE, 2006.

Tanacredi, John T., Mark L. Botton, and David Smith. *Biology and Conservation of Horseshoe Crabs.* New York: Springer, 2009.

Liz Swain

Hospital wastes *see* **Medical waste.**

Household waste

Household waste is commonly referred to as garbage or trash. As the population of the world expands, so does the amount of waste produced. Generally, the more automated and industrialized human societies become, the more waste they produce. For example, the industrial revolution introduced new manufactured products and new manufacturing processes that added to household solid waste and industrial waste. Modern consumerism and the excess packaging of many products also contribute significantly to the increasing amount of solid waste.

The amount of household waste produced in the United States each year equates to about thousands of pounds of trash for each American. About 40 percent of this total is paper and paper products. Much of the waste comes from packaging materials. Other types of waste produced by consumers are durable goods such as tires, appliances, and furniture, while other household solid waste is made up of nondurable goods such as paper, disposable products, and clothing. Many of these items could be recycled and reused, so they also can be considered a nonutilized resource.

In less industrialized times and even today in many developing countries, households and industries disposed of unwanted materials in bodies of water or in land dumps. However, this practice creates undesirable effects such as health hazards and foul odors. Open dumps serve as breeding grounds for disease-carrying organisms such as rats and insects. As the first world became more alert to environmental hazards, methods for waste disposal were studied and improved. Today, however, governments, policy makers, and individuals still wrestle with the problem of how to improve methods of waste disposal, storage, and recycling.

In 1976 the U.S. Congress passed the Resource Conservation and Recovery Act (RCRA) in an effort to protect human health and the environment from hazards associated with waste disposal. In addition, the Act aims to conserve energy and natural resources and to reduce the amount of waste Americans generate. Further, the RCRA promotes methods to manage waste in an environmentally sound manner. The Act covers regulation of solid waste, hazardous waste, and underground storage tanks that hold petroleum products and certain chemicals.

Most household solid waste is removed from homes through community garbage collection and then taken to landfills. The garbage in landfills is buried, but it can still produce noxious odors. In addition, rainwater can seep through landfill sites and leach out pollutants from the landfill trash. These are then carried into nearby bodies of water. Pollutants can also contaminate groundwater, which in turn leads to contamination of drinking water.

In order to fight this problem, sanitary landfills were developed. Clay or plastic liners are placed in the ground before garbage is buried. This helps prevent water from seeping out of the landfill and into the surrounding environment. In sanitary landfills, each time a certain amount of waste is added to the landfill, it is covered by a layer of soil. At a predetermined height the site is capped and covered with dirt. Grass and trees can be planted on top of the capped landfill to help prevent erosion and to improve the look of the site. Sanitary landfills are more expensive than open pit dumps, and many communities do not want the stigma of having a landfill near them. These factors make it politically difficult to open new landfills. Landfills are regulated by state and local governments and must meet minimum requirements set by the U.S. Environmental Protection Agency (EPA). Some household hazardous wastes such as paint, used motor oil, or insecticides can not be accepted at landfills and must be handled separately.

Incineration (burning) of solid waste offers an alternative to disposal in landfills. Incineration converts large amounts of solid waste to smaller amounts of ash. The ash must still be disposed of, however, and it can contain toxic materials. Incineration releases smoke and other possible pollutants into the air. However, modern incinerators are equipped with smokestack scrubbers that are quite effective in trapping toxic emissions. Many incinerators have the added benefit of generating electricity from the trash they burn.

Composting is a viable alternative to landfills and incineration for some biodegradable solid waste. Vegetable trimmings, leaves, grass clippings, straw, horse manure, wood chippings, and similar plant materials are all biodegradable and can be composted. Compost helps the environment because it reduces the amount of waste going into landfills. Correct composting also breaks down biodegradable material into a nutrient-rich soil additive that can be used in gardens or for landscaping. In this way, nutrients vital to plants are returned to the environment. To successfully compost biodegradable wastes, the process must generate high enough temperatures to kill seeds or organisms in the composted material. If done incorrectly, compost piles can give off foul odors.

Families and communities can help reduce household waste by making some simple lifestyle changes. They can reduce solid waste by recycling, repairing rather than replacing durable goods, buying products with minimal packaging, and choosing packaging made from recycled materials. Reducing packaging material is an example of source reduction. Much of the responsibility for source reduction is with manufacturers. Businesses need to be encouraged to find smart and cost-effective ways to manufacture and package goods in order to minimize waste and reduce the toxicity of the waste created. Consumers can help by encouraging companies to create more environmentally responsible packaging through their choice of products. For example, consumers successfully pressured McDonald's to change from serving their sandwiches in nonbiodegradable Styrofoam boxes to wrapping them in biodegradable paper.

Individual households can reduce the amount of waste they send to landfills by recycling. Paper, aluminum, glass, and plastic containers are the most commonly recycled household materials. Strategies for household recycling vary from community to community. In some areas materials must be separated by type before collection. In others, the separation occurs after collection.

Recycling preserves natural resources by providing an alternative supply of raw materials to industries. It also saves energy and eliminates the emissions of many toxic gases and water pollutants. In addition, recycling helps create jobs, stimulates development of more environmentally sound technologies, and conserves resources for future generations. For recycling to be successful, there must be an end market for goods made from recycled materials. Consumers can support recycling by buying "green" products made of recycled materials.

Composting is another alternative waste disposal option that is increasingly gaining favor. In Edmonton, Alberta, Canada, a municipal composting operation that is the size of fourteen National Hockey League rinks composts about 250,000 tons (226,800 metric tons) of household waste each year. In the United States, San Francisco (California) and Seattle (Washington) mandate the composting of food and yard waste.

Battery recycling is also becoming increasingly common in the United States and is required by law in many European countries. In 2001 a nonprofit organization called Rechargeable Battery Recycling Corporation (RBRC) began offering American communities cost-free recycling of portable rechargeable batteries such as those used in cell phones, camcorders, and laptop computers. These batteries contain cadmium, which is recycled back into other batteries or used in certain coatings or color pigments.

Resources

BOOKS

Grossman, Elizabeth. *High Tech Trash: Digital Devices, Hidden Toxics, and Human Health.* Washington DC: Shearwater Books, 2007.

Rodgers, Heather. *Gone Tomorrow: The Hidden Life of Garbage.* New York: New Press, 2006.

Royte, Elizabeth. *Garbage Land: On the Secret Trail of Trash.* Boston: Back Bay Books, 2006

Teresa G. Norris

HRS *see* **Hazard ranking system.**

Hubbard Brook Experimental Forest

The Hubbard Brook Experimental Forest is located in West Thornton, New Hampshire. It is an experimental area established in 1955 within the White Mountains National Forest in New Hamphire's central plateau and is administered by the U.S. Forest

Service. Hubbard Brook was the site of many important ecological studies beginning in the 1960s that established the extent of nutrient losses when all the trees in a watershed are cut.

Hubbard Brook is a north temperate watershed covered with a mature forest, and it is still accumulating biomass. In one early study, vegetation cut in a section of Hubbard Brook was left to decay while nutrient losses were monitored in the runoff. Total nitrogen losses in the first year were twice the amount cycled in the system during a normal year. With the rise of nitrate in the runoff, concentrations of calcium, magnesium, sodium, and potassium rose. These increases caused eutrophication and pollution of the streams fed by this watershed. Once the higher plants had been destroyed, the soil was unable to retain nutrients.

Early evidence from the studies indicated that total losses in the ecosystem because of the clear-cutting were a large number of the total inventory of species. The site's ability to support complex living systems was reduced. The lost nutrients could accumulate again, but erosion of primary minerals would limit the number of plants and animals sustained in the area.

Another study at the Hubbard Brook site investigated the effects of forest cutting and herbicide treatment on nutrients in the forest. All of the vegetation in one of Hubbard Brook's seven watersheds was cut and then the area was treated with the herbicides. At the time the conclusions were startling: deforestation resulted in much larger runoffs into the streams. The pH of the drainage stream went from 5.1 to 4.3, along with a change in temperature and electrical conductivity of the stream water. A combination of higher nutrient concentration, higher water temperature, and greater solar radiation because of the loss of forest cover produced an algal bloom, the first sign of eutrophication. This signaled that a change in the ecosystem of the watershed had occurred. It was ultimately demonstrated at Hubbard Brook that the use of herbicides on a cut area resulted in their transfer to the outgoing water.

Hubbard Brook Experimental Forest continues to be an active research facility for foresters and biologists. Most current research focuses on water quality and nutrient exchange. The Forest Service also maintians an acid rain monitoring station, and conducts research on old-growth forests. The results from various studies done at Hubbard Brook have shown that mature forest ecosystems have a greater ability to trap and store nutrients for recycling within the ecosystem. In addition, mature forests offer a higher degrees of biodiversity than do forests that are clear-cut.

See also Aquatic chemistry; Cultural eutrophication; Decline spiral; Experimental Lakes Area; Nitrogen cycle.

Resources

BOOKS

Allaby, Michael, and Richard Garratt. *Temperate Forests (Biomes of the Earth)*. New York: Chelsea House Publications, 2006.

Herring, Margaret J., and Sarah Greene. *Forest of Time: A Century of Science at Wind River Experimental Forest*. Corvallis: Oregon State University Press, 2007.

Moran, Emilio F., and Elinor Ostrom, eds. *Seeing the Forest and the Trees: Human-Environment Interactions in Forest Ecosystems*. Cambridge, MA: MIT Press, 2005.

OTHER

Center for International Forestry Research (CIFOR). "Center for International Forestry Research." Indonesia Headquarters Web site. http://www.cifor.cgiar.org (accessed November 11, 2010).

Linda Rehkopf

Hudson River

Starting at Lake Tear of the Clouds, a two-acre (0.8-ha) pond in New York's Adirondack Mountains, the Hudson River runs 315 miles (507 km) to the Battery on Manhattan Island's southern tip, where it meets the Atlantic Ocean. Although polluted and extensively dammed for hydroelectric power, the river still contains a wealth of aquatic species, including massive sea sturgeon (*Acipenser oxyrhynchus*) and short-nosed sturgeon (*A. brevirostrum*). The upper Hudson is fast-flowing trout stream, but below the Adirondack Forest Preserve, pollution from municipal sources, paper companies, and industries degrades the water. Stretches of the upper Hudson contain so-called warm water fish, including northern pike (*Esox lucius*), chain pickerel (*E. niger*), smallmouth bass (*Micropterus dolomieui*), and largemouth bass (*M. salmoides*). These latter two fish swam into the Hudson through the Lake Erie and Lake Champlain canals, which were completed in the early nineteenth century.

The Catskill Mountains dominate the mid-Hudson region, which is rich in fish and wildlife, though dairy farming, a source of runoff pollution, is strong in the region. American shad (*Alosa sapidissima*), historically the Hudson's most important commercial fish, spawn on the riverflats between Kingston and Coxsackie. Marshes in this region support snapping turtles (*Chelydra serpentina*) and, in the winter, muskrat (*Ondatra zibethicus*) and mink (*Mustela vison*). Water chestnuts (*Trapa natans*) grow luxuriantly in this section of the river.

Deep and partly bordered by mountains, the lower Hudson resembles a fiord. The unusually deep lower river makes it suitable for navigation by ocean-going vessels for 150 miles (241 km) upriver to Albany. Because the river's surface elevation does not drop between Albany and Manhattan, the tidal effects of the ocean are felt all the way upriver to the Federal Lock and Dam above Albany. These powerful tides make long stretches of the lower Hudson saline or brackish, with saltwater penetrating as high as 60 miles (97 km) upstream from the Battery.

The Hudson contains a great variety of botanical species. Over a dozen oaks thrive along its banks, including red oaks (*Quercus rubra*), black oaks (*Q. velutina*), pin oaks (*Q. palustris*), and rock chestnut (*Q. prinus*). Numerous other trees also abound, from mountain laurel (*Kalmia latifolia*) and red pine (*Pinus resinosa*) to flowering dogwood (*Cornus florida*), together with a wide variety of small herbaceous plants.

The Hudson River is comparatively short. More than eighty American rivers are longer than it, but it plays a major role in New York's economy and ecology. Pollution threats to the river have been caused by the discharge of industrial and municipal waste, as well as pesticides washed off the land by rain. From 1930 to 1975, one chemical company on the river manufactured approximately 1.4 billion pounds (635,000 kg) of polychlorinated biphenyls (PCBs), and an estimated 10 million pounds (4.5 million kg) a year entered the environment. In all, a total of 1.3 million pounds (590,000 kg) of PCB contamination allegedly occurred during the years prior to the ban, with the pollution originating from plants at Ford Edward and Hudson Falls. A ban was put in place for a time prohibiting the possession, removal, and eating of fish from the waters of the upper Hudson River. A proposed cleanup was designated, to proceed by means of a 40-mile (64.4-km) dredging and sifting of 2.65 million cubic yards (202607037 m³) of sediment north of Albany, with an anticipated yield of 75 tons (68 metric tons) of PCBs.

In February of 2001 the U.S. Environmental Protection Agency (EPA), having invoked the Superfund law, required the chemical company to begin planning the cleanup. The company was given several weeks to present a viable plan of attack, or else face a potential $1.5 billion fine for ignoring the directive in lieu of the cost of cleanup. The cleanup cost, estimated at $500 million was presented as the preferred alternative. The engineering phase of the cleanup project was expected to take three years of planning and was to be scheduled after the offending company filed a response to the EPA. The company responded within the allotted time frame in order to placate the EPA, although the specifics of a drafted work plan remained undetermined, and the company refused to withdraw a lawsuit filed in November of 2000, which challenged the constitutionality of the so-called Superfund law that authorized the EPA to take action. The river meanwhile was ranked by one environmental watchdog group as the fourth most endangered in the United States, specifically because of the PCB contamination. Environmental groups demanded also that attention be paid to the issues of urban sprawl, noise, and other pollution, while proposals for potentially polluting projects were endorsed by industrialists as a means of spurring the area's economy. Among these industrial projects: the construction of a cement plant in Catskill where there is easy access to a limestone quarry; and the development of a power plant along the river in Athens, which generated controversy, stemming from the industrial asset afforded by development along the river versus the advantages of a less-fouled environment. Additionally, the power plant, which threatened to add four new smokestacks to the skyline and to aggravate pollution, was seen as potentially detrimental to tourism in that area. Also in recent decades, chlorinated hydrocarbons, dieldrin, endrin, DDT, and other pollutants have been linked to the decline in populations of the once common Jefferson salamander (*Ambystoma jeffersonianum*), fish hawk (*Pandion haliaetus*), and bald eagle (*Haliaeetus leucocephalus*).

The precipitous environmental state of the Hudson River in the 1960s prompted folk singer and activist Pete Seeger to found an organization called Hudson River Sloop Clearwater. An important facet of the groups was the construction of the sailboat Clearwater. The group's efforts helped galvanize resolve to clean up the river.

See also Agricultural pollution; Dams (environmental effects); Estuary; Feedlot runoff; Industrial waste treatment; Runoff.

Resources

BOOKS

DuLong, Jessica. *My River Chronicles: Rediscovering the Work that Built America; A Personal and Historical Journey*. New York: Free Press, 2009.

Dunwell, Frances F. *The Hudson: America's River*. New York: Columbia University Press, 2008.

Lewis, Tom. *The Hudson: A History*. New Haven, CT: Yale University Press, 2007.

David Clarke

Human ecology

Human ecology may be defined as the branch of knowledge concerned with relationships between human beings and their environments. Among the disciplines contributing seminal work in this field are sociology, anthropology, geography, economics, psychology, political science, philosophy, and the arts. Applied human ecology emerges in engineering, planning, architecture, landscape architecture, conservation, and public health. Human ecology, then, is an interdisciplinary study that applies the principles and concepts of ecology to human problems and the human condition. The notion of interaction—between human beings and the environment—is fundamental to human ecology, as it is to biological ecology.

Human ecology as an academic inquiry has disciplinary roots extending back as far as the 1920s. However, much work in the decades prior to the 1970s was narrowly drawn and was often carried out by a few individuals whose intellectual legacy remained isolated from the mainstream of their disciplines. The work done in sociology offers an exception to the latter (but not the former) rule; sociological human ecology is traced to the Chicago school and the intellectual lineage of Robert Ezra Park, his student Roderick D. Mackenzie, and Mackenzie's student Amos Hawley. Through the influence of these men and their school, human ecology, for a time, was narrowly identified with a sociological analysis of spatial patterns in urban settings (although broader questions were sometimes contemplated).

Comprehensive treatment of human ecology is first found in the work of Gerald L. Young, who pioneered the study of human ecology as an interdisciplinary field and as a conceptual framework. Young's definitive framework is founded upon four central themes. The first of these is interaction, and the other three are developed from it: levels of organization, functionalism (part-whole relationships), and holism. These four basic concepts form the foundation for a series of field derivatives (niche, community, and ecosystem) and consequent notions (institutions, proxemics, alienation, ethics, world community, and stress or capacitance). Young's emphasis on linkages and process set his approach apart from other synthetic attempts in human ecology, which some viewed as largely cumbersome classificatory schemata. These attempts were subject to harsh criticism because they tended to embrace virtually all knowledge, resolve themselves into superficial lists and mnemonic "building blocks," and had little applicability to real-world problems.

Generally, comprehensive treatment of human ecology is more advanced in Europe than it is in the United States. A comprehensive approach to human ecology as an interdisciplinary field and conceptual framework gathered momentum in several independent centers during the 1970s and 1980s. Among these have been several college and university programs and research centers, including those at the University of Gothenburg, Sweden, and, in the United States, at Rutgers University and the University of California at Davis. Interdisciplinary programs at the undergraduate level were first offered in 1972 by the College of the Atlantic (Maine) and The Evergreen State College (Washington). The Commonwealth Human Ecology Council in the United Kingdom, the International Union of Anthropological and Ethnological Sciences' Commission on Human Ecology, the Centre for Human Ecology at the University of Edinburgh, the Institute for Human Ecology in California, and professional societies and organizations in Europe and the United States have been other centers of development for the field.

Dr. Thomas Dietz, President of the Society for Human Ecology, defined some of the priority research problems that human ecology addresses in his testimony before the U.S. House of Representatives Subcommittee on Environment and the National Academy of Sciences Committee on Environmental Research. Among these, Dietz listed global change, values, post-hoc evaluation, and science and conflict in environmental policy. Other human ecologists would include in the list such items as commons problems, carrying capacity, sustainable development, human health, ecological economics, problems of resource use and distribution, and family systems. Problems of epistemology or cognition such as environmental perception, consciousness, or paradigm change also receive attention.

Our Common Future, the report of the United Nation's World Commission on Environment and Development of 1987, has stimulated a new phase in the development of human ecology. A host of new programs, plans, conferences, and agendas have been put forth, primarily to address phenomena of global change and the challenge of sustainable development. These include the *Sustainable Biosphere Initiative* published by the Ecological Society of America in 1991 and extended internationally; the United Nations Conference on Environment and Development; the proposed new United States National Institutes for the Environment; the Man and the Biosphere Program's Human-Dominated Systems Program; the report of the National Research Council Committee on Human Dimensions of Global Change and the associated National Science Foundation's Human Dimensions of Global Change Program; and green plans published by the governments of Canada, Norway, the Netherlands, the United Kingdom, and Austria. All of these programs call for an integrated, interdisciplinary approach to complex problems of human-environmental relationships. The next challenge for human ecology will be to digest and steer these new efforts and to identify the perspectives and tools they supply.

Resources

BOOKS

Curry, Patrick. *Ecological Ethics: An Introduction.* Cambridge, UK: Polity Press, 2006.

Des Jardins, Joseph R. *Environmental Ethics: An Introduction to Environmental Philosophy.* Belmont, CA: Wadsworth Publishing, 2005.

Jeremy Pratt

Humane Society of the United States

The largest animal protection organization in the United States, the Humane Society of the United States (HSUS) works to preserve wildlife and wilderness, save endangered species, and promote humane treatment of all animals. Formed in 1954, the HSUS specializes in education, cruelty investigations and prosecutions, wildlife and nature preservation, environmental protection, federal and state legislative activities, and other actions designed to protect animal welfare and the environment.

Major projects undertaken by the HSUS in recent years have included campaigns to stop the killing of whales, dolphins, elephants, bears, and wolves; to help reduce the number of animals used in medical research and to improve the conditions under which they are used; to oppose the use of fur by the fashion industry; and to address the problem of pet overpopulation.

The group has worked extensively to ban the use of tuna caught in a way that kills dolphins, largely eliminating the sale of such products in the United States and Western Europe. It has tried to stop international airlines from transporting exotic birds into the United States. Other high priority projects have included banning the international trade in elephant ivory, especially imports into the United States, and securing and maintaining a general worldwide moratorium on commercial whaling.

The HSUS Companion Animals section works on a variety of issues affecting dogs, cats, birds, horses, and other animals commonly kept as pets, striving to promote responsible pet ownership, particularly the spaying and neutering of dogs and cats to reduce the tremendous overpopulation of these animals. The HSUS works closely with local shelters and humane societies across the country, providing information, training, evaluation, and consultation.

Several national and international environmental and animal protection groups are affiliated and work closely with HSUS. One of them is the Humane Society International (HSI), which works abroad to fulfill the HSUS's mission and to institute reform and educational programs that will benefit animals.

In addition, the National Association for Humane and Environmental Education is the youth education division of the HSUS, developing and producing periodicals and teaching materials designed to instill humane values in students and young people.

The Center for Respect of Life and the Environment (CRLE) works with academic institutions, scholars, religious leaders and organizations, arts groups, and others to foster an ethic of respect and compassion toward all creatures and the natural environment. Its quarterly publication, *Earth Ethics*, examines such issues as earth education, sustainable communities, ecological economics, and other values affecting the human relationship with the natural world. The Interfaith Council for the Protection of Animals and Nature (ICPAN) promotes conservation and education mainly within the religious community, attempting to make religious leaders, groups, and individuals more aware of their moral and spiritual obligations to preserve the planet and its myriad life forms.

HSUS has been quite active, hard-hitting, and effective in promoting its animal protection programs, such as leading the fight against the fur industry. It accomplishes its goals through education, lobbying, grassroots organizing, and other traditional, legal means of influencing public opinion and government policies.

With about eleven million members or constituents and an annual revenue of over $218 million, HSUS, headquartered in Washington, DC, is considered the largest and one of the most influential animal protection groups in the United States and, perhaps, the world.

Resources

OTHER

Center for Respect of Life and Environment. "Home Web page." http://center1.com/ (accessed October 13, 2010).

The Humane Society of the United States. "Home Web page." http://www.humanesociety.org/ (accessed October 13, 2010).

Interfaith Council for the Protection of Animals and Nature. "Home Web page." http://www.icpanonline.org/ (accessed October 13, 2010).

ORGANIZATIONS

The Humane Society of the United States, 2100 L Street, NW, Washington, DC, USA, 20037, (202) 452–1100, http://www.hsus.org

Lewis G. Regenstein

Humanism

A perspective or doctrine that focuses primarily on the interests, capacities, and achievements of human beings. This focus on human concerns has led some to conclude that human beings have rightful dominion over the earth and that their interests and well-being are paramount and take precedence over all other considerations. *Religious humanism*, for instance, generally holds that God made human beings in His own image and put them in charge of His creation. *Secular humanism* views human beings as the source of all value or worth. Some environmentally-minded critics of humanism assert that anthropocentric (human-centered) thinking can be detrimental to other animal species and the environment.

Human-powered vehicles

Finding easy modes of transportation seems to be a basic human need, but finding easy *and clean* modes is becoming imperative. Traffic congestion, overconsumption

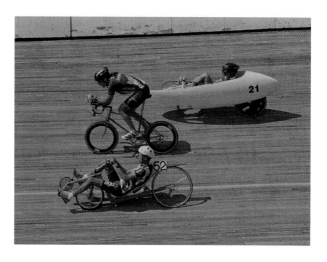

Competitors in the Human-powered Vehicles British Open Championships in Leicester, United Kingdom. *(©Nigel Farrow/Assignments Photographers/CORBIS)*

of fossil fuels, and air pollution are all direct results of automotive lifestyles around the world. The logical alternative is human-powered vehicles (HPVs), perhaps best exemplified in the bicycle, the most basic HPV. New high-tech developments in HPVs are not yet ready for mass production, nor are they able to compete with cars. Pedal-propelled HPVs in the air, on land, or under the sea are still in the expensive, design-and-race-for-a-prize category. But the challenge of human-powered transport has inspired a lot of inventive thinking, both amateur and professional.

Bicycles and rickshaws comprise the most basic HPVs. Of these two vehicles, bicycles are clearly the most popular, and production of these HPVs has surpassed production of automobiles in recent years. The number of bicycles in use throughout the world is roughly double that of cars; China alone contains approximately 270 million bicycles, or one-third of the total bicycles worldwide. Indeed the bicycle has overtaken the automobile as the preferred mode of transportation in many nations. There are many reasons for the popularity of the bike: It fulfills both recreational and functional needs, it is an economical alternative to automobiles, and it does not contribute to the problems facing the environment.

Although the bicycle provides a healthy and scenic form of recreation, people also find it useful in basic transportation. In the Netherlands, bicycle transportation accounts for 30 percent of work trips and 60 percent of school trips. One-third of commuting to work in Denmark is by bicycle. In China, the vast majority of all trips there are made via bicycle.

A surge in bicycle production occurred in 1973, when in conjunction with rising oil costs, production doubled to 52 million per year. Soaring fuel prices in the 1970s inspired people to find inexpensive, economical alternatives to cars, and many turned to bicycles. Besides being efficient transportation, bikes are simply cheaper to purchase and to maintain than cars. There is not a need to pay for parking or tolls, expensive upkeep is not required, and high fuel costs are eliminated.

The lack of fuel costs associated with bicycles leads to another benefit: Bicycles do not harm the environment. Cars consume fossil fuels and in so doing release more than two-thirds of the United States' smog-producing chemicals. Furthermore, they are considered responsible for many other environmental ailments: depletion of the ozone layer through release of chlorofluorocarbons (CFCs) from automobile air conditioning units; cause of cancer through toxic emissions; and consumption of the world's limited fuel resources. With human energy as their only requirement, bicycles have none of these liabilities.

Nevertheless, in many cases—such as long trips or traveling in inclement weather—cars are the preferred form of transportation. Bicycles are not the optimal choice in many situations. Thus, engineers and designers seek to improve on the bicycle and make machines suitable for transport under many different conditions. They are striving to produce new human-powered vehicles—HPVs that maximize air and sea currents, that have reasonable interior ergonomics, and that can be inexpensively produced. Several machines designed to fit this criteria exist.

As for developments in human-powered aircraft, success is judged on distance and speed, which depend on the strength of the person pedaling and the lightness of the craft. The current world record holder (for straight distance and duration) is Greek Olympic cyclist Kanellos Kanellopoulos who flew *Daedalus 88*. *Daedalus 88* was created by engineer John Langford and a team of Massachusetts Institute of Technology (MIT) engineers and funded by American corporations. Kanellopoulos flew *Daedalus 88* for three hours, fifty-three minutes, and thirty seconds across the Aegean Sea between Crete and Santorini, a distance of 74 miles (119 km), on April 23, 1988. The craft averaged 19.0 miles per hour (30.6 kph) and flew 15 feet (4.6 m) above the water. Upon arrival at santorini, however, the sun began to heat up the black sands and generate erratic shore winds and *Daedalus 88* plunged into the sea. It was a few yards short of its goal, and the tailboom of the 70-pound (32-kg) vehicle was snapped by the wind. But to cheering crowds on the beach, Kanellopoulos rose from the sea with a victory sign and strode to shore.

The International Human Powered Vehicle Association (IHPVA) acts as the sanctioning and regulatory body for new world records in human powered land, water, and air vehicles. Various records for HPVs are found on its Web page.

In the creation of a human-powered helicopter, students at California Polytechnic State University (in San Luis Obispo) had been working on perfecting one since 1981. In 1989 they achieved liftoff with Greg McNeil, a member of the United States National Cycling Team, pedaling an astounding 1.0 horsepower. The graphite epoxy, wood, and Mylar craft, *Da Vinci III*, rose 7 inches (17.7 cm) for 6.8 seconds. But rules for the $10,000 Sikorsky prize, sponsored by the American Helicopter Society, stipulate that the winning craft must rise nearly 10 feet (3 m) and stay aloft one minute.

On land, recumbent vehicles, or recumbents, are wheeled vehicles in which the driver pedals in a semi-recumbent position, contained within a windowed enclosure. The world record was set in 2008 by Canadian Sam Whittingham during the World Human Powered Speed Challenge held outside of Battle Mountain, Nevada in an HPV named *Varna Diablo III*. Whittingham pedaled a peak speed of 82.3 miles per hour (132.5 kph).

Unfortunately, the realities of road travel cast a long shadow over recumbent HPVs. Crews discovered that they tended to be unstable in crosswinds, distracted other drivers and pedestrians, and lacked the speed to correct course safely in the face of oncoming cars and trucks. On July 19, 2009, *Varna Tempest* set a new IHPVA record recognized for a human powered vehicle when pedaled by Sam Whittingham, at Romeo, Michigan, to cover a distance of 56.29 miles (90.60 km) in an hour.

In the sea, being able to maneuver at one's own pace and be in control of one's their vehicle—as well as being able to beat a fast retreat undersea—are the problems faced by HPV submersible engineers. Human-powered submarines (subs) are not a new idea. The Revolutionary War (1775–1783) created a need for a bubble sub that was to plant an explosive in the belly of a British ship in New York Harbor. (The naval officer, breathing one-half hour's worth of air, failed in his night mission, but survived).

The special design problems of modern two-person HP-subs involve controlling buoyancy and ballast, pitch, and yaw (to turn by angular motion about the vertical axis); reducing dra; increasing thrust; and positioning the pedaler and the propulsor in the flooded cockpit (called "wet") in ways that maximize air intake from scuba tanks and muscle power from arms and legs.

Depending on the design, the humans in HP-subs lie prone, foot to head or side by side, or sit, using their feet to pedal and their hands to control the rudder through the underwater currents. Studies by the U.S. Navy Experimental Dive Unit indicate that a well-trained athlete can sustain 0.5 horsepower for ten minutes underwater.

On the surface of the water, fin-propelled watercraft—lightweight inflatables that are powered by humans kicking with fins—are ideal for fishermen whom maneuverability, not speed, is the goal. Paddling with the legs, which does not disturb fish, leaves the hands free to cast. In most designs, the angler sits on a platform between tubes, his feet in the water. Controllability is another matter, however. In open windy water, the craft is at the mercy of the elements in its current design state. Top speed is about 50 yards (46 m) in three minutes.

Finally, over the surface of the water, the first human- powered hydrofoil, *Flying Fish*, with national track sprinter Bobby Livingston, broke a world record in September 1989 when it traveled 328 feet (100 m) over Lake Adrian, Michigan, at 18.5 miles per hour (29.8 kph). A vehicle that pedaled like a bicycle, resembled a model airplane with a two-blade propeller and a 6-foot (1.8-m) carbon graphite wing, *Flying Fish* sped across the surface of the lake on two pontoons.

Resources

BOOKS

Angus, Colin. *Beyond the Horizon: The First Human-Powered Expedition to Circle the Globe*. Birmingham, AL: Menasha Ridge Press, 2009.

Herlihy, David V. *Bicycle: The History*. New Haven, CT: Yale University Press, 2004.

Smith, Stevie. *Pedaling to Hawaii: A Human-Powered Odyssey*. Woodstock, VT: Countryman Press, 2006.

OTHER

Wired.com. "World's Fastest Cyclist Hits 82.3 MPH." http://www.wired.com/autopia/2008/09/worlds-fastest/ (accessed October 15, 2010).

ORGANIZATIONS

International Human Powered Vehicle Association, Post Office Box 357, Cutten, CA, USA 95534-0357, (877) 333-1029, http://www.ihpva.org/home/

Stephanie Ocko
Andrea Gacki

Humus

Humus is essentially decomposed organic matter in soil. Humus can vary in color but is often dark brown. Besides containing valuable nutrients, there are many other benefits of humus: It stabilizes soil mineral particles into aggregates, improves pore space relationships and aids in air and water movement, aids in water holding capacity, and influences the absorption of hydrogen ions as a pH regulator.

Hunting and trapping

Wild animals are a potentially renewable natural resource. This means that they can be harvested in a sustainable fashion, as long as their birthrate is greater than the rate of exploitation by humans. In the sense meant here, "harvesting" refers to the killing of wild animals as a source of meat, fur, antlers, or other useful products, or it refers to the outdoor sport. The harvesting can involve trapping, or hunting using guns, bows and arrows, or other weapons. From the ecological perspective, it is critical that the exploitation is undertaken in a sustainable fashion; otherwise, serious damages are caused to the resource and to ecosystems more generally.

Unfortunately, there have been numerous examples in which wild animals have been harvested at grossly unsustainable rates, which caused their populations to decline severely. In a few cases this caused species to become extinct—they no longer exist anywhere on earth. For example, commercial hunting in North America resulted in the extinctions of the great auk (*Pinguinnis impennis*), passenger pigeon (*Ectopistes migratorius*), and Steller's sea cow (*Hydrodamalis stelleri*). Unsustainable commercial hunting also brought other species to the brink of extinction, including the Eskimo curlew (*Numenius borealis*), northern right whale (*Eubalaena glacialis*), northern fur seal (*Callorhinus ursinus*), grey whale (*Eschrichtius robustus*), and American bison or buffalo (*Bison bison*).

Fortunately, these and many other examples of overexploitation of wild animals by humans are regrettable cases from the past. Today, the exploitation of wild animals in North America is undertaken with a view to the longer-term conservation of their stocks, that is, an attempt is made to manage the harvesting in a sustainable fashion. This means that

Hunter using a birdcall in a blind in the Sacramento Valley, California. Hunting blinds are often located in harvested rice fields, which are then flooded to attract waterfowl. *(Ron Sanford/Photo Researchers, Inc.)*

trapping and hunting are much more closely regulated than they used to be.

If harvests of wild animals are to be undertaken in a sustainable manner, it is critical that harvest levels are determined using the best available understanding of population-level productivity and stock sizes. It is also essential that trappers and hunters respect harvest quotas and that illegal exploitation (or poaching) does not compromise what might otherwise be a sustainable activity. The challenge of modern wildlife management is to ensure that good conservation science is sensibly integrated with effective monitoring and management of the rates of exploitation.

Ethics of trapping and hunting

From a strictly ecological perspective, sustainable trapping and hunting of wild animals is no more objectionable than the prudent harvesting of timber or agricultural crops. However, people have widely divergent attitudes about the killing of wild (or domestic) animals for meat, sport, or profit. At one end of the ethical spectrum are people who see no problem with the killing of wild animals as a source of meat or cash. At the other extreme are individuals with a profound respect for the rights of all animals, and who argue that killing any sentient creature is ethically wrong. Many of these latter people are animal-rights activists, and some of them are involved in organizations that undertake high-profile protests and other forms of advocacy to prevent or restrict trapping and hunting. In essence, these people object to the lethal exploitation of wild animals, even under closely regulated conditions that would not deplete their populations. Most people, of course, have attitudes that are intermediate to those just described.

Trapping

The fur trade was a very important commercial activity during the initial phase of the colonization of North America by Europeans. During those times, as now, furs were a valuable commodity that could be obtained from nature and could be sold at a great

Floodwaters from Hurricane Katrina pour through a levee along Inner Harbor Navigaional Canal near downtown New Orleans, LA., August 30, 2005, a day after Katrina passed through the city. *(AP Photo/Vincent Laforet, Pool)*

of Mexico, all four of these storms reached Category 5 on the Saffir-Simpson scale.

Hurricane Katrina, which made landfall three times during August 2005, was by far the most damaging and deadly of the 2004 and 2005 storms. Katrina caused more than $60 billion in damage, making it the most expensive natural disaster in U.S. history. Katrina's storm surge, arriving at an already high tide, reached 30 feet (9.14 m) and was the highest ever recorded in the United States. Strong winds in excess of 140 mph (225 kph) also whipped waves on top of the surge to more than 60 feet (18 m) in height. Extensive damage ranged from Louisiana to the Florida panhandle. The approach of Katrina as a Category 5 hurricane prompted a mandatory evacuation of New Orleans, a city largely below sea level. Despite warnings, many residents remained in the city either by choice or necessity. Most of the city was flooded when heavy rains and the storm surge caused levees between New Orleans and Lake Pontchartrain to fail after the storm had passed. In September 2010, five years after the storm and subsequent flooding, estimates of the number of people killed by the storm varied from 1,800 to 3,600 people. Many bodies were unidentified or remain unclaimed. Significant storm damage is still visible along the Gulf Coast from Texas to the Florida panhandle. Repairs to about 350 miles (563 km) of levees, along with installation of a renovated pumping system, continue as of 2010. Engineers estimate completing the levee repairs and renovations before the 2011 hurricane season. Both environmental and engineering experts argue that renovation and restoration of Louisiana's wetlands, currently lost at a rate of approximately 24,000 acres (9,712 ha) per year, will be required to provide adequate protection from the strongest categories of hurricanes. Such storms normally carry a substantial storm surge that can be blunted only by wetlands and barrier islands.

In 2008 Hurricane Ike killed more than six hundred people and caused massive flooding and homelessness in the Caribbean before crossing the Gulf to strike Galveston Island off the Texas coast near Houston. Although it struck the United States as a Category 2 storm, Ike was such a large storm—with a high storm surge more characteristic of a Category 4 or

Category 5 storm—that it caused tremendous flooding and damage. Hurricane Ike became the third costliest hurricane to strike the United States behind Hurricanes Katrina and Andrew (1992).

Global warming and hurricanes

Climate scientists and climate models vary in their predictions about the impact of global warming on hurricanes. Studies released late in 2009 suggested that global warming would result in fewer, but more intense, hurricanes. Although 2004 and 2005 produced a number of highly destructive hurricanes, there were also four years during the same decade (2000 to 2009) when no hurricane made landfall in the United States. Atlantic hurricane formation is also strongly correlated to a number of climate and weather factors that influence hurricane formation and strength. For example, El Niño events are historically associated with decreased hurricane activity in the Atlantic Ocean and Caribbean Sea.

Because of warmer water temperatures and other factors, forecasters predicted that the 2010 U.S. hurricane season, which ended on November 30, would be one of the most active seasons on record. Mathematical models estimated that fifteen to twenty named storms would develop during the 2010 season. Eight storms we predicted to reach hurricane strength, with three developing into major hurricanes. Hurricane Alex, a Category 2 storm eventually making landfall along the coast of Mexico about 100 miles (160 km) south of Brownsville, Texas, became the first Atlantic basin hurricane to form in the month of June in fifteen years.

See also Tornado and cyclone.

Hutchinson, George E.

1903–1991
American ecologist

Born January 30, 1903, in Cambridge, England, Hutchinson was the son of Arthur Hutchinson, a professor of mineralogy at Cambridge University, and Evaline Demeny Shipley Hutchinson, an ardent feminist. He demonstrated an early interest in flora and fauna and a basic understanding of the scientific method. In 1918, at the age of fifteen, he wrote a letter to the *Entomological Record and Journal of Variation* about a grasshopper he had seen swimming in a pond. He described an experiment he performed on the insect and included it for taxonomic identification.

In 1924 Hutchinson earned his bachelor's degree in zoology from Emmanuel College at Cambridge University, where he was a founding member of the Biological Tea Club. He then served as an international education fellow at the Stazione Zoologica in Naples from 1925 until 1926, when he was hired as a senior lecturer at the University of Witwatersrand in Johannesburg, South Africa. He was apparently fired from this position two years later by administrators who never imagined that in 1977 the university would honor the ecologist by establishing a research laboratory in his name.

Hutchinson earned his master's degree from Emmanuel College *in absentia* in 1928 and applied to Yale University for a fellowship so he could pursue a doctoral degree. He was instead appointed to the faculty as a zoology instructor. He was promoted to assistant professor in 1931 and became an associate professor in 1941, the year he obtained his U.S. citizenship. He was made a full professor of zoology in 1945, and between 1947 and 1965 he served as director of graduate studies in zoology. Hutchinson never did receive his doctoral degree, though he amassed an impressive collection of honorary degrees during his lifetime.

Hutchinson was best known for his interest in limnology, the science of freshwater lakes and ponds. He spent most of his life writing the four-volume *Treatise on Limnology*, which he completed just months before his death. The research that led to the first volume—covering geography, physics, and chemistry—earned him a Guggenheim Fellowship in 1957. The second volume, published in 1967, covered biology and plankton. The third volume, on water plants, was published in 1975, and the fourth volume, about invertebrates, appeared posthumously in 1993.

The *Treatise on Limnology* was among the nine books, nearly one hundred and fifty research papers, and many opinion columns that Hutchinson penned. He was an influential writer whose scientific papers inspired many students to specialize in ecology. Hutchinson's greatest contribution to the science of ecology was his broad approach, which became known as the "Hutchinson school." His work encompassed disciplines as varied as biochemistry, geology, zoology, and botany. He pioneered the concept of biogeochemistry, which examines the exchange of chemicals between organisms and the environment. His studies in biogeochemistry focused on how phosphates and nitrates move from the earth to plants, then animals, and then back to the earth in a continuous cycle. His holistic approach influenced later environmentalists when they began to consider the global scope of environmental problems.

hydrocarbons. Hydrocarbons are environmentally important for several reasons. First, hydrocarbons give off greenhouse gases, especially carbon dioxide, when burned and are important contributers to smog. In addition, many aromatic hydrocarbons and hydrocarbons containing halogens are toxic or carcinogenic.

Hydrochlorofluorocarbons

The term hydrochlorofluorocarbon (HCFC) refers to halogenated hydrocarbons that contain chlorine (Cl) and fluorine (F) in place of some hydrogen atoms in the molecule. They are chemically related to the chlorofluorocarbons (CFCs), but have less chlorine. A total of fifty-three HCFCs and CFCs are possible.

While HCFCs and CFCs have been commercially important since the mid-twentieth century, by the 1980s scientists realized the HCFCs and CFCs had a negative effect on the environment. Their growing significance has resulted from increasing concerns about the damage being done to stratospheric ozone (O_3) by CFCs.

Significant production of the CFCs began in the late 1930s. At first, they were used almost exclusively as refrigerants. Gradually, they were used in other applications that developed, especially as propellants and blowing agents. By 1970, the production of CFCs was growing by more than 10 percent per year, with a worldwide production of well over 662 million pounds (300 million kg) of one family member, CFC–11.

Environmental studies began to show, however, that CFCs decompose in the upper atmosphere.

Chlorine atoms produced in this reaction attack ozone molecules, converting them to normal oxygen (O_2). Because stratospheric ozone provides protection for humans against solar ultraviolet radiation, this finding was a source of great concern. In 1989, the Montreal Protocol, an international agreement to phase out the production of CFCs, went into effect. Every member nation of the Untied Nations has signed the Montreal Protocol.

Research to develop substitutes for CFCs began. The problem was especially severe in developing nations where CFCs are widely used in refrigeration and air-conditioning systems. Countries like China and India refused to take part in the CFC-reduction plan unless developed nations helped them switch over to an equally satisfactory substitute.

Scientists soon learned that HCFCs were a more benign alternative to the CFCs. They discovered that compounds with less chlorine than the amount present in traditional CFCs were less stable and often decomposed before they reached the stratosphere. By mid–1992, the U.S. Environmental Protection Agency (EPA) had selected eleven chemicals that they considered to be possible replacements for CFCs. Nine of those compounds are HFCs and two are HCFCs.

Both HCFCs and HFCs were developed as replacements for ozone-depleting CFCs, but it was later discovered that HCFCs also contribute to ozone depletion. CFCs, HCFCs, and HFCs all contribute to climate change. These compounds can also break down in the atmosphere into the exceptionally stable compounds such as trifluoroacetic acid (TFA) ($C_2HF_3O_2$), which have the potential to accumulate in surface waters. Computer models have shown that nearly all of the proposed substitutes will have at least some slight

CHF_3	HFC-23
$CHCl_2CF_3$	HCFC-123
CH_2FCClF_2	HCFC-133b
CH_3CHClF	HCFC-151a

Hydrochlorofluorocarbons. *(Reproduced by permission of Gale, a part of Cengage Learning)*

effect on the ozone layer or the greenhouse effect. In fact, the British government considered banning one possible substitute for CFCs, HCFC-22, almost as soon as the compound was developed. In addition, HCFC–123, was found to be carcinogenic in rats. In 1992, amendments to the Montreal Protocol, which currently has 191 participating countries, initiated the phaseout of HCFCs as part of the U.S. Clean Air Act (CAA) because of their ability to break down ozone and contribute to global warming. As of 2020, HCFC-22 will no longer be produced, but existing HCFC-22 can be used and recycled for maintenance of existing systems. In 2007 an international agreement accelerated the complete elimination of HCFCs, although developing nations were given until 2030 to attain this. Countries, such as the United States and China, who had previously resisted these changes agreed with the accelerated phaseout. The EPA lists HFC–134a and a blend of HFCs termed HFC-410a as substitutes for banned refrigerants.

See also Aerosol; Air pollution; Air pollution control; Air quality; Carcinogen; Ozone layer depletion; Pollution; Pollution control.

Resources

BOOKS

Anderson, Stephen O., and K. Madhava Sarma. *Protecting the Ozone Layer: The United Nations History*. London: Earthscan Publications, 2005.

Gillespie, Alexander. *Climate Change, Ozone Depletion, and Air Pollution*. Leiden, The Netherlands and Boston, MA: Nijboff/Brill, 2005.

Intergovernmental Panel on Climate Change. *Safeguarding the Ozone Layer and the Global Climate System: Special Report of the Intergovernmental Panel on Climate Change*. Cambridge, UK: Cambridge University Press, 2005.

Martins, John. *Ultraviolet Danger: Holes in the Ozone Layer*. New York: Rosen Publishing Group, 2006.

OTHER

United Nations Environment Programme (UNEP). "Ozone Secretariat." http://ozone.unep.org/ (accessed October 8, 2010).

David E. Newton

Hydrogen

The lightest of all chemical elements, hydrogen has a density about one-fourteenth that of air. It has a number of special chemical and physical properties.

For example, hydrogen has the second-lowest boiling and freezing points of all elements. The combustion of hydrogen produces large quantities of heat, with water as the only waste product. From an environmental standpoint, this fact makes hydrogen a highly desirable fuel. Energy experts predict that hydrogen-based energy technologies will replace some of the energy currently produced by fossil fuel technologies.

Hydrogeology

Sometimes called groundwater hydrology or geohydrology, this branch of hydrology is concerned with the relationship of subsurface water and geologic materials. Of primary interest is the saturated zone of subsurface water, called groundwater, which occurs in rock formations and in unconsolidated materials such as sands and gravels. Groundwater is studied in terms of its occurrence, amount, flow, and quality. Historically, much of the work in hydrogeology centered on finding sources of groundwater to supply water for drinking, irrigation, and municipal uses. More recently, groundwater contamination by pesticides, chemical fertilizers, toxic wastes, and petroleum and chemical spills have become new areas of concern for hydrogeologists.

Hydrologic cycle

The natural circulation of water on earth is called the hydrologic cycle. Water cycles from bodies of water, via evaporation to the atmosphere, and eventually returns to the oceans as precipitation, runoff from streams and rivers, and groundwater flow. Water molecules are transformed from liquid to vapor and back to liquid within this cycle. On land, water evaporates from the soil or is taken up by plant roots and eventually transpired into the atmosphere through plant leaves; the sum of evaporation and transpiration is called evapotranspiration.

Water is recycled continuously. The molecules of water in a glass used to quench your thirst today, at some point in time may have dissolved minerals deep in the earth as groundwater flow, fallen as rain in a tropical typhoon, been transpired by a tropical plant, been temporarily stored in a mountain glacier, or quenched the thirst of people thousands of years ago.

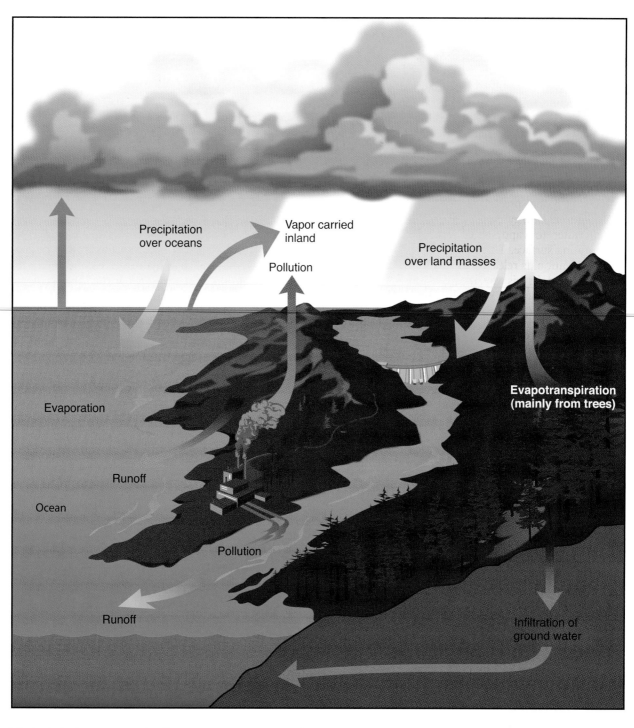

The hydrologic or water cycle. *(Reproduced by permission of Gale, a part of Cengage Learning)*

The hydrologic cycle has no real beginning or end but is a circulation of water that is sustained by solar energy and influenced by the force of gravity. Because the supply of water on earth is fixed, there is no net gain or loss of water over time. On an average annual basis, global evaporation must equal global precipitation. Likewise, for any body of land or water, changes in storage must equal the total inflow minus the total outflow of water. This is the hydrologic or water balance.

At any point in time, water on earth is either in active circulation or in storage. Water is stored in icecaps, soil, groundwater, the oceans, and other bodies of water. Much of this water is only temporarily stored. The residence time of water storage in the atmosphere is several days and is only about 0.04 percent of the total freshwater on earth. For rivers and streams, residence time is weeks; for lakes and reservoirs, several years; for groundwater, hundreds to thousands of years; for oceans, thousands of years; and for icecaps, tens of thousands of years. As the driving force of the hydrologic cycle, solar radiation provides the energy necessary to evaporate water from the earth's surface, almost three-quarters of which is covered by water. Nearly 86 percent of global precipitation originates from ocean evaporation. Energy consumed by the conversion of liquid water to vapor cools the temperature of the evaporating surface. This same energy, the latent heat of vaporization, is released when water vapor changes back to liquid. In this way, the hydrologic cycle globally redistributes heat energy as well as water.

Once in the atmosphere, water moves in response to weather circulation patterns and is transported often great distances from where it was evaporated. In this way, the hydrologic cycle governs the distribution of precipitation and, hence, the availability of fresh water over the earth's surface. About 10 percent of atmospheric water falls as precipitation each day and is simultaneously replaced by evaporation. This 10 percent is unevenly distributed over the earth's surface and, to a large extent, determines the types of ecosystems that exist at any location on earth and likewise governs much of the human activity that occurs on the land.

The earliest civilizations on earth settled in close proximity to fresh water. Subsequently, and for centuries, humans have been striving to correct, or cope with, this uneven distribution of water. Historically, we have extracted stored water or developed new storages in areas of excess, or during periods of excess precipitation, so that water could be available where and when it is most needed.

Understanding processes of the hydrologic cycle can help one develop solutions to water problems. For example, it is known that precipitation occurs unevenly over the earth's surface because of many complex factors that trigger precipitation. For precipitation to occur, moisture must be available and the atmosphere must become cooled to the dew point, the temperature at which air becomes saturated with water vapor. This cooling of the atmosphere occurs along storm fronts or in areas where moist air masses move into mountain ranges and are pushed up into colder air. However, atmospheric particles must be present for the moisture to condense upon, and water droplets must coalesce until they are large enough to fall to the earth under the influence of gravity.

Recognizing the factors that cause precipitation has resulted in efforts to create conditions favorable for precipitation over land surfaces via cloud seeding. Limited success has been achieved by seeding clouds with particles, thus promoting the condensation-coalescence process. Precipitation has not always increased with cloud seeding and questions of whether cloud seeding limits precipitation in other downwind areas is of both economic and environmental concern.

Parts of the world have abundant moisture in the atmosphere, but it occurs as fog because the mechanisms needed to transform this moisture into precipitation do not exist. In dry coastal areas, for example, some areas have no measurable precipitation for years, but fog is prevalent. By placing huge sheets of plastic mesh along coastal areas, fog is intercepted, condenses on the sheets, and provides sufficient drinking water to supply small villages.

Total rainfall alone does not necessarily indicate water abundance or scarcity. The magnitude of evapotranspiration compared to precipitation determines to some extent whether water is abundant or in short supply. On a continent basis, evapotranspiration represents from 56–80 percent of annual precipitation. For individual watersheds within continents, these percentages are more extreme and point to the importance of evapotranspiration in the hydrologic cycle.

Weather circulation patterns responsible for water shortages in some parts of the world are also responsible for excessive precipitation, floods, and related catastrophes in other parts of the world. Precipitation that falls on land, but that is not stored, evaporated, or transpired, becomes excess water. This excess water eventually reaches groundwater, streams, lakes, or the ocean by surface and subsurface flow. If the soil surface is impervious or compacted, water flows over the land surface and reaches stream channels quickly. When surface flow exceeds a channel's capacity, flash flooding is the result. Excessive precipitation can saturate soils and cause flooding no matter what the pathway of flow. For example, in August 2010 there were catastrophic floods and mudslides in the Gansu Province of China that resulted in nearly 1,800 fatalities, along with extensive property damage. Geologists and land management specialists argue that the effects of heavy rainfall were exacerbated by the removal of natural forest cover and various development programs by the Chinese government, including nearby hydroelectric power projects.

Although floods and mudslides occur naturally, many of the pathways of water flow that contribute to such occurrences can be influenced by human activity. Any time vegetative cover is severely reduced and soil exposed to direct rainfall, surface water flow and soil erosion can degrade watershed systems and their aquatic ecosystems.

The implications of global warming or greenhouse effects on the hydrologic cycle raise several questions. The possible changes in frequency and occurrence of droughts and floods are of major concern, particularly given projections of population growth. Global warming can result in some areas becoming drier, whereas others may experience higher precipitation. Globally, increased temperature will increase evaporation from oceans and ultimately result in more precipitation. The pattern of precipitation changes over the earth's surface, however, cannot be fully predicted at the present time.

The hydrologic cycle influences nutrient cycling of ecosystems, processes of soil erosion and transport of sediment, and the transport of pollutants. Water is an excellent liquid solvent; minerals, salts, and nutrients become dissolved and transported by water flow. The hydrologic cycle is an important driving mechanism of nutrient cycling. As a transporting agent, water moves minerals and nutrients to plant roots. As plants die and decay, water leaches out nutrients and carries them downstream. The physical action of rainfall on soil surfaces and the forces of running water can seriously erode soils and transport sediments downstream. Any minerals, nutrients, and pollutants within the soil are likewise transported by water flow into groundwater, streams, lakes, or estuaries.

Atmospheric moisture transports and deposits atmospheric pollutants, including those responsible for acid rain. Sulfur and nitrogen oxides are added to the atmosphere by the burning of fossil fuels. Being an excellent solvent, water in the atmosphere forms acidic compounds that become transported via the atmosphere and deposited great distances from their original site. Atmospheric pollutants and acid rain have damaged freshwater lakes in the Scandinavian countries and terrestrial vegetation in Eastern Europe. Once pollutants enter the atmosphere and become subject to the hydrologic cycle, problems of acid rain have little chance for resolution. However, programs that reduce atmospheric emissions in the first place provide some hope.

An improved understanding of the hydrologic cycle is needed to better manage water resources and the environment. Opportunities exist to improve the global environment, but better knowledge of human impacts on the hydrologic cycle is needed to avoid unwanted environmental effects.

See also Estuary; Leaching.

Resources

BOOKS

Brustaert, W. *Hydrology: An Introduction.* Cambridge, UK: Cambridge University Press, 2005.

Chang, M. *Forest Hydrology.* 2nd ed. Boca Raton, FL: CRC Press, 2006.

Darrigol, Olivier. *Worlds of Flow: A History of Hydrodynamics from the Bernoullis to Prandtl.* Oxford: Oxford University Press, 2005.

Fierro, Pedro, and Evan K. Nyer. *The Water Encyclopedia: Hydrologic Data and Internet Resources.* Boca Raton, FL: CRC/Taylor & Francis, 2007.

Gat, Joel. *Isotope Hydrology: A Study of the Water Cycle.* London: : Imperial College Press, 2010.

Hendriks, Martin R. *Introduction to Physical Hydrology.* Oxford: Oxford University Press, 2010.

Pinder, George Francis, and Michael Anthony Celia. *Subsurface Hydrology.* Hoboken, NJ: Wiley-Interscience, 2006.

Wood, Paul J., David M. Hannah, and J. P. Sadler. *Hydroecology and Ecohydrology: Past, Present, and Future.* Chichester, UK: Wiley, 2007.

OTHER

National Aeronautics and Space Administrations (NASA). "NASA Global Hydrology and Climate Center." http://www.ghcc.msfc.nasa.gov/ (accessed October 26, 2010).

U.S. Government; science.gov. "Water and Hydrology." http://www.science.gov/browse/w_119H.htm (accessed October 26, 2010).

United States Department of the Interior, United States Geological Survey (USGS). "Hydrodynamics." http://www.usgs.gov/science/science.php?term = 1678 (accessed October 26, 2010).

United States Environmental Protection Agency (EPA). "Water: Ground Water: Hydrogeology." http://www.epa.gov/ebtpages/wategroundwater hydrogeology.html (accessed October 26, 2010).

Kenneth N. Brooks

Hydrology

The science and study of water, including its physical and chemical properties and its occurrence on earth. Most commonly, hydrology encompasses the study of the amount, distribution, circulation, timing, and quality of water. It includes the study of rainfall, snow accumulation and melt, water movement over and through the soil, the flow of water in saturated, underground geologic materials (groundwater), the flow of water in channels (called streamflow), evaporation and transpiration, and the physical, chemical, and biological characteristics of water. Solving problems concerned with water excesses, flooding, water shortages, and water pollution are in the domain of hydrologists. With increasing concern about water pollution and its effects on humans and on aquatic ecosystems, the practice of hydrology has expanded into the study and management of chemical and biological characteristics of water.

Hydroponics

Hydroponics is the practice of growing plants in water as opposed to soil. It comes from the Greek terms *hydro* ("water") and *ponos* ("labor"), implying "water working." The essential macro- and micro-nutrients (i.e., trace nutrients) needed by the plants are supplied in the water.

Hydroponic methods have been used for more than 2,000 years, dating back to the Hanging Gardens of Babylon. More recently, they have been used by plant physiologists to discover which nutrients are essential for plant growth. Unlike soil, where nutrient levels are unknown and variable, precise amounts and kinds of minerals can be added to deionized water, and removed individually, to find out their role in plant growth and development. During World War II hydroponics was used to grow vegetable crops by U.S. troops stationed on some Pacific islands.

Hydroponics is becoming a more popular alternative to conventional agriculture in locations with low or inaccessible sources of water or where land available for farming is scarce. For example, islands and desert areas such as the American Southwest and the Middle East are prime regions for hydroponics. Plants are typically grown in greenhouses to prevent water loss. Even in temperate areas where fresh water is readily available, hydroponics can be used to grow crops in greenhouses during the winter months.

Two methods are traditionally used in hydroponics. The original technique is the water method, where plants are supported from a wire mesh or similar framework so that the roots hang into troughs that receive continuous supplies of nutrients. A recent modification is a nutrient-film technique (NFT), also called the nutrient-flow method, where the trough is lined with plastic. Water flows continuously over the roots, decreasing the stagnant boundary layer surrounding each root, and thus enhances nutrient uptake. This provides a versatile, light-weight, and inexpensive system. In the second method, plants are supported in a growing medium such as sterile sand, gravel, crushed volcanic rock, vermiculite, perlite, sawdust, peatmoss, or rice hulls. The nutrient solution is supplied from overhead or underneath holding tanks either continuously or semi-continuously using a drip method. The nutrient solution is usually not reused.

On some Caribbean Islands such as St. Croix, hydroponics is being used in conjunction with intensive fish farms (e.g., tilapia), which use recirculated water (a practice that is more recently known as aquaponics). This is a "win-win" situation because the nitrogenous wastes, which are toxic to the fish, are passed through large greenhouses with hydroponically-grown plants such as lettuce. The plants remove the nutrients and the water is returned to the fish tanks. There is a sensitive balance between stocking density of fish and lettuce production. Too high a ratio of lettuce plants to fish results in lower lettuce production because of nutrient limitation. Too low a ratio also results in low vegetable production, but this time as a result of the buildup of toxic chemicals. The optimum yield came from a ratio of 1.9 lettuce plants to 1.0 fish. One pound (0.45 kg) of feed per day was appropriate to feed 33 pounds (15 kg) of tilapia fingerlings, which sustained 189 lettuce plants and produced nearly thirty-three hundred heads of lettuce annually. When integrated systems (fish-hydroponic recirculating units) are compared to separate production systems, the results clearly favor the former. The combined costs and chemical requirements of the separate production systems was nearly two to three times greater than that of the recirculating system to produce the same amount of lettuce and fish. However, there are some drawbacks that must be considered: disease outbreaks in plants or fish; the need to critically maintain proper nutrient (especially trace element), plant, and fish levels; uncertainties in fish and market prices; and the need for highly-skilled labor. The integrated method can be adapted to grow other types of vegetables such as strawberries, ornamental plants such as roses, and other types of animals such as shellfish. Some teachers have even incorporated this technique into their classrooms to illustrate ecological as well as botanical and culture principles.

Some proponents of hydroponic gardening make fairly optimistic claims and state that a sophisticated unit is no more expensive than an equivalent parcel of farmed land. They also argue that hydroponic units (commonly called "hydroponicums") require less attention than terrestrial agriculture. What follows are some examples of different types of successful hydroponicums: a person in the desert area of southern California has used the NFT system for over eighteen years and grows his plants void of substate in water contained in open cement troughs that cover 3

Since refugia survived the past dry-climate phases, they have traditionally supplied the plants and animals for the restocking of the new-growth forests when wet conditions returned. Modern deforestation patterns, however, do not take into account forest history or biodiversity, and both forest refugia and more recent forests are being destroyed equally. For the first time in millions of years, future tropical forests that survive the present mass deforestation episode could have no species-rich centers from which they can be restocked.

See also Biotic community; Deciduous forest; Desertification; Ecosystem; Environment; Mass extinction.

Resources

BOOKS

Holzman, Barbara A. *Tropical Forest Biomes.* Westport, CT: Greenwood Press, 2008.

PERIODICALS

Stone, Roger D. "Tomorrow's Amazonia: as Farming, Ranching, and Logging Shrink the Globe's Great Rainforest, the Planet Heats Up." *The American Prospect* 18 (September 9, 2007): A2.

Weaver, Andrew J. and C. Hillaire-Marcel. "Global Warming and the Next Ice Age." *Science* 304 (2004): 400–402.

Neil Cumberlidge

Impervious material

As used in hydrology, this term refers to rock and soil material that occurs at the earth's surface or within the subsurface that does not permit water to enter or move through them in any perceptible amounts. These materials normally have small-sized pores or have pores that have become clogged (sealed), which severely restrict water entry and movement. At the ground surface, rock outcrops, road surfaces, or soil surfaces that have been severely compacted would be considered impervious. These areas shed rainfall easily, causing overland flow or surface runoff which pick up and transport soil particles and cause excessive soil erosion. Soils or geologic strata beneath the earth's surface are considered impervious, or impermeable, if the size of the pores is small and/or if the pores are not connected.

Improvement cutting

Removal of physically distorted (e.g., crooked and forked) trees or diseased trees from a forest in which tree diameters are 5 inches (13 cm) or larger. In forests where tree diameters are smaller, the same process is called cleaning or weeding. Both have the objective of improving species composition, stem quality and/or growth rate of the forest. Straight, healthy, vigorous trees of the desired species are favored. By discriminating against certain tree species and eliminating trees with cavities or insect problems, improvement cuts can reduce the variety of habitats and thereby diminish biodiversity. An improvement cut is often an initial step to prepare a neglected or unmanaged stand for future harvest.

See also Clear-cutting; Forest management; Selection cutting.

Inbreeding

Inbreeding occurs when closely related individuals mate with one another. Inbreeding may happen in a small population or due to other isolating factors; the consequence is that little new genetic information is added to the gene pool. Thus recessive, deleterious alleles become more plentiful and evident in the population. Manifestations of inbreeding are known as *inbreeding depression*. A general loss of fitness often results and may cause high infant mortality, lower birth weights, fecundity, and longevity. Inbreeding depression is a major concern when attempting to protect small populations from extinction.

Incidental catch *see* **Bycatch.**

Incineration

As a method of waste management, incineration refers to the burning of waste. It helps reduce the volume of landfill material and can render toxic substances non-hazardous, provided certain strict guidelines are followed. There are two basic types of incineration: municipal and hazardous waste incineration.

Municipal waste incineration

The process of incineration involves the combination of organic compounds in solid wastes with oxygen at high temperature to convert them to ash and gaseous products. A municipal incinerator consists of a series of unit operations that include a loading area under slightly negative pressure to avoid the escape of odors,

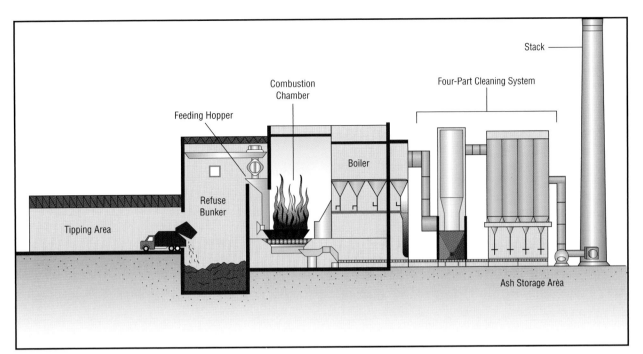

Diagram of a municipal incinerator. *(Reproduced by permission of Gale, a part of Cengage Learning)*

a refuse bin that is loaded by a grappling bucket, a charging hopper leading to an inclined feeder and a furnace of varying type—usually of a horizontal burning grate type—a combustion chamber equipped with a bottom ash and clinker discharge, followed by a gas flue system to an expansion chamber. If byproduct stream is to be produced either for heating or power generation purposes, then the downstream flue system includes heat exchanger tubing as well. After the heat has been exchanged, the flue gas proceeds to a series of gas cleanup systems which neutralizes the acid gases (sulfur dioxide and hydrochloric acid, the latter resulting from burning chlorinated plastic products), followed by gas scrubbers and then solid/gas separation systems such as baghouses before dischargement to tall stacks. The stack system contains a variety of sensing and control devices to enable the furnace to operate at maximum efficiency consistent with minimal particulate emissions. A continuous log of monitoring systems is also required for compliance with county and state environmental quality regulations.

There are several products from a municipal incinerator system: items that are removed before combustion such as large metal pieces; grate or bottom ash (which is usually water-sprayed after removal from the furnace for safe storage); fly (or top ash), which is removed from the flue system generally mixed with products from the acid neutralization process; and

finally the flue gases that are expelled into the environment. If the system is operating optimally, the flue gases will meet emission requirements, and the heavy metals from the wastes will be concentrated in the fly ash. (Typically these heavy metals, which originate from volatile metallic constituents, are lead and arsenic.) The fly ash typically is then stored in a suitable landfill to avoid future problems of leaching of heavy metals. Some municipal systems blend the bottom ash with the top ash in the plant in order to reduce the level of heavy metals by dilution. This practice is undesirable from an ultimate environmental viewpoint.

There are advantages to municipal waste incineration. The waste volume is reduced to a small fraction of the original. Reduction is rapid and does not require semi-infinite residence times in a landfill. For a large metropolitan area, waste can be incinerated on site, minimizing transportation costs. The ash residue is generally sterile, although it may require special disposal methods. By use of gas clean-up equipment, discharges of flue gases to the environment can meet stringent requirements and be readily monitored. Incinerators are much more compact than landfills and can have minimal odor and vermin problems if properly designed. Finally, some of the costs of operation can be reduced by heat-recovery techniques such as the sale of steam to municipalities or electrical energy generation.

There are disadvantages to municipal waste incineration as well. Generally, the implementation cost is high and is escalating as emission standards change. For example, an incineration plant proposed for Durham, Ontario, Canada, will, if approved, cost an estimated $275 million to build and require about $15 million annually to keep it running. Permitting requirements are becoming increasingly more difficult to obtain. Supplemental fuel may be required to burn municipal wastes, especially if yard waste is not removed prior to collection. Certain items such as mercury-containing batteries can produce emissions of mercury that the gas cleanup system may not be designed to remove. Continuous skilled operation and close maintenance of process control is required, especially since stack monitoring equipment reports any failure of the equipment which could result in mandated shut down. Certain materials are not burnable and must be removed at the source. Traffic to and from the incinerator can be a problem unless timing and routing are carefully managed. The incinerator, like a landfill, also has a limited life, although its lifetime can be increased by capital expenditures. Finally, incinerators also require landfills for the ash. The ash usually contains heavy metals and must be placed in a specially designed landfill to avoid leaching.

Incineration of municipal waste is controversial. For example, as of April 2010, an incineration-based power plant near Toronto, Ontario, was approved by the provincial government. The plan was actively opposed by those who would be nearby the plant and those downwind. Opposition centered around the possible health hazards of the emissions if inhaled and when toxic particulates settle on soil or water.

Hazardous waste incineration

For the incineration of hazardous waste, a greater degree of control, higher temperatures, and a more rigorous monitoring system are required. An incinerator burning hazardous waste must be designed, constructed, and maintained to meet Resource Conservation and Recovery Act (RCRA) standards. An incinerator burning hazardous waste must achieve a destruction and removal efficiency of at least 99.99 percent for each principal organic hazardous constituent. For certain listed constituents such as polychlorinated biphenyl (PCB), mass air emissions from an incinerator are required to be greater than 99.9999 percent. The Toxic Substances Control Act requires certain standards for the incineration of PCBs. For example, the flow of PCB to the incinerator must stop automatically whenever the combustion temperature drops below the specified value; there must be continuous monitoring of the stack for a list of emissions; and scrubbers must be used for hydrochloric acid control.

Recently medical wastes have been treated by steam sterilization, followed by incineration with treatment of the flue gases with activated carbon for maximum absorption of organic constituents. The latter system was installed at the Mayo Clinic in Rochester, Minnesota, as a model medical disposal system.

See also Fugitive emissions; Solid waste incineration; Solid waste volume reduction; Stack emissions.

Resources

BOOKS

Auyero, Javier, and Débora Alejandra Swistun. *Flammable: Environmental Suffering in an Argentine Shantytown.* New York: Oxford University Press USA, 2009.
Deverell, William, and Greg Hise. *Land of Sunshine: An Environmental History of Metropolitan Los Angeles.* Pittsburgh, PA: University of Pittsburgh Press, 2006.
Meuser, Helmut. *Contaminated Urban Soils.* New York: Springer, 2010.

Malcolm T. Hepworth

Incineration, solid waste *see* **Solid waste incineration.**

Indicator organism

Indicator organisms, sometimes called bioindicators, are plant or animal species known to be either particularly tolerant or particularly sensitive to pollution. The health of an organism can often be associated with a specific type or intensity of pollution, and its presence can then be used to indicate polluted conditions relative to unimpacted conditions.

Tubificid worms are an example of organisms that can indicate pollution. Tubificid worms live in the bottom sediments of streams and lakes, and they are highly tolerant of sewage. In a river polluted by wastewater discharge from a sewage treatment plant, it is common to see a large increase in the number of tubificid worms in stream sediments immediately downstream. Upstream of the discharge, the number of tubificid worms are often much lower or almost absent, reflecting cleaner conditions. The number of tubificid worms also decreases downstream, as the discharge is diluted.

Pollution-intolerant organisms can also be used to indicate polluted conditions. The larvae of mayflies live in stream sediments and are known to be particularly sensitive to pollution. In a river receiving wastewater discharge, mayflies will show the opposite pattern of tubificid worms. The mayfly larvae are normally present in large numbers above the discharge point; they decrease or disappear at the discharge point and reappear farther downstream as the effects of the discharge are diluted.

Similar examples of indicator organisms can be found among plants, fish, and other biological groups. Giant reedgrass (*Phragmites australis*) is a common marsh plant that is typically indicative of disturbed conditions in wetlands. Among fish, disturbed conditions may be indicated by the disappearance of sensitive species like trout, which require clear, cold waters to thrive.

The usefulness of indicator organisms is limited. Although their presence or absence provides a reliable general picture of polluted conditions, they are often little help in identifying the exact sources of pollution. In the sediments of New York Harbor, for example, pollution-tolerant insect larvae are overwhelmingly dominant. However, it is impossible to attribute the large larval populations to just one of the sources of pollution there, which include ship traffic, sewage and industrial discharge, and storm runoff.

The U.S. Environmental Protection Agency (EPA) is working diligently to find reliable predictors of aquatic ecosystem health using indicator species. The EPA has developed standards for the usefulness of species as ecological indicator organisms. A potential indicator species for use in evaluating watershed health must successfully pass four phases of evaluation. First, a potential indicator organism should provide information that is relevant to societal concerns about the environment, not simply academically interesting information. Second, use of a potential indicator organism should be feasible. Logistics, sampling costs, and timeframe for information gathering are legitimate considerations in deciding whether an organism is a potential indicator species or not. Third, enough must be known about a potential species before it may be effectively used as an indicator organism. Sufficient knowledge regarding the natural variations to environmental flux should exist before incorporating a species as a true watershed indicator species. Lastly, the EPA has set a fourth criterion for evaluation of indicator species. A useful indicator should provide information that is easily interpreted by policy makers and the public, in addition to scientists.

Additionally, in an effort to make indicator species information more reliable, the creation of indicator species indices are being investigated. An index is a formula or ratio of one amount to another that is used to measure relative change. The major advantage of developing an indicator organism index that is somewhat universal to all aquatic environments is that it can be tested using statistics. Using mathematical statistical methods, it may be determined whether a significant change in an index value has occurred. Furthermore, statistical methods allow for a certain level of confidence that the measured values represent what is actually happening in nature. For example, a study was conducted to evaluate the utility of diatoms (a kind of microscopic aquatic algae) as an index of aquatic system health. Diatoms meet all four criteria mentioned earlier, and various species are found in both fresh and salt water. An index was created that was calculated using various measurable characteristics of diatoms that could then be evaluated statistically over time and among varying sites. It was determined that the diatom index was sensitive enough to reliably reflect three categories of the health of an aquatic ecosystem. The diatom index showed that values obtained from areas impacted by human activities had greater variability over time than diatom indices obtained from less disturbed locations. Many such indices are being developed using different species, and multiple species in an effort to create reliable information from indicator organisms. As more is learned about the physiology and life history of indicator organisms and their individual responses to different types of pollution, it may be possible to draw more specific conclusions.

See also Algal bloom; Nitrogen cycle; Water pollution.

Resources

OTHER

United States Department of the Interior, United States Geological Survey (USGS). "Freshwater Ecosystems." http://www.usgs.gov/science/science.php?term=419 (accessed October 10, 2010).

United States Environmental Protection Agency (EPA). "Ecosystems: Aquatic Ecosystems: Freshwater Ecosystems." http://www.epa.gov/ebtpages/ecosaquaticecosystfreshwaterecosystems.html (accessed October 10, 2010).

United States Environmental Protection Agency (EPA). "Ecosystems: Ecological Monitoring: Environmental Indicators." http://www.epa.gov/ebtpages/ecosecologicalmoni environmentalindicators.html (accessed October 10, 2010).

Terry Watkins

Indigenous peoples

Cultural or ethnic groups living in an area where their culture developed or where their people have existed for many generations. Most of the world's indigenous peoples live in remote forests, mountains, deserts, or arctic tundra, where modern technology, trade, and cultural influence are slow to penetrate. Many had much larger territories historically but have retreated to, or been forced into, small, remote areas by the advance of more powerful groups. Indigenous groups, also sometimes known as native or tribal peoples, are usually recognized in comparison to a country's dominant cultural group. In the United States the dominant, non-indigenous cultural groups speak English, have historic roots in Europe, and maintain strong economic, technological, and communication ties with Europe, Asia, and other parts of the world. Indigenous groups in the United States, on the other hand, include scores of groups, from the southern Seminole and Cherokee to the Inuit and Yupik peoples of the Arctic coast. These groups speak hundreds of different languages or dialects, some of which have been on their continent for thousands of years. Their traditional economies were based mainly on small-scale subsistence gathering, hunting, fishing, and farming. Many indigenous peoples around the world continue to engage in these ancient economic practices.

It is often difficult to distinguish who is and who is not indigenous. European-Americans and Asian-Americans are usually not considered indigenous even if they have been in the United States for many generations. This is because their cultural roots connect to other regions. On the other hand, a German residing in Germany is also not usually spoken of as indigenous, even though by any strict definition she or he *is* indigenous. This is because the term is customarily reserved to denote economic or political minorities—groups that are relatively powerless within the countries where they live.

Historically, indigenous peoples have suffered great losses in both population and territory to the spread of larger, more technologically advanced groups, especially (but not only) Europeans. Hundreds of indigenous cultures have disappeared entirely just in the past century. In recent decades, however, indigenous groups have begun to receive greater international recognition, and they have begun to learn effective means to defend their lands and interests—including attracting international media attention and suing their own governments in court. The main reason for this increased attention and success may be that scientists and economic development organizations have recently become interested in biological diversity and in the loss of world rain forests. The survival of indigenous peoples, of the world's forests, and of the world's gene pools are now understood to be deeply interdependent. Indigenous peoples, who know and depend on some of the world's most endangered and

biologically diverse ecosystems, are increasingly looked on as a unique source of information, and their subsistence economies are beginning to look like admirable alternatives to large-scale logging, mining, and conversion of jungles to monocrop agriculture.

There are probably between 4,000 and 5,000 different indigenous groups in the world; they can be found on every continent (except Antarctica) and in nearly every country. The total population of indigenous peoples amounts to between 200 million and 600 million (depending upon how groups are identified and their populations counted) out of a world population just over 6.69 billion. Some groups number in the millions; others comprise only a few dozen people. Despite their worldwide distribution, indigenous groups are especially concentrated in a number of "cultural diversity hot spots," including Indonesia, India, Papua New Guinea, Australia, Mexico, Brazil, Democratic Republic of the Congo, Cameroon, and Nigeria. Each of these countries has scores, or even hundreds, of different language groups. Neighboring valleys in Papua New Guinea often contain distinct cultural groups with unrelated languages and religions. These regions are also recognized for their unusual biological diversity.

Industrial economies involved in international trade consume tremendous amounts of land, wood, water, and minerals. Indigenous groups tend to rely on intact ecosystems and on a tremendous variety of plant and animal species. Indigenous lands hold some of the richest remaining reserves of natural resources worldwide. Frequently state governments claim all timber, mineral, water, and land rights in areas traditionally occupied by tribal groups. In Indonesia, Malaysia, Burma (Myanmar), China, Brazil, Democratic Republic of the Congo Cameroon, and many other important cultural diversity regions, timber and mining concessions are frequently sold to large or international companies that can quickly and efficiently destroy an ecological area and its people.

Many indigenous communities, including various Arctic peoples, support international policy measures to reduce pollution, curb greenhouse gas emissions, and combat global climate change. Arctic indigenous leaders have spoken about the impact of warmer winters and less dense ice packs on life in the Arctic region. These indigenous groups assert that climate change threatens not only their local ecosystem, but also their way of life.

Resources

BOOKS

Johnston, Alison M. *Is the Sacred for Sale?: Tourism and Indigenous Peoples.* London, UK, and Sterling, VA: Earthscan, 2006.

OTHER

World Health Organization (WHO). "Health of Indigenous Peoples." http://www.who.int/entity/mediacentre/factsheets/fs326/en/index.html (accessed October 10, 2010).

World Health Organization (WHO). "Indigenous Population." http://www.who.int/topics/health_services_indigenous/en (accessed October 10, 2010).

Mary Ann Cunningham

Indonesian forest fires

For several months in 1997 and 1998, a thick pall of smoke covered much of Southeast Asia. Thousands of forest fires, burning simultaneously on the Indonesian islands of Kalimantan (Borneo) and Sumatra, are thought to have destroyed about 8,000 square miles (20,000 sq km) of primary forest, or an area about the size of New Jersey. The smoke generated by these fires spread over eight countries and 75 million people, covering an area larger than Europe. Hazy skies and the smell of burning forests could be detected in Hong Kong, nearly 2,000 miles (3,200 km) away. The air quality in Singapore and the city of Kuala Lumpur, Malaysia, just across the Strait of Malacca from Indonesia, was worse than any industrial region in the world. In towns such as Palembang, Sumatra, and Banjarmasin, Kalimantan, in the heart of the fires, the air pollution index frequently passed 800, twice the level classified in the United States as an air quality emergency that is hazardous to human health. People drove their automobiles with the headlights on, even at noon. They groped along smoke-darkened streets unable to see or breathe normally. The damage from the fires in 1997 and 1998 is estimated at about $9 billion, affecting tourism, environment, health, and other activities.

At least 20 million people in Indonesia and Malaysia were treated for illnesses such as bronchitis, eye irritation, asthma, emphysema, and cardiovascular diseases. It is thought that three times that many who could not afford medical care went uncounted. The

A firefighter battles a blaze in the sub-district of Ilir Barat outside Palembang city, South Sumatra, in September 2006.
(© STRINGER/INDONESIA/Reuters/Corbis)

number of extra deaths from this months-long episode is unknown, but it seems likely to have been hundreds of thousands, mostly elderly or very young children. Unable to see through the thick haze, several boats collided in the busy Straits of Malacca, and a plane crashed on Sumatra, killing 234 passengers. Cancelled airline flights, aborted tourist plans, lost workdays, medical bills, and ruined crops are estimated to have cost countries in the afflicted area several billion dollars. Wildlife suffered as well. In addition to the loss of habitat destroyed by fires, breathing the noxious smoke was as hard on wild species as it was on people. At the Pangkalanbuun Conservation Reserve, weak and disoriented orangutans were found suffering from respiratory diseases much like those of humans.

Geographical isolation on the 16,000 islands of the Indonesian archipelago has allowed evolution of the world's richest collection of biodiversity. Indonesia has the second largest expanse of tropical forest and the highest number of endemic species anywhere. This fact makes destruction of Indonesian plants, animals, and their habitat of special concern. The dry season in tropical Southeast Asia has probably always been a time of burning vegetation and smoky skies. Farmers practicing traditional slash-and-burn agriculture start fires each year to prepare for the next growing season. Because they generally burn only a hectare or two at a time, however, these shifting cultivators often help preserve plant and animal species by opening up space for early successional forest stages. Globalization and the advent of large, commercial plantations, however, have changed agricultural dynamics. There is now economic incentive for clearing huge tracts of forestland to plant oil palms, export foods such as pineapples and sugar cane, and fast-growing eucalyptus trees. Fire is viewed as the only practical way to remove biomass and convert wild forest into domesticated land. Although it can cost the equivalent of $200 to clear a hectare of forest with chainsaws and bulldozers, dropping a lighted match into dry underbrush is essentially free.

In 1997 to 1998, the Indonesian forest was unusually dry. A powerful El Niño/Southern Oscillation (ENSO) weather pattern caused the most severe droughts in fifty years. Forests that ordinarily stay green and moist even during the rainless season became tinder dry. Lightning strikes are thought to have started many forest fires, but many people took advantage of the drought for their own purposes. Although the government blamed traditional farmers for setting most of the fires, environmental groups claimed that the biggest fires were caused by large agribusiness conglomerates with close ties to the government and military. Some of these fires were set to

cover up evidence of illegal logging operations. Others were started to make way for huge oil-palm plantations and fast-growing pulpwood trees.

Neil Byron of the Center for International Forestry Research was quoted as saying that "fire crews would go into an area and put out the fire, then come back four days later and find it burning again, and a guy standing there with a petrol can." According to the World Wildlife Fund (WWF), thirty-seven plantations in Sumatra and Kalimantan were responsible for a vast majority of the forest burned on those islands. The plantation owners were politically connected to the ruling elite, however, and none of them was ever punished for violation of national forest protection laws. Indonesia has some of the strongest land-use management laws of any country in the world, but these laws are rarely enforced. In theory, more than 80 percent of its land is in some form of protected status, either set aside as national parks or classified as selective logging reserves where only a few trees per hectare can be cut. The government claims to have an ambitious reforestation program that replants nearly 1.6 million acres (1 million hectares) of harvested forest annually, but when four times that amount is burned in a single year, there is not much to be done but turn it over to plantation owners for use as agricultural land.

Aquatic life, also, is damaged by these forest fires. Indonesia, Malaysia, and the Philippines have some of the richest coral reef complexes in the world. More than 150 species of coral live in this area, compared with only about thirty species in the Caribbean. The clear water and fantastic biodiversity of Indonesia's reefs have made it an ultimate destination for scuba divers and snorkelers from around the world. However, soil eroded from burned forests clouds coastal waters and smothers reefs.

Perhaps one of the worst effects of large tropical forest fires is that they may tend to be self-reinforcing. Moist tropical forests store huge amounts of carbon in their standing biomass. When this carbon is converted into carbon dioxide (CO_2) by fire and released to the atmosphere, it acts as a greenhouse gas to trap heat and cause climate change. All the effects of human-caused global climate change are still unknown, but stronger climatic events such as severe droughts may make further fires even more likely. Alarmed by the magnitude of the Southeast Asia fires and the potential they represent for biodiversity losses and global climate change, world leaders have proposed plans for international intervention to prevent recurrence. Fears about imposing on national sovereignty, however, have made it difficult to come up with a plan for how to cope with this growing threat.

The fires have continued each year since 1997. In 2002 the Association of Southeast Asian Nations (ASEAN) countries signed an agreement termed the ASEAN Agreement on Transboundary Haze Pollution. This agreement focuses on the countries taking measures to control haze and preventing the haze from affecting other countries. Despite the agreement, the fires and haze continue each year during the dry season, with burning being particularly rampant in 2006. As of 2010, Indonesia had not ratified the agreement, meaning that the control provisions are not binding. Even with attempts at regional agreements over the fires, the problems persist. The haze in Southeast Asia seems almost an annual event, and 2008 was no different, when in April the growing number of forest fires from farming activities in Sumatra created concern in Malaysia.

Resources

BOOKS

Glover, David, and Timothy Jessup. *Indonesia's Fires and Haze: The Cost of Catastrophe*. Ottawa: International Development Research Centre, 2006. http://epe.lac-bac.gc.ca/100/200/301/idrc-crdi/indonisia_fire_haze-e/index.html

Gunn, Angus M. *Encyclopedia of Disasters: Environmental Catastrophes and Human Tragedies*. Westport, CT: Greenwood Press, 2008.

Jayachandran, Seema. *Air Quality and Early-Life Mortality: Evidence from Indonesia's Wildfires*. Cambridge, MA: National Bureau of Economic Research, 2008.

World Bank. *Indonesia and Climate Change Current Status and Policies (Indonesia dan perubahan iklim: status terkini dan kebijakannya)*. Jakarta: The World Bank, 2007.

William P. Cunningham

Indoor air quality

An assessment of air quality in buildings and homes based on physical and chemical monitoring of contaminants, physiological measurements, or psychosocial perceptions. Indoor air quality measures can also refer to the air in types of transportation such as airplanes, trains, buses, and passenger cars. Factors contributing to the quality of indoor air include lighting, ergonomics, thermal comfort, tobacco smoke, noise, ventilation, and psychosocial or work-organizational factors such as employee stress and satisfaction. Sick building syndrome (SBS) and building-related illness (BRI) are responses to indoor air pollution commonly described by office workers. Most symptoms are nonspecific; they progressively worsen during the week, occur more frequently in the afternoon, and disappear on the weekend.

Poor indoor air quality (IAQ) in industrial settings such as factories, coal mines, and foundries has long been recognized as a health risk to workers and has been regulated by the U.S. Occupational Safety and Health Administration (OSHA). The contaminant levels in industrial settings can be hundreds or thousands of times higher than the levels found in homes and offices. Nonetheless, IAQ in homes and offices has become an environmental priority in many countries, and federal IAQ legislation has been introduced in the U.S. Congress for the past several years. However, none has yet passed, and currently the U.S. Environmental Protection Agency (EPA) has no enforcement authority in this area. The EPA and the National Institute for Occupational Safety and Health (NIOSH) have developed a Building Air Quality Action plan with guidelines for improving and maintaining IAQ in public and commercial buildings.

Importance of IAQ

The prominence of IAQ issues has risen in part due to well-publicized incidents involving outbreaks of Legionnaires' disease, Pontiac fever, SBS, multiple chemical sensitivity, and asbestos mitigation in public buildings such as schools. Legionnaire's disease, for example, caused twenty-nine deaths in 1976 in a Philadelphia hotel due to infestation of the building's air conditioning system by a bacterium termed *Legionella pneumophila*. This microbe affects the gastrointestinal tract, kidneys, and central nervous system. It also causes the non-fatal Pontiac fever.

IAQ is important to the general public for several reasons. First, individuals typically spend the vast majority of their time—80–90 percent—indoors. Second, an emphasis on energy conservation measures, such as reducing air exchange rates in ventilation systems and using more energy efficient but synthetic materials, has increased levels of air contaminants in offices and homes. These newer buildings are more airtight and have fewer cracks and openings so minimal fresh air enters such buildings. Low ventilation and exchange rates can result in increased indoor levels of carbon monoxide (CO), nitrogen oxides (NO_x), ozone (O_3), volatile organic compounds, bioaerosols, and pesticides and maintain high levels of second-hand tobacco smoke generated inside the building. Thus, many contaminants are found indoors at levels that greatly exceed outdoor levels. Third, an increasing number of synthetic chemicals—found in

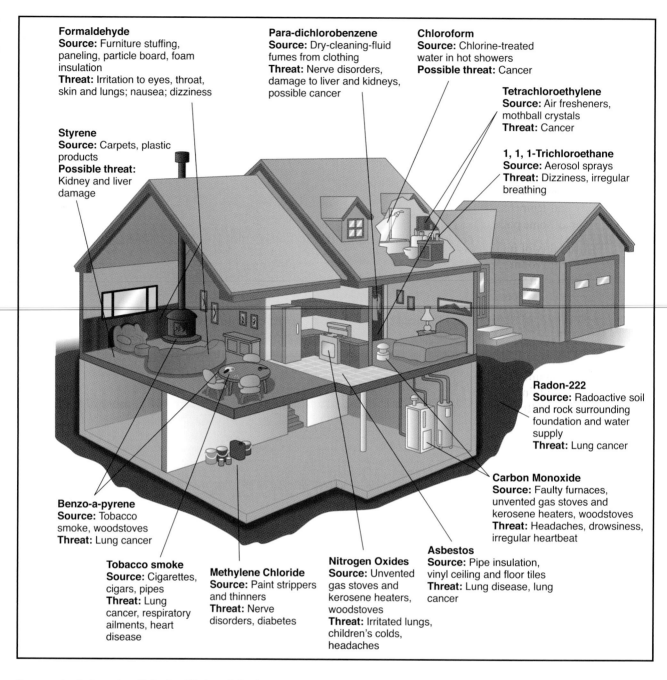

Formaldehyde
Source: Furniture stuffing, paneling, particle board, foam insulation
Threat: Irritation to eyes, throat, skin and lungs; nausea; dizziness

Para-dichlorobenzene
Source: Dry-cleaning-fluid fumes from clothing
Threat: Nerve disorders, damage to liver and kidneys, possible cancer

Chloroform
Source: Chlorine-treated water in hot showers
Possible threat: Cancer

Tetrachloroethylene
Source: Air fresheners, mothball crystals
Threat: Cancer

1, 1, 1-Trichloroethane
Source: Aerosol sprays
Threat: Dizziness, irregular breathing

Styrene
Source: Carpets, plastic products
Possible threat: Kidney and liver damage

Radon-222
Source: Radioactive soil and rock surrounding foundation and water supply
Threat: Lung cancer

Carbon Monoxide
Source: Faulty furnaces, unvented gas stoves and kerosene heaters, woodstoves
Threat: Headaches, drowsiness, irregular heartbeat

Benzo-a-pyrene
Source: Tobacco smoke, woodstoves
Threat: Lung cancer

Tobacco smoke
Source: Cigarettes, cigars, pipes
Threat: Lung cancer, respiratory ailments, heart disease

Methylene Chloride
Source: Paint strippers and thinners
Threat: Nerve disorders, diabetes

Nitrogen Oxides
Source: Unvented gas stoves and kerosene heaters, woodstoves
Threat: Irritated lungs, children's colds, headaches

Asbestos
Source: Pipe insulation, vinyl ceiling and floor tiles
Threat: Lung disease, lung cancer

Some major indoor air pollutants. *(Wadsworth Inc.)*

building materials, furnishing, cleaning and hygiene products—are used indoors. Fourth, studies show that exposure to indoor contaminants such as radon (Rn), asbestos, and tobacco smoke pose significant health risks. Fifth, poor IAQ is thought to adversely affect child development and lower productivity in the adult population. Demands for IAQ investigations of problem buildings have increased rapidly in recent years, and a large fraction of buildings are known or suspected to have IAQ problems.

Indoor contaminants

Indoor air contains many contaminants at varying but generally low concentration levels. Common contaminants include radon and radon progeny from the entry of soil gas and groundwater; from concrete and other mineral-based building materials; tobacco smoke from cigarette and pipe smoking; formaldehyde from polyurethane foam insulation and building materials; volatile organic compounds (VOCs) emitted from binders and resins in carpets, furniture, or building materials, as well as VOCs used in dry cleaning processes and as propellants and constituents of personal use and cleaning products, such as hair sprays and polishes; pesticides and insecticides; carbon monoxide, nitrogen oxides, and other combustion productions from gas stoves, appliances, and vehicles; asbestos from high temperature insulation; and biological contaminants including viruses, bacteria, molds, pollen, dust mites, and indoor and outdoor biota. Many or most of these contaminants are present at low levels in all indoor environments. High humidity and temperatures can result in increased concentration of some pollutants.

The quality of indoor air can change rapidly over time and from room to room. There are many diverse sources that emit various physical and chemical forms of contaminants. Some releases are slow and continuous, such as out-gassing associated with building and furniture materials, while others are nearly instantaneous, like the use of cleaners and aerosols. Many building surfaces demonstrate significant interactions with contaminants in the form of sorption-desorption processes. Building-specific variation in air exchange rates, mixing, filtration, building and furniture surfaces, and other factors alter dispersion mechanisms and contaminant lifetimes. Most buildings employ filters that can remove particles and aerosols. Filtration systems do not effectively remove very small particles and have no effect on gases, vapors, and odors.

Outdoor air enters indoor spaces via three different methods termed infiltration, natural ventilation, and mechanical ventilation. In infiltration, air enters through cracks and openings around windows, flooring, doors, and walls. Natural ventilation is achieved by opening windows and doors. Mechanical ventilation can be accomplished through the use of outdoor-vented fans or air handling systems that utilize fans and ducts to remove air from inside and to introduce outdoor air inside. By removing indoor air and replacing it with outdoor air, it is possible to remove contaminants or pollutants that have built up inside. Ventilation and air exchange units designed into the heating and cooling systems of buildings are designed to diminish levels of these contaminants by dilution. In most buildings, however, ventilation systems are turned off at night after working hours, leading to an increase in contaminants through the night. Though operation and maintenance issues are estimated to cause the bulk of IAQ problems, deficiencies in the design of the heating, ventilating, and air conditioning (HVAC) system can cause problems as well. For example, locating a building's fresh air intake near a truck loading dock will bring diesel fumes and other noxious contaminants into the building.

Health impacts

Exposures to indoor contaminants can cause a variety of health problems. Depending on the pollutant and exposure, health problems related to IAQ may include non-malignant respiratory effects, including mucous membrane irritation, allergic reactions, and asthma; cardiovascular effects; infectious diseases such as Legionnaires' disease; immunologic diseases such as hypersensitivity pneumonitis; skin irritations; malignancies; neuropsychiatric effects; and other non-specific systemic effects such as lethargy, headache, and nausea. In addition indoor air contaminants such as radon, formaldehyde (CH_2O), asbestos, and other chemicals are suspected or known carcinogens. There is also growing concern over the possible effects of low level exposures on suppressing reproductive and growth capabilities and impacting the immune, endocrine, and nervous systems.

Solving IAQ problems

Acute IAQ problems can be greatly eliminated by identifying, evaluating, and controlling sources of contaminants. IAQ control strategies include use of higher ventilation and air exchange rates, and the use of lower emission and more benign constituents in building and consumer products (including product use restriction regulations), air cleaning and filtering, and improved building practices in new construction. According to the EPA, the maximum acceptable amount of radon in a building is four picocuries per liter of air (pCi/L). Radon may be reduced by inexpensive sub-slab ventilation systems, which consist of a system of pipes and fans that originates below the home to collect the radon and vent it to the outside air. Installation of a heat recovery ventilator (HRV) can increase ventilation to decrease radon levels. There are also systems that pump air inside to increase the air pressure in the lower levels of a building preventing radon from entering. New buildings could implement a day of bake-out, which heats the building to temperatures over 90°F (32°C) to drive out volatile organic compounds. Filters to remove ozone, organic compounds, and sulfur (S) gases may be used to condition incoming and recirculated air. Copy machines and other emission sources should have special ventilation systems. Building designers, operators, contractors, maintenance personnel, and occupants are recognizing that healthy buildings result from combined and continued efforts to control emission sources, provide adequate ventilation and air cleaning, and good maintenance of building

Under INFOTERRA, participating nations designate institutions to be national focal points, such as the Environmental Protection Agency (EPA) in the United States. Each national institution chosen as a focal point prepares a list of its national environmental experts and selects what it considers the best sources for inclusion in INFOTERRA's international directory of experts.

INFOTERRA initially used its directory only to refer questioners to the nearest appropriate experts, but the organization has evolved into a central information agency. It consults sources, answers public queries for information, and analyzes the replies. INFOTERRA is used by governments, industries, and researchers in 177 countries.

Resources

ORGANIZATIONS

UNEP-Infoterra/USA, 1200 Pennsylvania Ave., NWMC3404T, Washington, DC, USA 20460, (202) 566-0544, library-infoterra@epa.gov

Linda Rehkopf

Injection well

Injection wells are used to dispose waste into the subsurface zone. These wastes can include brine from oil and gas wells, liquid hazardous wastes, agricultural and urban runoff, municipal sewage, and return water from air-conditioning. Recharge wells can also be used for injecting fluids to enhance oil recovery, injecting treated water for artificial aquifer recharge, or enhancing a pump-and-treat system. If the wells are poorly designed or constructed, or if the local geology is not sufficiently studied, injected liquids can enter an aquifer and cause groundwater contamination. Injection wells are regulated under the Underground Injection Control Program of the Safe Drinking Water Act.

See also Aquifer restoration; Deep-well injection; Drinking-water supply; Groundwater monitoring; Groundwater pollution; Water table.

Inoculate

To inoculate involves the introduction of microorganisms into a new environment. Originally the term referred to the insertion of a bud or shoot of one plant into the stem or trunk of another to develop new strains or hybrids. These hybrid plants would be resistant to botanic disease or they would allow greater harvests or range of climates. With the advent of vaccines to prevent human and animal disease, the term inoculate has come to represent injection of a serum to prevent, cure, or make immune from disease.

Inoculation is of prime importance in that the introduction of specific microorganism species into specific macroorganisms may establish a symbiotic relationships where each organism benefits. For example, the introduction of mycorrhiza fungus to plants improves the plants' ability to absorb nutrients from the soil.

See also Symbiosis.

Insecticide *see* **Pesticide.**

Integrated pest management

Integrated pest management (IPM) is a newer science that aims to give the best possible pest control while minimizing damage to human health or the environment. IPM means either using fewer chemicals more effectively or finding ways, both new and old, that substitute for pesticide use.

Technically, IPM is the selection, integration, and implementation of pest control based on predicted economic, ecological, and sociological consequences. IPM seeks maximum use of naturally occurring pest controls, including weather, disease agents, predators, and parasites. In addition, IPM utilizes various biological, physical, and chemical control and habitat modification techniques. Artificial controls are imposed only as required to keep a pest from surpassing intolerable population levels which are predetermined from assessments of the pest damage potential and the ecological, sociological, and economic costs of the control measures. Farmers have come to understand that the presence of a pest species does not necessarily justify action for its control. In fact, tolerable infestations may be actually desirable, providing food for important beneficial insects. This change in farming practices has occurred due to both the accumulation of pesticides in the environment and the negative effects of these chemicals.

The introduction of synthetic organic pesticides such as the insecticide DDT, and the herbicide 2,4-D (half the formula in Agent Orange) after World War II began a new era in pest control. These products were followed by hundreds of synthetic organic fungicides, nematicides (affects nematodes), rodenticides (affects rodents), and other chemical controls. These chemical materials were initially very effective and very cheap. Synthetic chemicals eventually became the primary means of pest control in productive agricultural regions, providing season-long crop protection against insects and weeds. They were used in addition to fertilizers and other treatments. However, IPM has applications that go well beyond farming and has been applied in other settings, such as in schools and museums. IPM in these buildings focuses on eliminating food, shelter, and water for pests and results in decreased exposure to potentially harmful chemical pesticides.

The success of modern pesticides led to widespread acceptance and reliance upon them, particularly in the United States. Of all the chemical pesticides applied worldwide in agriculture, forests, industry, and households, one-third to one-half were used in the United States. Herbicides have been used increasingly to replace hand labor and machine cultivation for control of weeds in crops, in forests, on the rights-of-way of highways, utility lines, railroads, and in cities. Agriculture consumes perhaps 65 percent of the total quantity of synthetic organic pesticides used in the United States each year.

For more than a decade, problems with pesticides have become increasingly apparent. Significant groups of pests have evolved with genetic resistance to pesticides. The increase in resistance among insect pests has been exponential, following extensive use of chemicals in the last forty years. Ticks, insects and spider mites (nearly four hundred species) are now especially resistant, and the creation of new insecticides to combat the problem is not keeping pace with the emergence of new strains of resistant insect pests. Despite the advances in modern chemical control and the dramatic increase in chemical pesticides used on U.S. cropland, annual crop losses from all pests appear to have remained constant or to have increased. Losses caused by weeds have declined slightly, but those caused by insects have nearly doubled. The price of synthetic organic pesticides has increased significantly in recent years, placing a heavy financial burden on those who use large quantities of the chemicals. As farmers and growers across the United States realize the limitations and human health consequences of using artificial chemical pesticides, interest in the alternative approach of IPM grows.

IPM aims at management rather than eradication of pest species. Since potentially harmful species will continue to exist at tolerable levels of abundance, the philosophy now is to manage rather than eradicate the pests. The ecosystem is the management unit. Every crop is in itself a complex ecological system. Spraying pesticides too often, at the wrong time, or on the wrong part of the crop may destroy the pests' natural enemies ordinarily present in the ecosystem. Knowledge of the actions, reactions, and interactions of the components of the ecosystems is requisite to effective IPM programs. With this knowledge, the ecosystem is manipulated in order to hold pests at tolerable levels while avoiding disruptions of the system.

The use of natural controls is maximized. IPM emphasizes the fullest practical utilization of the existing regulating and limiting factors in the form of parasites, predators, and weather, which regulate the pests' population growth. IPM users understand that control procedures may produce unexpected and undesirable consequences, however. It takes time to change over and determination to keep up the commitment until the desired results are achieved.

An interdisciplinary systems approach is essential. Effective IPM is an integral part of the overall management of a farm, a business, or a forest. For example, timing plays an important role. In order to effectively manage pest populations, monitoring of the crops for signs of infestation and for the status of the pest population is key to successful IPM. Knowledge regarding the life cycles of pests allows for more effective control. Certain pests are most prevalent at particular times of the year. By altering the date on which a crop is planted, serious pest damage can be avoided. Some farmers simultaneously plant and harvest, since the procedure prevents the pests from migrating to neighboring fields after the harvest. Others may plant several different crops in the same field, thereby reducing the number of pests. The variety of crops harbor greater numbers of natural enemies and make it more difficult for the pests to locate and colonize their host plants. In Thailand and China, farmers flood their fields for several weeks before planting to destroy pests. Other farmers turn or till the soil, so that pests are brought to the surface and die in the sun's heat.

The development of a specific IPM program depends on the pest complex, resources to be protected, economic values, and availability of personnel. It also depends upon adequate funding for research and to train farmers. Some of the techniques are complex, and expert advice is needed. However, while it is difficult to establish absolute guidelines, there are general guidelines that can apply to the management of any pest group.

Growers must analyze and continually monitor the pest status of each of the reputedly injurious organisms and establish economic thresholds for the real pests. The economic threshold is, in fact, the population level, and is defined as the density of a pest population below which the cost of applying control measures exceeds

Integrated pest management

the losses caused by the pest. Economic threshold values are based on assessments of the pest damage potential and the ecological, sociological, and economic costs associated with control measures. A given crop, forest area, backyard, building, or recreational area may be infested with dozens of potentially harmful species at any one time. For each situation, however, there are rarely more than a few pest species whose populations expand to intolerable levels at regular and fairly predictable intervals. Key pests recur regularly at population densities exceeding economic threshold levels and are the focal point for IPM programs.

Farmers must also devise schemes for lowering equilibrium positions of key pests. A key pest will vary in severity from year to year, but its average density, known as the equilibrium position, usually exceeds its economic threshold. IPM efforts manipulate the environment in order to reduce a pest's equilibrium position to a permanent level below the economic threshold. This reduction can be achieved by deliberate introduction and establishment of natural enemies (parasites, predators, and diseases) in areas where they did not previously occur. Natural enemies may already occur in the crop in small numbers or can be introduced from elsewhere. Certain microorganisms, when eaten by a pest, will kill it.

Newer chemicals show promise as alternatives to synthetic chemical pesticides. These include insect attractant chemicals, weed and insect disease agents, and insect growth regulators or hormones. A toxin produced by a pathogen, *Bacillus thuringiensis* (Bt), has proven commercially successful when the Bt gene that produces the toxin is introduced into the plant genome, thus conferring resistance to pests affected by the toxin. Since certain crops have a built-in resistance to pests, pest-resistant or pest-free varieties of seed, crop plants, ornamental plants, orchard trees, and forest trees can be used. Growers can also modify the pest environment to increase the effectiveness of the pest's biological control agents, to destroy its breeding, feeding, or shelter habitat or otherwise render it harmless. These measures include crop rotation, destruction of crop harvest residues and soil tillage, and selective burning or mechanical removal of undesirable plant species and pruning, especially for forest pests. Another method of ensuring low populations of pests is the planting of pest-free rootstocks when possible.

Although nearly permanent control of key insect and plant disease pests of agricultural crops has been achieved, emergencies will occur, and IPM advocates acknowledge this. During those times, measures should be applied that create the least ecological destruction.

Growers are urged to utilize the best combination of the three basic IPM components: natural enemies, resistant varieties, and environmental modification. However, there may be a time when pesticides may be the only recourse. In that case, it is important to coordinate the proper pesticide, the dosage, and the timing in order to minimize the hazards to non-target organisms and the surrounding ecosystems.

Pest management techniques have been known for many years and were used widely before World War II. They were deemphasized by insect and weed control scientists and by corporate pressures as the synthetic chemicals became commercially available after the war. Now there is a renewed interest in the early control techniques and in new chemistry.

Reports detailing the success of IPM are emerging at a rapid rate as thousands of farmers yearly join the ranks of those who choose to eliminate chemical pesticides. Sustainable agricultural practice increases the richness of the soil by replenishing the soil's reserves of fertility. IPM does not produce secondary problems such as pest resistance or resurgence. It also diminishes soil erosion, increases crop yields, and saves money over time. Organic foods are reported to have better cooking quality, better flavor, and greater longevity in the storage bins. And with little to no pesticide residue, the amount of chemicals ingested from food produced through IPM or organic methods are low or non-existent. The U.S. Department of Agriculture (USDA) has no certification for IPM-grown crops, comparable to their program for organically grown foods, but some areas are working to define IPM and label foods as IPM-grown.

See also Sustainable agriculture.

Resources

BOOKS

Cuperus, Gerrit W., G. S. Dhaliwal, and Opender Koul. *Integrated Pest Management: Potential, Constraints and Challenges.* Wallingford, UK: CABI, 2004.

Dwivedi, S. C., and Nalini Dwivedi. *Integrated Pest Management and Biocontrol.* Jaipur, India: Pointer Publishers, 2006.

Horne, Paul A., and Jessica Page. *Integrated Pest Management for Crops and Pastures.* Collingwood, Victoria, Australia: CSIRO Publishing, 2007.

Jepson, Paul C., Linda J. Brewer, and Susan B. Jepson. *Integrated Pest Management Resource Guide.* Corvallis, OR: Oregon State University Extension Service, 2006.

Kogan, M., and Paul C. Jepson. *Perspectives in Ecological Theory and Integrated Pest Management.* Cambridge: Cambridge University Press, 2007.

Koul, Opender, and Gerrit W. Cuperus. *Ecologically Based Integrated Pest Management.* Wallingford, UK: CABI, 2007.

Peshin, Rajinder, and A. K. Dhawan. *Integrated Pest Management.* Dordrecht, Netherlands: Springer, 2009.

OTHER

United States Environmental Protection Agency (EPA). "Pesticides: Pest Management: Integrated Pest Management (IPM)." http://www.epa.gov/ebtpages/pestpesticidemanagintegratedpestmanagementipm.html (accessed August 21, 2010).

Liane Clorfene Casten

Intergenerational justice

One of the key features of an environmental ethic or perspective is its concern for the health and well-being of future generations. Questions about the rights of future people and the responsibilities of those presently living are central to environmental theory and practice and are often asked and analyzed under the term *intergenerational justice*. Most traditional accounts or theories of justice have focused on relations between contemporaries: What distribution of scarce goods is fairest or optimally just? Should such goods be distributed on the basis of merit or need? These and other questions have been asked by thinkers from Greek philosopher Aristotle (384–322 BCE) to American philosopher John Rawls (1921–2002). Recently, however, some philosophers have begun to ask about just distributions over time and across generations.

The subject of intergenerational justice is a key concern for environmentally minded thinkers for at least two reasons. First, human beings now living have the power to permanently alter or destroy the planet (or portions thereof) in ways that will affect the health, happiness, and well-being of people living long after they are all dead. One need only think, for example, of the radioactive wastes generated by nuclear power plants that will be intensely "hot" and dangerous for many thousands of years. No one yet knows how to safely store such material for a hundred, much less many thousands, of years. Considered from an intergenerational perspective then, it would be unfair—that is, unjust—for the present generation to enjoy the benefits of nuclear power, passing on to distant posterity the burdens and dangers caused by society's (in)action.

Second, people not only have the power to affect future generations, but they *know* that they have it. And with such knowledge comes the moral responsibility to act in ways that will prevent harm to future people. For example, since people know about the health effects of radiation on human beings, having that knowledge imposes upon them a moral obligation not to needlessly expose anyone—now or in the indefinite future—to the harms or hazards of radioactive wastes. Many other examples of intergenerational harm or hazard exist: global warming, topsoil erosion, disappearing tropical rain forests, depletion and/or pollution of aquifers, among others. But whatever the example, the point of the intergenerational view is the same: the moral duty to treat people justly or fairly applies not only to people now living, but to those who will come later. To the extent that society's actions produce consequences that may prove harmful to people who have not harmed (and in the nature of the case cannot harm) people today is, by any standard, unjust. And yet it seems quite clear that the present generation is in many respects acting unjustly toward distant posterity. This is true not only for harms or hazards bequeathed to future people, but the point applies also to deprivations of various kinds.

Consider, for example, the present generation's profligate use of fossil fuels. Reserves of oil and natural gas are both finite and nonreplaceable; once burned (or turned into plastic or some other petroleum-based material), a gallon of oil is gone forever. Every drop or barrel used now is therefore unavailable for future people. As Wendell Berry observed, the claim that fossil fuel energy is *cheap* rests on a simplistic and morally doubtful assumption about the *rights* of the present generation: "We were able to consider [fossil fuel energy] 'cheap' only by a kind of moral simplicity: the assumption that we had a 'right' to as much of it as we could use. This was a 'right' made solely by might. Because fossil fuels, however abundant they once were, were nevertheless limited in quantity and not renewable, they obviously did not 'belong' to one generation more than another. We ignored the claims of posterity simply because we could, the living being stronger than the unborn, and so worked the 'miracle' of industrial progress by the theft of energy from (among others) our children."

And that, Berry adds, "is the real foundation of our progress and our affluence. The reason that we are a rich nation is not that we have earned so much wealth—you cannot, by any honest means, earn or deserve so much. The reason is simply that we have learned, and become willing, to market and use up in our own time the birthright and livelihood of posterity."

These and other considerations have led some environmentally minded philosophers to argue for limits on present-day consumption, so as to save a fair share of scarce resources for future generations. John Rawls, for instance, constructs a *just savings principle* according to which members of each generation may consume no more than their fair share of scarce resources. The main difficulty in arriving at and applying any such principle lies in determining what counts as fair share. As the number of generations taken into account increases, the share available to any single generation then becomes smaller; and as the number of generations approaches infinity, any one generation's share approaches zero.

Other objections have been raised against the idea of intergenerational justice. These objections can be divided into two groups, which can be called conceptual and technological. One conceptual criticism is that the very idea of intergenerational justice is itself incoherent. The idea of justice is tied with that of reciprocity or exchange; but relations of reciprocity can exist only between contemporaries. Therefore, the concept of justice is inapplicable to relations between existing people and distant posterity. Future people are in no position to reciprocate. Therefore, people now living cannot be morally obligated to do anything for them. Another conceptual objection to the idea of intergenerational justice is concerned with rights. Briefly, the objection runs as follows: Future people do not (yet) exist; only actually existing people have rights, including the right to be treated justly. Therefore, future people do not have rights which those in the present have a moral obligation to respect and protect. Critics of this view counter that it not only rests on a too-restrictive conception of rights and justice, but that it also paves the way for grievous intergenerational injustices.

Several arguments can be constructed to counter the claim that justice rests on reciprocity (and therefore applies only to relations between contemporaries) and the claim that future people do not have rights, including the right to be treated justly by their predecessors. Regarding reciprocity: Because it is acknowledged in ethics and recognized in law that it is possible to treat an infant or a mentally disabled or severely retarded person justly or unjustly, even though they are in no position to reciprocate, it follows that the idea of justice is not necessarily connected with reciprocity. Regarding the claim that future people cannot be said to have rights that require recognition and respect today: one of the more ingenious arguments against this view consists of modifying John Rawls's imaginary *veil of ignorance*. Rawls argues that principles of justice must not be partisan or favor particular people but must be blind and impartial. To ensure impartiality in arriving at principles of justice, Rawls suggests imagining an original position in which rational people are placed behind a veil of ignorance

wherein they are unaware of their age, race, sex, social class, economic status, etc. Unaware of their own particular position in society, rational people would arrive at and agree upon impartial and universal principles of justice. To ensure that such impartiality extends across generations, one need only *thicken* the veil by adding the proviso that the choosers be unaware of the generation to which they belong. Rational people would not accept or agree to principles under which predecessors could harm or disadvantage successors.

Some critics of intergenerational justice argue in technological terms. They contend that existing people need not restrict their consumption of scarce or nonrenewable resources in order to save some portion for future generations. They argue that substitutes for these resources will be discovered or devised through technological innovations and inventions. For example, as fossil fuels become scarcer and more expensive, new fuels—gasohol or fusion-derived nuclear fuel—will replace them. Thus people need never worry about depleting any particular resource because every resource can be replaced by a substitute that is as cheap, clean, and accessible as the resource it replaces. Likewise, people need not worry about generating nuclear wastes that they do not yet know how to store safely. Some solution is bound to be devised sometime in the future.

Environmentally minded critics of this technological line of argument claim that it amounts to little more than wishful thinking. Like Charles Dickens's fictional character Mr. Micawber, those who place their faith in technological solutions to all environmental problems optimistically expect that "something will turn up." Just as Mr. Micawber's faith was misplaced, so too, these critics contend, is the optimism of those who expect technology to solve all problems, present and future. Of course such solutions may be found, but that is a gamble and not a guarantee. To wager with the health and well-being of future people is, environmentalists argue, immoral.

There are of course many other issues and concerns raised in connection with intergenerational justice. Discussions among and disagreements between philosophers, economists, environmentalists, and others are by no means purely abstract and academic. How these matters are resolved will have a profound effect on the fate of future generations.

Resources

BOOKS

Des Jardins, Joseph R. *Environmental Ethics: An Introduction to Environmental Philosophy*. Belmont, CA: Wadsworth Publishing, 2005.

Kuipers, Theo A. F. *General Philosophy of Science: Focal Issues*. Handbook of the philosophy of science. Amsterdam: Elsevier/North Holland, 2007.

Page, Edward. *Climate Change, Justice and Future Generations*. Cheltenham, UK: Edward Elgar, 2006.

Terence Ball

Intergovernmental Panel on Climate Change (IPCC)

The Intergovernmental Panel on Climate Change (IPCC) was established in 1988 as a joint project of the United Nations Environment Programme (UNEP) and the World Meteorological Organization (WMO). The primary mission of the IPCC is to bring together the world's leading experts on the earth's climate to gather, assess, and disseminate scientific information about climate change, with a view to informing international and national policy makers. The IPCC has become the highest-profile and best-regarded international agency concerned with the climatic consequences of greenhouse gases, such as carbon dioxide (CO_2) and methane (CH_4), byproducts of fossil fuel combustion.

The IPCC was established partly in response to Mexican chemist and Nobel Laureate Mario Molina's

(1943–) 1985 documentation of chemical processes which occur when human-made chemicals deplete the earth's atmospheric ozone (O_3) shield. Ozone layer depletion results in increased levels of ultraviolet radiation reaching the earth's surface, producing a host of health, agricultural, and environmental problems. Molina's work helped to persuade most of the industrialized nations to ban chlorofluorocarbons (CFCs) and several other ozone-depleting chemicals. It also established a context in which national and international authorities began to pay serious attention to the global environmental consequences of atmospheric changes resulting from industrialization and reliance on fossil fuels.

Continuing to operate under the auspices of the United Nations and headquartered in Geneva, Switzerland, the IPCC is organized into three working groups and a task force, and meets about once a year. The first group gathers scientific data and analyzes the functioning of the climate system with special attention to the detection of potential changes resulting from human activity. The second group's duty is to assess the potential socioeconomic impacts and vulnerabilities associated with climate change. It is also charged with exploring options for humans to adapt to potential climate change. The third group focuses on ways to reduce greenhouse gas emissions and to stop or reduce climate change. The task force is charged with maintaining inventories of greenhouse emissions for all countries.

IPCC Chairman Rajendra Pachauri delivers a keynote speech at the thirty-second session of the Intergovernmental Panel on Climate Change (IPCC) in Busan, South Korea, in October 2010. The conference opened for a four-day run with some 400 officials from 194 IPCC member nations attending. (© STF/epa/Corbis)

Internalizing costs

Private market activities create so-called externalities. An example of a negative externality is air pollution. It occurs when a producer does not bear all the costs of an activity in which he or she engages. Since external costs do not enter into the calculations producers make, they will make few attempts to limit or eliminate pollution and other forms of environmental degradation.

Negative externalities are a type of market defect all economists believe is appropriate to try to correct. Milton Friedman refers to such externalities as "neighborhood effects" (although it must be kept in mind that some forms of pollution have an all but local effect). The classic neighborhood effect is pollution. The premise of a free market is that when two people voluntarily make a deal, they both benefit. If society gives everyone the right to make deals, society as a whole will benefit. It becomes richer from the aggregation of the many mutually beneficial deals that are made. However, what happens if in making mutually beneficial deals there is a waste product that the parties release into the environment and that society must either suffer from or clean up? The two parties to the deal are better off, but society as a whole has to pay the costs. Friedman points out that individual members of a society cannot appropriately charge the responsible parties for external costs or find other means of redress.

Friedman's answer to this dilemma is simple: Society, through government, must charge the responsible parties the costs of the clean-up. Whatever damage they generate must be internalized in the price of the transaction. Polluters can be forced to internalize environmental costs through pollution taxes and discharge fees, a method generally favored by economists. When such taxes are imposed, the market defect (the price of pollution which is not counted in the transaction) is corrected. The market price then reflects the true social costs of the deal, and the parties have to adjust accordingly. They will have an incentive to decrease harmful activities and develop less environmentally damaging technology. The drawback of such a system is that society will not have direct control over pollution levels, although it will receive monetary compensation for any losses it sustains. However, if the government imposed a tax or charge on the polluting parties, it would have to place a monetary value on the damage. In practice, this is difficult to do. How much for a human life lost to pollution? How much for a vista destroyed? How much for a plant or animal species brought to extinction? Finally, the idea that pollution is all right as long as the polluter pays for it is unacceptable to many people.

Government has tried to control activities with associated externalities through regulation, rather than by supplementing the price system. It has set standards for specific industries and other social entities. The standards are designed to limit environmental degradation to acceptable levels and are enforced through the Environmental Protection Agency (EPA). They prohibit some harmful activities, limit others, and prescribe alternative behaviors. When market actors do not adhere to these standards, they are subject to penalties. In theory, potential polluters are given incentives to reduce and treat their waste, manufacture less harmful products, and develop alternative technologies. In practice, the system has not worked as well as it was hoped in the 1960s and 1970s, when much of the environmental legislation presently in force was enacted. Enforcement has been fraught with political and legal difficulties. Extensions on deadlines are given to cities for not meeting clean air standards and to the automobile industry for not meeting standards on fuel economy of new cars, for instance. It has been difficult to collect fines from industries found to have been in violation. Some companies simply declare bankruptcy to evade fines. Others continue polluting because they find it cheaper to pay fines than to develop alternative production processes.

Pollution credit trading schemes, or a trade in pollution permits, are an alternative means of regulation. The government issues a number of permits that altogether set a maximum acceptable pollution level. Buyers of permits can either use them to cover their own polluting activities or resell them to the highest bidder (typically other polluters who have expended all of their allotted credits). The price of pollution is determined by market demand for unused credits. Credit schemes encourage polluters to internalize the environmental costs of their activities and invest in longer-term pollution-reduction solutions instead of buying pollution credits. Several nations have established pollution trading schemes for different kinds of pollutants such as acid-rain causing sulfur, carbon emissions linked to climate change, and pollutants released into waterways.

See also Environmental economics.

Resources

BOOKS

Babe, Robert E. *Culture of Ecology: Reconciling Economics and Environment.* Toronto, Canada: University of Toronto Press, 2006.

Eisler, Riane Tennenhaus. *The Real Wealth of Nations: Creating a Caring Economics.* San Francisco, CA: Berrett-Koehler, 2007.

Tietenberg, Thomas H. *Emissions Trading: Principles and Practice.* Washington, DC: Resources for the Future, 2006.

PERIODICALS

Kemfert, Claudia, et al. "The Environmental and Economic Effects of European Emissions Trading." *Climate Policy* 6 (2006): 441–456.

OTHER

Asia-Pacific Emissions Trading Forum. "Emissions Trading." http://www.aetf.emcc.net.au/HTML/whatis.html (accessed October 10, 2010).

United States Environmental Protection Agency (EPA). "Economics: Financing: Pollution Prevention Financing." http://www.epa.gov/ebtpages/econfinancing pollutionpreventionfinancing.html (accessed October 10, 2010).

Alfred A. Marcus
Marijke Rijsberman

Internal costs *see* **Internalizing costs.**

International Atomic Energy Agency

The first decade of research on nuclear weapons and nuclear reactors was characterized by extreme secrecy, and the few nations that had the technology carefully guarded their information. In 1954, however, that philosophy changed, and the United States, in particular, became eager to help other nations use nuclear energy for peaceful purposes. A program called "Atoms for Peace" brought foreign students to the United States for the study of nuclear sciences and provided enriched uranium to countries wanting to build their own reactors, encouraging interest in nuclear energy throughout much of the world.

But this program created a problem. It increased the potential diversion of nuclear information and nuclear materials for the construction of weapons, and the threat of nuclear proliferation grew. The United Nations created the International Atomic Energy Agency (IAEA) in 1957 to address this problem. The agency had two primary objectives: to encourage and assist with the development of peaceful applications of nuclear power throughout the world and to prevent the diversion of nuclear materials to weapons research and development.

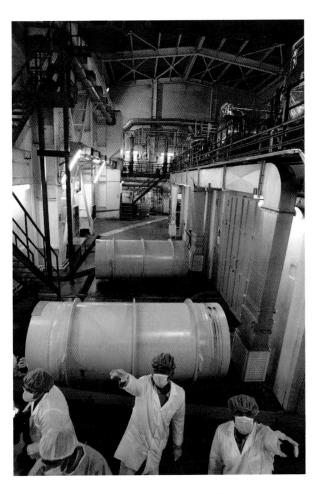

Technicians of the International Atomic Energy Agency inspect the site of the Isfahan uranium conversion plant in central Iran in 2007. (© Abedin Taherkenareh/epa/Corbis)

In first decade of the IAEA's existence, several nations independently signed bilateral non-proliferation treaties, choosing not to involve the IAEA in the negotiations or inspections process of those treaties. The 1970 nuclear non-proliferation treaty more clearly designated the IAEA's responsibilities for the monitoring of nuclear material, however, and its involvement in international non-proliferation discussions grew.

Today the agency is an active division of the United Nations, and its headquarters are in Vienna. The IAEA operates with a staff of more than 2,300 workers and a budget of about $396 million. To accomplish its goal of extending and improving the peaceful use of nuclear energy, the IAEA conducts regional and national workshops, seminars, training courses, and committee meetings. It publishes guidebooks and manuals on related topics and maintains the International Nuclear

Information System, a bibliographic database on nuclear literature that includes more than 2 million records. The database is made available on magnetic tape to its 151-member states.

The IAEA also carries out a rigorous program of inspection. In a typical year, IAEA activities include conducting safety reviews in a number of different countries, assisting in dealing with accidents at nuclear power plants, providing advice to nations interested in building their own nuclear facilities, advising countries on methods for dealing with radioactive wastes, teaching nations how to use radiation to preserve foods, helping universities introduce nuclear science into their curricula, sponsoring research on the broader applications of nuclear science, and aiding the implementation of nuclear medicine technologies in developing regions.

The IAEA and its former Director General Mohamed ElBaradei were awarded the Nobel Peace Prize in 2005.

Resources

BOOKS

Diehl, Sarah J., and James Clay Moltz. *A Handbook of Nuclear Weapons and Nonproliferation.* Warwick, UK: Pentagon Press, 2005.

OTHER

British Broadcasting Corporation (BBC). "Profile: International Atomic Energy Agency (IAEA)." BBC Monitoring Service. http://news.bbc.co.uk/1/hi/world/europe/country_profiles/2642835.stm (accessed October 12, 2010).

ORGANIZATIONS

International Atomic Energy Agency, P.O. Box 100, Wagramer Strasse 5, Vienna, Austria, A-1400, (413) 2600-0, (413) 2600-7, official.mail@iaea.org

David E. Newton

International Convention for the Regulation of Whaling (1946)

The International Whaling Commission (IWC) was established in 1949 following the inaugural International Convention for the Regulation of Whaling, which took place in Washington, D.C., in 1946. Many nations have membership in the IWC, which primarily

sets quotas for whales. The purpose of these quotas is twofold: They are intended to protect the whale species from extinction while allowing a limited whaling industry. In recent times, however, the IWC has come under attack. The vast majority of nations in the commission have come to oppose whaling of any kind and object to the IWC's practice of establishing quotas. Furthermore, some nations—principally Iceland, Japan, and Norway—wish to protect their traditional whaling industries and are against the quotas set by the IWC. With two such divergent factions opposing the IWC, its future is as doubtful as that of the whales.

Since its inception, the commission has had difficulty implementing its regulations and gaining approval for its recommendations. In the meantime whale populations have continued to dwindle. In its original design, the IWC consisted of two sub-committees, one scientific and the other technical. Any recommendation that the scientific committee put forth was subject to the politicized technical committee before final approval. The technical committee evaluated the recommendation and changed it if it was not politically or economically viable; essentially, the scientific committee's recommendations have often been rendered powerless. Furthermore, any nation that has decided an IWC recommendation was not in its best interest could have dismissed it by simply registering an objection. In the 1970s this gridlock and inaction attracted public scrutiny; people objected to the IWC's failure to protect the world's whales. Thus in 1972 the United Nations Conference on the Human Environment voted overwhelmingly to stop commercial whaling.

Nevertheless, the IWC retained some control over the whaling industry. In 1974 the commission attempted to bring scientific research to management strategies in its New Management Procedure. The IWC assessed whale populations with finer resolution, scrutinizing each species to see if it could be hunted and not die out. It classified whales as either initial management stocks (harvestable), sustained management stocks (harvestable), or protection stocks (unharvestable). Although these classifications were necessary for effective management, much was unknown about whale population ecology, and quota estimates contained high levels of uncertainty.

Since the 1970s, public pressure has caused many nations in the IWC to oppose whale hunting of any kind. At first, one or two nations proposed a whaling moratorium each year. Both pro- and anti-whaling countries began to encourage new IWC members to vote for their respective positions, thus dividing the commission. In 1982 the IWC enacted a limited moratorium on commercial whaling, to be in effect from 1986 until 1992. During that time it would thoroughly assess whale stocks and afterward allow whaling to resume for selected species and areas. Norway and Japan, however, attained special permits for whaling for scientific research: They continued to catch approximately 400 whales per year, and the meat was sold to restaurants. Then in 1992—the year when whaling was supposed to have resumed—many nations voted to extend the moratorium. Iceland, Norway, and Japan objected strongly to what they saw as an infringement on their traditional industries and eating customs. Iceland subsequently left the IWC, and Japan and Norway have threatened to follow. These countries intend to resume their whaling programs. Members of the IWC are torn between accommodating these nations in some way and protecting the whales, and amid such controversy it is unlikely that the commission can continue in its present mission.

Although the IWC has not been able to marshall its scientific advances or enforce its own regulations in managing whaling, it is broadening its original mission. The commission may begin to govern the hunting of small cetaceans such as dolphins and porpoises, which are believed to suffer from overhunting. The group has become concerned over the probable extinction of the vaquita (porpoise) unless immediate action is taken. There are at most 150 animals left of this species, which is only found in the upper Gulf of California off Mexico. The IWC recently held its sixtieth annual and associated meetings in Santiago, Chile, in June 2008. Here a key issue was a proposal by Denmark for an annual strike limit of 10 humpback whales for the period 2008–2012 for West Greenland. Although the Scientific Committee had agreed that this strike limit would not harm the population, there was concern that the need for the additional whales was not satisfactorily documented, and the proposal was defeated. There was discussion over a South Atlantic Whale Sanctuary. Additionally special permit whaling programs that allowed 551 Antarctic minke whales to be taken in 2007 and smaller numbers of common minke, sei, Bryde's and sperm whales from the North Pacific as with all special permit whaling deeply divides the commission. There were strong statements both in favor and against lethal research programs. There were additional worries about the endangered western North Pacific gray whale, whose feeding grounds coincide with oil and gas operations off Sakhalin Island.

Resources

BOOKS

Annual Report of the International Whaling Commission. Cambridge: International Whaling Commission, 2007.

Darby, Andrew. *Harpoon: Into the Heart of Whaling.* Cambridge, MA: Da Capo Press, 2008.

Freestone, David, Richard Barnes, and David M. Ong. *The Law of the Sea: Progress and Prospects.* Oxford: Oxford University Press, 2006.

Gillespie, Alexander. *Whaling Diplomacy: Defining Issues in International Environmental Law.* Cheltenham , UK: Elgar, 2005.

Kaschner, Kristin. *Competition between Marine Mammals and Fisheries: Food for Thought* Washington, DC: The Humane Society of the United States, 2004.

Morikawa, Jun. *Whaling in Japan: Power, Politics, and Diplomacy.* New York: Columbia University Press, 2009.

UK Department for Environment, Food and Rural Affairs. *The International Whaling Commission: The Way Forward.* London: DEFRA, 2008.

Washington College of Law. *Ocean and Fisheries Law.* Washington, DC: American University, Washington College of Law, 2006.

David A. Duffus
Andrea Gacki

International Council for Bird Preservation *see* **BirdLife International.**

International Environmental Legislation and Treaties

International environmental laws exist in numerous forms. Treaties, including conventions, are the most important and prominent source of international environmental law. A treaty is an instrument that two or more countries have agreed upon and ratified. A treaty contains agreed-upon provisions designed to govern the actions of each member nation. While international treaties may cover a wide array of topics—from nuclear arms reduction to trade agreements—environmental law has generated more international treaties than any other area of specialization. As of 2010, countries had agreed to more than 1,000 bilateral or multilateral treaties that either address environmental issues directly or contain provisions that touch on environmental issues.

Customary international law and subsidiary sources of law, such as decisions by international courts and tribunals, are also sources of international environmental law. Customary international law involves

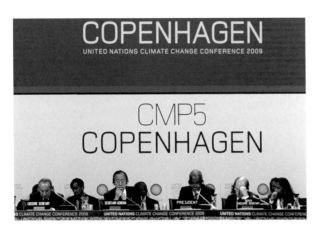

Executive-Secretary of the UN Climate Conference Yvo de Boer (L) and UN Secretary-General Ban Ki-Moon attend the plenary session at the end of the 2009 CMP5 UN Climate Change Conference. *(Oliver Morin / AFP / Getty Images)*

examining the practices between or among states. Pervasive customary actions and rules may be deemed to have become a binding part of international law. Customary international law is often difficult to determine and typically lies on a careful fact-based analysis of state behavior.

Decisions by international courts and tribunals may also become binding or persuasive rulings on environmental issues. International courts and tribunals that rule on environmental issues include the International Court of Justice, the European Court of Justice, and the World Trade Organization (WTO) Dispute Settlement Board, or a tribunal established to govern a specific convention or treaty, such as the International Tribunal of the Law of the Sea. The role of treaties and customary international law may be closely related to these subsidiary sources of law, because courts and tribunals may rely on treaty interpretation or findings of international custom to issue rulings. Once a court or tribunal issues a decision regarding a particular area of law, however, that decision should influence similar future conduct by nations.

Despite the prevalence of international environmental treaties and other regulations, enforcement of international environmental law remains a major issue. Enforcement of international environmental law involves two aspects: determining a nation's compliance with its obligations and imposing sanctions on non-compliant nations. Determining compliance in many situations is a simple task. For example, a nation may easily determine the source of an illegal discharge of water pollution from a neighboring nation.

Compliance management relies on oversight of activities and a system that encourages compliance through benefits or dissuades non-compliance through punitive measures. In some situations compliance determination is rather simple. For example, a nation may easily determine that a neighboring nation is not in compliance with a transboundary pollution treaty when polluted water flows across the border. In other situations, however, compliance determinations are more difficult. Determining whether a nation has met agreed upon standards for the emission of carbon dioxide or other greenhouse gases poses serious issues for the international community.

Often the issue of compliance relies upon self-reporting of activities by nations. Self-reporting does not necessarily produce reliable results, however, because of dishonest reporting by some nations or inaccurate reporting by nations that do no have the technical ability to monitor and report accurately. The Kyoto Protocol, however, established a Compliance Committee to monitor and ensure compliance with the agreement's greenhouse gas reduction targets. The enforcement branch of the Compliance Committee reviews allegations of non-compliance by examining expert reports from a variety of sources. International courts and tribunals may be able to determine compliance in certain situations, too. International pressure and the desire of nations to conform to accepted environmental norms also shape compliant behavior.

Typically, when a nation is non-compliant with its obligations under an environmental treaty, sanctions are imposed under the terms of the treaty. Sanctions are designed to impose punitive measures on non-compliant nations and encourage future compliance by the offending state and, by way of example, other states. Countermeasures, a form of self-help actions undertaken by other states in response to the non-compliant state, are often used in international law. A common example is the imposition of tariffs in response to another nation that imposes tariffs in violation of a trade agreement. In international environmental law, however, countermeasures are not a common sanction, because such reciprocal actions would undermine the treaty and further damage the environment. Environmental treaties and conventions, therefore, typically include mechanisms for the imposition of bilateral or collective sanctions against a non-compliant member nation.

Sanctions may include membership sanctions, which affect the non-compliant nation's membership standing in the organization governing the treaty. Membership sanctions may include the loss of voting rights, removal from committees, or expulsion from the organization. Expulsions are rare under multilateral treaties, however, because such actions often weaken the membership organization and grant the non-compliant nation free license to continue its objectionable behavior.

Bilateral or multilateral sanctions are the most common form of coercive compliance mechanisms. Bilateral and multilateral sanctions often impose civil penalties, including fines and injunctions, on non-compliant treaty member nations. Bilateral sanctions involve the imposition of sanctions by one nation against another nation under the terms of a treaty. Under the terms of most treaties, bilateral sanctions typically require the nation seeking sanctions to go before a tribunal, such as the WTO Dispute Settlement Board, to seek authorization to impose sanctions. Multilateral sanctions usually are imposed by a treaty's governing organization.

Over the past several decades, the international community has adopted several notable, multilateral environmental treaties, including the Convention on Biological Diversity (CBD), Convention on International Trade in Endangered Species (CITES), UN Convention to Combat Desertification (UNCCD), UN Framework Convention on Climate Change (UNFCCC), Kyoto Protocol to the United Nations Framework Convention on Climate Change (Kyoto, 1997), and Montreal Protocol of the Vienna Convention (Montreal). Although many of these treaties have been effective in accomplishing their environmental goals, many policymakers lament the cost of treaty compliance. Developing nations often do not have the financial resources to comply with the requirements of such treaties. A recent movement has focused on creating mechanisms that allow developed nations to cover the cost of implementation in developing nations.

Resources

BOOKS

Bodansky, Daniel, Jutta Brunnée, and Ellen Hey. *The Oxford Handbook of International Environmental Law*. Oxford: Oxford University Press, 2007.

Goyal, Anupam. *The WTO and International Environmental Law: Towards Conciliation*. New Delhi: Oxford University Press, 2006.

Kiss, Alexandre Charles, and Dinah Shelton. *Guide to International Environmental Law*. Leiden, Netherlands: Martinus Nijhoff Publishers, 2007.

Kiss, Alexandre Charles, and Dinah Shelton. *International Environmental Law*. Ardsley, NY: Transnational Publishers, 2004.

Louka, Elli. *International Environmental Law: Fairness, Effectiveness, and World Order*. Cambridge: Cambridge University Press, 2006.

OTHER

Environmental Law & Policy Center. http://www.elpc.org/ (accessed November 10, 2010).

United Nations Environment Programme (UNEP). "Environmental Law and Conventions (DELC)." http://www.unep.org/dec/ (accessed November 10, 2010).

Joseph P. Hyder

International Geosphere-Biosphere Programme (U.N. Environmental Programme)

A major effort to organize research on important, worldwide scientific questions such as climate change was begun in the early 1980s. Largely through the efforts of scientists from two U.S. organizations, the National Aeronautics and Space Administration (NASA) and the National Research Council, a proposal was developed for the creation of an International Geosphere-Biosphere Programme (IGBP). The purpose of the IGBP was to help scientists from around the world focus on major issues about which there was still too little information. Activity funding comes from national governments, scientific societies, and private organizations.

IGBP was not designed to be a new organization. Instead, it was conceived of as a coordinating program that would call on existing organizations to attack certain problems. The proposal was submitted in September 1986 to the General Assembly of the International Council of Scientific Unions (ICSU), where it received enthusiastic support.

Within two years, more than twenty nations agreed to cooperate with IGBP, forming national committees to work with the international office. A small office, administered by Harvard oceanographer James McCarthy, was installed at the Royal Swedish Academy of Sciences in Stockholm.

IGBP has identified research questions and existing programs that fit the program's goals and developed new research efforts. Because many global processes are gradual, a number of IGBP projects are designed with time frames of ten to twenty years and beyond.

Resources

BOOKS

Allen, John. *The Biospheres: A Memoir by the Inventor of Biosphere II.* Santa Fe: Synergetic Press, 2008.

Bolin, Bert. *A History of the Science and Politics of Climate Change: The Role of the Intergovernmental Panel on Climate Change.* New York: Cambridge University Press, 2008.

Mackenzie, Fred T. *Carbon in the Geobiosphere: Earth's Outer Shell.* Dordrecht, Netherlands: Springer, 2006.

David E. Newton

International Institute for Sustainable Development

The International Institute for Sustainable Development (IISD) is a nonprofit organization that serves as an information and resources clearinghouse for policy makers promoting sustainable development. IISD aims to influence decision-making worldwide by assisting with policy analysis, providing information about practices, measuring sustainability, and building partnerships to further sustainability goals. It serves businesses, governments, communities, and individuals in both developing and industrialized nations. IISD's stated aim is to "create networks designed to move sustainable development from concept to practice." Founded in 1990 and based in Winnipeg, Canada, IISD is funded by foundations, governmental organizations, private sector sources, and revenue from publications and products.

IISD works in seven program areas. The Business Strategies program focuses on improving competitiveness, creating jobs, and protecting the environment through sustainability. Projects include several publications and the EarthEnterprise program, which offers entrepreneurial and employment strategies. IISD's Trade and Sustainable Development program works on building positive relationships between trade, the environment, and development. It examines how to make international accords, such as those made by the World Trade Organization, compatible with the goals of sustainable development. The Community Adaptation and Sustainable Livelihoods program identifies adaptive strategies for drylands in Africa and India, and it examines the influences of policies and new technology on local ways of life. The Great Plains program works with community, farm, government, and industry groups to assist

communities in the Great Plains region of North America with sustainable development. It focuses on government policies in agriculture, such as the Western Grain Transportation Act, as well as loss of transportation subsidies, the North American Free Trade Agreement (NAFTA), soil salination and loss of wetlands, job loss, and technological advances. Measurement and Indicators aims to set measurable goals and progress indicators for sustainable development. As part of this, IISD offers information about the successful uses of taxes and subsidies to encourage sustainability worldwide. Common Security focuses on initiatives of peace and consensus- building.

IISD's Information and Communications program offers several publications and Internet sites featuring information on sustainable development issues, terms, events, and media coverage. This includes *Earth Negotiations Bulletin*, which provides online coverage of major environmental and development negotiations (especially United Nations conferences), and IISDnet, with information about sustainable development worldwide. IISD also publishes books, monographs, and discussion papers, including *Sourcebook on Sustainable Development*, which lists organizations, databases, conferences, and other resources. IISD produces five journals that include *Developing Ideas*, published bimonthly both in print and electronic form, and featuring articles on sustainable development terms, issues, resources, and recent media coverage. *Earth Negotiations Bulletin* reports on conferences and negotiation meetings, especially United Nations conferences. IISD's Internet journal, */linkages/journal/*, focuses on global negotiations. Its reporting service, Sustainable Developments, reports on environmental and development negotiations for meetings and symposia via the Internet. IISD also operates IISDnet, an Internet information site featuring research, new trends, global activities, contacts, information on IISD's activities and projects, including United Nations negotiations on environment and development, corporate environmental reporting, and information on trade issues.

Resources

BOOKS

Voigt, Christina. *Sustainable Development as a Principle of International Law: Resolving Conflicts between Climate Measures and WTO Law.* Leiden, Netherlands: Martinus Nijhoff Publishers, 2009.

OTHER

Sustainable Development Institute. http://www.susdev.org/ (accessed October 12, 2010).

Sustainable Development Research Institute. http://
www.sdri.ubc.ca/ (accessed October 12, 2010).

ORGANIZATIONS

International Institute for Sustainable Development, 161
Portage Avenue East, 6th Floor, Winnipeg, Manitoba,
Canada R3B 0Y4, (204) 958-7700, (204) 958-7710,
info@iisd.ca, http://www.iisd.org.

Carol Steinfeld

International Joint Commission

The International Joint Commission (IJC) is a
permanent, independent organization of the United
States and Canada formed to resolve trans boundary
ecological concerns. Founded in 1912 as a result of
provisions under the Boundary Waters Treaty of 1909,
the IJC was patterned after an earlier organization, the
Joint Commission, which was formed by the United
States and Britain.

The IJC consists of six commissioners, with three
appointed by the president of the United States, and
three by the governor-in-council of Canada, plus support personnel. The commissioners and their organizations generally operate free from direct influence or
instruction from their national governments. The IJC
is frequently cited as an excellent model for international dispute resolution because of its history of successfully and objectively dealing with natural resources
and environmental disputes between friendly countries.

The major activities of the IJC have dealt with
apportioning, developing, conserving, and protecting
the binational water resources of the United States
and Canada. Some other issues, including transboundary air pollution, have also been addressed by the
commission.

The power of the IJC comes from its authority to
initiate scientific and socio-economic investigations,
conduct quasi-judicial inquiries, and arbitrate disputes.

Of special concern to the IJC have been issues
related to the Great Lakes. Since the early 1970s, IJC
activities have been substantially guided by provisions
under the 1972 and 1978 Great Lakes Water Quality
Agreement plus updated protocols. For example, it is
widely acknowledged, and well documented, that environmental quality and ecosystem health have been substantially degraded in the Great Lakes. In 1985 the
Water Quality Board of the IJC recommended that

states and provinces with Great Lakes boundaries
make a collective commitment to address this communal problem, especially with respect to pollution. These
governments agreed to develop and implement remedial action plans (RAPs) toward the restoration of
environmental health within their political jurisdictions. Forty-three areas of concern have been identified
on the basis of environmental pollution, and each of
these will be the focus of a remedial action plan.

An important aspect of the design and intent of
the overall program, and of the individual RAPs, will
be developing a process of integrated ecosystem management. Ecosystem management involves systematic,
comprehensive approaches toward the restoration and
protection of environmental quality. The ecosystem
approach involves consideration of interrelationships
among land, air, and water, as well as those between
the inorganic environment and the biota, including
humans. The ecosystem approach would replace the
separate, more linear approaches that have traditionally been used to manage environmental problems.
These conventional attempts have included directed
programs to deal with particular resources such as
fisheries, migratory birds, land use, or point sources
and area sources of toxic emissions. Although these
non-integrated methods have been useful, they have
been limited because they have failed to account for
important inter-relationships among environmental
management programs, and among components of
the ecosystem.

Resources

OTHER

United States Environmental Protection Agency (EPA).
"International Cooperation: Border Issues:
Transboundary Pollution." http://www.epa.gov/
ebtpages/inteborderissuestransboundarypollution.html
(accessed October 12, 2010).
United States Environmental Protection Agency (EPA).
"International Cooperation: Water: Transboundary."
http://www.epa.gov/ebtpages/intewatertransboundary.
html (accessed October 12, 2010).
United States Environmental Protection Agency (EPA).
"Water: Water Pollution: Transboundary Pollution."
http://www.epa.gov/ebtpages/watewaterpollutiontrans
boundarypollution.html (accessed October 12, 2010).

ORGANIZATIONS

International Joint Commission, 2000 L Street, NW, Suite
615, Washington, DC, USA 20440, (202) 736-9024,
(202) 632-2007, http://www.ijc.org

Bill Freedman

International Primate Protection League

Founded in 1974 by Shirley McGreal, International Primate Protection League (IPPL) is a global conservation organization that works to protect non-human primates, especially monkeys and apes (chimpanzees, orangutans, gibbons, and gorillas).

IPPL has 30,000 members; branches in the United Kingdom, Germany, and Australia; and field representatives in thirty-one countries. Its advisory board consists of scientists, conservationists, and experts on primates, including the world-renowned primatologist Jane Goodall, whose famous studies and books are considered the authoritative texts on chimpanzees. Her studies have also heightened public interest and sympathy for chimpanzees and other nonhuman primates.

IPPL runs a sanctuary and rehabilitation center at its Summerville, South Carolina, headquarters, which houses two dozen gibbons and other abandoned, injured, or traumatized primates who are refugees from medical laboratories or abusive pet owners. IPPL concentrates on investigating and fighting the multi-million dollar commercial trafficking in primates for medical laboratories, the pet trade, and zoos, much of which is illegal trade and smuggling of endangered species protected by international law.

IPPL's work has helped to save the lives of thousands of monkeys and apes, many of which are threatened or endangered species. For example, the group was instrumental in persuading the governments of India and Thailand to ban or restrict the export of monkeys, which were being shipped by the thousands to research laboratories and pet stores across the world.

IPPL often undertakes actions and projects that are dangerous and require a good deal of skill. In 1992, its investigations led to the conviction of a Miami, Florida, animal dealer for conspiring to help smuggle six baby orangutans captured in the jungles of Borneo. The endangered orangutan is protected by the Convention on International Trade in Endangered Species of Wild Fauna and Flora (CITES), as well as by the United States Endangered Species Act. In retaliation, the dealer unsuccessfully sued McGreal, as did a multi-national corporation she once criticized for its plan to capture chimpanzees and use them for hepatitis research in Sierra Leone.

One of IPPL's victories occurred in April 2002. In 1997 Chicago O'Hare airport received two shipments from Indonesia, each of which contained more than 250 illegally imported monkeys. Included in the shipments were dozens of unweaned baby monkeys. After several years of pursuing the issue, the U.S. Fish and Wildlife Service and the U.S. Federal prosecutors charged the LABS Company (a breeder of monkeys for research based in the United States) and several of its employees, including its former president, on eight felonies and four misdemeanors.

IPPL publishes *IPPL News*, online news updates, and periodic emails alerting members of events, volunteer and fundraising opportunities, and issues that affect primates worldwide.

ORGANIZATIONS

International Primate Protection League, P.O. Box 766, Summerville, SC, USA 29484, (843) 871-2280, (843) 871-7988, info@ippl.org, http://ippl.org.

Lewis G. Regenstein

International Register of Potentially Toxic Chemicals (U. N. Environment Programme)

The International Register of Potentially Toxic Chemicals is published by the United Nations Environment Programme (UNEP). Part of UNEP's Chemicals Branch and the UN's three-pronged Earthwatch program, the register is an international inventory of chemicals that threaten the environment. Along with the Global Environment Monitoring System and INFOTERRA, the register monitors and measures environmental problems worldwide. Information from the register is routinely shared with agencies in developing countries. Such countries have long been the toxic dumping grounds for the world, and they still use many chemicals that have been banned elsewhere. Environmental groups regularly send information from the register to toxic chemical users in developing countries as part of their effort to stop the export of toxic pollution.

Resources

BOOKS

Rapp, Doris. *Our Toxic World: A Wake Up Call*. Buffalo, NY: Environmental Research Foundation, 2004.

Walker, C.H. *Principles of Ecotoxicology*. Boca Raton, FL, and London, UK: CRC, 2006.

OTHER

United Nations System-Wide EarthWatch. "Toxic Chemicals." http://earthwatch.unep.net/toxicchem/index.php (accessed October 19, 2010).

ORGANIZATIONS

International Register of Potentially Toxic Chemicals, Chemin des Anémones 15, Châtelaine, Switzerland, CH–1219, +41-22-979 91 11, +41-22-979 91 70, chemicals@chemicals.unep.ch

ORGANIZATIONS

International Society for Environmental Ethics, Center for Environmental Philosophy, University of North Texas, 115 Union Circle #310980, Denton, TX, USA 76203-5017, (940) 565-2727, http://www.cep.unt.edu/ISEE.html

Nicole Beatty

International Society for Environmental Ethics

The International Society for Environmental Ethics (ISEE) is an organization that seeks to educate people about the environmental ethics and philosophy concerning nature. An environmental ethic is the philosophy that humans have a moral duty to sustain the natural environment. It attempts to answer how humans should treat other species (plant and animal), use Earth's natural resources, and place value on the aesthetic experiences of nature.

The society is an auxiliary organization of the American Philosophical Association, with hundreds of members in more than twenty countries. Many of ISEE's current members are philosophers, teachers, or environmentalists.

ISEE publishes a quarterly newsletter available to members and maintains an online archive of its publications. Of special note is the ISEE Bibliography, an ongoing project that contains one of the world's largest databases of works on environmental ethics with records from journals such as *Environmental Ethics*, *Environmental Values*, and the *Journal of Agricultural and Environmental Ethics*.

The ISEE Syllabus Project maintains a database of university-level course offerings in environmental philosophy and ethics, based on information from two-year community colleges and four-year state universities, private institutions, and master's- and doctorate-granting universities.

Resources

BOOKS

Curry, Patrick. *Ecological Ethics: An Introduction*. Cambridge, UK: Polity Press, 2006.

Des Jardins, Joseph R. *Environmental Ethics: An Introduction to Environmental Philosophy*. Belmont, CA: Wadsworth Publishing, 2005.

Newton, Lisa H., et al. *Watersheds 4: Ten Cases in Environmental Ethics*. Belmont, CA: Wadsworth Publishing, 2005.

International trade in toxic waste

Just as computers, cars, and laundry soap are traded across borders, so too is the waste that accompanies their production. In the United States alone, industrial production accounts for at least 500 million lb (230 million kg) of hazardous waste a year. The industries of other developed nations also produce waste. Although some of it is disposed within national borders, a portion is sent to other countries where costs are cheaper and regulations less stringent than in the waste's country of origin.

Unlike consumer products, internationally traded hazardous waste has begun to meet local opposition. In some recent high-profile cases, barges filled with waste have traveled the world looking for final resting places. In at least one case, a ship may have dumped about ten tons of toxic municipal incinerator ash in the ocean after being turned away from dozens of ports. In recent years national and international bodies have begun to voice official opposition to this dangerous trade through bans and regulations.

Typically a manufacturing facility generates waste during the production process. The facility manager pays a waste-hauling firm to dispose of the waste. If the landfills in the country of origin cost too much, or if there are no landfills that will take the waste, the disposal firm will find a cheaper option, perhaps a landfill in another country. In the United States, the shipper must then notify the Environmental Protection Agency (EPA), which then notifies the State Department. After ascertaining that the destination country will indeed accept the waste, American regulators approve the sale.

Disposing of the waste overseas in a landfill is only the most obvious example of this international trade. Waste haulers also sell their cargo as raw materials for recycling. For example, used lead-acid batteries discarded by American consumers are sent to Brazil where factory workers extract and resmelt the lead. Though the lead-acid alone would classify as hazardous, whole batteries do not. Waste haulers can ship these batteries overseas without notification to Mexico,

deterioration of Canada's impending endangered species legislation. Their United States-based operation has built a reputation in the research field working with government agencies to ensure that whale-watching on the eastern seaboard does not harm the whales.

Resources

BOOKS

Chadwick, Douglas H. *The Grandest of Lives: Eye to Eye with Whales.* San Francisco: Sierra Club Books, 2006.

OTHER

University of California Museum of Paleontology. "Marine Vertebrates: Whales and Dolphins." http://www.ucmp.berkeley.edu/mammal/cetacea/cetacean.html (accessed October 12, 2010).

ORGANIZATIONS

International Wildlife Coalition, 70 East Falmouth Highway, East Falmouth, MA, USA 02536, (508) 548-8328, (508) 548-8542.

David Duffus

Intrinsic value

Intrinsic value describes attributes of an entity that may not be calculable in market-based monetary valuations. However, intrinsic values may nevertheless be valuable. The northern spotted owl (*Strix occidentalis caurina*) for example, has no instrumental or market value; it is not a means to any human end, nor is it sold or traded in any market. But, environmentalists argue, utility and price are not the only measures of worth. Indeed, they say, some of the things humans value most—truth, love, respect—are not for sale at any price, and to try to put a price on them would only tend to cheapen them. Such things have "intrinsic value."

Similarly, environmentalists say, the natural environment and its myriad life-forms are valuable in their own right. Wilderness, for instance, has intrinsic value and is worthy of protecting for its own sake. To say that something has intrinsic value is not necessarily to deny that it may also have instrumental value for humans and non-human animals alike. Deer, for example, have intrinsic value; but they also have instrumental value as a food source for wolves and other predator species.

See also Shadow pricing.

Introduced species

Introduced species (also called invasive species) are those that have been released by humans into an area to which they are not native. These releases can occur accidently, from places such as the cargo holds of ships. They can also occur intentionally, and species have been introduced for a range of ornamental and recreational uses, as well as for agricultural, medicinal, and pest control purposes.

Introduced species can have dramatically unpredictable effects on the environment and native species. Such effects can include overabundance of the introduced species, competitive displacement, and disease-caused mortality of the native species. Numerous examples of adverse consequences associated with the accidental release of species or the long term effects of deliberately introduced species exist in the United States and around the world. Introduced species can be beneficial as long as they are carefully regulated. Almost all the major varieties of grain and vegetables used in the United States originated in other parts of the world. This includes corn, rice, wheat, tomatoes, and potatoes.

The kudzu vine, which is native to Japan, was deliberately introduced into the southern United States for erosion control and to shade and feed livestock. It is, however, an extremely aggressive and fast-growing species, and it can form continuous blankets of foliage that cover forested hillsides, resulting in malformed and dead trees. Other species introduced as ornamentals have spread into the wild, displacing or outcompeting native species. Several varieties of cultivated roses, such as the multiflora rose, are serious pests and nuisance shrubs in fields and pastures. The purple loosestrife, with its beautiful purple flowers, was originally brought from Europe as a garden ornamental. It has spread rapidly in freshwater wetlands in the northern United States, displacing other plants such as cattails. This is viewed with concern by ecologists and wildlife biologists since the food value of loosestrife is minimal, while the roots and starchy tubes of cattails are an important food source to muskrats. Common ragweed was accidently introduced to North America, and it is now a major health irritant for many people.

Introduced species are sometimes so successful because human activity has changed the conditions of a particular environment. The Pine Barrens of southern New Jersey form an ecosystem that is naturally acidic and low in nutrients. Bogs in this area support a number of slow-growing plant species that are adapted to these conditions, including peat moss, sundews, and pitcher plants. But urban runoff, which contains fertilizers, and wastewater effluent, which is high in both nitrogen and phosphorus, have enriched the bogs; the waters there have become less acidic and shown a gradual elevation in the concentration of nutrients. These changes in aquatic chemistry have resulted in changes in plant species, and the acidophilus mosses and herbs are being replaced by fast-growing plants that are not native to the Pine Barrens.

Zebra mussels were transported by accident from Europe to the United States, and they are causing severe problems in the Great Lakes. They proliferate at a prodigious rate, crowding out native species and clogging industrial and municipal water-intake pipes. Many ecologists fear that shipping traffic will transport the zebra mussel to harbors all over the country. Scattered observations of this tiny crustacean have already been made in the lower Hudson River in New York.

Although introduced species are usually regarded with concern, they can occasionally be used to some benefit. The water hyacinth is an aquatic plant of tropical origin that has become a serious clogging nuisance in lakes, streams, and waterways in the southern United States. Numerous methods of physical and chemical removal have been attempted to eradicate or control it, but research has also established that the plant can improve water quality. The water hyacinth has proved useful in the withdrawal of nutrients from sewage and other wastewater. Many constructed wetlands, polishing ponds, and waste lagoons in waste treatment plants now take advantage of this fact by routing wastewater through floating beds of water hyacinth.

The reintroduction of native species is extremely difficult, and it is an endeavor that has had low rates of success. Efforts by the U.S. Fish and Wildlife Service to reintroduce the endangered whooping crane into native habitats in the southwestern United States were initially unsuccessful because of the fragility of the eggs, as well as the poor parenting skills of birds raised in captivity. The FWS then devised a strategy of allowing the more common sandhill crane to incubate the eggs of captive whooping cranes in wilderness nests, and the fledglings were then taught survival skills by their surrogate parents. Such projects, however, are extremely time and labor intensive; they are also costly and difficult to implement for large numbers of most species.

Due to the difficulties and expense required to protect native species and to eradicate introduced species, there are not many international laws and policies that seek to prevent these problems before they begin. Thus customs agents at ports and airports routinely check luggage and cargo for live plant and animal materials to prevent the accidental or deliberate transport of non-native species. Quarantine policies are also designed to reduce the probability of spreading introduced species, particularly diseases, from one country to another.

There are similar concerns about genetically engineered organisms, and some have argued that their creation and release could have the same environmental consequences as some introduced species. Both the Food and Drug Administration and the Environmental Protection Agency (EPA) impose some controls on the field testing of bioengineered products, as well as on their cultivation and use.

Conservation policies for the protection of native species are now focused on habitats and ecosystems rather than single species. It is easier to prevent the encroachment of introduced species by protecting an entire ecosystem from disturbance, and this is increasingly well recognized both inside and outside the conservation community.

See also Bioremediation; Endangered species; Fire ants; Gypsy moth; Rabbits in Australia; Wildlife management.

Resources

BOOKS

Mooney, Harold A. *Invasive Alien Species: A New Synthesis.* Washington, DC: Island Press, 2005.

OTHER

United Nations System-Wide EarthWatch. "Invasive Species." http://earthwatch.unep.net/emergingissues/biodiversity/invasivespecies.php (accessed November 6, 2010).

Usha Vedagiri
Douglas Smith

Inversion *see* **Atmospheric inversion.**

Iodine 131

Iodine 131 is a radioactive isotope of the element iodine. During the 1950s and early 1960s, iodine-131 attached to atmospheric water droplets and dust was a major public health concern because of open testing of atomic weapons. Along with cesium-137 and strontium-90, iodine-131 was one of the three most abundant isotopes found in the fallout from the atmospheric testing of nuclear weapons. These three isotopes settled to the earth's surface and were ingested by cows, ultimately affecting humans by way of dairy products. In the human body, iodine-131, like all forms of that element, tends to concentrate in the thyroid, where it may cause cancer and other health disorders. The 1986 Chernobyl nuclear reactor explosion in the Ukraine is also known to have released large quantities of iodine-131 into the atmosphere.

See also Radioactivity.

Ion

An atom that has gained or lost electrons. Ions are electrically charged. Positive ions result from negatively charged electrons and are called *cations* because when charged electrodes are placed in a solution containing ions, the positive ions migrate to the cathode (negative electrode). Negative ions (those that have gained extra electrons) are called *anions* because they migrate toward the anode (positive electrode). Environmentally important cations include the hydrogen ion (H^+) and dissolved metals. Important anions include the hydroxyl ion (OH^-) as well as many of the dissolved ions of nonmetallic elements.

See also Ion exchange; Ionizing radiation.

Ion exchange

The process of replacing one ion that is attached to a charged surface with another. An example of ion exchange is the exchange of cations bound to soil particles. Soil clay minerals and organic matter both have negative surface charges that bind cations. In a fertile soil the predominant exchangeable cations are Ca^{2+}, Mg^{2+} and K^+. In acid soils Al^{3+} and H^+ are also important exchangeable ions. When materials containing cations are added to soil, cations leaching through the soil are retarded by cation exchange.

Ionizing radiation

High-energy radiation with penetrating competence such as x rays and gamma rays, which induces ionization in living material. Molecules are bound

together with covalent bonds, and generally an even number of electrons binds the atoms together. However, high-energy penetrating radiation can fragment molecules resulting in atoms with unpaired electrons known as "free radicals." The ionized free radicals are exceptionally reactive, and their interaction with the macromolecules (DNA, RNA, and proteins) of living cells can, with high dosage, lead to cell death. Cell damage (or death) is a function of penetration ability, the kind of cell exposed, the length of exposure, and the total dose of ionizing radiation. Cells that are mitotically active and have a high oxygen content are most vulnerable to ionizing radiation.

See also Radiation exposure; Radiation sickness; Radioactivity.

Iron minerals

The oxides and hydroxides of ferric iron (Fe(III)) are important minerals in many soils, and are important suspended solids in some fresh water systems. Common oxides and hydroxides of iron include goethite, hematite, lepidocrocite, and ferrihydrite.

These minerals tend to be very finely divided and can be found in the clay-sized fraction of soils, and like other clay-sized minerals, are important adsorbers of ions. At high pH they adsorb hydroxide (OH^-) ions creating negatively charged surfaces that contribute to cation exchange surfaces. At low pH they adsorb hydrogen (H^+) ions, creating anion exchange surfaces. In the pH range between 8 and 9 the surfaces have little or no charge. Iron hydroxide and oxide surfaces strongly adsorb some environmentally important anions, such as phosphate, arsenate, and selanite, and cations such as copper, lead, manganese, and chromium. These ions are not exchangeable, and in environments where iron oxides and hydroxides are abundant, surface adsorption can control the mobility of these strongly adsorbed ions.

The hydroxides and oxides of iron are found in the greatest abundance in older highly weathered landscapes. These minerals are very insoluble; during soil weathering they form from the iron that is released from the structure of the soil-forming minerals. Thus, iron oxide and hydroxide minerals tend to be most abundant in old landscapes that have not been affected by glaciation, and in landscapes where the rainfall is high and the rate of soil mineral weathering is high. These minerals give the characteristic red (hematite or ferrihydrite) or yellow-brown (goethite) colors to soils that are common in the tropics and subtropics.

See also Arsenic; Erosion; Ion exchange; Phosphorus; Soil profile; Soil texture.

Irradiation of food *see* **Food irradiation.**

Irrigation

Irrigation is the method of supplying water to land to support plant growth. This technology has had a powerful role in the history of civilization. In arid regions sunshine is plentiful and soil is usually fertile, so irrigation supplies the critical factor needed for plant growth. Yields have been high, but not without costs. Historic problems include salinization and water logging; contemporary difficulties include immense costs, spread of water-borne diseases, and degraded aquatic environments.

One geographer described California's Sierra Nevada as the "mother nurse of the San Joaquin Valley." Its heavy winter snowpack provides abundant and extended runoff for the rich valley soils below. Numerous irrigation districts, formed to build diversion and storage dams, supply water through gravity-fed canals. The snow melt is low in nutrients, so salinization problems are minimal. Wealth from the lush fruit orchards has enriched the state.

By contrast, the Colorado River, like the Nile, flows mainly through arid lands. Deeply incised in places, the river is also limited for irrigation by the high salt content of desert tributaries. Still, demand for water exceeds supply. Water crossing the border into Mexico is so saline that the U.S. government built a desalinization plant at Yuma, Arizona. Colorado River water is imperative to the Imperial Valley, which specializes in winter produce in the rich, delta soils. To reduce salinization problems, one-fifth of the water used must be drained off into the growing Salton Sea.

Salinization and water logging have long plagued the Tigris, Euphrates, and Indus River flood plains. Once fertile areas of Iraq and Pakistan are covered with salt crystals. Half of the irrigated land in the western United States is threatened by salt buildup.

Some of the worst problems are degraded aquatic environments. The Aswan High Dam in Egypt has greatly amplified surface evaporation, reduced nutrients to the land and to fisheries in the delta, and has contributed to the spread of schistosomiasis via water snails in irrigation ditches. Diversion of drainage away from the Aral Sea for cotton irrigation has severely lowered the shoreline, and threatens this water body with ecological disaster.

In the United States, spray irrigation in the High Plains is lowering the Ogallala Aquifer's water table, raising pumping costs. Kesterson Marsh in the San Joaquin Valley has become a hazard to wildlife because of selenium poisoning from irrigation drainage. The U.S. Bureau of Reclamation has invested huge sums in dams and reservoirs in western states. Some question the wisdom of such investments, given the past century of farm surpluses, and argue that water users are not paying the true cost.

Irrigation still offers great potential, but only if used with wisdom and understanding. New technologies may yet contribute to the world's ever-increasing need for food.

See also Climate; Commercial fishing; Reclamation.

Resources

BOOKS

Svendsen, Mark. *Irrigation and River Basin Management: Options for Governance and Institutions.* Wallingford, UK, and Cambridge, MA: CABI Publishing, 2005.

OTHER

United States Department of the Interior, United States Geological Survey (USGS). "Irrigation." http://www.usgs.gov/science/science.php?term=607 (accessed October 19, 2010).

Nathan H. Meleen

Farm irrigation system. *(Tish1/Shutterstock.com)*

Island biogeography

Island biogeography is the study of past and present animal and plant distribution patterns on islands and the processes that created those distribution patterns. Historically, island biogeographers mainly studied geographic islands—continental islands close to shore in shallow water and oceanic islands of the deep sea. In the last several decades, however, the study and principles of island biogeography have been extended to ecological islands such as forests and prairie fragments isolated by human development. Biogeographic "islands" may also include ecosystems isolated on mountaintops and landlocked bodies of water such as Lake Malawi in the African Rift Valley. Geographic islands, however, remain the main laboratories for developing and testing the theories and methods of island biogeography.

otherwise extinct species of plant or animal) scraps of mainland ecosystems in which little changed—or closed systems mainly driven by evolution. That view began to change radically in 1967 when Robert H. MacArthur and Edward O. Wilson published *The Theory of Island Biogeography*.

In their book, MacArthur and Wilson detail the equilibrium theory of island biogeography—a theory that became the new paradigm of the field. The authors proposed that island ecosystems exist in dynamic equilibrium, with a steady turnover of species. Larger islands—as well as islands closest to a source of immigrants—accommodate the most species in the equilibrium condition, according to the theory. MacArthur and Wilson also worked out mathematical models to demonstrate and predict how island area and isolation dictate the number of species that exist in equilibrium.

Equilibrium theory

Until the 1960s, biogeographers thought of islands as living museums—relict (persistent remnant of an

Dispersion

The driving force behind species distribution is dispersion—the means by which plants and animals

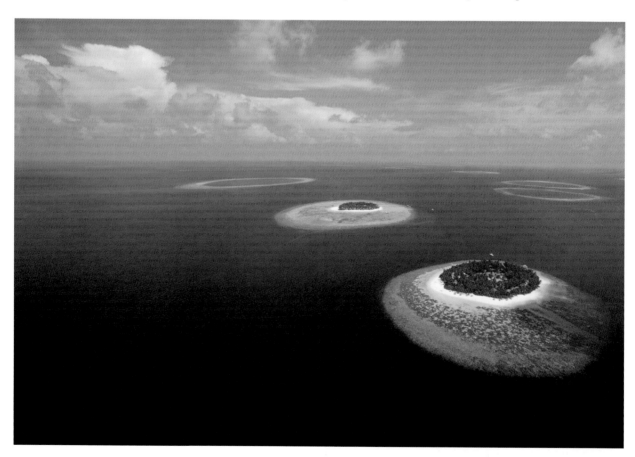

Aerial view of the Maldives Islands in the Indian Ocean. The islands are fringed with coral reefs and separated by deep water. Ring-shaped coral atolls are also seen. *(Alexis / Photo Researchers, Inc.)*

actively leave or are passively transported from the source area. An island ecosystem can have more than one source of colonization, but nearer sources dominate. How readily plants or animals disperse is one of the main reasons equilibrium will vary from species to species.

Birds and bats are obvious candidates for anemochory (dispersal by air), but some species normally not associated with flight are also thought to reach islands during storms or even normal wind currents. Orchids, for example, have hollow seeds that remain airborne for hundreds of kilometers. Some small spiders, along with other insects like bark lice, aphids, and ants (collectively knows as aerial plankton), often are among the first pioneers of newly formed islands.

Whether actively swimming or passively floating on logs or other debris, dispersal by sea is called thallasochory. Crocodiles have been found on Pacific islands 600 miles (950 km) from their source areas, but most amphibians, larger terrestrial reptiles, and, in particular, mammals, have difficulty crossing even narrow bodies of water. Thus, thallasochory is the medium of dispersal primarily for fish, plants, and insects. Only small vertebrates such as lizards and snakes are thought to arrive at islands by sea on a regular basis.

Zoochory is transport either on or inside an animal. This method is primarily a means of plant dispersal, mostly by birds. Seeds ride along either stuck to feathers or survive passage through a bird's digestive tract and are deposited in new territory.

Anthropochory is dispersal by human beings. Although humans intentionally introduce domestic animals to islands, they also bring unintended invaders, such as rats.

Getting to islands is just the first step, however. Plants and animals often arrive to find harsh and alien conditions. They may not find suitable habitats. Food chains they depend on might be missing. Even if they manage to gain a foothold, their limited numbers make them more susceptible to extinction. Chances of success are better for highly adaptable species and those that are widely distributed beyond the island. Wide distribution increases the likelihood a species on the verge of extinction may be saved by the rescue effect, the replenishing of a declining population by another wave of immigration.

Challenging established theories

Many biogeographers point out that isolated ecosystems are more than just collections of species that can make it to islands and survive the conditions they encounter there. Several other contemporary theories of island biogeography build on MacArthur and Wilson's theory; other theories contradict it.

Equilibrium theory suggests that species turnover is constant and regular. Evidence collected so far indicates MacArthur and Wilson's model works well in describing communities of rapid dispersers who have a regular turnover, such as insects, birds, and fish. However, this model may not apply to species who disperse more slowly.

Proponents of historical legacy models argue that communities of larger animals and plants (forest trees, for example) take so long to colonize islands that changes in their populations probably reflect sudden climactic or geological upheaval rather than a steady turnover. Other theories suggest that equilibrium may not be dynamic, that there is little or no turnover. Through competition, established species keep out new colonists; the newcomers might occupy the same ecological niches as their predecessors. Established species may also evolve and adapt to close off those niches. Island resources and habitats may also be distinct enough to limit immigration to only a few well-adapted species.

Thus, in these later models, dispersal and colonization are not nearly as random as in MacArthur and Wilson's model. These less random, more deterministic theories of island ecosystems conform to specific assembly rules—a complex list of factors accounting for the species present in the source areas, the niches available on islands, and competition between species.

Some biogeographers suggest that every island—and perhaps every habitat on an island—may require its own unique model. Human disruption of island ecosystems further clouds the theoretical picture. Not only are habitats permanently altered or lost by human intrusion, but anthropochory also reduces an island's isolation. Thus, finding relatively undisturbed islands to test different theories can be difficult.

Since the time of naturalists Charles Darwin and his colleague, Alfred Wallace, islands have been ideal "natural laboratories" for studying evolution. Patterns of evolution stand out on islands for two reasons: island ecosystems tend to be simpler than other geographical regions, and they contain greater numbers of endemic species, plant, and animal species occurring only in a particular location.

Many island endemics are the result of adaptive radiation—the evolution of new species from a single lineage for the purpose of filling unoccupied ecological niches. Many species from mainland source areas simply never make it to islands, so species that can immigrate find empty ecological niches where once they faced competition. For example, monitor lizards immigrating to several small islands in Indonesia found the niche for large predators empty. Monitors on these islands evolved into Komodo Dragons, filling the niche.

Conservation of biodiversity

Theories of island biogeography also have potential applications in the field of conservation. Many conservationists argue that as human activity such as logging and ranching encroach on wild lands, remaining parks and reserves begin to resemble small, isolated islands. According to equilibrium theory, as those patches of wild land grow smaller, they support fewer species of plants and animals. Some conservationists fear that plant and animal populations in those parks and reserves will sink below minimum viable population levels—the smallest number of individuals necessary to allow the species to continue reproducing. These conservationists suggest that one way to bolster populations is to set aside larger areas and to limit species isolation by connecting parks and preserves with wildlife corridors.

Islands with the greatest variety of habitats support the most species; diverse habitats promote successful dispersal, survival, and reproduction. Thus, in attempting to preserve island biodiversity, conservationists focus on several factors: the size (the larger the island, the more habitats it contains), climate, geology (soil that promotes or restricts habitats), and age of the island (sparse or rich habitats). All of these factors must be addressed to ensure island biodiversity.

Resources

BOOKS

Cox, C. Barry, and Peter D. Moore. *Biogeography: An Ecological and Evolutionary Approach*. New York: Sinauer Associates, 2010.

Lomolino, Mark V., Brett R. Riddle, Robert J. Whittaker, and James H. Brown. *Biogeography, Fourth Edition*. New York: Springer, 2010.

Losos, Jonathan, and Robert E. Ricklefs. *The Theory of Island Biogeography Revisited*. Princeton, NJ: Princeton University Press, 2009.

ORGANIZATIONS

Environmental Protection Agency (EPA), 1200 Pennsylvania Avenue, NW, Washington, DC, USA 20460, (202) 272-0167, public-access@epa.gov, http://www.epa.gov

Darrin Gunkel

ISO 14000: International Environmental Management Standards

ISO 14000 refers to a series of environmental management standards that were adopted by the International Organization for Standardization (ISO) in 1996 and are implemented by businesses worldwide, including businesses themselves, their legal representatives, environmental organizations and their members, government officials, and others.

What is the ISO and what are ISO standards?

The International Organization for Standardization (ISO) is a private (nongovernmental) worldwide organization whose purpose is to promote uniform standards in international trade. Its members are elected representatives from national standards organizations in 163 countries. The ISO covers all fields involving promoting goods, services, or products and where a member body suggests that standardization is desirable, with the exception of electrical and electronic engineering, which are covered by a different organization called the International Electrotechnical Commission (IEC). However, the ISO and the IEC work closely together.

Since the ISO began operations in 1947, its Central Secretariat has been located in Geneva, Switzerland. Between 1951 (when it published its first standard) and 2010, the ISO issued over 10,000 standards. Standards are documents containing technical specifications, rules, guidelines, and definitions to ensure equipment, products, and services perform as specified.

The ISO series is a set of standards for quality management and quality assurance. The standards apply to processes and systems that produce products; they do not apply to the products themselves. Further, the standards provide a general framework for any industry; they are not industry-specific. A company that has become registered under ISO has demonstrated that it has a documented system for quality that is in place and consistently applied. ISO 14000 standards apply to all kinds of companies whether large or small, in services or manufacturing.

The impetus for the ISO 14000 series came from the United Nations Conference on the Environment and Development (UNCED), which was held in Rio de Janeiro in 1992 and attended by representatives of over one hundred nations. One of the documents resulting from that conference was the *Global Environmental Initiative*, which prompted the ISO to develop its ISO 14000 series of international environmental standards. The ISO's goal is to insure that businesses adopt common internal procedures for environmental controls including, but not limited to, audits. It is important to note that the standards are process standards, not performance standards. The goal is to ensure that businesses are in compliance with their own national and local applicable environmental laws and regulations. The initial standards in the series include numbers 14001, 14004, and 14010-14012; all of them adopted by the ISO in 1996.

Provisions of ISO 14000 Series standards

ISO 14000 sets up criteria pursuant to which a company may become registered or certified as to its environmental management practices.

Central to the process of registration pursuant to ISO 14000 is a company's Environmental Management System (EMS). The EMS is a set of procedures for assessing compliance with environmental laws and company procedures for environmental protection, identifying and resolving problems, and engaging the company's workforce in a commitment to improved environmental performance by the company.

ISO 14001 series can be divided into two groups: guidance documents and specification documents. The series sets out standards against which a company's EMS will be evaluated. For example, it must include an accurate summary of the legal standards with which the company must comply, such as permit stipulations, and relevant provisions of statutes and regulations, and even provisions of administrative or court-certified consent judgments. To become certified, the EMS must: (1) include an environmental policy; (2) establish plans to meet environmental goals and comply with legal requirements; (3) provide for implementation of the policy and operation under it including training for personnel, communication, and document control; (4) set up monitoring and measurement devices and an audit procedure to insure continuing improvement; and (5) provide for management review. The EMS must be certified by a registrar who has been qualified under ISO 13012, a standard that predates the ISO 14000 series.

The ISO 14004 series is a guidance document that gives advice that may be followed but is not required. It includes five principles, each of which corresponds to one of the five areas of ISO 14001 listed above.

ISO 14010, 14011, and 14012 are auditing standards. For example, 14010 covers general principles of environmental auditing, and 14011 provides guidelines for auditing of an EMS. ISO 14012 provides guidelines for establishing qualifications for environmental auditors, whether those auditors are internal or external.

Plans for additional standards within the ISO 14000 Series standards

The ISO is considering proposals for standards on training and certifying independent auditors (called registrars) who will certify that ISO 14000-certified business have established and adhere to stringent internal systems to monitor and improve their own environmental protection actions. Later the ISO may also establish standards for assessing a company's environmental performance. Standards may be adopted for use of eco-labeling and life cycle assessment of goods involved in international trade.

Benefits and consequences of ISO 14000 Series standards

A company contemplating obtaining ISO 14000 registration must evaluate its potential advantages as well as its costs to the company.

ISO 14000 certification may bring various rewards to companies. For example, many firms are hoping that, in return for obtaining ISO 14000 certification (and the actions required to do so), regulatory agencies such as the U.S. Environmental Protection Agency (EPA) will give them more favorable treatment. For example, leniency might be shown in less stringent filing or monitoring requirements or even less severe sanctions for past or present violations of environmental statutes and regulations.

Further, compliance may be merely for good public relations, leading consumers to view the certified company as a good corporate citizen that works to protect the environment. There is public pressure on companies to demonstrate their environmental stewardship and accountability; obtaining ISO 14000 certification is one way to do so.

The costs to the company will depend on the scope of the EMS. For example, the EMS might be international, national, or limited to individual plants operated by the company. That decision will affect the costs of the environmental audit considerably. National and international systems may prove to be costly.

Nevertheless, a company may realize cost savings. For example, an insurance company may give reduced rates on insurance to cover accidental pollution releases to a company that has a proven environmental management system in place. Internally, by implementing an

EMS, a company may realize cost savings as a result of waste reduction, use of fewer toxic chemicals, less energy use, and recycling.

ISO 14000 has potential consequences with respect to international law as well as international trade. ISO 14000 is intended to promote a series of universally accepted EMS practices and lead to consistency in environmental standards between and among trading partners. Some developing countries such as Mexico are reviewing ISO 14000 standards and considering incorporating their provisions within their own environmental laws and regulations. On the other hand, some developing countries have suggested that environmental standards created by ISO 14000 may constitute nontariff barriers to trade, in that costs of ISO 14000 registration may be prohibitively high for small- to medium-size companies.

Companies that implemented various revisions of ISO have learned to view their companies' operations through the lenses of quality of management and environmental quality. ISO 14000 has the potential to lead to two kinds of cultural changes. First, within the corporation, it has the potential to lead to consideration of environmental issues throughout the company and its business decisions ranging from hiring of employees to marketing. Second, ISO 14000 has the potential to become part of a global culture as the public comes to view ISO 14000 certification as a benchmark connoting good environmental stewardship by a company.

Resources

OTHER

United States Environmental Protection Agency (EPA). "International Cooperation: Voluntary Standards: ISO 14000." http://www.epa.gov/ebtpages/intevoluntary standiso14000.html (accessed October 19, 2010).

Paulette L. Stenzel

Isotope

Atoms of the same element (number of protons) with different numbers of neutrons. Isotopes have the same atomic number, but differ in atomic mass. Atoms consist of a nucleus, containing positively charged particles (protons) and neutral particles (neutrons), surrounded by negatively charged particles (electrons). Isotopes of an element differ only in the number of neutrons in the nucleus and hence in atomic

weight. Isotopes vary in stability and some may be highly unstable and undergo relatively rapid radioactive decay. An element may have several stable and radioactive isotopes, but most elements have only two or three isotopes that are common. Also, for most elements, the radioactive isotopes are only of concern in material exposed to certain types of radiation sources. Carbon has three important isotopes with atomic weights of 12, 13, and 14. C-12 is stable and represents 98.9 percent of natural carbon. C-13 is also stable and represents 1.1 percent of natural carbon. C-14 represents an insignificant fraction of naturally occurring carbon, but it is radioactive and important because its radioactive decay is valuable in the dating of fossils and ancient artifacts. It is also useful in tracing the reactions of carbon compounds.

See also Nuclear fission; Nuclear power; Radioactivity; Radiocarbon dating.

Itai-itai disease

The symptoms of Itai-itai disease were first observed in 1913 and characterized between 1947 and 1955; it was 1968, however, before the Japanese Ministry of Health and Welfare officially declared that the disease was caused by chronic cadmium poisoning in conjunction with other factors such as the stresses of pregnancy and lactation, aging, and dietary deficiencies of vitamin D and calcium. The name arose from the cries of pain, "itai-itai" (ouch-ouch), by the most seriously stricken victims—older Japanese farm women. Although men, young women, and children were also exposed, 95 percent of the victims were post-menopausal women over fifty years of age. They usually had given birth to several children and had lived more than thirty years within 2 miles (3 km) of the lower stream of the Jinzu River near Toyama.

The disease started with symptoms similar to rheumatism, neuralgia, or neuritis. Then came bone lesions, osteomalacia, and osteoporosis, along with renal disfunction and proteinuria. As it escalated, pain in the pelvic region caused the victims to walk with a duck-like gait. Next, they were incapable of rising from their beds because even a slight strain caused bone fractures. The suffering could last many years before it finally ended with death. Overall, an estimated 199 victims have been identified, of which 162 had died by December 1992.

The number of victims increased during and after World War II as production expanded at the Kamioka

Mine owned by the Mitsui Mining and Smelting Company. As 3,000 tons of zinc-lead ore per day were mined and smelted, cadmium was discharged in the wastewater. Downstream, farmers withdrew the fine particles of flotation tailings in the Jinzu River along with water for drinking and crop irrigation. As rice plants were damaged near the irrigation inlets, farmers dug small sedimentation pools that were ineffective against the nearly invisible poison.

Both the numbers of Itai-itai disease patients and the damage to the rice crops rapidly decreased after the mining company built a large settling basin to purify the wastewater in 1955. However, even after the discharge into the Jinzu River was halted, the cadmium already in the rice paddy soils was augmented by airborne exhausts. Mining operations in several other Japanese prefectures also produced cadmium-contaminating rice, but afflicted individuals were not certified as Itai-itai patients. That designation was applied only to those who lived in the Jinzu River area.

In 1972 the survivors and their families became the first pollution victims in Japan to win a lawsuit against a major company. They won because in 1939 Article 109 of the Mining Act had imposed strict liability upon mining facilities for damages caused by their activities. The plaintiffs had only to prove that cadmium discharged from the mine caused their disease, not that the company was negligent. As epidemiological proof of causation sufficed as legal proof in this case, it set a precedent for other pollution litigation as well.

Despite legal success and compensation, the problem of contaminated rice continued. In 1969 the government initially set a maximum allowable standard of 0.4 parts per million (ppm) cadmium in unpolished rice. However, because much of the contaminated farmland produced grain in excess of that level, in 1970, under the Foodstuffs Hygiene Law this was raised to 1 ppm cadmium for unpolished rice and 0.9 ppm cadmium for polished rice. To restore contaminated farmland, Japanese authorities instituted a program in which, each year, the most highly contaminated soils in a small area were exchanged for uncontaminated soils. Less contaminated soils were rehabilitated through the addition of lime, phosphate, and a cadmium sequestering agent, EDTA.

By 1990 about 10,720 acres (4,340 ha), or 66.7 percent, of the approximately 16,080 acres (6,510 ha) of the most highly cadmium contaminated farmland had been restored. In the remaining contaminated areas where farm families continued to eat homegrown rice, the symptoms were alleviated by treatment with massive doses of vitamins B1, B12, D, calcium, and various hormones.

See also Bioaccumulation; Environmental law; Heavy metals and heavy metal poisoning; Mine spoil waste; Smelter; Water pollution.

Resources

BOOKS

Lippmann, Morton, ed. *Environmental Toxicants: Human Exposures and Their Health Effects.* Hoboken, NJ: Wiley-Interscience, 2006.
Rapp, Doris. *Our Toxic World: A Wake Up Call.* Buffalo, NY: Environmental Research Foundation, 2004.

OTHER

United States Environmental Protection Agency (EPA). "Human Health: Toxicity." http://www.epa.gov/ebtpages/humatoxicity.html (accessed October 12, 2010).
United States Environmental Protection Agency (EPA). "Pollutants/Toxics: Soil Contaminants: Cadmium." http://www.epa.gov/ebtpages/pollsoilccadmium.html (accessed October 12, 2010).

Frank M. D'Itri

IUCN—The World Conservation Union

Founded in 1948, the International Union for the Conservation of Nature (IUCN) works with governments, conservation organizations, and industry groups to conserve wildlife and approach the world's environmental problems using "sound scientific insight and the best available information." Its membership, currently over 1,000 member organizations, comes from 140 countries and includes over 200 government agencies and over 800 non-governmental organizations (NGOs). IUCN exists to serve its members, representing their views and providing them with the support necessary to achieve their goals. Above all, IUCN works with its members "to achieve development that is sustainable and that provides a lasting improvement in the quality of life for people all over the world." IUCN's three basic conservation objectives are: (1) to secure the conservation of nature, and especially of biological diversity, as an essential foundation for the future; (2) to ensure that where the earth's natural resources are used, this is done in a wise, equitable, and sustainable way; (3) to guide the development of human communities toward ways of life that are both of good quality and in enduring harmony with other components of the biosphere.

IUCN is one of the few organizations to include both governmental agencies and nongovernmental organizations. It is in a unique position to provide a neutral forum where these organizations can meet, exchange ideas, and build partnerships to carry out conservation projects. IUCN is also unusual in that it both develops environmental policies and then implements them through the projects it sponsors. Because the IUCN works closely with, and its membership includes, many government scientists and officials, the organization often takes a conservative, pro-management, as opposed to a preservationist, approach to wildlife issues. It may encourage or endorse limited hunting and commercial exploitation of wildlife if it believes this can be carried out on a sustainable basis.

IUCN maintains a global network of almost 11,000 scientists and wildlife professionals who are organized into six standing commissions that deal with various aspects of the union's work. There are commissions on Ecosystem Management; Education and Communication; Environmental, Economic, and Social Policy; Environmental Law; Protected Areas; and Species Survival. These commissions create action plans, develop policies, advise on projects and programs, and contribute to IUCN publications, all on an unpaid, voluntary basis.

IUCN publishes an authoritative series of Red Data Lists, describing the status of rare and endangered wildlife. Each volume provides information on the population, distribution, habitat and ecology, threats, and protective measures in effect for listed species. The Red-Data-Lists concept was originated in the mid-1960s by the famous British conservationist Sir Peter Scott, and the series now includes a variety of publications on regions and species. Other titles in the series of "Red Data Lists" include *Dolphins, Porpoises, and Whales of the World*; *Lemurs of Madagascar and the Comoros*; *Threatened Primates of Africa*; *Threatened Swallowtail Butterflies of the World*; *Threatened Birds of the Americas*; and books on plants and other species of wildlife, including a series of conservation action plans for threatened species.

Other notable IUCN works include *World Conservation Strategy: Living Resources Conservation for Sustainable Development* and its successor document *Caring for the Earth—A Strategy for Sustainable Living*; and the *United Nations List of Parks and Protected Areas*. IUCN also publishes books and papers on regional conservation, habitat preservation, environmental law and policy, ocean ecology and management, and conservation and development strategies.

Every four years the IUCN publishes a detailed analysis of its Red List. The most recent analysis, released in July 2009 and titled *Wildlife in a Changing World—An Analysis of the 2008 IUCN Red List of Threatened Species*, examines all 44,838 species on the IUCN Red List and presents the results by groups of species, geographical regions, and habitats. It shows that life on Earth remains under serious threat. Nearly one-third of amphibians, more than 12 percent of birds, and nearly 25 percent of mammals are threatened with extinction. The report also details trends of extinction risk in groups of species, and these trends reveal that birds, mammals, amphibians, and corals show a continuing decline, with a particularly rapid decline among coral species. An analysis of bird and mammal species used for food and medicine shows that these species are much more threatened than other birds and mammals. Habitat analyses also paint a troubling picture. For example, marine species are experiencing significant losses due to pollution, coastal development, over-fishing, invasive species, and climate change. At least 17 percent of sharks and rays, 85 percent of marine turtles, 27 percent of reef-building corals, and 27.5 percent of marine birds are threatened.

Resources

BOOKS

Freyfogle, Eric T. *Why Conservation Is Failing and How It Can Regain Ground*. New Haven: Yale University Press, 2006.

Ladle, Richard J. *Biodiversity and Conservation: Critical Concepts in the Environment*. London: Routledge, 2009.

OTHER

International Union for Conservation of Nature (IUCN). "The IUCN Red List of Threatened Species." http://www.iucnredlist.org/ (accessed October 2, 2010).

IUCN - International Union for Conservation of Nature. Switzerland Headquarters Website. http://www.iucn.org (accessed October 2, 2010).

ORGANIZATIONS

IUCN—The International Union for Conservation of Nature, Rue Mauverney 28, Gland, Switzerland, 1196, +41 (22) 999-0000, +41 (22) 999-0002, mail@ iucn.org, http://www.iucn.org.

Lewis G. Regenstein

Ivory-billed woodpecker

The ivory-billed woodpecker (*Campephilus principalis*) is one of the rarest birds in the world and is considered by most authorities to be extinct in the United States. The last confirmed sighting of ivory-bills was in Cuba in 1987 or 1988. Though never common, the ivory-billed woodpecker was rarely seen in the United States after the first years of the twentieth century. Some were seen in Louisiana in 1942, and since then, occasional sightings have been unverified. Interest in the bird rekindled in 1999, when a student at Louisiana State University claimed to have seen a pair of ivory-billed woodpeckers in a wilderness preserve. Teams of scientists searched the area for two years. No ivory-billed woodpecker was sighted, though some evidence made it plausible the bird was in the vicinity. By mid-2002, the ivory-billed woodpecker's return from the brink of extinction remained a tantalizing possibility, but not an established fact.

The ivory-billed woodpecker was a huge bird, averaging 19–20 inches (48–50 cm) long, with a wingspan of over 30 inches (76 cm). The ivory-colored bills of these birds were prized as decorations by Native Americans. The naturalist John James Audubon found ivory-billed woodpeckers in swampy forest edges in Texas in the 1830s. But by the end of the nineteenth century, the majority of the bird's prime habitat had been destroyed by logging. Ivory-billed woodpeckers required large tracts of land in the bottomland cypress, oak, and black gum forests of the Southeast, where they fed off insect larva in mature trees. This species was the largest woodpecker in North America, and they preferred the largest of these trees, the same ones targeted by timber companies as the most profitable to harvest. The territory for breeding pairs of ivory-billed woodpeckers consists of about 1.15 square miles (3 square km) of undisturbed, swampy forest, and there was little prime habitat left for them after 1900, for most of these areas had been heavily logged. By the 1930s one of the only virgin cypress swamps left was the Singer Tract in Louisiana, an 80,000-acre (32,375-ha) swathe of land owned by the Singer Sewing Machine Company. In 1935 a team of ornithologists descended on it to locate, study, and record some of the last ivory-billed woodpeckers in existence. They found the birds and were able to film and photograph them, as well as make the only sound recordings of them in existence. The Audubon Society, the state of Louisiana, and the U.S. Fish and Wildlife Service tried to buy the land from Singer to make it a refuge for the rare birds. But Singer had already sold timber rights to the land. During World War II, when demand for lumber was particularly high, the Singer Tract was leveled. One of the giant cypress trees that was felled contained the nest and eggs of an ivory-billed woodpecker. Land that had been virgin forest then became soybean fields.

Few sightings of these woodpeckers were made in the 1940s, and no reports exist for the 1950s. But in the early 1960s, ivory-billed woodpeckers were reportedly seen in South Carolina, Texas, and Louisiana. Intense searches, however, left scientists with little hope by the end of that decade, as only six birds were reported to exist. Subsequent decades yielded a few individual sightings in the United States, but none were confirmed.

In 1985 and 1986, there was a search for the Cuban subspecies of the ivory-billed woodpecker. The first expedition yielded no birds, but trees were found that had apparently been worked by the birds.

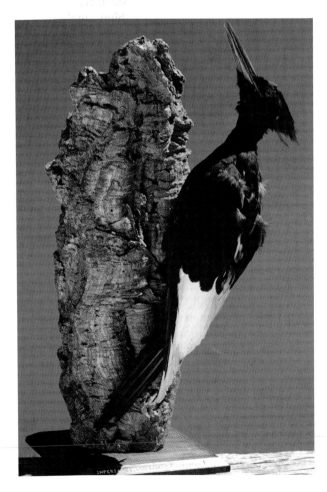

Stuffed ivory-billed woodpecker. (*Campephilus principalis*)
(© Grant Heilman Photography / Alamy)

restore the trees, but it will not restore the array of organisms that were present in the original forest.

Large scale logging activities have had a largely negative impact on the local economies in exporting regions because whole logs are shipped to Japan for further processing. Developed countries such as the United States and Canada, which in the past harvested timber and processed it into lumber and other products, have lost jobs to Japan. Indigenous cultures that have thrived in harmony with their forest homelands for centuries are displaced and destroyed. Provision has not been made for the survival of local flora and fauna, and provision for forest re-establishment has thus far proven inadequate. As resentment has grown in impacted areas, and among environmentalists, efforts have emerged to limit or stop large-scale timber harvesting and exporting.

Although concern has been voiced over all large-scale logging operations, special concern has been raised overharvesting of tropical timber from previously undisturbed primary forest areas. Tropical rain forests are especially unique and valuable natural resources for many reasons, including the density and variety of species within their borders. The exploitation of these unique ecosystems will result in the extinction of many potentially valuable species of plants and animals that exist nowhere else. Many of these forms of life have not yet been named or scientifically studied. In addition, overharvesting of tropical rainforests has a negative effect on weather patterns, especially by reducing rainfall.

Japan is a major importer of tropical timber from Malaysia, New Guinea, and the Solomon Islands. Although the number of imported logs has declined in recent years, this has been matched by an increase in imported tropical plywood manufactured in Indonesia and Malaysia. As a result, the total amount of timber removed has remained fairly constant. An environmentalist group called the Rainforest Action Network (RAN) has issued an alarm concerning recent expansion of logging activity by firms affiliated with Japanese importers. The RAN alleges that: "After laying waste to the rain forests of Asia and the Pacific islands, giant Malaysian logging companies are setting their sights on the Amazon. Major players include the WTK Group, Samling, Mingo, and Rimbunan Hijau." The RAN claims that "the same timber companies in Sarawak, Malaysia, worked with such rapacious speed that they devastated the region's forest within a decade, displacing traditional peoples and leaving the landscape marred with silted rivers and eroded soil."

One large Japanese firm, the Mitsubishi Corporation, has been targeted for criticism and boycott by the RAN, as one of the world's largest importers of timber. The boycott is an effort to encourage environmentally-conscious consumers to stop buying products marketed by companies affiliated with the huge conglomerate, including automobiles, cameras, beer, cell phones, and consumer electronics equipment. Through its subsidiaries, Mitsubishi has logged or imported timber from the Philippines, Malaysia, Papua New Guinea, Bolivia, Indonesia, Brazil, Chile, Canada (British Columbia and Alberta), Siberia, and the United States (Alaska, Oregon, Washington, and Texas). The RAN charges that "Mitsubishi Corporation is one of the most voracious destroyers of the world's rain forests. Its timber purchases have laid waste to forests in the Philippines, Malaysia, Papua New Guinea, Indonesia, Brazil, Bolivia, Australia, New Zealand, Siberia, Canada, and even the United States." The Mitsubishi Corporation itself does not sell consumer products, but it consists of 190 interlinked companies and hundreds of associated firms that do market to consumers. This conglomerate forms one of the world's largest industrial and financial powers. The Mitsubishi umbrella includes Mitsubishi Bank, Mitsubishi Heavy Industries, Mitsubishi Electronics, Mitsubishi Motors, and other major components.

The Mitsubishi Corporation countered criticism by launching a program "to promote the regeneration of rain forests . . . in Malaysia that plants seedlings and monitors their development." The corporation formed an Environmental Affairs Department, one of the first of its kind in Japan, to draft environmental guidelines, and coordinate corporate environmental activities. In the words of the Mitsubishi Corporation Chairman, "A business cannot continue to exist without the trust and respect of society for its environmental performance." Mitsubishi Corporation launched a program to support experimental reforestation projects in Malaysia, Brazil, and Chile. In Malaysia, the company cooperated with a local agricultural university, under the guidance of a professor from Japan. About 300,000 seedlings were planted on a barren site. Within five years, the trees were over 33 feet (10 m) in height and the corporation claimed that they were "well on the way to establishing techniques for regenerating tropical forest on burnt or barren land using indigenous species." However, in spite of management's declaration of increased concern for the environment, Mitsubishi Paper Mills Ltd.—a part of the Mitsubishi conglomerate—was implicated in a 2008 scandal involving several Japanese corporations that used deceptive advertising regarding their paper products. The companies involved inflated the percentage of recycled paper being used in various consumer products, as well as lied about the sources of woodchips used to produce their paper goods. The offending companies were found to be using woodchips derived from old growth forests in Tasmania—contradicting previous assertions that they (the paper companies) had only used wood products from approved sources, such as secondary forests (i.e., non-old-growth) and tree plantations. Mitsubishi Paper Mills Ltd. thereafter pledged to discontinue the use of woodchips and other wood products derived from old growth forests.

Resources

BOOKS

Tacconi, Luca. *Illegal Logging: Law Enforcement, Livelihoods and the Timber Trade*. London: Earthscan Publications, 2007.

OTHER

Centers for Disease Control and Prevention (CDC). "Logging." http://www.cdc.gov/niosh/topics/logging/ (accessed November 8, 2010).

International Tropical Timber Organization (ITTO). "International Tropical Timber Organization." Japan Headquarters Website. http://www.itto.or.jp (accessed November 8, 2010).

Bill Asenjo

K

Kapirowitz Plateau

The Kapirowitz Plateau, a wildlife refuge on the northern rim of the Grand Canyon, has come to symbolize wildlife management gone awry, a classic case of misguided human intervention intended to help wildlife that ended up damaging the animals and the environment. The Kapirowitz is located on the Colorado River in northwestern Arizona, and is bounded by steep cliffs dropping down to the Kanab Canyon to the west, and the Grand and Marble canyons to the south and southeast. Because of its inaccessibility, according to naturalist James B. Trefethen, the Plateau was considered a "biological island," and its deer population "evolved in almost complete genetic isolation."

The lush grass meadows of the Kapirowitz Plateau supported a resident population of 3,000 mule deer (*Odocoileus hemionus*), which were known and renowned for their massive size and the huge antlers of the old bucks. Before the advent of Europeans, Paiute and Navajo Indians hunted on the Kapirowitz in the fall, stocking up on meat and skins for the winter. In the early 1900s, in an effort to protect and enhance the deer population of the Kapirowitz, the federal government prohibited all killing of deer, and even eliminated the predator population in the area. As a result, the deer population exploded, causing massive overbrowsing, starvation, and a drastic decline in the health and population of the herd.

In 1893, when the Kapirowitz and surrounding lands were designated the Grand Canyon National Forest Reserve, hundreds of thousands of sheep, cattle, and horses were grazing on the Plateau, resulting in overgrazing, erosion, and large-scale damage to the land. On November 28, 1906, President Theodore Roosevelt established the 1 million-acre (400,000-ha) Grand Canyon National Game Preserve, which provided complete protection of the Kapirowitz 's deer

population. By then, however, overgrazing by livestock had destroyed much of the native vegetation and changed the Kapirowitz considerably for the worse. Continued pasturing of over 16,000 horses and cattle degraded the Kapirowitz even further.

The Forest Service carried out President Roosevelt's directive to emphasize "the propagation and breeding" of the mule deer by not only banning hunting, but also natural predators as well. From 1906 to 1931, federal agents poisoned, shot, or trapped 4,889 coyotes (*Canis latrans*), 781 mountain lions (*Puma concolor*), 554 bobcats (*Felis rufus*), and twenty wolves (*Canis lupus*). Without predators to remove the old, the sick, the unwary, and other biologically unfit animals, and keep the size of the herd in check, the deer herd began to grow out of control and lost those qualities that made its members such unique and magnificent animals. After 1906, the deer population doubled within ten breeding seasons, and by 1918 (two years later), it doubled again. By 1923, the herd had mushroomed to at least 30,000 deer, and perhaps as many as 100,000 according to some estimates.

Unable to support the overpopulation of deer, range grasses and land greatly deteriorated; and by 1925, 10,000 to 15,000 deer were reported to have died from starvation and malnutrition. Finally, after relocation efforts mostly failed to move a significant number of deer off of the Kapirowitz, hunting was reinstated, and livestock grazing was strictly controlled. By 1931, hunting, disease, and starvation had reduced the herd to under 20,000. The range grasses and other vegetation returned, and the Kapirowitz began to recover. In 1975 James Trefethen wrote, "the Kapirowitz today again produces some of the largest and heaviest antlered mule deer in North America."

In the fields of wildlife management and biology, the lessons of the Kapirowitz Plateau are often cited (as in the writings of naturalist Aldo Leopold) to demonstrate the valuable role of predators in maintaining the balance of nature (such as between herbivores and the plants they consume) and survival of the fittest. The experience of the Kapirowitz shows that in the absence of natural predators, prey populations (especially ungulates) tend to increase beyond the carrying capacity of the land, and eventually the results are overpopulation and malnutrition.

See also Predator-prey interactions; Predator control.

Resources

BOOKS

Bolen, Eric, and William Robinson. *Wildlife Ecology and Management.* New York: Benjamin Cummings, 2008.

Leopold, A. *A Sand County Almanac.* New York: Oxford University Press, 1949.

OTHER

United Nations System-Wide EarthWatch. "Exceeding carrying capacity." http://earthwatch.unep.net/emergingissues/demography/carryingcapcity.php (accessed October 12, 2010).

Lewis G. Regenstein

Kennedy Jr., Robert

1954–
American environmental lawyer

Robert "Bobby" Kennedy Jr. had a very controversial youth. Kennedy entered a drug rehabilitation program, at the age of twenty-eight, after being found guilty of possession of heroin following his arrest in South Dakota. He was sentenced to two years probation and community service.

Clearly the incident was a turning point in Kennedy's life. "Let's just say, I had a tumultuous adolescence that lasted until I was twenty-nine" he told a reporter for New York magazine, which ran a long profile of Kennedy in 1995, entitled "Nature Boy." The title refers to the passion that has enabled Kennedy to emerge from his bleak years as a strong and vital participant in environmental causes.

A Harvard graduate and published author, Kennedy serves as chief prosecuting attorney for a group called the Hudson Riverkeeper (named after the famed New York river). Kennedy also serves as president of Waterkeeper Alliance. Kennedy previously served as a senior attorney for the Natural Resources Defense Council. Kennedy, who earlier in his career

served as assistant district attorney in New York City after passing the bar, is also a clinical professor and supervising attorney at the Environmental Litigation Clinic at Pace University School of Law in New York.

While Kennedy appeared to be following in the family's political footsteps, working, for example, on several political campaigns and serving as a state coordinator for his uncle and former Senator Ted Kennedy's 1980 presidential campaign, it is in environmental issues that Kennedy Jr. has found himself. He has worked on environmental issues across the Americas and has assisted several indigenous tribes in Latin America and Canada in successfully negotiating treaties protecting traditional homelands. He is also credited with leading the fight to protect New York City's water supply, a battle that resulted in the New York City Watershed Agreement, regarded as an international model for combining development and environmental concerns.

Opportunity was always around the corner for a young, confident, and intelligent Kennedy. (A number of his relatives achieved great success in the political arena.) After Harvard, Kennedy Jr. earned a law degree at the University of Virginia. In 1978, the subject of Kennedy's Harvard thesis—a prominent Alabama judge—was named head of the Federal Bureau of Investigation (FBI). A publisher offered Kennedy Jr. money to expand his previous research into a book, published in 1978, called *Judge Frank M. Johnson Jr.: A Biography.*

Kennedy Jr. did a publicity tour that included TV appearances, but the reviews were mixed. In 1982, Kennedy Jr. married Emily Black, a Protestant who later converted to Catholicism. Two children followed: Robert Francis Kennedy III and Kathleen Alexandra, named for Kennedy Jr.'s aunt Kathleen, who died in a plane crash in 1948.

The marriage, however, coincided with Kennedy Jr.'s fall out through drug addiction. In 1992, Kennedy Jr. and Emily separated, and a divorce was obtained in the Dominican Republic. In 1994, Kennedy Jr. married Mary Richardson, an architect, with whom he would have two more children.

During this time Kennedy Jr. emerged as a leading environmental activist and litigator. Kennedy is quoted as saying: "To me...this is a struggle of good and evil between short-term greed and a long-term vision of building communities that are dignified and enriching and that meet the obligations of future generations. There are two visions of America. One is that this is just a place where you make a pile for yourself and keep moving. And the other is that you put down roots and build communities that are examples to the rest of humanity." Kennedy goes on: "The environment cannot be separated from the economy, housing, civil rights. How we distribute the goods of the earth is the best measure of our democracy. It's not about advocating for fishes and birds. It's about human rights."

Among his accomplishments, Kennedy Jr. is an experienced falconer and white-water rafter. His efforts with Riverkeeper to restore the Hudson River led him to be named a "hero of the planet" by Time.com. From 2005 to 2010, he has advocated strongly for wind power. While he flirted with running for the U.S. Senate, he renounced that intention early in 2008.

Resources

BOOKS

Cronin, John, and Robert F. Kennedy Jr. *The Riverkeepers: Two Activists Fight to Reclaim Our Environment as a Basic Human Right.* New York: Scribner, 1999.

ORGANIZATIONS

Riverkeeper, Inc., 20 Secor Road, Ossining, NY, USA, 10562, (800) 21-RIVER, info@riverkeeper.org, http://www.riverkeeper.org

Kepone

Kepone ($C_{10}Cl_{10}O$) is an organochlorine pesticide that was manufactured by the Allied Chemical Corporation in Virginia from the late 1940s to the 1970s. Kepone was responsible for human health problems and extensive contamination of the James River and its estuary in the Chesapeake Bay. It is a milestone in the development of a public environmental consciousness, and its history is considered by many to be a classic example of negligent corporate behavior and inadequate oversight by state and federal agencies.

Kepone is an insecticide and fungicide that is closely related to other chlorinated pesticides such as dichlorodiphenyltrichloroethane (DDT) and aldrin. As with all such pesticides, Kepone causes lethal damage to the nervous systems of its target organisms. A poorly water-soluble substance, it can be absorbed through the skin, and it bioaccumulates in fatty tissues from which it is later released into the bloodstream. It is also a contact poison; when inhaled, absorbed, or ingested by humans, it can damage the central nervous system as well as the liver and kidneys. It can also lead to neurological symptoms such as tremors, muscle spasms, sterility, and cancer. Although the manufacture and use of Kepone is now banned by the Environmental Protection Agency (EPA), organochlorines have long half-lives, and these compounds, along with their residues and degradation products, can persist in the environment over many decades.

Allied Chemical first opened a plant to manufacture nitrogen-based fertilizers in 1928 in the town of Hopewell, on the banks of the James River in Virginia. This plant began producing Kepone in 1949. Commercial production was subsequently begun, although a battery of toxicity tests indicated that Kepone was both toxic and carcinogenic and that it caused damage to the functioning of the nervous, muscular, and reproductive systems in fish, birds, and mammals. It was patented by Allied in 1952 and registered with federal agencies in 1957. The demand for the pesticide grew after 1958, and Allied expanded production by entering into a variety of subcontracting agreements with a number of smaller companies, including the Life Science Products Company.

In 1970, a series of new environmental regulations came into effect, which should have changed the way wastes from the manufacture of Kepone were discharged. The Refuse Act Permit Program and the National Pollutant Discharge Elimination Program (NPDES) of the Clean Water Act required all dischargers of effluents into United States waters to register their discharges and obtain permits from federal agencies. At the time these regulations went into effect, Allied Chemical had three pipes discharging Kepone and plastic wastes into the "gravelly run", a tributary of the James River, about 75 miles (120 km) north of Chesapeake Bay.

A regional sewage treatment plant, which would accept industrial wastes, was then under construction but not scheduled for completion until 1975. Rather than installing expensive pollution control equipment for the interim period, Allied chose to delay. They adopted a strategy of misinformation, reporting the releases as temporary and unmonitored discharges, and they did not disclose the presence of untreated Kepone and other process wastes in the effluents. The Life Science Products Company also avoided the new federal permit requirements by discharging their wastes directly into the local Hopewell sewer system. These discharges caused problems with the functioning of the biological treatment systems at the sewage plant; the company was required to reduce concentrations of Kepone in sewage, but it continued its discharges at high concentrations, violating these standards with the apparent knowledge of plant treatment officials.

During this same period, an employee of Life Science Products visited a local Hopewell physician, complaining of tremors, weight loss, and general aches and pains. The physician discovered impaired liver and nervous functions, and a blood test revealed an astronomically high level of Kepone—7.5 parts per million. Federal and state officials were contacted, and the epidemiologist for the state of Virginia toured the manufacturing facility at Life Science Products. This official reported that "Kepone was everywhere in the plant," and that workers wore no protective equipment and were "virtually swimming in the stuff." Another investigation discovered seventy-five cases of Kepone poisoning among the workers; some members of their families were also found to have elevated concentrations of the chemical in their blood. Further investigations revealed that the environment around the plant was also heavily contaminated. The soil contained 10,000 to 20,000 ppm of Kepone. Sediments in the James River, as well as local landfills and trenches around the Allied facilities, were just as badly contaminated. Government agencies were forced to close 100 miles (161 km) of the James River and its tributaries to commercial and recreational fishing and shellfishing.

In the middle of 1975, Life Science Products finally closed its manufacturing facility. It has been estimated that since 1966, it and Allied together produced 3.2 million pounds (1.5 million kg) of Kepone and were responsible for releasing 100,000 to 200,000 pounds (45,360–90,700 kg) into the environment. In 1976, the Northern District of Virginia filed criminal charges against Allied, Life Science Products, the city of Hopewell, and six individuals on 1,097 counts relating to the production and disposal of Kepone. The indictments were based on violations of the permit regulations, unlawful discharge into the sewer systems, and conspiracy-related to that discharge.

The case went to trial without a jury. The corporations and the individuals named in the charges negotiated lighter fines and sentences by entering pleas of "no contest." Allied ultimately paid a fine of $13.3 million, although their annual sales reach was $3 billion. Life Science Products was fined $4 million, which it could not pay because of lack of assets. Company officers were fined $25,000 each, and the town of Hopewell was fined $10,000. No one was sentenced to a jail term. Civil suits brought against Allied and the other defendants resulted in a settlement of $5.25 million to pay for cleanup expenses and to repair the damage that had been done to the sewage treatment plant. Allied paid another $3 million to settle civil suits brought by workers for damage to their health.

Environmentalists and many others considered the results of legal action against the manufacturers of Kepone unsatisfactory. Some have argued that these results are typical of environmental litigation.

It is difficult to establish criminal intent beyond a reasonable doubt in such cases; and even when guilt is determined, sentencing is relatively light. Corporations are rarely fined in amounts that affect their financial strength, and individual officers are almost never sent to jail. Corporate fines are generally passed along as costs to the consumer, and public bodies are treated even more lightly, since it is recognized that the fines levied on public agencies are paid by taxpayers.

Today, the James River has been reopened to fishing for those species that are not prone to the bioaccumulation of Kepone. Nevertheless, sediments in the river and its estuary contain large amounts of deposited Kepone, which is released during periods of turbulence. Scientists have published studies which document that Kepone is still moving through the food chain and the ecosystem in this area, and Kepone toxicity has been demonstrated in a variety of invertebrate test species. There are still deposits of Kepone in the local sewer pipes in Hopewell; these continue to release the chemical, endangering treatment plant operations and polluting receiving waters.

Usha Vedagiri
Douglas Smith

Kesterson National Wildlife Refuge

One of a dwindling number of freshwater marshes in California's San Joaquin Valley, Kesterson National Wildlife Refuge achieved national notoriety in 1983 when refuge managers discovered that agricultural runoff was poisoning the area's birds. Among other elements and agricultural chemicals reaching toxic concentrations in the wetlands, the naturally occurring element selenium was identified as the cause of falling fertility and severe birth defects in the refuge's breeding populations of stilts, grebes, shovelers, coots, and other aquatic birds. Selenium, lead, boron, chromium, molybdenum, and numerous other contaminants were accumulating in refuge waters because the refuge had become an evaporation pond for tainted water draining from the region's fields.

The soils of the arid San Joaquin Valley are the source of Kesterson's problems. The flat valley floor is composed of ancient sea bed sediments that contain high levels of trace elements, heavy metals, and salts.

The sun sets over a man-made wetland at the San Luis National Wildlife Refuge near Gustine, California. *(AP Photo/Marcio Jose Sanchez)*

But with generous applications of water, this sun-baked soil provides an excellent medium for food production. Perforated pipes buried in the fields drain away excess water—and with it dissolved salts and trace elements—after flood irrigation. An extensive system of underground piping, known as tile drainage, carries wastewater into a network of canals that lead to Kesterson Refuge, an artificial basin constructed by the Bureau of Reclamation to store irrigation runoff from central California's heavily-watered agriculture. Originally a final drainage canal from Kesterson to San Francisco Bay was planned, but because an outfall point was never agreed upon, contaminated drainage water remained trapped in Kesterson's twelve shallow ponds. In small doses, selenium and other trace elements are not harmful and can even be dietary necessities. But steady evaporation in the refuge gradually concentrated these contaminants to dangerous levels.

Wetlands in California's San Joaquin valley were once numerous, supporting huge populations of breeding and migrating birds. In the past one-half

century drainage and the development of agricultural fields have nearly depleted the area's marshes. The new ponds and cattail marshes at Kesterson presented a rare opportunity to extend breeding habitat, and the area was declared a national wildlife refuge in 1972, one year after the basins were constructed. Eleven years later, in the spring of 1983, observers discovered that a shocking 60 percent of Kesterson's nestlings were grotesquely deformed. High concentrations of selenium were found in their tissues, an inheritance from parent birds who ate algae, plants, and insects—all tainted with selenium—in the marsh.

Following extensive public outcry the local water management district agreed to try to protect the birds. Alternate drainage routes were established, and by 1987 much of the most contaminated drainage had been diverted from the wildlife refuge. The California Water Resource Control Board ordered the Bureau of Reclamation to drain the ponds and clean out contaminated sediments, at a cost of well over $50 million. However, these contaminants, especially in such large volumes and high concentrations, are difficult to contain, and similar problems could quickly emerge again. Furthermore, these problems are widespread. Selenium poisoning from irrigation runoff has been discovered in atleast nine other national wildlife refuges, all in the arid west because it appeared at Kesterson. Researchers continue to work on affordable and effective responses to such contamination in wetlands, an increasingly rare habitat in this country.

Resources

BOOKS

Harris, T. *Death in the Marsh*. Washington, DC: Island Press, 1991.

PERIODICALS

Claus, K. E. "Kesterson: An Unsolvable Problem?" *Environment* 89 (1987): 4–5.

Harris, T. "The Kesterson Syndrome." *Amicus Journal* 11 (Fall 1989): 4–9.

Marshal, E. "Selenium in Western Wildlife Refuges." *Science* 231 (1986): 111–112.

Tanji, K.; A. Láuuchli; and J. Meyer. "Selenium in the San Joaquin Valley." *Environment* 88 (1986): 6–11.

ORGANIZATIONS

Kesterson National Wildlife Refuge, c/o San Luis NWR Complex, 340 I Street, PO Box 2176, Los Banos, CA, USA, 93635, (209) 826-3508

Mary Ann Cunningham

Ketones

Ketones belong to a class of organic compounds known as carbonyls. They contain a carbon atom linked to an oxygen atom with a double bond ($C = O$). Acetone (dimethyl ketone) is a ketone commonly used in industrial applications. Other ketones include methyl ethyl ketone (MEK), methyl isobutyl ketone (MIBK), methyl amyl ketone (MAK), isophorone, and diacetone alcohol.

As solvents, ketones have the ability to dissolve other materials or substances, particularly polymers and adhesives. They are ingredients in lacquers, epoxies, polyurethane, nail polish remover, degreasers, and cleaning solvents. Ketones are also used in industry for the manufacture of plastics and composites and in pharmaceutical and photographic film manufacturing. Because they have high evaporation rates and dry quickly, they are sometimes employed in drying applications.

Some types of ketones used in industry, such as methyl isobutyl ketone and methyl ethyl ketone, are considered both hazardous air pollutants (HAP) and volatile organic compounds (VOC) by the Environmental Protection Agency (EPA). As such, the Clean Air Act regulates their use.

In addition to these industrial sources, ketones are released into the atmosphere in cigarette smoke and car and truck exhaust. More "natural" environmental sources such as forest fires and volcanoes also emit ketones. Acetone, in particular, is readily produced in the atmosphere during the oxidation of organic pollutants or natural emissions. Ketones (in the form of acetone, beta-hydroxybutyric acid, and acetoacetic acid) also occur in the human body as a by-product of the metabolism, or break down of fat.

Resources

PERIODICALS

Kirk-Othmer. "Ketones." In *Kirk-Othmer Encyclopedia of Chemical Technology*. 5th ed. New York: John Wiley & Sons, 2004.

OTHER

United States Environmental Protection Agency (EPA). "Pollutants and Toxics: Ketones." http://www.epa.gov/ebtpages/pollchemicketones.html (accessed October 23, 2010).

ORGANIZATIONS

American Chemical Society, 1155 Sixteenth St. NW, Washington, DC, USA, 20036, (202) 872-4600, (202) 872-4615, (800) 227- 5558, help@acs.org, http://www.chemistry.org

Paula Anne Ford-Martin

Keystone species

Keystone species have a major influence on the structure of their ecological community. The profound influence of keystone species occurs because of their position and activity within the food chain. In the sense meant here, a "major influence" means that removal of a keystone species would result in a large change in the abundance, and even the extirpation (i.e., the local extinction) of one or more species in the community. This would fundamentally change the structure of the overall community in terms of species composition, productivity, and other characteristics. Such changes would have substantial effects on all of the species that are present, and could allow new species to invade the community.

The original use of the word "keystone" was in architecture. An architectural keystone is a wedge-shaped stone that is strategically located at the summit of an arch. The keystone serves to lock all other elements of the arch together, and it thereby gives the entire structure mechanical integrity. Keystone species play an analogous role in giving structure to the "architecture" of their ecological community.

The concept of keystone species was first applied to the role of certain predators (i.e., keystone predators) in their community. More recently, however, the term has been extended to refer to other so-called strong interactors. This has been particularly true of keystone herbivores that have a relatively great influence on the species composition and relative abundance of plants in their community.

Keystone species directly exert their influence on the populations of species that they feed upon, but they also have indirect effects on species lower in the food web. Consider, for example, a hypothetical case of a keystone predator that regulates the population of a herbivore. This effect will also, of course, indirectly influence the abundance of plant species that the herbivore feeds upon. Moreover, by affecting the competitive relationships among the various species of plants in the community, the abundance of plants that the herbivore does not eat will also be indirectly affected by the keystone predator. Although keystone species exert their greatest influence on species with which they are most closely linked through feeding relationships, their influences have ramifications throughout the food web.

Ecologists have documented the presence of keystone species in many types of communities. The phenomenon does not, however, appear to be universal in that keystone species have not been identified in many ecosystems.

Predators as keystone species

The term "keystone species" was originally used by the American ecologist Robert Paine to refer to the critical influence of certain predators. His original usage of the concept was in reference to rocky intertidal communities of western North America, in which the predatory starfish (*Pisaster ochraceous*) prevents the mussel (*Mytilus californianus*) from monopolizing the available space on rocky habitats and thereby eliminating other, less-competitive herbivores and even seaweeds from the community. By feeding on mussels, which are the dominant competitor among the herbivores in the community, the starfish prevents these shellfish from achieving the dominance that would otherwise be possible. This permits the development of a community that is much richer in species

than would occur in the absence of the predatory starfish. Paine demonstrated the keystone role of the starfish by conducting experiments in which the predator was excluded from small areas using cages. When this was done, the mussels quickly became strongly dominant in the community and eliminated virtually all other species of herbivores. Paine also showed that once mussels reached a certain size they were safe from starfish predation. This prevented the predator from eliminating the mussel from the community.

Sea otters (*Enhydra lutris*) of the west coast of North America are another example of a keystone predator. This species feeds heavily on sea urchins when these invertebrates are available. By greatly reducing the abundance of sea urchins, the sea otters prevent these herbivores from overgrazing kelps and other seaweeds in subtidal habitats. Therefore, when sea otters are abundant, urchins are not, and this allows luxurious kelp "forests" to develop. In the absence of otters, the high urchin populations can keep the kelp populations low, and the habitat then may develop as a rocky barren ground. Because sea otters were trapped very intensively for their fur during the eighteenth and nineteenth centuries, they were extirpated over much of their natural range. In fact, the species had been considered extinct until the 1930s, when small populations were discovered off the coast of California and in the Aleutian Islands of Alaska. Thanks to effective protection from trapping, and deliberate reintroductions to some areas, populations of sea otters have now recovered over much of their original range. This has resulted in a natural depletion of urchin populations, and a widespread increase in the area of kelp forests.

As of late 2010, the sea otter was still listed as endangered under the U.S. Endangered Species Act. The endangered status remains in place in spite of population counts that have generally demonstrated a slow but steady rise in sea otter numbers. Sea otters are grouped into southern and northern regions along the western coast of North America. Southern sea otters (also called California sea otters) have seen their numbers grow from approximately 1,300 in 1985 to a record high of just over 3,000 in 2007. The 2010 count, led by the U.S. Geological Survey (USGS), was just over 2,700 southern otters. In order to delist the southern sea otter from endangered status, the three-year population average must exceed 3,090 otters.

Herbivores as keystone species

Some herbivorous animals have also been demonstrated to have a strong influence on the structure and productivity of their ecological community. One such example is the spruce budworm (*Choristoneura fumiferana*), a moth that occasionally irrupts in abundance and becomes an important pest of conifer forests in the northeastern United States and eastern Canada. The habitat of spruce budworm is mature forests dominated by balsam fir (*Abies balsamea*), white spruce (*Picea glauca*), and red spruce (*P. rubens*). This native species of moth is always present in at least small populations, but it sometimes reaches very high populations, which are known as irruptions. When budworm populations are high, many species of forest birds and small mammals occur in relatively large populations that subsist by feeding heavily on larvae of the moth. However, during irruptions of budworm most of the fir and spruce foliage is eaten by the abundant larvae; and after this happens for several years, many of the trees die. Because of damages caused to mature trees in the forest the budworm epidemic collapses, and then a successional recovery begins. The plant communities of early succession contain many species of plants that are uncommon in mature conifer forests. Eventually, however, another mature conifer forest redevelops, and the cycle is primed for the occurrence of another irruption of the budworm. Clearly, spruce budworm is a good example of a keystone herbivore, because it has such a great influence on the populations of plant species in its habitat, and also on the many animal species that are predators of the budworm.

Another example of a keystone herbivore concerns snow geese (*Chen caerulescens*) in salt marshes of the western Hudson Bay. In the absence of grazing by flocks of snow geese this ecosystem would become extensively dominated by several competitively superior species, such as the salt-marsh grass (*Puccinellia phryganodes*) and the sedge (*Carex subspathacea*). However, vigorous feeding by the geese creates bare patches of up to several square meters in area, which can then be colonized by other species of plants. The patchy disturbance regime associated with goose-grazing results in the development of a relatively complex community, which supports more species of plants than would otherwise be possible. In addition, by manuring the community with their droppings, the geese help to maintain higher rates of plant productivity than might otherwise occur. In recent decades, however, large populations of snow goose have caused severe damages to the salt-marsh habitat by over-grazing. This has resulted in the development of salt-marsh barrens in some places, which may take years to recover.

Plants as keystone species

Some ecologists have also extended the idea of keystone species to refer to plant species that are extremely influential in their community. For example, sugar maple (*Acer saccharum*) is a competitively superior species that often strongly dominates stands of forest in eastern North America. Under these conditions most of the community-level productivity is contributed by sugar maple trees. In addition, most of the seedlings and saplings are of sugar maple. This is because few seedlings of other species of trees are able to tolerate the stressful conditions beneath a closed sugar-maple canopy.

Other ecologists prefer to not use the idea of keystone species to refer to plants that, because of their competitive abilities, are strongly dominant in their community. Instead, these are sometimes referred to as "foundationstone species." This term reflects the facts that strongly dominant plants contribute the great bulk of the biomass and productivity of their community, and that they support almost all herbivores, predators, and detritivores that are present.

Resources

BOOKS

Akcakaya, H. Resit. *Species Conservation and Management: Case Studies*. New York: Oxford University Press, 2004.

Arsuaga, Luis, et al. *Chosen Species: The Long March of Human Evolution*. Malden, MA: Blackwell Publishing, 2005.

Askins, Robert. *Saving Biological Diversity: Balancing Protection of Endangered Species and Ecosystems*. New York: Springer, 2008.

Carroll, Sean B. *Remarkable Creatures: Epic Adventures in the Search for the Origins of Species*. Boston: Houghton Mifflin Harcourt, 2009.

Ellis, Richard. *No Turning Back: The Life and Death of Animal Species*. New York: Harper Perennial, 2005.

Grant, Peter R., and B. Rosemary Grant. *How and Why Species Multiply: The Radiation of Darwin's Finches*. Princeton, NJ: Princeton University Press, 2008.

Mackay, Richard. *The Atlas of Endangered Species*. Berkeley: University of California Press, 2008.

Sax, Dov; John Stachowicz; and Steven Gaines. *Species Invasions: Insights into Ecology, Evolution, and Biogeography*. Sunderland, MA: Sinauer Associates, 2005.

Bill Freedman

Killer bees *see* **Africanized bees.**

Kirtland's warbler

Kirtland's warbler (*Dendroica kirtlandii*) is an endangered species and one of the rarest members of the North American wood warbler family. Its entire breeding range is limited to a seven-county area of north-central Michigan. The restricted distribution of the Kirtland's warbler and its specific niche requirements have probably contributed to low population levels throughout its existence, but human activity has had a large impact on their numbers over the past hundred years.

The first specimen of Kirtland's warbler was taken by Samuel Cabot in October 1841 and brought on ship in the West Indies during an expedition to the Yucatan. However, this specimen went unnoticed until 1865, long after the species had been formally described. Charles Pease is credited with discovering Kirtland's warbler. He collected a specimen on May 13, 1851, near Cleveland, Ohio, and gave it to his father-in-law, Jared P. Kirtland, a renowned naturalist. Kirtland sent the specimen to his friend, ornithologist Spencer Fullerton Baird, who described the new species the following year and named it in honor of the naturalist.

The wintering grounds of Kirtland's warbler is the Bahamas, a fact which was well-established by the turn of the century, but its nesting grounds went undiscovered until 1903, when Norman Wood found the first nest in Oscoda County, Michigan. Every nest found since then has been within a 60 mile (95 km) radius of this spot.

In 1951, the first exhaustive census of singing males was undertaken in an effort to establish the range of Kirtland's warblers as well as its population level. Assuming that numbers of males and females are approximately equal and that a singing male is defending an active nesting site, the total of 432 in this census indicated a population of 864 birds. Ten years later another census counted 502 singing males, indicating the population was over 1,000 birds. In 1971, annual counts began, but for the next twenty years these counts revealed that the population had dropped significantly, reaching lows of 167 singing males in 1974 and 1987. In the early 1990s, conservation efforts on behalf of the species began to bear fruit and the population began to recover. By 2001 the annual census counted 1,085 singing males or a total population of over 2,000 birds.

The first problem facing this endangered species centers on its specialized nesting and habitat requirements. The Kirtland's warbler nests on the ground, and its reproductive success is tied closely to its selection of young jack pine trees as nesting sites. When the jack pines are 5 to 20 feet (1.5–6 m) tall, at an age of between eight and twenty years, their lower branches are at ground level and provide the cover this warbler needs. The life cycle of the pine, however, is dependent on forest fires, as the intense heat is needed to open the cones for seed release. The advent of fire protection in forest management reduced the production of the number of young trees the warblers needed and the population suffered. Once this relationship was fully understood, jack pine stands were managed for Kirtland's warbler, as well as commercial harvest, by instituting controlled burns on a fifty-year rotational basis.

The second problem is the population pressures brought to bear by a nest parasite, the brown-headed cowbird (*Molothrus ater*), which lays its eggs in the nests of other songbirds. Originally a bird of open plains, it did not threaten Kirtland's warbler until Michigan was heavily deforested, thus providing it with appropriate habitat. Once established in the warbler's range, it has increasingly pressured the Kirtland's population. Cowbird chicks hatch earlier than other birds and they compete successfully with the other nestlings for nourishment. Efforts to trap and destroy this nest parasite in the warbler's range have resulted in improved reproductive success for Kirtland's warbler.

A 2007 census of Kirtland's warbler suggested that there were slightly fewer than 5,000 birds remaining. The species remains endangered but the population has shown encouraging growth since the year 2000.

See also Deforestation; Endangered Species Act (1973); Rare species; Wildlife management.

Resources

BOOKS

Alderfer, Jonathan K. *National Geographic Complete Birds of North America*. Washington, DC: National Geographic, 2006.
Audubon, John James. *Birds of America*. Philadelphia: J. B. Chevalier, 1842.

OTHER

U.S. Government; science.gov. "Birds." http://www.science.gov/browse/w_115A1.htm (accessed October 13, 2010).

Eugene C. Beckham

Krakatoa

The explosion of this triad of volcanoes on August 27, 1883, the culmination of a three-month eruptive phase, astonished the world because of its global impact. Perhaps one of the most influential factors, however, was its timing. It happened during a time of major growth in science, technology, and communications, and the world received current news accompanied by the correspondents' personal observations. The explosion was heard some 3,000 miles (4,828 km) away, on the Island of Rodriguez in the Indian Ocean. The glow of sunsets was so vivid three months later that fire engines were called out in New York City and nearby towns.

Krakatoa (or Krakatau), located in the Sunda Strait between Java and Sumatra, is part of the Indonesian volcanic system, which was formed by the subduction of the Indian Ocean plate under the Asian plate. A similar explosion occurred in AD 416, and another major eruption was recorded in 1680. Now a new volcano is growing out of the caldera, likely building toward some future cataclysm.

This immense natural event, perhaps twice as powerful as the largest hydrogen bomb, had an extraordinary impact on the solid earth, the oceans, and the atmosphere, and demonstrated their interdependence. It also made possible the creation of a wildlife refuge and tropical rain forest preserve on the Ujung Kulon Peninsula of southwestern Java.

Studies revealed that this caldera, like Crater Lake, Oregon, resulted from Krakatoa's collapse into the now empty magma chamber. The explosion produced a 131-foot (40-m) high tsunami, which carried a steamship nearly 2 miles (3.2 km) inland, and caused most of the fatalities resulting from the eruption. Tidal gauges as far away as San Francisco Bay and the English Channel recorded fluctuations.

The explosion provided substantial benefits to the young science of meteorology. Barometers all around Earth recorded the blast wave as it raced toward its antipodal position in Columbia, and then reverberated back and forth in six more recorded waves. The distribution of ash in the stratosphere gave the first solid evidence of rapidly flowing westerly winds, as debris encircled the equator over the next thirteen days. Global temperatures were lowered about 0.9°Fahrenheit (0.5°C), and did not return to normal until five years later.

An ironic development is that the Ujung Kulon Peninsula was never resettled after the tsunami killed most of the people living there. Without Krakatoa's explosion, the population would have most likely grown significantly, and much of the habitat there would likely have been altered by agriculture. Instead, the area is now a national park that supports a variety of species, including the Javan rhino (*Rhinoceros sondaicus*), one of Earth's rarest and most endangered species. This park has provided a laboratory for scientists to study nature's healing process after such devastation.

In 1927 debris began to appear above the ocean surface above the collapsed caldera of Krakatoa, and Anak Krakatoa was thus born. Significant pyroclastic flows cascaded down the right flank of Anak Krakatoa in 1952. The latest cycle of eruptions began in April 2008. Continuous releases of hot gases, rocks, and lava have prompted officials to recommend a 4.8-mile (3-km) exclusion zone around the island.

See also Mount Pinatubo; Mount St. Helens; Volcano.

Resources

BOOKS

Gunn, Angus M. *Encyclopedia of Disasters: Environmental Catastrophes and Human Tragedies*. Westport, CT: Greenwood Press, 2008.

Tagawa, Hideo. *The Krakataus: Changes in a Century since Catastrophic Eruption in 1883*. Kagoshima, Japan: Kagoshima: University, 2005.

Winchester, Simon. *Krakatoa: The Day the World Exploded: August 27, 1883*. Paw Prints, 2008.

Nathan H. Meleen

Krill

Marine crustaceans in the order Euphausiacea. Krill are zooplankton, and most feed on microalgae by filtering them from the water. In high latitudes, krill may account for a large proportion of the total zooplankton. Krill often occur in large swarms and in a few species these swarms may reach several hundred square meters in size with densities over 60,000 individuals per square meter. This swarming behavior makes them valuable food sources for many species of whales and seabirds. Humans have also begun to harvest krill for use as a dietary protein supplement.

Krutch, Joseph Wood

1893–1970
American literary critic and naturalist

Through much of his career, Krutch was a teacher of criticism at Columbia University and a drama critic for *The Nation*. But then respiratory problems led him to early retirement in the desert near Tucson, Arizona. He loved the desert and there turned to biology and geology, which he applied to maintain a consistent, major theme found in all of his writings, that of the relation of humans and the universe. Krutch subsequently became an accomplished naturalist.

Readers can find the theme of man and universe in Krutch's early work, such as in *The Modern Temper* (1929) and in his later writings on human-human and human-nature relationships, including natural history—what Rene Jules Dubos described as "the social philosopher protesting against the follies committed in the name of technological progress, and the humanist searching for permanent values in man's relationship to nature." Assuming a pessimistic stance in his early writings, Krutch despaired about lost connections, arguing that for humans to reconnect, they must conceive of themselves, nature, "and the universe in a significant reciprocal relationship."

Krutch's later writings repudiated much of his earlier despair. He argued against the dehumanizing and alienating forces of modern society and advocated systematically reassembling—by reconnecting to nature—"a world man can live in." In *The Voice of the Desert* (1954), for instance, he claimed that "we must be part not only of the human community, but of the whole community." In such books as *The Twelve Seasons* (1949) and *The Great Chain of Life* (1956), he demonstrated that humans "are a part of Nature...whatever we discover about her we are discovering also about ourselves." This view was based on a solid anti-deterministic approach that opposed mechanistic and behavioristic theories of evolution and biology.

His view of modern technology as out of control was epitomized in the automobile. Driving fast prevented people from reflecting or thinking or doing anything except controlling the monster: "I'm afraid this is the metaphor of our society as a whole," he commented. Krutch also disliked the proliferation of suburbs, which he labeled "affluent slums." He argued in *Human Nature and the Human Condition* (1959) that "modern man should be concerned with achieving the good life, not with raising the [material] standard of living."

An editorial ran in *The New York Times* a week after Krutch's death: Today's younger generation, it read, "unfamiliar with Joseph Wood Krutch but concerned about the environment and contemptuous of materialism," should "turn to a reading of his books with delight to themselves and profit to the world."

Resources

BOOKS

Krutch, J. W. *The Desert Year.* New York: Viking, 1951.

Gerald L. Young

Kudzu

Pueraria lobata or kudzu, also jokingly referred to as "foot-a-night" and "the vine that ate the South," is a highly aggressive and persistent semi-woody vine introduced to the United States in the late nineteenth century. It has since become a symbol of the problems possible for native ecosystems caused by the introduction of exotic species. Kudzu's best-known characteristic is its extraordinary capacity for rapid growth, managing as much as 12 inches (30.5 cm) a day and 60 to 100 feet (18–30 m) a season under ideal conditions. When young, kudzu has thin, flexible, and downy stems that grow outward as well as upward, eventually covering virtually everything in its path with a thick mat of leaves and tendrils. This lateral growth creates the dramatic effect, common in southeastern states such as Georgia, of telephone poles, buildings, neglected vehicles, and whole areas of woodland being enshrouded in blankets of kudzu. Kudzu's tendency toward aggressive and overwhelming colonization has many detrimental effects, killing stands of trees by robbing them of sunlight and pulling down or shorting out utility cables. Where stem nodes touch the ground, new roots develop that can extend 10 feet (3 m) or more underground and eventually weigh several hundred pounds. In the nearly ideal climate of the Southeast, the prolific vine easily overwhelms virtually all native competitors and also infests cropland and yards.

A member of the pea family, kudzu is itself native to China and Japan. Introduced to the United States at the Japanese garden pavilion during the 1876 Philadelphia Centennial Exhibition, kudzu's broad leaves and richly fragrant reddish-purple blooms made it seem highly desirable as an ornamental plant in American gardens. It now ranges along the eastern seaboard from Florida to Pennsylvania, and westward to Texas. Although hardy, kudzu does not tolerate cold weather and prefers acidic, well-drained soils and bright sunlight. It rarely flowers or sets seed in the northern part of its range and loses its leaves at first frost.

For centuries, the Japanese have cultivated kudzu for its edible roots, medicinal qualities, and fibrous leaves and stems, which are suitable for paper production.

After its initial introduction as an ornamental plant, kudzu also was touted as a forage crop and as a cure for erosion in the United States. Kudzu is nutritionally comparable to alfalfa, and its tremendous durability and speed of growth were thought to outweigh the disadvantages caused for cutting and baling by its rope-like vines. But its effectiveness as a ground cover, particularly on steeply-sloped terrain, is responsible for kudzu's spectacular spread. By the 1930s, the United States Soil Conservation Service was enthusiastically advocating kudzu as a remedy for erosion, subsidizing farmers as well as highway departments and railroads, with as much as $8 an acre to use kudzu for soil retention. The Depression-era Civilian Conservation Corps also facilitated the spread of kudzu, planting millions of seedlings as part of an extensive erosion control project.

Kudzu also has had its unofficial champions, the best-known of whom is Channing Cope of Covington, Georgia. As a journalist for Atlanta newspapers and a popular radio broadcaster, Cope frequently extolled the virtues of kudzu, dubbing it the "miracle vine" and declaring that it had replaced cotton as "king" of the South. The spread of the vine was precipitous. In the early 1950s, the federal government began to question the wisdom of its support for kudzu. By 1953, the Department of Agriculture stopped recommending the use of kudzu for either fodder or ground cover. In 1982, kudzu was officially declared a weed.

Funding is now directed more at finding ways to eradicate kudzu or at least to contain its spread. Continuous over-grazing by livestock will eventually eradicate a field of kudzu. Repeated applications of defoliant herbicides can kill or limit many kudzu vines, but some have developed a resistance to common herbicides. Stubborn patches may take five or more years to be completely removed. Controlled burning is usually ineffective and attempting to dig up the massive root system is generally an exercise in futility, but kudzu can be kept off lawns and fences (as an ongoing project) by repeated mowing and vigorous pruning.

A variety of new uses are being found for kudzu, and some very old uses are being rediscovered. Kudzu root can be processed into flour and baked into breads and cakes; as a starchy sweetener, it also may be used to flavor soft drinks. Medical researchers investigating the scientific bases of traditional herbal remedies have suggested that isoflavones found in kudzu root may significantly reduce craving for alcohol in alcoholics. Eventually, derivatives of kudzu may also prove to be useful for treatment of high blood pressure. Methane and gasohol have been successfully produced from kudzu, and kudzu's stems may prove to be an economically viable source of fiber for paper production and

other purposes. The prolific vine has also become something of a humorous cultural icon, with regional picture postcards throughout the south portraying spectacular and only somewhat exaggerated images of kudzu's explosive growth. Many fairs, festivals, restaurants, bars, rock groups, and road races have borrowed their name and drawn some measure of inspiration from kudzu, poems have been written about it, and kudzu cookbooks and guides to kudzu crafts are readily available in bookstores.

Kudzu is only one of several aggressive, invasive species found in the southern United States. Air potato, *Dioscorea bulbifera*, a native of Africa and India is more aggressive in warm climates than kudzu. The U.S. Department of Agriculture and several individual states have banned the importation, sale, and planting of various types of invasive species. They have also worked to educate the public about the environmental problems caused by nonnative, invasive plants.

Resources

BOOKS

May, Suellen. *Invasive Terrestrial Plants*. New York: Chelsea House Publications, 2006.

PERIODICALS

Withgott, Jay. "Are Invasive Species Born Bad?" *Science* 305 (2004): 1,100–1,101.

OTHER

U.S. Government; science.gov. "Harmful Invasive Species." http://www.science.gov/browse/w_115A15.htm (accessed October 13, 2010).
United States Department of the Interior, United States Geological Survey (USGS). "Invasive Species." http://www.usgs.gov/science/science.php?term = 602 (accessed October 13, 2010).

Lawrence J. Biskowski

Kwashiorkor

Kwashiorkor is one of many severe protein energy malnutrition disorders that are a widespread problem among children in developing countries. The word's origin is in Ghana, where it means a deposed child, or a child that is no longer suckled. The disease usually affects infants between one and four years of age who have been weaned from breast milk to a high starch, low protein diet. The disease is characterized by lethargy, apathy, or irritability. Over time, the individual will experience delayed growth processes, both physically and mentally. Kwashiorkor occurs most often among children living in developing countries where food insecurity occurs because of famine, drought, or conflict.

Kwashiorkor results in amino acid deficiencies, which inhibit protein synthesis in all tissues. The lack of sufficient plasma proteins, specifically albumin, results in systemic pressure changes, ultimately causing generalized edema (swelling because of fluid accumulation). The edema, however, often masks wasted muscle tissue, severely reduced body fat, and overall dehydration. Hair becomes dry and brittle, and the color usually fades. Skin lesions often resemble either flaking paint or burns, and both leave the child vulnerable to infection. Children with kwashiorkor usually have little energy or appetite, and it is often difficult to persuade them to eat.

Eventually, the liver swells with stored fat because there are no hepatic proteins being produced for digestion of fats. Kwashiorkor additionally results in reduced bone density and impaired renal function. If treated early on in its development, the disease can be reversed with proper dietary therapy and treatment of associated infections. If the condition is not reversed in its early stages, prognosis is poor.

See also Sahel.

Kyoto Protocol/Treaty

In the mid-1980s, a growing body of scientific evidence linked man-made greenhouse gas emissions to global warming. In 1990, the United Nations General Assembly issued a report that confirmed this link. The United Nations Framework Convention on Climate Change (UNFCC), an accord signed by various nations at the Earth Summit in Rio de Janeiro, Brazil, in 1992, committed industrialized nations to stabilizing their emissions at 1990 levels by 2000.

In December 1997, representatives of 160 nations met in Kyoto, Japan, in an attempt to produce a new and improved treaty on climate change. Major differences occurred between industrialized and still-developing countries with the United States perceived, particularly by representatives of the European Union (EU), as not doing its share to reduce emissions, especially those of carbon dioxide.

The outcome of this meeting, the Kyoto Protocol to the United Nations Framework Convention on Climate Change (UNFCCC), required industrialized nations to reduce their emissions of carbon dioxide, methane, nitrous oxide, hydrofluorocarbons, sulfur

Members of Japanese environmental groups stage a demonstaration in downtown Kyoto to celebrate the Kyoto Protocol, the world's most far-reaching environmental treaty. *(Kazuhiro Nogi /AFP/ Getty Images)*

dioxides, and perfluorocarbons below 1990 levels by 2012. The requirements would be different for each country and would have to begin by 2008 and be met by 2012. There would be no requirements for the developing nations. Whether or not to sign and ratify the treaty was left up to the discretion of each individual country.

Global warming

The organization that provided the research for the Kyoto Protocol was the Intergovernmental Panel on Climate Change (IPCC), set up in 1988 as a joint project of the United Nations Environment Programme (UNEP) and the World Meteorological Organization (WMO). In 2001, the IPCC released a report, "Climate Change, 2001." Using the latest climatic and atmospheric scientific research available, the report predicted that global mean surface temperatures on earth would increase by 2.5 to 10.4°Fahrenheit (1.5–5.9°C) by the year 2100, unless greenhouse gas emissions were reduced. This warming trend was seen as rapidly accelerating, with possible dire consequences to human society and the environment. These accelerating temperature changes were expected to lead to rising sea levels, melting glaciers and polar ice packs, heat waves, droughts and wildfires, changes in agricultural production, and a profound and deleterious effect on human health and well-being.

Some of the effects of these temperature changes may already be occurring. According to the Intergovernmental Panel on Climate Change's (IPCC) "Fourth

Assessment Report" in 2007, global surface temperatures increased by an average of $1.33 \pm 0.32°$ Fahrenheit ($0.74 \pm 0.18°C$) during the twentieth century. Sea ice is melting in both the Antarctica and the Arctic. The world's glaciers are melting at an alarming rate with a mean thickness loss of approximately 46 feet (14 m). The sea level is rising at three times its historic rate. Florida is already feeling the early effects of global warming with shorelines suffering from erosion, with dying coral reefs, with saltwater polluting the fresh water sources, with an increase in wildfires, and with higher air and water temperatures. In Canada, forest fires have more than doubled since 1970, water wells are going dry, lake levels are down, and there is less rainfall.

Controversy

Since its inception, the Kyoto Protocol has generated a great deal of controversy. Richer nations have argued that the poorer, less-developed nations are getting off easy. The developing nations, by contrast, have argued that they will never be able to catch up with the richer nations unless they are allowed to develop with the same degree of pollution as that which let the industrial nations become rich in the first place.

Another controversy rages between environmentalists and big business. Environmentalists have argued that the Kyoto Protocol does not go far enough, while petroleum and industry spokespersons have argued that it would be impossible to implement without economic disaster.

In the United States, the controversy has waxed especially high. The Kyoto Protocol was signed under the administration of President Bill Clinton, but was never ratified by the Republican-dominated U.S. Senate. Then, in 2001, President George W. Bush, backed out of the treaty, saying it would cost the U.S. economy $400 billion and 4.9 million jobs. Bush unveiled an alternative proposal to the Kyoto accord that he said would reduce greenhouse gases, curb pollution, and promote energy efficiency. But critics of his plan argued that by the year 2012, the Bush plan would have increased the 1990 levels of greenhouse gas emissions by more than 30 percent.

Soon after the Kyoto Protocol was rejected by the Bush administration, the EU criticized the action. In particular, Germany was unable to understand why the Kyoto restrictions would adversely affect the American economy, noting that Germany had been able to reduce their emissions without serious economic problems. The Germans also suggested that President Bush's program to induce voluntary reductions was politically motivated and was designed to prevent a drop in the unreasonably high level of consumption of greenhouse gases in the United States, a drop that would be politically damaging for the Bush administration. In rejecting the Kyoto Protocol, President Bush claimed that it would place an unfair burden on the United States. He argued that it was unfair that developing countries such as India and China should be exempt.

Politics has always been at the forefront of this debate. The IPCC has provided assessments of climate change that have helped shape international treaties, including the Kyoto Protocol. However, the Bush administration, acting at the request of ExxonMobil, the world's largest oil company, and attempting to cast doubts upon the scientific integrity of the IPCC, was behind the ouster in 2002 of the IPCC chairperson Robert Watson, an atmospheric scientist who supported implementing actions against global warming.

The ability of trees and plants to fix carbon through the process of photosynthesis, a process called *carbon or C sequestration*, results in a large amount of carbon stored in biomass around the world. In the framework of the Kyoto Protocol, C sequestration to mitigate the greenhouse effect in the terrestrial ecosystem has been an important topic of discussion in numerous recent international meetings and reports. To increase C sequestration in soils in the dryland and tropical areas, as a contribution to global reductions of atmospheric CO_2, the United States has promoted new strategies and new practices in agriculture, pasture use, and forestry, including conservation agriculture and agroforestry. Such practices should be facilitated particularly by the application of article 3.4 of the Kyoto Protocol covering the additional activities in agriculture and forestry in the developing countries and by appropriate policies.

Into the future

In June 2002, the fifteen member nations of the EU formally signed the Kyoto Protocol. The ratification by the fifteen EU countries was a major step toward making the 1997 treaty effective. Soon after, Japan signed the treaty, and Russia was expected to follow suit.

To take effect, the Kyoto Protocol must be ratified by fifty-five countries, but these ratifications have to include industrialized nations responsible for at least 55 percent of the 1990 levels of greenhouse gases. As of 2002, over seventy countries had already signed, exceeding the minimum number of countries needed. If Russia signs the treaty, nations responsible for over 55 percent of the 1990 levels of greenhouse gas pollution will have signed, and the Kyoto Protocol will take effect.

Before the EU ratified the protocol, the vast majority of countries that had ratified were developing countries. With the withdrawal of the United States, responsible for 36.1 percent of greenhouse gas emissions in 1990, ratification by industrialized nations was crucial. For example, environmentalists hope that Canada will ratify the treaty as it has already committed compliance.

Although the Bush administration opposed the Kyoto Protocol, saying that its own plan of voluntary restrictions would work as well without the loss of billions of dollars and without driving millions of Americans out of work, the Environmental Protection Agency (EPA), under its administrator Christine Todd Whitman, sent a climate report to the United Nations in 2002 detailing specific, far-reaching, and disastrous effects of global warming upon the American environment and its people. The EPA report also admitted that global warming is occurring because of man-made carbon dioxide and other greenhouse gases. However, it offered no major changes in administration policies, instead recommending adapting to the inevitable and catastrophic changes.

Although the United States was still resisting the Kyoto Protocol in mid–2002, and the treaty's implications for radical and effective action, various states and communities decided to go it alone. Massachusetts and New Hampshire enacted legislation to cut carbon emissions. California was considering legislation limiting emissions from cars and small trucks. Over 100 U.S. cities had already opted to cut carbon emissions. Even the U.S. business community, because of their many overseas operations, was beginning to voluntarily cut back on their greenhouse emissions.

As of 2010, 187 nations had signed and ratified the Kyoto Protocol. The Kyoto Protocol, however, will expire at the end of 2012. The international community has engaged in negotiations to produce a successor greenhouse gas reduction treaty. Delegates to the 2007 United Nations Climate Change Conference in Bali, Indonesia, adopted a roadmap to reach a successor agreement in 2009. Delegates to the 2009 United Nations Climate Change Conference in Copenhagen, Denmark, however, failed to reach consensus on a comprehensive climate change agreement.

Resources

BOOKS

Bolin, Bert. *A History of the Science and Politics of Climate Change: The Role of the Intergovernmental Panel on Climate Change.* New York: Cambridge University Press, 2008.

Cowie, Jonathan. *Climate Change: Biological and Human Aspects.* Cambridge: Cambridge University Press, 2007.

Dessler, Andrew Emory, and Edward Parson. *The Science and Politics of Global Climate Change: A Guide to the Debate.* Cambridge, UK: Cambridge University Press, 2006.

Giddens, Anthony. *Politics of Climate Change.* Cambridge, UK: Polity Press, 2009.

OTHER

United Nations, Secretariat of the United Nations Framework Convention on Climate Change (UNFCCC). "Introduction to the UNFCCC and its Kyoto Protocol." http://unfccc.int/files/press/backgrounders/application/pdf/unfccc_and_kyoto_protocol.pdf (accessed November 8, 2010).

United Nations, Secretariat of the United Nations Framework Convention on Climate Change (UNFCCC). "Kyoto Protocol." *United Nations Framework Convention on Climate Change.* http://unfccc.int/kyoto_protocol/items/2830.php (accessed November 8, 2010).

United Nations, Secretariat of the United Nations Framework Convention on Climate Change (UNFCCC). "The Kyoto Protocol." http://unfccc.int/files/press/backgrounders/application/pdf/fact_sheet_the_kyoto_protocol.pdf (accessed November 8, 2010).

ORGANIZATIONS

IPCC Secretariat, c/o World Meteorological Organization, 7bis Avenue de la Paix, C.P. 2300, CH- 1211, Geneva, Switzerland, 41-22-730-8208, 41-22-730-8025, ipcc-sec@wmo.int, http://www.ipcc.ch

Douglas Dupler